International Conference on Engineering Technology (ETIC 2016)

MATEC Web of Conferences Volume 97 (2017)

Ho Chi Minh City, Vietnam
5 - 6 August 2016

Editors:

M.A.B. Abdullah
S.Z. Abd Rahim
M.E. Muhammad Suandi
M.N. Mat Saad
M.F. Ghazali

ISBN: 978-1-5108-3607-5

Printed from e-media with permission by:

Curran Associates, Inc.
57 Morehouse Lane
Red Hook, NY 12571

Some format issues inherent in the e-media version may also appear in this print version.

Printed by Curran Associates, Inc. (2017)

For additional information, please contact EDP Sciences – Web of Conferences at the address below.

EDP Sciences – Web of Conferences
17, Avenue du Hoggar
Parc d'Activité de Courtabœuf
BP 112
F-91944 Les Ulis Cedex A
France

Phone: +33 (0) 1 69 18 75 75
Fax: +33 (0) 1 69 28 84 91

contact-edps@webofconferences.org

Additional copies of this publication are available from:

Curran Associates, Inc.
57 Morehouse Lane
Red Hook, NY 12571 USA
Phone: 845-758-0400
Fax: 845-758-2633
Email: curran@proceedings.com
Web: www.proceedings.com

TABLE OF CONTENTS

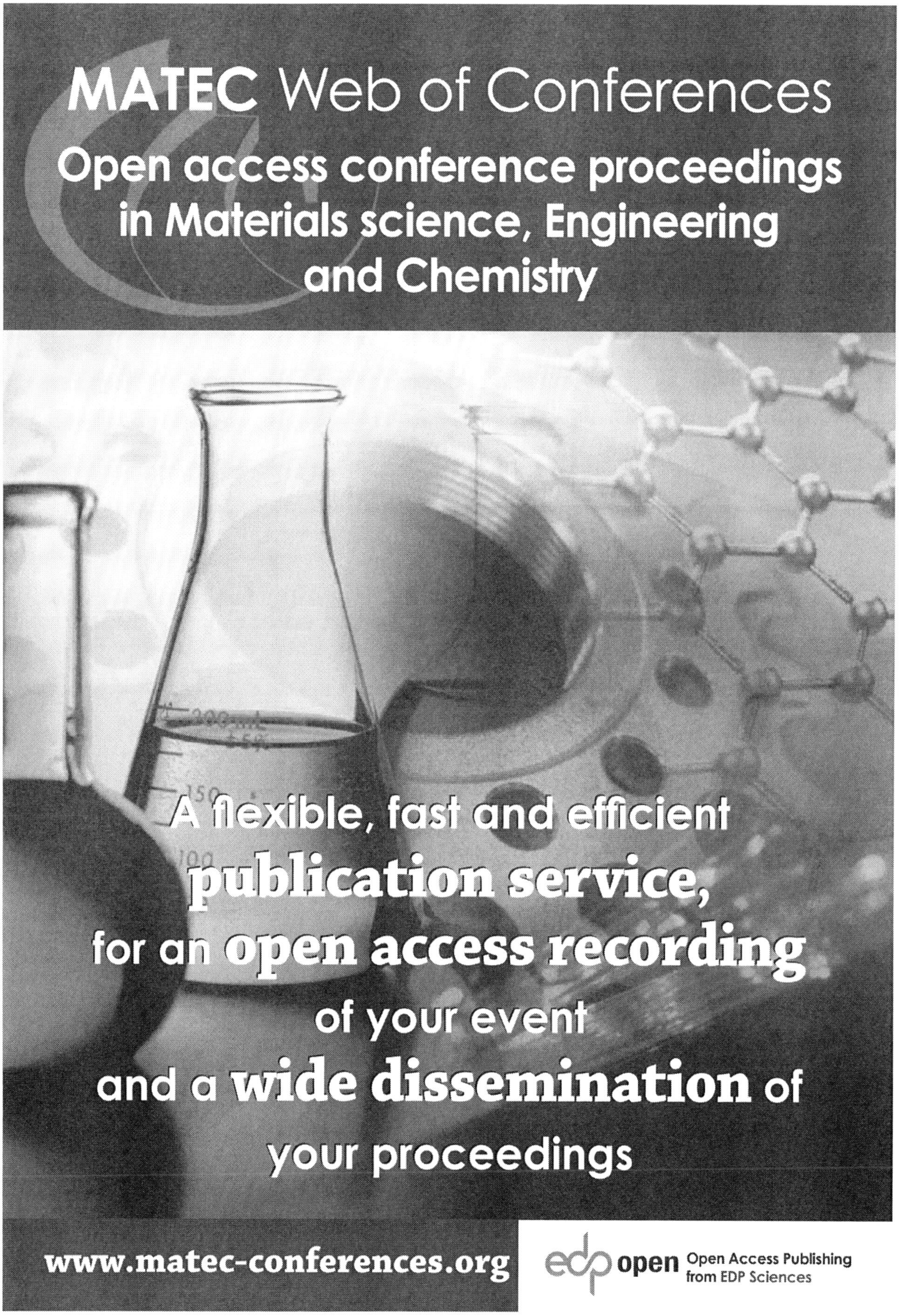
MATEC Web of Conferences
Open access conference proceedings in Materials science, Engineering and Chemistry
A flexible, fast and efficient publication service, for an open access recording of your event and a wide dissemination of your proceedings
www.matec-conferences.org
edp open Open Access Publishing from EDP Sciences

edp sciences

MATEC Web of Conferences

Statement of Peer review

In submitting conference proceedings to MATEC Web of Conferences, I certify to the Publisher that I adhere to the **Policy on Publishing Integrity** of the journal in order to safeguard good scientific practice in publishing.

1. All articles have been subjected to peer review administered by the proceedings editors.
2. Reviews have been conducted by expert referees, who have been requested to provide unbiased and constructive comments aimed, whenever possible, at improving the work.
3. Proceedings editors have taken all reasonable steps to ensure the quality of the materials they publish and their decision to accept or reject a paper for publication has been based only on the merits of the work and the relevance to the journal.

Title, date and place of the conference

Engineering Technology International Conference 2016 (ETIC2016)
5-6 August 2016, Ho Chi Minh City, Vietnam

Proceedings editor(s):

List of Editors:
Associate Professor Dr. Mustafa Al Bakri Abdullah
Dr. Shayfull Zamree Bin Abd Rahim
Mr. Mohd Effendi Muhammad Suandi
Mr. Mohd Nasir Mat Saad
Dr. Mohd Fathullah bin Ghazali

Date and editor's signature

Associate Proffesor Dr. Mohd Mustafa Al Bakri Abdullah
Faculty of Engineering Technology (FETech)
Universiti Malaysia Perlis (UniMAP)
Pejabat Pos Besar, 01000,
Kangar, Perlis, Malaysia

03-Jan-2017

Mathematical Derivation of Switching Angles of Multilevel Voltage Source Inverter based on Alternative Phase Opposition Disposition (APOD)

Mohd Alif Ismail[1,*], *Mohd Junaidi* Abdul Aziz[2], *Muhd Hafizi* Idris[1], *Noor Fazliana* Fadzail[1], *Mohd Saufi* Ahmad[1], *Melaty* Amirruddin[1], *Farrah Salwani* Abdullah[1], and *Nurhakimah* Mohd Mukhtar[1]

[1]PPK Sistem Elektrik, Universiti Malaysia Perlis
[2]Faculty of Electrical Engineering, Universiti Teknologi Malaysia

Abstract. Modular structured multilevel inverter is very useful for electrical application especially in high voltage and high power applications. The main function of this multilevel inverter is to produce multilevel AC output voltage from several separate DC sources. This project is to derive a newmathematical formulation of multilevel voltage source inverter switching instants. The proposed method for this project is based on the sinusoidal natural sampling PWM (SPWM) by comparing several modified modulation signal with a triangular carrier signal. This resulting intersection points between this modulation and carrier signal become the switching instants of the PWM pulses. Derivation also based on Alternative Phase opposition disposition (APOD). A cascaded multilevel inverter is selected as a topology for this project due to major advantages compare with other topology. The derived formula is analyzed by using MATLAB simulation software. It is found that the results that use the derived formula are almost identical to simulation result.

1 Introduction

The Multilevel Voltage Source Inverter (MVSI) has drawn tremendous interest in the power industry. Multilevel inverter is an array of capacitors voltage sources and power semiconductor switches which will generate an output voltage of stepped waveform [1]. This inverter is commonly use in application of high voltage and high power. It also has a few advantages. It is easier to build a high voltage and high power inverter with the multilevel structure. This is due to the way in which devices voltage stresses are controlled in the structure. The power rating can be increase by increasing the number of voltage levels in the inverter without requiring higher ratings on individual devices.

Besides that, the multilevel voltage source inverter also has other advantages such as

* Corresponding author: alifismail@unimap.edu.my

possibility to construct more cheap applications that does not need galvanic isolation such as adjustable speed drives. Furthermore, multilevel voltage source inverter also can improve harmonics performance where it can reduce switching losses. This is very important during application of high power as increased power losses can complicate thermal management system design. One of the famous techniques that normally be applied to the multilevel inverter is sinusoidal pulse width modulation (SPWM) [4].

2 Methodology

This project is focused on sinusoidal natural sampling PWM (SPWM) since SPWM is the most popular modulation method. The SPWM for multilevel voltage source inverter is based on the intersection between triangular carrier signals with a modified sinusoidal signal. The switching angle of the PWM pulses is come from the points of intersection. As far as the carrier signals alignment is concerned, three carrier disposition method are suggested which are Alternative Phase Opposition Disposition (APOD), Phase Opposition Disposition (POD) and Phase Disposition (PD) [3].

For this project, it only focuses on derivation of the switching angle equations for APOD Scheme. The topology that had been selected for this project is Cascaded Multilevel Inverters (CMI) because CMI have several advantages compare to other topologies [3].

2.1 Mathematical derivation of the switching angles equations for Alternative Phase Opposition Disposition (APOD) scheme.

The switching angle point is presented by α and this point is obtained from intersection between carrier signal and modulation signal. To illustrate the principle of the proposed scheme, a five level inverter had being used in this project. Figure 1 shows the output of PWM pulses generate from intersection between modulation signal and carrier signal for APOD scheme. For this five level output, the switching point are obtain from intersection between four modulation signals that is $S_1(k)$and $S_2(k)$ $(0 < t < \pi)$, $S_3(k)$ and $S_4(k)$ $(\pi < t < 2\pi)$with single triangular carrier c(k). Signal $S_2(k)$ is actually similar with $S_1(k)$ and only shifted down by the amplitude of triangular carrier signal Ac.

The PWM pulses for level 1 output signal is obtain from the comparison between modified sinusoidal signal, $S_1(k)$ and carrier signal, c(k), while for level 2 output signal is obtain from comparison between modified sinusoidal signal, $S_2(k)$ and carrier signal, c(k). The comparison is designed such that if modified sinusoidal signal, $S_1(k)$ is greater than carrier signal, c(k), a pulse-width for level 1 is generated and if the modified sinusoidal signal, $S_2(k)$ is greater than carrier signal, c(k), a pulse-width for level 2 is generated. So, if there is no intersection, then output signal remain at 0. As a result, a multilevel output voltage Vout is produce. Figure 2 show that the k_{th}rising edge is defined as the intersection of the negative slope carrier, $c^{-}(k)$, positive slope carrier $c^{+}(k)$ and two set of modulating signals $S_1(k)$ and $S_2(k)$.

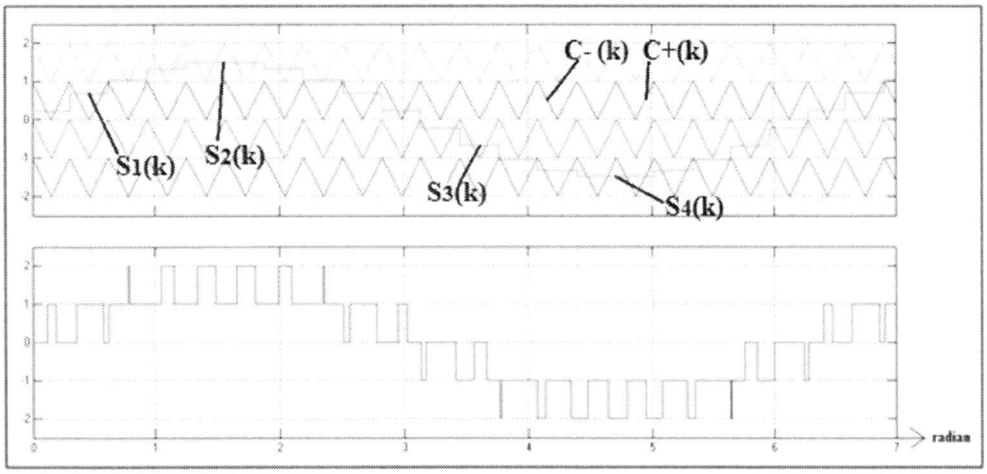

Fig. 1. The PWM pulse generate from intersection between modulation Signal and carrier signal for Alternative Phase Opposition Dispossion (APOD) scheme.

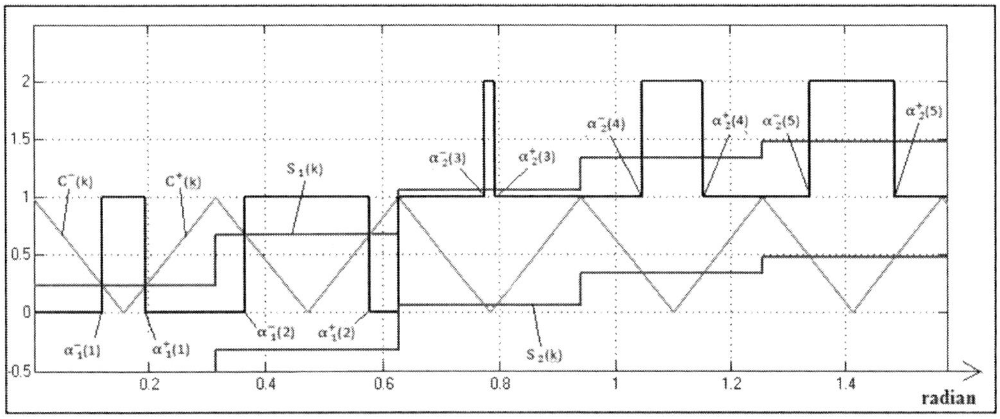

Fig. 2. Intersection between modified sinusoidal modulation signals and single carrier signal with it symbol.

2.1.1 First step : Find the carrier signal equation,c(k).

To find equation for C_1^-(negative slope at level 1), basic equation need to be use that is y = mx + c where y = C_1^- and x = α.

For $C_1^-(1)$,
$m = \frac{-A_C}{\frac{T_C}{2}}$, $c = A_C$, So, equation for the C_1^- (1) is $C_1^-(1) = \frac{-A_C}{\left(\frac{T_C}{2}\right)} \propto +A_C$

For $C_1^-(2)$,
$m = \frac{-A_C}{\frac{T_C}{2}}$, $c = 3A_C$, So, equation for the C_1^- (2) is $C_1^-(2) = \frac{-A_C}{\left(\frac{T_C}{2}\right)} \propto +3A_C$

General equation for C_1^- is
$C_1^-(k) = \frac{-A_C}{\left(\frac{T_C}{2}\right)} \propto_1^- + (2k-1)A_C$Where k = 1,2,3,4..

By using similar method as above, the equation for C_I^{+}(positive slope) can be obtain.
General equation for $C_I^{+}$is

$C_1^{+}(k) = \frac{A_C}{\left(\frac{T_C}{2}\right)} \propto_1^{\mp} - (2k-1)A_C$ Where k = 1,2,3,4..

2.1.2 Second step : Find the modified sinusoidal modulation signal equation, s(k).

To obtain the equation for S_1(k), the middle point for every level of our sample point need to be use . By using basic equation for sinusoidal equation, y= A sin(wt), equation for S_1(k) can be obtain.

For this equation,

$$\omega = \frac{2\pi f_o}{f_c}$$

Where *fo* is the frequency of modulation signal and *fc* is the frequency of carrier signal.

So, for $S_1(1)$ $\quad S_1(1) = A_m \sin\left(\frac{\pi x f_o}{f_c}\right)$

For $S_1(2)$ $\quad S_1(2) = A_m \sin\left(\omega + \frac{\pi x f_o}{f_c}\right)$

For $S_1(3)$ $\quad S_1(3) = A_m \sin\left(\omega(2) + \frac{\pi x f_o}{f_c}\right)$

General equation for S_1(k)

$S_1(k) = A_m \sin\left(\omega(k-1) + \frac{\pi x f_o}{f_c}\right)$ Where k = 1,2,3,4....

To find the equation S_2(k), it is similar like S_1(k) but the different is only this signal is shift down from S_1(k) signal by the magnitude of Ac. But for ($\pi < t < 2\pi$) period, S_3(k) and S_4(k) will intersect with carrier signal. S_3(k) signal is same with S_1(k) but the different is only this signal is shift up from S_1(k) signal by the magnitude of Ac and S_4(k) is different from S_1(k) that is shift up by the magnitude of 2Ac.

So, the general equation for S_2(k)

$$S_2(k) = A_m \sin\left(\omega(k-1) + \frac{\pi x f_o}{f_c}\right) - A_C$$

So, the general equation for S_3(k)

$$S_3(k) = A_m \sin\left(\omega(k-1) + \frac{\pi x f_o}{f_c}\right) + A_C$$

And the general equation for S_4(k)

$S_4(k) = A_m \sin\left(\omega(k-1) + \frac{\pi x f_o}{f_c}\right) + 2A_C$ Where *k* = 1,2,3,4....

2.1.3 Third step : Find the switching angle equation for Pulse Width Modulation(PWM) switching signal.

To obtain the switching angle,α equation for PWM signal , the method of intersection between two equation need to be use and the intersection point between equation of carrier signal and equation of modulation signal represent the value of switching angle, α.

For C_1^{-}(k) intersection with S_1(k) at ($0 < t < \pi$),

$$C_1^{-}(k) = S_1(k)$$

$$\frac{-A_C}{\left(\frac{T_C}{2}\right)} \propto_1^- (k) + (2k-1)A_C = A_m \sin\left(\omega(k-1) + \frac{\pi \cdot f_o}{f_c}\right)$$

$$A_C\left[\frac{-\propto_1^- (k)}{\left(\frac{T_C}{2}\right)} + (2k-1)\right] = A_m \sin\left(\omega(k-1) + \frac{\pi \cdot f_o}{f_c}\right)$$

$$\frac{-\propto_1^- (k)}{\left(\frac{T_C}{2}\right)} = -(2k-1) + \frac{A_m}{A_C}\sin\left(\omega(k-1) + \frac{\pi \cdot f_o}{f_c}\right) \propto_1^- (k)$$

$$= \frac{T_C}{2}\left[(2k-1) - \frac{A_m}{A_C}\sin\left(\omega(k-1) + \frac{\pi \cdot f_o}{f_c}\right)\right]$$

By using similar method as above, the equation for$C_1^-(k)$ intersection with $S_2(k)$ at ($0 < t < \pi$), $C_1^+(k)$ intersection with $S_1(k)$ at ($0 < t < \pi$), $C_1^+(k)$ intersection with $S_2(k)$ at ($0 < t < \pi$), $C_1^-(k)$ intersection with $S_3(k)$ at ($\pi < t < 2\pi$), $C_1^-(k)$ intersection with $S_4(k)$ at ($\pi < t < 2\pi$), $C_1^+(k)$ intersection with $S_3(k)$ at ($\pi < t < 2\pi$), and for $C_1^+(k)$ intersection with $S_4(k)$ at ($\pi < t < 2\pi$) can be obtain.Lastly, it can be summerize that the switching angle equation for PWM signal are derived like below:

i. Negative slope ($0 < t < \pi$)

$$\propto_1^- (k) = \frac{T_C}{2}\left[(2k-1) - \frac{A_m}{A_C}\sin\left(\omega(k-1) + \frac{\pi \cdot f_o}{f_c}\right)\right]$$

$$\propto_2^- (k) = \frac{T_C}{2}\left[(2k) - \frac{A_m}{A_C}\sin\left(\omega(k-1) + \frac{\pi \cdot f_o}{f_c}\right)\right]$$

ii. Positive slope ($0 < t < \pi$)

$$\propto_1^+ (k) = \frac{T_C}{2}\left[(2k-1) + \frac{A_m}{A_C}\sin\left(\omega(k-1) + \frac{\pi \cdot f_o}{f_c}\right)\right]$$

$$\propto_2^+ (k) = \frac{T_C}{2}\left[(2k-2) + \frac{A_m}{A_C}\sin\left(\omega(k-1) + \frac{\pi \cdot f_o}{f_c}\right)\right]$$

iii. Negative slope ($\pi < t < 2\pi$)

$$\propto_1^- (k) = \frac{T_C}{2}\left[(2k-2) - \frac{A_m}{A_C}\sin\left(\omega(k-1) + \frac{\pi \cdot f_o}{f_c}\right)\right]$$

$$\propto_2^- (k) = \frac{T_C}{2}\left[(2k-3) - \frac{A_m}{A_C}\sin\left(\omega(k-1) + \frac{\pi \cdot f_o}{f_c}\right)\right]$$

iv. Positive slope ($\pi < t < 2\pi$)

$$\propto_1^+ (k) = \frac{T_C}{2}\left[(2k) + \frac{A_m}{A_C}\sin\left(\omega(k-1) + \frac{\pi \cdot f_o}{f_c}\right)\right]$$

$\propto_2^+ (k) = \frac{T_C}{2}\left[(2k+1) + \frac{A_m}{A_C}\sin\left(\omega(k-1) + \frac{\pi \cdot f_o}{f_c}\right)\right]$Where k = 1,2,3,4..

3 Results

After the equation of switching angle are carried out, verification need to be done by comparing the result that come from the derived equation and the result that come out from simulation. If the errors between both results are very small or equal to zero, than the derived equation is verified and accepted.

3.1 Comparison between Matlab simulation results and result from derive equation.

Table 1 and Table 2 below show that the different between simulation result and result from derive equation. It shows that percentage error between both result are equal to zero. It also shows that the equation are verified and accepted.

Table 1. Comparison Result for ($0 < t < \pi$)

	Simulation (radian)	Derived Equation(radian)	% ERROR
α1(1)	0.1202	0.1202	0.000
α1(1)	0.1939	0.1939	0.000
α1(2)	0.3643	0.3643	0.000
α1(2)	0.5782	0.5782	0.000
α2(3)	0.7759	0.7759	0.000
α2(3)	0.7949	0.7949	0.000
α2(4)	1.0467	1.0467	0.000
α2(4)	1.1524	1.1524	0.000
α2(5)	1.3381	1.3381	0.000
α2(5)	1.4894	1.4894	0.000
α2(6)	1.6522	1.6522	0.000
α2(6)	1.8035	1.8035	0.000
α2(7)	1.9892	1.9892	0.000
α2(7)	2.0949	2.0949	0.000
α2(8)	2.3467	2.3467	0.000
α2(8)	2.3657	2.3657	0.000
α1(9)	2.5634	2.5634	0.000
α1(9)	2.7773	2.7773	0.000
α1(10)	2.9476	2.9477	0.000
α1(10)	3.0214	3.0214	0.000

Table 2. Comparison result ($\pi < t < 2\pi$)

	Simulation (radian)	Derived Equation(radian)	% ERROR
α1(11)	3.1784	3.1785	0.000
α1(11)	3.4189	3.4189	0.000
α1(12)	3.5627	3.5627	0.000
α1(12)	3.6630	3.6629	0.000
α2(13)	3.7794	3.7794	0.000
α2(13)	4.0746	4.0745	0.000
α2(14)	4.1369	4.1369	0.000
α2(14)	4.3454	4.3454	0.000
α2(15)	4.4739	4.4739	0.000
α2(15)	4.6368	4.6368	0.000
α2(16)	4.7880	4.7880	0.000
α2(16)	4.9509	4.9509	0.000
α2(17)	5.0794	5.0794	0.000
α2(17)	5.2878	5.2878	0.000
α2(18)	5.3502	5.3502	0.000
α2(18)	5.6453	5.6453	0.000
α1(19)	5.7619	5.7618	0.000
α1(19)	5.8620	5.8621	0.000
α1(20)	6.0059	6.0059	0.000
α1(20)	6.2463	6.2463	0.000

4 Discussion

After result has been verified, there show that simulation result is identical to the result that obtains from the derived equation. Based on Table 1 and Table 2, It shows the PWM output signal that obtain from derived equation and matlab simulation. It is exactly same and zero percentage error. This derived equation will make more easier for us because it can be use in digital purpose. It is well know that the digital technique had many advantages compared to analog technique.

5 Conclusion

The objectives for this project are to derive a new mathematical equation that defines the switching instants of sinusoidal pulse width modulation (SPWM) for multilevel voltage source inverter. This derive equation also was verified by comparing with MATLAB simulation. From the comparison, both results that come from simulation and from derived equation is exactly same. It gives almost zero percentage error. It means that the derived equation are verified and accepted.

References

1. J. Aziz, Z. Salam, "*An alternative PWM scheme for multilevel voltage source inverter.* Power Engineering Conference, 2003. PECon 2003. Proceedings. National. 83-87, (2003)
2. J.A. Aziz, Z. Salam, *A PWM strategy for the modular structured multilevel inverter suitable for digital implementation.* Power Electronics Congress, 2002. Technical Proceedings. CIEP 2002. VIII IEEE International. 160-164, (2002)

3. S. Zainal, J. Aziz, S.S. Ahmed, *Single carrier PWM scheme for cascaded multilevel voltage source inverter.* Power Electronics and Drive Systems, 2003. PEDS 2003. The Fifth International Conference. IEEE. **1**, 406-410, (2003)
4. V.K. Chinnaiyan, J. Jerome, J. Karpagam, T. Suresh, *Control techniques for multilevel voltage source inverters.* 2007 International Power Engineering Conference (IPEC 2007). IEEE. 1023-1028, (2007)
5. X. Xu, Y. Zou, K. Ding, F. Liu, *Cascade multilevel inverter with phase-shift SPWM and its application in STATCOM.* Industrial Electronics Society, 2004. IECON 2004. 30th Annual Conference of IEEE. **2**, 1139-1143, (2004)

DOI: 10.1051/matecconf/20179701002

The Influence of Salt Water on Chloride Penetration in Geopolymer Concrete

Like Novia Halim[1,2,*], *Januarti Jaya* Ekaputri[1,2], and *Triwulan*[1,2]

[1]Civil Engineering Department, Faculty of Civil Engineering and Planning, Institut Teknologi Sepuluh Nopember (ITS), Sukolilo, Surabaya, Indonesia
[2]Konsorsium Riset Geopolimer Indonesia (KORIGI), Laboratorium Beton dan Bahan Bangunan, 60111 Kampus ITS Sukolilo Surabaya, Indonesia

Abstract. This paper presents the influence of chloride ion penetration in geopolymer concrete. Fly ash as based material for geopolymer concrete was used in this mixture. Fly ash was mixed with sodium hydroxide (NaOH) 8 M and sodium silicate (Na_2SiO_3) as the alkali solution. The sizes of cylindrical specimens were prepared with a diameter of 100 mm and 200 mm high. Some specimens were immersed in salt water at a concentration of 3.5%, and other control specimens were cured in tap water for 30, 60, 90, and 120 days. The mechanical properties were determined with compressive test which was conducted at 28, 30, 60, 90 and 120 days. Some durability tests were performed for porosity, chloride penetration, and pH measurement. It was found that geopolymer concrete has higher compressive strength than concrete made with Ordinary Portland cement (OPC). However, chloride penetration in geopolymer concrete is higher than OPC. The pH measurement showed that geopolymer concrete has less pH than OPC concrete. The porosity of concrete has been found to influence chloride penetration and pH of concrete.

1 Introduction

Concrete is one of the most widely used construction materials in the world. In addition, ordinary Portland cement (OPC) is the main binding material in concrete. The consumption of energy during manufacturing of Portland cement releases a large amount of greenhouse gases into the atmosphere causing global warming. The cement industry is responsible for about 6-8% of all CO_2 emissions, since the production of one ton of Portland cement emits approximately one ton of CO_2 into the atmosphere [1, 2].

The previous studies showed geopolymer cement is one of the solutions to replace the Portland cement in construction [3, 4]. Fly ash is one of the construction materials, which is used commonly to produce geopolymer concrete [5]. Fly ash-based geopolymer concrete has many significant benefits such as environmental friendly, high-performance material solution, and economic benefits. Fly ash-based geopolymer concrete is also resistant to corrosion [6].

* Corresponding author: like.halim77@gmail.com

DOI: 10.1051/matecconf/20179701002

Durability of geopolymer concrete in salt water is affected by its permeability. It is the major factor for determining the long-term durability of concrete in marine environment. Denser concrete will lead against destructive agents to penetrate and flow through the pores [7]. However, the chloride environment sometimes can increase the compressive strength of the geopolymer concrete [8]. According to some previous research, fly ash-based geopolymer concrete has better durability than Portland-based concrete in aggressive environment such as sulphate, acid, and fire [9-13]. Nevertheless, there are only few literatures discussing about the influence of salt water on chloride penetration in geopolymer concrete. This paper discusses factors affecting the strength geopolymer concrete in salt water.

2 Materials

2.1 Fly ash

Class F fly ash was obtained from the fertilizer factory, PT. Petrokimia Gresik Indonesia. The composition of fly ash is listed in Table 1.

Table 1. Chemical composition of fly ash (%) by XRF.

SiO_2	Al_2O_3	Fe_2O_3	TiO_2	CaO	MgO	Cr_2O_3	K_2O	Na_2O	SO_3	Mn_2O_3
48.47	26.05	12.54	0.92	5.18	2.77	0.02	1.66	0.47	1.05	0.19

2.2 Alkaline activator

Activators used in the mixture are NaOH and Na_2SiO_3. NaOH solution was prepared by dissolving flakes form in distilled water. Concentration of NaOH used is 8M. Sodium silicate was obtained from PT. Kasmaji Inti Utama (PTKIU), Indonesia. Sodium silicate contains 18.5% of Na_2O, 36.4% of SiO_2, and 45.1% of H_2O. The mass ratio of Na_2SiO_3 to NaOH was kept at 2.5.

2.3 Aggregates

Coarse aggregate and fine aggregate were collected from PT. Surya Beton Indonesia. The densities of coarse aggregate and fine aggregate are 2.6 g/cm^3 and 2.7 g/cm^3 respectively.

2.4 Superplasticizer

Polycarbonic Acid Salts used in the mixture was obtained from PT. Sika Indonesia. It was applied to the mixture to ensure the workability of fresh concrete. The dosage of SP was 2% by weight of fly ash.

3 Methods

The fresh concrete was cast in cylindrical molds with a diameter of 100 and 200 mm high. One day after casting, the concrete was demolded and cured at room temperature for one day before immersing into tap water for 28 days. After that, the immersion was continued in two conditions which were tap water and salt water. All specimens were cured for 30, 60, 90, and 120 days. Fig. 1 shows the sequence of treatment and testing.

	The specimens were immersed in tap water				The specimens were immersed in salt water				
Treatment	Mixing	Open the mold and immerrsed the specimens in normal water	Three specimens were kept at room temperature	Test of concrete compressive strength at 28 days	Some specimens were immersed into salt water	Compressive strength test	Compressive strength test	Compressive strength test	Compressive strength test
Days	0	1	27	28	30	60	90	120	180
Cylinder concrete						√	√	√	√

Fig. 1. The treatment and testing of cylindrical concrete specimens.

4 Test procedures

The mechanical properties of concrete were determined by compressive strength test according to ASTM C 39 [15]. Porosity test for the specimens at 28 days (initial condition) and those which were cured in salt water for 30, 60, 90, and 120 days. The tests for chloride penetration and pH were only measured for geopolymer concrete and OPC, which were immersed in salt water, containing 3.5% of NaCl solution. Chloride ion penetration test and pH measurement were selected from samples at three points as described in Fig. 2. Three points are determined with some distance from a concrete surface at a depth of 0 cm, 4 cm, and 8 cm.

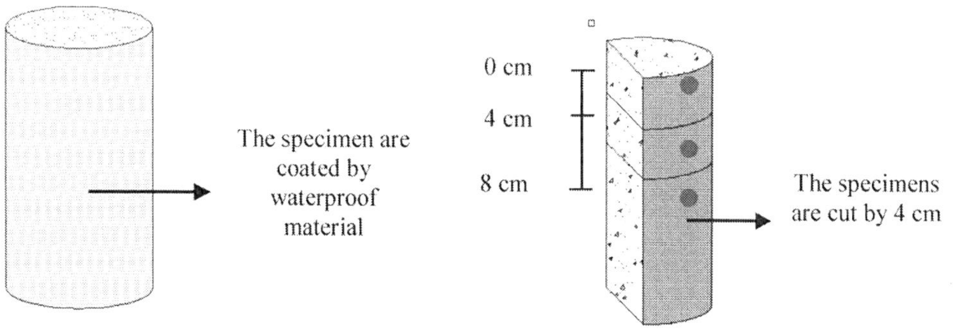

Fig. 2. The model of sample for chloride penetration and pH measurement.

DOI: 10.1051/matecconf/20179701002

5 Results and discussions

5.1 Compressive strength

OPC concrete was used as a control specimen to compare its compressive strength with geopolymer concrete. Table 2 and Fig. 3 showed that geopolymer concrete has higher compressive strength than OPC in salt water. It is proved that salt water influences the strength of geopolymer concrete. In addition, geopolymer concrete has also higher compressive strength in tap water than OPC.

Fig. 3 shows that the compressive strength of geopolymer and OPC concrete in tap water were decreased at 90 and 120 days. However, the compressive strength of geopolymer concrete decreases less significant than OPC concrete. Compressive strength of OPC specimen which was immersed in salt water reduces drastically from 59.99 MPa at 90 days to 45.21 MPa at 120 days. This proves that the geopolymer concrete is more resistant than OPC concrete in aggressive environment. It has a good agreement with Ekaputri [14]. The durability of structure was decreased from the first year (2005) when research was conducted and after 20 years of investigation . Ekaputri [14] calculated the safety factor to obtain the structure performance in sea water and it was found that the protection of structure has been declined year by year. Bayuaji, et al [8] found that geopolymer concrete cured in sea water provides higher compressive strength than the other curing system using fresh water and room temperature. Giasuddin [16] explained that there is abundance of Na^+ in the surrounding medium in salt water curing. It is very likely that Na^+ as well as other cations (Ca^{2+}) in the sample are less prone to leaching. These excess cations in the surrounding curing solution and in the sample will help to promote reaction mechanism and geopolymer formation.

Table 2. The comparison for compressive strength in geopolymer concrete and OPC concrete.

Type of Concrete		Compressive Strength				
	Curing (days)	0	30	60	90	120
	Age (days)	28	58	88	118	148
Geopolymer	Salt water	51.42	65.49	68.33	70.39	65.19
	Tap water	51.42	57.76	68.07	66.27	65.66
OPC	Salt water	46.61	52.28	60.14	59.99	45.21
	Tap water	46.61	48.35	57.76	55.69	52.64

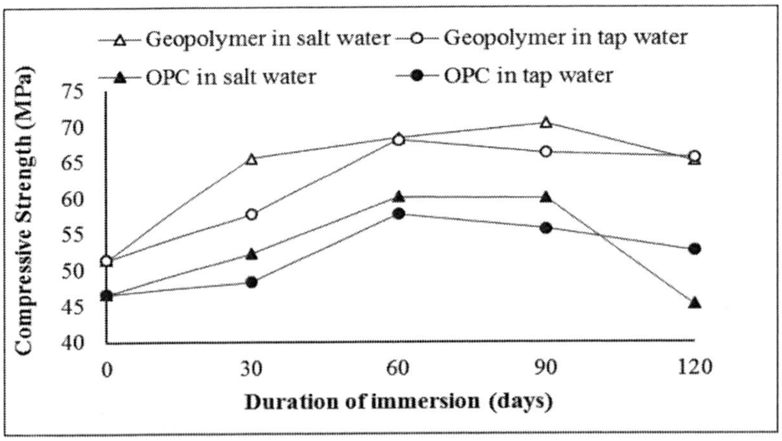

Fig. 3. The comparison for compressive strength in geopolymer and OPC concrete.

Another investigation showed that geopolymer concrete has a good development of compressive strength in aggressive environment for a long term [9]. During the geopolimerisation process, silica and alumina in fly ash dissolve with NaOH to form sodium-alumina-silicate, which contributes to the strength of geopolymer concrete. Salt water which containing sodium ions from NaCl keeps the ions in geopolymer concrete during geopolimerization [12]. It causes the compressive strength of geopolymer concrete higher than OPC concrete. The compressive strength on OPC was decreased because portlandite ($Ca(OH)_2$) leaks out from OPC concrete. Then, this compound bonds with NaCl to form calcium chloride ($CaCl_2$) which then reacts with aluminate in the cement. As a result, this reaction forms Friedle salt ($3CaO.Al_2O_3.CaCl_2.10H_2O$) which fills the pores. Chloride in salt water can cause cracks and accumulation of salt crystal as well [17]. It also happened on the compressive strength of geopolymer concrete which was immersed in salt water. The strength was decreased at the age of 118 to 148 of days for 0.61 MPa because calcium oxide in the fly ash will react with NaCl to form calcium silicate chloride ($Ca_2SiO_3Cl_2$). In addition, the calcium content in the fly ash is not as much as in the cement and this compound is very easy to react with other compounds.

5.2 Porosity

Porosity measurement on the cylindrical specimen was obtained from total porosity (Pt). The measurement of porosity test was conducted after the specimens were immersed in salt water and tap water at specific age.

Table 3 shows that OPC concrete has better durability in tap water than in salt water. It is caused by the porosity of OPC concrete in salt water is greater than in tap water. If it is compared to geopolymer concrete, it shows that total porosity of OPC concrete is less than geopolymer in salt water.

In wet curing, sodium leaks out of concrete to the water. It causes the geopolymer concrete become more porous. It was proved by the result of pH measurement listed in Table 4. Before the concretes were soak in the water, the pH of immersion water was 7.9 with a concentration of solution was 266 ppm. The pH and concentration increased day by day until 10.4 and 1283 ppm respectively at 10 days. It is likely can be proposed that moist curing may be more recommendable than wet curing.

Table 3. Porosity test of geopolymer concrete and OPC concrete

Type of Concrete		Total Porosity Pt (%)				
	Curing (days)	0	30	60	90	120
	Age (days)	28	58	88	118	148
Geopolymer	Salt water	20.10	20.56	21.71	21.19	20.67
	Tap water	20.10	20.40	23.94	22.38	23.58
OPC	Salt water	14.20	18.27	22.38	20.00	21.15
	Tap water	14.20	22.83	20.42	22.28	20.48

Table 4. The result of pH measurement and concentration of immersion water

Days of immersion	pH	Concentration of solution Ppm
1	7.9	266
2	9.6	383
3	9.7	461
4	10.1	593
5	10.0	685
6	10.1	763
7	10.1	801
8	10.2	958
9	10.4	1096
10	10.4	1283

5.3 Chloride penetration

The measurement of chloride penetration was intended to determine the depth of concrete that has been damaged by chloride ions. Fig. 4 illustrates that the chloride penetration of geopolymer concrete is larger than OPC. For instance, geopolymer concrete which was immersed in salt water for 30 days contained chloride penetration at 0 cm, 4 cm, and 8 cm from the surface were 0.15%, 0.13%, and 0.12% respectively. Whereas, OPC concrete has 0.10% at 0 cm, 0.07% at 4 cm, and 0.03% at 8 cm. This indicates that there is a correlation between porosity and chloride penetration. The greater the total porosity, the more chloride ion penetrate to concrete. In the previous study [9, 18], the geopolymer concrete was kept at room temperature before it was immersed in salt water. In this research, the concrete was directly soaked into normal water for 28 days. It causes the bond of Si-Al in the concrete becomes weaker because its greater pores. Olivia and Nikraz [18] said that the chloride concentration was large in geopolymer concrete because there is no C_3A in fly ash, which could enable a chloride binding mechanism to minimize direct chloride ion penetration of the concrete paste. However, the use of fly ash can slow the effects of chloride and CO_2 into concrete [19].

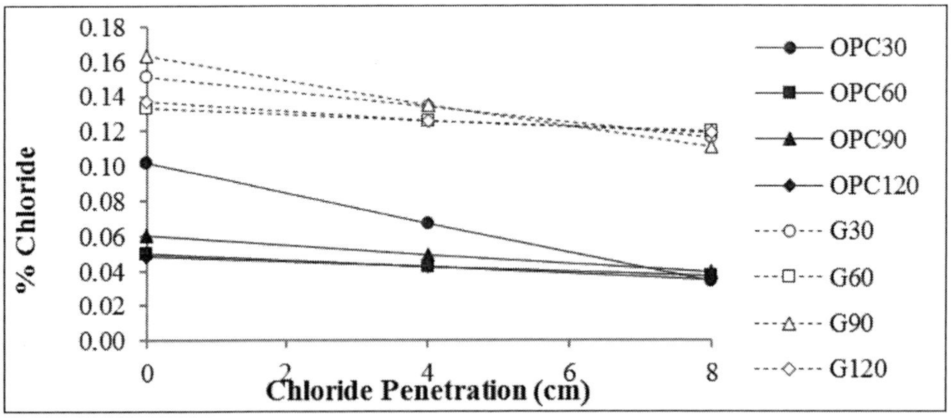

Fig. 4. Chloride penetration in geopolymer and OPC concrete.

5.4 pH measurement

The measurement was conducted for geopolymer and OPC concrete in salt water. It was intended to determine the correlation between pH value and chloride penetration. Generally, geopolymer concrete has a high pH because of the reaction which occurred when fly ash was mixed with alkaline solution. Fig. 5 shows the result of pH measurement for the different of penetration depth at 0 cm, 4 cm, and 8 cm from the concrete surface.

Fig. 5 shows that pH in geopolymer concrete is less than OPC concrete. pH value is influenced by porosity. The greater the total porosity, the higher sodium is leaching. This is proved by the measurement of initial pH on geopolymer concrete is 11.24 before it is cured in normal water for 28 days. After that, the specimens were immersed in salt water. pH value decreases because the leaching of sodium from geopolymer concrete. It is proved with the pH measurement that is shown in Table 4. In addition, pH value of geopolymer concrete and OPC ranged between 10.32 until 11.70. It means that both concrete are in alkaline condition and good for corrosion resistance.

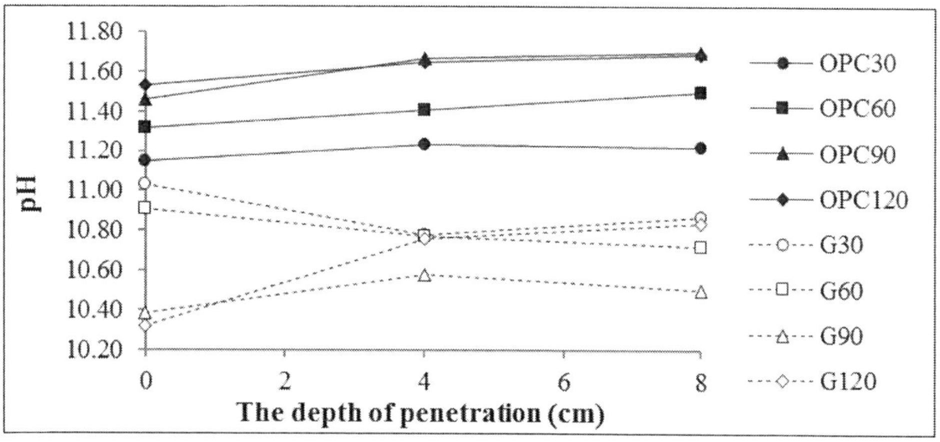

Fig. 5. pH measurement in geopolymer concrete and OPC concrete.

6 Summary

Geopolymer concrete has higher compressive strength than OPC in salt water. It proves that geopolymer concrete is durable in aggressive environment because there is abundance of Na^+ in salt water. Na^+ as well as other cations (Ca^{2+}) in the sample are less prone to leaching. These excess cations in the surrounding curing solution and in the sample which help to promote reaction mechanism and geopolymer formation. Porosity of OPC concrete in salt water is larger than in tap water. It shows that OPC concrete has a good durability in tap water. If it is compared with geopolymer concrete, it shows that the total porosity of OPC concrete is smaller than geopolymer in salt water. In wet curing, sodium leaks out of concrete to the water. It causes the geopolymer concrete become more porous. The chloride penetration of geopolymer concrete is higher than OPC concrete. This indicates that there is a correlation between porosity and chloride penetration. Greater porosity increases chloride penetration into geopolymer concrete. pH value in geopolymer concrete and OPC ranged between 10.32 until 11.70. It means that both concrete are alkaline and good for corrosion resistance.

References

1. McCaffrey, *Climate Change and the Cement Industry* (GCL Magazine, London, 2002)
2. J. Davidovits, Rev. Geopolymer Inst., 1 (2013)
3. A.M. Mustafa Al Bakri, H. Kamarudin, M. Binhussain, I. Khairul Nizar, A.R. Rafiza, Y. Zarina, Rev. Adv. Mater. Sci., **30**, 90 (2011)

4. Z.F. Farhana, H. Kamarudin, Azmi Rahmat, A.M. Mustafa Al Bakri, Aust. J. Basic Appl. Sci., **7**, 230 (2013).
5. Information on http://www.geopolymer.org/applications/global-warming
6. K. Zerfu, J.J Ekaputri, Mater. Sci. Forum, **841**, 162 (2016)
7. D.V Reddy, J.B Edouard, K. Sobhan, S.S Rajpathak, J. Mater. Civ. Eng., **25**, 1465 (2011)
8. R. Bayuaji, M.S Dharmawan, B. Wibowo, N.A Husin, S. Subektie, J.J Ekaputri, Mech. Mater., **754**, 400 (2015)
9. P. Chindaprasirt, W. Chalee, Constr. Build. Mater., **63**, 303 (2014)
10. M.M Hossain, M.R Karim, M.K Hossain, M.N Islam, M.F.M Zain, Constr. Build. Mater., **93**, 95(2015)
11. Kupwade-Patil, K. and Allouche, E.N., J. Mater. Civ. Eng., **25**, 1465 (2013)
12. F.U.A Shaikh, (2014), Adv. Concr. Constr., **2**, 109 (2014)
13. Singh N, Vyass S, Pathak RP, Sharma P, Mahure NV, Gupta SL, Int. J. Eng. Innov. Tech., **3**, 2277(2013)
14. J.J. Ekaputri, *Proceedings Fourth International Conference on Forensic Engineering*, (2008)
15. ASTM C 39 / C 39M-03 (2003), *Standard Test Method for Compressive Strength of Cylindrical Concrete Specimens*, United Stated.
16. Giasuddin, M. Haider, Sanjayan, G. Jay, P.G. Ranjith, Fuel, **107**, 34 (2013)
17. T. Ishida, S. Miyahara, T. Maruya, J. Adv. Concr. Tech., **6**, 287(2008)
18. Olivia, Monita, H.R. Nikraz. *Proceedings of Concrete Conference,* (2011).
19. S. Dhawan, S. Bhalla, B. Bhattacharjee, *International Symposium on Advanced Science and Technology in Experimental Mechanics*, (2014)

The Effect of Pulp Industrial Waste as Chemical Admixture to Compressive Strength of Fly Ash Based Alkali Activated Materials

Andrie Harmaji[1,*], *Aishah* Mahyarni Imran[1], and *Bambang* Sunendar[1]

[1]Advanced Material Processing Laboratory, Faculty of Industrial Technology, Institut Teknologi Bandung, Indonesia

Abstract. Black liquor is a toxic by-product from industrial pulp manufacture. It contains sodium hydroxide that can be used as precursor activator for alkali activated material, which is an aluminosilicate material that can be prepared from thermal activation of solid material containing alumina and silica as precursor and alkali activator solution. In this work, alkali activated mortar was prepared by mixing fly ash as main precursors, aggregate, followed by addition of activator solution containing sodium hydroxide solution and waterglass, and chemical admixture which is lignin or black liquor. The best compressive strength was 34.40 MPa achieved in addition of 10 wt% of black liquor to alkali activated mortar. X-ray diffraction demonstrated the formation of albite in mortars, indicating that geopolymerization have been successfully formed. FTIR spectra showed the presence of siloxo and sialate peaks which commonly found in geopolymerization.

1 Introduction

Industrial paper manufacture involves a procedure known as the Kraft process, where wood is converted into wood pulp and then into paper. However, this process produces a toxic byproduct referred as black liquor. This primarily liquid mixture of pulping residues (like lignin and hemicellulose) and inorganic chemicals from the Kraft process (sodium hydroxide and sodium sulfide, for example) is toxic and until the invention of recovery boilers in the early 20th century was often simply released into. Recovery boilers allowed paper manufacturers to reuse and recover inorganic chemicals and extract energy from the pulping residues. Where recovery boiler capacity is limited and a bottleneck in the pulp mill the lignin in the black liquor may be extracted. Lignin used by plants for structure, water transport, and defense, and it is the second most abundant biopolymer on Earth after cellulose [1]. Due to presence of sodium hydroxide in black liquor and lignin, this material can be used as chemical admixture for alkali activated materials.

* Corresponding author: harmaji.a@gmail.com

DOI: 10.1051/matecconf/20179701003

In this work, chemical admixture combined with fly ash, which is also by product material containing high silica and alumina, was used in production of alkali activated materials. The activation of fly ash with alkali activator forms an inorganic binder through geopolymerization. Silica and alumina were reacted with alkali activator to generate aluminosilicate materials with CaO facilitated the setting of the materials. In this work, fly ash was utilized as precursor in geopolymerization to make alkali activated material (AAM). Fly ash that has reacted after geopolymerization will not leach out to the surrounded environment. Alumina and silica consisted in fly ash were first reacted with alkali activator solution and then it was combined with aggregate to make alkali activated mortars. In order to enhance the mechanical properties of fly ash-based AAM, chemical admixture in form of lignin, and black liquor were added to geopolymeric body up to 10% (wt/wt). The effect of various chemical admixtures toward macroscopic and microscopic properties of resulting AAM was also evaluated.

2 Materials

Most of the materials used in this experiment were originally come from industrial waste. Fly ash was obtained from Suralaya Coal Fired Power Plant in Banten. The oxides composition of fly ash was determined using X-Ray Fluorescence. The fly ash contains SiO_2 (52.30 %), Al_2O_3 (26.57 %), CaO (6.00 %), Fe_2O_3 (7.28 %), and some metal oxides. Cimalaka sand in saturated surface-dry (SSD) was used as fine aggregate for making mortar phase. NaOH 12M solution was made by mixing 480g of solid NaOH with 1 litre of aquadest. Alkali activator solution was made by mixing water glass solution and sodium hydroxide (NaOH) 12M. The alkali activator was stored at least 4 hours before used in geopolymerization [2]. Lignin and black liquor were kindly provided by Center of Pulp and Paper, West Java, Indonesia

3 Experimental procedure

Alkali activated mortar was made by mixing fly ash, aggregate, alkali activator solution and lignin or black liquor as chemical admixture. All components were mixed using Hobart^{TM} mixer to form slurry. Once the slurry has been mixed thoroughly and reached proper consistency, it was then tested using Flow Test Table conformed to ASTM C192 to study its workability. The slurry was filled into cubical mould 5cm x 5cm x 5cm. The mortars were cured under ambient (Room Temperature). After few hours, through hydration the mortars hardened and were put out from the mould. The samples were stored at room temperature until it perfectly set. Ratio of alkali activator to binder used was kept constant at 0.5.

The composition of various mortars is shown in Table 1. All composition ratios indicated in that table are in weight unit. The ratio of activator: binder: filler was kept 1 : 2 : 3 [3]. After 28 days, the samples were tested by using Universal Testing Machine (UTM) conformed to ASTM C-39 to obtain the maximum compressive strength data. Chemical composition analysis was performed by using X-Ray diffraction (XRD) and Fourier Transform infrared spectroscopy (FTIR). XRD provided the information of phase identification of a crystalline material of resulting mortars using Philips Diffractometer PW1710. FTIR used to obtain an infrared spectrum of the absorption or emission of a solid, liquid or gas from the broken mortars using (FTIR) Prestige 21 Shimadzu.

Table 1. Mix design of alkali activated materials.

Code	Act:Prec: Aggregate	Curing Method	Admixture
Control	1:2:3	Ambient	-
FAL	1:2:3	Ambient	Lignin
FABL	1:2:3	Ambient	Black Liquor

4 Results and discussion

4.1 Compressive strength test result

The values of compressive strength are shown in Figure 1. Control samples demonstrated average compressive strength of 26.40 MPa, while FAL and FABL showed the increased values of compressive strength which is 30.33 MPa and 34.40 MPa, respectively. It shows that the addition of lignin and black liquor increase the compressive strength from sample without admixture. This is due to high concentration of NaOH used in making black liquor (14M), which means there are addition of Na^+ in the solution is sufficient to bind silicate in fly ash. Samples containing black liquor showed denser mortars than others.

Table 2. AAM mix design, workability and compression strength results.

Code	Workability Result (cm)	Compressive Strength (MPa)
Control	15	26.40
FAL	19	30.33
FABL	20	34.40

4.2 Workability test result.

Workability of FAL (19 cm) and FABL (20 cm) was higher than Control (15 cm), thus makes the slurry of FAL and FABL easier to pour into mould because of extra liquid presents in chemical admixture. It makes lignin and black liquor might be sufficient as water reducer chemical admixture also.

 DOI: 10.1051/matecconf/20179701003

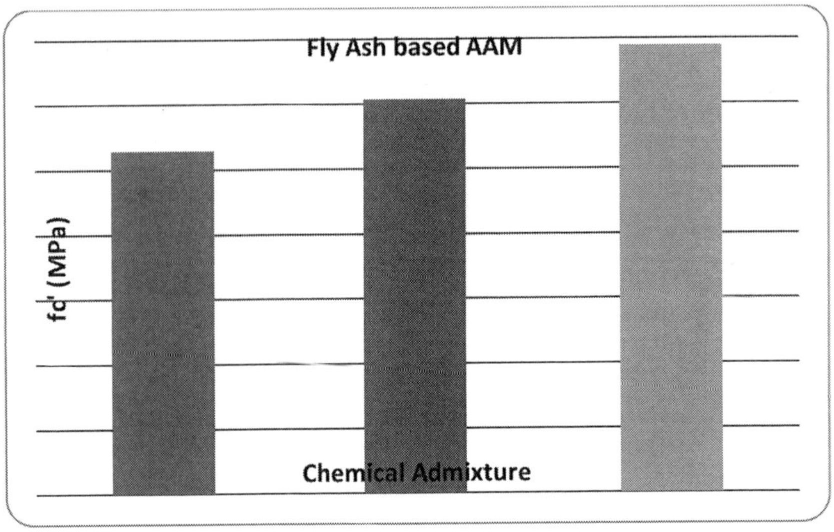

Fig. 1. The effect of various chemical admixture of fly ash based AAM towards compresive strength (fc').

4.3 XRD test result

XRD diffractograms which according to Joint Committee on Powder Diffraction Standards (JCPDS) are presented in Figure 2. demonstrated the presence of quartz (SiO_2) (Peak Q, JCPDS No. 46-1045), Hematite (Fe_2O_3) (Peak H, JCPDS No. 33-0664), both presents in fly ash precursor and alkali activated mortars. Alkali activated mortars has albite ($NaAlSi_3O_8$) (Peak A, JCPDS No. 20-0572), which is only found in alkali activated materials. The formation of albite indicates the reaction between fly ash precursors and alkali activator (sodium hydroxide and water glass) has successfully occurred. Silica and alumina contents in fly ash dissolved in alkali activator and then reacted with water glass as source of Na^+ ion and silica. The reaction is similar with geopolymerization, resulted in aluminosilicate compounds as shown by the presence of albite that has similar formula with alkali activated material (Mn [-(SiO_2)z-AlO_2]n. wH_2O) and base peak line hump XRD pattern has shifted as shown in Figure 3. It means that resulting mortars have more semi crystalline area than fly ash or in the other words fly ash is less amorphous than product which is in agreement with the work of Skavara [4].

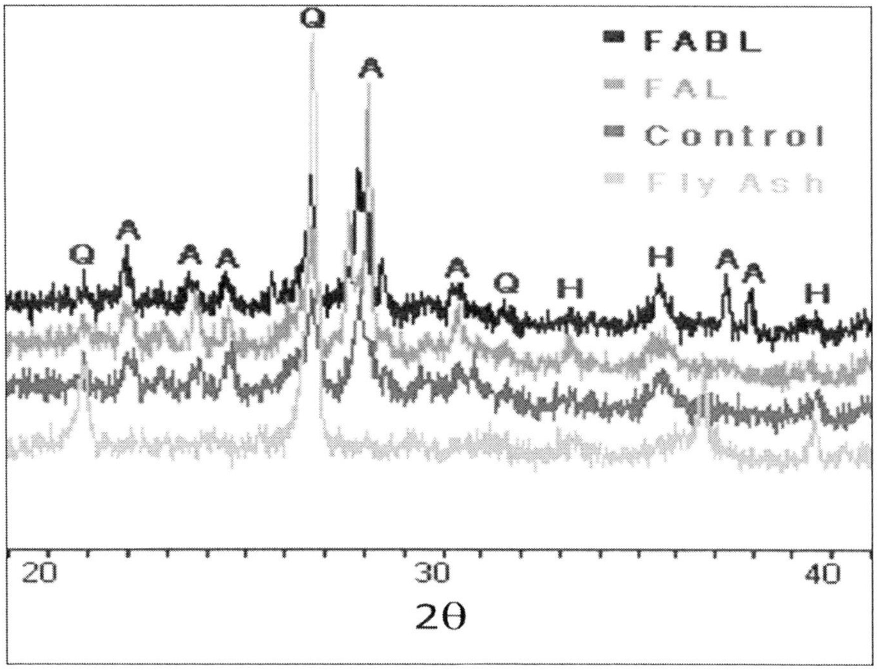

Fig. 2. XRD patterns of various fly ash-based AAM.Peaks A=albite, Q=quartz and H=hematite.

4.4 FTIR result

FTIR results of various alkali activated mortars are shown in Table 3, while Figure 3 shows FTIR spectra of Control, FAL and FABL samples. Peak around 470 cm^{-1} originates from bending vibration Si-O-Si bonding. The highest peak of Si-O-Al and Si-O-Si are shown by absorption at 570-780 cm^{-1} and 1010-1030 cm^{-1} indicating formation of albite, a typical compound of alkali-activated material in these variants as the result of alumina and silica dissolution in fly ash.

Table 3. FTIR results summary of various alkali activated mortars.

Code	Stretching –OH (cm-1)	Bending H-O-H (cm-1)	Stretching Si-O-Si (cm-1)	Symmetric Vibration Si-O-Al (cm-1)		Bending Vibration Si-O-Si (cm-1)
Control	3448.72	1641.42	1012.63	775.38	578.64	472.56
FAL	3450.65	1641.42	1010.70	775.38	578.64	449.41
FABL	3448.72	1641.42	1020.34	775.38	578.64	472.56

DOI: 10.1051/matecconf/20179701003

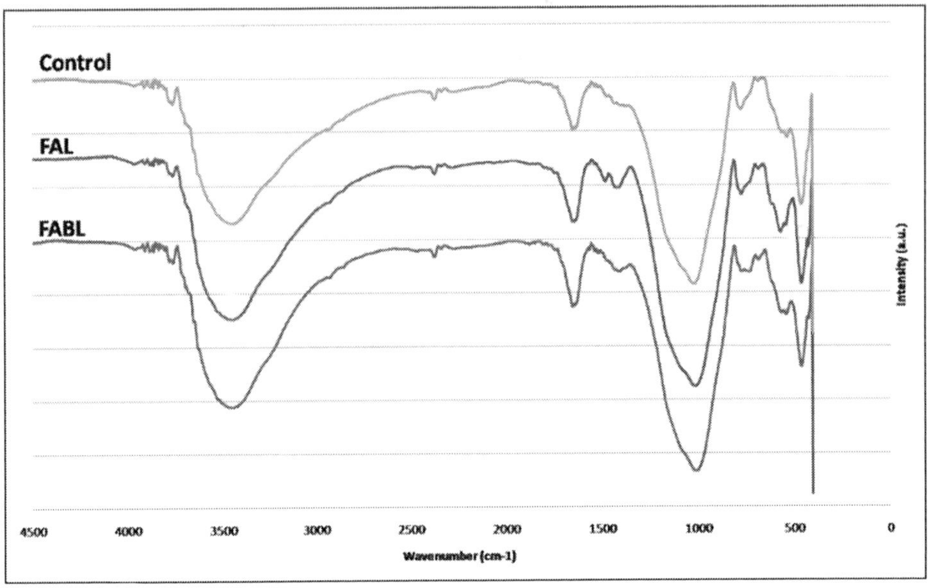

Fig. 3. FTIR spectra of various AAM samples.

The similar spectra in Criado works and stated that the formation of solidification product marked with wavelength number reduction in similar bonding in fly ash based alkali-activated material [5]. Broad peak around 3400 cm^{-1} indicates the presence of O-H bending vibration. FTIR spectra support the XRD results, showing that geopolymerization has been occurred as demonstrated by the presence of functional groups.

5 Conclusions

Alkali activated mortars have been successfully prepared from fly ash, aggregate, alkali activator solution, and chemical admixture followed by curing. The compressive strength of resulting mortars can be varied by changing the chemical admixture, where the addition of black liquor in AAM resulting in highest compressive strength (34.40 MPa). Thus, lignin and black liquor was a potential chemical admixture to be developed for alkali activated materials.

References

1. J.G Linger, D.R Vardon, M.T Guarnieri, E.M Karp, G.B Hunsinger, M.A Franden, C.W Johnson, G. Chupka, T.J. Strathmann, P.T. Pienkos, G.T. Beckham, *Proceedings of National Academy of Sciences of the United States of America*, (2014)
2. D. Hardjito, S.S. Fung, Modern Applied Science, **4**, 44 (2010)
3. P.H. Simatupang, R. Hanafi, B. Sunendar, I. Imran, I. Pane, J. Adv. Mat. Res., **450**, 281(2012)
4. F. Skavara, T. Jilek, L. Kopecky, Ceram. Silik., **49**,195(2005)
5. M. Criado, A. Palomo, A. Fernandez-Jimenez, P.F.G. Banfill, Rheol Acta, **48**, 447(2009)

 DOI: 10.1051/matecconf/20179701004

A Study on Marble-Based Geopolymer

Ta-Wui Cheng[1,*], Yung-Chin Ding[1], Wei-Hao Lee[1], and Po-chang Lu[1]

[1]Institute of Mineral Resources Engineering, National Taipei University of Technology, Taipei City, Da'an Dist., 106, Taiwan, R.O.C.

Abstract. Cement manufacturing is one of the highest carbon dioxide emission industries. It releases 880 kg carbon dioxide for every ton of cement produced. In Taiwan, the carbon dioxide emission from cement industry is more than 10 million tons per year. The development of low carbon dioxide emission green cement can be a solution to replace part of OPC for civil applications and reduce carbon dioxide emission. Geopolymer consists of a polymeric Si–O–Al framework, similar to zeolites which can be formed by mixing aluminosilicate materials, sodium hydroxide and sodium silicates. Geopolymers have advantage of high compressive strength, fire resistance, low shrinkage, acid resistance and low thermal conductivity. In this study, marble-based geopolymer was studied by mixing marble powder, blast furnace slag (BFS) and alkali solution. Physical/mechanical properties of marble-based geopolymer prepared with different marble/BFS weight ratio and SiO_2/Na_2O mole ratio were tested. The experimental results indicated that the compressive strength of marble-based geopolymer can reach 44 MPa within 120 minutes setting time. The microstructure of geopolymeric green cement contains C-S-H gel and aluminosilicate in the framework. According to the experiment results, marble-based geopolymer has the potential to be used as a new civil engineering construction material.

1 Introduction

According to International Energy Agency IEA/OECD statistics in 2015, energy related carbon dioxide emissions was over 248.7 million tonnes in Taiwan and ranked No. 22 in the world's [4]. Therefore, the reduction of carbon dioxide emissions has become a very important issue to alleviate the problem. It is well known that the production of OPC emits large quantity of carbon dioxide [1]. Cement manufacturing is one of the highest carbon dioxide emission industries. It releases 880 kg carbon dioxide for every ton of cement produced. The annual global cement production has reached 2.8 billion tonnes [5]. The OPC production contributes approximately 5–8 % of global CO_2 emissions [6].
Geopolymer is a three-dimensional framework aluminosilicate material [2], and is an inorganic non-metallic material newly developed in recent years. In comparision with Portland cement production, the carbon dioxide emission of geopolymer is extremely low. It has very good characteristics, such as early compressive strength, durability, fire resistance, immobilization of heavy metals, etc. [3]. Previous studies have presented when

* Corresponding author : twcheng@ntut.edu.tw

DOI: 10.1051/matecconf/20179701004

using geopolymeric technology to make geopolymer could reduce over 80 % carbon dioxide emission. In this study, marble and blast furnace slag (BFS) were used as raw materials to fabricate marble-based geopolymer under the activation of alkaline solutions of various molar ratios at ambient condition. The characteristics of marble-based geopolymer in workability (viscosity, setting time), compressive strength and ^{29}Si NMR analysis were presented in this study.

2 Experimental

The marble samples used in this study were collected from a marble mine in eastern Taiwan. The BFS samples were produced from China Steel in southern Taiwan. The chemical composition of marble and BFS were analyzed by X-ray fluorescence (XRF), as listed in Table 1.The major composition in marble and blast furnace slag was calcium oxide, however, the blast furnace slag also contains 27.7 % of SiO_2. Alkali solution was prepared by mixing NaOH and Sodium silicate, and the SiO_2/Na_2O molar ratio was controlled at 0.96, 1.28, and 1.40. The experimental procedure and test parameters were shown in Figure 1 and Table 2.

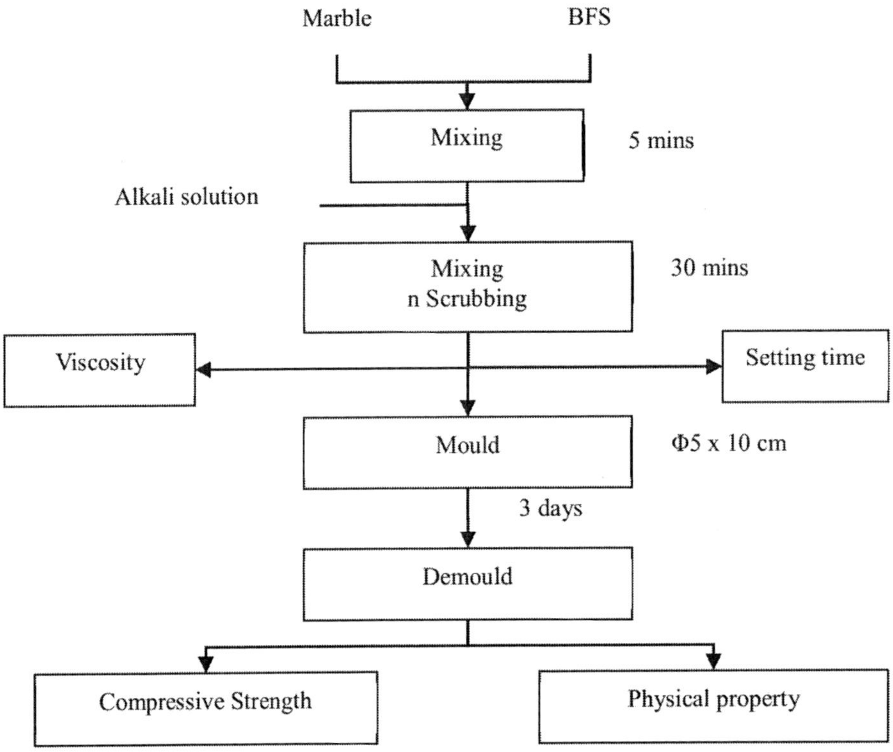

Fig. 1. Overall experiment procedures

Table 1. Marble & Blast furnace slag Chemical Composition

Composition / wt. %	SiO_2	CaO	Al_2O_3	Fe_2O_3	L.O.I.	Other
Marble	1.3	63.9	1.3	0.2	28.0	5.3
Blast furnace slag	27.7	57.9	1.2	0.4	-	2.8

DOI: 10.1051/matecconf/20179701004

Table 2. Experiment Parameters

Experiment No.	Solid Materials wt.%		Alkali Solution	Liquid / Solid wt.% ratio	Viscosity (mPa·s)
	Marble	BFS	SiO_2/Na_2O		
M5S5	50	50	1.28	0.4	2900
M6S4	60	40			2000
M7S3	70	30			1100
M7S3-LS040					1100
M7S3-LS038				0.38	2200
M7S3-LS035				0.35	5500
M7S3A096			0.96	0.4	6700
M7S3A128			1.28		1100
M7S3A140			140		1800

3 Results and Discussion

3.1 Effect of marble/BFS ratio

The effect of marble/BFS ratio on setting time and compressive strength of cement mortar are illustrated in Figure 2. The initial and final setting time increases as the amount of BFS decreases. Because BFS has higher reactivity than marble, the setting time of geopolymer paste can be increased by increasing the amount of BFS in the marble BFS mixture. The compressive strength of marble-based geopolymer was also affected by the mixing ratio of marble and BFS. As shown in Figure 2, the compressive strength of cement mortar increases as the amount of BFS increase. However, micro cracks on geopolymer specimen was observed if the amount of BFS is over 40 %, and thus, resulted in the compressive strength reduction of marble-based geopolymer. According to the experiment results, while using marble and blast furnace slag at the Wt% ratio of 7:3 to form geopolymer, the final setting time of geopolymer paste was around 120 mins, and the viscosity was 1100 mPa • s. The compressive strength can reache 42 MPa after curing for 28 days.

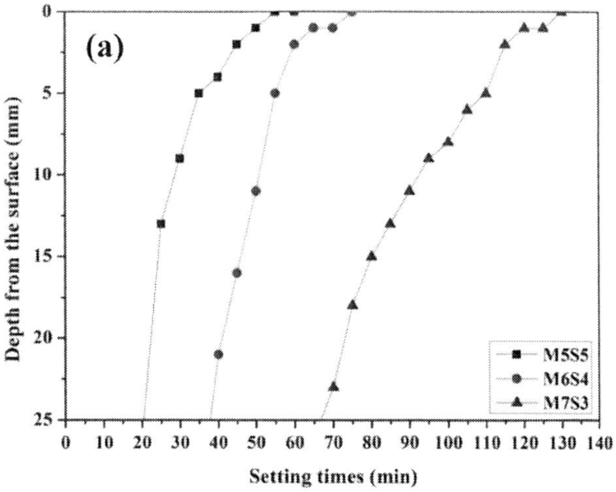

DOI: 10.1051/matecconf/20179701004

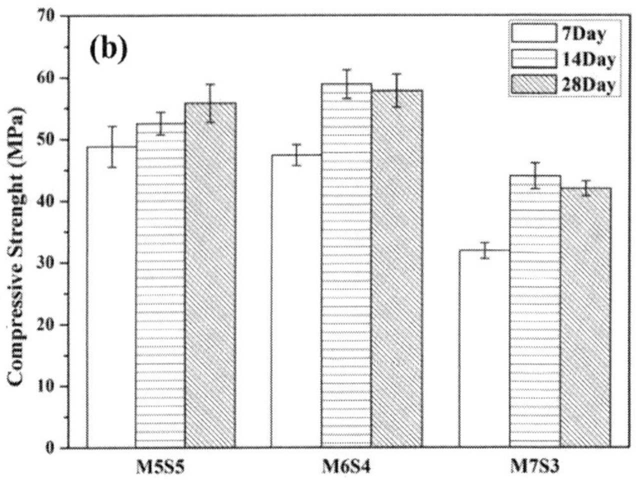

Fig. 2. Effect of marble/BFS ratio for (a) Setting time (b) Compressive strength

(Alkali Solution SiO_2/Na_2O=1.28, L/S = 0.4)

3.2 Effect of Alkali solution/solid materials ratio (L/S ratio)

The effect of L/S ratio on setting time and compressive strength are shown in Figure 3. It was found that the initial and final setting time of geopolymers were strongly affected by the liquid/solid ratio, when the amount of alkali solution increased, the viscosity of geopolymer paste rapidly reduced, and caused the setting time increased. The compressive strength of marble based geopolymer was also affected by the alkali solution/solid materials ratio. Geopolymer compressive strength increased while decreasing L/S ratio. The optimal parameter for higher compressive strength result was the test samples of M7S3-LS035, but the initial and final setting time of M7S3-LS035 and M7S3-LS038 was too fast, their workability were unsatisfactory.

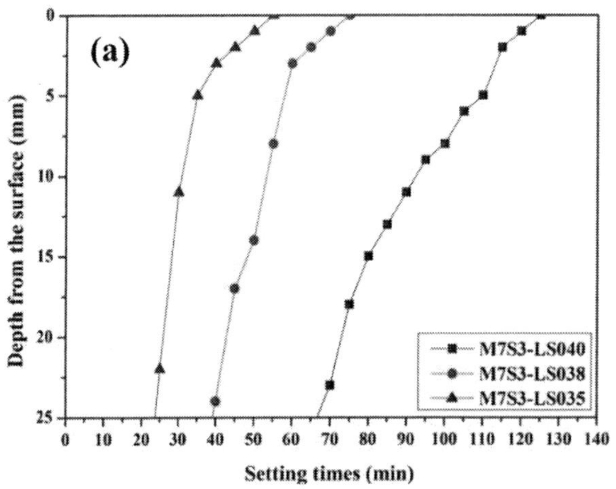

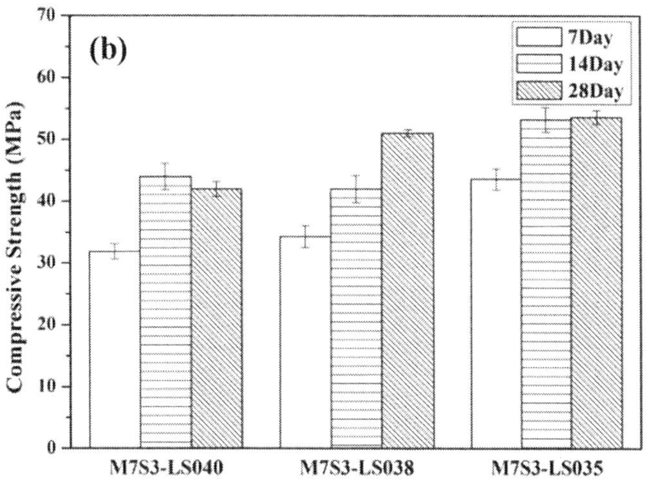

Fig. 3. The effect of Alkali solution/solid materials ratio for (a) Setting time (b) Compressive strength (Alkali Solution SiO_2/Na_2O=1.28, Marble / BFS wt.% ratio = 70:30)

3.3 Effect of SiO_2/Na_2O molar ratio

The effect of SiO_2/Na_2O molar ratio for setting time and compressive strength are shown in Figure 4. As SiO_2/Na_2O molar ratio was higher or lower than 1.28, the initial and final setting time was reduced. Generally, the compressive strength increases as the curing days extends. After curing for 28 days, the compressive strength of all samples can reach 40 MPa and above. By considering the workability of geopolymer, using alkali solution with SiO_2/Na_2O molar ratio=1.28 is the optimal value.

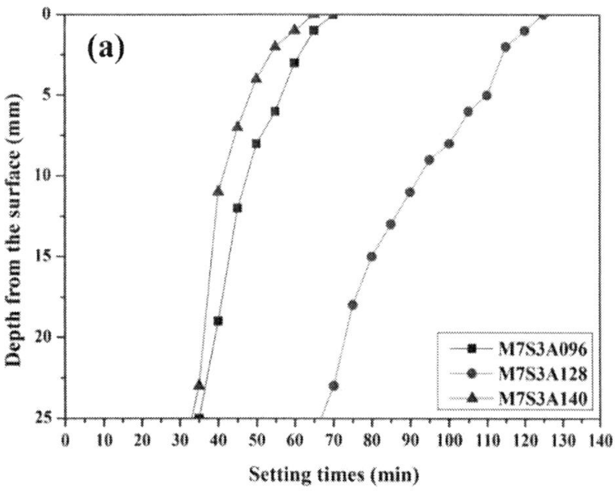

DOI: 10.1051/matecconf/20179701004

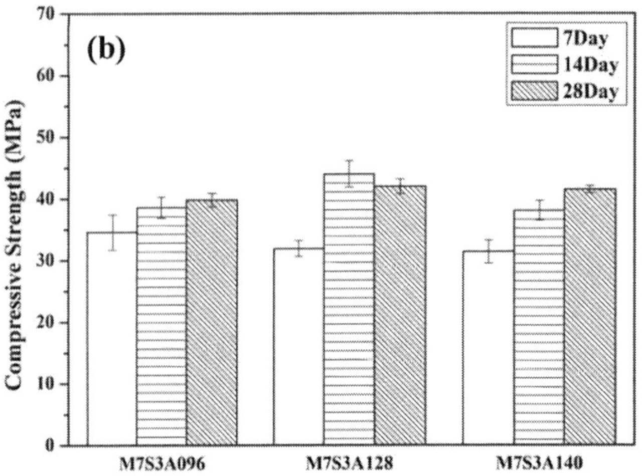

Fig. 4. Alkali solution different SiO_2/Na_2O molar ratio – (a) setting time (b) compressive strength. Marble / BFS wt.% ratio = 70:30, Liquid / Solid wt.% ratio = 0.4.

3.4 NMR analysis

After 28 days curing, marble-based geopolymer (M7S3) sample was analyzed using Nuclear Magnetic Resonance Spectroscopy (NMR). By comparing with traditional coal fly ash-based geopolymer, as shown in Figure 5, it can be found that the Q_4^4 in fly ash based geopolymer was higher than that of in marble-based geopolymer. This is because the major element in coal fly ash is Si, after geopolymeric reaction, coal fly ash can dissolve large amount of Si for forming Q_4^4. In marble-based geopolymer system, the main Si sources is BFS or sodium silicate, therefore only 15% Q_4^4 was measured in the marble based geopolymer.

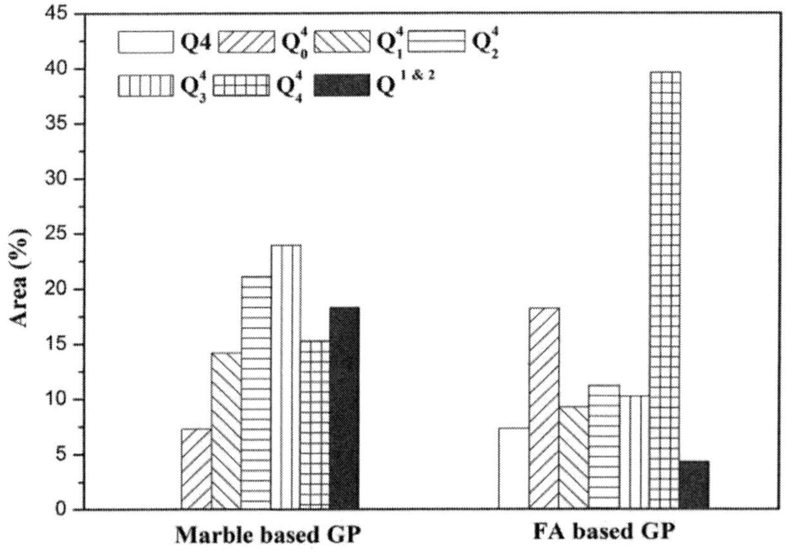

Fig. 5. ^{29}Si NMR analyze result Marble based GP: Marble / BFS wt.% ratio = 70:30, Alkali Solution SiO_2/Na_2O=1.28, Liquid / Solid wt.% ratio = 0.4.FA based GP: Coal fly ash / BFS wt.% ratio = 70:30, Alkali Solution SiO_2/Na_2O=1.28, Liquid / Solid wt.% ratio = 0.4.

4 Conclusions

Marble and BFS based geopolymer was successfully developed in this study. After curing for 28 days, the compressive strength of marble-based geopolymer can reachs 42~60 MPa. The viscosity of geopolyme paste can be controlled between 1100 ~ 6700 mPa s, and the final setting time can be controlled between 40~120 min. It is believed that the fabrication of marble-based geopolymer has great potential for engineering application.

References

1. D.L.Y. Kong, J.G. Sanjayan, Cem. Concr. Compos., **30**, 986 (2008)
2. I. Lecomte, C. Henrist, M. Li´egeois, F. Maseri, A. Rulmont, R. Cloots, J. Eur. Ceram. Soc., **26**, 3789 (2006)
3. P. Duxson, A. Fernandez Jimenez, J.L. Provis, G.C. Lukey, A. Palomo, J.S.J. Van Deventer, J. Mater. Sci., **42**, 2917-2933 (2006)
4. IEA/OECD Key World Energy Statistics, (2015)
5. M. Schneider, M. Romer, M. Tschudin, H. Bolio, Cem. Concr. Res., **41**, 642 (2011)
6. J.S.J. Van Deventer, J.L. Provis, P. Duxson, Miner. Eng., **29**, 89 (2012)

The Effect of Temperature Curing on Geopolymer Concrete

Triwulan[1,2,*], *Januarti Jaya* Ekaputri [1,2] and *Nur Fadlilah* Priyanka [1]

[1] Department of Civil Engineering, Faculty of Civil Engineering and Planning
ITS Surabaya, Indonesia, 60111.
[2]Konsorsium Riset Geopolimer Indonesia (KORIGI), Lab Beton dan Bahan Bangunan ITS, Surabaya, Indonesia

Abstract .This study presents curing temperature influences to behaviors of geopolymer concrete. The concrete was fly ash based, with the ratio of fly ash to alkaline activator was 65%:35%. Ratio coarse aggregate to fine aggregate was 60%:40%, ratios $Na_2SiO_3/NaOH$ were 1.5:2 and 2.5. A polycarboxilate superplasticizer with ratio was 2% to fly ash weight. Form of the test specimens were cylinder with diameter of 10 cm and 20 cm high. Steam curing conducted were at 40°C, 60°C and 80°C for 24 hours. The control specimens were treated at normal moist curing. Tests conducted were compressive strength, tensile strength, porosity and elasticity modulus. The results indicated that at the elevated temperature the compressive strength increases as well as tensile strength but decreases the closed porosity of specimens. The elasticity modulus had a similar tendency.

1 Introduction

This paper presents the influence of temperature curing to behaviors of geopolymer concrete. It is already known that manufacturing process of Ordinary Portland Cement (OPC) produce CO_2 as pollutants [1]. Since it is a main substantial in concrete industries, an innovative eco-friendly material is necessary to be considered. Recently, geopolymer concrete is introduced as one of the solutions. It is formed by using source materials containing reactive silica (SiO_2) and alumina (Al_2O_3) activated with alkali solutions. The source of materials can be obtained abundantly, such as fly ash (FA), calcined clay, metakaolin or granulated blast furnace slag (GBFS) [1,2]. Researches on geopolymers mostly were application of class F fly ash mixed with activator alkali to produce high strength geopolymer paste [4]. In general, mixing class F fly ash and activator alkali with mass ratio of Na_2SiO_3 to NaOH in the range of 1.5-2.0 resulted in high compressive strength of geopolymer concrete. In this case, the consentration of NaOH was eight molars [4]. In some countries the geopolymer concrete are applied as structure members [5]. It is normal in concrete that curing factors have an effect on the compressive strength, as well as

* Corresponding author: triwulan_marwan@yahoo.com

 DOI: 10.1051/matecconf/20179701005

for geopolymer concrete. At elevated temperature of curing, the strength of concrete has tendency to increase.

2 Materials and experiments

The concrete was based on fly ash. The mass ratio fly ash to alkali activator was 65% : 35%, mass ratio of coarse aggregate to fine aggregate was 60% :40%, and mass ratio Na_2SiO_3 to NaOH were varied at 1.5, 2, 2.5. Superplasticizer used was polycarboxilate based with a ratio of 2% by fly ash weight. In this paper, geopolymer concrete were prepared for cylindrical specimens with a diameter of 10 cm and 20 cm high. After casting, the specimens were steam cured at temperature varied at 40°C, 60°C and 80°C. Some tests were performed for compressive strength, tensile strength and elasticity modulus. Some specimen controls were treated at room temperature.

3 Experimental results

3.1 Material analysis.

The results XRD analysis of fly-ash is shown in Figure 1 and Table 1, and XRF result is illustrated in Table 2.

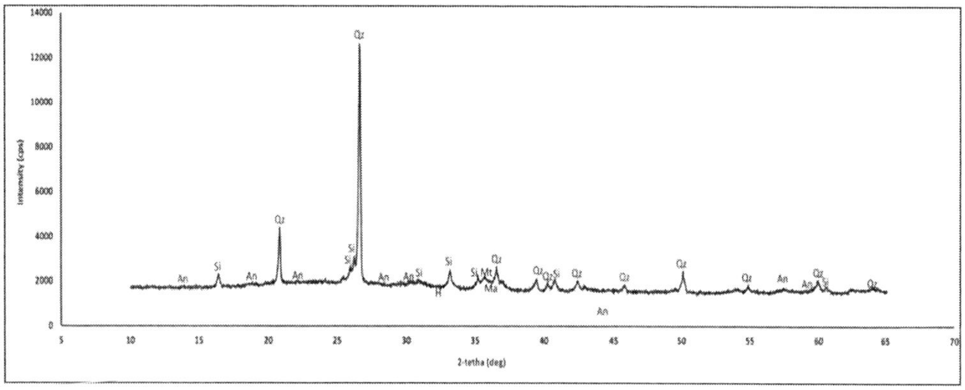

Fig. 1. XRD analysis of fly-ash.

Table 1: Chemical compounds of fly ash from XRD analysis.

Code	Mineral	Chemical Formula
1	Quartz low	SiO_2
2	Alumunium Oxide	Al_2O_3
3	Magnemite	TiO_2
4	Hematite HP, iron(III) oxide	Fe_2O_3
5	Anorthite, sodian	$Na_{0.45}Ca_{0.55}Al_{1.55}Si_{2.45}$

As shown in Figure 1 and Table 1 the main mineral of fly-ash are quartz, aluminum oxide, iron oxide and anorthite.

DOI: 10.1051/matecconf/20179701005

As shown in Table 2, the total quantity of three main oxides ($SiO_2 + Fe_2O_3 + Al_2O_3$) of fly-ash is more than 70%. Therefore, the material is classified as pozzolanic material according to ASTM C618-2012.

Table 2. Chemical composition of fly ash by XRF analysis.

No.	Oxide Contain	Percentage (%)
1.	SiO_2	48,47
2.	Al_2O_3	26,05
3.	Fe_2O_3	12,54
4.	TiO_2	0,92
5.	CaO	5,18
6.	MgO	2,77
7.	Cr_2O_3	0,02
8.	K_2O	1,66
9.	Na_2O	0,47
10.	SO_3	1,05
11.	Mn_2O_3	0,19

3.2 Compressive strength

The specimens were tested for compressive strength at 28 days. The results is shown in Figure 2.

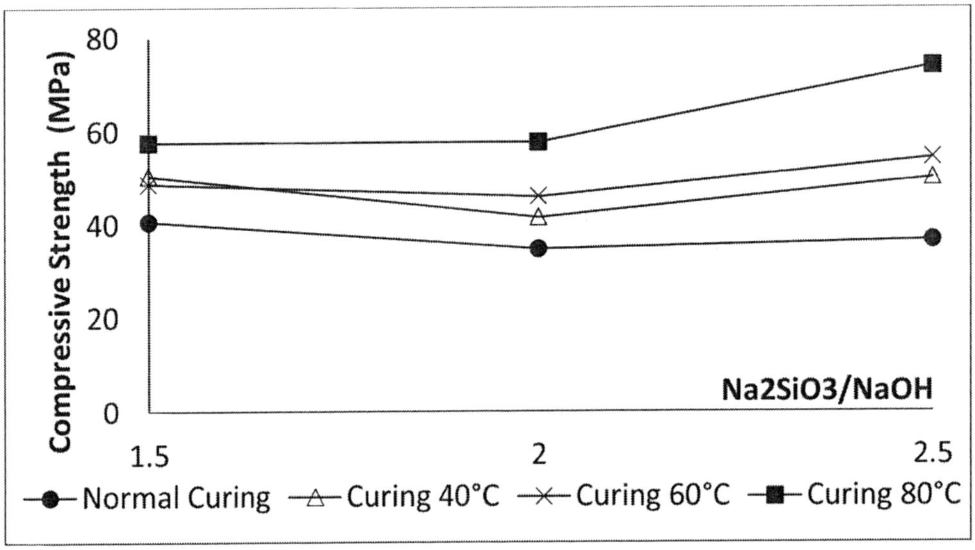

Fig. 2. Influence of curing temperature to the strength.

It is clearly as illustrated in Figure 2 that increasing of curing temperatures improves mechanical strength. The highest compressive strength is achieved at a temperature of 80°C. Increase of strength at elevated temperature is contributed by acceleration of geopolymerization reaction [1]. Another study explained that the maximum of compressive strength obtained with the curing temperature of 60-80°C with a duration of 6-12 hours [5]. The increasing of compressive strength due to the formation of the crystal lattice at 2Θ angle in the mids of 5°-14°. Increasing of curing temperature even increases compressive strength of the concrete with GGBTS (Ground granulated Blas Furnace Slag) as substitution [6]. With the normal curing there were formation of crystall quarts, mullite and magnetite on the 2Θ angle in the mids of 28°-35° [7]. In this research, geopolymer-based fly ash was more appropriate with a mixture of alkaline activators Na_2SiO_3 and NaOH than only with NaOH or Na_2SiO_3 .However, with the increasing mass ratio of Na_2SiO_3 to NaOH the increasing of compressive strength relatively small. The highest compressive strength was achieved at a ratio of Na_2SiO_3/ NaOH = 2.5, equal to the results of previous study [8]. It seems that contribution of soluble silica from alkaline activator takes part in the geopolymerization process in improving the ratio of Si to Al. It significantly effects to developing of mechanical properties of concrete [9]. Other study proved that the influence of curing temperature changed the formation of the crystal [10, 11] , but at longer duration of curing time causes less impact in the formation of crystals.

3.3 Split tensile strength

The specimens were tested for splitting tensile strength at 28 days. The results are shown in Figure 3.

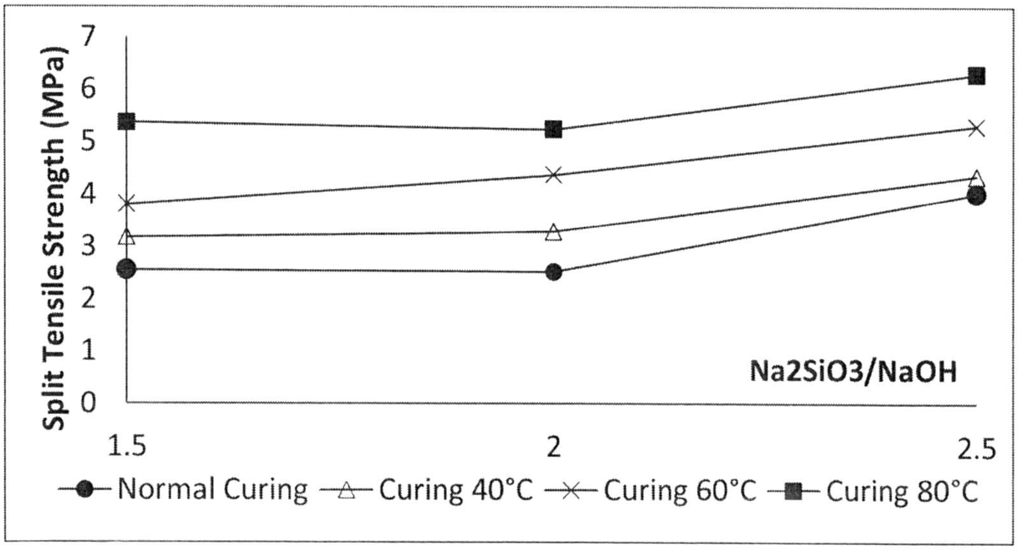

Fig. 3. Influence of curing temperature to the split tensile strength.

Illustration in Figure 3 reveals that the effect of curing temperature in split tensile strength has an equal tendency to the strength. The increasing of temperature curing increases the split tensile strength as well. The highest split tensile strength was achieved at a temperature of 80°C. The rising of curing temperature even increases the split tensile strength for concrete with GGBFS (Ground Granulated Blas Furnace Slag) as substitution [12]. The highest influence ratio of Na_2SiO_3/NaOH on split tensile strength was achieved on the ratio 2.5, similar to the results of previous study [13].

DOI: 10.1051/matecconf/20179701005

3.4 Closed porosity

The specimens were measured to obtain closed porosity 28 days. The results are shown in Figure 4.

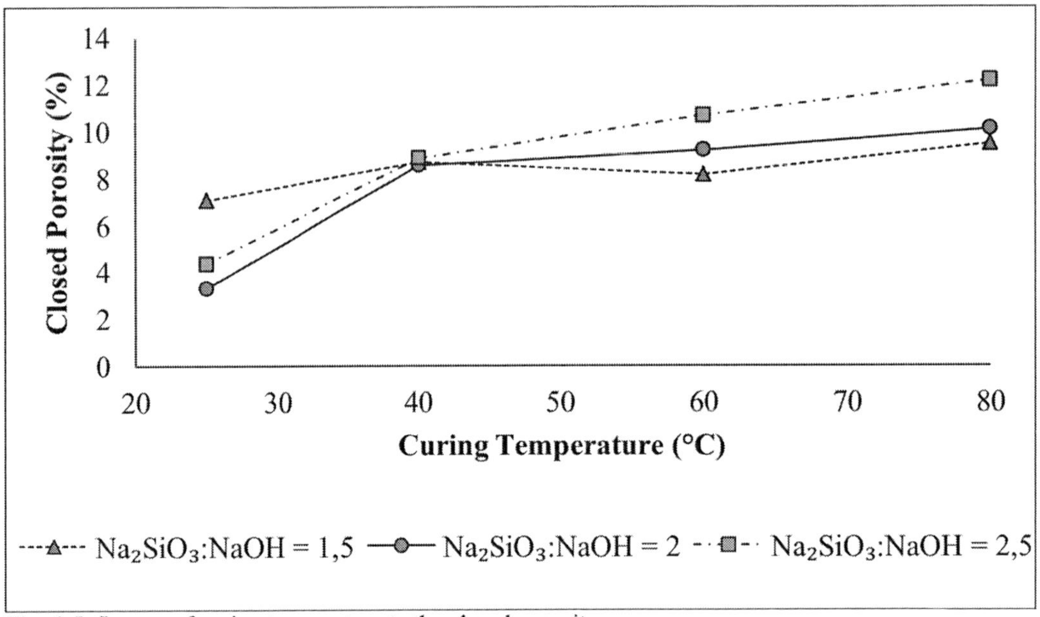

Fig. 4. Influence of curing temperature to the closed porosity.

The influences of increasing curing temperature to the closed porosity have equal tendency to the results of compressive strength. The increasing of curing temperature increases the percentage of closed porosity as well.

3.5 Elasticity modulus

The mixtures were tested for determine the modulus of elasticity. The result is provided in Figure 5.

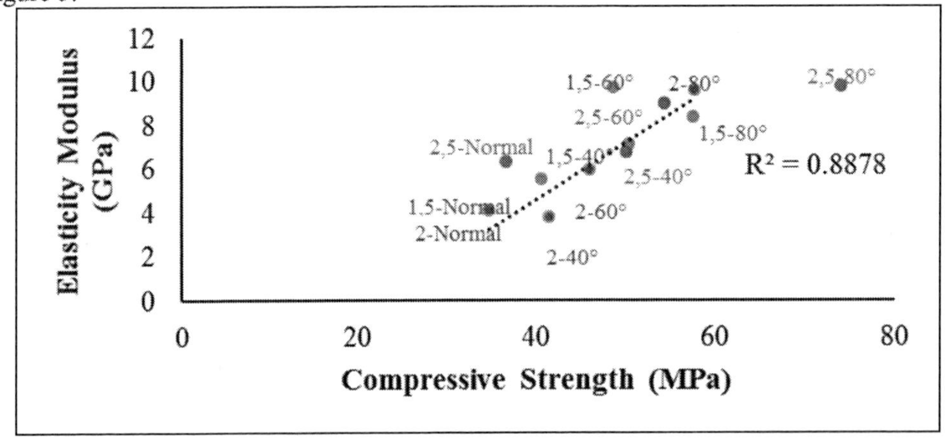

Fig. 5. Graph of elasticity modulus for all mixtures.

Notes: There are two signs on the chart (numbers – letters = **x-y**) or (numbers – numbers = **x-y**).
x = are the ratio of Na_2SiO_3/NaOH, **y** = are the curing conditions

Due to the limitations of existing equipment, each mixture was analyzed for one specimen at 28 days. It is not simple to discuss all existing graph. However it can be explained that increasing temperature of curing promotes the elasticity modulus. The results were corresponding to the results for compressive strength and split tensile strength.

4 Conclusions

Curing temperature has a significant effect to increase compressive strength and split tensile strength. The curing temperature enhances the percentage of closed porosity and Elastic Modulus as well. The maximum ratio of Na_2SiO_3 to NaOH to increase the optimum mechanical properties is 2.5. In this composition, the increase of compressive strength is up to 100%, split tensile strength increased up to 56%, and modulus of elasticity increased up to 53%. At a ratio of Na_2SiO_3 to NaOH = 2 the increase of compressive strength is up to 65%, split tensile strength increased up to 100%, and the increase of modulus of elasticity was up to 133. For the ratio of Na_2SiO_3 to NaOH = 1.5 the increase of compressive strength up to 42.5%, the split tensile strength was up to 101%, and the increase of the modulus of elasticity was up to 51

References

1. J. Davidovits, *Geopolymer: Chemistry & Applications* (Institut Geopolymere, France, 2008).
2. S. Kumar, R. Kumar, S.P. Mehrotra, J. Mater. Sci., **45**, 607 (2010)
3. D. Hardjito, E.W. Steenie, M.J.S Dody, B. Vijaya Rangan, ACI Material Journal,**6**, 467 (2004)
4. J.J. Ekaputri, Triwulan, S. Junaedi, H. Fansuri, R. Bayuaji, Materials Science Forum **803**, 63 (2015)
5. Frantisek Skvara, Tomas Jilek, Lubomir Kopecky , Ceram. Silik., **49**, 195 (2005)
6. J.J. Ekaputri, K. Maekawa, T. Ishida, Mater. Sci. Forum, **857**, 305 (2016)
7. Tanakorn Phoo-ngernkham, Akihiro Maegawa, Naoki Mishima, Shigemitsu Hatanaka, Prinya Chindaprasirt, Constr. Build. Mater., **91**, 1 (2015)
8. Triwulan, P. Wigestika , J.J. Ekaputri, *International Conference on Civil Engineering Research* , (2016)
9. M. Rowles, B. O'Connor, J. Mater. Chem., **13**, 1161 (2003)
10. E. Arioz, O. Arioz, O. M. Koçkar, Int. J. Chem. Eng. Appl., **496,** 423 (2013)
11. S. Apodaca Garcia, S.P. Arredondo Rea, J.M. Gomez Soberon, J.L. Almaral Sanchez, R. Corral Higuera, Int. J. Mater. Sci. Eng., **2**, 50 (2015)
12. S. Chithra, G. Dhinakaran, International Journal of ChemTech Research, **6(2)**, 1516 (2014)
13. M.D.J. Sumajouw, B.V. Rangan, (Curtin University of Technology, 2006)

DOI: 10.1051/matecconf/20179701006

Properties of Palm Oil Fuel Ash (POFA) Geopolymer Mortar Cured at Ambient Temperature

Monita Olivia[1,2,*], *Lora* Mona Tambunan[1] and *Edy* Saputra[3]

[1]Department of Civil Engineering, Faculty of Engineering, Universitas Riau, Pekanbaru, Riau 28293, Indonesia.
[2]Konsorsium Riset Geopolimer Indonesia (KORIGI), Laboratorium Beton dan Bahan Bangunan Kampus ITS Sukolilo Surabaya 60111, Indonesia.
[3]Department of Chemical Engineering, Faculty of Engineering, Universitas Riau, Pekanbaru, Riau 28293, Indonesia.

Abstract. Geopolymer material needs high temperature curing to produce good microstructure, high strength, and durable product. However, curing at ambient temperature is more preferable and practical in application for cast in situ geopolymer. In order to allow curing at ambient temperature, the geopolymer is mixed with mineral additives that has high calcium content such as slag, Ordinary Portland Cement (OPC) and high calcium fly ash. In this study, the Ordinary Portland Cement (OPC) was added in the Palm Oil Fuel Ash (POFA) geopolymer mortar to induce setting and hardening at ambient temperature. Setting time, compressive strength and porosity of the POFA geopolymer mortar were measured. The OPC was added into the geopolymer at dosages of 0%, 20%, 25%, 30%, 35%, and 40%. The alkaline activator used was a combination of NaOH (16M) and sodium silicate with a ratio of 2.5 by mass. The POFA geopolymer mortar were cast and cured at ambient temperature. Results show that addition of 35% OPC increased the setting time by 99.44%, increased the compressive strength of mortar by 95.46% and decreased the porosity by 5.27% at 28 days. It can be concluded that inclusion of the OPC could improve the setting and final strength of the geopolymer material.

1 Introduction

Geopolymer is a binder from material contains silica and alumina and activated by alkaline solution. The geopolymer paste usually needs high temperature curing to accelerate geopolymerization and activate silica to produce strong and dense matrix. Kovalchuck, et al. [1] stated that geopolymer paste hardens very slowly at ambient curing; therefore, this type of concrete is usually subjected to high temperature curing. It was found that high temperature curing in the range of 60-90^{0}C is necessary to complete the amorphous aluminosilicate process [2]. Heat will accelerate the reaction in a geopolymer mixture. However, curing of material at high temperature, it is impractical in application since it needs

* Corresponding author: monita1306@yahoo.com

special methods or equipments. In this case, geopolymer product is considered costly related due to high temperature curing.

Additives that have high CaO content such as slag, high calcium fly ash and OPC were reported included to accelerate geopolymer curing at ambient temperature. Slag contains high CaO, hence Deb et al. [3] and Nath and Sarker [4] suggested of using slag to increase reactivity and hydration process, reduce setting time, assist workability and increase early strength properties of geopolymer concrete at ambient temperature. Furthermore, instead of slag, the commercial OPC that contains more CaO than slag (approximately of 65%) was found practical to decrease setting time, and reduce the curing time and temperature. In one study, it was found that the optimum mixture for curing at ambient temperature consisted of 40% alkaline activator and addition of 5% OPC to replace fly ash (by total binder) could produce mortar with compressive strength more than 50 MPa [5]. Inclusion of OPC to the mixture was stated could improve setting time, mechanical properties and durability of the fly ash geopolymer [6].

Previous studies about geopolymer material containing Ordinary Portland Cement (OPC) as aluminosilicate source and additive were reported. Initially, the geopolymer was mixed with the OPC to produce OPC-geopolymer composite with stronger properties than the OPC paste alone [7]. In other study, the OPC was added in the geopolymer to produce mortar with the highest strength of 43.1 MPa at room temperature [8]. Approximately 5-12% of the OPC (by cement weight) was included in the geopolymer paste mixture only to assist the geopolymerization process by generating heat during hydration [9]. The mixtures were also cured at $60^{O}C$ for 24 hours. It was reported that the inclusion of OPC in the mixtures could reduce setting time, porosity and water absorption with the increase of OPC content. There was also improvement in compressive strength and elastic modulus since the OPC could modify the microstructure and pore of the geopolymer paste. Another study that investigate the effect of using OPC in high calcium geopolymer mixture at different curing regime (membrane curing, temperature curing and wet curing) was reported by the same research group. The geopolymer samples cured in tempeture curing ($40^{O}C$ for 24 hours) showed better physical properties and high early compressive strength [10].

Utilization of source materials that rich in silica and alumina content has driven usage of agro-industrial biomass such as Palm Oil Fuel Ash (POFA). POFA as geopolymer main source material in Southeast Asia has become popular due to abundant supply from palm oil plantations. POFA is produced from combustion of empty fruit bunch and palm oil kernel. Some studies were carried out to investigate the suitability of POFA as geopolymer's main material and suggested that POFA is a potential source material for the geopolymer due to high silica content [11,12]. Furthermore, local POFA from Riau Province could be used to produce gepolymer concrete with compressive strength of 22.58 MPa [13]. Although curing of geopolymers at ambient temperature is important in application, there are still few published studies on POFA geopolymer cured in ambient temperature using the OPC to accelerate curing. The current study aims to investigate the effect of using OPC at various percentage in setting time, compressive strength, and porosity of the POFA geopolymer cured at ambient temperature.

2 Material and methods

Palm Oil Fuel Ash (POFA) from Dumai, Riau, was used in this study. Ordinary Portland Cement (OPC) type I was used to accelerate curing at ambient temperature. The chemical composition the Ordinary Portland Cement (OPC) and POFA is presented in Table 1.

Table 1. Chemical composition of OPC and POFA.

Components	Percentage [%]	
	Ordinary Portland Cement (OPC)*	Palm Oil Fuel Ash (POFA)**
SiO_2	20.92	54.53
Al_2O_3	5.49	15.90
Fe_2O_3	3.78	3.86
MgO	-	1.21
CaO	65.21	7.49
Na_2O	-	0.09
K_2O	-	4.31
MnO	-	0.07
P_2O_5	-	4.90
SO_3	-	0.87
Cu	-	0.02
Zn	-	0.01

*Salain 2009, **PSD Geologi Bandung

Alkaline solution used was a combination of sodium silicate (Na_2SiO_3) and sodium hydroxide (NaOH). NaOH concentration was 16M and the ratio of NaOH/sodium silicate used was 2.5. The alkaline activator dosage was based on the recommendations from the previous study [14]. Sodium silicate with a SiO_2/Na_2O ratio of 2.19 in a 53.7% liquid and density of 1.56 was mixed with the dissolved sodium hydroxide (Na_2O) 99% wt. Fine aggregates from Kampar river in saturated surface dry condition with fine modulus of 2.83 and specific gravity of 2.65 were mixed with binders. Superplasticizer contains napthlene sulphonated formaldehyde from Sika was included in the mixture to improve workability. The OPC was included as 0%, 20%, 25%, 30%, 35%, and 40% in the mixtures GP1, GP2, GP3, GP4, GP5 and GP6, respectively.

Table 2. Mixture composition.

Mixes	Quantity [kg/m^3]					SP	w/s
	Sand	POFA	OPC	SS	NaOH		
GP1 (OPC0)	1364	495.73	0	330.49	165.24	7.44	0.57
GP2 (OPC20)	1364	495.73	99.15	330.49	165.24	7.44	0.57
GP3 (OPC25)	1364	495.73	123.93	330.49	165.24	7.44	0.57
GP4 (OPC30)	1364	495.73	148.72	330.49	165.24	7.44	0.57
GP5 (OPC35)	1364	495.73	173.51	330.49	165.24	7.44	0.57
GP6 (OPC40)	1364	495.73	198.29	330.49	165.24	7.44	0.57

OPC = OPC Percentage as additive, R = ratio of Na_2SiO_3/NaOH, SS = Sodium Silicate, SP = superplasticizer, w/s = water/solid

The mortar specimens were cast in 50x50x50mm mould and cured at room temperature. Those specimens were demoulded after 24 hours and placed in the lab until testing date. The measured properties were setting time, compressive strength (at 7 and 28 days), density and porosity.

3 Results and discussion

3.1 Setting time

Table 3 shows the initial and final setting times measured for POFA geopolymer mortar at various OPC content. It can be seen that the initial and final setting time of mix GP1 (OPC 0%) mortar at ambient temperature was much longer than other mixture specimens (GP2-GP6). The POFA geopolymer mixture certainly needs high temperature curing to assist fast geopolymerization process, and curing at ambient temperature delay the paste activation. It can be observed that the setting time for all POFA geopolymer mortars decreased with the increase of the OPC percentage in the mixtures. Both mixes GP5 and GP6 that have shown a rapid increase of setting time than the control mix (GP1). This was confirmed in the previous study that stated fly ash geopolymer mixtures with calcium from the OPC as an additive yielded fast setting concrete [5], which has been also observed for the POFA geopolymer mortar in this research. It was stated that the calcium content is responsible to increase solidification rate and rapid hardening by providing extra nucleation sites for precipitation of dissolved species in geopolymer system [15]. An increase in setting time rate with calcium additive will be beneficial for the POFA geopolymer mortar as repair binding material or plaster in situ.

Table 3. Initial setting and final setting of the geopolymer POFA at different OPC percentages.

Mix	OPC Percentages [%]	Initial setting time [min]	Final setting time [min]
GP1	0	2860	3900
GP2	20	62	240
GP3	25	41	135
GP4	30	24	105
GP5	35	16	90
GP6	40	13	75

3.2 Compressive strength

Figure 1 shows the compressive strength of POFA geopolymer mortar at various OPC content cured at ambient temperature. At 28 days, the compressive strength of the mortar was in the range 11.73 to 22.93 MPa. It can be seen that the compressive strength increased with concrete age for all mixes. A gain in compressive strength of the mortar continued until GP5 (OPC content of 35%), before it decreased for mix GP6 (OPC content of 40%). The compressive strength of GP5 has increased by 95.46% compared to the GP1 (control mix). This could be due to the increase amount of calcium in the geopolymer mixes that form additional Calcium Silicate Hydrate (CSH) in the matrix. The increase percentage of CaO up to 35% has altered microstructure of geopolymer gels that could finally enhance the physical, mechanical and properties. The increase in compressive strength was related to the OPC and pozzolanic reactions to form additional CSH in the matrix [5, 7, 9]. However, when the OPC content was 40%, the compressive strength started to decline. An increase in OPC content enhanced strength development of the geopolymers, but excess calcium caused rapid setting time, and precipitation of aluminosilicate gel altogether with CSH at the very early stages, resulting in lower strength mortars.

Chemical composition of POFA determine the type of geopolymer product and its interaction with CaO in the system. For example, POFA used in the previous study by Salih et al. [16, 17] consists of SiO_2, CaO and small content of Al_2O_3. Thus, the POFA geopolymer mix produced more CSH than aluminosilicate product due to higher calcium content than the aluminium content. This low Al content lead to slow setting and low strength development of the POFA geopolymer. In this research, the POFA used has relatively higher Al_2O_3 (15.90%) than CaO content (7.90%). Hence, the aluminosilicate product could be more dominant than the CSH gels in the geopolymer network. Inclusion of OPC with a reasonably high CaO content (65.21%) in the mix certainly increased CSH amount, setting and strength development in the POFA geopolymer mortar cured at ambient temperature. This is also probably due to the OPC that generates internal heat inside the specimens. The exothermic heat could enhance further formation of geopolymeric gel and subsequently will improve the mechanical strength of the geopolymers. Similar observation was reported by Moon et al. [18].

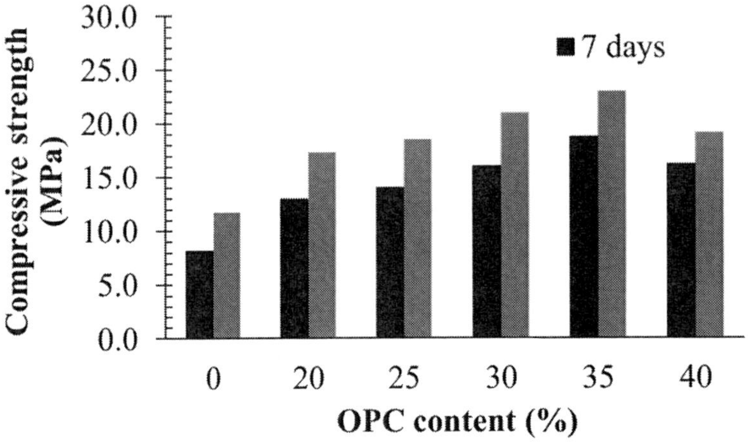

Fig. 1. Compressive strength at 7 and 28 days due to variation of OPC content.

3.3 Porosity

Porosity for all mixtures is presented in Figure 2. There was a decline trend of the mortar, showing a slight change in porosity. The porosity values were in the range of 20.73-22.09%. It can be seen that the OPC mortar (control mix GP1) had the highest porosity (22.09%) compared to other geopolymer mixes. The OPC content in the mortar had a good impact for porosity since it increased the calcium content in the geopolymer to produce CSH gels. An increase of calcium content could increase in specimen compressive strength due to increase in the amount of geopolymer product within each sample. On the other hand, the porosities of other mixes were 20-73-21.66%, respectively. Inclusion of the OPC has definitely increased hydration products and amount of CSH in the mixtures. The CSH products developed and filled the pores and thus reduced the porosity of the mortars, as can be seen for mix GP5 that has shown a decreased in porosity by 5.27%. Pangdaeng et al. [10] reported similar trend for porosity of the fly ash geopolymer mortar using 5-15% of OPC cured at ambient temperature. It was stated that an extra participation of CSH from the calcium (OPC) improve the microstructure and decrease porosity, indicating by dense and compact matrix.

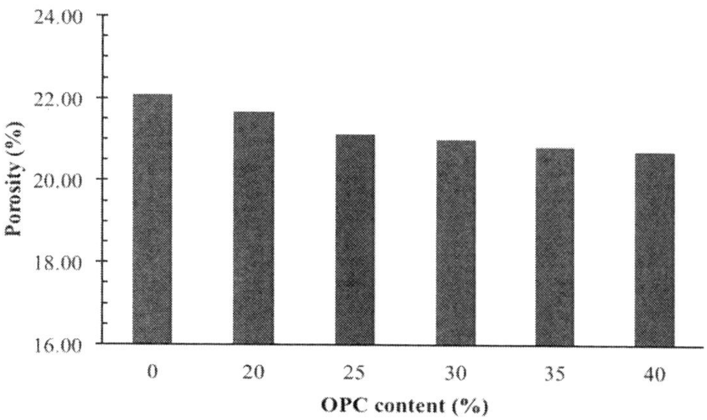

Fig. 2. Porosity at 7 and 28 days due to variation of OPC content.

The effect of inclusion of OPC to assist POFA geopolymer mortar curing at ambient temperature was investigated in this research. Previous study stated that calcium during geopolymerization had two roles [6]. The first role is to aid aluminosilicate dissolution from fly ash particles. The second role of calcium is as counter-balancing cation in the geopolymer pore structure by incorporating calcium in the geopolymer network system. Another role of calcium in the geopolymer mix was observed to increase internal heat from the exthermic reaction of OPC hydration that could assist ambient curing condition [19]. In this research, the optimum combination obtained was using NaOH 16M concentration, ratio of $NaOH/Na_2SiO_3$ of 2.5, using 35% OPC as additive that will produce mortar with compressive strength of 22.93 MPa and porosity of 20.82%. It can be summarized that inclusion of the OPC could improve the setting, final strength and porosity of the POFA geopolymer mortar.

4 Conclusions

The present study investigated the use of various percentage of OPC in POFA geopolymer for curing at ambient temperature. The OPC mortars (GP1) was used as control specimens. It could be seen that there was a decrease in setting time, an increase in compressive strength and a decrease in porosity of POFA geopolymer at 7 and 28 days. The OPC addition in the geopolymer system increased the calcium content in the mixture and produced Calcium Silicate Hydrate (CSH). The additional CSH improved mictrostructure and strength development of the geopolymer. In this research, the optimum combination obtained was using NaOH 16M concentration, ratio of $NaOH/Na_2SiO_3$ of 2.5, using 35% OPC as additive that will produce mortar with compressive strength of 22.93 MPa and porosity of 20.82%. It can be summarized that the OPC inclusion will improve setting, final strength and porosity of the POFA geopolymer mortar.

The authors would like to acknowledge the Geopolymer Research Group Universitas Riau, all member of Final Year Project Group Batch 2 and 3, and the Structural Engineering and Concrete Laboratories, Faculty of Engineering, Universitas Riau.

References

1. G. Kovalchuck, A. Fernandez Jimenez, A. Palomo, Fuel, **86**, 315 (2007)
2. D. Hardjito, S.E. Wallah, D.M.J. Sumajouw, V. Rangan, ACI Mater. J., **101**, 467 (2004)
3. P.S. Deb, P. Nath, P.K. Sarker, Mater. Des., **62**, 32 (2014)
4. P. Nath, P.K. Sarker, Constr. Build. Mater., **66**, 163 (2014)
5. P. Nath, P.K. Sarker, Cem. Concr. Compos., **55**, 205 (2015)
6. G.M. Canfield, J. Eichler, K. Griffith, J.D. Hearn, J. Mater. Sci., **49**, 5922 (4014)
7. J. Tailby, K.J.D. MacKenzie, Cem. Concr. Res., **40**, 787 (2010)
8. A. Nazari, H. Khammohammadi, M. Amini, H. Hajiallahyari, A. Rahimi, Mater. Des., **41**, 43 (2012)
9. T. Phoo ngernkham, P. Chindaprasirt, V. Sata, S. Pangdaeng, T. Sinsiri, Int. J. Min. Met. Mater., **20**, 214 (2013)
10. S. Pangdaeng, T. Phoo ngernkham, V. Sata, P. Chindaprasirt, Mater. Des., **53**, 269 (2014)
11. M.A. Salih, A.A.A. Ali, N. Farzadnia, Constr. Build. Mater., **64**, 591 (2014)
12. M.J.A. Mijarsh, M.A.M. Johari, Z.A. Ahmad, Constr. Build. Mater., **52**, 473 (2014)
13. M. Olivia, A. Kamaldi, I.R. Sitompul, I. Diyanto, E. Saputra, Mater. Sci. Forum, **803**, 110 (2015)
14. D. Hardjito, S.E. Wallah, D.M.J. Sumajouw, B.V. Rangan, ACI Mater. J., **101**, 467 (2004)
15. W.K.W. Lee, J.S.J. van Deventer, Colloids Surf., A **211**, 115 (2002)
16. M.A. Salih, A.A.A. Ali, N. Farzadnia, Constr. Build. Mater., **65**, 592 (2014)
17. M.A. Salih, N. Farzadnia, A.A.A. Ali, R. Demirboga, Constr. Build. Mater., **93**, 289 (2015)
18. J. Moon, S. Bae, K. Celik, S. Yoon, K. Kim, K.S. Kim, P.J.M. Monteiro, Cem. Concr. Compos., **53**, 2118 (2014)
19. T. Suwan, M. Fan, N. Braimah, Constr. Build. Mater., **114**, 297 (2016)

MATEC Web of Conferences **97**, 01007 (2017) DOI: 10.1051/matecconf/20179701007
ETIC 2016

On The Physico-Mechanics, Thermal and Microstructure Properties of Hybrid Composite Epoxy-Geopolymer for Geothermal Pipe Application

Ira Firawati[1,*], *Jasruddin*[1], and *Subaer*[1]

[1]Laboratorium Fisika Material, FMIPA, Universitas Negeri Makassar Jalan Daeng Tata Raya, Makassar 90223, Indonesia

Abstract. The objective of this study is to determine the effect of epoxy resin on the physico-mechanics, thermal and microstructure properties of geopolymers hybrid composites for geothermal pipe application. Hybrid composite epoxy-geopolymers pipes were produced through alkali activation method of class-C fly ash and epoxy resin. The mass of epoxy-resin was varied relative to the mass of fly ash namely 0% (SPG01), 5% (SPG02), 10% (SPG03), 15% (SPG04), and 20% (SPG05). The resulting materials were stored in open air for 28 days before conducting any measurements. The densities of the product composites were measured before and after the samples immersed in boiling water for 3 hours. The mechanical strength of the resulting geothermal pipes was measured by using splitting tensile measurement. The thermal properties of the pipes were measured by means of thermal conductivity measurement, differential scanning calorimetry (DSC) and fire resistance measurements. The chemical resistance was measured by immersing the samples into 1M H_2SO_4 solution for 4 days. The microstructure properties of the resulting materials were examined by using x-ray diffraction (XRD) and Scanning Electron Microscopy-Energy Dispersive Spectroscopy (SEM-EDS). The results of this study showed that hybrid composite epoxy-geopolymers SPG02 and SPG03 are suitable to be applied as geothermal pipes.

1 Introduction

The needs for energy consumption is increasing rapidly every year and forcing new efforts to search for new and renewable energy resources [1]. One of the energy resources which readily available in many countries, including Indonesia is geothermal. Kamojang in West Java Province is one of the geothermal sites in Indonesia, which releases hydrothermal fluid with a temperature between 180 °C to 220 °C [2]. The hydrothermal fluid moves upward and reaches the surface of the earth after passing the drill of impermeable stones [3].

* Corresponding author: ira.firawati28@gmail.com

The hydrothermal fluid requires heat and chemical corrosion resistant pipes known as geothermal pipes [4]. Geopolymer is inorganic polymer which has good mechanical properties, excellent fire and heat resistant and able to stand on strong chemical attacks [5-8]. Geopolymers can be synthesized from pure aluminosilicate minerals such as kaolinite and clays or industrial waste such as fly ash, furnace slag and red mud [9,10].

Fly ash is an inorganic material comprising of active SiO_2 and Al_2O_3 species as a result of coal burning in power plants and has been used as geopolymers raw material for many years [10, 11, 12]. Fly ash is abundantly available and categorized as dangerous waste and hence requires careful and thorough handling. Data from the Ministry of Environment shows that in 2006, Indonesia produced 52.2 tons fly ash per day and bottom ash as high as 5.8 tons per day. This figure is increasing in recent years [13, 14].

Geopolymers can be fabricated as a hybrid composite through the addition of certain organic material into the geopolymers paste such as commercial epoxy-resin [15, 16]. The used of phenolic resin and Mekpo hardener for example was found to improve the heat resistance of producing geopolymers. Beside that epoxy-resin has been used in space craft industries to develop fire resistance exterior composites [16].

2 Experimental

Geopolymers were produced from class C-fly ash activated with alkali solution and mixed manually until the mixture was homogenous. Epoxy-resin (Union®) was added to the geopolymer paste by varying its mass relative to the mass of fly ash. The samples destination were SPG01 (0% epoxy-resin), SPG02 (5% epoxy-resin), SPG03 (10% epoxy resin), SPG04 (15% epoxy-resin), and SPG05 (20% epoxy resin). The mixture was molded to produce pipe shaped hybrid composite samples having an outer diameter of 4.3 cm and inner diameter of 3.3 cm. The samples were cured at 70°C for 2 hours before remolded. The produce composites were then stored in open air for 28 days before any measurements were performed [17]. The splitting tensile measurements were conducted by using Tokyo Testing Machine (TTM) with a maximum load of 1000 kN. The chemical resistance of the sample was measured by immersing the sample into 1M H_2SO_4 solution for 4 days, the length of the time was chosen to be enough for measurement. Thermal properties of the samples were examined through thermal conductivity measurement, differential scanning calorimetry (DSC) (Perkin Elmer DSC 4000 series), and fire resistance by using torcher flame 1500°C for 30 minutes. The microstructures of the raw and produced materials were examined by using Mini-Flex II x-ray diffraction (XRD) and Tescan SB3 Scanning Electron Microscopy-Energy Dispersive Spectroscopy (SEM-EDS).

3 Results and discussion

Figure 1 shows examples of hybrid composites, epoxy-geopolymer produced in this research with a length of 9.0 cm and prepared for various measurements. It can be seen that the surface of all samples is smooth, shining and free from apparent pores.

DOI: 10.1051/matecconf/20179701007

Fig. 1. Samples of hybrid composites, epoxy-geopolymer pipes

Table 1 shows the density of the samples before and after immersing them into boiling water for 3 hours. It can be seen that the density of hybrid composite epoxy resin-geopolymer is lower than pure geopolymers (SPG01). It was also observed that there were no physical damage or cracks occured on all samples after 3 hours in boiling water although their densities were slightly increased except for sample SPG05.

Table 1. The density, splitting tensile strength and thermal conductivity of hybrid composites, epoxy-geopolymer samples

Sample	**Density (g/cm^3)**		**Splitting Tensile Strength (MPa)**	**Thermal Conductivity (W/m)**
	Before soaked in boiling water	**After soaked in boiling water**		
SPG01	2.13	2.24	2.56	0.34
SPG02	1.96	2.02	3.33	0.32
SPG03	1.96	1.95	2.22	0.31
SPG04	1.88	1.92	2.22	0.31
SPG05	1.91	1.87	1.91	0.33

The splitting tensile strength of the samples was measured horizontally by using the following equation [18].

$$P = \frac{2F}{d \, x \, L} \tag{1}$$

Where, F is applied force (N), d is the thickness of the pipe (mm), and L is the length of the pipe (mm).

The measuring results showed that best splitting tensile strength of the samples is 3.33 MPa for sample SPG02 with the addition of 5% epoxy resin relative the mass of fly ash. The splitting tensile strength of the rest of the samples decreases as the mass of epoxy resin increase. These results were expected since epoxy resin is an organic material with low mechanical strength. Similar measurement has been conducted by [19] and achieved the splitting tensile strength for their samples 1.10 MPa.

The thermal conductivity (k) measurements were performed by using a steady state method with the following equation [8],

 DOI: 10.1051/matecconf/20179701007

$$k = \frac{Q \, x \, l}{A \, x \, \Delta T} \tag{2}$$

Where, Q is heat energy (W), l is length (mm), A is cross section (mm^2), and ΔT is temperature difference.

Table 1 shows that the thermal conductivity of all samples ranges from 0.31 to 0.33 W/mK which means that the ability of the material to conduct heat is low. The thermal conductivity of geopolymer can be tailored depending on raw material, chemical composition of alkali, type and size of aggregate (minerals) and the volume of pores [10].

The acid resistance was examined by soaking the sample into 1 M H_2SO_4. Figure 2 shows the picture of the sample after 4 days. Thin layer of epoxy resin appeared to peel out from SPG04 and SPG05 while other samples remain physically unchanged. These results indicate that the maximum addition of epoxy-resin into the network of geopolymer in order to stand the acid attack was 15%.

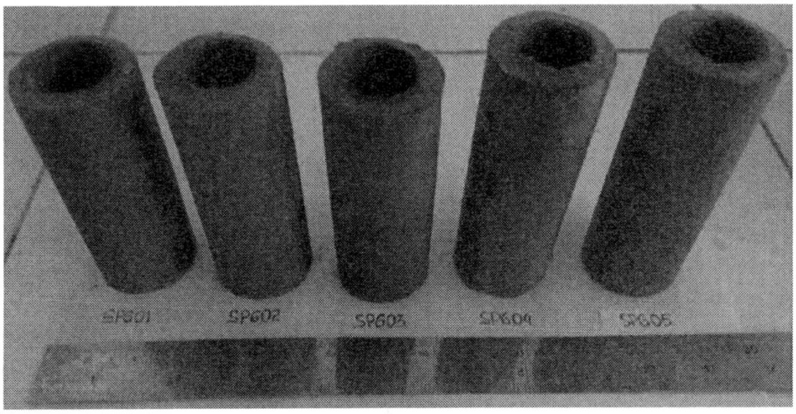

Fig. 2. The condition of hybrid composites, epoxy-geopolymer after soaking in 1 M H_2SO_4 solution for 4 days

Differential Scanning Calorimetry (DSC) measurements were performed to study the nature and the ability of the composites to absorb or release heat from its network. The measurements were conducted at a temperature range between 30 to 400°C with a heating rate 20°C/minute. Figure 3 is the results of DSC measurements for all samples showing a single broad exothermic curve. The peaks of exothermic curve and the enthalpy (ΔH) decrease as the mass of epoxy resin increase. This indicates that the presence of epoxy resin reduces the ability of the composites to inhibit heat and fire.

 DOI: 10.1051/matecconf/20179701007

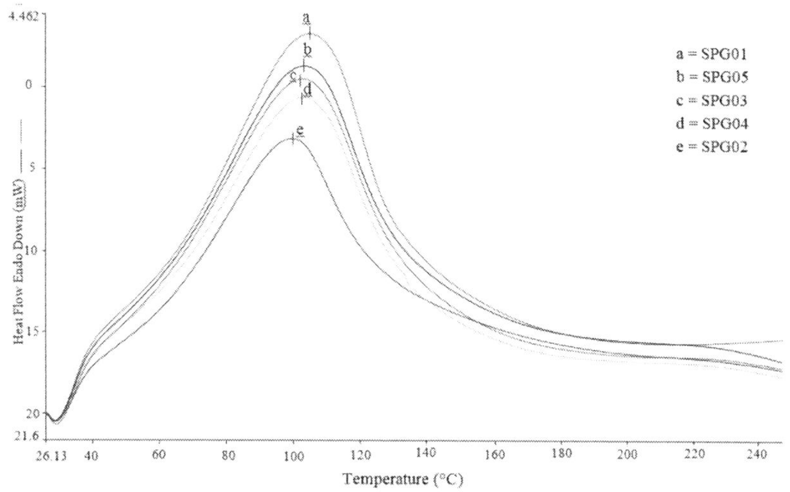

Fig. 3. DSC results of as-prepared hybrid composite epoxy-geopolymers

Table 2. The result of DSC of as-prepared hybrid composite epoxy- geopolymers

Sample	Peak (^{0}C)	Peak Height (mW)	ΔH (J/g)
SPG01 (a)	108.80	-18.5412	-157.089
SPG05 (b)	103.43	-16.3115	-133.647
SPG03 (c)	102.75	-15.6946	-123.202
SPG04 (d)	102.72	-15.2520	-121.717
SPG02 (e)	99.67	-15.2252	-108.911

Figure 4 shows the results of fire resistant measurement of the composite. The measurements were performed by using a torcher flame reaching 1500°C in a few minutes. The surface of the samples were exposed to fire for 30 minutes and a thermocouple was inserted into the inner diameter of the sample to register the temperature inside the pipe. The results showed that sample SPG01, SPG02, and SPG03 were able to resist the temperature between 200 – 800°C without burning or significant physical damage. Sample SPG04 and SPG05 which contain 15% and 20% of epoxy-resin were cought on fire and released the smoke at temperature around 400°C. Figure 5 shows the photograph of the samples after fire resistance measurements. It can be seen that sample SPG01 which did not contain epoxy resin formed a new phase on its surface after 30 minutes exposed to fire. Sample SPG03 which contain 10% epoxy-resin was found to be the best composite produced for high temperature application. This sample showed to be excellent as fire and chemical resistance.

DOI: 10.1051/matecconf/20179701007

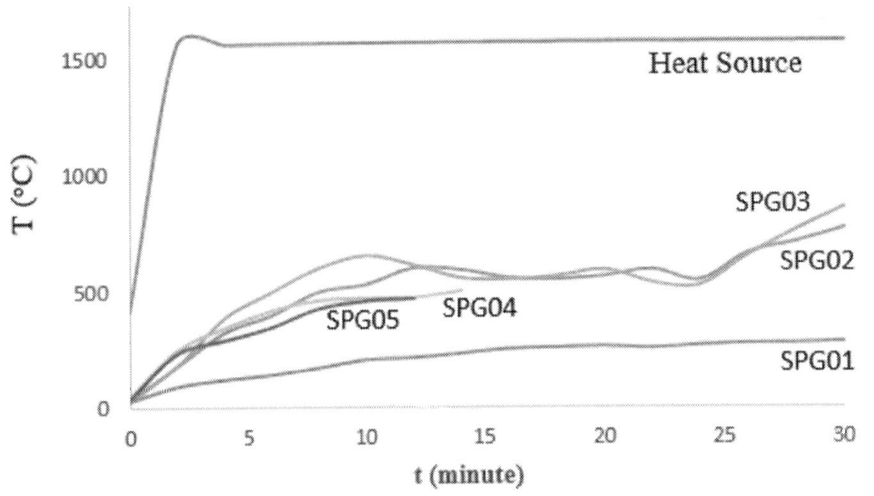

Fig. 4. Temperature versus time during fire resistance measurement of hybrid composites, epoxy-geopolymer

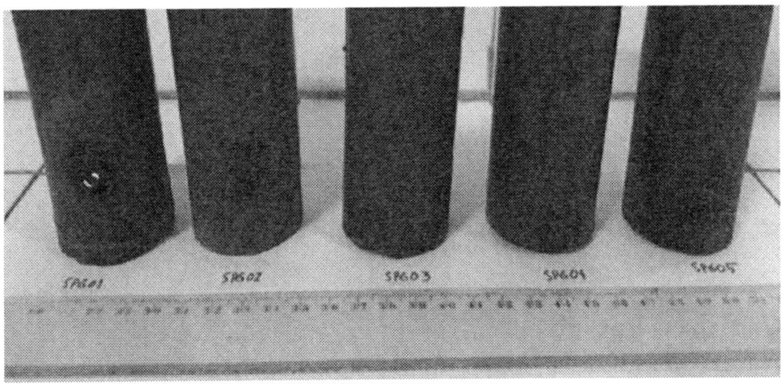

Fig. 5. Photograph of samples surface after fire testing

XRD examinations were performed for as prepared samples and after the samples were soaked in 1M H_2SO_4 solution for 4 days. This was intended to examine the formation of any new compound (phase) due to strong acid attack on the surface of the samples. Diffractogram of the samples before and after soaking in the acid solution is shown in figure 6 (A) and (B), respectively. It can be seen clearly that XRD results did not show any discernible differences between the samples before and after acid attack. This result indicates although few layers of epoxy peeled out from the surface of the sample, the structure of polymeric network of hybrid composite epoxy-geopolymer did not change due to the intrusion of H_2SO_4 solution, at least during the time scale of measurement.

DOI: 10.1051/matecconf/20179701007

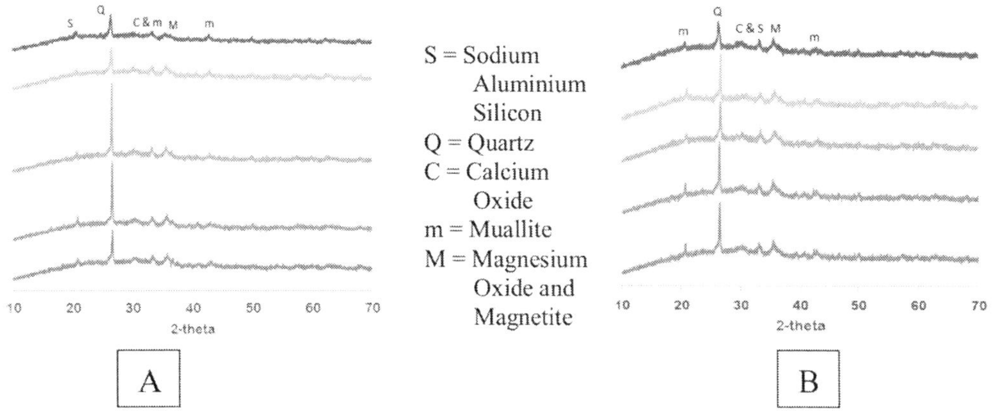

Fig. 6. The diffractogram of hybrid composite epoxy-geopolymer; (A) before, (B) after soaking in 1 M H_2SO_4 solution for 4 days

Microstructure examination of the samples before and after acid attack was performed by using SEM-EDS. Figure 7 shows SEM micrographs for as-prepared samples. The micrographs shows that epoxy resin create a strong mechanical bond with the geopolymer matrix. The morphology of the composites also shows the presence of substantial unreacted fly ash particles which is normally observed on geopolymers made from fly ash. Figure 8 shows the surface morphology of the sample after immersing in 1M H_2SO_4 solution for 4 days. It can be seen that the attack of strong acid did not form any new phase on the surface of the sample. The surface morphology of all samples remains similar to those of as-prepared sample. This indicates that at the time scale of the measurement, H_2SO_4 did not affect the integrity of geopolymer structure. Further study needs to be conducted by immersing the samples into a strong acid solution for much longer time.

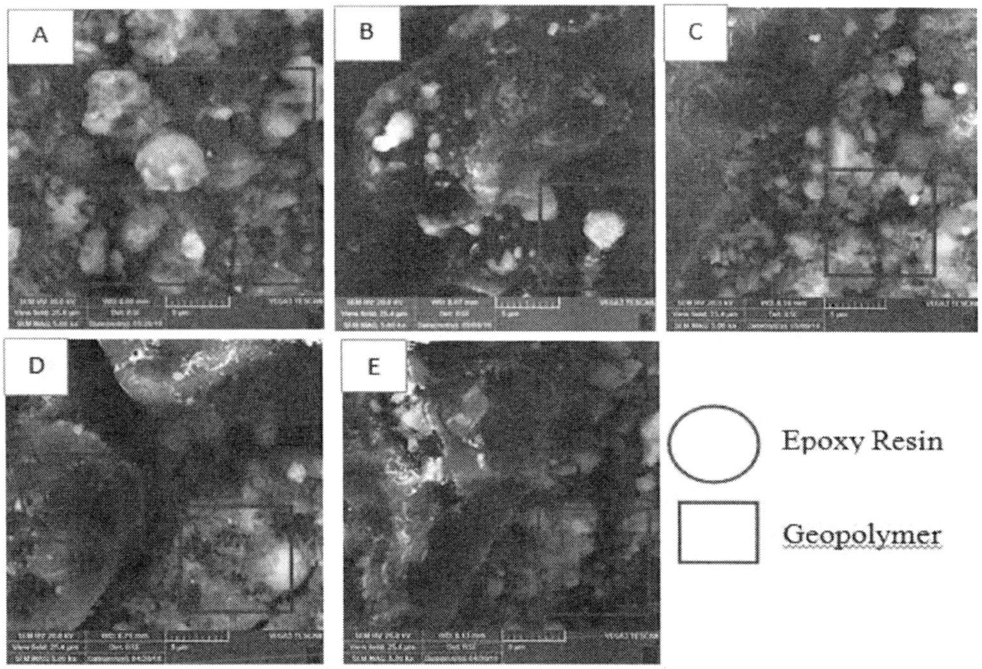

Fig. 7. SEM image of hybrid composite epoxy-geopolymer before soaking in 1 M H_2SO_4 solution for 4 days (A)SPG01 (B)SPG02 (C)SPG03 (D)SPG04 (E)SPG05

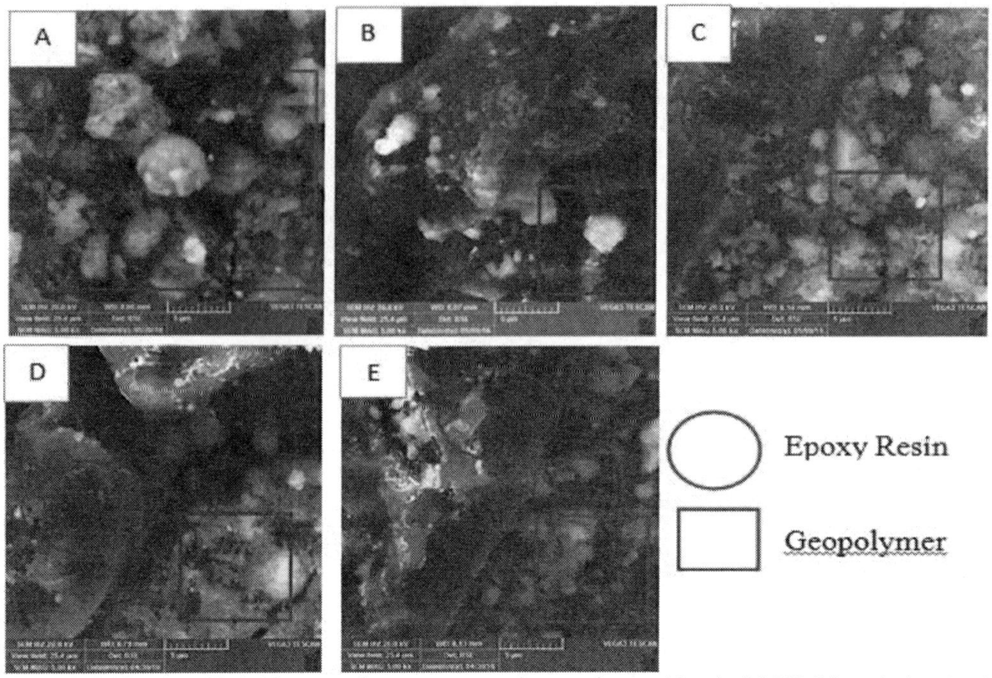

Fig. 8. SEM image of hybrid composite epoxy-geopolymer after soaking in 1 M H_2SO_4 solution for 4 days (A)SPG01 (B) SPG02 (C)SPG03 (D)SPG04 (E)SPG05

4 Conclusion

Hybrid composite epoxy resin-geopolymers have been successfully produced and characterized. The splitting tensile strength of the composite reached 3.33 MPa for the sample contain 5% epoxy resin. The fire resistance, thermal conductivity and DSC measurements showed that (figure 3) the thermal properties of the composites were suitable to be applied as geothermal pipe. The microstructure examinations for as-prepared samples and after immersing the samples into 1M H_2SO_4 solutions for 4 days showed that the produced composites remain stable and did not show the formation of a new phase or physical damage. The results of this study indicate that hybrid composite epoxy resin-geopolymer offer a high potential to be applied as geothermal pipe.

References

1. ESDMMAG. *Eksplorasi Panas Bumi; Potensi Energi Baru Terbaharukan; Upaya Mengentaskan Kemiskinan Energi.* Media Komunikasi Kementerian Energi dan Sumber Daya Mineral (2012)
2. D. Yudiantoro, S. Yuwono, I. Takashima, D. Ishiyama, Y. Kamah, J. Hutabarat, Indonesian Journal on Geoscience, **8** (2013)
3. S. Suparno, Universitas Indonesia, Jakarta (2009)
4. Bastarina, Annisa Destavia, *Energi Panas Bumi A Present from the Heart of The Earth.* Universitas Indonesia. **1** (2008)
5. A. Haris, A. Irhamsyah, J.J. Ekaputri, Mater. Sci. Forum., **841**, 170 (2016)
6. V. Sreevidya, R. Anuradha, D. Dinakar, R. Venkatasubramani, Int. J. Eng. Sci. Res. Technol., **4** (2012)

7. Luhar, Salmabanu, Urvashi Khandelwal, Salmabanu Luhar, Int. J. Eng. Res. Ind. Appl., **8**, 17 (2015)
8. A. Fernandez Jimenez, A. Palomo, Cem. Concr. Res., **35**, 1984 (2005)
9. I. Nikolic, D. Durovic, R. Zejak, L. Karanovic, M. Tadic, D. Blecic, V.R. Radmilovic, J. Serb. Chem. Soc., **78**, 851 (2013)
10. Subaer, *Pengantar Fisika Geopolimer*. Makasar: Badan Penerbit UNM. (2014)
11. Y. Yusuf, Z. Zuki, G. Refnita, *Pengaruh Penambahan Abu Terbang (Fly Ash) Terhadap Kuat Tekan Mortar Semen Tipe PCC Serta Analisis Air Laut yang Digunakan untuk Perendaman,* Prosiding SEMIRATA. **1** (2013)
12. K. Diharjo, B. Kusharjanta, R.A.P. Tarigan, A.R. Andhika, *Pengaruh Kandungan dan Ukuran Serbuk Genteng Sokka terhadap Ketahanan Bakar Komposit Geopolimer.* Rekayasa Mesin, **4**, 27-34 (2013)
13. J. Setiaka, I. Ulfin, N. Widiastuti, *Adsorpsi Ion Logam Cu (II) dalam Larutan pada Abu Dasar Batubara Menggunakan Metode Kolom.* Prosiding Tugas Akhir. (2010)
14. A. Lasryza, D. Sawitri, *Pemanfaatan Fly Ash Batubara Sebagai Adsorben Emisi Gas CO Pada Kendaraan Bermotor.* Jurnal Teknik Pomits. **1** (2012)
15. G. Roviello, L. Ricciotti, C. Ferone, F. Colangelo, R. Cioffi, O. Tarallo, Materials, **6**, 3943 (2013)
16. C.M. Thomson, Jr. Smith, J.W. Connell, P.W. Hergen rother, *Flame Retardant Epoxy Resin*. NASA Research Centre. (2007)
17. B.S. Nasional, *Tata Cara Perhitungan Struktur Beton untuk Bangunan Gedung.* **SNI 03-2847-2002** (2002)
18. SNI (Standar Nasional Indonesia). *Metode Kuat Tarik Belah Beton.* **SNI 03-2492-2002** (2001)
19. A. Elkadi, J. Van Mier, Int. J. Fract., **1**, 55 (2004)

DOI: 10.1051/matecconf/20179701008

Development of Nano TiO_2–Geopolymer Functional Composite as Antifouling Bricks

Nurul Kusuma Wardani[1,*], *Andi* Ikhsan Maulana[1], *Nurfadilla*[1], *Subaer*[1]

[1]Department of Physics, FMIPA, Makassar State University, Makassar 90233, Indonesia

Abstract. The purpose of study is to examine the ability of nano TiO_2 – geopolymer functional composite as antifouling bricks. The samples were synthesized through alkali-activation method at 70^0C for 1 hour by mixing metaclay with TiO_2 nanoparticles and activated with sodium silicate solution. There were two series of samples produced, namely, GT_A with addition of 2% nanoTiO_2 and GT_B with addition of 4% nano TiO_2 relative to the mass of metaclay. The samples were immersed in water and in 1M H_2SO_4 solution for 4 days to examine the resistance of composites in hars environment. The x-ray diffraction (XRD) was performed to examine the chemical compositions of the samples before and after environmental test. The morphology of the samples surfaces was examined by using Scanning Electron Microscopy (SEM) coupled with energy dispersive spectroscopy (EDS). Based on this study, sample GT_A shows its excellent properties as antifouling bricks. The addition of nano TiO_2 was found to improve the quality of geopolymers as a high performance bricks.

1 Introduction

Brick is one of construction materials that have been known by people around the world, both in rural and urban areas. Conventionally, manufacture of bricks need to be burned and dried (traditional bricks) and it may cause air pollution in the surrounding environment [1-2,19]. Traditional bricks as a raw material in the building are vulnerable to weathering both chemically and mechanically, brittle and sensitive to the salting reaction, and directly related to the presence of micro-organisms deposit on the building [3-5, 20-21]. Biofouling can be defined as the accumulation of micro-organisms, plants, and animals which are attached to either temporary or permanent on building facades [6]. One of the promising material which has abundant precursor and easy to produce is geopolymer bricks.

Geopolymer was developed by Davidovits in the late 1978, synthesized through polycondensation of aluminosilicate minerals (e.g., metakaolin, clay, fly ash) with a solution of alkali (e.g., sodium hydroxide and sodium silicate) [7- 9]. Many researchers use clay as raw material for manufacturing geopolymer for bricks application [10-11]. The most important advantages of geopolymer bricks are resistance to acid attack, high early strength, long-term durability, good fire resistance, low manufacturing energy consumption, and low CO_2 emission, which make these novel material to be categorized as "green material"

* Corresponding author: nurulkusumaa@yahoo.co.id

[11-14]. Recently, the latest innovation that provides added value to the geopolymer bricks is to incorporate anti-fungal or anti biofouling material.

Nanomaterials are increasingly developed due to its size (1-100 nanometers) in at least one dimension. Their physical and chemical properties differ from those of bulk materials (controlling matter at atomic level). It can be used as prevention of building materials from organic contamination. Titanium dioxide (TiO_2) is one of the most used material to realize self-cleaning and its general use is attributed to its main features: high catalysis efficiency and chemical stability, inexpensiveness, compatibility with traditional construction [15]. The advantage of using TiO_2 as follows: (i) protective or anti corrosion coating mechanism; (ii) thermal control; (iii) easy to clean; (iv) anti bacterial coating for work surfaces; (v) anti grafitti coating for building and structure, (vi) low-cost [16, 17]. Furthermore, the most importance of nano-TiO_2 is its inert properties, corrosion resistance due to light or chemicals and non-toxic [18].

This study utilizes clay-based geopolymer as the raw material forming the matrix of geopolymer and nano TiO_2 as antifouling agent. This study was carried out to determine the quality and capability of functional composite nano TiO_2 - geopolymer as antifouling bricks.

2 Experimental

This study is a development research that lead to the development of Nano TiO_2 – geopolymer as antifouling bricks through alkaline-activation method. Metaclay was produced from calcined clay at 750^0C for 6 hours as a source of aluminosilicate and TiO_2 (anatase) as antifouling agent (filler). The alkaline solution for geopolymerization was a mixture of sodium hydroxide (NaOH), sodium silicate ($Na_2O.3SiO_2$) and distilled water (H_2O) with a certain concentration. Geopolymer bricks was produced by mixing the basic ingredients of the metaclay with nano TiO_2 and activated with an alkaline solution to obtain a homogeneous paste.

Geopolymer paste was poured into the block-shaped glass mold with size 20 cm x 8 cm x 4 cm and then cured in an oven at 70^0C for 1 hour. After the age of 24 hours, the samples were removed from the mold and tested after 3 days. After the age of 7 days, the samples were immersed in water solution and sulfuric acid (H_2SO_4 1 M). The physical properties (pH levels, bulk density, mass, and colour change) were monitored daily. Durability test have been performed to the geopolymer bricks containing Nano TiO_2, which consist of acid attack and environment test. Samples which have been tested were characterized by XRD to examine the structure of the samples. SEM characterization was used to study the structure and morphology of the sample before and after testing.

3 Results and discussions

After the age of 7 days, Nano TiO_2-Geopolymer samples were tested. Each sample was immersed in solutions of sulfuric acid (H_2SO_4 1 M) and water for 4 days. Table 1 below shows the value of bulk density of each sample before and after testing. Sample geopolymer without nano-TiO_2 was produced by Riska [22] with an average composite strength 58.44 MPa. This research focused on the mechanical strength.

DOI: 10.1051/matecconf/20179701008

Table 1. The Apparent Porosity and Bulk Density before and After Testing

Sample ID	Bulk density (gr/cm^3)			
	Acid Attack Test		**Environmental Test**	
	Before	**After**	**Before**	**After**
Sample GT_A	1.68	1.24	1.65	1.58
Sample GT_B	1.64	0.76	1.68	1.18

Figure 1 shows the appearance of (a) GT_A (b) GT_B before acid attack and environment test of each sample.

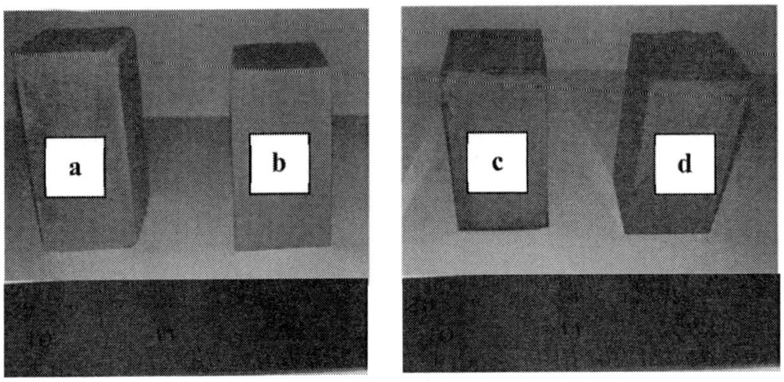

Fig. 1. (i) The appearance of : (a) MC + 2% $NanoTiO_2$; (b) MC + 4% $NanoTiO_2$ before acid attack, (ii) The appearance of (c) MC + 2% $NanoTiO_2$; (d) MC + 4% $NanoTiO_2$ before environmental test

The appearances of each sample before and after testing in strong acid solution 1 M H_2SO_4 and water for 4 days were shown clearly in Figure 2. It can be explained that there are physical change and densification after immersing in acid solution.

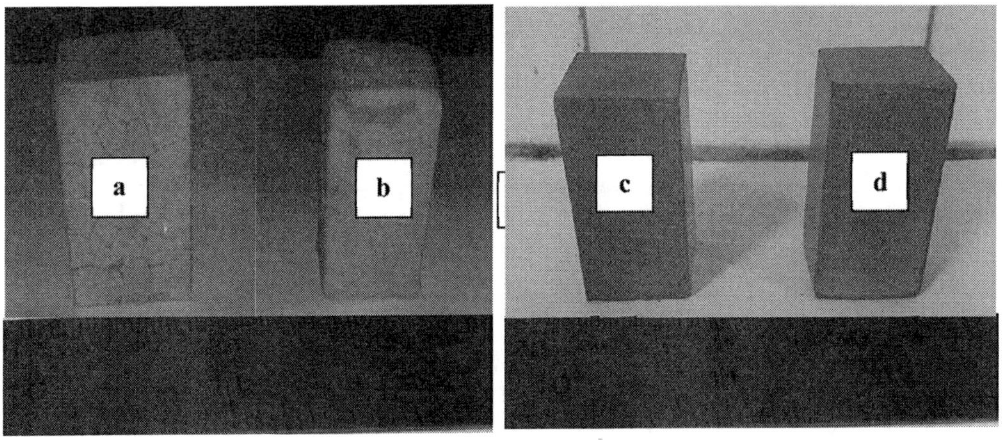

Fig. 2. The appearence of: a) MC + 2% Nano TiO_2; (b) MC + 4% Nano TiO_2 after acid attack, (ii) The appearence of: (c) MC + 2% Nano TiO_2; (b) MC + 4% Nano TiO_2 after environment test

In this study, clay is used as raw material due to its abundant availability in South Sulawesi, Sidrap Regency, Indonesia. Table 2 shows the composition of clay raw material. Based on XRF analysis, it can be shown that the main oxide composition of clay were SiO_2, Al_2O_3, and Fe_2O_3 followed by TiO_2 and K_2O.

Table 2. Composition of Clay as Raw Material Based on XRF Result

Compound	wt (%)	Element	wt (%)
SiO_2	66.140	Si	30.920
Al_2O_3	16.310	Al	8.630
Fe_2O_3	14.780	Fe	10.330
TiO_2	1.310	Ti	0.780
K_2O	0.860	K	0.718
CaO	0.240	Ca	0.172
MnO	0.151	Mn	0.117
ZrO_2	0.018	Zr	0.063
BaO	0.032	Ba	0.028
SrO	0.023	Sr	0.019
Nb_2O_5	0.018	Nb	0.013
MoO_3	0.012	Mo	0.008
ZnO	0.010	Zn	0.008
Y_2O_3	0.007	Y	0.006
In_2O_3	0.005	In	0.004

Figure 3a shows the XRD pattern of meta-clay that has a certain crystalline structure with the highest peak lies between 25^0– 27^0 (2θ). The XRD analysis shows that the main compositions of meta-clay were SiO_2 in the form of *quartz,* magnetite and hematite. Figure 3b shows the characterization results of the Nano TiO_2 and all the peaks can be indexed as anatase form of TiO_2 (the average particle size is 12 nm).

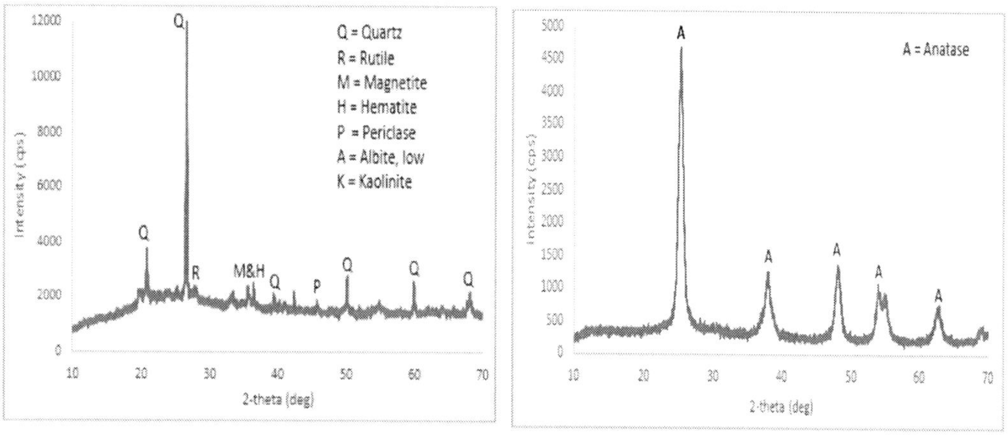

Fig. 3. Phase characterization of (a) metaclay; (b) nano TiO_2

Figure 4 is the result of phase characterization of sample namely GT_A with addition of 2% nano TiO_2 relatively to its precursor mass and GT_B with addition of 4% nano TiO_2. The diffraction pattern shows the sample maintain its level of crystallinity.

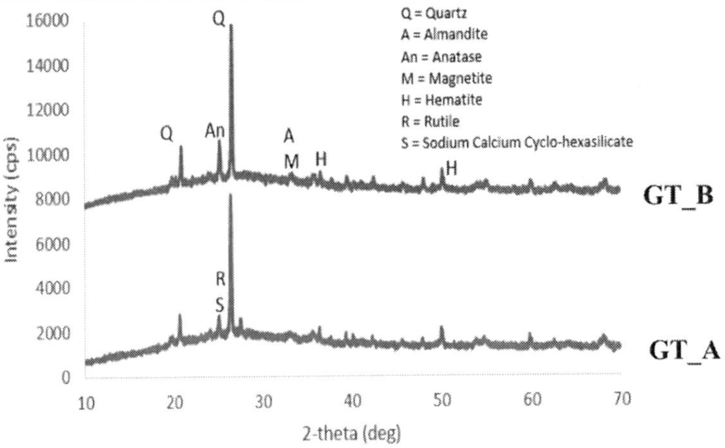

Fig. 4. Phase characterization of (a) GT_A (b) GT_B as prepared samples

Phase characterization shows that the most dominant peaks both of GT_A and GT_B lies between 20^0-30^0 (2θ) which consist of quartz, almandite, magnetite, hematite, rutile, sodium calcium-cyclo hexasilicate joining with anatase. Overall, there is no significant change due to the addition of nano TiO_2 in geopolymer structure. Table 2 shows the chemical composition of the phase of nano TiO_2-geopolymer (in wt%).

Table 3. Chemical composition of the sample in Figure 4 based on XRD measurement

Phase	Sample	
	(GT_A) wt (%)	**(GT_B) wt (%)**
Quartz	72	75
Anatase	8.2	10.6
Rutile	7.2	7
Magnetite	3.4	3.1
Hematite	5.2	5.1
Sodium calcium cyclo-hexasilicate	4	-
Almandine	0.3	-

Figure 5a shows the diffraction pattern of the sample after acid attack and Fig.5b shows the diffraction pattern of the sample after environmental test.

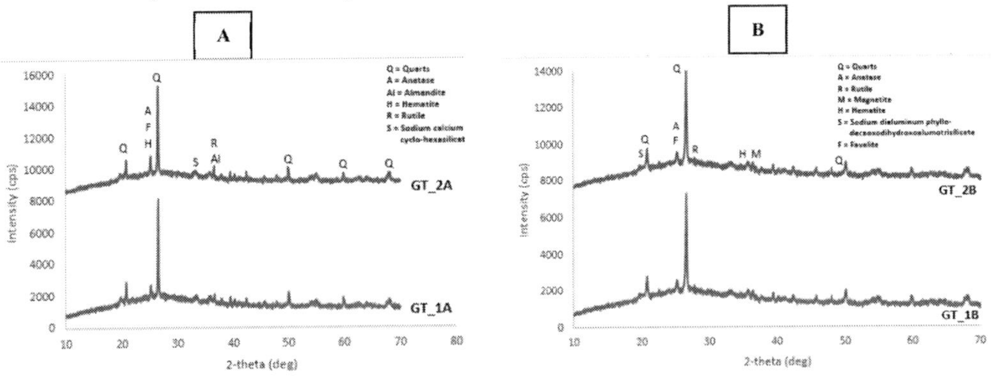

Fig. 5. Phase Characterization of sampel (a) GT_A (b) GT_B after acid attack and XRD pattern of sampel (a) GT_A (b) GT_B after environmental test.

Phase characterization shows that the most dominant peaks both of GT_A and GT_B after acid attack lies between 27^0-30^0 (2θ) which consist of quartz, almandine, magnetite, hematite, rutile, sodium calcium-cyclo hexasilicate and anatase. Figure 5b shows that all

 DOI: 10.1051/matecconf/20179701008

samples consist of quartz, anatase, rutile, hematite, with addition of sodium dialuminium-phyllo decaoxodihydroxo allumotrisilicate and favalite. Similarly, its diffractogram shows that the most dominant peaks after environment test lies between 27^0-30^0 (2θ).

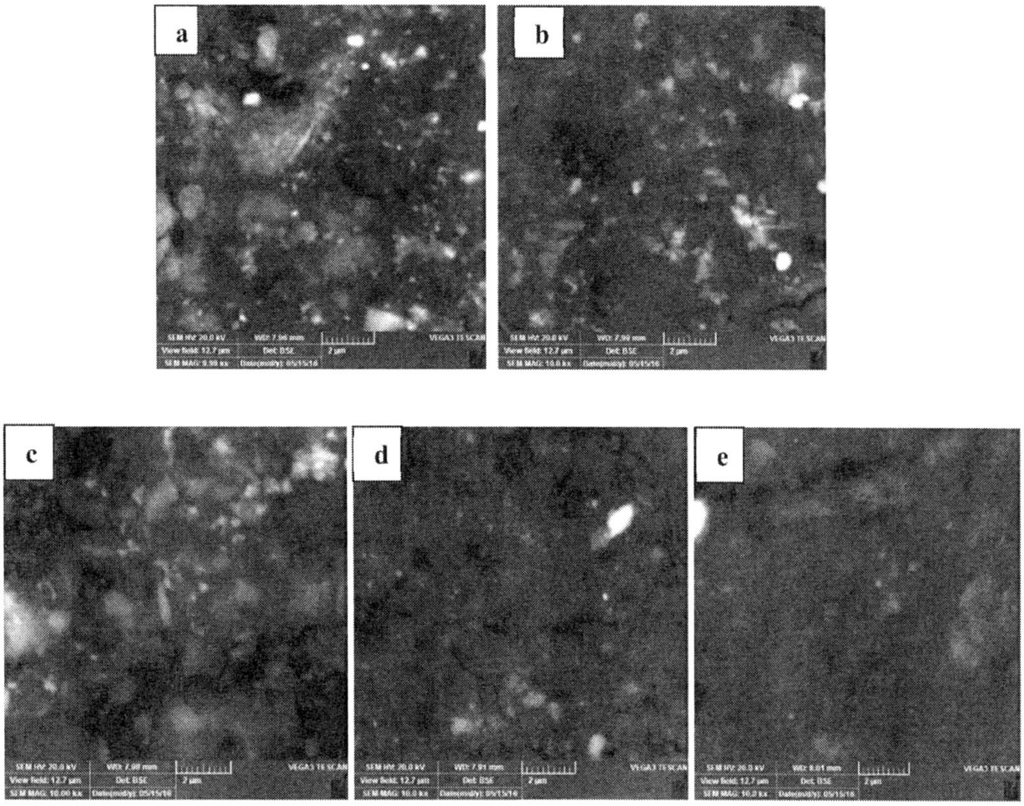

Fig. 6. SEM images of (a) GT_A; (b) GT_B after acid attack; (c) GT after environment testing

Fig.6 clearly shows the SEM image of geopolymers surface displaying the presence of nano TiO_2 represented with white spots. The microstructure of samples, (i) before testing is shown in Fig. 6a and 6b; (ii) after acid attack is shown in Fig. 6c and 6d; (iii) and after environmental test is shown in Fig.6e. It is well known that nano TiO_2 with particle size below 500 nm, can strongly bonded in geopolymer although it was attacked by strong acid solution and hars enviromental [20]. In addition, the most commonly used oxide is the anatase form of TiO_2, which for practical reason of safety and easily for handling and encapsulated in an inorganic matrix [16-19]. The chemical composition of each sample before and after acid attack testing is presented in Table 3.

Table 4. Chemical composition of Nano-TiO_2 – Geopolymer before and after acid attack test based on EDS result

Element	Compound norm.	Comp. C (wt%)		
		Before	After Acid Attack Test	After Environment Test
Sodium	Na_2O	12.10	12.17	11.81
Magnesium	MgO	0.93	1.15	1.31
Aluminium	Al_2O_3	21.78	22.31	24.50
Silicon	SiO_2	53.40	52.15	52.13
Potassium	K_2O	0.38	0.38	0.81
Titanium	TiO_2	4.90	4.70	3.36
Iron	FeO	6.52	6.58	6.26

4 Summary

Functional composite of nano TiO_2 - geopolymers have been succesfully synthesized from a mixture of meta-clay with nano TiO_2 through alkali-activation method at 70 ^{0}C for 1 hour. The addition of nano TiO_2 improved the quality of geopolymer to resist acid attack and hars environment. Based on this research, sample containing 2% nano TiO_2 shows excellent properties as antifouling bricks.

References

1. T.H. Tran, A. Govin, R. Guyonnet, P. Grosseau, C.L.E. Garcia Diaz, D. Damidot, O. Deves, B. Ruot, Int. Biodeterior. Biodegrad., **70**, 31 (2012)
2. K.S. Mohan, T. Srivastava, Biochem Tech., **2**, 210 (2010)
3. L. Graziani, M. D'Orazio, Coatings., 357 (2015)
4. G.V. Gomez de Saravia, Rev. Metal Madrid., 49 (2003)
5. L. Zhang, Constr. Build. Mater., **47**, 643 (2013)
6. M.V. Graham, N.C. Cady, Coatings., **4**, 37 (2014)
7. Assaedi, Journal of Asian Ceramic Societies, **4**, 19 (2016)
8. M.T. Faheem, Key Eng. Mater., **594**, 406 (2014)
9. S. Subaer, A.Haris, N.Nurhayati, A.Irhamsyah, J.J Ekaputri, Mater. Sci. Forum., **841**, 170 (2016)
10. M. Faheem, M.A. Bakri, R. Binhussain, Adv. Mater. Res., **626**, 878 (2013)
11. A.M. Mustafa Al Bakri, H. Kamaruddin, H. Mohamed, C.M. Ruzaidi, A.R. Rafiza, M.T. Faheem, A.M. Izzat, Aust. J. Basic Appl. Sci., **5**, 824 (2011)
12. N. Essaidi, B.Samet, S. Baklouti, S. Rossignol, Ceram.-Silik., **3**, 251 (2013)
13. Nasser, E.M. Mohson, Int. J. Miner., Metall. Mater., **21**, 609 (2014)
14. S. Thokcom, P. Ghosh, S. Ghosh, Asian Research Publishing Network, **4**, 28 (2009)
15. S. Sabihuddin, Int. J. Eng. Res. Ind. Appl., 92-95 (2014)
16. R. Vijayalaksmi, V. Rajendran, Arch. Appl. Sci. Res., 1183-1190 (2012)
17. Gupta, Tripathi, Chin. Sci. Bull., **56**, 1639 (2011)
18. A.M. Ferrari, M. Pini, P. Neri, F. Bondioli, Coatings. 232-245 (2015)
19. K.J.D Mackenzie, Mat. & Sci. Eng., 1-8
20. A.Allahvredi, F.Skvara, Ceram.-Silik. **49**, 225 (2005)

DOI: 10.1051/matecconf/20179701008

21. X.N Thang, D. Kroisova, P. Louda, O. Bortnovsky, P. Prokopcakova, P. Zdobinska, P. Kejzlar, *Environment and Chemical Resistant of Geopolymer Composites*. The 7th International Conferene, TEXSCI (2010)
22. A.Riska, *Pengaruh Suhu dan Waktu Curing terhadap Sifat Mekanik dan Struktur Mikro Geopolimer Berbasis Lempung Laterite*. Skripsi. (2016)

Development of Hybrid Composite Rice Husk Ash (RHA)–Geopolymer for Bricks Bearing Buildings

Andi Ikhsan Maulana[1,2,*], *Nurul* Kusuma Wardani[1,2], *Syamsidar* D[1,2], *Ariani*[1], *Nurfadilla*[1,2], and *Subaer*[1,2]

[1]Material Physics Laboratory, FMIPA, Makassar State University, Makassar 90223, Indonesia
[2]Konsorsium Riset Geopolimer Indonesia (KORIGI), Laboratorium Beton dan Bahan Bangunan, Kampus ITS Sukolilo Surabaya, Indonesia, 6011

Abstract. The aim of this study is to synthesize hybrid composite rice husk ash (RHA)-geopolymer as materials for bricks bearing buildings application. Class-C fly ash was used as raw material and rice husk ash (RHA) taken directly from the field, washed and dried at 100°C for 4 hours. Geopolymers were synthesized through alkali activation method at 60°C for 1 hour with its molar oxide ration of $SiO_2/Al_2O_3 = 3.0$, $Na_2O/SiO_2 = 0.2$, and $H_2O/Na_2O = 10$. There were three series of samples produced by mixing fly ash with addition of 5%, 10% and 15% RHA (relative to the mass of fly ash). The samples were tested after 7 days. The highest compressive strength were 26 MPa which is reached by samples containing 10% RHA. The fire resistance measurement showed that the addition of RHA improved fire retardancy of geopolymers. The samples were immersed in 1 M H_2SO_4 solution for 4 days for acid resistance examination. The X-Ray Diffraction was performed to examine the chemical compositions of the samples before and after the test took place. Scanning Electron Microscopy (SEM) was performed to examine the morphology of these samples surfaces. The addition of 10% RHA in geopolymer showed its excellent properties in terms of mechanical strength, thermal properties, and acid resistance. The study suggests that hybrid composite RHA-geopolymer and can be developed as bricks bearing buildings.

1 Introduction

The advancement of material science and structural technology has fulfilled our needs on better and safer buildings. Nowadays, the main materials used for buildings or other structural products are bricks and Portland cement. Traditional bricks used for buildings normally have low mechanical strength, low heat, fire and strong chemical resistance. On the other hand, the needs to substitute the use of Portland cement have been increasing in recent decades. The production of Portland cement has been recognizing as one of the main contributor of greenhouse effect [1-3]. One of the promising environmentally friendly materials is geopolymers [4].

* Corresponding author: andi.ikhsan.maulana@gmail.com

Geopolymer was first introduced by Davidovits in 1978 as inorganic polymers based on aluminosilicate minerals activated with alkali solution [4]. Geopolymers has been recognized as fire and acid resistant material, and show high early compressive strength similarly with Portland cement [2-6]. The excellent properties of geopolymers have placed these materials suitable to be used as structural materials [6]. Fly Ash (FA) has been used to produce geopolymers for many applications including concrete and bricks. FA produced by some coal-burning industries particularly in Indonesia is categorized as class-C. FA is an industrial waste rich with silica and alumina and its production is increasing every year causing serious environmental problems. Utilizing this waste as geopolymers precursor will turn FA from dangerous material into green and high quality material [1-2, 7-8].

Rice Husk Ash (RHA) is a one of the main source of silica (SiO_2) [4, 9,10]. This material is suitable to be used as fillers in the production of Portland cement or geopolymers. A mixture between RHA and FA will result in organic-inorganic hybrid composite [11]. The use of rice husk ash in bricks-making can provide several advantages such as improved compressive strength, reduce material costs, reduce environmental impact of waste materials, and reduce carbon dioxide emissions [1, 7, 10].

Bearing bricks is a model of bricks preparation that can be locked to each other. This model can reduce the use of cement that binds the bricks in the cavity of the bricks. The advantage gained by this model are; (a) reducing the use of cement, (b) no building frame is needs. Figure 1 shows that the appearance of bricks. This brick has 3 holes on it and Figure 1c shows that these bricks bind by cement through the holes. There is one hole that not filled by cement. The function of the center hole is to reduce the heat transfer and to act as wave and vibration absorbance.

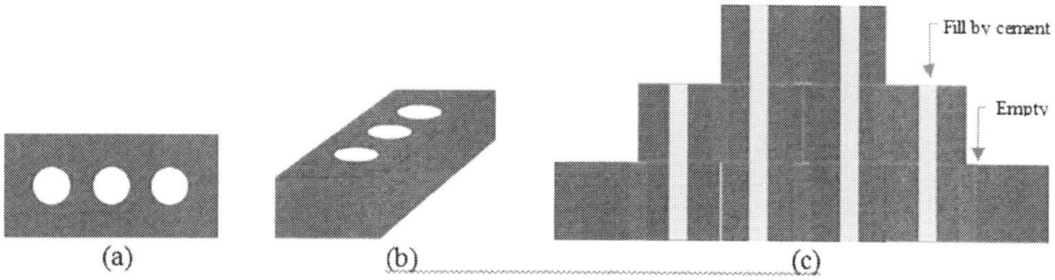

Fig. 1. Model of bricks: (a) from the top; (b) in 3D and (c) formation and cement filling

2 Experimental

This study is a development research that leads to the development of bricks bearing building model using class-C Fly Ash (FA) added with Rice Husk Ask (RHA) through alkali-activation method. RHA was washed and oven dried at 100°C for 4 hours. A mixture of sodium hydroxide (NaOH), sodium silicate ($Na_2O.3SiO_2$), and distilled water H_2O with a certain concentration was used as an alkaline solution. Geopolymer was synthesized by mixing fly ash with variation of 5%, 10%, and 15% RHA relative to the mass of fly ash.

Geopolymer paste was poured into the beam-shaped glass mold and then cured at 60°C for 1 hour. After the age of 24 hours, the samples were removed from the mold and tested after 7 days Mechanical properties of samples were performed to investigate its compressive strength. Thermal measurement was performed to investigate the resistance of samples ability from fire and heat. The chemical resistance of the samples was measured by immersing the sample into the sulphuric acid solution (1 M H_2SO_4) for 4 days. Samples which have been tested were chacterized by XRD to determine the compounds and crystalline structure. SEM-EDS to determine the morphological structure and its containing element of starting material and samples.

3 Results and discussion

Table 1 shows the magnitude of samples bulk density and apparent porosity measured by using Archimedes method. The magnitude of bulk density of the samples decrease as the content of RHA increase. Similarly, the apparent porosity also decreased as the RHA content increase. Similar result was reported by Safiuddin et. al. [12] who found that the porosity of their samples decrease as the the RHA content increase.

Table 1. The density and porosity of sample hybrid composite rice husk ash-geopolymer

Sample	Density (g/cm^3)	Porosity (%)
FA-ASP 5%	2.08	13.70
FA-ASP 10%	2.07	6.59
FA-ASP 15%	2.03	6.16

In this study, RHA is used as filler because it has high silica content. Fig. 2 shows the XRD characterization of rice husk ash with the highest peak lies between $17^0 – 20^0$ (2θ). The XRD analysis indicates that the main composition of were SiO_2 (97 wt%) minerals in the form of tridymite and crystobalite, and 3 wt% TiO_2 (Rutile).

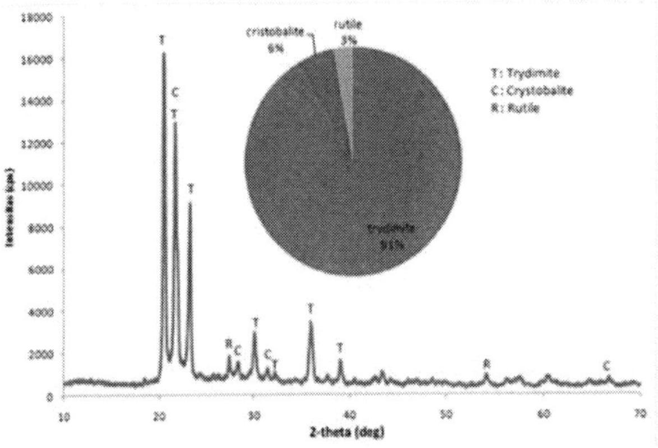

Fig. 2. X-Ray diffractogram of RHA

XRD diffractograms of fly ash as raw material shows the most dominant peak lies between $20^0–30^0$ (2θ) which consist of quartz, magnesium dialuminate, mullite, calcium peroxide and magnetite. The XRD pattern also shows that fly ash is a mixture of amorphous and crystalline phases. The main oxides in FA is quartz (32 wt%) and magnetite (29 wt%).

DOI: 10.1051/matecconf/20179701009

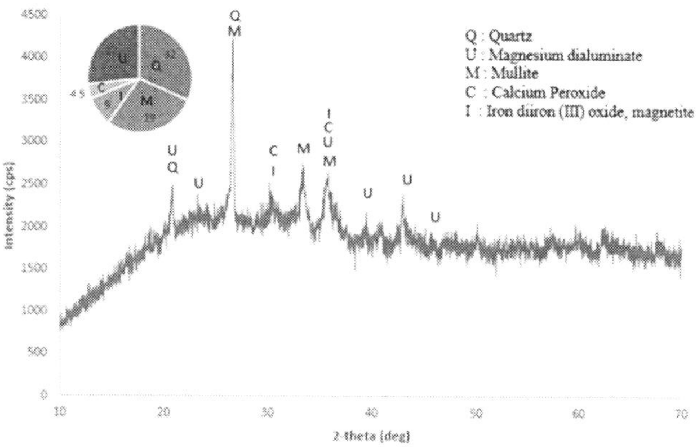

Fig. 3. X-Ray diffractogram of fly ash

Fig. 4 shows the diffraction pattern of three samples (with 5%, 10%, and 15% variation of RHA) hybrid composite rice husk ash-geopolymer with the highest peak lies between 20^0–30^0 (2θ). It shows that the addition of RHA can increase the peaks of trydimite and cristobalite.

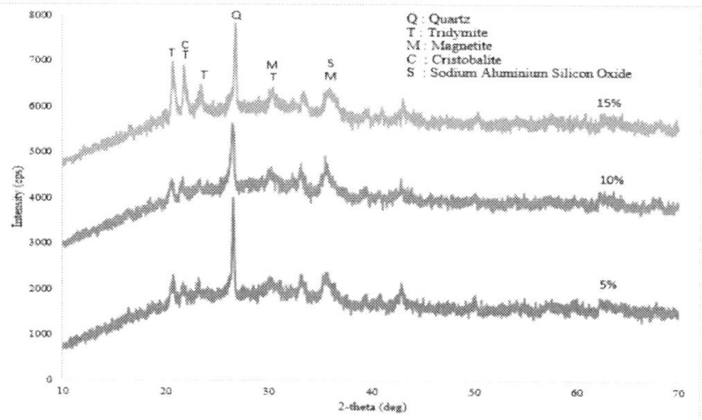

Fig. 4. X-Ray diffractogram of hybrid composite rice husk ash-geopolymer with RHA variation of 5%, 10%, and 15%

Fig. 5 shows the morphology of the samples with addition of RHA. It can be seen that there are some FA and RHA particles which did not fully react in alkaline solutions. In addition, it was also found that micro-cracks appeared on the surface of the samples which is categorized as secondary cracks caused by polishing and vacuum during the preparation of the sample for SEM-EDS examination.

 DOI: 10.1051/matecconf/20179701009

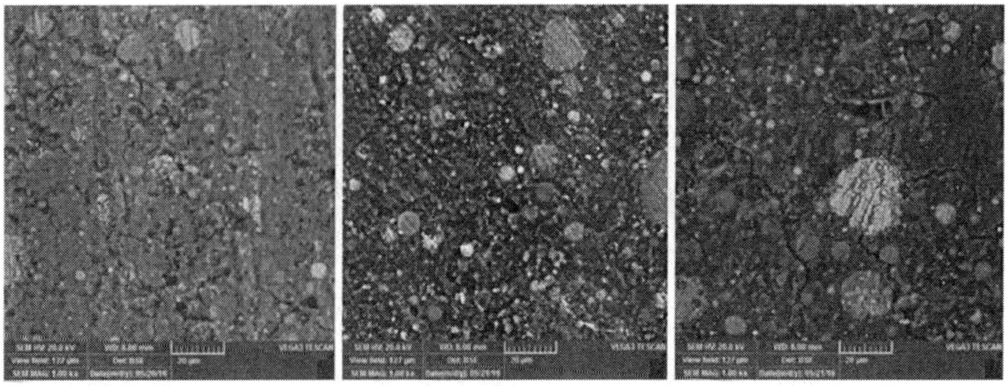

(a) (b) (c)

Fig. 5. SEM image of geopolymer with the addition of: (a) 5%; (b) 10%; and (c) 15% of RHA

Table 2 shows that the compressive strength of the sample contain 15% RHA was declined. The highest compressive strength was 26. 21 MPa obtained on the sample containing of 10% RHA. Nurfadilla [8] found that geopolymers made from FA have compressive strength about 23.4 MPa. The addition of RHA into geopolymer structure was found to increase the magnitude of compressive strength up to 10%. The presence of RHA will increase Si/Al molar ratio of the starting material, the density of Si-O-Si chain and improve the strength of geopolymers [13].

Table 2. 7 days compressive strength of hybrid composite RHA-geopolymers

Sample	Compressive Strength (MPa)
FA-ASP 5%	26.00
FA-ASP 10%	26.21
FA-ASP 15%	25.59

Fig. 6a shows that one side of the sample was exposed to fire up to 1540°C. The temperature of the sample was measured by using a thermocouple placed around 2 cm from the fire source for 30 minutes. It was observed that the sample did not burn, release smoke and showed significant physical damage after testing as shown in Figure 7(b).

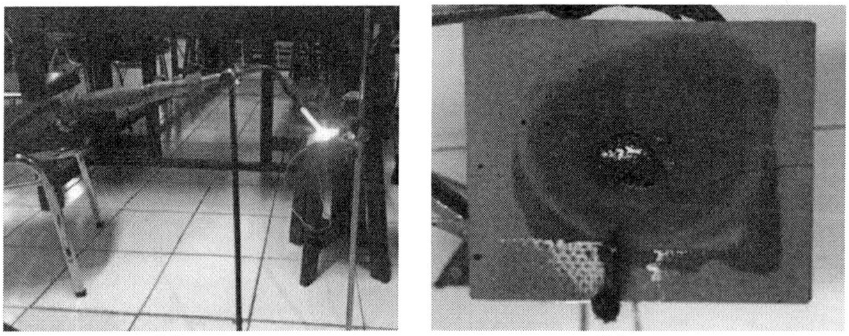

(a) (b)

Fig. 6. (a) Fire resistance measurement (b) sample condition after measurement

Fig. 7 shows the result fire resistance measurement of the resulting composites. The temperature different between fire source and 2 cm from the source reaching 1038°C. This graph shows that these materials can be applied as heat and fire shielding materials including bricks as fire is the most devastating calamity in many buildings. The results of study was similar to those reported by Irfanita et. al. [14] who also found that geopolymers made from FA were able to resist fire up to 1000°C. Kaloari et. al. [5] also investigated the potential of geopolymers as coating materials and the addition of RHA were able to increase the thermal properties of geopolymers.

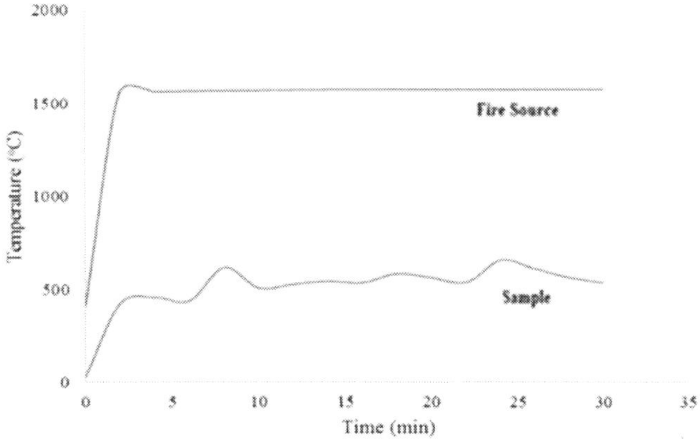

Fig. 7. The graph of temperature vs time for hybrid composite geopolymer containing 10% RHA

1 M H_2SO_4 solution was used for acid resistance measurement. The samples were soaked in the solution for 4 days. Fig.8 showed the appearance of the samples after 4 days in acid solution. It was observe the color change of the samples due to strong acid attack on the surface of the samples.

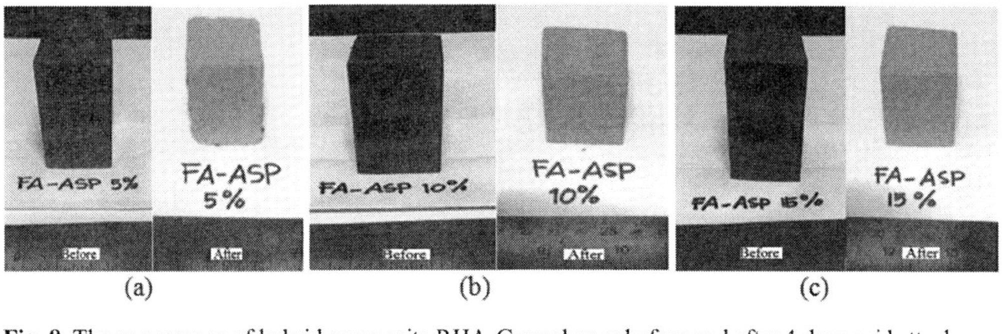

Fig. 8. The appearance of hybrid composite RHA-Geopolymer before and after 4 days acid attack

Table 3 shows the bulk density of the sample before and after 4 days in acid solution. It shows that the density of samples decreased. It indicates that the structural change in geopolymers network or the formation of new phases. Figure 9 is the cross section of the sample after acid attack showing the depth penetration of H_2SO_4 on the surface of the sample. The addition of RHA into geopolymers also improves the resistance of the material in high acidity [15].

DOI: 10.1051/matecconf/20179701009

Table 3. The density of hybrid composites RHA-geopolymer before and after acid attack

Sample	**Density (g/cm^3)**	
	Before Acid Attack	**After Acid Attack**
FA-ASP 5%	2.08	1.90
FA-ASP 10%	2.07	1.92
FA-ASP 15%	2.03	1.90

Fig. 9. The cross section of the sample after acid attack

Fig. 10 shows the diffraction patterns of hybrid composite RHA-geopolymer after acid attack test. It can be seen the formation of gypsum mineral due to the reaction between H_2SO_4 and CaO of fly ash. It was found that the composite with 15% showed the highest peak of gypsum phase. Composite contained 10% RHA showed the lowest peak of gypsum phase indicating that this composition is the most resist composite in acid attack.

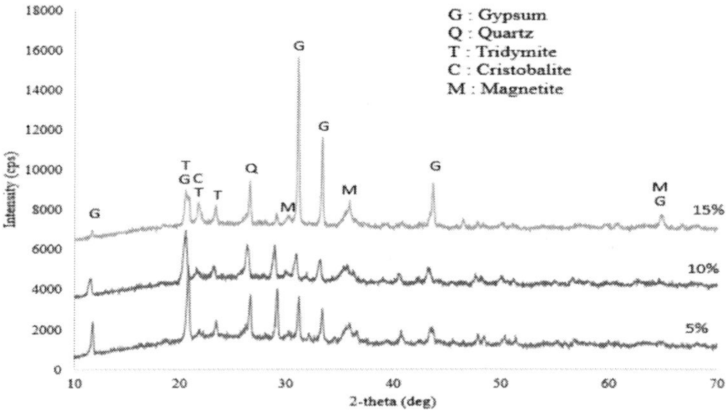

Fig. 10. The diffractogram of hybrid composite RHA-geopolymer contain of 5%, 10%, and 15% RHA after 4 days acid attack

Figure 11 shows the morphology of the RHA-geopolymers with the addition of 5% to 15% that has been soaked in 1 M H_2SO_4 solution for 4 days. The micrographs showed the different phases of the sample after acid attack. It can be seen the physical damage as well as the change of chemical composition of the area which were in contact with H_2SO_4. The x-ray diffraction results showed that the reaction between H_2SO_4 and CaO in fly ash formed gypsum ($CaSO_4(H_2O)_2$) minerals. It is well known that the presence of substantial gypsum minerals in the network of geopolymers will decrease its mechanical strength. This research finding suggest that hybrid composite RHA-geopolymer has a potential to be used as a raw material for bricks bearing building.

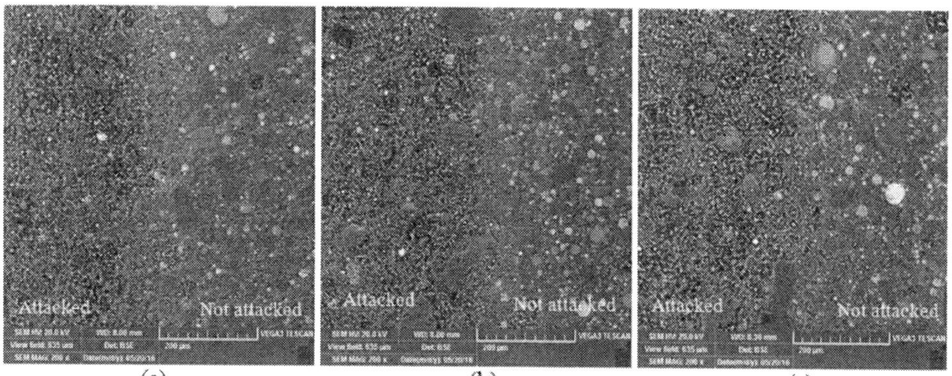

(a) (b) (c)

Fig. 11. SEM image of hybrid composite geopolymer with the addition of (a) 5% RHA, (b) 10% RHA, and (c) 15% RHA after acid attacked

4 Summary

Hybrid composite rice husk ash (RHA)-geopolymers based on class-C fly ash has been synthesized through alkali activation method at 60^0C for 1 hour. The addition of 10% RHA improved the compressive strength, fire and acid resistance of geopolymers. The results of this study suggest that hybrid composite RHA-geopolymers is suitable to be applied as material for bricks bearing building.

References

1. P. Turgut, J. Clean. Prod., **37**, 361 (2012)
2. X. Guo, H. Shi, W.A. Dick, Cem. Concr. Compos., **32**, 142 (2010)
3. L.N. Assi, E. Deaver, M.K. ElBatanouny, P. Ziehl, Constr. Build. Mater., **112**, 807 (2016)
4. C.L. Hwang, T.P. Huynh, Constr. Build. Mater., **101**, 1 (2015)
5. R.M. Kaloari, A. Haris, Mater. Sci. Forum., **841**, 79 (2016)
6. Z. Pan, J.G. Sanjayan, Cem. Concr. Compos., **34**, 261 (2012)
7. C.A. Jeyasehar, G. Saravanan, A.K. Ramakrishnan, S. Kandasamy, Asian J. Civ. Eng. BHRC., **14**, 797 (2013)
8. Nurfadilla, M. Dzulkifli, F. Ramli, Mater. Sci. Forum. **841**, 21 (2016)
9. V. Saraswathy, H.W. Song, Constr. Build. Mater., **21**, 1779 (2007)
10. P. Chindaprasirt S. Rukzon, Constr. Build. Mater., **22**, 1601 (2008)
11. C. Ferone, G. Roviello, F. Colangelo, R. Cioffi, O. Tarallo, Appl. Clay Sci., **73**, 42 (2013)
12. M. Safiuddin, J.S. West, K.A. Soudki, Cem. Concr. Compos., **32**, 708 (2010)
13. J. He, Y. Jie, J. Zhang, Y. Yu, G. Zhang, Cem. Concr. Compos., **37**, 108 (2013)
14. R. Irfanita, A. Ansar, Subaer, Mater. Sci. Forum., 841 (2016)
15. Subaer, Pengantar Fisika Geopolimer. Makassar: BP Universitas Negeri Makassar, (2012)

The Properties of Nano Silver (Ag)-Geopolymer as Antibacterial Composite for Functional Surface Materials

Armayani. M[1,2,*], *Muhammad* Akbar Pratama [1,2], and *Subaer*[1,2]

[1]Laboratorium Fisika Material, FMIPA, Universitas Negeri Makassar Jalan Daeng Tata Raya, Makassar 90223, Indonesia
[2]Konsorsium Riset Geopolimer Indonesia (KORIGI), Laboratorium Beton dan Bahan Bangunan, Kampus ITS Sukolilo Surabaya, Indonesia, 6011

Abstract. The purpose of this research was to produce and characterize nano silver (Ag)-geopolymer composite for functional surface materials. Geopolymer matrix was synthesized through alkali activation of metakaolin and nano silver was added into geopolymers paste with a mass of 0, 0.5 g, 1 g, 1.5 g and 2 g keeping the mass of metakaolin constant. The mixture was cured at 70°C/1 hour and stored for 7 days before conducting any measurements. The structure of the resulting composite was examined by using Rigaku Mini Flex II x-ray diffraction (XRD). Scanning Electron Microscopy (SEM) coupled with Energy Dispersive Spectroscopy (EDS) was used to examine the morphology of the composite surface as well as the capability of the composite to isolate the growth of bacteria. The thermal properties of composites in terms of their working temperature and enthalpy were examined by using Perkin Elmer Differential Scanning Calorimetry (DSC). The heat resistance of composite was observed through calcination at 750°C for 18 hours. The results indicate that the resulting composites were able resist up 750°C. SEM examinations showed that nano Ag-geopolymer composites were effectively restraining the growth of bacteria. It is suggested that nano Ag-geopolymer composites are suitable for functional surface applications such as floor and wall, kitchen ware utensils, hospital instruments, art and decoration materials.

1 Introduction

Nanotechnology plays important role in the advancement of science and technology. This technology has already find a place in a wide range of physics, chemistry, biology, and medical applications. Nano silver (Ag) is one of the attractive nano materials for antibacterials application. Nano silver has been applied as antimicrobial in public area such as train stations and elevators. It is a well-known fact that silver ions and silver-based compounds are highly toxic to microorganisms which include 16 major species of bacteria [1]. Silver nanoparticles with a size between 1 – 100 nm show unique physical, chemical and biological properties which is very much different when in bulk size [2-4]. Nanoparticles have vast surface area relative to volume and therefore minuscule amounts of silver nanoparticles can lend antimicrobial effects to hundreds of square meters of its host material [5].

As antibacterial, the activity of nano silver is influenced by the type and number of bacteria colony as well as time contact between nano silver and bacteria [6]. The size of nano silver will determine the penetration process on the particles into bacteria. Silver's antimicrobial is based on the interaction between free silver ions with specific bacterial cell

* Corresponding author: armha_physics09@yahoo.com

structure. The silver ions block the enzymatic respiration chain, Ion perak memblokir rantai respirasi enzimatik, cause destruction of the cell membrane (lysis), or to prevent mitosis of these organisms [7].

Composite nano silver-geopolymers can be applied to cover a wall in which Ag nanoparticles will prevent the formation of biofilm, or the wall of bacteria cell and therefore unable to split [8]. This study was conducted to examine the potential of nano Ag-geopolymer composites to be applied as functional surface, arts dan decoration materials which have unique color and structure [9], as well as material that inhibit the growth of bacteria.

2 Research Method

Composite nano silver (Ag)-geopolimers were synthesized through alkali activation of metakaolin. The resulting geopolymer paste was added with silver nanoparticle up to 2 g keeping the mass of metakaolin constant. The mixture was stirred manually and poured into polycarbonate container and cured at 70°C for 1 hour. Silver nanoparticle was prepared from $AgNO_3$ precursor by using the bioreduction method of the extract of catappa (*Terminalia catappa*) leaves for 4 hours.

The resulting composites were stored in open air for 14 day before commencing any measurements. X-ray diffraction technique was performed to observe the structure of as-prepared and calcined composites at 750°C. Samples for Scanning Electron Microscopy (SEM) was prepared and stored in oper air to examine the presence of any bacteria on the surface of composites. Differential Scaning Calorimetry (DSC) was performed to study the working temperature and the entalhy of the resulting composites.

3 Results and discussion

Figure 1(a) is the morphology of silver nanoparticles prepared through bioreduction method of $AgNO_3$. The SEM image shows that the particle size of silver below 100 nm and with this size silver nanoparticles have important optical properties, mechanics, and toxicity [10]. Figure 1(b) is the SEM image of metakaolin particles used in this study. The image showed that metakaolin consist of flakes particles similar to those of kaolin.

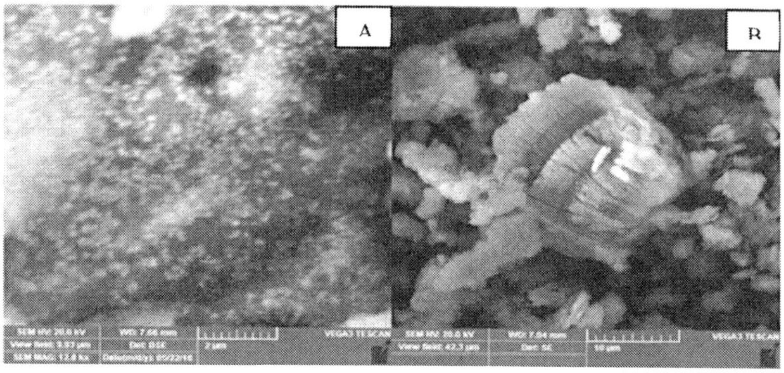

Fig. 1. SEM images of (a) silver nanoparticles, (b) metakaolin

DOI: 10.1051/matecconf/20179701010

Table 1 shows the chemical compositions of metakaolin based on EDS measurement.

Table 1. Chemical composition of Silver nanoparticles and metakaolin

Element	Nano-perak (Ag)		Metakaolin	
	Atom C [at.%]	Comp C [wt %]	Atom C [at.%]	Comp. C [wt.%]
Silver	89.88	96.06	-	-
Magnesium	1.64	0.66	0.19	0.37
Iron	0.95	0.68	0.52	1.82
Copper	3.78	2.38	0.24	0.74
Silicon	0.39	0.23	15.73	45.99
Aluminium	-	-	19.95	49.49
Sodium	-	-	0.27	0.41
Titanium	-	-	0.19	0.72
Potassium	-	-	0.20	0.45

The EDS results showed that the produced silver nanoparticles contain 89.88 wt% silver and other elements derived from catappa leaves. Metakaolin consists of 45.99 wt% SiO_2 and 49.49 wt% Al_2O_3. Because the weight of SiO_2 is lower than Al_2O_3 the synthesis of geopolymers requires higher sodium silicate solution.

Figure 2(a) shows the diffractogram of silver nanoparticles showing almost single phase. By using Debye-Scherer equation it was found that the particle size of silver nanoparticles was 21.3 nm. Figure 2(b) is the diffractogram of metakaolin showing the amorphous nature with some minor minerals such as quartz, anatase and dialuminium silicate oxide.

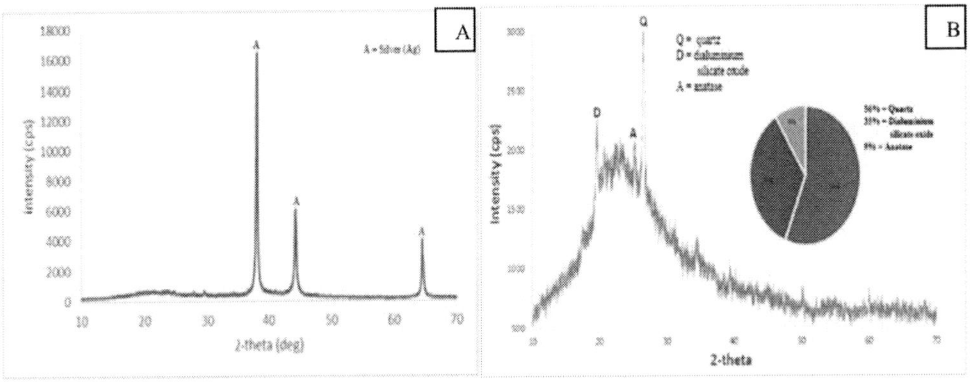

Fig. 2. Diffractograms of (a) silver nanoparticle, (b) metakaolin

Figure 3 shows diffractogram of the resulting composites contain different amounts of nano silver.

It can be seen that the peak of silver is at 38° 2θ and increase as the amount of silver increase. The presence of silver did not influence the overall amorphous structure of geopolymers made from metakaolin.

 DOI: 10.1051/matecconf/20179701010

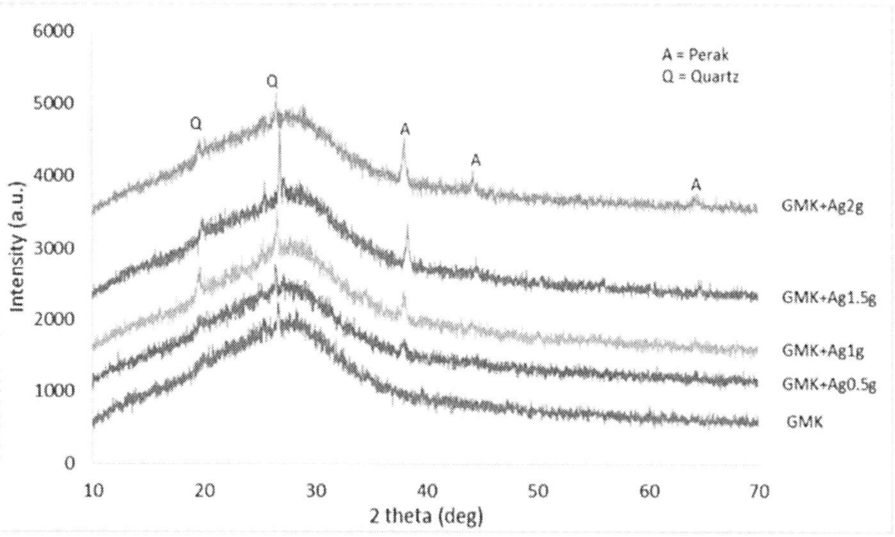

Fig. 3. Diffractogram of composite nano Ag-geopolimers having different amount of Ag

The thermal properties of the resulting composites were studied by means of differential scanning calorimetry (DSC). Figure 4(a) shows DSC result of geopolymer without silver nanoparticle. The measurements were conducted at temperature range from 30°C to 400°C with a heating rate 20°C/minutes and sample mass of 20 mg. Geopolymer is an exothermic material and at temperature around 100°C water from the interior of geopolymer start to evaporate and results in the first peak at the exothermic curve (figure 4.a).

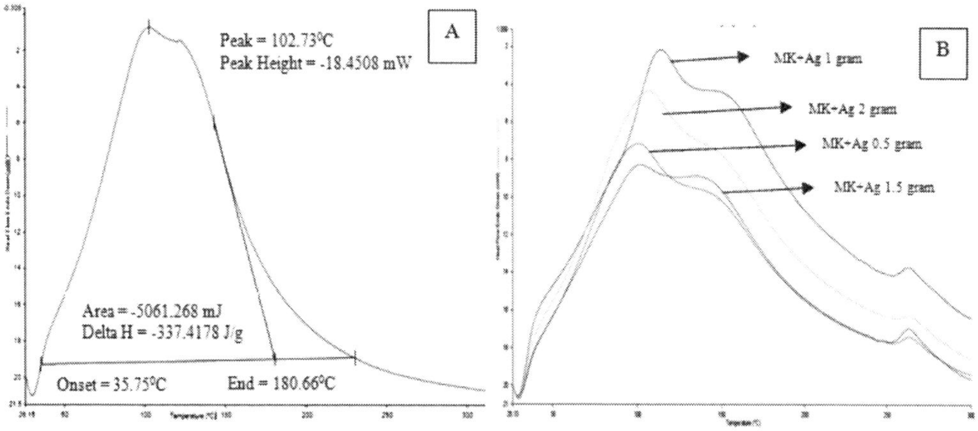

Fig. 4. DSC results of (a) pure geopolymer, (b) Nano silver-geopolymers

The presence of silver nanoparticle in the network of geopolymers change the exothermic profile of this material as can be seen from figure 4 (b). The enthalpy of the composites increase as the mass of silver increase as can be seen in table 2. The results also showed that the working temperature of the composite reach a value around 200°C.

Table 2. DSC results of pure geopolymers and nano silver-geopolymers composites

Sample	Range Temperature (C^0)	Peak Endoterm (C^0)	Peak Exoterm (C^0)	Enthalpy (J/g)
GMK	55.24 – 207.77	-	140.98	317.91
GMK+Ag 0,5 gram	50.27 – 205.97	-	100.89	189.09
GMK+Ag 1 gram	74.82 – 215.55	-	114.34	213.69
GMK+Ag 1,5 gram	44.28 – 206.87	-	101.90	201.03
GMK+Ag 2 gram	56.20 – 194.44	-	106.98	230.90

Figure 5 shows the SEM images of composite nano Ag-geopolymers. Figure 5(a) shows the typical morphology of geopolymers surface made from metakaolin. This sample did not contain silver nanoparticles. Figure 5 (b,c,d, and e) shows the morphology of composites surfaces which contain 0.5 g, 1.0 g, 1.5 g, and 2.0 g of nano silver and have been exposed to open air before coating with Pd/Au for several days. The images show that on the surfaces of all samples contain certain types of bacteria. This indicates that the presence of nano silver on the surface of the samples effectively captured and inhibited the growth of bacteria. Yaohui et.al. [11] reported that nano silver effectively capture 650 types of bacteria.

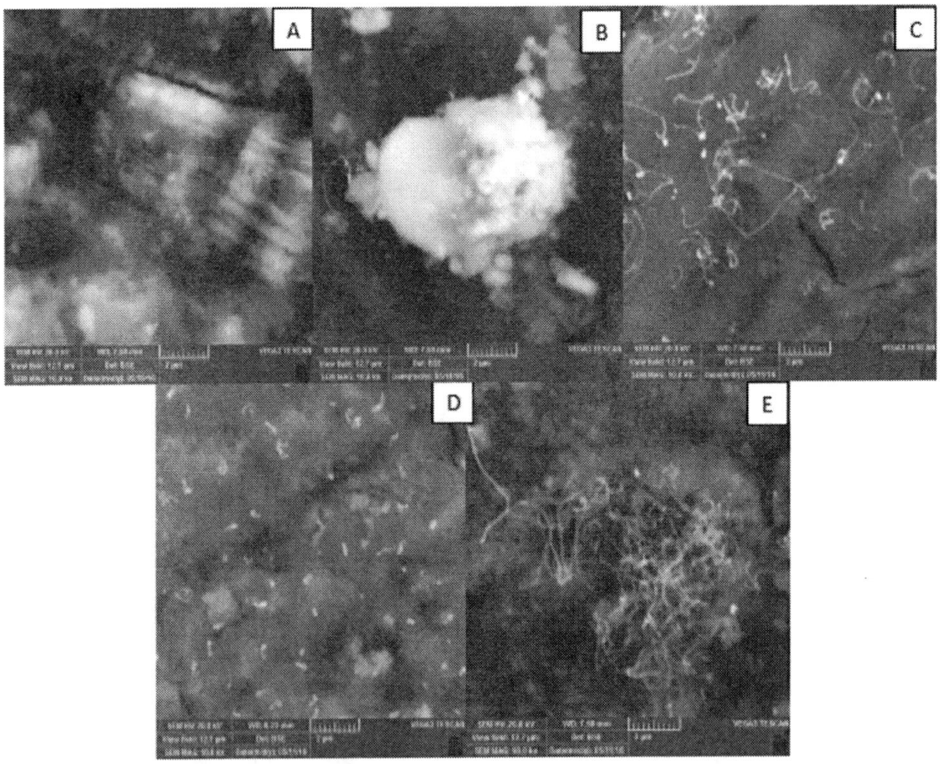

Fig. 5. SEM images of nano silver-geopolymers (a) GMK, b). GMK+Ag 0,5 g, c). GMK+Ag 1g, d.) GMK+Ag 1,5 g, and e). GMK+Ag 2 g. Notice the presence of certain bacteria on the surface of each sample.

Figure 6 show that the presence of Nano silver in the network of geopolymers did not reduce the ability of geopolymers to retards heat. Although substantial cracks were observed on the surface of all samples after heat resistance measurement but the integrity of the samples remains strong. The occurrence of cracks was a result of dry shrinkage of the samples which start to take place at temperature 100°C [12, 13]. Table 3 shows the bulk density of the sample before and after heat resistance measurement. The reduction of densities of the samples after heat measurements due to substantial cracks on the surface of the samples.

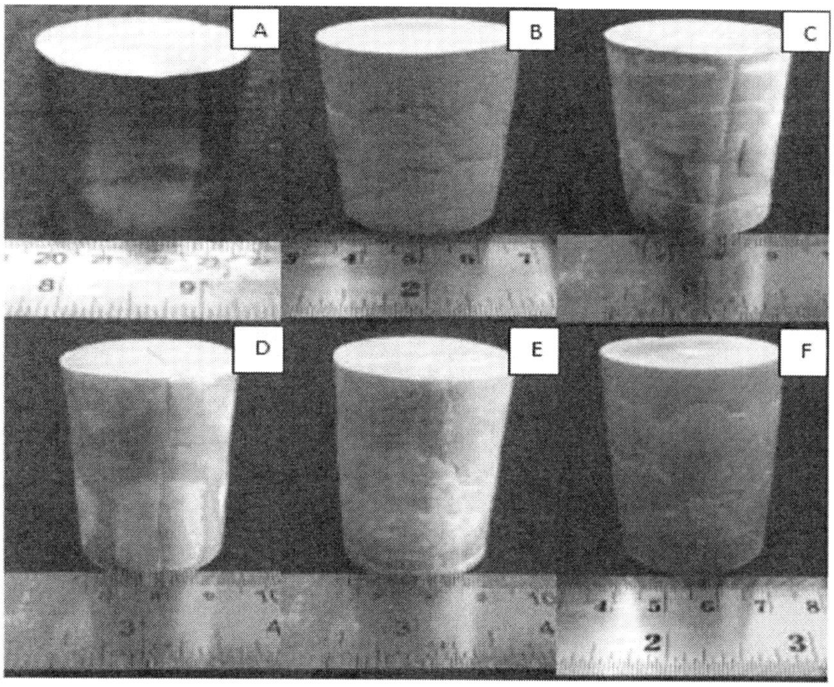

Fig. 6. The appearance of nano silver-geopolymers after heat resistance measurement. (a) as-prepared sample (b) 0 g Ag, (c) 0.5 g Ag, (d) 1.0 g Ag, (e) 1.5 g Ag, and (f) 2.5 g Ag.

Table 3. Bulk density of composite Nano silver-geopolymers before and after heat resistance measurement

Sample	Density as-prepared sample (g/cm^3)	Density after heat resistance measurement (g/cm^3)
GMK	1,71	1,58
GMK + Ag 0.5 g	1,76	1,66
GMK + Ag 1 g	1,94	1,67
GMK + Ag 1.5 g	1,74	1,60
GMK + Ag 2 g	1,75	1,57

4 Summary

Functional composite nano silver-geopolymer based on metakaolin have succesfully produced and characterised. Silver nanaparticles was produced through bioreduction method (solution of catappa leaves) of $AgNO_3$. The composites were able to inhibit certain types of bacteria which indicate that the presence of nano silver works effectively as microbial agent. The resulting composites also showed excellent properties as heat resistance material. The result of this study suggest that functional composite nano silver-geopolymers can be applied as antimicrobial material.

References

1. S. Prabu, E.K. Poulose, Int. Nano Lett., (2012)
2. J. Palanivelu, M.M. Kunjumon, A. Suresh, A. Nair, Dr. Ramalingam, J. Pharm. Sci. Res., **7**, 690 (2015)
3. J.R. Morones, J.L. Elechiguerra, A. Camacho, K. Holt, J.B. Kouri, J.T. Ramirez, M.J. Yacaman, Nanotechnology, **16**, 2346 (2005)
4. K. Burridge, J. Johnston, T. Borrman, J. Mater. Chem., **21**, 734 (2011)
5. T. Theivasanthi, M. Alagar, Anti-bacterial Studies of Silver Nanoparticles. arXiv preprint arXiv (2011)
6. N. Beyth, Y.H. Haddad, A. Domb, W. Khan, R. Hazan, J. Evidence-Based Complementary Altern. Med., (2015)
7. Herman, Raumedic Infect-Protect., (2010)
8. B. Domenech, M. Munoz, D.N. Muraviev, J. Macanas, Polymer-Silver Nanocomposites as Antibacterial materials. Microbial Pathogens and Strategies for combating them: Science, Technology and Education. (2013)
9. G. Grimal, A.F. Sculptor and Painter, Potential utilizations in Art and Decoration Georges Grimal. Geopolymer Institute. (2006)
10. Y. Masakke, Sulfikar, M. Rasyid, Biosynthesis of Silver Nanoparticles Using Methanol Extract of Mangosteen Leaves (Garcinia Mangostana L.). Journal Sainsmat. IV (2014)
11. Y. Lv, H. Liu, Z. Wang, L. Hao, J. Liu, Y. Wang, G. Du, D. Liu, J. Zhan, J. Wang, Polym. Adv. Technol., **19**, 1455 (2008)
12. Subaer. Pengantar Fisika Geopolimer. Direktorat Jenderal Pendidikan Tinggi. (2012)
13. Z. Zuhua, Y. Xiao, Z. Huajun, C .Yue, Journal of Applied Clay Science. **43**, 218 (2009)

The Effect of Si/Al on Mechanical Properties and Fracture Behavior of Stainless Steel Mesh/Cr_p Reinforced Geopolymer Composites

Jingkun Yuan[1,*], *Peigang* He[1] , and *Dechang* Jia[1]

[1]School of Materials Science and Engineering, Harbin Institute of Technology, P. O. Box 433 Harbin 150001, P. R. China

Abstract. In this study, a series stainless steel mesh/Cr_p reinforced geopolymer composites with different Si/Al molar ratio (N) were designed and prepared, where $N = 1.75$, 2 and 2.25, respectively. The effect of Si/Al molar ratio in the geopolymer matrix on mechanical properties and fracture behavior of the geopolymer composites were investigated. The microstructure of geopolymer became more compact when Si/Al increased from 1.75 to 2, which was beneficial to the improvement of geopolymer's mechanical properties. And continuing to rise to 2.25 for Si/Al, the completely curing of geopolymer composites required more time compared with lower Si/Al, which can be attributed to the different microstructure and chemical composition caused by the different Si/Al. The optimum Si/Al molar ratio was about 2 at which the composites samples present the best mechanical properties with the flexure strength of 115.3 MPa and elastic modulus of 11.0 GPa, respectively. The results of fracture behavior suggested that geopolymer composites with N is 2.25 displayed the behavior characteristics of metal materials, which can be attributed to a poor integrated condition in interface between reinforcements and geopolymer matrix.

1 Introduction

While traditional design and evaluation approaches are based on the principle of maximizing the economic efficiency and include quality, cost, and time, the new approach of 'sustainable construction' emphasizes the importance of reducing the environmental impact of buildings and infra-structures [1]. Therefore, the reduction of CO_2 emissions and energy consumption in the cement industry has recently become a contentious issue, which corresponds to approximately 7% of the total emissions into the earth's atmosphere. In an effort to limit the environmental impact of concrete materials, there is increasing interest in the development an environmental-friendly material. As an outstanding candidate, geopolymer can provide comparable performance to traditional cementitious binders in a range of applications, but with the added advantage of significantly reduced greenhouse emissions [2-7].

* Corresponding author: yuanjingkun@126.com

Geopolymers are a class of cementitious aluminosilicate binder materials synthesized by a geopolymerization reaction between alumino-silicate oxides and alkali-metal silicate solutions under alkaline conditions, with partially or fully amorphous polymeric structures containing tetrahedral SiO_4 and AlO_4 randomly distributed along the polymeric chains [8-11]. The researches on the geopolymerization mechanism in this system are attracting more and more attentions, and the process may be summarized briefly as follows: dissolution, rearrangement, condensation and re-solidification [12-18]. The structure and mechanical properties of geopolymers are affected by several parameters such as chemical composition and reactivity of the raw materials, nature and concentration of the activating solution, curing conditions. The Si/Al molar ratio has a significant effect on the structural and mechanical performance. An optimum Si/Al ratio increases the strength of the resulting geopolymer. Apart from the ideal ratio, some losses may occur in the mechanical properties of the material due to the change of the geopolymer structure [19].

In this paper, a series stainless steel mesh/Cr_p reinforced geopolymer composites with different Si/Al molar ratio (N) were designed and prepared, where N = 1.75, 2 and 2.25, respectively. The effect of Si/Al molar ratio in the geopolymer matrix on mechanical properties and fracture behavior of the geopolymer composites were investigated.

2 Experiments

2.1 Material and synthesis

Geopolymers resin with composition of Si/Al = N, 1.75, 2 and 2.25, respectively, were obtained by mixing metakaolin powder with potassium silicate solution. The metakaolin was prepared by calcining kaolin at 800 °C for 2 h and the potassium silicate solution was prepared by dissolving a certain amount of amorphous silica (Shanghai Dixiang Indus, China) into KOH (Tianjin Fuchen Indus., China) solution stirring for 48 h in order to dissolve the silica completely. Then the metakaolin powder and chromium powder (40% mass ratio of metakaolin) were gradually added into the potassium silicate solution under ultrasonic-assisted high-shear mixer. To prepare composite samples, seven stainless steel mesh layers (Ø 70 mm) were impregnated with the geopolymer resin and laid on top of each other. The lamination structure was formed by vacuum-bag technique and pressed at ~3 MPa for 1 h, followed by curing 3 d at 60°C. After demoulding the hardened geopolymer composites were further cured at the same temperature for another 2 d. The major ingredient of the stainless steel mesh is $Cr_{0.19}Fe_{0.7}Ni_{0.11}$. The Si/Al molar ratio was adjusted by adding extra amorphous silica, and the amount of amorphous silica was calculated according to the Si/Al ratio.

2.2 Characterization

The phases of geopolymer composites were characterized using an X-ray diffractometer (XRD: Rigaku, D/MAX-2200VPC, Tokyo, Japan) with CuKa radiation at a scan rate of 4 °/min. Flexural strength measurements were conducted on specimens (4 mm×3 mm×36 mm) using a three-point-bending fixture on an Instron-500 testing machine with a span length of 30 mm at a crosshead speed of 0.5 mm/min. Tensile strength measurements were carried out on a Work of fracture was calculated by the area between load curves and X axis in the load/displacement curves till the load dropped to 90% of the maximum load. Six specimens were tested under each test condition. Fracture morphology of the geopolymer composites were investigated by a scanning electron microscope (SEM, 30 KV, Quanta 200 FEG, FEI Co., USA).

 DOI: 10.1051/matecconf/20179701011

3 Results and discussion

3.1 Phase composition

Fig. 1 provided XRD patterns of geopolymer composites with different Si/Al ratio. As can be seen, all the samples show typical amorphous character, i.e., a broad amorphous hump at ~28° 2θ [20], which suggested that the different Si/Al ratio did not changed the nature of the geopolymer matrix. With the rise of Si/Al, the amount of SiO_2 in the composites decreased gradually. The results of XRD also indicated that the stainless steel mesh and chromium powder were not involved in the reaction with geopolymer matrix [21].

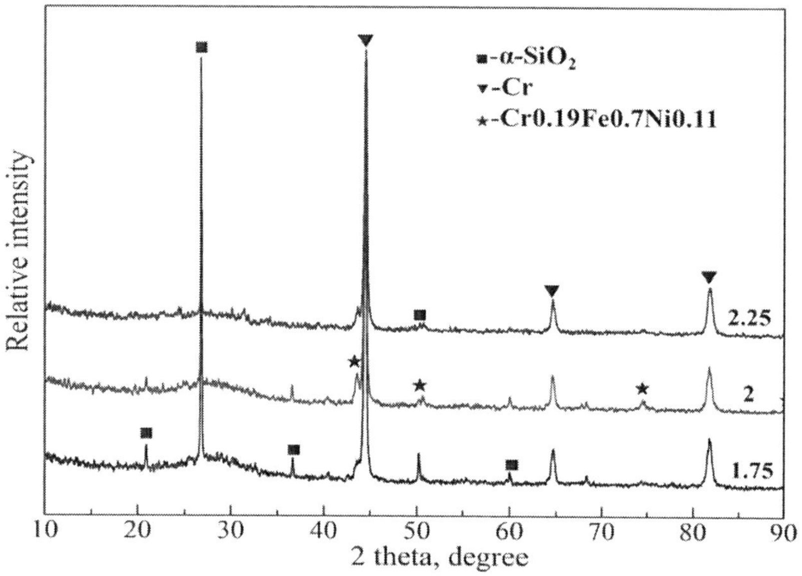

Fig. 1. XRD patterns of geopolymer composites with different Si/Al ratio.

3.2 Microstructure

Fig. 2 displayed the surface morphologies of geopolymer composites with different Si/Al ratio. The surface morphology of geopolymer composites became much more compact and smoothly when N increased from 1.75 to 2. When N =1.75, the viscosity of geopolymer slurry was relatively low, which favored the escape of the water in the initial stage of curing process, which can be explained the pore in the samples surface (Fig. 2c). Meanwhile, more surface water and pore water will loss in the curing process due to the loose microstructure. However, a geopolymer slurry with a moderate viscosity of was obtained when the $N = 2$ compared with that of 1.75 [22, 23]. Therefore, more uniform and compact microstructure was observed in the samples surface (Fig. 2b). When N came to 2.25, there were significant difference with N = 1.75 or 2. After 3 d curing, the geopolymer gel has not completely solidified, which was attributed to the presence of massive potassium silicate [24]. There were many cracks in the samples surface with N = 2.25, which may be caused by the extrusion in the demoulding process.

DOI: 10.1051/matecconf/20179701011

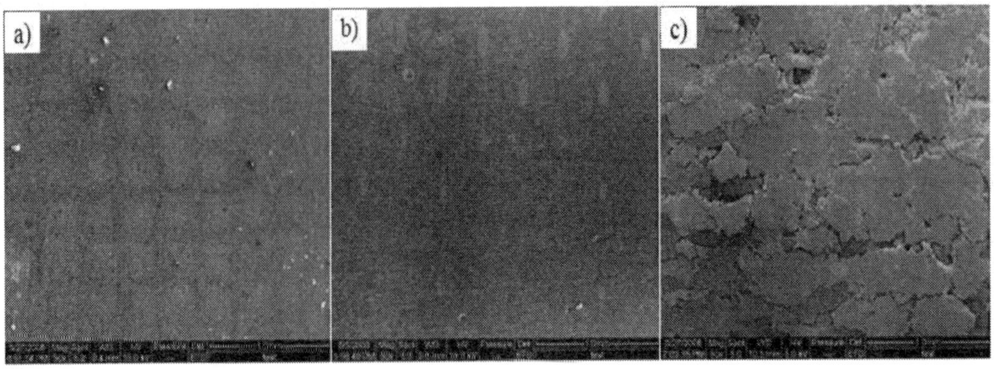

Fig. 2. Surface morphologies of geopolymer composites with different Si/Al ratio, a) 1.75, b) 2 and c) 2.25

3.3 Mechanical properties

The Si/Al ratio has a great influence on the mechanical properties of the resulting geopolymer composites, as shown in Fig. 3. With the increase in Si/Al ratio, flexural strength, elasticity modulus, compressive strength and work of fracture of the composites showed similar trends, i.e., increasing first, reaching the highest value at $N = 2$, and then decreasing. Take flexural strength for example (Fig .4): it increased from 96.3 MPa for composites with $N = 1.75$ to the highest of 15.8 MPa for composites with $N = 2$. And almost all the mechanical properties reached the peak value when $N = 2$. However, there was a significant declined in performance when $N = 2.25$. As shown in Fig. 4 d, a mass of geopolymer matrix flaked away. The surface of stainless steel mesh was still smooth and no corrosion in the alkaline environment [21]. This might result from the weak interfacial strength between stainless steel mesh and geopolymer matrix [25]. In addition, microscopic defects such as interfacial gaps or the cracks in the surface also caused the loss of mechanical properties of geopolymer composites [26].

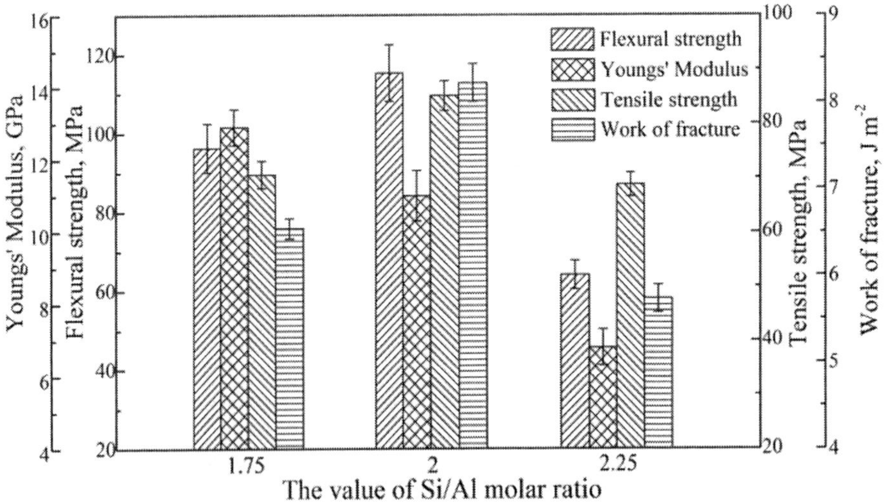

Fig. 3. Mechanical properties of geopolymer composites with different Si/Al ratio, a) 1.75, b) 2 and c) 2.25.

 DOI: 10.1051/matecconf/20179701011

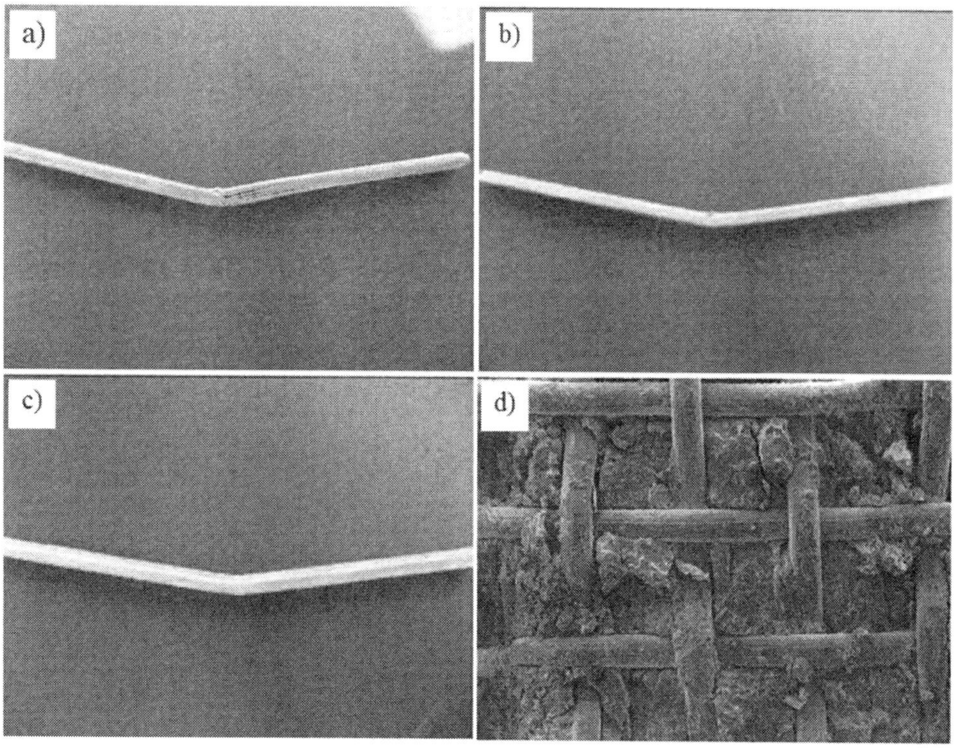

Fig. 4. The samples after three-point bending tests, a) 1.75, b) 2, c) 2.25 and d) the surface morphology with $N = 2.25$.

3.4 Fracture behavior

Fig. 5. shows the load-displacement curves of geopolymer composites with different Si/Al molar ratio. Unlike $N = 2.25$, the samples with $N = 1.75$ or 2 showed a non-linear elastic increase of load before reaching the maximum load values. Then, the load decreased slowly with the displacement. The reason may be attributed to the bridging and sliding after debonding and pulling-out of the stainless steel mesh from the weak geopolymer matrix [27]. Meanwhile, the presence of chromium powder also was beneficial to enhancement of composites' toughness. However, the samples with $N = 2.25$ displayed a fracture behavior as similar with metal materials, which indicated that there was little interfacial strength between stainless steel mesh and geopolymer matrix. The change of work of fracture also suggested the optimum Si/Al ratio should be 2.

 DOI: 10.1051/matecconf/20179701011

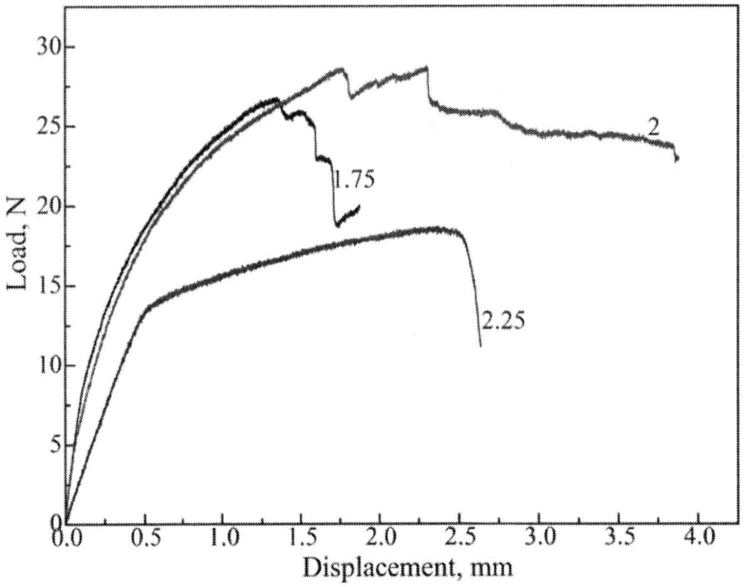

Fig. 5. Fracture behavior of geopolymer composites with different Si/Al ratio.

4 Summary

In this paper, the effect of Si/Al molar ratio on the performance of stainless steel mesh/Cr_p reinforced geopolymer composites was investigated. The results indicated that Si/Al molar ratio had a significant impact on the final microstructure and mechanical properties of resulting geopolymer composites. The microstructure of geopolymer became more compact when Si/Al increased from 1.75 to 2, and the corresponding mechanical properties also reached the peak value. However, when Si/Al came to 2.25, the mechanical properties of composites declined. The results of fracture behavior suggested that the samples with $N =$ 2.25 displayed a fracture behavior as similar with metal materials, which different with $N =$ 1.75 and 2. The reason can be ascribed to different interfacial strength between reinforcements and geopolymer matrix, and composites with $N = 2$ has an optimum interfacial strength compared with $N = 1.75$ and 2.25.

This work was supported by the Fundamental Research Funds for the Central Universities (Grant no. HITNSRIF20165), and the National Natural Science Foundation of China (NSFC, Nos. 51372048, 51502052, 51321061 and 51225203).

References

1. K.H. Yang, J.K. Song, K.I. Song, J. Clean. Prod., **39**, 265 (2013)
2. P. Hlavacek, V. Smilauer, F. Skvara, L. Kopecky, R. Sulc, J. Eur. Ceram. Soc., **35,** 703 (2015)
3. A. Favier, G. Habert, J.B.D.E.d. Lacaillerie, N. Roussel, Cem. Concr. Res., **48**, 9 (2013)
4. Z. Zhang, X. Yao, H. Zhu, Appl. Clay Sci., **49**, 1 (2010)
5. V.F.F. Barbosa, K.J.D. Mackenzie, Mater. Res. Bull., **38**, 319 (2003)
6. F. Slaty, H. Khoury, H. Rahier, J. Wastiels, Appl. Clay Sci., **104**, 229 (2015)
7. W.D.A. Rickard, L. Vickers, A.V. Riessen, Appl. Clay Sci., **73,** 71 (2013)
8. J. Davidovits, J. Therm. Anal., **37**, 1633 (1991)
9. J. Davidovits, J. Therm. Anal., **35,** 429 (1989)

10. K. Komnitsas, D. Zaharaki, A. Vlachou, G. Bartzas, M. Galetakis, Adv. Powder Technol, **26**, 368 (2015)
11. B.H. Mo, Z. He, X.M. Cui, Y. He, S.Y. Gong, Appl. Clay Sci., **99**, 144 (2014)
12. J.L. Provis, J.S.J.V. Deventer, *Geopolymers : Structures, Processing, Properties and Industrial Applications* (Elsevier, United Kingdom, 2009)
13. X. Yao, Z. Zhang, H. Zhu, Y. Chen, Thermochim Acta, **493**, 49 (2009)
14. L. Weng, K.S. Crentsil, J. Mater. Sci., **42,** 2997 (2007)
15. P.S. Singh, M. Trigg, I. Burgar, T. Bastow, Mater. Sci. Eng. A Struct., **396**, 392 (2005)
16. M.R. Rowles, B.H. O'Connor, J. Am. Ceram. Soc., **92**, 2354 (2009)
17. I. Ozer, S.S. Uzun, Ceram. Inter., **41**, 10192 (2015)
18. K. Komnitsas, D. Zaharaki, V. Perdikatsis, J. Mater. Sci., **42**, 3073 (2006)
19. G. Gorhan, G. Kurklu, Compos. Part B-Eng., **58**, 371 (2014)
20. P. Duxson, A.F. Jimenez, J.L. Provis, G.C. Lukey, A. Palomo, J.S.J.V. Deventer, J. Mater. Sci., **42**, 2917 (2006)
21. P.G. Allison, C.A. Weiss, J.R.D. Moser, A.J. Diaz, O.G. Rivera, S.S. Holton, Compos. Part B-Eng., **78**, 131 (2015)
22. P. Duxson, G.C. Lukey, J.S.J.V. Deventer, J. Non Cryst Solids, **352,** 5541 (2006)
23. P. Duxson, G.C. Lukey, J.S.J.V. Deventer, Langmuir, **22**, 8750 (2006)
24. A. Hajimohammadi, J.L. Provis, J.S.J.V. Deventer, Cem. Concr. Res., **41**, 210 (2011)
25. A. Castel, S.J. Foster, Cem. Concr. Res., **72**, 48 (2015)
26. J. Yuan, P. He, D. Jia, S. Yan, D. Cai, L. Xu, Z. Yang, X. Duan, S. Wang, Y. Zhou, Ceram. Inter., **42**, 5345 (2016)
27. Q. Zhao, B. Nair, T. Rahimian, P. Balaguru, J. Mater. Sci., **42**, 3131 (2007)

Effect of Alkaline Activator to Fly Ash Ratio for Geopolymer Stabilized Soil

Muhammad Sofian Abdullah[1,2,*], and *Fauziah* Ahmad[1]

[1]School of Civil Engineering, Universiti Sains Malaysia (USM), Engineering Campus, 14300 Nibong Tebal, P. Pinang, Malaysia
[2]Center of Excellence Geopolymer and Green Technology (CEGeoGTech), School of Materials Engineering, Universiti Malaysia Perlis (UniMAP), Kangar, Perlis, Malaysia.

Abstract. Geopolymer technology have been developed and explored especially in the construction material field. However, lack of research related to geopolymer stabilized soil. In this research, the utilization of geopolymer has been investigated to stabilize the soil including the factors that affecting the geopolymerization process. Unconfined compressive test (UCT) used as indicator to the strength development and hence evaluating the performance of geopolymer stabilized soil. This paper focusing on the effect of fly ash/alkaline activator ratio, Na_2SiO_3/NaOH ratio and curing time on geopolymer stabilized soil. A various mix design at different fly ash/alkaline activator ratio, Na_2SiO_3/NaOH ratio were prepared and cured for 7 and 28 days. Molarity and the percentage of geopolymer to soil were fixed at 10 molar and 8 percent respectively. Then, the UCT tests were carried out on 38mm diameter x 76mm height specimens. The highest strength obtained at the fly ash/alkaline activator ratio 2.5 and Na_2SiO_3/NaOH ratio 2.0 at 28 days curing time.

1 Introduction

Infrastructures such as road, highways, slope, and embankment are built on top of the soil. In fact some of them are built by cutting and trimming the soil (i.e; slope and embankment). Thus, the strong and stable soil is an important part of the infrastructure construction. However, in many construction cases, to find the soil that naturally high strength and high durability is very difficult and the soil need to be treated. Since a decade, the treatment technologies and methods are well developed by researchers and some of them are well implemented to stabilize the soil before the constructions begin. One of the recent material technology (since 1970's) is geopolymer. This material developed by Davidovits [1] and gives a number of benefits to the construction material field especially for the building structures. Since then, extensive research on geopolymer material are conducted to understand the behaviour, properties and strength at various conditions.

* Corresponding author : msofian78@gmail.com

Looking at the advantages of this material, the research on suitability of this material for soil stabilization is presented.

In general, geopolymer produced from the reaction between raw material (i.e; fly ash) and alkaline activator. The termed 'geopolymers' comes when the inorganic polymeric material synthesized in a manner similar to thermosetting organic polymers. For this reason, these materials names geopolymer [2]. Scientifically, geopolymer is the reaction of a solid aluminosilicate with a highly concentrated aqueous alkali hydroxide or silicate solution to produces a synthetic alkali aluminosilicate material [3]. Fly ash (residue from coal combustion) is used finely divided residue resulting from combustion of coal and has been used as cement replacement for the recent years [4]. Sodium silicate (Na_2SiO_3) and sodium hydroxide (NaOH) solutions is commonly used to make an alkaline activator [5,6].

A lot of factors contribute to the strength development of geopolymer stabilize soil such as types of soils, types of raw materials, solid to liquid ratio, sodium hydroxide to sodium silicate ratio, temperature, and curing time [2], [7-11]. The influence of solid to liquid ratio and sodium hydroxide to sodium silicate ratio are highlighted in this paper. Some researchers stated that optimum compressive strength can be achieved at solid to liquid ratio in between 2.0 and 3.0 [12, 13]. However, the ratio can be varies depend on its application. It should be noted that, the compressive strength is optimum at the ideal solid to liquid ratio. For geopolymer, some researchers discovered that a geopolymer strength up to 70 MPa could be achieved when the mixture is formulated at ratio 2.0 [12]. Different curing time also give an impact to the strength of geopolymer. Geopolymerisation process highly influence by curing time. Researchers reported the strength increased as the curing time increased [11-14].

The objective of this paper is to present the effect of various fly ash to alkaline activator ratio, Na_2SiO_3/NaOH Ratio and curing time on the compressive strength of geopolymer. The result obtained will be used as a design mix for soil stabilization.

2 Experimental setup

2.1 Material selection

Fly ash is an industrial by product resulting from coal combustion. This raw material collected from Tenaga Nasional Berhad (TNB) Manjung Power Station, Telok Rubiah, Lumut, Perak, Malaysia. The compositions of this fly ash are suitable for geopolymerization process. The activator used in this study is combination between sodium silicate (Na_2SiO_3) and sodium hydroxide (NaOH). NaOH was prepared by mixing sodium hydroxide pellets (97–99 % purity) with distilled water at 10M concentration. Untreated soil used collected from Kampung Padang Durian, Pendang, Kedah, Malaysia. After classification test the soil fall under coarse grain sand.

2.2 Sample preparation

Series of sample at various fly ash/alkaline activator ratio and Na_2SiO_3/NaOH ratio were prepared, then curing for 7 and 28 days at room temperature. The percentage of geopolymer to soil was fixed at 8 percent (8%). The uniaxial compressive test (UCT) used to determine the shear strength of each sample. Details mix design tabulated in Table 1.

DOI: 10.1051/matecconf/20179701012

Table 1. Mix design geopolymer stabilize soil.

Sample	**Fly-Ash / Activator ratio**	**Na_2SO_3 / NaOH ratio**
1-1	2.0	2.0
1-2	2.0	2.5
1-3	2.0	3.0
2-1	2.5	2.0
2-2	2.5	2.5
2-3	2.5	3.0
3-1	3.0	2.0
3-2	3.0	2.5
3-3	3.0	3.0

First, soil and fly ash were dry mixed together to allow the fly ash distributed uniformly. Meanwhile Na_2SO_3 were mix together with NaOH at design ratio to form alkaline activator (AA) and rest for 30 minutes for complete reaction. After that, the AA is poured into fly-ash and soil and mixed together until the AA mixed evenly. Then, the 38mm diameters x 76 mm height sample were prepared. The process continued at different design mix. The sample then test under UCT to obtain the maximum shear strength.

3 Results and discussion

3.1 Soil classification

Particle size distribution (PSD) test used to classify type of soil for this study. A set of sieve varies from 2 mm to 0.063 mm opening sorting according to British Standard (BS). Soil sample was sieving for 10 minutes to allow the soil particle distribute accordingly. The particle distribution curve is presented in Figure 1. From figure 1, soil use for this study are classify as Coarse Sand (CS).

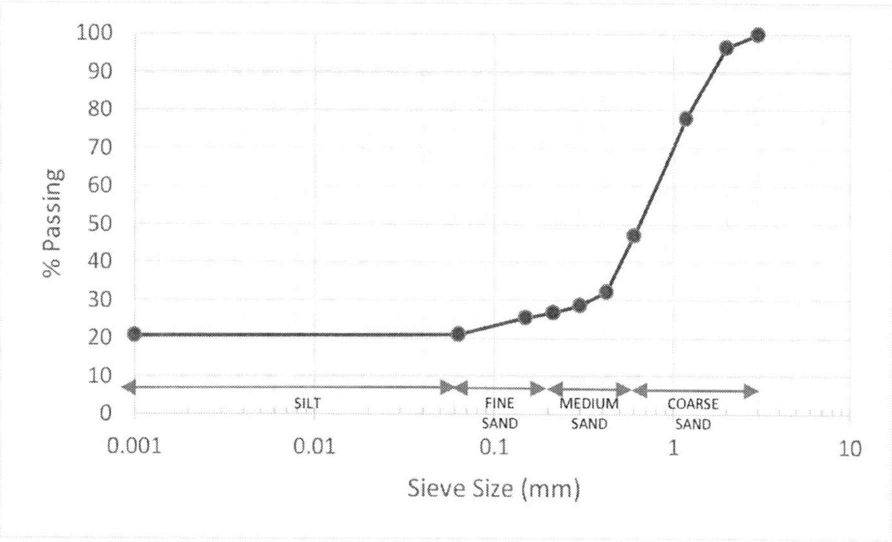

Fig. 1. Particle size distribution curve.

3.2 Chemical composition of fly ash

X-ray fluorescence (XRF) test was used to determine chemical composition of fly ash. The main chemical composition of fly ash is silica oxide SiO_2, followed by Fe_2O_3, CaO, and Al_2O_3. The XRF result is tabulate in Table 2 below.

Table 2. Chemical composition of fly ash using XRF.

Chemical	SiO_2	Al_2O_3	Fe_2O_3	TiO_2	CaO	MnO	CuO	K_2O	P_2O_5	SO_3	SrO
wt.%	38.8	14.7	19.48	1.02	18.1	0.16	0.041	1.79	0.054	1.50	0.11

Percentage of calcium oxide (CaO) is 18.1% and the summation of SiO_2, Al_2O_3, and Fe_2O_3 is 72.98%. Referring to fly ash standard classification in ASTM C618, this fly ash classified as Class F.

3.3 Standard compaction test

Standard compaction test is used to determine the optimum moisture content (OMC) and maximum dry density (MDD) of soil. OMC and MDD results are used to prepare the specimen for uniaxial compressive test (UCT). Compaction curve from the test is shown in Figure 2. From the curve, the OMC and MDD are 14.7 % and 1.78 Mg/m^3 respectively.

 DOI: 10.1051/mateconf/20179701012

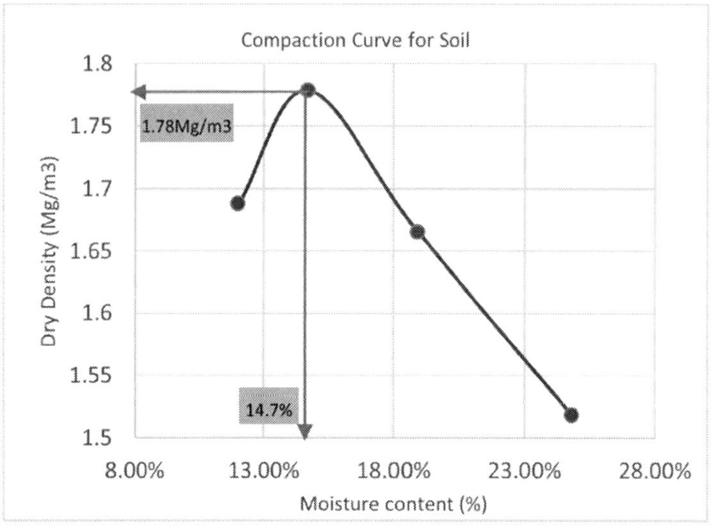

Fig. 2. Compaction curve.

3.4 Uniaxial compressive strength test

All samples were test under uniaxial compressive test (UCT) to determine the shear strength for each sample. UCT test used to indicate the best mix design for fly ash-geopolymer stabilize soil. The result then plot in Figure 3 and Figure 4. From both figures, the highest shear strength obtained at sodium silicate (Na_2SiO_3) to sodium hydroxide (NaOH) ratio of 2.0 and a fly ash to activator ratio of 2.5. There are significant increase of shear strength at various solid to liquid ratio and activator ratio. At 7 days curing time, the different between the highest strength (0.395MPa) and the lower strength (0.215MPa) is about 46% increment whereby at 28 days curing the different between the highest strength (0.645MPa) and the lower strength (0.337MPa) is about 48% increment.

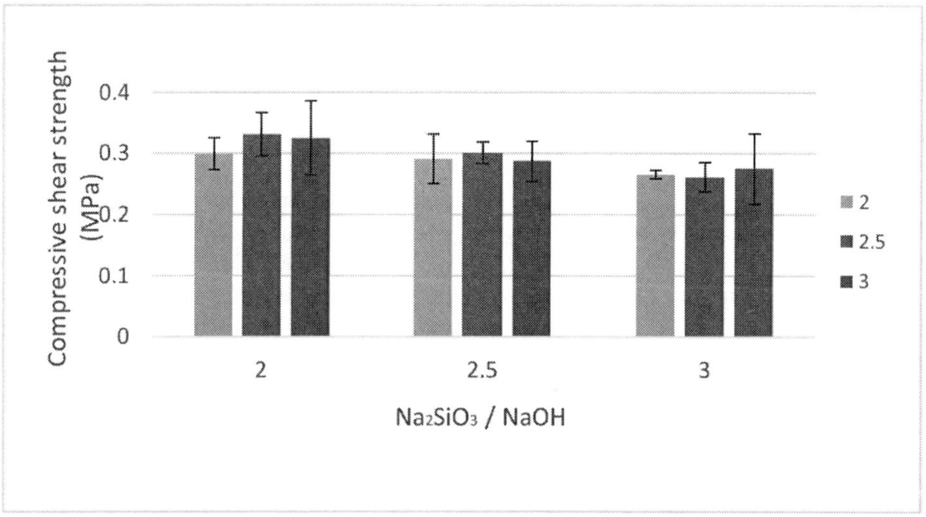

Fig. 3. Compressive strength of various proportions of activators at 7 days curing.

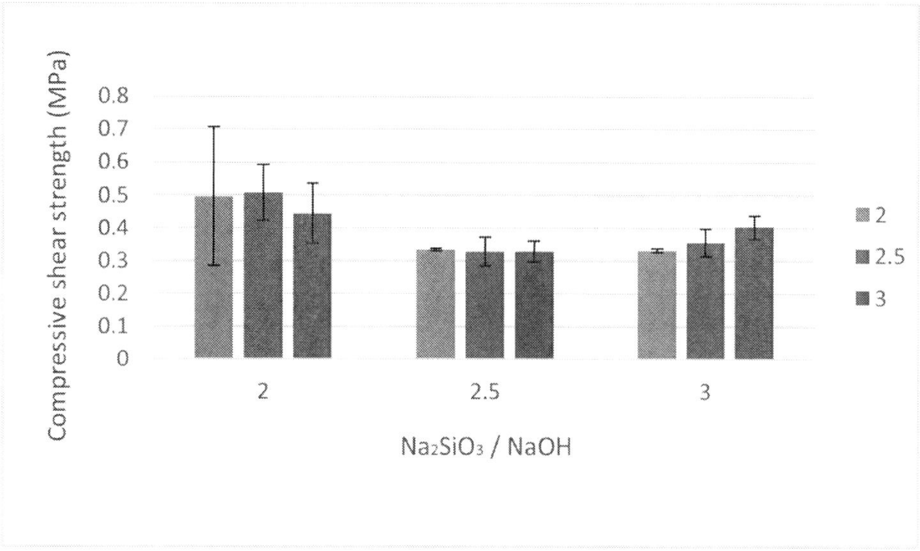

Fig. 4. Compressive strength of various proportions of activators at 28 days curing.

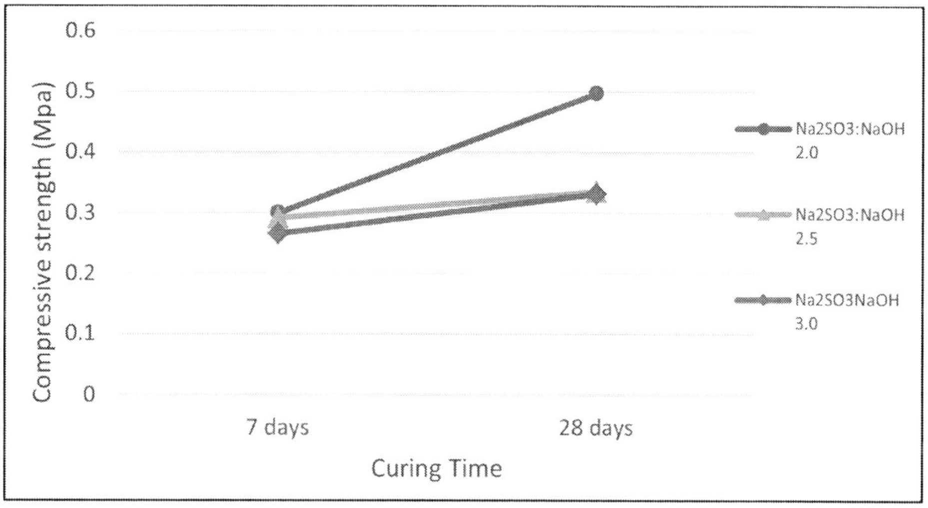

Fig. 5. Effect of Na_2SO_3 / NaOH ratio at different curing time.

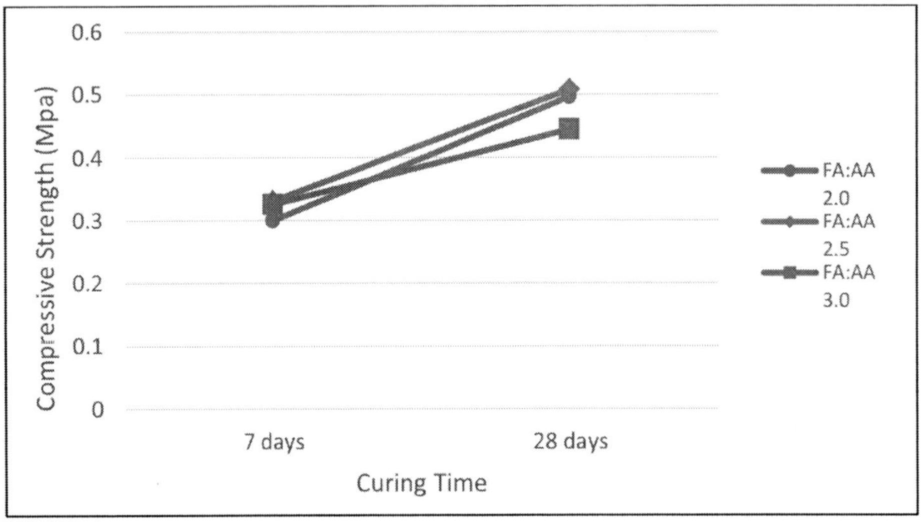

Fig. 6. Effect of FA / AA ratio at different curing time.

Figure 5 and Figure 6 shows the effect of Na_2SiO_3 / NaOH ratio and effect of FA / AA ratio at different curing time respectively. From both figures, there is significant strength increment from day 7 to day 28 of curing time. The different ratio of Na_2SiO_3 / NaOH gives a significant impact compare to the different ratio of FA / AA. At the ratio Na_2SiO_3 / NaOH of 2.0, there are 54% increment from day 7 to day 28 of curing. This is showing that geopolymerization process occurs during the curing time. This result shows the curing time plays an important role for strength development in geopolymer stabilized soil.

4 Conclusions

The aim of this paper is to present the effect of fly ash/alkaline activator ratio, Na_2SiO_3/NaOH ratio and curing time for geopolymer stabilized soil. From the experimental results, the following conclusions have been made:
1) The highest strength determine at the ratio of fly ash/alkaline activator and Na_2SiO_3/NaOH 2.5 and 2.0 respectively.
2) Curing time plays an important role where the highest strength determines at 28 days curing.

References

1. P.J. Davidovits, *Geopolymer 2002 Conference*, (2002)
2. D. Khale, R. Chaudhary, J. Mater. Sci., **42**, 729 (2007)
3. P. Duxson, A.F. Jiménez, J.L. Provis, G.C. Lukey, A. Palomo, J.S.J. Deventer, J. Mater. Sci., **42**, 2917 (2006)
4. A. Gonzalez, R. Navia, N. Moreno, Waste Manag. Res., **27**, 976 (2009)
5. T.P. Ngernkham, A. Maegawa, N. Mishima, S. Hatanaka, P. Chindaprasirt, Constr. Build. Mater., **91**, 1 (2015)
6. D. Hardjito, *Studies on Fly Ash-Based Geopolymer Concrete*, (Curtin University of Technology, 2005)
7. R.A. Mozumder, A.I. Laskar, Comput. Geotech., **69**, 291 (2015)
8. M. Zhang, H. Guo, T.E. Korchi, G. Zhang, M. Tao, Constr. Build. Mater., **47**, 1468

(2013)

9. P. Sukmak, S. Horpibulsuk, S.L. Shen, P. Chindaprasirt, C. Suksiripattanapong, Constr. Build. Mater., **47**, 1125 (2013)
10. D. Hardjito, S.E. Wallah, D.M.J. Sumajouw, B.V. Rangan, Advances in Geopolymer Science and Technology, **6**, 88 (2007)
11. W.K. Part, M. Ramli, C.B. Cheah, Constr. Build. Mater., **77**, 370 (2015)
12. M.M.A. Abdullah, H. Kamarudin, M. Hussain, I.K. Nizar, A.R. Rafiza, Y. Zarina, Adv. Mater. Res., **328**, 1475 (2011)
13. N. Cristelo, S. Glendinning, T. Miranda, D. Oliveira, R. Silva, Constr. Build. Mater., **36**, 727 (2012)
14. N. Cristelo, S. Glendinning, L. Fernandes, and A.T. Pinto, Constr. Build. Mater., **29**, 167 (2012)

The Properties of Nano TiO_2-Geopolymer Composite as a Material for Functional Surface Application

D. Syamsidar[1,2,*], *Nurfadilla*[1,2] , and *Subaer*[1,2]

[1]Department of Physics, FMIPA, Makassar State University, Makassar 90233, Indonesia
[2]Konsorsium Riset Geopolimer Indonesia (KORIGI) Laboratorium Beton dan Bahan Bangunan Kampus ITS Sukolilo Surabaya, Indonesia, 6011

Abstract. The aim of this study is to examine the properties of Nano TiO_2-geopolymer as a material for functional surface applications such as walls, floors, bench top, arts and decoration materials. Class-C fly ash and metakaolin were used as raw materials to produce geopolymers pastes (binder). Geopolymers were synthesized through alkali activation method cured at 50°C for 2 hours using molar oxide ratios of SiO_2/Al_2O_3 = 3.0, Na_2O/SiO_2 = 0.2, and H_2O/Na_2O = 10. Nano TiO_2 was added into geopolymers paste at different concentration namely 0 wt%, 5wt%, 10wt% and 15wt % relative the weight of fly ash or metakaolin). The measurements were commenced after the samples aged 7 days. The samples made from fly ash were immersed in 1 M H_2SO_4 solution for 3 days for acid resistance examination. The self-cleaning properties of the composites were observed by immersing the sample into red clays solution. The X-Ray Diffraction (XRD) was performed to examine the structure and phase of the samples before and after acid resistance measurement. Scanning Electron Microscopy (SEM) was performed to examine the surface morphology of the resulting composites. The measurements results showed that Nano TiO_2–geopolymers composite can be applied as functionally surface materials.

1 Introduction

The quality of building materials can be valued not only by their mechanical, but also durability and structural performance. Structural application refers to application that requires mechanical performance (e.g. strength, stiffness and vibration damping ability) in material [1]. The essential requirements consist of low density, corrosion resistance, and ability to resist high temperature. Functional or hybrid composite based on geopolymer are currently investigated due to its excellent properties such as durability, mechanical, and thermal properties [2].

Geopolymer, is a green and eco-friendly material, developed as a promising material because of its high mechanical strength, heat and fire resistance, good surface and

* Corresponding author : sidarsyams@gmail.com

resistance to acid attack [3,4]. In addition, geopolymers have been recognized as excellent binder material and with an addition of certain aggregate such as organic or inorganic fibres or particles will form high performance composites. The advantages of producing composite are to prevent the presence and development of cracks and to enhance the mechanical strength, particularly flexural strength, of geopolymer matrix [4]. Recently, composites based on geopolymer have been studied for producing high performance and functionally surface materials such as walls and floors, bench top, and other applications [3-5,6].

Nanomaterial is one of the promising material because of its size (1-100 nanometres) produced for wide range applications. Metal oxide such as TiO_2 is a semiconductor material which has photo-catalyst properties [7-10] and it is one of the most practical nanoparticles which offer a promising platform for different applications, e.g. high performance catalysis, photo-catalysis, self-cleaning and electro-catalysis systems [8-9, 11-13]. Titanium dioxide (TiO_2) is one of the most used materials to realize self-cleaning and its general use is attributed to high catalysis efficiency and chemical stability [14-17].

This study used Class-C fly ash and metakaolin as raw materials for producing geopolymers pastes and nano TiO_2 as filler. The study was carried out to determine the properties of the resulting composite such as mechanical strength, self-cleaning ability, and resistance to acid attack.

2 Experimental

This research is aimed at developing geopolymer composites made from fly ash and metakaolin added with TiO_2 nanoparticles as filler for functional surface. Fly ash is derived from Bosowa Power Plant Jeneponto, South Sulawesi and it is categorized as class-C. The molar oxide ratios of the starting materials were SiO_2/Al_2O_3 = 3.0, Na_2O/SiO_2 = 0.2 and H_2O/Na_2O = 10.0. This composition will result in geopolymer type poly (sialate-siloxo) according to Davidovits's terminology. The concentration of TiO_2 (Anatase phase) was varied relative to the weight of fly ash or metakaolin, namely 0%, 5%, 10% and 15%. TiO_2 nanoparticle was prepared through ball milling process for 10 hours. The mixture between geopolymer paste and TiO_2 was stirred manually, poured in various moulds and cured at 50^oC for 2 hours. The resulting materials were stored in open air for 7 days before commencing any measurements. The compressive strength was measured by using Tokyo Testing Machine (TTM). X-ray diffraction (XRD) measurements were conducted to study the structure and phase of the starting and the resulting composites. The acid resistance of the material was measured by immersing the samples into 1M H_2SO_4 solutions for 3 days. The self-cleaning ability of the samples was observed by immersing the samples into red clay solution for several minutes. The surface morphology of the samples was examined by using scanning electron microscopy (SEM).

3 Results and discussion

The chemical compositions of fly ash and metakaolin used in this study were examined by using Energy Dispersive Spectroscopy (EDS) as shown in table 1.

DOI: 10.1051/matecconf/20179701013

Table 1. Chemical composition (wt%) of fly ash and metakaolin (EDS result).

	Al_2O_3	CaO	FeO	K_2O	MgO	Na_2O	SiO_2	SO_3	TiO_2
Fly ash	22.25	13.95	23.41	1.39	6.20	0.99	30.10	0.51	1.20
Metakaolin	50.88	-	1.14	0.89	0.29	0.18	46.46	-	0.16

Table 1 shows that the content of CaO is more than 10wt% and this make the time setting of geopolymer paste is much less than 30 minutes and made difficult to produce in large volume. In contrast, the time setting of geopolymer paste made from metakaolin exceeded 2 hours and therefore much easier to produce in large scale.

Figure 1 show photographs of the example of the composites produced in this study. It can be seen that the surface of the composite is smooth and shiny, free from apparent pores or air bubbles. The densities of the composite with TiO_2 addition of 0 wt%, 5wt%, 10wt%, and 15wt% relative to the weight of fly ash were 2.10 g/cm^3, 2.16 g/cm^3, 2.18 g/cm^3 and 2.20 g/cm^3, respectively.

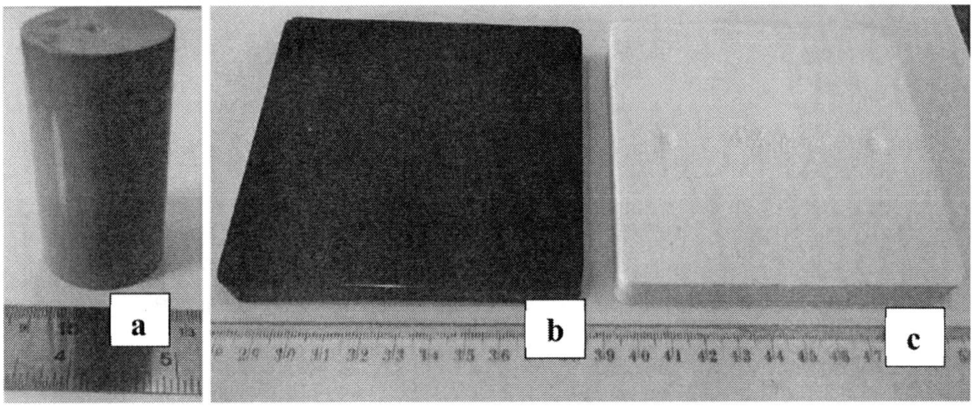

Fig. 1. Photographs of geopolymers; (a) & (b) Fly Ash + TiO2 nanoparticle, (c) Metakaolin + TiO_2 nanoparticle.

The compressive strength of the composites were measured for samples having cylindrical in shape with a length of 4.00 cm and 2.00 cm in diameter. Table 2 shows the compressive strength of the samples contain different concentration of TiO_2. It can be seen that the magnitude of the compressive strength increases as the weight of TiO_2 increase up to 10% and start to decrease at the addition of 15wt% TiO_2. Data for mechanical characterisations of nano TiO_2-geopolymers composite made from metakaolin unfortunately did not available at the time of writing this paper.

Table 2. The average of 7 days compressive strength of Composite TiO_2-Geopolymers made from fly ash.

Sample ID	Compressive Strength (MPa)
HDS 0% TiO_2	25.79
HDS 5% TiO_2	28.98
HDS 10% TiO_2	39.65
HDS 15% TiO_2	37.56

 DOI: 10.1051/matecconf/20179701013

Figure 2(a) shows typical diffractogram of fly ash and metakaolin. The crystallinity level and chemical compositions of fly ash and metakaolin are very much different. Metakaolin is x-ray amorphous and considered as pure aluminosilicate although contain some minor phases such as Quartz and Anatase. Figure 2(b) shows diffractogram of TiO_2 nanoparticle used in this study.

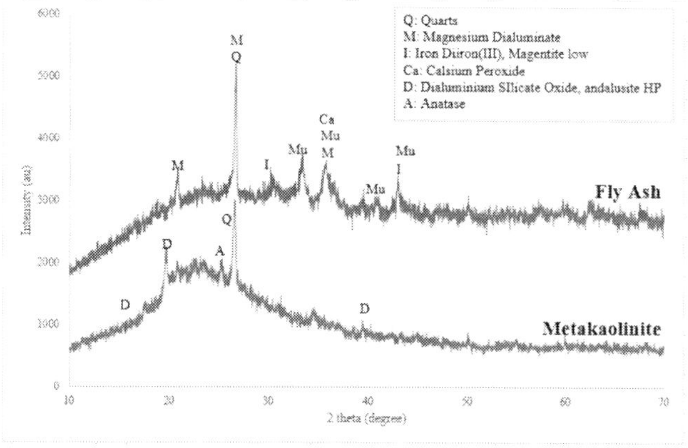

(a)

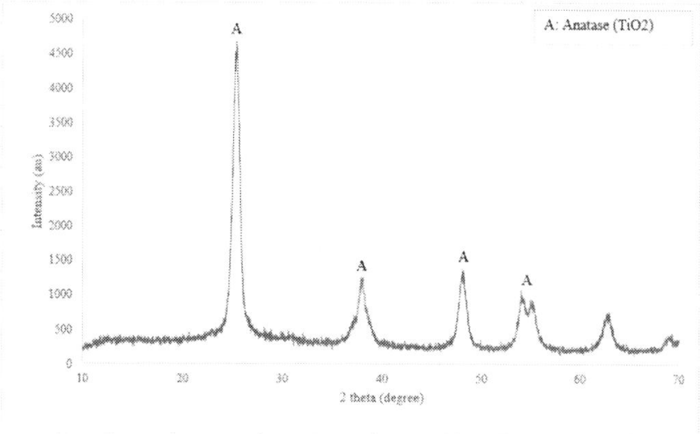

(b)

Fig. 2. Diffractogram of (a) Metakaolin and Fly Ash (b) TiO_2 nanoparticle.

The average size of TiO_2 nanoparticle was about 62 nm as calculated by using Debye-Scherer equation. Previous research which utilizes vanadium-doped TiO_2 nanoparticles has identical anatase phase with an average crystal size of 10-20 nm [16].

Diffractogram of as-prepared composite TiO_2-geopolimer based on fly ash and metakolin is shown in Figure 3. The peak of TiO_2 increase as the concentration of TiO_2 increase in the network of geopolymers. Figure 3 also shows that there was no new phase formed in the structure of geopolymer due to the addition of nano TiO_2 indicating that TiO_2 did not react with other fly ash or metakaolin constituents.

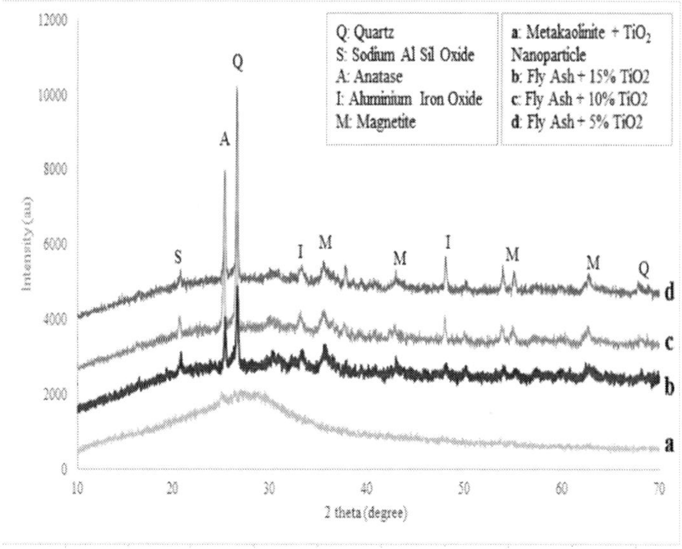

Fig. 3. Diffractogram of TiO_2-geopolymers composite made from fly ash containing 5 wt%, 10wt%, 15wt% of TiO_2 along with TiO_2-geopolymer made from metakaolin.

Figure 4 shows SEM micrographs of composite nano TiO_2-geopolymer produced from fly ash and metakaolin. Substantial cracks on the surface of composites made from fly ash occurred due to polishing and water loss during evacuation in coating and SEM chambers. This kind of cracks is known as secondary cracks [17]. Cracks development in composite made from metakaolin did not as extensive as those of fly ash. The micrographs also show good mechanical bonds between geopolymer matrix and TiO_2 nanoparticles.

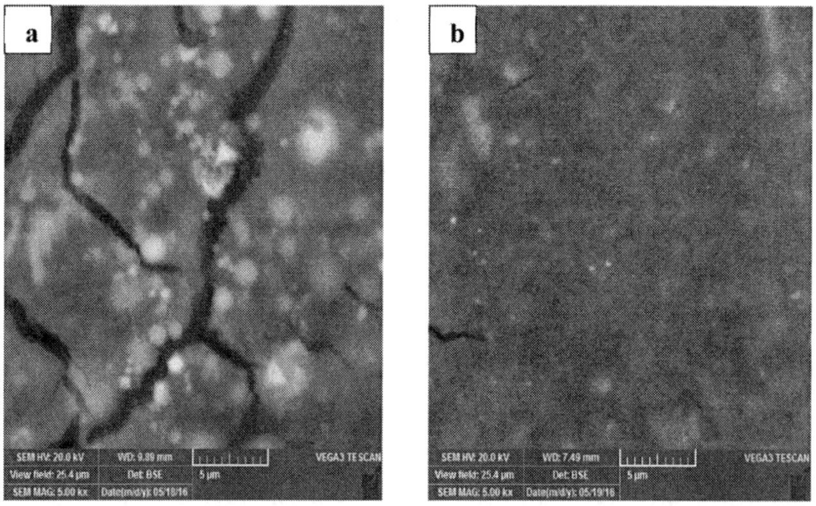

Fig. 4. SEM micrographs of composite TiO_2-geopolymers made from (a) fly ash and (b) metakaolin. The composites contain 10% of TiO_2.

The composites made from fly ash which contains 0%, 5% and 15% of TiO_2 were subjected to acid resistance measurement by using 1M H_2SO_4 solution for 3 days. Figure 4 shows the diffractogram of the sample after 3 days in H_2SO_4 solution. The diffractogram revealed that H_2SO_4 react strongly with CaO in fly ash forming gypsum ($CaSO_4.2H_2O$)

crystal. The sample containing 5wt% TiO2 showed the highest formation of gypsum. The presence of excess gypsum in the interior of geopolymers will reduce its mechanical strength [18].

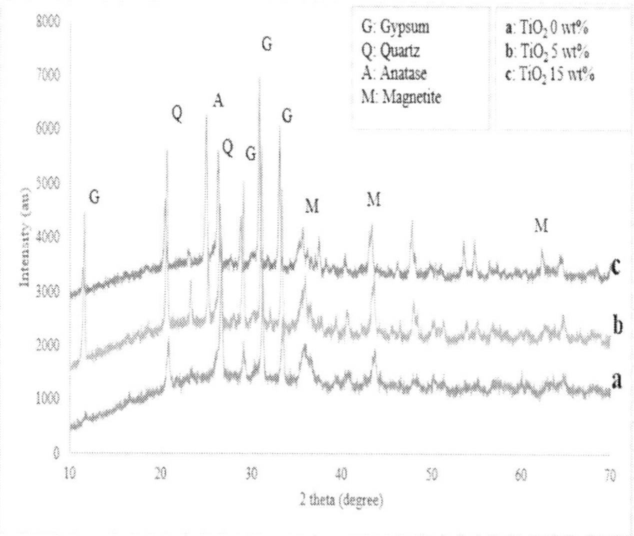

Fig. 5. Diffractogram of samples with 0%, 5% and 15% TiO_2 after 3 days acid attack.

SEM results on the composite samples made from fly ash after 3 days immersed in H_2SO_4 solutions is shown in Figure 6. The micrographs confirmed the formation of gypsum crystals in geopolymer network as observed in the diffractogram figure 5. Samples contain 5% TiO_2 showed the largest formation of gypsum crystal. This result suggests that geopolymers produced from fly ash containing high CaO will form gypsum crystals when in contact with H_2SO_4 solution and hence deteriorating the mechanical strength of geopolymers.

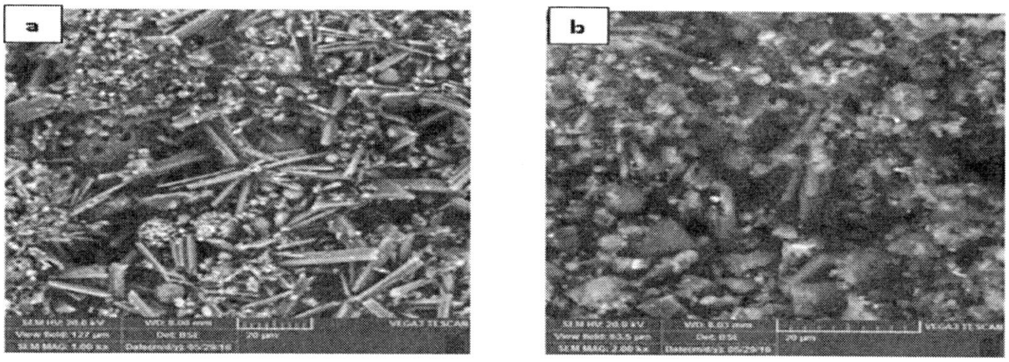

Fig. 6. SEM micrographs of composites made from fly ash after 3 days in H_2SO_4 solution showing the growth of gypsum crystals (a) 5% TiO_2, (b) 10% TiO_2.

 DOI: 10.1051/matecconf/20179701013

The addition of TiO_2 nanoparticles in the network of geopolymers result in functional composite which have a self-cleaning ability. It is well known that TiO_2 nanoparticle is an excellent photocatalyst material particularly when it size below 100 nm. The self-cleaning ability of the composites made from fly ash and metakaolin were examined by immersing the sample into red clay solution for several minutes as in Figure 7. The sample was then taken out and it was found that the surface of the composite remains clean and no dirt or clay particulates remain on its surface.

Fig. 7. Functional composite TiO_2-geopolymer made of fly ash exhibited self-cleaning ability.

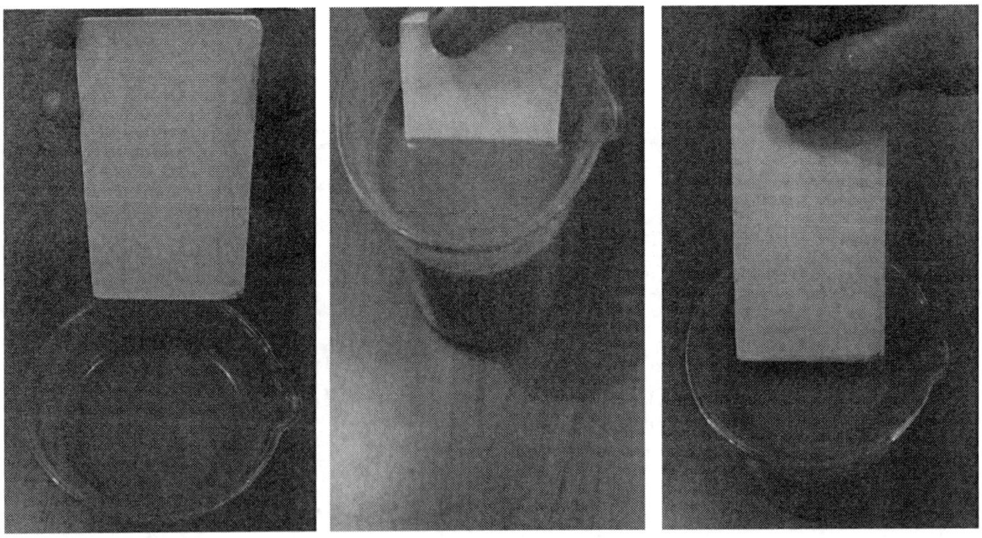

Fig. 8. Functional composite TiO_2-geopolymer made of fly ash exhibited self-cleaning ability.

4 Conclusions

Nano TiO_2-geopolymer composites based on fly ash and metakaolin have been successfully produced through alkali activation method. The mechanical strength of the composites (made from fly ash) increase with the increase of TiO_2 concentration up to 10wt%. The composite made from high CaO fly ash suffer from the formation of substantial gypsum crystal when in contact with H_2SO_4 solution. The nano TiO2-geopolymers composite has a potential to be applied as functionally surface material and exhibit self-cleaning properties.

References

1. D.L. Chung, *Composite Material: Science and Applications* (Springer, New York, 2004)
2. A. Natali, S. Manzi, M.C. Bignozzi, Procedia Eng., **21**, 1124 (2011)
3. W. Zhao, Y. Wang, X. Wang, D. Wu, Ceram. Inter., **42**, 6329 (2015)
4. A.R. Sakulich, Sustainable Cities and Society, **1**, 196 (2011)
5. I. Phummipan, S. Horpibulsuk, P. Sukmak, A. Chinkulijiniwat, A. Arulrajah, S. Shen, Road Mater. Pavement., **17**, 1 (2015)
6. R.H. Kupaei, U.J. Alengaram, M.Z. Jumaat, Electronical Journal of Structural Engineering, **13(1)**, 1 (2013)
7. K. Sobolev, I. F. Vivian, *The 4th International fib Congress*, (2014)
8. K. Ghavani, C. Rodrigues, S. Paciornik, Asean Journal of Civil Engineering (Building and Housing), **4(1)**, 1 (2003)
9. M.S. Omima, A.A. Saleh, M.F. Mostafa, A.W.A.A. Magd, Int. J. Photoenergy, **2008**, 1 (2007)
10. M. Mashiana, M.C. Kenneth, M. Yusoff., Solid State Sci. Tech., **21**, 23 (2013)
11. D. Yang, H. Liu, Z. Zheng, Y. Yuan, J.C. Zhao, E.R. Waclawik, X. Ke, H. Zhu, JACS, **131(49),** 17885 (2009)
12. E.E.B. Moushoul, Y. Mansourpanah, K. Farhadi, M. Tabatabaei, Chem. Eng. J., **283**, 29 (2015)
13. S. Sabihuddin, Int. J. Eng. Res. Appl., **4(3)**, 92 (2014)
14. R. Vijayalaksmi, V. Rajendran, Arch. Appl. Sci. Res., **4(2)**, 1183 (2012)
15. Gupta, Tripathi, Chin. Science Bulletin, **56**, 1639 (2011)
16. N.M. Thuy, D.Q. Van, L.T.H. Hai, Nanomater. Nanotechnol., **2**, 1 (2012)
17. Subaer, *Pengantar Fisika Polimer* (D. P. Tinggi, Jakarta, 2012)
18. Y. Jun, J.E. Oh.,Materials, **8**, 3053 (2015)

The Potential of Laterite Soils Deposit South Sulawesi as a Precursor for Na-Poly (Ferro-Sialate) Geopolymers

Subaer[1,*], *Nursyamsih* Amalia[1] , *Nurfadilla*[1] , *Andi* Riska[1] ,*Sulfiana*[1] , and *Fitria Pebriyanti* San[1]

[1]Material Physics Laboratory, FMIPA, Makassar State University, Makassar 90233, Indonesia

Abstract. The main objectives of this study was to investigate the potential of lateritic soils deposit South Sulawesi, Indonesia as a precursor for Na-poly(ferro-sialate) geopolymers. The samples of laterite soils were taken from three different regions, namely Sidrap, Bone, and Gowa regency. The soil was clean, grounded, sieves 200 mesh, and dehydroxylated at 750°C for 2 hours. The x-ray fluorescence (XRF) and energy dispersive spectroscopy (EDS) were used to examine the chemical compositions of the soils. The geopolymers was synthesized through alkali activation method by adjusting the molar oxide ratios of $SiO_2/(Al_2O_3+Fe_2O_3)$, Na_2O/SiO_2 and H_2O/Na_2O in accordance with the chemical compositions of the soils. The functional groups of the resulting geopolymers were examined by using Fourier Transform Infra-Red (FTIR). The structure and phase of the resulting material were studied by using x-ray diffraction (XRD). The surface morphology of geopolymers was studied by using scanning electron microscopy (SEM). The mechanical strength of the materials was examined through compressive strength measurement. The results of this study showed that high strength Na-poly (ferro-sialate) geopolymers were successfully produced and characterized.

1 Introduction

Laterite is one of a residual soils rich in iron and aluminum formed through weathering processes in tropical areas and subtropical climates [1,2]. Lateritic or laterites soils with a chemical designation of Fe_2O_3-Al_2O_3-SiO_2-H_2O, are derived from kaolinite in which a high proportion of Al^{3+} is replaced by Fe^{2+} or Fe^{3+}. Since iron and aluminum oxides are prominent in lateritic soils, and with the seasonal fluctuation of the water table, these oxide result in the reddish to brown in color that is seen in laterite soils [3].

Laterite have been used extensively in construction of dams, embankment as well as buildings, and other environmentally friendly materials [4,5,6]. Laterite incorporation in structural concrete elements will likely reduce the cost of construction of buildings significantly. On the other hand, the use of these materials has some tendencies to reduce environmental problems [7].

* Corresponding author: jzubayir@yahoo.com

Over the last three decades, geopolymers have been intensively studied as a new form of inorganic polymer material that could substantially substitute for conventional or ordinary Portland cement, plastics, ceramics-composites and many mineral-based products. Geopolymers are a subset of the broader class of alkali-activated binders, and for considerations, it appears to be a potential alternative to the classis hydraulic binders produced at low temperatur [8, 9]. The defining characteristic of a geopolymer is that the binding phase comprises an alkali aluminosilicate gel, with aluminium and silicon linked in a three-dimensional tetrahedral gel framework that is relatively resistant to dissolution in water [16,17, 8]. Research has shown that geopolymers may be readily synthesised through alkali-activation of inexpensive and pure starting materials such as kaolinitic clays [19,20,21], as well as waste products such as fly ash and furnace slag [18,22,23,24,25].

The use of laterite soils as a raw material for geopolymers was patented in 2012 by Davidovits, et.al.,[26] and designated as ferro-aluminosilicate type with an empirical formula of $(Ca,Na,K)\cdot(FeO)_x\cdot(SiOAlO)_{1-x}\cdot(SiO)_y$, where x is a less than or equal to 0.5, y is a value between 0 and 25. Several studies followed the utilization of aluminosilicates minerals rich with Fe in the production of geopolymers [8, 27].

South Sulawesi is one of the provinces in Indonesia where deposit of lateritic soils can be easily found. This research was conducted to investigate the properties of lateritic soils from 3 different sites in South Sulawesi, namely Sidrap, Bone, and Gowa regencies, as potential raw materials for poly-(ferro-sialate) or [Na] [- Fe-0-Si-0-Al-0-] type of geopolymers.

2 Experimental methods

This study was conducted to develop South Sulawesi lateritic soils as a raw material for poly-(ferro-sialate) geopolymers. The examples of lateritic soils were taken from three different location; Sidrap, Bone, and Gowa regencies. The examples of soils were clean, immersed in water for 24 hours, dried, grounded and sieved 200 mesh. The soils was dehydroxylated at 750°C for 2 hours and subjected to x-ray fluorescence (XRF) or energy dispersive spectroscopy (EDS) examinations to examine its chemical compositions.

Geopolymers were synthesized by varying the molar ratio of $SiO_2/(Al_2O_3+Fe_2O_3)$, Na_2O/SiO_2, and H_2O/Na_2O in accordance with the amount of SiO_2, Al_2O_3 and Fe_2O_3 available in lateritic soils samples. Sodium silicate solution was used as activator added with NaOH pellets to increase its pH up to 13. The geopolymers gel was molded in accordance with the measurements requirement, cured at 70°C for 1 hour. The resulting samples were stored in open air for 28 days.

The functional group of the geopolymers paste was examined by means of Fourier Transform Infra-Red (FTIR). X-ray diffraction (XRD) was used to study the phase and crystallinity of the starting and the resulting materials. Scanning Electron Microscopy-Energy Dispersive Spectroscopy (SEM-EDS) was used to study the surface morphology and chemical compositions of the resulting materials. Compressive strength was measured to examine the mechanical properties of the samples.

3 Results and discussion

Figure 1 shows the dehydroxylated laterite soils produced in this research. The soils from Gowa regency contain high amount of Fe_2O_3 and appear to be dark brown.

 DOI: 10.1051/matecconf/20179701014

Fig. 1. Dehydroxylated laterite soils from: (a) Sidrap, (b) Bone, and (c) Gowa

Table 1 shows the chemical composition of the soils from three different sites measured by using x-ray fluorescence (XRF). The soils contain Fe_2O_3 between 10 – 30wt% indicating the characteristics of lateritic soils. The soils taken from Gowa regency comprise 30.42wt% of Fe_2O_3 and give dark brown in color.

Table 1. Chemical composition of laterite soils.

Compound	Sidrap (wt%)	Bone (wt%)	Gowa (wt%)
SiO_2	47.85	50.10	30.68
Fe_2O_3	14.79	16.12	30.42
Al_2O_3	32.74	22.90	30.94
CaO	0.240	2.41	-
TiO_2	1.31	2.33	2.92
K_2O	0.864	1.09	-
MnO	0.151	0.366	1.21
SrO	0.0227	0.0732	-
ZrO_2	0.0814	0.0679	0.0931
BaO	0.032	0.046	-
Nb_2O_5	0.0185	0.0226	0.0265
MoO_3	0.0122	0.0153	-
ZnO	0.0110	0.0150	-
Y_2O_3	0.0077	0.0085	-
In_2O_3	0.0053	0.0077	0.0052

Figure 2 shows diffractogram of Dehydroxylated lateritic soils from three different sites. Laterite from Sidrap regency contain high wt% of quarzt, while albite mineral was the main species found in laterite Bone regency. Iron oxide in the form of hematite was the main constituent of Gowa lateterite.

 DOI: 10.1051/matecconf/20179701014

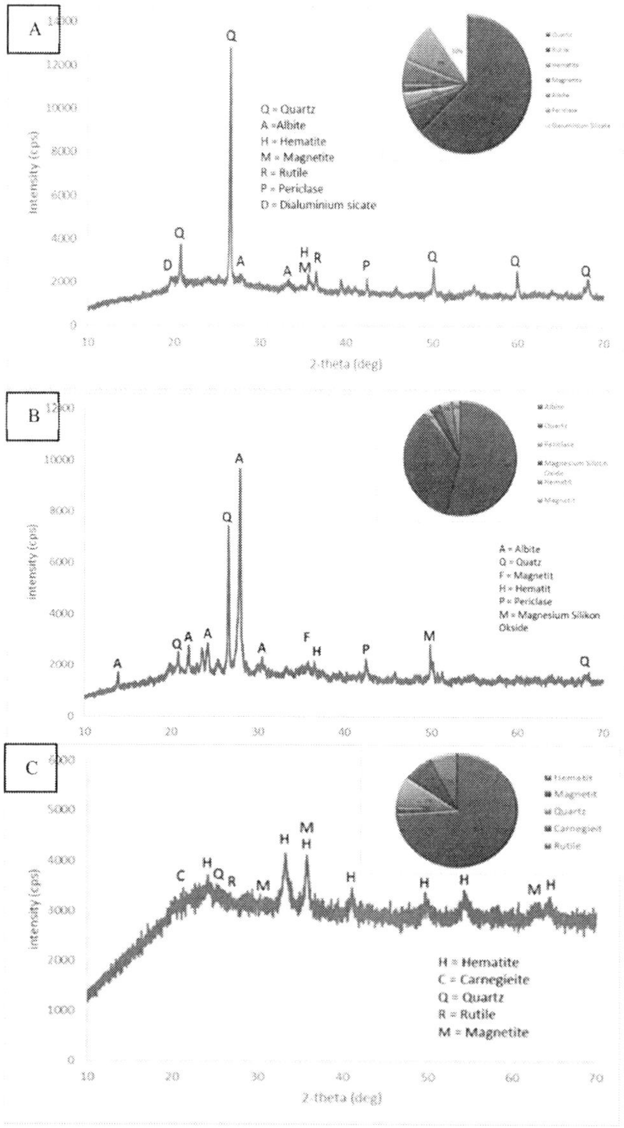

Fig. 2. Diffractogram patterns of laterite soils from (a) Sidrap, (b) Bone, and (c) Gowa.

Figure 3 shows SEM images of dehydroxylated soils from three different sites. The morphology of laterite soils particles from Sidrap shows the shape of kaolinite minerals as this soil contains high percentage of kaolin. Leterite soils from Bone and Gowa contain high percentage of albite and hematite with irregular particles shape.

DOI: 10.1051/matecconf/20179701014

Fig. 3. SEM images of laterite soils from a) Sidrap, b) Bone, and c) Gowa.

Figure 4 displays the IR spectra of product geopolymers made from three different laterite soils. The broad bands appearing at 2316-3351cm^{-1} were due to stretching vibrations of O–H and H–O–H [11]. The high content of iron oxides of lateritic soil from Gowa (figure 3C) results in slightly different IR bands. The peaks characteristics of water are almost absent indicating that the laterites based inorganic polymers have relatively low water content compared to the standard metakaolin based geopolymers. The band at 1389 cm-1 was attributed as Al–OH or Si–O vibration. The vibration around 1000 cm^{-1} was due to a strong band for Si–O–Si stretching vibration and it is the fingerprint of geopolymerisation [12-14]. The Si–O–Al bond was observed between 400 cm^{-1} – 875 cm^{-1} indicating the reaction between alkali solution and solid Al–O sheet in the lateritic clays. The bands at 442.35 cm^{-1} was identified as Si–O–Fe or Si–O stretching modes. Vibration band of Si–O– Fe was also found around 468.9 cm^{-1} [15]. The presence of Si–O–Fe vibrations confirmed that the produced geopolymers were categorized as poly(ferro-sialate).

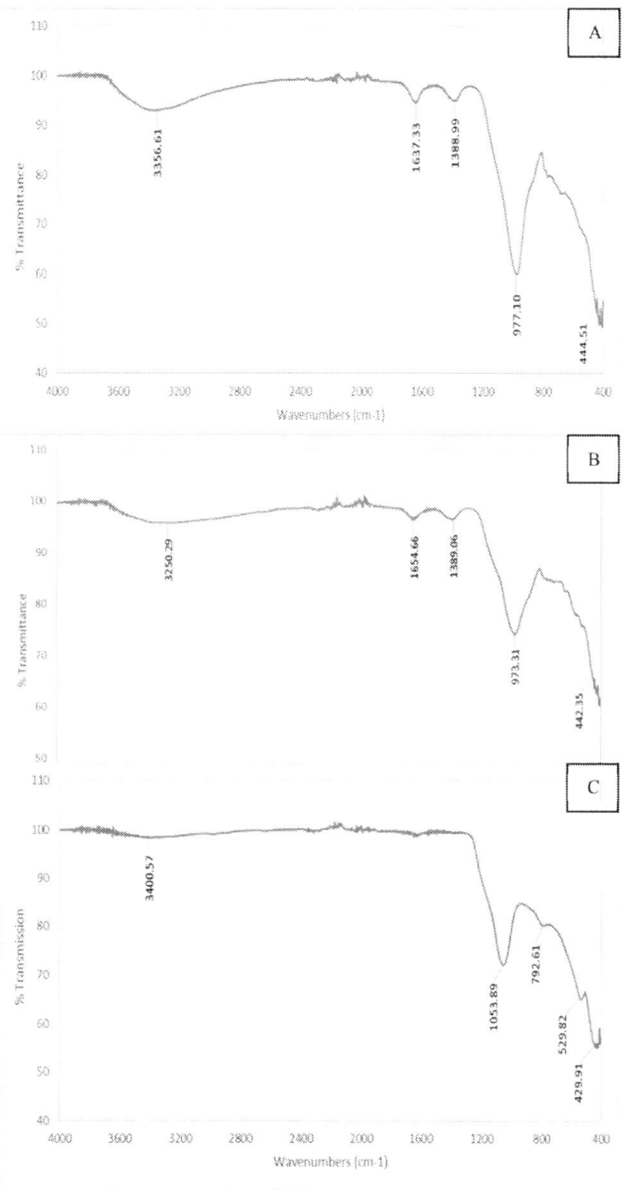

Fig. 4. IR spectra of geopolymer pastes made from laterite soils (a) Sidrap, (b) Bone, and (c) Gowa.

Figure 5 shows x-ray diffractogram of geopolymer pastes produced by using lateritic soils from (a) Sidrap, (b) Bone, and (c) Gowa. Diffaractogram of geopolymer made from Sidrap laterite soils was dominated by almandite and sodium aluminium silicate phases. sampel lanah Sidrap yang menunjukkan fase almandite dan sodium aluminium Silicate yang terbentuk dari reaksi antara aktivator dan bahan dasar. Aegirine phase, (Na)(Fe-Si-Al-O), was observed in geopolymer made from Bone laterite. Sodium iron silicate was observed in geopolymer made from Sidrap and Gowa laterites. The results confirmed that lateritic soils from three different locations produced ferro-sialate type of geopolymers.

DOI: 10.1051/matecconf/20179701014

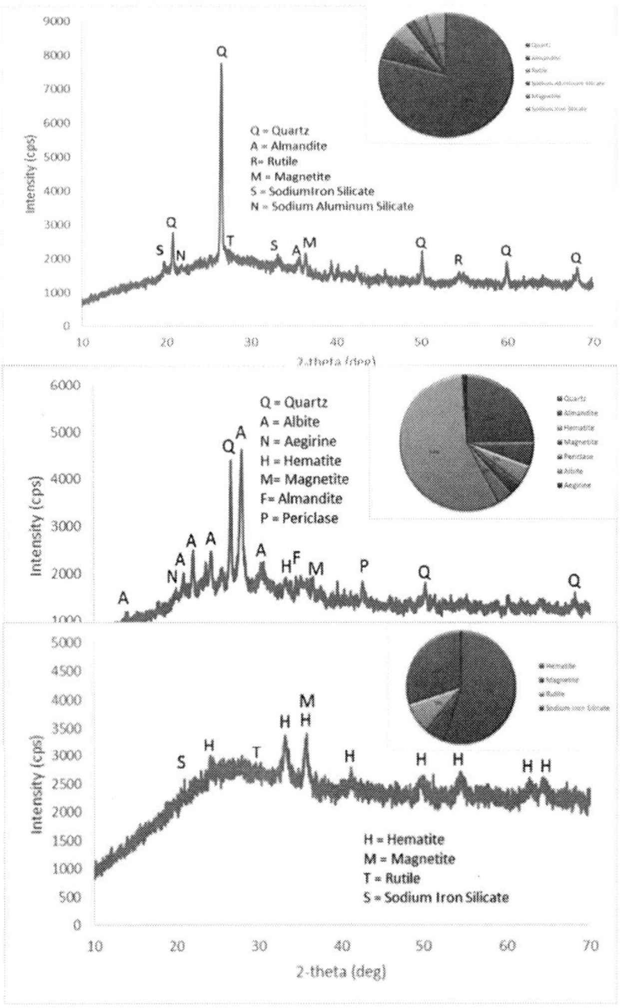

Fig. 5. Diffractogram patterns of geopolymers made from laterite (a) Sidrap, (b) Bone, and (c) Gowa.

SEM examination results of geopoplymer pastes made from lateritic soils is shown in figure 6. The surface of geopolymers made from Sidrap and Gowa laterites suffered from white efflorescence and produced long and sharp crystals of sodium carbonate ($Na_2CO_3(H_2O_7)$) as can be seen in figure 6(a) and (b). SEM image of geopolymer made from Gowa laterite showed the presence of substantial amount of iron oxide particles (bright color).

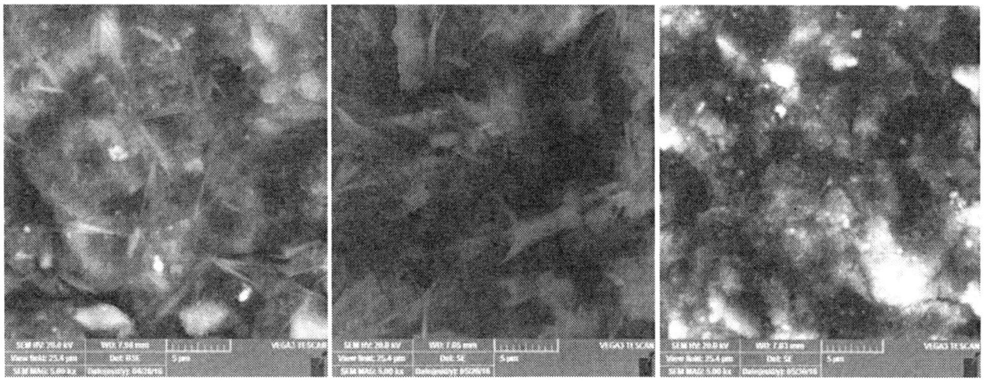

Fig. 6. SEM images of geopolymers paste made from laterites soils (a) Sidrap, (b) Bone, and (c) Gowa

Figure 7 shows the examples of geopolymer pastes made from laterite soils (a) Sidrap, (b) Bone, (c) Gowa for compressive strength measurements. The samples were produced by using acrylic mould and resulting a silinder shape of specimens 4.0 cm in length and 2.0 cm in diameter. The compressive strength measurement was conducted after the samples aged 28 days.

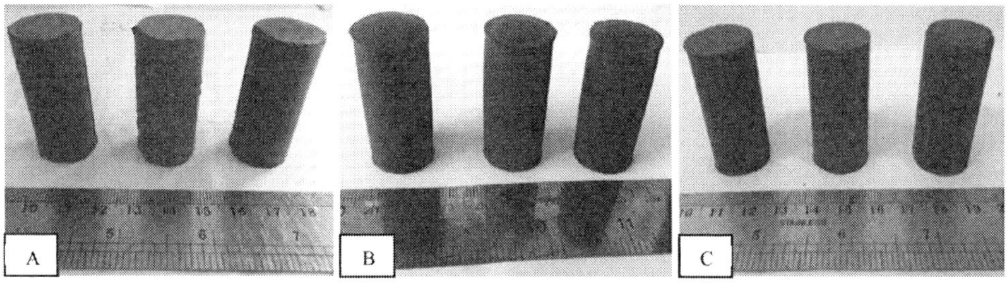

Fig. 7. Examples of geopolymer pastes made from laterite soils (a) Sidrap, (b) Bone, (c) Gowa for compressive strength measurements

Table 2 shows the average magnitude of compressive strength of geopolymer pastes produced from different laterites soils. The value of compressive strength of iron rich-geopolymers produced in this study is similar to those made from metakaolin geopolymer [10]. The compressive strength of Ca-poly(ferro-sialate) type geopolymer is between 60 – 90 MPa [26]. The difference can be attributed to the type of alkali activator as well as the concentration of iron oxides in lateritic soils used.

Table 2. The average of compressive strength of geopolymers made from laterite soils

Laterite soils	Compressive Strength (MPa)
Sidrap	43.24
Bone	44.88
Gowa	39.91

4 Summary

Iron rich-geopolymers or Na-poly(ferro-sialate) based on three different lateritic soils as raw materials have been successfully produced. The content of iron oxides (hematite and magnetite) of the lateritic soils range from 10 – 30wt%. The presence of functional groups of Si-O bands and Si-O-Al bands as well as the presence of (Na) (Fe-Si-Al-O) phase confirmed the geopolymerisation of lateritic soils used in this study. The compressive strength of the produced was comparable to those made from metakaolin.

References

1. I.E. Kalu, E.E. Jossou, F. Jacob, A.I. Ilboudo, A.F. Adedeji, W.O. Soboyejo, Inter. J. Eng. Technol., **5**, 87 (2015)
2. T.H. Ko, ScientificWorldJournal, **1** (2014)
3. P.N. Lemougna, U.F.C. Melo, E. Kamseu, A.B. Tchamba, Sustainability, **3**, 293 (2011)
4. I.A. Oyediran, J. Okosun, Provessional Paper, **60**, 287 (2012)
5. J.O. Ukpata, M.E. Ephraim, G. A. Akeke, J. Eng. Appl. Sci., **7**, 81 (2012)
6. E.A. Adam, Compressed Stabilized Earth Block Manufacture in Sudan (UNESCO, Paris, 2001)
7. J.O. Ukpata, M.E. Ephraim, J. Eng. Appl. Sci., **7**, 324 (2012)
8. W. Hajjaji, S. Andrejkovicova, C. Zanelli, M. Alshaaer, M. Dondi, J.A. Labrincha, F. Rocha, Mater. Des., **52**, 648 (2013)
9. F. Skvara, L. Kopecky, J. Nemecek, Z. Bittnar, Ceramics, **50**, 208 (2006)
10. Subaer, Pengantar Fisika Geopolimer (Direktorat Jendral Pendidikan Tinggi, Jakarta, 2015)
11. Z. Yahya, M.M. Abdullah, K. Hussin, K.N. Ismail, R.A. Razak, A.V. Sandu, Materials, **8**, 2227 (2015)
12. A. Allahverdi, E.N. Kani, Int. J. Civ. Eng., **7**, 154 (2009)
13. A.M.M. Bakri, H. Kamarudin, M.B. Hussain, I.K. Nizar, A.R. Rafiza, Y. Zarina, Adv. Meter. Sci., **30**, 90 (2012)
14. E. Arioz, O. Arioz, O. M. Kockar, Inter. J. Chem. Eng. Appl., **4**, 423 (2013)
15. P.S. Nayak, B. K. Singh, Bull. Mater. Sci., **30**, 235 (2007)
16. K.J.D. Mackenzie, What are these things called geopolymers? A physico-chemical perspective,(2003)
17. C.A. Rees, J.L. Provis, G.C. Lukey, J.S.J.V. Deventer, Langmuir, **23**, 8170 (2007)
18. C. Shi, P.V. Krivenko, D. Roy, Alkali-activated Cements and Concretes (Taylor &Francis, London, 2006)
19. J. Davidovits, U.S Patent No.4,349,386 (1982)
20. H. Rahier, B.V. Melle, M. Bieseman, J. Wastiels, X. Wu, J. Mater. Sci., **31**, 71(1996)
21. H. Rahier,W. Simons, B.V. Melle, M. Biesemans, J. Mater. Sci., **32**, 2237(1997)
22. A. Palomo, M.W. Grutzeck, M.T. Blanco, Cem. Concr. Res., **29**, 1323 (1999)
23. J.G.S.V. Jaarsveld, J.S.J.V. Deventer, L. Lorenzen (1997), Miner. Eng., **10**, 659 (1997)
24. J.G.S.V. Jaarsveld, J.S.J.V. Deventer, (1999), Cem.Concr. Res., **29**, 1189 (1999)
25. A.R. Brough, A. Atkinson, Cem.Concr. Res., **32**, 1 (2002)
26. J. Davidovits, F. Davidovits, R. Davidovits, Patent No. WO/2012/056125/2012 (2002)
27. E.A. Obonyo, E. Kamasue, P.N. Lemougna, A.B. Tchamba, U.C. Melo, C. Leonelli, Sustainability, **6**, 5535 (2014)

DOI: 10.1051/matecconf/20179701015

Synthesis and Characterization of Functional Composite Carbon-Geopolymers for Precast Panel Application

Kharisma Noor Afifah[1,2,*], *Asrianti* Bt.Sunardi[1,2], *Resky* Irfanita[1,2], and *SUBAER*[1,2]

[1]Laboratorium Fisika Material, JurusanFisika, UniversitasNegeri Makassar, Indonesia
Jl. Daeng Tata Raya, Makassar, Indonesia
[2]Konsorsium Riset Geopolimer Indonesia (KORIGI) Laboratorium Beton dan Bahan Bangunan Kampus ITS Sukolilo Surabaya, Indonesia, 6011

Abstract. The purpose of this study is to examine the influence of carbon (C) particles as filler (aggregate) in the production of geopolymers functional composite for possible precast panel application. Geopolymers was synthesized through alkali activation of metakaolin added with carbon particles relative to the mass of metakaolin. The mixture was cured at 70°C for 2 hours and the resulting composites were stored in open air for 28 days. The bulk density and the apparent porosity of the composites were measured by using Archimedes method. The thermal properties of the samples was examined by using thermal conductivity measurement and differential scanning calorimetry (DSC). The microstructure characterization of the samples were performed by using x-ray diffraction (XRD) and Scanning Electron Microscopy-Energy Dispersive Spectroscopy (SEM-EDS).

1 Introduction

Geopolymer is a new type of inorganic polymers synthesized through alkali activated Aluminosilicate materials at temperature below 100°C [1]. This new inorganic material consists of polymeric chains of –Si–O–Al– similar to zeolite but its network is amorphous [2,3]. Geopolymers is well known for its high early strength, low shrinkage, acid, corrosion, and fire resistance [3,4]. Nowadays, geopolymers can be synthesized from kaolin, clays, fly ash, red mud and other aluminosilicate materials [1].

This study utilized metakaolin as a raw material to produce geopolymers paste. Metakaolin is dehydroxilated product of kaolinite mineral at temperature around 750°C [2,5]. Geopolymers produced from metakaolin has already applied in many areas due to its excellent physical and mechanical properties [8]. Similarly, geopolymers synthesized from

* Corresponding author: Knoorafifah@gmail.com

fly ash has also been applied as green concrete in many areas such as precast floor panels, airport runways, and pedestrians paths [4,5,9,12].

Functional composite is a material comprising two or more different starting materials, engineered to gain high performance material [1]. In this study, geopolymers were synthesized by adding carbon (C) particles as aggregate. The use of carbon fiber and particulates as filler (aggregate) of geopolymers have studied intensively in recent years. The addition of carbon fibers or particulates was found to improve the compressive, flexural, and tensile strength as well as heat resistance of geopolymers. The effectivity of carbon fibers and carbon nanoparticles with geopolymers matrices were studied at 750°C and revealed that the carbon fibers or nanoparticles did not react with geopolymers constituents [8,10,11].

2 Experimental method

Geopolymers paste were synthesized through alkali activation of metakaolin. The composition of the starting materials were adjusted to achieved the molar ratio of SiO_2/Al_2O_3 = 3.0, Na_2O/SiO_2 = 0.25, and H_2O/Na_2O = 10. Carbon particles were produced by burning coconut shell, grounded and shieved 200 mesh. The quality of carbon was examined by x-ray diffraction. The mass of carbon particles was varied from 1.0%, 1.5%, and 2.5% relative to the mass of metakaolin. The mixture between geopolymer paste and carbon particles was stirred manually and poured into plastic mold and cured at 70°C for 2 hours. The resulting composites were aged 14 days before conducting any measurements.

The bulk density and porosity of the resulting composites were measured by using Archimedes method. The flexural strength of the composites were measured by using three bending flexural measurement. The thermal properties of the materials were studied by using thermal conductivity measurement and Differential Scanning Calorimetry (DSC). The chemical composition and microstructure characters of the resulting materials were examined by using X-ray Diffraction (XRD) and Scanning Electron Microscopy-Energy Dispersive Spectroscopy (SEM-EDS).

3 Results and discussion

Fig. 1 shows diffractogram of metakaolin and carbon particles used this study. Both metakaolin and carbon particles have amorphous structures. Metakaolin contain quartz which is originated form kaolinite mineral.

 DOI: 10.1051/matecconf/20179701015

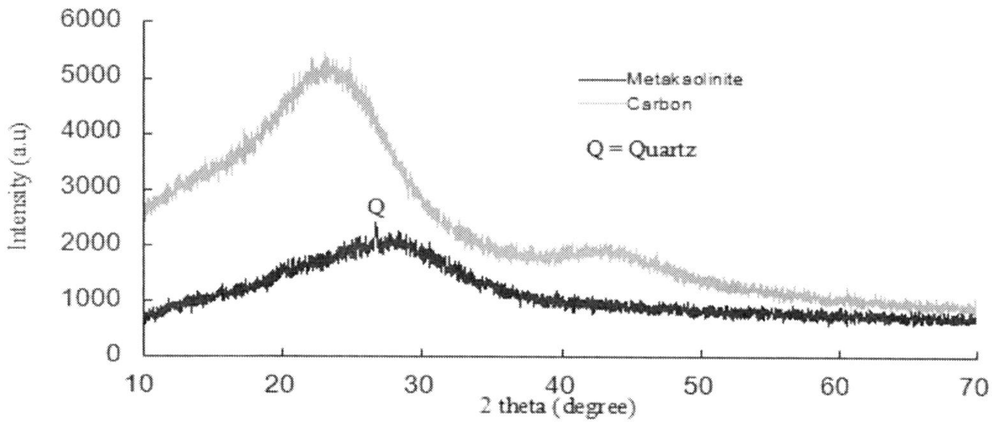

Fig. 1. Diffractogram of metakaolin and carbon particles.

Fig. 2 shows example of carbon-geopolymers composites produced in this study. Sample (a) contain 1.5% carbon, and sample (b) contain 2.5% carbon relative to the mass of metakaolin. Scratch on the surface of sample (B) was originated from defect on the surface of mold.

Fig. 2. Carbon-geopolymers composite (a) 1.5%, (b) 2.5% of carbon relative to the mass of metakaolin.

The maximum mass of of carbon particles added into geopolymers paste was 2.5% relative to the mass of metakolin. The addition of more carbon into the geopolymers paste keeping constant the concentration of alkali solution will be difficult to mix and results in low strength composite. The bulk density and apparent porosity of the resulting composites were determined by Archimedes method. Table 1 shows the density, the apparent porosity as well as electrical conductivity of the composites. The density and the apparent porosity of the samples revealed that the resulting composites were light and the incorporation of carbon did not increase the volume of pores in the bulk of geopolymers. The thermal and electrical conductivity was measured since carbon is a material with good thermal and electrical properties. The addition of carbon into geopolymers matrices was found to increase the thermal conductivity and it was higher than those made from standard metakaolin geopolymers [4].

DOI: 10.1051/matecconf/20179701015

Table 1. Bulk density, apparent porosity, thermal and electrical conductivity of produced composites.

No.	Sample	Density (g/cm^3)	Apparent Porosity (%)	Thermal Conductivity (W/mK)	Electrical Conductivity (S/cm)
1	Geo-C 1.0%	1.62	8.72	0.77	9.47E-07
2	Geo-C 1.5%	1.63	6.67	0.78	1.21E-06
3	Geo-C 2.5%	1.65	8.20	0.83	1.37E-06

The addition of carbon was found to increase the electrical conductivity of the composites but their magnitudes were still in the range of insulator materials [7].

Fig. 3 shows the DSC results of the composites. The measurement was conducted at temperature range 30-400°C with a heating rate of 20°C/minutes. The addition of carbon particles reduced the heat flow and the enthalpy of the composites. The enthalpy of composite contain 1.0% carbon was 315.5406J/g, 278.2725J/g for 2%, and 259.851 J/g for 2.5% carbon. This results are in good agreement with thermal conductivity measurements which showed that the addition of carbon increased the thermal conductivity of the composites.

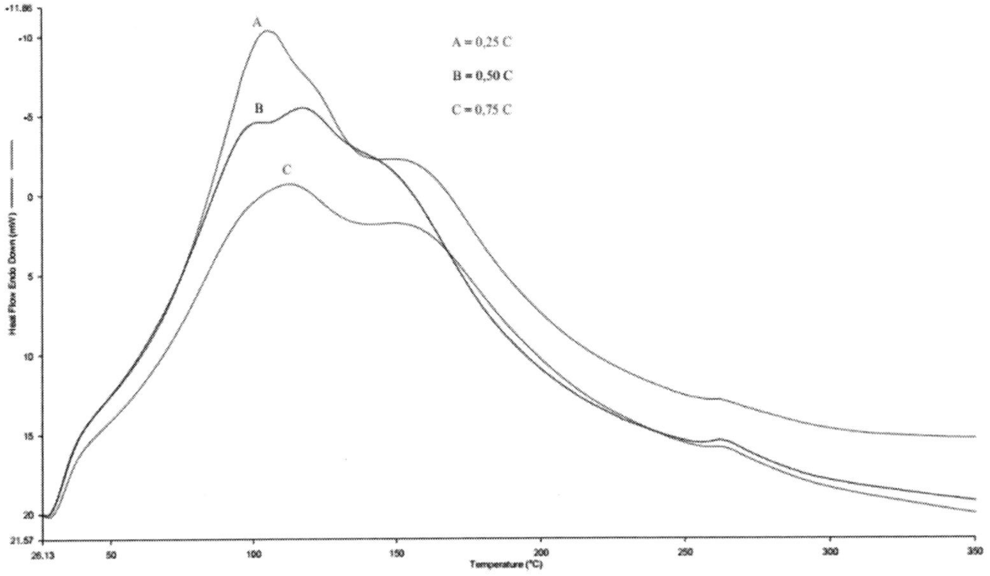

Fig. 3. DSC results of carbon-geopolymer composites with different concentration of carbon.

Diffractogram patterns of the resulting composites is shown in Fig. 4 which showed typical amorphous structure of geopolymers made from metakaolin. The reaction between alkali solution with aluminisilicate species in geopolymers formed a phase of sodium dialuminum phyllo-decaoxodihdroxoalumotrisilicate, $NaAl_2(AlSi_3)O_{11}$, The addition of carbon was found did not affect the amorphous structure of geopolymers and a small hump observed between 40 – 45° 2θ was due to the amorphous structure of carbon.

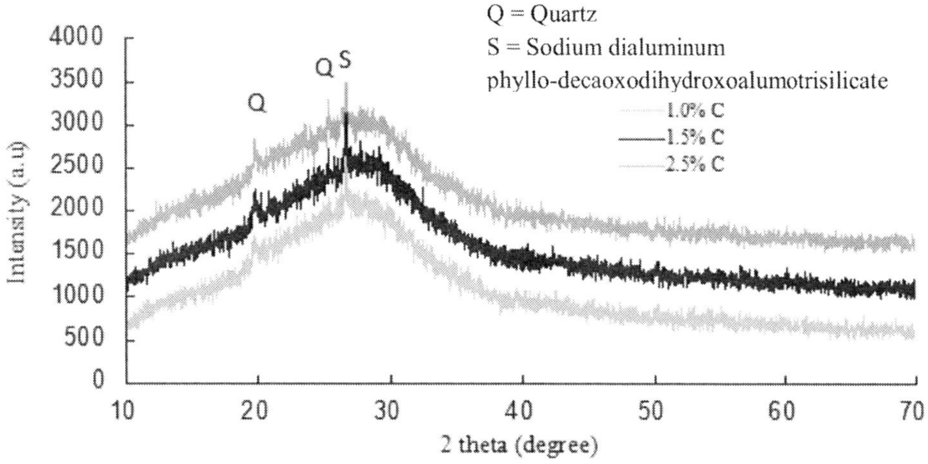

Fig. 4. Diffractogram patterns of carbon-geopolymer composites.

The resulting composites were calcined at 750°C to examine the formation of new phases due to high temperature and reaction between carbon particles with other constituent of geopolymers. Fig. 5 shows diffractogram of calcined composites in which the peaks of amorphous structure of geopolymers shifted to lower 2θ indicating restructure of geopolymers network [4]. Besides that, calcination at 750°C caused the loss of $NaAl_2((AlSi_3)O_{11}$ phase from the network of geopolymers.

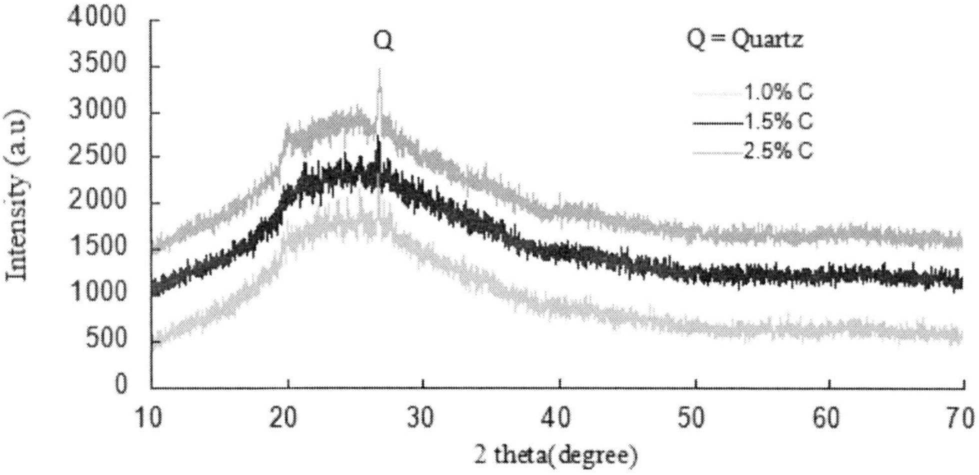

Fig. 5. Diffractogram of calcined carbon-geopolymer composites

Fig. 6 shows SEM images of the surface of the samples. The images showing the interface between carbon particles and geopolymers matrices. The bond between carbon and geopolymers matrix become weaker as the concentration of carbon increase as can be seen in Fig. 6C which shows the propagation of cracks around the interfacial zone between carbon particle and the matrix. Figure 6A and 6B showed that the bond between carbon particle and the matrix is strong and there is no sign of cracks development. This suggest that the addition of excess carbon into the geopolymers matrix will affect the strength of the resulting composites.

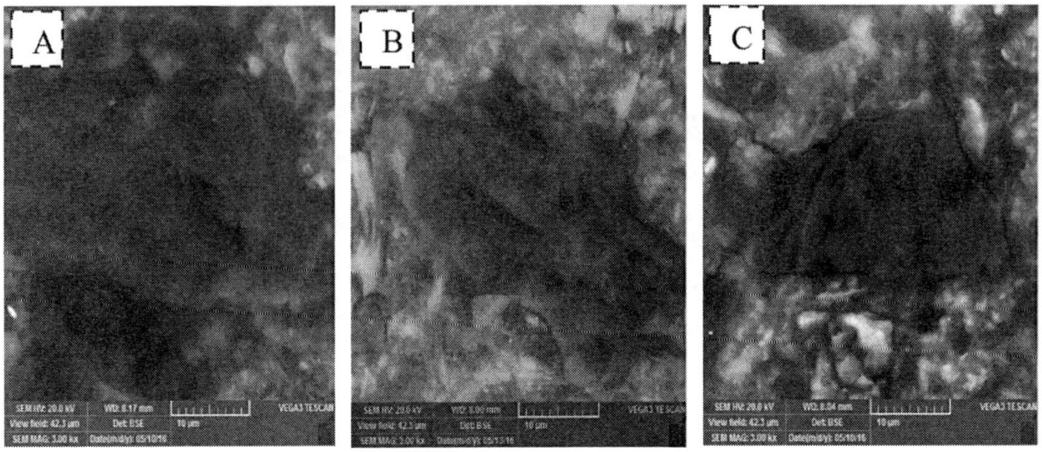

Fig. 6. SEM images of carbon-geopolymer composites contain (a) 1.0% (b) 1.5% (c) 2.5% of carbon particles.

4 Conclusion

The synthesis and characterizations of functional composite carbon-geopolymers have been conducted. The geopolymers matrix was synthesized through alkali activation of metakaolin and carbon was derived from burning coconut shell. The resulting composites reveal good surface quality, amorphous structure and good thermal and electrical conductivity for precast panel application. The maximum content of carbon which result in good quality of composite was 1.5%. SEM investigation showed that the bond between carbon and the matrix for samples containing 1% and 1.5% of carbon was strong and did not show any cracks opening.

References

1. J. Davidovits, Geopolimer Poly(sialat)/Poly(sialat-siloxo) Mineral Matrices for Composite Minerals, in ICCM VI & ECCM 2, Elsevier Applied Science, London, **1** (1987)
2. A. Shvarzman, The Effect of Dehydroxylation/Amorphization Degree on Pozzolanic Activity of Kaolinite, Cement Concrete Res., **33** (2002)
3. J.G.S. Van Jaarsveld, J.S.J. van Deventer, Part I. Theory and Applications, Miner. Eng., **10** (1996)
4. Subaer, Pengantar Fisika Geopolimer, Direktorat Jenderal Pendidikan Tinggi (2012)
5. M.A. Soleimani, R. Naghizadeh, A.R. Mirhabibi, F. Golestanifad, Iranian Journal of Materials Science & Engineering, **9** (2012)
6. M.L. Granizo, M.T. Blanco-Valera, A. Palomo, J. Mater. Sci., **35** (2000)
7. P. Duxson, A. Fernandez-Gimenez, L. Provis, G.C. Lukey, A. Palomo, J.S.J Van Deventer, J. Mater. Sci., **42** (2007)
8. R. Bligh, T. Glasby, Development of geopolymer precast floor for the global change institute at university of Queensland, www.asec2014.org (2013)

9. P. Garces, L.G. Andion, G. Gatala, Characterization of Porland Cement Concrete with Carbon Material Admixture, American Concrete Institute, **234** (2006)
10. A.K.M.S. Rahman, NanoFiber Reinforcement of a geopolymer matrix for improved composite materials mechanical performance, Colorado State University. *Dissertation* (2015)
11. Jorena, Menentukan Energi Gap Semikonduktor Silikon Melalui Pengukuran Resistansi Bahan pada Suhu Beragam, Universitas Sriwijaya (2009)
12. K. P. Puji, O. Tri, N. Lutfiyati, P. Gontjang, Uji Konduktivitas Listrik pada CaCO3 dan arang kayu dengan metode Four Point Probe, Fisika Laboratorium, Lab.Material (2015)

DOI: 10.1051/matecconf/20179701016

A Study of Hybrid Composite Hydroxyapatite (HA)-Geopolymers as a Material for Biomedical Application

Saleha[1,*], *Andi* Riska[1], *Ainun* Nisa Makmur[1], *Fathurrakhman* Ansar[1], and *Subaer*[1]

[1]Laboratorium Fisika Material, Jurusan Fisika, Universitas Negeri Makassar, Indonesia Jl. Daeng Tata Raya, Makassar, Indonesia

Abstract. The main purpose of this research is to study the physical properties and microstructure characters of hybrid composites HA-geopolymers as a material for biomedical application. Hybrid composite HA–geopolymers were produced through alkaline activation method of metakaolin as a matrix and HA as the filler. HA was synthesized from eggshell particles by using a precipitation method. The addition of HA in metakaolin paste was varied from 0.5%, 1.0%, 1.5%, and 2.0% relative the weight of metakaolin. FTIR was used to examine the absorption bands the composites. X-ray diffraction (XRD) was used to study the crystal structure of the starting and the resulting materials. Scanning Electron Microscopy-Energy Dispersive Spectroscopy (SEM-EDS) was used to investigate the surface morphology of the composites. The thermal properties of the samples was examined by means of Differential Scanning Calorimetry (DSC). Capacitance measurement was conducted to investigate the bioactive properties of HA. The study results suggest that hybrid composite HA-geopolymers has a potential to be applied as a biomedical such as biosensor material.

1 Introduction

Hybrid composite is a versatile material comprising organic–inorganic materials. This type of hybrid composite has been intensively studied including in the area of geopolymers research for various purposes [1]. One of the organic biomaterials which attract a lot of attentions lately is hydroxyapatite (HA). The chemical formula of HA is $Ca_{10}(PO_4)_6(OH)_2$) and it is a compound that formed the constituent of tissues in the human body such as bones and teeth [2]. HA which can be applied as synthetics bones synthesize from materials containing high calcium, such as cows bone, seashell and eggshell [3]. Eggshell for instance consist of 94% $CaCO_3$, 1% $MgCO_3$, 1% $CaPO_4$ [4]. Recently, HA was synthesized by using sol-gel or precipitation method [5, 6]. HA possesses biocompatible properties and has a reasonable high compressive strength [7]. However, HA is a brittle material like

* Corresponding author: saleha.alimuddin@gmail.com

ceramics hindering this material from wide range applications and mostly used as a biomaterial for low pressure application [8]. The properties of HA can be improved by combining with other material and form a composite like structure.

Hule, et.al., [9] reported that the nanocomposite hydroxyapatite-etilen-vinyl acetate (HAp-EVA) has a flexural strength and elongation which higher than a single particle of hydroxyapatite, and it can be used as matrix of bones regeneration. Chia-Jung, et.al., [10] showed that silicate polymer nanocomposite which synthesized by using HA as an aggregate changed the structure, modulus, strength and toughness of HA [10].

This research was conducted to synthesize and examine the properties of hybrid composite made from geopolymers and HA. Metakaolin, an amorphous mineral produced from hydroxylated kaolin at temperature between 600 -750°C, was used as inorganic material to produce geopolymers paste which act as a binder (matrix) and hydroxyapatite (HA) which synthesized from eggshell as organic fillers. The geopolymers paste produced was categorized as poly(sialate-siloxo) [11].

2 Experimental

Hybrid composite HA–geopolymers were produced through alkaline activation of metakaolin. HA was synthesized from eggshell by using a precipitation method. The eggshell was grounded and sieved to obtain fine particles. The resulting particles was calcined at 750°C for 2 hours followed by precipitation process by using Na3PO4, NH3, and HCl solutions. The resulting material was dried at 110oC for 12 hours and then calcined again at 750oC to obtain hydroxyapatite particles. The composites were produced by added geopolymer paste with HA varying the weight percentage of HA 0.5%, 1.0%, 1.5% and 2.0% relative to the weight of metakaolin. The mixtures were stirred manually, molded and then cured at 60°C for 1 hour.

The functional groups of hybrid composite HA-geopolymers were studied by means of Fourier Transform Infra Red (FTIR). The phase and the structure of the starting materials and produced hybrid composite HA-geopolymers were examined by using X-ray Diffraction (XRD). Scanning Electron Microscopy-Energy Dispersive Spectroscopy (SEM-EDS) was used to study the morphology of the surface of the sample as well as the interface between geopolymer matrix and HA particles. The thermal properties of the samples was studied by using Differential Scanning Calorimetry (DSC). Capacitance measurement was conducted to investigate the bioactive properties of the composites.

3 Results and discussion

Fig. 1 showes the representative samples produced in this study. Figure 1(a) shows the appearance of geopolymers without HA. Fig. 1(b) shows hybrid composite of HA-geopolymers having different compositions of HA.

 DOI: 10.1051/matecconf/20179701016

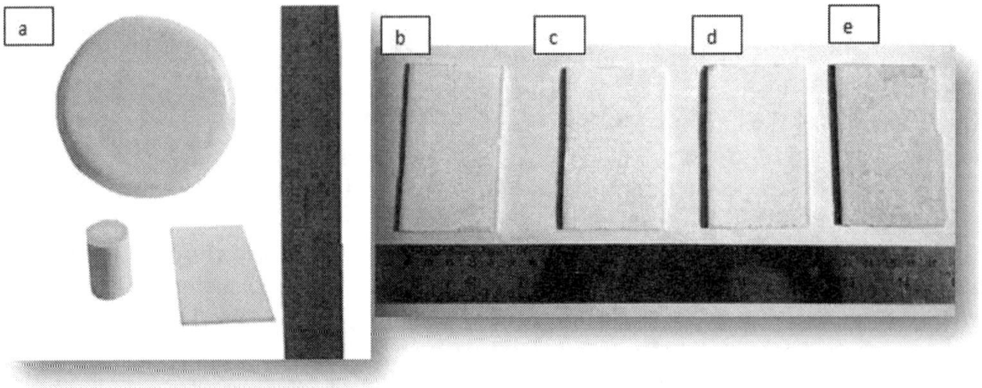

Fig. 1. Hybrid composite HA-Geopolymers (a) geopolymer without HA; (b) 0.5 % HA; (c) 1.0 % HA; (d) 1.5 % HA, (e) 2.0 % HA.

Fig. 2 shows the IR bands of pure geopolymer (0% HA) and geopolymer added with 2% of HA. The IR spectrum of pure geopolymer is typical with a band at 3398 cm^{-1} and 1649 cm^{-1} which is due to absorbed atmospheric water. A strong band for Si–O stretching was found at 977 cm^{-1} which is a bit lower than that of metakaolin (1050 cm^{-1}). This is due to tetrahedra condensation of geopolymer. The Si –O–Al bond was observed between 680 cm^{-1} – 875 cm^{-1} indicating the reaction between alkali solution and solid Al-O sheet in metakaolin. The absorption bands of CaO was found at 1381 cm^{-1} for geoplymer without HA, and at 1437 cm^{-1} and 1458 cm^{-1} for geopolymers with 2wt% and 4% HA, respectively. The shift of Cao bands is due to the high content of CaO in HA.

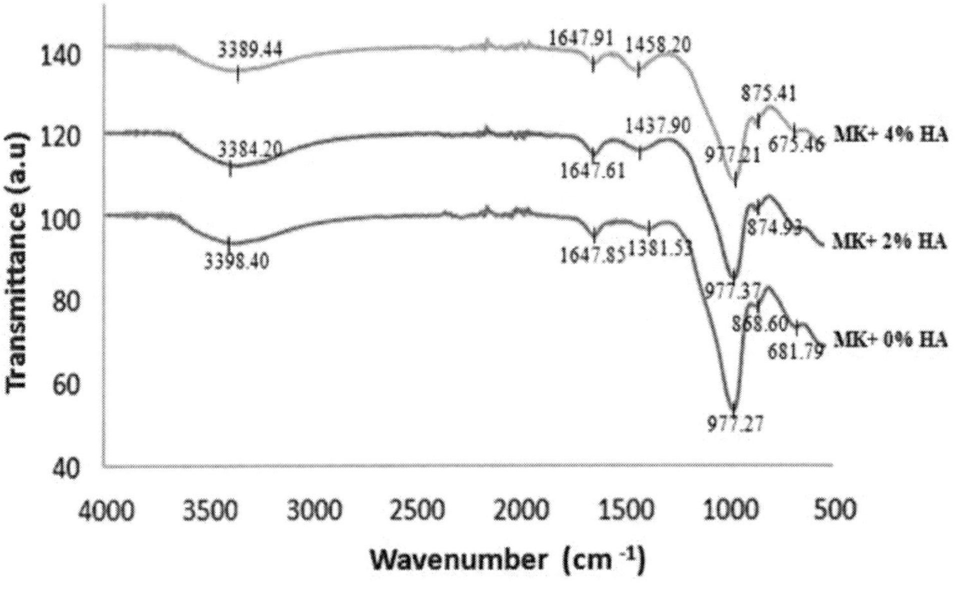

Fig. 2. FTIR Spectra of geopolymer (0%HA) and hybrid HA-geopolymers.

Fig. 3(a) shows the diffractogram of hydroxyapatite precursor and alkali activated hydroxyapatite. The diffraction patterns show that HA has a high level of crystallinity alongside with CaO and $Ca(OH)_2$ crystals. Fig. 3(b) shows the SEM image of the activated hydroxyapatite which formed a well-structured crystals. Based on the XRD examination,

DOI: 10.1051/matecconf/20179701016

the content of CaO and $Ca(OH)_2$ increase while the weight percentage of HA decrease. This indicates that HA reacts with alkaline solution. Fig. 3(b) shows the SEM image of hydroxyapatite activated with alkaline solution. These results indicate that HA and geopolymer paste will form a strong chemical bond as they both can be activated by alkaline solution and hence creates hybrid II composite.

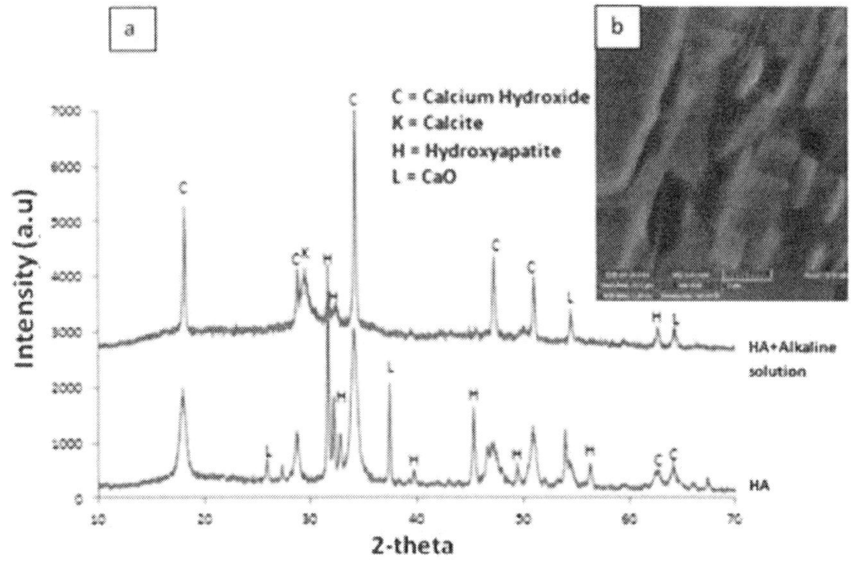

Fig. 3. (a) Diffractogram of HA and alkali activated HA, (b) SEM image of alkali activated HA.

Table 1 shows the phase composition of hydroxyapatite (wt %) measured with XRD. It can be seen that the eggshell produced in this research reached 47 wt% of HA and 46 wt% $Ca(OH)_2$. The alkali activated HA produced a material that rich with $Ca(OH)_2$ and $CaCO_3$, reducing the wt% of HA and CaO.

Table 1. The composition of HA and alkali activated HA (wt%).

Phase	Sample	
	HA	HA+Alkaline solution
$Ca_5(PO_4)_3(OH)$	47,0	6,9
$CaCO_3$	0,3	33
CaO	6,6	0,2
$Ca(OH)_2$	46,0	60

Fig. 4 shows the diffractogram of hybrid composites HA-geopolymers along with the diffractogram of HA and metakaolin. A broad hump of diffraction patterns between 20 – 40° 2θ indicate the amorphous nature of geopolymers. The addition of 2wt% HA relative to the weight of metakaolin did not change the amorphous phase of geopolymers. Quartz mineral originated from metakaolin remains on the network of geopolymer although its peak slightly decreases as the content of HA increases.

DOI: 10.1051/matecconf/20179701016

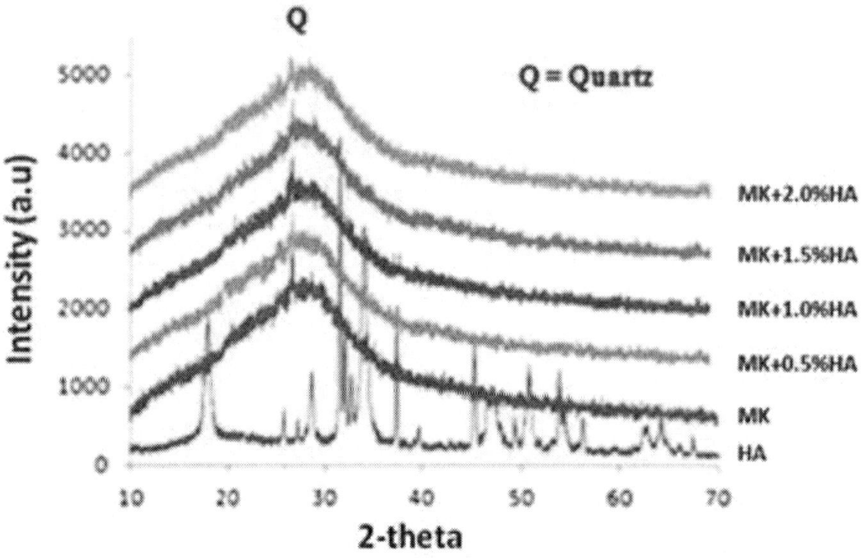

Fig. 4. Diffractogram patterns of HA, metakaolin and four samples of hybrid composites.

Fig. 5 shows the SEM images of metakaolin, HA, and hybrid composites produced in this research. The metakaolin particles (5a) appear to retain the flake shape of kaolinite crystallinity. The morphology of HA particles tend to form a hexagonal shape with a size between 5 – 10 μm similar to that reported in [12, 13, 14]. HA tends to form agglomerates with an average size of crystals in the range of 15-50 nm [15] as well as irregular morphology [16]. Fig. 5 (c), (d), (e), and (f), show the morphology of hybrid composite HA-Geopolymers. It can be seen that the matrix of geopolymer form a strong chemical bond with HA particles confirming the formation of hybrid type II. Fig. 5(d) and (e) show some unreacted of metakaolin particles during the polymerization.

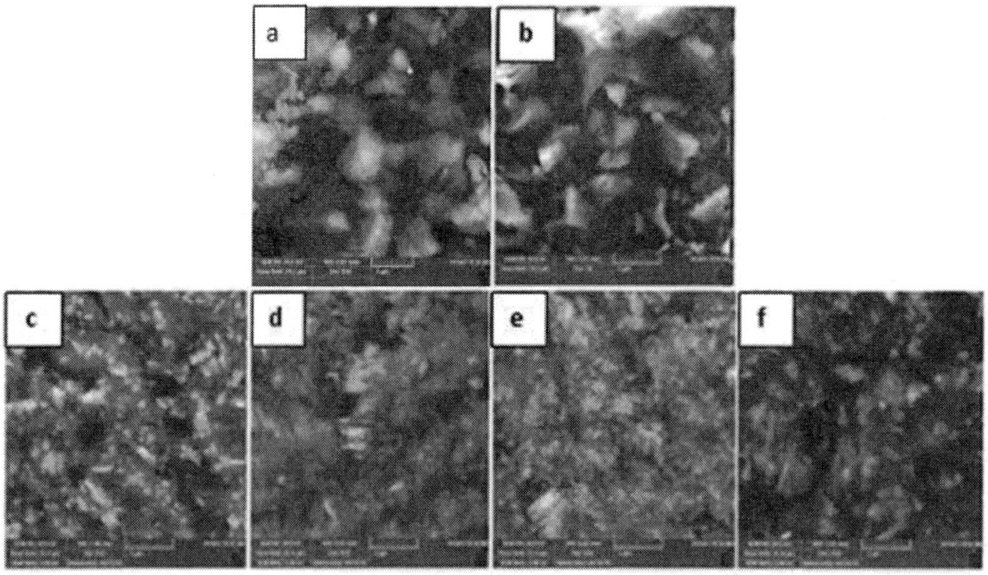

Fig. 5. SEM images (a) metakaolin, (b) HA, (c) metakaolin + HA 0.5%, (d) metakaolin + HA 1.0%, (e) metakaolin + HA 1.5%, (f) metakaolin + HA 2.0%.

Fig. 6 shows the results of DSC measurements for as-prepared hybrid composite HA-geopolymers (a) and hybrid composite HA-geopolymer cured at 60°C/1 hour. The measurements were conducted at the temperature range between 30 – 400°C with a heating rate 20°C/minute. The presence of HA in geopolymer network reduced the exothermic peak as well as the enthalpy of the composites before and after curing (Table 2). It was also observed that the exothermic peak and enthalpy of pure geopolymer were higher than those of geopolymer with HA addition. This results point out that the capacity of geopolymer to absorb or release heat is higher than that of hybrid composite HA-geopolymers.

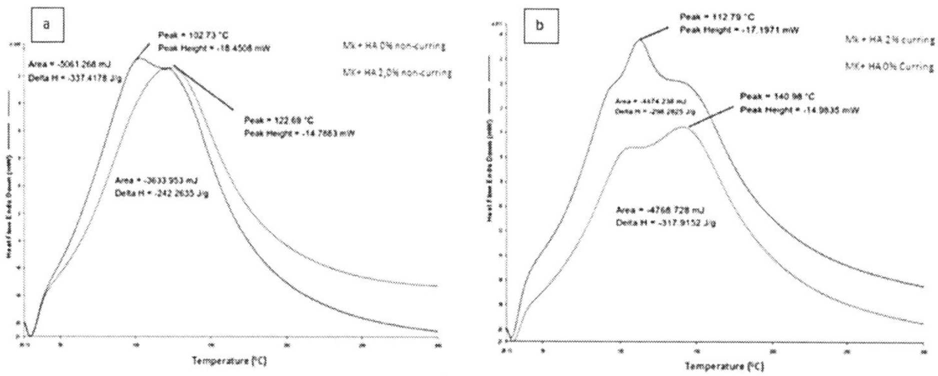

Fig. 6. DSC results of (a) as-prepared samples, (b) cured samples.

Table 2. DSC analysis of pure geopolymer and hybrid composite HA-geopolymer.

Nama Samples	Onset(°C)	Peak (°C)	End(°C)	Delta H (J/g)
Pure geopolymer paste	35.75	102.73	180.66	337.4178
Pure Composite (HA 2%)	60.17	122.69	185.65	242.2635
Cured geopolymer	55.24	140.98	207.77	317.9152
Cured Composite (HA 2%)	58.70	112.79	171.29	298.2825

Fig. 7 shows that the dielectric properties of hybrid composites as a function of wt% HA and the thickness of the sample conducted by using capacitance measurement. The results indicate that the magnitude of samples capacitance is directly proportional to the weight of HA and hence the thickness of the samples. This results prove that the hybrid composite HA-geopolymer exhibit a bioactive properties [17, 18]. The results also suggest that use of geopolymers to bind HA do not reduce the bioactive properties of HA and therefore can be applied as biosensor [19, 20].

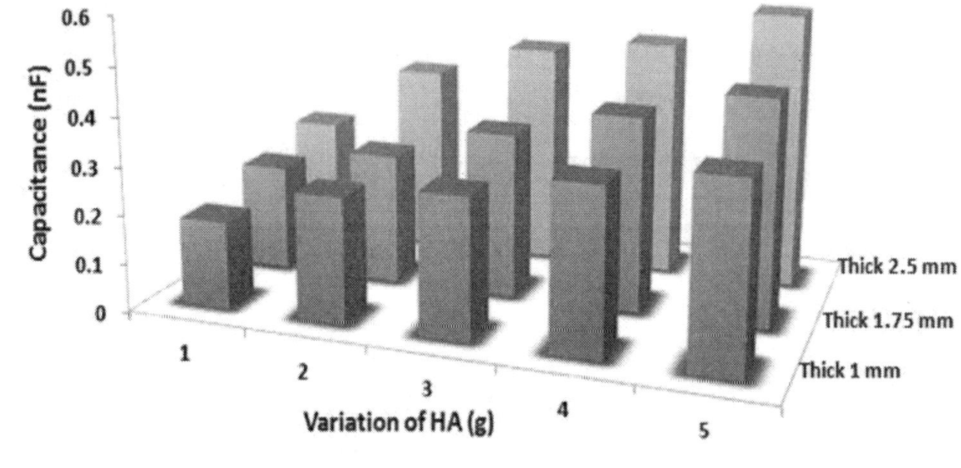

Fig. 7. The relation between capacitance capacitive and the variation of HA.

4 Conclusion

Hybrid composites hydroxyapatite (HA)- geopolymers based on metakaolin have been successfully produced. The FTIR and XRD analysis shows that the addition of HA did not affect the overall absorption bands and the amorphous phase of geopolymer. SEM examination reveal good chemical bond between HA particles with geopolymer matrix indicating strong mechanical properties. DSC results indicate that the presence of HA reduce the capacity of geopolymer to absorb or release heat from the interior of geopolymers. The dielectric properties of hybrid composite HA-geopolymers based on capacitance measurement showed that this composite has a potential to be applied as biosensor material.

References

1. R. Giuseppina, R. Laura, F. Claudio, C. Francesco, C. Raffaele, T. Oreste, *Materials*, **6** (2013)
2. S. Vikramjit, M. Neeraj, Internatinonal Journal of Science and Engineering Investigations, **1** (2012)
3. S.T. Wahyudi, S.U. Dewi, A. Anggraeni, K. Dahlan, Akhmaloka, M.A. Zulfikar, R. Hertadi, Adv. Mat. Res., **896** (2014)
4. S.M. Fabio, O.R. Patrik, M.T. Celia, A.S.S. Marcos, Food Sci. Technol. (Campinas), **27** (2007)
5. M.S. Mazrooei, M.H. Fathi, Ceram. Int., **38**, 2012.
6. I. Mobasherpour, M.S. Heshajin, A. Kazemzadeh, M. Zakeri, J. Alloys. Compd., **430** (2007)
7. P. Triyunda, B.W. Haris, R.Z. Andi, Solids and Structures, **1** (2012)
8. J. Luan, S. Wang, Z. Hu, L. Zhang, Curr. Org. Synth., **9** (2012)
9. A. Hule, J. Pochan, Darrin, Mrs. Bull., **32** (2007)
10. W. Chia Jung, K.G. Akhiles, Gaharwar, Materials, **3** (2010)
11. Subaer, *Pengantar Fisika polimer.* Jakarta: DP2M Dikti (2007)
12. K.B. Ravi, E.J.P. Gerrard, F. Derek, American Journal of Materials Science, **4** (2013)

13. G. Taglieri, C. Mondelli, V. Danieli, E. Pusceddu, G. Scoccia, Advances in Materials Physics and Chemistry, **4** (2014)
14. Taglieri, C. Mondelli, V. Danieli, E. Pusceddu, G. Scoccia, Advances in Materials Physics and Chemistry, **3** (2013)
15. P. Hui, S.L. Meena, G. Singh, R.D. Agarawal, S. Prakash, Journal of Minerals and Materials Characterization and Engineering, **8** (2010)
16. C.M. Luis, L.R. Geysy, C.M. Raphaella, Materials Sciences and Applications, **3** (2012)
17. P.N. Lim, E.Y. Teo, B. H0, B.Y. Tay, E.S. Thian, J. Biomed. Mater. Res. A., **101A** (2013)
18. G. Tetteh, A.S. Khan, R.M. Delaine Smith, G.C. Reilly, I.U. Rehman, J. Mech. Behav. Biomed. Mater., **39** (2014)
19. I. Sporysh, E. Shynkaruk, O. Lysko, A. Synkaruk, Mater. Sci. Eng., B, **169** (2010)
20. H. Li, Y. Liu, L. Luo, Y. Tan, Q. Zhang, K. Li, Mater. Sci. Eng., C, **59** (2015)

The Effect of Different Ratio Bottom Ash and Fly Ash Geopolymer Brick on Mechanical Properties for Non-loading Application

Laila Mardiah Deraman[1,*], *Mohd Mustafa Al Bakri* Abdullah[2,1], *Liew* Yun Ming[3], *Kamarudin* Hussin[2,1], *Wan* Mastura Wan Ibrahim[1], and *Andrei* Victor Sandu[3,1]

[1]Center of Excellence Geopolymer and Green Technology (CEGeoGTech), Universiti Malaysia Perlis (UniMAP), Malaysia
[2]Faculty of Engineering Technology, Universiti Malaysia Perlis (UniMAP), Perlis, Malaysia
[3]Gheorghe Asachi Technical University of lasi, Faculty of Materials Science and Engineering, lasi, Romania

Abstract. This paper studies the finding of strength and water absorption of geopolymer bricks using bottom ash and fly ash as a geopolymer raw material for non-loading application with minimum strength. The study has been conducted to produce bottom ash and fly ash geopolymer bricks by varying the ratio of fly ash-to-bottom ash, solid-to-liquid and sodium silicate (Na_2SiO_3)-to-sodium hydroxide (NaOH) in the mixing process. The compressive strength range between 3.8-4.5 MPa was obtained due to the minimum strength of non-loading application with 70°C curing temperature within 24 hours at 7 days of ageing. The optimum ratio selected of bottom ash-to-fly ash, solid-to-liquid and Na_2SiO_3-to-NaOH are 1:2, 2.0 and 4.0 respectively. The water absorption result is closely related to the amount of bottom ash used in the mix design.

1 Introduction

The huge demand from construction industry due to the increased population has entailed the need for sustainable building materials especially bricks [1]. Production of coal power plant leads to many environmental problems. Statistic shows in Malaysia, around 18 000 tons/day of coal needed by the Tanjung Bin Power Plant alone to generate electricity, as a result, large amounts of bottom ash and fly ash were produced as waste materials [2]. Hence, to overcome this problem, these waste materials will be useful when treated using geopolymerization methods to become a construction material. Bottom ash and fly ash can be utilized using inorganic polymers where it is formed by alkaline activation of alumina-silicate and alkaline solution through a polymerization process at ambient temperature to become a geopolymer because it has aluminium oxide-silicate [3]. Bottom ash and fly ash has been considered as pozzolanic materials and can be activated by using alkaline

* Corresponding author: lailamardiahderaman@gmail.com

activation, which is acting as a binder through a chemical polymerization reaction which was known as geopolymer [4]. This paper presents the strength and water absorption of bottom ash and fly ash geopolymer bricks with a different ratio. By using the waste product also will contribute to the environmental and economic benefits [3–5, 6].

2 Raw materials

2.1 Bottom ash

Bottom ash was obtained from Manjung Coal-Fired Power Station, Lumut, Perak, Malaysia. The chemical composition of bottom ash where's the content of SiO_2 and Al_2O_3 in the bottom ash is high which is about 43.2 % and 13.7 % respectively which is suitably used for geopolymer materials.

2.2 Fly ash

Fly ash (Class F ASTM) also obtained from Manjung Coal-Fired Power Plant, Perak Malaysia. The chemical composition of fly ash, which also contains highest Si and Al where it is suitably used for geopolymer materials.

2.3 Alkaline Solution

In this research, alkaline solution that was used is sodium silicate (Na_2SiO_3) and sodium hydroxide (NaOH). Sodium silicate is a liquid that consists 12 % of Na_2O and 30 % of SiO_2. The sodium hydroxides are originally in the form of a pellet with 98 % of purity, then being diluted with distilled water (12 M) [4].

3 Experimental method

The mix design of geopolymer bricks is shown in Table 1. Bottom ash and fly ash were firstly weighed and mixed together properly about 2 minutes based on the mixed design. Sodium silicate and sodium hydroxide were mixing together and the mixture was stirred for 2 minutes. Then, alkaline solution was added into bottom ash and fly ash to become a geopolymer paste. The molarity of NaOH was fixed about 12 M based on previous research done by Wan Mastura et al. [7] which is given the best compressive strength for geopolymer-based materials. The sized of brick are based on BS 3921: 1985 [8] where it is about (215×102.5×65) mm in work size. After a homogeneous mixture of geopolymer was obtained, it was placed into a machine that has been set following brick standard sized. The bricks were cured at temperature 70 °C for 24 hours and it was left until 7 days ageing.

 DOI: 10.1051/matecconf/20179701017

Table 1. The Mix Design of Bottom Ash and Fly Ash Geopolymer Brick.

<table>
<tr><th>Ratio of
Fly ash : Bottom Ash</th><th>Ratio of
Solid : Liquid</th><th>Ratio of
Na_2SiO_3 : NaOH</th></tr>
<tr><td>1 : 2</td><td rowspan="3">2.0</td><td rowspan="3">2.5</td></tr>
<tr><td>1 : 3</td></tr>
<tr><td>1 : 4</td></tr>
<tr><td rowspan="3">1 : 2</td><td>1.5</td><td rowspan="3">2.5</td></tr>
<tr><td>2.0</td></tr>
<tr><td>2.5</td></tr>
<tr><td rowspan="3">1:2</td><td rowspan="3">2.0</td><td>2.0</td></tr>
<tr><td>3.0</td></tr>
<tr><td>4.0</td></tr>
</table>

3.1 Testing

The strength of bottom ash and fly ash geopolymer brick was evaluated using compressive strength test. Compressive strength is tested by imposing the bricks to compression load until failure. The compressive strength tested based on ASTM C67-11 by using hydraulic Compressive Testing Machine VU 2000 at the load speed rate 0.6 $N/mm^2/s$ [9]. The effects of the compressive strength of geopolymer brick were only tested on their stretcher face area [7-9].The maximum load was recorded and the strength calculated by dividing the maximum load by the area of the surface subject to loading. In this research, three samples are tested at each ratio to evaluate the strength due to the non-loading application. The strength was recorded in N/mm^2 to the nearest 0.1 N/mm^2. Water absorption test was carried out based on ASTM C140 [10] for every age of curing specimens. Test specimen immersed in water at temperature 15 to 27 °C for 24 hours. After 24 hours, the specimens are removed and weighted and recorded as W_s (saturated weight). Then, the sample is dried in the oven at 110 °C for not less than 24 hours or until two successive weight at intervals of 2 hours show an increment of loss not greater than 0.2 % of the last previous determined weight of the specimen. The weights of dried specimen are recorded as W_d (oven-dry weight).

4 Results and discussion

Fig. 1 shows the compressive strength of ratio fly ash-to-bottom ash. From the data obtained, ratio 1:2 showed a compressive strength of 4.5 MPa which has reached the minimum strength for non-load bearing based on ASTM C129-11. The compressive strength of geopolymer bricks was decreased as the ratio of fly ash to bottom ash increased. Since bottom ash is not fineness and do not have a high surface area, it couldn't dissolve more silica and alumina thus can lead to the low compressive strength according to the properties.

 DOI: 10.1051/matecconf/20179701017

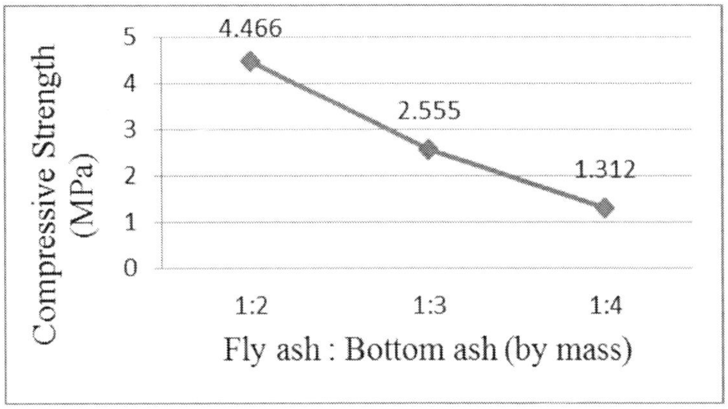

Fig. 1. The compressive strength of fly ash-to-bottom ash ratio.

Fig. 2 shows the water absorption of geopolymer brick. Ratio 1:2 showed the lowest readings of water absorption which is 4.3 % of water absorption, while the ratio of 1:3 and 1:4 are 4.4% and 4.5% respectively. The higher ratio showed more bottom ash used in the designed mixture. More bottom ash used leads to the higher percentage of water absorption. Since the properties of bottom ash are porous and irregular surface than fly ash which allows it to behave like a water reservoir that retains water, so, the percentage of water absorption is higher.

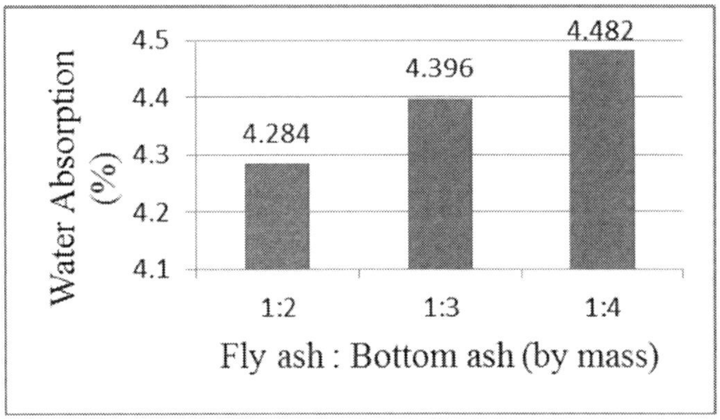

Fig. 2. The water absorption of fly ash to bottom ash ratio.

The compressive strength for effect solid-to-liquid ratio is shown in Fig. 3. After obtaining the optimum value compressive strength of fly ash to the bottom ratio (1:2), then the experiment continues with finding the optimum ratio of solid-to-liquid. Three ratios are selected based on the trial and error that has been done by other researchers which are 1.5, 2.0 and 2.5 [3, 7, 11]. At a ratio of 2.0, compressive strength showed 4.44 MPa where it is acceptable for minimum strength for the non-load bearing application. Highest compressive strength shows more fly ash used.

 DOI: 10.1051/matecconf/20179701017

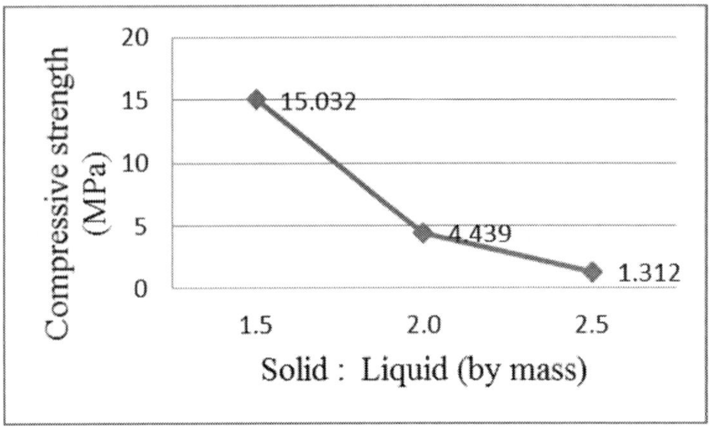

Fig. 3. The compressive strength of solid-to-liquid ratio.

Water absorption results with three different ratios of solid-to-liquid shown in Fig. 4. The lowest percentage resulting in solid-to-liquid 2.0 with value 5.18 %. Then, for solid-to-liquid ratio, 2.5 have the highest water absorption which is 6.96 % and the ratio of 1.5 has a value 5.16 %. The higher percentage of water absorption indicates to not fully reacted precursor with alkaline activator causing water retained in the pores especially in bottom ash because physical properties of bottom ash are porous and less pozzolanic.

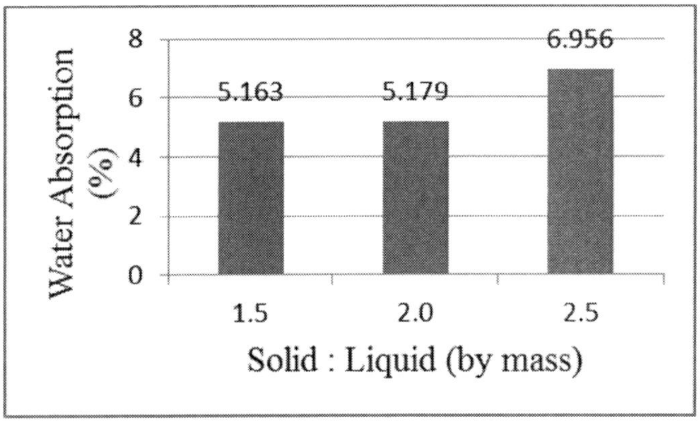

Fig. 4. The water absorption of solid to liquid ratio.

Experiments continue with finding the optimum ratio for sodium silicate (Na_2SiO_3)-to-sodium hydroxide (NaOH). Fig. 5 shows the graph results of ratio Na_2SiO_3-to-NaOH. The minimum compressive strength for non-load bearing indicates at a ratio of 4.0 which are 3.84 MPa. Then, the ratio of 2.0 and 3.0 has a value of 6.405 MPa and 6.19 MPa respectively. According to Zarina et al. (2015), when the Na_2SiO_3-to-NaOH ratio was more than 3.0 the compressive strength tends to decrease that may due to excessive alkali content which retarded the geopolymerization process. It occurred when Si-Al phase [12].

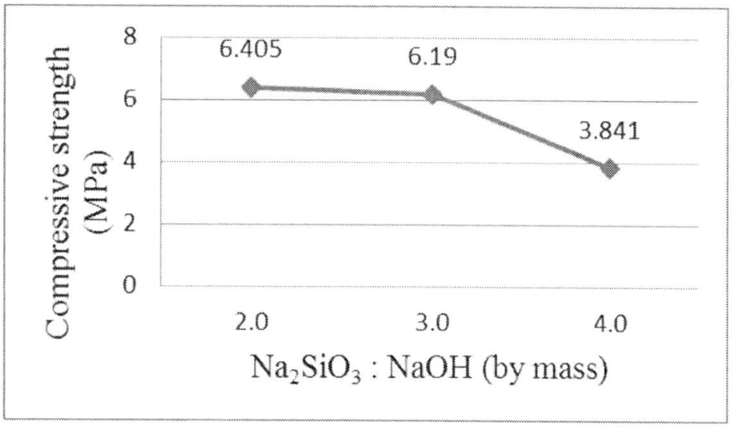

Fig. 5. The compressive strength of Na_2SiO_3-to-NaOH ratio.

Fig. 6 shows the graph percentage of water absorption with different Na_2SiO_3-to-NaOH ratios. The results show the lowest percentages of water absorption are obtained at 4.0 ratios which are 3.21 %. The highest percentage of water absorption shows at 3.0 ratio which is 5.85 % while ratio 2.0 calculated 5.30 % of water absorption. The pores in the geopolymer bricks may affect the absorption of water. Lower compressive strength shows lower in the absorption of water.

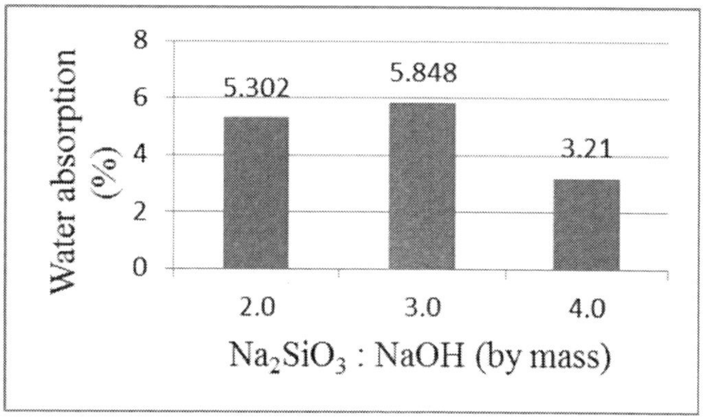

Fig. 6. The water absorption of Na_2SiO_3 to NaOH ratio.

5 Conclusion

According to the analysis of compressive strength results of bottom ash and fly ash geopolymer brick, the strength will increase with the increasing of fly ash. This may due to the size particle of fly ash, which is very fine and act like an ordinary Portland cement. The optimum ratio for bottom ash-to-fly ash, solid-to-liquid and Na_2SiO_3-to-NaOH ratio are 1:2, 2.0 and 4.0 respectively. The maximized used of bottom ash, the minimum strength for the non-load bearing is selected since it is not as widely used compared with fly ash or other geopolymer material. More bottom ash used will result higher percentage of water absorption due to the more surface area found and the pozzolanic of bottom ash also can retain water.

This study was supported by Center of Excellence Geopolymer and Green Technology (CEGeoGTECH) UniMAP and School of Materials Engineering, UniMAP.

DOI: 10.1051/matecconf/20179701017

References

1. J.M. Bennet, M. Sudhakar, C. Natarajan, European International Journal of Science and Technology, **2** (2013)
2. A. Muhardi, M. Aminaton, K.K. Kassim, M.A. Mahir, Electron J. Geotech. Eng., **15** (2010)
3. Y. Zarina, A.M. Mustafa Al Bakri, H. Kamaruddin, I. Khairul Nizar, A.R. Rafiza, Advance Material Science, **34** (2013)
4. J.C. Swanepoel, C.A. Strydom, Appl. Geochem., **17** (2002)
5. M.T. Muhammad Fahem, A.M. Mustafa Al Bakri, C.M. Ruzaidi, H. Kamaruddin, A.M. Izzat, A. Alida, Engineering Materials, **594** (2014)
6. K.S. Gan, P.C. Chui, CV978 Report, School of Civil and Environmental Engineering, Nanyang Technological University, Singapore (2003)
7. W.I Wan Mastura, A.M. Mustafa Al Bakri, A.V. Sandhu, H. Kamaruddin, I.G. Sandu, I. Khairul Nizar, A.K. Aeslina, M. Bnhussain, Rev. Chim.-Bucharest, **65** (2014)
8. BS 3921 - Spesifications of Clay Bricks, (1985)
9. American Society for Testing and Material, ASTM C67-11 (2011)
10. American Society for Testing and Material, ASTM C140-12 (2011)
11. A.M. Mustafa Al Bakri, H. Kamarudin, M. Bnhussain, I. Khairul Nizar, A.R Rafiza, Y. Zarina, J. Eng. Technol. Res., **3** (2011)
12. Y. Zarina, A.M. Mustafa Al Bakri, H. Kamarudin, I. Khairul, R. Rafiza, S. Andrei, Materials, **8** (2015)

Natural Fiber Filament Wound Composites: A Review

Suriyati Mohamed Ansari[1,*], *Che Mohd Ruzaidi* Ghazali[2,3], and *Kamarudin* Husin[1,2]

[1]Center of Excellence Geopolymer and Green Technology (CEGeoGTech), Universiti Malaysia Perlis (UniMAP), Malaysia
[2]Faculty of Engineering Technology, Uniciti Alam Campus, Universiti Malaysia Perlis (UniMAP), 02100, Sungai ChuChuh, Padang Besar, Perlis, Malaysia
[3]School of Fundamental Science, Universiti Malaysia Terengganu, 21030 Kuala Terengganu, Terengganu

Abstract. In recent development, natural fibers have attracted the interest of engineers, researchers, professionals and scientists all over the world as an alternative reinforcement for fiber reinforced polymer composites. This is due to its superior properties such as high specific strength, low weight, low cost, fairly good mechanical properties, non-abrasive, eco-friendly and bio-degradable characteristics. In this point of view, natural fiber-polymer composites (NFPCs) are becoming increasingly utilized in a wide variety of applications because they represent an ecological and inexpensive alternative to conventional petroleum-derived materials. On the other hand, considerable amounts of organic waste and residue from the industrial and agricultural processes are still underutilized as low-value energy sources. This is a comprehensive review discussing about natural fiber reinforced composite produced by filament winding technique.

1 Introduction

In the past few decades, research and engineering interest has been shifting from monolithic materials to fiber-reinforced polymeric materials. Fiber reinforced polymers (FRP) are composite materials comprised of a polymer matrix combined with high-strength fibers such as glass, aramid and carbon. These composite materials such as carbon and glass fiber reinforced plastics has dominated the aerospace, leisure, automotive, constructions and sporting industries. Although these composites materials exhibit excellent mechanical properties, they also give rise to environmental pollution due to their non-degradability [1, 2]. Glass fibers are the most widely used material to reinforced plastics due to their low cost compare to other materials and have fairly good mechanical properties. However, these fibers have serious drawback as indicated in Table 1 where the properties of glass fiber were compared with natural fibers. On the other hand, the neutrality properties of natural fibers towards carbon dioxide is particularly attractive. Burning of substances derived from fossil products such as petroleum releases enormous amounts of carbon dioxide into the

*Corresponding author: suriyati.mohamed91@gmail.com

atmosphere. Thus, this phenomenon is believed to be the root cause of the greenhouse effect and by extension the world's climate changes [3].

Built upon those phenomenon, several attempts have been made to use natural fiber composites in place of glass mostly in non-structural applications. As a result, a good number of automotive components previously made with glass fiber composites are now being manufactured using environmental friendly composites [3, 4]. Currently, plenty research materials are being generated on the potential of cellulose based fibers as reinforcement for plastics. However, all researchers that been working in the area of natural fibers and their composites agreed that these renewable abundantly available materials have several feebleness such as; poor wettability, incompatibility with some polymeric matrices and high moisture absorption by the fibers [5].

Table 1. Comparison between natural and glass fibers

	Natural fibers	Glass fibers
Density	Low	Twice that of natural fibers
Cost	Low	Low, but higher than NF
Renewability	Yes	No
Recyclability	Yes	No
Energy consumption	Low	High
Distribution	Wide	Wide
CO_2 neutral	Yes	No
Abrasion to machines	No	Yes
Health risk when inhaled	No	Yes
Disposal	Biodegradable	Non-biodegradable

A composite is a structural material that consist of two or more combined constituents that are attached together at microscopic level and not soluble in each other. One constituent is called the reinforcing phase and the one in which it is embedded is called the matrix. The reinforcing phase material may be in the form of fibers, particles, or flakes. The matrix phase materials are generally continuous [6]. In case of that, fiber-matrix adhesion has become the first and most important problem during achieving a good natural fiber reinforced composites. The role of the matrix in a fiber reinforced composite is to transfer the load to the stiff fibers through shear stresses at the interface. A good bond between the polymeric matrix and the fibers was required in this process. Poor adhesion at the interface means that the full capabilities of the composite cannot be exploited and leaves it vulnerable to environmental attacks that may weaken it, thus reducing its life span. Poor mechanical properties of the natural fiber reinforced polymer composites maybe due to insufficient adhesion between hydrophobic polymers and hydrophilic fibers. However, these properties can be improved by [7, 8] physical treatments (cold plasma treatment, corona treatment) and chemical treatment (maleic anhydride, organosilanes, isocyanates, sodium hydroxide, permanganate and peroxide).

2 Mechanical and physical properties of natural fiber

The mechanical and physical properties of natural fibers are very important for industrial applications and can contribute to the use of natural fibers in numerous application and can contribute to the use of natural fibers in numerous applications. Thus, various researchers have studied the physical and mechanical properties of natural fiber-reinforced polymer composites [9-12]. Lower density leads lower-weight structures in the automotive industry and aerospace applications [13-16]. Mechanical properties such as tensile properties, flexural properties and impact strength are strongly affected by fiber content, as shown in Figure 2 [17]. The tensile properties of jute oil palm fiber hybrid composites are enhanced substantially with increased jute compared with oil palm epoxy composites.

On the other hand, Venkateshwaran et al. [18] studied the mechanical and water absorption properties of banana/sisal-reinforced hybrid composites using fiber length and weight percentage as the main variables. They reported that hybridizations of sisal fibers with banana/epoxy composites of up to 50% by weight enhance the mechanical properties and degrade the water absorption properties of these fibers. The overall tensile and flexural properties of natural fiber-reinforced polymer hybrid composites are highly dependent on the aspect ratio, moisture absorption tendency, morphology and dimensional stability of the fibers used.

Zampaloni et al. [19] concluded that both tensile and flexural strength that is very similar to the 40% by weight flax and hemp polypropylene systems contrast the tensile strength is higher and the flexural strength is almost doubled when compared against the coir and sisal systems. Beside that, the effect of weaving patterns and random orientation on the mechanical properties of banana, kenaf, and banana/kenaf fiber-reinforced hybrid polyester composites was studied by Alavudeen et al. [20]. The result reported in this study showed major improved tensile properties compared to the twill type in all the fabricated composites. Moreover, the maximum increase in mechanical strength was observed in the plain woven hybrid composites rather than in randomly oriented composites. Natural fibers have been used to reinforce various polymer matrices. These fibers include bamboo, flax, coir [21-23] softwood, sisal, banana, coir, jute, abaca oil palm (fruit), kenaf oil palm, spider silk and rice husk [24-27].

3 Mechanical characterization of filament wound tubes

There is a considerable research that was carried out in order to characterize the mechanical properties of filament wound composites. In most of studies, mechanical constants and characteristics are determined by means of several mechanical tests and analytical studies. The effect of multi axial filament winding on tubular structures of three different winding configurations was studied by P. Mertiny [28]. This study was conducted by comparing the data of [$\pm$45, $\pm60_2$] and [$\pm$30, $\pm60_2$] with [$\pm60_3$], so called baseline data, under constant ratios of biaxial loads. In this work, two types of failures are distinguished that is functional failure, where leakage of test fluids takes place but the structure still carries load, and structural failure, where the specimen can no longer carry any load. The experimental hoop and axial stresses at both functional and structural failure was evaluated by the researchers, derived the elastic constants for pure hoop and axial loading conditions and studied the failure envelopes. From the research, it can be found that the failure modes depend strongly on applied stress ratio, and matrix damage can be minimized using [$\pm$30, $\pm60_2$] configuration. On the other hand, configurations of [$\pm$45, $\pm60_2$] and [$\pm$30, $\pm60_2$] showed higher functional and structural strength compared to the baseline configuration under hoop-to-axial load ratios smaller than 1, due to the reduced axial strain in $\pm$60 covers; whereas the baseline configuration performed best under pure hoop loading condition.

Soden [29] carried out a study on the effect of winding angle on the mechanical properties of filament wound composites that are subjected to uniaxial and biaxial stresses. Biaxial loading (hoop and axial loadings for lined and unlined specimens at different winding angles and uniaxial compression is applied to tubular specimens. During testing, electrical resistance strain gages are attached to some specimens and the data obtained are analyzed in order to determine elastic constants. Thus, stress-strain curves are plotted using the strain data from the strain gages and the axial stress together with hoop stress data evaluated from the burst tests. In this work also, the variation of the failure stresses ate different tests, effect of the wall thickness on the compressive properties, effect of stress concentrations through the gauge length and comparing the stress envelopes in the work and literature for different winding angles were studied.

Ellyin [30] studied the behavior of multidirectional filament wound glass fiber/epoxy tubular structures under biaxial loading The aim of this study is to investigate the effects of the rate and ratio of biaxial loading on failure strength, damage accumulation and monotonic stress-strain behavior of glass reinforced epoxy tubes. In order to achieve that, different hoop to axial stress ratios are employed in testing, then the failures are studied and biaxial failure envelopes of stress and strain are developed and interpreted. Thus, from the study, it can be concluded that strength and stiffness are functions of biaxial stresses, failure modes and damage accumulations at failure are dependent on biaxial stress ratio, and linear elastic behavior of the tubes can only be seen at relatively low temperatures.

Beside that, the fatigue behavior of glass fiber reinforced $\pm55°$ filament wound composite tubes was investigated experimentally by N. Tarakcioglu, L. Gemi and A. Yapici [31]. The specimens that particularly produced from E-glass fiber with epoxy resin are tested at an open ended internal pressure apparatus and the fatigue tests are performed according to the ASTM standard D2992. In this study, six different stress levels based on different percentages of the static strength of the specimen were applied at one frequency. Three damage mechanism are observed during the testing. These mechanisms are identified as crazing, leakage and final failure followed by the number of cycles to these predetermined levels are observed and recorded. From the results, it was shown that there is an analogy between macro and micro damage stages of the specimens. Thus, it can be concluded that at higher stress levels, the final failure of the specimen occurs just after the leakage whereas for low stress levels, there is much more cycles in between leakage and final failure stages.

Another study was carried out by J. Bai, P. Seeleuthner, and P. Bompard [32] which focus in the mechanical behavior of $\pm55°$ filament wound glass fiber/epoxy resin tubes. In this study, the tube was made of six plies of $\pm55°$ winding angle. Under different loading conditions, a series of mechanical tests were performed which is pure tensile loading, pure internal pressure and combined loading. The work mainly discussed on the damage mechanisms under these conditions. It is stated that the damage and failure process can be described by three steps that is initiation of the damage process by microcracking, delamination between the different plies, and the development and coalescence of cracks and delamination in different plies. In this study also, it can be observed that matric cracks occur perpendicular to the tensile direction at that zones free of fibers whereas at the zones where fiber volume fraction is low, microcracks propagated around fiber bundles. At the zone of high fiber volume fraction, the microcracks propagated at the fiber/matrix interfaces.

The analysis of filament wound cylindrical shells under combined centrifugal, pressure and axial loading was studied by P.M. Wild and G.W. Vickers [33]. Both of these researchers developed an analytical procedure to asses the stresses and deformations of filament wound structures under different loading types. The procedure developed is mainly based on classical laminated plate theory, and models both plane stress and plain strain states of a cylindrical shell having a number of sub-layers, which are cylindrically orthotropic. As a result, it can be concluded that when there is no axial loading, benefit of wind angle variation is more significant. On the other hand, if axial loads are present, the benefits of winding angle variation are more considerable under the last ply criterion.

4 Kenaf fiber filament wound tubes

Kenaf is natural reinforcement fiber in which its capability need to be explore so that it can replaced and compared with other common commercial reinforcement materials. The unique advantages which is high strength to weight ratio of composite materials have been supported by the excellent characteristics of the synthetic fibers. The domination of carbon and e-glass fibers are commonly majoring materials in the aeronautic, automotive and industrial sector. However, serious drawbacks are reported from the usage of these fibers such as non-recyclable, high energy required for manufacturing process, health risks when inhaled and non-biodegradable [34]. Thus, as a solution, any materials that are being considered "green" are very important to many researchers and engineers. Several natural fibers have been used and excellent results have been recorded for various non-load bearing applications. Kenaf with scientific name (Hibiscus cannabinus L) as an example of natural fibers that shows a lot of promises as a reinforcement material and used to live among a variety weather conditions [35,36]. Kenaf is gaining attention as a reinforcement material among researchers to develop a product.

Alkbir [37] analyzed the effect of geometry on crashworthiness parameters of natural fiber kenaf. By using hexagonal tubes and hand lay-up process, kenaf fiber mat was formed. By this, they concluded that kenaf hexagonal tubes produces a variation results and indicated that the angle of kenaf tubes reinforced composite hexagonal tubes affected the crashworthiness parameters. On the other hand, an experimental study on the dynamic mechanical property of hybridized Kenaf/PALF-reinforced HDPE composites were studied by Aji et al. [38]. They mentioned that the presence of the kenaf fiber helps reduce the loss modulus in the dynamic mechanical properties. Yousif et al. [39] conduct an experiment on investigating the effect of untreated and alkali treated kenaf fiber on flexural properties of epoxy composites. From the result, it can be found that reinforcement of epoxy with treated kenaf fibers increased the flexural strength of the composite about 36% whereas for untreated fiber, there was improvement about 20%. They also stated that the increamental results are mainly due to the high improvement of the chemical treatment (NaOH) on the interfacial adhesion of the fibers and the porosity of the composites which prevented the debonding, detachments or pull out of fibers.

Mokhtar et al. [40] focus a study on experimental analysis of kenaf filament wound tubes under axial compression load. Firstly, axial compression test was performed as an early indication to identify the performance of kenaf filament wound composite tubes. Comparisons have been done towards basalt, e-glass and carbon tubes using polyester as a resin. The axial compression test of kenaf/polyester and kenaf/epoxy tubes were conducted with different winding angles involved which are 45°, 55°, 65° and 75°. From the testing it can be seen that 45° kenaf/epoxy tubes generated the higher compressive strength followed by other winding angle in the ascending order. The layer strength identification has been conducted in 55 winding angle sample in which indicate the increment layer of winding is uniform between one, two and three layers in ascending orders. The comparison between the different reinforcement materials show carbon tubes produced the higher compressive strength followed by e-glass, basalt and kenaf. Kenaf/epoxy recorded 38.7% lower than the e-glass tubes. Kenaf/epoxy tubes were observed to identify the improvement from kenaf/polyester tubes and results showed at least 22% increment have been generated. From this experiment, it can be concluded that kenaf presence as a reinforcement material was successfully combine as composite system under axial compressive load as well as lead to the promising indication to be introduced in low bearing application.

Another study on kenaf reinforced composite was studied by S. Misri et al. [41] where the study focusing on the torsional behavior of filament wound kenaf yarn fiber reinforced unsaturated polyester composite hollow shafts, with a specific focus on the maximum torsion capacity of the composite hollow shaft for different winding angles and aluminum reinforcement. The conventional filament winding machine was modified and added to a new resin bath mechanism in order to produce a new natural fiber composite hollow shaft using kenaf yarn fiber reinforced with unsaturated polyester resin. It can be found that the torsion capacity is significantly affected by changing the winding angle and the presence of aluminum in the static torque test capacity properties. The torsion test results were obtained in comparison with different orientation between 90° and 45° winding angle together with aluminum reinforcement. The results showed that the 45° orientation filament and aluminum reinforcement in the hollow kenaf composite has the highest value for torsion strength and the longest rotation to crack composite. Finite element analysis (FEA) using Abaqus software 6.10 was carried out and showed a good agreement with the experimental results.

M.H. Zamri et al. [42] investigate on development of green pultruded composites using kenaf fiber: influence of linear mass density on weathering performance. This research reported the performance of pultruded kenaf fiber reinforced composites after exposed to natural weathering condition up to 200 days. Three different linear mass density (Tex) of kenaf yarn fiber that is 1400, 2200 and 3300 g/km were used as reinforcement material in producing the pultruded composite samples. Fro each composite, moisture content over exposure time and the changes of mechanical properties such as flexural and compression properties were recorded. The result showed that sample made of 3300 linear mass density recorded higher moisture absorption as compared to sample with 1400 linear mass density. Throughout the exposure period, the flexural and compression properties consistently decreased as the absorption of moisture into the composites had weakened the interfacial bonding between kenaf fiber and the matrix resin. However, the pultruded composites that utilized the kenaf fibers with lower fiber linear mass density proved to have better mechanical properties compared to the pultruded composites with a higher number of fiber linear mass density used. Since the degradation started at the point where the kenaf fiber is directly exposed to the environment, the protective layer such as surfing veil which normally used in pultrusion industry is recommended and expected to slow sown the degradation process when exposed to natural environment.

5 Conclusion

Natural fiber reinforced filament wound composite appear to have very bright future for wide range of applications. These filaments wound composite have beneficial properties such as low density, less expensive, and reduced solidity when compared to synthetic filament wound composite products. Using natural fibers as reinforcement for polymeric composites introduced positive effect on the mechanical behavior of polymers. These bio-composite materials with various interesting properties may soon be competitive with the existing conventional filament wound product. However, the present low level of production restricts them for to be applied in industrial application. In addition, its hydrophilic properties make the real challenge to design the product which can be a good candidate for various applications. Thus, further research and improvement should be conducted so that these fully environmental friendly composite can easily be manipulated and can give benefit to all mankind and environmental issues.

We would like to appreciate Center of Excellence Geopolymer and Green Technology (CEGeoGTech), School of Materials Engineering, Universiti Malaysia Perlis and Faculty of Engineering Technology, Universiti Malaysia Perlis.

References

1. A. Gr, M, Sanjay, B. Yogesha, Carbon, **30** (2014)
2. M. Thiruchitrambalam, A. Alavudeen, A. Athijayamani, N. Venkateshwaran, A.E. Perumal, Mater. Phys. Mech., **8** (2008)
3. H. Larbig, H. Scherzer, B. Dahlke, Poltrock, J. Cell. Plast., **34** (1998)
4. A Leao, R. Rowell, N. Tavares, *Applications of Natural Fibers in Automotive Industry in BrazilThermoforming Process*, In: 4th International conference proceedings. Cairo, Egypt: Plenum press, (1997)
5. A. Vazguez, J. Riccieri, L. Carvalho, Polym. Composite, **20**,1 (1999)
6. K. Autar, Kaw, *Mechanics of Composite Materials*, 2nd ed, CRC Press Taylor & Francis Group, (2006)
7. S. Luo, A. Netravali, Polym. Composite, **20**, 3 (1999)
8. I. Racz, H. Hargitai, Int. J. Polym. Mater., **47** (2000)
9. R. Mohjoub, J.M. Yatim, A.R.M. Sam, S.H. Hashemi, Constr. Build. Mater., **55** (2014)
10. P. Herrera Franco, A. Valadez Gonzalez, Compos. A: Appl. Sci. Manuf., **35** (2004)
11. S. Ochi, Mech. Mater., **40** (2008)
12. S.M. Sapuan, A. Leenis, M. Harimi, Y.K. Beng, Mater. Des., **27** (2006)
13. J. Datta, P. Kopczynska, Ind. Crops. Prod., **74** (2015)
14. C. Alves, P. Ferrao, A. Silva, L. Reis, M. Freitas, L. Rodrigues, J. Cleaner Prod., **18** (2010)
15. T. Gurunathan, S. Mohanty, S.K. Nayak, Compos. A: Appl. Sci. Manuf., **77** (2015)
16. B. Yousif, S.T. Lau, S. McWilliam, Tribol. Int., **43** (2010)
17. Y. El Shekeil, S.M. Sapuan, M. Jawaid, O. Al Shuja'a, Mater. Des., **58** (2014)
18. N. Venkateshwaran, A.E. Perumal, A. Alavudeen, M. Thiruchitambalam, Mater. Des., **32** (2011)
19. M. Zampaloni, F. Pourboghrat, S. Yankovich, B. Rodgers, J. Moore, L. Drzal, Compos. A: Appl. Sci. Manuf., **38** (2007)
20. A. Alavudeen, N. Rajini, S. Karthikeyan, M. Thiruchitrambalam, N. Venkateshwaran, Mater Des, **66** (2015)
21. A.V. Vuure, J. Baets, K. Wouters, K. Hendrickx, Mater. Lett., **149** (2015)
22. F. Arrakhiz, M. E. Achaby, A. Kakou, S. Vaudreuil, K. Benmoussa, R. Bouhfid, Mater. Des., **37** (2012)
23. A.C. Manalo, E. Wani, N.A. Zukarnain, W. Karunasena, K.T. Lau, Compos. B: Eng., (2015)
24. A, Bledzki, P. Franciszczak, Z. Osman, M. Elbadawi, Ind. Crops. Prod., **70** (2015)
25. N, Defoirdt, S. Biswas, I.D. Vriese, J.V. Acker, Q. Ahsan, L. Gorbatikh, Compos. A: Appl. Sci. Manuf., **41** (2010)
26. H.Y. Cheung, M.P. Ho, K.T. Lau, F. Cardona, D. Hui, Compos. B: Eng., **40** (2009)
27. E. Ayswarya, K.V. Francis, V. Renju, E.T. Thachil, Mater. Des., **41** (2012)
28. P. Mertiny, F. Ellyin, A. Hothan, Compos. Sci. Technol., **4** (2004)
29. P.D. Soden, R. KIthing, P.C. Tse, Y. Tsavalas, Compos. Sci. Technol., **46** (1992)
30. F. Ellyin, M. Carroll, D. Kujawski, A.S. Chiu, Compos. A: Appl. Sci. Manuf., **28** (1997)
31. N. Tarakcioglu, L. Gemi, A. Yapici, Sci. Technol., **65** (2004)
32. J. Bai, P. Seeleuthner, P. Bompard, Compos. Sci. Technol., **57** (1997)
33. P.M. Wild, G.W. Vickers, Compos. A: Appl. Sci. Manuf., **28**, (1997)
34. H.Y. Cheung, M.P. Ho, K.T. Lau, F. Cordona, D. Hui, Compos. B: Eng., **40**, 7 (2009)
35. A. Ashori, J. Harun, J. Raverty, M.N.M. Yusoff, Polym. Plast. Technol. Eng., **45**, 1 (2006)

DOI: 10.1051/matecconf/20179701018

36. I.S. Sapuan, S.M. Zainudin, E.S.K. Abdan, International Journal of Mechanical and Materials Engineering, **4**, 3 (2009)
37. M.F.M. Alkbir, S.M. Sapuan, M.R. Ishak, Mater. Des., **60** (2014)
38. I.S. Zainudin, S.M. Sapuan, A. Khalina, M.D. Khairul, Polym. Plast. Technol. Eng., **51**, 2 (2012)
39. Yousif, B.F. Shalwan, A. Chin, C.W. Ming, Mater. Des., **40** (2012)
40. Mokhtar, I. Yahaya, M.Y. Kadir, M.R.A. Kambali, Polym. Plast. Technol. Eng., **52**, 11 (2013)
41. S. Misri, S.M. Sapuan, Z. Leman, M.R. Ishak, Mater. Des., **65** (2015)
42. M.H. Zamri, M.R. Osman, H.M. Akil, M.H.A. Shahidan, Z.A. Mohd Ishak, J. Cleaner Prod., **125** (2016)

Properties and Behavior of Geopolymer Concrete Subjected to Explosive Air Blast Loading: A Review

Nurul Aida Mohd Mortar[1,2,*], *Kamarudin* Hussin[1,2], *Rafiza* Abdul Razak[1,2], *Mohd Mustafa* Al Bakri Abdullah[1,2], *Ahmad Humaizi* Hilmi[3], and *Andrei Victor* Sandu[4,1]

[1]Center of Excellence Geopolymer and Green Technology (CEGeoGTech), Universiti Malaysia Perlis (UniMAP), Malaysia.
[2]Faculty of Engineering Technology, Universiti Malaysia Perlis (UniMAP), P.O Box 77, D/A Pejabat Pos Besar, 01000 Kangar, Perlis, Malaysia.
[3]School of Manufacturing Engineering, Kampus Tetap Pauh Putra, 02600 Arau, Perlis, Malaysia
[4]Gheorghe Asachi Technical University of lasi, Faculty of Materials Science and Engineering, lasi, Romania.

Abstract. The severe damage to civilian buildings, public area, jet aircraft impact and defense target under explosive blast loading can cause a huge property loss. Most of researcher discusses the topics on design the concrete material model to sustain againts the explosive detonation. The implementation of modern reinforcement steels and fibres in ordinary Portland cement (OPC) concrete matrix can reduce the extreme loading effects. However, most researchers have proved that geopolymer concrete (GPC) has better mechanical properties towards high performance concrete, compared to OPC. GPC has the high early compressive strength and high ability to resist the thermal energy from explosive detonation. In addition, OPC production is less environmental friendly than geopolymer cement. Geopolymer used can lead to environmental protection besides being improved in mechanical properties. Thus, this paper highlighted on an experimental, numerical and the analytical studies cause of the explosive detonation impact to concrete structures.

1 Introduction

Geopolymer is an alternative material as to replace the used of ordinary Portland cement (OPC). The side effect of using OPC is increasing the percentage of carbon dioxide release in the air surrounding. Nowadays, geopolymer concrete is widely used in constructing buildings namely as beam, slab, girder, column and wall. These main structures must be mechanically design to withstand high loads and external impact. Researchers have reported geopolymer concrete is higher compressive strength and can resist temperature up to 800 °C [1, 2]. The significant finding is that the use of sodium silicate as alkaline activator increases the compressive strength, potentially due to the high concentration of silica and alumina [3]. The existing of silica to form geopolymer chains is compulsory for strong bonds within the

* Corresponding author: nurulaida@unimap.edu.my

polymer chain, generating subsequently a hardened structure [4]. The strength of GPC is also caused by the nature of geopolymer concrete that is self-compacting, avoid creating air traps as well as eliminating pour inside while curing [5, 6].

Most of the previous researches, explosive blasting only discussed on OPC concrete instead of geopolymer concrete. Many researchers have been exploring the reinforcing and mixing addition, such as steel, fiber, carbon and nylon into the concrete mold [7-11]. Recent developed concrete materials enhanced mechanical properties and durability to strengthen itself [12]. As known, concrete has a brittle properties materials. Then, the testing under blast loads was conducted to ensure the reliability of concrete structure to be proof against blast [13]. Limiting the damage concrete structures also aggressively design to protect the civilian buildings in a war or terrorist attack area [14].

An explosive detonation was conducted on air blast, blast in confined space [15] and, consequently, direct contact with blast load [9, 16]. Blast in confined space will trigger more energy concentration while the pressure behavior in air blast is low frequency blast vibration. It is because air overpressure waves radiate outward to the air surrounding then lower the energy waves concentration [17]. The extreme energy from detonation can create impulsive impact and in consequence, many researchers proposed the high strength of the concrete structures [12-15, 26]. Therefore, this paper reviews on the geopolymer concrete performance towards explosive blast loading.

2 Experimental model on blast events.

In the past studies, most of the experimental set up on concrete blasting was conducted through air blast conditions [8, 18, 19]. The critical considered parameter in experimental before setting up the blast events are standoff distance, mass of explosive, types of explosive and concrete model material. Figure 1 shows the experimental setup from previous study which is investigating on deformation of reinforced concrete (RC) beams under blast [20].

Fig. 1. Test set up for air blast loading [20]

Furthermore, a pressure sensor transducer also locates by other researcher to measure the pressure exerted from the blast. Then, the amount of energy from the explosion hit the concrete beam can be obtained by using the pressure sensor. The maximum deflection and residual deflection are computed through the measurement Linear Variable Differential Transformer (LVDT) [21, 22].

3 Explosive selections

Velocity of detonation (VOD) and density of explosives was differed through their physical state as solids, liquids or gases [23]. Solids explosive cause more damage after explosion. The detonation of solid explosives may give rise shock wave under pressure up to 30000 MPa and temperatures about 3000- 40000C [14]. The types of secondary solid explosives that is trinitrotoluene (TNT) [24-26] and ammonium nitrate/ fuel oil (ANFO) [11] mostly used in concrete blast events.

The commercial explosive also been used in blast experiment, although the velocity of detonation is fewer than TNT and ANFO. Table 1 shows the different type of Emulex explosive as commercial explosive in the industry according to its capability effect.

Table 1. Types of Emulex as commercial explosive [27]

Emulex Explosive	Application	Velocity of detonation (m/s)	Density (g cm^{-3})
Emulex® 100 (Pipe Charge for Pre-Splitting)	For general surface blasting. Intended for column charging and priming. Without aluminum powder.	4500	1.05 - 1.15
Emulex® 150 (Surface & Underwater Blasting)	Specialised in surface and underwater blasting. Use as both column and bottom charge. Added with aluminum to provide higher explosive energy.	4,500 - 5,400	1.12 - 1.24
Emulex® 180 (underground/tu nnelling blasting	Formulated for underground blasting and tunneling needs. Intended for column charging or priming. Posses superb explosion energy to enhance pull factor.	4,500- 5,700	1.10 - 1.25

4 Standoff Distance

Theoretically, the standoff distance may vary according to material use for blasting as refer to Unified Facilities Criteria-DoD Minimum Antiterrorism Standards for Bulidings, UFC 4-010-01,31 July 2002 [28]. The mass of explosive is directly proportional to standoff distance. There are a research that performed a blast test with 0.51kg mass of TNT with 400mm standoff distance, 34kg and 1675mm [11], respectively. Some researcher also conducted an experiment that evaluate the level of performance of reinforce concrete slab by varying the explosive weight and standoff distance [29]. The effective distance can be obtained by calculating using the formula [8] as below:

$$Z = R/\sqrt[3]{W}\,, \tag{1}$$

where R is the standoff distance and W is the mass of explosive.

DOI: 10.1051/matecconf/20179701019

5 Materials and structural element types

Concrete is a brittle construction materials. It can provide effective resistance to fire and shock wave during blast. The high resistance also provided by reinforced concrete structures which are important in structure close to explosion sources. There are several types of fiber reinforcement use in concrete enhancing strength as shown in Table 2. Commonly, the main role of fiber reinforcement imparts more resistant to impact loading, thus, increase ductility and reduce permeability of concrete elements [30].

Table 2. Classification of fibres [4]

Types of Fiber	Descriptions
Steel Fiber	Different shapes and dimensions, also microfibres
Glass Fiber	In cement matrices used only as alkali-resistant (AR) fibres
Synthetic Fiber	Fibres made with different materials: polypropylene, polyethylene and polyolefin, polyvinyl alcohol (PVA)
Carbon Fiber	Pitch and polyacrylonitrile (PAN) fibres

A previous research conducted a tests on six 2.74 m x 0.2 m x 0.2 m simply supported RC beams with two 16 mm diameter rebars. The results show that the fibre reinforced beam has high effectiveness in increasing the blast resistance compared to non-fibre reinforcement [31]. Moreover, research has been carried out for improving the blast resistance by employing fibre reinforced concrete. The result of blast test is shown in Table 3 was computed on crack width, length of deflection and mass loss in percentage for different reinforcement. The studies indicated that the variation of the reinforcement provided a different level of damage and performance.

Table 3. Performance of reinforced concrete after blast detonations [32]

Specimen (Ø 10)	Crack width	Deflection (mm)	Mass loss (%)
Steel bars reinforced concrete	16	82	30
Fibre reinforced concrete (FRC)	NA	NA	50
FRC + steel bars	2	5	0
Reinforced concrete (Glass fibre reinforced polymer bars)	2	9	5
RC (steel bars) + GFRP laminate	3	8	8

6 Numerical analysis of concrete structures

6.1 Pressure behaviour during explosion

The tremendous energy from explosive detonation generates higher heat stress and shock pressure wave in the atmosphere. Figure 2 shows the pressure behaviour for air blast. Before the shock wave reach, the atmospheric pressure, P_a at t_A, then fluctuate until the peak pressure, P_{so} after explosion. The pressure drop to P_a at t_A+t_D, next achieve P_{so}', the negative pressure peak before return again at atmospheric pressure [14].

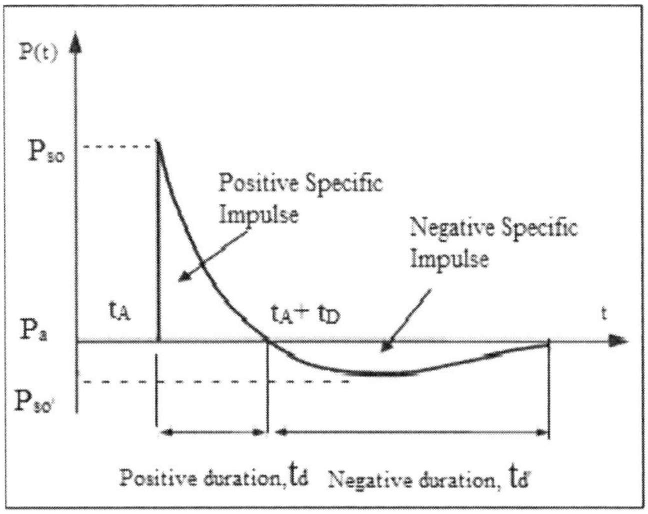

Fig. 2. Air blast pressure- time profile [14]

Theoretically, it is very inconsistent to determine the air blast pressure wave, depending on the environment condition. Due to that, Baker proposed equation to express the pressure as simplify the process [33].

$$P_{so}(t) = P_{so}\,(1 - \frac{t}{t_0})e^{\frac{-t}{t_0}} \quad (2)$$

where t is duration of the pressure wave travel to the given location, t_o is the time from peak pressure to atmospheric pressure. The impulse generates by positive pressure also can be obtained using its integral to time.

$$I_{so} = \int_{t_a}^{t_a+t_o} P_{so}(t)dt \quad (3)$$

A researcher proposed the empirical equations for shock wave calculation by collecting data from experimental and numerical [34].

$$P_{so} = 1.4072Z^{-1} + 0.554Z^{-2} - 0.0357Z^{-3} + 0.000625Z^{-4}\ (0.1 \leq Z \leq 0.3)$$
$$P_{so} = 0.619Z^{-1} - 0.033Z^{-2} + 0.213Z^{-3}\ (0.3 \leq Z \leq 1)$$
$$P_{so} = 0.066Z^{-1} - 0.405Z^{-2} + 0.329Z^{-3}\ (1 \leq Z \leq 10) \quad (4)$$

where Z is the scaled distance,expressed by $Z = \frac{R}{\sqrt[3]{W}}$, R is the standoff distance and W is the mass of explosive.

DOI: 10.1051/matecconf/20179701019

6.2 Strain-rate effect

The strain rate effect is very critical factors for a concrete structural response. Concrete dynamic behaviour also influenced the strain rate effect. The Dynamic Increase Factor (DIF), Compressive Dynamic Increase Factor (CDIF) and Tensile Dynamic Increase Factor (TDIF) empirical equations [35] have been suggested to calculate the effect on tensile and compressive strength.
For concrete compression,

$$\text{CDIF} = \frac{f_c}{f_{cs}} = \left(\frac{\dot{\varepsilon}}{\dot{\varepsilon}_{cs}}\right)^{1.206\,\alpha} \text{ for } \dot{\varepsilon} \le 30\text{s}^{-1}$$

$$\text{CDIF} = \frac{f_c}{f_{cs}} = \gamma_s\left(\frac{\dot{\varepsilon}}{\dot{\varepsilon}_{cs}}\right)^{\frac{1}{3}} \text{ for } \dot{\varepsilon} > 30\text{s}^{-1} \qquad (5)$$

where f_c is the dynamic compressive strength at strain rate $\dot{\varepsilon}$, f_{cs} is the static compressive strength at $\dot{\varepsilon}_{cs}$; $\log\gamma = 6.156\alpha - 0.49$; $\alpha = 1/(5 + 3f_{cu}/4)$ and f_{cu} is the static cube compressive strength in MPa.

For concrete tension,

$$\text{TDIF} = \frac{f_t}{f_{ts}} = \left(\frac{\dot{\varepsilon}}{\dot{\varepsilon}_{ts}}\right)^{\delta} \text{ for } \dot{\varepsilon}_{td} \le 1\text{s}^{-1}$$

$$\text{TDIF} = \frac{f_c}{f_{ts}} = \beta = \left(\frac{\dot{\varepsilon}}{\dot{\varepsilon}_{ts}}\right)^{\frac{1}{3}} \text{ for } \dot{\varepsilon}_{td} > 1\text{s}^{-1} \qquad (6)$$

where f_t is the dynamic tensile strength at strain rate $\dot{\varepsilon}$ in the range of 10^{-6} to 160 s^{-1}, f_{ts} is the static tensile strength at $\dot{\varepsilon}_{ts}$; $\log\beta = 6\delta - 2$; $\delta = 1/(1 + 8f'_c/f'_{co})$; f'_c is the static uniaxial compressive strength of concrete and f'_{co} is taken as 10MPa.

While the DIF relationship for steel reinforcement formulated as below[36].

$$\text{DIF} = \left(\frac{\dot{\varepsilon}}{10^{-4}}\right)^{\alpha} \text{ , take as } \alpha = 0.074 - 0.040 f_y/414 \qquad (7)$$

where f_y is steel yield strength in MPa.

6.3 Analytical approach for explosive blasting

An analytical analysis are performing by using the computer simulation software to illustrate the real and programming of explosive detonation. Typically, researcher carried out the simulation on explosive blasting using computer program is summarize in Table 4.

Table 4. List of computer software for blast simulation [14]

Software	Ability	Author/Vendor
LS-DYNA	Structural response+ CFD (Couple analysis)	Livermore Software Technology Corporation (LSTC)
BLASTX	Blast prediction, CFD code	SAIC
CTH	Blast prediction, CFD code	Sandia National Laboratories
FEFLO	Blast prediction, CFD code	SAIC
FOIL	Blast prediction, CFD code	Applied Research Associates, Waterways Experiment Station
SHARC	Blast prediction, CFD code	Applied Research Associates,Inc
DYNA3D	Structural response+ CFD (Couple analysis)	Lawrence Livermore National Laboratory (LLNL)
ALE3D	Couple analysis	Lawrence Livermore National Laboratory (LLNL)
Air3D	Blast prediction, CFD code	Royal Military of Science College, Cranfield University
CONWEP	Blast prediction (empirical)	US Army Waterways Experiment Station
AUTODYN	Structural response+ CFD (Couple analysis)	Century Dynamics
ABAQUS	Structural response+ CFD (Couple analysis)	ABAQUS Inc

From the past study, the test result of concrete damage after blasting are consulted by using LS DYNA and the accuracy of the analytical model is provided by comparing the results on analytical and experimental as shown in Figure 3.

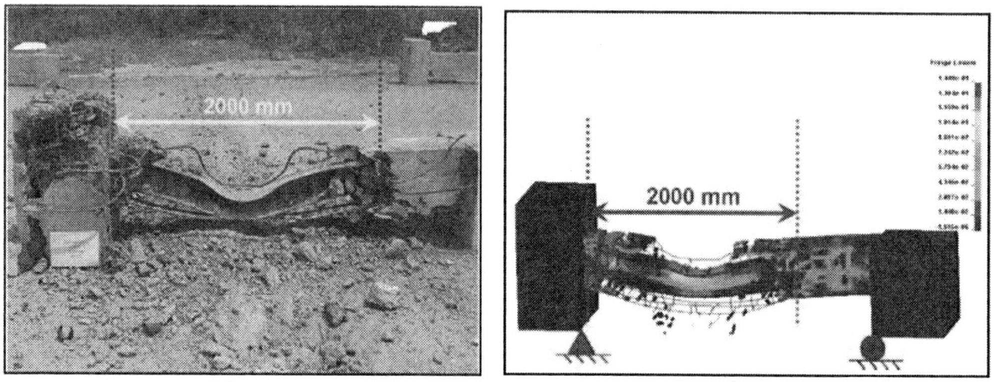

(a) (b)

Fig. 3. Comparison of damage between experimental (a) and analytical (b)[16]

Based on the studies have been done, LS-DYNA and AUTODYN commonly used as it is commercial numerical hyrocodes [18, 24, 25, 37]. A 2D finite element model was simulated by using LS-DYNA for modelling fiber reinforcement concrete and predict the observed damage [8]. Next, the numerical simulation of the dynamic response and residual axial capacity of composite columns was also investigated by using LS-DYNA [9]. While, the previous researcher studied using LS-DYNA to simulate the prestressed reinforced concrete beam under blast loads [11]. In contrast, according to the past research, AUTODYN was suggested to simulate the blast wave propagation in 2D simulation [37].

7 Summary

Design consideration in concrete withstands the shock wave from the explosion is very important. Hence, it can minimize and reduce the risk and damage impact from the detonation. Definitely, public buildings and defense area may lead to safety as to prevent attack from terrorist or war. Therefore, the design of high strength concrete may result in high cost consumption, in the meantime, it can be decrease by using the geopolymer concrete as well as reduce pollution problems.

References

1. I.H. Aziz, M.M. Al Bakri Abdullah, H. Cheng Yong, L. Yun Ming, K. Hussin, E.A. Azimi, Key Eng. Mat., **34** (2015)
2. H.y. Zhang, V. Kodur, L. Cao, S.l. Qi, Procedia Eng. **71**, 153 (2014)
3. L.N. Assi, E. Deaver, M.K. ElBatanouny, P. Ziehl, Constr. Build. Mater. **112**, 807 (2016)
4. O.A. Abdulkareem, M. Ramli, Modern Applied Science, **9**, 61 (2015)
5. E. Arioz, O. Arioz, O.M. Kockar, Procedia Eng., **42**, 100 (2012)
6. H.Y. Zhang, V. Kodur, S.L. Qi, L. Cao, B. Wu, Constr. Build. Mater., **55**, 38 (2014)
7. P.A. Buchan, J.F. Chen, Woodhead Publishing, 269 (2010)
8. S.j. Yao, D. Zhang, F.y. Lu, W. Wang, X.g. Chen, Eng. Fail. Anal., **62**, 103 (2016)
9. A.M. Coughlin, E.S. Musselman, A.J. Schokker, D.G. Linzell, Int. J. Impact Eng., **37**, 521 (2010)
10. D. Kakogiannis, F. Pascualena, B. Reymen, L. Pyl, J. M. Ndambi, E. Segers, Fire Safety J., **57**, 69 (2013)
11. Z.S. Tabatabaei, J.S. Volz, J. Baird, B.P. Gliha, D.I. Keener, Int. J. Impact Eng., **57**, 70 (2013)
12. A.M. Brandt, Compos. Struct., **86**, 3 (2008)
13. M.A. Yusof, N. Mohamad Nor, A. Ismail, N. Choy Peng, R. Mohd Sohaimi, M.A. Yahya, Adv. Mater. Sci. Eng., **2013**, (2013)
14. T. Ngo, P. Mendis, A. Gupta, J. Ramsay, Electronic Journal of Structural Engineering **7**, 76 (2007)
15. C.F. Zhao, J.Y. Chen, Y. Wang, S.J. Lu, Theor. Appl. Fract. Mec., **61**, 12 (2012)
16. K.C. Wu, B. Li, K.C. Tsai, J. Constr. Steel. Res., **67**, 602 (2011)
17. C. Wu, M. Lukaszewicz, K. Schebella, L. Antanovskii, J. Loss. Prevent. Proc., **26**, 737 (2013)
18. W. Chen, H. Hao, S. Chen, Mater. Des., **65**, 662 (2015)
19. X. Lin, Y.X. Zhang, P.J. Hazell, Mater, Desi, **56**, 620 (2014)
20. B. Yan, F. Liu, D. Song, Z. Jiang, Eng. Fail. Anal., **51**, 9 (2015)
21. J. Li, C. Wu, H. Hao, Mater. Des, **82**, 64 (2015)
22. G. Carta, F. Stochino, Eng. Struct., **49**, 306 (2013)
23. J. P. Agrawal, *High energy materials: propellants, explosives and pyrotechnics* (John Wiley & Sons, 2010)

24 G. Thiagarajan, A.V. Kadambi, S. Robert, C.F. Johnson, Int. J. Impact. Eng. **75**, 162 (2015)
25. L. Chernin, M. Vilnay, I. Shufrin, Int. J. Mech. Sci., **106**, 331 (2016)
26. Y. Shi, M.G. Stewart, Struct. Saf., **53**, 13 (2015)
27. Z.X. Zhang, *Explosives and Detonators," in Rock Fracture and Blasting* (Butterworth-Heinemann, 179, 2016)
28. J.P. Hanus, R.W. Welch, *Applying UFC 4-010-01 in Baghdad, Iraq* (Structures Congress 2005@ sMetropolis and Beyond, 1, 2005)
29. P.F. Silva, B. Lu, Compos. Part B-Eng., **38**, 523 (2007)
30. P.A. Buchan, J. F. Chen, Compos. Part B-Eng., **38**, 509 (2007)
31. C.A. Ross, M. Purcell, E.L. Jerome, Building to Last, 673 (1997)
32. C.P. Pantelides, T.T. Garfield, W.D. Richins, T.K. Larson, J.E. Blakeley, Eng. Struct. **76**, 24 (2014)
33. W.E. Baker, *Explosions in air* (University of Texas Press, 1973)
34. J. Henrych, R. Major, *The dynamics of explosion and its use* (Elsevier Amsterdam, 1979)
35. Y. Shi, M.G. Stewart, Int. J. Impact Eng., **85**, 5 (2015)
36. L.J. Malvar, Aci Mater J -American Concrete Institute, **95**, 609 (1998)
37. X. Q. Zhou, V. A. Kuznetsov, H. Hao, J. Waschl, Int. J. Impact Eng., **35**, 1186 (2008)

 DOI: 10.1051/matecconf/20179701020

The Effect of High Temperature on Compression Strength of Geopolymer Paste

Ridho Bayuaji[1,*], *Sigit* Darmawan[1], *Nur Ahmad* Husin[1], *Boedi* Wibowo[1], *Srie* Subekti[1], *Januarti Jaya* Ekaputri[1], *Mohd Mustafa Al Bakri* Abdullah[2], and *Totok* Ruki Biyanto[3]

[1]Civil Engineering Department, Institut Teknologi Sepuluh Nopember, Indonesia
[2]Center of Excellence Geopolymer and Green Technology (CEGeoGTech), Universiti Malaysia Perlis (UniMAP), Malaysia
[3]Physic Engineering Department, Institut Teknologi Sepuluh Nopember, Indonesia

Abstract. This research has a purpose to confirm the resistance of geopolymer concrete towards high temperature. This high temperature becomes the issues in material structure defense due to the danger of fire. Therefore, this research is done on geopolymer paste, the main binding ingredient from fly ash as local industry waste material. The method that is used in order to achieve the objective is done experimentally in laboratory with the geopolymer testing paste constructed of type C fly ash. Fly ash was obtained from PT. Petrokimia Gresik. The activator alkali sodium hydroxide (NaOH) was designed 6, 10, and 14 Molar. The dimension of specimen shape was cylinder 2.5 cm × 5 cm. The maintenance of specimen done in a room temperature (27±3°C) until testing period 28 days and proceeded by test item burning process on 200°C and 600°C for 4 hours. The testing of the test item that is done includes the pressure strength test according to SNI-1974-1990. The result of this research shows that the pressure strength of geopolymer paste with NaOH 6, 10, and 14 molar at 200°C temperature has a better pressure strength than room temperature geopolymer paste.

1 Introduction

Population growth in Indonesia and the acceleration of infrastructure are not balanced, especially in urban areas. It raises unawareness, one of which is the threat of fires in the building. Fires often cause damage to elements of the structure. Research on the building on the Makasar mall temperature recorded due to the fire can reach 1400°C.

On the other hand, the current global research directed toward the mindset of green environment, how the reduction of primary material partly or entirely of building materials and structures that are cementitious. Portland cement concrete as the main adhesive ingredient reaps a lot of attention because of greenhouse gas emissions (carbon dioxide) generated in the cement production process. The combustion process requires energy to the transformation temperatures of up to 1400°C, for 1 ton of cement produces 0.55 tonnes of CO_2 [1].

* Corresponding author: bayuaji@ce.its.ac.id

One of the innovations to reduce the 100% cementitious material is the technology of geopolymer concrete, in which its binding material consists of material with pozzolan characteristics, generally have significant silicate oxide substance and alumina oxide[2]. Examples of geopolymer concrete raw materials derived from coal combustion waste (fly ash). Utilization of industrial waste material and non industry are the other reason why geopolymer concrete worth the continued research [3-10] to support environmental researches.

This research focuses on those two main things, the threat of fire and waste utilization by formulation of the problem, to which extent the effect of temperature towards the strength of geopolymer concrete paste as the primary binding material for geopolymer concrete.

2 Methodology

Research method that is used to answer this research is by experimental laboratory. Fly ash used in this study as the basic material for the manufacture of geopolymer from PT. Petrokimia, Gresik. To determine the chemical composition contained in the fly ash, tests will be done in XRD (X-Ray Diffraction) and XRF (X-Ray Fluorescence). Activators for alkaline solution used is sodium silicate (Na_2SiO_3) and sodium hydroxide (NaOH) with molarity 6, 10 and 14. For the manufacture of test specimens using a cylinder with a diameter of 2.5 cm × 5 cm.

After the specimen is manufactured it will be tested as, firing test done by furnace for 4 hours in temperature of 200°C, 600°C, and 1000°C then the binder is rested until it is cooled off for the next testing process.

Test on the test object is the compressive strength test with the standard ISO-1974-1990. The age compressive strength test binder geopolymer concrete is done at the age of 28 days. For each test compressive strength, used 3 specimens of each combustion temperature condition.

3 Result and discussion

Some of the results from fly ash testing were presented on Figure 1, Table-1 and Table-2: This test is used to determine the elemental composition analysis in the ingredients. From the test result of XRD is known that fly ash used in this study is amorphous SiO_2 + Ca (FeO_3) 58% so as to the fly ash that is used is a type C. CaO content of fly ash used in this study is 10.74%, according to ASTM C 618-84 fly ash which has a CaO content of more than 10% are classified into type C fly ash.

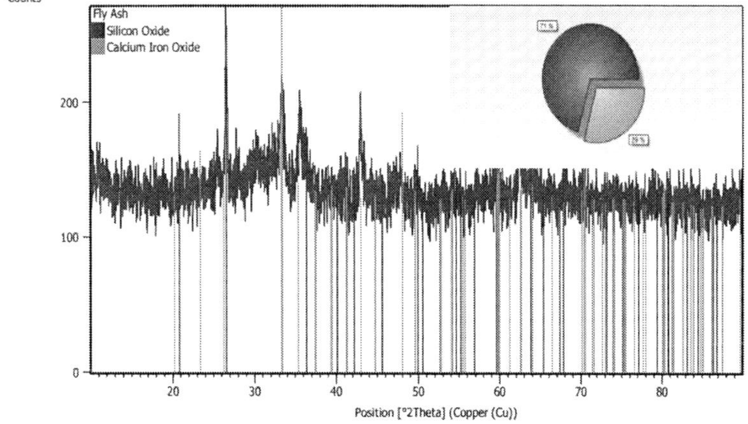

Fig. 1. X-Ray Difraction Test Result

 DOI: 10.1051/matecconf/20179701020

Table 1. Fly Ash content based on XRD test

Score	Compound Name	Displacement [°2Th.]	Scale Factor	Chemical Formula
38	Silicon Oxide	0.124	0.992	$Si\ O_2$
20	Calcium Iron Oxide	-0.171	0.467	$Ca\ (\ Fe\ O_3\ )$

Table 2. XRF Test Result

Component	iO_2	Al_2O_3	CaO	MgO	Na_2O	K_2O	TiO_2	$Ca(FeO_3)$	SO_3	P_2O_5	MnO	SrO
Percentage (%)	18,32	6,74	10,74	2,28	0,27	1,2	0,68	20,62	0,94	0,17	0,1	0,51

Compressive strength testing is conducted on Material and Building Structure Laboratory, FTSP Civil Engineering ITS. Here are the results of compressive strength testing. For the test result of the geopolymer concrete are described on Figure 2 to Figure 4. From figure 2, it is known that the average compressive strength of geopolymer paste test specimens treated with the room temperature compressive strength will increase if molar activator is enlarged. The increase in 6 molar to 10 molar by 9% and 10 molar to 14 molar increased 13%. The highest compressive strength at room temperature is 14 molar. From figure 3, it is known that the average compressive strength of geopolymer paste test specimens treated with burnt 200°C temperature compressive strength will have different patterns with the normal temperature. At 6 molar to 10 molar compressive strength decreased by 7% and 10 molar to 14 molar increased 3%. The highest compressive strength a temperature of 200°C was 14 molar. From figure 4, it is known that the average compressive strength of geopolymer paste test specimens treated with burnt 600°C temperature compressive strength will increase if molar activator is enlarged. At 6 molar test specimens were broken and could not be tested and 10 molar to 14 molar increased 68%. The highest compressive strength 600°C temperature is 14 molar. The highest compressive strength on geopolymer paste was caused the heat effect during curing system. The heat curing generated the water was expelling from the geopolymer matrix, causes discontinuous nano-pores within matrix, which increases the strength of geopolymers[11].

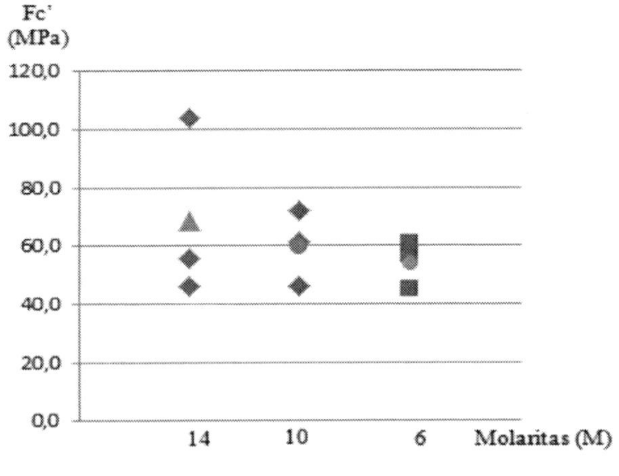

Fig. 2. Compression strength of room temperature geopolymer paste from different molarities

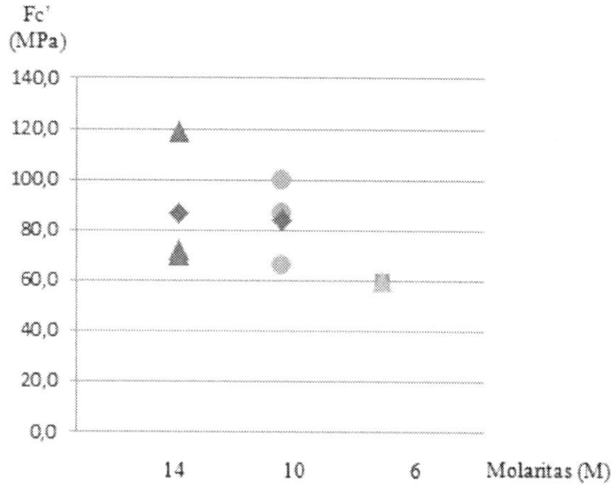

Fig. 3. Compression strength of 200°C geopolymer paste from different molarities

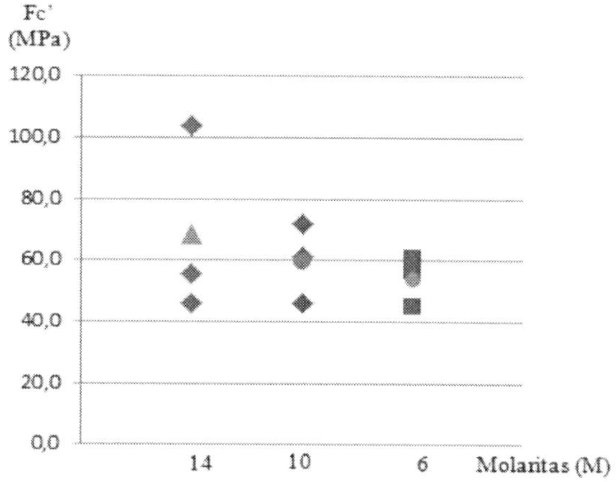

Fig. 4. Compression strength of 6000C geopolymer paste from different molarities

4 Conclusion

From the research with cylindrical test specimen, it can be concluded that, geopolymer paste (6, 10 and 14 molar) at normal temperature resulted the higher the molar would have increased compressive strength, the axis also lowered if the molar is enlarged. However, unlike UPV experiment, the result of experiment on 10 molar UPV is greater than 14 molar. Compressive strength of geopolymer paste (6, 10, and 14 molar) at a temperature of 200° C increased, decreased at a temperature of 600° C. The highest compressive strength is the specimen with 14 molar and burned at a temperature of 200° C (14-200).

The authors would like to thank RISTEK-DIKTI and LPPM ITS for provided financial assistance to complete this laboratory research grants. In addition, the support of Laboratory of Testing Materials of Diploma Civil Engineering Program ITS is greatly appreciated.

References

1. J. Davidovits, J. Therm. Anal., **37**, 1633 (1991)
2. J. Davidovits, Word Resource Review **6.2**, 263 (1994)

3. R. Bayuaji, M. Nuruddin, S. Francis, J. Ekaputri, S. Junaedi, H. Fansuri, Mater. Sci. Forum, **803**, 49 (2015)
4. R. Bayuaji, M. Sigit Darmawan, B. Wibowo, N. Ahmad Husin, S. Subekti, J.J. Ekaputri, Appl. Mech. Mater., 400 (2015)
5. H. Fansuri, A. Fatmawati, W.P. Utomo, W. Supriadi, R. Bayuaji, Mater. Sci. Forum, (2016)
6. A. Kusbiantoro, N. Rahman, S.C. Chin, R. Bayu Aji, Mater. Sci. Forum,**16** (2016)
7. L.Y. Dwijayanti, J.J. Ekaputri, R. Bayuaji, Mater. Sci. Forum, **140** (2016)
8. J.J. Ekaputri, S. Junaedi, R. Aji, Mater. Sci. Forum, **63** (2014)
9. S. Mane, H. Jadhav, Magnesium, **1**, 5 (2012)
10. W. Supriadi, R. Bayuaji, R. Burhan, H. Fansuri, Mater. Sci. Forum, **841,** (2016)
11. J. Davidovits, *Chemistry of Geopolymeric Systems, Terminology* (Proceedings of 99 International Conference," eds. Joseph Davidovits, R. Davidovits & C. James, France, 1999

Review on Various Types of Geopolymer Materials with the Environmental Impact Assessment.

Hariz Zain[1,*], *Mohd Mustafa Al Bakri* Abdullah[1,2], *Kamarudin* Hussin[1,2], *Nurliyana* Ariffin[1,2], and *RIDHO* Bayuaji[3]

[1]Faculty of Engineering Technology, Universiti Malaysia Perlis (UniMAP), Perlis, Malaysia
[2]Center of Excellence Geopolymer and Green Technology (CEGeoGTech), Universiti Malaysia Perlis (UniMAP), Malaysia
[3]Civil Engineering Department, Institut Teknologi Sepuluh Nopember, Indonesia

Abstract. The development of green technology in the construction industry since 10 years ago is something to be proud of Malaysia. Several alternative geopolymer materials were invented in Malaysia such as fly ash, POFA, kaolin, metakaolin, and dolomite based geopolymer materials to achieve sustainable development especially in the building and construction sector. Those alternative materials are very important to replace the application of OPC, which is said to be the main cause of global warming. A review on the content of the chemical differences with the environmental impact resulting from the production of geopolymer is carried out in this study. In conclusion, fly ash based geopolymer material showed the best performance in terms of aluminosilicate content and also it is the best practice in the environmental protection applications for the moment. However, when compared with the OPC, fly ash geopolymer concrete was still able to reduce the effects of global warming potentials, but it is rather gave a negative impact on some aspects of the environment such as abiotic depletions, human toxicity, freshwater ecotoxicity, terrestrial ecotoxicity and acidification.

1 Introduction

The development of green technology in the construction industry since 10 years ago is something to be proud of Malaysia. It also involves and feedback on initiatives in the construction and development of geopolymer. The increment of environmental awareness in recent years around the world including Malaysia has led to a positive assessment of the environmental impact for building materials, in addition to their technical properties. As geopolymer concretes don't contain Portland cement and the powder binder used is typically an industrial waste or a minimally-processed natural material, they can have lower carbon dioxide emissions than classic concrete and be presented as environmental friendly. However, despite the release of carbon become the main criteria used in the assessment of environmental impacts, there are also other aspects that affect the environment, such as fresh

* Corresponding author: harizhijaureen@gmail.com

water and marine ecotoxicity, human toxicity, ozone depletion, acidification, and eutrophication.

Geopolymer concrete had been introduced to reduce the environmental impacts. It is also showed good properties such as high compressive strength, low creep, good acid resistance and low shrinkage [1]. In addition, integrating green building materials into building projects can help to reduce the environmental impacts associated with the extraction, transport, processing, fabrication, installation, reuse, recycling, and disposal of these building industry source materials [2]. Attempts are being made to reduce CO_2 emissions and lower the energy consumption. One of the alternatives to produce more environmentally friendly concrete is to replace the amount of OPC in concrete with by-product materials such as fly ash in the form of blended cement [3]. However, the major drawbacks of blended cements are low early strength. Another way to have environmentally friendly concrete, which can lower CO_2 emission, is the development of inorganic alumino-silicate polymer, called geopolymer, synthesized from materials of geological origin or by-product materials such as fly ash that is rich in silicon and aluminum [4]. This paper reviews the comparisons of selected types of geopolymer based materials with their environmental impact assessment.

2 The types of geopolymer materials

Fly Ash. Fly ash is the residue from the combustion of coal which is widely available worldwide and lead to the anthropogenic pollution. Thus, fly ash-based geopolymer concrete is a good alternative to overcome the abundance of fly ash. Fly ash is rich in silicate and alumina, hence it reacts with alkaline solution to produce aluminosilicate gel that binds the aggregate to produce good concrete. In fly ash-based geopolymer concrete, the silica and the alumina present in the source materials are first induced by alkaline activators to form a gel known as aluminosilicate. The common materials used as alkaline solution in producing fly ash-based geopolymer are sodium silicate and potassium hydroxide [5].

Usually either of this material was mixed with sodium hydroxide to produce the alkaline solution and the molarity (M) of alkaline solution is 7 to 10 M [5]. Abdul Aleem and Arumairaj [27] used to mix fly ash with 10 M alkaline solution while Kawade et al. [28] varied the molarity of sodium silicate solution like 12 M, 14 M and 16 M.The alkaline solution was prepared a day before it is mixed with fly ash. Then, the materials are mixed together with fine aggregate and coarse aggregate to form concrete and curing process been done. To produce higher strength of geopolymer, the optimum sodium silicate to sodium hydroxide ratio was in the range of 0.67 to 1.00. Meanwhile, the concentration of NaOH is in between 10 and 20 M will give small effect on the strength [5-6].

Setting time for geopolymer depend on many factors such as composition of alkaline solution and ratio of alkaline liquid to fly ash by mass. However, the curing temperature is the most important factor for geopolymer. As the curing temperature increases, the setting time of concrete decreases [7]. During curing process, the geopolymer concrete experience polymerization process. Due to the temperature increment, polymerization become more rapid and the concrete can gain 70% of its strength within 3 to 4 hours of curing period [8].

Palm oil fuel ash (POFA). The large production of palm oil in Malaysia has made it become the second largest oil palm producer in the world on 2010, where 18.6 million metric tonnes of crude palm oil has been made [9]. However, waste from the palm oil industry also abundantly produced which caused criticism and complaint. Waste such as palm fibers, nut shells, palm kernel and empty fruit bunches are the solid waste that can be obtained from palm oil mill. Furthermore, these wastes were incinerated in boilers and from this process two types of oil palm ashes were produced which is boiler ash and palm oil fuel ash (POFA).

Boiler ash can be obtained from the burning of oil palm fiber and palm kernel shells in the boiler where it consists of clinker and ash [10]. POFA is by-product from power plant that generate electricity which used palm fiber, shell and empty fruit bunches as fuel and burnt at 800 – 1000 °C [11]. The production of boiler ash and POFA was estimated more than 4 million tonnes/year in Malaysia only [12]. In order to reduce environmental problems, palm oil ash has been utilized in many applications such as raw material for geopolymer composite, cement replacement in production of concrete, wastewater treatment and air purifier in cleaning atmospheric contaminants [9]. Compressive strength of cement paste containing pozzolan materials is contributed by hydration reaction, packing effect and pozzolanic reaction. Hydration reaction is the chemical between Portland cement and water as pozzolanic reaction silica compound and calcium hydroxide [11]. Moreover, packing effect is a proper arrangement of small particles which fill the voids and contribute to the increment of compressive strength [29,30]. Researchers Tangpagasit et al. [31] have to find the compressive strength due to packing effect and pozzolanic reaction use of insoluble material.

Kaolin. Kaolin is a fine clay, rich in kaolinite and used in ceramics. Frequently used aluminosilicate sources are of kaolinite, fly ash, callcined kaolin, and chemically synthesized kaolin. Geopolymers are synthesized by polycondensation below 100 °C at ambient pressure in an alkaline solution. In terms of past literatures, effects of calcined kaolin at high temperatures (800-900 °C) towards properties of post obtained geopolymer have not been elaborately discussed [13].

Kaolin, most versatile white mineral that possess outstanding properties such as chemically unreactive over a wide range of pH and good covering powder when used as a pigment or extender. Secondary kaolins that are fine had been used as gloss materials due to their smaller particle size. Whereas platy kaolin particles, provides excellent coating opacity and printability [14]. Kaolin is also soft, non-abrasive, poor heat and electric conductor. The compressive strength of geopolymer cement paste samples increases with increasing the calcination temperature of kaolin between 600 and 700 °C but decreases above 700°C [16].

Kaolin composes of kaolinite as its core mineral component with handful of secondary minerals, including anatase, quartz, dickite, halloysite and nacrite. In strong alkaline environments, dissolution rate and behaviour of quartz and kaolin differs as quartz has lower reactivity that kaolin. Structural characteristic of dehydrated halloysite, dickite, and nacrite differ while having similar chemical composition as kaolinite. Kaolinite has a sheet structure while halloysite has a tubular structure. In terms of hydroxyl groups, kaolinite has two out of the three hydroxyl groups contribute to the bond while, bonding between layers of dickite involves the collaboration of all inner-surface hydroxyl groups. It has been studied that various hydroxyl groups result in varying dehydroxylation behaviour, which may impact the reactivity of kaolin. Thus, it can be concluded that these secondary minerals present in kaolin will affect the reaction process and final properties of geopolymer. Initial research on how secondary minerals content in kaolin affects the geopolymers should be investigated due to the various types and quantity of mineral composition in kaolin [15].

Metakaolin is a dehydroxylated form of the clay mineral kaolinite. An aluminosilicate material such as kaolinite can be dissolved in an alkali-silicate solution to form a rock hard brittle ceramic. Using the proper chemistry, one can attain a high strength material that can set as little as a few hours at room temperature [16]. During the last few decades, fly ash, slag, kaolinite, mine tailings, etc. are used as raw materials to synthesize geopolymers. Among them kaolinite is the most common raw material due to its relatively purer components. In its raw form, kaolin can react and form a fully hardened geopolymer [15-16].

However, metakaolin is often used due to increased reactivity over raw kaolinite. Metakaolin obtain from calcination or dehydroxylation of kaolin clay at 500-900 °C. This removes chemically bonded water and change a large portion of the octahedral coordinated aluminum which is found in kaolin to four and five-fold configuration [16]. Kong et al. [32] reported that S/L ratio of 0.8 gave nearly optimum strength and provided good workability. Higher S/L ratio than 0.8 had very low workability and deteriorated the properties of the paste produced. Liew et al. [33] activated metakaolin pastes with alkali activation solution at S/L ratios, by mass, ranging between 0.40 and 1.20. The alkali activation solution was Na_2SiO_3/NaOH with different ratios. The results of bulk density and compressive strength showed that the S/L of 0.8 gave the highest values at Na_2SiO_3/NaOH ratio of 0.20. Weng et al. [34] studied different specific surface areas of metakaolin activated with sodium silicate and NaOH They concluded that higher specific surface area of metakaolin powders were characterized by quicker setting time, higher compressive strength and more homogeneous microstructure.

Dolomite. Calcium carbonate and calcium-magnesium carbonate in the form of limestone, dolomite, marl, chalk, and Oyster shell are one of the most widely utilized non-metallic materials in the industrial world. The largest use of limestone or calcium carbonate is in the cement industry where it is used as a source of CaO and also in the concrete industry where it is used as the primary coarse aggregate. Following the cement industry, the second largest user would be the lime industry [17].

Dolomite is relatively soft and easily crushed to a fine powder, which is used as agricultural lime ('aglime') by farmers to reduce soil acidity and also to adjust magnesium deficiencies. Study by Patel and Shah [35] reveals that value of cohesion is observed to be decreasing while angle of internal frication angle is increasing with increases in number of joints. The Normal stress is found to be decreasing as numbers of joints are increasing. Strength of jointed rock is dependent on the direction of applied loading with respect to orientation of joints [35]. In jointed rock specimen the failure is observed in terms of hair cracks surrounding the jointed rock area where as in unjointed specimen the failure is observed in terms of broken pieces of specimen. The strength of the rock specimen jointed by microfine cement is higher than the unjointed specimen. The load carrying of vertical cut specimen is higher than the horizontal cut specimen and also with increase in number of horizontal cut the load carrying capacity of specimen decreases. Observed that shear angle was dependent on confining pressure and the spacing of joint in the specimen [35].

Dolomite is equally good as limestone in neutralizing soil acidity but magnesium is also an important element itself as a plant nutrient [18]. Dolomite is a common rock-forming mineral. It is a calcium magnesium carbonate with a chemical composition of $CaMg(CO_3)_2$. It is the primary component of the sedimentary rock known as dolostone and the metamorphic rock known as dolomitic marble. Limestone that contains some dolomite is known as dolomitic limestone. The mineral dolomite crystallizes in the trigonal-rhombohedral system. It forms white, tan, gray, or pink crystals. Dolomite is a double carbonate, having an alternating structural arrangement of calcium and magnesium ions. It does not rapidly dissolve or effervesce (fizz) in dilute hydrochloric acid as calcite does. Crystal twinning is common [19].

3 Chemical composition (XRF)

Based on Table 1, fly ash, metakaolin and kaolin consists highest SiO_2 and Al_2O_3 which is very important in geopolymerization process. POFA contains high SiO_2 but it has low Al_2O_3 content and almost of its weight which is 21.6% was lost in ignition. Dolomite shows highest CaO content which is 23.0% and highest in MgO which is 17.2% but it has quite low SiO_2 which is only 15.37%.

Table 1. The review of chemical composition of various geopolymer materials using XRF.

Chemical composition	Fly ash [20]	POFA [9]	Kaolin [21]	Metakaolin [22]	Dolomite [23]
SiO_2	52.11	51.18	52.00	55.90	15.37
Al_2O_3	23.59	4.61	35.00	37.20	1.69
Fe_2O_3	7.39	3.42	1.00	1.70	0.51
TiO_2	0.88		0.90	2.40	0.015
CaO	2.61	6.93	<0.05	0.11	23.00
MgO	0.78	4.02	0.70	0.24	17.20
K_2O	0.80	5.52	2.00	0.18	0.195
Na_2O	0.42	0.06	0.05	0.27	0.013
SO_3	0.49			0.02	
P_2O_5	1.31			0.17	0.019
Loss in ignition		21.6		0.80	

The two main constituents of geopolymers are source materials and alkaline liquids. Source should be rich in silicon (Si) and aluminum (Al). These could be natural minerals such as kaolinite and clays. Alternatively, fly ash, silica fume, slag, rice-husk ash, red mud, and metakaolin, among others, could be used as source materials as byproduct materials [5].

As reported by Kroehong et al. [11], total content of SiO_2, Al_2O_3 and Fe_2O_3 of 2 types of POFA geopolymers were 56.9% and 58.6%, respectively. For insoluble material, the main chemical composition is SiO_2, which is approximately 93-94%.

Al_2O_3 composition is high in kaolin, metakaolin and fly ash but it is less in POFA and dolomite. Heah et al. [21] reported that Al_2O_3 composition was 35% in kaolin geopolymer while Jabiulla and Neeraja [22] reported that Al_2O_3 composition was 37.2% in metakaolin geopolymer.

4 Environmental impacts of geopolymer

Based on Table 2, the best performance and less environmental impact geopolymer material goes to fly ash based because it has optimum SiO_2 and Al_2O_3 composition and it promotes waste reduction by reusing waste products from coal mining industry. Moreover, fly ash based geopolymer doesn't deplete natural resources like metakaolin, kaolin and dolomite based geopolymer materials.

Table 2. Various types of geopolymer based with their sources

Various types of geopolymer based.	Highest chemical composition	Sources of raw materials
Fly ash [20]	SiO_2 & Al_2O_3	Waste product from coal mining
Metakaolin [9]	SiO_2 & Al_2O_3	Natural resources
Kaoline [21]	SiO_2 & Al_2O_3	Natural resources
POFA [22]	SiO_2	Waste product from oil palm industry
Dolomite [23]	CaO, MgO & SiO_2	Natural resources

The use of fly ash which would otherwise be ending up in landfills in geopolymer concrete further proves that it is an environmentally friendly material. The reduced greenhouse gas emission by using Australian geopolymer products is estimated to be 44 – 64% compared with that of OPC [24].

The reduction in CO_2 emissions for geopolymer system is due to the use of minimum processed natural minerals and industrial wastes to form the binding agents. The process of using this waste material as a component to produce the binder helps to mitigate the environmental problems and provides new environmentally friendly green concrete [25].

Table 3 shows the result of the environmental impact of various geopolymer materials and as reported by Habert et al. [26], the main environmental impacts that might be affected by the three types of geopolymer materials are global warming potentials and human toxicity. The environmental impacts are evaluated according to the baseline method of CML01 by Habert et al. and they evaluated 10 environmental impacts such as abiotic depletion, global warming, ozone layer depletion, fresh and marine water ecotoxicity, terrestrial ecotoxicity, human toxicity, eutrophication, acidification and photochemical oxidation.

Table 3. Environmental impact of various geopolymer materials. Adopted from Habert et al. [26].

Environmental Impacts	**Fly Ash**	**Metakaolin**	**Blast Furnace Slag Granulated**
Abiotic depletions (kg Sb eq.)	1.95×10^{-2}	3.43×10^{-2}	1.21×10^{-2}
Global warming potentials (kg CO_2 eq.)	2.51	4.32	1.25
Ozine layer depletions (kg CFC-11 eq.)	2.43×10^{-8}	2.59×10^{-7}	2.45×10^{-8}
Human toxicity (kg 1,4-DB eq.)	5.01×10^{-1}	1.74	3.99×10^{-1}
Freshwater ecotoxicity (kg 1,4-DB eq.)	3.17×10^{-2}	5.69×10^{-2}	2.10×10^{-1}
Marine ecotoxicity (kg 1,4-DB eq.)	2.97×10^{-3}	1.80×10^{-4}	5.20×10^{-2}
Terresterial ecotoxicity (kg 1,4-DB eq.)	4.48×10^{-4}	1.38×10^{-2}	3.35×10^{-3}
Photochemical oxydation (kg C_2H_4 eq.)	6.62×10^{-4}	2.26×10^{-3}	8.39×10^{-4}
Acidification (kg SO_2 eq.)	1.92×10^{-2}	2.42×10^{-2}	4.85×10^{-3}
Eutrophication (kg PO_4^{3-} eq.)	1.06×10^{-3}	1.87×10^{-3}	6.77×10^{-4}

The environmental impact of these materials are calculated and compared with a 100% OPC concrete displaying an equivalent mechanical strength. For each type of geopolymer concrete, a mean impact and a standard deviation are evaluated. The impact of standard OPC concrete is considered as a reference and set at a 100% value and relative values are calculated for each mix-design [26].

This study used the Life Cycle Assessment methodology to carry out a detailed environmental evaluation of the production of geopolymer concrete. The results show that the production of most standard types of geopolymer concrete has a slightly lower impact on global warming than standard OPC concrete. However they also reveal that the production of geopolymer concrete has a higher environmental impact regarding other impact categories than global warming. This is due to the heavy effects of the production of the sodium silicate solution [26].

Geopolymer concrete made from fly ashes or granulated blast furnace slags based require less of the sodium silicate solution in order to be activated. They therefore have a lower environmental impact than geopolymer concrete made from pure metakaolin. However, when the production of fly ashes and granulated blast furnace slags is taken into account during the life cycle assessment (using either an economic or a mass allocation procedure), it appears that geopolymer concrete has a similar impact on global warming than standard concrete [26].

A concrete made with OPC and displaying an equivalent mechanical strength shall contain 354 kg m^3 of cement [26]. The environmental impact of these mix-designs is presented in Table 4. It is obvious that the use of a sodium silicate solution is responsible for the major part of the environmental impact in the case of geopolymer concrete.

Table 4. Environmental impacts for a standard fly ash based geopolymer concrete. Adopted from Habert et al. [26].

Environmental Impacts	OPC concrete	OPC	Fly ash	Geopolymer concrete	Na Silicate	NaOH Powder
Abiotic depletions (kg Sb eq.)	0.61	5.64 x 10^{-1}	8.25 x 10^{-2}	1.19	7.44 x 10^{-1}	2.72 x 10^{-2}
Global warming potentials (kg CO_2 eq.)	305.9	299.1	2.14	168.5	117.8	3.71 x 10^{-1}
Ozone layer depletions (kg CFC-11 eq.)	8.74 x 10^{-6}	8.07 x 10^{-6}	1.37 x 10^{-6}	1.39 x 10^{-5}	9.08 x 10^{-6}	2.28 x 10^{-6}
Human toxicity (kg 1,4-DB eq.)	18.90	14.26	6.44 x 10-1	105.4	82.75	15.84
Freshwater ecotoxicity (kg 1,4-DB eq.)	2.52	1.46	7.18 x 10-2	27.01	21.84	3.98
Marine ecotoxicity (kg 1,4-DB eq.)	9.68 x 10^{-3}	6.89 x 10^{-3}	7.86 x 10^{-2}	4.59 x 10^{-4}	3.42 x 10^{-4}	7.87 x 10^{-3}
Terresterial ecotoxicity (kg 1,4-DB eq.)	0.45	4.16 x 10^{-1}	6.84 x 10^{-3}	1.77	9.23 x 10^{-1}	7.72 x 10^{-1}
Photochemical oxydation (kg C_2H_4 eq.)	1.67 x 10^{-2}	1.51 x 10^{-2}	7.88 x 10^{-4}	3.65 x 10^{-2}	2.51 x 10^{-2}	7.67 x 10^{-3}
Acidification (kg SO_2 eq.)	0.45	4.09 x 10^{-1}	1.35 x 10^{-2}	0.82	5.37 x 10^{-1}	1.78 x 10^{-1}
Eutrophication (kg PO_4^{3-} eq.)	6.83 x 10^{-2}	6.13 x 10^{-2}	2.01 x 10^{-3}	7.96 x 10^{-2}	5.10 x 10^{-2}	1.34 x 10^{-2}

According to Habert et al. [26], when this type of geopolymer concrete is compared with hydraulic cement based standard concrete, it is obvious that this new type of binder allows for a strong reduction of the global warming potential. From 306 kg of equivalent CO_2 per m^3 for OPC based concrete, the geopolymer concrete releases only 169 kg of equivalent CO_2 per m^3, which represents a saving of 45%. However, it is interesting to note, that this value is not so different from CO_2 emission reduction reached with an improvement in cement technology efficiency, where a 50% of reduction can be achieved by using existing technologies.

Based on the study, the most vulnerable environmental impacts affected by the materials of fly ash geopolymer and OPC are global warming potentials, human toxicity and terrestrial ecotoxicity. Fly ash geopolymer concrete and OPC concrete are recorded to be causing the abiotic depletions which are limestones and other lithospheric resources while the least impact goes to fly ash since fly ash consumption into geopolymer processing will reduce the environmental pollution.

As reported by Habert et al. [26], the most impactful material for global warming potential is OPC concrete with 305.9 kg CO_2 eq and OPC with 299.1 kg CO_2 eq GWP levels.

DOI: 10.1051/matecconf/20179701021

Geopolymer concrete at least cut the global warming impact into half which is 168.5 kg CO_2 eq. The least impact of this study goes to fly ash and NaOH powder.

If the comparison have been made between OPC concrete and fly ash based geopolymer concrete, OPC concrete shows more global warming potential than fly ash based geopolymer concrete since it cut almost half of the OPC concrete CO_2 emission from its life cycle. But, fly ash based geopolymer concrete gives some other environmental impacts rather than OPC concrete but the impacts are not so serious and the impacts are listed as abiotic depletions 1.19 kg Sb eq., human toxicity 105.4 kg 1,4-DB eq., freshwater ecotoxicity 27.01 kg 1,4-DB eq., terresterial ecotoxicity 1.77 kg 1,4-DB eq. and acidification 0.82 kgSO_2 eq .

5 Conclusion

From the comparison review between fly ash, metakaolin, kaolin, POFA and dolomite based geopolymer materials, fly ash based geopolymer material shows the highest performance in terms of SiO_2 and Al_2O_3 composition and it is the best option for environmental protection and sustainability because it apply waste reduction by reusing coal mining waste products to produce geoplymer. From the comparison review between fly ash concrete and OPC concrete, fly ash concrete based still shows positive results to the environment because it produced almost half CO_2 emission lower than OPC concrete during its production, but fly ash concrete based gave some low negative impacts to the environment such as abiotic depletions, human toxicity, freshwater ecotoxicity, terresterial ecotoxicity and acidification. The impact still can be controlled and further study is needed to overcome those little impacts to make sure that geopolymer is ready to empower green technology development for the sustainable future.

This work was supported by the Center of Excellence of Geopolymer and Green Technology (CEGeogtech) and Universiti Malaysia Perlis (UniMAP) under the supervision of Assoc. Prof. Dr. Mohd Mustafa Al Bakri Abdullah.

References

1 G.P. Lodeiro, Cem. Concr. Res., 175-183 (2007)
2 N.M. Mohamed. J. Am. Sci., **6**, 1381(2010)
3 N.B. Singh, B. Middendorf, Cem. Int., **6**, 76 (2008)
4 S. Ahmari, X. Ren, V. Toufigh, L. Zhang, Constr. Build. Mater., **35**, 718 (2012)
5 M.M.A. Abdullah, H. Kamarudin, H. Mohammed, I. Khairul Nizar, A.R. Rafiza, Y. Zarina, J. Eng. Technol. Res., **3**, 1 (2011)
6 P. Chindaprasirt, T. Chareerat and V. Siricicatnanon, Cem. Concr. Compos., **29**, 224 (2007)
7 N.V. Chanh, A.B. Trung, D.V. Tuan. Recent Research Geopolymer Concrete. The 3rd ACF International Conference - ACF/VCA, 1-5 (2008)
8 D.L.Y. Kong, J. G. Sanjayan, Cem. Concr. Compos., **30**, 986 (2008)
9 Y. Zarina, M.M.A. Abdullah, H. Kamarudin, I. Khairul Nizar, A.R. Rafiza, Rev. Adv. Mater. Sci., **34**, 37 (2013)
10 V. Subramaniam, M.A. Ngan, C. Yuen May, N.S. Nik Meriam, American Journal of Environmental Sciences, **4**, 310 (2008)
11 W. Kroehong, T. Sinsiri, Jaturapitakkul. Procedia Eng., **14**, 361 (2011)
12 A.R. Mohamed, K.T. Lee, N.M. Noor, N.F. Zainudin, Procedia Eng., **125**, 804 (2015)
13 M.R. Wang, D.C. Jia, P. He, Y. Zhou, Mater. Lett., **64**, 2551 (2010)
14 K. Rissa, T. Lepisto, K. Yrjola, Prog. Org. Coat., **55**, 137 (2006)

15 S. Ramasamy, H. Kamarudin, M.M.A. Abdullah, C.M. Ruzaidi Ghazali, A.V. Sandu, M. Binhussain, Rev. Adv. Mater. Sci., **42**, 83 (2015)
16 M.A. Soleimani, R. Naghizadeh, A.R. Mirhabibi, F. Golestanifard, Iranian Journal of Materials Science & Engineering, **9**, 43 (2012)
17 B.D. Wheeler, The Rigaku Journal, 16 (1999)
18 S. Gunasekaran, G. Anbalagan, Bull. Mater. Sci., **30**, 339 (2007)
19 A. Kranjc, Acta Carsologica, **35**, 40 (2006)
20 M.M.A. Abdullah, H. Kamarudin, M. Binhussain, I. Khairul Nizar, A.R. Rafiza, Y. Zarina, Rev.Adv.Mater. Sci., **30**, 90 (2012)
21 C.Y. Heah, M.M.A. Abdullah, M. Binhussain, M. Luqman, I. Khairul Nizar, C.M Ruzaidi, Y.M. Liew, Phys. Procedia, **22**, 305 (2011)
22 S. Jabiulla, D.D. Neeraja, International Journal for Technological Research In Engineering, 1309 (2015)
23 M.A.A. Mohammed, A. Salmiaton, W.A.K.G. Wan Azlina, M.S. Mohamad Amran, Y.H. Taufiq Yap, Journal of Energy, 1 (2013)
24 B.C. McLellan, R.P. Williams, J. Lay, A. Van Riessen, G.D. Corder, J. Clean. Prod., **19**, 1080 (2011)
25 S. Joshi, M. Kadu, International Journal of Environmental Science and Development, **3**, 417 (2012)
26 G. Habert, J.B. 'Espinose de Lacaillerie, N. Roussel, J. Clean. Prod., **19**, 1229 (2011)
27 M.I. Abdul Aleem, P.D. Arumairaj, Indian J. Sci. Technol., **5**, 2299 (2012)
28 U.R. Kawade, P.A. Salunkhe, S.D. Kurhade, Int. J. Innov. Res. Sci. Eng. Technol., 135 (2014)
29 A. Goldman, A. Bentur, Cem. Concr. Res., **23**, 962 (1993)
30 G.C. Isaia, A.L.G. Gastaldini, R. Moraes, Cem. Concr. Compos., **25**, 69 (2003)
31 W. Tangchirapat, C. Jaturapitakkul, P. Chindaprasirt, Constr. Build. Mater., **23**, 2641 (2009)
32 L.Y. Kong Daniel, G. Sanjayan Jay, Kwesi Sagoe Crentsil. Cem. Concr. Res., **37**, 1583 (2007)
33 Y.M. Liew, A.M. Mustafa Al Bakri, M. Bnhussain, M. Luqman, I. Khairul Nizar, C.M. Ruzaidi, Constr. Build. Mater., **37**, 440 (2012)
34 L. Weng, Kwesi Sagoe Crentsil, T. Brown, S. Song, Mater. Sci. Eng. B, **117**, 163 (2005)
35 M.B. Patel, M.V. Shah, International Journal of Science Technology & Engineering, **1**, 114 (2015)

Review on Characterization and Mechanical Performance of Self-cleaning Concrete

Siti Norsaffirah Zailan[1,*], *Norsuria* Mahmed[1] , *Mohd Mustafa Al Bakri* Abdullah[1] , *Andrei* Victor Sandu[3] , and *Noor Fifinatasha* Shahedan[1]

[1]Centre of Excellence Geopolymer and Green Technology (CEGcoGTech), School of Materials Engineering, Universiti Malaysia Perlis (UniMAP), 01007, P. O Box 77, D/A Pejabat Pos Besar, Kangar, Perlis, Malaysia
[2]Faculty of Engineering Technology, Universiti Malaysia Perlis (UniMAP), P. O Box 77, D/A Pejabat Pos Besar, Kangar, Perlis 01000, Malaysia
[3]Gheorghe Asachi Technical University, Faculty of Materials Science and Engineering, D. 61A Mangeron Str., 700050, Iasi, Romania

Abstract. Self-cleaning concrete is an effective alternative to provide cleaner environment which contribute to sustainability and towards a green environment. It is in accordance with the requirements of environmental issues on huge energy consumption and air pollution from carbon dioxide (CO2) emissions. Photocatalyst in self-cleaning concrete accelerates the decomposition of organic particulates, hence pollution could be reduced through photocatalytic degradation of gaseous pollutants. Mechanical performances of self-cleaning concrete were improved by adding photocatalytic materials. Self cleaning abilities were evaluated in the photocatalytic activity test under UV light and photocatalytic degradation of gaseous pollutant was measured by depollution test. This review aims to give an overview about the characteristics of photocatalytic materials and mechanical performances of self-cleaning concrete.

1 Introduction

The demand usage of concrete increases as cities grow. However, the utilization of concrete causes pollution to the environment, hence give negative impacts on the environment. Quality of air becomes an important issue for public health and the environment. Thus, recently, many research having a great concern in the development of cementless concrete in order to reduce CO_2 emission that caused by concrete dust [1]. Geopolymer concrete with self-cleaning behavior is an innovative construction material that had been introduced to reduce this problem. In geopolymer concrete, the role of the binder is replaced by fly ash, which is cementitious materials as ordinary Portland cement (OPC) due to high content of alumina and silicate, whereas alkaline solution act as an activator which is a combination of sodium hydroxide (NaOH) and sodium silicate (Na_2SiO_3) [2].

Self-cleaning concrete has an ability to clean their surfaces by itself. Self-cleaning technology has been derived from nature. The surface of self-cleaning inspired by the

* Corresponding author: sitinorsaffirah@yahoo.com

nanostructures of a lotus leaf which based on superhydrophobic effect, which residues are washed off by rain, thus carrying away dirt and debris. This is due to the photocatalytic actions with the addition of photocatalyst onto the concrete [3]. Photocatalytic reaction uses the energy from ultraviolet rays to oxidize organic compounds. This reaction is able to accelerate the process of natural oxidation and faster pollutant decomposition. Photocatalytic have two primary benefits which are kept surface free of dirt and ensure a cleaner environment. Photocatalytic materials is not only apart, having a wide range of applications, but it also offers various benefits, which include reduction in maintenance cost and time (process of cleaning), thereby energy was saved in cooling the buildings by reducing the temperature [4].

Previous research reported about the presence of voids in concrete, which is about 1-2% that leads to decrement of compressive strength [5]. Thus, nanoparticles photocatalyst would fill the voids between cement particles, then smaller pores are produced in order to increase the strength, modify the properties and improve durability, thermal, mechanical, and electrical properties of cementitious materials. Nanoparticles are commonly used in the research as photocatalyst is silicon oxide (SiO_2), zinc oxide (ZnO), aluminium oxide (Al_2O_3) and titanium oxide (TiO_2) [5]. In the previous study [6], concrete containing of titanium dioxide with white cement are prepared, while titanium dioxide is added in relatively small amounts of cement and was doped with silver (Ag) to improve the early resistance of the concrete. Many research activities are currently conducted about applications of nano materials on construction materials. Recently nano silica (nano-SiO_2) and nano titanium (nano-TiO_2) are reported that can improve the mechanical properties of mechanical and durability properties of concrete which causes an improvement of the refined porous structure of concrete, whereas resistance to chloride penetration increased [7].

2 Processing of Self-cleaning concrete

Mix Design. The study of fly ash and photocatalyst was mixed in different proportions and used to produce self-cleaning concrete. Recently in the previous experiments [5], nano-TiO_2 powder was used as the photocatalyst, while OPC and white ordinary Portland cement were used as the cementitious materials which act as a binder. TiO_2-modified cement are made of ordinary white cement, water and different contents of TiO_2 nanoparticles (0%, 0.5%, 1%, 1.5%, 2%, 2.5% and 3%) for self-cleaning and testing for setting time. Other than that, sand with standard size was added for compressive and flexural strength tests. The same method was practice of another reported research [8], where there only differs in step of addition the nanoparticles onto the modified concrete. The samples dispersed TiO_2 powder in ethanol followed by ultrasonic vibration for 15 min for TEM observations.

Other research reported that, the crude titanium dioxide was later added to the cement powder in the preparation process of the cement plates [9]. Previous researchers state that, ordinary Portland cement and fly ash class F (low calcium fly ash) were used as the cementitious materials to produce geopolymer concrete. The amount of fly ash used was 25% from weight of cement. Other than that, natural river sand was used as fine aggregate and were washed before using it. Cement, water, sand and fly ash were mixed at binder ratio 0.4. The amount of photocatalyst in the mixture was varied with ratios of 1%, 3% and 5% of the cement weight and all mixtures (single, binary and ternary) are remaining constant [10].

Mixing Process. W. Shen *et al.* [8] proposed mixing cement with various nano particles TiO_2 contents (0%, 2.5%, 5.0% and 7.5%) by using a ball mill with a diameter of 500 mm for 30 min. Besides that, other researchers who have used TiO_2 as photocatalyst in their study was added powdered TiO_2 to the dry cement before mixing, then subsequent water was added. Then, cementitious paste was mechanically stirred at 140 rpm with 62 planetary motions per minute for 30 s. The paste are then will be cast into molds with dimension 8.6

DOI: 10.1051/matecconf/20179701022

cm-diameter Petri dish. The duration for curing process are 24 hours and was placed at room temperature at 20°C under 70% relative humidity (RH) and, 100% RH for the next 27 days. Before the characterization and the photocatalytic tests were conducted, samples were aged at least three months [11].

The same method of mixing was applied by D. Feng *et. Al* [12] but the TiO_2 nanoparticles and cement were soaked in water and vibrated ultrasonically for 30 min in order to achieve better distribution before mixing process. On the other hand, P. Duan *et. al* [13] reported that the purpose of vibration is to remove the entrained air bubbles. However, other researchers [9] have stated that the different amount of ratio were added to raw cement and mixed mechanically by using porcelain mortar which are then be cast in silicone mold (20 mm × 20 mm × 6 mm) and curing for 28 days at room temperature. Then, homogenous powders that obtained were mixed with water (water-to-cement ratio w/c = 0.48). The author believes that excessive additional of water will creates additional voids which known as capillary pores and give a big influence on the strength and permeability of the concrete.

3 Characterization materials in self-cleaning concrete

X-ray Fluorescence (XRF). Previous research has studied about the characteristics of the raw materials that use in geopolymer concrete by using X-ray Fluorescence (XRF). Based on the chemical composition, these previous raw materials that have been used in geopolymer concrete contain high content in Silica (Si), Alumina (Al), Calcium (Ca) and Ferum (Fe) which indicates the performance of geopolymer materials are good enough to be applied as cementitious materials [13]. According to M. Mustafa *et al.* [14], fly ash that commonly used as a binder in current research have two types, which are Class C (high calcium fly ash) and Class F (low calcium fly ash). Fly ash are categorized under class C type if the composition analysis of CaO content are more than 10%, while class F type of fly ash which contain less than 10% of the CaO. Fly ash class C type are commonly being used as replacement of cement in construction materials.

X-ray Diffraction (XRD). Previous published research has studied the characterization of raw material (TiO_2) that has been used as a photocatalytic application by using X-ray Diffraction (XRD) [15]. According to P. Krishnan *et al.* [15], crystal structure deterioration is a possible reason for the efficiency for photocatalytic of titania to decrease. However, the XRD patterns before and after 400 hours of UV exposure demonstrates stability of the TiO_2 crystal structure due to nearly identical of TiO_2 photocatalyst. The author suggests that the deterioration of crystal structure would not occur within the investigated duration because it may cause decreased photocatalysis of TiO_2.

According to previous research has found that XRD pattern measured for raw cement had no impact on the phase composition. The published research revealed that the presence of additional peaks which assigned to the anatase and rutile phases when Titania photocatalyst was added to the raw cement [9]. The narrower anatase peak indicates higher crystallinity that occurred in photocatalyst. Additionally, it was applied in high temperature of calcination which caused a partial phase transformation from anatase to rutile.

Scanning Electron Microscopy (SEM). According to Hui Li *et. al* [16], the microstructures between the cement mortar mixed with the nano-particles and the plain cement mortar are observed by Scanning Electron Microscope (SEM). The result showed that the pores was filled up with nano-Fe_2O_3 and nano-SiO_2 and the $Ca(OH)_2$ compound are reduced between the hydrate. From the SEM observation, nano-particles also act as activator either than filler in order to improve the microstructure of the cement paste, where nano-particles were spread uniformly.

Another research [6] reported that, surface of prepared cement samples were analyzed using SEM as shown in Figure 1. SEM images obviously show the surface of pure white cement particle were closely packed while even distributions of TiO_2 are highlighted in circles on the surface of the sample cement slab with TiO_2 which have a ratio 0.1 and 0.2. The photocatalyst are well distributed on the surface. The author also reported that there is some dust and small cracks are being observed on the surface of pure cement. However, P. Duan *et al.*[13] reported that the addition of nano-TiO_2 enhances the geopolymer formation and microstructures become denser with fewer crack.

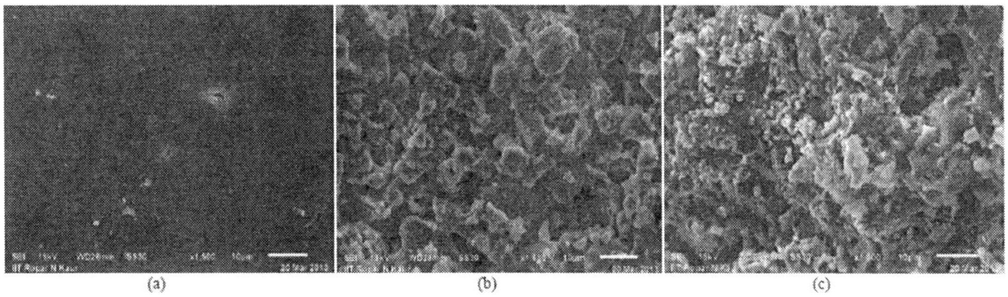

Fig. 1. SEM images at 1500x of (a) Cement slab made up of white cement only (b) Cement slab made-up of TiO2 to white cement ratio of 0.1 (c) Cement slab made-up of TiO2 to white cement ratio of 0.2 [6]

4 Mechanical properties on self-cleaning concrete

Compressive Strength. Previous research that study on the mechanical properties TiO_2-modified cementitious materials state that the partial addition of TiO_2 nanoparticles into cement improved the mechanical properties of the cement mortar [5]. On the other hand, compressive strength increases with increasing the amount of TiO_2 nanoparticles by 1%. This is because pozzolanic activity of TiO_2 in hydrating cement attributed to the improvement of compressive strength of the modified concrete. There is a research on the effect of nanomaterials as cement replacement on physical properties of concrete, reported that the compressive strength of concrete and cement by using various nano products that commonly used in published research as additives and admixtures was reduced as in Table 1 [17].

From the other reported research on the mechanical properties, state that compressive strength increase up to 1% replacement of cement by TiO_2. The main reason increasing of compressive strength is because of voids of mortar was filled with small size titanium dioxide, which leads to increase the strength. In addition, the cement particles covered by the excess amount of TiO_2 which disrupts the water cement reaction, hence the decrease the strength on further increment. As in Figure 2 shows a comparison of compressive strength with the replacement of cement by titanium dioxide (1%, 2%, 3%) that cured in 28 days. The author suggests to extend this study on various properties of concrete by changing the particle size of titanium dioxide and various grades of concrete [10].

Table 1. Compressive strength test of cement cube (MPa) [17]

Days	Normal Concrete Mortar	Nano Alumina Mortar	Nano Titania Mortar	Nano Silica Mortar
7	22.069	16.050	12.049	7.356
14	32.102	20.731	18.725	16.719

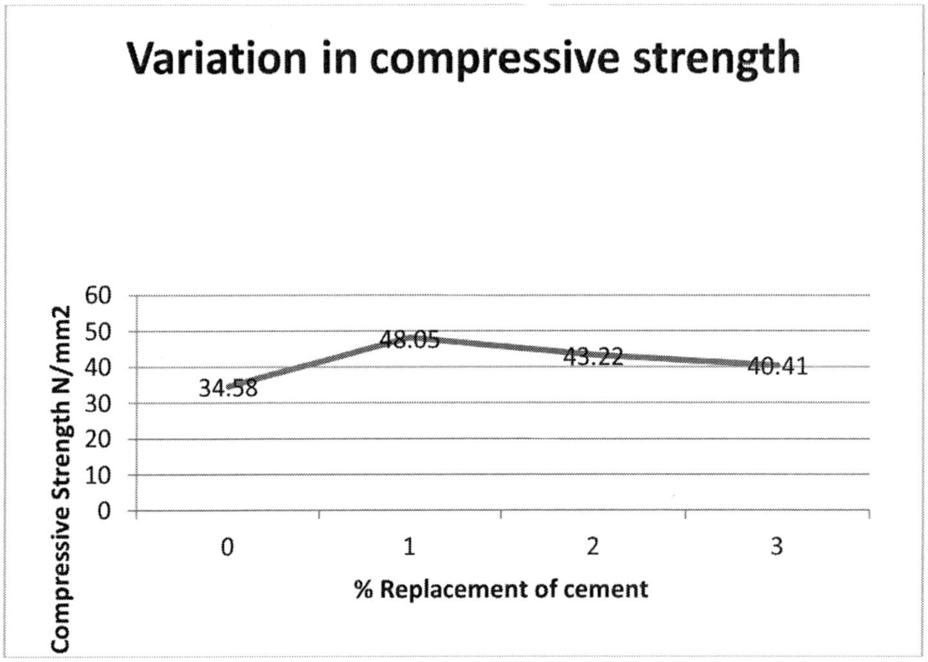

Fig. 2. Comparison of compressive strength with the replacement of cement by titanium dioxide (Days 28) [10]

Flexural Strength. From the previous research, flexural strength is conducted for measurement of the tensile strength of the concrete to resist failure in bending [5]. It was reported that the increases of flexural strength with increasing the amount of TiO_2 nanoparticles up to 1% due to the reaction between TiO_2 nanoparticles and calcium hydroxide are fast. These reactions were made in the cement hydration in order to form the denser structure. Published research that study on the compressive, tensile and flexural strength of cement mortar containing nanoparticles state that the mechanical properties of Al_2O_3 nanoparticles (1% and 3%) are better than ordinary cement mortar [7]. The results of flexural strength are lower than compressive strength results. However, the results with the addition of photocatalytic materials (nano-TiO_2 and nano-SiO_2) are less significant increased.

5 Self-cleaning test

Photocatalytic testing. A self-cleaning test was conducted to assess the abilities of self-cleaning of the material. Previous research reported that photo degradation of methylene blue (MB) solution was studied in order to evaluate the photocatalytic properties of self-cleaning on the surface of concrete under UV radiation. As shown in Figure 3, the color of the solution indicates a photocatalytic effect of self-cleaning concrete on the degradation of MB blue with different percentage of Titania as photocalyst by exposing time. The concrete with high content of TiO_2 has obvious effect on the degradation of MB blue, which has a higher degradation ratio than the lower TiO_2 content. This shows that nano-TiO_2 particles have the potential for as a photocatalyst for building materials [8]. Other than the degradation of MB, there are other published research reported that performances of self-cleaning mortars were verified by monitoring the discoloration of organic dyes (Rhodamine B (RhB) [18].

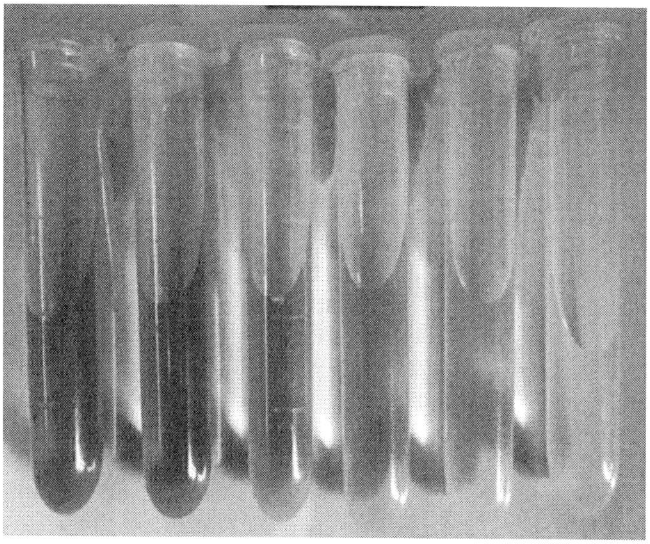

Fig. 3. Photocatalytic effect of the self-cleaning concrete on the MB solution with the exposing time [8]

Depollution testing (NOx Oxidation test). A NOx oxidation test was conducted to evaluate depollution effects of the materials. Jun Chen *et al.* [19] used a stirred flow reactor to measure the photodegradation of organic compounds on the photocatalytic material surface regarding the degradation of VOCs. Photocatalytic activity of pure samples of TiO_2 was observed three to ten times greater than the cementitious sample with 3% catalyst that was prepared.

The same test was reported by P. Krishnan *et al.* [15] but differ from machine that used to test photocatalytic degradation of gaseous pollutant. In these researches, a continuous flow reactor was modified and used for photocatalytic NOx removal test, while the concentration of NOx concentration was measured by a Chemiluminescence NOx analyzer. On the other hand, the photocatalytic degradation efficiency was not increased even dosage of TiO_2 increased from 2% to 4% in the mortar.

6 Summary

The current knowledge shows that the influence of adding photocatalyst in the cementitious material is essential for enhances the mechanical properties of self-cleaning concrete. As a conclusion, nanoparticles of photocatalyst are a potential material to be as filler or replacing part of cement in order to improve the performances of self-cleaning behavior. Based on this review, further study on development of geopolymer concrete of self-cleaning applications can be conducted.

References

1. M.B. Karakoc, F. Kantarci, I. Turkmen, R. Demirboga, M.M. Maras, M.U. Toprak, Renewable Energy Research and Applications (ICRERA), 2013 International Conference on IEEE, (2013)
2. Sharma, K. Sanjay, Himmi Gupta, *Development of Paver Block by Using Foundry Sand Based Geopolymer Concrete* (2015).
3. Banerjee, Swagata, Dionysios D. Dionysiou, Suresh C. Pillai, Appl. Catal. B-Environ., **176,** 396 (2015)
4. A. Khitab, M. Alam, H. Riaz, S. Rauf, Adv. Life Sci., **1**, 47 (2014)
5. R. Khataee, V. Heydari, L. Moradkhannejhad, M. Safarpour, S. W. Joo, J. Nanosci. Nanotechnol, **13**, 5109(2013)
6. Dahlan, Dahyunir., *Effect of Aging Time on the Synthesis of Fe-doped TiO2 Thin Films by Spin Coating Method* (2016).
7. S. Rao, P. Silva, J. De Brito, Constr. Build. Mater, **96**, 508(2015)
8. Shen, Weiguo, J. Clean Prod., **82**, 762(2015)
9. M. Janus, J. Zatorska, A. Czyzewski, K. Bubacz, E. Kusiak Nejman, A. Morawski, Appl. Surf. Sci, **330,** 200 (2015)
10. Khushwaha, A., Rachit Saxena, Shilpa Pal., J. Civ. Eng. Manag, **2**, 482(2015)
11. A.H. Aissa, E. Puzenat, A. Plassais, J.M. Herrmann, C. Haehnel, C. Guillard, Appl. Catal. B-Environ, **107,** 1(2011)
12. D. Feng, N. Xie, C. Gong, Z. Leng, H. Xiao, H. Li, X. Shi, Ind. Eng. Chem. Res., **52**, 11575(2013)
13. P. Duan, C. Yan, W. Luo, W. Zhou, Constr. Build. Mater., **106**, 115 (2016)
14. Mohd Mustafa Al Bakri, Abdullah, *Geopolymerization technology for soil stabilization application* (2014).
15. P. Krishnan, M.H. Zhang, L. Yu, H. Feng, Constr. Build. Mater, **44**, 309 (2013).
16. H. Li, H.G. Xiao, J. Yuan, J. Ou, Compos. Part B-Eng, **35,** 185 (2004)
17. Bastos, Guillermo, Materials, **9,** 972(2016)
18. E. Jimenez Relinque, J.R. Rodriguez Garcia, A. Castillo, M., Castellote, Cement Concrete Res., **71,** 124 (2015)
19. Chen, Jun, Chi-sun Poon, Build. Environ., **44,** 1899 (2009)

Review on Adsorption of Heavy Metal in Wastewater by Using Geopolymer

Nurliyana Ariffin[1,2,*] , *Mohd Mustafa Al Bakri* Abdullah[1,2] , *Mohd Remy Rozainy* Mohd Arif Zainol[3,2] , *Mohd Fared* Murshed[3] , *Hariz*-Zain[1,2] , *Meor Ahmad* Faris[1,2] , and *RIDHO* Bayuaji[4]

[1]Faculty of Engineering Technology, Universiti Malaysia Perlis (UniMAP), Perlis, Malaysia
[2]Centre of Excellent Geopolymer and Green Technology, Universiti Malaysia Perlis (UniMAP), Perlis, Malaysia
[3]School of Civil Engineering, Engineering Campus, Universiti Sains Malaysia (USM), 14300, Nibong Tebal, Pulau Pinang, Malaysia
[4]Civil Engineering Department, Institut Teknologi Sepuluh Nopember, Indonesia

Abstract. Rapid development of advanced technology in Malaysia gave impact increasing in the accumulation of heavy metal every day in our daily life through wastewater. Long term exposure of human bodies to heavy metals susceptible to receives various infections and diseases. From an environmental and economic perspective, adsorption is acceptable process that can be applied in wastewater treatment. However, usage of activated carbon most acknowledged and costly adsorbents lead people to find an alternative to activated carbon. Several studies of physical properties of geopolymer make them gain attention to replace an activated carbon in the treatment of heavy metal. This paper review adsorption of heavy metal by using geopolymer.

1 Introduction

Nowadays, management of waste and quality of water is two of the important proposition in human life. Accretion of technologies in urbanization and industrialization lead to increase in percentage accumulation of waste all around the world and release of heavy metal in the water streams. These harmful heavy metals are an output from varieties activities such as industrial, waste disposal, agricultural and others. Accumulation of heavy metal in wastewater streams acts against human bodies and cause to death. Technologies used in the treatment of wastewater include chemical precipitation, ion-exchange, adsorption, membrane filtration, coagulation-flocculation, flotation and electrochemical [1-3].

Adsorption is one of a common treatment applied in the heavy metal removal involves adsorbents such as undoubtedly activated carbon. However, production and regeneration of activated carbon are very costly and tend to encourage people to find another alternative [4, 5]. Recently, studies of low cost adsorbent are intensively gain an attention to the scientist

* Corresponding author: anurliyana.arifin@gmail.com

DOI: 10.1051/matecconf/20179701023

were usually waste product from industrial, agricultural and food production which produced abundantly [6].

Currently, geopolymer gain new interest to treat wastewater and replacing the conventional adsorbent use which normally used in the production of cement. First identification of geopolymeric materials as an alternative to conventional Portland cement and concrete composites in 1979 [7]. Geopolymer can be viewed as an economic zeolite which produced by reacting solid aluminosilicate with highly alkali hydroxide [8]. A unique structure of geopolymer which good in mechanical, chemical and thermal making them as alternatives for a variety of application [9]. The purpose of this review is to evaluate the potential of geopolymer used to remove dangerous heavy metal in wastewater treatment in the recent studies.

2 Current technologies available for treatment methods of heavy metal

There a diverse of technologies in the treatment of wastewater in term of metal removal that are divided into physical, chemical and biological process such as metal precipitation, ultra filtration, biological systems, oxidation, solvent extraction electrolytic processes, ion exchange, membrane filtration and adsorption. Physical and chemical treatment is more costly compared to biological treatment in removing metal however biological treat are lack effectiveness and timely process. Advantages and disadvantages of physical and chemical current treatment were simplified as in Table 1.

Table 1. Advantage and disadvantages of current treatment technologies for heavy metal involving physical and chemical processes [10].

Methods	Advantages	Disadvantages
Oxidation	Rapid process for toxic pollutants removal	High energy costs and formation of by products
Ion exchange	Good removal of a wide range of heavy metals	Adsorbents requires regeneration or disposal
Membrane filtration technologies	Good removal of heavy metals	Concentrated sludge production and expensive
Adsorption	Flexibility and simplicity of design, ease of operation and insensitivity to toxic pollutants	Adsorbents require regeneration
Coagulation/flocculation	Economically feasible	High sludge production and formation of large particles.
Electrochemical treatment	Rapid process and effective for certain metal ions	High energy costs and formation of large particles.
Ozonation	Applied in gaseous state; alteration of volume	Short half life
Photochemical	No sludge production	Formation of by-product
Irradiation	Effective at lab scale	Require a lot of dissolved O_2
Electrokinetic coagulation	Economically feasible	High sludge production
Fentons reagents	Effective and capable of treating variety of wastes and no energy input necessary to activate hydrogen peroxide	Sludge generation
Biological treatment	Feasible in removing some metals	Technology yet to be established and commercialized

3 Adsorption

Adsorption process gained is a promising method for a long term treatment and economically proven. In the process, heavy metals can be removing and minimizing the heavy metals even at a low concentration enhances the application of adsorption as one practical treatment [11]. Moreover, adsorption process tender to produce high quality treated effluent and offers flexibility in design and operation. Adsorbents used can be regenerate by suitable desorption process because of its sometime reversible characteristic[1].

Parameters that should considered in this treatment include the characteristic of physical and chemical of the adsorbents and adsorbate, temperature, contact time, pH and the concentration of adsprbate in liquid. There is no specific mechanism in the adsorption process but sorption isotherms are used to explain to describe the mechanism occurred on how adsorbate ions interact on the surface of the adsorbent. Table 2 showed adsorption model that are usually used to prove the reaction between adsorbate and adsorbent.

Table 2. Adsorption models of the single-component system [10, 11].

Types of mechanism	Equations	Nomenclature
a) Adsorption isotherms i) Langmuir isotherms	$q_e = \frac{q\ max\ bCa}{1+bCa}$	q_e is equilibrium metal sorption capacity, *Ce* is equilibrium solute concentration in solution, *q max* and *b* are Langmuir constants related to maximum sorption capacity (monolayer capacity) and bonding energy of adsorption.
ii) Freundlich isotherms	$q_e = K_f C_e^{1/n}$	K_f is a biosortion equilibrium constant, q_e is the sorption capacity, *n* is a constant indicative of biosorption intensity.
b) Adsorption kinetics i) Pseudo-first order	$\frac{dqt}{dt} = k_1 (q_e - q_t)$	q_e and q_t are the sorption capacity at equilibrium and at time *t*, k_1 is the rate constant of pseudo -second
ii) Pseudo-second order	$\frac{dqt}{dt} = k (q_e - q_t)^2$	q_e and q_t are the sorption capacity at equilibrium and at time *t*, *k* is the rate constant of pseudo -second
c) Thermodynamics Parameters	$K_c = \frac{CA}{Ce}$ $\Delta G^o = Rt\ln K_c$ $\ln ke = \frac{\Delta S^\circ}{R} - \frac{-\Delta H^\circ}{RT}$	K_c is the equilibrium constant, C_A is the solid phase concentration at equilibrium, C_e is the equilibrium concentration T(K) is the adsolute temperature, *R* is the gas constant (8.314 J/mol K), ΔG^o is the Gibbs free energy ΔH^o is the entalphy change, ΔS^o is the entrophy change

DOI: 10.1051/matecconf/20179701023

4 Heavy Metal

Heavy metal poses impact to serious environmental problem due to their characteristic that accumulates in living organism, causing various disease and disorder which in the Table 3 [12]. Heavy metal that can be found in the landfill leachate such as iron, aluminium, arsenic, cadmium, copper, iron, manganese, mercury, nickel, silver and zinc. Heavy metal are not easy to break down and easy to enrich in different ambient mediums. It has been proven that heavy metal of leachate may lead to secondary pollution [13].

Table 3. Sources and effects of heavy metals.

Heavy Metal	Sources	Effects
Copper	Water pipes, Copper water heaters, Frozen freens and canned greens using copper to produce an ultra green color, alcoholic beverages from copper brewery equipment, instant gas hot water heaters, hormone pills, pesticides, insecticides, fungicides, copper cooking pots.	Mental disorder, anaemia, arthritis, hypertension, nausea/vomiting, hyperactivity, schizophrenia, insomnia, autism, stuttering, postpartum psychosis, inflammation and enlargement of liver, heart problem, cystic fibrosis.
Nickel	Effluents of silver refineries, electroplating, zinc base casting and storage battery industries.	Dermatis, myocarditis, encephalopathy, pulmonary fibrosis, cancer of lungs, nose and bone, headache, dizziness, nausea and vomiting, chest pain, rapid respiration.
Chromium	Steel and textile industry	Skin rashes, respiratory problems, haemolysis, acute and renal failure, weakened immune systems, kidney and liver damage, alteration of genetic material, lung cancer, pulmonary fibrosis.
Lead	Industries such as mining, steel, automobile, batteries and paints. Pollutant arising from increasing industrialization.	Nausea, encephalopathy, headache and vomiting, learning difficulties, mental retardation, hyperactivity, vertigo, kidney damage, birth defects, muscle weakness, anorexia, cirrhosis of the liver, thyroid dysfunction, insomnia, fatigue, degeneration of motor neurons, schizophrenic-like behaviour.
Mercury	Industries like chloro-alkali, paints, pulp and paper, oil refining, rubber processing and fertilizer, batteries, dental fillings adhesive, fabric softeners, drugs, thermometers, fluorescent light tube and high intensity street lamp, pesticides, cosmetic and pharmaceuticals.	Tremors, birth defects, kidney damage, nausea, loss of hearing or vision, gingivitis, chromosome damage, mental retardation, tooth loss, seizures, cerebral palsy, blindness and deafness, hypertonia – muscle rigidity, minamata disease.

5 Geopolymer

Geopolymer or alkali-activated aluminosilicate is a diverse group of ceramic-like materials produced by geosynthetic reaction of aluminosilicate minerals in the presence of an alkali solution at low temperature (less than 100 degree celcius) [14-16]. It consists of a polymeric silicon-oxygen-aluminium framework with alternating silicon and aluminium tetrahedral joined together in three directions by sharing all the oxygen atoms [17]. Fly ash, dolomite, expanded clay, natural zeolite, kaolinite and many are example of geopolymers. Geopolymers had interesting properties that are encouraged to investigated and developed. The major nature of geopolymers are tendency to drastically decrease the mobility of most heavy metals ions contained within the geopolymeric structure, quick compressive development of strength, resistance to acid and fire, fat setting, low permeability and good resistance to freeze-thaw cycles [18-20].

Theoretically, any alkali can be used in geopolymerization reactions; however most of the studies have focused on the effect of sodium (Na+) and potassium (K+) ions. Both NaOH and KOH could be used in the activation process, but the extent of dissolution was higher when NaOH is used; this is due to the smaller size of Na+ which can be better stabilize the silicate monomers and dimmers present in the solution, enhancing the minerals dissolution rate [8]. Rios conclude that geopolymer treated with NaOH is better than treated geopolymer by KOH [21].

Fly ash. Fly ash is a familiar materials used as geopolymerized adsorbents. It is a complex and abundant anthropogenic material which by product of coal combustion in thermal power plants [22]. Fly ash is a waste substance that are easily to found and had been reported that could be effective for removal of heavy metals. However, it is shows lower adsorption capacity unless it is treated or activated called as geopolymer. Fly ash based geopolymer are growing in commercial potential due to their reliance on cheap and readily available waste fly ash as the main feedstock. There is a significant trend in recycling of waste materials and converting it to usable and valuable materials [23]. One of these materials is coal fly ash. The disposal of the large amount of fly ash has become a serious environmental and economic problem [8]. One approach to deal with fly ash is converting to geopolymer which not only effective to heavy metal removal but also help to reduce the waste accumulation.

Dolomite. Dolomite has been a subject of interest for over six decades. Dolomite material is available very cheaply and in abundance around the world (important occurrences are found in India, Indonesia, Turkey and China). Dolomite which has similar properties to limestone is sometimes known as magnesium-limestone in industry. Its crystal structure consists of alternative layers of magnesium and calcium carbonate [24]. Therefore, dolomite had a characteristic to remove heavy metals as effective as limestone.

6 Factor affecting adsorption of heavy metal by geopolymeric adsorbents

Some factors are influence in the process of adsorption of heavy metal into adsorbents such as pH, adsorbent dosage, initial concentration, contact time and temperature [25]. Removal efficiency of heavy metals can be described as following equation [19]:

$$\text{Removal efficiency } (\%) = \frac{(Co-Ct)}{Co} \, x \; 100 \tag{1}$$

 DOI: 10.1051/matecconf/20179701023

Where Co (mg/L) is the initial concentration and Ct (mg/L) is the concentration at time t (min).

Adsorption amount at time t, qt (mg/g) and at equilibrium, qe (mg/g) are determined using following equations :

$$qt = \frac{(Co - Ct)V}{m} \tag{2}$$

$$qe = \frac{(Co - Ct)V}{m} \tag{3}$$

Effect of pH: The surface charge of the solution, degree of ionization, and the adsorbate species influenced by the pH of the solution. Most of the metal adsorbed increase with the increasing of pH of the solution until certain point and followed by reduction if further increasing of pH. The equation of pH can be written as :

$$\text{pH} = \text{pka} - \log \frac{[AH]}{[A-]} \tag{4}$$

where, [A-] and [AH] represent the concentration of deprotonated and protonated surface groups and the equilibrium constants pKa resembled the carbonyl groups.
Study by al-Zboon show that adsorption efficiency increase from 1% to 90.66% as pH of the solution increase from 1 to 5 and slightly decrease at pH 6 [18]. Therefore, pH 5 is indicated as zero point charge.

Effect of adsorbent dosage: Dose of adsorbent also is one of the main points to determine the capacity uptake of heavy metals by adsorbents. Usually, increase in the dose of adsorbents will increase in the adsorbed capacity until its reach a limit. If further increase the dose, the adsorption capacity will be constant. Wang study adsorbent loading from 0.5 g until 2.0 g, removal of copper increase with increase of adsorbent loading [5].

Effect of initial concentration: Adsorption dosage gain a strong effect by initial concentration of heavy metals. Generally, adsorption capacity increased with the increased initial concentration of heavy metals. Playing as important driving force, initial concentration influence in overcome all mass transfer resistance between solid and aqueous phases. Several studies have shown that removal efficiency of heavy metal is concentration dependent and there exist decreasing trend if further increase initial concentration [26].

Effect of contact time: The interaction of functional group between the solution and the surface of adsorbent result in the adsorption capacity if adsorbate into adsorbent. Specific time needed to maintain equilibrium interaction therefore the adsorption process undergo completion. Cadmium removal using zeolite based geopolymer achieve equilibrium contact time at 7 hours [3]. However, fly ash-based geopolymer for lead removal gain equilibrium contact time at 120 min and thereafter, its remain constant [18].

Effect of temperature: The natures of the processes either exothermic or endothermic are depends on the adsorption equilibrium that affected by the temperature used. The uptake capacity of the adsorbates increases with the rises of the temperatures. This happens due to the enlargement of pores and activation of the sorbent surface. Research by Javadian shows removal of cadmium was increase with temperature range from 25°C to 45°C [3].

7 Summary

The role of geopolymers in the wastewater treatment has been investigated a few years ago. Among other geopolymeric adsorbents, fly ash is the most interested in the removal of heavy metal because of they are reliance on cheap and readily available waste. Adsorption capacity of heavy metals adsorbed usually modeled by Langmuir and Freundlich isotherm. Factors that influence of adsorption capacity are pH, adsorbent dosage, initial concentration, contact time and temperature. Normally, increase of dosage, contact time and initial concentration will increase the uptake capacity to a certain point before constant. However, some materials and adsorption may be different. From reviews, it can be conclude that usage of geopolymeric adsorbent as alternative to costly activated carbon are effective to remove heavy metal from wastewater. Most of research focused on the industrial wastewater instead of improvement in the leachate treatment. Moreover, there has been limited research of mixture of geopolymeric adsorbents.

This work was financially supported by the Centre of Excellence Geopolymer and Green Technology, Universiti Malaysia Perlis, Malaysia.

References

1. F. Fu, Q. Wang, Journal of Environmental Management, **92** , 407 (2011)
2. T.W. Cheng, M.L. Lee, M.S. Ko, T.H. Ueng, S.F. Yang, Applied Clay Science, **56**, 90 (2012)
3. H. Javadian, F. Ghorbani, H.A. Tayebi, S. H. Asl, Arabian Journal of Chemistry, **8.6**, 837 (2015)
4. O.E.A. Salam, N.A. Reiad, M. M. ElShafei, Journal of Advance Technology, **2.4**, 297 (2011).
5. S. Wang, L. Li, Z. H. Zhu, J. Hazard Mater., **139.2**, 254 (2007)
6. C.G. Lee, M.K. Song, J.C. Ryu, C. Park, J.W. Choi, S.H. Lee, Chemosphere **153**, 1 (2016)
7. M. Jo, L. Soto, M. Arocho, J.S. John, S. Hwang, Constr. Build. Mater., **93**,1097 (2015)
8. M.S. Al Harahsheh, K.A. Zboon, L. AL Makhadmeh, M. Hararah, M. Mahasneh, J. Environ. Chem. Eng., **3**, 1669 (2015)
9. A. Gharzouni, L. Vidal, N. Essaidi, E. Joussein, S. Rossignol, Mater. Des., **94**, 221 (2016)
10. M. Ahmaruzzaman, Adv. Colloid Interface Sci., **166**, 36 (2011)
11. S.N.A. Abas, M.H.S. Ismail, M.L. Kamal, S. Izhar, Applied Sciences Journal **28**, 1518 (2016)
12. T. Motsi, N.A. Rowson, M. J. H. Simmons, Int. J. Miner Process, **92**, 42 (2009)
13. H.H. Liu, S.X. Sang, Environ. Monit. Assess., **165**, 349 (2010)
14. J.W. Phair, J.S.J.V. Deventer, J.D. Smith, Appl. Geochem., **19.3**, 423 (2004)
15. N. Waijarean, S. Asavapisit, K. Sombatsompop, Constr. Build. Mater., **50**, 486 (2014)
16. V. Nikolic, M. Komljenovic, Z. Bascarevic, N. Marjanovic, Z. Miladinovic, R. Petrovic, Constr. Build. Mater., **94**, 361(2015)
17. Y. Ge, Y. Yuan, K. Wang, Y. He, X. Cui, J. Hazard. Mater., **299**, 711 (2015)
18. K. Al Zboon, M.S. Al Harahsheh, F.B. Hani, J. Hazard. Mater., **188.1**, 414 (2011)
19. T. Luukonen, M. Sarkkinen, K. Kemppainen, J. Ramo, U. Lassi, Appl. Clay Sci., **119**, 266 (2016)
20. S. Andrejkovicova, A. Sudagar, J. Rocha, C. Patinha, W. Hajjaji, E.F.D. Silva, A. Velosa, F. Rocha, Appl. Clay Sci., **126**, 141 (2016)

DOI: 10.1051/matecconf/20179701023

21. C.A.R.R., C.D. Williams, C.L. Roberts, Fuel, **88.8**, 1403 (2009)
22. Z.T. Yao, X.S. Ji, P.K. Sarker, J.H. Tang, L.Q. Ge, M.S. Xia, Y.Q. Xi : A, Earth-Science Reviews, **141**, 105-121 (2015)
23. L. Li, S. Wang, Z. Zhu, J. Colloid. Interface Sci., **300.1**, 52 (2006)
24. A.B. Albadarin, C. Mangwandi, A.A. H. Al Muhtaseb, G.M. Walker, S.J. Allen, M.N.M. Ahmad, Chem. Eng. J., **179**, 193-202 (2012)
25. A.P. Lim, A. Z. Aris, Rev. Environ. Sci. Biotechnol., **13,** 163 (2014)
26. J.G.S.V. Jaarsveld, J.S.J.V. Deventer, G.C. Lukey, Chem. Eng. J., **89.1**, 63 (2002)

Synthesis and Characterization of Fly Ash Geopolymer for Water Absorbent Material

Ufafa Anggarini[1,*], *Ndaru Candra* Sukmana[2], and *Fandi Angga* Prasetya[1]

[1]Departement of Chemical Engineering, Universitas Internasional Semen Indonesia, Indonesia
[2] Departement of Engineering Management, Universitas Internasional Semen Indonesia, Indonesia

Abstract. Geopolymer as a water absorbent material has been synthesized from fly ash. This research aims to determine the ability of geopolymer to save water content with variations of NaOH molar ratio. In this research, the synthesis of geopolymer was conducted by setting NaOH molar ratio at 3, 4, 5, 6 and 7 M. The resulting material characterization was done by using XRD, FTIR and SEM in order to characterize the geopolymer structures. Water absorption capacity was measured by immersing the geopolymer specimens in water for 24 hours. The results show that the optimum molar ratio of NaOH was 3 M, which water absorption capacity was 11.10%wt.

1 Introduction

Porous material of ceramic is a potential material to be used as an absorbent (water absorber)[1]. Water absorber is very useful, especially as a planting medium where the landmass area is limited. Porous ceramics prepared from clay minerals show high water adsorption capacity, high adsorption rate and slow water release rate due to their unique microporous structure [2]. Another new advantage of using porous ceramics is that such material is produced environmentally friendly without the need for combustion processes at high temperatures. Geopolymer is one of the best materials to prepare a porous ceramic material. Synthesis of geopolymer is generally carried out by reacting an alkali silicate solution with a solid aluminosilicate such as metakaolin, fly ash and so forth under alkaline conditions. Thokchom et al. (2009) state that the less alkali specimens the geopolymer contains, the higher porosity and water absorption the geopolymer possess. In this study, geopolymer was synthesized by using fly ash as a base material and alkaline solution and sodium silicate as additional materials. The synthesis was carried out at temperature of 25°C involving variations of NaOH molar ratio. The product was characterized by X-Ray Diffraction (XRD), Fourier Transform Infrared (FTIR) and Scanning Electron Microscopy (SEM). The measurement of maximum water absorption capacity was conducted by immersing the samples of geopolymer in distilled water for 24 hours.

* Corresponding author: aufafa.anggarini@uisi.ac.id

2 Experimental section

2.1 Material and instrumentation

The materials used in this research were fly ash, the source of aluminium and silica, which is the waste from coal combustion at Steam Power Plant (PLTU) Paiton, Na2SiO3, NaOH pellet (99%, Merck) as alkali activator, distilled water and KCl. The analysis X-ray Diffraction (XRD) was conducted to identify the chrystal phase of geopolymer by using Cu-Kα radiation. FTIR spectrum of geopolymer samples were analyzed to determine the presence of KCl inside geopolymer matrix. Scanning Electron Microscope (SEM) images were analyzed to identify the morphology of geopolymer. The water storage capacities of geopolymer samples were measured by immersing geopolimer matrix in distilled water for 24 hours at 25°C.

2.2 Procedure

The fly ash refers to industrial waste collected from a power plant in Paiton, Indonesia which is used as aluminosilicate source. This material was prepared by shieving with sieve no. 325 and drying at 85°C for 24 hours. NaOH solution was prepared by dissolving NaOH pellet in distilled water, continued with 24 hours storage. Alkali solution was prepared by mixing sodium silicate and NaOH solution. In this research, the variations of NaOH molar ratio were 3, 4, 5, 6 and 7 M. Fly ash, as the starting material, was mixed with alkali solution, then KCl salt was added into geopolymer paste. This mixture was casted and cured at 85°C for 24 hours.

2.3 The measurement of water storage capacity

Geopolymer specimen was weighed and determined as initial weight (wo). These samples were then immersed in destilled water for 24 hours at 25°C. The saturated dry surface was weighed as final weight (wf). Water absorption capacity was calculated and reported as increase in weight percentage.

3 Result and discussion

Geopolymer has been synthesized using fly ash and alkali solution. Abdullah, et al (2011) stated that geopolymer containing silica and alumina will react with alkali solution to produce chain and ring that contain three dimensional polymer. The existance of three dimensional ring in geopolymer can be used for water and nutrient storage for plantation aplication. Okada, et al (2009) stated that geopolymers are good candidates for water absorbers due to their microstructure.

3.1 Effect of KCl content toward the structure of geopolymer

In this research, the synthesis of geopolymer was followed by adding 3% wt of KCl as plant nutrition. Potassium chloride is the most widely used potassium fertilizer because it contains more potassium than any other source do and this material has no significant pollution impact on water. The FTIR spectrum of geopolymer synthesized under varying NaOH molar ratio is shown in Figure 1.

DOI: 10.1051/matecconf/20179701024

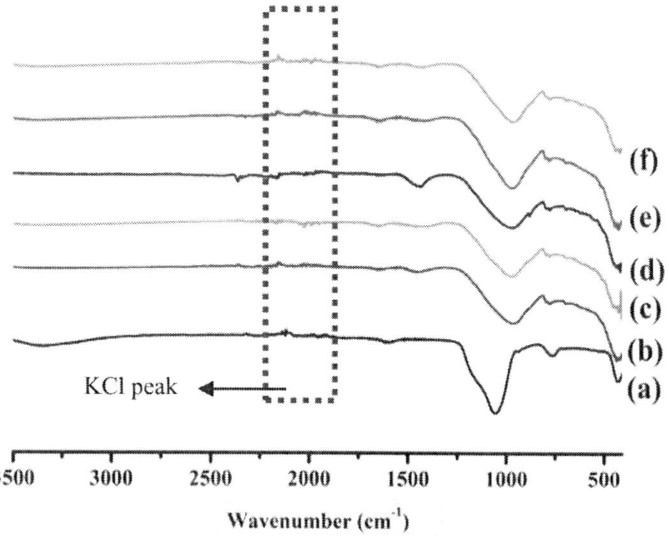

Fig. 1. FTIR spectrum of KCl (a), Geopolymer with NaOH molar ratio variation : 3M (b), 4M (c), 5M (d), 6M (e), and 7M (f)

KCl sample was investigated by using FTIR and the result shows high peak at wavenumber of 1096 cm-1. FTIR spectrum of all samples geopolymer shows the presence of KCl mineral and this result indicates that KCl does not react with geopolymer matrix and can be used for plant nutrition.

3.2 Effect of NaOH molar ratio toward the structure of geopolymer

The crystal phase of geopolymer samples were investigated by using XRD. Geopolymerization reaction can form phases, such as quartz, mullite, magnetite and sodalite. Figure 2 shows the XRD pattern of fly ash and geopolymer samples. The diffractogram of fly ash can be seen in Figure 2(a), while quartz was the phase most widely found at peak (2θ) of 26.72°, 39.49° and 50.49°. Another phase found in fly ash sample was mullite shown at peak (2θ) of 16.48°, 23.47°, 31.15°, 33.2°, 35.29°, 40.89°, 42.55°, 49.52°, 53.66° and 57.56°. The diffractogram of geopolymer samples synthesized from fly ash exhibits transformation from fly ash diffractogram. XRD pattern of geopolymer shows the presence of hump peak at (2θ) of 11°-40° indicating the formation of geopolymer phase which has been done. However, the quartz phase found in XRD pattern of geopolymer samples at (2θ) of 27° indicate the presence of unreacted fly ash.

 DOI: 10.1051/matecconf/20179701024

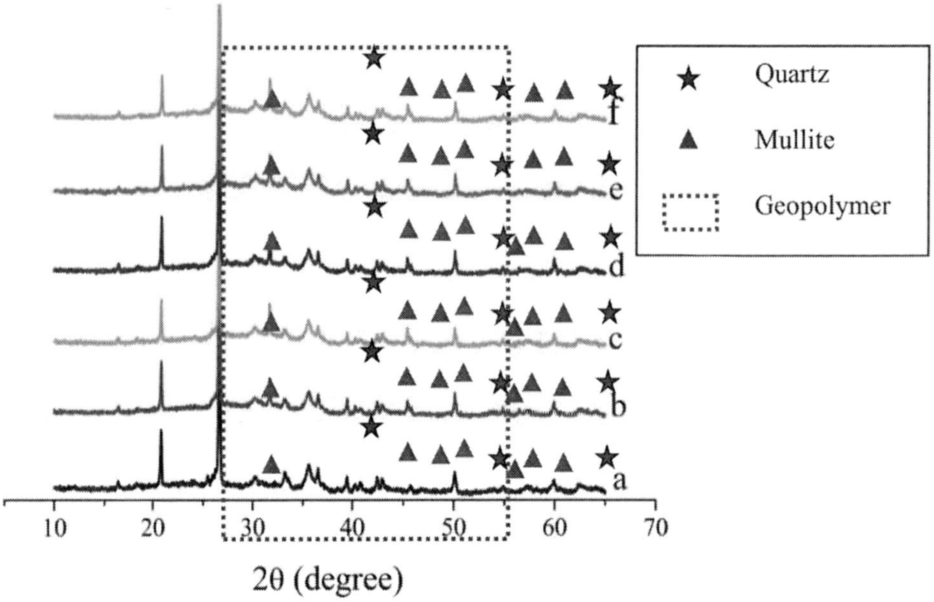

Fig. 2. The XRD pattern of (a) fly ash and geopolymer with NaOH molar ratio variation of (b) 3M, (c) 4M, (d) 5M, (e) 6M and (f) 7M

3.3 The Effect of NaOH molar ratio toward the morphology of geopolymer

The morphology of geopolymer samples was characterized by using SEM as seen in Figure 3. The particle size of geopolymer varied with the NaOH molar ratio. The particle size of geopolymer synthesized with 3M NaOH solution is smaller than those synthesized with more concentrated NaOH solution. Hence, the higher NaOH molar ratio used to synthesize geopolymer, the bigger its particle size. Thokchom, et al (2009) stated that as the NaOH molar ratio increases, the reactivity of geopolymerization will also increase. As the particle size become smaller, geopolymer will certainly has higher surface area and more enhanced absorption capacity. Geopolymer with high porosity consists of agglomerated fine particles and less platy shape [1]. Geopolymer with 3M NaOH produces fine particle agglomerate, as seen in Figure 3(a). This result is supported by water absorption capacity measurement data, which shows that geopolymer synthesized with 3M NaOH, the most dilute NaOH solution used, absorb more water than the others do.

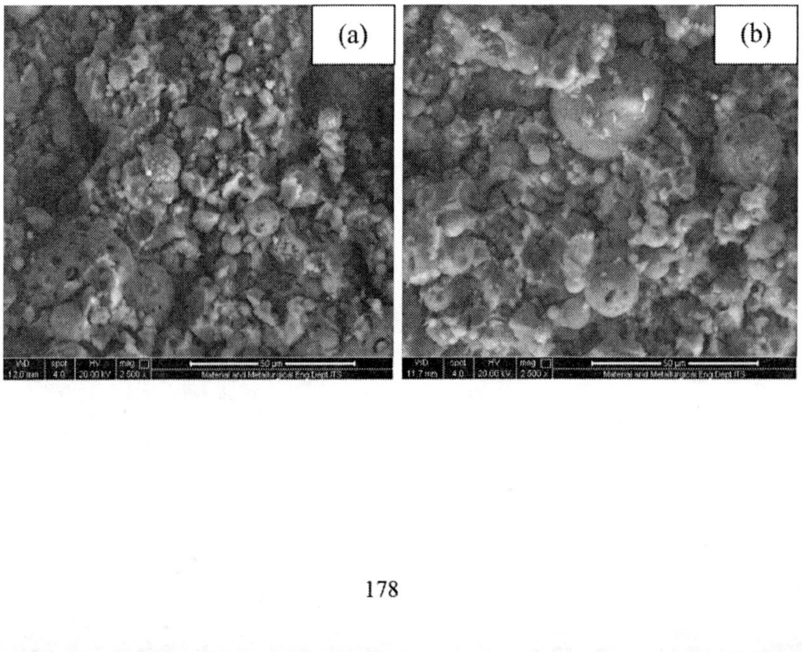

DOI: 10.1051/matecconf/20179701024

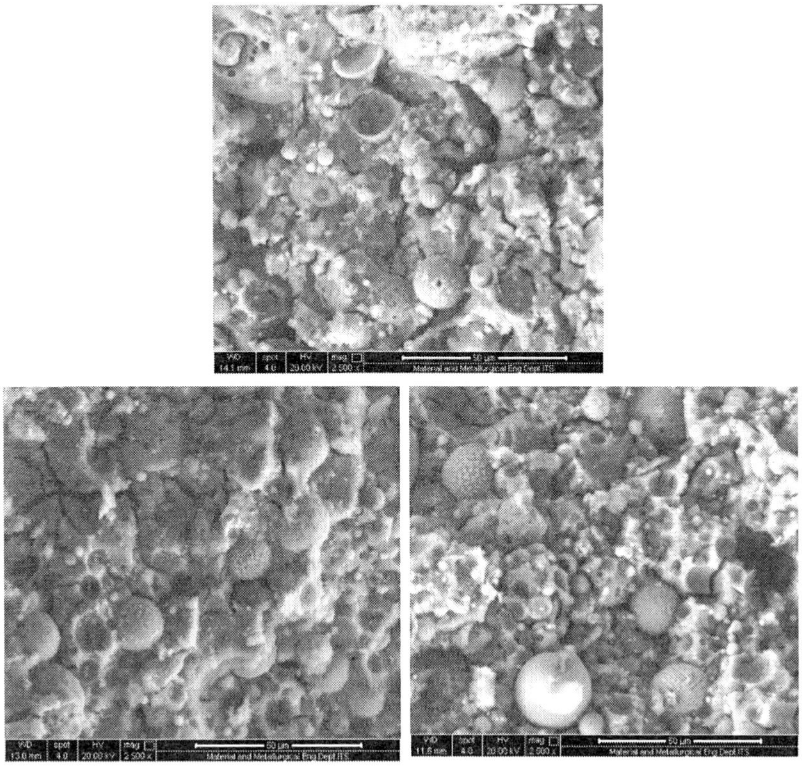

Fig. 3. SEM images of geopolymer samples with variation NaOH molar ratio of (a) 3M, (b) 4M, (c) 5M, (d) 6M and (e) 7M

3.4 Water absorption capacity measurement

Figure 4 indicates the corellation between NaOH molar ratio variation and water storage capacity of geopolymer samples. The results show that the activity of water storage material increases with the decrease of NaOH molar ratio. Thockhom, et al (2009) stated that NaOH used as catalyst in geopolymerization reaction will increase higher alkali content in the mix.

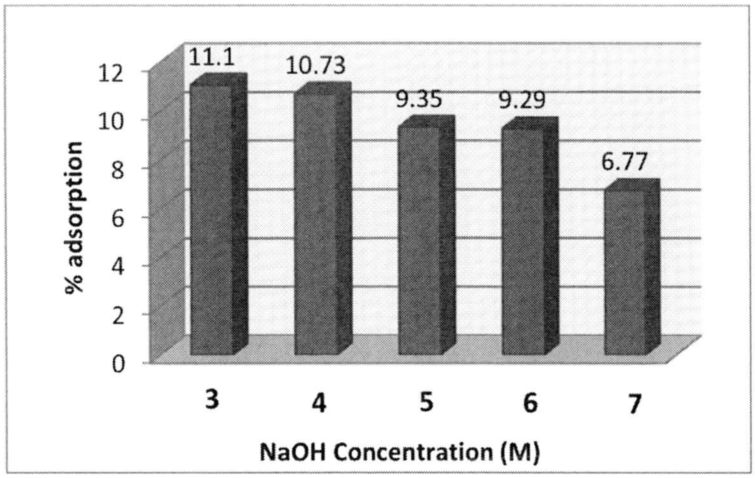

Fig. 4. Water absorption capacity of geopolymer samples with NaOH molar ratio variation

Moreover, the reactivity of geopolymer will also increase as fly ash is used, so that denser microstructure of geopolymer will be accomplished. This result is appropriate with the XRD result showing that denser microstructure of geopolymer was achieved by using NaOH solution which molar ratio is higher. Geopolymer synthesized with 3M NaOH produces a material with higher pore formation and allows more water molecules to penetrate the geopolymer specimen.

Summary

Geopolymer has been synthesized from fly ash as an aluminosilicate source with NaOH molar ratio variations of 3, 4, 5, 6 and 7 M. The XRD and SEM results show that the lower NaOH molar ratio, the higher microstructure pores and the lower particle size of geopolymer. Moreover, geopolymer specimen synthesized with lower NaOH molar ratio shows higher water storage capacity.

The authors wish to thank LPPM UISI for providing Hibah Riset Bersaing UISI, contract number 03/KP.01/10-01/02.16 date February 19th, 2016

References

1. K.Okada, A.Ooyama, T.Isobe, Y. Kameshima, A. Nakajima, K.J. MacKenzie, J. Eur. Ceram. Soc., **29**, 1917 (2009)
2. K. Okada, S. Matsui, T. Isobe, Y. Kameshima, A. Nakajima, Journal of Ceram. Intern., **34**, 345 (2008)
3. S. Thokchom, P. Ghosh, S. Ghosh, J. Eng. Appl. Sci. **4**, 28 (2009)
4. M.M.A. Abdullah, K. Hussin, M. Bnhussain, K. N. Ismail, W.M.W. Ibrahim, Int. Journal of Pure and App. Sci and Technol., **6**, 35 (2011)
5. U. Rattanasak, P. Chindaprasirt, Miner. Eng., **22**, 1073 (2009)
6. J. Davidovits, J. Therm. Anal., **37**, 1633 (1991)
7. Skvara, Frantisek, Kopecky, Lumboir, Nemecek, Jiri, Bittnar, Zdenek, Ceram.-Silikaty, **50**, 208 (2006)
8. Xu, Hua, Deventer, S.J. Van, Ind. Eng. Chem. Res., **42**, 1698 (2003)
9. Mishra, Anurag, Choudhary, Deepika, Jain, Namrata, Kumar, Manish, Sharda Nidhi, and Dutt, Durga, ARPN Journal of Engineering and Applied Sciences, **3**, 14 (2008),
10. Xu, Hua, Deventer, S.J. Van, Ind. Eng. Chem. Res., **42**, 1698 (2003)

MATEC Web of Conferences **97**, 01025 (2017) DOI: 10.1051/matecconf/20179701025
ETIC 2016

Correlation between Compressive Strength and Fire Resistant Performance of Rice Husk Ash-Based Geopolymer Binder for Panel Applications

Mohd Salahuddin Mohd Basri[1,3,*], *Norkhairunnisa* Mazlan[1,2], *Faizal* Mustapha[1,3], and *Mohamad* Ridzwan Ishak[1,2,3]

[1] Department of Aerospace Engineering, Faculty of Engineering, University Putra Malaysia, 43400 UPM Serdang, Selangor, Malaysia
[2] Institute of Tropical Forestry and Forest Products (INTROP), Universiti Putra Malaysia, 43400 UPM Serdang, Selangor, Malaysia
[3] Aerospace Manufacturing Research Center (AMRC), Universiti Putra Malaysia, 43400 UPM Serdang, Selangor, Malaysia

Abstract. Panel structures which are mainly used as insulation materials should possess high fire resistance characteristic. In addition, their mechanical requisites for walls and doors such as compressive strength must not be unduly compromised. Rice husk ash (RHA) was used as an aluminosilicate source and two factors namely RHA/AA ratio and NaOH concentration were analyzed using statistical tool to study the effect of both factors on the compressive strength. Surface morphology and fire resistant behavior of four selected samples based on their compressive strength (brittle, semi-brittle, ductile, and semi-ductile samples) were studied to determine the correlation between compressive strength and fire resistant performance of those selected samples. Results showed that RHA-based geopolymer sample recorded high compressive strength above 28 MPa when its RHA/AA ratio and NaOH concentration were high ranging from 0.7 to 0.8 and 12M to 14M, respectively. Brittle geopolymer sample (GS) with low Si/Al ratio shows high compressive strength together with high degree of geopolymerization. Ductile GS in comparison, shows low compressive strength irrespective of its degree of geopolymerization. Semi-ductile GS showed the best fire resistant properties with a maximum non-exposed surface temperature of only 50°C after 50 minutes (after it was exposed to a direct fire with temperature of 900°C) followed by semi-brittle and brittle GS.

* Corresponding author: salahuddin_basri@yahoo.com

1 Introduction

Rice husks (RHs) is a byproduct of paddy rice processing. It is a good source of renewable energy and weight approximately 20% of the harvested paddy. In 2008, about 661 million tons of rice were produced worldwide yieldingly 132 million tons of RHs [1]. In Malaysia, approximately 3.6 million tons of RH was produced annually [2]. RHs is one of the largest available biomass resources and the low cost of RH [3, 4] may potentially reduce the costs of geopolymer through reducing cost of production compared by using other resin systems available in the market such as polyimides, cyanide esters, silicone, epoxy, polyurethane and acrylic resins.

Geopolymer is known for its good mechanical properties and high early-aged strength. The potential advantages of RHA-based geopolymer material, in enhancing its mechanical and thermal properties, are mainly associated with RHA which having high silica content. Compared to fly ash or other aluminosilicate sources, RHA contained the highest amount of silica which is approximately 85 to 95% and relatively low alumina content which is approximately 0.5 to 2.0%. Higher loading of RHA will result in higher silica content which consequently leads to high Si/Al ratio. Therefore, high compressive strength can be achieved as proved by Komnitsas et al. [5] and Silva et al. [6]. Although high Si/Al ratio proved to provide higher compressive strength, further increase in Si/Al ratio has a tendency to reduce the compressive strength as studied by Songpirijakij et al. [7].

Some existing materials for wall panels in the market have poor fire resistant properties which made from plastic based. Promoting the use of environmental-friendly and energy consumption materials as an alternative for the plastic materials can be considered. Applying geopolymer binder as part of the panel materials tends to improve the fire resistance properties of the panel. Even though the main criteria of some panel structures should possess high fire resistance characteristic, their mechanical requisites for walls and doors must not be sacrificed too. Moreover, intensive research study on the relationship between compressive strength and fire resistant properties of geopolymer for fire resistant panel materials are not well established yet.

Therefore, the objective of this study were to elucidate the correlation between compressive strength and fire resistant performance of different geopolymer material behavior, namely, brittle and ductile, and to study the effect of two factors namely RHA/AA ratio and NaOH concentration on the compressive strength.

2 Materials and experimental details

2.1 Design of Experiment.

Experiment was statistically designed and analyzed using Minitab software. Three replications were considered with a total number of 30 experimental runs. RHA/AA ratio and NaOH concentration were chosen as factors. Other factors, such as the ratio of AA solution, curing temperature and curing time were kept constant at 5.5, 50°C, and 7 days, respectively.

2.2 Raw materials.

Rice husk ash (RHA) provided by Maerotech Sdn. Bhd and was use as received. The RHA was then ground using planetary mill Pulverisette 4. Then, follow by sieve and the produced particle size were tested under particle size distribution test.

2.3 Sample preparation.

Geopolymer was first prepared by mixing the sodium hydroxide and sodium silicate solutions at designated ratios. The AA solution was mixed with RHA and the mixture stirred gently for 30 seconds before using mechanical stirrer for another 30 minutes until the solution become homogenous. For compressive strength test, the mixture was poured in a mould with a dimension of (12.7 x 12.7 x 25.4) mm. For fire resistant test, the mixture was casted in mould with dimension of (100 x 100 x 10) mm. The samples were then cured in an oven for 24 hours followed by 6 days at room temperature for complete curing.

2.4 Compressive strength test.

The compressive test was conducted by using the Instron 3382 Floor Model Universal Testing System in accordance with ASTM D695 standard. The specimen was placed between the compressive plates and load was applied at constant speed of 1.30 ± 0.13 mm/min.

2.5 Fire resistant test.

The fire retardant tests were conducted by using the ThermaCam S500 (infrared camera). Distance between infrared camera and sample, (60 cm) and between sample and blow, (7 cm) were kept constant throughout the test. Geopolymer samples were exposed to direct fire with temperature around 900°C for at least 20 minutes or until equilibrium temperature was reached. Temperature on the back of the samples was plotted as a function of time.

2.6 Morphology analysis.

Experimental data were analyzed using MINITAB 7. Scanning electron microscopy (SEM) was performed using JEOL JSM-6400 to analyze the microstructure of selected geopolymer coating.

3 Result and discussion

3.1 Raw materials.

Size of RHA particles were in the range of 3.4 to 59.5 μm, with an average size aggregate of approximately 22.8 μm. The chemical composition of RHA is shown in Table 1. SiO2 was found to be the major constituent in RHA.

Table 1. Chemical composition of RHA

Component	SiO_2	PdO	Al_2O_3	Fe_2O_3	CaO	K_2O	Cr_2O_3	MnO	NiO	CuO	ZnO
Mass %	87.4	6.0	3.0	1.49	1.40	0.49	0.27	0.19	0.068	0.045	0.035

Sodium silicate solution used comprised of H2O = 30.3%, Na2O = 11.9% and SiO2 = 57.8%. Sodium hydroxide (NaOH) pellets with 97% purity was used and five different concentrations were prepared using NaOH pellets dissolved in de-ionized water.

DOI: 10.1051/matecconf/20179701025

3.2 Effect of factor on compressive strength.

Figure 1 illustrates the contour plot on effect of RHA/AA ratio and NaOH concentration on compressive strength of geopolymer binder. In geopolymerization process, use of high concentration of NaOH promotes greater dissolution of RHA and improves bonding between particles. As can be seen from Figure 1, it is clear that compressive strength increased with increasing the value of both factors which are NaOH concentration and RHA/AA ratio. Higher compressive strength of more than 28.00 MPa was observed in the range of 0.7 to 0.8 for RHA/AA ratio and 12M to 14M of NaOH concentration. Geopolymer binder contained NaOH solution with concentration of 10M and S/L of 0.85, is sufficient to achieve remarkable compressive strength up to 34.60MPa. It was reported that high concentration of 12M alkaline activator solution in FA-based geopolymer will promote rapid dissolution and hydrolysis, thus increasing compressive strength [8]. However, the compressive strength value starts to decrease as the S/L ratio reduces from 0.85 to 0.25, even though the NaOH concentration is maintained at 10M. This indicates that high volume of water presence during preparation of alkaline solution which affects the geopolymerization process. In addition, the S/L ratio for geopolymer based on RHA was found to be nearly similar to the S/L geopolymer based metakaolin which is 0.8 [8]. The low S/L ratio indicated that the source materials require higher liquid demand due probably to its finer particle size and difference in particle shape. However, increased further the NaOH concentration, lead to the reduction of the compressive strength at any S/L ratio. This is due to the formation of aluminosilicate gel precipitation at the early stages of the mixing process and hence decrease the compressive strength property of the geopolymer binder [9].

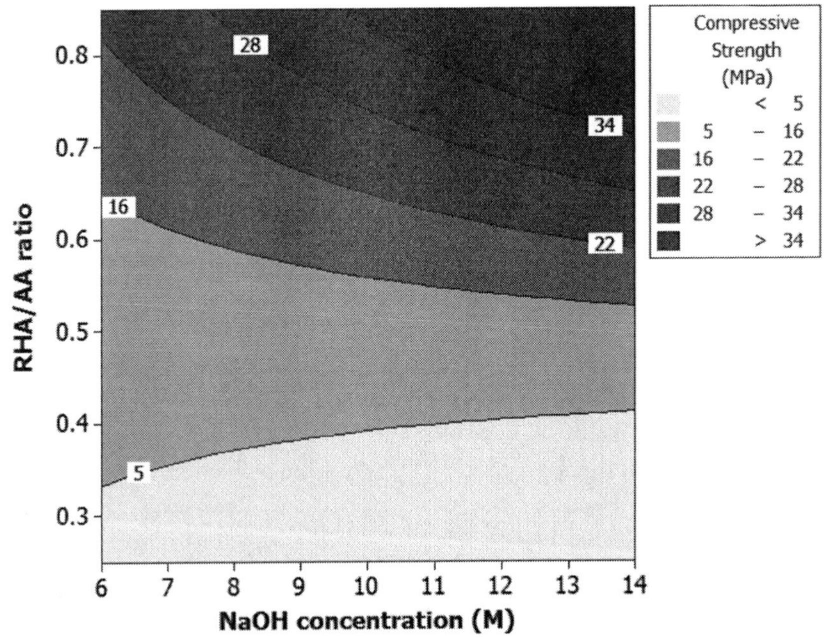

Fig. 1. Contour plot of the effect of RHA/AA ratio and NaOH concentration on compressive strength.

Table 2 also indicates the design matrix for the factors and all responses for 30 experimental runs. Experiment were conducted based on five factors which are the ratio of AA solution, RHA/AA ratio, NaOH concentration, curing temperature, and curing time.

However, only RHA/AA ratio and NaOH concentration were showed in Table 2 since other factors namely the ratio of AA solution, curing temperature, and curing time were kept constant for each sample at 5.5, 50°C, and 7 days, respectively. In order to easily determine the relationship between RHA/AA ratio, NaOH concentration, and compressive strength, contour plot in Figure 1 was generated based on the data shown in Table 2.

Table 2. Experimental results of compressive strength test

Sample	RHA/AA ratio	NaOH Concentration (M)	Compressive strength (MPa)	Sample	RHA/AA ratio	NaOH Concentration (M)	Compressive strength (MPa)
S01	0.40	8	0.80	S16	0.40	12	0.64
S02	0.70	12	31.07	S17	0.70	8	22.52
S03	0.70	12	33.55	S18	0.55	10	15.34
S04	0.70	8	23.20	S19	0.55	6	16.12
S05	0.40	8	0.89	S20	0.85	10	30.97
S06	0.40	12	0.48	S21	0.25	10	0.05
S07	0.55	10	14.39	S22	0.55	14	20.29
S08	0.70	12	33.32	S23	0.85	10	34.60
S09	0.55	10	13.82	S24	0.55	6	15.53
S10	0.40	12	0.65	S25	0.25	10	0.04
S11	0.55	10	15.10	S26	0.55	14	17.78
S12	0.70	8	25.49	S27	0.85	10	30.19
S13	0.40	8	1.03	S28	0.25	10	0.04
S14	0.55	10	13.31	S29	0.55	14	21.90
S15	0.55	10	14.88	S30	0.55	6	15.64

3.3 Surface morphology and geopolymer properties.

Study on the surface morphology was performed on four selected samples of geopolymer binder based on their compressive strength properties. The samples were S23 and S7 which exhibited brittle and semi-brittle behavior, whereas S28 and S5 which were exhibited ductile and semi-ductile behavior. Figure 2 shows SEM images of four geopolymer samples (S28, S5, S7 and S23) with different material behavior subsequent to compressive test. Sample S28 showed that its amorphous matrix was homogenous with no obvious crystalline precipitates present. The material initiated no cracks and had large plastic deformation range (which enabled it to stretch a few times its original length). Sample S5, S7, and S23 initiated minor to major cracks due probably to the presence of voids that initiate formation of cracks in the structure of the geopolymer binder. During plastic deformation, presence of voids will increase in number and grow bigger. Finally, the sample will fracture and form cracks due to the growth and coalescence of the voids [10]. Sample S5 had a homogenous amorphous matrix with very minimal unreacted RHA particles which were bonded in the matrix. Although no void can be seen on the surface of sample S5, there were microvoids present which initiated the cracks. Cracks formed in sample S5 were fewer and narrower compared to those in sample S7.

Sample S7 and S23, as shown in Figure 2 (c) and (d), had a non-homogeneous amorphous matrix with some of the RHA particles remaining unreacted but were bonded in the matrix. However, the unreacted particles had only minor effect on compressive strength since they were not acting as filler [11]. Compared to sample S7, the higher number and larger size of voids in S23 led to wider and higher number of cracks formed. Although S23 had a higher number and larger size of voids compared to the other three samples, it still exhibited the highest compressive strength. However, this sample failed under compression load at a rate of four times faster compared to that in S7. The wide range of behavior shown in geopolymer binder suggests wide option potentials in industrial applications.

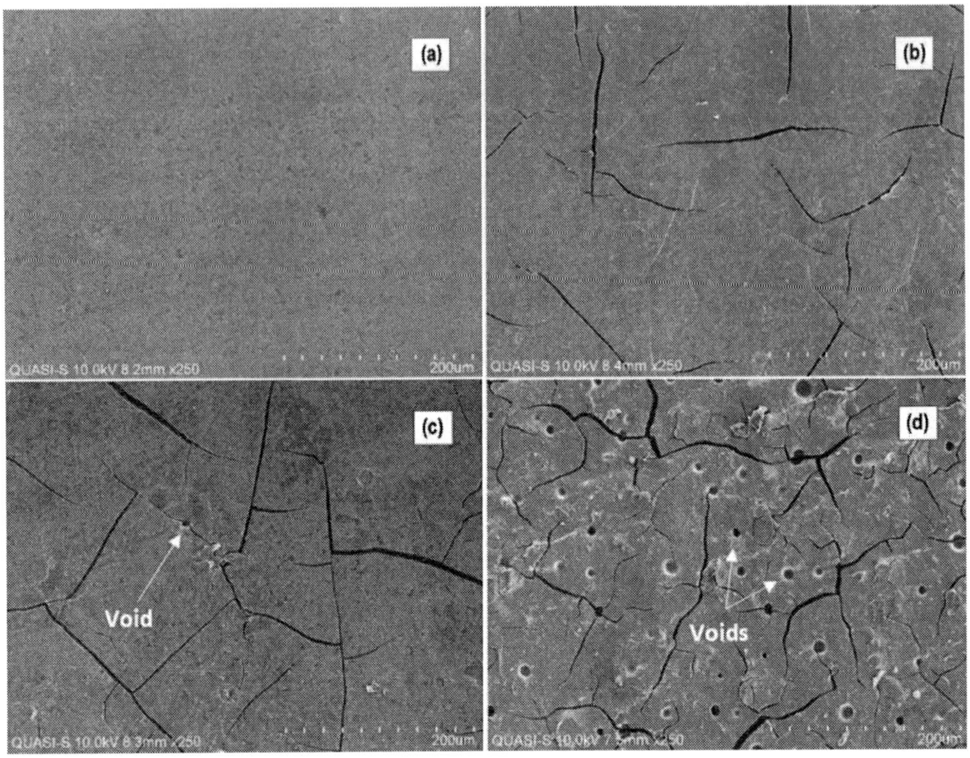

Fig. 2. The SEM micrographs of RHA-based geopolymer after compressive test; (a) S28 – ductile, (b) S5 - semi-ductile, (c) S7 - semi-brittle, and (d) S23 – brittle.

Semi-brittle material such as S7 possessed lower compressive strength compared to the brittle sample S23. The behavior of the former material was due to its lower viscosity caused by a higher water content and Si/Al ratio as shown in Figure 3. A semi-ductile material, such as S5, possessed higher compressive strength due to lower water content and Si/Al ratio compared to fully ductile material S28.

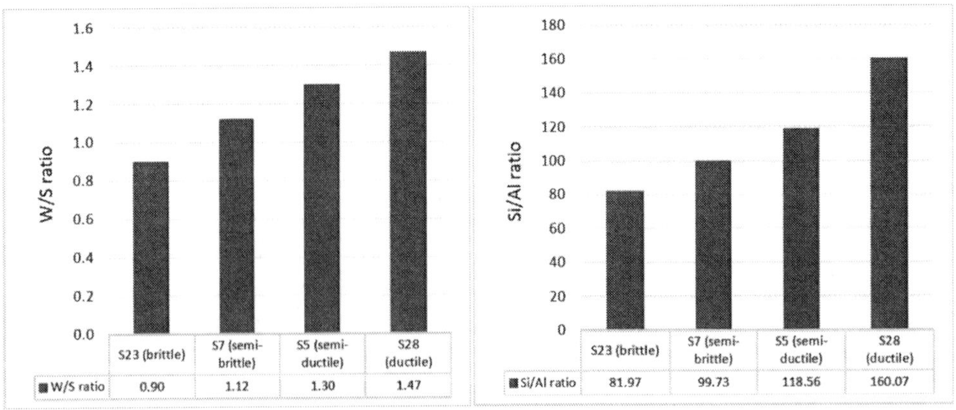

Fig. 3. W/S ratio and Si/Al ratio in selected brittle and ductile samples.

 DOI: 10.1051/matecconf/20179701025

3.4 Fire resistant performance.

Sample S28 which exhibited ductile behavior has poor compressive strength and fire resistant behavior as shown in Figure 4 (a). One of the major reasons for this is the high water content. When the sample was heated rapidly to a high temperature, part of its water content, in the form of bound or absorbed water, evaporated. Pressure developed simultaneously in the pores of the material vacated by the evaporation. According to Luna-Galiano et al. [12], evaporated water molecules supposedly travel from the hot fire-exposed surface to the inner part of the material which was cooler. Due to rapid evaporation rate, however, the material pores expanded faster resulting in a weak matrix structure and the release of the evaporated water from the material. Simultaneously, the intumescent process was not able to take place and the geopolymer surface could not crystallize in time. With the failure of the geopolymer system, heat and fire were able to travel through the binder matrix and created a hole.

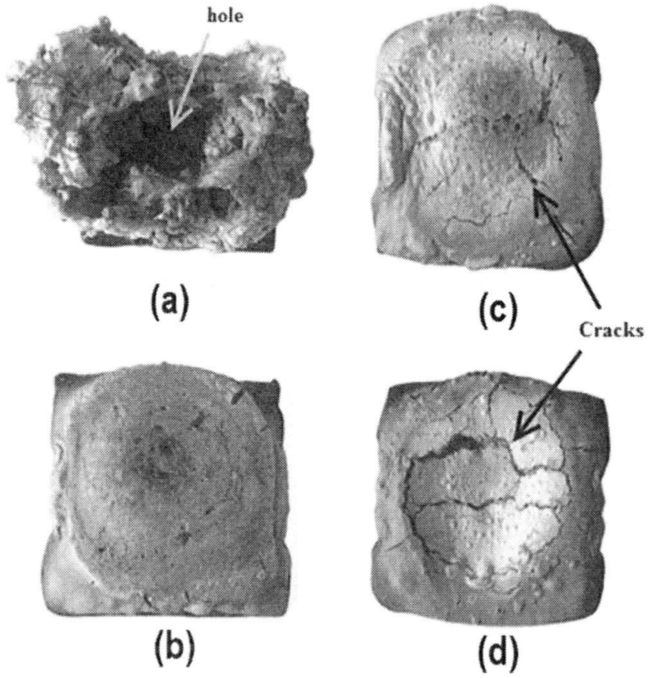

Fig. 4. Images of RHA-based geopolymer samples (a) S28, (b) S5, (c) S7, and (d) S23 after fire resistant test.

Sample S5 (see Figure 4 (b)) which shows semi-ductile behavior produced the best fire resistant properties with a maximum non-exposed surface temperature of only 50°C after 50 minutes (see Figure 5). There were no cracks or holes formed after the fire test which indicated that the sample exhibited a proper intumescent process. A portion of the water content evaporated when exposed to the fire, and generated pressure in the pores of the material and probably in the microvoids. Since crystallization of the geopolymer surface and intumescent process successfully took place in the sample, the evaporated water molecules were transported to cooler area of the material resulting in very low temperature at the non-exposed area which has been identified previously by Luna-Galiano et al. [12].

Sample S7 (see Figure 4 (c)) which showed semi-brittle behavior also possessed good fire resistant properties. This sample had similar process as that in the semi-ductile S5. The only difference was the presence of small voids in S7 which initiated cracks when the voids

DOI: 10.1051/matecconf/20179701025

and pores expanded under pressure created by evaporated water. The surface cracks directly in contact with the fire allowed heat to enter and penetrate to cooler area in the material thus raising the non-exposed surface temperature.

Sample S23 as shown in Figure 4 (d) which exhibited brittle behavior also displayed similar process as S7 and S5. However, due to more voids present in the geopolymer material and wider crack openings generated during the fire test, a higher amount of heat can penetrate and dissipate in the material resulting in higher non-exposed surface temperature of around 85°C after 50 minutes. With further propagation of surface cracks, the exposed geopolymer structure began to collapse especially at the center of the material.

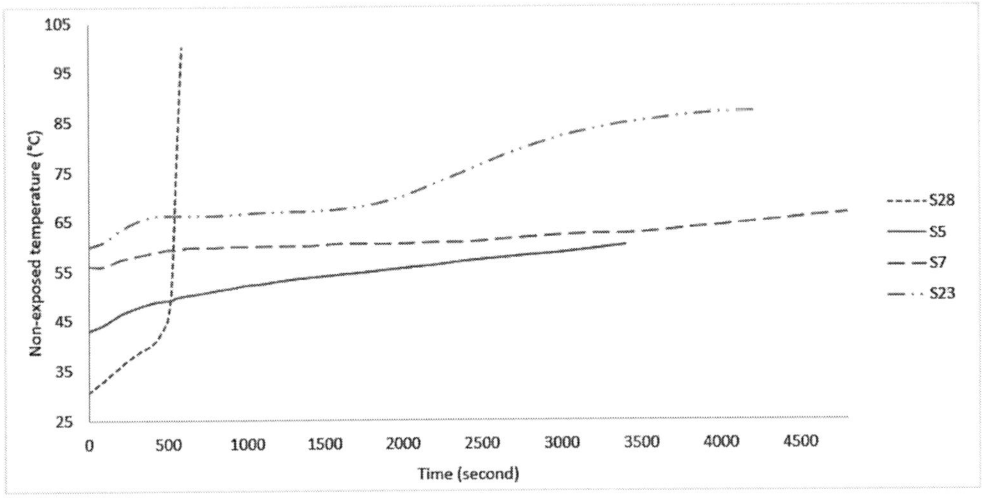

Fig. 5. Time-temperature curves of selected geopolymer coating samples.

4 Conclusion

RHA-based geopolymer sample recorded high compressive strength above 28 MPa when its S/L ratio was within range of 0.7 to 0.8, and its NaOH concentration within 12M to 14M. Brittle geopolymer sample (GS) with low Si/Al ratio (88.95) was high in compressive strength (33.55 MPa) and also showed high degree of geopolymerization. Ductile GS in comparison registered low compressive strength (0.04 MPa) irrespective of its degree of geopolymerization. Ductile GS with very low compressive strength (0.04 MPa) did not show any cracks while non-ductile GS (semi-ductile, semi-brittle, and brittle) initiated minor to major cracks. Brittle GS had higher amount of unreacted RHA particles compared to ductile ones. Water content within the GS had a major effect on its fire resistant properties. Semi-ductile GS showed the best fire resistant properties with a maximum non-exposed surface temperature of only 50°C after 50 minutes (when exposed to a direct fire with temperature of 900°C) followed by semi-brittle and brittle GS. Ductile GS however failed as fire resistant panel material due to very high amount of water which not allowed intumescent process and surface crystallization of the sample to take place.

Appreciation is given to Universiti Putra Malaysia (UPM) for funding this research study under the Geran Putra – IPS (9466800), Research University (RUGS) Grant (93771000 & 9393700) and Fundamental Research Grant Scheme (FRGS) sponsored by the Ministry of Education (5524464).

References

1. A. Belonio, in: Rice husk gas stove: handbook., Appropriate Technology Center, Department of Agricultural Engineering and Environmental Management, College of Agriculture, Central Philippine University, Iloilo City, Philippine (2005).
2. K.T. Lee, A.M. Mohtar, N.F. Zainudin, S. Bhatia, A.R. Mohamed, Fuel, **84**, 143 (2005)
3. R. Han, D. Ding, Y. Xu, W. Zou, Y. Wang, Y. Li, L. Zou, Bioresour. Technol., **99**, 2938 (2008)
4. E.B. Oyetola, M. Abdullahi, Leonardo El. J. Pract. Technol., **8**, 58 (2006)
5. K. Komnitsas, D. Zaharaki, Miner. Eng., **20**, 1261 (2007)
6. P.D. Silva, K. Sagoe-Crenstil, V. Sirivivatnanon, Cem. Concr. Res., **37**, 512 (2007)
7. S. Songpiriyakij, T. Kubprasit, C. Jaturapitakkul, P. Chindaprasirt, Constr. Build. Mater., **24** , 236 (2010)
8. Z. Zuhua, Y. Xiao, Z. Huajun, C. Yue, Appl. Clay Sci., **43**, 218 (2009)
9. F.A. Memon, M.F. Nuruddin, S.A.D.A.Q.A.T.U.L.L.A.H. Khan, N.A.S.I.R. Shafiq and T.E.H.M.I.N.A Ayub, Journal of Engineering Science and Technology, **8**, 44 (2013)
10. M. Oyane, T. Sato, K. Okimoto, S. Shima, J. Mech. Work. Technol., **4**, 65 (1980)
11. H. Xu, J.S.J. Van Deventer, Int. J. Miner. Process., **59**, 247 (2000)
12. Y. Luna Galiano, A. Cornejo, C. Leiva, L.F. Vilches C. Fernandez Pereira, Mater. Construcc., **65** (319), e059 (2015)

 DOI: 10.1051/matecconf/20179701026

Effect of Variability of Fly Ash Obtained from the Same Source on the Characteristics of Geopolymer

Antoni[1,*], *Juan* Satria[1], *Agung* Sugiarto[1], and *Djwantoro* Hardjito[1]

[1]Department of Civil Engineering, Petra Christian University, Indonesia

Abstract. Geopolymer properties is notably influenced by the quality of fly ash used as the precursor material. Fly ash obtained from the same source may have varying physical and chemical characteristics. The characteristics are influenced by, among other, the method and temperature of burning, the degree of grinding and the quarry sources of the coal. In this study, high calcium fly ash from a single power plant was investigated, to analyze the effect of its variability on the characteristics of geopolymer mortar. Fly ash samples were collected at ten different sampling periods. Three mixture compositions, with different sodium silicate solution to sodium hydroxide ratios, were prepared to produce geopolymer mortars. From the results, it was found that there are broad variation of fly ash chemical properties, however only small variation on the fineness and LOI was discovered. The setting time and compressive strength of geopolymer are influenced by the fly ash particle size, pH value and chemical content. Higher level of calcium oxide content increases the compressive strength, while finer particle size could have effect on the later age increase of the compressive strength. Faster setting time is also correlated with the higher pH value of the fly ash, with the tendency of flash setting occurrence for geopolymer using fly ash of high pH level.

1 Introduction

Fly ash based geopolymer concrete has been researched extensively due to the availability of fly ash in many regions. Fly ash, as a waste product from boiler burning process in factory and power plant, consequently has very broad quality range, depending on the variability of the coal quality and the burning process. Fly ash quality and its suitability to be used as precursor material in making geopolymer is determined by the source of the fly ash, its chemical composition, and physical shape and size of the fly ash. The presence of unburned coal, shown by the loss on ignition (LOI), is also a significant factor in classifying fly ash quality, as fly ash with higher LOI would absorb the alkaline activator in a higher rate, and reduce the geopolymer reaction [1,2].

* Corresponding author: antoni@petra.ac.id

In ordinary concrete technology, fly ash is classified into the low calcium and high calcium fly ash, to differentiate the hydration reaction and pozzolanic reaction with the calcium hydroxide. Thus, this classification is also applicable in the fly ash-based geopolymer concrete technology. Low calcium fly ash causes more predictable setting time of geopolymer, due to its low calcium content [3-5]. Low calcium fly ash geopolymer commonly has long setting time and requires heat for curing. On the other hand, high calcium fly ash creates challenges, when it is used for making geopolymer concrete [6-8]. Higher strength is typically obtainable when using high calcium fly ash, however the occurrence of rapid setting may hinder its application in large volume.

The authors' previous studies [9,10] have found that fly ash obtained from different power plants in Java island, Indonesia, produced different properties of the resulting geopolymer mortar. Setting time and compressive strength are affected by the physical and chemical properties of the fly ash used. The physical properties of fly ash are revealed by its particle size distribution, particle shape and the presence of contaminant in the fly ash, while the chemical properties are shown by the chemical compositions and LOI. The type of alkaline activator used in making the geopolymer significantly affects the properties of geopolymer. Fly ash obtained from one source stood out by having the strength of resulting geopolymer of almost double than those obtained from other sources, while maintaining the same mixture composition.

As the coal used in a power plant may be obtained from several quarries and the power plant may only interested in obtaining the highest energy, the quality of the fly ash produced could differs significantly between shipments. The difference also includes the calcium content in the fly ash.

In the current study, the suitability of fly ash obtained from one high quality source as precursor for making geopolymer mortar is investigated. Variations of fly ash quality, collected from different sampling time, are studied. The resulting fresh and hardened properties of the geopolymer mortar are correlated with physical and chemical properties of fly ash.

2 Experimental methods

Materials. Fly ash samples for making geopolymer were obtained from a power plant located in Paiton, East Java, Indonesia. It was reported as a good precursor for making good geopolymer [10]. A combination of chemical grade sodium hydroxide (NaOH) and sodium silicate solution was used as the alkaline activator. Sodium silicate solution was composed of 16.27% Na2O, 38.23% SiO2 and 45.5% H2O. Sand was obtained from Lumajang quarry and conditioned to conform to standard sand gradation of ASTM C778 [11].

Mixture and Testing. The following mixture composition was used to produce geopolymer mortar, i.e. sand to fly ash mass ratio of 2, distilled water to fly ash ratio of 0.25, 8M NaOH concentration, and sodium silicate liquid to sodium hydroxide solid ratio (S/N) were at 2, 2.5 and 3, respectively. Mixing was carried out using hand drill and performed in small batches to produce three 50 mm cubes, as it was found that there was tendency of rapid setting. Vicat setting time measurement was conducted on geopolymer paste at room condition [12]. Paste mixture was the same mixture proportion as the mortar, only without sand. The mortar mixture was cast in the 50 mm cube mould for compressive strength testing at 7, 14 and 28 days. After casting, the mould was covered with plastic sheet, and placed in the oven at 60°C for 24 hours. After demolding, the specimens were cured in laboratory room condition until the day for compressive testing. Each data represents the average result from three compressive strength tests.

DOI: 10.1051/matecconf/20179701026

2 Results and discussion

Fly Ash Variability. Ten samples of fly ash were collected within the period of four months, from July to October 2015, with one to two weeks intervals. Fly ash from Paiton power plant in East Java is considered as good quality fly ash with very low LOI. Table 1 shows the pH and fineness of ten samples of fly ash, as the proposed rapid indicators for the quality of the fly ash. Values of pH were measured from solution of 80 gram distilled water and 20 gram fly ash. Table 2 shows the chemical composition of selected fly ash samples as measured from XRF test. LOI values were consistently less than 1%. Fly ash samples have medium to high content of CaO, ranges from 9.40% to 20.42%. It was found that the pH values correlate with the content of calcium oxide (CaO) and magnesium oxide (MgO) compounds. The physical properties was measured only by sieving the fly ash passing sieve #325 (smaller than 44 μm). It was found that fly ash used in this study has a consistent fineness, with the exception of FA3, that has 76% particle size finer than 44 μm.

Setting Time. Setting times of geopolymer paste using four different fly ash samples as measured using Vicat needle are shown in Fig. 1. Higher alkali activator ratio (S/N) causes longer setting time as the higher content of sodium silicate solution increases the wetness of the mixture. It is shown that the high pH values of more than 11.0 correlate with the occurrence of fast setting. FA4 with pH = 11.8 produces the fastest setting time of geopolymer paste at about 10 minutes. Different behavior was shown on FA2 and FA10, with only 0.1 pH value difference, whereby the setting times of geopolymer paste were very contrast. FA2 causes significantly longer setting time than FA10. Davidovits [13] classifies the pH range of fly ash of 8 to 11 that may cause rapid setting, while high possibility of flash setting may occur when pH value of fly ash is more than 11. With the exception of FA2, this study confirms the classification. However, the setting behavior could be affected also by other factors besides pH, especially for the ones with pH values about 11.

The results also showed that for geopolymer paste without any flash setting tendency, the higher the S/N ratio, the longer the setting time. But, for those with flash setting tendency, the S/N ratio does not really affect the setting time.

Table 1. pH value and particle fineness of the fly ash samples

	FA1	FA2	FA3	FA4	FA5	FA6	FA7	FA8	FA9	FA10
pH	10.9	11.1	10.4	11.8	10.6	10.8	10.6	11.7	11.4	11.2
Passing sieve #325 (44μm) (%, mass)	84	88	76	88	80	84	84	88	92	84

DOI: 10.1051/matecconf/20179701026

Table 2. XRF results for selected fly ash samples

No.	Compound	Chemical Composition (%, by mass)			
		FA2 (11.1)	FA3 (10.4)	FA4 (11.8)	FA5 (10.6)
1	SiO_2	43.74	43.36	32.47	42.26
2	Al_2O_3	22.03	29.74	14.92	24.43
3	Fe_2O_3	14.68	7.33	16.50	12.91
4	TiO_2	1.28	1.00	0.71	1.01
5	CaO	9.40	13.30	20.42	11.19
6	MgO	4.33	1.80	7.95	3.69
7	Cr_2O_3	0.14	0.00	0.14	0.01
8	K_2O	1.55	0.42	1.32	0.80
9	Na_2O	1.56	1.88	2.92	1.85
10	SO_3	0.53	0.40	1.88	0.91
11	Mn_2O_3	0.15	0.14	0.18	0.24
12	LOI	0.80	0.60	0.43	0.44
$SiO_2+Al_2O_3+Fe_2O_3$		80.45	80.43	63.89	79.60
SiO_2/Al_2O_3		1.99	1.46	2.18	1.73

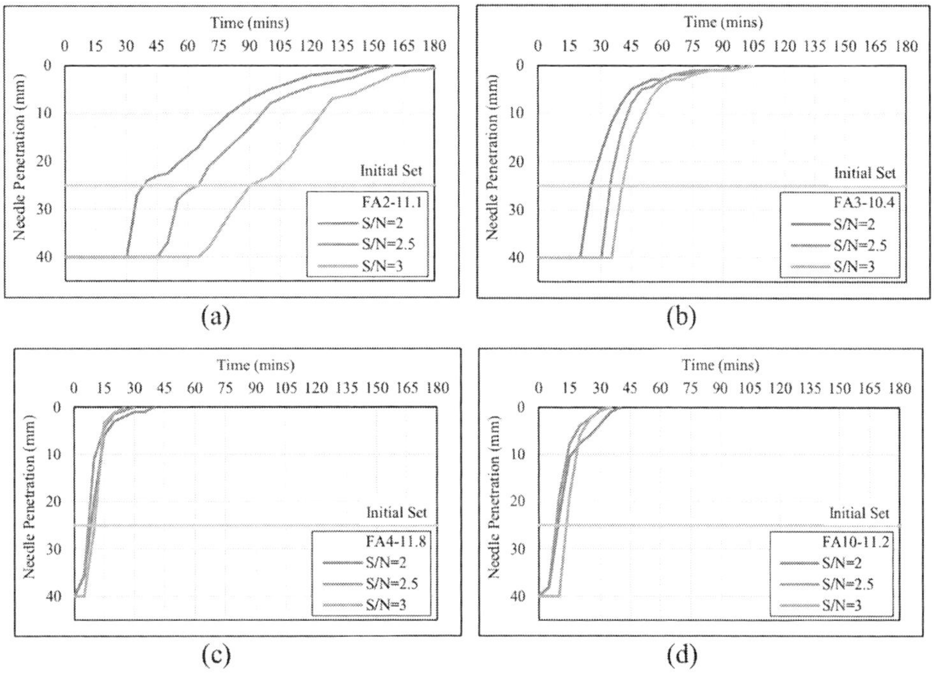

Fig. 1. Setting time of geopolymer paste using four different fly ash samples, (a) FA2-pH 11.1, (b) FA3-pH 10.4, (c) FA4-pH 11.8 and (d) FA10-pH 11.2

Compressive Strength. The effect of fly ash variability and S/N ratio on the compressive strength of geopolymer mortar are shown in Fig. 2. The series was sorted based on the pH values of the fly ash, from the lowest to the highest. The results show that all geopolymer samples have initial strength at 7 days higher than 55 MPa. Most samples show the increase of compressive strength with age, although the rates vary with the variability of fly ash. It was found that there was a large increase of compressive strength between 14 and 28 days for those with FA2, FA4, FA8, and FA9; while there was only slight increase of strength for the other fly ashes. The highest compressive strength of 85.0 MPa was achieved by mortar with FA4, with S/N ratio of 2.5, at 28 days. The compressive strength of geopolymer mortar normally reaches a plateau at later age, after thermal curing,

DOI: 10.1051/matecconf/20179701026

as indicated by Chindaprasirt et al. [7]. The increase at later age strength suggest that there could be a hydration reaction takes place beside the geopolymer reaction. The increase of compressive strength with age for high calcium fly ash geopolymer was also stated by Gou et al. [6] and Nurwidayati et al. [14], but no further discussion was made. The effect of different S/N ratios on the compressive strength does not show any single trend.

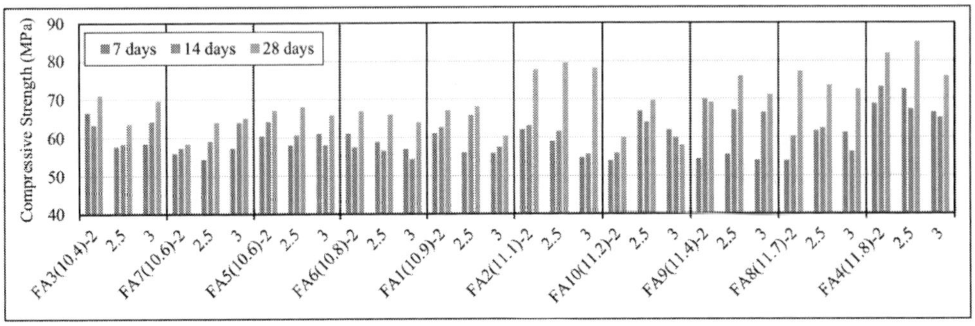

Fig. 2. Effect of variability of fly ash and S/N ratio on compressive strength of geopolymer mortar

Strength Gain. The increase of the compressive strength from 7 to 28 days with the rapid indicators of the fly ash is shown in Fig. 3. Lower compressive strength values show the 7 days strength and linked with the higher values for the 28 days strength. From the figure, it is shown that the higher increase of compressive strength from 7 to 28 days occurs to the geopolymer mortar with higher pH value fly ashes. The trend is true with exception of FA10, showing only small increment. Fig 3(b) shows the effect of material fineness with the strength increment. There is a good correlation between finer materials with the increase of compressive strength. This condition shows that there could be an additional hydration reaction in the geopolymer mortar due to the high calcium content in the fly ash. The bigger strength gain is also attributed by the finer particle size of the fly ash.

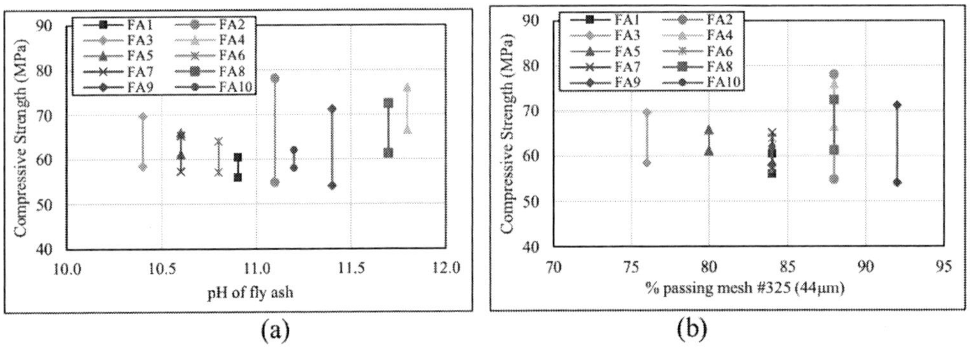

Fig. 3. Effect of fly ash properties on the increase of compressive strength from 7 to 28 days for geopolymer mixture with S/N = 3, (a) pH effect and (b) particle fineness.

3 Conclusions

Fly ash obtained from one power plant with various sampling time was used to made geopolymer mortar. The following findings was obtained:

There is a large variation, in term of the chemical properties, of the fly ash collected from the same source in various sampling time. This may cause large variation in the compressive strength and setting time of the resulting geopolymer mortar. The high CaO content in the fly ash can increase the compressive strength due to additional hydration reaction that may take place. However, the higher content of CaO may increase the possibility of flash setting time occurrence.

On the physical properties, i.e. the particle size and LOI, it was found that the fly ash studied has good and relatively constant properties. This may be atttributed as the source of excellent compressive strength of resulting geopolymer mortar, higher than 50 MPa, with the range of 58 to 85 MPa for 28 days compressive strength.

Two rapid indicators, namely pH value and % passing sieve #325, can be good indicators to assess the quality of the fly ash, given the suitability of the fly ash as source material for making geopolymer has been confirmed.

The authors acknowledge the financial support from Ministry of Research, Technology and Higher Education, Indonesia, in this research study.

References

1. J. van Jaarsveld, J. van Deventer, G. Lukey, Mater. Lett. **57**, 1272 (2003)
2. E. Diaz, E. Allouche, S. Eklund, Fuel **89**, 992 (2010)
3. D. Hardjito, C. Chua, C. Ing, Modern applied science, **2**, 3 (2008)
4. C. Montes, E. Allouche, Coal Combustion and Gasification Products, **4**, 1 (2012)
5. S.E. Wallah, Civil Engineering Dimension **12**, 73(2010)
6. X. Guo, H. Shi, W. Dick, Cement Concrete Comp., **32**, 142 (2010)
7. P. Chindaprasirt, J. Mater. Civ. Eng., **23**, 264 (2010)
8. S. Pangdaeng, Mater. Design., **53**, 269 (2014)
9. Antoni, S. Wijaya, D. Hardjito, Mater. Sci. Forum., **841**, 90 (2016)
10. Antoni, S. Wijaya, D. Hardjito, Mater. Sci. Forum., **841**, 98 (2016)
11. ASTM C778. Standard Specification for Standard Sand. ASTM International, (2002)
12. ASTM C191, Standard Test Method for Time of Setting of Hydraulic Cement by Vicat Needle, ASTM International, (2004)
13. J. Davidovits, Geopolymer Chemistry and Applications, Institut Geopolymere, France, 390-394 (2008)
14. R. Nurwidayati, Mater. Sci. Forum., **841** (2016)

DOI: 10.1051/matecconf/20179701027

Optimizing High Performance Self Compacting Concrete

Yonathan Raymond A[1,*], *Ekaputri* Januarti Jaya[1,2] , and *Triwulan*[2]

[1]Civil Engineering Department, Faculty of Civil Engineering and Planning, Institut Teknologi Sepuluh Nopember (ITS), Sukolilo, Surabaya, Indonesia
[2]Konsorsium Riset Geopolimer Indonesia (KORIGI), Laboratorium Beton dan Bahan Bangunan, Kampus ITS Sukolilo Surabaya, Indonesia, 60111

Abstract. This paper's objectives are to learn the effect of glass powder, silica fume, Polycarboxylate Ether, and gravel to optimizing composition of each factor in making High Performance SCC. Taguchi method is proposed in this paper as best solution to minimize specimen variable which is more than 80 variations. Taguchi data analysis method is applied to provide composition, optimizing, and the effect of contributing materials for nine variable of specimens. Concrete's workability was analyzed using Slump flow test, V-funnel test, and L-box test. Compressive and porosity test were performed for the hardened state. With a dimension of 100x200 mm the cylindrical specimens were cast for compressive test with the age of 3, 7, 14, 21, 28 days. Porosity test was conducted at 28 days. It is revealed that silica fume contributes greatly to slump flow and porosity. Coarse aggregate shows the greatest contributing factor to L-box and compressive test. However, all factors show unclear result to V-funnel test.

1 Introduction

Condensed Silica Fume is now commonly used as a cementitious supplementary material in concrete [1]. It improves the strength of concrete and also the durability of concrete [2]. Using waste glass powder as supplementary material for cement is quite popular worldwide due to the increased disposal costs and environmental concern. It contains silica based material in amorphous form, and increase the strength of concrete [2-4]. It was also found promising to be added in Self compacting concrete mixture [5].
Previous study showed that using silica fume and glass powder as cement replacement could be used to gradually increase strength of concrete paste [6,7]. Self Compacting Concrete is well known for its workability, which can provide many advantages such as lower the rates of human error, doesn't need any compacting activities, can fill through narrow gaps, etc [8-13]. Using this previous study as reference, High Performance Self Compacting Concrete should be developed to increase the productivity and quality for the construction. It is complicated to optimize the mix design using trial and error, since it is time and energy consuming, therefore using statistic method is helpful to obtain more result

* Corresponding author: rayonathan@hotmail.com

with less trial and error, since it is time and energy consuming, therefore using statistic method is helpful to obtain more result with less trial and error.

Fractional factorial designs was known as money and time saver statistical approach. Despite being widely and effectively used, this method become unwieldy in cost and time the number and variables is large, two designs for same experiments may resulting different results, and the interpretation of the experimental result with a large number of factors may be difficult due to lack of clear design and analysis guidelines. Genichi Taguchi of Japan proposed an innovative method by simplified and standardized fractional factorial designs, thus he able to overcome the limitation of fractional factorial designs. [14]

This paper discusses about the optimizing and the contributing factors of high performance self compacting concrete using Taguchi's single response method as the data analysis.

2 Material

Aggregates. The densities of coarse aggregate (±1.2 cm) and fine aggregate are 2.74 g/cm3 and 2.72 g/cm3 respectively.

Superplasticizer. There were two kind of superplasticizer used in the mixture to achieve the workability of concrete, which are Polycarboxylate Ether and Polyhydroxy Carbon Salts. Polycarboxylate Ether was used in the controlled factors as shown in Table 1. Polyhydroxy Carbon Salts was used as additional superplaticizer with 0.2% from binders weight.

Binders. Ordinary Portland Cement (OPC) with specific gravity of 3.14 gr/cm3. Silica Fume with specific gravity of 2.105 gr/cm3. Waste ground glass with specific gravity of 2.381 gr/cm3. Ground glass need to be pulverized before applied as cement replacement, after that particle size distribution was conducted. Using glass powder with size in average of 25 μm which its distribution as shown in Fig.1.

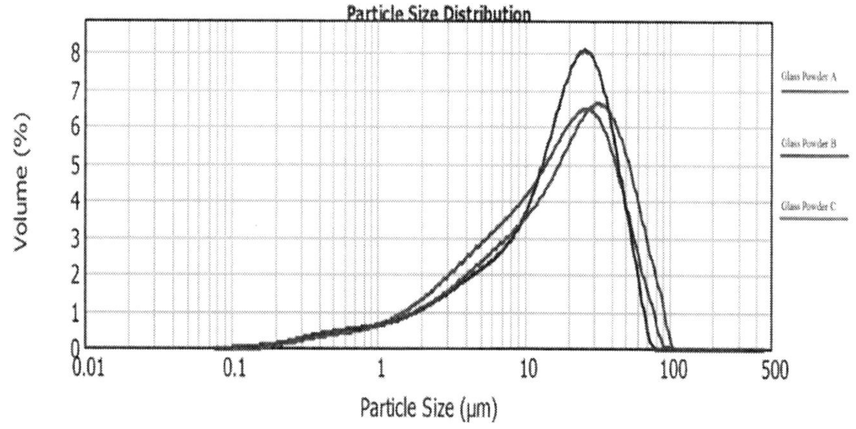

Fig. 1. Distribution of Glass Particles

DOI: 10.1051/matecconf/20179701027

3 Methods

The composition of binders were determined from previous research [6,7]. Using Taguchi analysis method, the factors, levels, degrees of freedom, and orthogonal array need to be determined to generate the mix design [14,15] as shown in Table 1, Table 2, Table 3, and Table 4 respectively. Using glass powder (10%, 15%, 20% from total weight of cement plus glass powder), silica fume (20%, 40%, 60% from total weight of glass powder plus silica fume), Polycarboxylate Ether (0.8%, 1%, 1.2% from binder's weight), and coarse aggregate (45%, 50%, 55% from total volume of aggregate) as the controlled factors.

Table 1. Factors and Level

Codes	Controlled Factors	Level 1	Level 2	Level 3
		(%)	(%)	(%)
A	Coarse Agregate	45	50	55
B	Glass Powder	10	15	20
C	Silica Fume	20	40	60
D	*Polycarboxylate Ether*	0.8	1	1.2

Table 2. Degree of Freedom

Factors	DF
A	2
B	2
C	2
D	2
Total	8

Table 3. Orthogonal Array L9 (3^4)

Codes	A	B	C	D
R1	1	1	1	1
R2	1	2	2	2
R3	1	3	3	3
R4	2	1	2	3
R5	2	2	3	1
R6	2	3	1	2
R7	3	1	3	2
R8	3	2	1	3
R9	3	3	2	1

Table 4. Mix Design

Codes	Cement (kg/m^3)	Glass Powder (kg/m^3)	Silica Fume (kg/m^3)	Water ($Liter/m^3$)	*Polycarboxylate Ether* ($liter/m^3$)	*Polyhydroxy Carbon Salts* ($Liter/m^3$)	Fine Agregate (kg/m^3)	Coarse Agregate (kg/m^3)
R1	612.75	54.47	13.62	170.21	5.11	1.36	897.6	739.8
R2	569.85	60.34	40.22	167.6	6.29	1.34	897.6	739.8
R3	527.27	52.73	79.09	164.77	7.43	1.32	897.6	739.8
R4	607.73	40.52	27.01	168.81	7.61	1.35	816	822
R5	570.06	40.24	60.36	167.67	5.04	1.34	816	822
R6	532.79	106.56	26.64	166.5	6.25	1.33	816	822
R7	608.53	27.05	40.57	169.04	6.35	1.35	734.4	904.2
R8	569.63	80.42	20.1	167.54	7.55	1.34	734.4	904.2
R9	532.5	79.88	53.25	166.41	5	1.33	734.4	904.2

Fresh concretes were tested for slump flow test, V-funnel test, and L-box test [16]. After that, the fresh concrete was casted in cylindrical molds with a diameter of 100 and 200 mm high [17]. One day after casting, concretes were cured in water (wet curing system) until the age of 3, 7, 14, 21, and 28 days to be tested for compressive strength. One day before compressive test, the rough top surface of concrete was cut to obtain smooth and solid surface. The compressive test is corrected using correction factor, since length of cylinder may be less than initial condition [18,19]. Porosity test conducted at the age of 28 days.

4 Results and discussions

Concrete's test result is shown by Table 5.

Table 5. Test Results

Codes	Compressive Strength on 28 Days	Slump	V-Funnel	L-Box	Total Porosity
	(MPa)	(cm)	(s)		(%)
R1	56.34	79.5	8.17	0.83	17.93
R2	53.2	72.5	12.3	0.79	15.51
R3	40.98	69.5	7.1	0.83	20.89
R4	42.93	71	11.8	0.76	16.11
R5	45.99	68.5	14	0.73	24.57
R6	42.76	80.5	8.3	0.82	18.5
R7	32.16	73.5	12.94	0.78	29.02
R8	22.35	79	37.37	0.8	21.95
R9	31.82	71	8.81	0.79	22.71

 DOI: 10.1051/matecconf/20179701027

Taguchi data analysis is applied for Slump flow test, V-funnel test, L-box test, Compressive test, and Porosity test with 85% of confidence level. Analysis result is shown by Table 6.

Table 6. Taguchi's data result (single response)

Test	Coarse Aggregate	Glass Powder	Silica Fume	*Polycarboxylate Ether*	Estimated optimum Condition
Slump flow (cm)	55% **	10%	20% *	1% *	78.778 ± 2.040 cm
L-Box	55% *	15% *	40% *	1% **	0.760 ± 0.030
V-Funnel (s)	45%	20%	40% **	1%	-
Compressive Strength (MPa)	45% *	10%	40% **	0.8% *	54.037 ± 5.496 MPa
Porosity (%)	45% *	15% **	40% *	1.2% *	14.274 ± 0.677 %
Note: (*) Significant Factor					
(**) Pooled Factor					

Slump Flow. It is found that Silica Fume contributes greatly to slump flow with contribution percentage 88.88% followed by Polycarboxylate Ether with contribution percentage 5.74%. Applying 55% of coarse aggregate, 10% of glass powder, 20% of silica fume, and 1% of Polycarboxylate Ether to achieve optimum slump flow with 78.8 ± 2.04 cm.

This result agreed with previous study [5-7] that Silica Fume absorb more water than glass powder. Adding more silica fume resulted in fresh concrete more condensed. In addition, adding glass powder shows better workability. It was noted that in R6 and R8 bleeding occurred, which contains higher glass powder. It is proved that adding glass powder cannot reduce bleeding.

V-Funnel. It is found that no factors have significant effect on the V-funnel. Which means that there are other factors or other levels that may have significant effect on V-funnel that were not estimated during the experiments. The optimum condition could not be estimated, however the composition can still be used.

L-Box. It is found that coarse aggregate contributes greatly to L-box with contribution percentage 27.44% followed by silica fume with contribution percentage 26.16%, and then glass powder with contribution percentage 25.47%. Applying 55% of coarse aggregate, 15% of glass powder, 40% of silica fume, and 1% of Polycarboxylate Ether to achieve optimum L-box with 0.76 ± 0.03.

This result agreed with previous study [8,9] that composition of mortar plays an important role in making self compacting concrete.

Compressive Strength. It is found that coarse aggregate has the biggest effect on the compressive strength with contribution percentage 76.35% followed by Polycarboxylate

Ether with contribution percentage 14.42%. Applying 45% of coarse aggregate, 10% of glass powder, 40% of silica fume, and 0.8% of Polycarboxylate Ether to achieve optimum compressive strength with 54.037 ± 5.496 MPa.

This result agreed with previous study [6,7] that the use of Glass powder and Silica Fume can improves the strength because of Glass Powder is more reactive than Silica Fume, with recommendation amount 10-15% Glass Powder replacement to Cement, and 40% of Silica Fume replacement to Glass Powder. By using 45% of coarse aggregate, the compressive strength increases due to the smaller void or porous reducing in the concrete.

There is bubble occurred when mixing the concrete, especially on R8 where badly bubbled. It has not been confirmed what the cause of the bubble. Which lead to cutting the surface of cylinders before tested.

Porosity. It is found that silica fume has the biggest effect on the porosity with contribution percentage 50.23% followed by coarse aggregate with contribution percentage 44.82%, and then Polycarboxylate Ether with contribution percentage 4.34%. Applying 45% of coarse aggregate, 15% of glass powder, 40% of silica fume, and 1.2% of Polycarboxylate Ether to achieve optimum total porosity with 14.274 ± 0.677%.

Glass powder and silica fume have smaller particle size than cement and also lowering coarse aggregate's percentage, the optimum composition used high percentage of smaller materials to achieves the lowest percentage of porous. Which means by lowering the porous of concrete could increase the compressive strength [20,21]. It was found that by using high range water reducer and microsilica could decreased porosity [2], but by increasing high range water reducer could also decrease compressive strength.

5 Summary

It is revealed that silica fume shows the greatest contribution in Slump flow and Porosity test. However, coarse aggregate shows more contribution in L-box and Compressive test. Applying 55% of coarse aggregate, 10% of glass powder, 20% of silica fume, and 1% of Polycarboxylate Ether can achieve optimum slump flow at 78.8 ± 2.04 cm. Optimum L-box test result at 0.76 ± 0.03 can be achieved by applying 55% of coarse aggregate, 15% of glass powder, 40% of silica fume, and 1% of Polycarboxylate Ether. Optimum compressive strength at 54.037 ± 5.496 MPa can be achieved by applying 45% of coarse aggregate, 10% of glass powder, 40% of silica fume, and 0.8% of Polycarboxylate Ether. Applying 45% of coarse aggregate, 15% of glass powder, 40% of silica fume, and 1.2% of Polycarboxylate Ether can achieve optimum total porosity at 14.274 ± 0.677%.

Using 40% of silica fume by glass powder weight is recommended. In addition, 10-15% of glass powder is also recommended. Since the density of silica fume is less than glass powder, it influence the performance of fresh state of concrete. Hence, bleeding of concrete can be avoided.

Applying 1% of Polycarboxylate Ether is recommended to achieve better workability. Furthermore, to obtain less porous in concrete, it is recommended to use superplasticizer up to 1.2% because it can produce denser concrete. In consequences, it decrease the strength.

It is proved that controlling the volume of coarse aggregate plays an important role to produce high performance concrete in hardened state. Less coarse aggregate leads to compacted concrete.

References

1. E.H. Kadri, R. Duval, S. Aggoun, S. Kenai, ACI Struct. J., **106(2)**, 107 (2009)

2. K. Wille, A.E. Naaman, G.J. Parra Montersinos, ACI Struct. J., **108(1)**, 46 (2011)
3. J.M. Khatib, E.M. Negim, H.S. Sohl, N. Chileshe, Eur. J. Appl. Physiol., **4(4)**, 173 (2012)
4. A. Shayan, A. Xu, Cem. Concr. Res., **36(3)**, 457 (2006)
5. M. Liu, Constr. Build. Mater., **25 (2)**, 919 (2010)
6. H. Limantono, J.J. Ekaputri, Triwulan, Key Eng. Mater. ,**673**, 37 (2016)
7. J.J. Ekaputri, H. Limantono, S. Susanto, Eddy, Triwulan, B. Abdullah, MM. AI, Key Eng. Mater., **673**, 83 (2016)
8. H. Okamura, M. Ouchi, J. Adv. Concr. Technol., **1(1)**, 5 (2003)
9. H. Brouwers, H. Radix, Cem. Concr. Res., **35(11)**, 2116 (2005)
10. S. Dhiyaneshwaran, P. Ramanathan, I. Baskar, R. Venkatasubramani, Jordan, J. Civ. Eng., **7(3)**, 342 (2013)
11. C.F. Ferraris, L. Brower, C. Ozyildirim, J. Daczko, Proceedings of the international symposium on high performance concrete, 398 (2000)
12. O. Gencel, C. Ozel, W. Brostow, G. Martinez Barrera, Mater. Res. Innov., **15(3)**, 216 (2011)
13. Z. Grdic, I. Despotovicc, G. Topliccic CCurccic, Facta universitatis-series: Architecture and Civil Engineering **6(2)**, 173 (2008)
14. R.K. Roy, *A Primer on the Taguchi Method* (Society of Manufacturing Engineers, U.S.A, 2010)
15. R.K. Roy, *Design of Experiments Using the Taguchi Approach: 16 Steps to Product and Process improvement* (John Wiley & Sons, New York, 2001)
16. F. Aslani, Mag. Concrete. Res., **65(15)**, 914 (2013)
17. ASTM C 192/C 192M – 02, *Standard practice for making and curing concrete test specimens in the laboratory* (ASTM International, West Conshohocken, 2002)
18. ASTM C 39/C 39 M – 01, *standart test method for compresive strength of cylindrical concrete specimen* (ASTM International, West Conshohocken, 2001)
19. ASTM C 42M – 03, *Standart test method for obtaining and testing drilled cores and sawed beams of concrete* (ASTM International, West Conshohocken, 2003)
20. M. Rashid, M. Mansur, J. Civ. Eng., **37(1)**, 53 (2009)
21. A. Logan, W. Choi, A. Mirmiran, S. Rizkalla, P. Zia, ACI Struct. J., **106(5)**, 413 (2009)

The Fatigue Crack Growth Rate Due to Single-Step Austempered Heat Treatment in Nodular Cast Iron

Andoko Andoko[1,*], and *Poppy* Puspitasari[1]

[1]Mechanical Engineering Department, Engineering Faculty, Universitas Negeri Malang, Indonesia

Abstract. Nodular cast iron comes to be the most used material in manufacturing industry. Recently, a number of engineering to enhance the mechanical properties of nodular cast iron have been being given including through the heat treatment called single-step austempered. In general, the process of single-step heat treatment is conducted through austenization at the temperature of 900°C with the holding time of 60 minutes and quenched at various austempered temperatures of 280, 310, and 340°C in the holding time of 60 and 120 minutes. The observation on fatigue crack behaviour was performed in comparison of stress R=0 and at 11-Hz frequency. From the result of the research, a graphic about relationship of crack length (a) and the number of cycles (N), and the graphic of the fatigue crack growth rate (da/dN) with the stress intensity factor of (Δk). In addition, it is found that the single-step austempered heat treatment at the austempered temperature at 340°C with the holding time of 120 minutes showed a slower fatigue crack growth rate in comparison to the temperature and time below. The slowness of the fatigue crack growth rate was caused by the higher percentage of the pearlite matrix phase in comparison to the ferrite matrix, nodular distance, and the relatively even nodule distribution.

1 Introduction

One of the weaknesses of cast iron is related to its brittleness, thus being liable to break easily particularly in accepting a dynamic loading, impact loading (sudden attack) or flexure loading. In response to these weaknesses, British Cast Iron Research Association (BCIRA) in 1948 mixed some elements such as cerium, copper, nickel and other alloyed elements into the cast iron in order to result in a round-shaped graphite cast iron. This endeavour was then followed by the emergence of some tests, one of which is by adding the magnesium into cast iron. Based on the result of metallographic test, it is known that the addition of magnesium into the cast iron is able to produce the round-shaped graphite cast iron and the result from the laboratory test reveals that nodular graphite in fact has tensile strength and other mechanical properties that can reach double – even its ductility is comparable to carbon steel. This finding was then patented in 1949, later remarked as the year of the emergence of Nodular

* Corresponding author: andoko.ft@um.ac.id

Cast Iron (also called as Ferro Casting Ductile or FCD) [1]. Further, to improve the mechanical and other physical properties of nodular cast iron (FCD), the heat treatment was given in which the result of this treatment process is known as Austempered Ductile Iron (ADI).

The heat treatment on FCD is performed through the heating process at the proper temperature with a number of holding times. Subsequently, FCD is processed for cooling at a lower temperature with certain speed. In other words, temperature and time are becoming two core parts in heat treatment. The heat treatment on FCD is aimed to improve its mechanical properties either in strength, hardness, ductility and toughness [2]. Also, it is to reduce the residual stress and wear resistance of cast materials given the heat treatment [3]. It has been recognized so long that Austempered Ductile Iron (ADI) has the high level of ductility, strength, wear resistance, toughness and machinability that can substitute the wrought steel in various applications [4]. ADI has been proven as a suitable material for the high abrasive wear resistance – even better than that of wrought steel. For these natures, ADI is able to substitute the wrought steel in any applications [5], some of which are used in Crankshaft, camshaft, and other agricultural devices.

2 Methodology

The specimens consist of three types of treatment which are single step austempered at 280°C, 310°C, and 340°C with holding time at 60 minutes and 120 minutes. The observation and analysis on the tested specimens have been conducted by observing the crack length (a) and the number of cycles (N) in the testing machine of fatigue. The results of the observation on the crack length and the number of cycles were recorded to be used in making a plot of correlation diagram between the crack length (a) and cycle number (N) and to measure the fatigue crack growth rate (da/dN) and the stress intensity factor (Δk). The test to observe the crack length and the number of cycles was performed at the stress ratio (R=0) with the stress level 1/3 yield strength at the frequency of 11 Hz.

3 Results and discussion

3.1 Fatigue crack growth test

Fatigue crack growth of specimens were observed using cyle (N) versus crack length graph and dA/dN versus Δk. There were two set of fatigue crack growth graphs for two different holding times with three different heating temperatures (280°C, 310°C, and 340°C).

 DOI: 10.1051/matecconf/20179701028

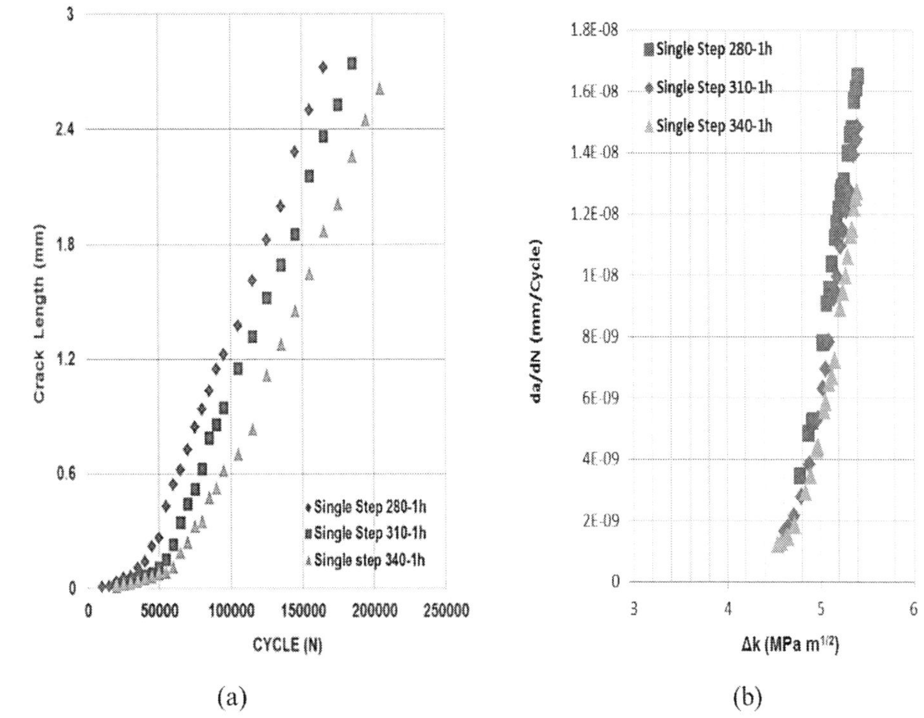

Fig. 1. The Graphic of Correlation between the crack length (a) and the number of cycles (N) b. The graphic of correlation between (da/dN) and (Δk)

Figure 1a shows the graphic of correlation between the crack length (a) and the number of cycles (N) in the single step with various austempered temperatures (280, 310 and 340°C) for the holding times of 60 minutes. Meanwhile, Figure 1b presents the graphic of correlation between the fatigue crack growth rate (da/dN) and the stress intensity factor (Δk) in single step with various austempered temperatures (280, 310 and 340°C) for the holding times of 60 minutes.

DOI: 10.1051/matecconf/20179701028

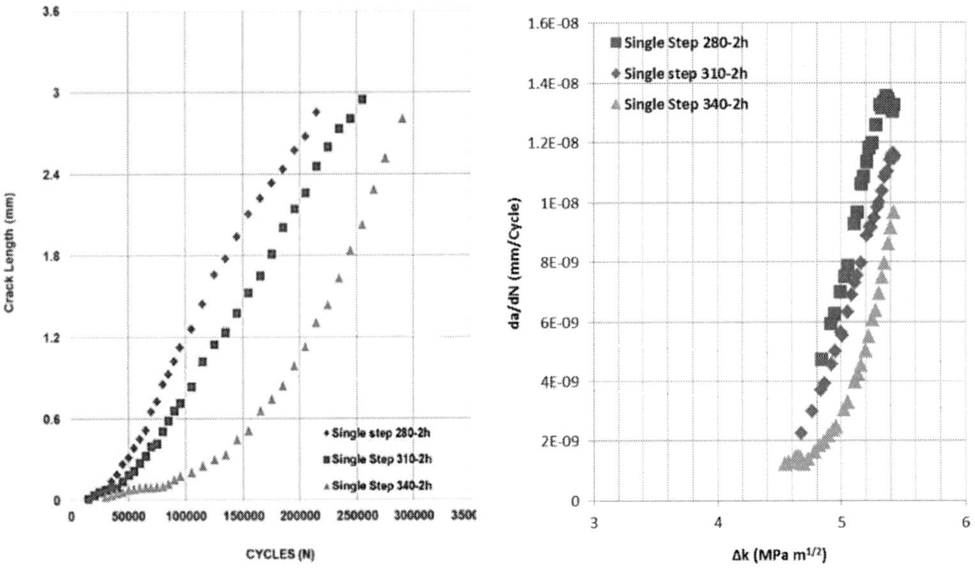

(a) (b)

Fig. 2. The graphic of correlation between the crack length (a) and the number of cycles (N). (b) The graphic of correlation between (da/dN) and (Δk)

Figure 2a shows a combination of the graphics of the correlation between the crack length (a) and the number of cycles (n) in single step with various austempered temperatures (280, 310 and 340°C) for the holding times of 60 minutes and 120 minutes on each temperature. Figure 2b, meanwhile, shows the combination of the graphic of the correlation between the fatigue crack growth rate (da/dN) and the stress intensity factor (Δk) in single step with various austempered temperatures (280, 310 and 340°C) for the holding times of 60 minutes and 120 minutes.

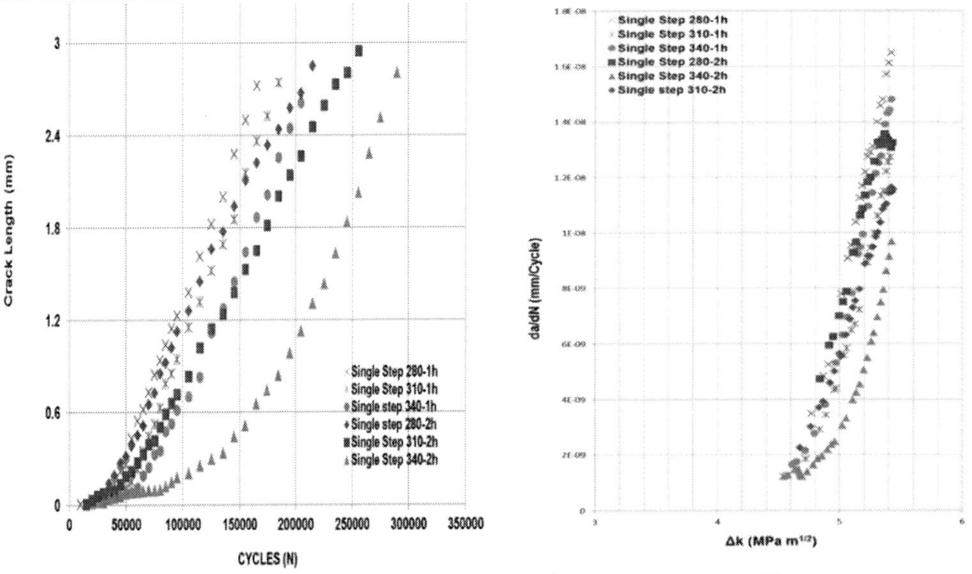

(a) (b)

Fig. 3. The graphic of correlation between the crack length (a) and the number of cycles (N). (b) the graphic of correlation between (da/dN) and (Δk)

As seen in that graphic combination (Figure 3), the best results for the crack length and the number of cycles were obtained at the austempered temperature at 340°C with the holding time of 120 minutes in comparison to other temperatures and holding times. It shows that the failure in material occurred approximately in 28,000, 46,300 and 70,200 cycles for R=0.1. In contrast, there was no any failure when the specimen was tested at the maximum voltage of 500 MPa with the ratio of R=0.1. For the ratio of R=0,5 with the application of maximum tension of 1100 MPa, the failure occurred around 32,000-43,000. The initiation of the fatigue crack exclusively occurred in pores present both on the surface and directly under the surface. Decohesion of the nodular graphite and the next initiation and the micro crack growth can cause the deflection of the dominant crack system. This also indicates a significant increase in the frequency of the nodular graphite along the crack paths as the tension increased and the further initiation (limited to the nodular graphite on crack tip the plastic zone) for the crack occurred to spread the crack tip that was dominant during the specimen period. Thus, the change in the cast micro caused by the heat treatment has caused the more fatigue level in the performance of the crack growth. This was in view of the lack of eutectic carbide and the relatively high number of austenite in micro.

The austempered temperature at 340°C with the holding time of 120 minutes at single step showed a microstructure in the form of the refined ausferrite with the remaining austenite phase added by the number, size and the evener graphite distribution. The content of the remaining austenite is fewer than that of the ferrite acicular phase. The longer the austempered time was, the more the ausferrite were formed. However, in the austempered time of 120 minutes with the austempered temperature at 340°C in single step the content of the ausferrite got fewer, then making the mechanical properties particularly the tensile strength and toughness resulted lower compared to the two step austempered. [6]

The specimen of the result of single step heat treatment showed a matrix formed by the coarse laths from acicular ferrite, the characteristics of upper part of bainite, and pieces of austenite with a number of remainders of ferrite around graphite elements. The characters of matrix and distribution of nodular graphite in general had a significant influence on fatigue. A previous research has shown that the fatigue strength of the result of heat treatment increased with the increase of the number of nodules, distribution, uniformity and the shape of nodule [7]. In addition, the fatigue strength was influenced by the nodular orientation and the distance between the relative nodular graphite towards the load direction. The pores and microshrinkages were also observed in the tested material in which they had a significant influence at the beginning of fatigue crack growth as shown in [8].

In single step, the crack path started growing due to the presence of stress concentration on the notch tip. The next fatigue crack growth occurred in a vertical direction towards the direction of cyclic loading. The fatigue crack growth was little bit different in the single step heat treatment. In the initial part, the crack tended to be in straight direction and in the following part; it slightly experienced a change in direction. This phenomenon indicates that the fatigue crack growth experienced by single step tended to have a delay. In this case, the very different performance of the fatigue crack growth is highly determined by nodular distribution, distance between nodules, the size of nodules and the phase of ferrite and pearlite matrix [9].

The fatigue crack path changed the behaviour controlled by the presence of nodular graphite. The crack growth always occurred in the interface of matrix graphite, while the nodular graphite was still not broken. This is the characteristic for both two single step treatments. The crack of nodular graphite spread mainly through pearlite, which, due to its refined flat composite, provided an advantageous condition for the crack growth.

As reported previously, the matrix of the result of heat treatment consisted of ferrite and pearlite. The crack always started in the graphite matrix and spread along the energy course commonly as an interface between laths ferrite and pearlite. According to [10] the interface matrix graphite is relatively irregular with a sharp angle that, in some cases, is the microcracks coming from the nodule. The main fore space enhances the value of the stress intensity factor in the spot of crack initiation. This then stimulates the growth of microcracks from the nodules towards the main space to the opposite direction of the growth with the general crack growth until it joins with the main crack. In this case, the common crack path continuously occurs in some other spots in all nodules in which the advantageous condition lies in the beginning of a new space that makes a process of growth, and repeats the process when facing other nodules. This mechanism then justifies the existence of crack branching as observed from the concurrence between the main crack and the nodular.

The secondary initiation of microcracks was observed near the main crack. When the small cracks simultaneously spread except the main cracks, the existing elastic energy for the growth from the main cracks certainly would be reduced, particularly in view of the creation of the larger crack surface; thus reducing the general level of crack growth. In some cases, it locally can cause a sudden crack. In addition, the orientation of the ferrite laths in matrix can influence the crack path. The space frequently spreads along the ferrite-austenite interface located more or less normal towards the applied load direction in which the small cracks cut through the bainite ferrite laths located in parallel towards the applied loading direction. On the other side, as some bainite ferrite laths commonly have the equal orientation, laths cluster is shaped then. In general, the orientation of each single group, however, is various.

Theoretically, the transformation of the temperature above 350°C can result in the coarser ausferrite matrix called upper ausferrite showing an acicular structure. Meanwhile, in the transformation of the lower temperature under 350°C, the more refined ausferrite matrix could be obtained called bottom ausferrite presenting a structure that highly resembled to the martensitic temper. For the austempered time of 60 minutes in the two-step austempered the highest toughness occurred at the temperature of 340°C. For the austempered time of 120 minutes, the highest toughness occurred at the temperature of 340°C and the lowest one was at the temperature of 280°C. This was in view of the formation of bottom ausferrite with few of the remaining austenite with higher percentage of acicular ferrite phase compared to the remaining austenite. At the temperature of 280°C the matrix structure formed was upper ausferrite (coarse ausferrite) resembling to the refined pearlite [9].

4 Summary

Based on the research results, it is observed that the single step heat treatment at the austempered temperature at 340°C with the holding time of 120 minutes showed a slower fatigue crack growth rate in comparison to the single step heat treatment at the temperature of 310, and 280°C with the holding time of 60 minutes. The slowness of the fatigue crack growth rate particularly is indicated by the higher percentage of pearlite matrix phase in comparison to the ferrite matrix, nodular distance, nodular size and the relatively even nodular distribution.

I would like to express my gratitude to Engineering Faculty, Universitas Negeri Malang which fully supports the opportunity to join AGS 2016 Conference

References

1. W. Siefer, K. Orths, T. Am. Fish. Soc., **78**, 382 (1970)
2. Andoko, Rudy Soenoko, P. Anindito, Yudy S. Irawan, International Journal of Applied Engineering Research, **9(5)**, 541 (2014)
3. AFS 115th Metalcasting Congress, (Schaumburg, IL USA, 2011)
4. O. Eric, D. Rajnovic, S. Zec, L. Sidjani, T. Milan, Mater. Charact., **57**, 211 (2006)

5. J. Zimba, D.J. Simbi, E. Navara, Cem. Concr. Compos., **25**, 643 (2003)
6. Andoko, P. Poppy, ARPN Journal of Engineering and Applied Sciences, **11(2)**, 863 (2016)
7. R.A. Gonzaga, P.M. Landa, A. Peres, P. Villanueva, Journal of Achievements in Materials and Manufacturing Engineering, **33(2)**, 150 (2009)
8. N. Costa, N. Machado, F.S. Silva, Ciencia & Tecnologia dos Materiais, **20**, 120 (2008)
9. Andoko, Rudy Sonoko, P. Anindito, Yudi S. Irawan, Aust. J. Basic Appl. Sci., **8(3)**, 277 (2014)
10. G.L. Greno, J.L. Otegui, R.E. Boeri, Int. J. Fatigue, **21**, 35 (1999)

DOI: 10.1051/matecconf/20179701029

Properties of $Mn_{0.4}Zn_{0.6}Fe_2O_4$ and $Mn_{0.6}Zn_{0.4}Fe_2O_4$ as Nanocatalyst for Ammonia Production

Poppy Puspitasari[1,3,*], *Andoko* Andoko[1], *Heru* Suryanto[1], *Puput* Risdanareni[2,3], *Januarti Jaya* Ekaputri[3,4]

[1]Mechanical Engineering Department, Engineering Faculty, Universitas Negeri Malang, Indonesia
[2]Civil Engineering Department, Engineering Faculty, Universitas Negeri Malang, Indonesia
[3]Konsorsium Riset Geopolimer Indonesia (KORIGI), Lab Beton dan Bahan Bangunan ITS, Surabaya, Indonesia
[4]Civil Engineering Department, Faculty of Civil Engineering and Planning, Institut Teknologi Sepuluh Nopember (ITS), Surabaya, Indonesia

Abstract. Ammonia synthesis requires high pressure and high temperature process. Unfortunately, the capital intensive cost resulting low yield of ammonia by using recent catalyst which is iron oxide. Therefore, manganese zinc ferrite as a soft ferrite material will be introduced as a new nanocatalyst to enhance the ammonia yield. As a new nanocatalyst for ammonia production, study of comparasion two different concentration of MnZn Ferrite is very important. This paper will compare the yield of ammonia by using two different nanocatalyst which are $Mn_{0.4}Zn_{0.6}Fe_2O_4$ and $Mn_{0.6}Zn_{0.4}Fe_2O_4$. Both were synthesized by sol-gel method and has been characterize by using FESEM (morphology), XRD (phase identification), EDX (elemental analysis) and TPR (oxide reduction). The ammonia was produce with and without magnetic field applied. The result shows that the ammonia yield is higher for $Mn_{0.4}Zn_{0.6}Fe_2O_4$ nanocatalyst than $Mn_{0.6}Zn_{0.4}Fe_2O_4$ by using magnetic field applied. 67.2% of yield has been achieved by using new nanocatalyst $Mn_{0.6}Zn_{0.4}Fe_2O_4$ and magnetic field applied at ambient environment.

1 Introduction

Ammonia is a chemical substance which has been used as an additive in various applications such as explosives, detergent and about 76% from the total production of ammonia has been applied in fertilizer industry. Unfortunately, the current production is only capable to generate 10-20% of ammonia yields with capital and energy intensive. In order to overcome these drawbacks, nanotechnology is seen as an excellent solution. By introducing nanocatalyst which is manganese zinc ferrite ($Mn_{0.4}Zn_0.6Fe_2O_4$ and $Mn_{0.6}Zn_{0.4}Fe_2O_4$) with the new concept of applied magnetic field, the catalytic activity can be induced and the yield can be enhanced.

* Corresponding author: poppy@um.ac.id

MnZn-ferrites are considered as one of the great important soft ferrite ceramic materials. This type of ceramics find extensive applications in electronic and telecommunication industry and catalyst. Spinel ferrites combine interesting soft magnetic properties with rather high electrical resistivities [1]. Because of the chemical composition and crystal structure of Mn–Zn ferrite, this material has a high initial permeability, saturation magnetization, and a relatively low eddy current losses compared with alloy cores.

Various processing techniques including conventional and nonconventional have been developed for the synthesis of ferrites. However, each one of these techniques has its specific limitations. Non-conventional techniques such as co-precipitation, thermal decomposition, sol-gel, and hydrothermal, self-propagating high temperature synthesis (SHS) and other wet chemical techniques were widely used [2-5]. The main disadvantage concerns economic and environmental. Therefore, the manufacturing of near fully dense manganese zinc ferrites through minimum number of steps with high magnetic properties will be the focus of this research.

The properties of manganese zinc ferrites depend on their microstructure. The grain size and porosity of the sample will affect the strength performance of the material. One of the main problems during conventional sintering of manganese zinc ferrites is the elimination of porosity. It is well known that the domains structure of magnetic materials is affected by crystal structure, saturation magnetization, magnetocrystalline anisotropy and magnetostriction as well as the size and shape of grains, porosity and crystal defects, which are determined by the processing route [6]. This works deals with the synthesis of manganese zinc ferrite ($Mn_{0.4}Zn_{0.6}Fe_2O_4$ and $Mn_{0.6}Zn_{0.4}Fe_2O_4$) and the ammonia synthesis by using magnetic induction (1Tesla). Both nanocatalyst will compare which one is the best nanocatalyst for ammonia synthesis due to their mole fraction difference.

2 Methodology

2.1 Nanocatalyst Preparation

The precursor for $Mn_{0.4}Zn_{0.6}Fe_2O_4$ and $Mn_{0.6}Zn_{0.4}Fe_2O_4$ were prepared by mixing $Mn(NO_3)_2.4H_2O$, $Zn(NO_3)_2.6H_2O$, and $Fe(NO_3)_3.9H_2O$ with of 65% HNO_3. As-prepared samples were stirred for 1 week and gradually heating until the gel formed at 70°C. The sample was then dried in an oven at 110°C for 24 hours and annealed at 700°C, 800°C, 900°C, 1000°C, and 1100°C in inert environment for 4 hours.

2.2 Nanocatalyst Characterization

The as-synthesized nanocatalysts were characterized by using X-Ray Diffraction (XRD) with CuKα and λ=1.5418Å, Field Emission Scanning Electron Microscopy (FESEM) SUPRA 35VP, Energy Dispersive X-ray Spectroscopy (EDX), and Temperature Program Reduction (TPR). XRD was used for the identification of phases, crystalline and crystallite size determination. The FESEM gives the microstructure evaluation on the sintered powder. While TPR gives the information about reduction temperature and hydrogen consumption of the samples. The samples of $Mn_{0.4}Zn_{0.6}Fe_2O_4$ labeled as sample 1 and the samples of $Mn_{0.6}Zn_{0.4}Fe_2O_4$ labeled as sample 2.

3 Result and Discussion

3.1 Phase Identification

Phase identification and lattice parameter were observed using XRD result. The single phase $Mn_{0.4}Zn_{0.6}Fe_2O_4$ and $Mn_{0.6}Zn_{0.4}Fe_2O_4$ obtained at 1100°C sintering temperature (Fig 1 and Fig 2). Intensity of major peak revealed the FWHM, d-spacing and 2 theta degree which also explained the crystallite structure in term of lattice parameter as shown at Table 1 and Table 2. The same value of a, b, c were described that lattice parameter of $Mn_{0.4}Zn_{0.6}Fe_2O_4$ and $Mn_{0.6}Zn_{0.4}Fe_2O_4$ are cubic.

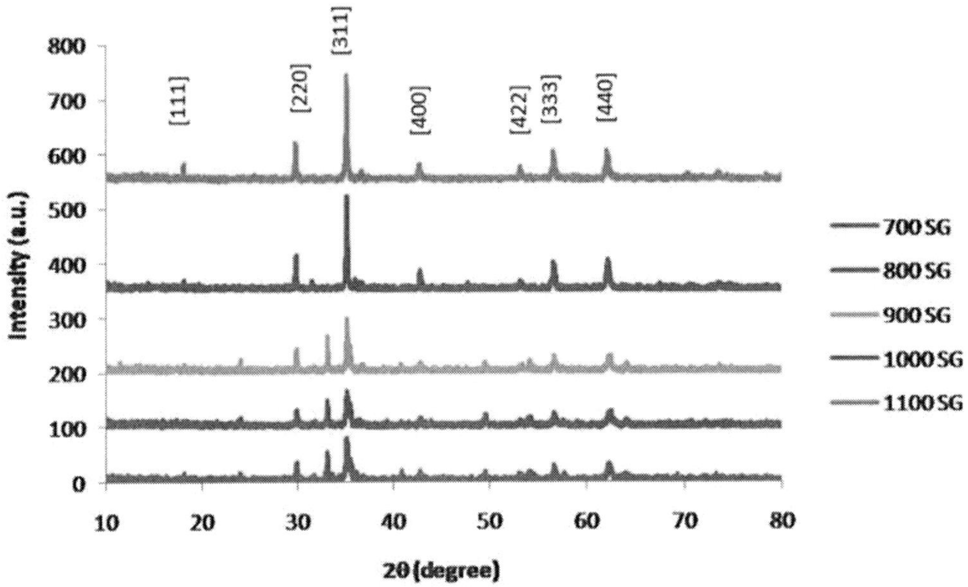

Fig. 1. Phase identification for $Mn_{0.4}Zn_{0.6}Fe_2O_4$ synthesized by sol gel method annealed at 700°C, 800°C, 900°C, 1000°C, and 1100°C.

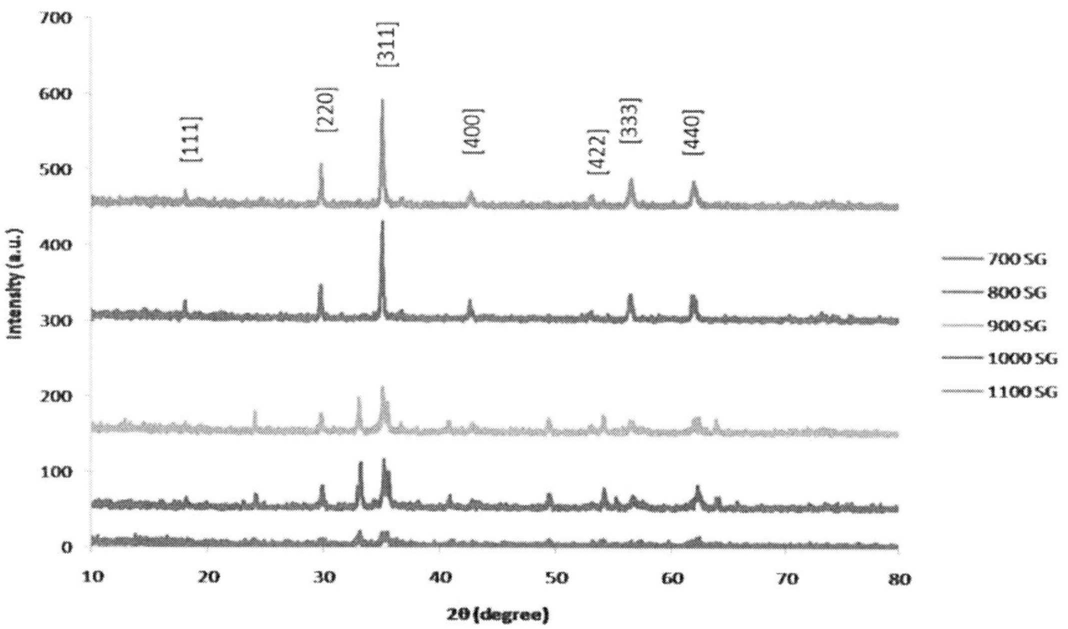

Fig. 2. Phase identification for $Mn_{0.6}Zn_{0.4}Fe_2O_4$ synthesized by sol gel method annealed at 700°C, 800°C, 900°C, 1000°C, and 1100°C.

Figure 1 shows the XRD pattern for $Mn_{0.4}Zn_{0.6}Fe_2O_4$ at five different temperatures. The highest crystallinity is $Mn_{0.4}Zn_{0.6}Fe_2O_4$ annealed at 1100°C. It shows the single phase nanoparticles at [311] plane according to the standard card SSNNN 89-6609 (C). The samples of 1-700, 1-800, and 1-900 are not single phase since there are another peak appearing as hematite peak. This is because the annealing temperature is not high enough to arrange the atoms to occupied at their own lattice. Whereas, at 1000°C the as-synthesized sample shows as single phase with [311] plane. By using the Scherer equation, the crystallite size can be obtain as mention at table I. The crystallite size is getting bigger since the annealing temperature increased. All sample for $Mn_{0.4}Zn_{0.6}Fe_2O_4$ shows the hexagonal structure according to the value of a=b=c. The smallest particle size is for sample 1-700 which is 70.8nm. Figure 2 shows the XRD pattern for $Mn_{0.6}Zn_{0.4}Fe_2O_4$ at five different annealing temperature. It shows at 1000°C and 1100°C, the as-synthesized samples have formed a single phase at [311] plane. At 700°C, the samples has not formed into the single phase nanoparticle. It happens because of impurities of the samples and the annealing temperature is not reach the suitable temperature for the ferrite to formed the single phase nanoparticles. The crystallite size can be obtain by using Scherer equation and the values are as mention at table 1 and table 2. Higher annealing temperature resulting bigger crystallite size.

DOI: 10.1051/matecconf/20179701029

Table 1. Value of Intensity, d-spacing, FWHM, crystallite size and lattice parameter of $Mn_{0.4}Zn_{0.6}Fe_2O_4$ synthesized by sol gel method.

Sample	Intensity	d-spacing	FWHM	Crystalite size (nm)	2theta	a	b	c
1-700	78.9	2.54	0.233	70.8	35.33	8.46	8.46	8.46
1-800	67.5	2.54	0.184	89.9	35.34	8.46	8.46	8.46
1-900	96.9	2.54	0.163	102.7	35.29	8.46	8.46	8.46
1-1000	157.0	2.54	0.131	130.8	35.30	8.46	8.46	8.46
1-1100	189.0	2.54	0.113	145.9	35.23	8.46	8.46	8.46

Table 2. Value of Intensity, d-spacing, FWHM, crystallite size and lattice parameter of $Mn_{0.6}Zn_{0.4}Fe_2O_4$ synthesized by sol gel method.

Sample	Intensity	d-spacing	FWHM	crystallite size (nm)	2theta	a	b	c
2-700	19.90	2.70	0.48	37.5	35.27	8.37	8.37	8.37
2-800	66.10	2.54	0.34	48.6	35.26	8.37	8.37	8.37
2-900	62.10	2.55	0.24	68.4	35.16	8.38	8.38	8.38
2-1000	122.00	2.55	0.18	89.3	35.13	8.50	8.50	8.50
2-1100	142.00	2.55	0.16	102.7	35.10	8.50	8.50	8.50

3.2 Morphology Analysis

Figure 3 shows the morphology of samples 1 with different annealing temperatures. At 700°C, the average particle size is 43nm and shows the agglomeration particle. At 800°C the average particle is 52nm. At 900°C the average particle size is 63nm. At 1000°C and 1100°C the samples were not in the range of nanoparticle size since the particle size is more than 100nm. Figure 4 shows the morphology of samples 2 at five different annealing temperature. The particle size for 2-700 is 34nm while 2-800 has average particle size 43nm. The other sample, 2-900 shows the average particle size 150nm, 2-1000 has bigger crystallite size which is 250nm and 2-1100 has average crystallite size 350nm. All nanoparticles show the agglomeration and increasing annealing temperature deal with the increasing of the crystallite size.

 DOI: 10.1051/matecconf/20179701029

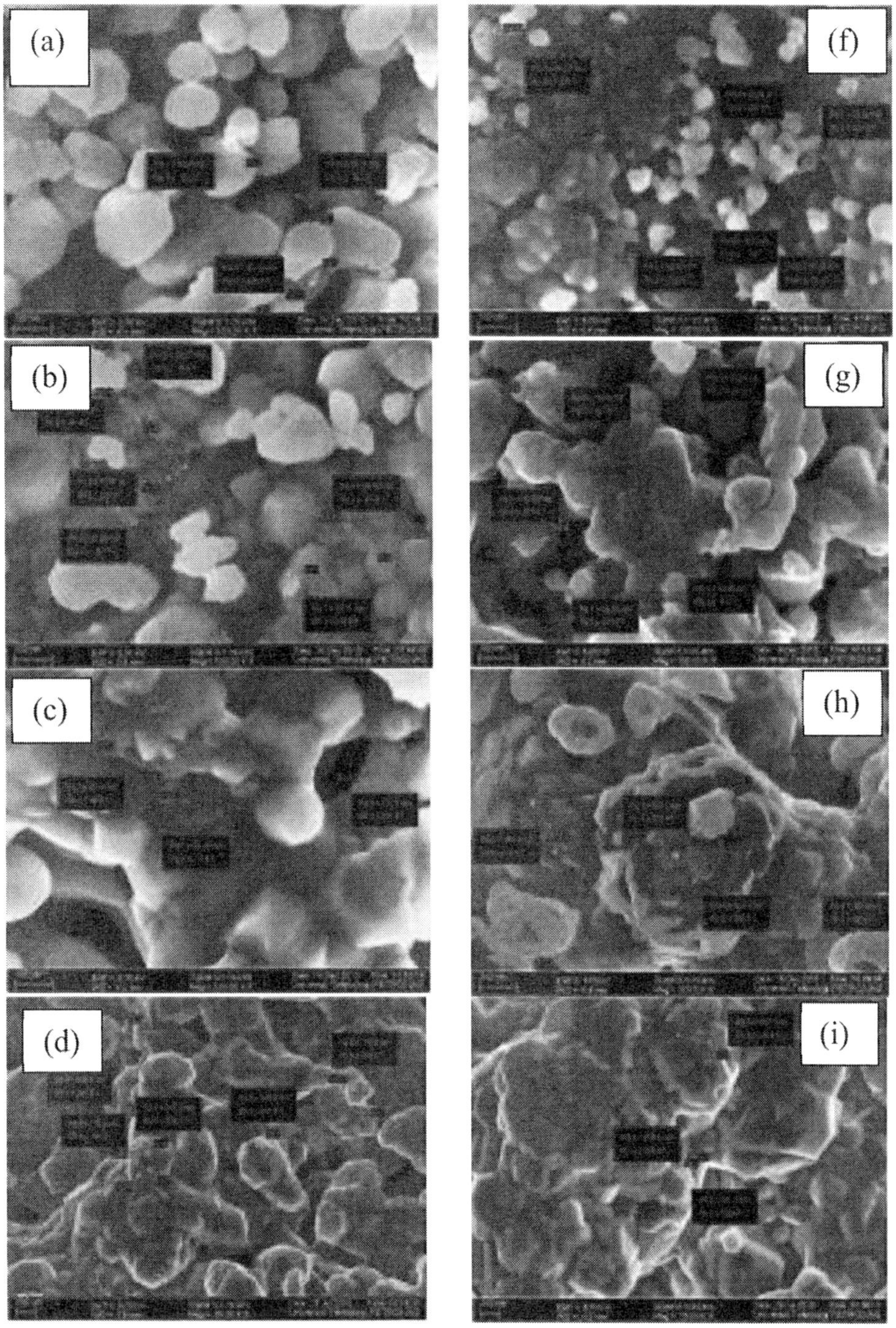

DOI: 10.1051/matecconf/20179701029

Fig. 3. Morphology of $Mn_{0.4}Zn_{0.6}Fe_2O_4$ synthesized by sol gel method and annealed at (a) 700°C, (b) 800°C, (c) 900°C, (d) 1000°C, and (e) 1100°C

Fig. 4. Morphology of $Mn_{0.6}Zn_{0.4}Fe_2O_4$ synthesized by sol gel method and annealed at (f) 700°C, (g) 800°C, (h) 900°C, (i) 1000°C, and (j) 1100°C

3.3 Oxide reduction

Temperature Programmed Reduction (TPR) was identify reduction of the oxide catalyst to the metallic state. The highest flow rate has been used in this process at 20 ccm/min.

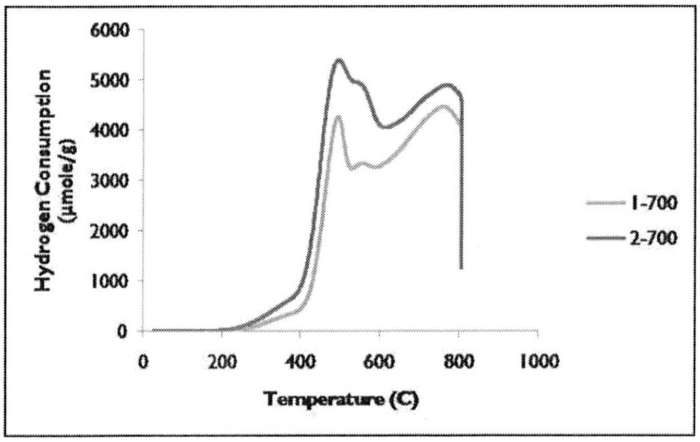

Fig. 5. Reduction profile of samples 1-700 and 2-700

Figure 5 shows the reduction profile for sample 1 and sample 2. It shows that sample 1 needs higher hydrogen consume to reduce the oxide to metallic state. Due to the XRD and FESEM result, sample 2 has smaller particle size (37.5nm) compared to sample 1 (70.5nm). Therefore, the percentage hydrogen consumption to reduce is higher for sample 2. Table 3 provides the information for percentage hydrogen and phase behavior at reduction process for samples 1 and 2.

DOI: 10.1051/matecconf/20179701029

Table 3. H_2-TPR process for samples 1 and 2 annealed at 700°C

Sample	$T_{max(1)}$	Hydrogen Consumption (µmole/g)	Reduction	$T_{max(2)}$	Hydrogen Consumption (%)	Reduction
1-700	478°C	3789.98	Fe_2O_3 → Fe_3O_4	734°C	4364.93	Fe_3O_4 → Fe
2-700	476°C	5030.52		767°C	4898.99	

The reduction of Fe_2O_3 to Fe_3O_4 occurs at 478°C for sample 1 and 476°C for sample 2. Other researcher found the reduction of Fe_2O_3 occurs at 348°C and complete metallic state obtain at 800°C [7]. The reduction from Fe_3O_4 to FeO happens at 670°C [7]. At this figure, sample 1 shows the middle peak at 559°C that can be speculated as the reduction temperature for Fe_3O_4 to FeO. Sample 2 shows at temperature 555°C the reduction process for Fe_3O_4 to FeO has been occurred. There has been mentioned that amount of zinc will decrease the hydrogen consumption for reduction process [8]. This can be seen from Figure 5 that zinc contain for sample 1 is higher than sample 2. Thus, the hydrogen consumption for reduction process at sample 1 is lower than sample 2. At 800°C all the oxide has been reduce completely to the metallic state.

3.4 Ammonia formation

The formation of ammonia was detected using Kjeldhal Method. The experiment was taken at ambient pressure under five different temperature which were 28°C , 68°C, 108°C , 148°C , and 188°C to measure the highest yield.

Figure 6 shows the value of ammonia yield using nanocatalyst $Mn_{0.4}Zn_{0.6}Fe_2O_4$. The highest yield 67.2% was obtain at room temperature (28°C). The yield was decreasing as the temperature increasing. This is due to the effect of currie temperature of nanocatalyst below 100°C [9].

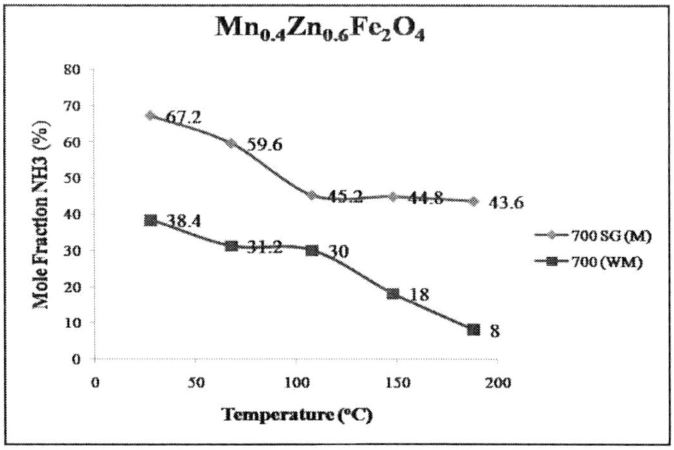

Fig. 6. Ammonia result by using nanocatalyst $Mn_{0.4}Zn_{0.6}Fe_2O_4$

DOI: 10.1051/matecconf/20179701029

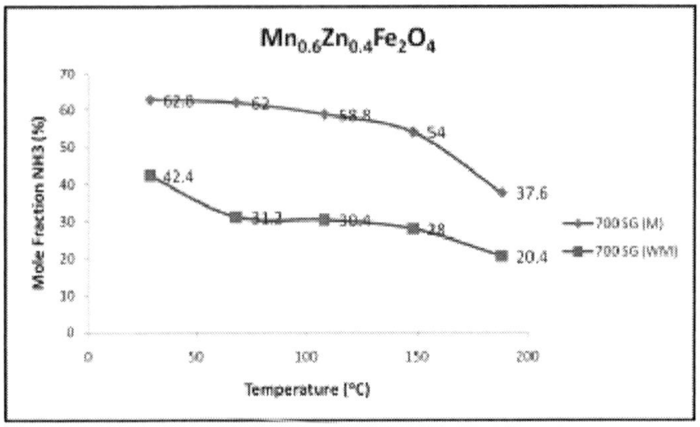

Fig. 7. Ammonia result by using nanocatalyst $Mn_{0.6}Zn_{0.4}Fe_2O_4$

Figure 7 shows the value of ammonia yield by using nanocatalyst $Mn_{0.6}Zn_{0.4}Fe_2O_4$. The highest yield 62.8% was obtain at room temperature (28°C). The yield was decreasing as the temperature increasing.

Both nanocatalyst was obtain the highest yield at room temperature and ambient pressure by applied magnetic field. It has been proven by other researcher that by using applied magnetic field, the yield will increase tremendously compared to the reaction with the absence of magnetic field [10]. This work has a similar condition with other researcher who has been obtain the higher yield of ammonia in the presence of magnetic field by using nanocatalyst $Mn_{0.8}Zn_{0.2}Fe_2O_4$ [11]. It was speculated that $Mn_{0.4}Zn_{0.6}Fe_2O_4$ posses better spin waves which enhanced the catalytic reaction. The interaction between homogeneous magnetic field in the vicinity of the active centers and magnetic moments of the gaseous particles. In the case H_2 and N_2, had elevated the production of ammonia in the ambient pressure and room temperature. It is because the electrons in a good alignment which increase the chemical reaction of the process [12]. The best nanocatalyst is $Mn_{0.4}Zn_{0.6}Fe_2O_4$ compared to nanocatalyst $Mn_{0.6}Zn_{0.4}Fe_2O_4$.

4 Summary

The new nanocatalysts from $Mn_{0.4}Zn_{0.2}Fe_2O_4$ and $Mn_{0.6}Zn_{0.4}Fe_2O_4$ have been used for ammonia synthesis. The $Mn_{0.4}Zn_{0.2}Fe_2O_4$ nanocatalyst shows higher ammonia yield (67.2%) as compared to the $Mn_{0.6}Zn_{0.4}Fe_2O_4$ nanocatalyst (62.6%). The excellent ammonia yield has been achieved at ambient pressure and room temperature by applying 1 Tesla applied magnetic field. The TPR result shows that the higher amount of zinc will decrease the temperature reduction of nanocatalyst. Activation energy shows the exothermic reaction for the $Mn_{0.4}Zn_{0.2}Fe_2O_4$ nanocatalyst while by using $Mn_{0.6}Zn_{0.4}Fe_2O_4$ nanocatalyst the reaction endothermic. From the XRD, FESEM, EDX, and ammonia result, the best nanocatalyst is $Mn_{0.4}Zn_{0.6}Fe_2O_4$ compared to $Mn_{0.6}Zn_{0.4}Fe_2O_4$. Room temperature and ambient pressure are the most conducive environment to synthesis ammonia because it is easy to control, safe and cost effective.

I would like to express my gratitude to Ministry of Research, Technology and Higher Education (RISTEK DIKTI) for the financial support under Penelitian Unggulan Perguruan Tinggi (PUPT) batch 2016.

References

1. Emad M. M. Ewais, Mahmoud M. Hessien, Abdel-Hady A. El-Geassy, J. Aust. Ceram. Soc., **44**, 57 (2008)
2. L. Nalbandian, A. Delimitis, V.T. Zaspalis, E.A. Deliyanni, D.N. Bakoyannakis, and E.N. Peleka, Micropor. Mesopor. Mat., **114**, 465 (2008)
3. P. Puspitasari, N. Yahya, N.A. Zabidi, N.A. Ahmad, International Journal of Applied Sciences, **11(7)**, 1199 (2010)
4. M.M Rashad, Mater. Sci. Eng. **127(2)**, 123 (2006)
5. Lai Zhenyu, Xu Guangliang, Liu Min, A. Ahniyaz, M. Yoshimura, J. Wuhan Univ. Technol. Mater. Sci. Ed., **23(2)**, 151 (2008)
6. S. Kazemi, G. Fantozzi, Mater. Sci. Eng. A, **442**(1), 496 (2006)
7. A. Khan, P.G. Smirniotis, J. Mol. Catal. A: Chem., **280(1)**, 43 (2008)
8. Meisheng Liang, Wenkai Kang, Kechang Xie, J. Nat. Gas Chem., **18**, 110 (2009)
9. S.R. Janasi, D. Rodrigues, F.J.G. Landgraf, Mater. Sci. Forum, **498**, 119 (2005)
10. N. Yahya, P. Puspitasari, K. Koziol, P. Guiseppe, *Ammonia Synthesis in Carbon and Oxide Nanostructure* (Springer, Berlin Heidelberg, 2010)
11. N. Yahya, P. Puspitasari, K. Koziol, N.A.M. Zabidi, M.F. Uthman, International Journal of Basic and Applied Sciences, **10 (1)**, 95 (2010)
12. R.Y. Hong, Y.J. Wu, B. Feng, Q.D. Di, H.Z. Li, B. Xi, Y.Z. Heng, D.G. Wei, J. Magn. Magn. Mat, **321(8)**,1106 (2009)
13. S. Arrhenius. Z., Phys.Chem, **4**, 226 (1889)
14. A.F Dethlefsen, E. Mariani, W. Wegscheider, R.J Haug, Physica E., **34**, 108 (2008)

MATEC Web of Conferences **97**, 01030 (2017) DOI: 10.1051/matecconf/20179701030
ETIC 2016

Flexural Test of Fly Ash based Geopolimer Concrete Beams

Nindyawati[1,2,*], *Sri Umniati* B.[1,2] , and *Puput* Risdanareni[1,2]

[1]Department of Civil Engineering, Universitas Negeri Malang, Malang, Indonesia
[2]Konsorsium Riset Geopolimer Indonesia (KORIGI), Lab Beton dan Bahan Bangunan ITS, Surabaya, Indonesia

Abstract. Fly ash is a by-product from the coal industry, which is widely available in Indonesia. Fly ash contains quite high silicate and alumina. Silica and alumina reacts with alkaline solution to produce alumina silicate gel which binds the aggregate to produce geopolymer concrete. Geopolymer concrete is introduced as an environmental concrete with high compressive strength. The use of geopolymer concrete beams is a solution to reduce the effects of greenhouse gases. This research uses experimental designs. The data are obtained from the testing of 4 pieces of reinforced geopolymer concrete beams and reinforced ordinary concrete beams with a / d of 1.11 and 2.24. The results are obtained from the maximum load that can be accepted by the beam. The results of this study are: (1) Geopolymer concrete cylinder has 26.78% higher compressive strength than ordinary concrete cylinders (2) Ordinary concrete beams can withstand 34.8% load higher compared to the geopolymer concrete beam (3) Reinforced ordinary concrete beams experience bending shear collapse while reinforced geopolymer concrete beam experience pure bending collapse.

1 Introduction

Concrete is the most popular construction material. The benefits are reduction in the use of cement to reduce air pollution and improve the propertiess of concrete. In the cement factory, every ton of Portland cement produces about one ton of CO2 in the atmosphere [1]. The replacement alternatives for cement are fly ash, silica fume, rise husk ash, metakaolin, or a combination of the above materials. Silicate and alumina are rich in fly ash and hence reacts with alkaline solution to produce alumina silicate gel that will be formed as geopolymer concretes.

Indonesia is a coal producing countryy. The coal is used in electricity generatingg fuel. One of the material produced by the burning power plants are fly ash. The abundant sources of fly ash can be utilized for the geopolymer concrete production.

Based on previous study by Rangan, the geopolymer concrete beams with widths of 200 mm, a height of 300 mm and a length of 3300 mm beam using reinforcement 12 mm by 4 units with 150 mm spacing between reinforcement, reinforcement ratio of 0.64% - 2.69% were made. The cylinders generated compressive strength of 550 MPa - 560 MPa while

* Corresponding author: nindyawati.ft@um.ac.id

beams generated ultimate momen of 168.7 KNm [2]. In addition, crack pattern of beams was highly influenced by shear span (a) and effective high (d) of the beam.

Geopolymer Concrete thin-Webbed T-Beams with a size of 250mm x 300mm and has a/d value of 1.5 and a/ d of 2 experience bending collapse behavior [1]. The maximum moment of geopolymer concrete beams with a/d of 1.5 and a/d of 2 are 211.97 KNm and 233.29 KNm.

This study is comparing the flexural load on geopolymer concrete beams and ordinary concrete beams. Variations a/d beams of =1.11 and a/d =2.24 was used. Hence, compressive strength of ordinary and geopolymer concrete was also compared.

2 Materials

Cement. Pozollan Portland Cement conforming to Standard Nasional Indonesia is used in the present investigation of specific gravity 3.12.

Fly ash. Fly ash used in this study was obtained from Paiton Power Corporation, East Java Indonesia and the specific gravity of fly ash is 2.14.

Aggregates . The locally available river sand of zone III was used as fine aggregate in the present in Tabel 1. The properties of aggregate used specific gravity of fine and coarse aggregate - 2.66 and 2.70, fineness modulus of fine and coarse aggregate - 2.43 and 6.71

Concrete. In this study, geopolymer concrete mix design compressive strength plan of 44 MPa was used. In addition, ordinary concrete using DOE (Degree Of Environment) with a compressive strength plan of 35 MPa. Adding margin value of 8,5 Mpa at DOE mix design calculation was conducted in order to adjust ordinary and geopolymer concrete compressive strength plan [3]. Futhermore, compressive strength plan of ordinary concrete was 43.5 MPa. The mix proportion of geopolymer and ordinary concrete is presented in Table 1.

Table 1. Mix Proportion of Concrete

Description	Quantity						
	Cylinder		*AD11		*AD22		
	*B	*G	*B	*G	*B	*G	
Cement	2,21	-	0,77	-	1,35	-	kg
Fly Ash	-	0,65	-	0,36	-	0,62	kg
Fine aggregates	2,84	0,44	0,99	0,24	1,73	0,42	kg
Coarse aggregates	4,49	0,88	1,57	0,48	2,74	0,84	kg
Water	1	-	0,35	-	0,61	-	kg
NaOH	-	0,08	-	0,04	-	0,07	kg
Na_2SiO_3	-	0,15	-	0,08	-	0,15	kg
Superplasticziser	-	0,01	-	0,007	-	0,01	kg

Description :
* B: Ordinary Concrete
* G: Geopolymer Concrete
* AD11: Beams with a / d = 1.11
* AD22: Beams with a / d = 2.24

Alkaline Liquid. Alkaline activators made the day before use by mixing NaOH with aquades thus concentrating 10M. The solution was mixed with Na_2SiO_3 with ratio between Na_2SiO_3 / NaOH is 2.

 DOI: 10.1051/matecconf/20179701030

Mixing and Casting. All the materials were measured, ie fly ash, gravel and sand. Fly ash, gravel and sand mixed into the mixer, and mix for 3 minutes. Then alkaline activator and superplasticziser liquid naphthalene sulphonate were added and stirred for 4 minutes until a homogeneous mixture blended. After the concrete mix homogeneous then the mixture can be removed from the mixer and moulded in a tub. Slump test was performed in accordance with (ASTM C 143-78). Fresh geopolymer concrete was moulded and compacted in the formwork using vibrator. This is treatment was conducted in order to produce good fresh concrete mixture and to reduce open pore of concrete.

Specimen Details. Four numbers of reinforced concrete beams with span of 800 mm, wide of 50 mm and hight of 100 mm were produced. Beams were placed at simple suppot with an effective span of 700 mm. The clear cover of the beam was 8 mm. The spesimens was produced from two types of concrete that are geopolymer concrete and ordinary concrete. The beams are designed as under reinforced concrete. Specimen details are presented at Table 2 and Fig 1.

Table 2. Specimen Details

Beams Code	Explanation	a/d	b	h	L	Ø reinforcement	Ø stirrups
			mm	mm	mm	mm	mm
BB11	Normal Concrete	1.11	50	100	800	4,4	2,5–25
BG11	Geopolimer Concrete	1,11	50	100	800	4,4	2,5–25
BB22	Normal Concrete	2,24	50	100	800	4,4	2,5–35
BG22	Geopolimer Concrete	2,24	50	100	800	4,4	2,5– 35

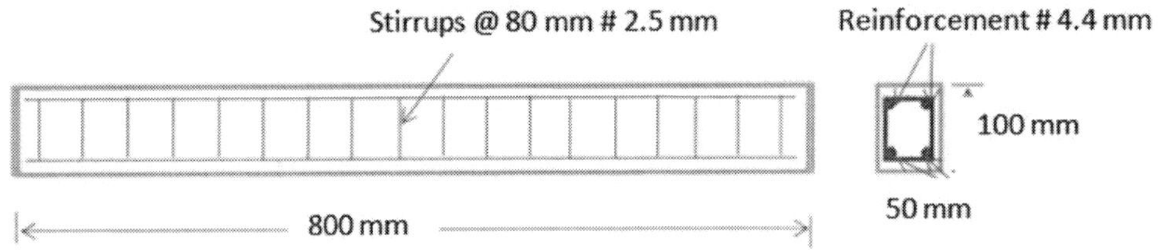

Fig. 1. Specimens Details

The beams were simply supported over a span of 700 mm. The load was applied on two point each: 100 mm (a/d = 1.1) and 200 mm (a/d = 2.2) away from the support. The beams were subjected to two-point loads under load control mode. The load was applied in increment 2 kN. Dial gauge was placed in the midspan beam bottom. The gauge was used for measuring deflections. Dial gauge of 0.001 mm readings are recorded at increment loads. The movement of the beam was observed carefully.

 DOI: 10.1051/matecconf/20179701030

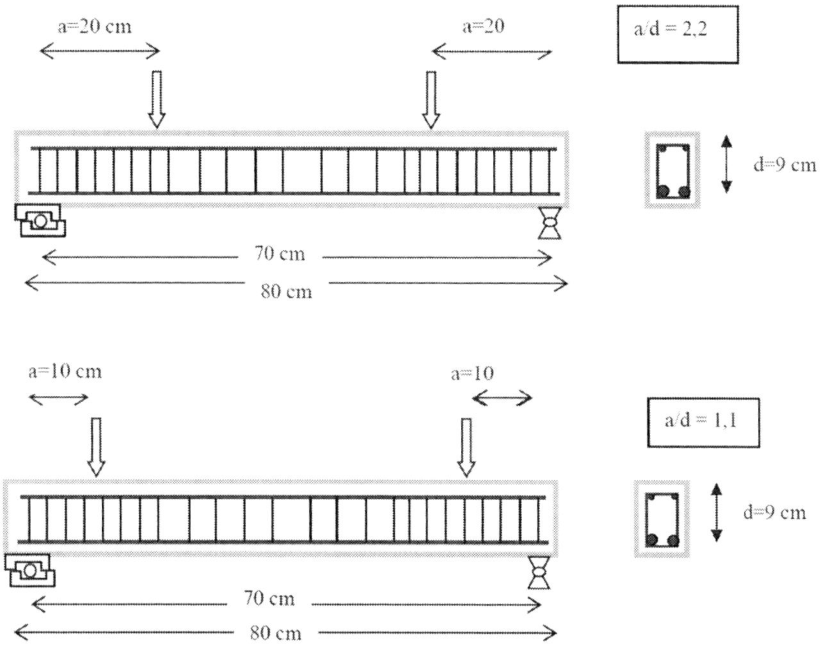

Fig. 2. Set Up Loading

3 Result and discussion

Compressive Strength of Concrete. Compressive strength of geopolymer and ordinary concrete was shown at figure 3. The results show that compressive strength of geopolymer concrete is higher than ordinary concrete. The average compressive strength of ordinary concrete cylinders is 24 Mpa. The average compressive strength of geopolymer concrete cylinders is 31 Mpa. Compressive strength of geopolymer concrete is 26.78% higher than ordinary concrete.

Geopolymer concrete and ordinary concrete compressive strength decreased from its plan. Geopolymer concrete compressive strength plan is 31 MPa while ordinary concrete compressive strength is 24.41 MPa. The first factors that predicted led the decreasing compressive strength of geopolymer concrete and ordinary concrete is the less flat of concrete surface specimens. The smoothing surface of concrete cylinder can improve the strength of concrete by 1 % - 21 % [4]. Porosity of concrete was also predicted to be another factor that led decreasing strength of concrete. Previous research stated that the higher porosity percentage of concrete, the lower compressive strength owned [9].

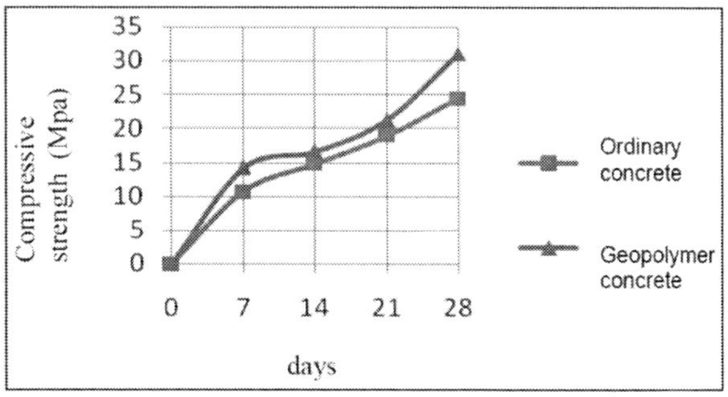

Fig. 3. Compressive strength of concrete in different age

DOI: 10.1051/matecconf/20179701030

Beams Deflection. Ultimate load and maximal deflection of beams was presented at table 3, while load versus deflection curve was illustrated at figure 4. BG11 and BG22 are expected to have higher bending force than BB11 and BB22 but this study revealed different result. It can be seen at table 3 that BB11 and BB22 which was produced from ordinary concrete has higher ultimate load capacity than BG 11 and BG 22 which was produce from geopolymer concrete. Beam type BB 22 has the highest ultimate load compare to other specimens with the value of 13 kN. In addition, beam BB22 has the smallest deflection with value of 0.16 mm.

Table 3. Ultimate Load and Maximum Deflection of Beams

Beams Code	Ultimate Load (kN)	Max. Deflection (mm)
BB11	11.6	0.094
BG11	8.6	0.06
BB22	13	0.182
BG22	5	0.16

At figure 4 it can be seen that ordinary concrete beam has higher flexural strength than geopolymer concrete beams. This result is little bit different from author expectation, geopolymer beams expected to higher flexural strength than ordinary concrete due to its high compressive strength. This unexpected result occurs because geopolymer concrete beams are more brittle than ordinary concrete beams.

The stiffness of ordinary and geopolymer concrete can be determined from its elasticity modulus value. Higher elasticity modulus value will deliver brittle characteristic of the concrete. This result has a good agreement with research conducted by Datatreya, that geopolymer concrete has higher modulus of elasticity (Ec) than ordinary concrete [6].

Differences load serviceability between BB11 and BG11 or BB22 and BG22 was also caused by the lower flexural strength of geopolymer concrete compared to ordinary concrete. Ordinary concrete beams have higher flexural strength than geopolymer concrete although geopolymer concrete has higher compressive strength [7]. Differences in the bending moment on geopolymer concrete beams and concrete beams was also caused by low capacity strength absorption of geopolymer concrete beams. The capacity strength absorption of geopolymer concrete beams affecting its moment result , the lower capacity strength absorption of the beam, the lower moment capacity owned [3].

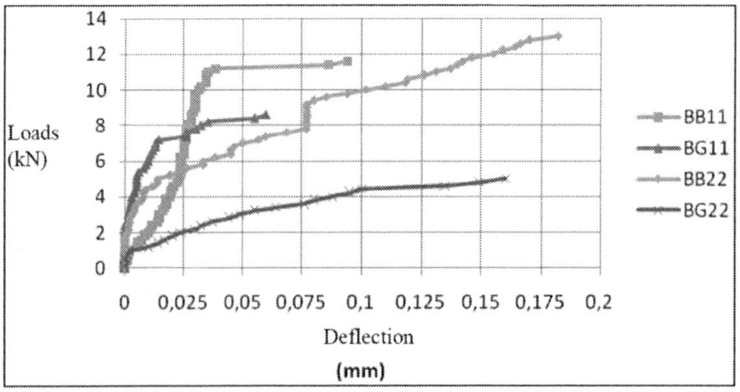

Fig. 4. Load and Deflection Curve

Crack Pattern. Crack patern of geopolymer and ordinary concrete beams was presented at figure 5 and 6. Pure flexural failure is characterized by a vertical crack which occurs only in the middle of the beam span. At figure 5 it can be seen that ordinary concrete beam with a/d variation of 1- 2.5 bending collapse mostly occurs than shear collapse. The bending collapse occurred because the bond between geopolymer binder with ageregat is more solid compared to cement ties with aggregate in ordinary concrete. This condition led geopolymer concrete to spalling easly than normal concrete. The solid bond between binder and agregate at geopolymer concrete led to the brittle characteristic of its concrete, thus the crack pattern of the beams only occurs at mid span . This result is apropriate with previous research by Sumanjauw which is stated that geopolymer concrete has high compressive strength than ordinary concrete [8].

Referring to the previous research, it can conclude that the BB11 and BB22 has bending shear collapse while BG11 and BG22 has pure bending collapse. This condition occurs because geopolymer has more solid bond with aggregate than ordinary concrete [6].

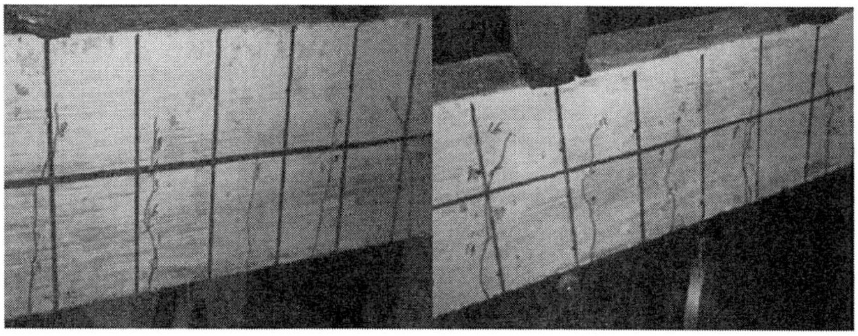

Fig.5. Crack pattern of Ordinary Concrete Beams (BB11 and BB22)

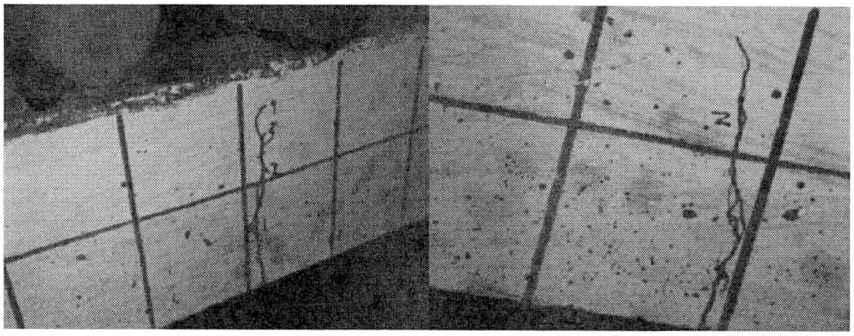

Fig.6. Crack pattern of Geopolymer Concrete Beams (BG11 and BG22)

4 Summary

From experimental investigation of study flexural test of fly ash based geopolymer concrete beams the following conclusion can be made :

1. Geopolymer cylinder concrete has 26.78% higher compressive strength than the ordinary concrete cylinder.
2. BB11 has 34.8% higher load capacity than BG11 while BB22 has 160% higher load than BG22. This condition occurs because the flexural strength of geopolymer concrete is lower than the ordinary concrete.

3. BB11 and BB22 experience bending shear collapse, while BG11 and BG 22 experience bending collapse because geopolymer concrete bond is more solid compared to ordinary concrete bond.

This work was financially supported by Kementrian Riset, Teknologi, dan Pendidikan Tinggi (Hibah Bersaing).

References

1. C. K. Madheswaran, P. S. Ambily, N. Lakshmanan, J. K. Dattatreya, S. A. Jaffer Sathik, ACI Mater. J., **11(1)**, 89 (2014)
2. V. B. Rangan, D. Sumanjouw, S. Wallah, D. Hardjito, 31st Conference on Our World In Concrete & Structure (2006)
3. J. Davidovits, Special Publication, **144**, 383 (1994)
4. K.D. Sushil, K.A. Sajal, Z.M. Abu, H.M. Mosaddek, D. Giash, *Effect Capping System on Compressive Strength of Concrete* (ICETCESD 2012)
5. Z.F. Farhana., H. Kamarudin, R. Azmi, A.M. Mustafa A.B, Key. Eng. Mater. **549-595** (2012)
6. J.K. Dattatreya, N.P. Rajamane, D. Sabitha, P.S. Ambily, M.C. Nataraja, International Journal of Civil And Structural Engineering, **2(1)**, 138 (2011)
7. E. Suryawangi, Y. Takarendehang, D. Hardjito, Jurnal Dimensi Pratama Teknik, **2(1)**, 1 (2013)
8. J.R. Yost, A. Radlinska, S. Ernst, M. Salera, N.J. Martignetti, Mater. Struct., **46(3)**, 449 (2013)
9. X. Chen, W. Huang, J. Zhou, Indian J. Eng. Mater. Sci., **19(6)**, 427 (2012)

Chemical and Physical Characterization of Fly Ash as Geopolymer Material

Puput Risdanareni[1,4,*] *Poppy* Puspitasari[2,4], *Ekaputri* Januarti Jaya[3,4]

[1]Department of Civil Engineering, Universitas Negeri Malang, Malang, Indonesia
[2]Departement of Mechanical Engineering, Universitas Negeri Malang, Malang, Indonesia
[3]Departement of Civil Engineering, Institut Teknologi Sepuluh Nopember, Surabaya, Indonesia
[4]Konsorsium Riset Geopolimer Indonesia (KORIGI), Lab Beton dan Bahan Bangunan ITS, Surabaya, Indonesia

Abstract. Research on finding suitable cement substitute material becomes massive due to environmental effect. Geopolymer as inorganic material is potential to be the smart solution to overcome global warming issue. Fly ash is a waste material rich in silica and alumina becomes popular raw material to produce geopolymer. The best properties ofgeopolymer paste come from the high quality of fly ash. Therefore, it is important to investigate various types of fly ash and geopolymer properties. Their chemical and physical properties characterized by XRF, pH value, XRD and SEM. The results showed that type of fly ash depended on amount of Si-based of Ca-based compound which consisted of spherical morphology. Geopolymer paste produced from the ash with different compound has bulky and irregular shape morphology. The pH value of each ash has also a correlation with the setting time of fresh paste.

1 Introduction

Recently, the utilization of cement in construction industry becomes a popular issue due to its environmental effect. Many studies revealed that in its production, portland cement exceeds high number of CO_2 to the atmosphere causing global warming [1, 2].Geopolymer, an inorganic material synthesized from high silica and alumina materials activated with alkali solution becomes one of the smart solutions to overcome global warming issue due to cement production [3]. Utilizing geopolymer material as construction material such as concrete proved can reduce global warming effect due to its cement use reduction.

Fly ash, waste material which is rich in silica and alumina derived from coal combustion process becomes popular raw material to produce geopolymer materials [4]. ASTM C618 classified fly ash into two categories based on its oxides content of Si, Al and Fe [5]. Fly ash is categorized as class F for its rich in silica content but fly ash class C is known to have high calcium content [5]. However, identifying fly ash character becomes more complicated due to its heterogeneous composition and its physical properties. The varying composition of fly ash is depending on type of coal, boiler type, operation condition, post combustion technology and power plant technology [6].

* Corresponding author: puput.risdanareni.ft@um.ac.id

DOI: 10.1051/matecconf/20179701031

Research investigation on fly ash characterization conducted by Nurwidayati revealed that mechanical properties of geopolymer paste is strongly influenced by physical and chemical characteristic of fly ash such as fineness, specific surface area and soluble content [7]. Other study by Tajunnisa concluded that similar mechanical properties of geopolymer paste were less influenced by ashes used from the same class of fly ash [8].

Several testing such as XRF, pH, XRD and SEM was conducted in order to identify physical and chemical character of fly ash. Using XRF test, chemical composition of fly ash that can be used to categorize them into class F or class C could be ascertain [9]. Research conducted by Antoni stated that degree of acid or pH from fly ash plays an important role in setting time of geopolymer paste. Higher pH level of fly ash induces faster setting time of geopolymer paste [10, 11]. XRD test was performed in order to discover crystalline mineral component of fly ash [12]. On the other hand, SEM is required to analyze the shape, surface and internal structure of fly ash particle [6, 12]. Previous research explained that chemical composition, morphology and crystalline mineral component of fly ash influenced mechanical and physical properties of its geopolymer paste such as compressive strength and setting time [13-15].

In general, it is summarized that chemical and physical characterization on fly ash are required to predict and control mechanical and physical properties of geopolymer paste. Therefore, the characterization data of fly ash is necessaryto make an accurate mix design of geopolymer paste and concrete [9].

2 Chemical characterization

2.1 XRF test

XRF test was conducted for chemical composition of fly ash [9]. The percentage of important compound such as silica, alumina, calcium and ferrite were determined with this testing. Furthermore, identifying percentage of calcium in fly ash, categorization fly ash as class F (low calcium) or C (high Calsium) referred to ASTM C618 were assessed. Research conducted by Antoni discovered that CaO content in fly ash plays important role in physical properties of geopolymer paste. High CaO content in fly ash induces geopolymer paste set rapidly [16]. XRF testing result of fly ashes from several power plants and cement industries is listed in Table 1 and Table 2.

Table 1. Chemical Composition of Fly Ash from Paiton, Rembang and TanjungJati, Indonesia [10]

Code	SiO	Al_2O_3	CaO	Fe_2O_3	K_2O	MgO	SO_3	LOI	$SiO_2+Al_2O_3+Fe_2O_3$
Y 11.2	30.78	17.87	15.47	15	1.32	6.45	1.32	0.49	63.65
J 10.6	38.24	15.28	9.15	24.28	0.78	5.19	0.61	3.9	77.8
P 10.3	51.03	25.13	5.49	9.66	1.58	3.25	0.51	1.44	85.82
Y 9.8	51.12	18.9	5.54	17.71	0.82	3.17	0.47	6.96	87.73
R 9.6	50.14	29.08	4.03	9.66	1.53	1.11	0.77	0.63	88.88

Table 2. Chemical Composition of Fly Ash from PT. Surya Beton and Semen Indonesia [8]

Code	SiO	Al_2O_3	CaO	Fe_2O_3	K_2O	MgO	Na_2O	P_2O_5	SO_3	SiO_2+Al_2O_3+ Fe_2O_3
SB1	37.45	24.49	15.73	11.79	1.13	0.32	0.99	2.52	0.80	73.73
SB2	38.76	29.43	4.05	18.43	0.88	0.01	0.88	2.19	0.67	86.62
SB3	39.16	31.03	10.26	11.60	0.46	0.72	0.68	2.27	0.70	81.79
SB4	39.11	30.92	10.04	11.93	0.47	0.62	0.69	2.30	0.69	81.96
SI1	41.17	34.48	1.89	11.54	1.05	1.74	0.96	2.04	0.65	87.19
SI2	34.65	22.55	15.24	12.52	1.31	6.82	1.58	3.23	1.17	69.72

Table 1 shows that fly ash type Y11.6 is categorized as class C due to its high calcium content (more than 10 %), while the rest of samples are categorized as class F because they contain silica oxide, alumina and ferrite more than 70 %.

Listed in Table 2, since its contain silica oxide, alumina and hematite less than 70%, fly ash SI 2 is categorized as class C, while the rest of the samples are categorized as class F. There is an interesting part of Table 2. Sample SB1, SB3, and SB4 have calcium content more than 10% as categorized as class C. However, they contain silica, alumina and hematite more than 70 % which meet criteria as class F.

The compressive strength test result of geopolymer paste produced from fly ash from Table 2 is presented at Figure 1.

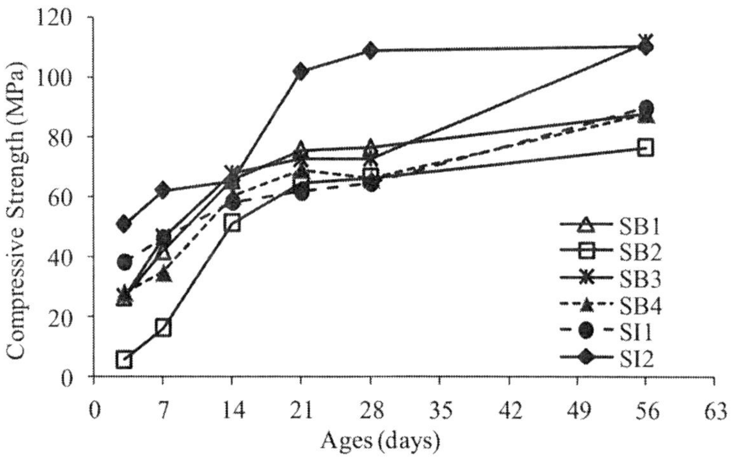

Fig. 1. Compressive Strenghtof Geopolymer Paste at Different Ages [8]

It shows that sample SI2 has high compressive strength although it contains less CaO as compared to SB1. Previous research stated that high CaO content in fly ash resulted high compressive strength on geopolymer paste [10]. This might occur due to ash's particle size and reactivity of silica content as stated by Fernandez that the key factors of fly ash potential reactivity are: the reactive silicacontent, the vitreous phase content and the particle size distribution [17].

 DOI: 10.1051/matecconf/20179701031

2.2 pH test.

Research conducted by Antoni revealed that pH value of fly ash influence the setting time of geopolymer paste [11]. The pH value was tested by dissolving fly ash into distilled water followed with pH measurement. Fly ash with pH value ranged from 8 to 11 tends to experience rapid setting [14]. In addition, pH value of fly ash has closely correlation with its CaO content. High CaO content in fly ash resulted in higher pH value [10].Correlation between pH value and Ca content of fly ash to initial setting time and compressive strength of geopolymer paste is presented at Table 3.

Table 3. The Effect pH [10, 11]

Sample Code	pH Level	Ca Content (% mass)	Initial setting time (minutes)	Compressive Strength (MPa)
Y 11.2	11.2	15.47	18	98
J 10.6	10.6	9.15	62	58
P 10.3	10.3	5.49	110	40
Y 9.8	9.8	5.54	120	47
R 9.6	9.6	4.03	42	58

High CaO content causes not only polymerization reaction but also hydration reaction thatimproves the compressive strength [10]. However, one sample containing less CaO (R9.6) shows rapid setting time. It might have correlation with its physical shape or other contaminant from the coal origin which is discussed later in this paper.

3 Physical characterization

3.1 XRD test.

X-Ray diffraction with CuKα radiation, Kα= 0.89 and wavelength λ = 1.54056 Å was used to determine the crystal structure [18]. Fig. 2 shows the XRD pattern of geopolymer compared between Indonesian Fly Ash and Japanese Fly Ash. The results show that both ash types have a same major peak of crystalline phase i.e quartz (SiO_2). Besides, Japanese fly ash also contained crystalline phase of hematite (Fe_2O_3), and mullite ($3Al_2O_3.2SiO_2$). In addition, Indonesian fly ash also contained srebrodolskite ($Ca_2Fe_2O_5$), magnetite (Fe_3O_4), anhydrite ($CaSO_4$) and lime (CaO) [8].

The second major peak presents that both of Fly Ash contain Fe-based material with different oxidation state (hematite and magnetite). It seems that Indonesian Fly Ash is easier to oxidize compared to Japanese Fly Ash. Indonesian Fly Ash also contains more $SiO_2$and CaO than Japanese Fly Ash although the major peak happened at the similar 2 theta degree at 20-30. Previous studies informed that quartz identified as peak phase of geopolymer crystal occured in the range degree of 20-30 with the intensity ranged 1200-1800 [8].

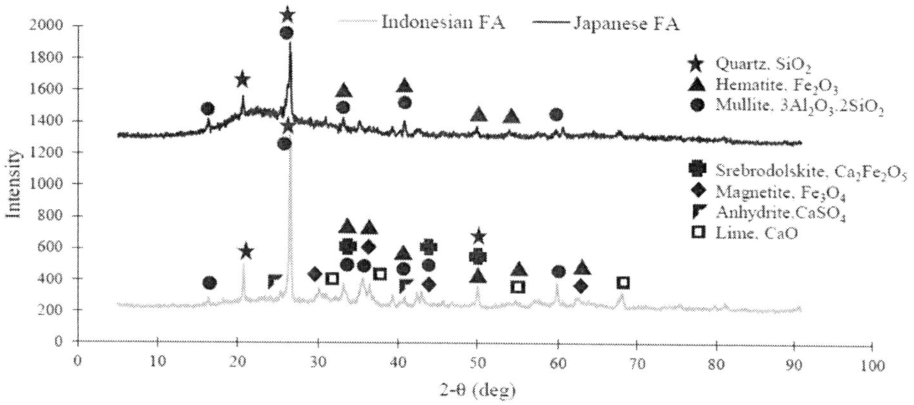

Fig.2. XRD Pattern of Indonesian and Japanese Fly Ash [8]

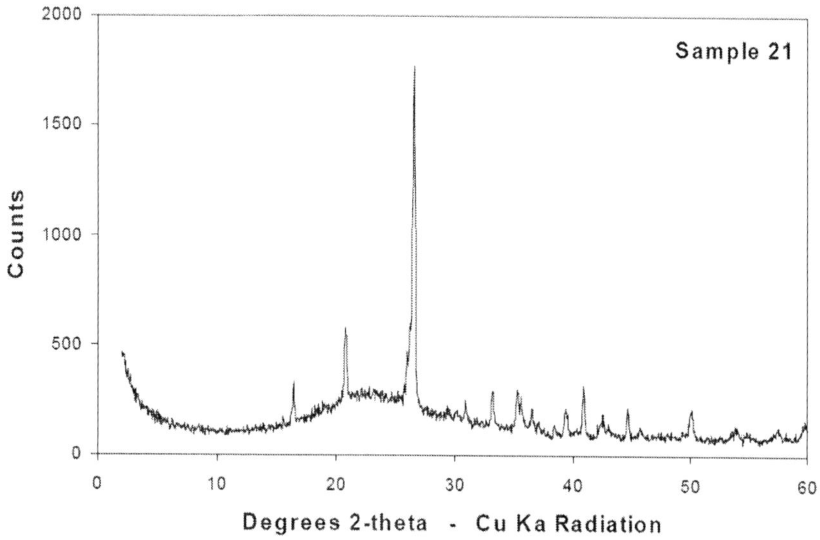

Fig. 3. XRD pattern of Australian Fly Ash [12]

Figure 3, illustrates the XRD pattern of Australian Fly Ash with major peak at 20-30 two theta degree. It was characterized using Philips X'pert Diffracto meter. The scans were run from 2 to 60° 2θ, with increments of 0.04° and counting time of 2 seconds per step [13]. The peaks consist of quartz, mullite, magnetite, hematite, calcium aluminate, and glass.

The previous research concluded that sharp major peak occur at small degree of 2 theta, that means the distance of particles is closer. This is due to large particle size and small surface area. Small surface area resulted less mechanical properties [7].

Australian, Indonesian and Japanese fly ash as presented in Figure 2 and Figure 3 was categorized as fly ash class F due to its high silica and alumina content. Based on their XRD test, all specimens perform a quartz as their peak phase. Therefore, it can be concludes that fly ash class F which has high silica and alumina content will perform quartz compound as their peak phase.

DOI: 10.1051/matecconf/20179701031

3.2 SEM test

Scanning Electron Microscope (SEM) was conducted in order to investigate the morphology of particles [17]. In addition, a morphology particle shape data can be used for predicting the mechanical and physical properties of material [12]. The results of SEM test for several fly ashes obtained from different sources are illustrated in Figure4 and Figure 5.

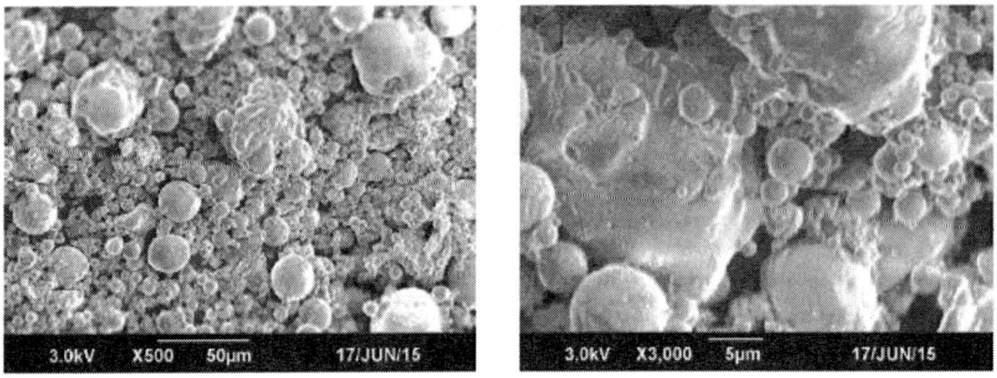

Fig.4. SEM results of Indonesian Fly Ash [8]

Fig. 5. SEM results of Japanese Fly Ash [8]

In Figure 4 and Figure 5, the shape of fly ash particles aregenerally spherical shape.Fly ash obtained from Indonesia power plant has irregular shape pattern (bulky) compared to Japanese fly ash. However, the result found that geopolymer paste produced from Indonesian fly ash has higher compressive strength compared to whose produced from Japanesefly ash.This condition occurs because the bulky particle shape of Indonesian fly ash has short particle distance compared to a homogeneous Japanesefly ash.Previous study by Fernandez stated that particles size distribution of fly ash play an important role in mechanical properties of geopolymer paste [17]. Hence, morphology of several fly ash and its geopolymer paste are illustrated at Figure 6 and Figure 7.

The morphology of fly ash and geopolymer paste showed in Figure 6 are clearly different in their shape. Fly ash obtained from Texas categorized as class C has spherical shape particles while geopolymer paste is in bulk structure. Spherical shape of fly ash seems heterogenic and has various distances between particles due to large difference of particle diameter.

In Figure 7, fly ash obtained from Tarong categorized as class F and its geopolymer paste shared a similar particle's morphology. Additionally, the coating layer at the surface of fly ash particles is occurs at morphology of geopolymer paste. This explains that geopolymerization reaction between fly ash and alkali solution only occurs at the surface of fly ash particles [13]. Therefore, it proves that the short distance between particles improves coated and linked form in geopolymer paste.

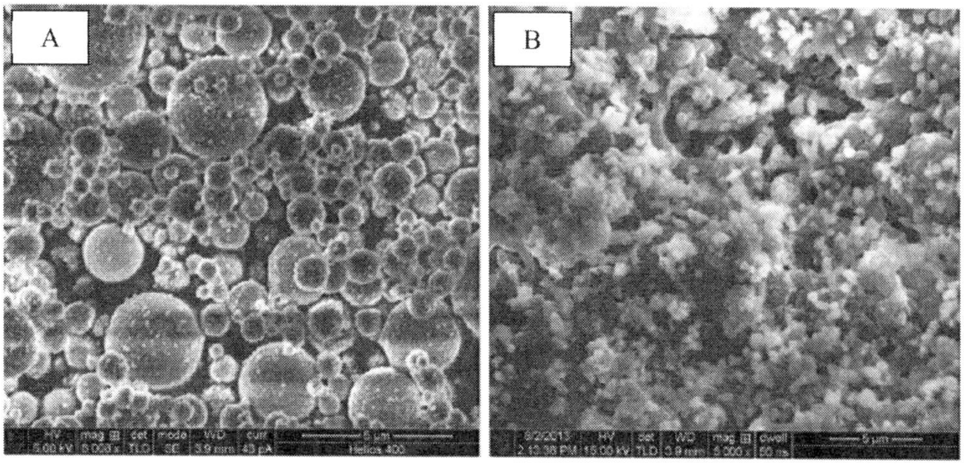

Fig. 6. Morphology of Texas Fly Ash (A) and Geopolymer Paste (B) [19]

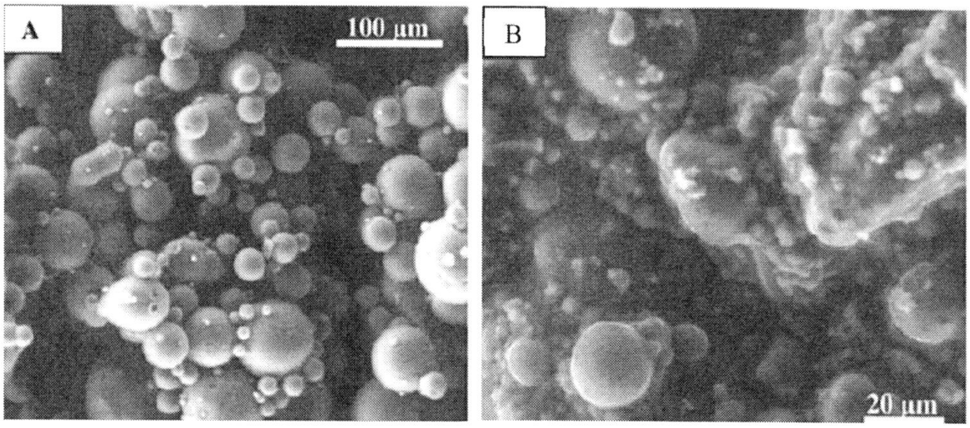

Fig.7. Morphology of TarongFly Ash (A) and Geopolymer Paste (B) [13]

The research conducted by Tajunnisa stated that more irregular shape particles required more water in mixing geopolymer paste[8].With the purpose of produce geopolymer paste, water content is represented by alkali solution. It is also concluded that concentration and composition of alkali solution plays an important role in developing mechanical properties of geopolymer paste [20-22].

4 Summary

The various type of fly ash resulted in different characterization results. Japanese fly ash consists of Ca-based compound and magnetite, Indonesian fly ash consists of Si-based compound and hematite, while Australian FA consists of quartz, hematite, magnetite, and

CaO. The pH value and CaO content of fly ash influences setting time of geopolymer paste. All diffraction patternsof fly ash showed the major peak at 20-30 two theta degree. The morphology of fly ash particles was mostly in spherical shape while geopolymer paste as their product was bulky irregular shape. Distance of particles contributed by their size is one of important factors to increase mechanical strength of geopolymer paste.

References

1 C.A. Hendriks, *Proceedings of the 4th International Conference on Greenhouse Gas ControlTechnologie* (Interlaken, Austria, 1998)
2 E. Gartner, Cem. Concr. Res., **34**, 1489 (2004)
3 J. Davidovits, *Proceeding First International Conference on Alkali Cement and Concrete* (1994)
4 D.Hardjito, *Studies on Fly Ash-Based Geopolymer Concrete* (Curtin Univ of Technology, 2005)
5 American Society for Testing and Materials, ASTM C618, *Annual Book of ASTM Standard,* (West Conshohocken, US, 2004)
6 B.G. Kutchko, A.G. Kim, Fuel, **85**, 2537 (2006)
7 R. Nurwidayati, M.B Ulum, J.J Ekaputri, Triwulan, P. Suprobo, Mater. Sci. Forum, **841**, 118 (2016)
8 Y. Tajunnisa, M. Sugimoto, T. Sato, J.J. Ekaputri, M.Shigeishi, *International Conference and Exhibition of Structural Faults and Repair 2016* (Edinburg, 2016)
9 J. Van Jaarsveld, J. Van Deventer, G. Lukey, Mater. Lett., **57**, 1272 (2003)
10 Antoni, S.W. Wijaya, D. Hardjito, Mater. Sci. Forum, **841**, 90 (2016)
11 Antoni, S.W. Wijaya, D. Hardjito, Mater. Sci. Forum, **841**, 98 (2016)
12 C.R. Ward, D. French, *WOCA* (Lexington, Kentucky, USA, 2005)
13 J. Duchesne, L. Duong, T. Bostrom, R. Frost, Waste Biomass Valori, **1**, 367 (2010)
14 J.J. Ekaputri, Triwulan, K.A. Priadana, T.E. Susanto, S. Junaedi, *The 2013 World Congress on Advances in StructuralEng and Mech (ASEM13)* (Jeju, Korea, 2013)
15 J.J. Ekaputri, M.B. Ulum, Triwulan, R. Bayuaji, T.E. Susanto, M.M.A.B. Abdullah, AMM, **754**, 320 (2015)
16 Antoni, R. Gunawan, D. Hardjito, AMM, **815**, 153 (2015)
17 A. Fernandez Jimenez, A. Palomo, Fuel, **82**, 2259 (2003)
18 N. Yahya, P. Puspitasari, J. Nano Res., **21**, 131 (2013)
19 A.R. Kotwall, Y.J. Kim, J. Hu, V. Sriraman, International Journal of Concrete Structures and Materials, **9**, 35 (2014)
20 P. Risdanareni, J.J Ekaputri, Triwulan, Mater. Sci. Forum **803**, 125 (2015)
21 A.N. Hashim, K. Hussin, N. Begum, M.M.A.B. Abdullah, K.A. Razak, J.J. Ekaputri, AMM, **754**, 300 (2015)
22 J.J. Ekaputri, Triwulan, S. Junaedi, H. Fansuri, R. Bayuaji, Mater. Sci. Forum, **803**, 63 (2015)

Experimental Study On Lateral Load Capacity of Bamboo RC Beam Column Joints Strengthened By Bamboo Mechanical Anchors

B. Sri Umniati[1,4,*], *Sri Murni* Dewi[2], Nindyawati[1,4], and *Gatot Adi* Susilo[3]

[1]Civil Engineering Department, Universitas Negeri Malang, Malang, Indonesia
[2]Civil Engineering Department, Brawijaya University, Malang, Indonesia
[3]Architectural Department, Institut Teknologi Nasional, Malang, Indonesia
[4]Konsorsium RisetGeopolimer Indonesia (KORIGI), Lab Beton dan Bahan Bangunan ITS, Surabaya, Indonesia

Abstract. In this paper, the prospective of bamboos which available abundantly especially in Indonesia as rebars and mechanical anchors are studied. And also the endurance of the bamboos mechanical anchors to withstand cyclic loading were observed. Nine classes of bamboos bar were evaluated: consist of 3 different anchors (0, 4 and 8 anchors) and 3 different compressive strength (19.19 MPa, 29.61 MPa and 37.96 MPa) means 3x3 parameters. The results show that the lateral load capacity increased significantly with the present of bamboo anchors specimens: 26.04 % for 4 anchors specimens (C2) and 25 % for the 8 anchors specimens (C3) compared to zero anchor specimens (C1). On the other hand, the compressive strength of concrete have no significant effects to the lateral load capacity. Overall it can be concluded that, bamboo can be used as mechanical anchorage to strengthen beam column joint.

1 Introduction

Many structures collapse during earthquakes due to severe damage to beam column connections, whereas, their other parts remain unaffected. Therefore, it is necessary to strengthen beam column joints by using mechanical anchorages in order to better the overall performance of concrete structures. This study aimed to find out whether bamboo mechanical anchor could be used as an anchorage in beam column joints to withstand earthquakes.

The use of bamboo as a significant alternative to steel bar in concrete structure is an interesting subject for research as it gives many benefits, for example in the matter of strength, galah bamboo has a tensile strength of 253 MPa, the same tensile strength as steel reinforcement of around 240 MPa to 400 MPa [1]. From the aspect of costs, bamboo is far cheaper than steel but has the same strength. Furthermore, it is an abundant and renewable resource found in all tropical regions of the world. Several species of bamboo have been used by the authors as bars in concrete such as 'petung' (Dendrocalamus asper), 'ori' (Bambusa

* Corresponding author: b.sri.umniati.ft@um.ac.id

DOI: 10.1051/matecconf/20179701032

blumeana) and 'tali' (Gigantochloa apus). These are 3 types of bamboo have proved to be highly recommended for replacing steel bars in concrete from previous study [2-7]. For this study, the bamboo was prepared as follows: 1) the bamboo was cut and split into the sizes required and then dried for about 28 days ; 2) when dry, it was painted with 2 coats of paint used for wood as water proofing with an interval 24 hours between coats ; 3) immediately after the second coat, the freshly painted bamboo was buried in sand. Sand adheres to the bamboo and improves the binding of the concrete to the bamboo and, as soon as the paint is dry, the bamboo is ready for use as bars in concrete [8].

Available data on structural damage due to earthquakes indicates that it is most prevalent in beam column joints, whereas, other parts are relatively unaffected with a few hairline cracks in those areas closest to beam column joints. These findings are supported by the author's previous experiments [4] where bamboo reinforcements to concrete frame, which had already been damaged by cyclical loads from a UTM axial test results indicated that these columns could bear axial loads of between 12 and 17.5 tons. These results are slightly lower than the theoretical axial load capacity [9] for the column sections of 18 tons, showing that within the structure of the columns, the bamboo reinforcements remained adhered to the concrete and that the column sections had functioned well. However, therefore, there are still room for improvement and further studies are required to into beam column joints to make them even better in order to ensure that the whole structure is stronger. Conventional beam column joints commonly use development hooks (are constructed from longitudinal steel bars bent into an 'L' shape at angle of 90°. But the weakness of this type, apart from being understrength, can also give rise to reinforcement congestion especially in exterior joints [10-13]. Thus the use of alternative methods such as mechanical anchors or 'headed bars' [10, 13] that are combined with vertically distributed reinforcements [12] to optimize structural strength as a whole. The aim of this study was (1) to discover the characteristics of the relationship between load (P) and displacement (Δ) on beam column joints under cyclic loads using quasi-cyclic loading test; (2) to find out the effect of concrete strength and the number of anchors installed on their capacity to withstand earthquakes loads; and (3) to draw up models of P-Δ relationships of beam column joints.

2 Experimental program

2.1 Specimens

The specimens tested were 'T' shaped interior beam column joints. The variables were concrete strength (variable A), and the total number of anchors installed (variable C). In this study, a group of a 3^2 factorial design [14] was employed. 9 specimens were made in this group. This group had 2 factors (A and C) at 3 levels (low, medium, and high), is shown in Table 1. Details of the specimens are presented in Figure 1.

Table 1. Specimens coding.

Number of anchors (headed bars)	Comp. strength [fc'] A1=19.19 MPa	Comp. strength [fc'] A2=29.61 MPa	Comp. strength [fc'] A3=37.96 MPa
C1 = 0	A1C1	A2C1	A3C1
C2 = 4 anchors	A1C2	A2C2	A3C2
C3 = 8 anchors	A1C3	A2C3	A3C3

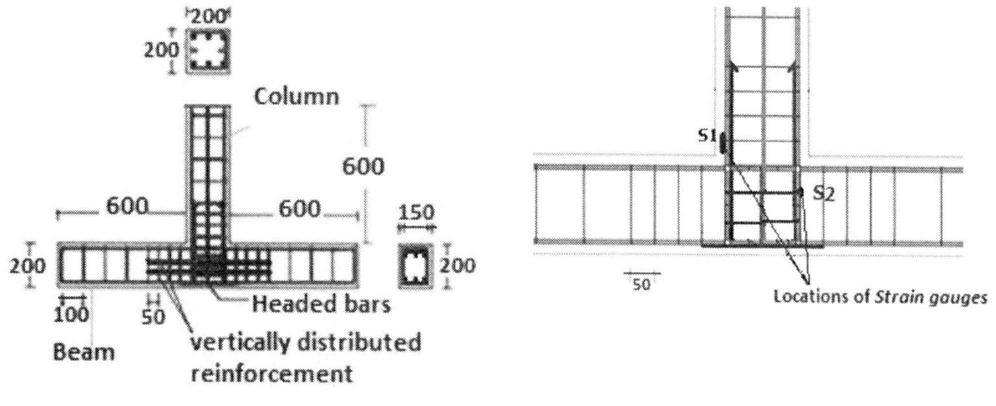

(a) (b)

Fig. 1. a). Detail of specimen (unit in mm); b). *Strain Gauges* locations

The bamboo reinforcement used in these experiments were made from 3-year-old Bambusa blumeana (Indonesian's name: 'Ori'), taken from the base up to 6 metres. The tension yield strength of the Ori bamboo used was 94.778 MPa and its ultimate yield strength was 124.26 MPa. The steel bars used for shear reinforcements were 6 mm in diameter with a yield strength of 419.17 MPa and an ultimate yield strength of 609.07 MPa. The parameters measured were (1) cyclical load, namely, the lateral load applied to the test subjects; and (2) deformations in test subjects due to cyclical loads in the form of lateral displacement of the column ends and bases, strain in bamboo bars and in shear reinforcements, and the form of cracks patterns.

2.2 Test set-up

The set-up for the instruments used to measure cyclical and monotonic loads, as well as the application of force are shown in Fig. 2.

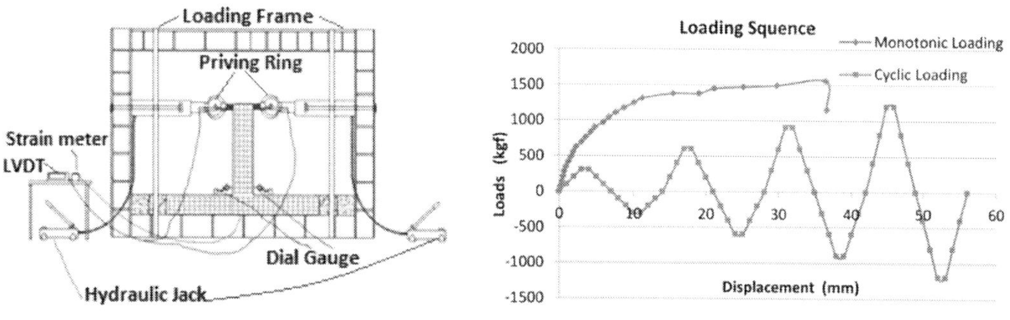

(a) (b)

Fig. 2. a). *Test Set-up* ; b). Loading Squence

2.3 Test procedure

The beam column joints were subjected to lateral cyclic loads. Load intervals were adjusted for the proving ring. The first interval reading was 3 x 100 kgf. For the second and third readings, force was applied twice and three times respectively. Due to page limits for

DOI: 10.1051/matecconf/20179701032

publication, only graphs showing the relationship of lateral loads (P) to displacements (Δ) are shown in this paper.

3 Results

The test results obtained from the cyclical loads applied enabled a graph to be plotted showing the relationship between the parameters measured for each load application. Each cycle load was plotted in one colour. The energy dissipations is computed by summing the curve area of each cycle (not included here). For test subjects A1C1, A1C2 and A1C3, the graphs of the relationship between lateral loads and lateral displacement shown Figure 3a, b and c.

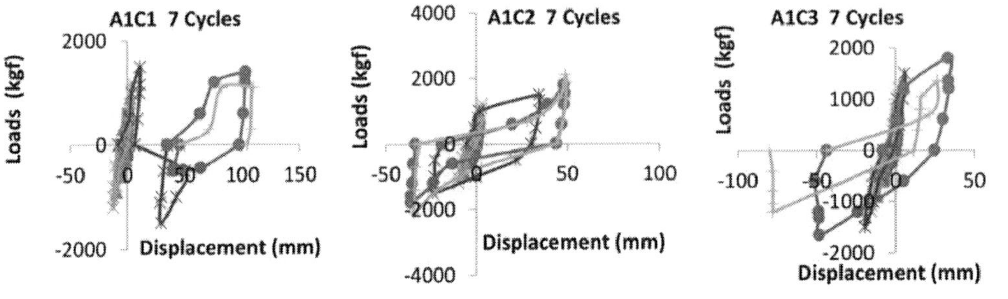

(a) (b) (c)

Fig. 3. The relationship between lateral load (P) and lateral displacement for test subjects a) A1C1; b) A1C2 and c) A1C3.

Figure 3 show the level of lateral displacement for each load application in concrete compressive strength A1 and the number of bamboo anchors installed for test subjects C1,C2 and C3 respectively. For test subjects A2C1, A2C2, an A2C3 the graphs of the relationship between lateral loads and lateral displacement are indicated in figure 4a, b, and c.

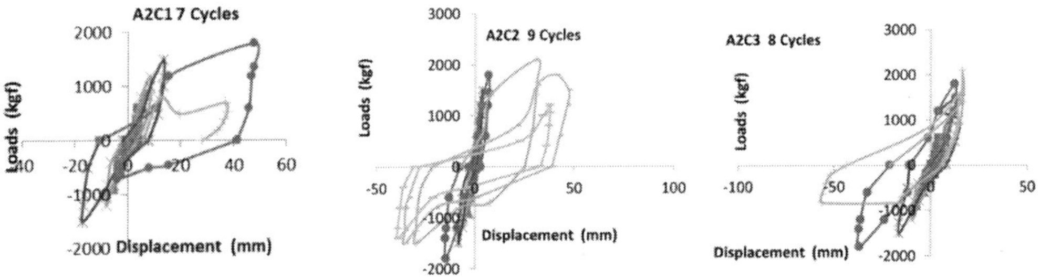

(a) A2C1 (b) A2C2 (c) A2C3

Fig. 4. The relationship between lateral load (P) and lateral displacement, for A2 test subjects

Figure 4 demonstrates lateral displacement levels for each lateral load application in concrete compressive strength A2 and the number of bamboo anchors installed in test subjects C1, C2 and C3 respectively. For test subjects A3C1, A3C2 and A3C3 the data for the relationship between lateral loads and displacement are indicated in Figure 5 a, b and c.

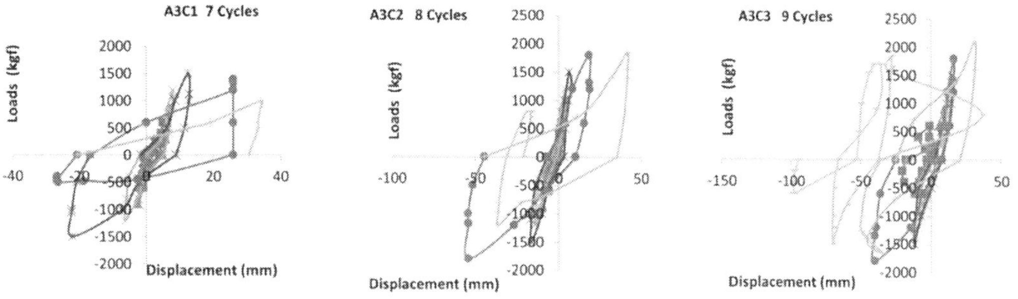

(a) A3C1 (b) A3C2 (c) A3C3

Fig. 5. The relationship between lateral load (P) and lateral displacement for A3 test subjects

The maximum compression and tension lateral loads for each test subjects, summarized in Table 2.

Table 2. Maximum loads, compression and tension.

Specimen	Maximum Compression Load [kgf]	Maximum Tension Load [kgf]
A1C1	1500	1500
A1C2	2100	2100
A1C3	1800	1650
A2C1	1800	1500
A2C2	2100	1500
A2C3	2100	1800
A3C1	1500	1300
A3C2	1850	1800
A3C3	2100	1800

3.1 Cracks pattern

In the test on the beam column joints using cyclical lateral loads, the form of the cracks that appeared in the test each of beam column joints after the test were finished was noted. The form of these cracks demonstrate the type of collapse that occurred in them. Figure 6 a, b and c show crack form for beam column joints as well as the total anchors installed for test subjects C1, C2 and C3 respectively for concrete compressive strength A1.

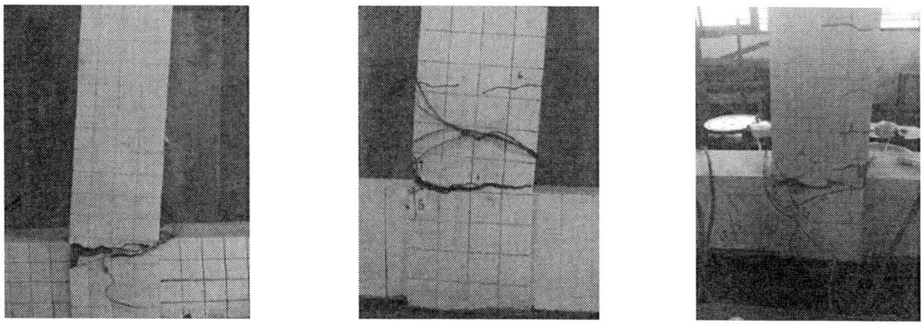

(a) (b) (c)

Fig. 6. Cracks pattern in a) A1C1; b) A1C2 and c) A1C3

Figure 7 a, b and c crack form in beam column joints, for concrete compressive strength A2 as well as bamboo anchors installed in C1, C2 and C3 respectively.

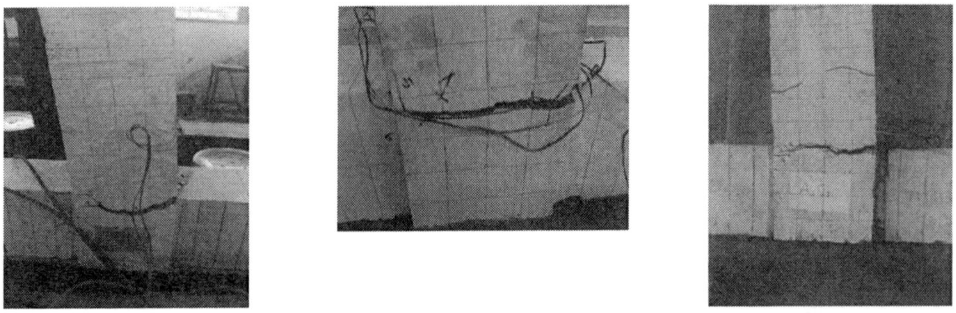

(a) (b) (c)

Fig. 7. The Cracks pattern in a) A2C1; b) A2C2; c) A2C3.

Figure 8 a, b and c crack form in beam column joints, test subjects C1, C2 and C3 respectively for A3 concrete compressive strength.

(a) (b) (c)

Fig. 8. The form of the cracks in test subjects a) A3C1; b) A3C2 an c) A3C3

4 Discussion of results

The above graphs demonstrate the relationship between lateral load and deformation in the tested beam column joints as well as maximum compression and tension lateral loads for each test subjects, summarized in Table 3 and Table 4.

Table 3. Maximum compression lateral loads for concrete compressive strength A1, A2 and A3 and for C1. C2 and C3 totals for bamboo anchors installed.

Number of anchors installed	**A1 [kgf]**	**A2 [kgf]**	**A3 [kgf]**	**Total [kgf]**	**Difference [%]**
C1= 0	A1C1=1500	A2C1=1800	A3C1=1500	ΣAiC1=4800	0
C2=4	A1C2=2100	A2C2=2100	A3C2=1850	ΣAiC2=6050	26.04167
C3=8	A1C3=1800	A2C3=2100	A3C3=2100	ΣAiC3=6000	25
Total	ΣA1Ci=5400	ΣA2Ci=6000	ΣA3Ci=5450		
Difference [%]	**0**	**11.11**	**0.93**		

Table 3 indicates that the highest P max. from all 3 concrete compressive strengths was achieved by the C2 (4 anchors) test subjects. Also, the highest P max. values were obtained by the A2 concrete compressive strength. The deformation, as shown in the graphs for the relationship between load (P) and displacement (Δ) and crack formation demonstrates that in test subjects of higher concrete compressive strength without anchors, this did not significantly affect the deformation that occurred. Whereas, for test subjects with anchors installed endurance to lateral loads increased, these test subjects became more ductile as proved by their withstanding repeated cycles of load applications and by their crack formation (without or with hairline cracks of column). Maximum load equation, namely, P max. = 1600 + 158.3 C – 13.54 C^2 with a coefficient regression of R^2 = 0.674 where C is amount of anchors installed. And the P max. values for tension are shown in Table 4.

Table 4. Maximum lateral load (tension) for concrete compressive strength A1, A2 and A3 and for C1, C2 and C3 totals for installed anchors, as well as the lowest difference levels.

Number of anchors installed	A1 [kgf]	A2 [kgf]	A3 [kgf]	Total [kgf]	Difference [%]
C1= 0	A1C1=1500	A2C1=1500	A3C1=1300	ΣAiC1=4300	**0**
C2=4	A1C2=2100	A2C2=1500	A3C2=1800	ΣAiC2=5400	**25.5814**
C3=8	A1C3=1650	A2C3=1800	A3C3=1800	ΣAiC3=5250	**22.09302**
Total	ΣA1Ci=5250	ΣA2Ci=4800	ΣA3Ci=4900	14950	
Difference [%]	**0**	**-8.57**	**-6.67**		

For the above tension values, the P max. was 1433 + 143.7 C – 13.02 C^2 with a regression coefficient of R^2 =0.517.

5 Summary

From research results and discussion, the following conclusions can be made:

1. The installation of bamboo mechanical anchors to beam column joints increased lateral load capacity by 26.04 % for compression lateral load (Table 2) and 25.58 % for tension lateral load (Table 3), and reduced deformation.
2. The test with the optimum number of bamboo anchors was A2 (fc'=29,61 MPa) with 4 anchors (C2).
3. The regression model for lateral load capacity for number of anchors installed variables, are as follows: Under compression lateral load, the P max was 1600 + 158.3 C – 13.54 C^2 with a coefficient regression of R^2 = 0.674 where C is amount of anchors installed.
4. The regression model for lateral load capacity for number of anchors installed under tension lateral loads the P max was 1433 +143.C - 13.02 C^2 with a regression coefficient of R^2 = 0.517 .

This research can be established with a grant of PHB DP2M Indonesian Directorate General of Higher Education, Universitas Negeri Malang (UM) and Brawijaya University (UB), Malang Indonesia.

DOI: 10.1051/matecconf/20179701032

References

1 Morisco, *Teknologi Bambu,* (MTBB UGM, Yogyakarta, 2005)
2 Karyadi, B.S. Umniati, Nindyawati, *Pengaruh Bentuk Pengekang di Jalur Gaya Tekan Penampang Balok Beton Bertulangan Bambu Terhadap Kapasitas Beban dan Lendutan Balok Pada Rasio Tulangan yang Berbeda,* (Fundamental Research Report, DP2M Funding, 2007)
3 Nindyawati, B.S. Umniati, *Portal Beton Bertulangan Bambu Yang Dikekang Di Jalur Gaya Tekannya, Sebuah Solusi Pembangunan Rumah Sederhana Tahan Gempa,* (PHB Report, DP2M Funding, 2009)
4 Nindyawati, B.S. Umniati, *Pengembangan Model Sambungan Balok Kolom Pada Struktur Portal Beton Terkekang Bertulangan Bambu Tahan Gempa,* (PHB Report, DP2M Funding, 2011)
5 B.S. Umniati, Karyadi, *Pengembangan Model Struktur Beton Bertulangan Bambu Tahan Gempa Sistim Ganda (Kombinasi Portal dan Dinding Geser) Untuk Pembangunan Rumah Sederhana Tahan Gempa pada Wilayah Gempa Tertinggi (Zona 6) di Indonesia*, (Hibah Strategis Nasional Research Report, DP2M Funding, 2009)
6 B.S. Umniati, Nindyawati, *Pengembangan Model Sambungan Balok Kolom Pada Struktur Portal Beton Terkekang Bertulangan Bambu Tahan Gempa*, (Year PHB Report, DP2M Funding, 2012)
7 B.S. Umniati, Nindyawati, *Pengembangan Model Sambungan Balok Kolom Pada Struktur Portal Beton Terkekang Bertulangan Bambu Tahan Gempa*, Year PHB Report, DP2M Funding, 2013)
8 J.A. Kankam, M.B. George, S.H. Perry, ACI Struct. J. **61** (1988)
9 T. Paulay, M.J.N. Priestley, *Seismic Design Of Reinforced Concrete And Masonry Building* (John Wiley & Sons Inc., New York, 1992)
10 S.C. Chun, O. Bohwan, S.H. Lee, J.C. Naito, ACI Struct. J., **579** (2009)
11 S.C. Chun, O. Bohwan, H.S. Lee, T.H.K. Kang, J.W. Wallace, ACI Struct. J., **102** (2007)
12 H.J. Lee, S.Y. Ying, ACI Struct. J., **329** (2009)
13 T.H. Kang, M. Shin, N. Mitra, J.F. Bonacci, ACI Struct. J., **106,** 868 (2009
14 D.C. Montgomery, *Design and Analysis of Experiments, 5th Edition,* (Jhon Wiley & Sons. Inc., New York, 2001)

Improving of Water Resistance of Asphalt Concrete Wearing Course Using Latex-Bitumen Binder

Henri Siswanto[1,2,*]

[1]The Department of Civil Engineering, The Faculty of Engineering, Universitas Negeri Malang, Jl. Semarang 5 Malang, Indonesia
[2]Konsorsium Riset Geopolimer Indonesia (KORIGI), Lab Beton dan Bahan Bangunan ITS, Surabaya, Indonesia

Abstract. It is well known that presence of water in a bituminous mix is a critical factor which can lead to premature failure of flexible pavements. This requires solutions one of which is to formulate an asphalt mix that has a high resistance to moisture and one way to do this is to mix latex with the asphalt mix. The purpose of this experimental study was to investigate the effect of water on Marshall stability of asphalt concrete wearing course (ACWC) made with a latex-bitumen binder. Latex-bitumen was mixed with aggregate and four levels of latex content were investigated in this study, namely, 0%, 2%, 4% and 6% respectively by weight of asphalt. Wet procces was used in the blending of mixtures. The procedure used to obtain the optimum binder contents conformed to the Marshall procedure (SNI 06-2489-1991). Six Marshall specimens at optimum binder content were prepared for each binder mix investigated. Three of six specimens from each group were tested under Marshall standards. The remaining specimens were tested by immersion in a bath at 60^0 C for 24 hours. The Marshall index of retained stability was used to evaluate the effect of water on the Marshall stability of ACWC. The results indicated that the addition of up to 4% latex to ACWC mix increased the retained Marshall stability, whereas the addition of latex above 4% decreased the retained stability of the mixture. The addition of 4% CRM significantly improved the retained stability of the mixture and was the best latex – ACWC mix.

1 Introduction

Many variables affect the degree of moisture damage which occurs in an asphalt concrete mixture and some of these variables are related to the materials forming hot mix asphalt (HMA) such as aggregate and bitumen. Others are related to mixture design and construction (air void levels, film thickness, permeability and drainage), environmental factors (temperature, pavement age, freeze–thaw cycles, and presence of ions in the water), traffic conditions and type and properties of the additives [1].

* Corresponding author: henri.siswanto.ft@um.ac.id

As in many tropical countries, the main causes of road deterioration in Indonesia are from high road surface temperatures, high rainfalls and high traffic loads [2]. Indonesia experiences frequent floods and these floods are a major cause of pavement damage by stripping/ separating the bitumen layers from the aggregate thus contributing to its deterioration.

Traffic and water are two key elements in the degradation of asphalt concrete pavements, water causes loss of adhesion at the bitumen–aggregate interface. This premature failure of adhesion is commonly referred to as stripping in asphalt concrete pavements. Strength is impaired since the mixture ceases to act as a coherent structural unit and loss of adhesion renders the cohesive resistance of the interstitial bitumen body useless. Water may enter the interface through diffusion across bitumen films and penetrate directly into partially coated aggregate. Futhermore, water can cause stripping through five different mechanisms, namely, detachment, displacement, spontaneous emulsification, pore pressure and hydraulic scour [3]. The solution of this problem is to obtain an asphalt mixture that has a high resistance to water.

Additives have been used to improve endurency of asphalt concrete pavements to permanent deformation, moisture damage, and fatigue or low-temperature cracks. There are a number of different additives available which can be introduced directly to the asphalt cement as a binder modifier, or can be added to the mixture with the aggregate [4].

One of the more promising additives is latex which improves its resistance to water. Research has shown that the addition of latex not only reduces asphalt's sensitivity to heat but also improves its binding and thus durability[5]. Latex improves the physical properties of asphalt by incresing its softening point and durability [6].

Asphalt concrete wearing course (ACWC) is widely used as road pavement in Indonesia and latex as an additive has the potential of increasing ACWC water resistance and thus its durability, therefore, the purpose of this experimental study was to investigate the effect of water on Marshall stability of asphalt concrete wearing course (ACWC) made with a latex-bitumen binder. The approach of this study was to find ways to improve the retained stability of ACWC. The higher the level of retained stability, the higher the water resistance of the mix.

2 Method

2.1 Materials

The materials used in this study were a 60/70 penetration bitumen, aggregate and latex. The latex used in this research was natural rubber of 60 % dry rubber content (DRC 60) available in Indonesia. The aggregate gradation used can be seen in Figure 1. The Indonesian National Standards (SNI 1737–1989) were used and all test results had to meet SNI standards.

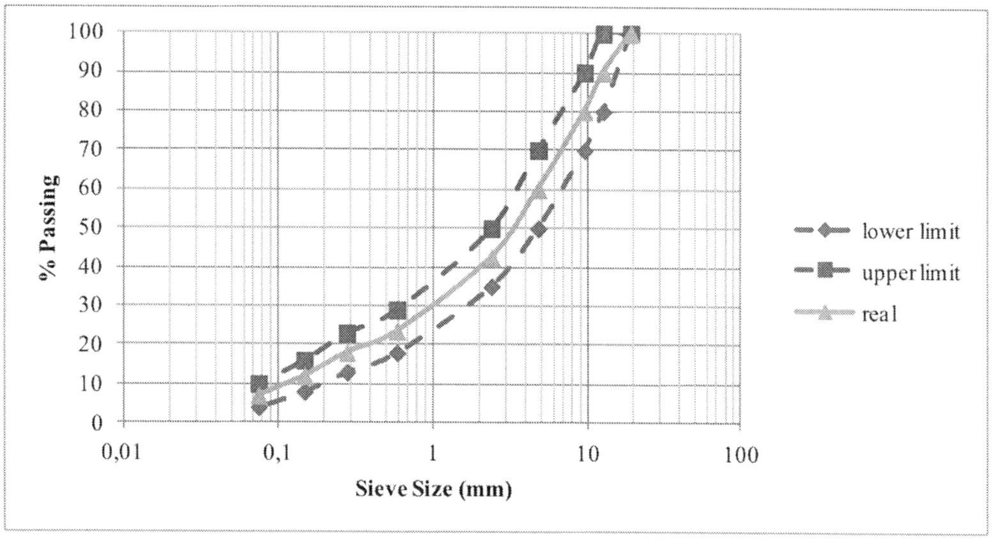

Fig. 1. The Grading Curves of Asphalt Concrete Wearing Course.

2.2 Blending latex and bitumen method

According to Road Note 36 Specification for the Manufacture and Use of Rubberised Bituminous road materials and binders, the method of blending rubber with penetration grade bitumen will vary in detail depending on the form of rubber to be used [7]. In general, the finer the particle size is, the more rapid the blending is. For binders harder than 300 pen., the binder should be heated to a temperature of 140-170°C depending on bitumen hardness, in a boiler the capacity of which will allow for at least a 50% increase in binder volume. The boiler should be fitted with a propeller-type stirrer or other means of agitation which will ensure that any material added to the binder will be drawn down into it. Latex should be added slowly in small quantities and allowed to remain on the surface for about 20 seconds before stirring it. The intensity of foaming should be gauged and the remainder of the latex added at such a rate to ensure that this is not too great for the capacity of the boilier used. After the addition of the latex, the binder should be allowed to cool below 130°C unless it is to be used immediately. Stirring should be continued for a minimum of 20 minutes after the addition of the rubber.

The procedure used to obtain the optimum binder contents was in accordance with the Marshall procedure. The binder contents investigated varied from 5.0 % to 7.0% of total mixture weight, in increments of 0.5 %. Three specimens were prepared for each binder content. For mixing purposes, the binder was heated to the temperature that produced a kinematics viscosity of 170±20 centistokes, for compaction the required viscosity was 280±30 centistokes.

Compaction was carried out by applying 75 blows to each face of the specimens, in accordance with the criterion for heavy traffic. The specimens were weighed in air and in water to determine the bulk specific gravity of the compacted mixes. At each binder content the following mix properties were determined; the specific gravity of the compacted mix, the void in the mineral aggregate (VMA), the voids in the mix (VIM) and the voids filled with the binder (VFB).

DOI: 10.1051/matecconf/20179701033

2.3 Testing of the specimens

Six Marshall specimens with optimum binder content were prepared for each binder type investigated. Four levels of latex content were investigated in this study, at 0 %, 2 %, 4 % and 6 % latex by weight of mix respectively. A wet process was used in the blending of mixtures. Three specimens from each group were tested under the Marshall standard to obtain Marshall stabilty data. The remaining specimens were tested following immersions in a bath at 60°C for 24 hours. The percentage that retained Marshall stability provided an indication of the effects of water on the mixes. A higher value indicated higher water resistance.

The retained Marshall stability was used to evaluate resistance to water as some researchers have used retained Marshall stability for evaluating resistance of asphalt mixtures to moisture [1, 8, 9, 10]. It is well known that presence of water in a bituminous mix is a critical factor which can lead to premature failure of flexible pavements.

3 Results and discussion

The mix made without latex (0%) had an optimum binder content of 6.5%. The results of the Marshall tests are summarized in Table 1.

Table 1. Summary of the Design Results of the Marshall Method Compared to Indonesian Specifications

Design Parameters	Unit	Value	Indonesian Specifications	
			Min	Max
Marshall Stability	kg	1121	800	-
Flow	mm	3.39	3	-
VIM	%	4.74	3.5	5.5
VMA	%	17.74	15	-
VFB	%	75.66	65	-
Marshall Quotient	kg/mm	330	250	-

The average Marshall stability and flow value was determined for each binder content. The parameters obtained were then plotted against binder content, namely; stability, flow, unit weight, VIM, VFB and VMA contents for each binder specimen. Optimum Binder Content (OBC) was determined from their graphs and showed in Figure 2.

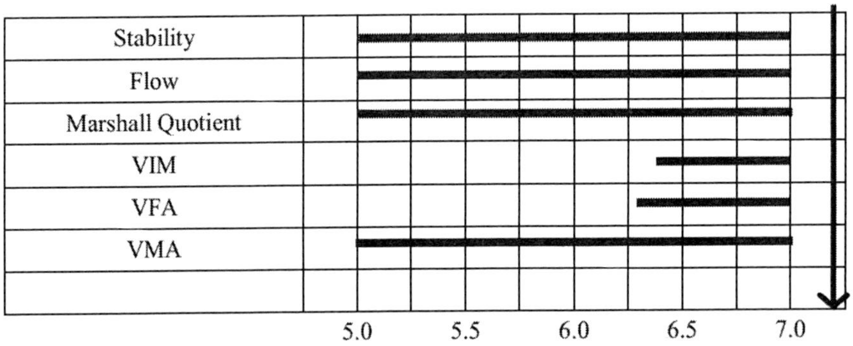

Fig. 2. Determination of Optimum Bitumen Content

All mixtures prepared had this optimum bitumen content. Fig. 3 shows an increase in latex increased stability from 0 % to 4 % which decreased when latex increased to 6 %. The mix with 4% latex had a stability level 13 % higher than that of the mix with no latex (0 %).

DOI: 10.1051/matecconf/20179701033

Adding latex to asphalt affects its softening, penetration index and bitumen film point increase so that the stability of the mix incresases. Stability of the mix is influenced by its components, namely agregate and asphalt binder.

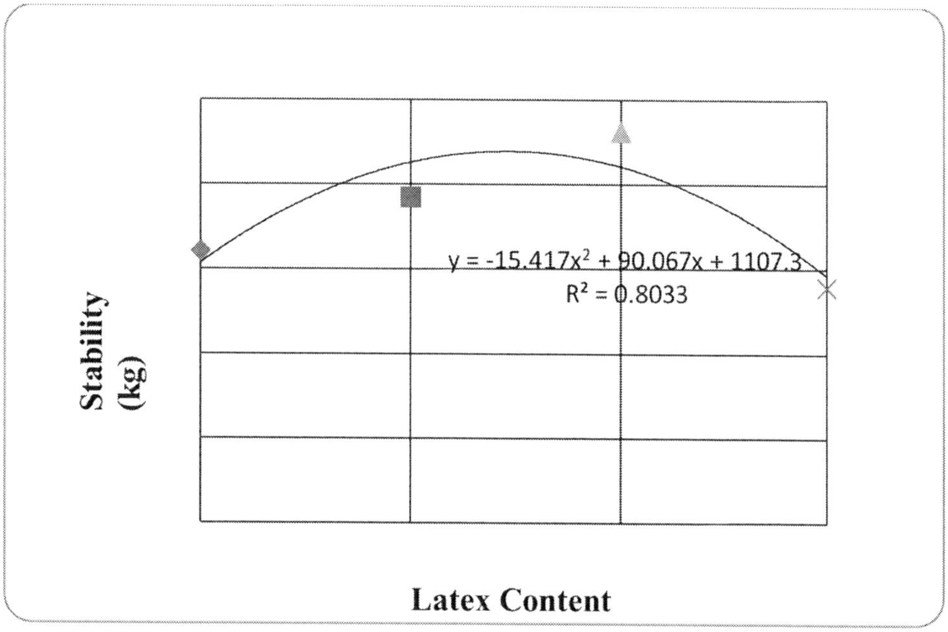

Fig. 3. Variations in the Marshall stabilities of ACWC against Latex Content by Marshall standard tests.

The Marshall immersion tests, Fig. 4, found that stability increased with the rise in latex levels from 0 % - 4 %, but decreased when the latex content was increased. Further the mix with 4 % latex had the best stability. Adding 4 % latex made the bitumen film more stable and the more durable. That mixtures with latex contents of over 4 % lost stability was possibly due to the fact this latex excessive and did not blend well.

Fig. 3 and Fig. 4 show that values of Marshall stability under immersion conditions were lower than those of the standard condition for each level of latex content. This was due to the effect of water on Marshall stability values. The absorbed water may have penetrated between the binder film and the aggregate surface causing a weakening in the bonding and adhesion between the aggregate particles, so that immersion of the mixtures decreased the Marshall stability.

Cohesion is developed in a mastic and is influenced by the rheology of the filled binder. The cohesive strength of a mastic is a function of the interaction between the asphalt cement and mineral filler, not just of the individual components alone. The cohesive strength of a mastic is weakened due to the presence of water through increased saturation and void swelling or expansion [11].

DOI: 10.1051/matecconf/20179701033

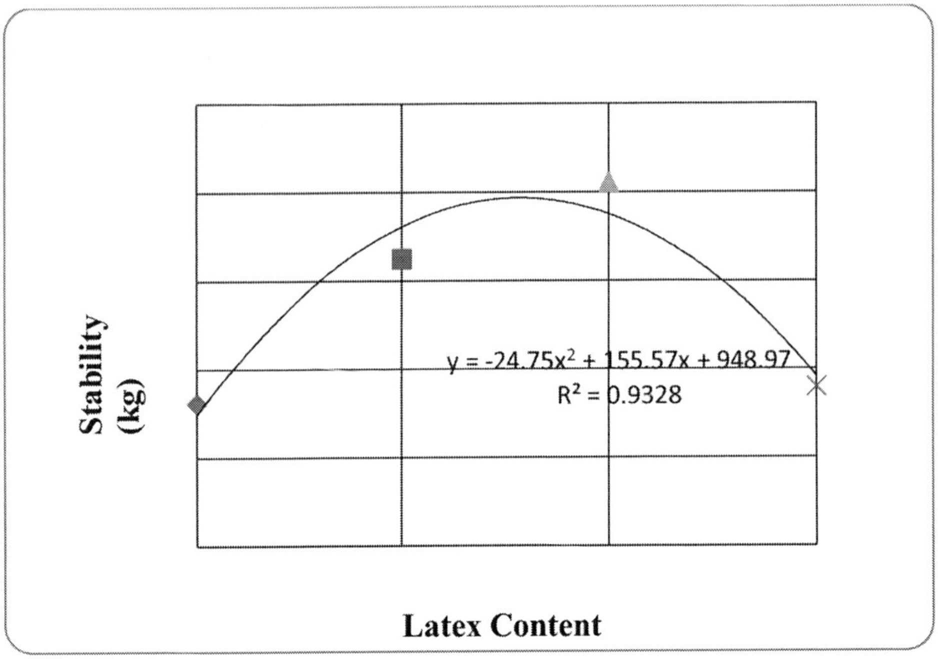

Fig. 4. Variations in the Marshall stabilities of ACWC between differing Latex Contents from Marshall immersion tests.

Fig. 5. Shows that retained stability increased with the increase in latex content from 0% to 4% but decreased with higer latex content. The addition of 2% or 4% latex appears to improve mix resistance to water significanltly as indicated by the retained stability. A smaller improvement is apparent when 6% latex is added.The mix with 4% latex had the highest retained stability value. This is attributed to the fact that the addition of 4% latex increased the resistance of the ACWC mix to water. In addition, the stability of the mix with no latex (0%) differed by 12% from that with 4%. However, there may be practical problems in producing a uniform blend at this low latex content which corerresponds to a dry rubber content of 2.4%.

Some studies on adding latex to asphalt mixes also showed that it improve the binder and the mixtures performance. The binder test results showed that all the asphat rubber (AR) binders had improved viscosity–temperature susceptibility [12]. By using asphalt rubber as a binder, the film thickness is increased to a value of 19–36 lm compared to the typical densegraded Hot Mix Asphalt (HMA) film thickness of about 9 lm [13].

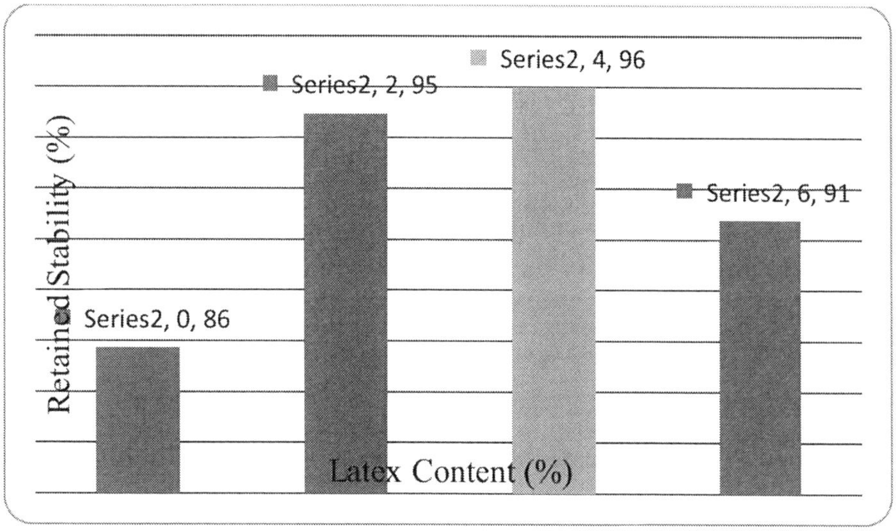

Fig. 5. Variations in ACWC retained Marshall stabilities with differing Latex Contents.

4 Conclusion

The results of this study lead to the following conclusions: That the Marshall immersion test results indicated that the addition of up to 4 % latex to ACWC mix increases the retained Marshall stability, that the addition of latex above 4 % decreased the retained stability of the mixture and that the addition of 4 % latex significantly improved mix resistance to water and was the best latex – ACWC mix.

References

1 K.D. Stuart, *A state of art report.* (Research Development and Technology, Turner-Fairbank Highway Research Center, 1990)
2 P. Rantetoding, National Symposium Highway Surfacing (1986)
3 C. Gorkem, B. Sengoz: Constr. Build Mater., **23**, 2227 (2009)
4 R. Roque, B. Birgisson, C. Drakos, G. Sholar, *Guidelines for use of modified binders.* (Florida Department of Transportation, Florida, 2005)
5 H. Siswanto, *The Research and Studies II Technological and Professional Skill Development Sector Project* (Bandung, 2006)
6 N. Vichitcholchai, J. Panmai, N. Na-Ranong, Rubberthai Journal, **1**, 39 (2012)
7 Transport Road Research Laboratory, *Road Note 36 Specification for the Manufacture and Use of Rubberised Bituminous road materials and binders* (HMSO, 1968)
8 A.E.A. Behiry, Ain Shams Eng. J., **4**, 353 (2013)
9 A. Hadidy, A. T. Hameed, TJES, **18**, 70 (2011)
10 A. Mishhadani, H.A. Baidhani, H.H. Zghair, TJES, **21**, 14 (2014)
11 Iowa State University, *Evaluation of hot mix asphalt moisture sensitivity using the Nottingham asphalt test equipment,* (IHRB Project TR-555, final report, 2010).
12 K.E. Kaloush, Constr. Build. Mater., **67**, 260 (2014)
13 Way G, Sousa J. Kaloush K, *International RILEM symposium on performance testing and evaluation of bituminous materials, PTEBM'03*, 543 (2003)

Fire Related Temperature Resistance of Fly Ash Based Geopolymer Mortar

R. Jeyalakshmi[1,*], *M.* Sivasakthi[2], *T.* Revathi[2], *N.P.* Rajamane[3]

[1]Professor, Department of Chemistry, SRM University, Kattankulathur, 603 203, India
[2]Research Scholar, Department of Chemistry, SRM University, Kattankulathur, 603 203 India
[3]Head, CACR, SRM University, Kattankulathur, 603203, India

Abstract. The study presented in this paper is on the effect of heat treatment on fly ash based geopolymer mortar synthesized from fly ash (Class F –Low lime) using alkaline binary activator solution containing sodium hydroxide (18 M) and sodium silicate solution (MR 2.0), cured at 80oC for 24 h. 7 days aged specimen heated at elevated temperature (200°C, 400°C, 600°C and 800°C) for the sustained period of 2hrs. The TGA/DTA analysis and thermal conductivity measurement as per ASTM C113 were carried out besides the compressive strengths. The thermal stability of the fly ash mortar at elevated temperature was found to be high as reflected in the observed value of f800°C/f30°C being more than 1 and this ratio was raised to about 1.3 with the addition of 2% Zirconium di oxide (ZrO_2). No visible cracks were found on the specimens with and without ZrO_2 when 800°C was sustained for 4 hrs in smaller specimens of size: 50 mm diameter x 100 mm height and in also bigger size specimens: 22 cm x 11 cm x 7 cm) specimens. TGA/DTA analysis of the geopolymer paste showed that the retention of mass was around 90%. The addition of ZrO2 improved thermal resistance. The micro structure of the matrix found to be intact even at elevated temperature that was evident from the FESEM studies.

1 Introduction

Geopolymers (GPs) are systems of inorganic binders proposed by Prof Davidovits in 1970's [1] have emerged as an alternative to Ordinary Portland cement, with superior strength, durability and environmental performances. Geopolymer technology is opt for field applications in light of today's burning issues, i.e. sustainability and need for reduction of greenhouse gas emissions from cement-based industries which is contributing to about 7% of world CO_2 emission. These inorganic alumino-silicate polymers can be synthesized from alkali activation of any material rich in silica and alumina. The reported investigations are on use of the precursor materials like metakaolin of geological origin or industrial by-products such as fly ash, blast furnace slag, red mud and rice husk ash and are activated using hydroxides of Na or K or in combination with silicates of Na or K. The

[*] Corresponding author: author@email.org

choice of the geopolymeric source materials (GSM) depends on availability, cost and type of applications. Extensive studies reported on GPs using metakaolin, fly ash and GGBS independently as sole precursors and tried to elucidate the kinetics, mechanisms and microstructure [2-7]. Geopolymers possess polymeric silicon-oxygen-aluminium three dimensional materials containing variety of amorphous to semi-crystalline phases, exhibit higher mechanical strength, much more excellent chemical, fire and heat resistances, compared to that of ordinary Portland cement [8-9]. Heat resistant properties of metakaoline based GPs were investigated and its thermal stability upto 1200-1400°C was reported. Heat resistant properties of alkali activated fly ash were investigated by several workers [10-15] and geopolymeric paste indicated high shrinkage as well as large changes in compressive strength at elevated temperature in the range 800-1200°C. To improve thermal resistance, alternate GSMs such as mine tailings, waste ceramics were tried and satisfactory thermo stable geopolymeric paste was obtained [16-17].

In the present study to achieve fire temperature resistance geopolymers, zirconium di oxide (ZrO_2), as a refractory additive was included in high alkali activation of fly ash geopolymerisation reactions. Thermal properties, TGA/DTG, thermal conductivity and thermal expansion studies were conducted and the results were presented/discussed here.

2 Materials and methods

2.1 Materials

Fly ash (FA) sample from a local Thermal Power Plant was used. To prepare activating solution, called AAS, the materials used were: industrial grade sodium silicate solution (15% Na_2O, 33% SiO_2 and 52% H_2O: molar ratio - SiO_2/ Na_2O=2.2, lye (50% NaOH) and deionized water. Mixture of sodium silicate and sodium hydroxide, as AAS was kept for 24 hrs in order to cool it to room temperature. AR grade ZrO_2 was used as an additive. Natural river bed sand (conforming to Grading Zone II of IS 383: 1970) used as fine aggregate had specific gravity of 2.58 and fineness modulus of 3.01.

2.2 Experimental procedures.

The materials were weighed accurately using digital balances. Sand and Fly Ash were first dry mixed for about 3 minutes in a digital mortar mixer till a uniform mix is achieved. Then, the AAS of required quantity was poured into it and the mixing continued for further 7 minutes until a uniform fresh mortar mix was achieved. The fresh mix was then placed into plastic cylindrical moulds of size 50mm diameter x 100mm height in three equal layers and compacted using a laboratory vibrator. After 24hrs of casting, the cylindrical moulds, containing geopolymer matrix, were kept for heat curing at 80°C for a period of 24hrs in a hot air oven. The specimens taken out of the oven were allowed to cool to room temperature before demoulding operations were carried out. Specimens were then heat treated in a high temperature furnace (capacity 1400°C, rate of heating 5°C/min) for 2 hrs at 200°C, 400°C, 600°C and 800°C. After the required high temperature is attained, the furnace was maintained for 1hr and then, it is switched off and the specimens were allowed to cool to room temperature. The compressive strength of the geopolymer mortars were measured using a CTM with the force applied at a rate of 2 KN/s. Compressive strength gains or losses of the calcined specimens were then computed. The results reported here are the average of 5 replicates.

DOI: 10.1051/matecconf/20179701034

Thermal Analysis (STA) [NETSCH 2500 Regulus] was used to study the thermal stability in the temperature range 30-1200°C in floating Air/Nitrogen (60 ml/min) with the heating rate of 10°C/min. Thermal conductivity of Mortar was examined by hot wire method according to the ASTM standard code C1113M-09 (Thermal conductivity apparatus - VBCC /TCF/1400°C-04) using the specimen size (22 cm X 11 cm x 7 cm). The required temperature was reached by the heating rate of 10°C per minute and the uniform temperature throughout the sample was maintained by holding the sample for 4 hrs at the same temperature.

Dilatometer measurements were conducted with specimens of size: 50mm diameter x 100 mm height in the stainless steel mould placed horizontally in the dilatometer tube. Co-efficient of thermal expansion at different temperatures were measured by push rod dilatometer method according to the ASTM standard E228-11(Computerized dilatometer - VBCC/DIL/1000°C /13/14). The specimens were heated upto 800°C, in the stepwise heating rate of 5°C per minute and allowed to reach thermal equilibrium at selected temperature before the length changes were recorded.

Scanning Electron Microscopy (Qunata) under vacuum mode using polished samples was utilised for microstructure characterization.

3 Results and discussion

3.1 Composition of Fly ash

Chemical composition of FA sample having sum of oxides SiO_2, Al_2O_3 and Fe_2O_3 exceeded 80% and low content of CaO (less than 5%) is in accordance with Class F defined in ASTM C 618. The total chemical composition of Fly ash, alkaline activator solution used in the sample preparation with geopolymer mortar are given in the Table1 and Table 2. Fly ash mix with Zirconium di oxide at 2% is designated as FAGP2z.

Table 1. Properties and chemical composition of Fly Ash (EDXRF analysis).

Physical Properties		**Chemical composition (wt. %)**	
Specific Gravity	2.15	CaO	2.09
Fineness (Blaine)	330 m^2/kg	SiO_2	47.55
Bulk density	1005 kg/m^3	Al_2O_3	33.45
Residue on 45 micron sieve	35 %	Na_2O	0.4
Cementing Efficiency Factor at 28 day	0.93	K2O	1.65
Loss on Ignition	1.4%	MgO	0.8
		Fe_2O_3	10.17

Table 2. Mix ID and composition of alkaline activating solution (AAS).

GPM mix ID	% FA	%ZrO_2	% FA	AAS Content (wt %)			Fly Ash /AAS ratio	FA: Sand
				Na_2O	SiO_2	H_2O		
FAGP1	100	0	100	21.5	12	66.5	0.55	1:2
FAGP2	100	0	100	26	21	53	0.60	
FAGP2z	100	2	100	26	21	53	0.60	

3.2 High temperature property

The compressive strengths of the mortar prepared at room temperature and at elevated temperature were measured. Strength development at different curing ages was followed and strength at 14days (f_R) has been taken as a reference for calculating the strength retention (f_T/f_R) at T temperature exposure. It may be generally said that the strength of alkali-activated fly ash is influenced by the modulus (MS) and the concentration of Na_2O in the alkaline activator, where the optimum performance was achieved at Na_2O concentration of 7 to 10% and at MS of 0.6 to 1. The curing conditions are the dominant factor controlling the strength of fly ash geopolymer. In our present study, it was observed that the specimens with 100% FA as source materials did not have any mechanical strength when AAS contained NaOH of 3.5M concentration; the addition of ceramic additive was also found to be ineffective. However, when the concentration of the NaOH solution in AAS was increased to 5M, (FAGP1) there was significant strength at ambient conditions; but, the test specimens lost most of their strengths after exposure to 800°C. The specimens (FAGP 2) with AAS containing NaOH solution of concentration 10M showed room temperature strength similar to that achieved with (FAGP1) 5M NaOH solution in AAS.
The relative increase or decrease in the compressive strength of each specimen as compared to its original strength before heating for the mortars as shown in Figure 1. The strength gain occurred with and without addition of zirconia, ZrO_2, with the increase of alkali concentrations at all the studied temperature range.

Visual observations showed that all specimens reacted well having smooth surface and showed no visible cracks in heated specimen. The reported literature on fly ash base geopolymer activated by hybrid activator in the similar reaction conditions showed that geopolymer paste, mortar and concrete strength level at ambient conditions are approximately same but under thermal treatment, paste underwent 73% residual loss and mortar did not retain any residual strength above 600C, even they split into two halves [15]. In our present study, the strength of geopolymer mortar specimens increased as temperature increased, attaining a peak strength at 200 °C. Subsequently, this strength was observed to deteriorate slightly for the remainder of the heating regime but maintaining the 80% strength gain to that of unheated mortar; this may be due to stronger thermal compatibility and/or sintering effect by unreacted fly ash particle. In general, the strength gain phenomenon after thermal treatment observed in small size specimens of cement or geopolymer matrixes are no longer evident in the larger size specimens. There is always greater difference in thermal gradient between hotter exterior and cooler interior of these larger specimens. Mostly thermal gradients increase with increasing size, which consequently induces thermal cracking.

DOI: 10.1051/matecconf/20179701034

The heat treatment of FAGP 2 specimens of size 22 cm x 11cm x 7 cm casted to 800°C for sustained period of 4 hrs did not show any visible surface changes.

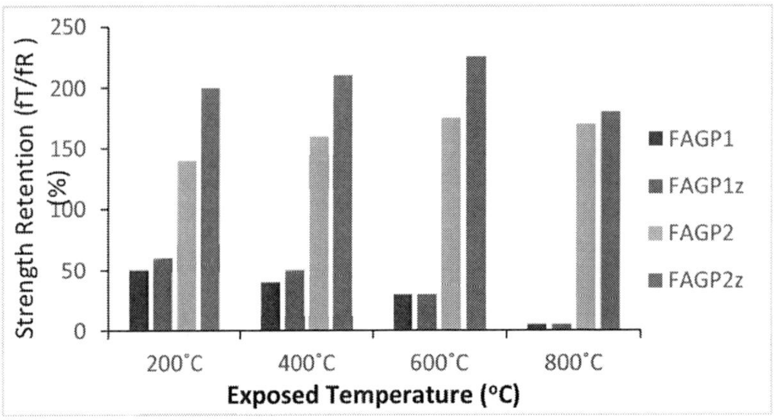

Fig. 1. Compressive Strength retention of geopolymer mortar after thermal treatment.

Fig. 2. Specimens before and after exposure to elevated temperature.

The integrity of matrix had not been affected (Figure 2), no peeling and even no cracks on the surface of the specimen after fire temperature. The water released from the geopolymerisation reaction trapped in the matrix is expelled at slower rate due to higher surface tension in the larger specimens.

3.3 Thermal analysis TGA/DTG

Figure 3 show the TGA analysis of geoplymer paste of the powedered specimens. In this TGA test, the mass loss was measured while the specimens were gradually exposed to increasing temperatures. Powdered specimens were used in TGA to ensure the achievement of thermal equillibrium during transient heating. TGA thermogram of mixes FAGP1 and FAGP2 showed that sharp decrease in weight before 250°C which is attributed to the loss of evapourable water content. After the initial rapid down, the rate of weight loss stabilised between 250°C to 780°C. After 780°C, little change in weight has occurred. After 1300°C exposure of geopolymer paste without addition of ZrO_2, the total weight loss was 12% only and retained was around 88% for both FAGP2 whereas with the addition of zirconia in FAGP2z, residual mass was found to be 90%. TG curve of FAGP2z was found to be smooth in nature. Similar TG analsis of ordinary portland cement paste showed that retention of mass was around 70% and visible crack of mortar specimen occurred around 400°C.

The decrease in weight at around 120°C indicted by the DTA endothermic curve is due to the dehydration of water molecule. The endothermic peak from the ambient temperature to 180°C and centred at approximately 67°C, may be due the dehydration of water correspondingly the weight loss of approximately 8% and is the dominant loss the whole heating regimen. The exothermic peak observed at 337°C and 686°C may be due to ordering of Ferric oxide present in the fly ash. The ZrO_2 addition does not alter the thermogram except the weight loss due to the evapouration of water absorbed on the surface during the reaction. DTA derivative curves were found to be very smooth indicate that there was no thermal reaction in the matrix with ZrO_2 till 1200°C.

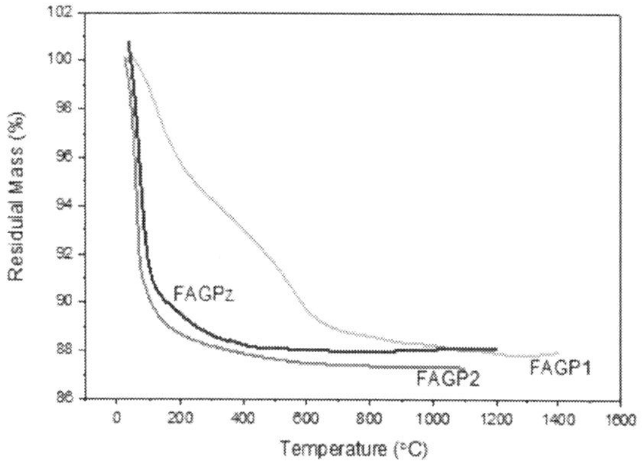

Fig. 3. TGA thermogram of the geopolymer paste.

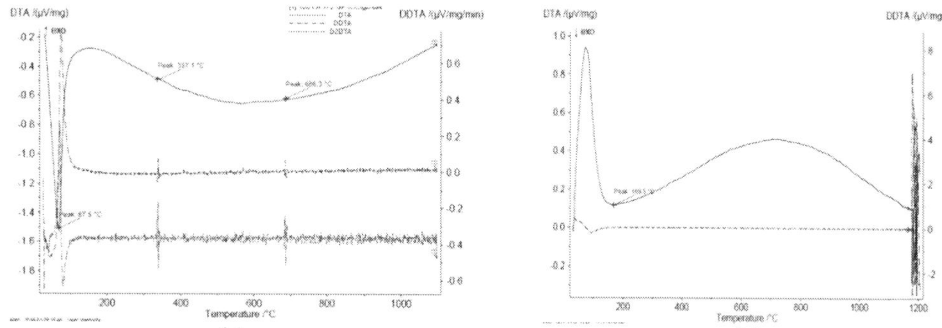

(a) (b)

Fig. 4. (a) DTA and its derivatives of FAGP2; (b) DTA and its derivatives of FAGP2z

3.4 Thermal conductivity and thermal expansion.

Thermal conducivity value of FAGP2 and FAGP2z at 80°C, and 800°C are about 1.7 and 0.3 W/m.K respectively and decreases with increasing tempeature. The thermal conductivity measurements exposed to temperature less than 120°C are only the function of conductivity of gel and the pores, at this temperature range most of the water evoporates, and hence pores in these can be assumed to contain only air.

The specimens cured at 80°C was used for dilatometric measurements which understand the thermal compatibility of the matrix and changing dimensions were noted after the attainment of thermal equilibrium. The coefficient of thermal expansion for the FAGP2z are found to be the range 0.9-1.0×10^{-6}/°C in the temperature 250-800°C. Further, it was about 2-2.5×10^{-6}/°C in the temperature below 200°C. Test data on Portland cement mortar

DOI: 10.1051/matecconf/20179701034

was found to be in the range 11-14.3×10^{-6}/°C in the 100-200°C and it was on higher value compared to that of geopolymer prepared.

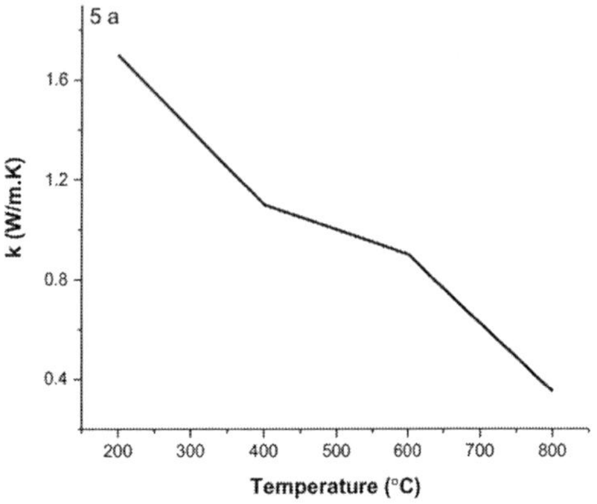

(a)

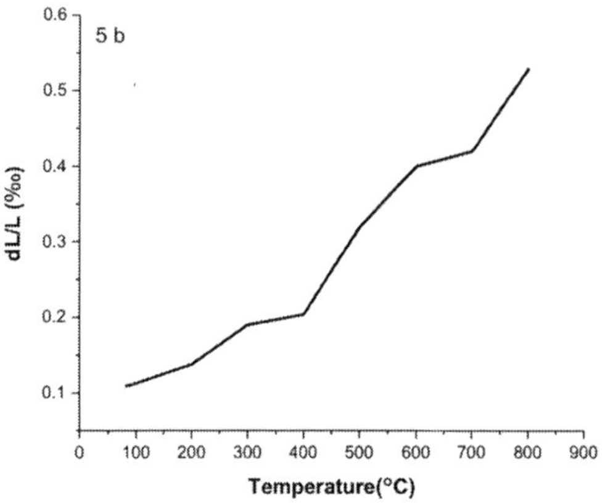

(b)

Fig. 5. (a) Thermal Conductivity at different temperature; (b) dL/Lo‰ of FAGP2 at different temperature.

The change in length for mortars after heating, the strain dL/Lo (**‰**) was calculated. Figure 5(b) is a representative curve of FAGP2z. Minor expansion were observed in the geopolymer as temperature increased from 80 to 800 with the range of values 0.109 dL/L **‰** to 0.53dL/L **‰** without any drastic shrinkage. Literature cited in the both fly ash and metakaolin based geopolymers, was appreciable shrinkage below 100°C due to release of water and as temperature rises, between 300-700°C dehydroxylation of gels causes shrinkage and second shrinkage at above 800°C due to structural densification. [18-19]. However, the studied geopolymers with and without addition of ZrO_2, has not shown any damage and negligible dimension change to the surface which may be due the lower adsorbed water on the surface of the gel.

3.5 SEM analysis

The Scanning Electron Microscopy of selected geopolymers before and after exposure are shown in the Figure 6, indicate distinct reaction product layers around the fly ash particle and the surface is more homogeneous with less porosity. At elevated temperature the rate of dissolution of the fly ash particle are enhanced and the condensation of alumino-silicate gel are obvious from the micrograph. The addition of zirconium di oxide after heating to 800°C shows compact and dense surface also the surface roughness has been reduced driven by surface tension thus resulting in good compressive strength.

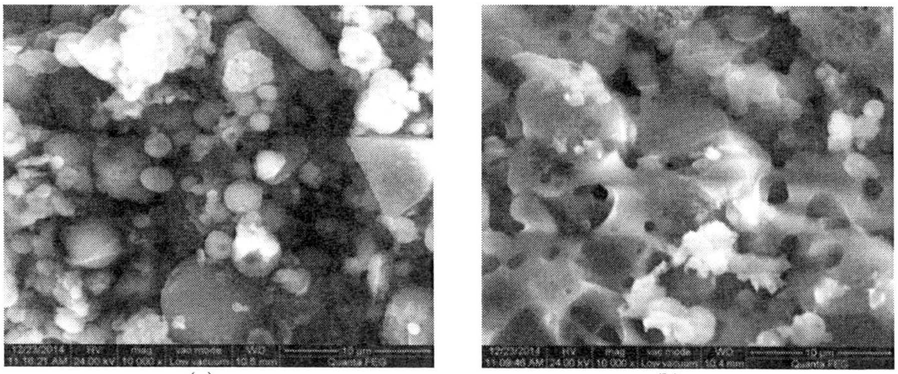

(a) (b)

Fig. 6. SEM photograph of FAGP2 before (a) and after (b) thermal exposure (800°C).

4 Conclusions

- In the current work geopolymer mortars prepared by using 100 % Fly ash activated by the sodium silicate solution with final MR of 0.8 ($SiO2/Na2O$ = 0.8) experienced strength gaining at elevated temperature heating from 200-800C without any crack or shrinkage.
- Addition of 2% zirconium di oxide improved the thermal stability further without undergoing any thermal reaction at elevated temperature.
- Higher thermal conductivity values of FAGP2 with low linear thermal expansion indicates the faster rate of heat dissipation which may protect the skeleton of alumino-silicate gel.

Authors gratefully acknowledge financial support from Department of Science and Technology (DST) under grand no DST/TSG/STS/2012/20.

References

1. Joseph Davidovits, *Geopolymer Chemistry and applications 3rd edition* (Geopolymer Institute)
2. M.W. Palomo, M.T. Grutzek, M.T. Blanco, Cem. Concr. Res., **29,** 1323 (1999)
3. J.L. Provis, J.S.J. van Deventer, *Geopolymers: Structures, Processing, Properties and Industrial Applications* (2009)
4. C. Shi, P.V. Krivenko, D.M. Roy, *Alkali-Activated Cements and Concrete* (Taylor & Francis, London and New York, 2006)
5. P. Duxson , A. Fernandez Jimenez, J.L.Provis , A.G.C. Lukey, Paloma, J.S.J. Van Deventer, J. Mater. Sci., **42**, 2917 (2007)
6. A. Fernanez Jimenez, A. palomo, M. Criado, Cem. Concr. Res., **35**, 120 (2005)
7. Chao Li, Henghu Sun, Longtu Li, Cem. Concr. Res., **40**, 1341 (2010)
8. Q. Li, H. Xu, F. Li, P. Li, L. Shen, J. Zhai, Fuel, **97**, 366 (2012)
9. M.M. Al Bakri, H. Mohammed, H. Kamarudin, I.K. Niza, Y. Zarina, J. Eng. Technol. Res., **3**, 1 (2011)
10. J. Davidovits, M. Davidovits, N . Davidovits, U.S. Patent No, 5,342,595 (1994)
11. V.F.F. Barbosa, K.J.D. MacKanzie, Mater. Res. Bull., **38**, 319 (2003)
12. Z. Pan, J.G. Sanjayan, B.V. Rangan, J. Mater. Sci., **44**, 1873 (2009)

DOI: 10.1051/matecconf/20179701034

13. T. Bakharev, Cem. Concr. Res., **36**, 1134 (2006)
14. D.L.Y. Kong, J.G. Sanjyayan, K.S. Crentsil, Cem. Concr. Res., **3**, 1583 (2007)
15. D.L.Y. Kong, J.G. Sanjyayan, Cem. Concr. Res., 40, 334 (2010)
16. Z. Sun, H. Cui, H. An, D. Tao, Y. Xu, J. Zhai, Q. Li, Constr. Build. Mater., **49**, 281 (2013)
17. M.A. Villaquiran Caicedo, R.M. de Gutierrez, S. Sulekar, C. Davis, J.C. Nino, Appl. Clay. Sci., **118**, 276 (2015)
18. J.L. Provis, C.Z. Young, P. Duxon , J.S. Van Devnter, Colloids. Surf. A Physicochem Eng. Asp., **336**, 57 (2009)
19. P. Duxon, G.C. Lukey, J.S.van Deventer, J. Non-Cryst. Solids, **352**, 5541 (2006)

An Investigation of Bond Strength of Reinforcing Bars in Fly Ash and GGBS Based Geopolymer Concrete

C. Boopalan[1,*], and *N.P.* Rajamane[2]

[1]PhD Scholar, Centre for Advanced Concrete Research (CACR), SRM University, Kattankulathur 603203, TN, India.
[2]Head, Centre for Advanced Concrete Research (CACR), SRM University, Kattankulathur 603203, TN, India.

Abstract. Geopolymers are amorphous aluminosilicate materials. Geopolymers are binders formed by alkali activation of Geopolymer Source Materials (GSM) using an alkaline activator solution. Concretes made using Geopolymer binders are excellent alternative to the Ordinary Portland Cement concretes from strength, durability, and ecological considerations. Especially, usage of industrial waste materials such as Fly Ash and Slags as GSMs considerably lower the carbon footprint of concrete and mitigate the damage due to the unscientific dumping/disposal of these materials. To use the Geopolymer concrete (GPC) for reinforced structural members, the composite action of reinforcing bars with Geopolymer concrete i.e. the bond behaviour should be well understood. This paper describes the bond behaviour of 12mm and 16mm dia. bars embedded in Fly ash and GGBS based Geopolymer concrete and conventional Portland Pozzolana cement concrete specimens investigated using the pull-out tests as per Indian Standard Code IS:2770(Part-I); the bond stresses and corresponding slips were found out. The bond stress increased with increase in compressive strength. The peak bond stress was found to be 4.3 times more than the design bond stress as per IS:456-2000. The Geopolymer concretes possess higher bond strength compared to the conventional cement concretes.

1 Introduction

Creating structures of modern world is quite impossible without concrete. Huge structures are constructed using cement concrete, which leads to the heavy usage of Portland cement. Concrete is the second most consumed substance on earth after water. Therefore, any improvement in concrete technology would influence the economy and ecology of the world over. The Portland cement, an important component of concrete, accounts for nearly 7% of the world's CO_2 emission [1], needed mining, manufacturing process, plant and machinery and transportation etc. The production of concrete involves utilization of billion

* Corresponding author: cgboopal@gmail.com

DOI: 10.1051/matecconf/20179701035

tons of natural resource materials such as lime stone, shale, clay, etc., which are actually non-renewable. Also the environment must be protected by preventing the unscientific dumping of industrial waste materials [2] such as Fly Ash (FA), Slags, Red mud etc., and in this regard, it is seen that the R&D efforts all over the world confirm that these materials have potential for converting them into binder for producing concrete. This way, the harmful environmental hazards due to these materials can be eliminated. In respect of the above environmental concerns, the Geopolymer based concretes, with much lower carbon footprint compared to that with Portland cement, show considerable promise for application in the construction industry [3].

Geopolymer concrete (GPC) utilizes Fly Ash and Blast Furnace Slag powder (generally known as GGBS), which are abundantly available as industrial waste materials as a result of the burning of coals (in Thermal Power Plants) and making of iron and steel (in Steel plants). These industrial by-products are good candidate materials [4] for creating eco-friendly binder called as geopolymers to function as cement in concretes. When compared to Ordinary Portland Cement concretes, higher compressive and tensile strengths can be achieved in GPCs which also have very low shrinkage and creep, besides being resistant to aggressive chemicals, acids, sulphates and corrosive chloride and salt water.

For the practical application of geopolymer concrete in structural members, the effectiveness of bond between the reinforcing bars and the geopolymer concrete should be established. In Reinforced concrete members, steel reinforcement receives its share of loads only from the surrounding concrete [5]. Therefore, the bond stress between concrete and steel must be ensured under various loads such as tension, compression, flexure etc., The bond action is mainly due to adhesive force of steel bar with the surrounding concrete and the mechanical interlocking of ribs on the deformed bars. This paper presents the experimental findings of bond strength obtained from pull-out test on reinforcing bars embedded in geopolymer concrete and Portland pozzolana cement concrete specimens. The bond stress versus various slip values for 12mm and 16mm diameter bars were studied and compared with geopolymer concrete and Portland pozzolana cement concrete specimens.

2 Geopolymers

In general, geopolymers are amorphous alumino silicate binder materials formed by alkali activation of silica and alumina containing powdery materials. The term 'geopolymer' was introduced in 1970's by French Scientist and Engineer Joseph Davidovits. The defining characteristics of a geopolymer is that the binding phase comprises an alkali alumino silicate gel, with Aluminium and silicon linked in a three dimensional tetrahedral gel frame work as described by sialate nomenclature of Aluminosilicate structure as represented in Fig. 1[1];

DOI: 10.1051/matecconf/20179701035

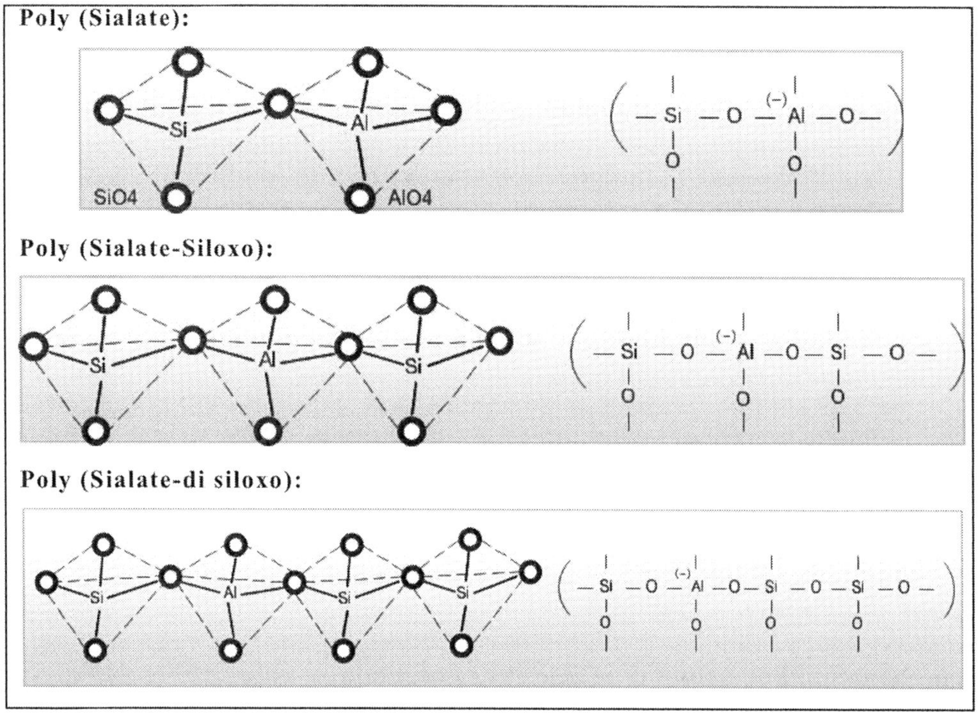

Fig. 1. Alumino Silicate Structure [1].

An alkaline liquid (NaOH or KOH) used to react with Silicon (Si) and Aluminium (Al) in a source material such as metakaolin, Fly Ash or Slags to produce gel binders. During the process of formation of geopolymer binder gel, alumino silicate materials are dissolved in alkali solution to form free SiO4 and AlO4 tetrahedral units and these units are linked alternatively. The source materials can be of geological origin i.e. naturally occurring clays are also used for geopolymerisation. Major efforts are being made worldwide to use waste materials such as fly ash, slags, red mud etc., instead of naturally occurring materials from resource conservation point of view.

2.1 Production of geopolymer concrete

The main difference between the geopolymer concrete and cement concrete is the geopolymer binder gel instead of cement paste. The geopolymer binder binds the coarse aggregate (stone jelly) and fine aggregate (sand) together to form the geopolymer concrete. Conventional methods used in the manufacture of cement concrete can also be utilized for geopolymer concrete. The alkaline liquid mixed with superplasticiser and the extra water if any, is added to the dry materials and mixed well. The fresh concrete could be handled up to 120 minutes without any setting and degradation. After moulding, the concrete can be heat cured or steam cured at 60°C for 24 hours [2]. Low calcium fly ash based geopolymer concrete can be cured in ambient condition with the addition of ground granulated blast furnace slag (GGBS) to reduce the energy needed in case of heat curing.

2.2 Properties of Geopolymer Concrete

Geopolymer concrete undergoes less creep, low shrinkage, sulphate resistance, corrosion resistance, acid resistance [4], fire resistance and no alkali-aggregate reaction. Experimental studies [6] have shown that the aggregate-binder interfaces are strong in geopolymer concretes than in the case of cement concretes. The stronger interface leads to superior mechanical properties.

Experimental investigations show that the geopolymer concrete can be well adopted for reinforced concrete structural members [3,5,7,8]. The behaviour of geopolymer concrete members such as columns and beams are similar to the reinforced concrete members. The basic assumption of reinforced concrete design of the structural members is the good composite action between the reinforcing bars and the surrounding concrete. Therefore, it is very important to study the bond behaviour of reinforcing bars. This paper presents some experimental investigations by conducting pull-out test of reinforcing bars embedded in geopolymer concrete (GPC) specimens and Portland pozzolana cement concrete (PPCC) specimens.

3 Bond strength

The mechanics of bond stresses ensure the reinforcement is solidly anchored to the surrounding concrete. Analysis and design of the reinforced concrete composite members are based on the assumption that no slippage will occur in the interface of steel and concrete. Bond stress is the shear stress at the reinforcing bar – concrete interface which by transferring load between the reinforcing bar and the surrounding concrete, modifies the steel stresses. This bond stress enables the two materials to form a composite member. The efficiency of a reinforced concrete structural member is based on the adequate composite action between the steel and concrete.

3.1 Bond stress distributions

Bond stress in reinforced concrete member arises due to two distinct situations:
1. The Anchorage bond of bars in tension or compression.
2. The Flexural bond of bars in flexural members.

3.2 Anchorage bond

The generation of bond stress is idealized as shown in Fig. 2:

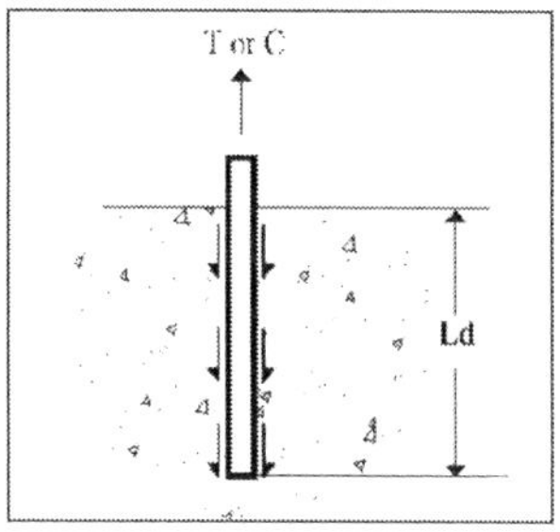

Fig. 2. Bar Anchorage.

A reinforcing bar must extend a length L_d beyond any section at which it is required to develop its full yield strength so that sufficient bond resistance is mobilized. Let τ_{ba} is the uniform bond stress along the anchorage length L_d in tension or compression,

$$\pi/4 \times d^2 \times f_s = \tau_{ba} \times \pi d \times L_d$$
$$L_d = (d\ f_s) / (4\tau_{ba}) \qquad (1)$$

where f_s=stress in steel; d=bar diameter.

3.3 Flexural Bond Stress

The magnitudes of bond stresses along the reinforcing bars are a function of the rate of change of tensile force in the reinforcing bars from section to section in flexural members. Change in the bending moment and the development of tension cracks cause the tensile force in the reinforcing bar to vary and induce variable bond stresses.

In flexural members, in regions, where the bending moment changes rapidly and the shear is consequently high, the tensile force in reinforcing bar necessarily under goes significant change in a short distance Δx as shown in Fig. 3, therefore large bond stresses called flexural bond stresses are required to equilibrate the difference in tension ΔT between the sections. This increased ΔT is capable of being transferred to the surrounding concrete without slip between the reinforcing bar and the concrete by the bond stresses between the reinforcing bar and the concrete.

Expressions for bond stresses are related to the shear, that is the measure of the rate at which the moment changes i.e. $V=\Delta M/\Delta x$,

$$\sum_0 \times \Delta x \times \tau_{bf} = \Delta T = \Delta M / jd$$
$$\tau_{bf} = V/\sum_0 jd \qquad (2)$$

where τ_{bf} = Flexural bond stress
$\sum_0$ = Surface area of a bar
jd = Internal lever arm

This paper deals with the anchorage type bond test in the form of pull out and comparison of mixes using this test is likely to be valid for flexural bond also.

DOI: 10.1051/matecconf/20179701035

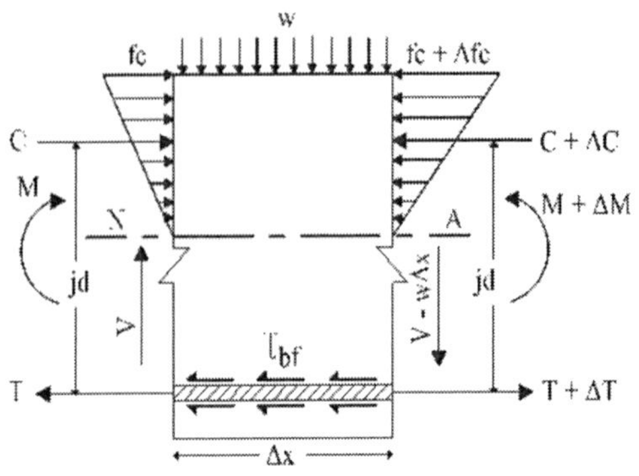

Fig. 3. Flexural Bond Stresses created by a Moment Gradient.

4 Investigation

The aim of this paper is to present some experimental findings on bond strength of embedded reinforcing bars of 12mm and 16mm dia. into geopolymer concrete cubes of 150x150x150mm size and compared with similar size cubes with Portland pozzolana Cement Concrete (PPCC). The effect of compressive strength of concretes on the bond stresses and the bond-slip relations were studied. The relation between slip and bond stresses were plotted and ultimate bond strength was found. The method of testing was carried out according to the Indian Standard Code IS: 2770(Part-I)-1997 [9].

4.1 Materials

Fly Ash and GGBS based geopolymer concrete mixes were used to make test specimens. The mixes were referred as GPC25g, with 25% GGBS and 75% FA and as GPC50g, with 50% GGBS and 50% FA contents. The Alkali Activator Solution (AAS) was prepared with the %Na_2O, %SiO_2& %H_2O for the mix GPC50g were 5.91%, 3.63% and 45.46% respectively and these values for mix GPC25g were 7.53%, 4.62% and 57.86% respectively. Typical PPCC mixes with 28% FA contents referred as PPCC35 and PPCC45 with W/C ratios 0.35 and 0.45 respectively were used to make test specimens for comparative study. The compressive strength levels of GPC and PPCC mixes are within the range of 2 to 7%. The GPC and PPCC mix proportions and details are summarized in Table 1&2.

Table 1. GPC Mix Proportion.

GPC Mix	GPC50g		GPC25g	
AAS / GSM	0.55		0.7	
	Wt Proportion	Content [Kg/m^3]	Wt Proportion	Content [Kg/m^3]
Fly Ash	0.5	212	0.75	308
GGBS	0.5	212	0.25	103
Sand	1.5	635	1.5	616
Coarse Aggregate	2.5	1059	2.5	1026
Density [kg/m^3]	2350		2340	
28day Strength [MPa]	55		45	

Table 2. PPCC Mix Proportion.

Mix ID	PPCC35		PPCC45	
W/C ratio	0.35		0.45	
	Wt Proportion	Content [kg/m^3]	Wt Proportion	Content [Kg/m^3]
Fly Ash	0.28	111	0.28	94
OPC	0.72	287	0.72	242
Sand	2.0	796	2.5	839
Coarse Aggregate	2.7	1075	3.2	1074
Density [kg/m^3]	2412		2402	
28day Strength [MPa]	56		41	

4.2 Specimens

The testing of bond in geopolymer concrete was carried out as per Indian standard Codal provision IS 2770(Part-I)-1997 [9]. Test specimens of 150x150x150 mm cubes with 12mm and 16mm dia. deformed bars conforming to IS 1786 with characteristic yield strength of 415MPa (Fig. 4). For each mix, three specimens for bond study and three 100x100x100 mm cubes for obtaining compressive strength were cast. The reinforcing bar was embedded vertically along the centre in each of cube specimen, and projected down by about 10 mm from the bottom face of the cube, and projected 650 mm upwards from the top surface of the cube for gripping the specimen while applying the load. The projecting lower end of the reinforcing bar was used to fix the dial gauge as shown in Fig. 5. The cubes were reinforced with a helix of 6 mm diameter plain mild steel reinforcing bar at 25 mm pitch. The test bar was aligned vertically at the time of compaction of the concrete. The geopolymer concrete specimens

DOI: 10.1051/matecconf/20179701035

were covered with gunny clothes for 3 days to avoid loss of AAS in the mix and allowed for ambient temperature curing.

Fig. 4. Specimens after Casting.

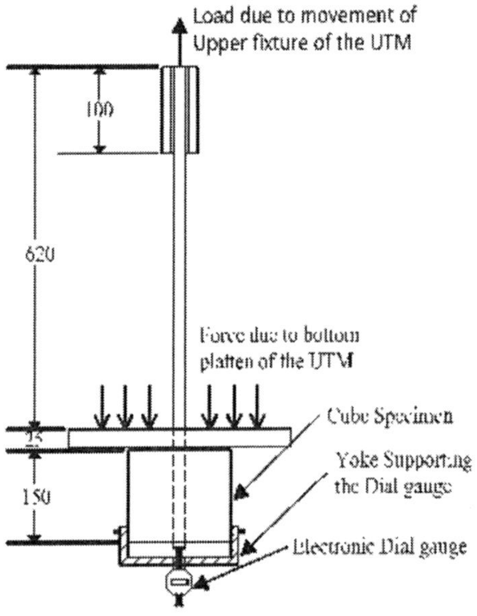

Fig. 5. Schematic Line Sketch of Test Setup.

4.3 Description of test method

The pull-out test setup shown in Fig. 6. The specimen was mounted in UTM of 40T capacity in such a manner that the bar was pulled axially from the cube. The top surface of cube was supported on plate with a hole at the centre to accommodate the reinforcing bar, which was projecting from the bearing surface. An electronic dial gauge of 0.001 mm least count was used for measuring the slip of the bar with respect to the free end of the bar.

The dial gauge was mounted on a suitable yoke, which was attached to the specimen. The pull-out action of the bar was indicated in the dial gauge

readings at different loads and loading was continued and readings were recorded until:
a) The yield point of reinforcing bar had reached.
b) The enclosing concrete failed or
c) A slip of 2.5mm occurred at the free end.

Fig. 6. Pull-out Test Setup

5 Bond test results

The average bond stress along the entire embedded length of the bar is obtained by dividing the applied load at the specified slip, by the surface area of the embedded length of bar and the expression is $\tau bd = P / (\pi dl)$; Where τbd = Bond stress in MPa; P = load in N; d = Dia of Reinforcing bar; l = Embedded length. The mean values of bond strengths on minimum of three specimens are used. The bond slip relationships for the concretes are shown in Table 3.

Table 3. Test Data on slip vs Bond Stresses.

	GPC		PPCC	
	GPC50g	**GPC25g**	**PPCC35**	**PPCC45**
AAS/GSM ratio (or) W/C for PPCC	0.55	0.7	0.35	0.45
28 days compressive strength, fc [MPa]	55	45	56	41
Slip value in mm	Test on 12mm dia deformed bar Bond stress value in MPa			
0.025	7.1	6.9	6.7	5.7
0.25	10.2	10.1	9.8	8.4
1.25	12.4	12.3	11.9	10.2
1.5	13.5	13.2	12.2	10.4
2.0	12.6	12.5	10.5	9.2
2.5	10.5	10.3	8.4	7.2
τ in IS 456:2000 0.45xSQRT(fc)	3.3	3.0	3.3	2.9
τ(Peak) / τ(IS 456)	4.1	4.4	3.6	3.6

 DOI: 10.1051/matecconf/20179701035

Slip value in mm	Test on 16mm dia deformed bar Bond stress value in MPa			
0.025	7.3	7.4	7.4	6.3
0.25	10.7	10.8	10.9	9.3
1.25	12.8	13.1	13.2	11.2
1.5	13.8	14.2	13.5	11.5
2.0	13.1	13.2	11.6	10.0
2.5	10.8	11.0	9.4	8.0
τ in IS 456:2000 0.45xSQRT(fc)	3.3	3.0	3.3	2.9
τ(Peak) / τ (IS 456)	4.2	4.7	4.0	4.0

6 Conclusion

From the test data, the following observations (Fig. 7 to 10) were made:

a) Increase of GGBS content increases the geopolymerisation and enhances compressive strength in GPC. In case of GPC with higher FA content, AAS/GSM ratio needs to be increased to enhance geopolymerisation and compressive strength. Bond stress increases with increase in compressive strength for both GPC and PPCC.

b) Bond stress in GPC is more than that of PPCC at similar strength levels. This can be due to the better adhesive bond of GPC with reinforcing bar.

c) The Bond stress increases marginally with bar diameter for both GPC and PPCC.

d) Peak Bond stress corresponding to slip for GPC were found to be higher than PPCC.

e) The design bond stress as per IS 456:2000 are presented in table. The peak bond stress obtained from investigation were found to be 4.3 times more than the design bond stress as per IS:456-2000 for GPC mixes and the same is 3.6 times more for PPCC mixes. Thus GPCs was found to possess higher bond strength compared to conventional Portland cement concretes.

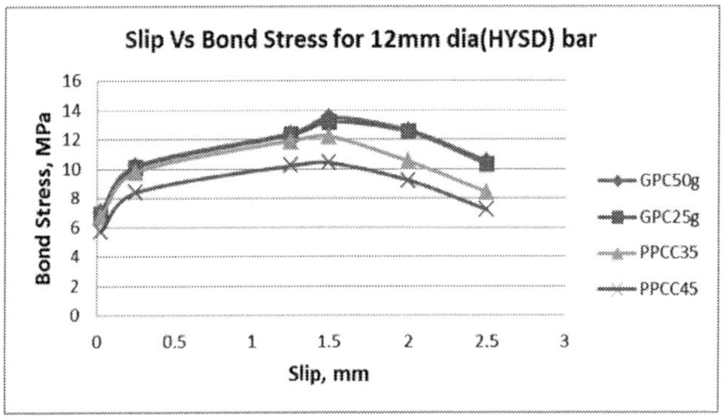

Fig. 7. Effect of Compressive strength on Bond Stress for GPC & PPCC.

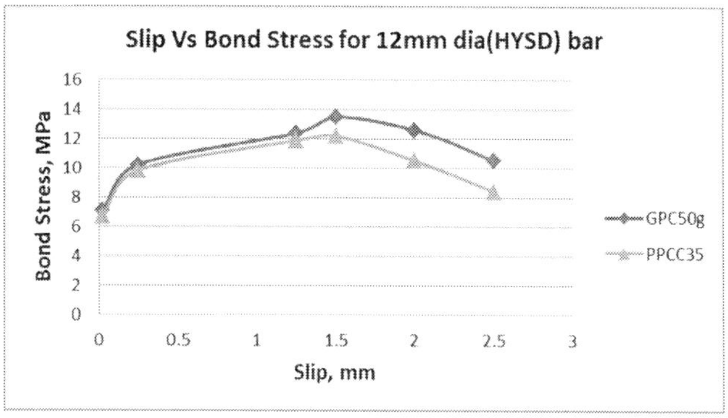

Fig. 8. Comparison of Bond strength of GPC and PPCC.

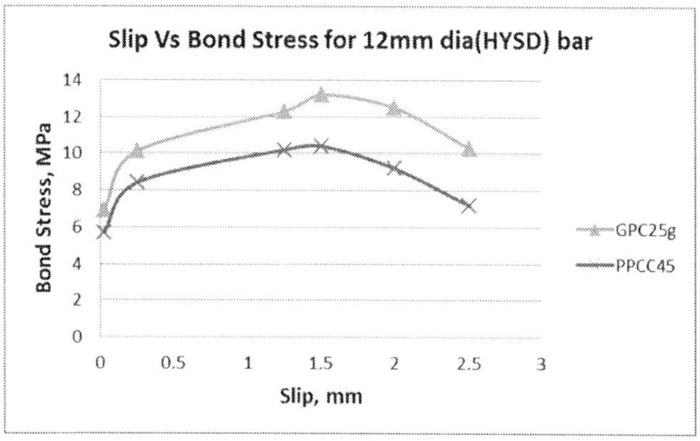

Fig. 9. Comparison of Bond strength of GPC and PPCC.

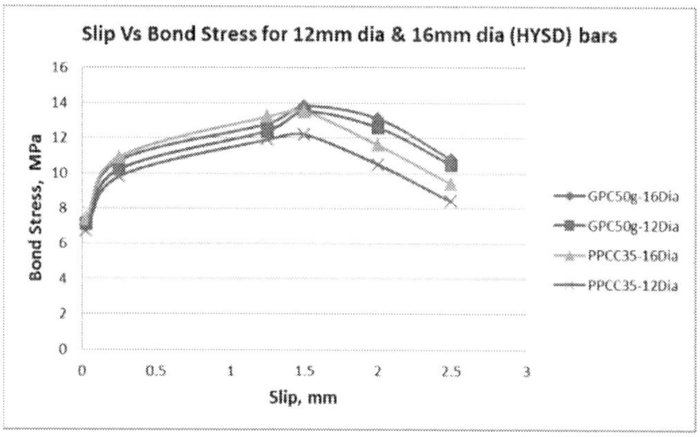

Fig. 10. Effect of bar diameter on Bond Strength for GPC & PPCC.

References

1. J. Davidovits, *Geopolymer Chemistry and Applications, Institute Geopolymer, Saint Quentin, France, 3rd Ed.* (2011)
2. B.V. Rangan, *ISTE-SRM International Journal of Civil Engineering* **1**, 1 (2015)

3. N.P. Rajamane, *Studies on Development of ambient temperature cured fly ash and GGBS based Geopolymer Concretes* (PhD thesis submitted to Visvesvaraya Technological University, Belgaum, Karnataka, 2013)
4. S.E. Wallah, B.V. Rangan, *Low-Calcium Fly Ash based Geopolymer Concrete-Long term properties* (Research Report GC2, Curtain University of Technology, Perth, Australia, 2006)
5. J.S. Kim, J.H. Park, International Journal of Civil, Environmental, Structural, Construction and Architectural Engineering, **9**, 96 (2015)
6. Lee, V. Deventer, Cem. Concr. Res., **34**, 195 (2004)
7. J.K. Dattatreya, N.P. Rajamane, D. Sabitha, P.S. Ambily, M.C.Nataraja, Int. J. Civ. Struct. Eng. Res., **2**,138 (2011)
8. D.M.J Sumajouw, D. Hardjito, S.E. Wallah, B.V. Rangan, J. Mater. Sci., **42**, 3124 (2017)
9. IS:2770, *Indian Standard methods of Testing Bond in Reinforced Concrete, Part I Pull-out test* (Bureau of Indian Standards, New Delhi, 1997)

The Effect of Corrosive Environment on Geopolymer Concrete Tensile Strength

Ridho Bayuaji[1,*], *Sigit* Darmawan[1], *Nur* Ahmad Husin[1], *Boedi* Wibowo[1], *Srie* Subekti[1], *Mohd Mustafa Al Bakri* Abdullah[2], *Totok* Ruki Biyanto[3]

[1]Civil Engineering Department, Institut Teknologi Sepuluh Nopember, Indonesia
[2]Center of Excellence Geopolymer and Green Technology (CEGeoGTech), Universiti Malaysia Perlis (UniMAP), Malaysia
[3]Physic Engineering Department, Institut Teknologi Sepuluh Nopember, Indonesia

Abstract. This study has the purpose to explore the potential of geopolymer concrete tensile strength in particular on the effects of corrosive environments. Geopolymer concrete, concrete technology used no OPC that has advantages, one of which is durability, especially for corrosive seawater environment. In addition, geopolymer concrete with polymerization mechanism does not require large energy consumption or an environmentally friendly concept. Geopolymer concrete in this study is using a type C fly ash from PT. International Power Mitsui Operation & Maintenence Indonesia (IPMOMI) Paiton. The type of alkaline activator used NaOH (14 molar) and Na_2SiO_3. Coarse and fine aggregate used are local aggregate. Geopolymer concrete molded test specimen with dimensions of (10 x 20) cm cylinder, further heating and without heating, then maintained at room temperature and seawater up to 28 days. Then to determine the mechanical properties, the tensile strength testing is done with reference. This result of study indicates the curing of geopolymer concrete at 60 ° C for 24 hours to raise the tensile strength of geopolymer concrete.

1 Introduction

The challenges to address the potential in Indonesia, which has sea area larger than the land area. It requires infrastructure that capable of supporting mobility between islands with the concept of sustainability. Normal concrete as a structural material currently still needs improvement to challenge the durability properties against seawater environment. Therefore, it is necessary to develop alternative structural materials that have better durability. In addition, community environmentalists have warned that one of the contributors of CO_2 is high enough portland cement industry [1].

Studies and research which address the challenges above are continued to be developed; the concept of the mechanism of pozzolan materials [2-4] partially substitute cement is to build extra compounds Calcium Silicate Hydrates (CSHs). These compounds are the source of strength in the concrete with the hydration process. On the other hand, geopolymer concrete is a green technology which is currently being developed [5-11] for the material

* Corresponding author: bayuaji@ce.its.ac.id

structure [12]. The main factor that becomes the appeal of geopolymer concrete is the polymerization process, the process of transforming material alumino-silicate oxide and alkaline activator into polysilicate product that has high mechanical strength without using cement. Moreover, in a previous study showed that the geopolymer concrete resistant to sulfuric and hydrochloric acid. Resistance properties are greatly influenced by the nature of the strong alkaline solution, treatment conditions, and the composition of the base material [13, 14]. Durability geopolymer concrete in corrosive environments has a compressive strength which does also have a good correlation with its tensile strength. As it is known that the normal concrete has a good and consistent correlation to compressive strength tensile strength. Therefore, this study has the purpose to explore the potential of geopolymer concrete tensile strength in particular on the effects of corrosive environments.

2 Methodology

Methodology to achieve the objectives of this study used an experimental method which is carried out in the Laboratory of Building Materials and Structures. Type alkaline activator used was NaOH (14 molar) and Na_2SiO_3 and fly ash are used come from PT IPMOMI Paiton with type C. Coarse aggregate used is local aggregates with a maximum diameter of 10 mm. Natural sand is used as a fine aggregate. The composition of geopolymer concrete building blocks to make 1 m^3 of fly ash is composed of 630 kg, 1,260 kg of gravel, sand 840 kg, 108 kg NaOH, Na_2SiO_3 162 kg, Superpalsticizier 18.9 kg, 12.6 kg of water. Tensile strength testing use standard ASTM C 496-94. Before being tested for tensile strength, geopolymer concrete specimen molded with dimensions of (10x20) cm cylinder, further treatment as described in Table 1.

Table 1. Curing Conditions on Geopolymer Concrete.

Combination	Information
1. Oven-Room (O-R)	After opening of the mold, the specimen cured at 60°C for 24 hours and subsequently cured at room temperature 27-30°C and an average humidity of 60-70%.
2. Oven-Sea (O-L)	After opening of the mold, the specimen cured at 60°C for 24 hours and subsequently cured in a corrosive environment with a temperature of 24-34°C and an average humidity of 50-92%.
3. Without Oven-Room (TO-R)	After opening the mold, then cured at room temperature 27-30°C and an average humidity of 60-70%.
4. Without Oven-Sea (TO-L)	After opening the mold, then cured in a corrosive environment with a temperature of 24-34°C and an average humidity of 50-92%.

3 Result and discussion

The test results obtained geopolymer concrete tensile strength is shown in Table 2. The compressive strength test was conducted on material testing laboratory ITS FTSP civil engineering diploma. The results of the testing of the strong sides of geopolymer concrete are as follows:

Table 2. Test Results of Geopolymer Concrete Tensile Strength.

No	Test Item Code	Curing Code	Test Strength (N)	Tensile Strength (N/mm^2)	Average (N/mm^2)
1	1A	O-R	146000	4.65	4.55
2	1B	O-R	149000	4.74	
3	1C	O-R	134000	4.27	
4	2A	O-L	136000	4.33	4.92
5	2B	O-L	150000	4.77	
6	2C	O-L	178000	5.67	
7	3A	TO-R	93000	2.96	3.16
8	3B	TO-R	105000	3.34	
9	3C	TO-R	100000	3.18	
10	4A	TO-L	101000	3.21	3.62
11	4B	TO-L	114000	3.63	
12	4C	TO-L	126000	4.01	

Tensile strength average of the test specimen cured with a temperature of 60°C for 24 hours and then placed at room temperature (O-R) have a strong sides that are lower than the specimen cured with a temperature of 60° C for 24 hours and then soaked in sea water (O-L). Immersion in sea water raises the tensile strength of geopolymer concrete at 8.15%. The specimens cured at room temperature (TO-R) has a tensile strength which is lower than the cured test specimens at room temperature and then cured in sea water (TO-L). Immersion in seawater raises the tensile strength as much as 14.43% of geopolymer concrete. The average compressive strength of the test specimen cured with a temperature of 60° C for 24 hours and then placed at room temperature (O-R) have a strong sides higher than specimens cured at room temperature (TO-R). Curing with a temperature of 60°C for 24 hours to raise tensile strength as much as 43.96% of geopolymer concrete. Average tensile strength of the test specimen cured with a temperature of 60°C for 24 hours and then immersed in sea water (O-L) have a tensile strength higher than specimens cured at room temperature and then cured in sea water (TO-L). Curing with a temperature of 60° C for 24 hours raises tensile strength as much as 36.07% of geopolymer concrete. Reddy et al [15] found similar trend in their experimental program were accelerated corrosion test in natural sea water.

4 Conclusion

This study suggests that the curing of geopolymer concrete at 60°C for 24 hours showed better tensile strength than cured at room temperature. Tensile strength of geopolymer concrete under the influence of chloride ion-containing environment of the sea showed better results than those cured at room temperature.

The authors would like to thank RISTEK-DIKTI and LPPM ITS for provided financial assistance to complete this laboratory research grants. In addition, the support of Laboratory of Material and Building Structure of Diploma Civil Engineering Program ITS is greatly appreciated.

DOI: 10.1051/matecconf/20179701036

References

1. J. Davidovits, World Resour. Rev., **6**, 263 (1994)
2. R. Bayuaji, R. Kurniawan, A. Yasin, H. A. Fatoni, F. Lutfi, *IOP Conference Series: Materials Science and Engineering*, 012014 (2016)
3. R. Bayuaji, Jurnal Teknologi, **75** (2015)
4. A.M. Neville, *Properties of Concrete* (Longman, 1995)
5. W. Supriadi, R. Bayuaji, R. Burhan, H. Fansuri, Mater. Sci. Forum, **841** (2016)
6. A. Kusbiantoro, N. Rahman, S.C. Chin, R. Bayu Aji, Mater. Sci. Forum, **841**,16 (2016)
7. H. Fansuri, A. Fatmawati, W. P. Utomo, W. Supriadi, R. Bayuaji, Mater. Sci. Forum, **841** (2016)
8. J. J. Ekaputri, Mater. Sci. Forum, **841** (2016)
9. L.Y. Dwijayanti, J.J. Ekaputri, R. Bayuaji, Mater. Sci. Forum, **841**,140 (2016)
10. M.S.D. Ridho Bayuaji, Boedi Wibowo, Nur Ahmad Husin, Srie Subekti, Januarti Jaya Ekaputri, Applied Mechanics and Materials, **754**, 400 (2015)
11. J.J. Ekaputri, M. Bahrul Ulum, B. Ridho, T.E. Susanto, M.M. Al Bakri Abdullah, Applied Mechanics and Materials, **754**, 320 (2015)
12. R. Bayuaji, M.F. Nuruddin, S. Francis, J. Ekaputri, S. Junaedi, H. Fansuri, Mater. Sci. Forum, **803**, 49 (2015)
13. A. Allahverdi, *International Cement Conference* (Iran, 2004)
14. D. Hardjito, S.E. Wallah, D. M.J. Sumajouw, B.V. Rangan, Australian Journal of Structural Engineering, **6**, 1 (2005)
15. D.V. Reddy, J.B. Edouard, K. Sobhan, J. Mater. Civ. Eng., **25**, 781 (2013)

Ultra-Low Power CMOS RF Mixer for Wireless Sensor Networks Application: A Review

S.A.Z Murad[1,*], *Muhammad M.* Ramli[1] , *A.* Azizan[1] , *M.N.M* Yasin[1] , and *I.S* Ishak[1]

[1]School of Microelectronic Engineering, University Malaysia Perlis, Pauh Putra Campus, 02600 Arau, Perlis, Malaysia Sciences

Abstract. This paper reviews ultra-low power CMOS RF mixer for wireless sensor networks (WSN) application. The study elaborates several types of mixer, mixer's performance parameters, the mixer's topologies and also the design technique to improve the mixer performance. The previously published works are compared and the best topology for low power mixer is discussed.

1 Introduction

Wireless sensor networks (WSN) are self-organizing network system which contains a large number of sensors using wireless communication technology [1-2]. In general, the WSN almost designed for data collecting, processing and transmitting in a wide range of applications such as military and national security, home automation, environmental monitoring, industrial control and so on. Brief in [3] the WSN is significant in military and national security because it produces an efficient system where it has a self-organizing and fault tolerance characteristics mostly in military command, control, intelligence, surveillance, reconnaissance and targeting systems. For home automation, as a rapid demand on advance technology, the WSN is able to provide a smart sensor nodes and actuators by burying the sensors in appliances like vacuum cleaners, refrigerators and micro-wave ovens [4]. These sensor nodes inside the devices is working through interactions with each other and with the external network via the Internet or Satellite [3] and it allows the end users to manage home devices easily either in local or remote [4]. In addition, the WSN also a well-used sensors in the environmental monitoring like in eco-systems monitoring, pollution monitoring (air, water, soil, noise and etc.) and forest exploration; irrigation systems; meteorological; agriculture and also medical applications [3, 5-6]. Fig. 1 shows the example of an overall system architecture supporting WSN applications which consist of RF transceiver and baseband processor.

Classically, the RF transceiver is built from several components such as low pass filter (LPF), mixer, power amplifier (PA), low noise amplifier (LNA) and voltage control oscillator (VCO). A mixer is an important component in a radio transceiver whose function is to translate the signal frequency to a higher or lower spectrum commonly generated by

* Corresponding author: sohiful@unimap.edu.my

 DOI: 10.1051/matecconf/20179701037

the multiplication of two signals [8]. Furthermore, the details of the mixer are discussed as section f

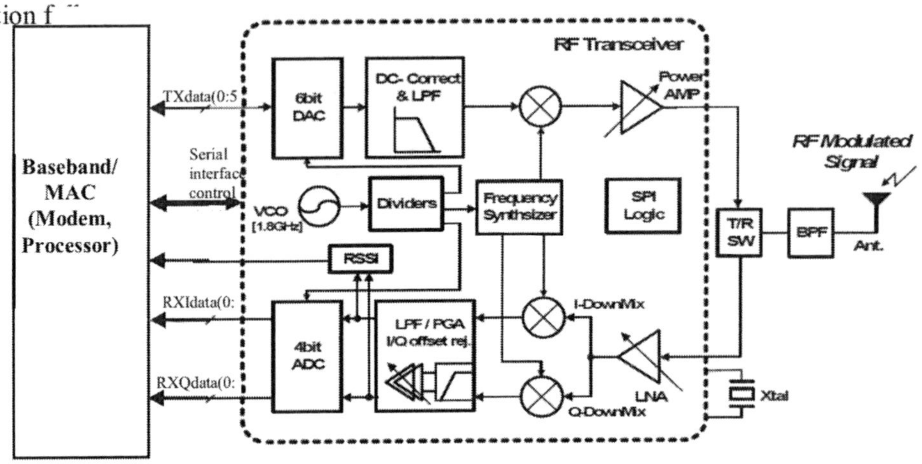

Fig. 1. Caption Example of an overall system architecture supporting WSN applications [7].

2 Mixer's Topology

The mixer circuit comprises a different topology which is suitable for a low power transmitters and receivers system. Based on previous researchers work there are three structures of mixer topologies which are a Gilbert-cell mixer, a bulk-driven mixer and a folded cascode mixer (FCM).

2.1 Gilbert-cell mixer

A Gilbert-cell is a conventional mixer whose frequently uses and most appropriate topology in designing of the active RF Mixer compared to other mixer's topologies. The Gilbert-cell topology [1,9] presents a several advantages such as attained a high linearity as well as conversion gain at low voltage operation. Besides that, this topology also has a good local oscillator (LO) to radio frequency (RF) and local oscillator (LO) to intermediate frequency (IF) port isolation in order to release the DC offset problems. Unfortunately, this structure is generated a large current that will affected by the noise of the mixer. The Gilbert-cell is includes of two different structures which are single-balance Gilbert-cell and double-balance Gilbert-cell such illustrated in Fig. 2. A single-balance mixer is a simple structure that proposed a moderate conversion gain and linearity compared to the double-balanced mixer which provided a high linearity and increases the conversion gain of the circuit system. Besides that, the double-balanced mixer also can improve the inherent isolation between an input port as well as an output port. Nevertheless, this topology produces a higher insertion loss but it can be gained by applying the source inductance structure [11].

 DOI: 10.1051/matecconf/20179701037

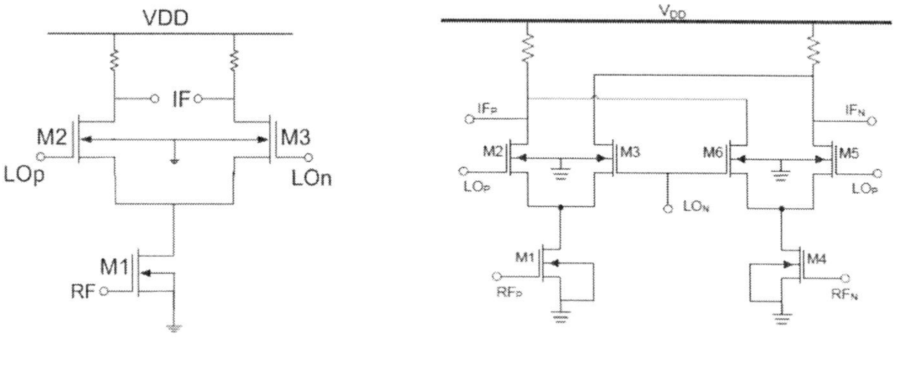

(a) (b)

Fig. 2. Conventional Gilbert-cell topology: (a) Single-balanced mixer, (b) Double-balanced mixer [11].

2.2 Bulk-driven mixer

In order to reduce the DC power consumption, the bulk-driven has been applied in designing the mixer circuit. This topology is formed by using a bulk terminal of transistor to be used as a controller terminal as well as RF and LO signals. However, this technique applies to depends on the target of performance before the bulk is connected to the required terminals [12]. Nevertheless, this topology also can produce a low voltage supply since the topology provided less stacked transistors. Fig. 3 illustrates the examples of bulk-driven topology that often used in mixer circuit design.

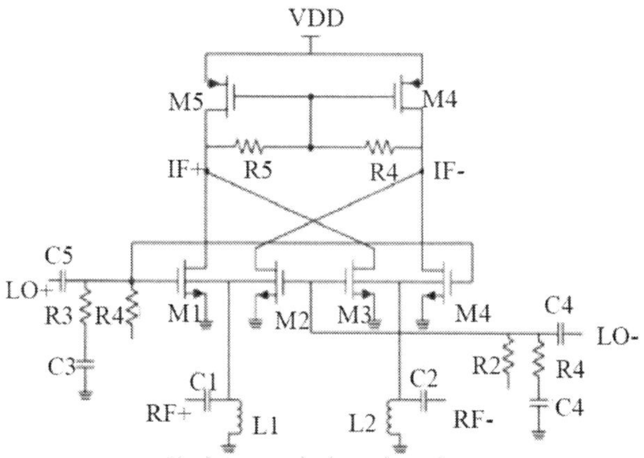

Fig. 3. Bulk-driven topology in conventional Gilbert cell mixer [13].

2.3 Folded cascode mixer

As reported in [22] the folded cascode mixer has been utilized for low power and low-voltage mixer. Ideally, the folded cascode topology is designed as Fig. 4 where the positive and negative RF input signals are differently injected into transconductance of a differential amplifier of M1 and M2. Meanwhile the IF output signals are generated from the switching stage and load stage. Essentially, the folded cascade obtained low supply voltage due to the fold configuration which has depressed the system's requirement for dc voltage [22]. Although, this fold configuration gives also a good noise performance because the current flow through the switching stage is smaller. However, the folded cascade topology is not easy in verifying the linearity of mixer circuit, but this matter can be solved by increasing the transconductance (gm) stage biasing current.

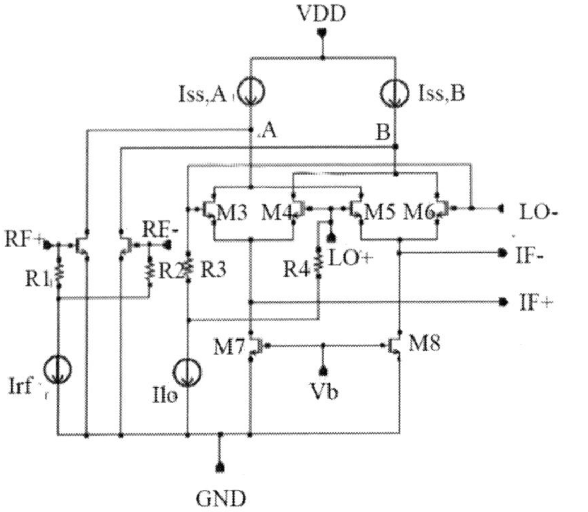

Fig. 4. Bulk- Folded cascode mixer topology [22].

3 Circuit design technique

A high-performances of the mixer circuit indicated by the parameters of the linearity as well as the conversion gain of the mixer core structure as depicted in Fig. 5. The mixer core consists of three major stages which are transconductance (gm) stage, switching stage and load stage. Therefore, there are several design techniques that can be applied in the enhancement of the linearity and also the conversion gain performances.

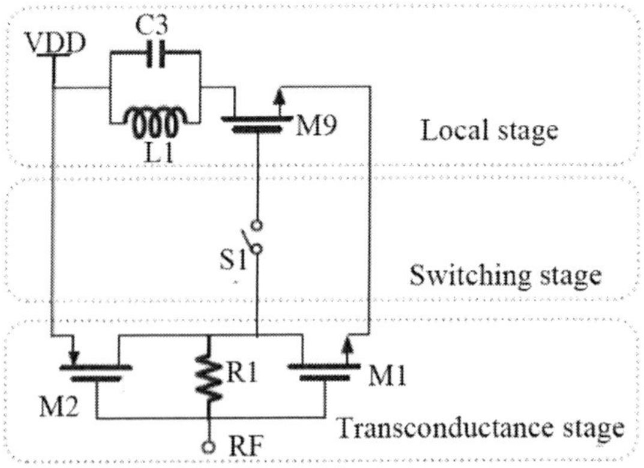

Fig. 5. Mixer core [23].

3.1 Self-bias

A self-bias technique proposed two advantages which are improving the conversion gain (CG) and also the linearity (IIP3) in the RF transconductance stage (gm). This technique is operated by the transistors alternately switched themselves using the IF input signals [15].

3.2 Current-reuse current-mirror

Current-reuse current-mirror technique is widely used consecutively to improve the gain and to reduce the power consumption of circuit systems. Fig. 6 shows the current-reuse current-mirror that has been modified from the traditional NMOS current-mirror structure. In brief, this technique can be reused the dissipated current at any stage to other stage because the total number of stacked transistor is not reduced same as the supply voltage [14, 16].

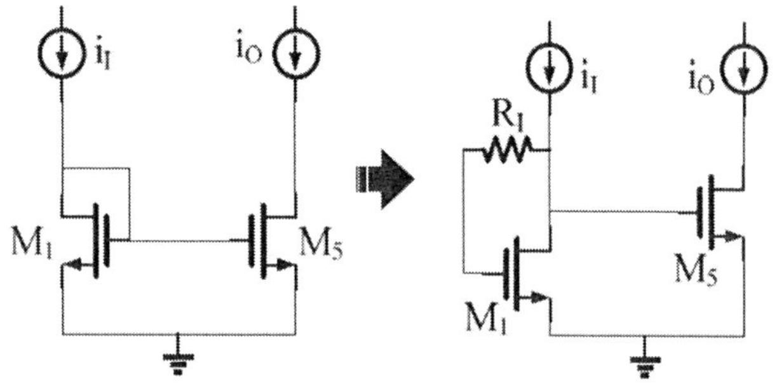

Fig. 6. The modification of traditional NMOS current-mirror to proposed current-mirror for current-reuse technique [2].

3.3 Current bleeding

The current bleeding technique is shown in Fig. 7 is particularly works as the injected current source as well as a part of the driving amplifier [2]. Commonly, this technique is constructed by connecting the RF input from the transconductance (gm) stage to the input of the bleeding stage. As reported in [17] the current is branched out in controlling the amount of a DC current through the transconductance (gm) stage and the switching stage independently. During this technique, the conversion gain of the mixer can be increase since this technique allowed a small amount of current flow through the switching stage besides reducing the gate-source capacitance (Cgs) of the switching transistor. Hence, this technique also improved the switching efficiency.

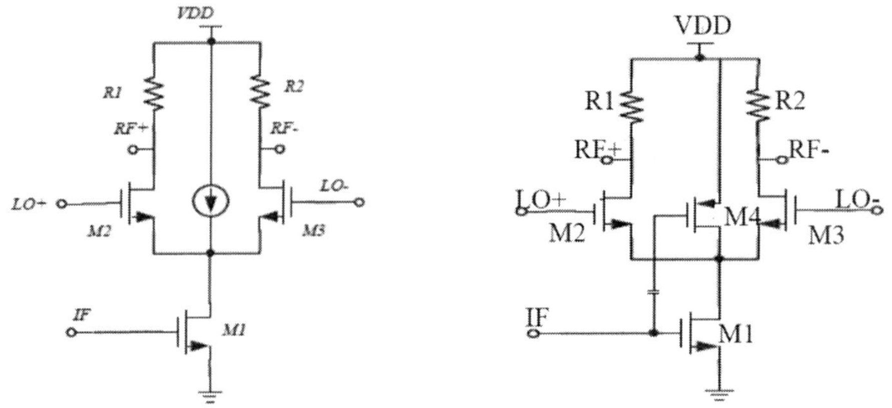

(a) Traditional (b) PMOS

Fig. 7. Current bleeding [5].

4 Summary of finding

Figures A high-performances and efficient, ultra-low power CMOS RF mixer becomes a crucial circuit block in transceiver system especially for wireless sensor networks (WSN) applications. The characteristics of low power are important parameter for mixer performance, however at the same time the high gain and good linearity also need be considered. Thus, to provide good performances of these parameters, the suitable mixer topology must be chosen. A Gilbert-cell topology with double-balanced structure proposed a several advantages which include a high conversion gain as well as high linearity. Besides that, this topology also can enhance the inherent isolation between an inputs and output port of the mixer. In contrast, this topology consumes large currents and it will affect to the noise of the mixer. However, this drawback can be improved by applying the current bleeding technique because this technique can reduce the currents in the mixer circuit system through the switching stage. On the other hand, the current-reuse current-mirror can also be used because it can dissipate the currents at any stage and reduced the power consumption in the mixer circuit.

Table 1 summarizes the performances of CMOS RF mixers that have been done in the previous works. The proposed mixers [1-2],[21-22] indicates a low power of RF mixer which is around 0.25 mW to 1.36 mW by using a low voltage operation of 1.0 V and also the current-reuse current-mirror technique. Due to the characteristics of conversion gain (CG) and linearity (IIP3), the proposed works of [18] and [24] consumes larger CG compared to the other works because the mixer circuit has been designed by using the current bleeding technique. Besides that, as shows in the work [14] and [18], the mixer circuit provides a good linearity (IIP3) and produces lowest noise figure (NF) amongst to the other works since the self-bias technique and has been used in designing the mixer circuit.

Table 1. The performance comparison of RF CMOS mixers.

Reference	[19]	[20]	[21]	[22]	[1]	[2]	[10]
Technology (μm)	0.18	0.13	0.18	0.18	0.18	0.18	0.18
V_{DD} (V)	1.8	1.0-1.2	1.2	1.0	1.0	1.0	1.0
Frequency (GHz)	2.4	2.4	2.4	2.4	2.4	2.4	2.4
Conversion gain (dB)	33	1	27.8	9.0	6.37	5.0	10.5
Noise figure (dB)	4.5	5.4	5.4	13.9	15.3	n/a	n/a
IIP3 (dBm)	-12	12-40	-23	2.2	n/a	-0.7	2.63
P1dB (dBm)	n/a	n/a	-30	-7.6	-10	-11	10.3
Current consumes (mA)	5.8	n/a	1.6	n/a	1.0	1.4	n/a
Power consumption (mW)	5.0	6.5-1.3	1.92	1.2	1.0	1.36	0.25

5 Conclusion

A review of an ultra-low power CMOS RF Mixer for wireless sensor network application has been discussed. The most popular topology is the Gilbert-cell mixer due to high isolation between the input port and output port and high conversion gain. The CMOS RF mixer can be designed in a low power by using a current-reused because it can dissipate the currents at any stage, thus reduce the power consumption.

This work was financially supported by the Fundamental Research Grant Scheme (9003-00387).

References

1. W. Chenjian, L. Zhiqun, S. Ge., J. of Semiconductor, **35**, 1 (2014).
2. S.A.Z. Murad, M.M. Shahimin, R.K. Pokharel, H. Hanaya, K. Yoshida, Micro. Opt. Technol. Lett., **54**, 923 (2012).
3. I.F. Akyildiz, W. Su, Y. Sankarasubramaniam, E. Cayirci., Computer Networks, **38**, 393 (2002).
4. E.M. Petriu, N.D. Georganas, D.C. Petriu, D. Makrakis, V.Z. Groza., IEEE Instrum. Meas. Mag., **3**, 31 (2000).
5. S.A.Z. Murad, M.M Shahimin, M.N. Md Isa, R.C. Ismail, N. Ahmad, M.N.M Yasin, Micro. Opt. Technol. Lett., **57**, 427 (2015).
6. M.L. Oliveira, P.C Rodrigues. J. Commun., **6**, 143 (2011).
7. J. Gutierrez, J.F Villa-Medina, A. Nieto, M.A Porta. IEEE Trans. Instrum. Meas., **63**, 166 (2014).
8. Information on httt://www.intechopen.com/books/wireless-sensor-network/

DOI: 10.1051/matecconf/20179701037

9. M. Voltti, T. Koivisto, E. Tiiliharju., *18th European Conference on Circuit Theory and Design*, (2007).
10. R. Mahmou, K. Faitah, International Journal of electronics and Communications (AEU), **68**, 883 (2014).
11. J.D. Chen, Z-M Lin, J.S Row, Analog Integr. Circ. S, **62**, 301 (2010).
12. Allen Gibson. *Design and Simulation of CMOS Active Mixers,* (University of Central Florida Orlando, Florida, 2011).
13. E. Abiri, M.R. Salehi, M. Bahrebar. *International Conference on Power Electronics, Systems and Applications,* (2012)
14. S. Dan, Z. Xiaolin. Chinese Journal of Aeronautics, **23**, 198 (2010)
15. L. Zhao, C. Wang. Springer Journal, Wireless Pers Communication, **80**, 277 (2015).
16. M.Kim, J. Lee, T. Yun, Electronic Letters, **48**, 1476 (2012)
17. S. He, C.E Saavedra, IEEE Trans. Micro. Theory Techn., **60**, 311 (2012)
18. G.H. Tan, R.M. Sidek, M.M. Isa, *IEEE Asia Pacific Conference on Circuits and Systems (APCCAS)*, (2014)
19. C. Choi, J. Choi, I. Nam., *IEEE 9th International Conference on ASIC (ASICON)*, (2011)
20. K. Ghosal, T. Anand, V. Chaturvedi, B. Amrutur. *19th IEEE International Conference on Electronics, Circuits and Systems (ICECS)*, (2012)
21. C.H. Li, C.Y. Hsu, C.N. Kuo. *IEEE 54th International Midwest Symposium on Circuits and Systems (MWSCAS)*, (2011).
22. M. Li, Z. Li, C. Wu, M. Zhang, J. Cao, Z. Wang, Adv. Mat. Res., **760**, 526 (2013).
23. G. Sun, Z. Li, C.Wu, M.Zhang, J.Cao, Z. Wang, Adv. Mat. Res., **760**, 516 (2013).

Development of Flat Slab – Column Interaction with Different Thickness

Ilya Binti Joohari[1,*], and *Norliyati* Binti Mohd Amin[2]

[1]School of Environmental Engineering, Universiti Malaysia Perlis, 02600 Arau, Malaysia
[2]Faculty of Civil Engineering, University Teknologi MARA, 40450 Shah Alam, Malaysia

Abstract. Flat slab – column connection has received much attention in recent years due to its simplicity of construction where beam is not required to support the slab. Flat slab has many usages in the construction field and has been investigated as a potential building material. Despite its many benefits, flat slab is easily subjected to punching shear failure. A simple approach to minimize punching shear failure is by increasing the slab thickness. It is generally accepted that the performance of flat slab is improved by using shear reinforcement; however, little attention has been paid to the selection of thickness. This research focuses on the crack behaviour of flat slab due to different slab thickness. Two specimens with different slab thickness have been constructed and tested to obtain the flat slab strength and crack pattern behaviour. Increasing the slab thickness was found to improve the strength of the flat slab and minimize the deflection and crack appearance on the slab.

1 Introduction

All A flat slab is a reinforced concrete slab that is supported directly by concrete columns without the use of beams [1]. It is widely used nowadays due to the many advantages it offers. One of the biggest contribution of flat slab is it allows faster construction [2]. It can be built faster because the formworks are simplified and reduced. Another highlight of using flab slab is it gives flexibility in room layout. It allows architect to introduce partition walls anywhere required and for owners to change the size of room layout. Flab slab also saves the building in height. It approximately saves 10% in vertical members, thus reducing foundation load [3].

Besides its various advantages, flat slab still has a weak point. It is susceptible to punching [4]. Punching shear is a type of failure when reinforced concrete slabs are subjected to localized forces. In flat slab structure, punching shear occurs at support column point. Punching shear can occur without any warning when the slab is subjected to extensive loading [5]. It can cause serious cracking and deflections.

To reduce the risk of punching shear in flat slab, a wide range of methods have been proposed. There are mainly two ways to increase the punching shear strength of concrete slabs which is by increasing the slab thickness in the vicinity of the column by providing a drop panel or a column head or second is by providing shear reinforcement [6].

* Corresponding author: ilya@unimap.edu.my

Conventionally, shear reinforcement can be in the form of stirrups or bent-up bars. However, these types of shear reinforcement have proven not to be very effective, especially in thin slabs (150-250 mm in overall thickness). Another down grade of this method is it is time consuming for construction. Recently, using steel fibers to strengthen the slab have been introduced [7]. Since it is quite new, it is rarely used in Malaysia. Thus, another practical and easy method is by increasing the slab thickness to increase its strength.

This research, based on experimental program done by other researchers on various methods, discusses the maximum shear strength of flab slab – column connection and the crack behavior for slab with different thickness. It also presents and compares the results for these two slabs.

2 Experimental work

This research is conducted to test the strength of flat slab-column connection when subjected to punching shear. A total of two flat slabs – column connection were fabricated for this research. The size of flat slab tested was 1m x 1m with thickness 100 mm for Sample 1. The column connected to the slab is of size 0.2 m x 0.2 m with height 0.15 m. The column is constructed just on top of the flat slab and exactly in the middle. The concrete grade used is C25. Steel bar of 7Y10 is used as the bottom reinforcement for both X and Y directions. The cover designed is 10 mm. Meanwhile, Sample 2 is fabricated with the exact same dimensions and material properties but only with different slab thickness which is 120 mm. No shear reinforcement is added to both samples. The laboratory test was focused on obtaining the maximum load value for both specimens and observing the very first crack pattern to the very last crack pattern.

The material used for this concrete is ordinary Portland cement (OPC), sand, aggregate and tap water. The ratio of concrete mix is 1:1: 2 which are for cement, sand and aggregate respectively. The size of sand is 2 mm and the size of aggregate is 16 mm. Both samples were concreted using the same concrete batch to ensure identical properties. The same concrete mix is also poured into a 150 mm x 150 mm x 150 mm cube mould to be tested for compressive strength of concrete. The design concrete strength is grade 25. Table 1 shows the material properties for both samples.

Table 1. Material Properties

Properties	
Type of cement	Portland
Size of sand	2 mm
Size of aggregates	16 mm
Ratio of concrete mix	1:1: 2
Water – cement ratio	0.5
Grade concrete	25 N/mm^2

The samples are loaded onto a support, ensuring the effective length of span to be 0.8 m by 0.8 m. Four concrete strain gauges are installed on the surface of the samples. The function of the strain gauges is to measure the amount of stresses generated. Two linear variable differential transformers (LVDT) are used to measure the deflections of the slab. The two LVDT's are placed at the bottom side and on one side of the slab respectively to measure vertical and horizontal displacement. Load increment of 2.5mm/min. is subjected on top of the column acting as vertical load. The samples are loaded continuously until failure.

DOI: 10.1051/matecconf/20179701038

3 Experimental result

Prior work has been documented that increasing the slab thickness with shear reinforcement improves the strength of flat slab column connection in terms of punching shear capacity. Bompa and One [1], for example, reported that the flat slab column connection with shear reinforcement improves the punching shear capacity. However, these studies have been focusing on the usage of shear reinforcement on large scale samples. In this study, the main focus is on the slab thickness of a smaller scale sample without the usage of shear reinforcement.

It is found that after testing, increasing the slab thickness even without shear reinforcement improves the strength, reduces the deflection and crack pattern on the flat slab surface. Sample 1 and Sample 2 were compared in terms of their maximum loading, deflection and crack pattern behaviour.

3.1 Load – Displacement Graph

Fig.1 shows the load – displacement graph which is obtained from the samples tested. The results of the load at first crack and ultimate failure load are tabulated in Table 2 and 3 respectively.

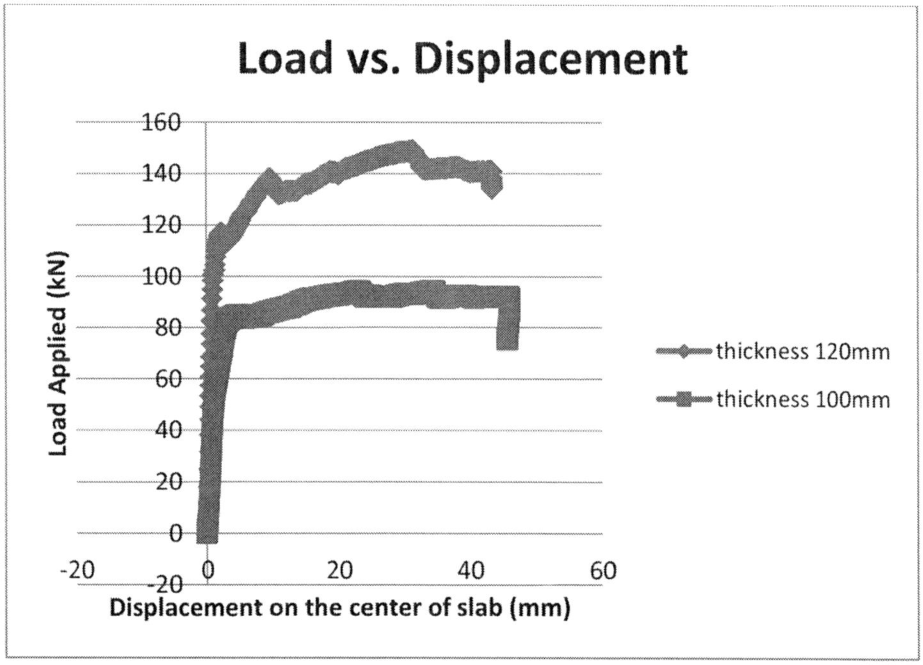

Fig. 1. Load vs. vertical displacement for Sample 1 and Sample 2

Table 2. Load at first crack

Name of Sample	Thickness of slab [mm]	Load at first crack [kN]
Sample 1	100	65
Sample 2	120	112

DOI: 10.1051/matecconf/20179701038

Table 3. Ultimate failure load

Name of sample	Thickness of slab [mm]	Ultimate failure load [kN]
Sample 1	100	93.92
Sample 2	120	148.81

The load applied on the structure causes the slab to deflect by hogging where the top part of the structure element is in compression and the bottom part is in tension. The tension failures in the structural element lead to the crack formation in the slabs.

It can be observed that Sample 2 has a higher load value at first crack compared to Sample 1 by 72.3%. The first crack for Sample 2 occurs at a load of 112 kN and at a displacement of 2.63 mm. 65 kN was the load value for Sample 1 at its first crack at a displacement of 1.96 mm. This shows that Sample 2 has a higher punching shear capacity for load at first crack. From the above discussion it is clear that an increase in slab thickness will result in a higher first crack load value.

From Table 3 it is observed that Sample 2 possess higher ultimate failure load compare to Sample 1. Sample 2 with thickness 120 mm show 148.81 kN to cause ultimate failure, which is a higher load in comparison with Sample 1. Sample 1 with slab thickness 100 mm shows lesser value which is 93.92 kN. This confirms that Sample 2 is superior to Sample 1 for load at ultimate failure. Sample 2 shows an increase of punching shear capacity compare to Sample 1. The increase is about 58.4% in punching shear strength for ultimate load failure.

3.2 Crack pattern behaviors

Both samples were tested by being loaded under a centric vertical load right on top of the column. Cracks propagations were recorded to study the effect of punching shear capacity with crack patterns developed. Both samples failed in punching shear. For illustration, photos of the crack patterns on the bottom face of Sample 1 and 2 are shown in Fig.2 respectively.

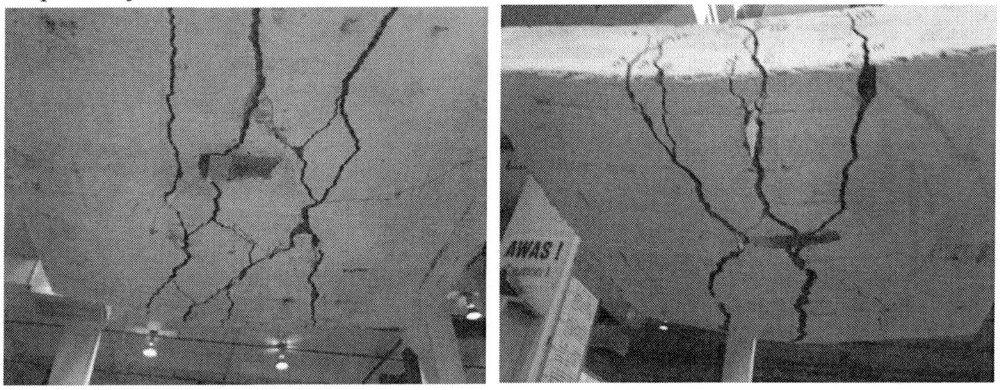

Fig. 2. Crack patterns for Sample 1 (left) and Sample 2 (right).

The crack development in both slabs followed a similar pattern at the early stage of loading. It is observed that the first crack originated at the center and gradually propagate radially towards the sides of the slab. The crack was firstly recorded at the side of slab was right at the center of the side at about 65 kN for Sample 1 and 112 kN for Sample 2. As the load applied was increased, the cracks became wider and more visible at the sides and bottom face of both samples. At the same time new cracks are formed. Besides longitudinal cracks, a circular crack area beneath the slab can be observed clearly after ultimate load failure.

Both samples failed in a brittle manner because no shear reinforcement was used. However, in terms of overall crack patterns, Sample 1 with slab thickness of 100 mm failed more severely compared to Sample 2 with slab thickness of 120 mm. A higher number of radial cracks were observed for sample 1 compare to sample 2. This can be concluded that increasing the slab thickness will increase punching shear capacity, thus decreasing crack formations.

4 Conclusion

Test results showed that all the specimens failed due to punching shear. The increase of slab thickness for the slab – column connection was effective and had improved and increased the maximum load applied and loads at first crack by 36.89% and 72.3% respectively. On the other hand, the test resulted in a decrease of 5.9% of deflection when using thicker slab.

In terms of crack, an increase in slab thickness has resulted in a decrease in the numbers of radial cracks. The surface failure of tested specimens' forms more severely overall crack patterns with thinner slab structure. Besides that, the distance between the column face and surrounding failure crack was larger in case for higher slab thickness.

Increasing slab thickness in design provides option to engineers to eliminate the use of shear reinforcement to increase punching shear capacity as installing shear reinforcement is more consuming for the project. In addition, increasing slab thickness allows designer to avoid using drop panels which can be a major architectural minus.

This work was financially supported by Universiti Malaysia Perlis.

References

1. D.V. Bompa, T. One, *Fib Symposium Tel-Aviv 2013*, (2013)
2. D.V. Bompa, T. One, *Technical University Of Cluj Napoca,* (2011)
3. D.V. Bompa,T. One, *Behaviour of Symmetric Flat-Slab Connections with Openings in the Control Perimeter* (Cluj-Napoca, Romania, 2010)
4. D.J.O. Cueva, I.N. Robertson, *Retrofit of Flat-Slab Column Connections Using CFRC Studs to Resist Punching Shear from Cyclic Loading* (Research Report, 2008)
5. R.B. Gomes, P.E. Regan, *Journal of Structural Engineering*, (1999)
6. H. Guan, Y.C. Loo, Can. J. Civ. Eng, **30**, 934 (2003)
7. M.A. Polak, *Ductility of Reinforced Concrete Flat Slab-Column Connections Department of Civil Engineering* (Blackwell Publishing, Canada, 2005)

DOI: 10.1051/matecconf/20179701039

Analytical Hierarchy Process for Natural Fiber Composites Automotive Armrest Thermoset Matrix Selection

M.U Rosli[1,*], *Mohd Riduan* Jamalludin[1], *C.Y.* Khor[1], *Muhammad Ikman* Ishak[1], *H.* Jahidi[1], *Norsyahadah* Yeop Wasir[1], *Wan Mohd* Faizal[1], *Wan Nur A'tiqah* Wan Draman[1], *Lailina* NM[1], *Ras Izzati* Ismail[1]

[1]Faculty of Engineering Technology, Universiti Malaysia Perlis (UniMAP), Level 1, Block S2, UniCITI Alam Campus, Sungai Chuchuh, 02100, Padang Besar, Perlis, Malaysia.

Abstract. The automotive industry is currently shifting to a 'green' outlook since that the popularity of natural fibers in composites plastics is accelerating in many areas and particularly the automotive industry. Nowadays, consumers are looking for vehicles more environmentally friendly and lighter in weight. For this reason, the engineers are now focusing to substitute the metal parts on utilizing the natural fiber composites. Selecting the right material in product development is a crucial decision. Imprecise decision can cause the product to be remanufactured and not in optimized condition. One of the methods that can be employed is Analytical Hierarchy Process (AHP). This paper illustrates the implementation of AHP method in order to select the most appropriate thermoset matrix for natural fiber composites automotive armrest. The selection is based on the weight reduction as the major aim of the study.

1 Introduction

In general, armrests are also found on chairs. In an automotive context especially in most modern and comfortable cars, an armrest is a function that provides the passengers a place to rest their arms. A wide armrest located between the back seats is part of interior design nowadays, if the central seating area is not in use, it can be folded out. Usually, armrests may be also equipped with other accessories such as cup holders and storage compartments. Some vehicles also offer the spot for controlller such as air-conditioner controls audio or window motors. Armrest also built into the door of the car, often forming part of the door pulling handle. Occasionally, one or two armrests may also be attached to each individual seat such as in MPVs and SUVs [1].

An armrest with a cup holder of a car as shown is Figure 1 which located at the back seat. In most design, bounded by soft polyurethane foam with an outer leather layer, the armrest has a steel frame inside [2]. The frame of the armrest is usually made by forming sheet metal which is formerly made by 2-mm-thick sheet metal which used to carry the load

* Corresponding author: uzair_rosli@yahoo.com.my

make it as the most vital part need to be focused. The steel frame should be replaced with plastics or composite materials due to the heavy weight and high manufacturing cost including punching, stamping and welding processes. By substituting with new material, the product's weight will be reduced as well as increase the production by the application of the injection molding manufacturing process.

Fig. 1. Armrest serves a cup holder function [3]

2 Material

Over the years, in European countries, the car manufacturers and suppliers for headliners, dashboards seat backs, door panels, and interior parts have been applied natural-fiber composites with thermoplastic and thermoset matrices widely [4]. The utilization of weight reduction, low cost natural fibers such as hemp, kenaf, jute, sisal, and flax put forward such advantages in return such as reductions in weight, cost and recyclability in various automotive interior and exterior parts. The manufacturing of natural-fiber composites consists of the use of either a thermoset or thermoplastic polymer binder system combined with the natural-fiber preform or mat. In applications of automotive industry, Polyester, Vinylester, and Epoxy resins are the key thermoset resins used in natural-fiber composites. Thermoset properties data is displayed in Table 1.

Table 1. Properties of Thermoset Polymers [4]

Property	Polyester resin	Vinylester resin	Epoxy
Density (g/cc)	1.2-1.5	1.2-1.4	1.1-1.4
Tensile strength (MPa)	40-90	69-83	35-100
Elastic modulus (GPa)	2-4.5	3.1-3.8	3-6
Compressive strength (MPa)	90-250	100	100-200
Elongation (%)	2	4-7	1-6
Cure shrinkage (%)	4-8	-	1-2
Izod Impact Notched (J/cm)	0.15-3.2	2.5	0.3
Water absorption (24h @ 20 c)	0.1-0.3	0.1	0.1-0.4
Cost (RM/KG) [5]	4.63-9.26/kg	9.26-18.52/kg	13.89-69.45/kg

In this paper, the material selection of thermoset matrix for natural fiber composites automotive armrest using Analytical Hierarchy Process (AHP) method is presented in results and discussion.

DOI: 10.1051/matecconf/20179701039

3 Results and discussion

Expert Choice software was utilized to select the most appropriate thermoset polymer between Polyester resin, Vinylester resin and Epoxy as depicted in Figure 2. Since that the aim was the weight and cost reduction, the property of density and cost was weighted as 18.6% or 0.186. The strength property including Elastic Modulus, Tensile Strength, Compressive Strength, Elongation, Izod impact and cure shrinkage was in second priority in this study with 9.6% or 0.096 and the least priority for water absorption with 5.7% or 0.057.

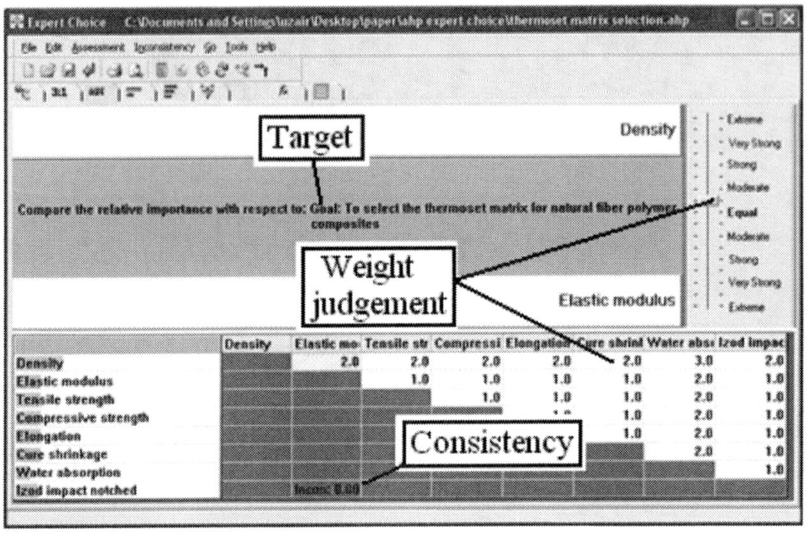

Fig. 2. Expert Choice software of AHP method

Figure 3 showed the result of each thermoset polymer alternatives with respect to each property. The weight for each alternative during the pair-wise comparison was based on the data in Table 1. The consistency analysis of the pair-wise was in best condition since the data came in numerical.

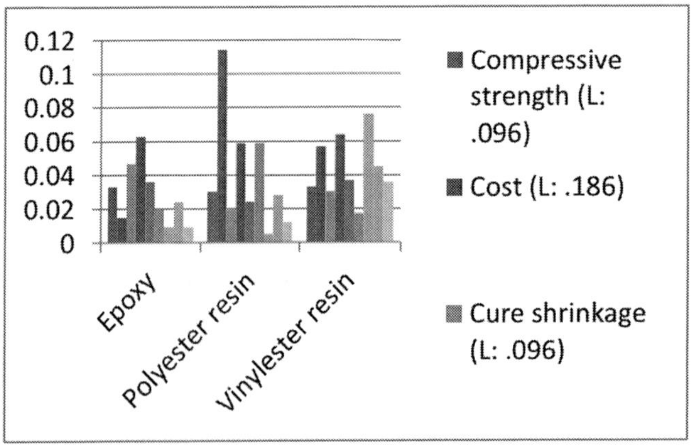

Fig. 3. Weight of each property and the thermoset alternatives

In the end result of the alternatives ranking, the Vinylester resin had the highest value (0.394 or 39.4%) among the other thermoset materials. The second highest was Polyester with (0.350 or 35.0%) and the lowest value or the last choice was Epoxy resin with only 0.256 (25.6%). Vinylester resin was preferred since it had the highest value among three alternatives.

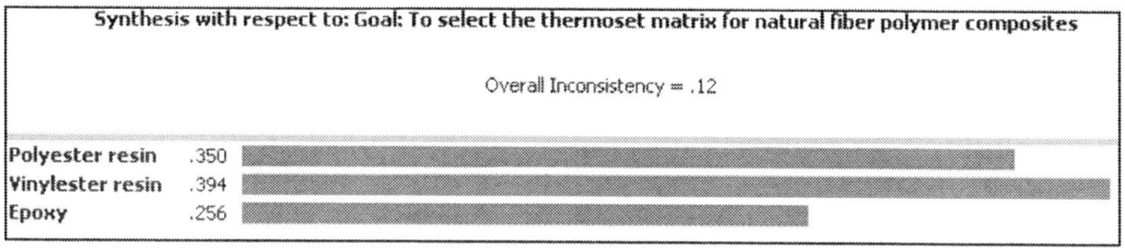

Fig. 4. Result of selection

As a new comer in the thermosetting resins family, Vinylester was a resin which usually formed by the result of reaction between epoxy resin and an ethylenically unsaturated monocarboxylic acid, with regularly used acids such as methacrylic and acrylic acid. The reaction product was then dissolved in a reactive solvent, such as styrene, to 35 - 45 percent content by weight [6]. Vinylester resins unite good thermal, excellent chemical resistance as well as mechanical properties, easeness of processes and quick cure characteristics of polyester resins. Compared with epoxies, Vinylester resin had enhanced moisture resistance when cured at room temperature. Despite of the similarity in their molecular structure to polyesters, however it differ in that the reactive sites were placed at the ends of the molecular chains which in position to absorb energy which results in a stronger material compared to polyesters [4].

4 Conclusions

As conclusion, in this study case of thermoset selection for natural fiber composites automotive armrest by the major aim of cost and weight reduction, Vinylester resin was discovered as the most appropriate material by implementing the AHP selection method. AHP could aid to evaluate the data and select the best alternative based on the criteria aspects of the decision. Application of AHP for material selection could improve the product's quality and product development process would be shortened.

The research was conducted under Fundamental Research Grant Scheme through project number 03-01-12-1108FR. The authors thank to Mr. Zaraq Shaifuddin Zamrah, Section Manager of Vendor Improvement and Education in Proton Automobile Company (Proton Sdn. Bhd) for providing the information and guidance in this research work.

References

1. R. Lueder, P. Allie, *Ergonomics Review: Armrest Design and Use*. An Ergonomics Review of the Literature for Steelcase Furniture (1997)
2. H.S. Park, X.P. Dang, Int. J. Automot. Techn., **12,** 83 (2011)
3. Information on http://nazakia.primuscore.com/category/sections/minivanssuvs/all-new-sportage
4. James Holbery, Dan Houston, Jom-Us, **58 (11)**, 80 (2006)
5. A. Brismarck, S. Mishra, T. Lampke, Natural fibres,Biopolymers and Biocomposites, 37, (2005)
6. D. Ray, B.K. Sarkar, A.K. Rana, N.R. Bose, Compos. Part A-Appl. S, **32**, 119 (2001)

DOI: 10.1051/matecconf/20179701040

Conceptual Design of Automotive Compressor for Integrated Portable Air Conditioning System

Muhammad Ikman Ishak[1,*], *C.Y.* Khor[1] , *Mohd Riduan* Jamalludin[1] , *M.U.* Rosli[1] , *Suhaimi* Shahrin[1] , *Norsyahadah* Yeop Wasir[1] , *M.S.* Zakaria[1] , *A.F.M.* Yamin[2] , *Noor Diyana* Dahlan[1] , and *Wan Nur A'tiqah* Wan Draman[1]

[1]Faculty of Engineering Technology, Universiti Malaysia Perlis, Level 1, Block S2, UniCITI Alam Campus, Sungai Chuchuh, 02100 Padang Besar, Perlis, Malaysia
[2]Faculty of Mechanical Engineering, Universiti Teknologi MARA, 13500 Permatang Pauh, Pulau Pinang, Malaysia

Abstract. This study introduces a new concept of portable air conditioner which integrated with some available components in automotive air conditioning system. This new idea intends to solve the storage problems as well as to reduce the price of current portable air conditioner since some devices could directly be used from the automotive air conditioning system. The primary emphasis of this study was on the modification of automotive compressor design so as the system may alternately be operated. The length of conventional compressor shaft is extended to place an additional clutch pulley, a drive plate and a clutch coil. The new concept particularly the shaft and pulley were analysed through slope deflection and computational finite element analyses. The result of engineering analyses exhibited that the new design of compressor shaft and clutch pulleys promote a low risk of failure as the data values recorded are lower than the critical value for each criterion investigated.

1 Introduction

An air conditioning system is a system used to acquire cooling effect. Generally, the system is used to provide human comfort and some to fulfill the industrial needs. The effect of external and internal heat load will increase the indoor air temperature that subsequently increases the room temperature [1, 2]. The mechanically-operated air conditioner may be classified into several groups such as individual room-typed, thermal storage, desiccant-based and many others including the portable-typed. Among the models of portable air conditioner that can be found are those of split system, single unit and evaporative cooler. The portable air conditioner is more demanding than the others as it provides minimal installation requirement, efficient energy consumption, economic space utilisation, ergonomic design and multi-functional system [3].

The use of portable air conditioner is not only limited for cooling a small room. It has also widely been used in outdoor travelling purposes. However, most portable air

* Corresponding author: ikman@unimap.edu.my

conditioners are heavy and thus giving difficulty for the users to move them from one place to another. The number of components used in the system is directly associated with its weight and price. Thus, one of the best ways to produce a light-weighted portable air conditioner is by reducing the number of components used. This may possibly be achieved by integrating the existing portable air conditioner with the automotive air conditioning system.

To the best of author's knowledge, the idea of integrating between both systems has not been investigated yet. Therefore, the overall aim of this study was to design a portable air conditioner which integrated with some suitable components from the automotive air conditioning system. An emphasis is placed on the conceptual design and analysis of new compressor design.

2 Design development

The main principle of automotive air conditioning system is to remove heat from the passenger compartment to ambient air. It is similar with that of portable air conditioner which is used to provide cooling effect for a limited space size. The only differences between both systems are the latter utilising larger component size and higher cooling capacity. The automotive air conditioning system consists of several components such as compressor, condenser, evaporator, receiver drier, expansion valve, hoses, tubing and magnetic clutch [4]. The system is driven by car engine via belt and magnetic clutch. The magnetic clutch which comprises clutch coil is operated by 12 V direct current. At certain rotating speed, the clutch will engage the pulley to drive the compressor. The compressor will compress refrigerant from the evaporator through suction line to a higher pressure so that the heat from passenger compartment could be removed to the outside environment.

It is noteworthy that in designing the proposed system, some modifications on the related parts of automotive air conditioning system are required. Those important parts are the compressor shaft, compressor inlet and condenser outlet. The new system also demands several additional components namely an evaporator, two flexible refrigerant lines, an electric motor and an adapter.

2.1 Compressor drive

The preferred working principle of the proposed design is to operate both automotive and integrated portable air conditioning systems alternately without interfering to one another. In order to achieve that, the length of compressor shaft is extended so that an additional pulley, drive plate and clutch coil could be installed. Moreover, an electric motor installed with another extra pulley is rigidly attached to the car chassis to drive the extended shaft portion via V-belt, whilst the clutch coil is operated by an external adapter. These modifications will enable the automotive and integrated portable air conditioning system to be driven separately by the car engine and electric motor, respectively.

The extended shaft portion is designed as a stepped cylinder where the shaft shoulders are helpful in carrying any thrust loads and axially locating shaft elements such as pulley and bearing. The shaft is idealised to be made of cold-drawn steel (ANSI 1020 steel alloy) and has a length of 226 mm which is 20% longer than the standard one. Besides, the diameter of additional pulleys that are attached to the compressor and electric motor shafts are 112 and 125 mm, respectively.

One may start to operate the integrated portable air conditioning system by turning on the electric motor and additional clutch coil adapter. The additional pulleys are then rotated and magnetic field will be generated. The clutch coil will pull the additional drive plate against the rotating pulley to drive the compressor. At the same time, the original drive plate will disengage and allow the compressor to be disconnected from the engine. If one wishes to turn on the automotive air conditioning system as before, they may reverse the process back to the original. Fig. 1 depicts the working principle of extended shaft portion elements in the proposed system.

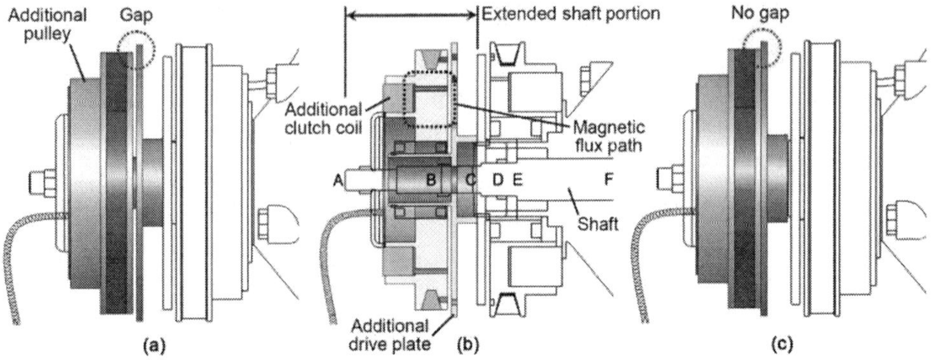

Fig. 1. The working principle of extended shaft portion elements. (a) There is a gap between the additional pulley and drive plate before the system starts. (b) As the electric motor and adapter are turned on, the additional clutch coil begins to generate magnetic field. (c) The rotating additional pulley pulls the additional drive plate to drive the compressor.

The magnitude of torque required to drive the compressor shaft by the electric motor is supposed to be equal or greater than that of car engine. It was thus determined that the cooling capacity of compressor is 6.04 kW for a space with the size of 29.5 m^2. This had been calculated based on American Society of Heating, Refrigerating and Air Conditioning Engineers (ASHRAE) handbook [5]. From thermodynamic analysis, it was evident that the power needed to run the system is 1.22 kW and therefore giving the value of torque imposed on the compressor shaft driven by the car engine is 5.83 Nm.

2.2 Ball valve

As the two systems will be alternately operated, it is therefore vitally important to control the flow of refrigerant from the evaporator to condenser such that both systems could be well performed. There are two ball valves are employed to direct the refrigerant to flow in the suction and liquid line connecting to the condenser and evaporator, respectively. The ball valve contains a hole that allows the refrigerant to flow through whenever it is in line with the inlet and outlet of the suction or liquid line.

2.3 Service port and schrader valve

The system also requires another two valves installed at both ends of detachable refrigerant line where one end is connected to a port at the condenser outlet and the other at the compressor suction inlet. The fitting at each port is equipped with a Schrader-typed valve. The valve comes with a depressor and installed at the end of the refrigerant lines. As the refrigerant line is connected to the port, the depressor pushes inward the valve stem to open the valve. The valve and refrigerant line are designed to be automatically closed by pushing back the depressor against the port to prevent refrigerant loss.

3 Engineering analysis

3.1 Slope deflection analysis

The slope deflection analysis was performed to find the maximum deflection of shaft body that probably generated while the shaft is operating. This is vitally important in shaft design as it may prevent excessive deflections which can eventually cause the edge-load bearings to failure. The method used in this analysis is Macaulay's or singularity function [6]. The analysis was conducted on the extended shaft body driven by the car engine and electric motor. The reaction forces imposed on the shaft due to ball bearing supports (at location B and E) for both drives were calculated from equilibrium force analysis. Fig. 2 illustrates all loadings applied on the shaft in free body diagram.

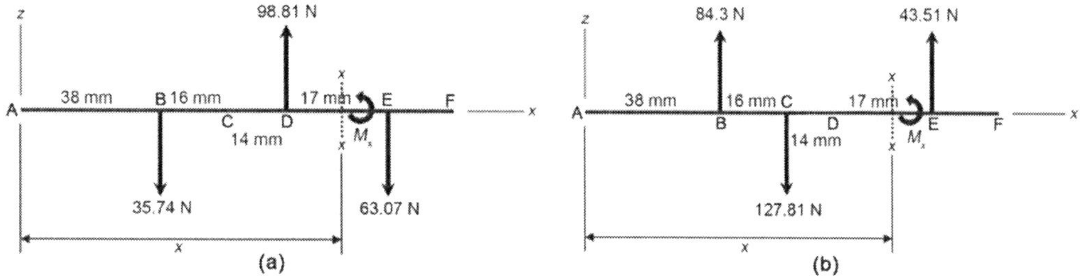

Fig. 2. The free body diagram of compressor shaft driven by (a) the car engine and (b) electric motor.

3.2 Computational finite element analysis

Another critical part of the proposed system which is the clutch pulley had been analysed via 3D finite element method. Both clutch pulleys driven by the car engine and electric motor together with the shaft body were designed using SolidWorks software. The clutch pulley and shaft were made of carbon steel and cold drawn steel alloy, respectively [7]. All materials were assumed to be isotropic, homogenous and linearly elastic [8-11]. For the car engine drive, a static torque of 5.83 Nm was applied onto the groove surface of pulley in the clockwise direction. Different magnitude of torque (8.18 Nm) was considered for the electric motor drive and applied at the same location. The contact surface between the clutch pulley and shaft was constrained in x, y and z directions to prevent any movement. The 3D models were then analysed using COSMOSXpress software to investigate the response of those parts towards the applied loading in terms of von Mises stress patterns.

4 Results and discussion

It was found in the slope deflection analysis that the electric motor drive (-0.41 μm) predicted 29% greater maximum deflection magnitude of the shaft as compared to the car engine drive (0.29 μm). Both drives seemed to share the same location where the maximum value was found which is towards the end of shaft body however occurred in different direction. The generated high value of deflection are likely to be due to a higher value of torque transferred from the electric motor (8.18 Nm) to the shaft body than that of car engine (5.83 Nm). Another possible clarification is that the additional clutch pulley connected to the electric motor has been installed and loaded slightly to the end of shaft body which leaving the shaft structure to be unevenly supported. This has resulted in a high bending moment generated at that particular area.

In terms of equivalent von Mises stress results, the clutch pulley driven by the electric motor was observed to bear a greater stress value than that of car engine. This finding was in agreement with the results of deflection analysis on the shaft body where the car engine drive recorded more favourable outcomes. Fig. 3 exhibits the von Mises stress peaked at 1.48 MPa in the electric motor drive and has no tendency to part failures since carbon steels are known can tolerate stresses up to 282.69 MPa [7]. The generated stress values were also under the yield strength of material used thus suggesting that the clutch pulleys appear to provide adequate stability.

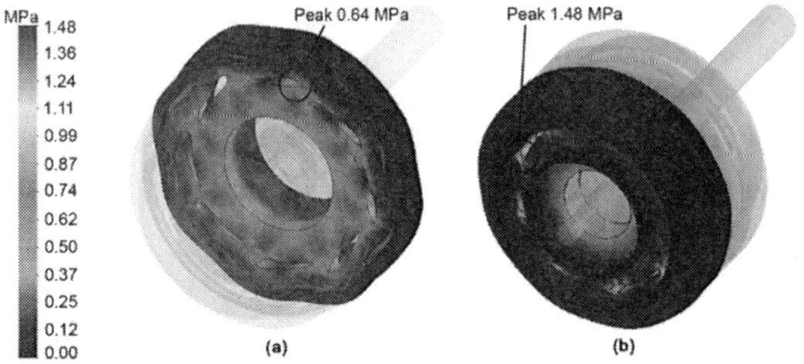

Fig. 3. Stress distribution within the clutch pulley driven by the (a) car engine and (b) electric motor.

5 Conclusions

The results of engineering analysis found that the extended compressor shaft satisfactorily sustains the bending moment loads in the slope deflection analysis for both drives. Both electric motor and car engine drives also provide adequate stability of the clutch pulleys in terms of mechanical stress outcomes obtained from the computational finite element analysis.

References

1. R. McDowall, *Fundamentals of HVAC systems* (Elsevier Science, 2007)
2. S. K. Wang, *Handbook of air conditioning and refrigeration* (McGraw-Hill Education, 2000)
3. Information on http://www.air-n-water.com
4. T. W. Birch, *Automotive heating and air conditioning* (Prentice Hall, 2005)
5. American Society of Heating, R. and A.-C. Engineers, *ASHRAE handbook: HVAC systems and equipment* (American Society of Heating, Refrigerating and Air-Conditioning Engineers, 2004)
6. F.P. Beer, E.R.J. Jr., J.T. DeWolf, *Mechanics of materials* (McGraw-Hill Education, 2005)
7. W.D. Callister, *Materials science and engineering: an introduction* (John Wiley & Sons, Inc., 2002)
8. M.I. Ishak, M.R. Abdul Kadir, E. Sulaiman, N.H. Abu Kasim, Int. J. Oral Maxillofac. Implants, **28**, 3 (2013)
9. M.I. Ishak, M.R. Abdul Kadir, E. Sulaiman, N.H. Abu Kasim, Int. J. Oral Maxillofac. Surg., **41** 9 (2012)
10. M.I. Ishak, M.R. Abdul Kadir, E. Sulaiman, N.H. Abu Kasim, Biomedical Engineering and Sciences (IECBES) (2010)
11. A. Omar, M.I. Ishak, M.N. Harun, E. Sulaiman, N.H. Abu Kasim, Applied Mechanics and Materials, **121** (2012)

Analysis of Signal Propagation in an Experiment Room with Epoxy Covered Floor for Wireless Sensor Network Applications

A. Harun[1,4,*], *S.A.Z. Murad*[1], *M.F.* Ramli[2,4], *L.M.* Kamarudin[2,4], *D.L.* Ndzi[3], *N.A.M.A* Hambali[1], *S.N.* Mohyar[1], and *M.N.* Isa[1]

[1]School of Microelectronic Engineering, Universiti Malaysia Perlis (UNIMAP), Perlis, Malaysia
[2]School of Computer and Communication Engineering, Universiti Malaysia Perlis (UNIMAP), Perlis, Malaysia
[3]School of Engineering, University of Portsmouth, Portsmouth, UK
[4]Center of Excellence for Advanced Sensor Technology (CEASTech), Universiti Malaysia Perlis, Malaysia

Abstract. As sensor applications combined with wireless network becoming more of an everyday applications, the optimal deployment becomes ever increasing important as that would be a key important factor in the trade-off between cost and link quality. This paper reports on the effect of epoxy covered floor on signal propagation characteristics in an experiment room. Microchip developed motes were used to measure signal propagation in an experiment room where sensors would be deployed extensively. The results show that the signal strength for 30 cm antenna height provides a significant margin with respect to signal noise floor. As for the 5 cm antenna height, there is still around 25 dB margin in average before the signal reaches noise floor. Analysis shows that the log-distance model is the best fit to the measured data. Free Space Loss model seemed to under estimate the overall performance of the signals. An important conclusion from this study is that wireless mote deployment must consider the margin between the two signals of antenna heights and the margin to noise floor to avoid link quality deterioration especially for sensitive data acquisition applications.

1 Introduction

It is unquestionable that sensor application coupled with wireless sensor network (WSN) has introduced tremendous amount of applications that are important and applicable in many everyday tasks and activities. As their use becomes more widespread, especially in time critical applications, so too is the need to ensure that they are reliable and can meet quality of service requirements. Detailed knowledge of wireless signal propagation within the specific environment is essential to ensure reliability in network performance [1].

* Corresponding author: aziziharun@unimap.edu.my

DOI: 10.1051/matecconf/20179701041

For specific wireless sensor deployment, such as in an experiment room, the link quality between the nodes is very important. Data are coming from various sensors located at multiple heights from the ground. Some sensors could be connected via long wires to wireless motes; however, some requires direct connection to wireless motes. To achieve the optimal deployment strategy for customized experiment room, it is necessary to exploit the entire spatial-temporal characteristics of propagation channel. In an effort to ensure the accuracy of indoor environment signal characterization, a comprehensive measurement activity has to be performed. All electromagnetic aspect of the channel has to be taken into consideration [2]. This paper reports on the effect of epoxy covered floor on signal propagation characteristic in an experiment room.

2 Literature Review

Application of WSN in data collection in customized application is one of the major uses that will have a profound impact on research communities. Until recently most studies in WSN have focused on the devices [3], protocols [4]-[6] and the network architecture [7]. Although there have been some studies focusing on signal propagation, such as in [8]-[11], specific signal propagation study for WSN network deployment in experimental room with epoxy covered floor has not been widely done. In an indoor propagation study, simple channel models such as the free space loss (FSL) given by Eq. (1) is often used.

$$L_{FSL} = -27.56 + 20\log_{10}(d) + 20\log_{10}(f) \quad (1)$$

The parameter f is the frequency in MHz; d is the distance between the isotropic transmitting and receiving antennas in meters.

Indoor propagation is very much associated with Log-distance model. Gay-Fernandez et al. [12] have shown that WSN channel can be modeled using such model. The log-distance model is described by Eq. (2) [13].

$$P_r(d) = P_{r_0} - 10\alpha\log_{10}(d) + X_\sigma \quad (2)$$

where $P_r(d)$ is the received power (in dBm) at a distance d (in meters) from the transmitter, P_{r_0} is the signal strength at 1 m antenna separation, α is the path loss exponent and $X\sigma$ represents a Gaussian random variable with zero mean and standard deviation of σ dB. Due to the short-range nature of wireless sensor nodes, proper study need to be performed to ensure careful network consideration to avoid high costing in network implementation [14][15].

The study reported in this paper uses RSSI for estimating the signal strength received at the receiver given a certain value of transmitted signal. It has been reported in [11] that the RSSI can be predicted and modeled based on average signal strength over the distance of radius centered at the receiver. The model is given by Eq. (3).

$$RSSI = -10n\log_{10}(d) + A \quad (3)$$

where n is the signal propagation constant, d is the distance between transmitting and receiving antennas and A is the average of received signal strength at 1.5 m radius. The maximum range of the mote, in this study, is defined as the length of the room to be measured which is at 13 m.

3 Equipments and measurements

This study has been performed using MICROCHIP mode utilizing Microchip manufactured RF chip which is ZIGBEE/IEEE802.15.4 compliant. The mote transmit in the 2.4 GHz – 2.5 GHz ISM band, however for this specific experiment the mote was programmed to transmit at 2.423 GHz. The mote uses supply voltage of 9.0 V and utilizes Omni-directional 8 dBi antenna. The noise floor for this mote is at -92 dBm.

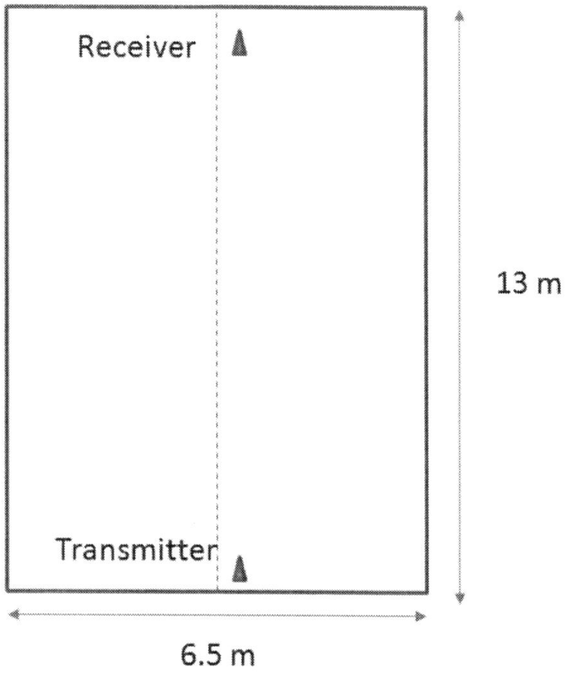

Fig. 1. Room dimension and setup of transmitter and receiver during measurement

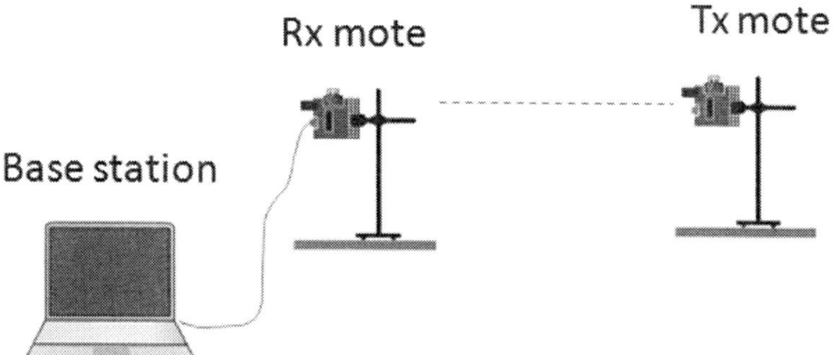

Fig. 2. Equipment setup for 30 cm antenna height

 DOI: 10.1051/matecconf/20179701041

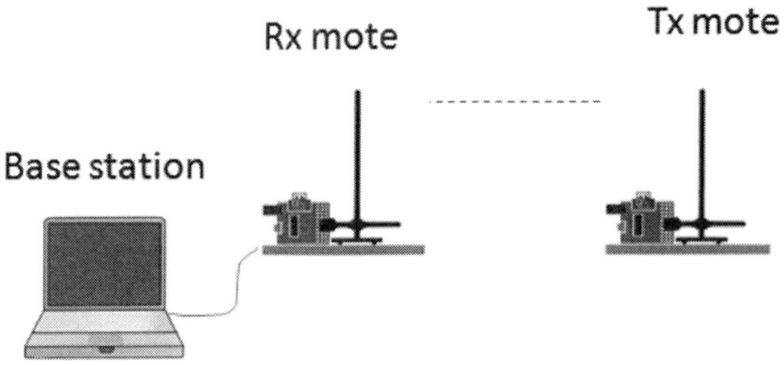

Fig. 3. Equipment setup for 5 cm antenna height

The measurements were performed in a room measuring 6.5 m by 13 m. The room is designed to be used for multiple types of sensor placements equipped with WSN motes to create network of WSN for real time data collection. The sensors are to be placed closed to the floor and around 25 cm to 30 cm all over the room for a customized application data collection. The room dimension is depicted as in Fig. 1. The maximum antenna separation was determined by the range available in the room which is limited by the room's width and length. During the study, the receiving nodes remained at a fixed position whilst the transmitting nodes were placed at specific distances from the receiver in a straight line. The experiments were carried out with antenna heights of 5 cm and 30 cm. The setting at 5 cm is to replicate placing motes on the floor for sensitive sensors requiring close proximity connection to mote while 30 cm setting is optimal mote placement for majority of the sensors. The setups are illustrated in Fig. 2 for 30 am antenna height and Fig. 3 for 5 cm antenna height.

4 Results and discussions

This section discusses the results obtained in indoor signal propagation measurement under the effect of epoxy coated floor. The results are presented based on the average RSSI values at various transmitting to receiving mote distances.

Fig. 4 compares the characteristic of signal propagation at 5 cm and 30 cm antenna heights. Signal profile with distance for both 5 cm and 30 cm antenna heights behaves almost similar for distance less than 2 m. However, after 2 m, signal profile for 5 cm antenna height shows greater power reduction that that for 30 cm antenna height. The difference in power level between the two antenna heights at 2 m until 4 m is almost 3 dB at the highest point. Signal strength for 30 cm antenna height shows an increasing trend between 5 m to 8 m distance while signal strength for 5 cm antenna height continues the reduction trend at this segment of the distance. Signal strength difference between the two antenna heights measures around 12 dB at the highest point. Nevertheless, as signal strength for 30 am antenna height shows decreasing trend after 8 m distance onward, signal strength for 5 cm antenna height starts the rebound and trending upward. At around 10 m distance signals at both antenna heights climb up for about 8 dB to 10 dB before starts showing downward trend again after around 11 m distance until signal strength for 30 cm antenna height reaches 56 dBm levels and signal strength for 5 cm antenna height reaches 64 dBm levels. Since the length of the room was limited, there are still about 28 dB margin before the signal strength hits the noise floor.

Even though the signal variation between the two antenna heights are not very distinct, there is still about 8 dB to 12 dB margin between them which implied having motes on the ground or very close to ground would reduce the signal strength significantly compared to having the motes at 30 am antenna height. As such, any signal strength reduction at transmitter should take this into consideration to avoid hitting noise floor and deteriorated signal quality.

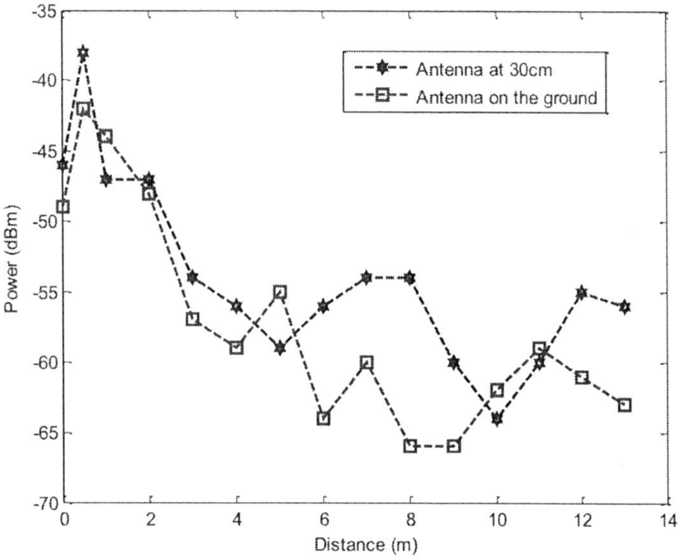

Fig. 4. RSSI variation over distance for both 30 cm antenna height and antenna on the ground (5 cm)

To model signal power variation with distance, the log-distance and FSL models were fitted to the data. Fig. 5 shows signal variation over distance for 5 cm antenna height fitted with log-distance and FSL models. The equation for log-distance model that describes the fitted model is given in the figure with root mean square error of 3.11. Based on the figure, FSL seems to have underestimated the signal strength for 5 cm antenna height on epoxy covered floor which resulted in bigger root mean square error of 11.99. Smaller variation in the signal power with distance contributed to low root mean square error for log-distance model.

Compared to results from 5 cm antenna height measurement, the signal decay with distance for 30 cm antenna height exhibited slightly larger variation in average thus contributed to higher root mean square error of 3.18 as depicted in Fig. 6.

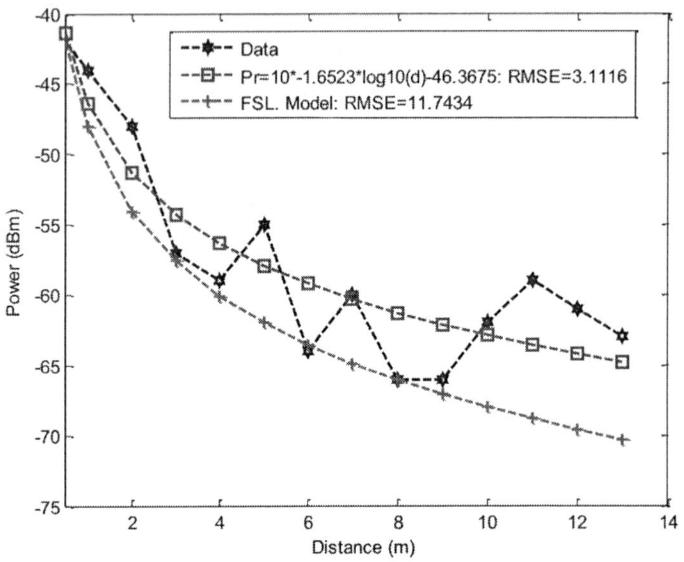

Fig. 5. 5 cm antenna height data with log-distance and FSL model fitting

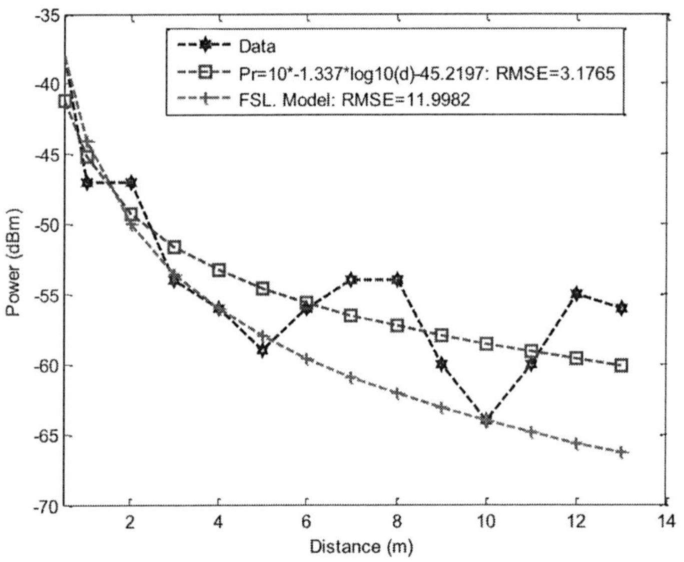

Fig. 6. 30 cm antenna height data with log-distance and FSL model fitting

It is apparent from the signal variations that both signals are undergoing severe attenuation throughout the distance in the room. Observable increase in signal strength for 30 cm antenna height at around 7.5 m to 8 m range was mainly contributed by first ground reflection occurred at 7.4 m range. Although the same reflection occurred in the signal propagation for 5 cm antenna height which occurred at 2 m range, the effect is insignificant due to the height of the antenna from ground. A summary of the log-distance model parameter values from fittings to the measured data is given in TABLE 1.

Table 1. Summary Of Fitted Log-Distance Model Parameter Values

Mote Type	Antenna height	α	P_{r_0}	RMSE (dB)	Site
Microchip	30 cm	1.33	-45.2	3.18	Epoxy floor
Microchip	5 cm	1.65	-46.3	3.11	Epoxy floor

Conclusion

This paper has presented a wireless propagation study performed in an epoxy covered room customized for wireless sensor motes deployment. The results show that the signal strength for 30 cm antenna height provides a significant margin with respect to signal noise floor. As for the 5 cm antenna height, there is still around 25 dB margin in average before the signal reaches noise floor. It is also observed that the 25 cm difference in antenna height contributed almost 12 dB margin between the two antenna heights. Although ground reflection contributed fairly to 30 am antenna height signals, the effect is not obvious for 5 cm antenna height signals.

Analysis shows that the log-distance model is the best fit to the measured data. FSL model seemed to under estimate the overall performance. An important conclusion from this study is that wireless mote deployment must consider the margin between the two signals of antenna heights and the margin to noise floor to avoid link quality deterioration especially for sensitive data acquisition project. Any power optimization attempt would also require close attention to the margins.

This work is supported financially by Ministry of Higher Education Malaysia (MOHE) under Fundamental Research Grant Scheme (9003-00387).

References

1. S. Puccinelli, M. Haenggi, IEEE Circuits Syst. Mag., **5**, 19 (2005)
2. T.H. Loh, L.R. Arnaut, *Experimental characterization of radiowave signal propagation for indoor UWM wireless communication*, *PIERS Online* **5(8)**, 2009.
3. H. Liu, Z. Meng, S. Cui, Intern. Conf. on Wireless Communications, Networking and Mobile Computing, 2344 (WiCom, Shanghai, 2007)
4. J. Wu, F. Dai, IEEE Trans. Mobile Comput., **4**, 259 (2005)
5. W.R. Heinzelman, A. Chandrakasan, H. Balakrishnan, *Energy- Efficient Communication Protocol for Wireless Microsensor Networks*, in Proceedings of the 33rd Hawaii International Conference on System Sciences **8**, (2000)
6. V. Kanakaris, D. Ndzi, K. Ovaliadis, International Journal of Wireless & Mobile Networks, **3**, (2011)
7. V. Handziski, A. Kopke, H. Karl, A. Wolisz, *A common wireless sensor network architecture?* In Proc. 1. GI/ITG Fachgesprach Sensornetze, Technical Report TKN-03-012 of the Telecoms. Networks Group, Technical University of Berlin, (2003)10-17
8. A. Harun, D. L. Ndzi, M. F. Ramli, A. Y. M. Shakaff, M. N. Ahmad, L. M. Kamarudin, A. Zakaria, Y. Yang, Prog. Electromagn. Res., **131**, 477 (2012)
9. David L. Ndzi, Azizi Harun, Fitri M. Ramli, Munirah L. Kamarudin, Ammar Zakaria, Ali Yeon Md. Shakaff, Mahmad N. Jaafar, Shikun Zhou, Rohani S. Farook, Elsevier Computer and Electronics in Agriculture, **105**, 83 (2014)

10. D.L. Ndzi, M.A. M. Arif, A.Y.M. Shakaff, M.N. Ahmad, A. Harun, L.M. Kamarudin, A. Zakaria, M.F. Ramli, M.S. Razalli, Prog. Electromagn. Res., **125**, 1 (2012)
11. J.S.C. Turner, L.M. Kamarudin, D.L. Ndzi, A. Harun, A. Zakaria, A.Y.M. Shakaff, A.R.M. Saad, S.M. Mamduh, *Modelling indoor propagation for WSN deployment in smart building*, Electronic Design (ICED), 2014 2nd International Conference, 19-21 (2014)
12. J.A. Gay Fernandez, M. Garcia Sanchez, I. Cuinas, A.V. Alejos, J.G. Sanchez, J.L. Miranda SierraProg. Electromagn. Res., **106**, 121 (2010)
13. W.C.Y. Lee, Mobile *Communications Engineering: Theory and Applications*, (McGraw-Hill, 1998)
14. A. Harun. A.Y.M. Shakaf, A. Zakaria, L.M. Kamarudin, M.N. Jaafar, D.L. Ndzi, "Wireless sensor network planning and deployment in tropical precision farming," Computational Intelligence, Modelling and Simulation (CIMSiM), 2011 Third International Conference on, pp. 346 – 350, 20-22 Sept 2011.
15. H. Liu, Z. Meng, and Y. Shang, "Sensor Nodes Placement for Farmland Environmental Monitoring Applications," 5th International Conference on Wireless Communications, Networking and Mobile Computing, Sep. 2009, pp1-4

Methacrylic Acid as Chemical Treatment on the Properties of Kapok Husk Filled Linear Low Density Polyethylene Eco-Composites

Nurul Fatin Syazwani Arshad[1,*], *Salmah* Husseinsyah[1], and *Lim Bee* Ying[1]

[1]School of Materials Engineering, Universiti Malaysia Perlis, Taman Muhibah 02600 Arau, Perlis, Malaysia.

Abstract. The effect of methacrylic acid (MAA) as chemical treatment on the properties of kapok husk (KH) filled linear low density polyethylene (LLDPE) has been studied. KH, as natural filler was added into LLDPE by using Brabender Plasticizer EC plus at temperature 160 °C. The addition of KH content decreased the tensile strength and elongation at break of the eco-composites, but modulus of elasticity increased. Crystallinity of LLDPE/KH eco-composites decreased with increasing of KH content. Chemical treatment by MAA altered the tensile properties and crystallinity of eco-composites. The treated LLDPE/KH eco-composites showed higher tensile strength, modulus of elasticity and crystallinity of eco-composites due to better interfacial interaction between LLDPE matrix and KH filler. This has proven by Scanning Electron Microscopic (SEM).

1 Introduction

Polymer eco-composites have become more attractive in either research field or industries field recently [1]. These phenomena occurred due to high market price of neat polymer over polymer eco-composites and environmental awareness. The using of natural fillers increasing importance compared to synthetic filler. Many researchers studied the effect of natural filler filled polymer eco-composites nowadays [2,3,4]. This is driven by processing of natural filler is much easier compared to synthetic filler instead of economic.

Linear low density polyethylene matrix has been used in many applications such as food packaging, grocery bags, and kitchen utensils [5]. Partly due to LLDPE is cheaper compare to other polyethylene types. LLDPE has significant number of short branches which enable its chains to slide against each other without being entangled upon elongation at break and this gives LLDPE has high tensile strength and high impact and puncture resistance. Due to increasing demands on the LLDPE, a few researchers have studied the incorporation of natural filler filled LLDPE eco-composites [6,7,8,9]. However, kapok husk filled LLDPE is a new studied broad for polymer eco-composites field. KH is one of the agriculture wastes in Malaysia which produced around 0.008 million tonnes annually of kapok product such as mattress, pillow, and clothes [10]. This waste could be reinforcing

* Corresponding author: fatinsyaz13@gmail.com

DOI: 10.1051/matecconf/20179701042

materials. The utilization does not only economic and abundance, it also will contribute to the waste disposal management as well to overcome the environmental problem.

However, the only disadvantage of polymer eco-composites is incompatibility between natural filler and polymer matrix which results in poor properties of the eco-composites. This encourages researchers to modify the properties of dispersion of natural filler into polymer matrix [11,12,13,14]. This is due to the reducing the hydrophilicity of natural filler by chemical treatments. In present work, LLDPE/KH eco-composites were developed and methacrylic acid (MAA) was used as chemical treatment towards KH to enhance the properties of the LLDPE/KH eco-composites. The effects of MAA on tensile properties, morphology and crystallinity of the LLDPE/KH eco-composites were studied.

2 Experimental

2.1 Materials.

Linear low density polyethylene (LLDPE) used as polymer matrix was supplied by PT Titan Petrokimia Nusantara, Indonesia with melt temperature 160°C and a density of 0.92g/cm^3. The kapok husk (KH) was obtained from Perlis, Malaysia. KH was cleaned and oven dried at 80 °C for 24 hours. Then, KH was crushed and grinded to become powder. The average particles size of KH was 63μm as determined by using Malvern Particle size analyser Instrument. MAA and ethanol was supplied by Sigma Aldrich (Penang, Malaysia).

2.2 Filler treatment.

MAA was dissolved in ethanol with amount of 3% of KH filler. KH filler was added slowly into the solution and it was stirred for 2 hours at room temperature. The treated KH was soaked for 12 hours. After that, it was filtered and oven dried at 80 °C for 24 hours.

2.3 Preparation of eco-composites samples.

The LLDPE/KH eco-composites were prepared by Brabender Plasticizer EC Plus at temperature 160°C. The total mixing LLDPE/KH eco-composites according to formulation in Table 1become homogenous was 7 minutes. After that, LLDPE/KH eco-composites were compressed by compression moulding machine GT-7014 A at temperature 160°C. The 2 mm thickness sheet eco-composites were discharged. The samples were cut into dumbbell shapes by using Wallace Dumbbell Cutter.

Table 1. Formulation of untreated and treated LLDPE/KH eco-composites

Materials	Untreated LLDPE/KH	Treated LLDPE/KH
LLDPE	100	100
KH (php)	10,20,30,40	10,20,30,40
MAA (%)	-	3

*php: part per hundred polymer

3 Characterization

3.1 Tensile test.

Tensile test was conducted according to ASTM D638 by using Instron Machine Model 5569 with 30 mm/min cross head speed. Tensile test were measured for10 identical samples for each eco-composites.The tensile strength, elongation at break and modulus of elasticity were authorized by software.

3.2 Morphological study.

Morphological study of fractured surface of LLDPE/KH eco-composites was carried out by scanning electron microscopic with JEOL JSM-6460LA series. The samples were mounted on an aluminium stubs and sputter coated with a thin layer of palladium.

3.3 Differential Scanning Calorimetric (DSC).

The DSC analysis was conducted to study the melting and crystallization behaviour of LLDPE/KH eco-composites. The 10 ± mg samples of eco-composites were analyzed by using Perkin Elmer DSC-7 analyzer. Samples were heated from 20 to 250 °C. The crystallinity (X_c) of eco-composites was calculated using the following equations:

$$X_c = \frac{\Delta H_f}{\Delta H°_f} \times 100\% \quad (1)$$

Where ΔH_f is the heat fusion of LLDPE/KH eco-composites, $\Delta H°_f$ heat fusion perfectly (100%) crystalline LLDPE ($\Delta H°_f$ = 288 J/g)

4 Results and discussion

4.1 Tensile properties.

The effect of KH content on tensile strength of untreated and treated LLDPE/KH eco-composites with MAA is shown in Fig. 1. The tensile strength of untreated eco-composites decreased with increasing of KH content. This is due to the incompatibility between LLDPE and KH. However, tensile strength of treated LLDPE/KH eco-composites improved compared to untreated eco-composites. It showed that the chemical treatment of KH by MAA promoted a better filler-matrix interfacial interaction due to the reduction of hydrophilic properties of KH filler. The highest improvement of treated LLDPE/KH eco-composites compared untreated eco-composites is at 20 php with 67.5% increment. This was attributed to esterification reaction of MAA with hydroxyl groups of KH.

DOI: 10.1051/matecconf/20179701042

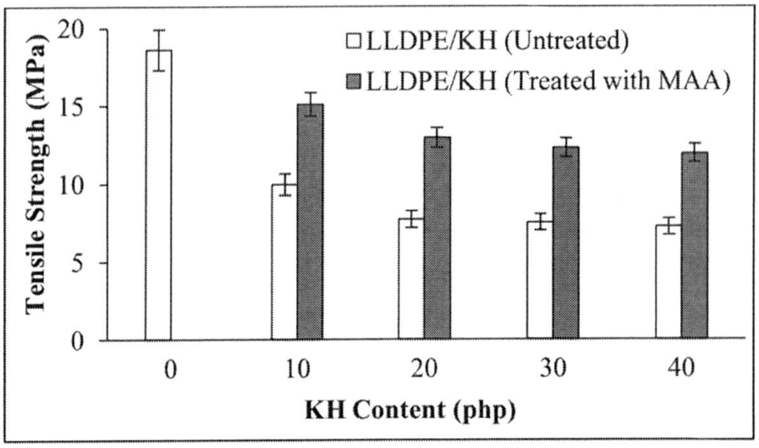

Fig. 1. Effet of KH content on tensile strength of untreated and treated LLDPE/KH eco-composites with MAA.

Fig. 2 demonstrates the effect of KH content on the elongation at break of untreated and treated LLDPE/KH eco-composites. The elongation at break of untreated eco-composites was decreased significantly with increasing of KH content. Nevertheless, treated LLDPE/KH eco-composites with MAA have lower elongation at break compared to untreated eco-composites. This is due to MAA treatment altered the KH filler by cooperating the OH group of KH and OH group of MAA. Thus, the treated eco-composites have better filler-matrix interaction and reduced the ductility of LLDPE/KH eco-composites simultaneously. Both untreated and treated eco-composites were dramatically decreased at 10 php KH content compared to neat LLDPE. Meanwhile the largest drop of elongation at break was found at 20 php treated eco-composites compared to untreated LLDPE/KH eco-composites.

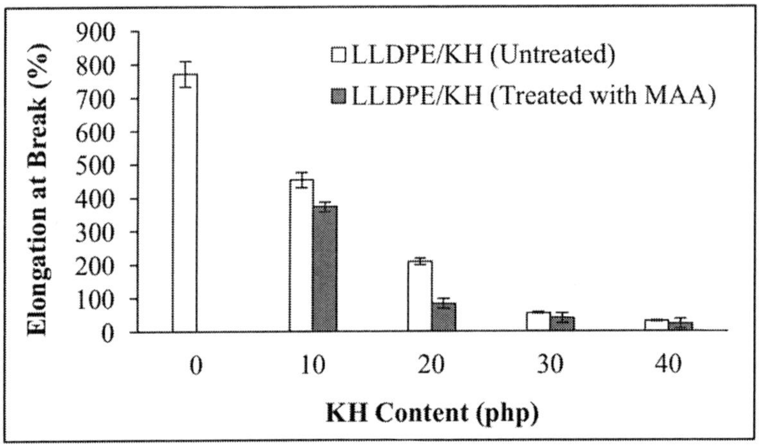

Fig. 2. Effect of KH content on elongation at break of untreated and treated LLDPE/KH eco-composites with MAA.

Fig. 3 presents the effect of KH content on modulus of elasticity for both untreated and treated LLDPE/KH eco-composites. The untreated and treated eco-composites were increased in modulus of elasticity with the increasing of KH content. However, treated LLDPE/KH eco-composites by MAA indicated higher modulus of elasticity compared to untreated eco-composites. The highest increment was 49.2% for treated eco-composites as compared to untreated LLDPE/KH eco-composites at 20 php. This is because the presence of MAA had reduced the hydrophilicity of KH filler through esterification reaction by created covalent bond on between KH filler and MAA. Thus, this reduced the deformability of the LLDPE/KH eco-composites and indicated decreasing in stiffness.

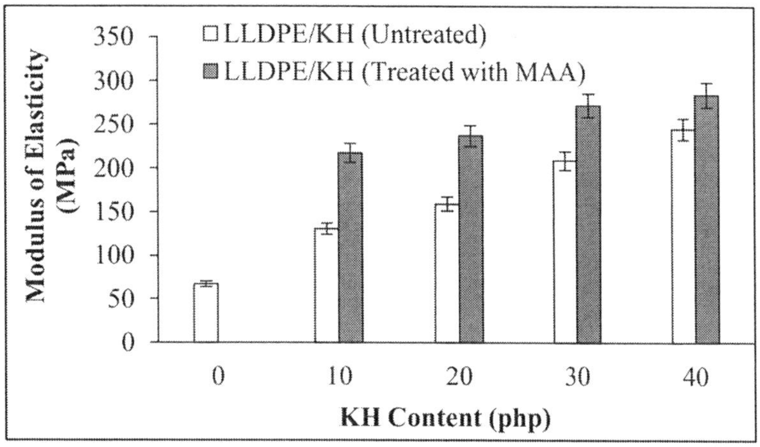

Fig. 3. Effect of KH content on modulus of elasticity of untreated and treated LLDPE/KH eco-composites with MAA.

4.2 Morphological study.

Fig 4 illustrates the SEM micrograph of tensile fracture surface of untreated and treated LLDPE/KH eco-composites by using MAA at 20 and 40 php, respectively. From the micrographs figures, treated KH filler had better dispersion within LLDPE compared to untreated eco-composites due to presence of MAA treatment. This was expected since the tensile strength of MAA treated LLDPE/KH eco-composites was higher compared to untreated LLDPE/KH eco-composites as discussed in section 4.2.1. This strongly verified that the effect of MAA had improved the filler-matrix interfacial interaction of the eco-composites.

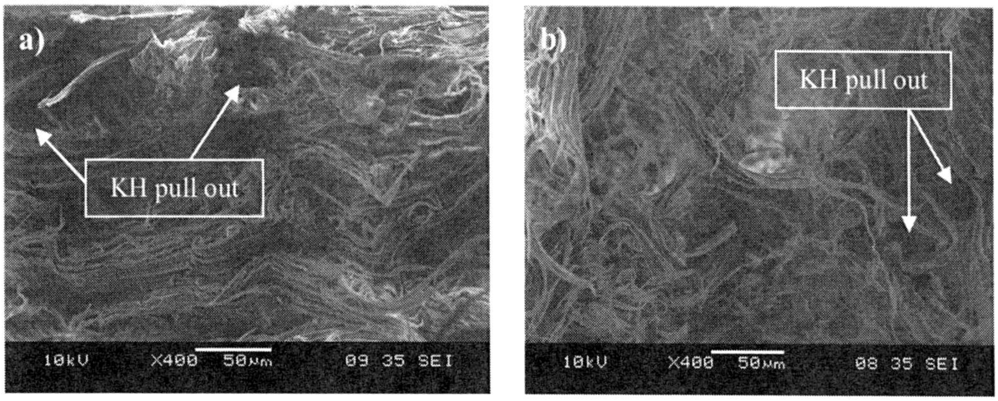

DOI: 10.1051/matecconf/20179701042

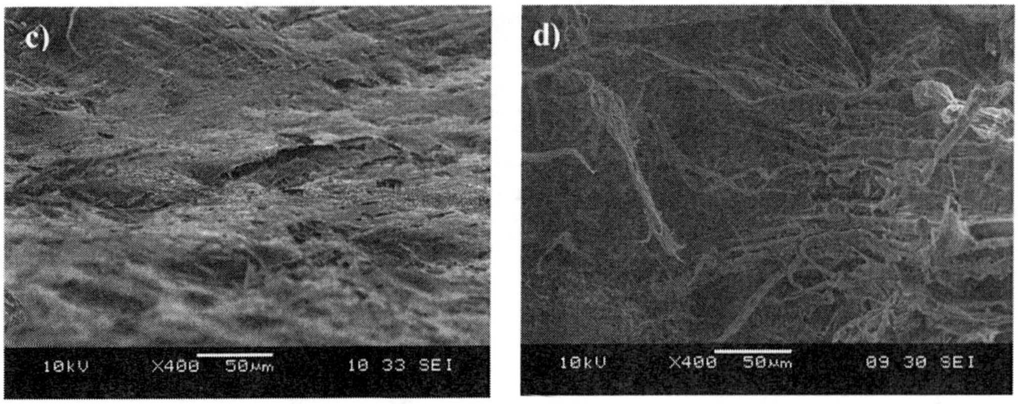

Fig. 4. SEM of tensile fracture surface: a) Untreated LLDPE/KH eco-composites at 20 php; b) Untreated LLDPE/KH eco-composites at 40 php; c) Treated LLDPE/KH eco-composites with MAA at 20 php; d) Treated LLDPE/KH eco-composites at 40 php.

4.3 Differential Scanning Calorimetry (DSC).

The DSC curve of the neat LLDPE, untreated and treated LLDPE/KH eco-composites with MAA at different KH content is illustrated in Fig. 5. Table 2 summarizes the DSC data for neat LLDPE, untreated and treated eco-composites. From the data, the T_m does not show any significant changes within untreated and treated eco-composites at different filler content. Meanwhile, the crystallinity of LLDPE/KH eco-composites decreased with increasing of KH content. This is due to interruption of KH in crystalline growth of LLDPE. However, the treated LLDPE/KH eco-composites had higher X_c compared to untreated eco-composites at similar KH content. This indicated that MAA was successfully attributed to the enhancement of the interfacial interaction between KH and LLDPE.

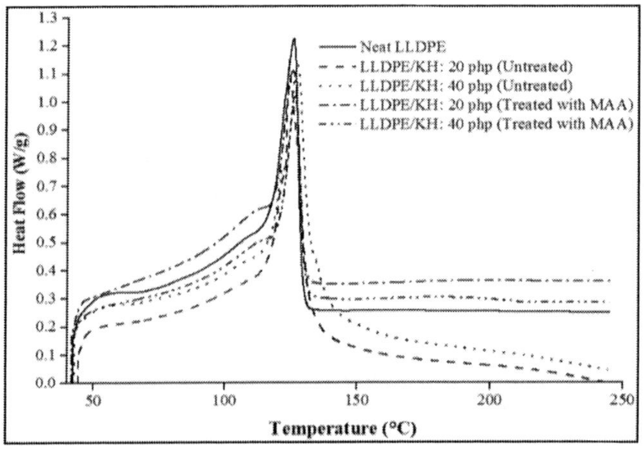

Fig. 5. DSC curves of neat LLDPE, untreated and treated LLDPE/KH eco-composites with MAA at different KH content.

Table 2. DSC data of untreated and treated LLDPE/KH eco-composites with MAA.

Materials	T_m (°C)	ΔH_f (J/G)	X_c
Neat LLDPE	126	82.6	28.9
LLDPE/KH: 20 php	128	60.1	20.9
LLDPE/KH: 40 php	127	56.7	19.7
LLDPE/KH: 20 php (Treated with MAA)	126	76.7	26.6
LLDPE/KH: 40 php (Treated with MAA)	126	69.1	24.0

5 Conclusion

The addition of KH had reduced the tensile strength, elongation at break and crystallinity of LLDPE/KH eco-composites. This was due to poor incompatibility of hydrophilic KH and hydrophobic LLDPE. However, treated LLDPE/KH with MAA has led to increasing of tensile strength and modulus of elasticity compared untreated eco-composites. The crystalline growth of eco-composites was improved after treated KH with MAA. This indicated that the treated KH with MAA was effective. The improvement of interfacial interaction of LLDPE and KH has been proven by SEM study.

References

1. N. Petchwattana, S. Covavisaruch, J. Bionic Eng., **10**, 110 (2013)
2. H. Essabir, M.E. Achaby, E.M. Hilali, R. Bouhfid, A. Qaiss, J. Bionic Eng., **12**, 129, (2015)
3. N. Sarifuddin, H. Ismail, Z. Ahmad, J. Phys. Sci., **24**, 97 (2013)
4. F. Zhou, G. Cheng, B. Jiang, Appl. Surf. Sci., **292**, 806 (2014)
5. S.C. Nwanonenyi, M.U. Obidiegwu, T.S. Onuchukwu, I.C. Egbuna, Int. J. Eng. Sci., **2**, 42 (2013)
6. N.S.A. Wahab, M.F. Omar, H.M. Akil, N.N. Zulkepli, M.M.A.B. Abdullah, A.V. Sandu, Mater. Plas., **52**, 1 (2016)
7. M. Sabetzadeh, R. Bagheri, M. Masoomi, Carbohydr. Polym., **119**, 126 (2015)
8. M.A. AlMaadeed, Z. Nogellova, I. Janigova, I. Krupa, Mater. Des., **58**, 209 (2014)
9. S. M. Mirmehdi, F. Zeinaly, F. Dabbagh, Composites Part B: Engineering, **56**, 137 (2014)
10. FAO, *Statistic of crop production in the world*. Food Agric. Organiz. (2014)
11. R. Narendar, K.P. Dasan, Composites Part B: Engineering, **56**, 770 (2014)
12. M.S.M. Rasidi, S. Husseinsyah, T.P. Leng, Bioresour. **9**, 2033 (2014)
13. S. Husseinsyah, A. Ismail, H. Osman, Appl. Mech. Mater., **754**, 187 (2015)
14. M.N.A. Lotfi, S. Husseinsyah, H. Osman, H. Ismail, Bioresour., **10**, 3409 (2015)

DOI: 10.1051/matecconf/20179701043

Microstructural Study of the Interfacial Transition Zone in Concrete using Optical Microscopy

Khairul Nizar Ismail[1,*], and *Tan* Let Hui[1]

[1]School of Environmental Engineering, Kompleks Pusat Pengajian Jejawi 3, Universiti Malaysia Perlis, 02600 Jejawi, Perlis, Malaysia.

Abstract. This paper presents an overview on the optical microstructural features in concrete. Microstructure is the small scale structure of a material, defined as the structure of a prepared surface of material as revealed by a microscope. The concrete specimens with the manipulation of water/cement ratio 0.3-0.7 (increment of 0.1) in the concrete mix design were used to study the optical microstructural features in concrete and to validate the existence of an interfacial transition zone (ITZ) in concrete. USB Digital Microscope is used to analyse the formation mechanism of microstructure in concrete where the optical microstructural images is analyse via respective curing periods of 1, 7, 28 and 56 days. The microstructure features are discussed with respect to their influence on the strength development of concrete.

1 Introduction

Concrete is the single most widely used material in the world, however, concrete behaviour is complex at it microscopic level. Microstructure can be observed by using a range of microscopy methods. The typical practiced are the optical microscopy and SEM analysis. It is crucial to select the microscopy methods by considering the length scale of the observations to study the microstructure of a material. Optical microscopy is applied to supplement the cement matrix composite test results. The advantage of using an optical microscope in concrete examination is that, it offers larger field of view [1].

Microscopic complex mass in concrete is view as a three-phase composite structure: aggregate, interfacial transition zone (ITZ), and matrix [2]. Interfacial transition zone (ITZ) is defines as a thin layer region of the cement paste around the aggregate particles and ITZ exists around both the coarse and fine aggregates [3]. The formation mechanisms consisted different microstructure of surrounding hydrated cement paste which is generated during casting of fresh concrete with a water/cement (w/c) ratio gradient develops around the aggregate particles. The properties of the microstructures: aggregate, cement bulk paste and particularly the w/c ratio of the mixture are significant factors that determining the concrete composite properties [4]. Zhang [5] carried out a research in determination of the

* Corresponding author: nizar@unimap.edu.my

microstructural features in concrete due to the effect of curing conditions on the manipulation of w/c ratio. It is however, w/c ratio and cement content have the limited effect on the microstructure in concrete; therefore, it is necessary to identify the modification of the microstructure towards the strength of concrete.

This research aims to describe the optical microstructure features in different concrete models in order to understand the strength development and performance properties of concrete. By understanding the microstructures of concrete is comprised of an interfacial zone of which the properties is differ from the bulk matrix and aggregates but the ITZ is not necessarily an intrinsic feature of concrete instead it depends on the degree of hydration; therefore, the water/cement ratio in the mix design is importance.

2 Materials and experimental methods

2.1 Materials

This research used:

i. Ordinary Portland cement (OPC - 42.5) (ASTM Type 1);

ii. Tap water is used for both the concrete mixes and curing. The temperature ranged between 20^0C- 27^0C. The water used should not contaminate.

iii. Coarse aggregate used was crushed granite stone with maximum size of 20mm and it properties is determined from the specific gravity test.

iv. Natural river sand with sand fraction passing through 600μm sieve was used and it properties are determined through the specific gravity test.

2.2 Mixture proportions

The concrete mixture proportion is design based on the British Standard on design of normal concrete mixes. Concrete with water/cement ratios of 0.30 to 0.70 with an increment of 0.10 was prepared. The design characteristic strength of the concrete for this research is 25MPa. The slump for w/c ratio = 0.30-0.40 is 10-30mm and the slump for w/c ratio = 0.50-0.70 is 30-60mm.

2.3 Curing condition

Selected curing ages of 1,7,28 and 56 days were used to analyze the microstructural features influences on the strength of concrete. One day curing age of this test was defined as the time spent for curing in water tank after removing the concrete specimens from the mold; before that, the concrete was in the process of molding and put in air for 24 h. The concrete casting process is carried out in accordance with BS EN 12390-2:2009, making and curing specimens for strength tests.

2.4 Concrete compressive test

The concrete compressive strength test was determined according to BS EN 12390-3:2009 at 1, 7, 28, and 56 curing ages. Total of 12 cubes samples were tested. The results obtained are the average compressive strength of the 3 cubes samples test results for each curing ages. The cubes are tested on the face perpendicular to the casting face.

2.5 Optical microstructure test

At the later stage, the microstructural test is carried out on each specimens when it achieved the desired curing age such that at 1, 7, 28 and 56 days. The specimens that used to undergo the optical microstructure test must be a thin flat surface so that to obtain a clearer images and is prepared from the specimens casted in cylinder molds and cutting with the concrete cutting machine. A USB Digital Microscope is used where the microscope camera attaches directly to the USB port of a computer, eyepieces are not required and the images are shown directly on the monitor.

3 Results and discussions

3.1 Effect of water/cement (w/c) ratio on compressive strength of normal concrete

The w/c ratio, largely determines the strength and durability of the concrete when it is cured properly. When the w/c ratio is high which mean the bigger amount of water in a concrete mix, the more dilute the cement paste will be and affects the compressive strength and leads to the porosity. The more the w/c ratio is increased; the strength of the resulting concrete is reduced. The results for 1, 7, 28 and 56 days curing ages with regard to test compressive strength for all the concrete cubes samples being computed in table and shown in Fig 1.

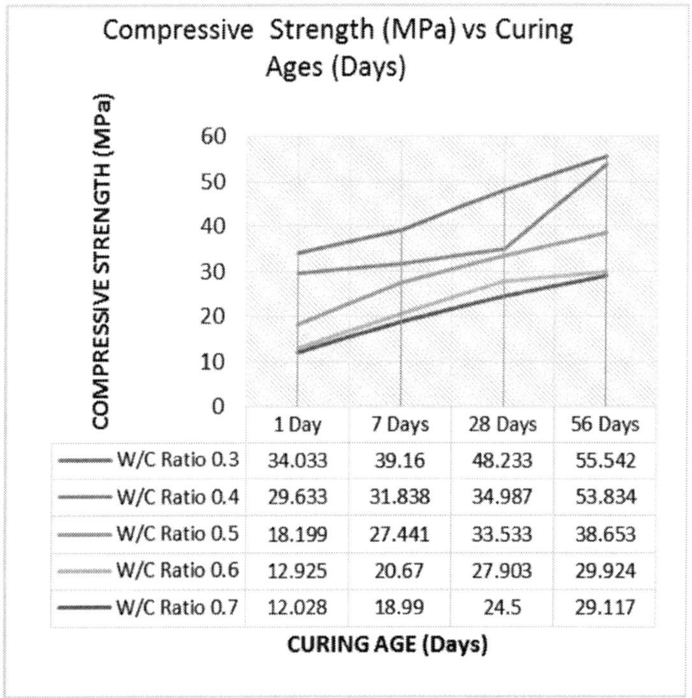

	1 Day	7 Days	28 Days	56 Days
W/C Ratio 0.3	34.033	39.16	48.233	55.542
W/C Ratio 0.4	29.633	31.838	34.987	53.834
W/C Ratio 0.5	18.199	27.441	33.533	38.653
W/C Ratio 0.6	12.925	20.67	27.903	29.924
W/C Ratio 0.7	12.028	18.99	24.5	29.117

Fig. 1. Graph of Compressive Strength (MPa) against Curing Ages (Days)

3.2 Optical microstructural analysis and development of interfacial transition zone in concrete for different curing ages

It can be observed that, the optical microstructure in normal concrete consisted of three distinct phases and boundaries. The three phases in concrete consisted of the aggregates phase, bulk cement paste and also the interfacial transition zone (ITZ). In addition, pores due to inefficient packing of concrete may also be observed using the optical microscope. The microstructural analysis images of concrete samples with water/cement ratio of 0.5 are shown in Fig. 2. The ITZ is presence due to the potentially effect from the inefficient packing between the aggregate surface and cement particles. However, the thickness of the ITZ layer will decrease or even disappeared in respond to it increasing in curing ages.

Fig. 2. Microstructural analysis images on concrete samples with water/cement ratio of 0.5.

Fig 3 shows the thickness (μm) of the interfacial transition zone (ITZ) against the curing ages (Days) for the five concrete samples with w/c ratio from 0.3-0.7. It was found that, as the w/c ratio is low, the thickness of the ITZ layer is significantly reduced and hence enhances the microstructure of the interface region. In addition, lower w/c ratio created thinner ITZ and help to increase the concrete strength since the microstructures are denser packed to each other. This is also justified w/c ratio of 0.3 has the highest compressive strength but the thinnest ITZ layer during the 28 days.

 DOI: 10.1051/matecconf/20179701043

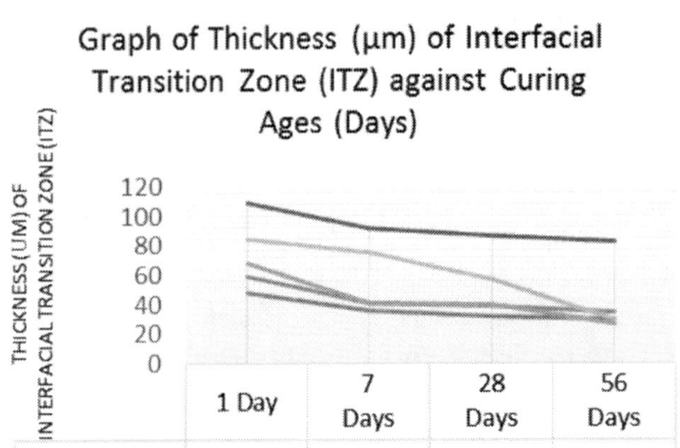

	1 Day	7 Days	28 Days	56 Days
W/C 0.3	47.569	35.001	31.306	29.21
W/C 0.4	58.546	40.782	38.753	33.956
W/C 0.5	68.44	40.356	39.635	25.971
W/C 0.6	83.865	75.499	56.514	29.21
W/C 0.7	108.793	91.571	86.654	81.777

Fig 3: Graph of thickness (µm) of the interfacial transition zone (ITZ) against the curing ages (Days) for the five concrete samples with w/c ratio from 0.3-0.7.

4 Summary

The ITZ is existed in the concretes for w/c ratio range from 0.3-0.7 and it is detectable by using the optical microscope. However, there is no constant trend can be observed for the existence of ITZ. It can be summarized that the ITZ thickness depends on w/c ratio. The higher w/c ratio resulted in thicker ITZ layer due to higher water concentration of the concretes and slower hydration process. From the optical microstructural analysis, it can be concluded that the formation mechanisms of the microstructure in concrete can be described as the continuous development of the microstructure with respective to its curing ages.

References

1. A.M. Brandt, N.D. Jozwiak, Z. Ranachowski, Advances in Civil Engineering Materials 1, **1**, 1, (2012)
2. L.S. Karen, R. Li, J. Pei, B. Xue, Constr. Build. Mater., **101**, 410, (2004)
3. B.D. Mason, H.B. Sarnes, D. Siamond, D.L. Wolch, Cem. Concr. Res., **32**, 1291, (2000)
4. J. Halbiniak, G. Prolopski, Cem. Concr. Res., **30**, 579 (2000)
5. Y.H. Zhang, Journal of Concrete and Structure, **3**, 146 (2010)

Preparation and mechanical performance of Cf-SiCf-(Al_2O_3p) reinforced geopolymer composites

Shu Yan[1,2], *Peigang* He[1,2,*], *Dechang* Jia[1,2], *Jinyan* Wang[1,2], *Zhihua* Yang[1,2], *Xiaoming* Duan[1,2], and *Yu* Zhou[1,2]

[1]Institute for Advanced Ceramics, Harbin Institute of Technology, Harbin 150080, China
[2]School of Materials Science and Engineering, Harbin Institute of Technology, Harbin 150080, China

Abstract. The C_f-SiC_f-(Al_2O_{3p})/KGP composites were successfully prepared by the method that mixing a kind of hybrid fiber preform pre-developed and geopolymer resin. The synergistic effects of the hybrid fibers and Al_2O_3 particles on the composites were investigated. Results show that all the composites showed typical amorphous structure. Mechanical properties of the C_f-SiC_f-(Al_2O_{3p})/KGP composites were improved remarkably due to the addition of the hybrid fibers and Al_2O_3 particles. The flexural strength and Young' s modulus of the C_f-SiC_f-Al_2O_{3p}/KGP composites with 1 mm length SiC_f were reached 62.4±2.1 MPa and 24.1±1.6 GPa, respectively. The C_f-SiC_f-(Al_2O_{3p})/KGP composites showed non-catastrophic behavior rather than the catastrophic failure.

1 Introduction

Geopolymers are an novel three-dimensional inorganic aluminosilicate polymers which synthesis by aluminosilicate source materials under highly alkaline conditions [1-4]. Geopolymers offer variety of characteristics and properties, such as low densities, low costs, easy processing and fire resistance, thus attract lots of industrial attentions [1-7]. However, the pure geopolymers showed disadvantages of both low mechanical performance and catastrophic fracture behavior, thus limits its applications as structural components [1-7]. In recent years, many studies have been explored on geopolymer matrix composites to overcome the disadvantages of itself low strength and brittle failure, including nanotube [8,9], graphene [10-11], particles [12-13], short fiber [14-16,20] and continuous fiber [17-19]. Ranjbar [19] evaluated fiber-matrix interaction and mechanical properties of two different fibers (polypropylene (PPF) and micro steel fibers (MSF)) reinforced fly ash based geopolymer. Results show that MSF has strong contact with the matrix because it tends to behave as a hydrophilic material, while PPF led to fiber-matrix debonding due to hydrophobic characteristics. Timakul et al [20] reported the addition of basalt fibers into the fly ash-based geopolymer can improve its compressive strength. Lin et

* Corresponding author: peiganghe@hit.edu.cn

DOI: 10.1051/matecconf/20179701044

al. [14, 21-22] prepared the short carbon fiber geopolymer matrix composites using sheet-like fiber preform, and improved the mechanical properties. The addition of α-Al_2O_3 particles into the composites could reduce the volume change of the composites and improve the mechanical properties, especially at high temperatures [22].

In our previous research, we have attempted short carbon fiber and SiC fiber reinforced geopolymer composites, respectively [21-22, 23]. The short carbon fiber reinforced showed good mechanical properties with non-brittle failure behavior [21]. The SiC fiber exhibited similar mechanical properties as carbon fiber, and much higher oxidation resistance [24], which makes it a prominent high-temperature reinforcing candidate. However, compared with the carbon fibers, the SiC fiber is much expensive. Thus, it inspired us to introduce C_f, SiC_f and Al_2O_3 particles into the geopolymer to to improve the mechanical properties.

In this study, we provided a simple method to prepare C_f-SiC_f-(Al_2O_{3p})/KGP composites. The synergistic effects of the hybrid fibers and Al_2O_3 particles on the composites were investigated. The microstructure and the mechanical properties of the composites were also studied.

2 Materials and experiments

The composition of the geopolymer matrix was SiO_2/Al_2O_3 = 4.0, K_2O/SiO_2 = 0.3 and H_2O/K_2O = 11 (mole ratio) according our previous work [17, 22]. The raw materials were KOH (Tianjin Guangfu Indus., China), silica sol (Jiangsu Xiagang, Indus., China), kaolin (95%, Fengxian Reagent Factory, China) and α-Al_2O_{3p}(Xiamen Fushixin Indus., China). The kaolin was heated treated at 800 °C for 2 h to get the metakaolin powder. Carbon fiber(Jilin Carbon Indus., China) used in this study has a diameter of 6-8μm and an average tensile strength of 3 GPa. SiC fiber (Jiangsu Sailifei Ceramic Fiber Indus., China) has a diameter of 13 μm and an average tensile strength of 1.5-1.6 GPa. The properties of Carbon and SiC fibers were shown in Table 1. SiC fiber used in this study was first treated at 370 °C in an air atmosphere for 2 h to remove the surface epoxy coating [25].

The alkaline solution was obtained by mixing KOH and silica sol for 3 days using magnetic stirring. Whereafter, the Al_2O_3 particle (the ratio of Al_2O_3/metakaolin was 0.75 wt.%) was added to geopolymer resin and mixed homogeneously. The composites were prepared by infiltrating geopolymer resin into the short carbon fiber and SiC fiber preform with the help of the ultrasonic vibration treatment, and stacked one by one to get a green sample with 20 layers. The carbon fiber content was 4.5 vol%, and fiber length was 7 mm. The SiC fiber content was also 4.5 vol%, and fiber length was 1, 3 and 5 mm. To remove the pores in the composites, degassing was applied at 70 °C for 3 days using a vacuum-bag technique after mechanical pressing.

Table 1. Properties of Carbon and SiC fiber

Brand	Density ($g \cdot cm^{-3}$)	Diameter (μm)	Tensile strength (GPa)	Modulus (GPa)	Elongation (%)
C_f-TX-3	1.76	6~7	≥3.0	210~230	≥1.4
SiC_f-SLFC1	2.36	13±0.5	1.5~1.6	140±10	1.3~1.4

The phases of C_f-SiC_f-(Al_2O_3) reinforced geopolymer composites were characterized using an X-ray diffractometer (XRD: Rigaku, D/MAX-2200VPC, Tokyo, Japan) with CuKa radiation at a scan rate of 4°/min. Mechanical properties of the composites in the x direction(in-plane and perpendicular to fiber axial direction) and z direction (laminate lay-up direction) were investigated. Flexural strength measurements were conducted on specimens(3×4×36 mm^3) using a three-point-bending fixture on an Instron-500 testing machine with a span length of 30 mm at a crosshead speed of 0.5 mm/min. Work of fracture was calculated by the area between load curves and x axis in the load/displacement curves till the load dropped to 90% of the maximum load. Five to six specimens were tested under each test condition. Surface micrographs of the C_f-SiC_f-(Al_2O_3) reinforced geopolymer composites were observed by optical microscope. Fractographs of the composites were observed by scanning electron microscopy.

3 Results and discussions

Fig. 1 shows the XRD patterns of the metakaolin, pure geopolymer matrix, C_f-SiC_f/KGP and C_f-SiC_f-Al_2O_{3p}/KGP composites. Metakaolin particles showed characteristic amorphous hump between 22° 2θ, and several reflections from quartz present introduced by the raw kaolin [17]. The pure geopolymer matrix displayed typical broad amorphous humps around 28° 2*θ*.The characteristic amorphous hump of the C_f-SiC_f/KGP composites was similar to the pure geopolymer matrix, but shifted to a lower position around 27°. The addition of C_f and SiC_f seems to retard the geopolymerization process of the matrix [25]. However, with the addition of Al_2O_{3p}, typical Al_2O_3 peaks in C_f-SiC_f-Al_2O_{3p}/KGP composites was observed, indicating the incorporation of Al_2O_{3p} may not participate in the geopolymerization process of the matrix.

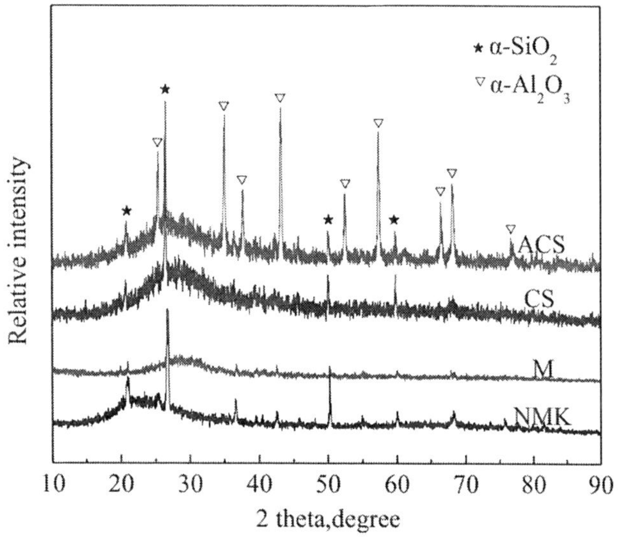

Fig. 1. XRD patterns of metakaolin, geopolymer matrix C_f-SiC_f /KGP and C_f-SiC_f-Al_2O_{3p}/KGP composites

Fig.2 shows the density data of the C_f-SiC_f-(Al_2O_{3p})/KGP samples. The density of the geopolymer matrix was 1.37 g·cm^{-3}. The addition of the C_f and SiC_f increased the density of the composites, while there was no obvious difference between the samples with different fiber length. The addition of the α-Al_2O_{3p} obviously increased the density of the composites. The density of the C_f-SiC_f/KGP and C_f-SiC_f-(Al_2O_{3p})/KGP were 2.08 g·cm^{-3} and 2.38 g·cm^{-3},respectively.

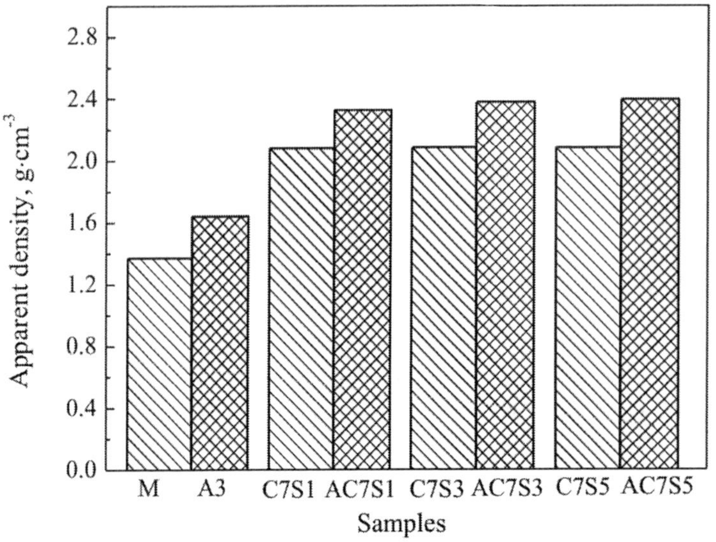

Fig. 2. the Density data of the C_f-SiC_f-(Al_2O_{3p})/KGP samples

With the help of the ultrasonic scattering treatment, it can effectively prevent fibers from too much aggregation. All the composites showed homogeneous microstructure, most short carbon fibers own a two-dimension in geopolymer matrix, as shown in Fig. 3.

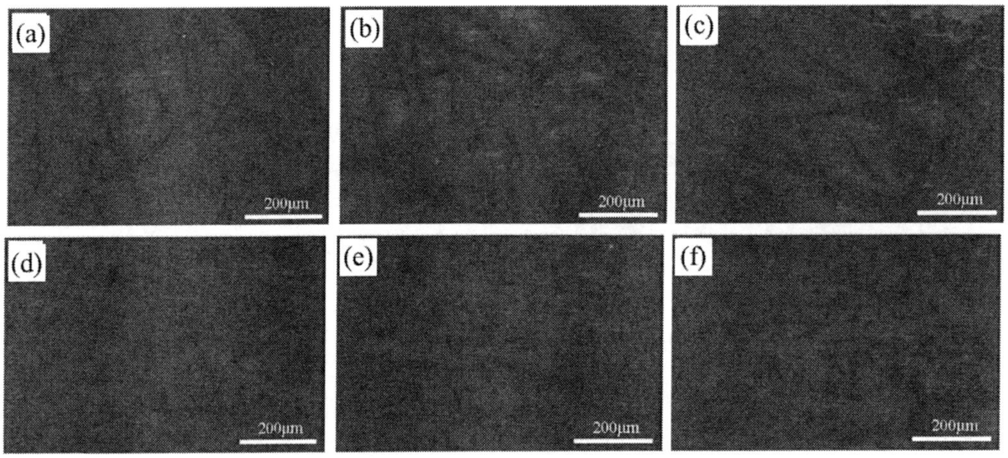

Fig. 3. Micrographs of the surface of C_f-SiC_f/KGP composites (a)-(c) Surface micrographs of C7S1, C7S3 and C7S5, (d)-(f) Cross-section micrographs of C7S1, C7S3 and C7S5

Table 2 shows the mechanical properties of the C_f-SiC_f-(Al_2O_{3p})/KGP composites. With the cooperation of hybrid fibers, the flexural strength and Young's modulus of the composites improved significantly. It should be pointed out that with Al_2O_{3p} addition, the flexural strength and Young's modulus of the corresponding composites were increased than before, indicating the significant strengthening effect from Al_2O_{3p}. There was no obvious difference of the mechanical properties between x direction and z direction. Fig.4 shows the mechanical properties of the C_f-SiC_f-(Al_2O_{3p})/KGP composites in x direction. The flexural strength for the composites ramped up first, reached the peak value when SiC fiber length was 1 mm, and then decreased with longer fiber length. The flexural strength of the C_f-SiC_f-Al_2O_{3p}/KGP composites was higher than the C_f-SiC_f/KGP samples, reached the peak value of 62.4±2.1 MPa for the AC7S1 sample, which was 6.4 times as high as that of geopolymer matrix (9.7±0.2MPa) and then decreased for the composite with the length of SiC_f fiber increased.

Table 2. Mechanical properties of the Cf-SiCf-(Al2O3p)/KGP composites

Samples	Flexural strength (MPa)		Young's modulus (GPa)		Fracture toughness ($MPa \cdot m^{1/2}$)	Work of fracture ($J \cdot m^{-2}$)
	x direction	*z* direction	*x* direction	*z* direction		
M	9.7±0.2	9.7±0.2	8.31±0.1	8.31±0.1	0.28±0.05	54.2±8.1
C7S1	55.6±1.1	60.8±1.3	15.7±1.3	19.0±0.5	1.43±0.14	3359.1±245.8
C7S3	41.3±2.7	50.2±4.5	13.3±0.7	10.4±0.7	1.95±0.17	2860.9±178.2
C7S5	41.8±2.3	42.5±1.0	12.5±1.4	8.2±0.3	1.46±0.11	2399.7±211.3
A3	20.2±0.7	20.2±0.7	9.53±0.9	9.53±0.9	0.48±0.08	48.7±7.2
AC7S1	62.4±2.1	60.9±3.4	24.1±1.6	27.8±0.4	1.18±0.11	3218.1±223.8
AC7S3	50.6±3.2	53.0±2.1	23.5±1.8	27.3±2.2	1.41±0.12	2788.5±163.9
AC7S5	50.1±4.3	47.2±2.3	21.9±1.2	22.3±1.6	1.15±0.04	2217.46±129.4

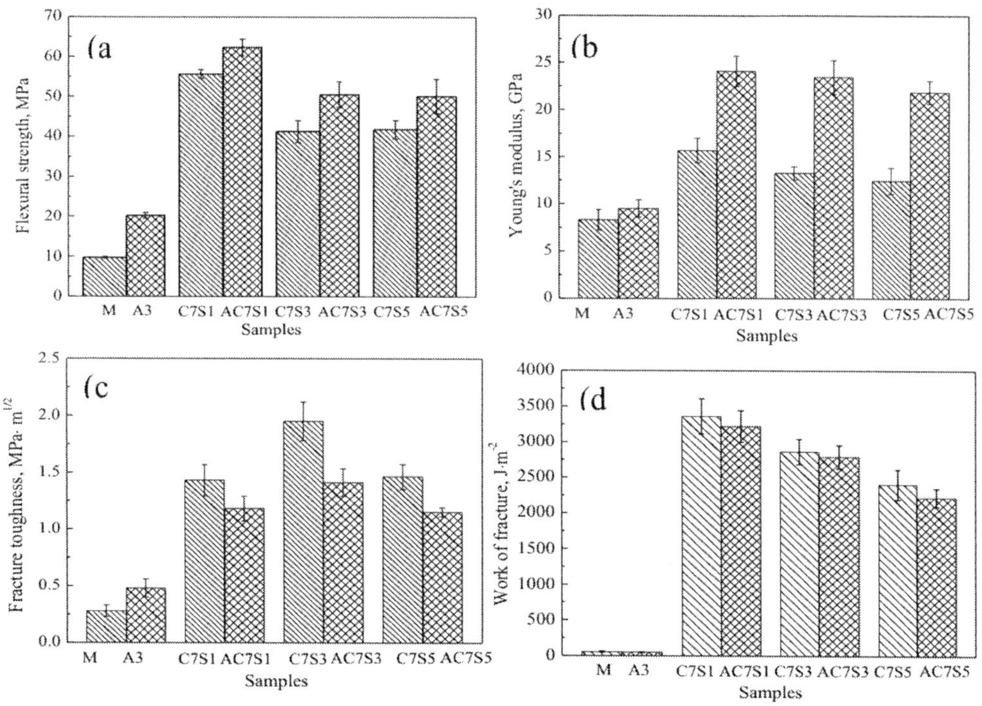

Fig. 4. Mechanical properties of the C_f-SiC_f-(Al_2O_{3p})/KGP composites in x direction (a) Flexural strength, (b) Young's modulus, (c) Fracture toughness, (d) work of fracture

DOI: 10.1051/matecconf/20179701044

Similarly, compared with the geopolymer matrix, the Young's modulus of the C_f-SiC_f-(Al_2O_{3p})/KGP composites showed increased with the incorporation of fiber and Al_2O_3 particles. the Young's modulus also reached the peak value of 24.1±1.6 GPa for AC7S1, which was 2.9 times as high as that of the geopolymer matrix (8.31±0.1 GPa),which shows a significant synergistic strengthening effect of the Al_2O_3 particles and fibers. With the increase in SiC fiber length, the Young's modulus of both the C_f-SiC_f/KGP and C_f-SiC_f-Al_2O_{3p}/KGP composites decreased.

While, the toughness effect of Al_2O_{3p} on the fracture toughness and work of fracture of the composites were not always positive. Fig. 4c and d show the fracture toughness and the work of fracture of the C_f-SiC_f-(Al_2O_{3p})/KGP composites. The fracture toughness of the C_f-SiC_f/KGP composites reached the peak value when SiC_f length was 3 mm, and then decreased for the composites with a longer fiber length of 5 mm. The addition of the Al_2O_{3p} increased the fracture toughness of the geopolymer matrix, from 0.28±0.05 to 0.48±0.08 $MPa \cdot m^{1/2}$, rising by 71.4%. However, compared with the C_f-SiC_f/KGP, within the SiC fiber was 1,3 and 5, the fracture toughness of the C_f-SiC_f-Al_2O_{3p}/KGP composites decreased by 17.4%,27.7% and 21.2% ,respectively. When fiber fiber was 1mm, the composite exhibited the highest work of fracture, 3359.1±245.8$J \cdot m^{-2}$, which was 61 times as high as that of pristine geopolymer. The work of fracture of the C_f-SiC_f-Al_2O_{3p}/KGP composites was much lower than the C_f and SiC_f reinforced ones. This may due to the addition of the Al_2O_{3p} may increase the viscosity of the slurry and bring some defects and pores, which was sensitive to cracks [22, 26].

Fig.5 shows the stress and displacement curves of C_f-SiC_f-(Al_2O_{3p})/KGP composites in x direction. The pure geopolymer matrix indicated a typical brittle failure mode, while all the composites showed non-catastrophic ones with the presence of both elastic region and nonlinear region. After stress reached the maximum values, the strength tended to reduce slowly and several significant steep drop-steps, especially for the C7S3 and AC7S3 samples, corresponding to interface debonding, fiber fracturing and pulling-out [23, 25].

The typical fractographs of the C_f-SiC_f-(Al_2O_{3p})/KGP composites were studied, as shown in Figs. 6. There was no obvious difference between the composites with and without Al_2O_3 particles. As for the composites, fiber fracturing and lots of fiber pulling-out could be seen on the fracture surface, which were due to the weak bonding strength between the fiber and the geopolymer matrix [14, 27]. They are all helpful to consume much higher fracture energy and enhance the fracture toughness of the composites [14, 27]. This ensured the effective toughening and could prevent the catastrophic fracture of the composites, which make the composites presents typical non-catastrophic behavior.

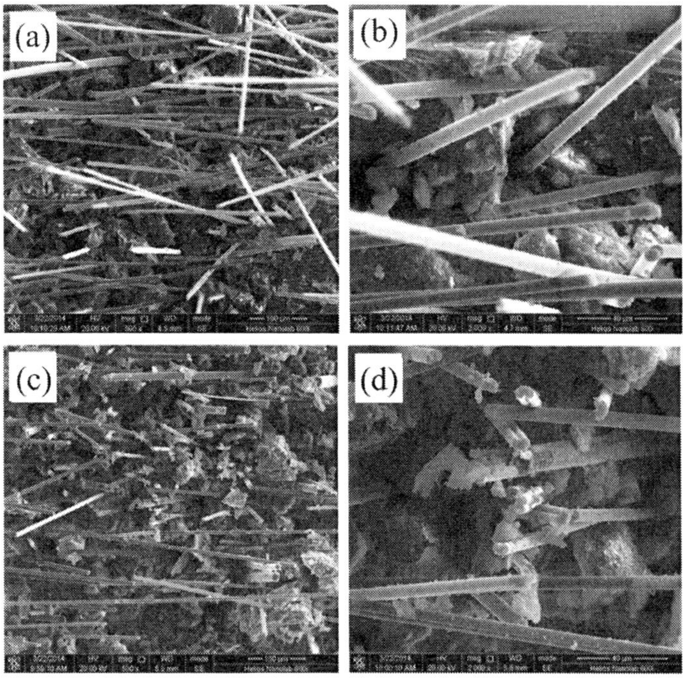

Fig. 5. Fractograph of the (a)-(b) C_f-SiC_f/KGP and (c)-(d) C_f-SiC_f-Al_2O_{3p}/KGP composites

4 Conclusions

In the present study, C_f-SiC_f-(Al_2O_{3p})/KGP composites with different silicon carbide fiber lengths and fixed carbon fiber lengths were prepared by the method that mixing a kind of hybrid fiber preform pre-developed and geopolymer resin. The synergistic effects of the hybrid fibers and Al_2O_3 particles on the on the microstructure and mechanical properties of the composites are studied.

The C_f-SiC_f-(Al_2O_{3p})/KGP composites showed typical amorphous structure and most short carbon fibers own a two-dimension in geopolymer matrix. The addition of the α-Al_2O_{3p} obviously increased the density of the composites. The density of the C_f-SiC_f/KGP and C_f-SiC_f-(Al_2O_{3p})/KGP were approximate 2.08 g·cm^{-3} and 2.38 g·cm^{-3}, respectively.

Mechanical properties of the C_f-SiC_f-(Al_2O_{3p})/KGP composites were improved remarkably due to the addition of the hybrid fibers and Al_2O_3 particles. The flexural strength and Young's modulus were enhanced further with the incorporation of Al_2O_{3p}. Compared with the geopolymer marix, the flexural strength and Young' s modulus of the C_f-SiC_f-Al_2O_{3p}/KGP composites with 1mm length SiC_f were increased 5.4 times and 1.9 times, respectively. All the C_f-SiC_f-(Al_2O_{3p})/KGP composites showed non-catastrophic behavior rather than the catastrophic failure. Fiber fracturing and lots of fiber pulling-out the fracture surface of the composites contributed to the enhancement of the toughness and work of fracture of the composites.

 DOI: 10.1051/matecconf/20179701044

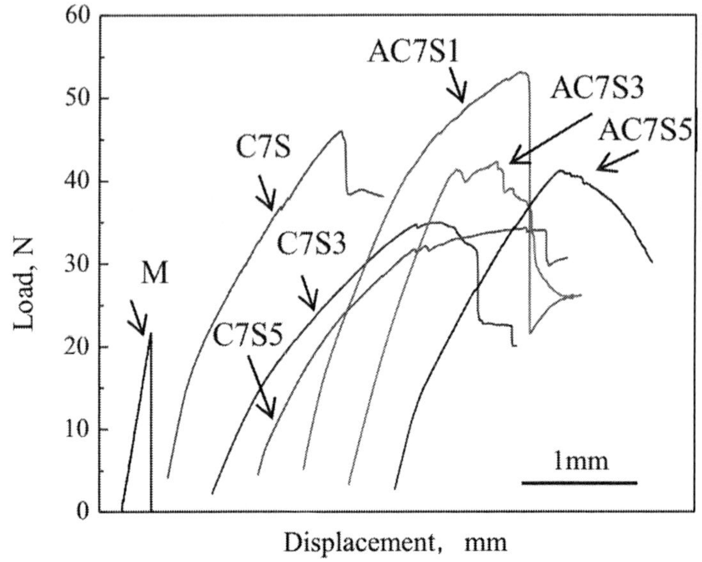

Fig. 6. Stress and displacement curves of SiCf/KGP composites in x direction.

This work was supported by the Fundamental Research Funds for the Central Universities (Grant no. HITNSRIF20165), and the National Natural Science Foundation of China (NSFC, Nos.51372048, 51502052, 51321061 and 51225203).

References

1 J. Davidovits, J. Therm. Anal., **37**, 1633 (1991)
2 K. Komnitsas, D. Zaharaki, Miner. Eng., **20**, 1261 (2007)
3 D. Khale, R. Chaudhary, Mater. Sci., **42**, 729 (2007)
4 K.J.D. MacKenzie, Ceram. Trans., **153**, 175 (2003)
5 M.M.A. Abdullah, K. Hussin, M. Bnhussain, K.N. Ismail, W.M.W. Ibrahim, Int. J. Pure Appl. Sci. Technol, **6**, 35 (2011)
6 J. Feng, R. Zhang, L. Gong, Y. Li, W. Cao, X. Cheng, Mater. Des., **65**, 529 (2015)
7 J Tailby, K.J.D. MacKenzie, Cem. Concr. Res., **40**, 787 (2010)
8 K.J.D. MacKenzie, M.J. Bolton, J. Mater. Sci., **44**, 2851 (2009)
9 H.M.M. Khater, H.A.A. Gawwad, Epitoanyag, 2, 38 (2015)
10 N. Ranjbar, M. Mehrali, M. Mehrali, U.J. Alengaram, M.Z. Jumaat, Cem. Concr. Res., **76,** 222 (2015)
11 M. Saafi, L. Tang, J. Fung, M. Rahman, F. Sillars, J. Liggat, X. Zhou, Smart Mater. Struct., **23**, 065006 (2014)
12 E. Kamseu, A. Rizzuti, C. Leonelli, D. Perera, J. Mater. Sci., **45**, 1715 (2010)
13 S.A. Bernal, J. Bejarano, C. Garzon, R. Mejía de Gutiérrez, S. Delvasto, E. D. Rodríguez, Compos. Part B: Eng., **43**, 1919 (2012)
14 T. Lin, D. Jia, P. He, M. Wang, D. Liang, Mater. Sci. Eng., A, **497**, 181 (2008)
15 Z. Yunsheng, S. Wei, Z. Li, X. Zhou, ACI Mater. J., **106**, 3 (2009)
16 V. Dhand, G. Mittal, K.Y. Rhee, S.J. Park, D. Hui, Composites Part B, 73, 166 (2015)
17 P. He, D. Jia, T. Lin, M. Wang, Y. Zhou, Ceram. Int., **36**, 1447 (2010)
18 Q. Zhao, B. Nair, T. Rahimian, P. Balaguru, J. Mater. Sci., **42**, 3131 (2007)
19 N. Ranjbar, S. Talebian, M. Mehrali, C. Kuenzel, H.S.C. Metselaar, M.Z. Jumaat, Compos. Sci. Technol., **122**, 73 (2016)
20 P. Timakul, W. Rattanaprasit, P. Aungkavattana, Ceram. Int., **42**, 6288 (2016)

21 T. Lin, D. Jia, M. Wang, P. He, D. Liang, Bull. Mater. Sci., **32**, 77 (2009)
22 T. Lin, D. Jia, P. He, M.R .Wang, Journal of Central South University of Technology, **16**, 881 (2009)
23 J. Yuan, P. He, D. Jia, S. Yan, D. Cai, L. Xu, Z. Yang, X. Duan, S. Wang, Y. Zhou, Ceram. Int., **42**, 5345 (2016)
24 B. Liang, Z. Yang, Y. Li, J. Yuan, D. Jia, Y. Zhou, Ceram. Int., **41**, 8868 (2015)
25 P. He, D. Jia, B. Zheng, S. Yan, J. Yuan, Z. Yang, X. Duan, J. Xu, P. Wang, and Y. Zhou, Ceram. Int., **42**, 12239 (2016)
26 H. Li, H. Xiao, J. Ou, Cem. Concr. Res., **34**, 435 (2004)
27 T. Lin, D. Jia, P. He, M. Wang, Mater. Sci. Eng., A **527**, 2404 (2010)

Optimization FSO Performance Under Strong Turbulence Effect By Enhanced New Carrier Data Transmission Technique

K.R. Ummul[1,*], *M.S.* Anuar[1] , *A.K.* Rahman[1] , *C.B.M.* Rashidi[1] , and *S.A.* Aljunid[1]

[1]School of Computer and Communication Engineering, Universiti Malaysia Perlis (UniMAP), Perlis, Malaysia

Abstract. This paper focus on mitigating the atmospheric turbulence effect in free space optical communication using a dual diffuser modulation (DDM) technique. The most deteriorate the FSOC is scintillation where it affected the wave front cause to fluctuating signal and ultimately receiver can turn into saturate or loss signal. DD approach enhances the detecting bit '1' and bit '0' and improves the power received to combat with turbulence effect. The performance focus on Signal-to-Noise (SNR) and Bit Error Rate (BER) by using the Kolmogorov's scintillation theory. The numerical result shows that the DD approach improves the range where estimated approximately 40% improvement under weak turbulence and 80% under strong turbulence.

1 Introduction

Free space optical communication is an attractive alternative over fiber optical communication where provides high bandwidth, fast-installation and high security [1].

However FSOC is suffering with the atmospheric turbulence which can lead the laser beam (a) beam wanders, (b) beam spreading and (c) scintillation. The beam wanders is refer to the condition beam deflected randomly through the changing refractive index oddities. Meanwhile beam spreading refers to conditions where the beam spread more than diffraction estimated predict. Lastly, for scintillation it affected the phase front of the beam can vary and resulting fluctuation irradiance or well known as intensity signal.

The combination of all these effect will cause both the spatial and temporal experience random fluctuations in refractive index of the variability of element factor such as temperature, pressure and wind variations along the optical propagation path through the channel [2-9].

* Corresponding author: ummulkhir@gmail.com

2 Dual-detection approach

The system employs two transmitters and On-Off Shift Keying (OOK) modulation as reference for conventional system. When the first transmitter sends binary '1', the second transmitter which is set in compliment condition will send the binary '0' in simultaneously and vice versa. Meanwhile at the receiver part, the signal will go through the subtractor for the differential detection process. Here we assume the ideal subtractor condition where no losses signal occurs during subtraction. Therefore, the signal output will become bit '1' for sending binary '1' and bit '-1' for sending binary '0'. This condition leads to modification on conventional OOK modulation particularly in improving signal threshold detection and reduce power loss.

3 Theory of analysis

Based on the Rytov theory, the field of the plane wave in random media is given [10],

$$E_o(\vec{r}) = A_o(\vec{r}) \exp\left(i\varphi_o(\vec{r})\right) \tag{1}$$

where $A_o(\vec{r})$ is amplitude of the laser beam in the atmosphere without the presence of turbulence. φ_o and $E_o(\vec{r})$are the phase and laser beam profile, respectively. However in the presence of turbulence, the scintillation effect is more dominant which change the refractive index and as a consequence lead to intensity fluctuations. Therefore this condition directly change the laser beam profile [10,11],

$$E_o(\vec{r}) = A_o(\vec{r}) \exp\left(i\varphi_o(\vec{r})\right) = E_o(\vec{r}) \exp(\Phi) \tag{2}$$

where $A_o(\vec{r})$ is amplitude of the laser in the turbulence atmospheric, ϕ is the exponential factor due to turbulence given as,

$$\begin{aligned} \Phi &= Ln\left(\frac{A(\vec{r})}{A_o(\vec{r})}\right) + i[\phi(\vec{r}) - \phi_o(\vec{r})] \\ &= \chi + iS \end{aligned} \tag{3}$$

where X represents the fluctuation of the log-amplitude of the field and S is the phase fluctuations. This equation predicts a fluctuation in intensity and phase of receiving signal due to turbulence is weak and assuming that the refractive index structure coefficient C_n^2 is symmetrical along the beam path, Rytov suggested a variance for the log-irradiance given as [6],

$$\sigma_{ln\,IR}^2 = \langle (ln\,I - \langle ln\,I \rangle)^2 \rangle = 1.23 C_n^2 k^{7/6} L^{11/6} \tag{4}$$

where k is wave number, $k=2\pi/\lambda$, L is propagation distance and I is intensity signal. However in practical FSO for plane wave in weak condition and for the symmetrical turbulence of the log-amplitude fluctuation is given by [9, 10],

$$\langle \chi^2 \rangle = 0.31 C_n^2 k^{7/6} L^{11/6} \tag{5}$$

By neglecting other source noises, we assume the only dominant noise source is atmospheric turbulence which gives the log-amplitude as,

$$\chi = Ln\left(\frac{A(\vec{r})}{A_o(\vec{r})}\right) = ln\left[\frac{A_o(\vec{r}) + A_n(\vec{r})}{A_o(\vec{r})}\right] = ln(1 + \varepsilon) \tag{6}$$

where $A_n(\vec{r})$ is the amplitude of noise and $\varepsilon = A_n(\vec{r})/A_o(\vec{r})$ is the ratio of the amplitude of noise to signal. Both SNR and BER are used to evaluate the quality of the communication systems. The BER performance depends on the average received power, the scintillation strength, receiver noise and threshold bits (1 or 0) detection [6]. Here the dual-detection approach is a modification from the conventional OOK detection that using direct detection where improves power received and eliminated the need of complex adaptive threshold. The SNR for case only dominant turbulence noise in term of mean signal and noise intensity I_o and $<I_n>$ given as:

$$SNR = \frac{I_O}{\langle I_n \rangle} = \frac{\langle A_o^2(\vec{r}) \rangle}{\langle A_n^2(\vec{r}) \rangle} = \frac{1}{\langle \varepsilon^2 \rangle} \quad (7)$$

So, the BER for OOK modulation can be written as [10]:

$$BER = \frac{1}{2} erfc\left(\frac{Q}{2}\right) = \frac{e^{(-\frac{Q^2}{2})}}{Q\sqrt{2\pi}} \quad (8)$$

where *erfc()* the complementary error function and Q is defined as,

$$Q = \frac{A_1 - A_0}{\sigma_1 + \sigma_0} \quad (9)$$

where A_1, σ_1 are the mean signal and noise at the receiver for a bit '1' respectively, and for a bit '0' the mean signal and noise are denoted as A_0, σ_0.

For conventional OOK bit detection,

$$A_1 =< A_o(\vec{r}) >, A_o = 0 \; and \; \sigma_1 + \sigma_0 \; = < A_o(\vec{r}) >$$

$$SNR = \frac{<A_o^2(\vec{r})>}{<A_n^2(\vec{r})>} = Q^2 \quad (10)$$

However for case dual-detection approach,

$$A_1 =< A_o(\vec{r}) >, A_o =< -A_o(\vec{r}) > \; and \; \sigma_1 + \sigma_0$$
$$= < A_o(\vec{r}) >$$

$$SNR = \frac{<2A_o^2(\vec{r})>}{<A_n^2(\vec{r})>} \quad (11)$$

For weak turbulence model, ε is very small thus (6) is given as [10]:

$$\chi = \ln(1 + \varepsilon) \; \approx \varepsilon \quad (12)$$

When turbulence is strong, the SNR and χ can be relating using Tailor series for function $f(x) = (e^x - 1)^2$ and can be approximately to:

$$SNR = \frac{1}{\langle \chi^2 + \chi^3 + \cdots \rangle} \cong \frac{1}{\alpha \langle \chi^2 \rangle} \quad (13)$$

4 Result and discussion

In this numerical result section, we use parameters wavelength 1550nm, $Cn^2 = 1 \times 10^{-15}$ for weak turbulence and $Cn^2 = 1 \times 10^{-14}$ for strong turbulence and range propagation 5km for all calculations. In Fig. 1 shows the SNR in dB response to turbulence strength. From the graph pattern, we can know that as the distance increase, the SNR also decrease. The SNR is influenced by turbulence strength. Therefore by using the dual-detection approach the SNR can improve approximately 3 dB or 17% compare to the conventional approach.

 DOI: 10.1051/matecconf/20179701045

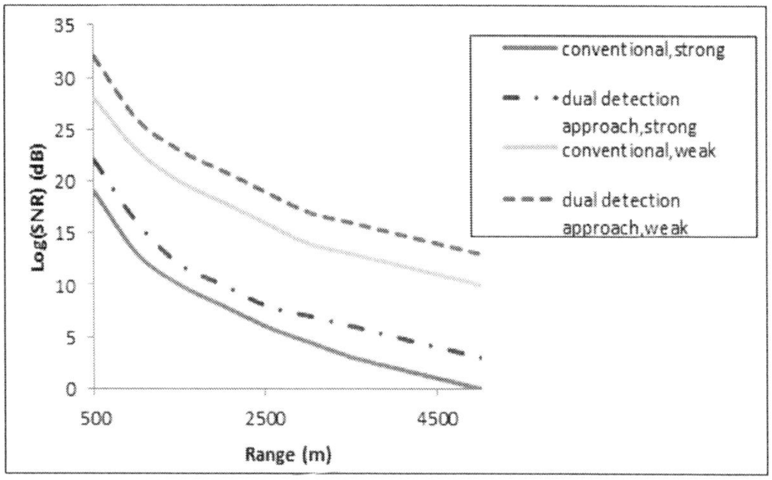

Fig. 1. SNR (dB) versus range propagation

In Fig. 2 shows the BER versus range propagation in response to weak turbulence. By using the standard communication for acceptable BER at 10-9, the conventional approach only reaches the range propagation 2.5km but the dual-detection approach can reach further to 3.5km which approximately 40% improvement.

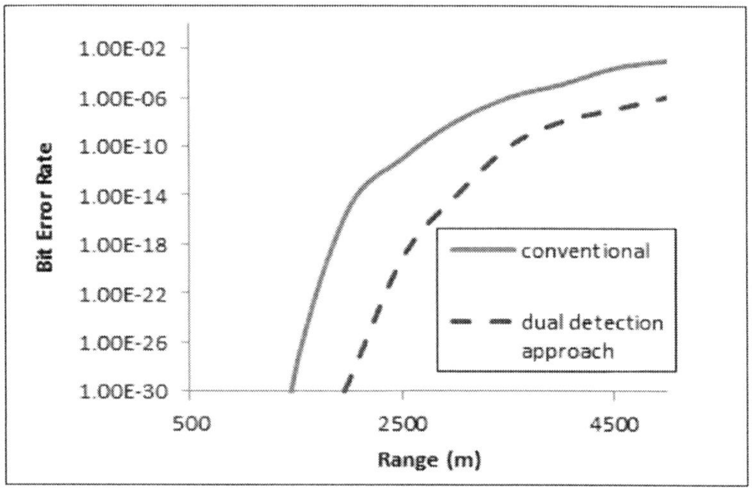

Fig. 2. BER versus range propagation for weak turbulence condition

In Fig. 3 shows the BER versus range propagation for strong turbulence condition. Here we assume the value of α is 2. In range of good quality communication (BER=10-9), the conventional approach only reaches the range propagation about 0.5km but the dual-detection approach still manage to reach 0.9km where about 80% improvement.

DOI: 10.1051/matecconf/20179701045

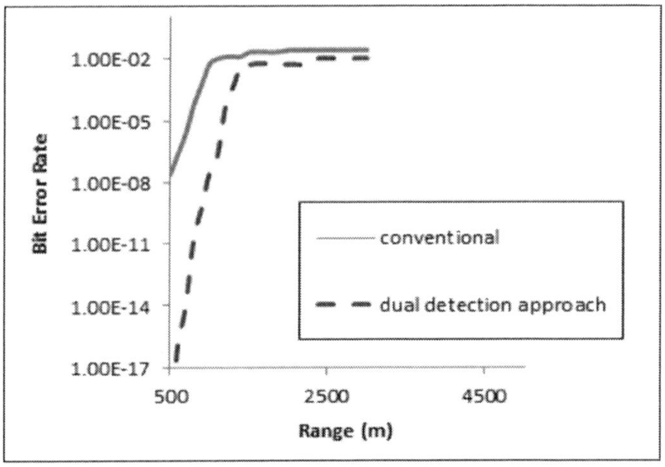

Fig. 3. BER versus range propagation for strong turbulence condition

Conclusion

In this paper, we focus on the scintillation effect deteriorate the performance of FSOC. We make a comparison performance in term of SNR and BER between two approach that are conventional (OOK-direct detection) and dual-detection. Analyses that carried out shows that the proposed dual-detection approach improve the margin SNR and BER. The range of propagation can be predicted improved 40% in weak turbulence and 80% in strong turbulence.

This work was financially supported by the Fundamental Research Grant Scheme (FRGS) in free space optic area research UniMAP.

References

1 I.I. Kim, R. Stieger, J.A. Koontz, C. Moursund, M. Barclay, P. Adhikari, J. Schuster, E. Korevaar, R. Ruigrok, C. DeCusatis, Opt. Eng., **37**, 3143 (1998)
2 A.K. Rahman, S.A. Aljunid, M.S. Anuar, H.A. Fadhil, Key Eng. Mater., **594**, 1027 (2014)
3 S. Trisno, I. Smolyaninov, S.D. Milnerb, C.C. Davis, *Optics & Photonics 2005*, (International Society for Optics and Photonics, **5892**, 589215, 2005)
4 M.C.R. Cordeiro, C.P. Colvero, J.P. von der Weid, *International Conference on Microwave and Optoelectronics,* 393 (2005)
5 M. Uysal, J. Li, M. Yu, IEEE Trans. Wireless Commun., **5**, 1229 (2006)
6 I.B. Djordjevic, B. Vasic, M.A. Neifeld, IEEE Photon. Technol. Lett., **18**, 1491 (2006)
7 M. Tariq, *Avionics, Fiber-Optics and Photonics Technology Conference, IEEE*, 21 (2007)
8 X. Zhu, J.M Kahn, IEEE Trans. Commun., **50**, 1293 (2002)
9 J. Li, M. Uysal, *Global Telecommunications Conference*, **5**, 2654 (2003)
10 X. Guoliang, Z. Xuping, W. Junwei, F. Xiaoyong, *Proc. SPIE, Optical Transmission, Switching, and Subsystems*, **5281**, 816 (2004)
11 L.C. Andrews, R.L. Phillips, C.Y. Hopen, *Laser Beam Scintillation with Applications*, (SPIE Optical Engineering Press, Bellingham, Washington, 2001)

DOI: 10.1051/matecconf/20179701046

Low Power CMOS Operational Amplifier with Integrated Common-Mode Feedback for Data Converter

S.A.Z Murad[1,*], *Muhammad M.* Ramli[1] , *A.* Azizan[1] , *M.N.M* Isa[1] , and *I.S* Ishak[1]

[1]School of Microelectronic Engineering, University Malaysia Perlis, Pauh Putra Campus, 02600 Arau, Perlis, Malaysia Sciences

Abstract. The development a high-performance design of analog circuits becomes increasingly challenging with the continuous trend towards reducing the voltage supply and low power consumption without neglecting the trade-off among other performance parameters. This paper presents the design and implementation of CMOS operational amplifier (op-amp) with integrated common-mode feedback (CMFB) circuit for data converter using 0.13-µm Silterra CMOS technology. The folded cascode topology is employed as a main op-amp design because it provides high gain and high bandwidth besides low power consumption. The simulation results indicate that the DC gain of 64.5 dB along 133.1 MHz unity gain bandwidth (UGB) is achieved for a 1 pF load capacitor. The slew rate of 22.6 V/µs, the phase margin (PM) of 68.4 ° with settling time of 72.4 ns are obtained. The power consumption of this op-amp is 0.3 mW through voltage supply of 1.8 V.

1 Introduction

Wireless Operational amplifier is a core element and integral part for most analog and mixed-signal systems. The behavioral of op-amp such as high gain, high input impedances, low output impedance, high bandwidth and fast settling makes this device often used amongst multiplicity of applications like in pipeline ADC. Traditionally, op-amp can be classified into several topologies which are telescopic, folded cascode, two-stage and gain-boosted [1]. Each topology has their own compensation, but they can be applied in any design op-amp circuit by considering the performance parameter.

A telescopic op-amp is a simple topology and provided a high gain as well faster performance [2, 3]. This topology is called as 'telescopic' because the cascades are attached between the voltage supplies with the transistor in the differential pair, occurs in a structure where each branch of the transistor connected directly together in a straight line [4]. Commonly the telescopic topology has a slighter swing because of lesser current legs and produces a small power consumption and low noise.

* Corresponding author: sohiful@unimap.edu.my

 DOI: 10.1051/matecconf/20179701046

In a folded cascode op-amp, the topology is normally customized from the telescopic op-amp and it issues higher gain and performance [5,6] compared to the telescopic because it consumes more currents legs. This topology is called 'folded cascode' because of small signal current is folded up or to down [1]. Generally, this op-amp allows the particular input common-mode level of being near to the voltage supplies as well as performing a high output swing, wide input common-mode range and preferably steering in low voltage supply circuits [4]. However, this topology contributed greater noise that effect from the more currents legs.

2 Design constraints

A design constraint is a key element in preparing the best performance of op-amp circuit. As a requirement of a high speed and high accuracy in pipeline ADC, there are numerous constraint parameters that should be considered such as gain, unity gain bandwidth, phase margin, slew rate and also output swing. Each parameter constraints are explained as follows.

2.1 Gain

Ideally, the gain is the product of the transconductance structure over the output resistances of the load structure that express as:

$$Gain = G_m \times R_{out} \quad (1)$$

where the gain extremely depends on the frequency of the input signal of an amplifier.

2.2 Unity gain bandwidth

The unity gain bandwidth state that the frequency at which the open loop gain of the amplifier is unity with the maximum capacitance at the output node. Thus the UBW is obtained as:

$$UBW = 2\pi f C_L \quad (2)$$

2.3 Phase margin

The purpose of phase margin (PM) is to determine the stability of the amplifier where the higher values of PM will allow the output signal to achieve a stable state without much swing. Noted that the PM is depends on the applications [1].

$$Phase\ margin\ (PM) = tan^{-1}\ [gm / (2\pi f C_L)] \quad (3)$$

2.4 Slew rate

Slew rate is defined as the rate of change in the output voltage that caused by a step change on the input. It can be determines by the output capacitance and the current across the output branch.

$$Slew\ rate = I_{out} / C_L \quad (4)$$

2.5 Output swing

This constraint relates to the output of the op-amp where the saturation voltage of load structure mainly defines the output swing of the op-amp. Commonly, most systems employing op amps require large voltage swings to accommodate a wide range of signal amplitude [1].

$$Output\ swing = (V_{DD} - V_{max/min}) / 2 \quad (5)$$

3 Design implementation

According to the op-amp specifications as presented in Table 1, the desired op-amp topology was determined. Folded cascode topology idyllically to be principle op-amp for this work since the design has been used in [7] for obtaining fast settling, high gain and high unity gain bandwidth besides low power consumption.

Table 1. Op-amp Specifications for Pipeline ADC.

Parameter	Value
Voltage supply	1.8 V
V_{in}	1.2 V
V_{out}	1.2 V
DC gain	> 70 dB
Phase margin	> 60°
Unity gain bandwidth	> 130 MHz

The architecture of folded cascode op-amp with common-mode feedback is illustrated as Fig. 1. Traditionally, the folded cascode op-amp has been designed by using a pair of PMOS type or NMOS type for the input range of op-amp through the limits of input common mode range [7]. For this work, the NMOS type is chosen to be an input of the differential amplifier since this input type can assign more large output gain compare to PMOS input type.

DOI: 10.1051/matecconf/20179701046

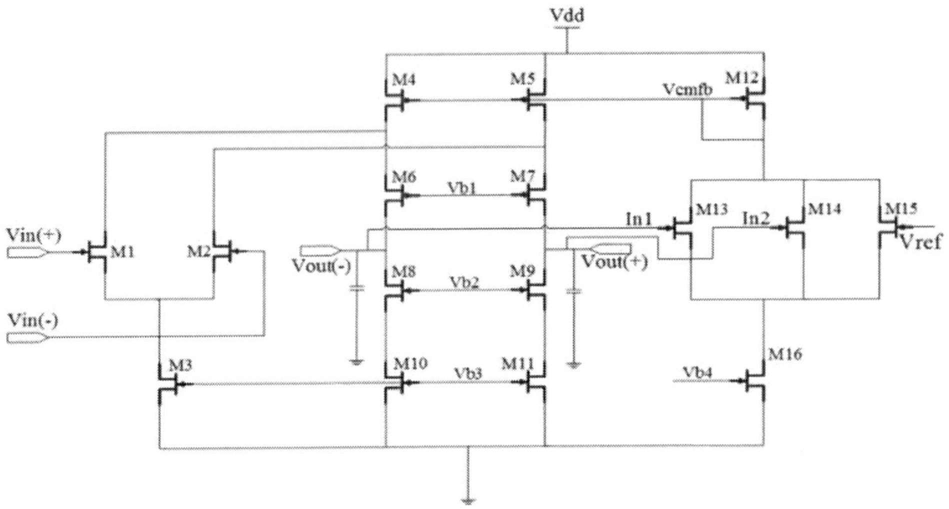

Fig. 1. Folded cascode op-amp with CMFB architecture.

As shown in Fig. 1, the folded cascode topology consists of two different structures which are NMOS differential amplifier (M1-M3) and folded cascode structure (M4-M11). The open loop voltage gain can be determined as:

$$A_v = g_m \times R_o \tag{6}$$

where g_m is a short-circuit transconductances of the output current gain across the transistor of M6 and R_o is the output impedance of folded cascode by looking into the drains of M6 and M8.

$$R_o = g_{m6}.r_{o6}(r_{o1}||r_{o4}) \ || \ (g_{m8}.r_{o8}.r_{o10}) \tag{7}$$

Therefore, the gain is expressed as:

$$A_v = g_m\{[g_{m6}.r_{o6}.(r_{o1}||r_{o4})] \ || \ [g_{m8}.r_{o8}.r_{o10}] \tag{8}$$

Meanwhile, the gain bandwidth of the folded cascode circuit is:

$$GBW = g_{m1} / C_L \tag{9}$$

where g_{m1} is a transconductance of M1 and C_L is the capacitance at the output node.

A CMFB circuit (M12-M16) is designed in order to fix the voltages at high impedances node to the desired voltage value of CMRR performance while ensuring the stability of common-mode voltage for fully differential op-amp [8]. As referred in Fig. 2, M12 is assigned to be a feedback to the folded cascode op-amp while M13 and M14 is an input of CMFB that attaches to the output of folded cascode op-amp and M15 represents as a reference voltage. This CMFB architecture is modified from the conventional amplifier as

DOI: 10.1051/matecconf/20179701046

in Fig. 1(a) in [9]. The modification is based on M12 because its need a stable voltage to fed the folded cascode op-amp. Subsequently, these two circuit blocks are implemented in 1.5-bit per stage pipeline ADC as shown in Fig. 2 in order to simulate the circuit performances. The result is discussed in section below.

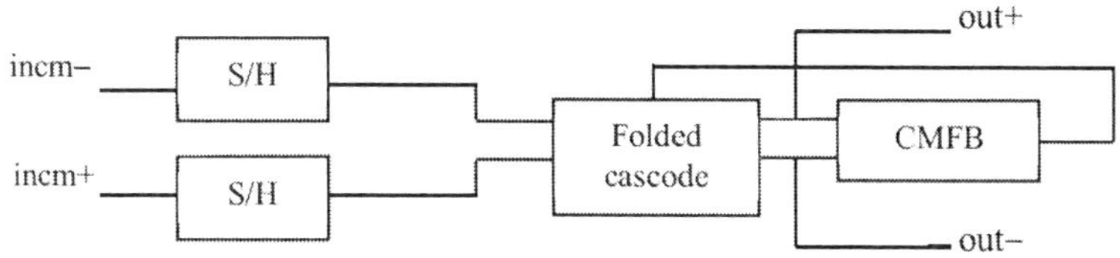

Fig. 2. Implementation circuit in 1.5-bit per stage pipeline ADC.

4 Results and discussion

The proposed folded cascode op-amp was designed using Cadence Software and implemented in 0.13-μm process technology with 1.8 V supply voltage whereas the simulation of circuit via Cadence Virtuoso spectre. The circuit design is analyzed in two methods which are AC analysis and Transient analysis. Fig. 3 shows the AC analysis result of the proposed circuit. The simulated DC gain demonstrated 64.5 dB with 68.4 degrees of phase margin (PM) as shown in Fig. 3 (a) and Fig. 3 (b), respectively. Meanwhile, the transient analysis result is depicted in Fig. 4. The slew rate performs 22.6 V/μs among 72.4 ns of settling times.

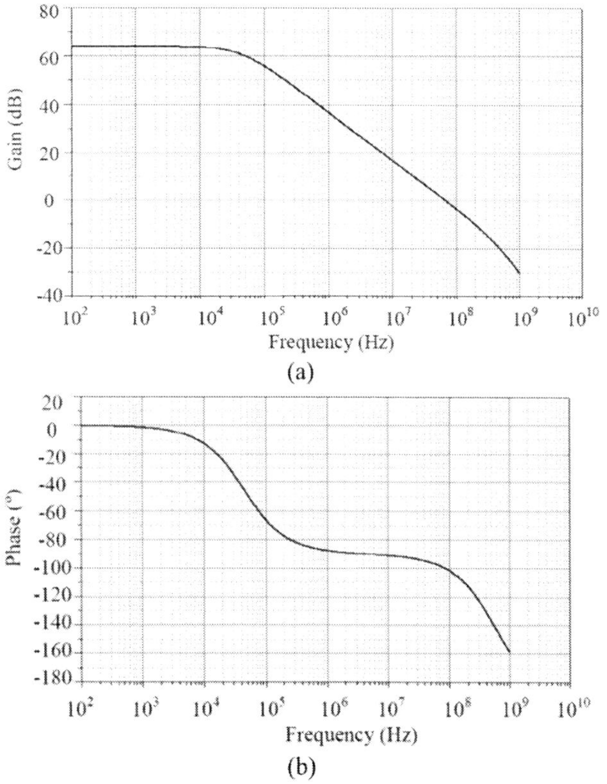

Fig. 3. AC analysis: (a) Gain, (b) Phase margin.

DOI: 10.1051/matecconf/20179701046

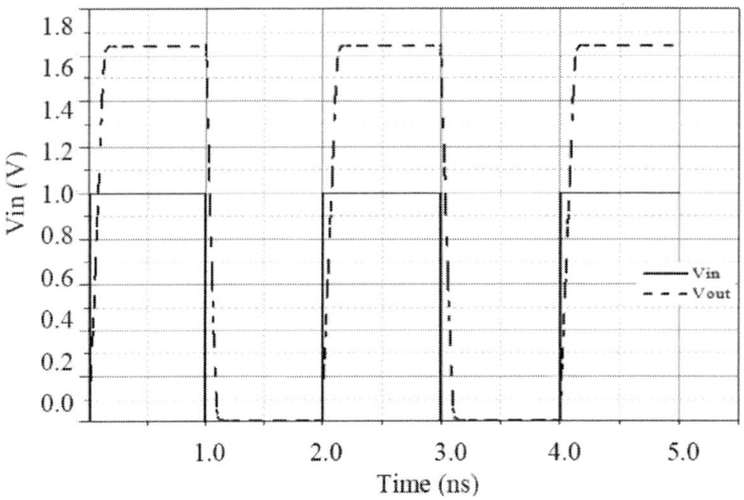

Fig. 4. Transient analysis.

Table 1 Fig. 5 illustrates the result for implementation of the circuit in 1.5 bit per-stage pipeline ADC. The performances of the circuit are referring to the process information as tabulated in Table 2.

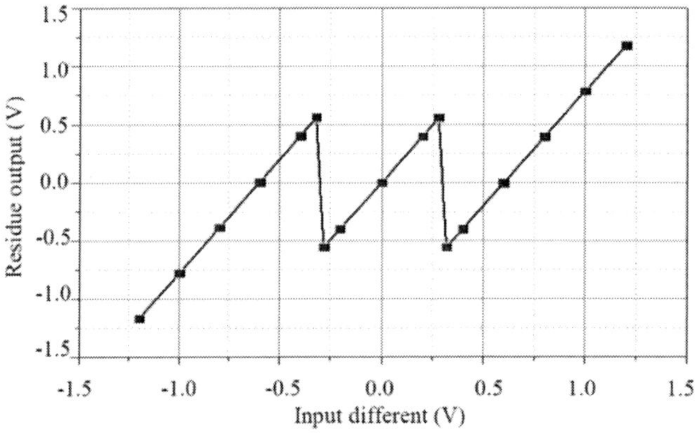

Fig. 5. Implementation results.

Table 2. Data converter information for 1.5-bit per stage pipeline ADC.

Vin	**B1**	**B0**	**DAC output**	**Residue output**
Vin > Vref/4	1	0	+Vref	0.18
-Vref/4 < Vin < Vref/4	0	1	0	1.0
Vin < -Vref/4	0	0	-Vref	2.4
Vin > Vref/4	1	0	+Vref	9.0

 DOI: 10.1051/matecconf/20179701046

The comparison for folded cascode op-amp performances between previous works is summarized as in Table 3. From the table, this circuit design attains low power consumption with smallest load capacitance; therefore the chip size can be reduced.

5 Conclusion

The design and implementation of CMOS op-amp with integrated common-mode feedback circuit for data converter using 0.13-μm Silterra technology is presented. The DC gain of the op-amp is obtained about 64.5 dB along with unity gain bandwidth (UGB) of 133.1 MHz for a 1 pF load and 68.4 degrees. The circuit in this work consumes a low power consumption which is around 0.3 mW at 1.8 V supply supply. The simulation result shows that the proposed design is suitable for data converter applications.

Table 3. Folded cascode op-amp performances.

Reference	**[10]**	**[11]**	**[12]**	**[6]**	**This wok**
Technology (μm)	0.18	0.18	0.13	0.13	0.13
V_{DD} (V)	1.8	1.8	3.0	1.8	1.8
DC gain (dB)	90.30	50.97	94.9	91.5	64.5
Unity gain Bandwidth (MHz)	700.7	944.0	41	714.5	133.1
Phase margin (°)	63.85	75.9	82.3	62.0	68.4
Slew rate (V/μs)	N/A	712.5	N/A	N/A	22.6
Settling time (ns)	N/A	N/A	6.17	40.0	72.4
Power consumption (mW)	3.24	172.8	11.0	9.0	0.3
Load capacitance (pF)	0.5	2.4	2.0	7.5	1.0

This work was financially supported by the Fundamental Research Grant Scheme (9003-00387).

References

1. Razavi, B. *Design of Analog CMOS Integrated Circuits*, (McGraw-Hill Higher Education, Singapore, 2001).
2. Nageshwarrao, D., Venkata Chalam, S. Malleswara Rao V.,International Journal of Electronic Engineering Research, **2**, 159 (2010).
3. S. Kant, O.P. Sahu, International Journal of Engineering Research & Technology, **1**, 1 (2012).
4. I.S. Ishak, S.A.Z. Murad, M.F. Ahmad, J. Eng. Appl. Sci., **11**, 20 (2016).
5. S.A.E. Ab Rahim, M.A. Ismail, A.I. Rahim, M.R. Yahya, A.F.A. Mat, *International Conference on Electronic Devices, Systems and Applications*, (2010)
6. X. Liu, J.F. Mcdonald J.F., *International Conference on IEE*, (2012).

7. W. Gu, W. Gao, World Acad. Sci Eng. Technol., **67**, 284 (2012).
8. I.S. Ishak, S.A.Z. Murad, M.F. Ahmad, S.C. Neoh, *Proceedings of Malaysian Technical Universities Conference on Engineering and Technology (MUCET)*, (2013)
9. M.F. Ahmad, S.A.Z. Murad, M.M. Shahimim, S.A.A. Rais, A.F. Hasan, Applied Mechanics and Materials, **446**, 992 (2014).
10. M.K. Hati, T.K. Bhattacharyya, *IEEE Computer Society Annual Symposium on VLSI*, (2011)
11. Z. Hao, F. Xiangning, S. Yutao, *International Conference on Wireless Communications and Signal Processing (WCSP2011)*, (2011).
12. S.A.E. Ab Rahim, I.M. Azmi, *International Symposium on Integrated Circuit*, (2011).

MATEC Web of Conferences 97, 01047 (2017) DOI: 10.1051/matecconf/20179701047
ETIC 2016

Ductile behavior of polyethylene fibre reinforced geopolymer composite

Shaikh Faiz Uddin Ahmed[1,*] and *Zammar* Ronnie[1,b]

[1] Dept. of Civil Engineering, Curtin University, Australia

Abstract. This paper presents the effects of various volume fractions of polyethylene (PE) fibres of 0.5%, 1%, 1.5% and 2% on tensile, flexure and compressive behavior of PE fibres reinforced geopolymer composites (PE-FRGC). Results show that the 1% PE fibre by volume is the optimum fibre volume fraction for fly ash geopolymer composite, which exhibit superior strain and deflection hardening behavior in uni-axial tension and three-point bending, respectively. Results also show that the compressive strength of the composites decreases with increase in volume fractions of PE fibre. The fibre surface examination of PE fibre in the gopolymer matrix using scanning electron microscope (SEM) and energy dispersive spectroscopy (EDS) revealed no significant damage of PE fibre in the alkaline geopolymer matrix.

1 Introduction

Geopolymer binder has been proposed as an alternative to ordinary Portland cement (OPC) binder in concrete and cement based construction materials. It exhibits comparable and even in some instances better mechanical, durability and fire resistance properties than its counterpart OPC binder. As a result real world application of geopolymer can be seen in many projects e.g. precast floor in a library building in University of Queensland, airport runway in Queensland, etc. With continuing research new emerging geopolymer technology will find more real world applications in future.

One of the inherent limitation of OPC concrete or geopolymer concrete is its extremely low tensile strength and crack resistance property. To overcome this limitation short fibres (metallic and polymeric) are added. Significant improvement in tensile strength and crack resistance are obtained in fibre reinforced geopolymer concrete (FRGC). However, the FRGC exhibits ductility after the peak load resulting in increase in crack width due to pullout of fibres or rupturing of fibres during strain softening or deflection softening range. The resulting FRGC material exhibit serious risk in the durability of the structure due to wide crack widths as aggressive chemicals or substances easily penetrates to cause corrosion of steel or deteriorate matrix of the FRGC. Recently, highly ductile strain hardening cement based composites exhibiting multiple fine cracks is developed and as well as its geopolymer based version e.g. [1-3]. The resulting strain hardening geopolymer

* Corresponding author: S.Ahmed@curtin.edu.au

DOI: 10.1051/matecconf/20179701047

composite and deflection hardening geopolymer composite exhibited superior ductility in terms of high tensile strain capacity and deflection at peak load in uni-axial tension and flexure, respectively [2-3]. Steel and polyvinyl alcohol (PVA) fibres are used to reinforced the above ductile geopolymer composites. Steel fibre is considered as high modulus high strength fibre and the PVA fibre is considered as low modulus low strength fibre, which result in high ultimate strength but low ductility and low ultimate strength but high ductility of the composite, respectively. Polyethylene (PE) fibre is another kind of polymeric fibre whose modulus is about twice the PVA and tensile strength is similar to steel fibre. Therefore, by reinforcing geopolymer matrix with PE fibre both high strength and high ductility can be achieved simultaneously. A number of research is reported on the ductility of PE fibre reinforced cement based composites [4-5]. However, no research so far is reported on geopolymer composite containing PE fibres. This paper presents experimental results on ductile behavior of PF fibre reinforced geopolymer composite containing four different volume fractions of PE fibre ranging from 0.5% to 2.0% in uni-axial tension, flexure and uni-xial compression. An optimum volume fraction of PF fibre to achieve the strain hardening and deflection hardening in uni-axial tension and flexure, respectively is also proposed. Microstructure of PF fibre reinforced geopolymer composite is also examined under scanning electron microscope to evaluate the effect of alkali of geopolymer matrix on the PE fibre degradation.

2 Materials, Mix Proportions, Casting and Testing Details

The fly ash used is originated from Gladstone power station in Queensland and satisfies ASTM class F classification. The chemical composition of fly ash is shown in Table 1. The activating solutions used are sodium silicate with a chemical composition of (wt.%): Na_2O = 14.7, SiO_2 = 29.4 and water = 55.9. The other characteristics of the sodium silicate solution are specific gravity = 1.53 g/cc and viscosity at 20°C = 400 cp. The sodium hydroxide solution is prepared from analytical grade sodium hydroxide pellets. The mass of the NaOH solids in the solution varied depending on the concentration of the solution expressed in terms of molar, M. In this study, the NaOH solution with a concentration of 8 M is considered and consisted of 8x40 = 320 g of NaOH solids per litre of the solution, where 40 is the molecular weight of NaOH. The sodium hydroxide solution is mixed with Na_2SiO_3 (sodium silicate) with the ratio of 0.4:1 and produced the alkali activator. The alkali activator solution is then used for the mixing of geopolymer composites.

In this study five series of mixes shown in Table 2 is considered. The first series is the unreinforced geopolymer paste considered as control. While the second to fifth series are geopolymer composites containing 0.5, 1.0, 1.5 and 2.0% PE fibre, respectively. The mixing is carried out in a Hobart Mixer. First fly ash and alkali activators are mixed for approximately 3 min and then the PE fibres are slowly added to the wet mix and continued mixing until the fibres are well dispersed in the mix. The geopolymer composite specimens were subjected to oven curing at 85°C immediately after casting, for 24 h. The specimens are then demolded after 24 h and stored in the laboratory in open air until the date of testing. A constant alkali activator to fly ash ratio of 0.4 is considered in all series. The properties of polyethylene (PE) fibre are shown in Table 3.

For each series, three 50mm cube specimens for compression, three plate specimens of 15x40x200 mm in dimension for uni-axial tension and three plate specimens of 15x40x400mm in dimension for three-point bending are cast. All specimens are tested after 7 days of curing using a universal testing machine under displacement control with a loading rate of 0.5 mm/min.

Table 1. Chemical compositions of class F fly ash

Compounds	SiO_2	Al_2O_3	Fe_2O_3	CaO	Na_2O	K_2O	MgO	P_2O_5	SO_3	TiO_2	MnO	LOI
Fly ash	51.11	25.56	12.48	4.3	0.77	0.7	1.45	0.885	0.24	1.32	0.15	0.57

Table 2. Mix proportions of geopolymer composites

Series	PE fibre (by Vol.)	Mix proportions by wt.		Superplasticizer (by wt. of fly ash)
		Fly ash	Alkali activator/ fly ash	
1(Control)	0 %	1	0.4	-
2	0.5 %	1	0.4	-
3	1.0 %	1	0.4	-
4	1.5 %	1	0.4	1%
5	2.0 %	1	0.4	2%

Table 3. Properties of Polyethylene (PE) fibre

Type of Fibre	Length (mm)	Diameter (mm)	Modulus of elasticity (MPa)	Fibre tensile Strength (MPa)	Density (gm/cm^3)	Elongation (%)
Polyethylene	12	0.012	73,000	2,580	0.97	-

3 Results and Discussions

The effect of various volume fractions of PE fibre on the compressive behavior of PE fibre reinforced geopolymer composite (PE-FRGC) is shown in Fig. 1. It can be seen that the control mix achieved about 80 MPa and the compressive strength of the composites decreased with increase in volume fractions of PE fibre. This gradual reduction in compressive strength can be attributed to the formation of entrapped air or pores inside the composite and it increased with increase in volume fractions of fibres. Similar reduction in compressive strength of polypropylene fibres reinforced geopolymer composite is also reported by Zhang et al. [6]. However, unlike the control geopolymer, the PE-FRGC did not exhibit such catastrophic failure. While the cube specimens of PE-FRGC indeed fractured they still maintained their overall shape and it was still possible for some compression force to be applied as evident from the softening tail of the compressive stress-axial strain behavior of various PE-FRGC shown in Fig. 1 and the post-peak axial deformation increases with increase in volume fractions of PE fibres.

DOI: 10.1051/matecconf/20179701047

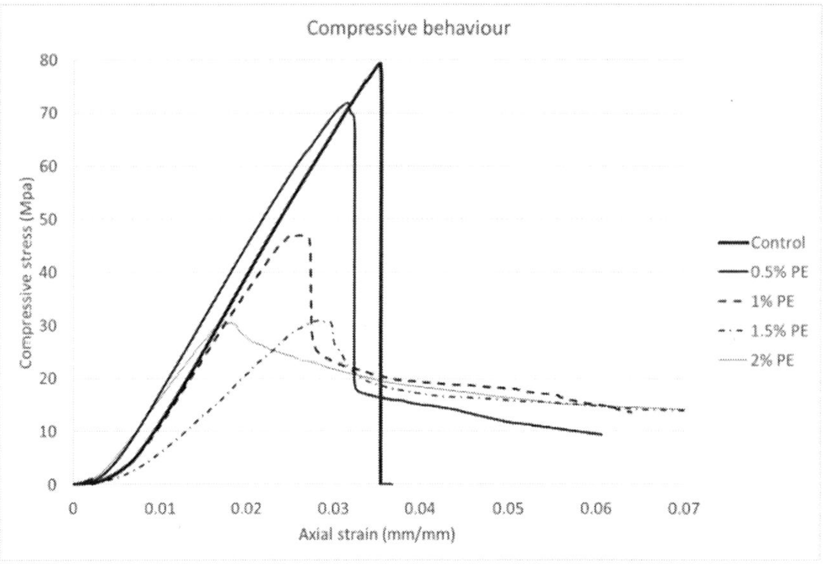

Fig. 1: Compressive stress vs. axial strain behavior of geopolymer composites.

The effects of various volume fractions of PE fibres on the tensile stress-stain behavior of geopolymer composites is shown in Fig. 2. The control specimen failed catastrophically under tension and hence its stress-strain curve was not possible to capture and is not shown in Fig. 2. It can be seen in Fig. 2 that all PE-FRGC exhibited strain hardening and multiple cracking behavior with composite containing 0.5% and 1% PE fibres exhibited exceptionally higher strain capacities from 4-5% at peak load. The number of multiple cracks and their spacing are also higher and lower, respectively compared to composites containing higher volume fractions of PE fibres e.g. at 1.5% and 2%. This can be attributed to the poor dispersion of PE fibre at higher volume fractions compared to lower volume fractions. Due to low density of PE fibre there is very likely possibility of formation of fibres balling during mixing which entrapped voids in the composite. These voids believed to be caused the stress concentration and propagated the crack formation as insufficient distributed fibres were present in the vicinity of the voids area.

By comparing with present composite containing 1% PE fibre, it can be seen that, similar level of strain capacity was achieved in PVA fibre reinforced geopolymer composite at 2% volume fractions in a PVA fibre reinforced geopolymer composite [2]. However, the ultimate tensile strength in present PE-FRGC containing 1% PE fibre is slightly lower than that of PVA fibre reinforced geopolymer composite at 2% volume fractions. The slightly lower tensile strength capacity of PE-FRGC than its PVA counterpart can be attributed to the half the volume fractions of PF fibre in the composite. Nevertheless, with half the volume fraction of fibre the PE-FRGC composite exhibited similar tensile ductility to its PVA fibre counterpart.

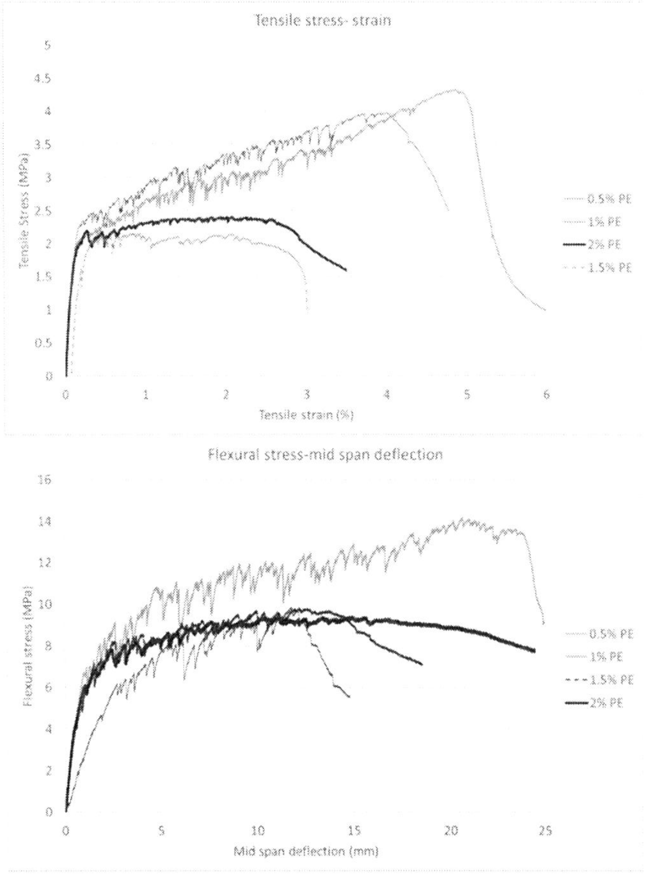

Fig. 2: Tensile strain hardening (left) and deflection hardening (right) behavior of PE-FRGC

The deflection hardening behavior of PE-FRGC under three point bending is also shown in Fig. 2. It can be seen in figure that the deflection hardening behavior under three point bending is very similar to the strain hardening behaviour observed in the same composites. The PE-FRGC containing 1% PE fibre exhibited the highest deflection capacity at peak load and the maximum flexural strength compared to other volume fractions. By increasing the volume fractions of PE fibre to 1.5% and 2% resulted in lower deflection capacity and flexural strength and can be attributed to the formation of fibre balling due to non-uniform dispersion of PE fibres at high volume fractions. It can also be seen that the ductility index which is defined as the ratio of deflection at peak load to that at first crack load of PE-FRGC containing 1% PE fibre is about 22, which is still high but lower than that obtained in the case of 2% (vol.) PVA fibre reinforced geopolymer composites reported in [3]. The lower ductility index obtained in this study can be attributed to the 50% more thickness of the flexure specimens, half the volume fraction of PE fibre and the 3-point bending test used in this study.

The effect of alkalinity of geopolymer on the degradation of PE fibre in the composite is also examined through SEM and EDS analysis. Fig. 3 shows the SEM image of individual PE fibre and that in the geopolymer matrix. The SEM image shows that the PE fibre is not damaged by the alkalinity of the geopolymer as evidenced by carbon peak obtained in the EDS analysis (Fig. 3). Fig. 3 also shows the evidence of adherence of geopolymer gel on

DOI: 10.1051/matecconf/20179701047

the PE fibre in the composite as indicated by four peaks of Si, Na, O and Al in EDS analysis.

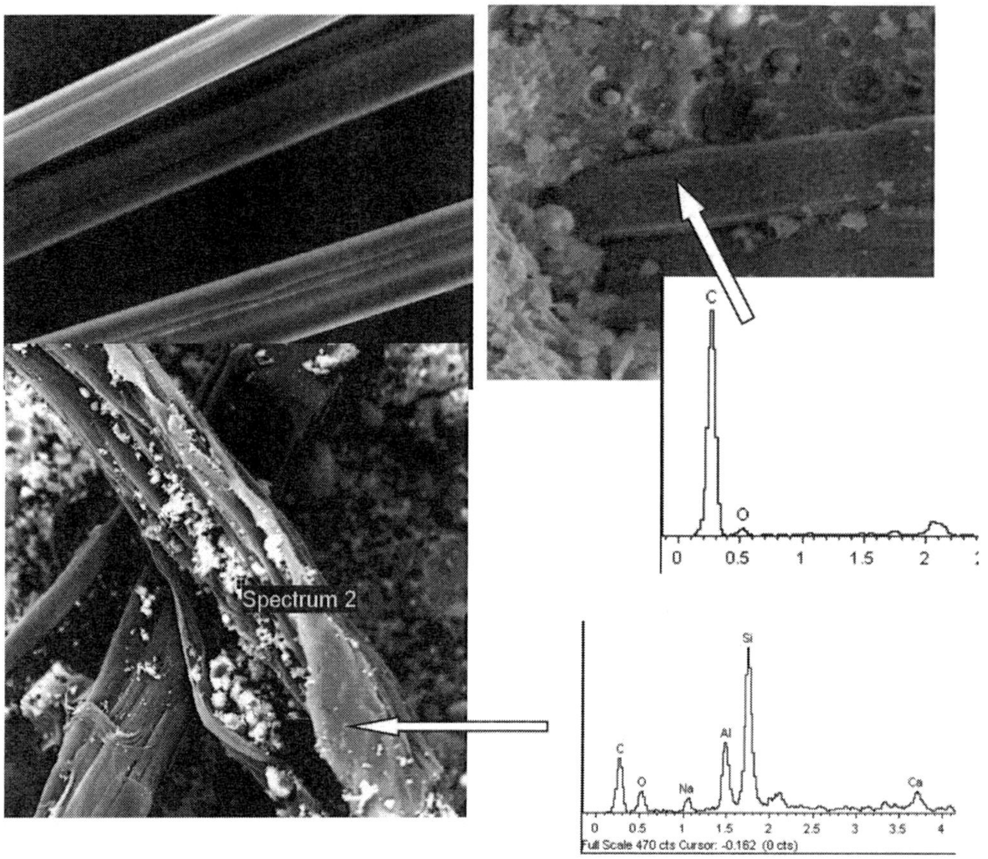

Fig. 3: SEM image of PE fibre surface as individual fibres (above-left) and no change of PE fibre surface in the geopolymer matrix (above-right and evidence of carbon peak for PE fibre in EDS insert). EDS showing the evidence of formation of geopolymer gel on the surface of PE fibres bundle (below-left) and EDS peaks in the insert.

4 Conclusion

Based on the limited variabilities of volume fractions of PE fibres and their effects on the tensile, flexure and compressive behavior of PE-FRGC, the following conclusions can be drawn:

- The compressive strength of the PE-FRGC decreases with increase in volume fractions of PE fibres due to formation of higher amount of entrapped air in the composite.
- The addition of PE fibres improves the ductility of PE-FRGC both in uni-axial tension and three-point bending. The PE fibre volume fraction of 1% is found to be the critical volume fraction for superior strain hardening and deflection hardening in uni-axial tension and three-point bending, respectively.

 DOI: 10.1051/matecconf/20179701047

- The uniform dispersion of PE fibres at 1% in the composite is believed to be the reason for such superior ductile behavior compared to higher fibre volume fractions.
- The examination of PE fibre surface through SEM and EDS analysis reveals no evidence of damage due to high alkalinity of geopolymer in the composite.

References

1. F.U.A. Shaikh, Mater. Des., **50**, 674 (2013)
2. B. Nematollahi, J. Sanjayan, F.U.A. Shaikh, J. Mater. Civ. Eng., (2015)
3. B. Nematollahi, J. Sanjayan, F.U.A. Shaikh, Constr. Build. Mater., **70**, 54 (2014)
4. S.F.U. Ahmed, M. Maalej, Constr. Build. Mater., **23(1)**, 96 (2009)
5. S.F.U. Ahmed, M. Maalej, P. Paramasivam, Constr. Build. Mater., **21(5)**, 1088 (2007)
6. Z. Zhang, Y. Xiao, H. Zhu, S. Hua, Y. Chen, J. Cent. South Univ. T., **16**, 49 (2009)

Variations of Ground-level Ozone Concentration in Malaysia: A Case Study in West Coast of Peninsular Malaysia

Nur Izzah Mohamad Hashim[1,*], Norazian Mohamed Noor[1]

[1]School of Environmental Engineering, University Malaysia Perlis (UniMAP), Kompleks Pusat Pengajian Jejawi 3, 02600 Arau, Perlis, Malaysia.

Abstract. Hourly ground ozone concentration, measured from the monitoring stations in the West Coast of Peninsular Malaysia for the period of 10 years (2003-2012) were used to analyse the ozone characteristic in Nilai, Melaka and Petaling Jaya. The prediction of tropospheric ozone concentrations is very important due to the negative impacts of ozone on human health, climate and vegetation. The mean concentration of ozone at the studied areas had not exceeded the recommended value of Malaysia Ambient Air Quality Guideline (MAAQG) for 8-hour average (0.06 ppm), however some of the measurements exceeded the hourly permitted concentration by MAAQG that is 0.1 ppm. Higher concentration of ozone can be observed during the daytime since ozone needs sunlight for the photochemical reactions. The diurnal cycle of ozone concentration has a mid-day peak (14:00-15:00) and lower night-time concentrations. The ozone concentration slowly rises after the sun rises (08:00), reaching a maximum during daytime and then decreases until the next morning.

1 Introduction

Ozone (O3) is a gas composed of three atoms of oxygen and occurs both in the Earth's upper atmosphere and at ground level [1]. It also can be beneficial "good ozone" or detrimental "bad ozone". In addition, ozone is also known as the greenhouse gas that has been the subject of the previous researches in the recent years [2,3]. Besides that, it regulates OH and NO3 radicals in the atmosphere during day and night-time respectively. It has been proved to be a serious air pollution problem for many countries all over the world [4] including Malaysia. In Malaysia, ground-level ozone was found to be one of the major pollutants since 2004 [5]. It is formed from chemical reactions among primary air pollutants, such as nitrogen oxides and volatile organic compounds in the presence of sunlight [6]. This air pollutant with high concentrations deserve the special attention due to capability of causing adverse health effect towards human health and on the environment [7] (WHO, 2003). O3 could irritate the respiratory system, reduce lung functions, and

* Corresponding author: nurizzah_hashim91@yahoo.com

aggravate asthma problems and chronic lung diseases [8] (EPA, 2008). Active children, active adults, and people with respiratory diseases are people groups particularly sensitive to ozone [9].

Thus, it is necessary and quite urgent to gain a good understanding of the characteristics of ozone pollution in Malaysia and it can be used as fundamental tool for forecast [10], hence mandatory public warnings and traffic restrictions can be planned [11]. In this study, 10 years of hourly ozone (O3) monitoring records from three stations i.e. Nilai, Melaka and Petaling Jaya were selected to be analysed. These three locations were chosen because there are situated in the west coast of peninsular Malaysia that are more developed than the East Coast region, hence the air quality are expected to be more deteriorated.

2 Material and methods

In this research, the measured records of hourly O3 and meteorological from three stations in Malaysia were chosen. In this research, three sites are selected namely Klang in Selangor, Muar in Johor and Seremban in Negeri Sembilan. The hourly O3 concentrations records from 2003 to 2012 were used due to the high level of O3 concentration were released or emitted in atmosphere during that particular time. All the hourly O3 concentration and meteorological data were provided by DoE and were regularly subjected to quality control.

2.1 Ozone Data Collection

Ten years of ground level ozone concentration from 2003 to 2012 were obtained from air quality monitoring in Nilai, Melaka and Petaling Jaya. The data were owned by Department of Environment (DOE) Malaysia, however are managed by Alam Sekitar Malaysia Sdn. Bhd. The ozone concentration was measured using Teledyne O3 Analyser Model 400A UV Absorption. The analyser uses a system based on the Beer-Lambert law for measuruing low ranges of O3 in ambient air.

2.2 Data Analysis

The hourly and diurnal data of ground level ozone concentration were examined visually using box and whisker plot. For the descriptive statistics, two tests were conducted i.e. i) measures of central tendency (mean, median) and; ii) measures of dispersion (range, standard deviation).

3 Results And Discussion

Table 1 and summarizes the descriptive statistics of ozone concentration for day and night-time in Nilai, Melaka and Petaling Jaya for day and night in Nilai, Melaka and Petaling Jaya respectively. The mean values for all areas during daytime and night-time were higher than the median. This indicated that the data are skewed to the right and that extreme events occurred. The mean concentration of ozone is recorded low than the recommended value of Malaysia Ambient Air Quality Guideline (MAAQG) that is 0.06 ppm. However, the maximum values recorded for all stations during daytime exceeded the stipulated value of MAAQG (0.06 ppm).

Figure 1 and Figure 2 shows the boxplot of ozone concentration for daytime and night-time at the studied areas. Obviously, the day concentration of ground-level ozone is much

DOI: 10.1051/matecconf/20179701048

higher compared to the night concentration. From Figure 1, for most of the years, Melaka gives the higher readings compared to Nilai and Petaling Jaya. However, the mean concentrations of ozone do not exceeded the recommended value of Malaysia Ambient Air Quality Guideline (MAAQG) that is 0.06 ppm. Though, it can be seen that for some of the years the ground-level concentration exceed the recommended value and this show that an extreme events occurred. The concentration of ground-level ozone during night- time (Figure 2) is far below the daytime concentration. Ozone need sunlight to undergo chemical reactions to be formed, hence, the concentration of O3 is higher during daytime. From Figure 2, Melaka recorded the highest concentration of O3 compared to Nilai and Petaling Jaya. Furthermore, Petaling Jaya shows that O3 concentration was more convergence and less variable compared to others. All the values are far below the MAAQG except for only 1 reading in Melaka (2009).

Table 1. Statistics of O_3 (ppm) in the Study Area for Daytime and Night-time

		Nilai	**Melaka**	**Petaling Jaya**
Daytime	Mean	0.0253	0.0277	0.0236
	Median	0.0244	0.0269	0.0223
	S.D	0.0092	0.0085	0.0097
	Min	0.0021	0.0030	0.0008
	Max	0.0665	0.0715	0.0657
Night-time	Mean	0.0074	0.0119	0.0039
	Median	0.0055	0.0112	0.0027
	S.D	0.0063	0.0067	0.0038
	Min	0.0000	0.0000	0.0000
	Max	0.0465	0.0690	0.0455

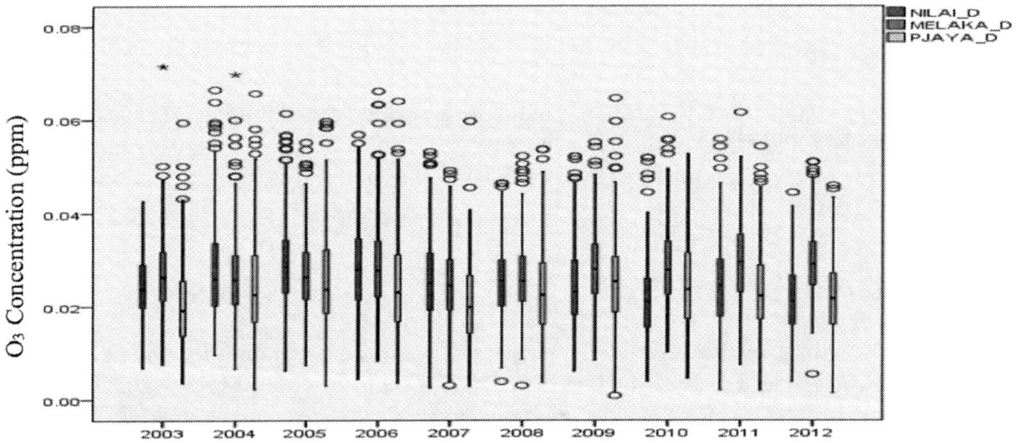

Fig. 1. Box plot (Day Record) for Nilai, Melaka and Petaling Jaya

DOI: 10.1051/matecconf/20179701048

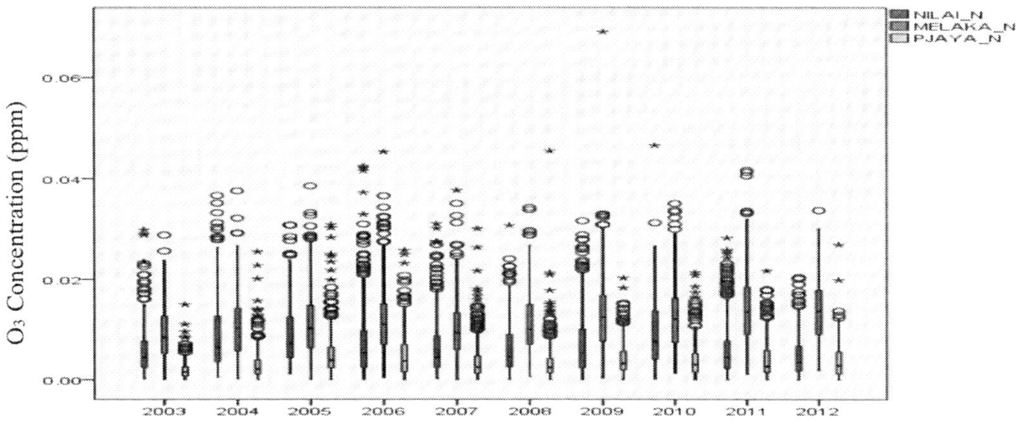

Fig. 2. Box plot (Night Record) for Nilai, Melaka and Petaling Jaya

The diurnal variations of O3 concentrations during the study period are graphically illustrated in Figure 3. The O3 diurnal variation of each site exhibited a similar pattern, but the magnitudes of the variations differed. The diurnal pattern of O3 for each site is characterized by a maximum concentration in the afternoon and a minimum during night-ime. This variation between day and night generally coincides with the amount of higher solar radiation intensity during the day, which is the favourable condition for powering photochemical reactions. In addition, the diurnal cycle of ozone are shaped like double waves. The diurnal cycle of ozone concentration reaches a peak during the middle of the day, while during the night-time, there was a lower ozone concentration [1]. Latif et al. (2012) stated that this is because the reaction of photochemical of O3 concentration such as VOC with NOx that were originated from transportation [12]. Ismail et al., (2016) proved that during the daytime, the ozone concentration slowly increases after the sun rises until it reached its maximum [13]. While, it slowly decreases until the next morning. The graphs above show that the increasing ozone concentration occurred from 08:00 to 13:00-14:00 due to the solar radiation [14]. From the boxplot, Nilai shows higher variations of diurnal ozone concentration compared to others. However, the highest peak of ozone concentration was recorded in Petaling Jaya at hour 13:00-14:00.

DOI: 10.1051/matecconf/20179701048

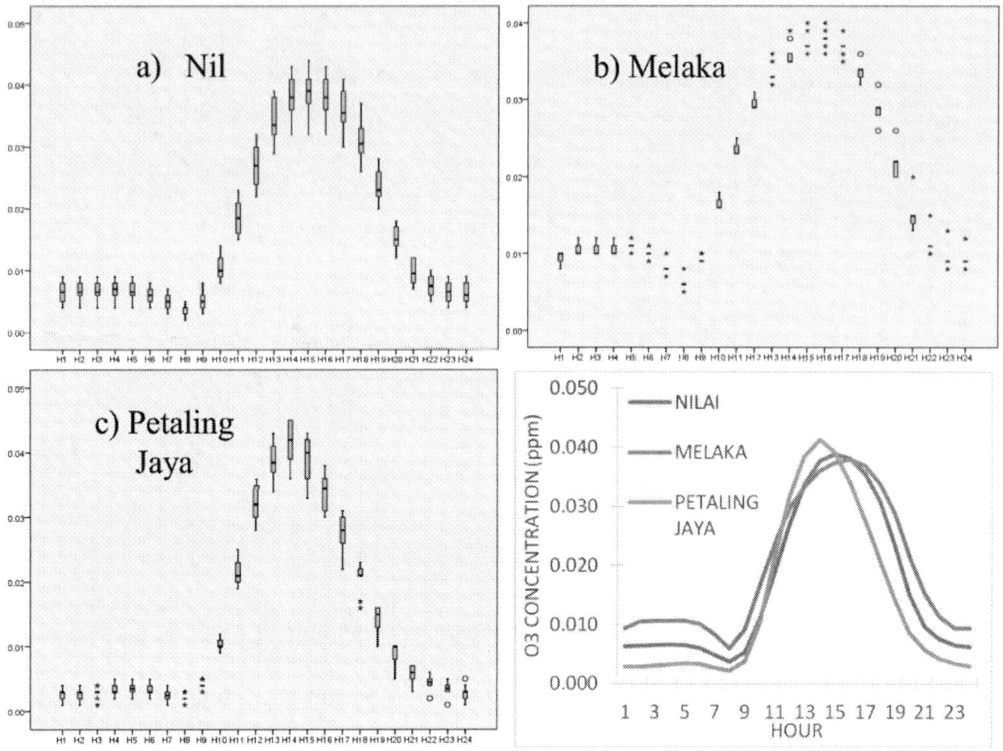

Fig. 3. Diurnal variations of O_3 concentrations

Conclusion

This study analyses the concentrations of O3 in the three areas in the West Coast of Peninsular Malaysia from 2003 to 2012. The descriptive statistics and the diurnal variations of the hourly O3 concentration in Nilai, Petaling Jaya and Melaka were observed and investigated. Higher concentration of ground level ozone during daytime was observed at all study areas due to higher solar radiation intensity during that time. Most of the time, Melaka recorded higher O3 concentration compared to other areas. However, the mean concentration of O3 in these three sites does not exceed the recommended value of MAAQG for 8-hour average that is 0.06 ppm. The diurnal variation of O3 concentration exhibited a mid-day peak at 14:00-15:00 and had lower night-time concentrations. The O3 concentration slowly rises after the sun rises (08:00), reaching a maximum during daytime and then decreases until the next morning. This pattern was majorly influenced by the activity of photochemical reactions of O3 during the daytime.

References

1. N.A. Ramli, N.D. Ghazali, A.S Yahaya, Malaysian Journal of Environmental Management **11**, 57 (2010)
2. M. Musa, A.A. Jemain, W.A.N. Zawiah, W.A.N Zin, Journal of Quality Measurement and Analysis, **9**, 9 (2013)
3. M.A. Barrero, J.O. Grimalt, L. Canton, Chemometr. Intell. Lab., **80**, 67 (2006)
4. C.M. Atkinson Palombo, J.A. Miller, J.R.C. Balling, Atmos. Environ. **40**, 7644 (2006)

5. Department of Environment (DOE) Malaysia (2012). Malaysia Environmental Quality Report 2012. Kuala Lumpur: department of Environment, Ministry of Sciences, Technology and Environment, Malaysia.
6. C. Duenas, M.C. Fernandez, S.J. Canete, Carretero, E. Liger, Chemosphere **61**, 1379 (2005)
7. Environmental Protection Agency. (2008). Ozone Pollution. Retrieved: 2nd August 2016. http://www.epa.gov/air/ozonepollution/pdfs/health.pdf
8. World Health Organization, Health aspects of air pollution with particulate matter, ozone and nitrogen dioxide. (Bonn, Germany, 2003)
9. E. Agirre, L.J.R Barron, Forecasting Model, 208-218 (2010)
10. S.I.V Sousa, F.G. Martins, M.C. Pereira, M.C.M. Alvim Ferraz, Chemosphere **64**, 1141 (2006)
11. S. Bordignon, C. Gaetan, F. Lisi, Stat. Methods. Appt., **11**, 227 (2002)
12. M.T. Latif, L.S. Huey, L. Juneng, Atmos. Environ., **61**, 434 (2012)
13. M. Ismail, S. Abdullah, F.S. Yuen, N.A. Ghazali, Environment Asia, **9**, 1 (2016)
14. A.G. Ulke, N.A.Mazzeo, Atmos. Environ., **32**, 1615 (1998)

 DOI: 10.1051/matecconf/20179701049

Screening for Direct Production of Lactic Acid from Rice Starch Waste by Geobacillus stearothermophilus

Balakrishnan Kunasundari[1,*], Mohamad Faizul Zulkeple[1], and Yi Peng Teoh[1]

[1]Faculty of Engineering Technology, Universiti Malaysia Perlis (UniMAP), P.O Box 77, D/A Pejabat Pos Besar Kangar, Perlis, 01000, Malaysia.

Abstract. Lactic acid recently became an important chemical where it is widely used in many industries such as food, cosmetic, chemical and pharmaceutical industry. The present study focuses on the screening for lactic acid production from rice starch waste using a thermophilic amylolytic bacterium, *Geobacillus stearothermophilus*. There is no information available on direct fermentation of lactic acid from rice starch waste using *G. stearothermophilus*. The effects of different parameters such as temperature, pH, incubation time, agitation speed, concentration of nitrogen and carbon sources on the lactic acid production were assessed. The highest concentration of lactic acid produced was 5.65 ± 0.07 g/L at operating conditions of 60°C, pH 5.5, 48 h, 200 rpm of agitation speed with 5% concentrations of both carbon and nitrogen source. The findings indicated that rice starch waste can be successfully converted to lactic acid by *G. stearothermophilus*.

1 Introduction

Lactic acid is considered to be one of the most important chemicals recently where it is widely used in many industries such as food, cosmetic, chemical and pharmaceutical. Besides that, the attraction for using it as a monomer in the manufacture of biodegradable and bio-compatible poly (lactic acid) or PLA has been increasing recently. It is an environmentally friendly and alternative to non-biodegradable plastics which are derived from petrochemicals. The worldwide demand on lactic acid is valued roughly to be around 130 000 to 150 000 (metric) tonnes per year. A viable approach for production of lactic acid includes fermentation of starch wastes using amylolytic lactic acid producing bacteria [1].

G. stearothermophilus is a thermophilic amylolytic bacterium that can convert starch directly to lactic acid. Most of the lactic acid producing microorganisms require hydrolysis of starch containing carbon sources to glucose before the fermentation process represents an advantage of using this particular bacterium. The other characteristics of this bacterium are facultative anaerobic and gram-positive from the phylum Firmicutes, with an optimum growth temperature between 50°C and 60°C, while pH of 6.2 and 7.5 [2, 3].

* Corresponding author: kunasundari@unimap.edu.my

Rice starch waste was used in this study which can be obtained during washing the rice. The colour of this liquid starch is milky and its texture is non-sticky. This raw material is rarely use by the researchers as the substrate in the fermentation of lactic acid because it is a municipal waste and available in a small volume only. However, when one considers the starch waste from rice industries which would be available in large quantities, then this substrate is an alternative option for production of lactic acid. So far to our knowledge, the potential of rice starch waste to be the substrate for fermentation are not being much investigated. Hence, this study aims to investigate lactic acid production from rice starch waste.

2 Materials and methods

Geobacillus stearothermophilus was used throughout this study. Growth profile construction and morphology identification of the bacterium were carried out. Fermentation for lactic acid production was conducted using mineral salt medium with rice starch waste as a sole carbon source by varying the process parameters according to the Plackett-Burman designs. Lactic acid produced from each experiment was quantified using Reflectometer.

2.1 Growth profile construction

G. stearothermophilus was maintained on nutrient agar. The culture was prepared by inoculating two loopfuls of the bacterium in 30 mL Nutrient broth (NB) of a 100 mL flask. The culture was agitated at 55°C and 100 rpm for 24 h. Three mililiter of culture was transferred into 97 mL of NB in a 250 mL of flask after 24 h of incubation time. Optical Density at 600 nm (OD_{600nm}) was determined by sampling at every 3 h interval for 48 h. At the same time, wet cell weight (WCW) and dry cell weight (DCW) were measured at every sampling point. The growth profile of *G. stearothermophilus* was plotted based on OD and DCW versus time for each sampling point [4].

2.2 Morphology identification

The morphology of the *G. stearothermophilus* was examined using Gram staining in which it can differentiate bacterial species into two large groups (Gram positive and Gram negative). This classification is based on the chemical and physical properties of bacteria's cell walls [4].

2.3 Fermentation process

Lactic acid production was carried in a batch fermentation using 250 mL flask with working volume of 100 mL. The inoculum was prepared as for growth profile construction and 3% of culture was transferred into flask containing 97 mL formulated media with 2 g/L KH_2PO_4, 2 g/L NaCl, 0.2 g/L $MgSO_4$, 0.05 g/L $MnSO_4$, 0.01 g/L $FeS0_4.7H_20$ and 12 g/L of $CaCO_3$. The amount of rice starch waste and yeast extract were varied according to the screening experiments. All experiments were conducted in triplicate to examine reproducibility and the average values are reported.

2.4 Lactic acid quantification

The analysis of fermentation product in this study was determined using Reflectometer. This equipment is functioned to detect the quantity of lactic acid that was produced. The sample strip was dipped into the fermented culture for 2 seconds before it was dried with filter paper. The presence of lactic acid was indicated when the colour of the sample strip turned purple from white. The detection time for each sample took 5 minutes to complete. The result was displayed on the screen of the equipment. The maximum concentration that can be detected by LA kit was 60 mg/l and the reading. Dilution was required for concentration of sample above that limit.

2.5 Screening for lactic acid production at various process parameters conditions

The Design-Expert® Version 7.1.5 software was used in this study for designing procedure of the Response Surface Methodology as shown in Table 1. By employing Plackett-Burman designs, different temperature, incubation time, pH, agitation speed, and different concentrations of carbon and nitrogen source were analyzed.

Table 1. Screening of the parameters using Plackett-Burman designs.

Run	Factor 1 (A) Temperature	Factor 2 (B) pH	Factor 3 (C) Time (h)	Factor 4 (D) Agitation (rpm)	Factor 5 (E) Carbon source	Factor 6 (F) Nitrogen source
1	50	5.5	24	200	1	10
2	60	7.5	24	200	5	10
3	60	7.5	24	100	1	10
4	60	5.5	24	100	5	5
5	60	5.5	48	200	5	5
6	60	5.5	48	200	1	10
7	50	7.5	48	100	5	10
8	50	5.5	24	100	1	5
9	60	7.5	48	100	1	5
10	50	7.5	48	200	1	5
11	50	7.5	24	200	5	5
12	50	5.5	48	100	5	10

3 Results and discussion

G. stearothermophilus is a thermophilic bacterium that can convert starch directly to lactic acid due to ability to produce α-amylase. Previously, the bacterium has been reported to produce lactic acid yield up to 81.2% when potato residues from food processing was used as a carbon source [2]. The present study focusing on the utilization of rice starch waste as an alternative substrate for potato waste. The growth profile was plotted in order to ensure the subsequent fermentation experiments were carried out using culture in log phase. The screening experiments at different process paramaters are important to determine the conditions that maximize the lactic acid production from rice starch waste by *G. stearothermophilus*.

3.1 Construction of growth profile

Figure 1 shows the growth profile of *G. stearothermophilus* where the optical density readings started to increase dramatically from 9 up to 27 h of incubation time. The increase in OD reading indicates that the cells were in exponential phase. The culture in this phase was used as inoculum for lactic acid experiments. After 27 h of incubation time, the OD reading at 600 nm started to decline until 48 h. This indicated that the dead cells of *G. stearothermophilus* were accumulating in the culture broth. Hence, culture incubated for more than 27 h was not suitable to be used as inoculum for fermentation process involving lactic acid production because the cell has reached its death stage.

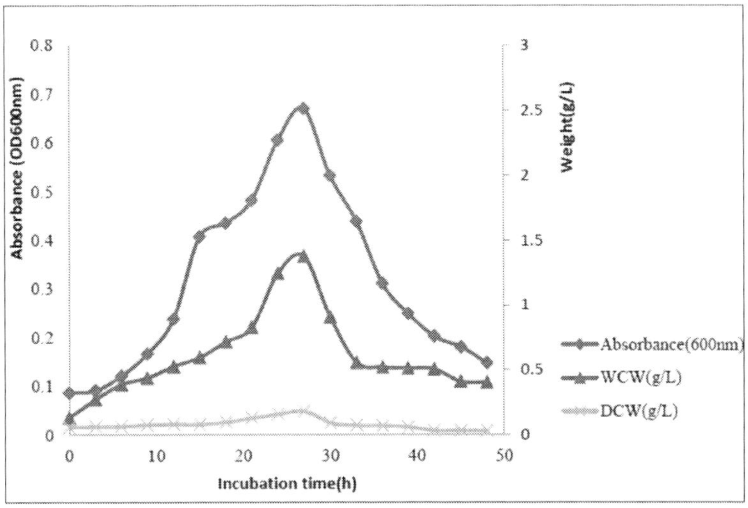

Fig. 1. Optical density (OD_{600nm}), Wet cell weight (WCW) and Dry cell weight (DCW) versus incubation time for 48 h at 55°C and 100 rpm.

3.2 Determination of *G. stearothermophilus* morphology

Literature search indicating that *G. stearothermophilus* is a rod-shaped, Gram-positive, spore-forming bacterium that able to grow either singly or in chains [2]. A Gram staining method was conducted followed by observation of the culture under the microscope.In general, the bacterium was found to be in rod-shape (Figure 2[A], [B] and [C] and staining procedure showed that it belongs to Gram-positive as reported previously [2]. Observation revealed the changes in cell morphology with increasing incubation time from 0 h to 48 h. At 0 h (Figure 2 [A]), the culture was determined to be shorther in size when compared to 9 h and 27 h (Figure 2 [B] and [C]). This is because the culture was in lag phase and required time to adapt to the growth medium. The culture was found to be longer in size at 9 h and 27 h while dispersed well in the medium which indicated they were in log phase. The bacterium was actively growing while the nutrients from the growth medium have been rapidly utilized. When the incubation time was increased to 48 h (Figure 2 [D]), the bacterial cells were observed to clump together which clearly pointing the culture were in death phase. The depletion of nutrients and the accumulation of metabolic waste products are the main reasons for the decline in the culture growth at 48 h.

DOI: 10.1051/matecconf/20179701049

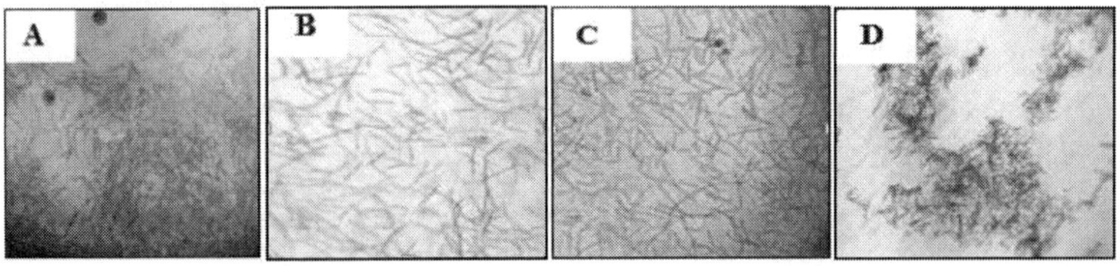

Fig. 2. Morphologies of *Geobacillus stearothermophilus* at [A] 0 h, [B] 9 h, [C] 27 h and [D] 48 h.

3.3 Screening for lactic acid production at various process parameters conditions.

The variation on concentrations of lactic acid were determined by the parameters used for each run. In total, there was 6 parameters tested and for each run the parameters was varied within specific range. From the result (Figure 3), run no.5 has produced highest concentration of lactic acid followed by run no.2 which was 5.65±0.07 g/L and 5.33±0.49 g/L respectively. Meanwhile, run no.8 that produced the lowest concentration of lactic acid with 0.75±0.21 g/L. Run no.5 that produced highest lactic acid concentration was conducted at 60°C, pH 5.5, 48 h of incubation, agitated at 200rpm, 5% of carbon and nitrogen source concentrations. The second highest was run no.2 which conducted with temperature at 60°C, pH 7.5, 24 h agitated at 200rpm, 5% of carbon source concentration and 10% nitrogen source concentration. It was reported that 37 g/L lactic acid was obtained in 20 h at 60°C using the same bacterium *G. stearothermophilus* when potato starch was used as a carbon source [2]. Our findings clearly demonstrated the potential of rice starch waste as a substrate and ability of *G. stearothermophilus* to convert this waste directly into lactic acid which is not the common feature of most of the lactic acid producers.

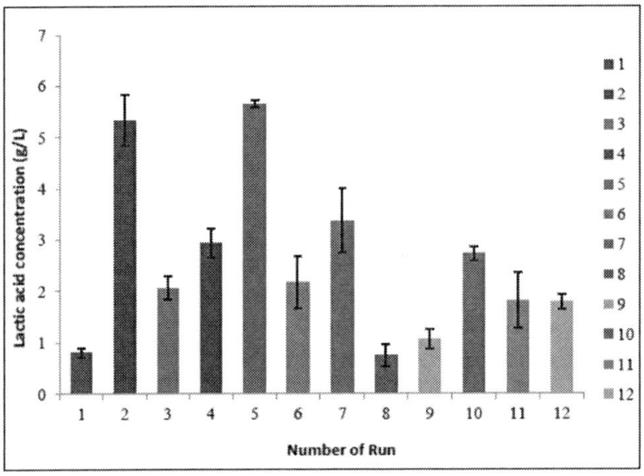

Fig. 3. Lactic acid concentration in each run.

3.4 Response Surface Methodology

The response surface methodology was used to evaluate the effects of the parameters that have been tested in this study. The three-dimensional plot is a graphical analysis of regression equation to analyze the optimum condition within the range that resulted in optimum response. The relationship between the response and the variables was envisioned by response surface developed concurring of two test variables. The effect of the temperature and concentration of carbon on the production of lactic acid from rice starch waste was shown in Figure 4 for 3-D surface. It shows a clear peak from 3-D response surface plot from the effect of temperature and concentration of carbon source. The peak for temperature was located at the area of 57.50°C to 60.00°C while 4.5 to 5% for carbon source. The 3-D graph demonstrates that the production of lactic acid increased gradually as the temperature increases up to 60°C while at the same time dependent on the concentration of carbon in the range 4 to 5%. An increased in temperature above 60°C and carbon source above 5% concentration did not effects the lactic acid production since the activity of the bacterium has stopped due to inactivation of the bacterial cells above its optimal temperature. In conclusion, DOE software was able to identify the variables that were significant for the production of lactic acid which was the temperature and carbon source concentration.

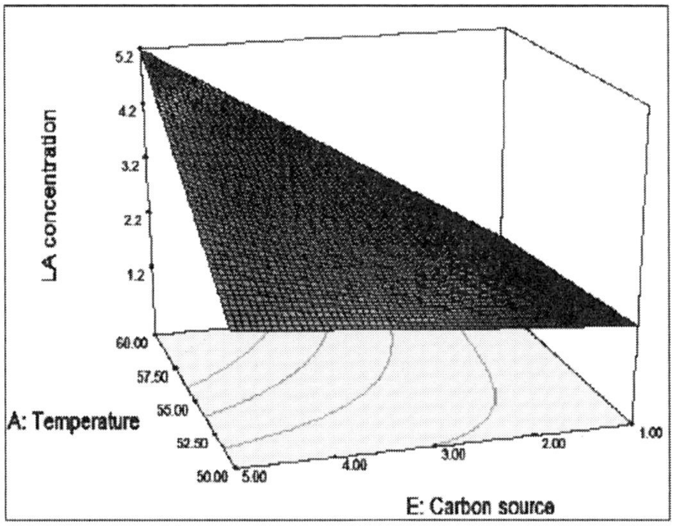

Fig. 4. The 3-D plot of temperature and carbon source interaction.

4 Conclusion

The six parameters was screened using Plackett-Burman design and there was 12 run in total for fermentation. From the result obtained, run no.5 showed the highest lactic acid concentration and followed by run no.2 with 5.65±0.07 and 5.33±0.49 respectively. However, further detail research on the optimization of lactic acid production by *G. stearothermophilus* using rice starch waste is important to prove it is an economically viable carbon source.

This study was supported by Fundamental Research Grant Scheme (FRGS), Grant No: 9003-00530[FRGS/1/2015/SG05/UNIMAP/02/4] entitled: Identification of New Cellulolytic Bacterial Strains from Tropical Mangrove Soil.

References

1. R. Palaniraj, P. Nagarajan, Int. J. ChemTech. Res., **4** , 1049, (2012)
2. M. Smerilli, M. Neureiter, S. Wurz, C. Haas, S. Fruhauf, W. Fuchs, J. Chem. Technol. Biotechnol., 1, (2015)
3. I. Pagana, R. Morawicki, T.J. Hager, J. Food. Sci. Tech., **49**, 641, (2013)
4. A.O Buyukkileci, S. Harsa, J. Chem. Technol. Biotechnol., **79**, 1036, (2004)

DOI: 10.1051/matecconf/20179701050

Solvent Extraction Characterization on Injection Molding of Mg Alloy

Nooraizedfiza Zainon[1,2,*] , *Salmah* Husseinshah[1] , *Mohd Afian* Omar[3] , and *Marina* Marzuki[2]

[1]Material Engineering School, Universiti Malaysia Perlis, Malaysia
[2]Manufacturing Engineering School, Universiti Malaysia Perlis, Malaysia
[3]Advance Materials Research Centre (AMREC), SIRIM Bhd., Kulim Hi-Tech Park, Kedah, Malaysia

Abstract. The effect of the leaching time and temperature on the solvent debinding process of Mg metal injection molding (MIM) green part has been investigated. In this study, both soluble binder, paraffin wax and stearic acid molecules were removed from the Mg MIM green part by immersing compact parts in heptane solution. Then, the solvent debinding rate has been investigated under the conditions of different leaching time and temperature. The weight loss percentages of paraffin wax and stearic acid were calculated and the pores structure was analyzed by scanning electron micrograph. The efficient diffusivity and activation energy of the soluble binder have also been studied. Result shows that the removal of soluble binder at high temperature is the fastest, but there are defects observed on the Mg MIM body. It was found that the total amount of binder removal was 68.24 wt.%, 75.56 wt.%, and 84.92 wt.% for leaching temperature of 40 °C, 60 °C, and 80 °C, respectively. The remaining binder will be removed during first cycle of sintering process.

1 Introduction

Solvent extraction is extensively employed with wax-polymer binder system. To allow handling after extraction process, the wax and polymer should be insoluble in each other. The solvent debinding can be accomplished by immersion in liquid solvent at low temperature and pressure (immersion debinding), by immersion in supercritical solvent at high temperature and pressure (supercritical debinding, or by contacting with saturated vapor of a liquid solvent vapor condensate on the surface of the specimens forming in a liquid film that extracts the soluble component of the binder [1]. Inadequate binder removal will lead to part defect such as bloating, blistering, surface cracking and large internal voids [2-3]. Solvent extraction is extensively employed with wax-polymer binder system. To allow handling after extraction process, the wax and polymer should be insoluble in each other. The solvent debinding can be accomplished by immersion in liquid solvent at low

* Corresponding author: nooraizedfiza@unimap.edu.my

temperature and pressure (immersion debinding), by immersion in supercritical solvent at high temperature and pressure (supercritical debinding, or by contacting with saturated vapor of a liquid solvent vapor condensate on the surface of the specimens forming in a liquid film that extracts the soluble component of the binder [4].

Several researchers had conducted a study on the solvent debinding with different influencing condition. Onbattuvelli et al. studied the effects of nanoparticle addition on the debinding of two different materials which is injection molded aluminum nitride (AlN) [5] and also silicon carbide (SiC) [6] compacts. The experiments varying the solvent debinding condition were performed on microscale and micro-nanoscale of injection molded compacts. The result shows that micro-nanoscale compacts has a slower solvent extraction of binder component compared to microscale compacts. Zaky et al., [7] studied on the debinding behavior of four different characteristics of paraffin wax. Data reveal that the binder system consisting 35 wt.% ethylene vinyl acetate, 62 wt.% of heavy paraffin wax, and 3wt.% stearic acid which extracted at 40 C for 5hours in the cyclohexane solvent, provides an excellent wax-based binder system. Previously, Zaky [8] proved that the higher solubility of n-hexane enhances the debinding rate. The efficiency diffusivity is proportional to leaching temperature/solvent viscosity. Subuki [9] studied on the debinding characteristics of the 316L stainless steel compacts with palm stearin as a wax-based binder system. The result shows that the binder will be removed rapidly for higher powder loading which presumably reflecting in reduction of binder. Ani et al. [10] studied on debinding process involving two stages which focuses for ceramic powder. They found that temperature and immersion time significantly affect binder weight loss during solvent debinding. Zaki et al. [11] successfully separate the paraffin wax grades by applying the solvent extraction technique at different extraction temperatures and different solvent feed ratios.

Kim compared the debinding kinetics between supercritical fluid (SCF) extraction and n-heptane solvent extraction the results proved that rapid debinding without defect formation was successfully performed by the injection molded ceramic bodies with skeleton pores of 68nm. The debinding rate of SCF extraction was 5 times higher than that of the n-heptane extraction. Severe defects appear for solvent when temperature is exceeding 318.5K. However, the debinding defects were invisible for SCF up to 338.5K [12]. Oliveira et al. investigated the influence of specimen dimension and solvent temperature on solvent debinding process. The green specimen with different bulk surface area ratio (As/V) were maintained in hexane at 20, 40, and 60°C. The results reveal that there are strong influences of both parameters on the debinding kinetics. Higher As/V ratio and temperature requires less time to debind [13]. Foudzi et al. investigated the effects of the rheological properties and debinding on the sintered part of the yttria stabilized zirconia. Solvent debinding was carried at the three temperature (50, 60, 70°C) using heptane solution. A large porous region was clearly identified at 70°C compared with 50°C. The powder loading of 41 vol.% debound at 60°C gave the best mechanical properties during the mechanical strength and hardness test with values of 417MPa and 876HV, respectively [14].

In our previous study, has been reported the effect of shear rate and temperature on rheology of Mg metal injection molding feedstock with multicomponent binder [15]. The result shows that the best powder loading was 65 vol.%. In this study, the heptane solution was used to dissolve paraffin wax and stearic acid. This study aims to investigate the effect of the leaching temperature and time on the debinding rate. The weight loss percentages of paraffin wax and stearic acid were calculated and the pores structure was analyzed by scanning electron micrograph. The efficient diffusivity and activation energy of the soluble binder will be also investigated.

2 Experimental

The magnesium powder with D_{50} of 22 µm supplied by MGH Industrial Co., Ltd, China was used. Multicomponent binder system comprising of paraffin wax (PW), high-density polyethylene (HDPE), and stearic acid (SA) obtained from Titan Polyethylene (M) Sdn. Bhd., Johor Bharu, Malaysia. 65 vol.% of powder loading with 35 vol.% multicomponent binder system (PW/HDPE/SA with fraction of 55/35/10) are used in this study. A Universal V4.0C TA Instruments DSC Q10 V8.2 Build 268 and Netzsch simultaneous TG/DTA analyzer (STA) model 409C were used to measure the melting point and decomposition temperature of the binders. Vertical MCP 100KSA injection molding machine manufactured by MCP, Germany was used to produce the molded green body. The solvent was heated by placing a glass container in a water bath manufactured by Tech-Lab Manufacturing. Heptane with molecular weight of 100 g/mol was added to the glass container to completely cover the samples. The glass container was covered with a watch glass to prevent evaporation of the heptane. The melting point of the binders used as the indicator for the selection of solvent temperature. Melting point for paraffin wax and stearic acid were 59.4 °C and 62.2 °C, respectively. The Mg sample were immersed into the solvent at different extraction temperature which is 40 °C, 60 °C, and 80 °C and at different periods of time ranging from 10 minutes to 300 minutes. The samples were placed at the center of the solvent bath and were removed at each time point for weight loss and gravimetric analysis. The leach specimens were dried in an oven at ±45 °C for 5 hours to completely remove the remaining heptane. The weight loss was determined by weighing dried samples. SEM was used to study the binder distribution and the development of pores resulting from the progressive removal of the paraffin wax and stearic acid during immersion in heptane. The fraction of the soluble binder remaining, F, can be calculated as Eq. 1.

$$F = 1 - [(w_i - w_t)/f_i w_i] \tag{1}$$

where, f_i is the initial weight fraction of the soluble binder, w_i is the initial mass of the compact and w_t is the mass of compact at immersion time, t.

Solvent debinding is a two stage process consisting of dissolution and diffusion [11]. Initially, solvent dissolves the paraffin wax and stearic acid and forming a porous surface. The solvent then penetrates into the pores by capillary action. This is followed by diffusion of dissolved paraffin wax out of the green body. The process can be expressed as Eq. 2 [7].

$$\mathrm{Ln}\left(\frac{1}{F}\right) = \frac{\mathrm{Det}\pi 2}{\psi 2} + K \tag{2}$$

where, F is the fraction of the remaining soluble, D_e is the effective diffusivity of soluble binder and solvent, t is time, ψ is shape factor and K represents the change in the mechanism controlling the debinding behavior. Injection molded green body for this study corresponds to shape factor, ψ of 1.9716. The shape factor is colume to surface area ratio and can be calculated using Eq. 3.

$$\Psi = \frac{\mathrm{HB(2L)}}{2(\mathrm{HB+B(2L)+(2L)H})} \tag{3}$$

where 2L, H, and B are the length, thickness, and breadth of the component, respectively.

DOI: 10.1051/matecconf/20179701050

The activation energy for the extraction of paraffin wax and stearic acid from the green molded part was expressed as equation 4.

$$D_e = D_o \exp(-Q/RT) \quad (4)$$

where, Q is the activation energy for the solvent extraction, R is the universal gas constant, T is the temperature in kelvin, D_e is the diffusion coefficient, and D_o is the maximum diffusion coefficient at infinite temperature. By plotting the graph of effective diffusivity, D_e against the 1/T, the activation energy can be acquired.

3 Results and discussions

The goal of solvent debinding is to take out the binder elements in a shortest time with the least impact on the green body. Fig.1 shows the effect of temperature and time on the amount of binder extracted during solvent debinding. From Fig.1, the total amount of binder extracted increases by increasing the extraction time ranging from 10 minutes to 300 minutes. The temperature dramatically influences the extraction rate at the first 30 minutes. This is due to the diffusion distance for the heptane and soluble binder which is short in the early stage of the debinding. The results clearly demonstrated that the speed removal of the paraffin wax and stearic acid occurs at higher leaching temperature of 80 °C. The total amount of binder removal was 68.24 wt.%, 75.56 wt.%, and 84.92 wt.% for leaching temperature of 40 °C, 60 °C, and 80 °C, respectively. At the first 30 minutes, it was observed that the binder extraction is faster at leaching temperature of 80 °C compared to temperature of 40 °C and 60 °C. This is due to high diffusivity resulted by the high temperature which finally increased debinding rate. However, too fast removal of paraffin wax and stearic acid result the molded part to distort and surface cracking. It was observed that leaching time at 40 °C and 60 °C provided a better shape retention compared to leach at 80 °C. This result indicates that leaching temperature of 60 °C was the most appropriate for leaching temperature. Ani [10] reported that the diffusion process is influenced by the microstructure layout of injected parts. Binders cannot be totally removed in case of dead ends since the porous channel in microstructures does not intersect with others.

Fig.2 shows the plots of ln (1/F) as a function of immersion time for solvent extraction of paraffin wax and stearic acid at different temperature. Slope of the graph represents the effective diffusivity of dissolution and diffusion stages which occur during solvent extraction process for Mg MIM molded part. From the Fig. 3, two different slopes can be observed which are associated with dissolution and diffusion stages. D_e of the soluble binder can be determined in both cases using equation 2 and summarize in Table 1. D_e value at the dissolution stage is higher than the diffusion stage. As temperature increase, the D_e increase at dissolution stage but no significant changes were recorded in diffusion stage. As a result, the dissolution stage is more efficient than diffusion when temperature increases. Similar result reported by Sotomayor et al. [16], in their study on debinding of ferritic and austenitic stainless steel. They reveal that the diffusion coefficient of paraffin wax during dissolution stage is almost 5.5 times higher than diffusion stage.

DOI: 10.1051/matecconf/20179701050

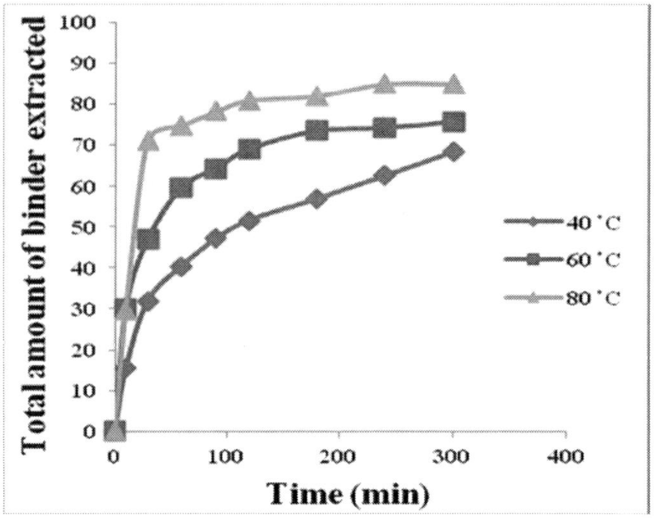

Fig. 1. Effect of extracted temperature and time on amount binder extracted of Mg MIM

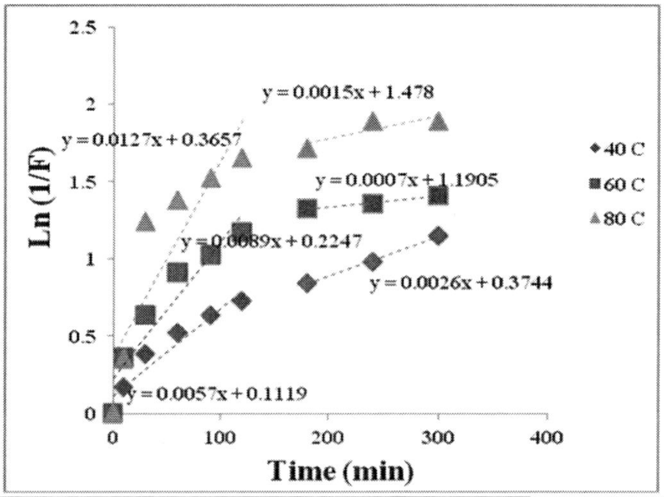

Fig. 2. Variation of ln (1/F) with immersion time at 40 °C, 60 °C, and 80 °C

Table 1. Efficient diffusivity of the soluble binder at different temperature

Temperature (°C)	Efficient diffusivity (De)	
	Dissolution stage	Diffusion stage
40	2.2453E-5	5.9085E-6
60	3.506E-5	2.7573E-6
80	4.727E-5	1.0241E-5

The activation energy can be calculated from the linear equation by plotting the graph of Ln D_e versus 1/T. According to Arrhenius equation as expressed in equation 4, the -Q/R

DOI: 10.1051/matecconf/20179701050

value represented by the slope of the linear curve. The activation energy for extraction the soluble binder from magnesium molded part was estimated to be 17.1584 KJ/mole. The activation energy of paraffin wax in heptane was 42 ± 5 KJ/mole [17]. This value was higher than the value obtained from this study. This is presumably due to different powder used and the present of the stearic acid in the soluble binder which may ease the extraction of the soluble binder form the system.

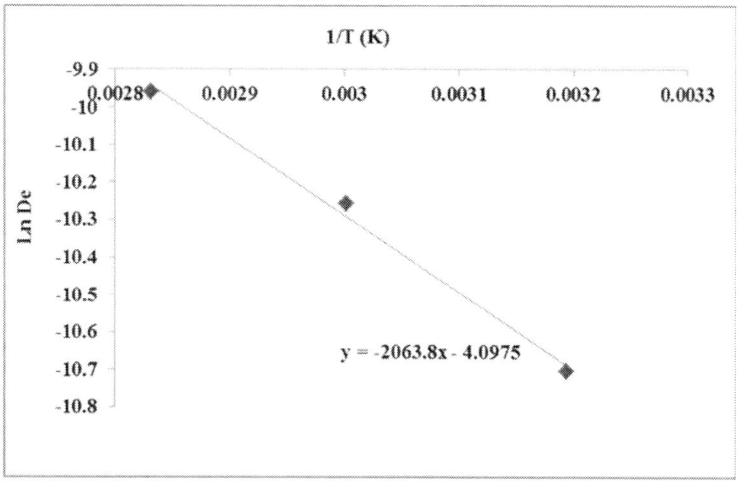

Fig. 3. The diffusion coefficient against 1/T of Mg MIM molded part

Fig.4 shows the microstructure of the molded part after solvent debinding at various times. Evaluation of the pore structure during solvent extraction at 60 °C was evaluated using SEM at fracture surface. At this temperature, the molded parts were fre free from defects. Fig.4 (a) to (d) shows the SEM of the samples at the center after 10, 60, 120, and 300 minutes leaching time, respectively. It reveals that pores with different sizes were formed. Some pore were within the binder represented by A in Fig.4 (a) and some were interparticle pores represented by B in Fig.4 (b), which indicate that all the binder component of PW, SA, and HDPE were mixed and interacted to a certain degree and that some PW and SA has been extracted. As the leaching time increases to 180 minutes as shown in Fig.4 (c), a large amount of binder leave spaces between the powder particles and increases the pore size which was labeled as C. While Fig.4 (d) shows the SEM of the sample after 300 minutes of leaching time. HDPE with irregular morphology was at the center region remained as interconnected capillary porosity inside the part which provided sufficient strength to hold the powder particles together prior to sintering process.

DOI: 10.1051/matecconf/20179701050

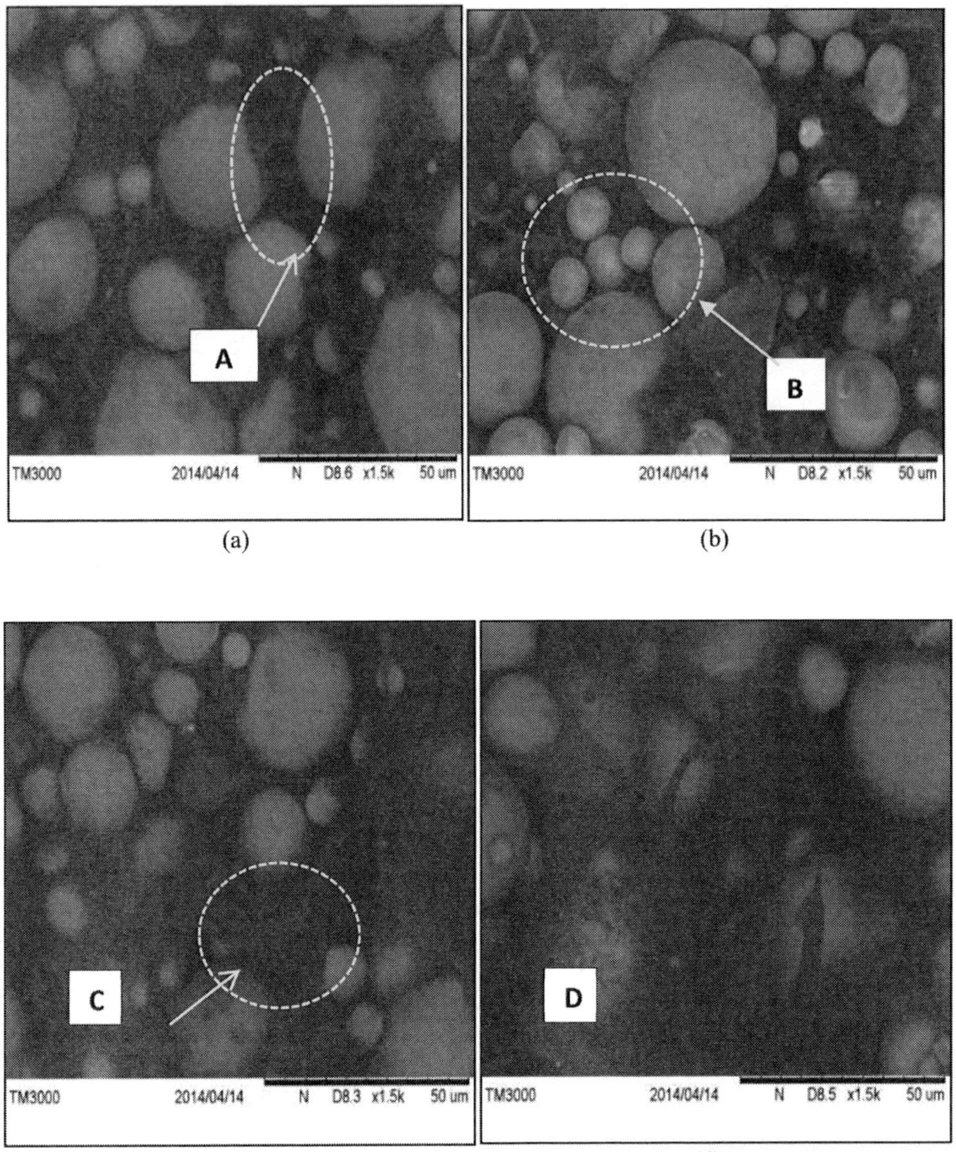

(a) (b)

(c) (d)

Fig. 4. The SEM on fracture surface Mg MIM molded part as a function of extraction time (a) 10 minutes, (b) 60 minutes, (c) 180 minutes, (d) 300 minutes.

4 Conclusion

The effect of leaching temperatures on the solvent debinding of paraffin wax and stearic acid in Mg MIM compact part were investigated. The result shows that the removal of soluble binder at high temperature is the fastest, but there are defects observed on the Mg MIM body. It was found that the total amount of binder removal was 68.24 wt.%, 75.56 wt.%, and 84.92 wt.% for leaching temperature of 40 °C, 60 °C, and 80 °C, respectively. Total amount of binder extracted has inverse relation to the diffusion coefficient of the soluble binder in the system with increasing time. The total amount of binder removed also increased, which reflects to the increasing of the diffusion distance and pore channels

DOI: 10.1051/matecconf/20179701050

extended to the inner region of the body. The temperature gives the significant impact on the removal rate during dissolution stage but not to diffusion stage. The efficient diffusivity for dissolution stage was much higher than the diffusion stage for approximately 4 times. Result also shows that the efficient diffusivity at 80 °C was higher than 60 °C and 40 °C due to of high diffusivity. The activation energy for extracting the soluble binder from magnesium molded part was 17.1584 KJ/mole. The SEM micrograph reveals more of pore developed after solvent debinding for 300 minutes.

This research was supported by Ministry of Higher Education (MOHE) for providing Fundamental Research Grant Scheme (FRGS) - No. 9003-00317.

References

1 R. M. German, A. Bose, Metal Powder Industries Federation (1997)
2 D.F. (Ed.) Heaney, *Handbook of metal injection molding*. Elsevier (2012)
3 V.A. Krauss, A.A.M. Oliveira, A.N. Klein, H.A. Al Qureshi, M.C. Fredel, J. Mater. Process Tech., **182**, 268 (2007)
4 M.A. Omar, R. Ibrahim, M.I. Sidik, M. Mustapha, M. Mohamad, J. Mater. Process Tech., **140**, 397 (2003)
5 V.P. Onbattuvelli, R.K. Enneti, S.J. Park, S.V. Atre, Int. J. Refract. Met. H., **36,** 183 (2013)
6 V.P. Onbattuvelli, R. Chinn, R.K. Enneti, S.J. Park, S.V. Atre, Ceram. Int., **40**, 13861 (2014)
7 M.T. Zaky, F.S. Soliman, A.S. Farag, J. Mater. Process. Tech., **209**, 5981 (2009)
8 J. Zaki, J.P. Mitchell, Curr. Dir. Psychol. Sci., **22**, 466 (2013)
9 I. Subuki, *Injection Moulding of 316L Stainless Steel Powder Using Palm Stearin Based Binder System* (Ph.D. Dissertation, Universiti Teknologi Mara, Malaysia, 2010).
10 S.M. Ani, A. Muchtar, N. Muhamad, J.A. Ghani, Ceram. Int., **40**, 2819 (2014)
11 M.T. Zaky, N.H. Mohamed, A.S. Farag, Fuel Process Technol., **92** , 2024 (2011)
12 S.W. Kim, J. Supercrit. Fluid., **51**, 339 (2010)
13 R.C.B. Oliveira, V. Soldi, M.C. Fredel, A.T.N. Pires, J. Mater. Process. Tech., **160**, 213(2005)
14 F.M. Foudzi, N. Muhammad, A.B. Sulong, H. Zakaria, Ceram. Int., **39**, 2665 (2013)
15 Z. Nooraizedfiza, H. Salmah, M.A. Omar, International Review of Mechanical Engineering, **7**, 1378(2013)
16 M.E. Sotomayor, A. Varez, B. Lavenfeld, Powder Technol., **200**, 30 (2010)
17 R.K. Enneti, T.S. Shivashankar, S.J. Park, R.M. German, S.V. Atre, Powder Technol, **228**, 14(2012)

Radiation Induced Grafting of Acrylate onto Waste Rubber: The Effect of Monomer Type

Siti Salwa M. Shirajuddin[1,*], *Che Mohd Ruzaidi* Ghazali[2,3], *Chantara* Thevy Ratnam[1], *Kamarudin* Husin[2], *Nor Azwin* Shukri[1], and *Nur Azillah Fatimah* Othman[1]

[1]Agensi Nuklear Malaysia, Bangi 43000 Kajang, Selangor, Malaysia
[2]Universiti Malaysia Perlis, 01000 Kangar, Perlis, Malaysia
[3]School of Fundamental Science, Universiti Malaysia Terengganu, 21030 Kuala Terengganu, Terengganu

Abstract. The effect of three different acrylate group monomers, namely n-butyl acrylate, methacrylic acid and tripropylene glycol diacrylate of radiation induced grafting onto waste rubber was studied. The electron beam accelerator operated at voltage of 2MeV was used to irradiate the waste rubber at 10 kGy and 100 kGy absorbed radiation dose, respectively. The formation of grafting was observed from the increase in the grafting yield and confirmed by Transformed Infra-Red Spectroscopy results. According to the result obtained, only tripropylene glycol diacrylate was selected to graft onto waste rubber. The carbonyl bond from acrylate groups was seen at 1726 cm^{-1} band which confirmed the presence of TPGDA in the polymer matrix. This indicates the successful preparation of the TPGDA-grafted waste rubber via radiation induced grafting techniques.

1 Introduction

The accumulation of waste rubber into the landfill has posing a serious threat to the health and environment. This is due to their durability and immunity to the biological degradation which makes the disposal and reprocessing of discarded tires is extensively difficult. With an increase in automobile use, the generation of waste rubber from the tire is also increase. In Korea, 28 million tons of scrap tires are produced annually (Ministry of Environment of Korea, 2011) while the United States generates approximately 300 million tons of scrap tires annually [1]. On the other hand, data from Malaysia Industrial Development Authority, MIDA shows that Malaysia generates about 150 000 tons of scrap tires every year. These statistics shows the vast number of scrap tires produced annually worldwide. The problems associated with the disposal of the waste rubber are also very clear if not recycle or discarded properly. Increasing concerns regarding to recycling waste rubber and the growing cost of discarded waste to the environment has led to a number of research on the utilization of waste rubber.

* Corresponding author: sitisalwa@nm.gov.my

However, waste rubber has low mechanical and structural properties which will cause a production of low quality product due to low properties of materials. In order to overcome this disadvantage, the modification techniques by using radiation induced grafting are proposed. Surface modification techniques shows an excellent performance in terms of engineering properties, such as mechanical properties, environmental stability, gas impermeability, flammability as well as adsorption properties [2, 3]. The grafted product obtained can be tailored into wide range of application especially for industrial application.

To date, there is a lot of research that has been done using radiation-induced grafting to modify the polymer properties such as Uwe Lappan *et al.* [4] conducted high energy radiation grafting of styrene onto fluoropolymers films and Przybytniak *et.al* [5] investigated functionalization of polymer surface by radiation-induced grafting of acryclic acid onto polyethylene, polypropylene and polystyrene. In 2014, Nor Azwin Shukri *et al.* [6] successfully grafted antimicrobial agent onto polyethylene film through radiation-induced graft copolymerization and grafted film showed a strong and stable antibacterial activity. Besides, numbers of study on the utilization of waste rubber from the discarded tires with various area of application have been reported; used as modifiers to asphalt paving mixture and used as an additive to Portland cement concrete in civil engineering application [1], used as a fillers in the composite blending [7] and used in high voltage outdoor insulation application [8]. Although studies could be found, concerning the application of radiation induced grafting to utilize waste rubber. At the author best knowledge, the incorporation of waste rubber with acrylate group n-butyl acrylate, methacrylic acid and tripropylene glycol diacrylate via radiation induced grafting is still scarce in the scientific literature. Moreover, no systematic research has been reported on the effect of grafting condition. Thus, this research is proposed to determine and to compare grafting work of various acrylate groups onto waste rubber. Comparison of these acrylate groups will be a guide for the use of suitable monomer in waste rubber for industrial application.

2 Methodology

2.1 Materials

Waste rubber from recycle tires used in this study was obtained from Sin Rubtech Consultancy Sdn. Bhd. The monomer used in this study is n-butyl acrylate (n-BA), tripropylene glycol diacrylate (TPGDA) and methacrylic acid (MAA). Solvents and chemical reagents were of laboratory grade and were used without further purification.

2.2 Radiation-induced grafting

The waste rubber samples were weighed and immersed in different concentration acrylates namely n-BA, MAA and TPGDA with concentration of 1% and 10%. The samples were then irradiated using electron beam accelerator operated at voltage of 2MeV to a dose of 10 kGy and 100 kGy respectively. The irradiated samples were placed in the water bath at 40 ^{0}C for the grafting reaction to occur. After 30 minutes, the samples were removed from the solution, cleaned by filtration and washed again repeatedly until no subsequent weight decrease. This is to remove the excess monomer and homopolymers. The weight of grafted sample was measured after drying in an oven at 60°C until constant weight was achieved.

The grafting yield is defined with the weight basis expression to confirm the grafted copolymer. The grafting percentage was calculated by the percentage weight gain of waste rubber after grafting process as defined below (Eq. 1)

$$GY(\%) = ((W_g - W_0)/W_0) \times 100 \quad (1)$$

Where, W_g and W_0 are denote the weights of grafted and un-grafted waste rubber samples, respectively.

2.3 Characteristic of waste rubber

The infrared of grafted waste rubber were obtained using Attenuated Total Reflectance-Fourier Transform Infrared Spectroscopy (ATR-FTIR) spectrometer (Mattson 6020-Galaxy Series) in transmittance mode at frequency in the range of 4000-400 cm^{-1}.

3 Results and discussion

The effect of absorbed radiation dose (kGy) on the grafting yield (%) of acrylate monomer onto waste rubber is shown in Figure 1. Grafting yield referred to the amount of monomer used was successfully grafted to the trunk polymer in percentage. In this study, grafting of MAA, n-BA and TPGDA monomer was done using post-irradiation method with doses of 10 kGy and 100 kGy. Figure 1 shows the study for radiation induced grating of MAA monomer and n-BA monomer was -0.99% and -1.345% of grafting yield for 10 kGy -1.46% and -1.575% of grafting yield for 100 kGy respectively. This clearly shows that there is no grafting occurs for MAA monomer and n-BA monomer since the grafting yield is in negative region for both of the monomer. In contrast, radiation induced grafting of TPGDA monomer for 10 kGy and 100 kGy is 12.285% and 6.34% of grafting yield respectively. Only TPGDA monomer was successfully grafted onto waste rubber via radiation induced grafting techniques.

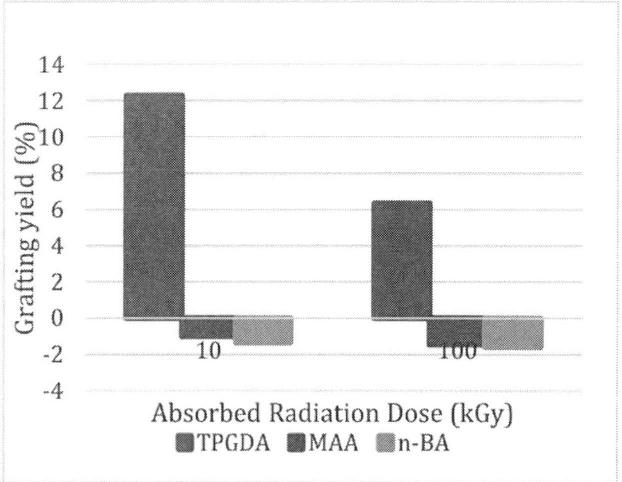

Fig. 1. Graph of absorbed radiation dose (kGy) versus grafting yield (%)

The effect of monomer concentration (%) on the grafting yield (%) of acrylate monomer onto waste rubber is shown in Figure 2. Grafting of MAA, n-BA and TPGDA was done using 1% and 10% of monomer concentration. Figure 2 shows the study for radiation induced grating of TPGDA at 1% concentration is 6.21% grafting yield. On the other hand, MAA and n-BA monomer shows no grafting occurs since both monomer of grafting yield was in negative region. Radiation induced grafting of 10% monomer concentration is 8.14% and 8.17% of grafting yield for TPGDA and MAA monomer respectively. Only TPGDA monomer was successfully grafted onto waste rubber via radiation induced grafting techniques at 1% and 10% of monomer concentration.

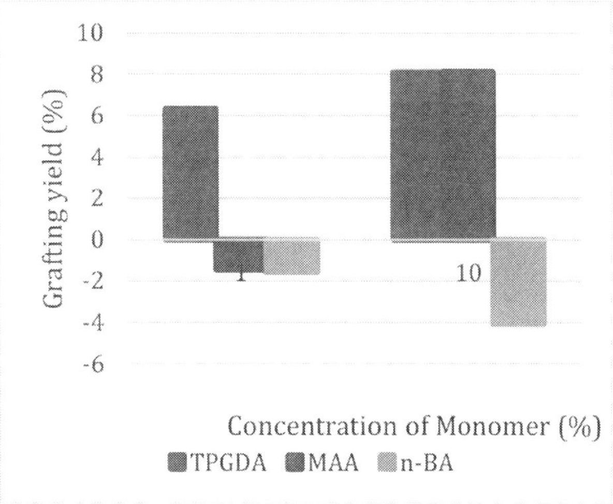

Fig. 2. Graph of concentration of monomer (%) versus grafting yield (%)

The radiation induced grafting of TPGDA onto waste rubber was further study with FTIR spectrometer and the FTIR spectrum of the sample is shown in Figure 3. The FTIR spectra shows a new band appear at 1726 cm-1 attributed to C=O carbonyl bond from the acrylate group from TPGDA monomer. This band clearly proved the TPGDA monomer was successfully grafted onto the waste rubber via radiation induced grafting.

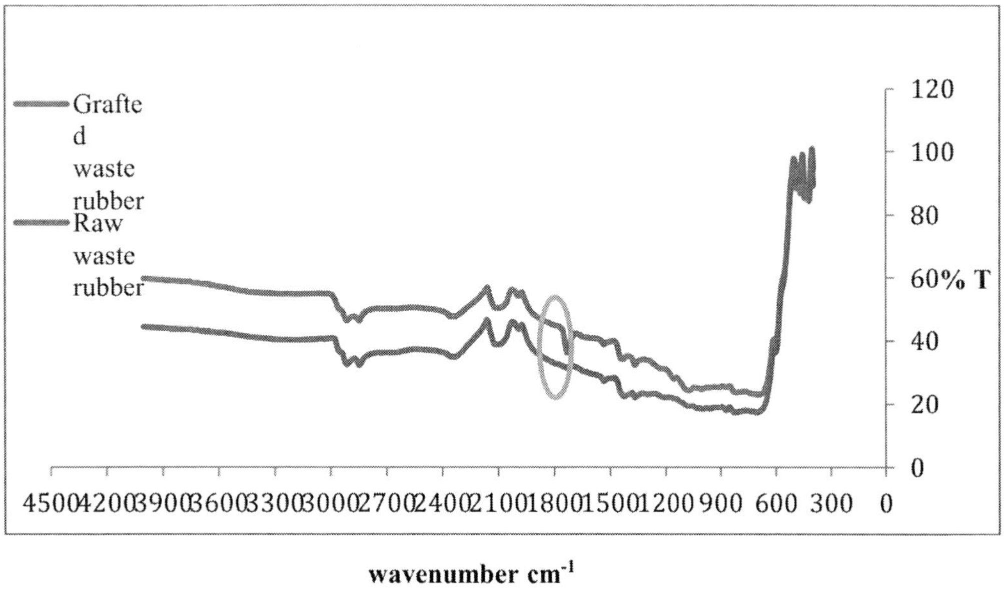

Fig. 3. FTIR spectra of un-grafted waste rubber and TPGDA grafted waste rubber

4 Conclusions

In conclusion, radiation induced grafting of MAA, n-BA and TPGDA monomer onto waste rubber was studied. Comparing the results of the effect of the absorbed radiation dose and monomer concentration, only TPGDA monomer has shown positive grafting yield. Meanwhile, for MAA and n-BA monomer did not show any grafting yield. This indicates that only TPGDA was successfully grafted onto the surface of waste rubber as proven by FTIR spectrometer. The carbonyl bond attributed to acrylate group from TPGDA monomer was confirmed in the polymer matrix. Thus, TPGDA was chosen as a suitable monomer for this system.

The authors wish to thank to Agensi Nuclear Malaysia and University Malaysia Perlis for the support and guidance throughout this study.

References

1. Q. Dong, B. Huang, X. Shu, Constr. Build. Mater., **48**, 116 (2013)
2. S. Shimada, Y. Takahasi, Y. Sugino, S. Hara, K. Yamamoto, J. Polym. Sci. B. Polym. Phys., **42**, 1705 (2004)
3. J.Y. Li, Y.M. Sun, H. Zeng, W.X. Xue, Y. Xiao, Q. Yu, J. Appl. Polym. Sci., **89**, 662 (2003)
4. U. Lappan, U. Geibler, U. Gohs, S. Uhlmann, Radiat. Phys. Chem., **79**, 1067 (2010)
5. G. Przybytniak, E.M. Kornacka, K. Mirkowski, M. Walo, Z. Zimek, Nukleonika, **53**, 89 (2008)
6. S. Nor Azwin, G. Zulkafli, O. Nor Azillah Fatimah, O. Siti Fatahiyah, W. Mat Uzir, Aust J. Basic Appl. Sci, **8**, 92 (2014)
7. C.T. Ratnam, S. Ramarad, M. Khalid, Noraini, J. Compos. Biodegradable Polym., **1**, 16 (2013)
8. A. Aman, M.M. Yaacob, M.A. Alsaedi, K.A. Ibrahim, Int. J. Elec. Power, **45**, 346 (2013)

Compressing Turbulence Effect in FSO using New Modulation Technique

Norizan Mohamed Nawawi[1,*], *Anuar* Mat Safar [1] , *Abdul* Rahman Kram[1], C.B.M. Rashidi[1] and *Junita* Mohd Nordin[1]

[1]Optical Research Group, Advanced Communication Engineering Centre of Excellence (ACE-COE), School of Computer & Communication Engineering, Universiti Malaysia Perlis (UNIMAP), Malaysia.

Abstract. Diffusers transmitter modulation to reduce the effect of atmospheric turbulence in free space optical communication system is described in this paper. This technique uses dual transmitters and dual receivers which are using differential mode for detection. The combination of these components produce superior modulation method especially to reduce scintillation index, to overcome signal detection with fix zero threshold and improve power received. In order to improve the performance of free space optical system, these three elements play an important role. The analysis result show that for receiving power dual diffuser modulation at 3 km distance propagation is 4.59dBm compared to conventional OOK using diffuser which only -7.6dBm and equivalent to 3dBm improvement or up to 40 percent better.

1 Introduction

Optical communication through free space is an attractive alternative over fiber optical communication where provides high bandwidth, fast-installation and high security [1]. However free space optic is suffering with the atmosphere disturbance which can lead the laser beam fading when reach the detector. Meanwhile, beam fading is a condition where the beam spread greater than diffraction estimated limit. Moreover, the front phase of the beam may varies and resulting intensity signal fluctuation. Taking account all these effect will result spatial and temporal random changes in refractive index of several elements such as temperature, pressure and wind along the optical propagation path [2-9].

Recently, phase screen diffuser which generates partial coherent beam has rising interest especially to improve the performance in laser communication system [10-13]. The original FSO employs the perfect coherence beam suffers some problems depends on weather condition. Propagation phenomena in the air such as scattering, absorption and turbulence spread the laser beam. In this paper we concern on atmospheric turbulence issue in the transmission system. Beam wander and redistribution of the intensity within the beam, phase-front distortion and beam broadening are three critical effects on the laser beam. The

* Corresponding author: norizan1123@gmail.com

DOI: 10.1051/matecconf/20179701052

scintillation or temporary redistribution of the intensity, is the outcome from the chaotic fluctuation an atmosphere and from thermal gradients within the optical channel. Consequently, these condition can cause signal fading that will degrade the FSO performance.

We demonstrate the alternative method in order to diminish these effect, which using the phase screen diffuser device. This technique creates a 'new' Gaussian beam characteristic which effectively propagates through atmospheric turbulence. Partially coherent beam can enhance the performance of FSO system by applying two diffusers at transmitter and at receiver employs reception in differential mode detection technique. The fundamental of partial coherent beam can be found in next section followed by demonstration of proposed system model and then result and discussion. Last but not least, the final section is for conclusion

2 Diffusers transmitter system model

The proposed system uses On Off Keying (OOK) modulation to modulate the signal as reference for conventional system. It employs two transmitter, when the transmitter A sends a bit '1', the transmitter B will send the bit '0' (transmitter B is set in compliment condition) in simultaneously and vice versa. Immediately the signals will go through the subtractor for the differential mode detection process. This occurred at receiver as shown in Figure 1. Subtractor were assume as an ideal subtractor where zero signal loss during this process. Therefore, the output signal is bit '1' for input signal binary '1' and bit '-1' for input signal binary '0'. Table 1 illustrate the binary transmitting procedure. Here we can see that, the modulation approach eliminated the need of adaptive threshold where the conventional OOK modulation depends on threshold detection to recognize the incoming signal received whether the bit '1' or '0'.

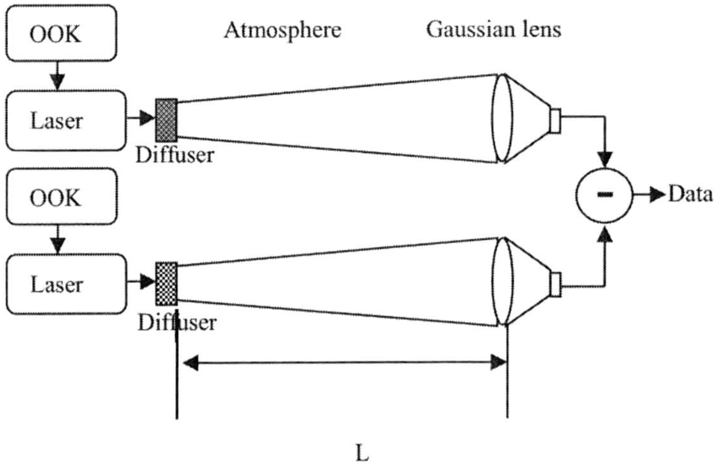

Fig. 1. Diffusers Modulation Setup

The conventional detection technique always misinterpretation by receiver where the noise alone capable to trigger as signal bit '1' (sending pulse) which also well known as 'false alarm' and sometimes the signal sending which contain the data is recognize as bit '0' (no pulse condition) when the signal is not reach the threshold value [17]. This condition is well known as 'miss detection' condition. Hence, the complex adaptive threshold has been suggested to overcome this problem. Whereas by employing dual diffuser modulation the detection signal is more precise with fix zero threshold. The incoming signal can be referenced to determine the input signal binary '1' or '0'. Additionally, we placed the low cost material phase screen diffuser at transmitter to combat the atmospheric turbulence effect more efficient then give the optimum performance for overall system. This combination creates the superior robust modulation in FSOC. According to Table 1, let say *Yn* is the received signal with zero threshold detection,

$$\left.\begin{array}{ll} Yn > 1 & \text{bit '1'} \\ Yn < 1 & \text{bit '0'} \end{array}\right.$$

The received signal can be written as:

$$Yn = 2\sqrt{E_b} + \frac{n_1 + n_2}{2} \tag{19}$$

where $\sqrt{E_b}$ is represent the average energy signal and n_1 and n_2 are representing the noise of photo detector 1 and 2 respectively, where n can be denoted as Additive White Gaussian Noise (AWGN) with zero mean and variance σ.

Table 1. Transmit and Receive bits using Diffusers Modulation Approach

Transmitter (Tx)		Receiver (Rx)		Subtraction Process	
Sending bit		Received bit		Rx A - Rx B	
Tx A	1	Rx A	1	Bit 1	1
	0		0		
Tx B (Compliment)	0	Rx B	0	Bit 0	-1

3 Partially coherent beam beam

Expressions for partially coherent beam are derived as same as a coherent beam [15] equation. The typical value for the diffuser correlation length (l_c^2) are 0.1, 0.01, 0.001 and 0.0001. Using the Kolmogorov spectrum and standard, extended Rytov theory on axis scintillation index for weak turbulence (inner scale l=0, Outer scale L=∞) partially coherent Gaussian-beam is given by,

$$\sigma_B^2 = 3.86\sigma_l^{\,2}\left\{\begin{array}{l} 0.4\left[\left(1+2\Theta_{ed}\right)^2 + 4(\Lambda_{ed})^2\right]^{5/12} \\ \times\left(\cos\left[\frac{5}{6}\tan^{-1}\left(\frac{1+2\Theta_{ed}}{2\Lambda_{ed}}\right)\right]\right) \\ -\left(\frac{11}{6}(\Lambda_{ed})^{5/6}\right) \end{array}\right\} \tag{1}$$

DOI: 10.1051/matecconf/20179701052

where σ_I^2 indicate the strength of intensity fluctuations and proportional to Rytov variance as

$$\sigma_R^2 = 1.23 C_n^2 k^{7/6} L^{11/6} \tag{2}$$

For weak fluctuation, it is less than 1 and for strong fluctuation it is greater than 1. C_n^2 is the refractive index structure constant that characterizes the strength of the index of refraction fluctuations. The typical C_n^2 value weak turbulence is 10^{-17} $m^{-2/3}$ and strong 10^{-12} $m^{-2/3}$.

3.1 Standard and effective beam parameters

The effect of diffuser on an optical beam at the receiver is characterized by replacing the standard beam parameter Θ_1, Λ_1 by effective beam parameter Θ_{ed}, Λ_{ed} define in term of N_s as follows

$$\Lambda_{ed} = \frac{\Lambda_0 N_S}{\Theta_0^2 + \Lambda_0^2 N_S} \tag{3}$$

and

$$\Theta_{ed} = \frac{\Theta_0}{\Theta_0^2 + \Lambda_0^2 N_S} \tag{4}$$

where

$$\Lambda_1 = \frac{\Lambda_0(L)}{\Theta_0^2(L) + \Lambda_0^2(L)} \tag{5}$$

and

$$\Theta_1(L) = \frac{\Theta_0(L)}{\Theta_0^2(L) + \Lambda_0^2(L)} \tag{6}$$

The initial Fresnel ratio $\Lambda_0(L)$ and the initial phase curvature $\Theta_0(L)$ are given by

$$\Lambda_0(L) = \frac{2L}{kw_0^2} \tag{7}$$

and

$$\Theta_0(L) = 1 - \frac{L}{F_0} \tag{8}$$

For all calculation, value for $F_o = \infty$, is refer to collimated. As a diffuser is added at the transmitter part and we can varies the value of correlation length, hence the output beam parameters also varies.

3.2 Turbulence scintillation index

For moderate to strong turbulence scintillation index is as above:

$$\sigma^2_{I,strong} = \exp\left\{ \frac{0.49\sigma_R^2}{\left[1+0.56\left(1+\Theta_{ed}\right)\sigma_R^{12/5}\right]^{7/6}} + \frac{0.51\sigma_R^2}{\left(1+0.69\sigma_R^{12/5}\right)^{5/6}} \right\} - 1 \tag{9}$$

3.3 Log intensity

Meanwhile the log intensity due to small-scale eddies is given by

$$\sigma^2_{lny}\left(D\right) = \frac{1.27\sigma_I^2 n_y^{-5/6}}{1+\dfrac{0.4n_y}{\Lambda_1+\Omega}} \tag{10}$$

where the corresponding cutoff frequency is

$$n_y = 3\left(\frac{\sigma_I}{\sigma_B}\right)^{12/5}\left(1+0.69\sigma_B^{12/5}\right) \tag{11}$$

Therefore the total log intensity due to large-scale and small-scale is

$$\sigma_I^2(D) = \exp\left[\sigma^2_{\ln x}(D) + \sigma^2_{\ln y}(D)\right] - 1 \tag{12}$$

4 Result and discussion

Fig. 2 shows the degradation of scintillation index when using the diffuser for strong turbulence. As we can see the scintillation index increase when the distance propagation increase. At short distance 2km, the scintillation index for coherent beam is 1.5 but with using the diffuser the scintillation index can be reduce to 1.2. However the effective of diffuser is not continuous where from the graph as we can see, above the distance propagation 3km it will turn to flat value. The effect of diffuser is more greatly effect in weak turbulence regime [18].

Fig. 3 illustrates the reduction of average intensity signal due to diffuser. The coherent beam (without diffuser) has a peak amplitude 0.045 at 1.4km but the partially coherent beam (with diffuser) at 0.8km. This is because the diffuser will expand the spot size beam and consequently reduce the mean intensity. Hence, as a result it reduce the power received at the receiver. However this effect can be reduce with using dual diffuser modulation approach where it can double up the magnitude power as shown in Fig. 4. As example at 3km distance propagation, the received power for coherent beam is -6.6dBm and for conventional OOK using diffuser is -7.6dBm but in diffusers modulation approach the received power is -4.59dBm. This show the improvement approximately 40 percent.

DOI: 10.1051/matecconf/20179701052

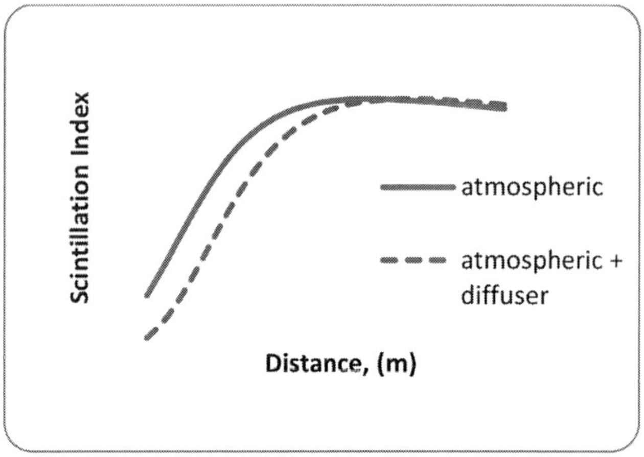

Fig. 2. Scintillation index versus distance for strong turbulence

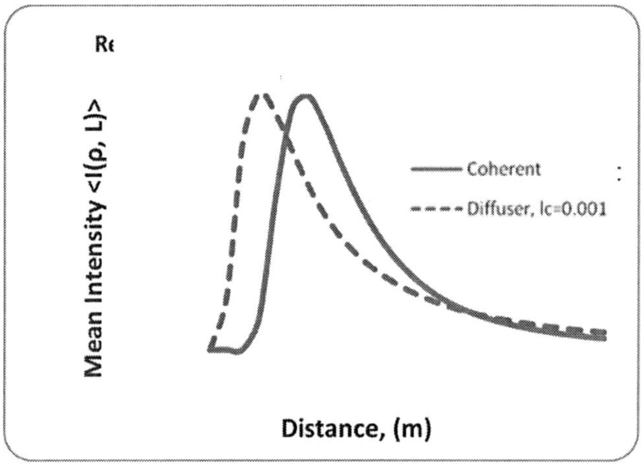

Fig. 3. Reduction of average intensity signal due to diffuser

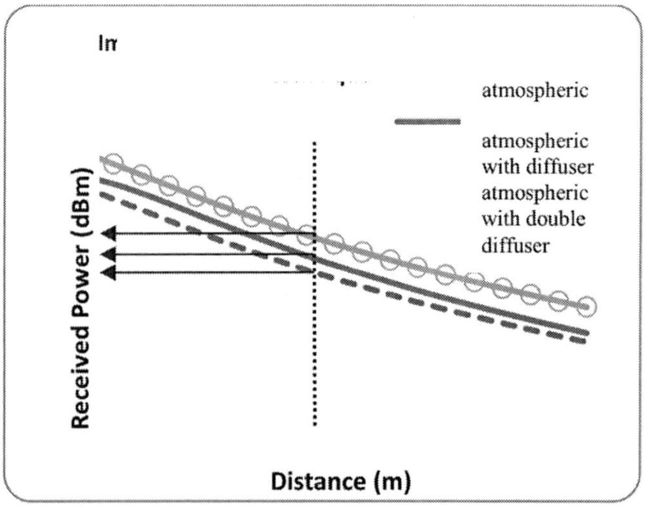

Fig. 4. Improvement of received power in Diffusers Modulation approach

5 Conclusion

In contribution for this paper, the diffusers modulation improve the performance of the free space optic communication in atmospheric turbulence channel. The BER in atmospheric are strongly influenced by scintillation index, power received and signal technique which employed the partially coherent beam method improved the power received up to 40 percent and BER distance 42 percent longer compared to conventional OOK. As a result, it create the robust modulation against the turbulence with fix zero threshold value.

References

1. I.I. Kim, R. Stieger, J.A. Koontz, C. Moursund, M. Barclay, P. Adhikari, J. Schuster, E. Korevaar, R. Ruigrok, C. DeCusatis, Opt. Eng., **37**, 3143 (1998)
2. S. Trisno, I. Smolyaninov, S.D. Milnerb, C.C. Davis, *Optics & Photonics 2005*, (International Society for Optics and Photonics, 589215, 2005)
3. N. Namazi, R. Burris, G. C.Gilbreath, *Optics & Photonics 2005*, (International Society for Optics and Photonics, 58920T, 2005)
4. M.C.R. Cordeiro, C.P. Colvero, J.P. von der Weid, *International Conference on Microwave and Optoelectronics*, IEEE, 393 (2005)
5. J. Li, M. Uysal, *Global Telecommunications Conference*, IEEE, **5**, 2654 (2003)
6. I.B. Djordjevic, B. Vasic, M.A. Neifeld, IEEE Photon. Technol. Lett., **18**, 1491 (2006)
7. M. Uysal, J. Li, M. Yu, IEEE Trans. Wireless Commun., **5**, 1229 (2006)
8. M. Tariq, *Avionics, Fiber-Optics and Photonics Technology Conference,* IEEE, 21 (2007)
9. X. Zhu, J.M. Kahn, IEEE Trans. Commun., **50**, 1293 (2002)
10. V.G. Sidorovich, V.V. Ragulsky, M.V. Vasil'ev, A.A. Leshchev, M.A. Sadovnikov, *High-Power Lasers and Applications,* International Society for Optics and Photonics, 191 (2002)
11. J.C. Ricklin, F.M. Davidson, J. Opt. Soc. Am. A, **19**, 1794 (2002)
12. T. Shirai, A. Dogariu, E. Wolf, *Directionality of some model beams propagating in atmospheric turbulence* (Optics Letters, submitted).
13. G. Gbur, E. Wolf, JOSA A, **19**, 8 (2002)
14. P. Raj, S. Sharma, P.C.S. Devara, J. Appl. Meteorol., **32**, 1161 (1993)
15. L.C. Andrews, R.L. Phillips, C.Y. Hopen, *Laser Beam Scintillation with Applications*, (SPIE Press, Bellingham, 2001)
16. J.C. Ricklin, S.M. Hammel, F.D. Eaton, *Atmospheric channel effectson free-space laser communication*, (SpringerScience, 2007)
17. H.E. Nistazakis, T.A. Tsiftsis, G.S. Tombras, IET Commun. **3**, 1402 (2009)
18. L.C. Andrews, R.L. Phillips, *Laser Beam Propagation through Random Media, 2nd ed.* (Bellingham, WA: SPIE Press., 2005)
19. O. Korotkova, L.C. Andrews, R. L. Phillips, *Speckle propagation through atmosphere: effects of a random phase screen at the source* (Proc. SPIE, 4821, 2000)

The Effects of Bottom Ash from MSWI Used as Mineral Additions in Concrete

Roshazita Che Amat[1,*], *Khairul Nizar* Ismail[1], *Norazian* Mohamed Noor[1] and *Norlia* Mohamad Ibrahim[1]

[1]School of Environmental Engineering, Universiti Malaysia Perlis, 02600 Arau, Perlis, Malaysia

Abstract. Municipal solid waste incinerators (MSWI) produce by products which can be classified as bottom and fly ashes. The bottom ash accounts for 85-90 % of the solid product resulting from MSW combustion. The aimed of the present work is to study the effect of replacing partial of bottom ash were manufactured. Fresh and hardened properties of the concrete were compared in order to study the suitable cement-bottom ash replacement. Bottom ash was found to have some reactivity, but without greatly affecting the hydration process of OPC at 10 % replacement. However at more than 10 % replacement, the addition of bottom ash greatly affected strength.

1 Introduction

Nowadays, Malaysia has experienced a problem of municipal solid waste (MSW) is growing more than 91 % due to increasing population and urbanization growth [1]. Municipal solid waste incinerator is the solution to release the pressure increasing of MSW. Nevertheless, by product of the incinerator were fly and bottom ash which are combined for disposal to landfill. The high cost of treatment, strategy of landfill space and also increased environmental awareness have prompted the alternative for uses the ash rather than disposal. Indeed, the recycling the waste in the concrete manufacturing is increasing due to the rising demand of infrastructures in developing countries.

Due to the construction demand was highly increased; human required a lot of concrete for construction. Concrete is used conventionally than any other man-made material in the world [3]. Even so, the use of cement as the production key ingredient is still happening. The traditional construction materials are being produced from the current natural resources. This is damaging the environment due to continuous query and depletion of pure resources [4].There is major drawbacks with respect to sustainability.

Bottom ash is mainly composed of Si, Fe, Ca, Al, K and others in oxides form and presents a similar composition of cement. In other countries, bottom ash was used as an aggregate substitute in road pavement [2]. This suggest that MSWI bottom ash could have pozzolonic or hydraulic behavior and its addition to concrete mix could have beneficial role in hydrated cement paste. This achievement could be a great sustainability as alternative material to cement.

* Corresponding author: roshazita@unimap.edu.my

2 Material and methods

Bottom ash from municipal waste incinerator from Langkawi Island, Malaysia was considered. Bottom ash was subjected to dying and grinding process in order to be suitable as mineral addition in concrete. Grinding of bottom ash was carried out to reduce their size particle to less than 75 micron. Concrete mix was designed for grade 20 with constant water cement ratio (w/c) of 0.5 was manufactured. The conventional concrete or 0% composes of cement, water, aggregates and sand were considered as a control mix without bottom ash addition. In order to study the effects of MSWI bottom ash on the properties of fresh and hardened concrete, other three mixes with the same composition as control mix were cast but with 10 %, 20 % and 30 % by volume replacement of cement with bottom ash as mineral addition.

Chemical composition of bottom ash and cement were determined directly by X-ray fluorescent (XRF) and were expressed as oxide percentage content. Workability of fresh concrete was measured by means of slump test. After 28 days curing, compressive strength was measured on 100 mm cubes. Water absorption was carried out in order to determine the percentage of water absorbed into the interconnectivity capillary pores in cement and bottom ash structure. To study the resistance to fire, the cube is initiated by burning the sample with heat rate and a predetermined time.

3 Results and discussion

3.1 Chemical composition

Table 1 below shows the Municipal solid waste contain a varies composition over time and differences between country to country, due to the variances in way of living, waste recycling process operations of a country and the ash content. This XRF reveal that several compound are present in the bottom ash. Table 1 shown the results chemical composition of cement and bottom ash for comparative purpose. Generally, the chemical and physical characterization of ash will depend on the compositions of the raw Municipal solid waste, the operational conditions, the type of incinerator and air pollution control system design. The chemical composition of cement and bottom ash shows that the major elements containing are Fe, Mg, Ca, Si, Al, K, Cl and Na. In addition, SiO_2, Al_2O_3, CaO, Fe_2O_3, Na_2O, K_2O are the common oxides found in ash.

The MSWI bottom ash could be reused as cement replacement materials since it contained SiO_2 which is one of the main building components in cement and concrete utilizations. However, the presence of alkalis content only limit from 0.3-1.2 % in cement. The presence of 2.83 % of K in bottom ash would influence the concrete strength [5].

DOI: 10.1051/matecconf/20179701053

Table 1. Chemical Composition of Cement and Bottom Ash

Substance	Compound	Cement	Bottom Ash
Aluminum	Al	2.90	4.07
Silicon	Si	13.40	9.59
Sulfur	S	3.43	2.45
Potassium	K	1.12	2.83
Calcium	Ca	73.47	63.94
Titanium	Ti	0.34	1.92
Vanadium	V	0.07	0.00
Chromium	Cr	0.035	0.12
Manganese	Mn	0.2	0.26
Iron	Fe	4.47	10.00
Copper	Cu	0.034	0.274
Strontium	Sr	0.051	0.41
Zirconium	Zr	0.025	0.12
Ruthenium	Ru	0.20	0.56

3.2 Workability of concrete

The effects on workability have been measured using slump test according to [6] and results summarized in Table 2 below. Similar range of slump (true slump) were obtained for both control concrete and concrete with addition of bottom ash.

Table 2. Result of Slump Test

Types of concrete	Slump (mm)	Type of slump
0 % BA	45	True Slump
10% BA	55	True Slump
20% BA	60	True Slump
30% BA	77	True Slump

The addition of bottom ash in mix leads to an increase in water/cement ratio which increases workability, and hence increases the flow. The very small particles and their angular surfaces increase the specific surface area of the mix, resulting in a higher water demand.

3.3 Water absorption

The test was carried out after completion of concrete immersed in water within a specified period. The test began with a weigh of concrete that was completed before the concrete curing in water. After completion of the mentioned concrete curing in water with the prescribed period the concrete sample will be taken and will be weighed for the second time. The difference between the original weights of the concrete used to assess the ability of concrete to absorb water. Results for the sample of concrete that has 0%, 10%, 20% and 30% converted to cement the bottom ash is shown in the Fig. 1. Subsequent tests were each mixed concrete has dried for 72 hours in an oven. Tests have been conducted on the concrete has cured to 28 days but the percentage varies replacement of cement with bottom ash which distributes to the development of a different result.

DOI: 10.1051/matecconf/20179701053

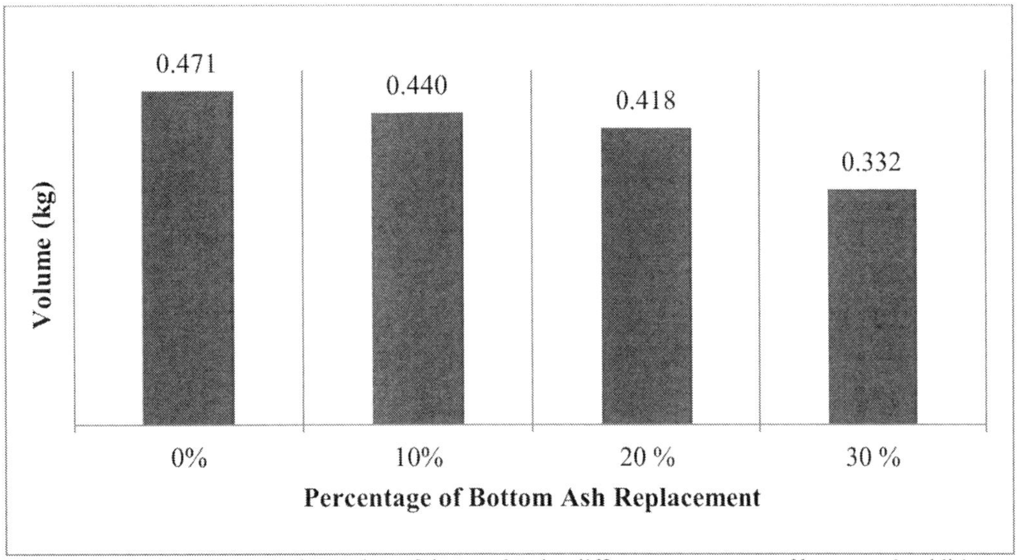

Fig. 1. Percentage of water absorption of three cubes by different percentage of bottom ash addition

3.4 Density

The density test was determined with water curing for 28 days and 7 days. Fig.2 shows the average density for the preservation of water and air curing results for 0 %, 10 %, 20 % and 30 % of the bottom ash with water curing for 28 days and 7 days. Sample of 30% indicates that the most significant difference was between 7 days and 28 days was a 72kN / m^3 showed. Sample 20 % showed a difference of as much as 43.4kN/ m^3, while the 10% sample also showed a change of 33kN/m^3 and 0% showed a difference of at least 14kN/m^3. Sample 10% was closest to the sample 0% and 20%, followed by neither the most comprehensive was 30%. Sample curing time actually affects the density of concrete by increasing the mass of concrete as cement to bond strongly absorbs water and acts response as a binder to the concrete.

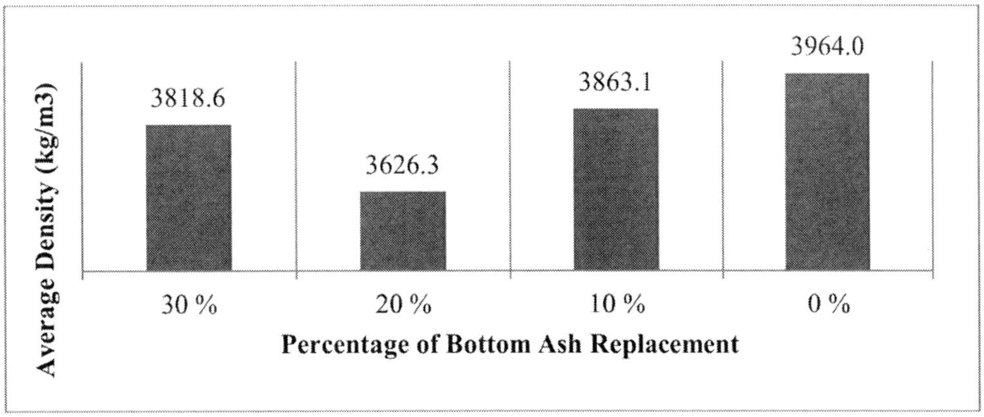

Fig. 2. Average of water curing and air curing density (kg/m^3) for 28 days

DOI: 10.1051/matecconf/20179701053

3.5 Compressive strength

Concrete strength was measured in accordance [7], the result of which strength can be found in Fig. 3. The concrete with only 10 % of bottom ash addition achieved the compressive strength more than 20 MPa. However, at higher levels of replacement, the difference in strength is not only due to the reduced quantity of cement, but also because of the large increased in water/cement ratio. Increase in water/cement ratio would increase porosity the free water to move around through a more porous matrix.

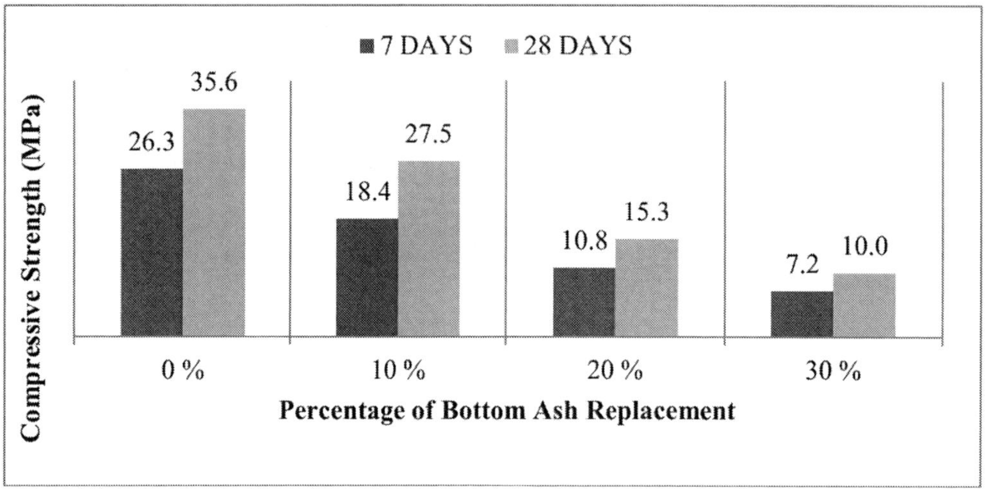

Fig. 3. Compressive strength by age of concrete

3.6 Fire resistance

Fire resistance tests showed strength will affect the heat or flame to move to the concrete cubes. The cubes was burned at constant temperature at 500 ° C for 2 hours each sample. Fig. 4 shows the result on concrete strength before and after fire resistance testing.

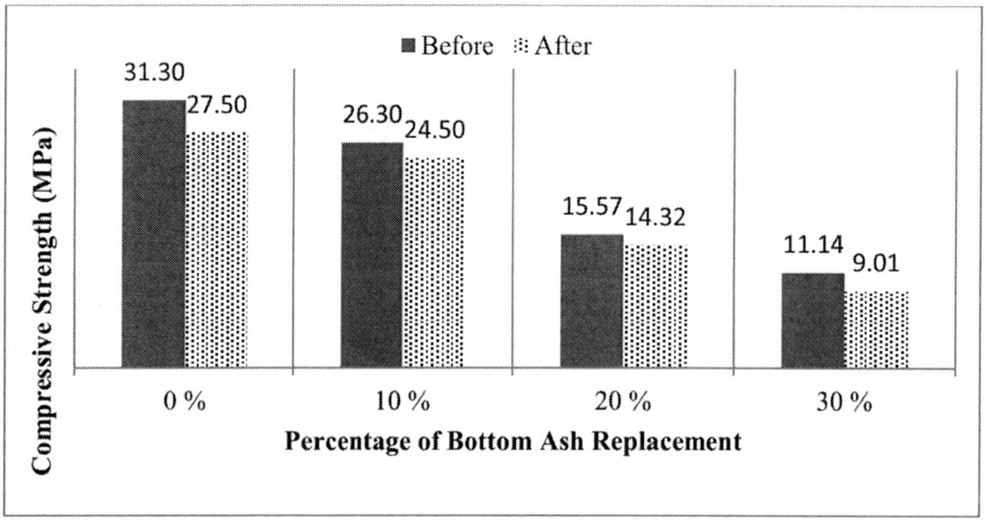

Fig. 4. Strength between sample before and after fire testing

After exposure with fire for 2 hours, control concrete reduced strength about 12 % and for concrete with addition of 10 % and 20 % have only 6-8 % strength reducing. 10% produced great quality concrete for optimum strength fire resistance.

Conclusion

Tests describeb in this paper shows that bottom ash from MSW incinerator was potentially attractive as mineral aditions for concrete. The bottom ass was reactive, but far less than cement. At low percentage of replacement (10 %) was not greatly affected the properties of concrete, but at more than 20 % there was a marked decrease performance. Bottom ash concret with replacement less than 20 % mixture met the requirement of 28-day compressive strength and the linear decrease of compressive strength after fire resistance testing.

This reserach was financially supported by Fundamental Research Grant Scheme from MOSTI and Universiti Malaysia Perlis.

References

1 Zamali Tarmudi, Mohd Lazim Abdullah, Abu Osman Md Tap, Jurnal Teknologi **51**, 1 (2009)
2 Izquierdo Maria, Vazquez Enric, Querol Xavier, Barra Marilda, Lopez Angel, *International Ash Utilization Symposium*, (Center for Applied Energy Research, 2001)
3 B. Lomborg, *The Skeptical Environmentalist: Measuring the Real State of the World*, (Cambridge University Press, Cambridge, 2006)
4 M. Safiuddin, M. A. Jumaat, R. Hashim, M. S. Islam, Int. J. Phys. Sci., **5**, 1952 (2010)
5 A.M. Neville, *Properties of Concrete* (Pearson, 2011)
6 British Standard 1983, *Testing Concrete: Method for Determination of Slump, BS 1881: Part 102 - 1983* (British Standard Online, 2014)
7 British Standard 1983, *Specification for Compression Testing Machines for Concrete, BS 1881 Part 115 - 1983* (British Standard Online, 2014)

DOI: 10.1051/matecconf/20179701054

Enhancing Growth and Yield of Grey Oyster Mushroom *(Plearotussajorcaju)* Using Different Acoustic Sound Treatments

Roshita Ibrahim[1,*], *Ahmad Azmil Irfan* Mohd Jamil[1] , *Sayed M. Zain* Hasan[2] , *Adzemi* Mat Arshad[2] , and *Zarina* Zakaria[1]

[1]Department of Chemical Engineering Technology, Faculty of Engineering Technology,Universiti Malaysia Perlis (UniMAP), Level 1,Block S2, Campus UnicitiAlam,02100 SgChucuh, Padang Besar, Perlis.
[2]School of Food Science and Technology, Universiti Malaysia Terengganu, 21030 Kuala Terenganu, Terengganu.

Abstract. Mushroom, as believed by many people, grows on specific time and condition as in the wild it grows after a heavy rain falls. The effects of lightning and thunderstorm may suppress the mychorrizal structure to grow and eventually forming fruiting body. This study was conducted to determine the effect of different acoustic sound treatments on the growth and yield of grey oyster mushroom (*Pleurotussajor-caju*). Five different acoustic sound treatments had been applied during spawning period which were thunder storm, hardcore music, soothing instrumental, Quranic recital and without any sound treatment applied which served as control. The parameters studied were mycelium growth rate, days of mycelium filled up the bags, days of pinhead emergence, days of fruiting body formation, total weight, percentage biological efficiency, pileus color and texture. There were significant differences ($P<0.05$) in the mycelium growth among different acoustic sound treatments where sound treated mushroom showed faster growth than control. Similar trend was also observed in yield (total weight and biological efficiency) where there were significant differences ($P<0.05$). However, there were no significant differences ($P>0.05$) observed in other parameters, such as pinhead emergence, fruiting bodies formation, pileus color and texture. In summary, treatments using different acoustic sound at 75 dB could be considered as better treatment to enhance the mycelium growth thus accelerate the mushroom cultivation process as well as increasing the mushroom productivity. This research could help farmers to grow and harvest their mushroom at specific time frame and fulfill customer's demand.

[*] Corresponding Author: roshita@unimap.edu.my

1 Introduction

Mushrooms are described as a macro-fungus with large fruiting bodies enough to be seen with naked eye and to be picked by hand [1]. Recently, agricultural field had shown popularity due to advances in the agricultural system technology. Mushroom market in Malaysia had shown high demands mostly caused by increasing in population and the advances in mushroom applications in both medicine and culinary. There are many methods to enhance the growth and yield of mushroom. Sound treatment had been proven to change the cell cycle [2]. This is based on different sound acoustic, sound intensities and sound frequencies. The difference in bio-acoustic sounds shows that mushroom interact with their selective sound waves. Various acoustic sounds were applied in this study which includes rock, soothing instrumental, explosion and religious. The effect of lighting and thunderstorm may encourage the response mechanism for the mushroom to grow and thus, forming fruiting bodies. Mushroom cultivation takes longer time due to its mycelium complicated growth. This affect the growers who tend to give up halfway due to extended cultivation period. Mushroom cultivation is a risky business. It needs a lot of effort and high cost to ensure a hygienic production. A whole mushroom house can be easily contaminated if there are diseases or the presence of outside contaminating agent.

2 Methodology

The first step is preparation of substrates. The substrates ingredients are mixed together in the ratio of 1:10:100 of calcium carbonate ($CaCO_3$), rice bran and rubber sawdust respectively. The media was ensured to be moist so that it will not disintegrate by adding a little amount of water at about 67% to 69%. The media of about 500g are filled into polyethylene bags. The bags were closed with cotton stuffed caps. The media bags were loaded into the shelf in autoclave followed by sterilization at 121°C for 30 minutes. The sterilized media bags were then be removed from autoclave and are allowed to be cooled in the laminar flow for 12 hours before inoculation process. The mushroom culture on corn that are used to inoculate the bags is about 10 g for each bag. To prevent anycontamination, the inoculation process will be conducted in aseptic way in the laminar flow. The bags were capped and arranged on the rack in the mushroom house for spawning process for 30 to 45 days.

2.1 Acoustic sound treatment

In every ten days intervals from the inoculation day, different acoustic sound treatments were applied to the mushroom bags. The mushroom bags were applied with four different acoustic sounds, which are, hardcore music, soothing instrumental, Quranic recital, and thunderstorm effect. Mushroom bags which are not been given any sound treatment were served as control. The reason for not applying the treatment everyday is to give time for the mushroom to rebuild the damaged tissues after the stress applied. The intensity of the sound was measured by using sound level meter (dB meter) which was kept constant at 75 dB at every treatment. The speaker are used to produce different acoustic sounds which are placed in a box together with the mushroom bags so that environmental condition are not being interfered.

2.2 Spawning stage

The spawning stage is the growth stage of mushroom mycelium. At this stage, the bags are arranged vertically with the purpose of accelerating of spawning process. Spawning process

 DOI: 10.1051/matecconf/20179701054

of mycelium takes 30 to 45 days from the date of inoculation to fill up the bags. Temperature, humidity and light are the basic needs for the growth and development of the mushrooms. The appropriate temperature and humidity were maintained at the range of 25-35°C and 75-85% respectively throughout the spawning process. To ensure the level of humidity are sufficient, the floor, wall and bags in the mushroom house were watered daily. Light intensity is only about 30% during this stage and the mushroom house are protected from dust and pest.

2.3 Mycelium growth rate

After the inoculation process are conducted, the mycelium growth rate are determined for every 5 days intervals. The mycelium growth are measured by using ruler in centimeter until the day mycelium filled up the bags (linear length). The mycelium growth rate (cm/day) are determined for each treatment of acoustic sounds and control.

2.4 Pin-head emergence

The cap of mushroom bags were opened to allow the pinhead emergence once mycelium filled up the bags. It takes about 4 to 5 days for the pinhead to emerge from the day the cap opened.

2.5 Fruiting body formation

The formation of fruiting body of the oyster mushroom are develop 3 to 5 days after pin-head emergence. The harvesting process required grower to pull the mushroom fruiting bodies from the substrate. After harvesting process done, the bags were closed back with the cap to give the mycelium time to accumulate the nutrient for the next fruiting cycle that takes around 10 to 14 days.

2.6 Yield of fruiting body

The fruiting body formation indicates the mushroom can be harvested. By using digital balance, the weight of the fruiting body are weighed. The number of the fruiting bodies were counted and recorded. The mushroom yield are expressed as the biological efficiency (BE) which is the total fresh weight of mushrooms divided by the dry weight of initial substrate.

2.7 Color analysis

The determination of the color properties of mushroom fruiting bodies are done with Minolta Chromometer using CIELAB Color parameter, L*, a*, b*. Three different points from the same surface of mushroom pileus were taken. The L*, a*, b* value were recorded. The L means monochrome scale expressing the amount of reflected light. L of 100 means pure white while L of 0 is matt black. There are two scales of colors, which are ‘a’ is red to green and ‘b’ is yellow to blue.

2.8 Texture analysis

The firmness of the mushrooms are measured by using texture analyzer TA.XTplus. Firstly, the sample are placed on the stage of texture analyzer. Probe P/2 are used to penetrate the samples at predetermined distance. Penetration test using 2 mm diameter probe are applied to puncture the mushroom cap from the top to a 5 mm depth. Texture Expert Software showed the texture profile of the penetration which include force, distance and time during the analysis process conducted. The maximum compression force are recorded.

2.9 Pileus diameter

By using a ruler, the diameter of fruiting body are taken. For each mushrooms that were harvested from the mushroom bags, the largest mushroom and the smallest mushroom were choosen for diameter measurement. The range of measurement are from each end side of the mushroom's pileus. The measurements were recorded.

2.10 Experimental design and statistical analysis

The experiment are designed based on Complete Randomized Design (CRD) with a control and 4 different acoustic sounds. Each treatment were having five replicates. The statistical analysis used are SAS version 16.0. The differences among the treatments were analyzed using one-way Analysis of Variance (ANOVA), considering significant at ($P<0.05$) using Tukey test.

3. Results and discussion

This section details out the results obtained and the discussion on the mycelium growth rate, the number of days taken for the mycelium to fill-up-the bag, pinhead emergence and fruiting bodies formation as well as the yield and the physical properties of the fruiting bodies harvested.

3.1 Mycelium growth rate

The result of mycelium growth rate on different sound treatment was presented in Figure 1. The graph shows that there were significant differences ($P<0.05$) among days of treatment. All the treated bags showed a faster mycelium growth compared to control. The fastest rate of mycelium reached 18.5 cm was when treated with thunder storm (28 days). The slowest rate of mycelium fill up the bags was when there is no sound treatment given to the bags (control). Control took 37 days to reach 18.5 cm. On the other hand, the treatment using hardcore music, soothing instrumental and quranic recital shared almost the same days of mycelium to reach 18.5 cm (30 days).

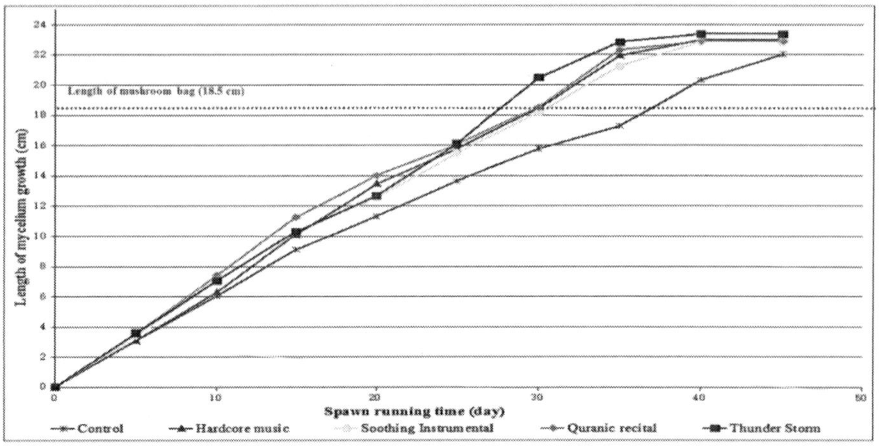

Fig. 1. The growth of mycelium subjected to different acoustic sound treatments during spawn running period.

Based on a research of mushroom growth using physical treatment which is done by [3], it was reported that control treatment took the longest time of 40.6 days to complete the spawning of 16 cm length of the mushroom bag. Faster spawning time were observed by other physically treated bags where high sound intensity treatment only took 23.6 days followed by bright light treatment which took 23.8 days, cold temperature treatment which took 24.2 days and electrical shock treatment which took 25.8 days [3]. All mushrooms species exhibited different mushroom growth pattern while *P. sajorcaju* showed a growth rate of 1.2 cm/day [4].He had also reported that there were significant effects of music on the germination of seeds when compared to untreated control plants.

3.2 Days of mycelium fill-up the bag, pinhead emergence and fruiting bodies formation

There were significant different ($P<0.05$) in the number of days taken for mycelium to filled up the bags among all the treatments (Figure 2). Sound treated bags show slightly shorter time compared to control, where thunder storm took 34 days followed by quranic recital (35 days), soothing instrumental (36 days) and hardcore music (37 days). Meanwhile, control took the longest time which is 42 days. However, there were no significant different ($P>0.05$) in pinhead emergence among different sound treatment. The time recorded for pinhead emergence are 3 to 5 days for sound treated bags, whereas for control it took 8 days for pinhead emergence. Pinhead emergence is an important factor in mushroom cultivation which indicate the initial stage of fruiting bodies formation. There were also no significant different ($P>0.05$) in fruiting bodies formation among different sound treatment. The time recorded for fruiting bodies formation are 8 days for sound treatment, and 9 days for control. Days for fruiting bodies formation was counted from the pinhead emergence until the mushroom pileus had properly developed.

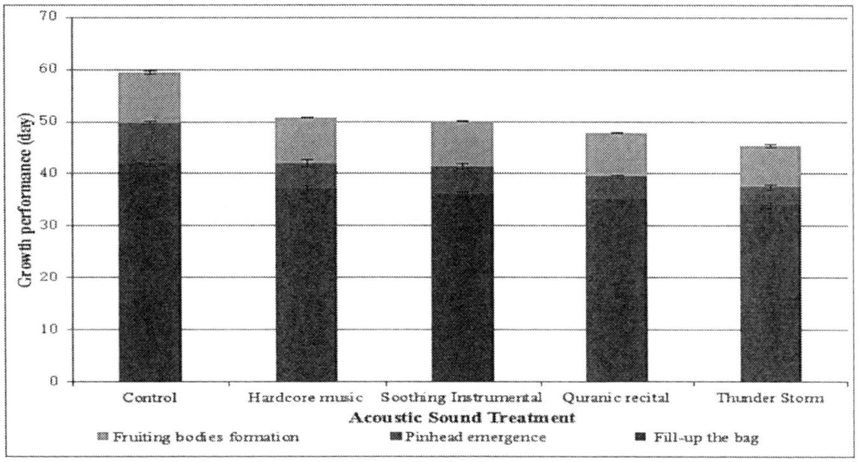

Fig. 2. The growth performance of grey oyster mushroom subjected to different sound treatments. Vertical bars represent standard errors.

A study that were done by [5] showed that there were significant different (P<0.05) in the number of days for mycelium to filled-up the bags among different sound intensity treatment. Based on his study, treatment with 65 dB and 75 dB showed significantly shorter time (36 days) followed by 85 dB and 95 dB (38 days), whereas for control (0 dB), it took the longest time (48 days). According to [6], sound waves may increase the quality of crops including mushrooms.

However, the result for mycelium to filled-up the bags in this study was relatively longer as compared to the result obtained from the study done by Shah *et al.* (2004) and Mondal*et al.* (2010) which only took 21 to 24 days. The slower growth of mycelium in this study is might due to the high temperature and low relative humidity in the mushroom house as compared to the ideal temperature and relative humidity for mushroom house (temperature : 28-30°C, relative humidity : 80-85%). The study done by [5] also showed that there were no significant differences (P>0.05) for pinhead emergence and fruiting bodies formation among different sound intensity treatment. The time recorded for pinhead emergence and fruiting bodies formation in his study were 3 to 4 days and 6 to 7 days respectively. A study done by [9] had reported that pinhead emergence appeared 5 days after spawning period. Meanwhile, the fruiting bodies formation recorded in this study was slower than the one reported by [8] which only took 2 to 3 days. The insignificant differences (P>0.05) for pinhead emergence and fruiting bodies formation is might due to the same mushroom species.

3.3 Yield

For total weight of fruiting bodies, there were significant differences (P<0.05) among all the treatments (Table 1). Treatment with hardcore music tends to show the highest yield (342.34 g), followed by thunder storm (313.23 g), soothing instrumental (299.92 g) and quranicrecital (281.89g). For control, it has the lowest yield which is only 147.54 g. For percentage of biological efficiency, the graph pattern is exactly the same as total weight (Table 1). Treatment with hardcore music shows the highest biological efficiency (68.47%), followed by thunder storm (62.65 %), soothing instrumental (59.98 %) and quranic recital (56.38 %). For control, it has the lowest biological efficiency which is only 29.51 %.

DOI: 10.1051/matecconf/20179701054

Table 1. The mean values for yield (total weight and biological efficiency) of grey oyster mushroom.

Acoustic Sound Treatments	Total Weight (g)	Biological Efficiency (%)
Control	147.54 ± 44.69^{b}	29.51 ± 8.94^{b}
Hardcore Music	342.34 ± 48.40^{a}	68.47 ± 9.68^{a}
Soothing Instrumental	299.92 ± 34.59^{a}	59.98 ± 6.92^{a}
Quranic Recital	281.89 ± 50.57^{a}	56.38 ± 10.11^{a}
Thunder Storm	313.23 ± 29.61^{a}	62.65 ± 5.99^{a}

Note : Values are means of 5 replicates. Means (n=5) ± standard deviation.
a-b : Values bearing the same superscript within the same column are not significant at 5% level (P>0.05).

Houand Mooneyham(1999)[6]had investigated the effect of Agri-wave technology on the plant meridian system to improve the yield and quality of plants. From his investigation, sound waves at frequencies of 0.1 kHz to 1 kHz and intensity of 70 dB for 3 hours from Plant Acoustic Frequency Technology (PAFT) generator significantly increased the yield of sweet pepper, cucumber and tomato by 30.05%, 37.10% and 13.20% respectively. Furthermore, the yield of lettuce, spinach, cotton, rice and wheat were increased by 19.60%, 22.70%, 11.40%, 5.70%, and 17.00% respectively. The audio treatment also increased the yield of edible mushrooms from 8.00% to 15.80% and also increased the fruit size from 2.40% to 43.30% respectively [6]. According to Royse *et al.* (2008), biological efficiency was determined by the ratio of fresh mushroom harvested per weight of dry substrate expressed in percentage. The study done by [5] had reported that for percentage of biological efficiency, there was no significant different (P>0.05) among all the sound intensity treatment. By comparing to this study, it shows that different type of acoustic sound treatment can produce different yield of grey oyster mushroom.

3.4 Physical properties

No significant differences (P>0.05) was observed in L*, a* and b* values among different sound treatment (Table 2). For L* value that indicates light intensity, all treatments show almost the same value (52.89-53.99). For a* value that indicates red to green color, all treatments show almost the same value (7.09-7.73).For b* value that indicates yellow to blue color, all treatments show almost the same value (9.87-11.38). There was no significant different (P>0.05) for the firmness of pileus among all the treatments (Table 2). The firmness values were in the range of 13.55 g to 15.00 g. Based on the study done by [5], there was no significant different (P>0.05) for the firmness of *Pleurotussajor-caju* among different sound intensity treatment, which the values were in the range of 117.31 g to 107.24 g. [3] also reported that there was no significant different (P>0.05) for the firmness of grey oyster mushroom among different physical treatment, which the values were in the range of 77.69 g to 61.19 g. The similarities of the firmness is might probably due to the same species of grey oyster mushroom that was used in this study.

Table 2. The mean values for physical properties (color and firmness) of grey oyster mushroom.

Acoustic Sound Treatments	Pileus Color			Firmness (g)
	Color L*	Color a*	Color b*	
Control	52.89 ± 8.01^a	7.62 ± 0.62^a	10.57 ± 1.94^a	13.89 ± 2.71^a
Hardcore Music	53.16 ± 1.88^a	7.29 ± 1.48^a	11.38 ± 1.18^a	14.22 ± 2.90^a
Soothing	53.99 ± 1.92^a	7.73 ± 0.51^a	9.87 ± 0.73^a	$13. \pm 2.18^a$
Quranic Recital	53.85 ± 3.27^a	7.28 ± 0.32^a	10.13 ± 0.68^a	14.89 ± 3.06^a
Thunder Storm	53.69 ± 1.59^a	7.09 ± 0.64^a	9.89 ± 1.32^a	15.00 ± 3.04^a

Note : Values are means of 6 replicates. Means (n=6) ± standard deviation.
a-b : Values bearing the same superscript within the same column are not significant at 5% level (P>0.05).

Color L* value represent the lightness of the color which range from 0 (black) to 100 (white), the a* value which range from -60 (green) to +60 (red), and the b* value which range from -60 (blue) and +60 (yellow) [3]. According to [5], no significant differences (P>0.05) were observed in L*, a* and b* values among different sound intensity treatment. The L* value obtained by [5] was in between 49.61 to 59.70 which is almost similar to the value reported by [3] which is 59.17. According to [3], the a* and b* values from mushroom cultivated through wood shaving was 7.79 and 9.84 respectively. The b* value that was also obtained by [5] had showed darker color which was 14.08 compared to this study which is in the range of 9.87 to 11.38. Firmness was defined as the force needed to break or penetrate the flesh. There were a huge differences of firmness values between this study and the study done by [5] and [3]. The reason is because of the condition of the grey oyster mushroom that were used in the firmness test. Firstly, the harvested mushrooms from the mushroom house were already aged. Secondly, all the mushroom's pileus were wet and slippery due to refrigerated storage. Both of these reasons had affected the firmness values of the mushrooms, and the real firmness value of grey oyster mushroom (*Pleurotussajor-caju*) could not be obtained.

4 Summary

In this study, there were significant differences (P<0.05) in the growth and yield of grey oyster mushroom treated with different acoustic sounds which were thunder storm, hardcore music, soothing instrumental and quranic recital. Each parameter shows differences in parameters analyzed. For example, treatment with thunder storm resulted in the best enhancement for the mycelium growth. Whereas, treatment with hardcore music was the best sound to enhance the yield of grey oyster mushroom. However, there were no significant differences (P>0.05) in other parameters. All the sound treatments did not show any differences in the number of days taken for pinhead emergence, fruiting body formation, as well as in the color and texture properties. Based on the results, it can be concluded that acoustic sound treatment could enhance the growth and yield of grey oyster mushroom, and each of different acoustic sound types have their own enhancement capabilities.

The authors would like to acknowledge all the staffs in the Department of Chemical Engineering Technology,UniMAPfor their helps and support.

DOI: 10.1051/matecconf/20179701054

References

1 S.T. Chang, P. Miles, Mycologist **6**, 64 (1992)
2 B.C. Wang, A. Sakanishi, Colloid Surface B, **12**, 89 (1998)
3 R. Ibrahim, K.M. Aziz, A.M. Arshad, S.M. Zain, Malays. Appl. Biol., **44**, 67 (2015)
4 K. Creath, G.E. Schwartz, J. Alt. Comp. Med., **10**, 113 (2004)
5 S.M. Firdaus, A.R. ShamsulBahri, M. Rahijan, AbdWahab A. Rahman, M. Rohana, R. Ibrahim, Advances in Life Science and Technology, 28 (2015)
6 T.Z. Hou, R.E. Mooneyham, Am. J. Chin. Med., **27**, 1 (1999)
7 R. Gush, *Mushworld, Oyster Mushroom Cultivation* (Mushroom Growers Handbook, 2004)
8 M. Dahmardeh, R. Hossienabadi, H. Safarpoor, J. Food Agri. Environ., **8**, 996 (2010)
9 S.R. Mondal, M.J. Rehana, M.S. Noman, S.K. Adhikary, *Comparative study on growth and yield performance of oyster mushroom (Pleurotusflorida) on different substrates* (Agrotechnology Discipline, Khulna University, Bangladesh 2010)

DOI: 10.1051/matecconf/20179701055

Effect of Different Curry Flavor Concentrations on the Fermentation Process and Lactic Acid Bacteria of Layang Scad (*Decapterus macrosoma*)

Roshita Ibrahim[1,*], *Kunasundari* Balakrishnan[1] , and *Norsaliha* Musa[1]

[1]Department of Chemical Engineering Technology, Faculty of Engineering Technology, UniCITI Alam Campus, Universiti Malaysia Perlis (UniMAP), 02100 Sungai Chuchuh, Padang Besar, Perlis, Malaysia

Abstract. For various socio-economic and technical factors, fermentation process is among the commonly used method for fish preservation in South-Eastern Asia. However, the fermented fish were not well accepted by some people due to its unpleasant odour and unique taste. In this study, layang scad (*Decapretus macrosoma*) has been used to study the effect of different curry flavor concentrations on the fermentation process and microflora of lactic acid bacteria. This study was carried out to determine the physical characteristics, microflora of lactic acid bacteria and sensory acceptability of the product. The fish were fermented for 10 days with different concentrations of curry flavor which were at 5%, 7.5%, 10% and control, no addition of curry flavor. The fermented fish were analyzed in terms of pH, firmness, lactic acid concentration and sensory evaluation. Over the fermentation period, pH and firmness values were decreased. Meanwhile, the concentration of lactic acid and colony-forming unit (CFU) of lactic acid bacteria were increased. The effect of fermentation by using different concentrations of curry flavor was observed and no significant different was found on CFU of lactic acid on MRS and M17 agar. Sensory evaluation showed that fermented fish with 10% concentration of curry flavor is the most acceptable by the sensory panelists as it has the highest acceptability score compared to other concentrations. It can be concluded that addition of 10% curry flavor in the formulation of fermented fish is the best in terms of physical characteristic, microbial quality and sensory acceptability.

1 Introduction

The cuisines of East Asia are usually based on the combination of rice, vegetables and proteins obtained mainly from aquatic living, fish [1]. Fish such as layang scad, (*Decapterus macrosoma*) contains a lot of protein that are required by human to maintain a healthy body. Fish is an important source of protein where it is rich in amino acids such as cystine, lysine, methionine, tryptophan and threonine [2]. Other than that, fish also contain microelements

* Corresponding author: roshita@unimap.edu.my

DOI: 10.1051/matecconf/20179701055

and macroelements like phosphorus, calcium, iodine and fluorine. Ismail (2005)[3] reported that fish also provide fats that act as energy sources, fat-soluble vitamins, and unsaturated fatty acids which have a hypo-cholesterolemic effect or anti-arteriosclerosis.

Fish, the most important source of proteins are easily spoiled after being caught, hence, need to be preserved. This is because the immune system of fish collapses after the fish died and bacteria are allowed to proliferate freely in the fish. Preservation of fish by applying the principles of engineering, chemistry and other branches of science is the method to increase the shelf life of fish products so that they can be stored longer because certain type of fish usually available only during certain seasons of the year and fresh fish spoils quickly [4]. Therefore, several methods of preservation of fish had been introduced such as salting, drying, cooling, freezing, canning and fermentation [4].

Preservation of food by fermentation does not only extend the shelf-life, it also improves the functional and nutritional properties of the food. Cleavage of proteins in food by microbial or indigenous proteases produced the bioactive peptides, substantially increases biological properties of the products [5]. Rajapakse et al. (2003)[6] stated that fermented food products are good source of amino acids and peptides. Fermentation process requires microorganisms with the enzymatic capacity like bacteria and yeast, specifically bacteria for lactic acid fermentation and yeast for ethanol fermentation [7].

Lactic acid bacteria (LAB) is the most frequently used microorganism in fermentation process of fish. They are important due to their metabolic capabilities and other physiological properties such as probiotic and substrate utilization properties. Silva et al. (2002)[8] reported that lactic acid bacteria have long been accepted as Generally Recognized as Safe (GRAS) to be consumed by Food and Drug Administration (FDA). A variety of LABs have been recognized from various fermentation products along with their function during or after fermentation process have increasingly been clarified [9].

2 Material and methods

This section details out the sample sources and preparations, the ingredients and procedures used in producing fermented fish and the analyses carried out which were physico-chemical analyses, the lactic acid micro-flora during fermentation and sensory evaluation of the fermented fish.

2.1 Sample sources and preparations

The samples for this project were layang scads that had commercial value for market which were obtained from a wet market in Kangar, Perlis. The length of the fish was about 150mm with width of 30mm and depth of 15mm. The fish were gutted and washed from dirt and blood using ice water. The fish were then toss-dried to remove the excess water and weighed using weighing scale.

2.2 Fermentation ingredients

Other fermentation ingredients that were used in this study were fine and coarse salt, palm sugar, roasted rice and curry flavor.

2.3 Production of fermented layang scad

Fresh layang scads were gutted and the intestines and fins were removed. The fish were thoroughly washed and drained then salted using fine and coarse table salt at 15% (w/w) and

DOI: 10.1051/matecconf/20179701055

kept in airtight container for 24 hours. The salted fish were then drained from the brine solution and coated with mixture of roasted rice, palm sugar and flavors at different concentrations. The fish were then kept in airtight containers at about 25°C according to different flavors concentration for fermentation process. The analysis of the physico-chemical properties of the fermented of the fermented fish were then carried out at 2 days intervals until day 10 (0, 2, 4, 6, 8 and 10 days). The degree of the fermentation process was determined mainly by the measurement of pH. Other physico-chemical properties measured were such as texture and lactic acid concentration. The sensory evaluation was also done after 6 days of fermentation to evaluate the acceptability of the fermented layang scad with different curry flavor concentrations.

2.4 Physico-chemical analyses:

The physico-chemical analyses involved the determination of the changes in pH and texture (firmness) during the fermentation process.

2.4.1 pH determination

The pH of the fish was measured on day 0 until day 10, with every 2 days intervals. This study was using pH meter to determine the pH data and the relationship between pH and formation of the lactic acid during fish fermentation. The pH meter was calibrated with buffer pH 7 and pH 4 for every different flavor concentrations.

2.4.2 Texture (Firmness) determination

The fish's nature was determined on day 0 until day 10, with every 2 days intervals. The firmness of the fish flesh was determined using Brookfield CT3 Texture Analyzer using TA10 stainless steel cylinder probe. The testing parameters were as follows:

Test mode: Compression
Pre-test speed: 1.0 mm/sec
Test speed: 0.5 mm/sec
Distance: 5.0 mm

2.5 Sample preparation for lactic acid bacteria studies

An amount of 10 g of fermented fish flesh was weighed and aseptically put into sterile food blender and mixed with 90 ml of sterile distilled water. The flesh was blended until homogenized for about 2 minutes in laminar flow and diluted up to 10^{-4} concentration. An amount of 0.1 ml of the sample was plated onto agar plate by using micropipette and spread on the agar plate. The same aqueous sample also was used for lactic acid determination.

2.5.1 Determination concentration of lactic acid

The samples were analyzed by dipping the lactic acid strip into the sample and inserted into the reflectoquant. The reading was taken and recorded.

2.5.2 Determination of lactic acid bacteria using De Man, Rogosa and Sharpe (MRS) agar

MRS media was prepared by dissolving 62 g of MRS agar in 1000ml of distilled water. It was then boiled to dissolve the medium completely. The dissolved media were then sterilized by autoclaving at 121 °C for 15 minutes. An amount of 0.1 ml of homogenized sample was plated onto De Man Rogosa and Sharpe (MRS) agar for isolation of lactobacillus species. Lactobacilli MRS Agar was developed by researchers De Man, Rogosa and Sharpe as an alternative non-selective media for the cultivation of fastidious lactobacilli. All the plates were then incubated for up to 3 days at 35°C in an anaerobic jar. The CFU of lactic acid on MRS media was counted and recorded.

2.5.3 Determination of lactic acid bacteria using M17 agar

M17 media was prepared by dissolving 50 g agar in 950 ml of distilled water and bring slowly to a boil. The mixture was sterilized by autoclaving at 121 °C for 15 minutes. It was then cooled and added with 50 ml of sterile 10 % lactose solution. The solution was then mixed thoroughly and dispended into petri dish. M17 agar was used for cultivation of the fastidious species *Streptococcus cremoris*, *Streptococcus lactis* and *Streptococcus diacetilactis*. An amount of 0.1 ml of homogenized sample was plated onto M17 agar before incubated for up to 3 days at 35 °C in an anaerobic jar. The CFU of lactic acid on M17 media was counted and recorded.

2.6 Sensory evaluation

The fermented fish were served to 30 respondents who were randomly selected among students and staffs of Universiti Malaysia Perlis. 7 point hedonic scales were used where score 1 indicating the most unacceptable, score 4 neither acceptable nor unacceptable and score 7 indicating the most acceptable. The respondents were presented with four fermented fish of different curry flavour concentrations and their evaluations were based on the attributes of colour, texture, aroma, taste and overall acceptability. The fermented fish were also served with rice, water and sensory evaluation form which were provided to each respondent to assist the process.

2.7 Data analysis

All the data obtained from the physico-chemical analyses were analysed by using one-way Analysis of Variance (ANOVA) and significant differences among the treatments were determined using Tukey Multiple Comparison Test at $p < 0.05$ from SAS software.

3 Results and discussions

This section reported on the results obtained which highlighted the changes in the physico-chemical properties, lactic acid bacteria studies and the sensory acceptability of fermented laying scads.

3.1 Changes in physico-chemical properties

Changes in pH of fermented layang scad with different curry flavor concentrations during fermentation period are shown in Fig. 1. The initial pH of all samples at day 0 was the same at pH 6.03. Generally the pH of the fermented fish with or without the addition of curry flavor showed a decreasing trend as the fermentation progress until day 8. This may be due to the release of free amino acids from proteins and large polypeptides. After day 8, the pH of all the samples started to increase. This might probably due to the beginning of putrefaction or spoilage process.

Among all the treatments, fermented fish with addition of curry flavour regardless its concentrations showed more drastic decrease in pH compared to control. This might probably be due to more production of lactic acid which results in lower pH values. However, among three concentrations of curry flavour the decrease in pH was observed to be much faster in the fermented fish with lower concentration (5%) compared to 10% concentration. The drop in pH value on the first few days of fermentation was also observed in the study of El Hag et al. (2012)[10].

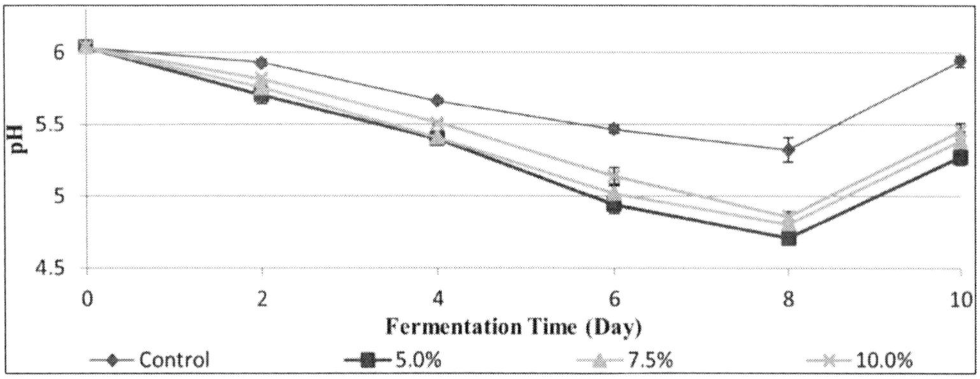

Fig. 1. pH of fermented layang scad with different concentrations of curry flavor during fermentation process. Vertical bars represent standard errors.

The changes in texture (firmness) of fermented layang scad with different curry flavor concentrations during fermentation are shown in Fig. 2. Higher addition of curry flavor resulted in decreased in the firmness of the fermented fish. Addition of curry flavor had a great influence on the textural properties of fermented fish. During fermentation, the samples with 5% curry had significantly ($p<0.05$) the most firm texture, whilst no significant difference in the samples between 7.5% and 10% concentrations. The development of firmness is related to pH reduction and moisture removal [11]. It was also reported that texture formation of fermented fish was closely associated to the rapid decline in pH, which induced conformational changes of gelation. Higher acidity in the fermented fish with the addition of curry flavor resulted in superior organoleptic properties.

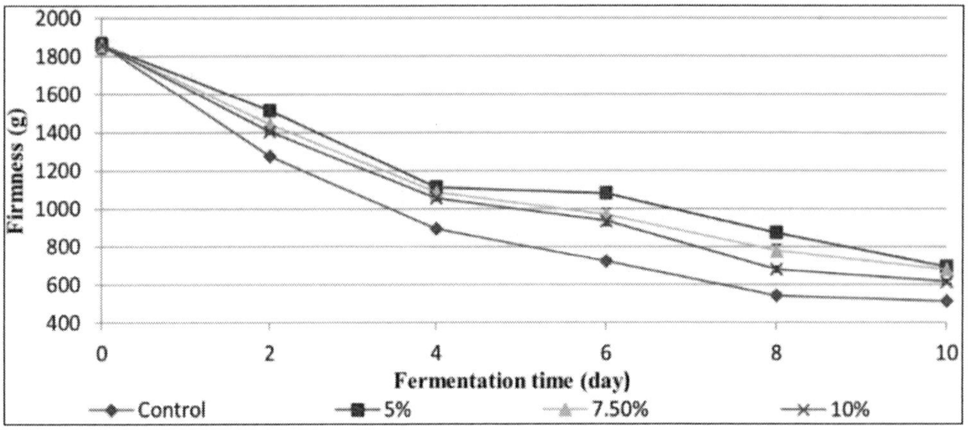

Fig. 2. The texture (firmness) of fermented layang scad with addition of different concentrations of curry flavor during fermentation process. Vertical bars represent standard errors.

The concentration of lactic acid on fermented layang scad with different curry flavor concentrations during fermentation are shown in Fig. 3.

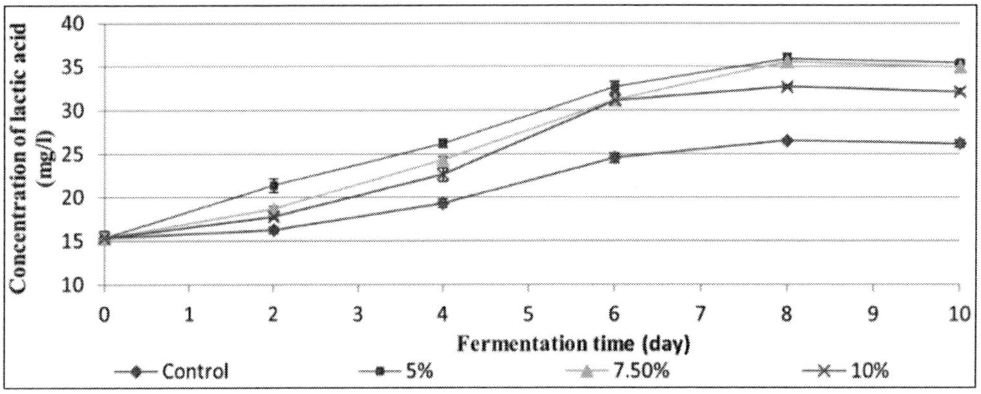

Fig. 3. Lactic acid concentration determined on fermented layang scad with addition of different curry flavor concentrations during fermentation process. Vertical bars represent standard errors.

The general trend of the concentration of lactic acid in fermented layang scad for all treatment was increasing during fermentation from day 0 until day 8. The concentration of lactic acid started to plateu off after day 8. Control fish showed the lowest concentration of lactic acid as compared to the curry flavoured fermented fish.

Based on Fig. 3, on day 0, there was no significant different in concentration of lactic acid of the fermented fish for all treatment. There were significant differences ($p<0.05$) among different curry concentration of lactic acid concentration on day 2 until day 10. The trends were the same for all treatments during the fermentation period where the 5% curry concentration showed the highest significant value as compared to other treatments.

The increasing of concentration of lactic acid were due to the decreasing in pH of fermented layang scad during fermentation progress. As the pH decreases the concentration will eventually increases as lactic acid bacteria prefer relatively acidic conditions, where the growth of spoilage and pathogenic bacteria were inhibited.

 DOI: 10.1051/matecconf/20179701055

3.2 The microflora of lactic acid bacteria

The determination of microflora of lactic acid bacteria on fermented layang scad with different curry flavour concentrations by using MRS agar are shown in Fig. 4.

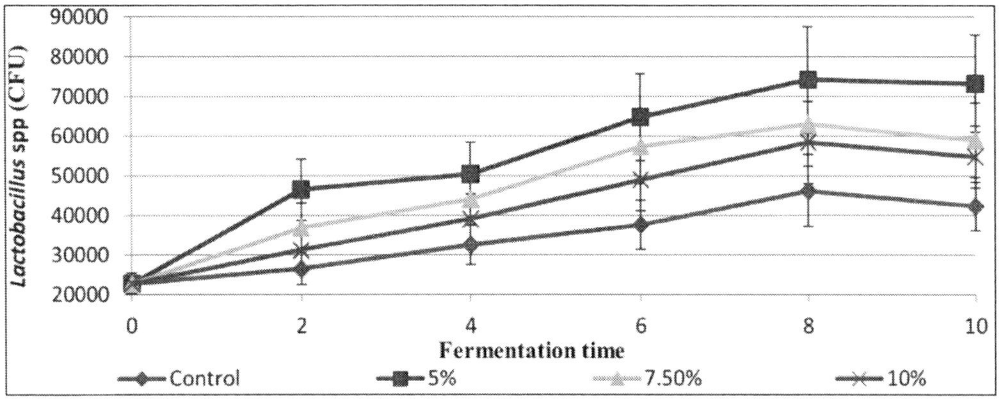

Fig. 4. CFU of lactic acid bacteria on fermented layang scad with addition of different curry flavour concentrations during fermentation process. Vertical bars represent the standard errors.

Generally the CFU of lactic acid bacteria of the fermented fish with or without addition of curry flavour showed an increasing trend as the fermentation progress. The highest CFU of *Lactobacillus* spp.was observed on day 8, for 5% curry flavour with value of 74167. After day 8, the CFU for all treatment started to decline. The increasing and decreasing CFU were due to the changes in pH of fermented layang scad during fermentation progress. As the pH decreases the CFU will eventually increases. Although there were high difference in number of CFU, there was no significant different ($p>0.05$) among all the treatments in terms of the CFU of *lactobacillus* in fermented layang scad.

The determination of microflora of lactic acid bacteria on fermented layang scad with different curry flavour concentrations by using M17 media during fermentation are shown in Figure 5.

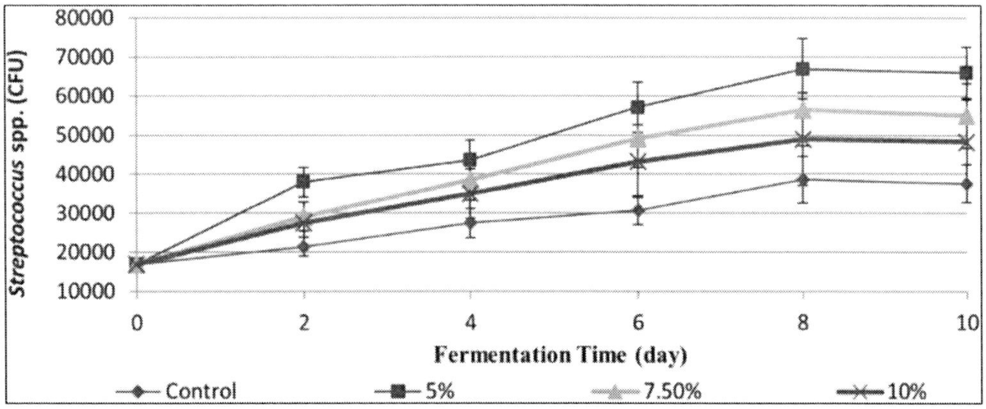

Fig. 5. CFU of lactic acid bacteria (*Streptococcus* spp.) on fermented layang scad with addition of different curry flavour concentrations during fermentation process. Vertical bars represent the standard errors.

Generally the CFU of streptococcus of the fermented layang scad with or without addition of curry flavour showed an increasing trend as the fermentation progress. The highest CFU was observed on day 8, for 5% curry flavour with value of 66990. After day 8,

DOI: 10.1051/matecconf/20179701055

the CFU for all treatment had a slight decline. Although there were high difference in number of CFU, there was no significant different (p>0.05) among all the treatments in terms of the CFU. The CFU for lactic acid bacteria of M17 agar for all treatment during the fermentation period were always lower than that of MRS agar. This showed that the fermented layang scad contained more *Lactobacillus* spp. than *Streptococcus* spp.

3.3 Sensory Acceptability

The sensory evaluation test of fermented layang scad with different concentrations of curry flavour was carried out by 30 panelist and the results of their acceptability are shown in Fig. 6.

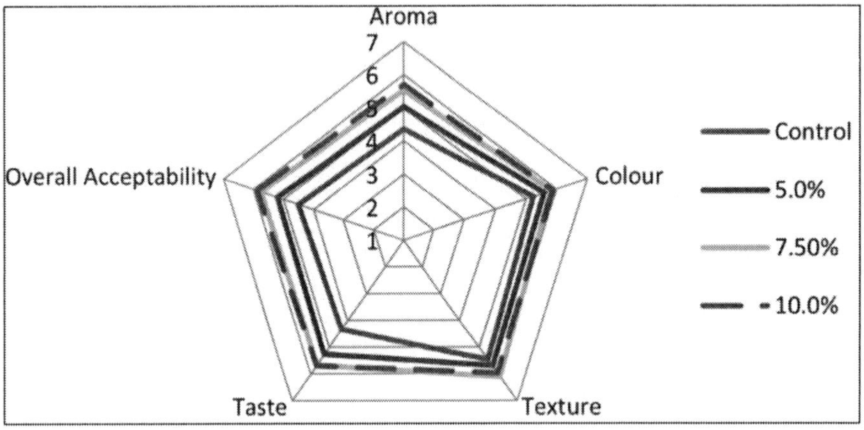

Fig. 6. Sensory acceptability of fermented layang scad with addition of different curry flavour concentrations.

4 Summary

In conclusion, the results of present study have indicated that the addition of curry flavor during fermentation of layang scad can give added value to the products in term of its texture, colour, aroma and taste. Over fermentation period, pH and firmness were decreased. Meanwhile, the lactic acid concentration and its CFU were increased. Since there was no chemical used in this process, this process would provide a safe product and also inhibit the growth of pathogenic microorganisms. Sensory evaluation showed that samples prepared using 10% curry concentrations are the most acceptable by the sensory panelists.

The addition of different concentrations of curry flavor during fish fermentation could also improve the sensory attributes of end product, and it can be concluded that the higher the concentration of curry used, the better the acceptance. The obtained results led us to conclude that the use of curry flavor could improve the preservation and fermentation of fish.

The authors would like to express their appreciations to Universiti Malaysia Perlis (UniMAP) for facilitating this project.

References

1. Ruddle, Kenneth, Naomichi Ishige, *On the origins, diffusion and cultural context of fermented fish products in Southeast Asia* (lobalization, Food and Social Identities in the Asia Pacific Region. Sophia University Institute of Comparative Culture, Tokyo, 2010)
2. Usydus, Zygmunt, Joanna Szlinder-Richert, Maria Adamczyk, Food Chem., **112**, 139 (2009)

3. Ismail, M. Hassan, Front Biosci., **10**, 1079 (2005)
4. Van Berkel, B.M.B. Boogaard, C. Heijnen, *Preservation of fish and meat*, (Technical Centre for Agricultural and Rural Co-operation, Wageningen (Países Bajos) (2004)
5. Steinkraus, H. Keit, Compr. Rev. Food Sci. F., **1**, 23 (2002)
6. N. Rajapakse, E. Mendis, W.K. Jung, J.Y. Je, S.K. Kim, Food Res. Int., **38**, 175 (2005)
7. Chilton, N. Stephanie, Jeremy P. Burton, Gregor Reid, Nutrients, **7**, 390 (2015)
8. J. Silva, A.S. Carvalho, P. Teixeira, P.A. Gibbs, Lett. Appl. Microbiol., **34**, 77 (2002)
9. Liu, Shan na, Ye Han, Zhi jiang Zhou, Food Res. Int., **44**, 643 (2011)
10. G.A. El Hag, B.Y. Abu Gideiri, M.E. Ali, I.M. Abu Zied, New York Science Journal, **5**, 32 (2012)
11. Riebroy, Siriporn, Soottawat Benjakul, Wonnop Visessanguan, LWT-Food Sci. Technol., **41**, 569 (2008)

Key Performance Indicators for Maintenance Management Effectiveness of Public Hospital Building

Mardhiah Farhana Omar[1,*], *Fazdliel* Aswad Ibrahim[2] , and *Wan* Mohd Sabki Wan Omar[1]

[1]School of Environmental Engineering, University Malaysia Perlis, 02600, Perlis, Malaysia
[2]Faculty of Engineering Technology, University Malaysia Perlis, 02100, Perlis, Malaysia

Abstract. Effectiveness of management in maintenance aspect holds the key element in influencing the performance of overall maintenance management. Similarly, public hospital building needs an effective maintenance management as this type of building in nature is one of the most complex issues in the field of maintenance. Improper building maintenance management adopted by the organization significantly will interrupt the overall operation of the building. Therefore, this paper is aim to identifying the key performance indicator (KPI) of effectiveness of maintenance management for the public hospital building. A total of 32 set of questionnaires were distributed to the maintenance manager for each hospital in the northern region of peninsular Malaysia by using self-administration strategy. The survey answer was analyzed by performing descriptive analysis in SPSS. Overall, the result of descriptive analysis shows that all the ten factors of effectiveness of maintenance management are accepted as KPI since the mean value is at least 3.93 which classified as important and significant. The most significant factor of effectiveness of maintenance management is task planning and scheduling with the mean score of 4.35. While less significant factor is identify as maintenance approach with the value of mean score is 3.93. The both results indicates that the management need to have well-structured planning for the maintenance works and also need to embrace the exact strategy of maintenance approach in order to achieved better overall performance of maintenance management. This study may draw a standard practice for the government in assessing the performance of public facilities in terms of maintenance management.

1 Introduction

Maintenance management of hospital buildings is one of the most complex issues in the field of maintenance [1]. As stated by [2], this scenario is occurred due to the fact that the complex nature of hospital buildings, the delicate mechanical and electrical system and inadequate maintenance budget which could affected the performance and operation of the

* Corresponding author: mardhifar@yahoo.com

hospital buildings besides the occurrence of numerous building defects. Building defects are becoming ordinary phenomena in the building maintenance services especially for the hospital buildings are over than 30 years old. Therefore, the longer a building is exposed to the atmosphere, the higher tendency for it to deteriorate [3]. Besides that, a common building design error is often occurred, usually in an effort to reduce the initial construction costs which project costing plays an important role in building design [4]. A research conducted by [1] found that the public hospital building in Nigeria is in very poor and deplorable conditions of structural and decorative disrepairs.

The Ministry of Health Malaysia (MOH) principally provides public healthcare in Malaysia and about 53 per cent of annual budget is allocated for health sector [5]. However, in 1997, the Government of Malaysia took the initiative to carry out a major privatization project for the provision, maintenance, and management of hospital support services (HSS) of public hospitals throughout the country [6]. As a result, the government has appointed concession companies (contractors) as a party to manage the maintenance management of public hospital. In Britain, for example, building maintenance activities account for 50 per cent of all annual construction activities. In the USA, 35 per cent of the overall turnover in the construction sector consists of renovation and modification projects [1]. The impact of building maintenance on quality and unit costs has to be understood. Many researchers are involved in the similar scenario of building maintenance. Essentially, the goal of the building maintenance organization especially in a hospital environment is to achieve zero defects in the hospital's physical operation.

The objective of the paper is to identify the key performance indicator for the effectiveness of maintenance management of the public hospital building. By this entire research will give information and data that more systematic and scientific towards improvement of building maintenance management which draw an assessment of the strengths and opportunities to improve maintenance management practices in our country.

2 The effectiveness of maintenance management

In order to identify the effectiveness of maintenance management factors, key performance indicator is the best strategy to measure the performances of the building. This statement is in line with [7] which affirmed that key performance indicator is a fundamental principle in maintenance management. In this study, the building is refers to public hospital. As proposed by [5], routine monitoring and supervision is one of the key elements to assess and identify the performance of organization in implementing plans, policies and procedures of maintenance management that relates to maintenance management. It will indicate the gap between the current performance and expected performance which could assist the organization to identify the opportunities for process improvement [5]. The proper supervision should be implemented in order to ensure that the maintenance work is performed in accordance to the work order conditions and based on standard maintenance procedures [8].

Besides that, routine monitoring and supervision need to be supported by task planning and working schedule. However, task planning and scheduling need to be provided and managed by first line maintenance foreman or supervisor as well as responsible in providing the feedback to the maintenance planner to ensure that the proposed job plans are relevant, efficient and effective [8]. The lack of planning and scheduling will definitely restrict a maintenance operation to meet its objective in delivering the optimum service as planned by the organization. As mentioned by [9], proper work planning can lead to high responsiveness, less unplanned work and less unplanned overtime with consequences of minimizing the overall maintenance cost. In addition, computerized maintenance management systems (CMMS) can be used to support the organization to automate the preventive planned maintenance (PPM) function and as an aid tool to control the maintenance inventories and material purchase for the purpose of maintenance works. CMMS is acknowledged as an effective tool that has capabilities of reporting the analysis that relates to overall maintenance management aspects quicker and accurate compared to than manual techniques. Other than that, accessibility and accuracy of information can provide more reliable and decisions in proper maintenance work [10]. Therefore, the above elements could be classified as individual factors because all the factors are concern about managerial aspect which influenced by a building maintenance leader.

The practice of maintenance approach is recognized as a factor that highly affects the maintenance performance. For instance, [9] mentioned that the reactive maintenance approach does not tackle the problem at the root level and always lead to repetitive failures. The distinction result between reactive and preventive planned maintenance (PPM) approach where the maintenance work planning program for preventive planned maintenance (PPM) approach drastically reduces errors in day-to-day operations, as well as increase the overall preparedness of plants in the case of an emergency. Moreover, the maintenance approach also will affect the spare part management. Mostly, spare parts are not available at what time required for routine maintenance, downtime is prolonged. In other hand, if too many parts are kept by the management, the utility absorbs excessive costs and overhead expenses due to of carrying the inventory [11]. Another aspect that related to maintenance approach is the type of maintenance work's contract, either it is conducted by in-house or apply an outsource strategy which award the project to the third party. Yet, for the outsource strategy, it still depend on how well the maintenance organizations manage the contract and task as well as "how much" benefit they gain out by applying the outsource strategy. The factor of maintenance approach, spare part management and outsourcing maintenance works could be summarized as a technical aspect as it involves top management who responsible to setting up the overall maintenance approach as well as the organization has to deal with third party for outsourcing maintenance works.

Policy deployment and organization is another recognized factor that influences the maintenance management effectiveness. As suggested by [9], the maintenance organization should consider this factor as one of organizational business strategy in order to improve the overall performance. Nevertheless, the organization also needs to have an appropriate human resource and effective human resource management to run the business efficiently as well to ensure the optimum performance. Furthermore, pointed out that human resource management program is one of the successful factor of maintenance management as the organization must acquire the right person to manage, plan, supervise and execute the maintenance works.

In order to support all the maintenance planning and execution, the allocation and control of budget play an important role in prompting the overall maintenance management. The elements that suggested by [9] in relation to financial are maintenance budget control, cost of monitoring the contractor, and cost control for labor and material aspect. For the purpose of improvise the organizational performance in terms of maintenance management, all the strategy, policy and managerial aspect need to be assessed for the purpose of continuous improvement. This practice is being adopted in Malaysia public hospital as it part of requirement for quality management system with the purpose of obtaining the quality ISO certificate as well as gauge the effectiveness of current practice of maintenance management [6]. The factors of policy deployment and organization, human resource management, financial aspect and continuous improvement could be summarized as an administration and organizational factor as all the factors are under the control of top management.

3 Research methodology

This study is a cross-sectional study where a set of questionnaires has been used as a medium to collect the related data in identifying the key performance indicators for effectiveness of maintenance management for the public hospital. The population for the study is public hospital in northern region of peninsular Malaysia. Overall, there are 32 public hospitals and maintenance manager for each hospital has been chosen to become as a respondent as they are involved in the overall stage of maintenance works and also responsible to monitor, supervise and reporting the maintenance works in the public hospital. Data collection for the current study was conducted based on self-administration method because it involved a short period of time, less expensive, gained quick answer and can clarify on the spot if there are any doubts from the respondents.

A set of questionnaires used in this study is based on 5-point likert-scale, which 1 is strongly disagree and 5 is strongly agree. The result obtained from the survey will be analyzed by performing descriptive analysis in SPSS. Therefore, for the purpose of data interpretation, this study adopted the mean scale developed by [2]. By adopting this mean scale, it enables this study to identify the key factors of effectiveness of maintenance management for the public hospital building. Therefore, this study will only consider mean scale of 3.50 and above as a key performance indicators as this mean scale is categorized as important and significant. The mean scales are as follows; 1.00-1.49 (Not Important and Significant), 1.50-2.49 (Not Important), 2.50-3.49 (Average), 3.50-4.49 (Important and Significant) and 4.50-5.00 (Very Important).

4 Result and finding

This study proposes three main components to be an indicator for effectiveness of maintenance management for the public hospital building, namely individual factors, maintenance aspects and administration and organizational factors. The individual factors consist of 4 factors, while maintenance aspects consider 3 factors and lastly, administration and organizational factors contain 3 factors; in total 10 factors. Figure 1 shows the overall score for the 10 factors of effectiveness of maintenance management. Overall, all the factors have scored at least in the range of 3.50 – 4.49 which indicates that the all factors are perceived as key performance indicators of effectiveness of maintenance management for the public hospital building.

The factor that notches the highest score is task planning and scheduling with the mean value of 4.35. This result demonstrates as very important in influencing the effectiveness of maintenance management as the maintenance work for the public hospital building involves technical complexity which the maintenance works need to be well-organized in order to elude any possibility for delay in completing the maintenance works. It is well acknowledged that the delay in construction works include the maintenance aspect will affect the function and daily routine of the building as well as contribute to the negative impact on overall maintenance management performance. In comparison, the lowest score of mean value is maintenance approach with mean value of 3.93 which classified as important and significant. The mean value of this factor is slightly different with previous studies which found that this factor is the common very important factor in prompting the effectiveness of maintenance management. However, this factor cannot be mistreated because the maintenance approach is a fundamental element in maintenance management and substantially determine overall path and direction of the organization in managing the building maintenance.

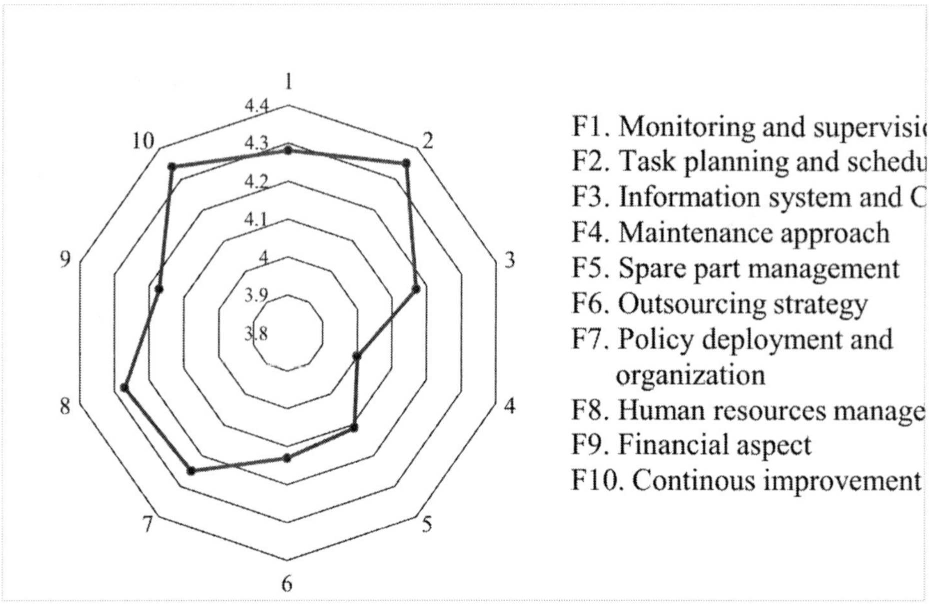

Fig. 1. Overall maintenance management performance ratings

The rest of the factors that consider in this study in identifying the effectiveness of maintenance management had men values of between 4.11 - 4.34 which fall under the classification of at least important and significant. The factor of continuous improvement has scored 4.34 of mean value. Therefore, the management is advised to conduct a maintenance performance assessment for the purpose of continuous improvement. Continuous improvement is significant in contributing to the effectiveness of maintenance management in the public hospital building as the management could achieve better performance for the following years and could setting up higher target of maintenance performance objective. While, the mean value for the factor of monitoring and supervision, human resource management and policy deployment and organization are classified as important and significant with the mean value of 4.28, 4.27 and 4.25 respectively. This result proves that the monitoring and supervision on the maintenance works particularly for the outsource maintenance works for the public hospital building is necessary and need to be performed efficiently in order to ensure the effectiveness of maintenance management. In addition, this result specifies that the top management plays an important role in the maintenance management for the public hospital building. This fact is significant as the top management is in the top organization hierarchy and holds the key in any organization. Thus, the mission and vision of the organization is mostly shaped by the top management which it will determine the direction of the organization for that period.

In addition, other factors that perceived as key performance indicator of effectiveness of maintenance management for the current study are information system and CMMS, financial aspect, outsourcing strategy, and spare part management. These factors notched between 4.11 – 4.17 of mean value which classified as important and significant. The effectiveness of maintenance management for the public hospital building could be better by embracing advance technology that is information system and CMMS. The adoption of technology software and hardware is needed in the modern world as it quicker and more effective which assists the organization to draw better alternative planning and result for the maintenance management. Correspondingly, the management on outsource strategy is apparent significant in improving the effectiveness of maintenance management. For certain maintenance works in the public hospital building in Malaysia is being awarded to the party. Thus, the selection of contractors and supervision on the overall maintenance work progress and quality are need to be more transparent in order to ensure the efficiency of the maintenance works as well as meet the organizational goals. Last but not least, overall organization mission and vision must be aligned with the overall allocation for the maintenance management purpose as the maintenance works involve high expenditures, implicate with complex management and deal with various technical and managerial issues which needs a maintenance expert to manage the maintenance works.

5 Conclusion

The ten factors of effectiveness of maintenance management proposed in this study seem that significantly contribute to the overall maintenance performance for the public hospital building in the northern region of peninsular Malaysia. The result implicates the needs of appointment the competent person to plan and scheduling the maintenance works efficiently. This is due to the fact that the building maintenance works deal with various technical complexity, policies, procedures, and managerial task which requires experienced person to manage the maintenance works. In addition, the competent person also important in figuring out and appointing the suitable contractors to perform the maintenance work for the outsource strategy. The selection of contractors for outsource strategy is important since the quality and work progress in totally depending on the third party, but still can be optimized if the correct approach is adopted by the management. The management via maintenance managers has to perform performance assessment and identify the gap between the current states of performance with the expected performance. By doing this practice, it will enable the administration and organization to maneuver new strategy for the purpose of continuous performance improvement for maintenance management in the public hospital buildings. Overall, this study finds that the three main components, namely, namely individual factors, maintenance aspects and administration and organizational factors has significant impact in determining the overall performance of maintenance management for the public hospital buildings.

References

1. M. Shohet, Facilities, **21**, 5 (2013)
2. O.A. Adenuga, JSDA, **14(1)**, 185 (2012)
3. O.M.A. Mydin, E3S Web of Conferences, **3**, (2014)
4. N. Ahzahar, N.A. Karim, S.H. Hassan, J. Eman, 2nd *International Building Control Conferences*, (2011)
5. N.H.A. Manaf, Int. J. Health Care Qual. Assur., **18(3)**, 204 (2005)
6. M. Ali, W. M. N. Mohammad, JFM, **7(2)**, 142 (2009)
7. A. Weber, R. Thomas, *Key performance indicators - Measuring and managing the maintenance,* IAVARA work smart, 1 (2005)
8. T. wireman, *Developing performance indicators for managing maintenance*, (Industrial Press, Inc., and New York, NY. 2005)
9. C. Cholasuke, R. Bhardwa, J. Antony, JQME, **10(1)**, 5 (2004)
10. A. Irajpour, A.F. Najafabadi, M.A Mahbod, M. Karimi, Math. Probl. Eng., 1 (2014)
11. P. Vaisnys, P. Contri, C. Rieg, M. Bieth, *Monitoring the effectiveness of maintenance programs through the use of performance indicators,* (Summary Report 2016)

Loop Mirror Multi-wavelength Brillouin Fiber Laser Utilizing Semiconductor Optical Amplifier and Fiber Bragg Grating

N. A. Idris[1,2,*], *N. A. M.* Ahmad Hambali[1,2], *M.H.A.* Wahid[1,2], *N. A.* Ariffin[1,2], *M. M.* Shahimin[1,2], and *M. A. M.* Azidin [1,2]

[1]School of Microelectronic Engineering, Universiti Malaysia Perlis (UniMAP), Pauh Putra, 02600 Arau, Perlis, Malaysia
[2]Semiconductor Photonics & Integrated Lightwave Systems (SPILS), School of Microelectronic Engineering, Universiti Malaysia Perlis, Pauh Putra, 02600 Arau, Perlis, Malaysia

Abstract. In this paper, the development of loop mirror multi-wavelength Brillouin fiber laser utilizing semiconductor optical amplifier and fiber Bragg grating is successfully demonstrated. A multi-wavelength BFL structure employs a single mode fiber that acts as the gain medium, while the fiber Bragg grating act to reflect the signals and semiconductor optical amplifier is used to amplify the signal are employed to produce multiple channels. The implementation of difference parameter for semiconductor optical amplifier driven current, Brillouin pump power, Brillouin pump wavelength and length of single mode fiber, plays an important role in producing the multiple channels. The highest number of channels which is 15 achieved once 9 km of SMF length, semiconductor optical amplifier current at 800mA and low Brillouin pump power at -12 dBm are utilized in the laser structure. Furthermore, the implementation of semiconductor optical amplifier and reflectivity of fiber Bragg grating in Brillouin fiber laser structure gives a better performance in producing the multiple channels.

1 Introduction

Throughout the years, multi-wavelength fiber laser has received much attention due to their numerous works with potential applications in optical fiber sensors [1], wavelength division multiplexing (WDM) systems [2], spectroscopy [3] and optical component testing [4].Various amplification methods have been proposed previously in producing the multi-wavelength output, such as semiconductor optical amplifier (SOA), Erbium-doped fiber amplifier (EDFA), and Raman amplifier [4].Typically, erbium-doped fiber amplifier (EDFA) is a favourable candidate in generating multiple channels which is has the capability of providing large gain, high saturation power and also low noise figure [5]. However, due to homogenous broadening mechanism from the erbium doped fiber which is

* Corresponding author: azuramalini@unimap.edu.my

caused by the competition between the laser modes and leads to fluctuating of output channels from laser system [5].

Thus, the low number of channels and unstable channels are produced. To overcome this problem, the inhomogeneous broadening characteristic that offered by SOA is the alternative way in producing a large number of channels [6]. Besides that, it minimizes the competition of the laser mode and has the ability to produce a stable multiple channels. The SOA has remarkable advantages, mainly due to compactness in terms of size, light weight, lower power consumption and also mass producible [6].

The Brillouin fiber laser (BFL) is one type of fiber laser that experienced a nonlinear effect known as Stimulated Brillouin Scattering (SBS). This nonlinear process involves the interaction between the intense pump light that being injected by the light source and acoustic wave in the optical fiber [7]. Through this process, a unique result is produced where the generated output signal propagate in backward direction and possess a downshifted frequency from the injected signal [7]. Based on the previous works, the generation of Brillouin semiconductor fiber laser associated with SOA and EDFA have been conducted. From this paper, the implementation of the EDFA as signal amplifier in producing multi-wavelength is restricted due to homogenous broadening effect. The improvement has been made by the use of SOA which able to produce a multi-wavelength in any wavelength of the outpurt power. However, only small peak of channels and low OSNR values are produced as output.

Thus, the loop mirror multi-wavelength Brillouin fiber laser (MWBFL) utilizing SOA and reflectivity of FBG has been studied to make an improvement in term or number of channels which is much highest that previous studies in [8]. In this paper, the performance of the BFL utilizing SOA and FBG was demonstrated. Implementation of SOA and FBG are resulted the better performance in producing the multiple channels. The FBG comprises of a special characteristic which is refractive index in the fiber core which leads the light ray to reflect back in generating a multiple channels [9]. Evidently, the highest number of channels is produced by proposed MWBFL structure utilizing SOA and FBG. At -12 dBm of Brillouin pump (BP) power and 1550 nm BP wavelength, 15 channels are produced. The variation of SOA driven current, BP wavelength and length of SMF affect the performance and generation of the channels.

2 Simulation and illustration of operating principle

The simulation design and structure illustration of loop mirror MWBFL utilizing SOA and FBG are shown in Fig 1 and Fig 2, respectively. Based on this structure, the external continuous wave (CW) laser plays the function as a Brillouin pump (BP) signal in the range of -12 dBm to 6 dBm. The SMF functions as Brillouin gain medium in the range of 7 km to 14 km length. In producing the multiple channels, reflection of FBG is the vital element in BFL structure. The FBG's has the characteristic of having central wavelength of 1550 nm. The 3-dBm of bidirectional optical coupler and optical circulator are used to extract the output and to prevent the signal from propagating in random direction [9], respectively. The BP power and BS signal undergoes signal amplification by SOA with driven current of 350 mA to 800 mA. The optical spectrum analyzer (OSA) is external elements that serve the purpose to measurement and observation output results.

DOI: 10.1051/matecconf/20179701057

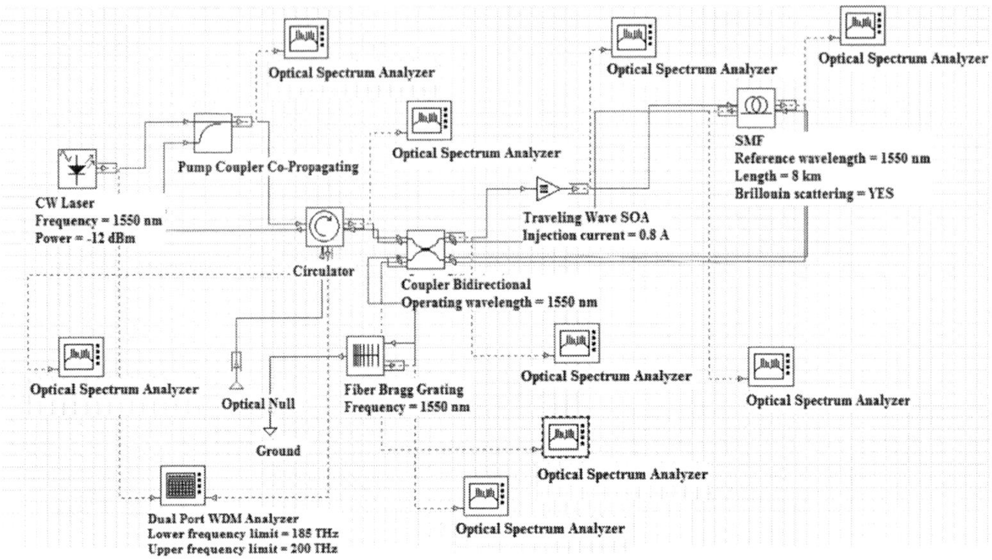

Fig. 1. The simulation design of loop mirror MWBFL utilizing SOA and FBG.

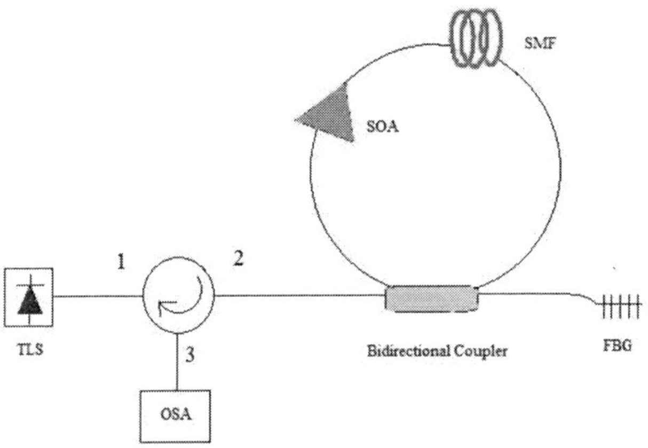

Fig. 2. The structure illustration of loop mirror MWBFL utilizing SOA and FBG.

The operation started by the injection of the BP signal from the external CW laser into the optical circulator where the signal is travelled from port 1 to port 2. Next, the BP signal passed through the bidirectional optical coupler and propagated into SOA. The amplified and stronger BP signals were produced in the SOA as it has a special feature to boost up the intensity of the optical signal [9]. Then, the BP signal is propagated into the SMF and experiences the SBS effect. After the Brillouin threshold condition in SMF is achieved, the BS signal known as channels is produced. The BP signal is then propagated to the FBG and reflected back in the loop mirror structure. Each of the generate BS signal act as a new pump to generate the next order of BS signals generation. After that the BS signal (channels) travelled from the bidirectional optical coupler and optical circulator (port 2 to port 3) and can be observed and measured through the OSA with bandwidth resolution of 0.01 nm.

3 Experimental result and discussion

Figure 3 shows the evolution of the number of channel plotted against the difference values of BP power. The BP power is injected into MWBFL structure in the range of -12 dBm to 6 dBm with 1550 nm of BP wavelength. The implementation of different SOA driven currents produced a different number of channels. At 800 mA and 3 dBm of BP power, 15channels are generated, followed by 700 mA, 650 mA and 350 mA which is produced about 6, 4 and 2 channels, respectively. This condition shows that the increment of the SOA driven current arise the number of channels produced due to sufficient energy. As the BP power increases to 6 dBm, the number of channels shows a slightly decrease due to several factor such as temperature, noise and the gain that shared among the channels to grow their peak power. This factor affected the performance in generating a number of channels.

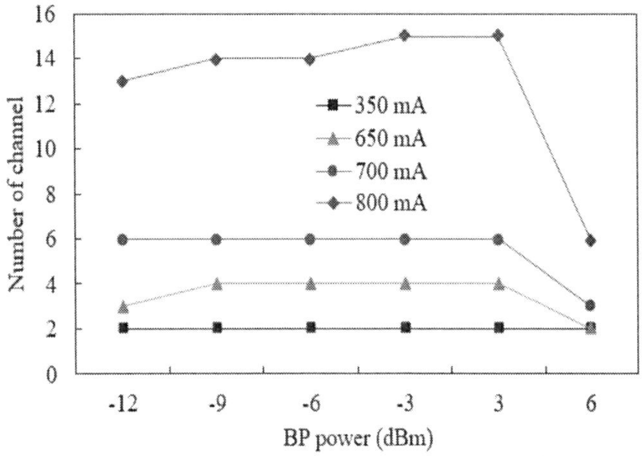

Fig. 3. The number of channels versus BP power at BP wavelength of 1550 nm with SOA driven current of 800 mA.

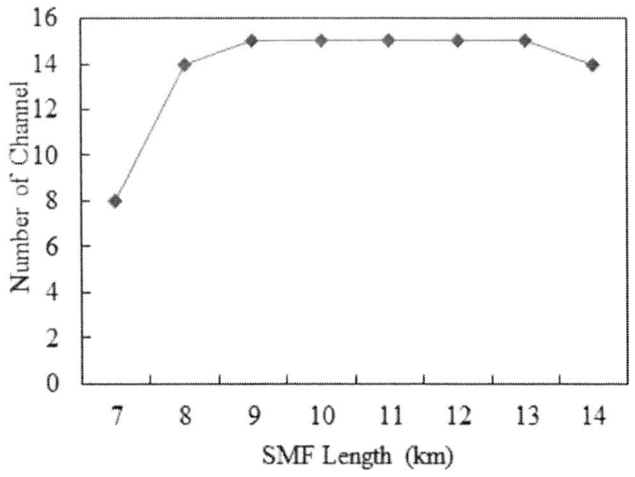

Fig. 4. The number of channels versus SMF length at -12 dBm of injected power and 800 mA of SOA driven current.

Next, the relationship between the numbers of channels versus SMF length and output spectra is plotted in Fig 4. The length of SMF is varied from 7 km to 14 km with the SOA driven current is fixed at 800 mA and BP power is injected at -12 dBm. From plotted graph, the MWBFL experienced an increasing number of channels as the length of the SMF is increases. When, the SMF length is increases to 14 km, it shown the decrement of number of channel. This factor is due to the attenuation in the fiber such as light absorption, scattering and radiation loss. Furthermore, the output spectra for different length of SMF with BP wavelength of 1550 nm and SOA driven current of 800 mA are shown in Figure 5.

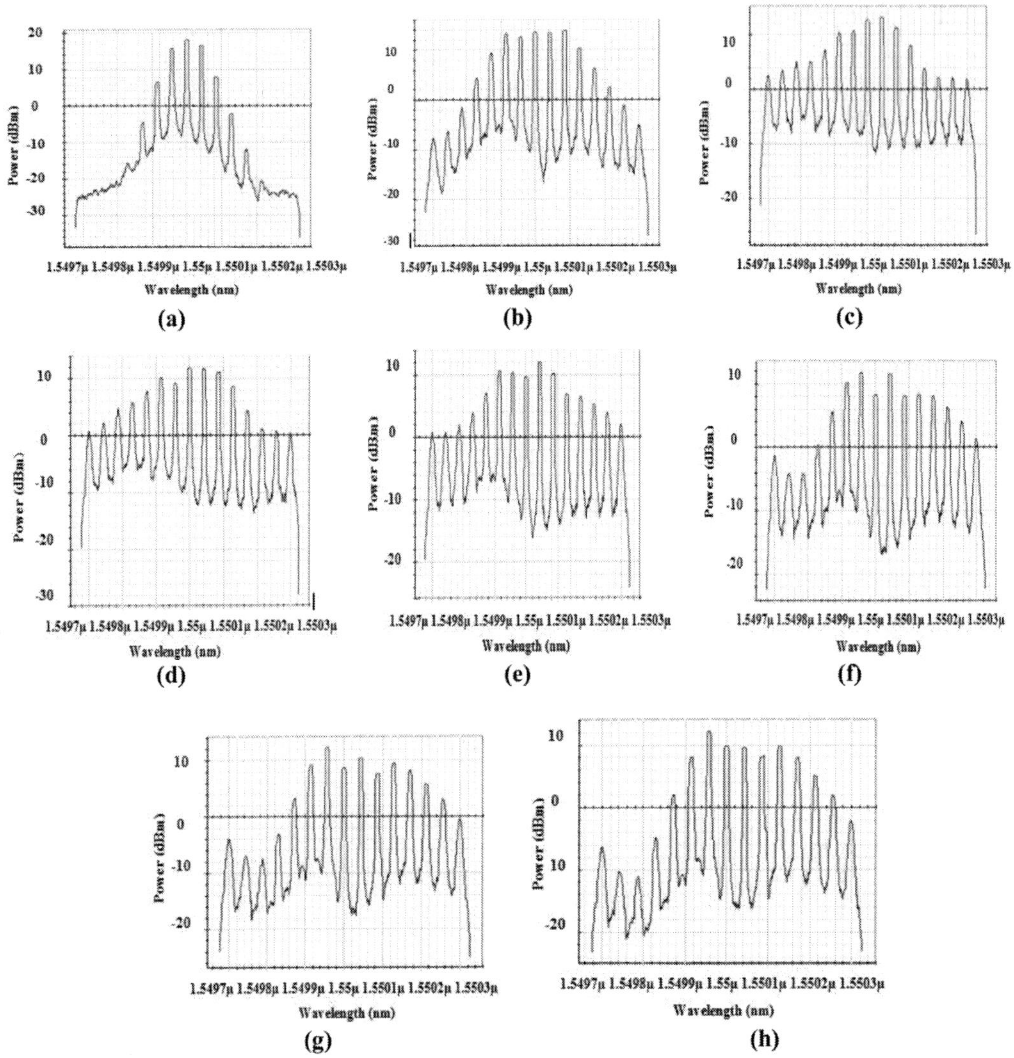

Fig. 5. The output spectra for different length of SMF with BP wavelength of 1550 nm and SOA driven current of 800 mA. (a) 7 km (b) 8 km (c) 9km (d) 10 km (e) 11 km (f) 12 km (g) 13 km and (h) 14 km.

Further analysis of the generation multi-wavelength BFL utilizing SOA and FBG is continued as shown in Fig 6. The BP wavelength is varied from 1530 nm to 1565 nm which enters the window transmission of conventional band (C- band). Meanwhile, the BP power and SOA driven current are fixed at -12 dBm and 800 mA, respectively. Based on Fig 6, 14 number of channels are produced in the range of 1530 nm to 1555 nm BP wavelength. The BP wavelength experienced a decrease in the number of channels at 1560 nm to 1565 nm.

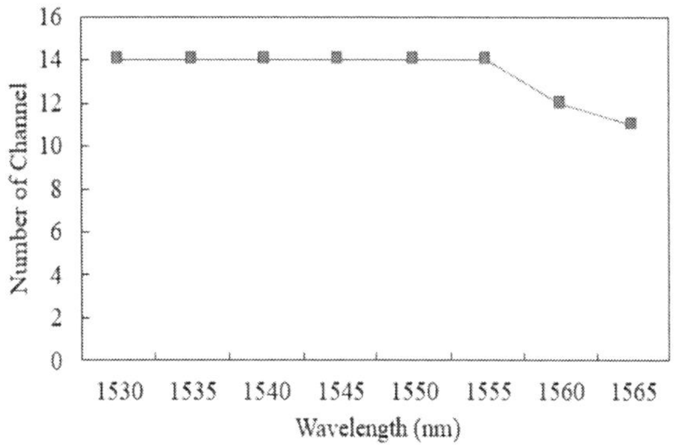

Fig. 6. The number of channels versus wavelength with BP power of -12 dBm, 800 mA of SOA driven current, 8 km of SMF and BP wavelength at 1550 nm.

4 Conclusion

As presented in this paper, the development of the loop mirror MWBFL utilizing SOA and FBG is successfully demonstrated in the C- and region which in the range of 1530 nm to1565 nm. The implementation of SOA and FBG plays the important role in producing multiple channels. At BP power of -12 dBm, SMF length of 9 km and SOA driven current of 800 mA, 15 number of channel are obtained. Besides that, 14 channels are produced due to the range of BP wavelength of 1530 nm until 1555 nm. Furthermore, it can be concluded that the performance of the loop mirror MWBFL is dependent on the characteristics of SOA and FBG to produce multiple channels and has great potential in DWDM system.

This work was fully supported by the Ministry of Higher Education, Malaysia under research grant #FRGS/9003-00532#

References

1. N.S. Shahabuddin, S.W. Harun H.Ahmad, J. Optoelectron. Adv. Mater., **3**, 1 (2009)
2. H. Ahmad, N.A. Hassan, M.H. Jemangin, S.W. Harun, M.Z. Zulkifli, J. Quantum Electron.,7, 602 (2011)
3. K.K. Qureshi, H.Y. Tam, Opt. Laser Technol., **44**, 1646 (2012)
4. S.F. Norizan , M.Z. Zulkifli ,Z.A. Ghani, S.W. Harun, H. Ahmad, Fiber Optics, **21**, 1 (2011)
5. A.H. Sulaiman, S. Shahi, S.W. Harun, H. Ahmad, Fiber Optics, **19**, 5 (2009)
6. H.C. Ooi, A.H. Sulaiman, K. Thambiratnam, M.Z. Zulkifli, S.W. Harun H. Ahmad, Microw. Opt. Technol., **50**, 20 (2008)
7. C. Wei, T.X. Bo, M.Z. Hu, X. Yang, Chin. Phys., **23**, 12 (2014)
8. A. Hayder, R.C. Lawrence, IEEE, **19**, 398 (2006)

Absorbance and Transmittance Capability of Mangoes, Grape and Orange in NIR Region

M. Ariff M. Hussin[1] , *M.Halim* A.Wahid[*,1], *N. Azura* M. Ahmad Hambali[1] , *M. Mukhzeer* Shahimin[2] , *N.Hafika* Hasanuddin[1] , and *M.Azarul* M. Azidin[1]

[1]Semiconductor Photonics & Integrated Lightwave Systems (SPILS), School of Microelectronic Engineering, Universiti Malaysia Perlis, Pauh Putra Main Campus, 02600 Arau Perlis, Malaysia.
[2]Department of Electrical and Electronic Engineering, Faculty of Engineering, National Defence University of Malaysia (UPNM), Kem Sungai Besi, 57000 Kuala Lumpur.

Abstract. This paper focuses on the usage of absorption and transmission properties in the near infra red (NIR) spectrum for quality monitoring of fruits particularly mangoes, grapes and oranges. These three fruits are classified into climacteric and non-climacteric, and analyzed in terms of its absorbance and transmittance capabilities. Experimental works are mainly performed in the study with some preliminary results in fruit ripeness detection. In comparison, climacteric fruits are less tendencies to absorb or transmit light. The non-climacteric fruits display more absorptive power and higher transmission ability. Grape populated as the best in this work due to its thinner and partially transparent skin. Absorbance in grapes starts as early as at 365 nm. Grapes also shows a slight increase in its spectra due to volume of the seeds. However, it does not observed in oranges because the seeds volume is smaller than the total volume of orange. Unriped mangoes give the smallest transmittivity.

1 Introduction

Generally, the NIR (near infrared) works because of specific organic molecules absorbing certain wavelengths of NIR light energies. The absorptions are directly correlated with the concentration of the organic molecules. This relationship must be scaled by a primary technique, otherwise known as the wet chemistry laboratory. Therefore, the NIR is dependent upon the wet chemistry methods to attain the linear relationship between the molecular absorptions and the actual constituent concentrations [1]. One advantage is that NIR can typically penetrate much farther into a sample than mid-infrared radiation. Near-infrared spectroscopy is, therefore, not a particularly sensitive technique, but it can be very useful in probing bulk material with little or no sample preparation [2].

The molecular overtone and combination bands seen in the NIR are normally very broad, leading to complex spectra and can be difficult to assign specific features to specific chemical components. Multiple variables calibration techniques are often employed to

* Corresponding author: mhalim@unimap.edu.my

extract the desired chemical information. Careful development of a set of calibration samples and application of multivariate calibration techniques is essential for near-infrared analytical methods [3].

Two major classifications of fruit ripeness, climacteric and non-climacteric, have been utilized to distinguish fruits on the basis of respiration and ethylene biosynthesis rates. Climacteric fruit, such as tomato, mango, avocado, banana, peaches, plums, and apples, are distinguished from non-climacteric fruits, such as strawberry, grape, and orange, by their increased respiration and ethylene biosynthesis rates during ripening [4]. Ethylene is not required for non-climacteric fruits ripening. In contrast, ethylene is necessary for the coordination and completion of ripening in climacteric fruits through analysis of inhibitors of ethylene biosynthesis and perception, and also in transgenic plants blocked in ethylene biosynthesis. It is also important to note that, though not nearly as well characterized in this regard, plant hormones, in addition to ethylene, are likely to influence climacteric fruits ripening [5]. Firmness decay, chlorophyll breakdown and carotenoid accumulation, controlled by ethylene, are major ripening events in mangoes. Pigment content and tissue structure affect the optical properties of the mesocarp, which can be measured nondestructively in the intact fruit by reflectance spectroscopy [6]. The source of light for this spectroscopy can be originated from the NIR.

Ideally mango is harvested at the preclimacteric and mature-green stage. The ripening stage is achieved in the postharvest time or phase. On the other hand, harvested fruits in a ripe condition are better quality as compare to preharvested. However, it lacks of lifetime which is important for long journey deliveries. Thus, identifying maturity level is critical in the relevant industries. NIR spectroscopy is one of the optical method to predict maturity state of fruits by observing its spectra [7]. This work is dedicated to evaluate the ripeness or maturity stage of mango, grape and orange by means of the NIR spectroscopy by looking at their spectrum.

2 Methodology

Observing light spectra that passes through fruits is the fundamental technique in this work. A spectrophotometer which is LAMBDA 950 that can emit ultra violet (UV), visible light (Vis) and near infra-red (NIR) is employed to measure the amount of light that a sample allows. Principally, a beam of light will go through a sample and the measurement is performed by collecting the intensity of light reaching the detector inside the LAMBDA 950. A simple technique yet, provides essential outcomes. The light beam consists of a stream of photons impinges on fruits analyte and being absorb partially. The analyte is firstly prepared from the fruits juice and placed in a transparent container that fits into the measurement slot of the spectrophotometer. The number of photons will reduce proportionally and thus making the intensity of light beam reduced as well. As a result, a spectra is produced.

Practically, there are three major steps that involved in the measurement. First, a blank container is used to measure the intensity of light (I_0) that passes through it. This light intensity determines the number of photons passes through per second. The container is filled with water since the actual sample is a liquid form which is obtained from blending process. Again, measurement is repeated and kept for baseline calibration. The filled container scatters light in a regular pattern that later will be superposed with the spectra of the fruits.

Secondly, measurement for the intensity of light (I) that passes through the sample (fruits) solution is conducted. In practical, the instrument does not measure the intensity of light but the power itself. The power is the energy generated per second, which is the product of the intensity (photons) and the energy per photon. In summary, energy generated per second is equal to the number of photon per second. In the last step, two quantities which are transmittance (T) and absorbance (A) are recorded from the output spectra of the spectrophotometer. Theoretically, the following equations can predict the experimental observation.

$$T = I/I_o \quad (1)$$

$$A= - Log_{10} T \quad (2)$$

The fraction of light in the original beam that passes through the sample and reaches the detector is being defined as transmittance. What is remain from the light, 1- T, is considered as the fraction of light being absorbed by the sample. In a lot of applications, it is common to relate the amount of absorbed light with the concentration of the absorbing molecules. Absorption spectra is more useful rather than the transmittance. If no light is absorbed, the absorbance will become zero, which means it has 100% transmittance. Each unit in absorbance corresponds with an order of magnitude in the fraction of transmitted light. For instance, if A is equal to 1, 10% of the light is being transmitted (T = 0.10) and the remaining which is 90% are absorbed by the sample.

3 Result and discussion

Three tested fruits in this work are mango (Chok Anam), red grape (Russian Seedless), and orange (Valencia). These fruits have difference visual, shape, color and also their characteristics. The fruits are taken from the local groceries and the percentage of its maturity is not properly defined by the supplier. To determine its maturity stage, spectra from LAMBDA 950 is solely analyzed and verified. A white light source sweeps from 200 nm to 1000 nm is used in the study. The absorbance and transmittance patterns can be seen in Fig. 1.

Fig.1 shows the absorbance spectrum for riped mango (purple) and unriped (blue), grape (black), and orange (red). These spectrum depict a huge difference between the three measured fruits. This is because both orange and grape are non-climacteric fruits while mango is a climacteric type of fruits. The highest value of absorbance goes to orange with the value of 4.13. The value is taken from the time it started to decrease and become constant which is in this case at 367 nm. The reference point is fixed at the starting point of 10. In grapes, the absorbance started to occur at 365 nm with the value of 3.54. The lowest value can be seen while doing the absorbance test for both mangoes. Aimed at unripe mango, the value of absorbance is around 0.04 while for ripe mango is 0.18 at 870 nm.

It can be summarized that orange starts to increase steadily from 500 nm and after it reaches about 850 nm, it starts to decrease and eventually becomes constant. In unripe fruit, the transmittance is about 0.06 percent. It means that most of the energy from the visible light source is being absorbed by the fruit as shown in Fig. 2. The highest value of absorbance is 3 at 850 nm. The unripe fruit has the least amount of sugar and water content, and less firmness compared to others making it to have more energy loss during transmittance. Orange tends to absorb more than grape because it has more free atoms and lower seeds volume, thus no bump is recorded. There is a slight increment for grapes in the range of 500-550 nm because the grape has seeds in it, thus making it to increase per short duration. Additionally, orange has more sugar and water contents.

 DOI: 10.1051/matecconf/20179701058

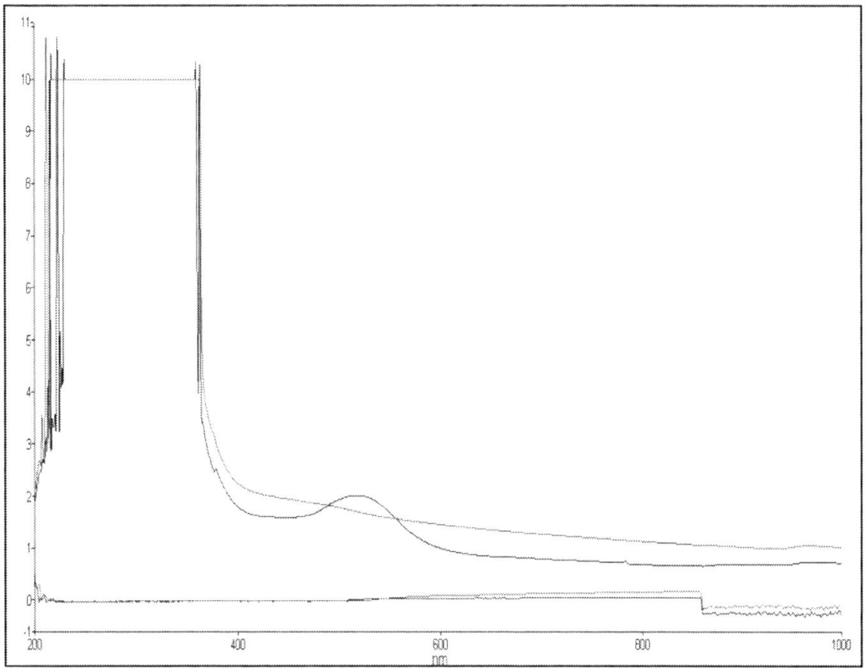

Fig. 1. Absorbance spectrum for grape (black) , mango (purple and blue), and orange (red).

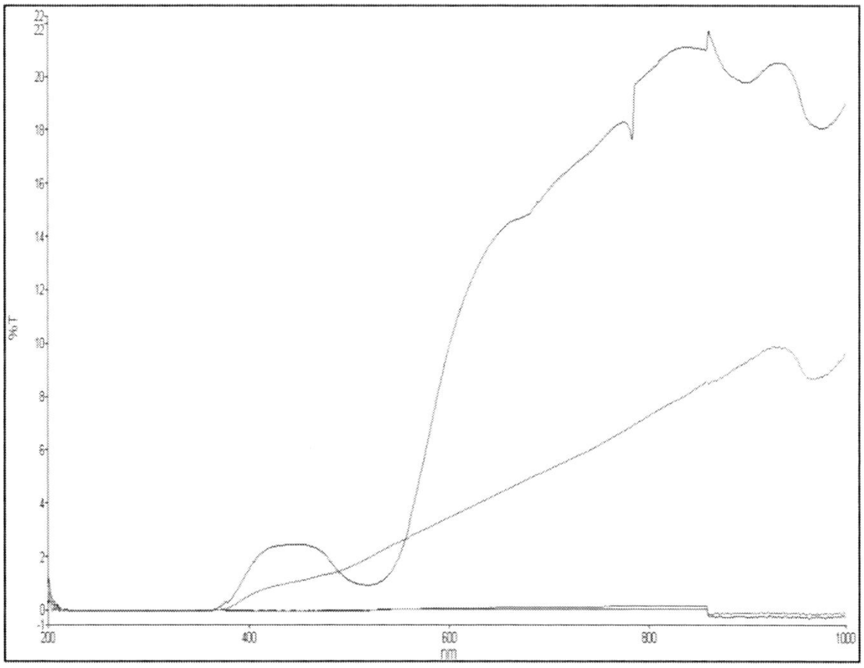

Fig. 2. Transmittance spectrum for grape (black), mango (purple and blue) and orange (red).

The transmittance spectrum sees grape as the highest with 22 percent leads orange by almost 15 percent. This is expected because the outer layer of the fruit itself can also give significant effect to the transmittance. Orange is a lot thicker skin than grape and mangoes. Environmental factor could also be one of the reasons why the value is different even though they are in the same type of non-climacteric fruits. Grapes are usually grown in high

DOI: 10.1051/matecconf/20179701058

ground places, cold weather, and covered with vinyl while oranges are mostly grown in low ground and the tree is bigger and takes more time to grow. Table 1 summarizes and compares the findings in this work. Grape can be concluded as the good absorbance and transmittance fruit.

Table 1. A summary of findings.

	CONDITION			
	Mango Unripe	**Mango Ripe**	**Grape**	**Orange**
Absorbance	0.04	0.18	3.54	4.13
Transmittance	0.0005	0.002	0.22	0.07
Transmittance Percentage (%)	0.05	0.2	22	7

4 Conclusion

This preliminary study concludes a major gap between climacteric and non-climacteric fruits in terms of absorptivity and transmittivity. Climacteric fruits are inclined to have a small value of absorbance and transmittance as compared to non-climacteric fruits because of ethylene contents. Near infra red (NIR) is suitable as a light source in investigating tropical fruits.

The authors would like to thank the School of Microelectronics Engineering, Universiti Malaysia Perlis (UniMAP) especially SPILS for their support in this work. Partial of this research is funded by FRGS 9003-00461 under Ministry of Higher Education, Malaysia.

References

1. B.M. Nicolai, K. Beullens, E. Bobelyn, A. Peirs, W. Saeys, K.I. Theron, J. Lammertyn, Postharvest Biol. Technol., **46**, 99 (2007)
2. B.S. Bhargava, M.E. Sumner, Commun. Soil Sci. Plant Anal., **18**, 581 (1987)
3. L.S. Magwaza, U.L. Opara, H. Nieuwoudt, P.J. Cronje, W. Saeys, B. Nicolai, Bioprocess Technology, **5(2)**, 425 (2012)
4. A.A. Kader, HortScience, **20**, 54 (1985)
5. J.M. Lelievre, A. Latche, B. Jones, M. Bouzayen, J.C. Pech, Physiol. Plant, **101**, 727 (1997)
6. P.E. Zerbini, M. Vanoli, A. Rizzolo, M. Grassi, R.M. De Azevedo Pimentel, L. Spinelli, A. Torricelli, Postharvest Biol. Technol., **101**, 58 (2015)
7. S.N. Jha, K. Narsaiah, P. Jaiswal, R. Bhardwaj, M. Gupta, R. Kumar, RJ. Sharma, Food Eng., **124**, 152 (2014)

Influence of Material Properties on the Fluid-Structure Interaction aspects during Molded Underfill Process

C.Y. Khor[1,*], *Muhammad Ikman* Ishak[1], *M.U.* Rosli[1], *Mohd Riduan* Jamalludin[1], *M.S* Zakaria[1], *A.F.M.* Yamin[2], *M.S.* Abdul Aziz[3], and *M.Z.* Abdullah[4]

[1]Faculty of Engineering Technology, Universiti Malaysia Perlis (UniMAP), Level 1, Block S2, UniCITI Alam Campus, Sungai Chuchuh, 02100, Padang Besar, Perlis, Malaysia.
[2]Faculty of Mechanical Engineering, Universiti Teknologi MARA, 13500 Permatang Pauh, Pulau Pinang.
[3]School of Mechanical Engineering, Universiti Sains Malaysia, Engineering Campus, 14300 Nibong Tebal, Penang, Malaysia.
[4]School of Aerospace Engineering, Universiti Sains Malaysia, Engineering Campus, 14300 Nibong Tebal, Penang, Malaysia.

Abstract. This paper presents the investigation of the effects of epoxy moulding compound' (EMC) viscosity on the FSI aspects during moulded underfill process (MUF). Finite volume (FV) code and finite element (FE) code were connected online through the Mesh-based Parallel Code Coupling Interface (MpCCI) method for fluid and structural analysis. The EMC flow behaviour was modelled by Castro-Macosko model, which was written in C language and incorporated into the FV analysis. Real-time predictions on the flow front, chip deformation and stress concentration were solved by FV- and FE-solver. Increase in EMC viscosity raises the deformation and stress imposed on IC and solder bump, which may induce unintended features on the IC structure. The current simulation is expected to provide the better understandings and clear visualization of FSI in the moulded underfill process.

1 Introduction

Transfer moulding is a well-known technique used for the integrated circuit (IC) encapsulation. The feeding of the epoxy-moulding compound (EMC) is transferred from the transfer pot to the mould cavity, which consists of the IC chip. The interaction between EMC and IC structures (i.e., silicon chip, solder bump, wire bonding, paddle and lead-frame) may induce unintended defects; hence, reduces packaging reliability. Although the IC encapsulation technology is quite mature, it is still not widely applied in moulded underfill (MUF) package. The encapsulation of the miniaturized IC package with a thinned silicon chip and the micro-bump has created the challenges to engineer to sustain the

* Corresponding author: cykhor@unimap.edu.my

DOI: 10.1051/matecconf/20179701059

package reliability. Therefore, the better understandings and clear visualization of the FSI phenomenon are important for the engineer and package designer, to eliminate the package defects and maintain the package reliability.

The FSI phenomenon in IC packaging was widely reported on the IC package that using wire bonding, lead frame and paddle. However, the investigations of the moulded underfill process for an IC package considering FSI aspect with MpCCI coupling method [1-3] are still lacking. Therefore, the FSI simulation using MpCCI coupling method was considered in the study on an IC package with dimensions of 9 mm × 9 mm × 0.7 mm, which consisting of a silicon chip (5 mm × 5 mm × 0.25 mm) and perimeter arrangement of solder bumps. FV-based and FE-based [4, 5] software were used to solve the fluid flow and structural analyses. EMC flow behaviour was modelled by Castro-Macosko model, which was incorporated into FV-solver by user-defined functions (UDFs) using C language. In the simulation, the flow-induced forces were transferred to FE-solver for simultaneous structural analysis and vice versa. In the current study, the EMC flow front profile, chip and solder bump stress were investigated through the FSI simulation. Three types of EMC [6-8] (e.g., Sumikon EME 6300HN, SUMITOMO 6300 HG) with different material properties, namely Case 1, 2 and 3 were considered in the analysis.

2 Viscosity models

The proper selection of viscosity model is crucial for the simulation of encapsulation process. In the simulation, the encapsulant material was assumed as generalized Newtonian fluid (GNF). The Castro-Macosko model [6, 9] was applied to predict the relationship between the viscosity and degree of polymerization and describe the viscosity of the encapsulant material as follows:

$$\eta(T,\dot{\gamma}) = \frac{\eta_0(T)}{1+\left(\frac{\eta_0\dot{\gamma}}{\tau^*}\right)^{1-n}}\left(\frac{\alpha_g}{\alpha_g-\alpha}\right)^{C_1+C_2\alpha}, \quad (1)$$

where

$$\eta_0(T) = B\exp\left(\frac{T_b}{T}\right). \quad (2)$$

B is an exponential-fitted constant, T_b is a temperature fitted-constant, n is the power law index, η_0 is the zero shear viscosity, α is the degree of conversion, α_{gel} is the degree of conversion at gel point, and τ^* is the parameter that describes the transition region between the zero shear rate and power law region of the viscosity curve.

Kamal's equation [6, 9] described the curing effect and integrated with the Castro-Macosko model in the current study. The rate of the chemical conversion of the compound in this model can be predicted as follows:

$$\frac{d\alpha}{dt} = (k_1 + k_2\alpha^{m_1})(1-\alpha)^{m_2}, \quad (3)$$

$$k_1 = A_1\exp\left(-\frac{E_1}{T}\right), \quad (4)$$

and

$$k_2 = A_2\exp\left(-\frac{E_2}{T}\right), \quad (5)$$

where A_1 and A_2 are Arrhenius pre-exponential factors, E_1 and E_2 are activation energies, m and n are reaction orders, and T is the absolute temperature.

3 Results and discussion

3.1 Deformation of silicon chip

Figure 1 shows the simulation results of the FV and FE analyses. The selected flow front advancement at two filling times are presented together with the chip deformation during IC encapsulation process. At filling time 1.4 s, the EMC flow front covered nearly 30% in the top space (upper stream) of the chip. Faster flow front at upper stream has induced the unstable forces acting on the silicon chip. Thus, the silicon chip deformed downwardly as depicted in Figure 1 (Column of FE-solver). Conversely, upward deformation was found at the chip edge that closer to inlet gate. This situation was attributed from the continuous inlet flow. At 3.2s, the filling of the mould cavity is nearly completed. The faster flow front totally covered the chip and resulted in the air trap beneath the chip [10]. The variations of the flow-induced forces on the chip caused the deformation around the middle region, which has no solder bump supported. Consequently, the chip edges deformed upwardly corresponded to the structural reactions.

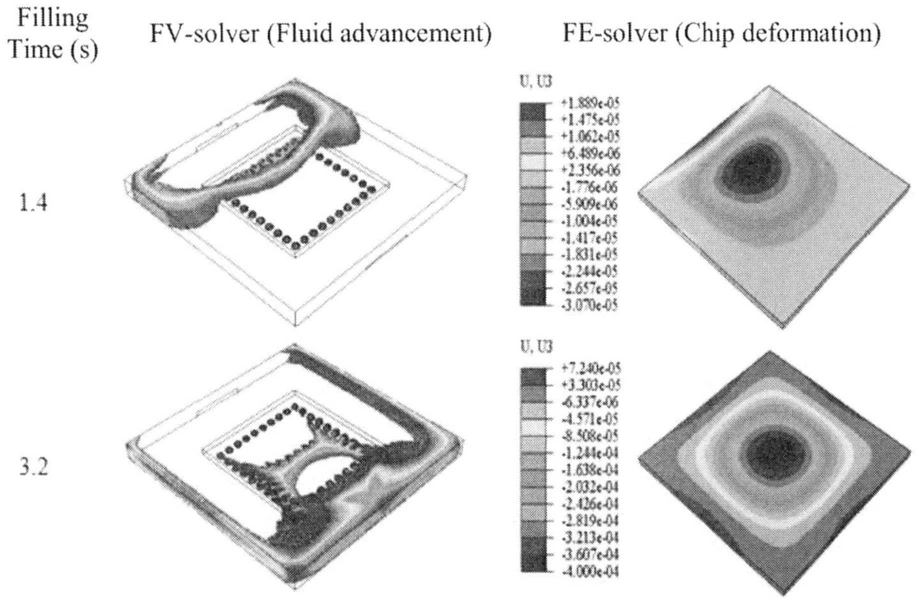

Fig. 1. FSI simulation results of FV-solver (Flow front advancement) and FE-solver (deformation, U in unit: mm)

3.2 Von Mises stress analysis

Aforementioned, the FSI has caused the deformation of the silicon chip. The unintended deformation may reduce the package reliability in the subsequent packaging process. During the encapsulation process, the silicon chip subjected to the stress due to the flow-induced forces. Figure 2 shows the stress distribution profile of the silicon chip at filling times, 1.4 s and 3.2 s. The stress distribution was corresponded to the EMC flow front and

 DOI: 10.1051/matecconf/20179701059

varied with the filling time. Figure 3 depicts the stress concentration on a solder bump. The predicted solder bump experienced higher stress concentration compared with the silicon chip during the process. High stress concentration may cause structure cracking or fracture, if exceeding the yield stress of the solder bump material. Therefore, the proper control, package design and material selection are important to maintain the package reliability. The moulded underfill process was investigated by using different EMC material properties. As illustrated in Figure 4, the IC structure (silicon chip and solder bump) of Case 3 experienced the highest stress. This situation was mainly contributed from the viscosity of the EMC used during the moulded underfill process. Success MUF process at low viscosity regime allows the EMC material flows easily and resulting low stress subjected to IC structure. The variations of the viscosity were found highest for Case 3 followed by Case 2 and Case 1. Case 3 has the highest viscosity at the shear rate regime >300 1/s compared to cases 1 and 2. Thus, the IC structure of Case 3 experienced highest stresses.

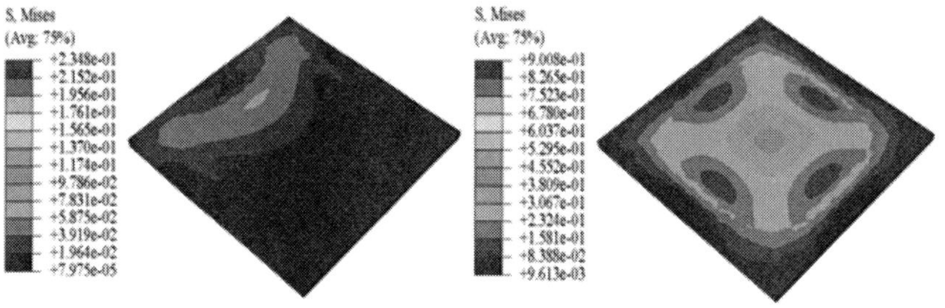

(a) 1.4s (b) 3.2s

Fig. 2. Stress concentration (Unit: MPa) at silicon chip.

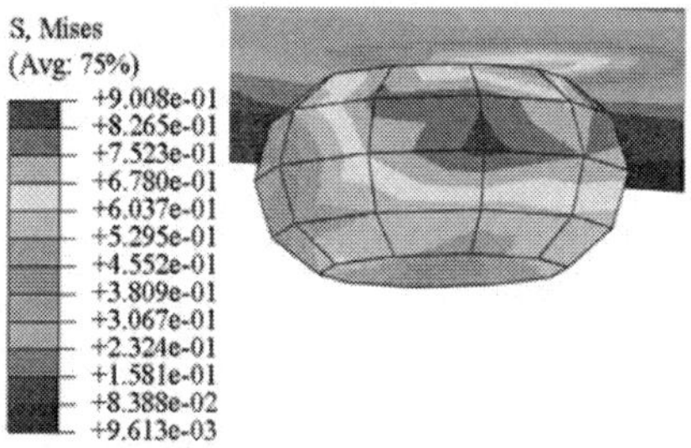

Fig. 3. Stress concentration at the solder bump.

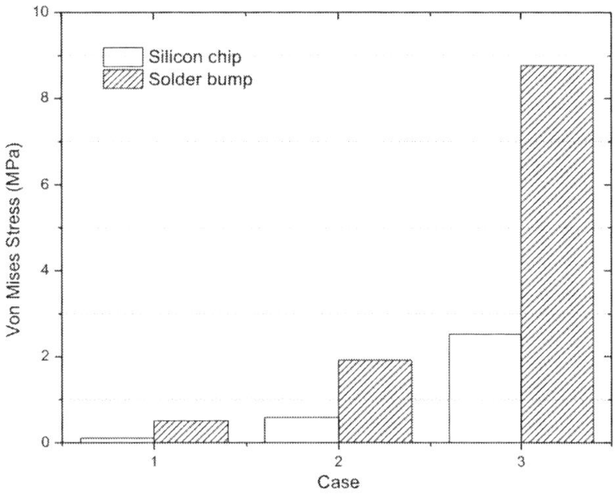

Fig. 4. Stress of silicon chip and solder bump for EMC 1, 2 and 3.

4 Conclusion

The investigation of the effects of material properties by considering the FSI aspects in MUF process has been carried out using finite volume and finite element codes, which coupled by MpCCI software. The FSI phenomenon and stress concentration during the MUF process were predicted through the simulation. The FSI methodology has been well validated. The increase in EMC viscosity raises the stress imposed on the IC structure and void in the package. The highest flow viscosity of Case 3 yield highest stress and void in the package. Extremely high stress and void will cause the deduction of package reliability in the subsequent manufacturing process. The maximum stress of solder bump and silicon chip for Case 3 were 8.8 MPa and 2.5 MPa. The proposed MUF simulation yields the clear visualization of FSI phenomenon and provides better understandings of the process to the engineers and package designers. The extension of this work will focus on the parametric investigations and development of analytical models for predicting the deformation and stress of the IC structure.

References

1. C.Y. Khor, M.Z. Abdullah, H.J. Tony Tan, W.C. Leong, D. Ramdan, Microelectron. Reliab., **52**, 241 (2012)
2. C.Y. Khor, M.Z. Abdullah, F. Che Ani, Microelectron. Eng., **88**, 3182 (2011)
3. C.Y. Khor, M.Z. Abdullah, W.C. Leong. IEEE Trans. Compon. Packag. Manuf. Technol., **2**, 604 (2012)
4. M.S. Abdul Aziz, M.Z. Abdullah, C.Y. Khor, A.M. Iqbal, F. Che Ani, Int. J. Heat Mass Transfer, **72**, 400 (2014)
5. C.S. Lau, M.Z. Abdullah, C.Y. Khor, Microelectron. Int., **30**, 151 (2013)
6. L. Nguyen, C. Quentin, W. Lee, S. Bayyuk, S.A. Bidstrup Allen, S.T. Wang, J. Electron. Packaging, **122**, 138 (2000)
7. J.H. Wu, A.A.O. Tay, IEEE Trans. Comp. Packag. Manuf. Technol. Part B: Adv. Packag., **21**, 65 (1998)

DOI: 10.1051/matecconf/20179701059

8. C.C. Pei and S.J. Hwang, Transaction of the ASME Journal of Electronic Packaging, **127**, 335 (2005)
9. D. Ramdan, M.Z. Abdullah, C.Y. Khor, W.C. Leong, W.K. Loh, C.K. Ooi, R.C. Ooi, IEEE Trans. Compon. Packag. Manuf. Technol., **2**, 1786 (2012)
10. C.Y. Khor, M.Z. Abdullah, M.F.M.A. Majid, M.A. Mujeebu, M.K. Abdullah, Z.M. Ariff, Heat Mass Transfer., **46**, 1315 (2010)

 DOI: 10.1051/matecconf/20179701060

Effect of Moisture Sorption Behavior on Color Quality of Cassia Alata Leaves

Norawanis Abdul Razak[1,*], *Abdul Razak Shaari*[1], *Nur Ain Mohd Shariff*[1], and *Lee Yit Leng*[1]

[1]Faculty of Engineering Technology, Universiti Malaysia Perlis (UniMAP), Campus of UniCITI Alam Sungai Chuchuh, Padang Besar, 02100 Perlis, Malaysia.

Abstract. The *Cassia alata* leaf is widely used for the medical uses as it is beneficial for treatment of skin diseases such as purities, eczema and allergy. The moisture sorption behavior included desorption and adsorption of *C. alata* leaves were investigated using standard static gravimetric method. The sorption isotherm was obtained at ambient temperature 24 °C and with the equilibrium relative humidity (ERH) range from 11.3 to 84.3 %. The equilibrium moisture content (EMC) of *C. alata* increased from 8.67 to 28.50 g water/dry matter with increasing ERH during desorption. It is similar to adsorption which the EMC increased from 21.28 to 32.85 g water/dry matter. The color quality of *C. alata* leaves were lower with increasing the ERH.

1 Introduction

An increase population's awareness about the negative effect on consuming synthetic medicine urge the community to find the alternative treatment using natural herb product. In order to perform good quality of herb, it is crucial to get appropriate moisture content of the herb during the drying process. High moisture content of the herbs would limit their shelf life and complicates their transport and storage [1]. Therefore, to reduce the moisture content, it is necessary to submit these herbs to a dehydration process. the moisture sorption isotherm are divided into two (adsorption) by addition of water to dry sample and (desorption) by removing the water from wet sample [2]. The study of moisture sorption is important to preserve the function of bioactive compounds constituent and maintain longer lifespan consumer.

In addition to moisture content, the color of the dehydrated herbs is another important quality factor, which is affected by the operation conditions. Food color usually is the first quality parameter evaluated by consumers and is critical in the acceptance of the products. In addition, color measurement is an objective parameter for the evaluation of quality changes during herb processing, and storage [3].

Cassia alata herb is traditionally used to treat a number of diseases such as constipation, skin problems, stomach pain, respiratory and digestive ailments [4]. The moisture content controlling has great influence on the quality of dehydrated *C. alata*. The removal of

* Corresponding author: norawanis@unimap.edu.my

moisture prevents the growth of microorganisms which deteriorate quality of herbs while while too low moisture content can badly influence the color, texture and even destroy nutrition contents of herbs. The colour measurement of food is commonly used in an alternate way to determine the quality variations as they are faster than a complete physicochemical examination. However, there is lack of information on the effects of moisture sorption behavior on the color changes of *C. alata* leaves. Thus, the aims of this study were to determine the moisture sorption isotherms and color changes of *C. alata* leaves at the various relative humidity range from 11.30 until 84.34 %. The hypothesis of this study is that the color quality of *C. alata* leaves decrease with the increasing equilibrium moisture content. Therefore, it has been expected to determine a suitable moisture content range of the leaves for preserving the color quality of *C. alata* during industrial scale production and storage.

2 Materials and methods

This section provided the preparation of material used and the methods that were utilized to analyze the samples.

2.1 Preparation of materials

The fresh Cassia alata leaves were collected randomly from Institute of Sustainable Agrotechnology (INSAT), which located at Sungai Chuchuh, Perlis. The samples were dried immediately using oven for 24 hours at 40 °C.

2.2 Development of moisture sorption isotherm

The static gravimetric method with different saturated salt solution (ranged from 11.30 until 84.34 %) was used to determine the equilibrium moisture content (EMC) of the samples. The saturated salt solutions were placed in the desiccators to give different range of relative humidity exposed to the herbs. The desiccators were placed at room temperature (24 °C ± 0.1 °C). About 2.0 to 2.5 gram of the fresh and dried samples for desorption and adsorption experiments, respectively, were prepared triplicate in each desiccator. The weight of the sample was recorded every 2 days by using analytical balance. The equilibrium of moisture content was considered to have reached when the weight difference for three consecutive weights was less than 1 mg.

Table 1. Saturated salt solutions used to establish the equilibrium moisture content (ERH) at different levels from 11.30 to 84.34 % (modified from Greenspan, 1977)

Saturated Salt Solutions	Relative Humidity at ambient temperature (24 °C)
Lithium chloride	11.30
Magnesium chloride	33.78
Sodium bromide	57.57
Sodium chloride	75.29
Potassium chloride	84.34

2.3 Determination of equilibrium moisture content

The equilibrium moisture content was measured using Equation 1.

$$EMC\ d.b = \frac{weight_{eq} - weight_{dm}}{weight_{dm}} \quad (1)$$

Where,

Weight $_{eq}$ is the final weight of the samples in g
Weight $_{dm}$ is the dry matter weight of the samples in g.

2.4 Determination of color quality

The determination of the color quality was based on lightness (L*), greenish to reddish (a*), and blue to yellowish (b*) color. The color quality of the sample were determined using colorimeter. The readings were recorded in triplicate.

3 Results and discussion

In this section, the pattern of the moisture sorption isotherms and the color changes of the materials were discussed in detail.

3.1 Moisture sorption isotherms of cassia alata leaves

The moisture sorption isotherms of the *Cassia alata* leaves were determined at ambient temperature (24 °C) and different relative humidity range from 11.30 to 84.34 %. Desorption and adsorption equilibrium moisture content (EMC) were shown in Table 2. In desorption experiment, the fresh *C. alata* leaves held lowest water content (8.67 g water/dry matter) at the lowest equilibrium relative humidity (ERH) (11.30 %). Whereas, the *C. alata* leaves obtained highest EMC (28.50 g water/dry matter) at the highest ERH (84.34 %). This finding was similar with work of Bejar et. al. [5], which also found that the dried leaves hold a large amount of water at high relative humidity.

For the adsorption, the EMC increase from 21.28 to 32.85 g water/dry matter as the ERH increased. This could be attributed to the adsorption water into the leaves as increasing ERH. The water moved from the high concentration of water to the lower concentration of water thus increasing in weight for adsorption slowly. This finding agreed with work of Abdul Razak et. al. [6], who also found the dried leaves with lower moisture content results in increment of moisture content after prolong storage.

3.2 Effect of equilibrium relative humidity (ERH) on the color quality of cassia alata leaves

The determination of the color quality is based on lightness (L*), greenish to reddish (a*), and blue to yellowish (b*) color. For L*, the scale is given as 0 which is black and 100 as white. For a*, negative value indicates greenness and positive value indicates reddish. Meanwhile for b*, negative value means blue and positive indicates that the yellowish. The effect of equilibrium relative humidity (ERH) on the color quality of *C. alata* leaves in desorption experiment is as shown in Table 3. The results showed that the L* values of the

DOI: 10.1051/matecconf/20179701060

leaves decreased as the ERH increased. The highest of L* was 54.09 at the lowest relative humidity (RH) while the lowest was 44.65 at 84.34 %RH.

Table 2. EMC values for desorption and adsorption of *Cassia alata* leaves at different ERH.

Desorption		Adsorption	
ERH (%)	**EMC (d.b)**	**ERH (%)**	**EMC (d.b)**
11.30	8.67 ± 0.28	11.30	21.28 ± 1.20
32.78	11.03 ± 0.36	32.78	22.26 ± 1.51
57.57	16.05 ± 0.81	57.57	26.57 ± 0.17
75.29	24.45 ± 0.45	75.29	30.12 ± 1.13
84.34	28.50 ± 0.19	84.34	32.85 ± 0.51

*Mean ± standard deviation. The samples were prepared in triplicate.

Table 3 shows an increase in a* values of the leaves with increasing the RH (-6.44, -6.29, -5.55, -2.25, 2.50 respectively). The green colors of leaves decreased as they were undergone desorption process. The change of color was probably due to degradation of chlorophyll. The a* value was also observed decreasing from green to reddish as the degradation of the chloropyll occurred [7]. The chloropyll was a main pigment of the leaves in order to retain the green color of the *C. alata* leaves with the humidity treatment. Besides that, Table 3 also showed the results of b* of *C. alata* at different relative humidity. The b* value decreased from 20.49 to 16.34 with increasing relative humidity.

Table 3. Effect of different relative humidity conditions for the color quality in *Cassia alata* leaves for desorption

Relative Humidity (%)	L*	a*	b*
11.30	54.09 ± 0.71	-6.44 ± 0.39	20.49 ± 0.38
32.78	53.16 ± 2.98	-6.29 ± 0.08	21.13 ± 1.48
57.57	51.39 ± 4.63	-5.55 ± 0.60	20.87 ± 0.42
75.29	48.91 ± 2.40	-2.25 ± 1.22	18.84 ± 1.30
84.34	44.64 ± 2.56	2.50 ± 1.25	16.34 ± 3.53

*Mean ± standard deviation. The samples were prepared in triplicate.

Table 4 shows the effect of equilibrium relative humidity (ERH) on the color quality of *C. alata* leaves in adsorption experiment. The highest L* value (52.56) was observed at 57.57 %RH while the lowest was 35.83 at 75.29 %RH. This finding was similar with work of Abdul Razak et. al. [6], which also found that the L* value of dried leaves decreased from 46.65 to 44.68 after 10 to 30 days storage. Nevertheless, an increase in a* values of the leaves was observed with increasing the relative humidity. The highest a* value was 4.01 at 84.34% RH while the lowest (-5.50) at 32.78% RH. The parameter of a* was important as it indicates the greenness of leaves and subsequently the quality of leaves. According to Rudra et. al. [8], the degree of greenness is an important characteristic for customers to determine the final dried products.

DOI: 10.1051/matecconf/20179701060

Table 4: Effect of different relative humidity conditions for color quality on *Cassia alata* leaves for adsorption

Relative Humidity (%)	L*	a*	b*
11.30	36.45 ± 2.75	-3.18 ± 0.19	13.75 ± 0.72
32.78	51.00 ± 3.34	-5.50 ± 0.55	25.64 ± 4.53
57.57	52.26 ± 1.74	-4.00 ± 0.59	25.21 ± 3.24
75.29	35.83 ± 1.22	0.19 ± 0.39	19.90 ± 3.28
84.34	46.01 ± 4.31	4.01 ± 0.20	18.85 ± 5.53

*Mean ± standard deviation. The samples were prepared in triplicate.

During the adsorption process, the leaves turned out brown color. This probably due to entering and removing of the water on the leaves itself. The color changes of *C. alata* could be attributed to the conversion of chlorophyll to pheophytins at high relative humidity and subsequently the leaves quality was affected in the presence of excessive water [7]. It appears that the b* value of *C. alata* was highest at 32.78 %RH and declining with increasing relative humidity However, the lowest b* value (13.75) was found at 11.30 %RH. This suggests that the carotenoid compound in *C. alata* can be sustained even at 32.78 %RH.

4 Summary

The study of moisture sorption isotherm is important to preserve the function of bioactive compounds constituent and maintain longer lifespan consumer. With higher relative humidity, the equilibrium moisture content of *Cassia alata* increased for both adsorption and desorption process. However, the color quality of *C. alata* leaves was lower with increasing relative humidity. This study suggests that the color quality of *C. alata* was the best at the relative humidity of 32.78. Low relative humidity was better in preserve the color quality of *C. alata* leaves.

This study was financially supported by the Ministry of Higher Education Malaysia (Fundamental Research Grant Scheme (FRGS9003-00520); the research facilities of the Institute of Sustainable Agrotechnology (INSAT) and Faculty of Engineering Technology, UniMAP.

References

1 P.B. Silva, C.R. Duarte, M.A.S. Barrozo, **8**, 62 (2015)
2 Z. Yan, M.J. Sousa-Gallagher, F.A.R. Oliveira, J. Food Eng., **86** , 342 (2008)
3 F. Pedreschi, J. Len, D. Mery, P. Moyano, Food Res. Int., **39**, 109 (2006)
4 M.A. Adiana, M.P. Mazura, J. Mol. Struct, **991,** 84 (2011)
5 A.K. Bejar, N.B. Mihoubi, N. Kechaou, Food Chemistry, **132**, 1728 (2012)
6 N. Abdul Razak, A.R. Shaari, A.F. Simbas, S. Abdullah, 5 (2014)
7 S. Gonda, L. Toth, G. Gyemant, M. Braun, T. Emri, G. Vasas, Phytochem. Anal., **23,** 88 (2012)
8 S.G. Rudra, H. Singh, S. Basu, U.S. Shivhare, J. Food Eng., **86**, 379 (2008)

DOI: 10.1051/matecconf/20179701061

Chemical Characterization of Pineapple Leaf Residue Chars generated by Controlled Combustion and by open burning

L.Y. Leng[1,*], *M. H. A.* Husni[2], *A. W.* Samsuri[2], *Norawanis* Abdul Razak[1], *N. M.* Lailina[1], and *Ras Izzati* Ismail[1]

[1]Faculty of Engineering Technology, Universiti Malaysia Perlis (UniMAP), Campus of UniCITI Alam Sungai Chuchuh, Padang Besar, 02100 Perlis, Malaysia.
[2]Department of Land Management, Faculty of Agriculture, Universiti Putra Malaysia, 43400 Serdang Selangor, Malaysia.

Abstract. This study was undertaken to compare the chemical characteristics of pineapple leave residues (PLRs) char generated by controlled combustion and by open burning. The properties of char generated by control combustion (CC) and open burning (OB) varied, due to differences in the production process. The total N, K and surface area of the char generated by CC were significantly higher than the OB. The results indicate that the CC process was better to be applied as a soil amendment than was the OB process.

1 Introduction

In Malaysia, nearly 70% of the pineapple crop is planted on peat [1]. Though relatively small compared to oil palm and rubber, the pineapple industry plays an important role in the country's socio-economic development of Malaysia.

Abundant of leave residues are produced in the field and open burning is commonly practiced to manage the plant residues in pineapple plantation. As a result, large amount of carbon (C), nitrogen (N) and greenhouse gaseous are emitted to the atmosphere from burning. Alternative way to manage the residues is converting the residues to char and applied as a soil amendment. The char is composed of large amount of stable aromatic carbon and sequester C if used as a soil amendment.

Furthermore, previous studies found out biochar is a good soil amendment as it can increase soil water holding capacity of sandy soil [2] and increase crop yields [3]. Field burning not only convert the biomass into volatile organic compounds, inorganic gas oxides but also chars. Approximate 4 to 6% of chars is produced during the open burning [4,5]. The application of biochar to the soil has been proposed to increase the stable nutrient pool. Control combustion is suggested a better way to retain the nutrient in the char rather than

* Corresponding author: yllee@unimap.edu.my

open burning. However, there is lack of information of the comparisons of chemical properties of pineapple leaf residue (PLR) chars from control combustion and open burning processes. This study was carried out to determine the chemical characteristics of PLR, chars generated from control combustion and field burning, to compare the characteristics of chars generated by control combustion and open burning.

2 Materials and methods

There were two types of chars being generated under different combustion conditions. The chemical characteristics of the chars varied under different processing and thus were analysed for pH, electrical conductivity (EC), macronutrient contents, cation exchange capacity (CEC) and BET surface area.

2.1 Open burning-generated char

The study site was the Peninsula Pineapple Plantation, Johor, Malaysia. Pineapple leaf samples were collected and air-dried for three months prior to char production under control combustion (CC). The chars were collected after the open burning (OB) conducted in a 6 ha area of the plantation [4]. The char samples from the open burning process were oven-dried, ground, sieved to pass a 2 mm sieve and kept in air-tight plastic containers prior to chemical analysis.

2.2 Control combustion-generated char

The pineapple leave residues (PLRs) were converted into chars in the laboratory by taking 90 g of air-dried leaf samples and combusted them for 3 hours at 340°C [6] in a Carbolite ELF 11/23 type furnace. The char samples from CC were processed similarly to the char samples from OB prior to chemical analysis.

2.3 pH and electrical conductivity (EC) of chars

The pH and EC content of leave and char samples were analyzed using a pH meter and a conductivity meter (Radiometer Analytical, ION check 30) by mixing a 1:10 (v/v) ratio of sample to water.

2.4 Total C and N analysis

Total nitrogen (N) was determined using a micro-Kjeldahl method [7] while Total C (%) was determined by a dry combustion method using a CT analyzer (LECO CR-412, Michigan, USA).

2.5 Total macronutrient analysis

Dry ashing was adopted for the determination of phosphorus (P), potassium (K), calcium (Ca) and magnesium (Mg).

DOI: 10.1051/matecconf/20179701061

2.6 Available P of Char analysis

Available P in the char samples was determined by Olsen method [8]. One g of char generated by CC and OB processes was weighed into the plastic vial. Then, 20 mL of the extracting solution was added and shaked for 30 minutes. Lastly, the extract was filtered using Whatman no.2 filter paper and the P in solution was determined using Auto Analyzer (AA).

2.7 Available K, Ca, Mg and cation exchange capacity (CEC) of char determination

Prior to analysis, char samples were washed 3 times with deionized water to remove soluble salts. Ammonium acetate shaking methods at pH 7 [9] was used to determine CEC of char samples. About 10 g of samples was weighed into centrifuge bottle, and added with 1 M ammonium acetate solution. Afterwards, it was shaken for 30 minutes at 180 rpm by using reciprocal shaker. After shaking, the bottle was centrifuged at 2500 rpm for 10 minutes. The supernatant was later filtered using Whatman no.2 filter paper, K, Ca and Mg in the supernatant was determined using AA and atomic absorption spectrophotometer (AAS). To remove free ammonium ion from the char, 100 mL of ethanol was added into the centrifuge bottle. Afterwards, it was shaken for 30 minutes, centrifuged, and the supernatant was discarded. This step was repeated for 2 more times. A 100 mL of 0.1 N potassium sulfate was added in the centrifuge bottle and shake for 30 minutes, centrifuged, then supernatant was filtered and send to AA for CEC determination.

2.8 BET surface area

The multipoint Brunauer, Emmett and Teller (BET) was used to measure surface area of char samples. Nitrogen adsorption technique at 77K with Automated Gas Sorption Analyzer (Autosorb-1, Quantachrome instruments version 2.01) was used for this purpose. The specific surface area was determined by application of BET analysis software available with the instrument.

2.9 Statistical analysis

An independent t-test was used to detect significant differences between chemical value of the char produced from the CC treatment and from the OB process. Statistical Analysis System (SAS) software version 9.1 was used for the statistical analysis.

3 Results and discussion

The chemical characteristic of the chars generated under CC and OB conditions were analysed and the two different processes will be compared for better quality in term of the nutrient content.

3.1 Macronutrient, pH and EC of char

The properties of char generated by CC and OB varied, due to differences in the production process. The properties of the char are shown in Table 1.0. Both CC and OB combustion processes show high pH (>9) and salinity (>7 mS/cm) in the char, and the high pH and salinity indicate the chars are good soil liming materials. Mineral ash produced during high heating rate in OB process resulted a significant higher pH in the char generated by OB. A significantly higher level of EC found for the CC process suggests that a higher content of metal salts was retained in the laboratory-produced char, probably due to slower heating rates [4].

Total C content under CC conditions (53.30%) was higher than OB conditions (42.19%). A controlled temperature and slow heating rates for the CC production condensed the C of PLR, thereby increasing the total C from the CC process. In contrast, high and uncontrolled temperatures during the OB in the field, would have produced a high level of mineral ash and resulted in more C weight loss [4]. This finding agrees with work of [10], who also found that controlled combustion could further increase the C content of the char that was produced.

The total N under CC conditions (2.60%) was significantly higher than under OB conditions (1.77%) (Table 1). This difference could be attributed to the volatilization of N compounds during open burning at high temperatures. However, N was not volatilized but underwent condensation in the CC process during the controlled combustion.

Table 1. pH, electrical conductivity (EC), total C, total N and C/N ratio of chars produced by controlled combustion (CC) and open burning (OB)

Parameter	CC	OB
pH	9.45^{b}	9.65^{a}
EC (mS/cm)	$12.27^{a} \pm 0.11$	$7.87^{b} \pm 0.87$
Total C (%)	$53.30^{a} \pm 0.70$	$42.19^{b} \pm 0.37$
Total N (%)	$2.6^{a} \pm 0.10$	$1.77^{b} \pm 0.03$

3.2 Total and available P, K, Ca, Mg concentrations and CEC of char

Total P, Ca, Mg and the relevant exchangeable nutrients as well under OB conditions were significantly higher than under CC conditions (Table 2). This difference could be attributed to parts of the PLRs were converted into ash during field burning at high heating rates and temperature and the presence of minerals in the ash residues, thereby increased the concentration of total P, Ca and Mg under CF conditions.

While the K concentration under OB conditions were significantly lower than under CC conditions, suggesting that K ions are highly mobile and vaporized at relatively high temperature during open burning in field [11]. The CEC of the char produced by both CC and OB processes was not significantly different and the values are consistent with the CEC of carbonized rice husk reported in literature [12]. The CEC of the chars tested in this study (15.97– 18.80 cmol(+)/kg) were greater by at least half than the CEC of Ultisols and Oxisols (3–7 cmol(+)/kg) in tropical area [13], except Histosols. The surface area of char made under OB process was found to be lower than under CC process (Table 2.0) suggests that the cell structure of char generated by OB was destroyed by rapid devolatilization rate of the PLRs burned at high heating rates. Higher surface area of char generated by CC enable it to adsorb and retain more nutrients.

Table 2: Available P, exchangeable K, Ca, Mg, CEC and surface area of pineapple leaf char produced by CC and OB processes

Parameter	CC	OB
Total P (%)	0.32^{b}	0.37^{a}
Total K (%)	$8.29^{a} \pm 0.06$	$5.17^{b} \pm 0.21$
Total Ca (%)	0.05^{b}	$0.93^{a} \pm 0.02$
Total Mg (%)	$0.35^{b} \pm 0.01$	$0.75^{a} \pm 0.02$
Exch. K (cmol(+)/kg)	$28.73^{a} \pm 0.84$	$10.84^{b} \pm 0.17$
Exch. Ca (cmol(+)/kg)	$1.71^{b} \pm 0.02$	$11.69^{a} \pm 0.05$
Exch. Mg (cmol(+)/kg)	$7.02^{b} \pm 0.28$	$15.88^{a} \pm 0.39$
CEC (cmol(+)/kg)	$18.8^{a} \pm 0.81$	$15.97^{a} \pm 0.86$
BET surface area (m^2/g)	5.46	2.08

4 Summary

The concentrations of total C, N and macronutrient contents were condensed and increased during conversion of biomass into chars. The chemical properties of PLRs chars varied depending on the combustion conditions. Higher concentrations of total N, K and surface area in the chars generated by CC process suggests the chars under CC process can be applied as a better soil amendment than OB process. However, the effect of PLRs chars on soil nutrient and plant growth should be examined in future research.

References

1. AGRIQUEST. Malaysian Agricultural Directory & Index 12th edition. Agriquest Sdn Bhd. (2009/2010)
2. K. Karhu, T. Mattila, I. Bergstrom, K. Regina, Ecosyst. Environ., **140**, 13 (2011)
3. F.P. Vaccari, S. Baronti, E. Lugato, L. Genesio, S. Castaldi, F. Fornasier, F. Miglietta, Eur. J. Agron., **34**, 231 (2011)
4. L.Y. Leng, M.H.A. Husni, A.W. Samsuri, Bioresource Technol., **102**, 10759 (2011)
5. G.D. Clay, F. Worrall, CJ. Environ. Manage., **92**, 675 (2011)
6. P.A. Jacinthe, C.D. Barton, S. Maharaj, Lal. Rattan, An evaluation of methodologies for assesing geogenic carbon in mine soils of the Eastern United States. In Soil Carbon Sequestration and the greenhouse effect, second edition, ed. R. Lal, R.F. Follett, 347-363. Soil Science Society of America, Inc, USA (2009)
7. J. M. Bremner, D. R. Keeney, Anal. Chim. Acta., **32**, 485 (1965)

8. S. Olsen, C. Cole, F. Wanatable, L. Dean, Estimation of available phosphorus in soils by extraction with sodium bicarbonate. USDA Circular Nr 939, US Government Printing Office, Washington, D.C, United States (1954)
9. J. Muramoto, I. Goto, M. Ninaki, Jpn. J. Soil Sci. Plant. Nutr., **63**, 210 (2009)
10. M.J. Antal, M. Gronli, Ind. Eng. Chem. Res., **42**, 1619 (2003)
11. J.E. Amonette, S. Joseph. Characteristics of biochar: Micro-chemical properties. In Biochar for environmental management: Science and Technology, ed. J. Lehman and S. Joseph, Earthscan, London (2009)
12. S.M. Haefele, C. Knoblauch, M. Gummert, Y. Konboon, S. Koyama. Black carbon (biochar) in rice-based systems: characteristics and opportunities. Springer Science and Business Media B.V. (2009)
13. R.D. Geoffrey, L.T. Emma. Soil Microbiology and Sustainable Crop Production, Springer, New York (2010)

Comparing Various Type of Natural Fibers as Filler in TPU: Mechanical Properties, Morphological and Oil Absorption Behavior

Nor Azwin Ahad[1,*], Nur Fatin Hidayu Mohd Radzi[1], Nor Suhaida Sadli[1] and Tan Jia Seng[1]

[1]School of Materials Engineering, Universiti Malaysia Perlis, Kompleks Pusat Pengajian Jejawi 2, 02600, Jejawi,Perlis, Malaysia.

Abstract. The idea of using natural fibers as filler in various polymers has been extensively studied. Various types of natural fibers and polymers have been identified and it can be varied according to the particular application and the two main composite materials will have advantages and disadvantages of each. However, natural fibers are usually selected as filler because it is readily available and environmentally friendly, inexpensive, non-toxic, biodegradable and still have good characteristics for a variety of uses. In this study, four types of natural fiber have been used which; coconut shell, coconut fiber, corn cob, and pineapple skin, as fillers in thermoplastic polyurethane (TPU). The mixing process conducted through melt mixing techniques. The percentage of TPU and natural fibers are 100/0, 95/5, 90/10 and 85/15. Different type of fiber will affect the mechanical properties of the composites and have been studied through tensile testing. It showed that the result for pineapple fiber at 5% was the highest and can also be related to the characterizations of this composite that have been studied via the SEM morphology. Swelling testing is also having been done to prove the absorbency ability by natural fiber composites in cooking oil and engine oil. Then it concluded that the pineapple fiber absorbed large amount of both oil compared to others.

1 Introduction

Natural fibers play an important role in developing high performance biodegradable composites which are the key material in solving nowadays environmental and 3R (Recycle-Reuse-Reduce) issues, because due to its low cost as well as easy to degrade [1]. The natural fibers are the one of popular resource that uses as reinforcement or filler in polymer matrix [2, 3]. Coconut husk and shell powder, pineapple fiber and corn cob are some of the sources of natural fibers that have potential to use as filler, which have been studied here and blended with TPU. There are several recent studies that combined the natural fiber and TPU, found it to produce some of the unique properties, such as TPU/kenaf by El-Shekeil et al. [4] and Janus & Patrycja [5]. They found that, kenaf fiber

* Corresponding author: norazwin@unimap.edu.my

were compatible with TPU from the view of mechanical properties. El-Shekeil et al. [4] studied TPU/cocoa pod husk and proved that, the tensile properties improved when the filler loading increased. Many researchers have been studied natural fiber in polymer composites and it's mechanical, morphological properties in order to investigate the compatibility of fiber/polymer composites. Due to that, we investigated the effect of four types of natural fiber as filler in TPU, with different content of filler (0%, 5%, 10% and 15%) on tensile strength and the morphology. Also, we determined the swelling percentage of the composite for the optimum content of filler in cooking oil and engine oil, to study the level of oil absorption by the composite of the four types of filler.

2 Experimental

The preparation of TPU composite with natural fibers ass filler in this study were done in polymer laboratory and material characteristics laboratory in School of Materials Enginering in Universiti Malaysia Perlis within 10 months schedule.

2.1 Materials

Polymer-based TPU in the pallet form was supplied from Duplas Marketing Sdn. Bhd. All natural fibers (pineapple, coconut husk, coconut shell and corncob) were collected from market, stall and village. They were cut into small size, washed and dried under direct sunlight for days to evaporate all the moisture. The dried fibers was grinded using grinding machine to produce fibers in the small size and were sieved using sieve shaker to get fine size between the range of 50 – 125 μm.

2.2 Preparation of TPU/ natural fiber composites

The composite were prepared by melt blending technique using laboratory internal mixer (RheoHaake 600p) at the weight percent of 100/0, 95/5, 90/10 and 85/15 for TPU and natural fiber. Blending was carried out at temperature 180°C, with a mixing speed of 60 rpm for 10 min [6]. Then it was pressed at 185°C under 45 MPa of pressure for about 2 min using hot press (Carver Laboratory Press) into thin sheet. The composite sheet then cut into dumbbell shape for tensile testing. The rest of the samples were prepared for swelling test.

2.3 Characterization of TPU/natural fibers composites

Tensile testing is to measure the tensile properties and was done using an Instron Testing Machine. The thin sheets of the sample were about 3 mm according to ASTM 638. Oil absorption test have been done to prove the ability of the composites to absorb cooking oil and engine oil.The samples with measurements of 20 mm x 20 mm x 1 mm were used as indicted by ASTM D570-98. The test have been observed and determined for 4 weeks until equilibrium at room temperature. The samples were weighted for everyday and calculated using the equation of;

$$\text{Oil absorption (\%)} = \frac{\text{intial weight} - \text{final weight}}{\text{final weight}} \times 100\% \qquad (1)$$

Scanning electron microscopy (SEM) was used to evaluate fiber's dispersion in the TPU matrix and to illustrate the fiber morphology of all the type of the fibers used. This testing

DOI: 10.1051/matecconf/20179701062

which using SEM HITACHI TM 300, was also to observe and investigate the morphology of the tensile fracture surface of the composites after tensile testing.

3 Results

The results will be presented here are on the tensile strength, oil absorption percentage and morphology of the composites that have been studied.

3.1 Tensile strength

The tensile strength of the TPU composites with various type of natural fiber can be inferred in Fig. 1. This figure illustrated the evolution of the tensile strength and a marked highest was observed at 5% of pineapple fiber, 10% of coconut husk, 15% of coconut shell and 10% of corncob, when we not compared it to the neat TPU sample. Also, we can consider these are the optimum filler loading for each type of natural fiber used in TPU. Due to various type of fiber, it can be seen that the tensile strength is affected by filler type, filler size, filler shape and their surface area [7]. It would also be controlled by the nature of the filler and matrix [8]. By comparing four natural fibers in this study, we may proved that pineapple fiber showed the highest tensile strength among others, although with low filler loading (5%) compared to other. Even, Kengkhetkit et al. [9] in their study resulted that the tensile strength show an increasing when filler at the range of 10% -to 30%. This may be due to the TPU, which have better interaction with hydroxyl groups on cellulosic fiber [10]. Beside, at higher loading of coconut shell (15%), the composite showed the highest tensile strength. This is due to the shape of coconut shell itself. The more loading are needed to make the distribution of filler become well enough and good interaction with the TPU matrix. Georgopoulos et al. [11] stated that, plant fibers are not easily dispersed in non-polar polymer due to their extensive intermolecular hydrogen-bonding. They also found in previous research, the decrease in tensile strength occurs upon loading the polymer with natural fiber. This fact also was supported by Myrath et al. [7]. They found that, the addition of natural fibers into the polymer, decreases tensile strength and this is the common observation with almost all filled polymer system. The decreasing in tensile properties of composites causes by the poor fiber dispersion. Consequently, the higher the cellulose content for each type of natural fiber (Table 1), the higher the tensile strength. Youssef et al. [1] and Georgopoulos et al. [11] proved that, high tensile strength and more obvious at the higher fiber loading of 10% corncob fiber, the value decreased. Referred to them, the decreasing may be due to decreasing of the internal bond. Differ to the coconut husk as filler (10%), resulting in an increasing of tensile strength and this is justified by Emmanuel et al. [3] which, with the enhancement of the interfacial adhesion and the ability to transfer stresses across the interface. Other possible reason is, some researchers proved that, most natural fibers start to lose strength at 160°C, which expected to cause poor mechanical properties in the final products [11]. This may be seen from the result of tensile strength.

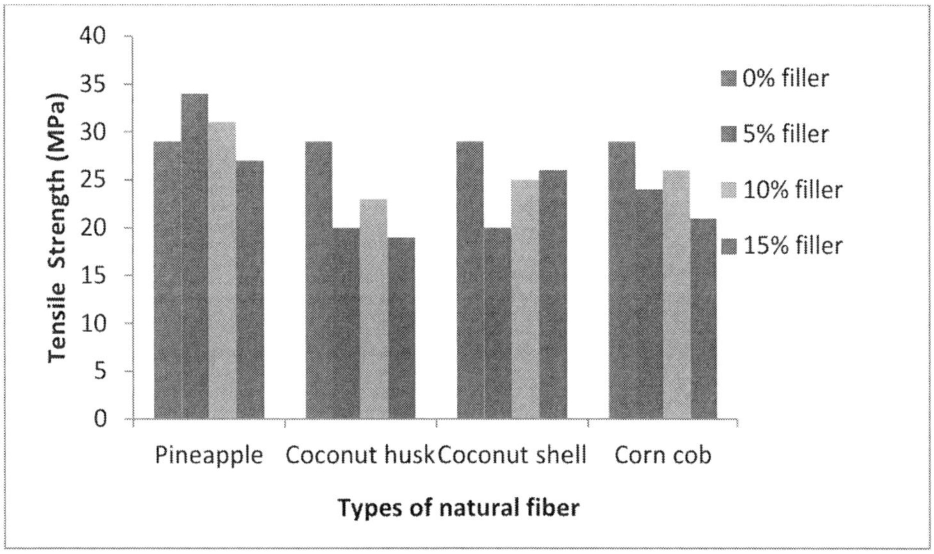

Fig. 1. Tensile strength for natural fiber/TPU composites at 0% - 15% of filler content.

3.2 Oil absorption behavior

The pattern of oil absorption after 4 weeks by the TPU composite with different types of natural fibers can be referenced in Fig. 2. All fibers showed higher oil absorption capacity for cooking oil compared to engine oil. Cooking oil has low viscosity so that the oil molecules can move easily to the pores of the fiber assembly and hence higher absorption capacity was observed. We may see that, coconut shell and corncob fiber have the lowest oil absorption capacity. Due to Marko et al. [12], several types of fibers have some drawback like oil absorption capacity due to its oleophobicity properties. Other reason is due to their shape, surface area and their nature. By looking to the shape and nature of filler, we can say that pineapple fiber and coconut husk contain long and fine fibrous structure which can easy access of the oil [9]. Coconut shell as we know, are in powder form with irregular shape and corncob contain only short fibers. Strong fiber and matrix interfacial adhesion also can help to diminish the oil penetration and avoiding the worsening of the mechanical performance of the composite, which we may relate to the tensile strength in previous discussion. Long fiber with hollow tubes structures may serves a valuable for oil absorption. Jayamol et al. [13] stated that, if the surface area is large, then there will be more area for absorption of oil molecules. Surface cavities are filled with oil molecules that have been absorbed. Furthermore, the percentage of oil absorption and swelling which obtained with different type of oil are determined by the ability of the oil to interact with the cellulosic fiber. The efficiency of the absorption process also depends on the packing density of the filler, but this is not yet to be determined in this study. From Emmanuel et al. [3] research, fiber assemblies with lower packing density are more porous and have a large hydraulic diameter of inter-fiber pores. The density of the oil used also influences the oil absorption capacity.

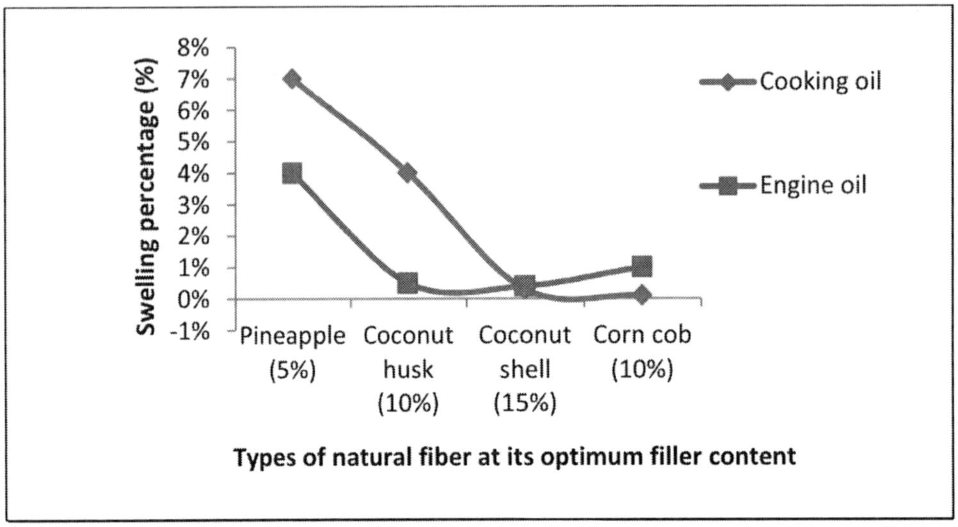

Fig. 2. Swelling percentage for natural fiber/TPU composites at optimum filler content, in cooking oil and engine oil after 4 weeks.

3.3 Morphology

Every natural fiber has its own surface morphology which decides the interfacial matrix-fiber adhesion and these will affected the mechanical/physical properties [14]. Cellulose content affects the mechanical properties of vegetable and fruits fiber [11]. Short fiber leads to higher composites strength, most probably due to better dispersion which gives anisotropic products. While particulate filler is dispersed into polymer melt and quite difficult to achieve a strong bond between particles and matrix. It is just because the poor wettability of the fine filler particles by the polymer melts. Since the particulate filler are very small, a high interfacial area with filler and the polar matrix. This area increases the worsening bonding between them and decreases the tensile strength. Irregular shape filler such as corncob [15] can caused to the decreases of strength of this composites. The cellulose content for all four types of natural fibers is presented in Table 1. Cellulose provides strength and stability to the cell walls and the fibers as a whole. The cellulose oriented in a fiber or bundle of fibers influences its properties [16]. Fig. 3 (a) showed the smooth surface of matrix and this described the compatibility between TPU matrix and pineapple fibers. Due to this good interfacial adhesion, it proved to the highest tensile strength [10]. Fig. 3 (b,c,d) showed the rough surface with the presences of fibers revealed less homogeneity [1]. The fibers also appear nearly to be free of the matrix material adhering to them. Also, could be seen, some cavities between fibers and the TPU component. The morphology of natural fibers reveals shapes in the form of long fiber, short fiber and also irregular particles.

Table 1. Cellulose content of natural fibers [17, 18].

Natural fiber:	Cellulose content (%):
Pineapple	73-81%
Coconut husk	32-43%
Coconut shell	34%
Corncob	31%

DOI: 10.1051/matecconf/20179701062

Fig. 3. Morphology for natural fiber/TPU composites at their optimum filler content; (a) Pineapple fiber at 5% loading (b) Coconut husk at 10% loading (c) Coconut shell powder at 15% loading and (d) corncob fiber at 10% loading.

At higher filler loadings, agglomerations of fibers caused poorer transfer of load and stresses by fiber agglomerates to matrix, which decreased the mechanical properties of composites. The composites which failed are mostly by fiber pull out from the matrix and indicate a poor interfacial bonding between the matrix and fibers [14]. Voids on the surface are also quite visible. Voids arise due to the entrapped air pocket in the matrix during composites processing [14].

4 Conclusions

Fruit and plant fiber are often seen as waste produced in the agricultural sector and can be transformed into profitable, useable products. The main characteristics of the most natural fibers however, are described in terms of their mechanical properties such as tensile strength. Among the four types of natural fibers been studied, pineapple at 5% filler content showed the highest strength and it proved by the composite's morphology. Good interaction between filler and matrix can be observed to the fiber break up and the surface roughness of the matrix. Different type of natural fiber represented different morphology and also displayed different tendency to absorb oil. Highest swelling percentage represented a highest level of TPU composite to absorb the both types oil with the incorporation of pineapple fibers.

DOI: 10.1051/matecconf/20179701062

References

1. A.M. Youssef, A. El Gendy , S. Kamel, Mater. Chem. Phys., **152**, 26 (2015)
2. G.J. Nilza, G.J. Virgo, V.E. Buchanan, Mater. Char., **59**, 1273 (2008)
3. M.F. Emmanuel, V.M. Correlo, J.F. Mani, R.L. Reis, Com. Sci. Tech., **78**, 56 (2013)
4. Y.A. El Shekeil, S.M. Sapuan, M.W. Algrafi, Mater. Des., **64**, 330 (2014)
5. D. Janusz, K. Patrycja, Indstl. Crops and Prods., **74**, 566 (2015)
6. A.A. Nor, S.H. Ahmad, N.M. Zain, Appl. Mech. Mater., **291**, 2654 (2013)
7. K. Myrath, H. Onggo A. Syampurwadi, J. Bio. Sci., **7 (2)**, 393 (2007)
8. H.M. da Costa, R.C.R Nunes, L.L.Y. Visconte, C.R.G. Furtado, Raw Mater. App., **54**, 242 (2001)
9. N. Kengkhetkit, T. Amornsakchai, Mater. Des., **55**, 292 (2014)
10. A. Elkhaoulani, F.Z. Arrakhiz, K. Benmoussa, R. Bouhfid, A. Qaiss, Mater. Des., **49**, 203 (2013)
11. Georgopolous, Tarantili, Augerinos, Andrepolous, Polym. Degr. Stab., **90**, 303 (2005)
12. L. Marko, M. Remskar, V. Ducman, F. Svegl, J. of Environ. Manage., **114**, 158 (2013)
13. G. Jayamol, S.S Bhagawan, S. Thomas, Comp. Sci. Tech., **58**, 1471 (1998)
14. R.P. Avinash, R.N. Jagtap, J. Mater. Environ. Sci., **6**, 902 (2015)
15. G. Junjie, A.H. Milford, Indstl. Crops Prods., **19**, 225 (2004)
16. N. Sena, A.R. Marco, A.M. Araujo, V.D. Souza, L.H.C. Mattaso, J.M. Marconcini, Indstl. Crops Prods., **43**, 529 (2013)
17. D. Rudi, K. Myrtha, A.S. Subyakto, D. Hermawan, A. Hadiyane, A. J.of Plant Sci., **15**, 42 (2016)
18. A.K. Bledzki, A.A. Mamun, J. Volk, Comp. Sci. Tech., **70** 840 (2010)

MATEC Web of Conferences **97**, 01063 (2017) DOI: 10.1051/matecconf/20179701063
ETIC 2016

Spatial Patterns of Public Facility Services based on Grid/Cell Supply Approach (Case Study: Elementary School in Malang, Surakarta & Banjarmasin City)

Deni Agus Setyono[1,*], *Septiana* Hariyani[1], *Denny* Dwi Cahyono[1], and *Masdar* Helmy[1]

[1]Urban and Regional Planning Department – Brawijaya University, MT. Haryono 167, Malang City, Indonesia

Abstract. Public facilities have various types and functions to support people activities. Provision of public facilities in Indonesia is carried out on the base of population size and spatial approach, which for the development of new facilities used the scale and capacity of services analysis on related facility. This study aims to measures service rates of elementary school in the Malang, Surakarta and Banjarmasin City based on supply (school avalaibility) aspect, and then to identify spatial pattern of the services. As previous, the measurement was done with the administrative boundary as analysis units but then the grid/cell approach were used to obtain accurate results. Based on the analysis result, service capacity of elementary school in the Malang City defined by the mash/grid/cell approach was 271,75% meanwhile in the Banjarmasin City was 535,68% and Surakarta City was 1.510,4%. These results means that school in these three cities can meet the demand with more than 3 – 15 times than existing population size. The results also illustrates that people in these three respective cities can access elementary school not only in their administrative boundary but also they access school everywhere of their city areas.

1 Introduction

The avalaibility of public facilities are quite important especially to enhance the urban growth and development. These urban facilities has functions that can support various activities that carried out by community. Hence provision of urban facilities adequately is necessary to improve the quality of city development and its community resources. One of the urban facilities that has important function is educational facilities which will be the

* Corresponding author: deni.setyono08@gmail.com

focus on this research, particularly for elementary schools. Procurement education in adequate sufficient is necessary to increase the quality of human resources. These research focused to identify the elementary schools or facilities service because there are national agenda in education sector named "*Wajib Belajar 9 Tahun*" whereas all young peoples in Indonesia are obliged to obtain education services especially in the elementary and junior high school level. The government tried to provide elementary school in various areas yet sometimes not supported by adequate planning procedure. The planning process must consider the existing education services by the number and distribution of elementary school in order to analyze the level to access school.

In another discussion, public access to school quite different in various areas, especially between the urban and rural area or the centre and fringe areas because the provision of facilities were generally concentrated on the activity centers. Particularly in rural areas, it is often represented have low access to the elementary school. These research is advanced research phase that conducted in aims to determine accessibility conditions of education facility services. In the early phase of research, we conducted the suitable method that can interpreted the service rates and accessibility conditions especially for elementary school. Related with those previous study, the accessibility of elementary school will be identified and analyzed using the cellular/mash approach in this part. This research conducted with 3 cities as area study namely Malang City, Surakarta City and Banjarmasin City which are indicated had similar characteristics as medium – large cities in Indonesia. The determination of these cities are expected to get the comprehensive results related elementary school service patterns in urban areas. To the best author's knowledge, there is no research that conducted to get spatial pattern of facility services especially for educational facilities in developing countries like Indonesia.

To define which method has valuable result then Moran's I was used. In mash/cellular approach, the calculation of service area focused on the built-up areas. Therefore author through this research want to give the comprehensive description about urban facilities services pattern in aim to devise a provision method of urban facilities.

2 Methods

The research was conducted in three cities that derived from different provinces, namely Malang City (East Java Province), Surakarta City (Central Java Province) and Banjarmasin City (South Kalimantan Province). Those three city has similar characteristic thus considered as study area in this research, which are as medium city based on population size, known as education city and also similar administrative division. Malang City has an administrave area about 110 km^2 and there are 851,298 inhabitants (BPS, 2016) live in this city. Malang City consist of 5 districts which are Blimbing, Kedungkandang, Klojen, Lowokwaru and Sukun District. Then, Surakarta City also known as Solo City also has similar characteristics with Malang City. The total city around 44 km^2 and total population reached 512,226 inhabitants in 2015 (BPS, 2016). Administratively, Surakarta City area also divided into 5 districts i.a Laweyan, Serengan, Pasar Kliwon, Jebres and Banjarsari District. And lastly is Banjarmasin City, which has administrative area about 98 km^2 and population size around 675,440 in 2015 (BPS, 2016). Banjarmasin City also divided by 5 districts, which are South Banjarmasin, East Banjarmasin, West Banjarmasin, Central

DOI: 10.1051/matecconf/20179701063

Banjarmasin and North Banjarmasin District. Overview of these three cities can be seen as the following map.

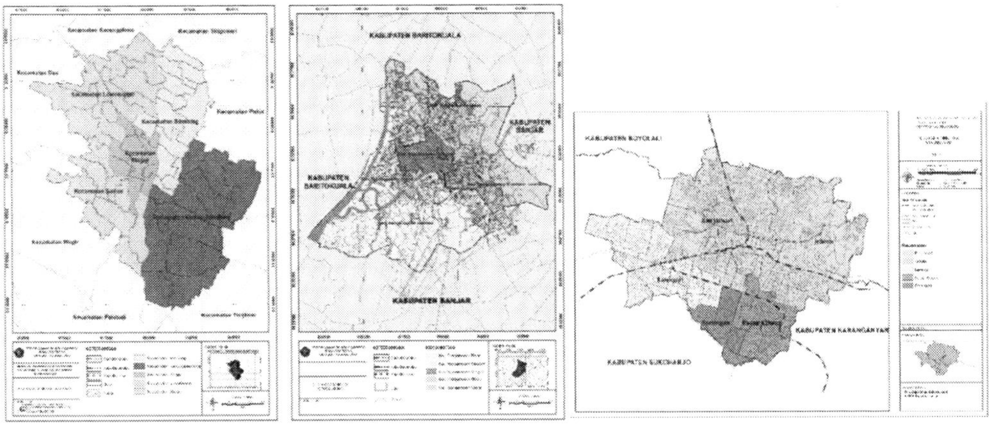

Fig. 1. Map of Administrative Area in Malang, Banjarmasin and Surakarta City.

Various educational facilities ranging from kindergarten until university are contained in these three city. Regarding the facility, the focus of this research aimed to elementary school as indicated relevance with national agenda on education sector "*Wajib Belajar 9 Tahun*". Development of elementary school quite important in order to meet the needs of educational services for all citizen and it is part of basic services on primary education development program. The provision of elementary facilities up to now based only form national standard with the main factor is population aspect and there is no emphasis with the local characteristics and conditions especially on spatial aspect.

Several procedures was used in this research. The first step was identification of school number and location in each city to get the existing spatial distribution characteristic. And another step was divide the city area using grid/cell/mash set with the cell size of 250 x 250 meter. Futhermore the range of services was identified using imaginary spheres with circular looking (layer) of each school location. And then assessment was carried out with overlay technique between service layers and grid/cell set layers as catchment process. Based on those process, it could be known how many school service layers that can be catched in each cell and then the total layer in every cell in each village boundary was calculated as mean value. These results can illustrates as average number of school that can serve or can be accessed by population and then the average number will determine the rate of elementary school service in each village. Moreover the result will be analyze using Moran's I to examine the links or relationship between area on spatial views. Based on author's knowledge, Moran's I analysis used to calculate spatial bounds, find a spatial model, etc. In this research, the authors tried Moran's I analysis as a method to compare spatial pattern of elementary school services.

3 Result and discussions

3.1 Spatial distribution of elementary school by existing units

This is early stage of the research, which is define the elementary school service rates based on the population size and existing number of school units that only located in each subdistrict or village boundaries. The spatial distribution of school can determine rate of service and shortages of facility in each city. Based on the data, Malang City has the highest number of elementary school which is 323 unit meanwhile Banjarmasin and Surakarta City has 252 & 275 unit. As further described, Kedungkandang District has the highest number of elementary school in Malang City reached 79 unit. While Banjarsari District has the most number of elementary school in Surakarta City with 82 unit and South District in the Banjarmasin City with 70 unit. These school number compared with the population size to get the basic rate of school services. Result of the calculation from this early stage served in Table 1.

Table 1. Average Service Rates of Elementary School in Malang, Banjarmasin and Surakarta City

District	Number of School Unit (Only Within Adm. Boundary)	Average Service Rates (%)
Blimbing	65	92.53
Kedungkandang	77	119.29
Klojen	48	94.80
Lowokwaru	62	90.58
Sukun	71	95.16
Malang City	**323**	**98.70**
Banjarsari	81	136.63
Jebres	54	89.62
Laweyan	53	113.61
Pasar Kliwon	55	145.94
Serengan	32	141.14
Surakarta City	**275**	**122.90**
West Banjarmasin	48	88.79
South Banjarmasin	72	111.89
Central Banjarmasin	40	98.55
East Banjarmasin	43	89.86
North Banjarmasin	49	84.20
Banjarmasin City	**252**	**95.67**

Source: Analysis Results, 2016

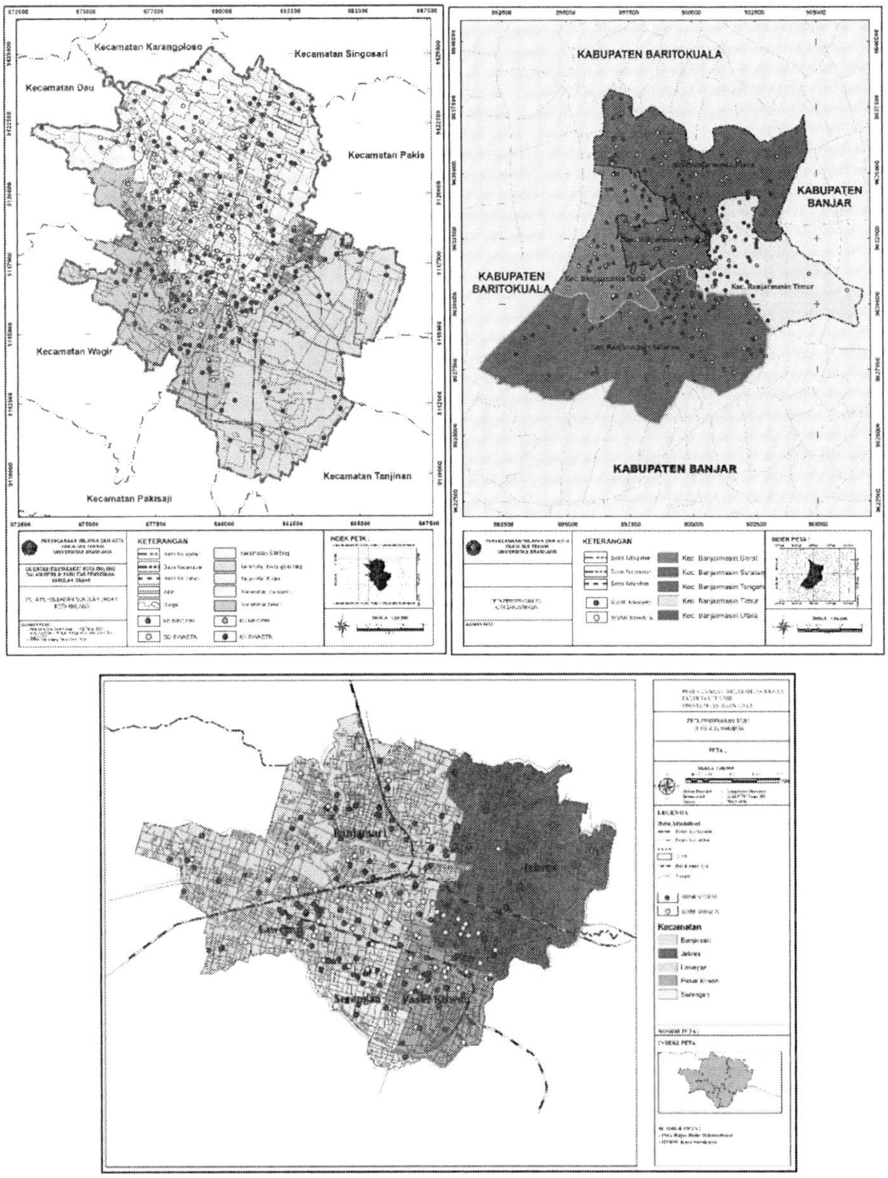

Fig. 2. Map of School Distribution in Malang, Banjarmasin and Surakarta City.

From the result (Table 1), it can be seen that the high average service rates was Surakarta City with 122.90% when the Malang and Banjarmasin City just has 98.70% & 95.67% respectively. In general, the rate value means elementary school in Surakarta City can meet the needs of facilities solely based on the population size (which can described that the good rate is more than 100%). And these can not be seen in the Malang and Banjarmasin City which are the service rates just below 100%. In detail, only Kedungkandang District which has high rate of school service reaches 119.29%. This is

DOI: 10.1051/matecconf/20179701063

same with Banjarmasin City, which is only South Banjarmasin District has the high school service rates as 111.89%. It is means that elementary school in these district can serves the demand of all population in the district areas. Another districts in Malang and Banjarmasin City has high rates but it still lower than 100%. So it can be indicated that all school in related districts can not meet demand of the population based on simple/basic calculation. In contrast, school in almost all districts in Surakarta City reaches high service rates, except in Jebres District. It means that avalaibility of school in this city can meet the demand of population. From this basic calcution or method shows that there is no specific spatial pattern to define the service rates.

3.2 Spatial distribution of elementary school by mash/cell catchment (MCC)

The another method that used in this research is Cell or Mash Catchment Approach which will observe and calculate service rates using grid/cell/mash set system. This method carried out from the conditions that people in the city can choose the elementary school not only near or around their home and they can choose various school in the their city areas. From this conditions, it is indicates that school service not limited by administrative boundaries and the service rates calculate not only facility within the areas but also facility from surrounding areas that can be accessed by people from related areas.

In the first phase, the city area divided by cell or mash set system. Then, the cell or mash set adjusted with land use and derived for the later phase only for cell/mash that contained the built-up area. This method was used to improve the concept of service rates calculation. The next phase is overlay process between school services layer and cell/mash set layer in aim to get number of service layers that can be catched in each cell. From this process, it will be known that every cells have different number of school layers and it interpretes that number of school that can be accesses people form those cell. Furthermore to get the final result, the value of each cell within village boundaries will be calculated as mean/average score as whole value for related villages. The result of services scale capacity based by cell or mash method can be seen in map as follows.

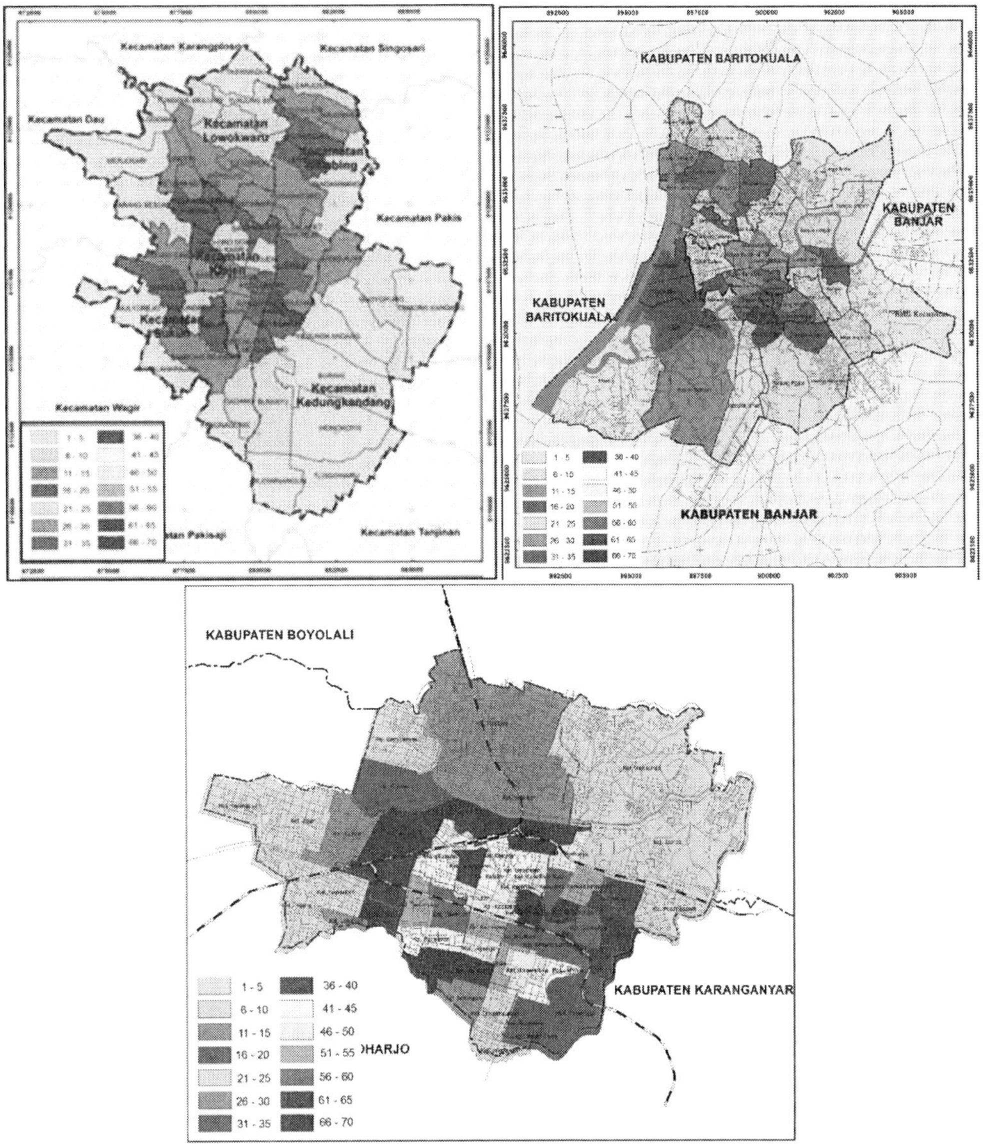

Fig. 3. Map of Elementary School Service Rates based on Mash/Cell Catchment

The mapping results shows number of elementary school that can serves or accessed by peoples in each village. It also can be seen the similarities that area with high value of school service rates in these three cities are villages that are located in the city centre. The highest number of schools that can be accessed by people from each village in Malang and Banjarmasin City was 35 – 40 units. While the Surakarta City has the higher number which are 65 – 70 units. It can illustrates that peoples in those villages can access or choose among 40 school freely if they have children in elementary school ages. From the mapping analysis results, it also described that villages in the peripheral or suburban areas has lower

DOI: 10.1051/matecconf/20179701063

number of school which average is 1 – 5 school units. These results will determine the elementary school service rates in each village. Rates of elementary school service that determine using mash or grid system method can be seen in Table 2.

Table 2. Average Service Rates of Elementary School in Malang, Surakarta and Banjarmasin City Using Mash/Cell/Grid Catchment Method

District	Number of School Units (Based on Service Layers)	Average Service Rates (%)
Blimbing	175	270.48
Kedungkandang	143	187.98
Klojen	272	569.21
Lowokwaru	122	177.65
Sukun	138	169.58
Malang City	**850**	**271.75**
Banjarsari	440	1611.40
Jebres	424	1752.63
Laweyan	300	988.67
Pasar Kliwon	401	1896.85
Serengan	232	1265.16
Surakarta City	**1797**	**1510.40**
West Banjarmasin	172	336.18
South Banjarmasin	235	468.61
Central Banjarmasin	336	1128.20
East Banjarmasin	172	354.05
North Banjarmasin	134	248.15
Banjarmasin City	**1049**	**535.68**

Source: Analysis Results, 2016

Based on Table 2, it is known that average school services rates using the mash/cell catchment method has higher than average rates of basic/existing method. The highest average rates located in Surakarta City that reached more than 1.510%. It is means that school in this city can servcs 15 times than the demand from existing population. Malang and Banjarmasin City has the lower average service rates which are 271,75% and 535,68% respectively. As it said before, areas (districts/villages) that has high rates of school service contentrated in the city centre. Klojen District has the highest service rates in Malang City which is 569,21%; then Pasar Kliwon as centre of Surakarta City and Central Banjarmasin also has the high rates of service which are 1.896,85% and 1.128,20% respectively.

These results considered more accurate compared with the basic or esixting method. In addition, the determination of cell/mash/grid in this research considered the land use characteristics in the process of of service rates measurement because it was unbuilt land there is no human activity especially residential activity. Therefore, Moran's I analysis was used to compare the accuracy level of measurement the elementary school service rates in Malang, Surakarta and Banjarmasin City.

3.3 Moran's I results

As it has been discussed, the use of Moran's I analysis is in order to compare the accuracy of the three methods that is measured the elementary school services capacity that has been done before. Moran's I analysis produced Moran's index that gained from spatial weight with GeoDa software. Moran's I analysis chosen because the characteristic of areas affected by activities on those area/characteristics of around so it is also used as a base assumptions in the measurement of elementary education services scale in this city. The measurement of three method that had been done and it obtained Moran index as follows.

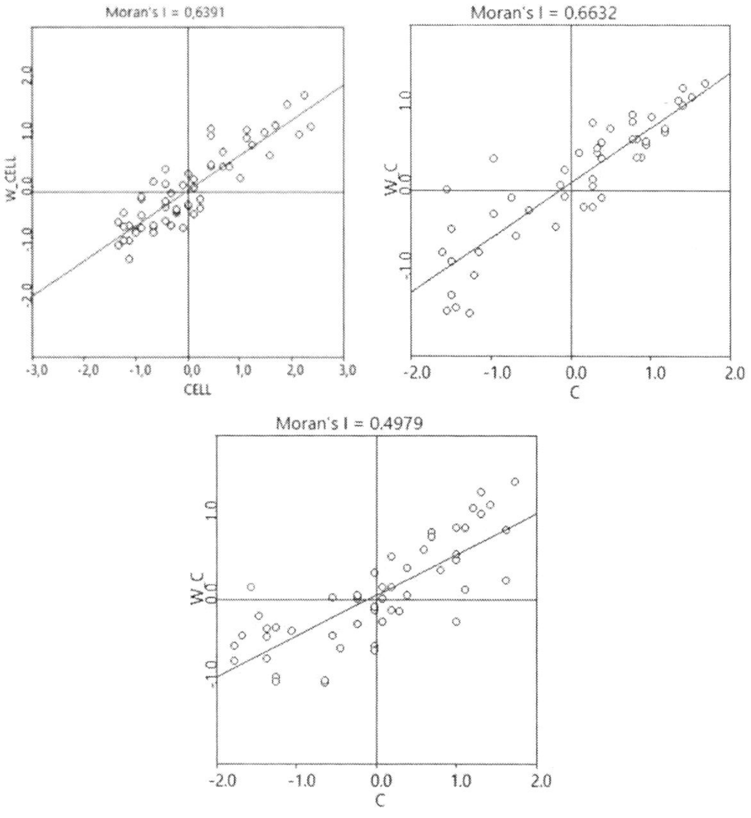

Fig. 4. Moran's I analysis results of service rates in Malang, Surakarta and Banjarmasin City

Based on the results of analysis, it can be seen the value of Moran's I analysis in the Malang and Surakarta City has high correlation index which are 0,6391 and 0,6632 respectively meanwhile Banjarmasin City has average correlation index that is 0,4979. The index value is indicated that spatial autocorrelation from the object studied is in the category of high correlation. These results strengthened with the LISA Cluster Map (another product of Moran's I analysis).

DOI: 10.1051/matecconf/20179701063

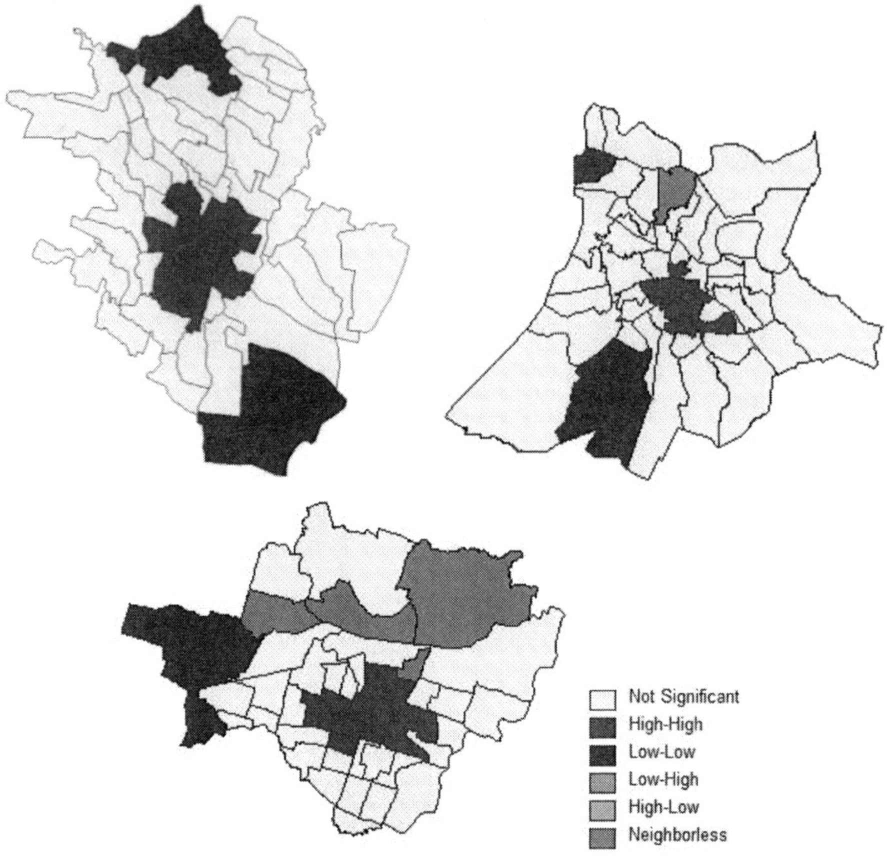

Fig. 5. LISA Cluster map of school service rates in Malang, Surakarta and Banjarmasin City.

Based on the Figure 5, it can be seen that villages in the centre areas has high-high correlation, both in the Malang, Surakarta and Banjarmasin City. These results has relevance with calculation results and it can be described the real condition in the field/city.

4 Conclusion

Based on research has been done to determine the method which has accurancy to determine elementary school service rates in Malang, Surakarta and Banjarmasin City using mash or cell or grid system approach. Based on the Moran's I analysis results, it can be said that interaction about access of elementary school is high mainly on urban centers. The result is also able to described the phenomena that happened in the urban areas especially in Indonesia which the parents are freely to choose education facilities for their children ignoring administrative boundaries and it is influenced by the availability of facilities around the city areas. And it is indicates different with rural cases, which is has the lower chance or opportunity to get or access the elementary school or public facility

DOI: 10.1051/matecconf/20179701063

freely. Topic about elementary school service pattern in rural areas will conducted in further research.

References

1 Minister of National Education RI, *Minister of National Education RI Policy No. 24, 2007 about Facilities and Infrastructure of Elementary, Junior High Schools and Senior high School Standarts* (Republik Indonesia, 2007)
2. National Indonesian Standart 03-1733-2004, *The Procedures for Planning a Residential Neighborhood in Urban Areas.*
3. R. Puspitasari, I. Susanto, *Prosiding of Mathematic and Mathematic Education National Conference* (Yogyakarta, 2011)
4. Statistic Agency of Banjarmasin City, *Banjarmasin City in Figure 2015 – 2016* (2015)
5. Statistic Agency of Malang City, *Malang City in Figure 2015 – 2016* (2015)
6. Statistic Agency of Surakarta City, *Surakarta City in Figure 2015 – 2016* (2015)

Application of Agent-Based Model Simulation for Tsunami Evacuation in Pacitan, Indonesia

Fadly Usman[1,*], *Keisuke* Murakami[2], *Agus* Dwi Wicaksono[1], and *Eko* Setiawan[3]

[1]Department of Urban and Regional Planning, Faculty of Engineering, University of Brawijaya, Malang, Indonesia
[2]Department of Civil and Environmental Engineering, University of Miyazaki, 1-1 Gakuen Kibanadai Nishi, Miyazaki 889-2192, Japan
[3]Faculty of Computer Science, University of Brawijaya, Malang, Indonesia

Abstract. The most significant measure in reducing tsunami disaster is an early evacuation. The aim of this study is to developments a dynamic evacuation model with using simulation of Agent-Based Modeling and Simulation (ABMS). The Shelter Plan Analysis is also conducted in this study to establish applicable evacuation places. All analysis in this study based on a geographic information system (GIS) data such as shelter location, evacuation networks, housing and other information. The shelter plan analysis is the methods to evaluate an applicable shelter. The important variable in this analysis is nearest accessibility based on the existing road from the residential area to the evacuation building. In order to set input data for this analysis, we conducted a questionnaire survey to the local people in Pacitan, which is a typical populated coastal city in Indonesia, located in East Java province. The agent-based model is according to multi criteria decision making (MCDM) for internal and external factor. The concept of MCDM was used to develop a weighted suitability decision matrix in a simulation's model. We also conducted the field survey in this city, especially focused on the condition of evacuation shelters and evacuation networks, in order to evaluate the shelter plan on GIS as well as to conduct the simulation of the agent-based model.

1 Introduction

This paper describes the development of an application of agent-based model simulation for tsunami evacuation in Pacitan City, Indonesia. Evacuation is a common strategy for handling emergency situations. Evacuation is a process in which threatened people displaced from dangerous areas to evacuation places, in order to reduce the health and life vulnerability of affected people. During disaster response, evacuation should be conducted accurately, and in a hurry.

As a result, preparation, testing and training of a proper evacuation plan required as a part of disaster preparedness. However, evacuation planning is a complex problem involving many behavioral and management facets (M. Saadatseresht et al. 2009). Sherali

* Corresponding author: fadlypwk@ub.ac.id, fadlypwkftub@gmai.com

et al. (1991) studied the evacuation plan for natural disaster such as hurricanes and floods with explicit consideration of the impact of shelter locations on evacuation time. Cova and Justin (2003) presented a network flow model for identifying optimal lane-based evacuation routing plans in a complex road network. A mixed-integer programming solver used to derive routing plans for sample networks. Pursals and Garzon (2009) considered the building evacuation problem and developed a model for selecting the proper routes for movement of people into a building during an emergency situation.

Mansourian et al. (2006) studied that most of the data required for emergency management and evacuation planning have a spatial component or location that represents a significant opportunity to utilize geographical information systems (GIS). Pal et al. (2003) have discussed the development of a traffic simulation methodology on GIS and spatial data, which could be used for emergency evacuation planning purposes. This methodology seeks to give optimal evacuation alternatives in the most efficient routes for evacuation of the population from the affected region. Mansourian et al. (2006) used spatial data infrastructure (SDI) concepts and Web-based GIS to facilitate disaster management. They used Web-based GIS to facilitate spatial data sharing among disaster management parties for better planning and decision-making.

ElDessouki (1998) had presented that to develop an evacuation plan. It is necessary to determine safe areas. Another significant job is involved determining the distribution of evacuees into the safe areas. Each evacuee understood where should they go and from which route. Wu and Zhang (2006) added to achieve a proper plan for evacuation places, a planner should consider the capacities of the safe areas and the distance to the safe areas as two main factors during the planning. In other words, in an evacuation plan, the distribution of evacuees into the safe areas should be based on the capacity of the safe areas. And at the same time, each evacuee should go to the nearest safe area. M. Saadatseresht et al. (2009) considered Multi Objective Evolutionary Algorithms (MOEA) and the geographical information system (GIS) for evacuation planning. In his study, he proposed a three-step approach the safe areas, is a spatial Multi Objective Optimization Problem (MOP), because the objective functions and data required for solving the problem has a spatial component.

The advantages of agent-based approach are capture emergent phenomena, provides a natural environment for the study of certain systems, and flexible, particularly in relation to the development of geospatial models (Andrew T. Crooks and Alison J. Heppenstall, 2012). The agent-based modeling approaches the complex system into individual component as an agent. Agent-Based Model and Simulation (ABMS) are a computational method and system, to understand the interacting agent behavior in an artificial environment (Sutikno and Murakami, 2011). The potential role of GIS in evacuation planning was explained by a number of studies. More specifically, there are few studies integrating GIS and multi objective techniques for evacuation planning (Balram and Dragicevic, 2006). For the problem of this research intends to present utilization of Multi Criteria Decision Making (MCDM) techniques and GIS as integrated frameworks for evacuation planning.

We added a consideration of household behavior into evacuation modeling in this study. The model developed on Repast Simphony toolkits, which can be used for early and the nearest network during evacuations processed. Both of analysis based on geographic information system (GIS) data such as shelter location, evacuation networks, housing, agent and other information. This system consists of two parts; (1) the modeling for the street, shelter and agents using geographical information system data in ArcGIS, and (2) analysis of disaster evacuation using agent-based model simulation. Furthermore, it is possible to investigate the appropriate evacuation route by the simulation.

2 Literature Review

Tsuji et al. (1995) reported that the last historical tsunami caused by the earthquake in Java Island, and it arrived the coast about 50 minutes after the main shock. The epicenter of this earthquake is different from the supposed earthquakes by Mardiatno et al. (2008) We estimate tsunami traveling time to Pacitan as approximately 21 minutes based on the shallow-water theory. Mardiatno et al. (2008) conducted tsunami inundation simulation on coastal area in Pacitan. For determining tsunami inundation area, we supposed possible maximum earthquake in this area as magnitude 8.0 from historical data. Located at the south coast of Java, Pacitan is facing a seismic gap between two tsunami-earthquake events, which are the 1994 (M 7.2) and the 2006 tsunami (M 7.7). (A. Muhari et al, 2012). The location of its hypothetic epicentre is 111.0°E and -9.5°S. Angle parameters are strike (θ) 276°, dip (δ) 10° and slip (λ) 95°. The other source parameters are length 120km, width 60.1 km, depth 10 km and dislocation 5.20 m. We conducted the field survey to get the spatial information of houses, household and settlement by using a digital camera with GPS, which used to specify the location on the Geographic Information Systems (GIS). This survey conducted to collect data about the street condition such as the width of street, pavement type, and level of service (LOS). LOS in this study is used to analyze the street condition by measuring traffic flow with number of vehicles and corresponding safe driving conditions. The local government of Pacitan City (BAPPEDA Pacitan, 2002) designated 44 evacuation shelters in case of a tsunami disaster as shown in Fig. 1.

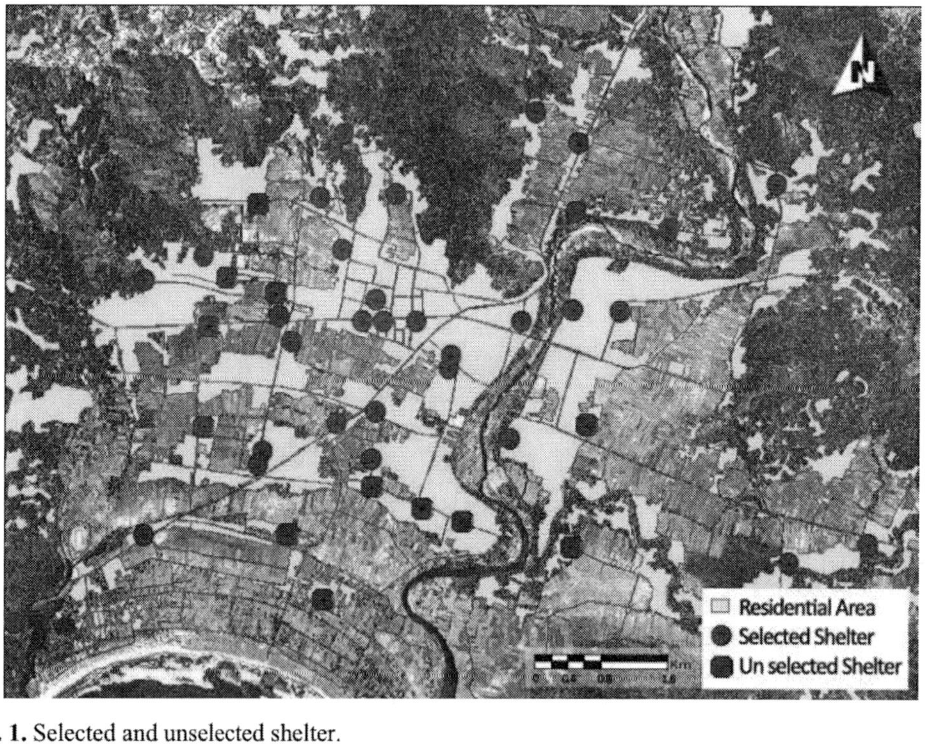

Fig. 1. Selected and unselected shelter.

It is assumed that around 5.4 km2 of the total of 10.8 km2 of residential areas in Pacitan City were inundated, and around 76% of them were inundated by water that was more than 1 m deep as seen on Fig. 2. These results occurred because the residential areas are located mainly on the low-land area in Pacitan City.

 DOI: 10.1051/matecconf/20179701064

Most of the evacuation shelters designated by the local government are the open-space areas located in the hilly terrain. Based on field survey and weighting analysis, this study found that 12 shelters (blue colour in Fig. 1) cannot be used as temporary evacuation places for several reasons. The main reason of unselected shelter is the poor of building quality and limited of building capacity (Usman and Murakami, 2011).

The distribution of selected and unselected shelter after scoring is shown on Fig. 1 which 32 facilities as temporal shelter shown in red-coloured dots. Based on several criteria such as nearest with inundated area and coverage of service area to the vulnerable area, 16 building facilities were selected as temporal shelter in this study (Fig 2).

Fig. 2. Shelter and inundated area.

This study was identified the location of mosques in Pacitan city as temporary evacuation places. Mosque is usually designed with specific criteria such as unique roof, simple shape and large capacity. Through field check, also found some public buildings such as hospitals, schools, hotels and government offices that can be supposed as the temporary evacuation place. Residential areas for this study include areas where there are houses, government offices, schools, shop-houses, market buildings, and other public facilities.

This study is organized two parts of analysis; the first part is an analysis by using the shelter coverage area which called Shelter Plan Analysis. The objective of this analysis is to identify the service area of the shelters based on the longest distance and limitation of evacuate time. In this study, evacuate time are grouped into two distances. Distance of 900m for 15 minutes of evacuates time, and the longest mileage is 1200m for 20 minutes of evacuates time. We set the average of walking speed is 1.0 m/s. The second part is an analysis by using Agent-Based Modeling and Simulation (ABMS). The objective of this part is to get the best routes in evacuation. The variables in ABMS are evacuee, street, and shelter.

DOI: 10.1051/matecconf/20179701064

3 Shelter plan analysis

Mardiatno et al. (2008) estimated that most of the residential areas in Pacitan city are inundated by the tsunami due to this supposed earthquake. The inundation areas more than 0.5 m depth spread to landward deeply as shown in Fig. 4. With considering above tsunami traveling time, this study set tsunami evacuation time as 15 and 20 minutes in order to investigate the maximum area that residents could evacuate to their shelters within the limitation of evacuation time. Fifteen-minute evacuation means an urgent situation that could be happened under the delay of evacuation warning.

Evacuation with vehicle sometimes causes a traffic jam, and it could disturb safety evacuation of others (Sutikno et al., 2011). This study supposes the evacuation by foot under tsunami evacuation. The walking speed defers depending on the physical condition of an evacuee such as age, sex, topographic condition, network condition and others. This study assumes that all residents evacuate with walking speed between 1.0 - 1.5 m/sec.

Evacuation speed of 1.0 meter/sec is speed of a dependent elderly during an evacuation and evacuation speed of 1.5 meter/sec is used as the average speed of the person together with a child during tsunami evacuation (Potangaroa, 2008). In this study, we choose the slowest of evacuation speed (1.0m/sec) to analyze the applicable shelter in the Pacitan city, using Network Analyst extension by ArcGIS. We are conducted two cases in shelter plan analysis; Case-1 consists of 11 shelters, where five shelters were not included in simulating according to an assessment of the evacuation building in the previous study (Usman and Murakami, 2011). Table 1 shows scoring of six-teen shelters in Fig. 2. And Case-2 in this study consists of 16 shelters which located on inundated area as seen on Fig 2.

Table 1. Scoring list of the Shelter.

No	Name	Building Level	Floor Area (m2)	Floor and Land Area (m2)	Scoring Distance	Scoring Capacity	Scoring Construction	Score
1	Masjid Al Muhajirin Perumnas	1	225	1 600	3	2	3	8
2	Masjid Ainul Yaqin	1	240	640	3	1	3	7
3	Masjid Kel Sidoharjo	1	180	1 091	2	2	1	5
4	Masjid Ds Sirnoboyo	1	110	160	3	1	1	5
5	SMK Bina Karya	2	1 700	5 112	3	3	2	8
6	Masjid LDII	2	500	1 250	3	2	3	8
7	MAN	2	350	4 557	3	2	3	8
8	Masjid Al Muhajirin	1	230	550	2	1	2	5
9	Masjid Perumnas	1	250	850	3	1	1	5
10	Masjid Pompes Al anwar	2	540	1 400	3	2	2	7
11	STKIP PGRI	2	700	2 902	3	2	2	7
12	SDN Ploso 1	2	520	980	3	2	2	7
13	Masjid Al Hikmah	1	245	430	3	1	2	6
14	PN Pacitan	2	940	4 200	3	3	3	9
15	SMA 1 PAcitan	2	1 800	9 000	3	3	2	8
16	SMK1 Pacitan	2	2 400	6 700	3	3	2	8

Fig. 3 shows the distributions of service area, which is defined as the maximum area in which residents can evacuate to the nearest shelter within each evacuation time. The service areas on each evacuation times are drowned in different colors. Table 2 shows the percentage of covered area and number of covered people in Pacitan city under the available condition of all allocated shelters and evacuation networks. As shown in Fig. 3, some areas are uncovered even in the case of 20 minutes of evacuation time.

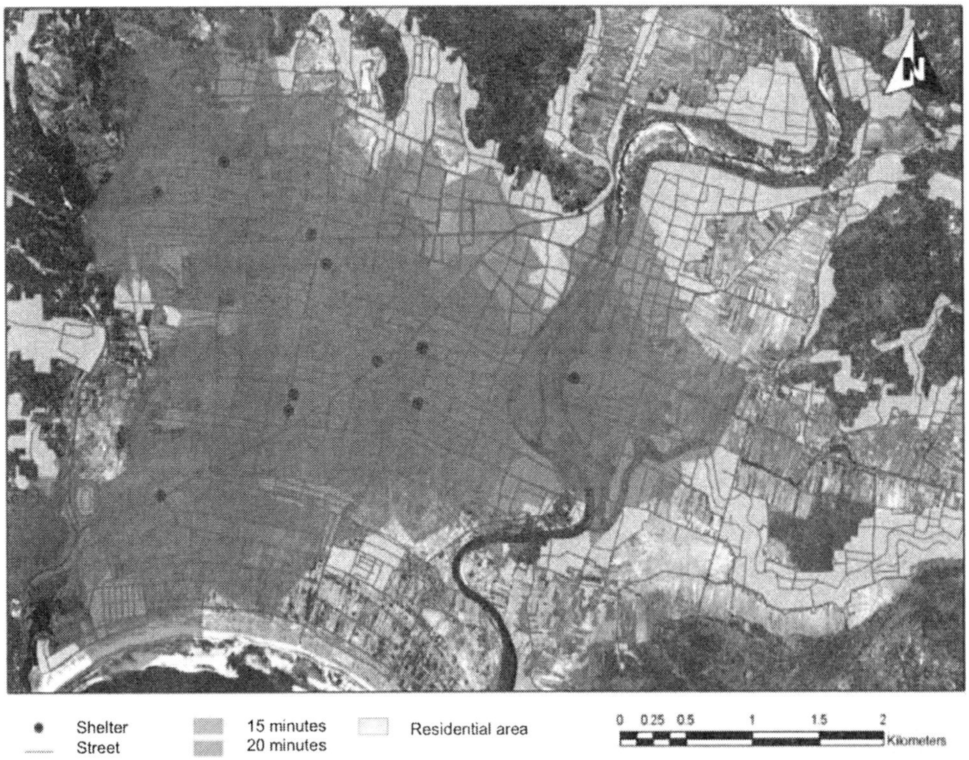

Fig. 3. Service area of Case-1 with 11 shelters.

With decreasing of evacuation time, covered area becomes smaller. In the case of 15 minutes of evacuation, which seems an urgent situation due to some troubles in the early-warning transmission system, only less than a half of residential area could be covered by current shelter allocation in case-1. About 73% of the residential area is covered by 11 shelters allocation in 20 minutes of evacuation. And only 61% of the residential area is covered in 15 minutes of evacuation time even in the new shelter allocation (Table 2).

Table 2. Percentage of coverage area in Case-1.

Evacuate Time	Coverage			Non coverage	
	Area	People	Percentage	Area	Percentage
15 mt	8.12	48,192	84.87%	1.45	15.13%
20 mt	9.21	54,661	96.26%	0.36	3.74%

This result shows that need more allocation shelter to cover all of service area for people evacuation. We use all of shelters in the vulnerable area in Case-2. These facilities which eliminated before could be used as a temporal shelter by improving structure quality and recommend expanding the capacity of shelter. Fig. 4 shows the service areas in Case-2.

In comparison with Fig. 3, the increase of the service areas can be observed in beach side and rural area in each evacuation time. Table 3 shows the percentage of the covered area in each village. About 96% of the residential area is covered by 16 shelters allocation in 20 minutes. On the other hand, only 84% of the residential area is covered in 15 minutes of evacuation time as shown on Table 3.

DOI: 10.1051/matecconf/20179701064

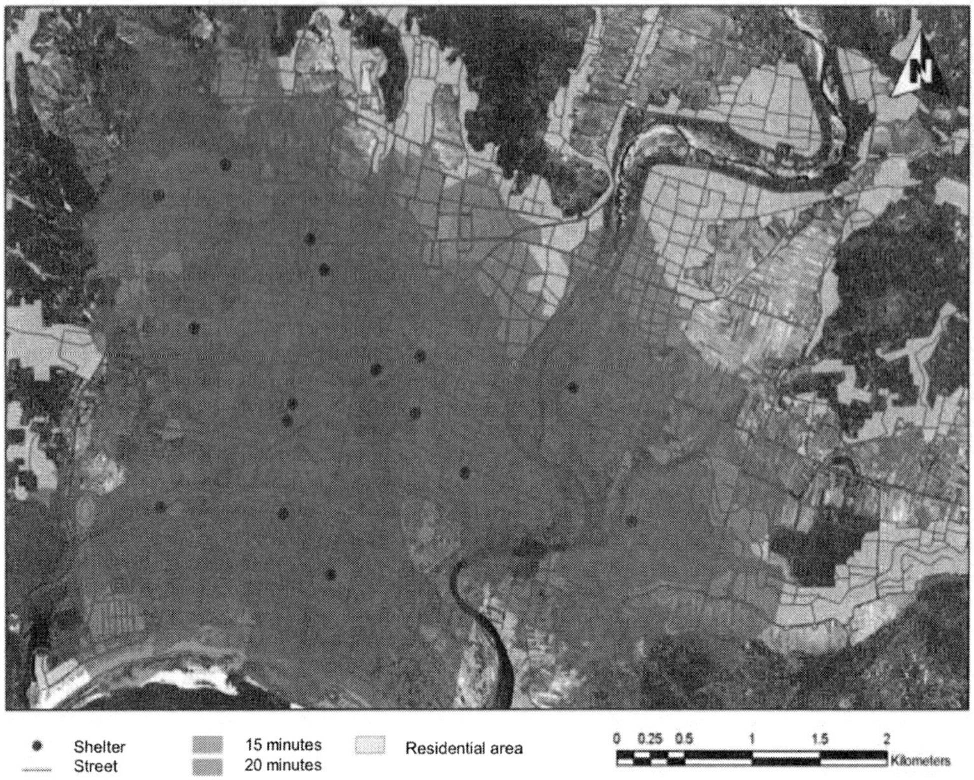

Fig. 4. Service area of Case-2 with 16 shelters.

Table 3. Percentage of coverage area in Case-2.

Evacuate Time	Coverage			Non coverage	
	Area	People	Percentage	Area	Percentage
15 mt	8.12	48,192	84.87%	1.45	15.13%
20 mt	9.21	54,661	96.26%	0.36	3.74%

4 Agent based model and simulation

The objective of agent-based model simulation is to verify the results that obtained through the Shelter Plan Analysis. In the analysis before, parameters of analysis are the nearest distance to determine coverage of service area. In the agent-based model, simulations carried out several approaches designed as closely as possible to the real condition and situation. This study is modifying the Agent-Based Model and Simulation (ABMS) of virtual city from RepasCity model, in order to develop agent-based for tsunami evacuation. The source codes of the models are developed by using Repast Simphony toolkit. Repast Simphony is designed to provide visual and tools for agent model design, agent behavior specification, model execution and result examination (North and Macal, 2007). Repast has multiple implementations in several program language and built-in adaptive features such as algorithm and regression. Repast is an open-source, cross platform, agent-based modeling and simulation toolkit.

In making a decision, an agent has their own condition called Internal Factors (IF). In this study, the Internal Factor of the agent is age that related to the speed of evacuation;

such width, street quality, and slope. All agents will choose a network based on the ability of each agent. Fig. 5 under is shown Multi Criteria Decision Making (MCDM) diagram for Internal Factor (IF) and External Factor (EF) in Agent-Based Model and Simulation. The movement of an agent is having unique condition and other variables that govern the movements of agents are street quality and building capacity as the External Factor in simulation. For example, a 60 years-old agent will choose the wide, flat, and smooth street. While other 20 years-old agents would like to choose any street as long as the paths are nearest to the shelter.

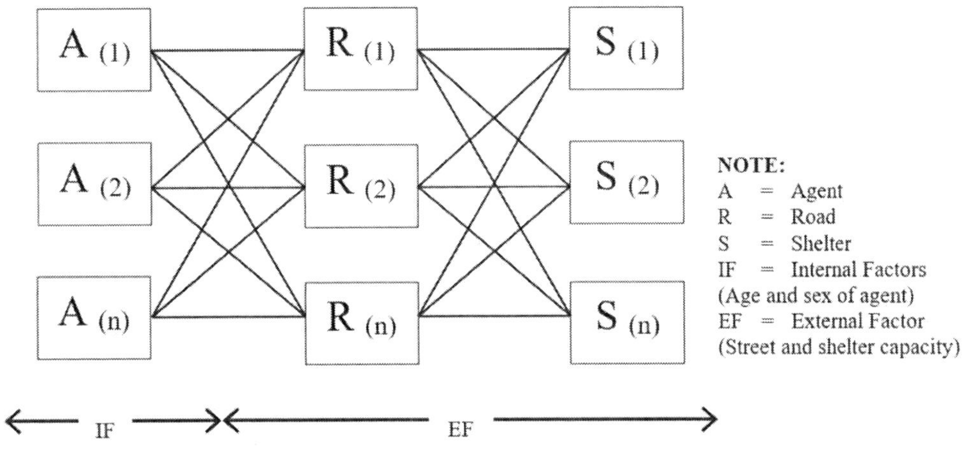

Fig. 5. Multi criteria hierarchy for agent-based model.

The development of tsunami evacuation consists three parts: construct context and projection, build environment, and define agent movement behaviour. A context is used to store a population of agent. Projection is used to define the relationship between each agent. Without any projection, agent would not be able to interact with each other (Sutikno and Murakami, 2011). Environment component in this simulation such as houses, shelter, road and junction layer are used to build an artificial environment of the model. Location data and properties of road network, shelter, building and agents are stored in GIS files, and then the agent behaviour is implemented as an algorithm in Java. The agent movement behaviour is defined to give agent's response to their environment when tsunami warning is announced. In this study is assumed that the individual households are treated as a self-supporting unit to making decision on the road to traffic environment and to the evacuation building. Evacuation modelling is a complex process, which is supposed to include many factors.

Table 4. List of scenario in agent based model and simulation.

No	Scenario	Shelter	Shelter Capacity	Building Collapse
1	Scenario-1	11	No	No
2	Scenario-2	16	No	No
3	Scenario-3	16	Yes	No
4	Scenario-4	16	Yes	Yes

In order to simply the model, this study assumes that the evacuees have to move to the nearest shelter without considering the capacity of shelter for Scenario-1 and 2. In Scenario-

DOI: 10.1051/matecconf/20179701064

choose different shelter, that nearest with evacuation shelter before. This study is divided speed of agent into 2 categories, elder (1.0m/sec), and young person (1.5m/sec) and number of agents in this simulation is 425 agents which representing 4 villages and 8,000 residents. In Scenario-4, we conduct a case with building collapse variable. In this scenario, the material of building collapse blocked network and cause some evacuee cannot escape from vulnerable area to the shelter. Table 4 shows the list of scenario conducted in this study.

4.1 Scenario 1

In Pacitan city, the local governments officially provide several shelters that could be used to place refugee in the case of the tsunami as shows on Fig. 1. This study is only evaluating the tsunami evacuation plans in prone residential base on inundated area as seen on Fig. 2. This study assumes that the evacuees are households that live in the vulnerable area. Their initial position is the center of their houses, and they start to evacuate from their house to the shelter after announcement. All of data are prepared by using ArcGIS toolkit in shape file data format. Fig. 6 shows location of 11 shelter and agents at initial condition.

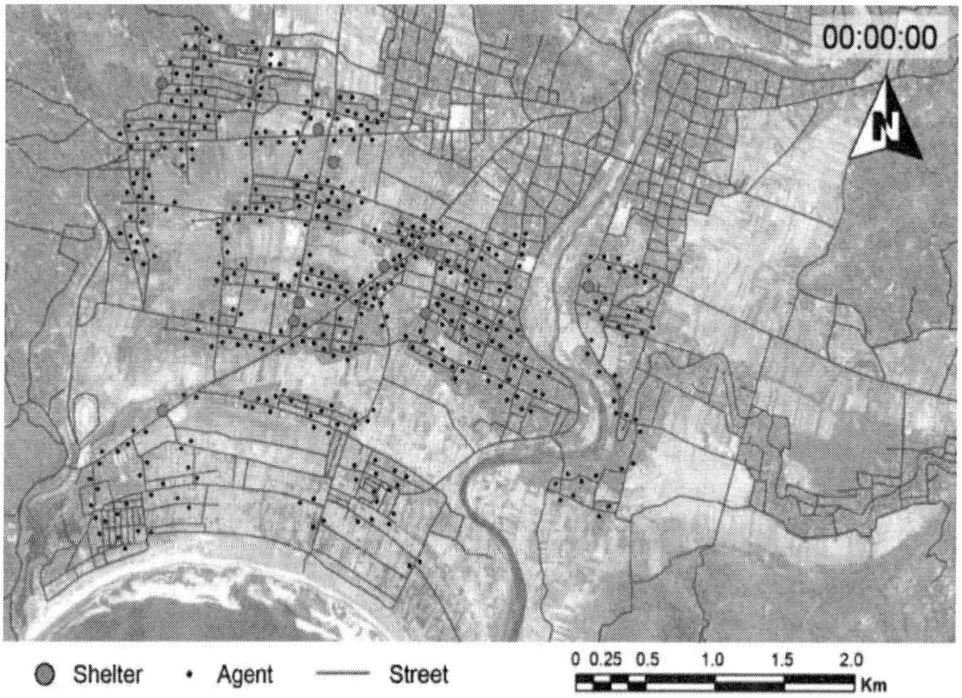

Fig. 6. Shelter and agents at initial condition in Scenario-1.

In Scenario-1, evacuation time is 27 minutes to evacuate all agents to the shelters. It is a long evacuation time because the required time for evacuation is about 15 minutes after the announcement of tsunami warning. Therefore, other scenarios were conducted to get better condition and faster than scenario-1 by add several shelters, which allocated at prone residential area.

4.2 Scenario 2

As the results obtained in the previous analysis in Shelter Plan Analysis, 11 evacuation places are not enough in the service area of building coverage and longer time of evacuation in Agent-Based Model and Simulation. In Scenario-2, simulation contains all shelter as seen in Fig. 2. The result in Agent-Based Model Simulation has similarities with the evaluation of Shelter Plan Analysis. In Shelter Plan Analysis, we found that coverage of the service area was increased significantly up to 96% in 20 minutes, and in the Agent-Based Model and Simulation, evacuate time is about 13.5 minutes only.

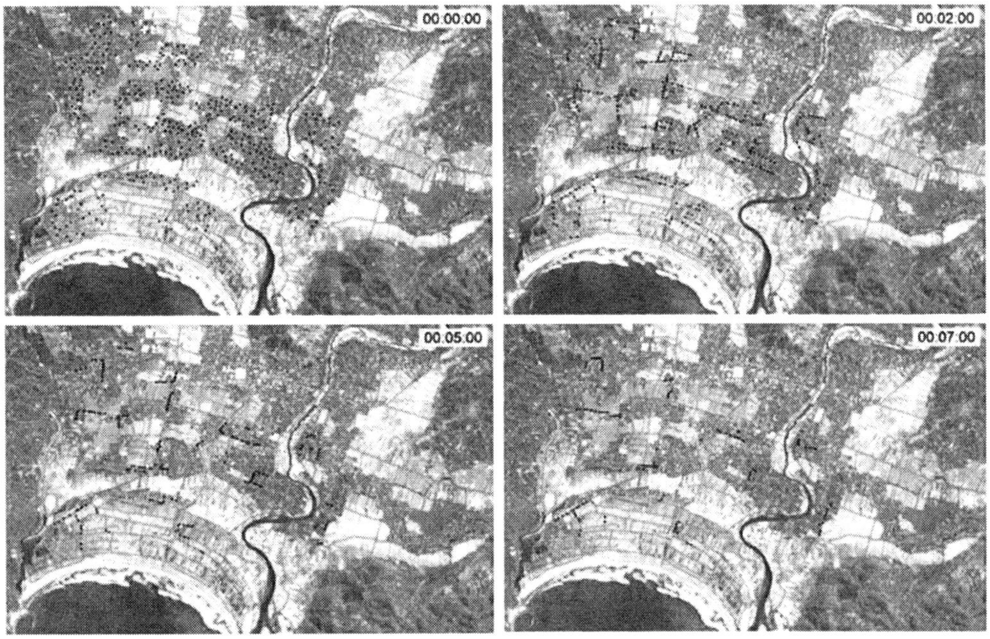

Fig. 7. Movement of agent at initial condition (0 minute), 2, 5 and 7 minutes in scenario-2.

In Scenario-2, evacuation time is 13.5 minutes which faster than Scenario-1 in evacuate all agents to the shelter. It is a better time to evacuate, because the required time for evacuation is about 15 minutes after the announcement of tsunami warning. Scenario-1 and 2 have not included the variable of building capacity. If the capacity of shelter is exceeded, agents require additional time to choose another nearby shelter. Fig. 7 shows movement of agent at initial condition, 2, 5 and 7 minutes.

4.3 Scenario 3

The behavior of agents will change when we conducted a parameter of building capacity. In Scenario-3, evacuation time is 19.5 minutes to evacuate all agents to the shelters. It is danger time to evacuate because the required time for evacuation is about 15 minutes after evacuee heard the announcement of tsunami warning. However, at the minute 13.5, agents have been evacuated to the shelter about 92%, that means only 8% need longer time to evacuate because exceeds the ability of the building capacity.

In Scenario-3, some agents move to another shelter after capacity of the nearest shelter was full. Fig. 8 shows that some agents move from shelter 10 to 9, from shelter 12 to 11 and from 14 to 13. Table 5 shows total agents, equivalent to the number of people,

 DOI: 10.1051/matecconf/20179701064

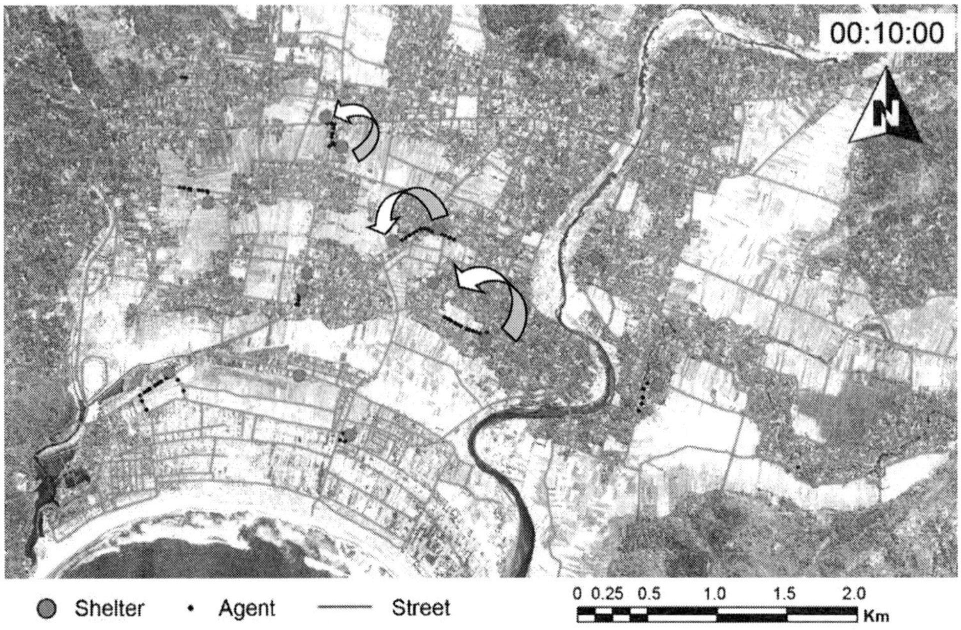

Fig. 8. Agents movement after 10 minutes in Scenario-3.

Table 5 shows that one agent representing twenty residents, so twenty agents representing about four hundred people. In Scenario-3 Shelter 10, 12 and 14 was filled full by the agent after 6 minutes, and other agent move to the nearest shelter from the previous shelter. This action requires a longer travel time for residents to reach the nearest shelter. Based on the phenomenon of agent movement in Scenario-3, a recommendation to the government is to increase the capacity of the evacuation building. This effort can be done to shorten the travel time.

Table 5. Total agents, building capacity and evacuate times.

No	Total Agent	Equivalent to the people	Building Capacity	Percentage (%)	Evacuate Time 1	Evacuate Time 2
1	26	520	800	65.0%	14	0
2	20	400	600	66.7%	10	0
3	28	560	600	93.3%	12	0
4	36	720	800	90.0%	14	0
5	22	440	500	88.0%	8	0
6	30	600	750	80.0%	13	0
7	26	520	600	86.7%	11	0
8	20	400	500	80.0%	8	0
9	47	940	1000	94.0%	9	16
10	20	400	400	100.0%	6	0
11	36	720	1000	72.0%	8	18
12	20	400	400	100.0%	7	0
13	26	520	600	86.7%	6	20
14	20	400	400	100.0%	7	0
15	27	540	600	90.0%	10	0
16	21	420				

 DOI: 10.1051/matecconf/20179701064

4.4 Scenario 4

According to Rakhman (2007) in his research said that many earthquakes such as in 2006 at Yogyakarta, Indonesia, have brought the serious houses damages. Reese et al. (2007) was reported that the dominant types of collapsed houses were Non-Permanent and Semi Permanent.

Fig. 9. Blocked network due to building collapse after earthquake and inundated area.

Heavy destruction of houses in a crowded residential area sometimes blocks evacuation networks, and also causes consumption of evacuation time. In agent-based model simulation, agents must follow road network data link from their houses to the evacuation building following the nearest possible path. In scenario-4, several networks were assumed damage due to the collapse of non-permanent houses, and damages of those houses could disturb safety evacuation within limited evacuation time. Fig. 9 shows the inundated areas in the case of suffering network damages.

The zoom in of blocked network is shown in Fig. 10. The number of shelters and agent behavior are same as in Scenario-3. However, in this scenario the agents could not through the damaged road by the collapse of non-permanent housing clusters along them. These conditions were assumed to be an unavailable network due to earthquake, and some agents will be trapped and unable to reach nearest evacuation building.

Fig. 11 shows movement of agent at 2, 7, 10 and 15 minutes. This figure can be seen that some agent cannot reach the evacuation building. In simulation, agents must follow the access road from their houses to the shelter following the nearest path. But in Scenario-4, some agents cannot find the street in front of their houses. About 5% of the agents trapped in front of a blocked network. In Scenario-4, evacuation time is 21.0 minutes to evacuate more than 95% agents to the applicable shelters.

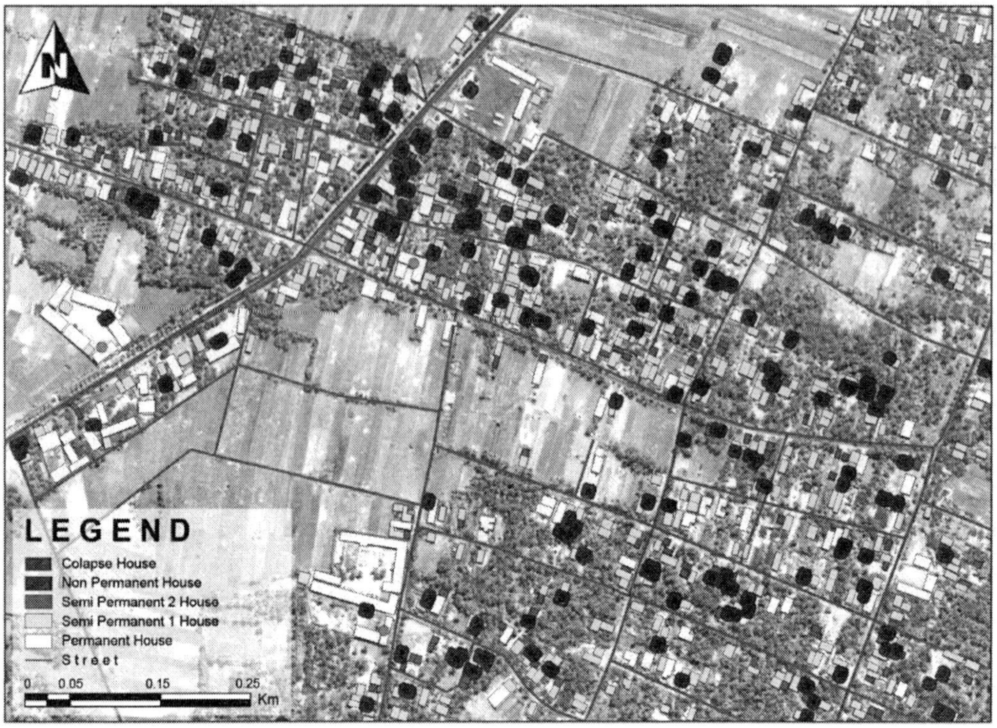

Fig. 10. Zoom in the vulnerable area in central of Pacitan city.

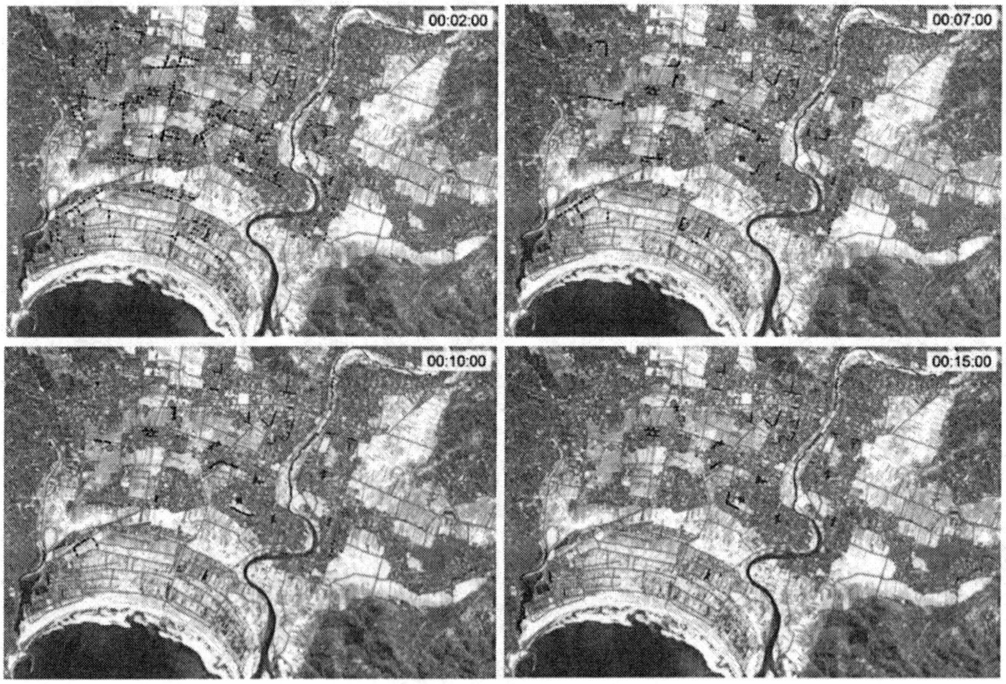

Fig. 11. Movement of agent at 2, 7, 10 and 15 minutes (scenario-4).

4.5 Comparison all scenarios

Figure 12 shows the comparison all of scenarios. Scenario-1 need more time to evacuate all agents to the shelter, Scenario-2 faster and good enough to evacuate the entire agent, but we conducted to add variable of building capacity then evacuate time become longer than before.

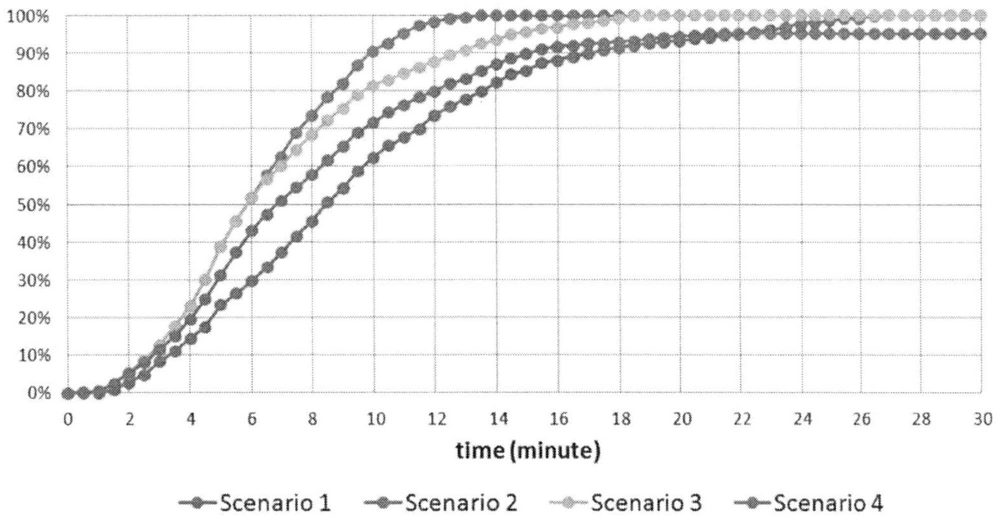

Fig. 12. Comparison all of scenarios.

Table 6 shows the percentage of household that can evacuate to the shelter in every step of evacuation time. This result was found that additional shelters in Scenario-2 provide a significantly result. Compared to the required time for evacuation in Scenario-1, evacuation time was decreased from 27 minutes to 13.5 minutes. That mean, Scenario-2 is 14 minutes faster than Scenario-1. Referring to the tsunami travel time based on the calculation, we estimate tsunami travel time to Pacitan as approximately 21 minutes.

Table 6. Total agents, building capacity and evacuate times.

Evacuation time (minutes)	Scenario 1 (%)	Scenario 2 (%)	Scenario 3 (%)	Scenario 4 (%)
5.0	23.41%	38.78%	38.78%	31.22%
10.0	62.44%	90.49%	81.46%	71.71%
13.5	79.76%	100.00%	92.44%	85.37%
15.0	87.32%		95.61%	89.76%
19.5	92.44%		100.00%	93.66%
21.0	94.15%			95.12%
25.0	98.54%			
27.0	100.00%			

However, it takes additional time to accommodate the evacuation process after the announcement of tsunami warning. Scenario-3 is only 19.5 minutes or less than 21 minutes. Better condition if evacuation time is calculated after the announcement of tsunami warning about 5-7 minutes after the people heard the tsunami warning. In this condition, Scenario-3 is still at the risk of evacuation time.

Evacuation time of scenario-4 is 21.0 minutes. This additional scenario was a scenario after an input variable of unavailable network due to building collapse after the earthquake.

Some agents need longer evacuation time to the temporal shelter. Around 5% evacuees trapped at non-permanent residential cluster and cannot reach the nearest shelter.

4.6 Comparison and similarities

Based on the results obtained from the previous analysis, in shelter plan analysis and agent-based model simulation, we found that both of analyses have the advantages of each analysis. Shelter plan analysis can provide the result about service area of the shelter based on street database by network analyst extension in ArcGIS. Expected result of this analysis can be obtained to determine the service area every evacuation building. The Agent-Based Model and Simulation (ABMS) is focused to agent, shelter and availability of street, so that the results obtained was different. The Agent-Based Model and Simulation (ABMS) is more accurate in determining the travel time of evacuation, while Shelter Plan Analysis is better in predict the service area of evacuation places. Table 6 shows comparison and similarity between the two analyses. In the Shelter Plan Analysis, distribution of the people is not required. However, in Agent Based-Model and Simulations, data of people distribution is needed as the agent to be simulated.

Table 7. Comparison and similarity of each analysis.

Analisis	Data				Measurement		
	street	shelter	house	people	distance	area	time
Shelter Plan	✓	✓	✓	×	✓	✓	×
Agent-based	✓	✓	✓	✓	✓	×	✓

From the discussion above, we recommend that to design an appropriate strategy related to planning the shelter it would be better if it begins with the Shelter Plan Analysis to determine the service area of each shelter. After analysed with Shelter Plan Analysis, continue the process with Agent-Based Model and Simulation to investigate the time required for evacuation.

With the consideration that evacuation with vehicle sometimes causes a traffic jam, and it could disturb safety evacuation of others, this study only supposes the evacuation by walking under tsunami evacuation. There are no agents ride any vehicles in the simulation, all of agent's speed based on walking speed.

5 Conclusion

This study applied both Shelter Plan Analysis and Agent-Based Model Simulation (ABMS) on GIS to establish an applicable strategy for tsunami evacuation plan. In order to set input data for this analysis, we conducted the questionnaire survey in Pacitan city, which is a typical populated coastal city in Indonesia.

From shelter plan analysis, this study revealed that service area of 16 shelters, which allocated on prone residential can serve up to 96% after upgrading several additional shelters. According to the result of Shelter Plan Analysis, we conducted the service area of current shelter allocation by Network Analyst on GIS. Agent-based model and simulation is used to verify the result of service area based on the speed of evacuation.

Furthermore, limited capacity of evacuation places to the service area was discussed throughout this study. Change of agent behavior in simulation shows that to reach another nearby shelter also requires extended time. These studies said that more additional shelters should be allocated in Pacitan to expand the service area even in the case of limited evacuation time. In addition, upgrading the quality of the shelter should be made, especially to upgrade the capacity of building and building strength. Unavoidable variable in the tsunami evacuation model of this study is the different of house quality, where poor quality of houses will block the network, and some agent trapped in that cluster area.

References

1. A. Muhari, M. Muck., S. Diposaptono, H. Spahn, Sci. Tsunami Haz., **31**, 244 (2012)
2. A.T. Crooks, A.J. Heppenstall, *Agent-Based Models of Geographical Systems: Introduction to Agent-Based Modelling* (Springer Science, London, 2012)
3. S. Balram, S. Dragicevic, *Collaborative Geographic Information Systems* (Idea Group Inc., New York, 2006)
4. T.J. Cova, P.J. Justin, Transport. Res. A-Pol., **37**, 579 (2003)
5. W.M. ElDessouki, *Some Developments in Transportation Network Analysis and Design with Application to Emergency Management Problems* (Ph.D. Dissertation, North Carolina State University, 1998)
6. J. Rakhman, Jurnal Permukiman, **3**, 4 (2008)
7. M.J. North, C.M. Macal, *Managing business complexity: discovering strategic solutions with agent-based modeling and simulation* (Oxford University Press, 2007)
8. A. Mansourian, A. Rajabifard, M.J. Valadan Zoej, I.P. Williamson, Comput. Geosci., **32**, 303 (2006)
9. D. Mardiatno, *Tsunami Risk Assessment Using Scenario- Based Approach, Geomorphological Analysis and Geographic Information System, A Case Study in South Coastal Areas of Java Island-Indonesia* (Dissertation, Faculty of Geo- and Atmospheric Sciences of the University of Innsbruck, 2008)
10. A. Pal, A.J. Graettinger, M.H. Triche, *Emergency evacuation modeling based on geographical information system data* (TRB 2003 Annual Meeting, 2003)
11. R. Potangaroa, *Proceeding from International Conference on Building Education and Research (BEAR) "Building Resillience"* (Auckland, New Zealand, 2008)
12. S.C. Pursals, F.G. Garzon, Eur. J. Oper. Res., **192**, 692 (2009)
13. Regional Development Agency of Pacitan, *Development planning of the settlement area in Pakistan city 2002-2012* (Final Report, Pemerintah Kabupaten Pacitan, Pacitan, 2002)
14. S. Reese, W.J. Cousins, Nat. Hazards Earth Syst. Sci., 573 (2007)
15. M. Saadatseresht, A. Mansourian, M. Taleai, Eur. J. Oper. Res., **198**, 305 (2009)
16. H.D. Sherali, T.B. Carter, A.G. Hobeika, Transport. Res. B-Meth., **25**, 439 (1991)
17. S. Sutikno, K. Murakami, A. Suharyanto, *Proceedings of the Twentieth International Offshore and Polar Engineering Conference* 599 (2010)
18. S. Sutikno, K. Murakami, *Proceeding of International Session in Conference on Coastal Engineering*, **2**, 26 (2011)
19. Y. Tsuji, F. Imamura, H. Matsutomi, C.E. Synolakis, P.T. Nanang, S. Harada, S.S. Han, K.I. Arai, B. Cook, Pure Appl. Geophys., **144**, 839 (1995)
20. F. Usman, K. Murakami, Journal of Japan Society of Civil Engineers, **67**, 595 (2011)
21. L.Y. Wu, X.S. Zhang, Comput. Oper. Res., **33**, 1226 (2006)

DOI: 10.1051/matecconf/20179701065

Investigate the Effectiveness of Seawall Construction using CADMAS Surf 2D

Fadly Usman[1,*] and *Supli* Effendi Rahim[2]

[1]Department of Urban and Regional Planning, Faculty of Engineering, University of Brawijaya, Malang, Indonesia.
[2]Agrotechnology Department, Agriculture Faculty, University of Palembang (UNPAL), 30139, Palembang, Indonesia.

Abstract. Tsunami that hit Japan in 2011 shocked the world. The earthquake at the coast of the Pacific Ocean was 9.0 on Richter scale leading to 10 meters of wave. The epicentre was reported to be off the coast of Oshika Peninsula, at the east coast of Tohoku, at a depth of 244 kilometres. The earthquake and tsunami caused more than 20,000 victims in six prefectures. In an effort to anticipate to the tsunami, Japan has long build walls around the beach directly face the Pacific Ocean—the walls are known as seawall. In this paper, facts show that the effectiveness of seawall in reducing the energy of tsunami is accompanied by tremendous destructive energy. Analysis in this research was done using a software named CADMAS Surf 2D with numerical analysis method. Observations were made using two variants of heights of seawall, two meters and three meters. Both are observed based on the heights of the wave coming from the Ocean, after passing through the seawall, and at some points of observation. The level of damage caused by the seawall was due to turbulence just behind the construction that resulted in crushing, scouring, and or destructive energy, the energy was even bigger after passing the seawall. Nevertheless, tsunami wave height and velocity decreased significantly after the seawall.

1 Introduction

The earthquake and tsunami was the most devastating natural phenomenon in human civilization. Both strike without warning, affecting all levels of society, in the form of losses of life, injury, property damage, and changes in socioeconomic structure. Several major earthquakes caused by tsunami, which in turn affects the condition of coastal and island communities [1].

Tsunamis can be generated by geophysical phenomena such as earthquakes, volcanoes, submarine landslides, and meteorite impacts. Historically, tsunamis in the world (1790 to 1990) are mostly generated by earthquakes (90.3 percent), volcanoes (6.4 percent), and landslides (3.3 percent) [2-3]. According to the Integrated Tsunami Data Base, at least

* Corresponding author: fadlypwk@ub.ac.id, fadlypwkftub@gmail.com

DOI: 10.1051/matecconf/20179701065

1,963 tsunamis have been recorded since 1628 to 2005 [4]. Earthquakes with a relatively close distance generate most tsunamis in tsunami-prone countries. As a result, the arrival of the tsunami varied from a few minutes to tens of minutes or a few hours. In such cases, the death toll from the tsunami waves can be very large, as seen in Aceh, which is located very close to the epicenter of the Indian Ocean, 2004. For other countries further away from the epicenter, the arrival times vary from one or two and up to ten hours. For example, Sri Lanka, India, and Thailand were struck by the 2004 tsunami around 2 and 2.5 hours after the quake. Some research shown that numerical simulations can predict the arrival time in such good enough way. History of the tsunami disaster showed that Indonesia is the country most vulnerable to tsunamis [5].

In this study, analysis was performed using numerical simulations to determine the behavior of the current wave hitting the seawall, when passing through it, and right after passing the walls. There are several important phenomena related to the failure of the seawall to halt tsunami, and this research report will show some of the facts related to the reduction rate of waves, decreased inundation, but with bigger destructive energy.

2 Literature review

Seawall is a huge wall built on the seafront, which is intended as a protector toward sea waves. Seawall to protect the city from tsunami waves is built high enough to prevent tsunami waves. In fact, seawall aimed as protector of the tsunami wave is one of the factors leading to the worst effect of tsunami itself. It can be seen from the massive damage despite being fortified by the tsunami retaining structures. One thing that must be understood to date is that the structure of any kind has not been able to withstand the enormous catastrophe of tsunami. Experience shows that buildings that surround the beach are directly destroyed when the tsunami comes (Fig. 1).

Fig. 1. Tsunami overtops seawall [6].

Seawall in the beginning is intended to withstand tsunami. The construction which is unsound or not in accordance with the magnitude of tsunami simply destroys it when the first wave comes before it reaches the buildings located around the coast. It becomes fatal when the tsunami moves with hydrodynamic force, a dynamic force that may increase from time to time. Therefore, when seawall is not able to withstand tsunami waves and then destroyed, then the collapse of the seawall will only increase the hydrodynamic force of the tsunami; and this is worse when seawall is built using concrete with large internal energy. Just imagine the accumulation of a large tsunami hydrodynamic force due to the collapse of the seawall, which makes the accumulation of tsunami wave to easily overtop and move quickly to settlements.

 DOI: 10.1051/matecconf/20179701065

In the last decade, the construction of seawalls continues to grow, and at least two major anti-tsunami seawalls are under construction. One in Kuji, a city in Iwate Prefecture once damaged by the tsunami and is scheduled to be completed soon. Seawall in Taro City is regarded as one of the most powerful anti-tsunami achievements in the world. It was built about 10 meters (30 feet) after the Sanriku quake in 1933 when the tsunami destroyed the village (Fig. 2).

Fig. 2. Collapsed seawall after being hit by tsunami [7].

3 Methode

This study uses CADMAS-Surf / 2D to perform numerical simulations. The equation used is composed of the continuity equation, the Navier-Stokes equation in the x-direction and the z direction as advection equations to discover water level. Last equation includes the F functions (x, z, t), which means the ratio of the volume of water in each cell numerically.

$$\frac{\partial \gamma_x u}{\partial x} + \frac{\partial \gamma_z w}{\partial z} = S_p \tag{1}$$

$$\lambda_v \frac{\partial u}{\partial t} + \frac{\partial \lambda_x uu}{\partial x} + \frac{\partial \lambda_z wu}{\partial z} = -\frac{\gamma_v}{\rho}\frac{\partial p}{\partial x} + \frac{\partial}{\partial x}\left\{\gamma_x v_e \left(2\frac{\partial u}{\partial x}\right)\right\} + \frac{\partial}{\partial z}\left\{\gamma_z v_e \left(\frac{\partial u}{\partial z} + \frac{\partial w}{\partial x}\right)\right\} - D_x u + S_u - R_x \tag{2}$$

$$\lambda_v \frac{\partial w}{\partial t} + \frac{\partial \lambda_x uw}{\partial x} + \frac{\partial \lambda_z ww}{\partial z} = -\frac{\gamma_v}{\rho}\frac{\partial p}{\partial z} + \frac{\partial}{\partial z}\left\{\gamma_z v_e \left(2\frac{\partial w}{\partial z}\right)\right\} + \frac{\partial}{\partial x}\left\{\gamma_x v_e \left(\frac{\partial w}{\partial x} + \frac{\partial u}{\partial z}\right)\right\} - D_z w + S_w - R_z - \gamma_v g$$

$$\gamma_v \frac{\partial F}{\partial t} + \frac{\partial \gamma_x uF}{\partial x} + \frac{\partial \gamma_z wF}{\partial z} = S_F \tag{3}$$

In the above equation, t mean time, x and z are horizontal and vertical coordinates, p is pressure, u and w means of horizontal and vertical velocity components. Then ρ is the fluid density, v is the summation of the molecular kinematic viscosity and eddy kinematic viscosity, g is the acceleration due to gravity. vγ is porosity, γx and γz is a component of air porosity, SF, Su and Sw is the source of wave generation, Dx and Dz is the coefficient for the sponge layer, and Rx and Rz is a resistance component due to the porosity on the x axis and z.

 DOI: 10.1051/matecconf/20179701065

Figure 3 is a schematic waveform flume for numerical simulations in the study. Flume has a length of 750 meters and height of 35 meters. Offshore depth wave is maintained at 10 meters as shown in Figure 3. The slope of the beach is 1/25. The size of the grid for the numerical simulation of the x and z direction are set respectively as 0.25 meter and 0.25 meter.

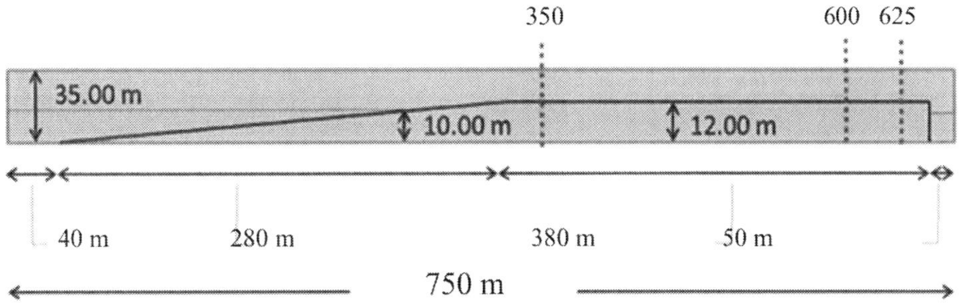

Fig. 3. schematic waveform flume for numerical simulations.

In the numerical simulation, it is assumed that the topography in numerical flume is in watertight condition, and the flatness of the topography is set as the original form so that the behavior of the wave before entering the beach will be the same as the original.

Based on reports from the height of the tsunami observed in the village of Lampon [8], numerical simulation in this study produces the type of bore waves like a tsunami with a height of 6 meters. This bore wave will produce a puddle with a height of around 3.5-5.0 meters on the shoreline. Simulation using CADMAS-Surf 2D takes time streak based on water level and velocity of fluid at the beginning of the boundary line so that it can generate tsunami waves with such profiles. This study assumes bore profile in offshore areas at first, and the velocity profile associated with the bore wave profile is obtained by using the equation below [9].

$$U = \frac{C\zeta}{H} = \zeta\sqrt{\frac{g(H+h)}{2H(H-\eta\zeta)}} \tag{4}$$

In this equation, U means the speed of an average water depth and g means gravitational acceleration. Which value of $H = h + \zeta$ means the total depth of the datum and ζ also means the height of temporal bore. And η is the coefficient derived from the ratio between the water depths early in the propagation area of total depth and set as 1.03 in this study.

4 Results and discussion

In this study, observations were made on water level and flow velocity was measured at three locations, 350 meters, 600 meters, and 625 meters respectively. The measurement points were chosen as a reference point for evaluating the effects of a reduction in the flow of tsunami inundation as shown in Figure 3 above.

Theory of the pressure profile of the impact of waves on the walls has been studied in detail by Peregrine [10]. It is said that violent impact of waves on the walls makes the speed and pressure to have far greater magnitude than those associated with ordinary wave propagation under gravity. This effect has been obtained by calculation of irrotational flow and to investigate the role of propagation and effects of trapped airwaves.

Figure 4 shows snapshot of the tsunami inundation in Cases 2 and 3 while Case 1 is a simple case where there are no obstructions on the topography of seawall. Reflection of flow can be observed on the ground slopes away from the wall when the flow overtops this embankment configuration. Reflections of this flow become greater at the measurement point of 350 meters after the waves hit the seawall.

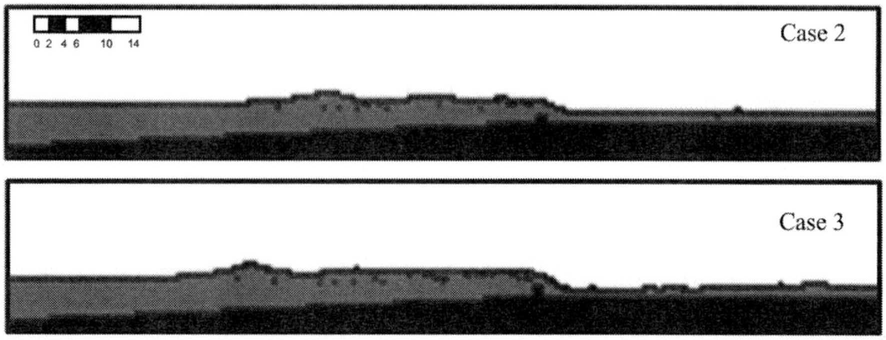

Fig. 4. Snapshot on Case 2 and 3 at t 60 seconds.

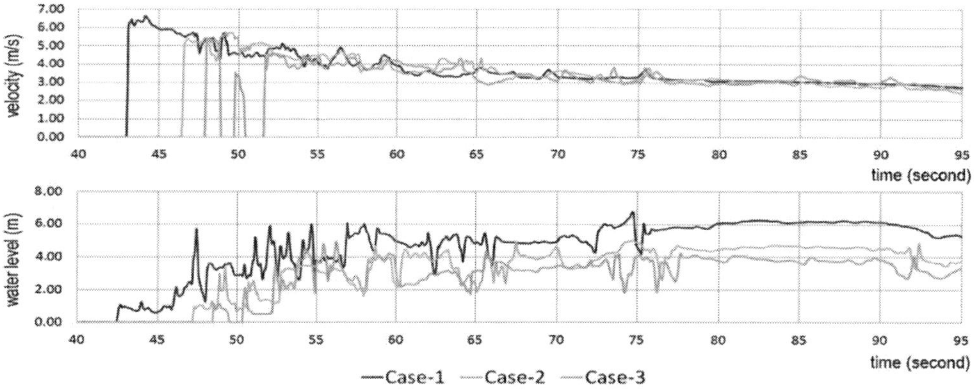

Fig. 5. Profile of maximum inundation and velocity of the wave at the measurement point of 350 meters, 600 meters, and 625 meters.

Reflection of flow at 600 and 625 meters respectively decreased approximately 1.00 meter compared to Case 1. Figure 5 shows the surface profile of the water and the speed at the location of 625 meters of the boundary input. Table 1 below shows the maximum value for the depth of flow and velocity obtained at several measuring points.

Table 1. The maximum inundation and velocity of the wave at the measurement point of 350 meters, 600 meters, and 625 meters.

Case	Wave height (m)			Velocity (m/s)		
	350m	600m	625m	350m	600m	625m
1	6.18	3.24	3.15	7.01	6.42	6.29
2	5.02	2.43	2.28	6.82	5.71	5.48
3	4.53	2.11				

Table 1 shows the maximum inundation and velocity of the wave at the measurement point of 350 meters, 600 meters, and 625 meters from the profile at Figure 5. By analysing the waves overtop the seawall, we can calculate the value of the Froude using the following equation.

$$Fr = \frac{v}{\sqrt{gh}} \tag{5}$$

Gravitation and wave height in this equation is used to calculate the speed for small amplitude on the long wave of the shallow water. It is obtained by assuming the hydrostatic pressure in the momentum equation. Hydrostatic assumption is no longer valid and the wave velocity depends on the wavelength.

Table 2. The Froude value at the measurement point of 350 meters and 600 meters

Case	350			600		
	v	h	Fr	v	h	Fr
2	6.82	5.02	0.96	5.71	2.43	1.16
3	5.91	6.18	0.75	5.45	2.11	1.19

Table 2 shows the value of the Froude at the measurement point of 350 meters and 600 meters. In this table, it is known that Fr 350 < 1 and Fr 600 > 1, which means the flow of the waves at 350 meters is lower than usual and at 600 is higher. The influence of the flow of this phenomenon directly affects the coverage area of inundation on the ground because the horizontal force transformation that occurs in the case of a seawall.

Figure 6 below shows the behaviour of waves a few moments before and after hitting the seawall. From the figure below, it can be seen that waves come up with great speed and energy, having retained the seawall and then transcend and turbulence occurs directly behind the seawall bottom.

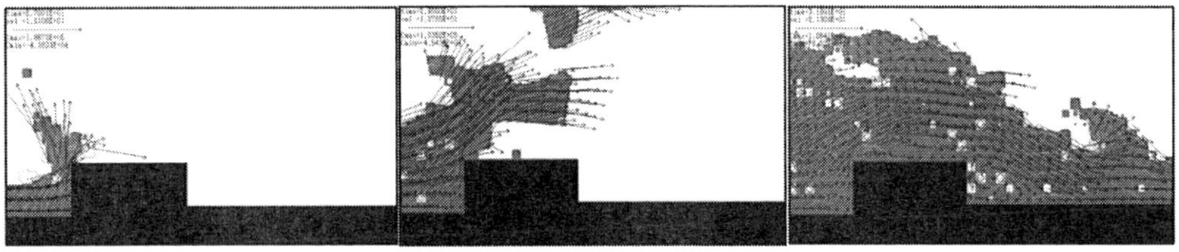

Fig. 6. Snapshot of tsunami waves hitting seawall.

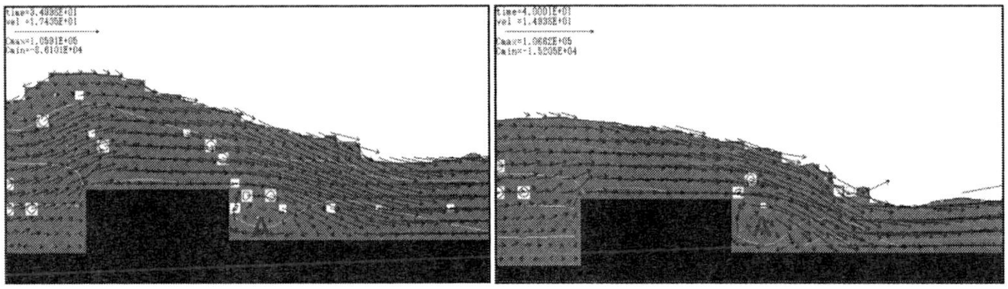

Fig. 7.

 DOI: 10.1051/matecconf/20179701065

Figure 7 shows the big volume of water that comes during the tsunami, and then passes the seawall, and causes continue turbulence at point A. Referring to Figure 2, it can be seen that the failure of seawall in Iwate prefecture is not because the seawall is damaged after the tsunami, but because the construction is toppled.

It is caused by erosion (scouring) on the sand foundation that the seawall is the toppled. In addition, the profile in Figure 8 shows that the velocity of the wave is going up and down with irregular frequency so that the destructive force behind the seawall becomes larger.

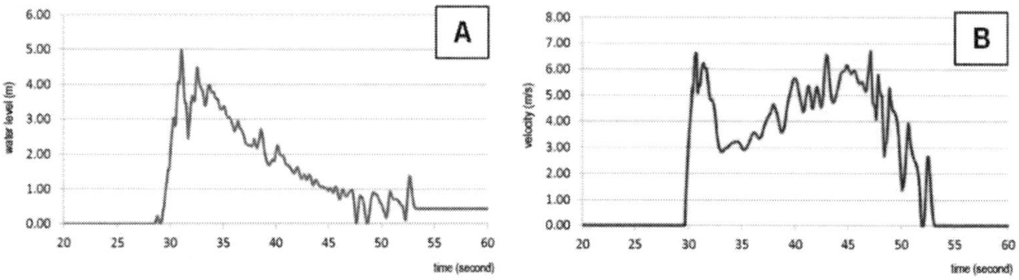

Fig. 8. Turbulence at the bottom of the seawall that could potentially.

One of the causes determining the amount of damage caused by the presence of seawall is that tsunami is a sea wave that moves with hydrodynamic power is a kind of force that may increase from time to time. When seawall built is not able to withstand the tsunami, it is actually going to add to the strength of hydrodynamic power.

If the seawall collapsed, the large internal energy adds to the hydrodynamic power making tsunami waves easily pass the seawall and quickly inundate settlements by bringing the seawall material.

5 Conclusion

Based on these results, it is known that the presence of seawall is capable of lowering the water level and reduces the speed of the wave at a distance of more than 300 meters from the construction of seawall. However, right behind the seawall, the opposite phenomenon occurs, the water level is getting higher yet the wave speed is not significantly reduced.

The large tsunami energy accumulates on the back of wave-retaining construction, where the dynamic hydrodynamic force is growing over time accompanied by the effects of turbulence. Turbulence in the incoming wave has resulted in erosion (scouring) at the bottom of the foundation, so the seawall is easily toppled a while after hit by the tsunami waves. According to the results of this research found that the seawall construction cannot reduce the effects of the tsunami. The existence of seawall construction adds horizontal force linear with greater energy which affected after tsunami hit the seawall.

References

1. Gajendra Prasad, Jack Rynn, Atu Kaloumaira, Sci. Tsunami Hazards, **18,** 35 (2000)
2. F. Imamura, H. Latief, N.T. Puspito, Journal of Natural Disaster Science, **22**, 25 (2000)
3. F. Imamura and H. Latief, *Proceedings of 2000 congress of Japan Society of Civil Eng.* (Tohoku Area, Moriyoka, Japan, 2000)
4. Tsunami Laboratory, *Integrated tsunami database for the world ocean,* (ICMMG SD, RAS, Novosibirks, Russia, 2005)

5. H. Latief, D.H. Natawidjaya, H. Sunendar, A.R. Gusman, Y. Tanioka, *Proceedings third international conference on urban earthquake engineering, Japan,* 259 (2006)
6. http://blogs.reuters.com/photographers-blog/2012/03/07/one-year-from-that-day/ (2014)
7. http://www.gettyimages.com/detail/news-photo/collapsed-sea-wall-stands-in-an-area-damaged-by-last-years-news-photo/151756230, (2014)
8. Y. Tsuji, F. Imamura, H. Matsutomi, S. Harada, Jumaidi, K. Arai, Pure. Appl. Geophys., **144**, 839 (1995)
9. Y. Fukui, S. Hidehiko, M. Nakamura, Y. Sasaki, Coast. Eng. J., **9**, 44 (1962)
10. D.H. Peregrine, Annu. Rev. Fluid Mech., **35**, 23 (2003)

Soil Loss Prediction on Mobile Platform Using Universal Soil-Loss Equation (USLE) Model

Supli Effendi Rahim[1,*], *Ahmad* Affandi Supli[2] , and *Nurhayati* Damiri[1]

[1]Agrotechnology Department, Agriculture Faculty, University of Palembang (UNPAL), 30139, Palembang Indonesia.
[2]School of Multimedia Technology and Communication, Universiti Utara Malaysia (UUM), 06010 Sintok Kedah, Malaysia.
[3]Agrotechnology Department, Agriculture faculty. University of Sriwijaya (UNSRI), 30862, Indralaya, Ogan Ilir, South Sumatra Indonesia.

Abstract. Indirect method for soil loss predictions are plentiful, one of which is Universal soil-loss equation (USLE) model. Available technology in mobile applications prompted the authors to develop a tool for calculating soil loss for many land types by transforming the USLE model into smart mobile application. The application is designed by using simple language for calculating each and every factor and lastly summing up the results. Factors that are involved in the calculation of soil loss are namely erosivity, erodibility, slope steepness, length of slope, land cover and conservation measures. The program will also be able to give its judgment for each of the prediction of soil loss rates for each and every possible land uses ranging from very light to very heavy. The application is believed to be useful for land users, students, farmers, planners, companies and government officers. It is shown by conducting usability testing using usability model, which is designed for mobile application. The results showed from 120 respondents that the usability of the system in this study was in "very good" classification, for three characteristics (ease of use, user satisfaction, and learnability). Only attractiveness characteristic that falls into "good" classification.

1 Introduction

Soil is a natural resource which can be utilized for agricultural purposes [1]. Two main functions of soil are namely as a nutrient sources for plants, and as the place where the roots of plants are anchoraged. According to Havlin et al. [2], soil is a dynamic material that always experiences changes both due to the inner materials contained in the soil, as well as due to the material coming from outside the soil. Those changes affect the productivity of soils. At the time where the productivity decreases, the damage of the soil is occurring.

* Corresponding author: sup_effendi@yahoo.co.id

Soil erosion is one of the important processes that cause soil damage. Erosion is simply the movement of material from one place to another by certain media such as running water, rainfall, wind, etc. Movement of soil from a place to another cause a number of unwanted impacts namely the openness of a land and the nutrients that are needed by plants disappear as part of the nutrients are eroded. Meanwhile, in places where the eroded soils that are sedimented cause disturbances of waterways or canals, if it is sedimented in rivers and lakes they may cause the decline in water surface quality [3, 4]. Nonetheless, Rahim [1] stated that sedimentation derived from erosion in upper parts can cause the increase of soil fertility in the lower parts of certain areas.

Knowing the rate of soil erosion in a given land is important for apart from knowing the amount of soil being eroded, the rate is useful in finding the solution for the problem through programs of prevention, rehabilitation and control. Such programs require data from soil erosion predictions. Soil erosion predictions can be conducted both direct and indirect methods. Predicting soil loss indirect means is difficult to do as it is time consuming. Indirect method for soil loss predictions are plentiful, one of which is USLE (Universal soil-loss equation). Apart from its weaknesses, the USLE model is widely used and so far it can produce reasonable data for land use planning.

According to Suripin [3], USLE is designed for long term prediction for soil erosion that is taking place in an area. The equation can also predict the soil erosion taking place in non-agricultural lands. However, the equation cannot be used to predict the sedimentation and cannot calculate sediments from gully erosion, river bank erosion and river bed erosion. The main reason why USLE is widely used is because the model is relatively simple and parameter inputs for the model are relatively easy to derive.

Prediction model for soil erosion using USLE is an empirical model that is taking into account a number of factors namely rainfall, soil characteristics and management of the land as well as of the plant. In a book written by Rahim [1], it is clearly described that the erosion rate prediction for an area can be conducted manually by calculating each and every factor that contribute to the eroding soils taking place in a given area.

The availability of mobile application that is most prevalent in the world nowadays prompted the authors to develop more quickly and practical in calculating soil loss that may occur in an area. This paper focuses on the development of a mobile application as a tool for calculating of factors that contributes to the soil erosion taking place in an area. This application is equipped with a very useful program, especially for the students of Faculty of Agriculture, Faculty of Technology – Civil engineering department, and Faculty of Forestry. This paper tries to describe the use mobile applications as a practical and fast tool for calculating soil erosion taking place in given areas.

The study foresees the need for the improvement in styles for calculating the soil erosion taking place in areas that is planned for certain land uses. The tool can offer recommendations for better management for a given land by knowing the factors that need proper correction and treatments as to reduce soil erosion that may occur during the utilization of the lands for various uses.

2 The framework for universal soil loss equation

Soil erosion model is the describing mathematically the processes of detachment, transportation and deposition of soil particles on the surface of a land. There exist many erosion models which were developed, namely USLE, and several other empirical models such as Revised Universal Soil Loss Equation (RUSLE), Modified Universal Soil-Loss Equation (MUSLE), which were developed from the concept of USLE itself. One of the

 DOI: 10.1051/matecconf/20179701066

physical model after USLE is called Griffith University Erosion System Template (GUEST) [5, 6].

Erosion model USLE has made possible planners to predict soil erosion taking place in any given land at certain slopes, at certain rainfall patterns, for a number of soil types and the implementation of conservation measures. USLE was designed for long term soil erosion prediction. The equation can also predict the soil erosion in many land types [7].

In calculating the hazard of soil erosion in a given land the factors need to considered are rainfall, length of slope, slope steepness, soil characteristics, land cover and conservations measures. Main factors for erosion to take place are rainfall and the existence of runoff. With the factors, the rate of erosion can be determined with USLE that was firstly developed by Wischmeier and Smith [8], as also narrated by Rahim [1].

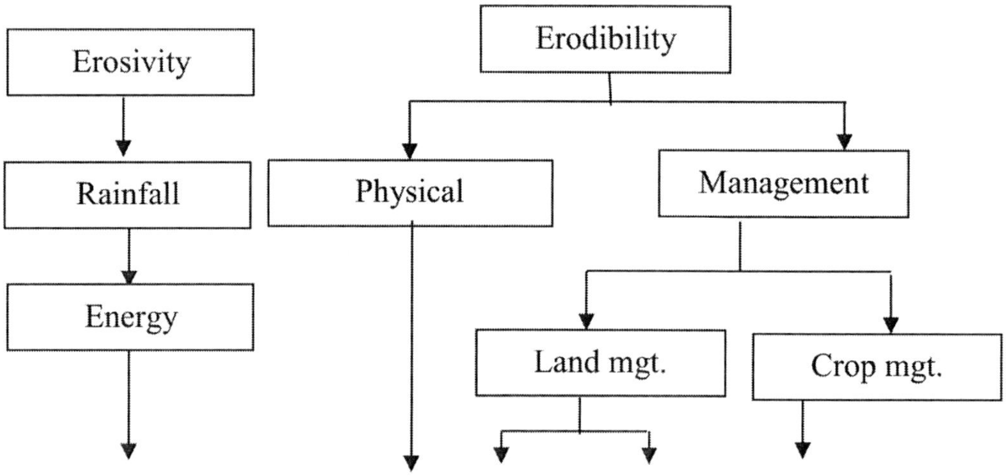

$$A = R \quad X \quad K \; X \; LS \; X \; P \; X \; C \tag{1}$$

where: A = Annual eroded soil (tonnes/ha)
R = Erosivity
K = Erodibiity
LS = Length and steepness of slopes
P = Conservation measures
C = Cropping management

In our study, erosivity factor is calculated using formula:

$$R = 6.119\,(RAIN)^{1.21}\,(DAYS)^{-0.47}\,(MAXP)^{0.53} \tag{2}$$

where: R = Average monthly Erosivity index
RAIN = Average monthly rainfall (cm)
DAYS = Average monthly rain days
MAXP = Maximum rainfall in 24 hours in the mentioned month

Erodibility factor is calculated using the following formula:

$$100\,K = (2.1\,M^{1.14})\,(12 - OM) + 4.20\,(s-2) + 3.23\,(p-3) \tag{3}$$

where:

DOI: 10.1051/matecconf/20179701066

K = Soil erodibility
OM = Organic matter (%)
S = soil structure code
P = permeability of soil code
M = M value can be also estimated if only known the soil texture
Slope length factor is defined mathematically as follows:

$$L = (l/22,1)^{m} \tag{4}$$

Slope Steepness is defined mathematically as:

$$S= (0.43 + 0.30s + 0.04s^{2}) /6.61 \tag{5}$$

where: S = actual slope steepness (%)

Land cover factor shows the overall effect of vegetation factor, soil surface condition, mulch, and land management which affect soil loss. The following table shows the values of C factor.

Spesific conservation measures have been made for the soils in many places. In Java such values have been determined.

3 Application for soil-loss prediction

Attempts to mobile applications for assessing land suitability conducted by Rahim et al. [9] have proved that such a simple and practical means to assess the suitability of the land is possible. In the current work, authors tried to use the principle to calculate the rate of soil loss from any given land uses. Since all land users always find it difficult to do when dealing with large amounts of data coming from the land quality of laboratory analysis of the soil and the availability of data Geo-physics and climate throughout Indonesia and in the world. What is important here is that much remains to be done when dealing with planning and monitoring of the soil loss rates of land uses.

Approach and soil loss calculation methods traditionally applied by land users no longer can be widely recommended. Indonesian government, in this case, the Department of Agriculture, can no longer be silent without giving a quick solution to provide a tool for planners to calculate rates of soil erosion rates of any given land. Authors, with the help of sponsors, will complete in one year or over a program that is friendly to the tablet, apple or windows that serve as a tool for calculating soil loss rates of given land uses in agricultural practices in general. This is very needed by farmers in remote areas of Indonesia.

Results of screenshot of the application can be seen at the figures below (see Figure 1 and Figure 2) :

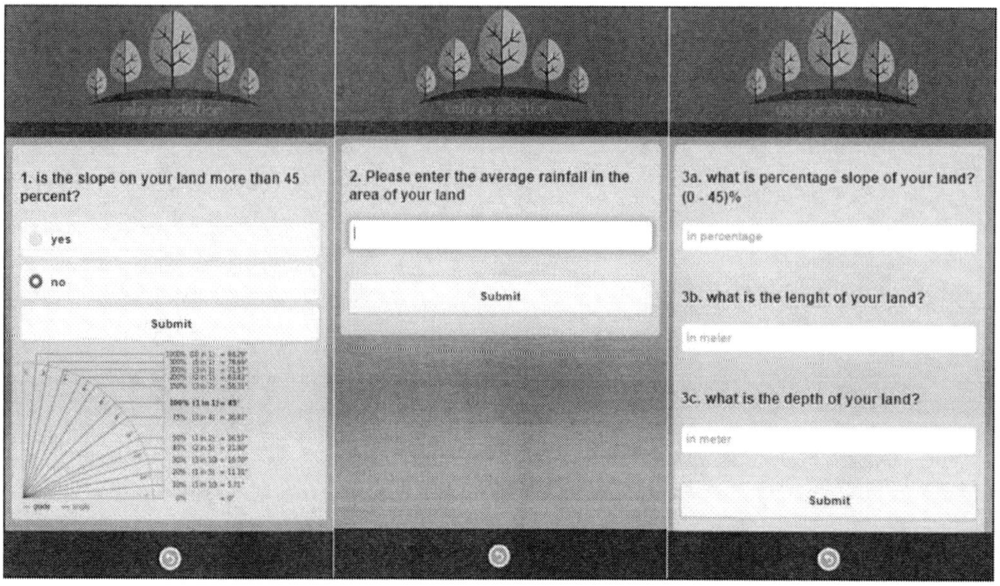

Fig. 1. The usle app questions.

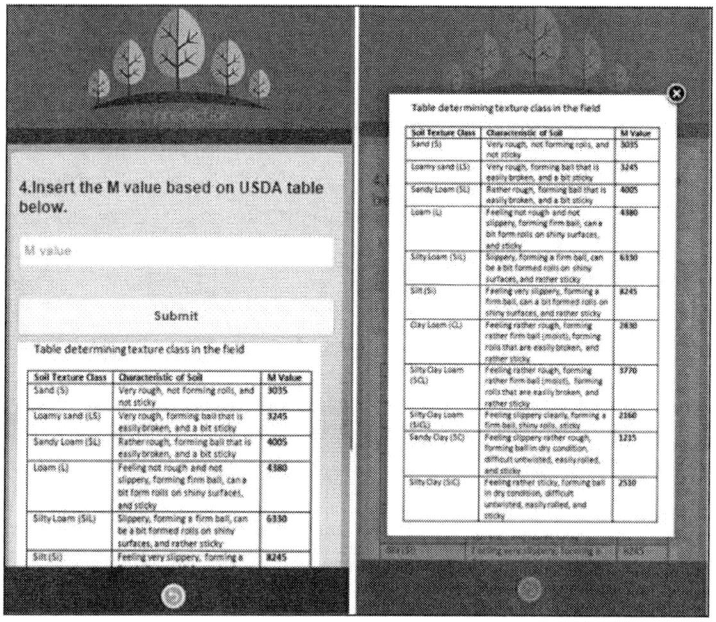

Fig. 2. Zooming feature.

The application is very straight forward and very easy to run in order to calculate the soil loss rates of users. The all variables in section 2 are all converted into 9 questions, which are easy to understand and have further detail description at several questions (see example in Figure 2), in a way to extract all information from a user. The zooming feature (Figure 2) is also provided in the app to see more clearly the additional description. Lastly, the result will be showed as a conclusion of soil loss rate (Figure 3). The conclusion for soil loss rate classification is given promptly (Figure 3) which shows factors that require treatments if the soil loss rate can be lowered to be allowable rates.

DOI: 10.1051/matecconf/20179701066

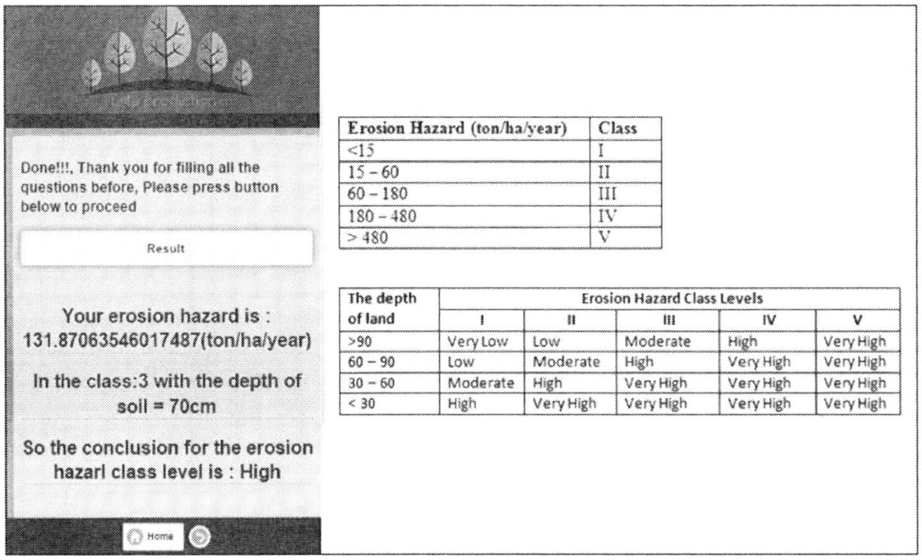

Erosion Hazard (ton/ha/year)	Class
<15	I
15 – 60	II
60 – 180	III
180 – 480	IV
> 480	V

The depth of land	Erosion Hazard Class Levels				
	I	II	III	IV	V
>90	Very Low	Low	Moderate	High	Very High
60 – 90	Low	Moderate	High	Very High	Very High
30 – 60	Moderate	High	Very High	Very High	Very High
< 30	High	Very High	Very High	Very High	Very High

Fig. 3. The result of soil loss.

4 Usability of the application

In order to measure the system's usability in this study, the authors adapt a model developed by [10]. This model is chosen due to the usability measurement is specifically designed for mobile application. The usability characteristic in this model is categorized into four characteristic: ease of use, user satisfaction, attractiveness, learnability. The authors then measure all variables using descriptive analysis in SPPS software. Beforehand, convenience sampling was chosen, in which the respondents were obtained from any members of the population who are conveniently available to provide it [11]. Furthermore, [12] reported that in order to determine the sample size, there exist rules of thumb that could be used:

i. Sample capacities larger than 30 and less than 500 are suitable for most research.
ii. For simple experimental research with tight experimental controls, successful research can be achieved with samples as small as 10 to 20 in size.

Therefore, all in all, this study managed to gather 120 responses to be involved in the experiment. The demography is exhibited in table 1 below:

Table 1. Demography of respondents' status.

		Frequency	Percent	Valid Percent	Cumulative Percent
Valid	lecturer	34	28.3	28.3	28.3
	employee	30	25.0	25.0	53.3
	students	18	15.0	15.0	68.3
	farmer	38	31.7	31.7	100.0
	Total	120	100.0	100.0	

All respondents were then required to fill the usability questionnaire that represents all the characteristics mentioned earlier. The questioner used the likert scale of 5 point (see Figure 4).

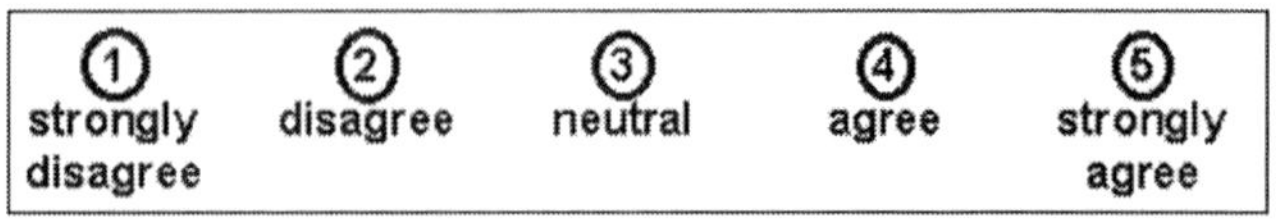

Fig 4. Linkert Scale 5-point

Then the all answers were analyzed by using descriptive analysis frequency. The calculation about mean, minimum, and maximum are analyzed. The results are shown as follows (see Table 2 and Table 3):

Table 2. Sub-characteristic Statistic Results

Descriptive Statistics

	N	Minimum	Maximum	Mean	Std. Deviation
Ease of use	120	2.00	5.00	4.0750	.60311
User satisfaction	120	2.00	5.00	3.9611	.65277
Attractiveness	120	2.00	5.00	3.1972	1.15186
Learnability	120	3.00	5.00	4.3375	.62565
Valid N (listwise)	120				

To interpret the results, the usage of five-point scale is used (see Table 3).

Table 3. The five-point Likert Scale.

Gap	Classification
1.00 – 1.80	Poor
1.81 – 2.61	Fair
2.62 – 3.41	Good
3.42 – 4.42	Very Good
4.43 – 5.23	Excellent

Based on Table 3, it can be clearly seen that almost all characteristics in table 2 reached in the range between 3.19 and 4.34 points. Thus, the attractiveness's characteristic (3.19) falls into good classification. Nevertheless, the results for the other characteristics (ease of use, user satisfaction and learnability) are included in very good classification, which have mean score 4.07, 3.96 and 4.33 respectively. It can be concluded that all respondents have very good perspective in usability of the mobile application in general.

5 Concluding remarks

A new solution for soil loss calculation is simple to run by changing the USLE model in smart mobile applications as the appropriate tool in lieu of soil loss calculation system which is manual. This application is designed with a simple language for each factor and also by utilizing Based System (RBS), an algorithm rules. Soil loss calculation in this paper is begun by calculating the erosivity factor, followed by erodibility, slope length, slope steepness, land cover and conservation measures. The program is also designed to give the user classifications of rates of calculated soil loss of a given land and suggesting factors that require further treatments as to lower any high rate of soil erosion so that it will become allowable one. Computer-based program has been designed and implemented as a means of calculating soil loss of any given land use. This program will be designed according to cross platform technology such as tablet, apple and windows. The program urgently need

DOI: 10.1051/matecconf/20179701066

officials to help them more easily manage their land for a better future. Last but not least, in terms of usability, the results showed that usability of the system are in "very good" classification for three characteristics: ease of use, user satisfaction, and learnability in which the all means of those are in the range between 3.96 and 4.33. Only the attractiveness dimension includes into "good" classification. In the future, this program will be designed according to the tablet as well as to other cross platform technology. The program is urgently needed by the users, farmers, companies, lecturers, students and government officials to help them more easily manage their land for a better future.

References

1. S.E. Rahim, *Pengendalian Erosi Tanah dalam Rangka Pelestarian Lingkungan Hidup* (Bumi Aksara, Jakarta, 2003)
2. J.L. Havlin, J.D.Beaton, S.L. Tisdale , W.L. Nelson, *Soil fertility and fertilizers: An introduction to nutrient management . Upper Saddle River, NJ* (Pearson Prentice Hall, 2005)
3. Suripin, *Pelestarian sumber daya tanah dan air* (Andi, 2002)
4. J.F. Boyle, A.J. Plater, C. Mayers, S.D. Turner, R.W. Stroud, J.E. Weber, J. Paleolimnol., **45**, 199 (2011)
5. A.A. Millward, E.M. Janet, Catena, **38**,109 (1999)
6. K. Subagyono, T. Vadari, R.L. Watung, F. Agus, *Land management for controlling soil erosion at microcatchment sale in Indonesia (No. H037205)* (International Water Management Institute, 2004)
7. K.G. Renard, G.R. Foster, G.A. Weesies, J.P. Porter, J. Soil Water Conserv., **46**, 30 (1991)
8. W.H. Wischmeier, D.D. Smith, *Predicting rainfall erosion losses-A guide to conservation planning* (1978)
9. S.E. Rahim, A.A. Supli, N. Shiratuddin, Nurhayati Damiri, Z.S. Mukti, *The International Joint Conference of Indonesia Malaysia Bangladesh Ireland* (Banda Aceh Indonesia, 2015)
10. A.A. Ali. *A framework for measuring the usability issues and criteria of mobile learning applications* (Doctoral dissertation, The University of Western Ontario, 2013)
11. R. Bougie, U. Sekaran, *Research methods for business: A skill building approach* (Wiley, London, 2010)
12. J.T. Roscoe, *Fundamental Statistics for the Behavioural Science* (New York, Holt, 1975)

Proposal for the risk management implementation phase in oil field development project by adding value on the refurbishment of critical equipment

Abdul Hamid[1,2,*], *Ishak* Bin Baba[1], and *Winardi* Sani[1]

[1]Faculty Technology Engineering Universiti Tun Hussein Onn Malaysia, 86400 Johor, Malaysia
[2]Faculty Engineering, Mechanical Engineering Department, Universitas Batam, 29464 Indonesia

Abstract. Refurbishment process is a conceptual stage in product life cycle. It is utilized in existing equipment in the field by adding value to recondition and repaired equipment. The main interest of this paper is to implement and design risk management implementation phase in oil field development project on the refurbishment of critical equipment in oil and gas industry. This paper is provided base on research and experiences in risk management and learned from practical team in industry which matched by an application in oil field development project in refurbishment of critical equipment. A framework of implementation phase for risk management in oil field development project in refurbishment critical equipment were reviewed and added value on communication skills of the project team to the stakeholder and organization, which support to external body and vice-versa. Risk management framework can be used for reference of refurbishment process with simply process and developed with same concept for the next wide development project in industry.

1 Introduction

Refurbishment process is a conceptual stage in product life cycle. Oil field equipment (OFE) such as tooling can be reconditioned after it is used in market. Performing the refurbishment process is to utilize the existing equipment in the field to determine the product cycle and sustainability [1]. Complete refurbishment of OFE equipment can eliminate waste along with the product used. These processes of refurbishment is similar to produce new OFE and depending of order scope, purchase order and contract but no limit for rental tool and equipment. Repair and remanufacture shall be complied with API spec 6A if it's an original equipment manufactured in accordance to the API international standard during lifecycle for continues service by user [2]. Many OFE Manufacturing companies in Batam Island Indonesia have operated refurbishment role and rent equipments and tools to support their organization survival and growth.

* Corresponding author: gn150003@siswa.uthm.edu.my

The traceability is important and shall be complied with standard requirement of API Spec 6A to increase profitability and compliance [2]. If it is not tracked properly and / or loose traceability, this will cause loosing an expensive drill and become cost prohibitive [3]. It is identified earlier to manage the traceability of the product during manufacturing process until delivered to field/hand over to client.

A risk is likelihood of specific consequence happening [4,5]. Refurbishment of OFE will potentially face more issue and risk in technical, quality, and economical uncertainty and more risk than new product scheme. Risk management is recognized as an integral part of a good management practice. Based on ISO 31000 as mentioned that a successful risk management initiative shall be proportionate to the all level of risk in organization, aligned with other corporate activities, comprehensive in its scope, embedded into routine activities and dynamic by being responsive to changing circumstances" [6].

2 Objectives

The aim of risk management is to obtain understanding by all parties in an organization and agreed on that can be risk really managed and improved through early detection and mitigation on product life cycle management such as remanufacture and refurbishment project. The objective of this paper is to propose risk management implementation phase in oil field development project by adding value on refurbishment product. The main interest of this paper is to review current method in the area of risk in refurbishment project and reflecting to risk and greenality.

3 Method approach

This paper is conducted based on research and experiences into risk management its potential to prevent and eliminate risk and learning from performance team in industry which matched by an application in project refurbishment in Batam manufacturing Indonesia.

3.1 Evaluating a risk management (Review and industry standard study)

OFE refurbishment requires more technical and economic uncertainty and risk [7]. Rush on delivery and limited time given is a risk. It is changing with project scope frequently and size of the project and the decision making process is hampered by the lack of clear in communication within client and manufacturer and definition can change a cost and profitability. Specific areas for risk management of conservation refurbishment project includes:

Design constrains. Varying an OFE and tools of level specification couple with restricted component and material selection has different risk implication [7]. A creativity Engineer and PM team to substitute material and focus on technical problem resolution rather than consider risk in systematic manner [8]. Poor design can cause of time delivery, cost and quality achievement and then expertise person is required to focused on resolution for project constrains [9]

Stakeholders and involvement external bodies. There are enhanced communication risk ranging from different area of expertise and between different corporate structure and culture and line communication between stakeholders may require different frame and task. The available ability of the professional team member with requisite experience or expertise at particular project time during project period. Invention from external bodies from various statutory bodies and any group it's potentially imposing expensive mandatory requirement or major restrictions upon the program work and this risk shall be evaluated at early project stage.

Manufacturing load/capacity. When a refurbishment OFE deal with project/stakeholder and sales with clearly scope and defined of area re-pair or remanufacturer than schedule and planning shall have capacity in plant. A risk can be identified by PM for any opportunity slot in and any project constrains face on high load capacity during the period of refurbishment schedule. Lack of communication and control the project has a risk to deliver it as per desire schedule and requirement.

Supply chain management (SCM). Several spare parts and material are quite difficult to supply with limited sources and lead time. High demand material and dependence on supplier or vendor make a manufacturer at a high risk [10]. Assessment to this SCM role for refurbishment OFE would be important to implement. The refurbishment project orientation in supply management which is a manufacturer face on adversity and deeper risk management culture and applied project management techniques to greater focus on risk management [10].

The OFE development project in the refurbishment critical equipment is a systematic process as suggested on API spec 6A for re-manufacturing annex H [2]. Figure 1 is the general Christmas tree assembly as critical equipment in the well field.

Fig. 1. Typical Christmas tree nomenclature as API license manufacturer [2].

Typical Christmas tree nomenclature as shown in figure 1 is used to control well production that consist of assembly of equipment including tubing head (bonnet), valves, tees, crosses, top connectors and chokes that attached to the uppermost of tubing head [2].

3.2 The risk management process for refurbishment project (Outline)

Generally, a project risk management (PRM) has been developed to associate the process model in literature [11, 12] :

Plan risk management. It is defining how to conduct risk management activities for refurbishment project. The risk management plan is vital to communicate with an agreement and knowledge supported by all stakeholders to ensure a risk management process is performed effectively over the project life cycle.

Identify risk. Listing all potential impact on the project [13] .It's a process of detection which a risk may be affecting and dangerous forwards the project. The objective is for the

PM to anticipate event occurred of the risk. Identifying risk is a typical continuous iterative process that cause the risk to exist and identified during project through its life cycle.

Qualitative and quantitative analysis. In practice these are method in nature to further analysis by assessing the probably and impact of the risk and used a numerical data collection. This process is to obtain risk assessment documentation [5].

Risk Prioritization. It can be conducted as long as measurement of priority system is used and taking from qualitative and quantitative risk list into account with rank.

Risk Response. To develop option and action to enhance opportunities and reduce/eliminate a thread to project objectives. This process is following the qualitative and quantitative risk analysis process. Selected the optimum risk response from each option is required to address the risk.

Monitoring risk. It can be monitored along the process in progress. Implementation of risk response plan, tracking the risk and evaluate with risk response. Evaluating implementation risk assessment effectiveness and review to issue a trigger impulse to support make a decision of significant change for risk assessment [12].

4 Result

4.1 Risk management method implementation (by adding value on communication skills)

The selected method was present from the possibility of refurbishment process of OFE for more focused to the method are used and effort eventually covers the risk management processes. Implementation phase on the risk management in the oil field development project on the refurbishment process can propose to use a framework as shown in Figure 2.

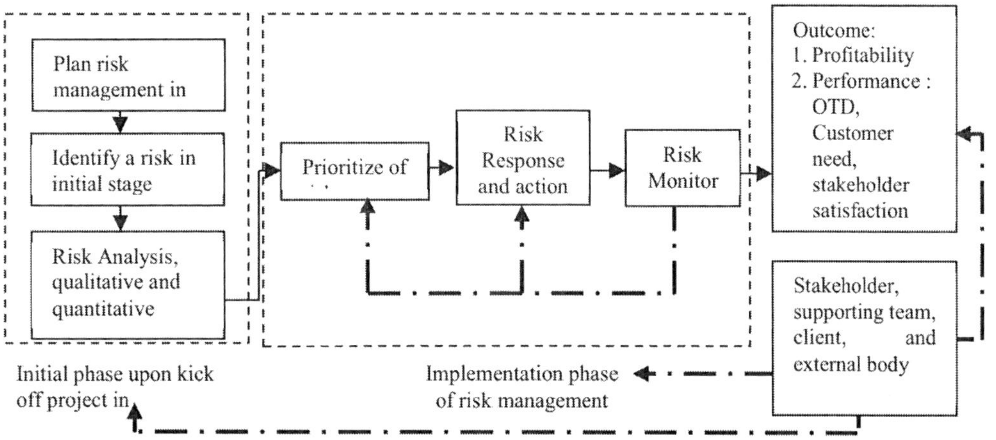

Fig. 2. Framework study to propose implementation phase for risk management oil field development project by adding value on refurbishment product.

Method of risk prioritization. This concept is to describe a probability and impact of risk. On implementation phase of risk management for development OFE development project in refurbishment with base/simple technique such pareto analysis and top 3 or 5 or 10 rank for the risk plan [14].

Method of Risk response and action. A checklist can be use following what an analysis count in risk assessment matrix. Risk assessment is used to measure any probability and impact on matrix grid[15]. It is required an action to prevent any missed and slip following the rank of risk and matrix grid.

Monitoring risk and feedback. There is a line communication in figure 2, were discussed with review literature and industry standard. Skill up communication is highly recommended to cover following aspect in refurbishment process to complete the task such as improved in engineering and technical aspect includes risk plan, identification, and analysis. Implement effective communication between PM and team to an internal and external organization and manufacturing include logistic and SCM and stakeholder.

5 Summary

The necessary implementation phase on risk management for OFE development project in refurbishment of critical equipment Christmas tree has been discussed with taking a literature, industry standard, and experimentally study. It is clear definition that risk can be manage and monitor by adding an action or correction by implement two way communication for stakeholders and PM team. A Risk management framework can be used for reference of refurbishment process with simple process and develop for the next project or development project in OFE manufacturer.

We would like to thank Faculty Engineering Technology UTHM and grateful for industry partner in Batam island and Universitas Batam Indonesia to significant support on this research project.

References

1. M. Richard, S. David, *Green Project Management*, (CRC Press NW, 2011)
2. C. API, M. Annex, O.F. *Specification for Wellhead and Christmas Tree Equipment*, (U.S. National, 2009)
3. S. Tiku, A.F. Veneruso, R.K. Etchells, M. Pecht, Oil Gas SCI. Technol. J., **60(4)**, 721 (2005)
4. A. Polytechnic, O. Adibo, Anal. Risk Manage. J., **5(29)**, 139 (2013)
5. A. Ahmed, B. Kayis, S. Amornsawadwatana, An International Journal, **14(1)**, 22 (2007)
6. Airmic, Alarm, Irm, Risk Manage. J., **7(1)**, 20 (2010)
7. J. Reyers, J. Mansfield, Struct. Surv. J., **19(5)**, 238 (2001)
8. J. Bowers, A. Khorakian, Eur. Innov. Manage. J., **17,** 25 (2014)
9. N.V. Thuyet, S.O. Ogunlana, P.K. Dey, Oil Gas. Int. Energy. Sect. Manage. J., **1(2)**, 175 (2007)
10. G.J.L. Micheli, E. Cagno, M. Zorzini, MRN, **31(11)**, 846 (2008)
11. P. Smith, *Managing risk proactively in product development projects*, 1 (2002)
12. Project Management Institute, *Project Management Body of Knowledge A Guide to the Project Management Body of Knowledge,* (2000)
13. Y.H. Park, Asian Qual. J., **11(1)**, 39 (2010)
14. J. Oehmen, B. Dick, U. Lindemann, W. Seering, *Proceedings DESIGN 2006, 9th International Design Conference*, (Dubrovnik, Croatia, 2006)
15. H. Zhang, M. Yongbo, *International Business and Management,* **23(1)**, 99 (2011)

 DOI: 10.1051/matecconf/20179701068

The Effect of HLRs on Nitrogen Removal by Using a Pilot-scale Aerated Steel Slag System

R. Hamdan[1,*], *Mohd. Razi* M.A.[2], and *Nurul Farhana* Zulkafli[1]

[1]Dept. of Civil Engineering Technology, Faculty of Engineering Technology, Universiti Tun Hussein Onn Malaysia 86400 Parit Raja, Batu Pahat, Johor, Malaysia.
[2]Department of Water and Environmental Engineering, Faculty of Civil and Environmental Engineering, Universiti Tun Hussein Onn Malaysia, Batu Pahat, Johor Malaysia

Abstract. Discharge from domestic wastewater treatment plant amongst the main sources of nitrogen pollution in the environment. However, to remove nitrogen conventionally in domestic wastewater require high cost and complex chemical treatment method. Vertical flow aerated rock filter emerged as one of attractive alternative wastewater treatment method due to simplicity and compactness of the system. However, the application is yet to be developed in warm climate countries in particular Malaysia. Therefore, this study was conducted to investigate the effect of hydraulic loading rate (HLR) to the performance of a pilot-scale Vertical Flow Aerated Rock Filter (VFARF) in removing nitrogen from domestic wastewater using pilot-scale VFARF systems with steel slag as the filter media. Furthermore, this study has been designed to focus on the effects of two HLRs; 2.72 and 1.04 m^3/m^3.day. Influent and effluent of the filter systems were monitored biweekly basis for 11 weeks and analyzed for selected parameters. Results from this study shows that the VFARF with HLR 1.04 m^3/m^3.day has performed better in terms of removal ammonium-nitrogen and TKN as the system able to remove 90.4 ± 6.9%, 86.2 ± 10.7%, whilst the VFARF with 2.72 m^3/m^3.day remove 87.4 ± 9.9%, 80 ± 11.7%, respectively. From the observation, it can be concluded that nitrogen removal does affect by HLR as the removal in lower HLR system was higher due to high DO level in the VFARF system with 1.04 m^3/m^3.day which range from 4.5 to 5.1 mg/L whilst the DO level was slightly lower in the VFARF system with 2.72 m^3/m^3.day in the range of 3.7 to 4.5 mg/L.

1 Introduction

Rock filter system has been widely used for more than 30 years ago in the United States to remove algal solids and associated BOD in effluents mainly from primary maturation ponds [1]. Conventionally, a rock filter is designed to remove algae suspended solids in pond effluent. The pond effluent undergoing further polishing treatment within the rock filter system is allowed to travel either horizontally or vertically through a submerged porous rock media bed. Whilst the liquid flows through the void spaces, the algae get

* Corresponding author: rafidahh@uthm.edu.my

trapped on the rock media surfaces. The use of rock filters for algae removal has been studied extensively at Eudora, Kansas; California, Missouri; and Veneta, Oregon [2]. Filtration is one of the method for the treatment of water and it is one of the oldest, simplest and widely used methods. It is the process of removing suspended solids from raw water by passing the water through a permeable fabric or porous bed materials. For large quantity of water, sand is generally used as the filter medium because it is cheaper and effective. Simplicity of filtration makes the process attractive for use in small communities and developing nations. Mara [3] also highlighted that the aerated rock filter system is an ideal treatment train for secondary treatment, because of their land area savings and potentially replacing the requirement for maturation ponds.

Rock filters has been used to treat effluents from domestic Wastewater Treatment Plant. From the previous study conducted by Hamdan (2010), an aerated system was introduced in rock filter system to remove ammonia nitrogen and phosphorus. Aeration also improve BOD and TSS removals. While, unaerated system primarily for algal suspended solids and organic matter removal, BOD [4]. Most rock filter operating systems were designed for horizontal flow with the rock bed placed at or near the effluent end of the final cell in the lagoon system. In general, vertical flow systems, such as ones located in Veneta, Oregon and West Monroe, Louisiana, perform better than horizontal flow systems.

Furthermore, previous study conducted by Hamdan [5], indicated that the filter using vertical flow aerated rock filter (VFARF) facilitate nitrogen removal better than the horizontal flow aerated rock filter (HFARF). The systems are also moderately inexpensive, have low energy requirements and do not require highly skilled personnel [6]. However, the VFARF system design is needed to be developed under warm climate as has been ssuggested in Hamdan and Mara [7]. Therefore, this study was conducted by using the VFARF system with low cost media which is steel slag. The purpose of this investigation was to evaluate the effect of different HLRs on ammonia nitrogen removal from domestic wastewater under Malaysian condition.

2 Materials and methods

2.1 Pilot-scale unit

Two identical vertical-flow aerated RF (VFARF) systems with 2.0 m height x 0.3 m diameter have been developed at the experimental station were operated in parallel at our experimental station at Wastewater Treatment Plant, Batu Pahat, Johor, Malaysia (1050'37.05"N, 102056'49.43"E). The WWTP comprises domestic wastewater from residential area consists of 15800 of population equivalent (P.E) about 90% and the rest 10% is from industrial wastewater. Wastewater after pre-treatment was pumped into primary sedimentation tank and then pumped into two parallel VFARF base using 15mm reinforced plastic pipework connected to a polyvinyl chloride (PVC) inlet strainer using peristaltic pumps with W77200-62 pump heads (Cole-Parmer Masterflex L/S Model 7524-40, USA). The Cole Parmer air flow meters were installed at the air inlet of the VFARFs and aerated using JUN Air Compressor (OF302-25B, Denmark). The pilot-scale VFARF systems have been operated at 1.04 m3/m3.day and 2.72 m3/m3.day and was maintained at 10 L/min of aeration rate using blast furnace slag media with 10-20 mm grain size.

2.2 Wastewater sampling and analysis

For this study, wastewater influent and effluent samples have been analyzed twice a week for the selected parameters including TKN, AN, DO, and nitrate, for both VFARF systems.

All the sampling, handling, preservation and laboratory analyses have been conducted according to *Standard Methods for Water and Wastewater Examinations* [8]. Laboratory analysis of collected samples were carried out at Wastewater Engineering Laboratory, Faculty of Civil and Environmental Engineering, UTHM. The VFARF performance have been monitored throughout this study. Final effluent quality of both VFARF systems then compared with Effluent Standard B of Malaysian Environmental Quality (Sewage) Regulations 2009.

3 Results and discussions

3.1 Total kjedhal nitrogen removal

Average concentrations of TKN in the influent was at 37.9 ± 17.4 mg NH_3- N/L. Effluent of the VFARF system average at 7.32 ± 5.1 mg NH3- N/L for 2.72 m^3/m^3.d whilst for HLR 1.04 m^3/m^3.d the average was at 4.9 ± 4.6 NH_3- N/L. TKN was consistently eliminated more efficiently in the VFARF system with HLR 1.04 m^3/m^3.d with highest value of removal efficiency. The average removal efficiencies in the VFARF system with HLR 2.72 m^3/m^3.d was at 80 ± 11.7% whereas their average removal efficiency in the VFARF system with lower HLR which is 1.04 m^3/m^3.d was at 86.2 ± 10.7 during the monitoring period as illustrated in Fig 1 represents these value. TKN concentration in the effluent was mainly in the form of ammonia [12]. As the HLR reduced, the removal efficiency was increased. According to Gerardi [12], the value for TKN is ranged from 8 to 35 mg NH_3- N/L. VFARF with HLR 1.04 and 2.72 m^3/m^3.d is within the limit. TKN removal were found to be relatively high after the wastewater was further treated in both VFARF filters. Results from this study shows that the VFARF with HLR 1.04 m^3/m^3.d has performed better in terms of removal TKN. However, both effluent systems however do not exhibit a significant different in terms of removal efficiencies. TKN removal were found to be relatively high after the wastewater was further treated in both filter.

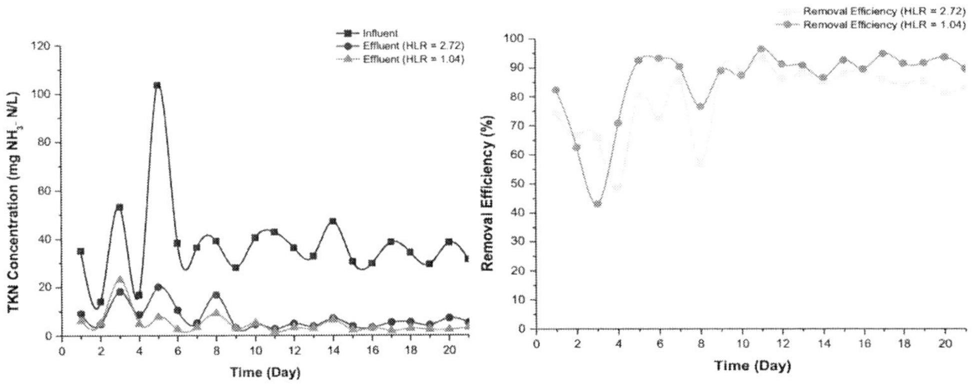

Fig. 1. The VFARF Influent and Effluent TKN Concentration and Their Removal Efficiency.

3.2 Ammonium nitrogen removal

Influent concentrations of AN entering the VFARF systems range from 4.90 to 33.41 mg NH_4- N/L. with an average of 25.3 ± 7.8 mg NH_4- N/L. Average effluent concentration of ammonium nitrogen (AN) for HLR 2.72 m3/m3. d was at 2.8 ± 1.5 mg NH4- N/L during this monitoring period whilst for HLR 1.04 m3/m3. d the average was at 2.2 ± 1.4 mg NH4- N/L, respectively (see Fig 2).

The concentration of ammonia nitrogen in the influent is higher than effluents in both VFARF systems. Ammonia nitrogen is mainly removed through nitrification process. The presence of ammonia nitrogen is due to incomplete nitrification process in the filter system. Nitrified wastewater is then can be further reduced in denitrification zone which is required anoxic condition. The highest removal efficiency is 97.69% over all sampling periods. Hence, the average removal efficiency was at 87.8 ± 9.9% for VFARF filter with HLR 2.72 m3/m3. d and 90.4 ± 6.9% for the lower HLR which is 1.04 m3/m3. d. The average removal efficiency performed slightly better in VFARF with lower HLR (1.04 m^3/m^3.d). This was attributable to the higher dissolved oxygen content, and a better nitrification rate [14]. Through the VFARF systems, ammonia nitrogen can be removed efficiently from domestic wastewater. In addition, VFARF system is more appropriate due to their ability in removing ammonia nitrogen.The low content of ammonia nitrogen shows the complete nitrification process. Effluents of ammonia nitrogen from both VFARF systems is complying with the effluent permissible limit for 'Standard B' from Environmental Quality (Sewage) Regulations 2009 (PU (A) 432) which is 20 mg/L. The value of ammonia nitrogen must follow standard prior discharge to river because higher ammonia nitrogen discharge to river will affect the aquatic life. As a conclusion, the VFARF system with HLR of 1.04 m^3/m^3.d eliminated AN more successfully than the VFARF system with HLR of 2.72 m^3/m^3.d as their capabilities in removing AN were consistently higher than the VFARF system with 2.72 m^3/m^3.d.

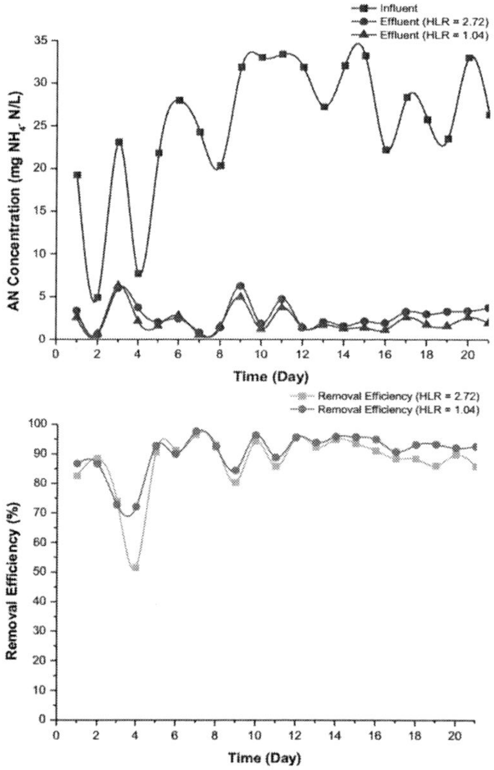

Fig. 2. The VFARF Influent and Effluent AN Concentration and Their Removal Efficiency

Furthermore, high concentration of nitrate-N in the VFARF effluent shows that nitrification process is taking place as AN has been converted to nitrate. An average concentrations of nitrate-N in the VFARF system effluent was 30.6 ± 14.5 mg NO_3-N/L and 33.0± 16.4 mg

NO3-N/L for HLR 2.72 m^3/m^3.d and 1.04 m^3/m^3.d, respectively as shown in Fig 3. From this observation, conversion of nitrate-N was slightly higher in the VFARF with 1.04 m^3/m^3.d HLR compared to the VFARF with 2.72 m^3/m^3.d HLR. This condition shows that nitrification rate was slightly higher in the system with higher nirate-N concentration in final effluent. Therefore, this observation then supported by nitrification rate data that was relatively higher in the VFARF system with HLR 1.04 m^3/m^3.d compared to VFARF system with HLR 2.72 m^3/m^3.d as illustrated in Fig. 4. Their average nitrification efficiency in the filter with HLR 2.72 m^3/m^3.d was at 80 ± 11.7% whereas their average nitrification efficiency in the filter with HLR 1.04 m^3/m^3.d was at 86.2 ± 10.7%. The rate of nitrification was found to be higher in filter with HLR 1.04 m^3/m^3.d. According to the previous study, the system achieved good nitrification and carbonaceous oxidation rate in treating domestic-strength synthetic wastewater by using vertically moving biofilm systems.

In addition, nitrification rate is also affected by dissolved oxygen (DO) level. It shows that the DO rate of the influent is more than 0.2 mg/1. The rate of the influent is between 0.5 to 2.5 mg/L. At lower DO than 1 mg/L, nitrification slows down in general. DO was found to be higher in the effluent of VFARF with 1.04 m^3/m^3.d whilst the lower DO was determined in the effluent of VFARF with 2.72 m^3/m^3.d. Average DO in effluent with 2.72 m^3/m^3.d and 1.04 m^3/m^3.d was at 4.1 ± 0.4 mg/L and 4.8 ± 0.3 mg/L, respectively. According to Gerardi [12], the nitrification rate was observed to improve at DO concentrations >3 mg/L. When DO drops to below 2 mg/L for an extended period, nitrification would be inhibited. The rates of nitrification are higher when the DO levels in the system are higher. According to the increasing level of DO in the systems, the VFARF pilot-scale system provides more favourable conditions for nitrification to occur as well as further removal of biodegradable organic matters. Hence, the VFARF pilot-scale is appropriate system to enhance the nitrogen removal mainly through nitrification process. As the organic matter in the VFARF influent were sufficiently lowered, the conditions will then allow nitrification to occur. In this environment, the growth of nitrifying bacteria becomes optimal as the nitrifying bacteria use DO and inorganic carbon to convert ammonia to nitrite and nitrate. The consumption of the DO is needed to oxidize ammonia to nitrite and later to nitrate. From this study, it was found that the DO level of the VFARF system with HLR 1.04 m^3/m^3.d. was significantly higher than VFARF system with HLR 2.72 m^3/m^3.d. Concentration of DO has a significant effect on the rates of nitrifier growth and nitrification in biological waste treatment systems [13].

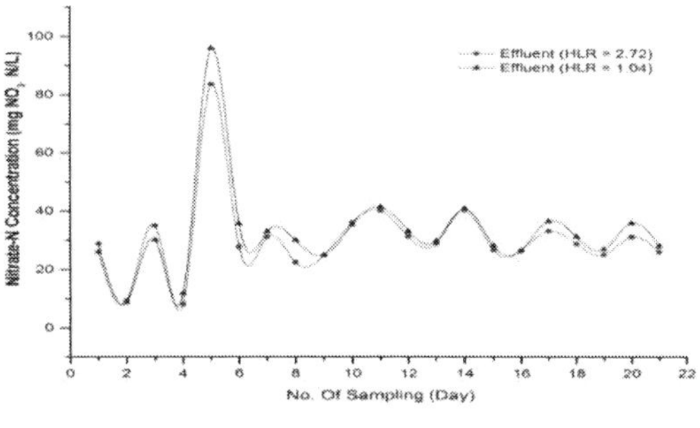

Fig. 3. The VFARF effluents Nitrate-N concentrations

 DOI: 10.1051/matecconf/20179701068

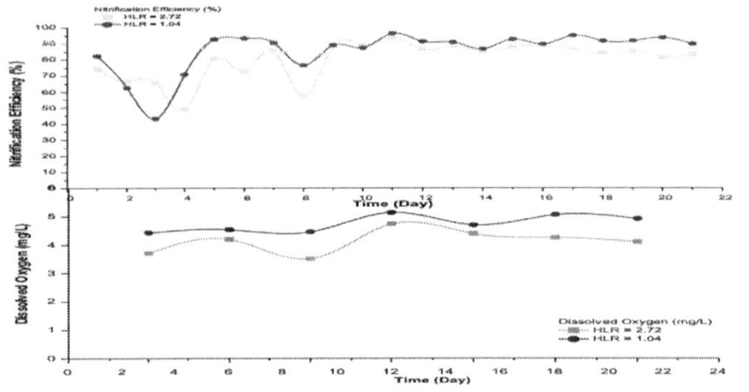

Fig. 4. Nitrification Rate of the VFARF and Their DO profile

4 Conclusion

The filter systems with HLR 1.04 m^3/m^3.d achieved a higher AN and TKN removal efficiency than filter with HLR 2.72 m^3/m^3.d. Nitrogen removal does affect by HLR as the removal in lower HLR system was higher due to high DO level in the VFARF system with 1.04 m^3/m^3.d which range from 4.5 to 5.1 mg/L whilst the DO level was slightly lower in the VFARF system with 2.72 m^3/m^3.d. in the range of 3.7 to 4.5 mg/L. It indicates that, the objective of this study is achieved. Furthermore, both effluents quality of the steel slag filter complies with Malaysian effluent discharge of standard B Environmental Quality (Sewage) Regulations 2009 (PU(A) 432). The application can be developed in warm climate countries in particular Malaysia.

References

1. W.J. O'Brien, R.E. McKinney, W.J. Martin, Wat. Sewage Works, **120(3)**, 66 (1973)
2. S.C. Reed, E.J Middlebrooks, R.W. Crites, *Natural Systems for Waste Management and Treatment 2nd Ed.*, (New York, USA, McGraw-Hill, 1988)
3. D.D. Mara, M.L. Johnson, Water Environ. J., **19(2)**, 143 (2006)
4. M.L. Johnson, Water Environ. J., **19(5)**, 143 (2005)
5. R. Hamdan, *Aerated Blast Furnace Slag Filters for Enhanced Nitrogen and Phosphorus Removal from Small Wastewater Treatment Plants,* (The University of Leeds, Ph.D Thesis 2010)
6. US Environmental Protection Agency, *Enhanced Nutrient Removal - Phosphorus, Onsite Wastewater Treatment Systems Technology Fact Sheet 8,* (Washington, DC, 2000)
7. R. Hamdan, D. Mara, Water Sci. Technol., **63(5)**, 841 (2011)
8. American Public Health Association, *Standard Methods for the Examination of Water and Wastewater, 21st ed.*, (Washington, DC, 2005)
9. M.A.A. Coetzee, M.M.P. Roux-Van, J. Badenhorst, Int. J. Environ. Res., **5(1)**, 119 (2011).
10. G.A. Ekama, M.C. Wentzel, Water Sci. Technol., **39**, 1 (2009)
11. M.S. Lee, A. Drizo, D.M. Rizzo, G. Druschel, N. Hayden, E. Twohig, Water Research, **44 (14)**, 4077 (2010)
12. K. M. Udert, T.A. Larsen, M. Biebow, W. Gujer, Water Res., **37**, 2571 (2003)
13. M.H. Gerardi, *Wastewater Bacteria,* (New Jersey: John-Wiley & Sons 2006)
14. B. Halling-Sorensen, S.E. Jorgensen, *The removal of nitrogen compounds from wastewater Volume 54*, (Elsevier, 1993)

MATEC Web of Conferences 97, 01069 (2017) DOI: 10.1051/matecconf/20179701069
ETIC 2016

Parameter Estimation and Verification of Unmanned Air Cushion Vehicle (UACV) System

Mohd Zamzuri Ab Rashid[1,*], *Hairol Nizam* Mohd Shah[1], *Md Nazri* Othman[1], *Esmail ali* Ali Alandoli[1], *Mohd Shahrieel* Mohd Aras[1], *Marizan* Sulaiman[1], *Mohd Faruq* Abdul Latif[2]

[1]Faculty of Electrical Engineering, Universiti Teknikal Malaysia Melaka, Hang Tuah Jaya, 76100 Durian Tunggal, Melaka, Malaysia.
[2] Faculty of Engineering Technology, Universiti Teknikal Malaysia Melaka, Hang Tuah Jaya, 76100 Durian Tunggal, Melaka, Malaysia.

Abstract. This project is mainly about the dynamic modelling and parameter estimation of Unmanned Air Cushion Vehicle (UACV). The purpose of developing mathematical model of the Unmanned Air Cushion Vehicle (UACV) is due to its under actuated nonlinearities where it has less input compared to the output required. This system able to maneuver over land, water and other surfaces either at certain speed or maintain at a stationary position. In order to model the UACV, the system is set to have two propellers which are responsible to lift the vehicle by forcing high pressure air under the system. The air inflates the "skirt" under the vehicle, causing it to rise above the surface while another two propellers are used to steer the UACV forward. UACV system can be considered as under actuated since it possess fewer controller inputs that its degree of freedom. The system's motions are defined by the six degrees of freedom which are; heaved, sway and surge. Another three components are rotational motions which can be elaborated as roll, pitch and yaw. The problem related to UACV is normally related to obtaining accurate parameters of the system to be included into the mathematical model of the system. This is due to the body inertia of the system during the static and moving condition. Besides, the air that flows into the UACV skirt to create the cushion causes imbalance and will affect the system stability and controllability. In this research, UACV need to be mathematically modelled using Euler-Lagrange method. Then, parameters of the system can be obtained through direct calculation and Solidworks software. The parameters acquired are compared and verified using simulation and experimental studies.

1 Introduction

This paper presents parameter estimation and verification of Unmanned Air Cushion Vehicle System (UACV). UACV is a type of Air Cushion Vehicle (ACV) where ACV is driven by human while UACV maneuver autonomously. UACV is a system that capable to move forward and reverse, turn left and right, and able to turn around 360 degrees by utilizing air

---

* Corresponding author: zamzuri@utem.edu.my

cushion that is generated underneath. It also can lift on underling surface and it is considered as advantages compared to other vehicles such as cars and ships. This UACV can be lifted by air flow through the air chamber between the deck, skirt and ground surface using a motor's propeller while another two propellers are used to provide air thrust which to control the movement of the UACV. In order to determined the parameters of the UACV system, a UACV has been designed and drafted using the SOLIDWORKS software. Then, the parameters of the UACV such as mass, mass moment of inertia, centre of gravity are generated based on the materials assigned on the system design. The parameters acquired then are inserted into the mathematical model developed earlier. The system model is simulated and compared with the experimental platform in order to verify the parameters acquired. The experimental setup is fabricated by utilizing materials like Perspex, aluminum, steel and plastic. The IMU sensor is used in this project to detect the movement of these UACV in surge, heave and also yaw. The objectives of this project are summarized as follows: (i) To design and acquire parameters via SolidWork software and develop of the laboratory scale of UACV, (ii) To verify the parameters of the UACV system in surge, heave and yaw by simulation and experimental studies.

2 Motivations

UACV system is a nonlinear system which has six degrees of freedom and quite challenging to control its movement. The challenge poses by this system has attracted various researchers to conduct and explore several fields such as system design, controller design, test the performance either using simulation or experimental studies. In this literature reviews, several past researches done by expertise have been thoroughly analyzed. In the paper by [1], the author had designed a lightweight UACV that has mass below 1.5kg and consists of the motor, batteries and other accessories. It was designed in a rectangular shape to ensure it has sufficient mounting points to allocate additional electronic and hardware and to maximize the surface area. The body of UACV was made by polystyrene. Two motors that were selected and used to lift the UACV are Johnson 400 that motors were secured on the top deck by hot glue and using 4 inches propeller. Next, in [2], the original air cushion vehicles with four cockpit controls were reduced to two, where turning rate and velocity were independently commanded. It was designed to cater its weight payload over surface of water. A throttle of this ACV was used to accelerate the system, while the steering wheel was to turn the ACV as resulting control system. In order to improve good handling, steering wheel deflection yield a turn rate proportional to the wheel angle. The author in [3] developed a high performance and compact a small size one seated ACV vehicle. The design combined a concept of bag skirt, peripheral jet and perforation was proposed for suitability in urban context.

However, in another paper done by [4] required novel sensory system for UACV where this task differed from conventional approach. This UACV was designed with the implementation of optic flow-based autopilot, a tiny magnetic compass and a micro-gyrometer. The two minimalist eyes were mounted sideway opposite to each other, and give the lateral Optic Flow (OF) to be measured. The paper in author [5] using the RCX controller that wired on the UACV and it will receive the signal from the computer using a USB tower. The program was transfered by the tower to the controller to operate the hovercraft system. The main purpose of the program was to have the light sensors recognize specific colors and to turn the fans ON and OFF so the UACV can move about. This author had two different programs that were written for their project. The first program was written to have the sensors followed a white path that was bordered in by two black strips on either side, which also guided it and for the second program was written to have the sensors to follow a single black strip of tape that would guide the UACV along its path.

Besides that in [6], based on the adaptive dynamic inversion control scheme, a neuralnetwork conjunction with a line-of-sight guidance law was adopted in this work for the under actuated UACV followed a prescribed path. Next, the author wrote this paper in [7] to develop a new algorithm that account for bounded sensor noise and bounded model uncertainty by the dynamic scheduling of multiple receding horizon control of the systems. Other than that, from the author from [8] was written about a dynamic second-order sliding mode control of the hovercraft vessel. This is the combination of the second-order sliding mode and the differential flatness property controller design technique. This kind of control proposed for the specification of a robust dynamic feedback multivariable controller. It was used to track earth coordinate position of a hovercraft vessel model.

From [9], the author had designed wing of the UACV. When UACV had reached a sufficient horizontal speed, it will lift the vehicle just off the surface as by the ground effect. Although the air cushion vehicles had same weight to helicopter, the UACV consumed less power than the helicopter. The system also can attain higher speed compare to mobile vehicle or ships. The UACV contained four control parameters of x, y and z-coordinate and orientation angle, much like standard as automobile. It was controlled with horizontal and vertical thrust and motion of rudders control the orientation. This system was open loop. It could be developed to be a closed loop by develop a feedback module in the form of monitoring of various operation of hovercraft. As to track the vehicle movements, it used a GPS guidance system.

3 System design

From the previous studies inside section II shown that most of the researches focused on the design and controller to control the movement either ACV or UACV systems. The system that is designed in this project consists of one propeller driven by a brushless motor to provide a lift thrust while another two propellers to provide a thrust for the system to move forward or during turning movement. Figure 1 shows the degree of freedom associated with this UACV system while Figure 2(a) – 2(d) show the isometric view, top view, side view and back view of the UACV system that have been drafted using the SOLIDWORKS software. Another figure, Figure 3(a) – 3(d) shows the experimental setup used in order to obtain more accurate information related with UACV system.

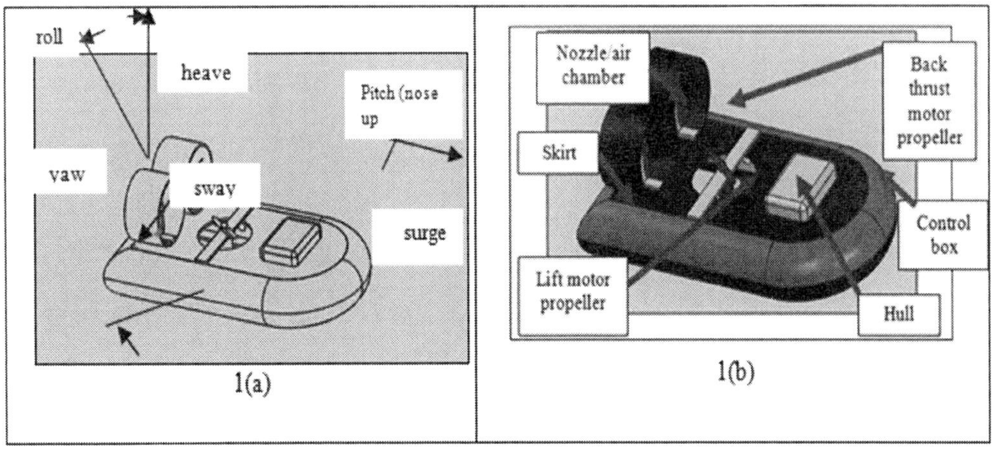

Fig. 1. (a) UACV degree of freedom, (b) UACV design

DOI: 10.1051/matecconf/20179701069

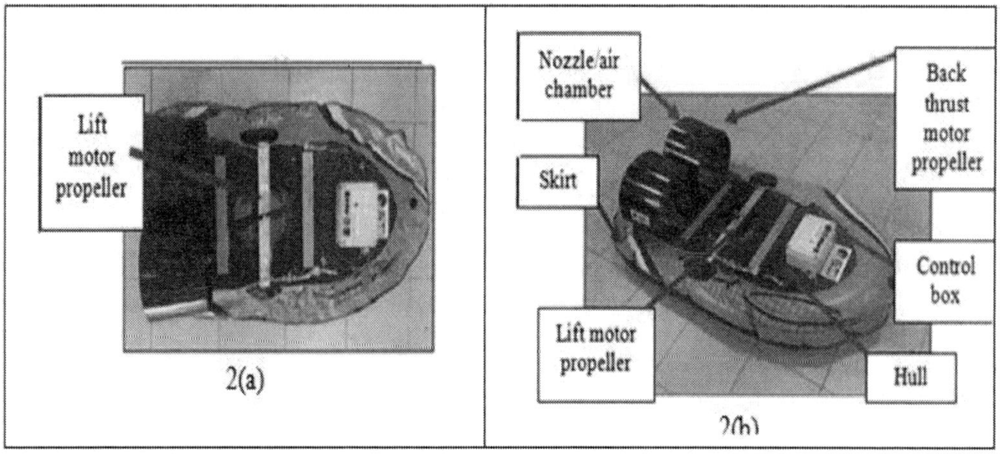

Fig. 2. (a) Top View of UACV Setup, (b) Isometric View of UACV

4 Results and discussion

Analysis have been done on the SOLIDWORKS design and the parameters of the UACV system have been obtained. Table 1 shows the parameters that have been acquired from the SOLIDWORKS sofware and moment of inertia (I) = 19.823kgm²

Table 1. Parameters of UACV

Parameters	Value	Unit
Mass, m	10.57	kg
Volume	0.032	m^3
Surface area	4.429	m^2
Center of mass	X = 0.500	m
	Y = 1.000	m
	Z = 0.400	m
Moment of inertia	I_{xx} =	kgm^2
	I_{yy} = 5.487	kgm^3
	I_{zz} =	kgm^4

The parameters obtained then are included into the mathematical model done by [13]. The state space is shown in Equation (3):

$$\dot{x} = Ax + Bu, \text{ and } y = Cx + Du$$

where:

$$A = \begin{bmatrix} 0 & 0 & 0 & 0.988 & 0 & 0 \\ 0 & 0 & 0 & 0 & 0.154 & 0 \\ 0 & 0 & 0 & 0 & 0 & 1 \\ -1 & 0 & 0 & 0 & 0 & 0 \\ 0 & -1 & 0 & 0 & 0 & 0 \\ 0 & 0 & -1 & 0 & 0 & 0 \end{bmatrix}, \quad B = \begin{bmatrix} 0 & 0 & 0 \\ 0 & 0 & 0 \\ 0 & 0 & 0 \\ 0.094 & 0 & 0 \\ 0 & 0.094 & 0 \\ 0 & 0 & -19.823 \end{bmatrix}$$

$$x = [x, y, z, r, p, \varphi]^T u = [\tau_u \quad 0 \quad \tau_r]$$

All the state space obtained has been included into the Matlab/SIMULINK and in order to prove simulation studies, a few experimental studies have been carried out and the outcomes of these experiments are elucidated in Figure 3 to Figure 4. The movement in straight line has a little bit vibration during the initial start but UACV moves steadily after certain period of time as shown in Figure 3(c).

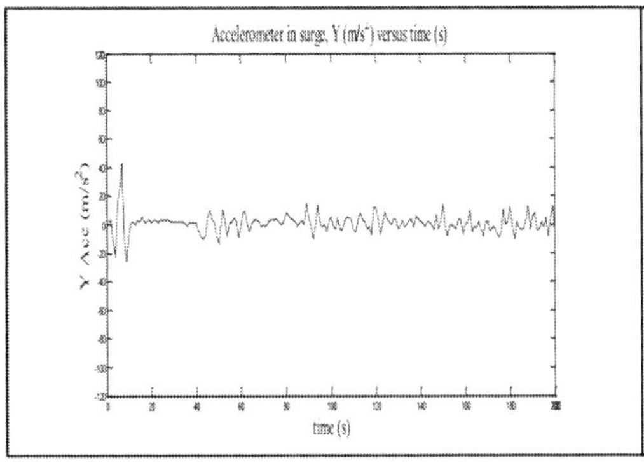

Fig. 3. (a) Acceleration in surge

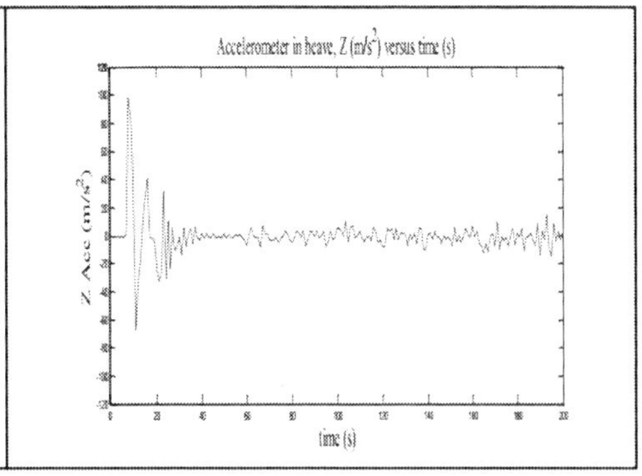

Fig. 3. (b) vertical lift (heave)

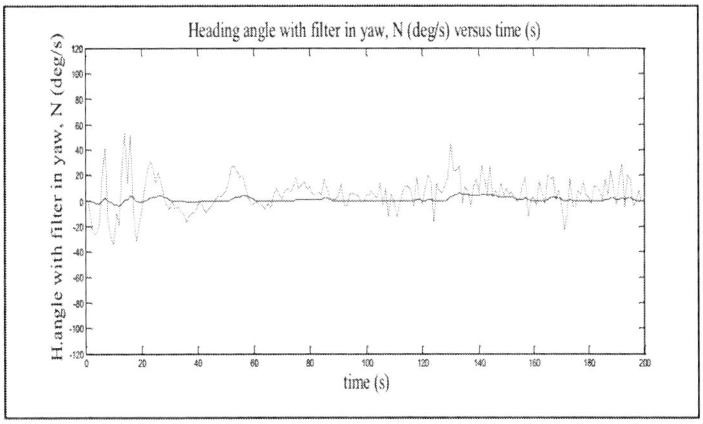

Fig. 3. (c). heading angle (yaw) during experimental straight motion

DOI: 10.1051/matecconf/20179701069

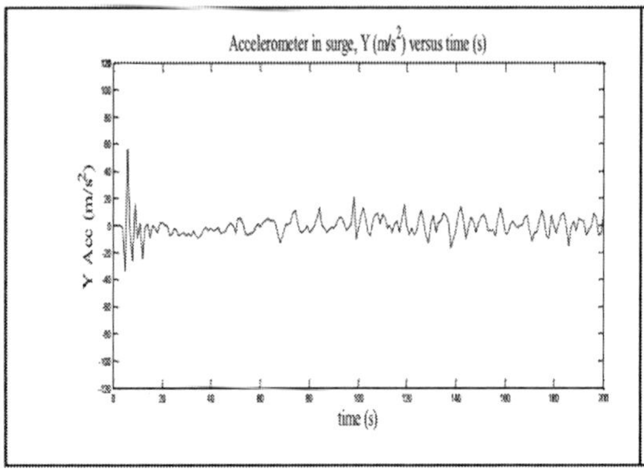

4(a)

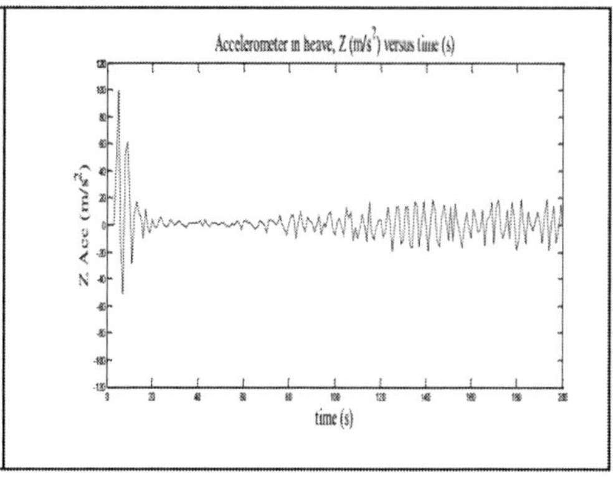

4(b)

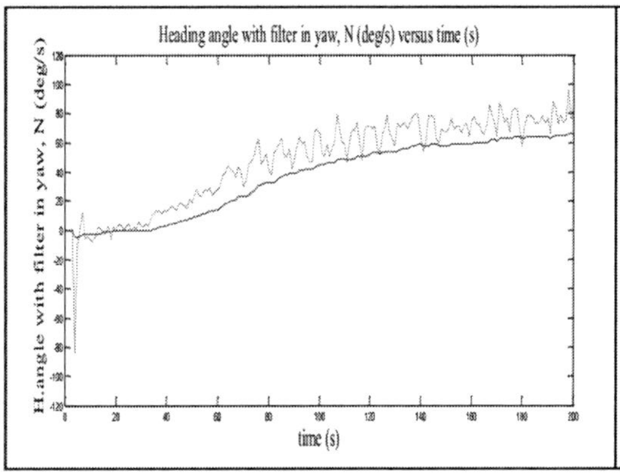

4(c)

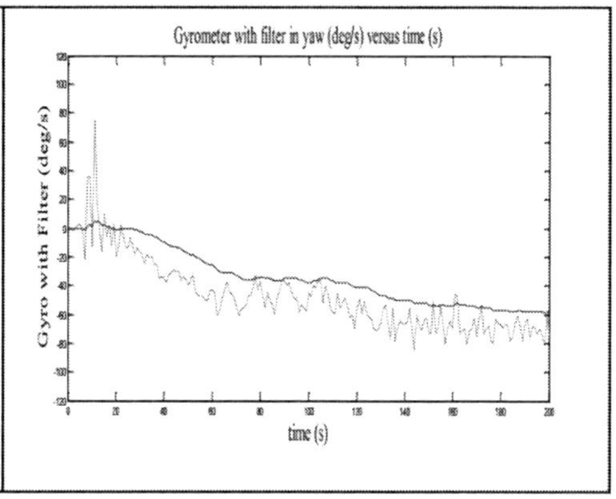

4(d)

Fig. 4. (a) Acceleration in surge, (b) vertical lift (heave), (c) heading angle (yaw) during right turn , (d) heading angle (yaw) during left turn

Figure 4(a) to Figure 4(c) show the movement of the UACV when it makes a right turn while 4(d) shows the movement of UACV when turn.

5 Conclusion

In conclusion, the UACV system has been developed and its parameters have been included into the mathematical model and simulation studies have been carried out. Then, the experimental studies have been conducted and the data obtained from the experimental studies are consistent with the simulation result.

The author want to express appreciation to Unmanned and Autonomous Research Group (UAIR), Centre of Excellence in Robotics and Industrial Automation (CERIA), Universiti Teknikal Malaysia Melaka (UTeM) and Ministry of Higher Education, Malaysia for sponsoring this project under RAGS grant no RAGS /1/2014/TK03/FKE/B00057.

References

1. M.Conyers, A.Walker, B.Warwick, Mechatronics Seng, 466 University of Victoria, Report the Second Phase of the Spring (2010)

DOI: 10.1051/matecconf/20179701069

2. R. Gran, G. Carpenter, R.W. Klein, *Computer Aided design of a Control System for a Hovercraft*. Decision and Control, *1982 21st IEEE Conference on*. 1347-1353 (1982)
3. S.M. Hein, H.C. Liaw, *International Conference on Advanced Intelligent Mechatronics,* 1516 (2013)
4. F.L. Roubieu, J. Serres, N. Franceschini, F. Ruffier, S. Viollet, *IEEE International Conference*, 1311 (2012)

River Debris Management System using Off-Grid Photovoltaic Module

Intan Mastura Saadon[1,*], *Herlanda* Windiarti[1] , *Emy Zairah* Ahmad[1] , and *Khalil Azha* Mohd Annuar[1]

[1] Department of Electrical Engineering Technology, Faculty of Engineering Technology, University of Technical Malaysia Melaka, Malaysia

Abstract. In Malaysia, Malacca River has long been the tourism attraction in Malacca. However, due to negligence, the river has been polluted by the litters thrown by tourists and even local residents, thus reflects a negative perception on Malacca. Therefore, this paper discusses about a fully automated river debris management system development using a stand-alone photovoltaic system. The concept design is to be stand alone in the river and automatically pull debris towards it for disposal. An off-grid stand-alone photovoltaic solar panel is used as renewable energy source connected to water pump and Arduino Uno microcontroller. The water pump rotates a water wheel and at the same time moves a conveyor belt; which is connected to the water wheel by a gear for debris collection. The solar system sizing suitable for the whole system is shown in this paper. The dumpster barge is equipped with an infrared sensor to monitor maximum height for debris, and instruct Arduino Uno to turn off the water pump. This system is able to power up using solar energy on sunny days and using battery otherwise.

1 Introduction

Malacca is a historical state in Malaysia and has been listed as UNESCO World Heritage Site since 7th July 2008. One of the tourism attractions in Melaka is the river cruise along Melaka River canal consequently, the cleanliness of the river is critical in order to preserve the tourism attraction. River debris that floats in the river is one of the major problems that occur in any water tourism attraction spots. It could be due to irresponsible visitors littering or the rain that washes the debris into the river. Besides affecting the tourism industry, polluted rivers also affect aquatic animals' habitat. To overcome this, an off-grid solar river debris management system prototype is developed and presented in this paper. It uses renewable energy for promoting environmental friendly technology in line with Malacca city slogan; City of Green Technology. Off-grid solar power is used to reduce maintenance and manpower besides being able to collect floating river debris automatically. Fig. 1 shows the concept design of this system. This concept will be developed based on the actual size upon the completion of the prototype

* Corresponding author : mastura@utem.edu.my

 DOI: 10.1051/matecconf/20179701070

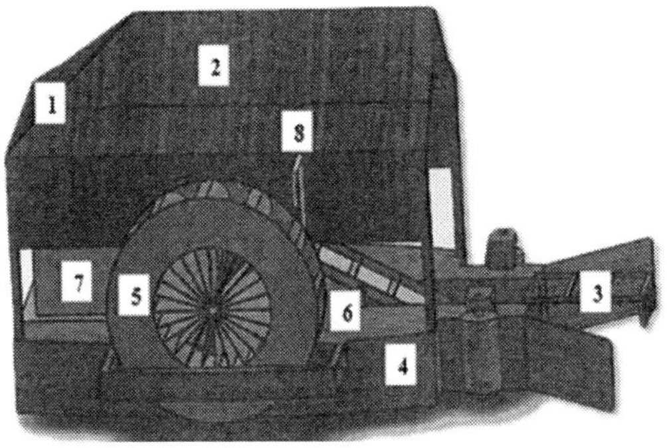

No	Parts
1	Solar Panel
2	Roof
3	Debris Raking
4	Base
5	Water Wheel
6	Conveyor
7	Dumpster Barge
8	Water Pump

Fig. 1. Side View of the Concept Design and the Parts Description

Figure 2 shows the electrical and electronic block diagram of the system. Solar charger controller will act as a regulator that controls the voltage in and out. Load is controlled from the solar charger controller and is supplied to Arduino Uno and the water pump. Water pump is controlled by a relay that is connected to Arduino, based on the input from IR sensor.

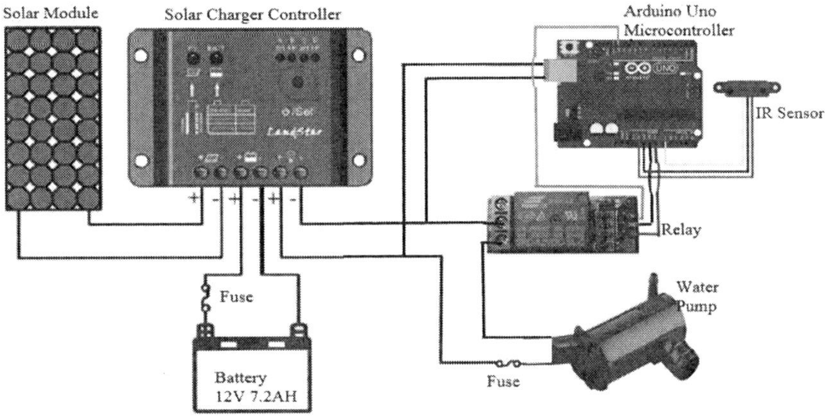

Fig. 2. Block Diagram of Electrical and Electronic Design for the System

1.1 Solar system design

1.1.1 Photovoltaic system

Photovoltaic (PV) is a system which converts the solar power into electrical energy. Sun is an endless source for PV system to produce direct current (DC) [1]. Photovoltaic effect is when light emitted on a two-layer semiconductor, it produces voltage between the layers. Produced output current can be used through external electrical circuit. Solar panels must be slanted and oriented at optimum angles so as to collect the maximum solar energy available in a specific region [2]. The orientation and the tilt of a solar panel strongly affect the amount of the collected yield. In Malaysia, tilt angle during dry season is 0° [3].

There are different types of silicon solar cells including mono-crystalline, thin-film, poly-crystalline silicon and amorphous. They differ in atomic structure, performance and comparisons as well as other aspects [4]. Amorphous is cheaper but there's a big difference in efficiency [5]. Mono-crystalline silicon has the highest efficiency compared to others [6], hence an off-grid mono-crystalline PV module is selected for this research. For off grid system, the DC output power from the battery will be used for the own system [7].

1.1.2 Battery

In small scale stand-alone system, choosing the suitable battery as power bank is important. The battery will store electrical energy produced by module during the day time and deliver it during cloudy days [8]. The battery charging will be controlled by charge controller. In stand-alone system, battery needs to supply higher current than PV array when it needs to start motors. A deep cycle battery is suitable to be used in PV system as it is maintenance free, low price and have longer life span [9]. It can store charge according to its capacity described in Amp-hour (AH).

1.1.3 Charger controller

Charger controller acts like a voltage regulator which is placed between batteries and PV module [8]. It controls the battery from over charging that weakens the bonds of electrolyte and over discharging that reduces the battery life span. There are few types of charger controller but mainly two types are widely used in solar system which is pulse-width-modulation (PWM) Controller and maximum power point tracking (MPTT) controllers [6]. The PWM charging efficiency ranges from 60% to 70%, and 75% to 85% while MPTT 90% to 93%. MPTT has excellent efficiency in charging but it is costly [10]. PWM controller is selected to be used in this project as it is cheaper and the efficiency of the controller is less considered compared to its price.

1.2 Water pump

Selecting water pump is based on the application of the pump and the amount of water can be pumped. The volume flow rate should be high [11]. This project uses a brushless motor submersible pool water pump. The pump specification is 6 watt, 12Vdc and water flow rate is 250L/hr made by plastic. A low powered dc pump is used to fit the size of this prototype which is sufficient to rotate the water wheel. In system development a higher power AC pump will be used but it has to consider an inverter for DC to AC conversion.

1.3 Analog distance infrared sensor

Infrared (IR) distance sensors are useful for measuring distances without actually touching a surface. Selection of an infrared sensor depends on the application, sensitivity and range of distance it can measure. Based on the size of this prototype, a 10-80 cm distance range infrared sensor is used at the dumpster barge. Distance sensor bounces IR off objects to determine how far away they are. It returns an analog voltage used to determine how close the nearest object is. The sensor measures with analog signal ranges between values of 0 to 1023 when observed by Arduino serial monitor [12].

2 Development of the prototype

Figure 3 shows the prototype developed for this system. Plywood is used as platform of this unit, supported by buoyant paddings to float the unit on water. Water wheel and conveyor belt is supported by an L-bar. When system is on, the water pumped is supplied with

DOI: 10.1051/matecconf/20179701070

sufficient power and is able to rotate the water wheel and this successfully collect all debris that float in the water.

Fig. 3. Prototype Developed for River Debris Management System

3 Results and analysis

3.1 PV system sizing

To determine the sizing of PV module, the total load should be determined and it is expressed as Watt hour per day (Wh/day). The formula used for PV module sizing and battery sizing are shown below:

$$Wh = Wp \; x \; PSH \; x \; Pr \tag{1}$$

$$I = Wh \,/\, V \tag{2}$$

Wh = units of energy produced, Wp = module peak watts, PSH = peak sun hour, Pr = performance ratio (0.60 – 0.70), I = battery current capacity (AH) and V = battery voltage.

For solar panel system sizing, the total load calculated is shown in Table 1. Load sizing for this prototype is calculated considering the water pump and Arduino which is 12Vdc and 1.0ampere, estimated to run for a maximum of 4 hours to do the analysis. From the table, the total load is 60Wh/day. Using the formula in Eq.1, considering 4 PSH and performance ratio of 0.7, the module peak watt obtained is 21.43Wp. The specifications of solar modules used in this project are 10W power, 12V voltage and 15A current; hence, to supply 21.43Wp, three sets of solar modules are used. The solar modules will be connected in parallel so that the output voltage is 12V, output current is 45A and output power is 30W.

Table 1. Total Load Power

Load	Quantity	Wattage (W)	Hours/day	Load Power (Wh/day)
Water pump	1	12	4	48
Others (system)	1	3	4	12
Total load	2	15	4	60

In battery sizing, the minimum capacity of battery current is determined using Eq.2. For this type of load, the battery used is 12V, so the minimum current capacity obtained is 5AH

DOI: 10.1051/matecconf/20179701070

and battery selected must be 5AH or above. Therefore, a 12V battery with capacity of 7.2AH is used.

3.2 PV module data collection and analysis

To analyze the solar module, data collection is made on the off-grid PV module from real condition test under the sun. The data collected are irradiance, temperature, open circuit voltage, short circuit current, voltage at maximum point, current at maximum point and battery charging voltage. The tilt angle is set to 15° angle for solar module and the results were recorded every 1 hour starting from 9.00am to 5.00pm.

To get the irradiance value, Seaward Solar 200R Irradiance Meter is used. It is placed on top of the module in order for it to read the data that has to be collected. While exposing the solar module under the sunlight, the temperature of the solar module is also recorded, by using Fluke 568 IR Thermometer. The fluke meter is shot behind the module to get the temperature. Besides that, the electric current that flows in the conductor is recorded using a clamp meter. The module is shorted to obtain the range of current. Power voltage and battery voltage values recorded using a multi meter. Results of the data collected for this off-grid PV solar system is recorded in Table 2.

Table 2. Data Collection for PV Module

Time	Tilt Angle , (°)	Solar Radiation , (W/m2)	Temperature , (°C)	Power Voltage, (Vmp)	Power Current , (Imp)	Power Point Watts, (W)	Battery Voltage , (V)
9.00am	15	807	39.2	12.04	0.07	0.84	11.42
10.00am	15	867	48.6	12.32	0.11	1.35	11.48
11.00am	15	896	54.9	12.32	0.16	1.97	11.56
12.00pm	15	912	61.4	12.34	0.13	1.60	11.64
1.00pm	15	1030	63.8	12.33	0.20	2.47	11.76
2.00pm	15	965	62.1	12.33	0.15	1.84	11.88
3.00pm	15	972	62.6	12.37	0.19	2.35	12.09
4.00pm	15	925	56.8	12.20	0.15	1.83	12.22

From Table 2, it can be said that the irradiance and temperature is low in the morning and increasing from time to time. The maximum irradiance is 1030W/m2 and maximum temperature is 63.8°C, both are seen at 1.00pm. This is because afternoon is the peak time to obtain high irradiance and the sky is cloudless. It can be concluded that the power produced is sufficient to charge the battery and run the system. The relationship between produced power voltage versus irradiance and temperature is shown in Figure 4 and Figure 5.

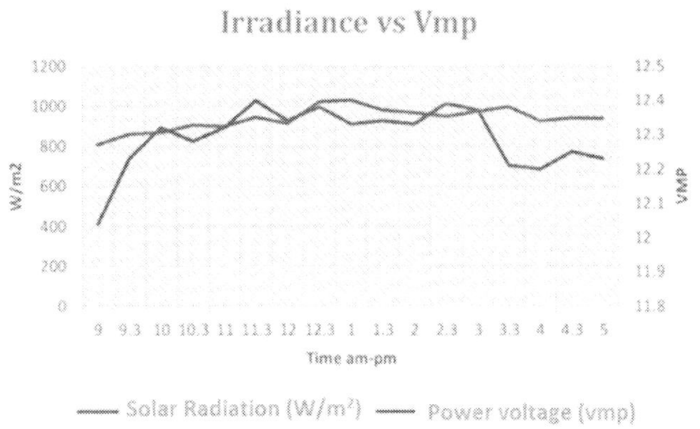

Fig. 4. Irradiance Effect on Voltage

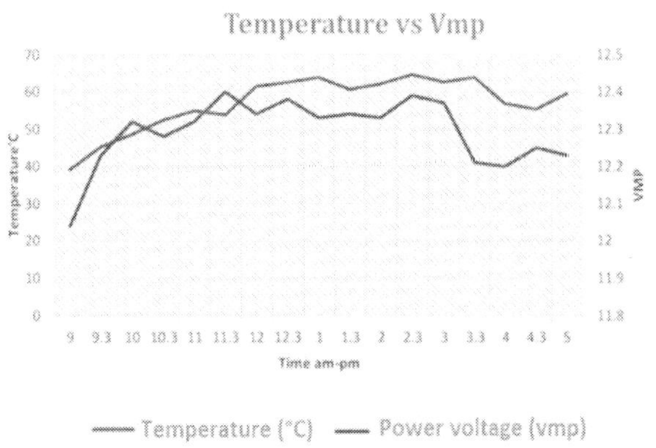

Fig. 5. Temperature Effect on Voltage

The voltage increases due to increasing of radiation and temperature but maintains until the evening. The voltage starts to drop from 3.00pm to 5.00pm. Solar energy intensity from sun which absorbed by solar module to produce energy. Irradiation may be blocked by cloud and shading factors. When the irradiance is blocked, the energy or volt produced by solar module decreases. The solar cell and panel materials absorb the solar irradiation and also absorb the heat. Temperature also plays an important role in producing energy. Due to this, the voltage starts to drop and the performance of solar panel reduces.

Figure 6 shows the battery charging time. The voltage increases steadily until it is fully charged ~12V. The continuous current required to charge the battery is 7.2AH by 4hours, which is 1.8 Ampere. The PV module continuous current is ~2.7A, and is sufficient to charge the battery.

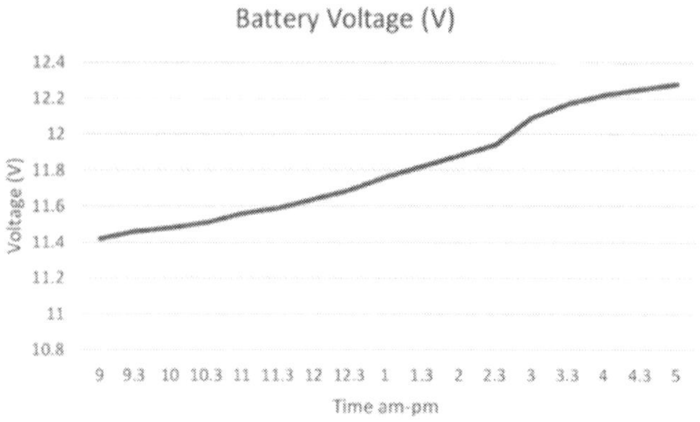

Fig. 6. Battery Charging Time

3.3 Water pump flow rate

The water pump is tested to check the flow rate per hour. The pump is turned on to supply 1000ml of water on the water wheel for three trials. The readings are shown in Table 3. From this table, the average reading is 13.9s or is 4,286ml/min. From this calculation, the flow rate is proven to be able to rotate the water wheel and at the same time create water current that brings river debris towards the system.

Table 3. Water Flow Rate by the Water Pump

Water Level	Reading 1 (s)	Reading 2 (s)	Reading 3 (s)	Average (s)
1000ml	13.53	14.03	14.11	13.90

3.4 IR Sensor on Dumpster Barge Sensitivity

. The dumpster barge prototype is set to collect trash up to 8cm. The distance sensor is placed above the dumpster barge, and is set to 18cm to indicate that the dumpster barge gas reached its maximum trash height and cutoff the water pump. A test is done by collecting sample trash into the dumpster barge to a maximum height, IR sensor detects and turns off the pump, and the time taken from empty dumpster barge to full dumpster barge is recorded in three readings as shown in Table 4. However, this depends on the type and quantity of trash. The time taken is consistent on all three readings which average is 1,139s or 19min.

Table 4. : Data for Full Dumpster Barge and Cutoff Water Pump

Trash Height	Reading 1 (s)	Reading 2 (s)	Reading 3 (s)	Average (s)
8cm	1,125	1,154	1,138	1,139

4 Summary

Automatic river debris management system is very important and should be placed at all rivers because it is environmentally safe and reduce man power. The prototype as presented in this paper is able to generate renewable energy to supply to the system. Besides that, the water wheel is capable to run the debris collection and moved into the dumpster barge by the conveyor belt. Moreover, the systems are automatically turned off when the dumpster barge is full, and avoid the barge from overflow. This system can be potentially developed into a larger scale, by considering a higher power for the water pump.

The authors would like to express deep appreciations to the Ministry of Higher Education Malaysia and Universiti Teknikal Malaysia Melaka (UTeM) for providing a PJP grant (PJP/2015/FTK(9A)/S01414), opportunity and necessary facilities to support this work.

References

1. F.Z. Zerhouni, M.H. Zerhouni, M.Zegrar, M.T. Benmessaoud, A.B. Stambouli, A. Midoun, Acta Polytech., **7**, 55-70 (2010)
2. T. Khatib, A. Mohamed, K. Sopian, Renewable Sustainable Energy Rev., **16**, 2864-2869 (2012)
3. T. Khatib, A. Mohamed, K. Sopian, Renewable Sustainable Energy Rev., (2012)
4. J. Balfour, M. Shaw, S. Jarosek, *Introduction to Photovoltaic,* (2013)
5. L.A. Dorbzanski, M. Musztyfaga, M. Giedroc and P. Pamek, J. Achv Mat and Man Eng., **55**, 307-315 (2012)
6. J. Balfour, M. Shaw, S. Jarosek, Advance Photovoltaic System Design (2013)
7. L.A. Dorbzanski, M. Musztyfaga, M. Giedroc, P. Pamek, J. Achv Mat and Man Eng., **53**, 7-13 (2012)
8. S. Sinha, A. Shukla, *From Sunlight to Electricity,* The Energy and Resources Institute. (2013)
9. A. Pradhan, S. M. Ali and P. Behera, Intl. J. of Sc. And Rsch. Publ., **2**, 1 (2012)
10. G.L. Kale and N.N. Shinde, Intl. J. of Rsch. Mech. Eng. Tech., **1**, 89 (2011)
11. A. Mohammedi, D. Rekioua and N. Mezzai, J. of Elec. Syst., **2**, 212 (2013)
12. Information on: https://www.adafruit.com/product/164

Optimization of Stiffened Square Plate Topographically under Uniform distributed Pressure

U.N. Eman[1,*], *H.M.S.* Firdaus[1], *M.Y.* Halyani[1], *Zulhasnizam* Hasan[1], and *R.F.* Farhana[2]

[1]Faculty of Engineering Technology, Technical University of Malaysia, Melaka, Hang Tuah Jaya, 76100, Durian Tunggal, Melaka, Malaysia.
[2]Armag Oil & Gas Academy, Kompleks Usahawan Terengganu, Kg. Kubang Jela, 21200 Kuala Terengganu, Terengganu, Malaysia.

Abstract. The structure of stiffened plate is widely used in engineering since 19th century and is currently used in many modern applications. When stiffened square plate was imposed to extremely high load, the thin plate structure will deform. This research is carried out to find the solution in optimizing the structure of stiffened square plate topographically. The structure of square plate is analysed by applying uniformly distributed load of 5 bar. Finite element analysis is carried out to identify the potential design in obtaining optimum solution in designing ultimate stiffened plate. Result shows that the performance of stiffened square plate structure can be improved approximately by 80% in terms of tension and compression using topography optimization of finite element analysis.

1 Introduction

Finite element analysis (FEA) is originally initiated for solving a solid mechanics problems by using numerical method to calculate the responses of a structure to combination of applied load such as pressure and force to a support. All function used should satisfy the boundary conditions of the existing problems [1]. It is also considered as a practicable technique for computer solution of complex engineering problem. Apart of FEA, optimization is a solution to a problem to obtain an optimum function for a particular design [2]. Optimization techniques is now view as the tool of the utmost importance for engineering in many industries such as in aerospace, manufacturing, automotive and chemical industry. Optimizations techniques can be classified as design optimization which includes shape optimization [8], topological optimization [10], topography optimization [9] and sizing optimization [11]. In this research, topography optimization is used to optimize the structure of stiffened square plate. A stiffened square plate is widely used in industry especially in ship structure [3, 4] and water tank base [5]. Stiffeners are used on a stiffened plate to resist highly directional loads while introduce multiple load paths that may provide protection against damage and crack growth under both compressive and tensile loads [12]. Ship hull structure received flexural compressive stress from water pressure [14], resulted

* Corresponding author: noor.eman91@gmail.com

in instability of it's structure. This situation will lead to plate deformation as well as increase the stress and strain of the plate. This paper explains the structure analysis of stiffened square plate under uniformly distributed pressure of 5 bar and topography optimization as tools in optimizing the design of stiffened square plate through topographic analysis. The advantage of the stiffener is to increase bending stiffness of the plate with minimum additional material, which makes these structures highly desirable for out-of-plane loads and destabilizing compressive loads. Optimized stiffened plate in developing inspection and correction of stiffened plate in the ship plating and typical web frame of a tanker is obtained by using this element.

2 Methodologies - modelling of stiffened plate

The size of the basic stiffened plate model is 300 mm x 300 mm x 5 mm as shown in Fig. 1. According to [13], a plate is considered as a thin plate structure if the ratio of width to length of the plate is maximum 0.1. Stainless steel which has elastic modulus of 210 GPa, tensile strength of 515 MPa and yield strength of 207 MPa is selected as material for the plate [6]. Topography analysis is carried out to obtain preliminary result in improving and optimizing plate performance [7]. The analysis is carried out by giving the uniformly distributed pressure of 5 bar to the square plate by using the inspire 2015 from Altair HyperWorks. The element size used for the analysis is obtained from calculation and graph convergence, which gives the value of 0.0060778m. Square mesh is choosing for this analysis because all of the models are in square shape. All 4 edges of the square plate are clamped.

Fig. 1. Basic Model of Stiffened Plate

Optimization of stiffened square plate are carried out based on the preliminary analysis that obtained from basic model to ensure the stiffened square plate are at optimum performance.The parameters that will change is types of stiffeners on the square plate. There are four designs that have been considered including the regular optimization that is widely used in ship structure plating and typical web frame of a tanker.

 DOI: 10.1051/matecconf/20179701071

3 Results and discussions

Fig. 2. Topography Analysis on Basic Model of Stiffened Plate

Fig. 2 shows stress contour for the basic model of stiffened plate. Dark color area shows critical section with maximum stress. Based on the result, stress concentration of deformation are located in the middle of the plate and at the fixed end that is not attached with stiffener. Colorless area shows area with minimum stress. Under the influence of the repeated stress, creep will happen at dark color area whereby plate becomes unstable and it leads to permanent plate deformation [15]. Therefore, stiffener must be present in supporting plate structure against load distribution in order to avoid brittle fracture. Four optimization designs are proposed in improving plate performance including regular optimization as shown by Fig. 3.

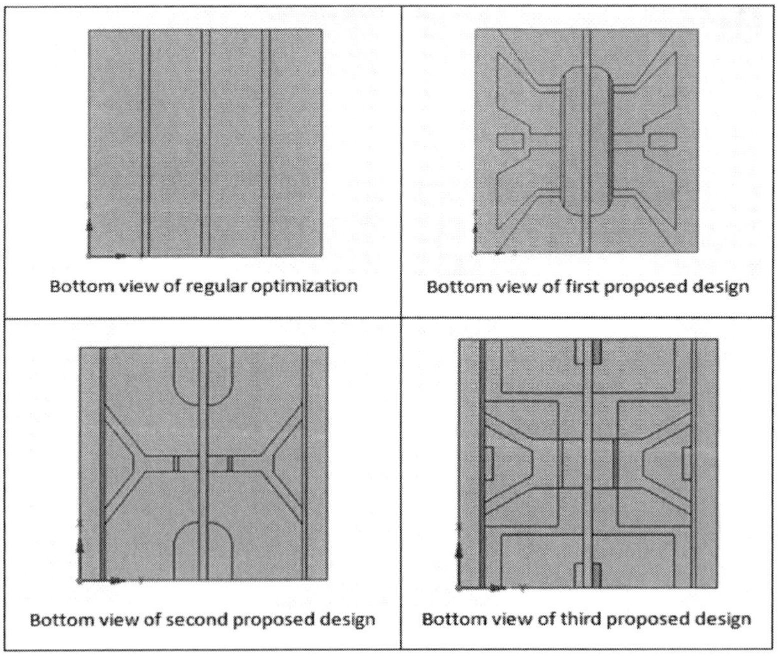

Fig. 3. Proposed Design of Stiffened Square Plate

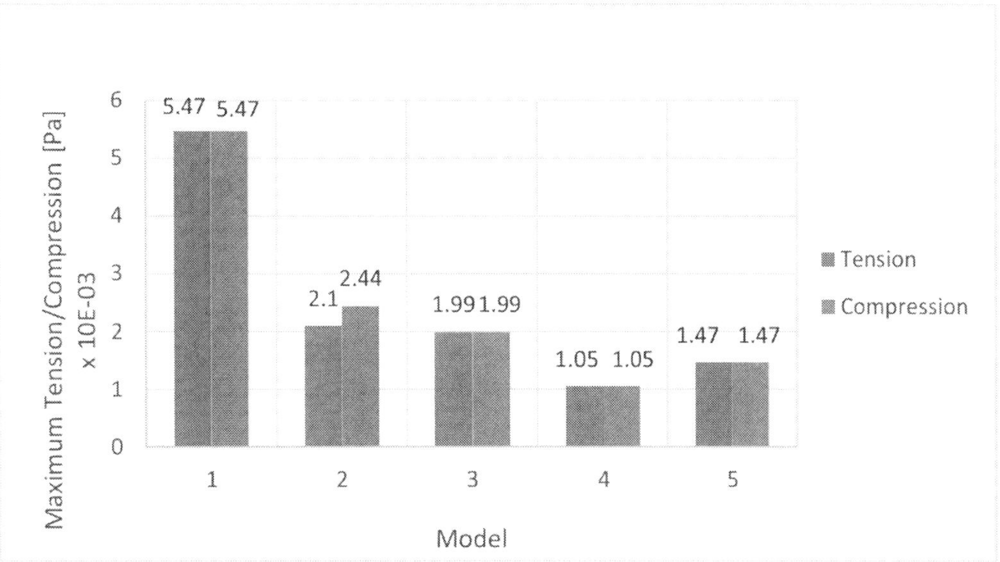

Fig. 4. Maximum Tension / Compression

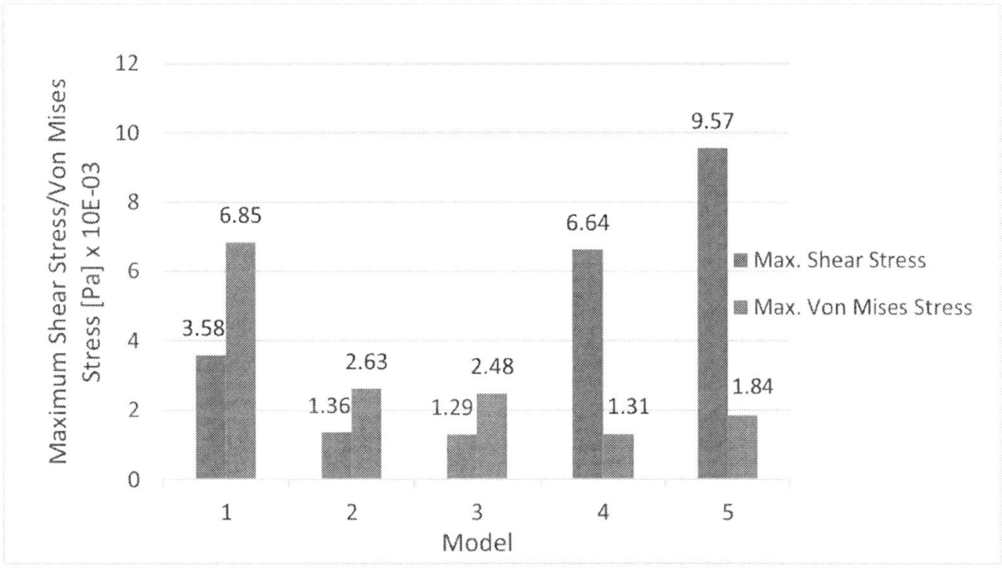

Fig. 5. Maximum Shear Stress / Von Mises Stress

 DOI: 10.1051/matecconf/20179701071

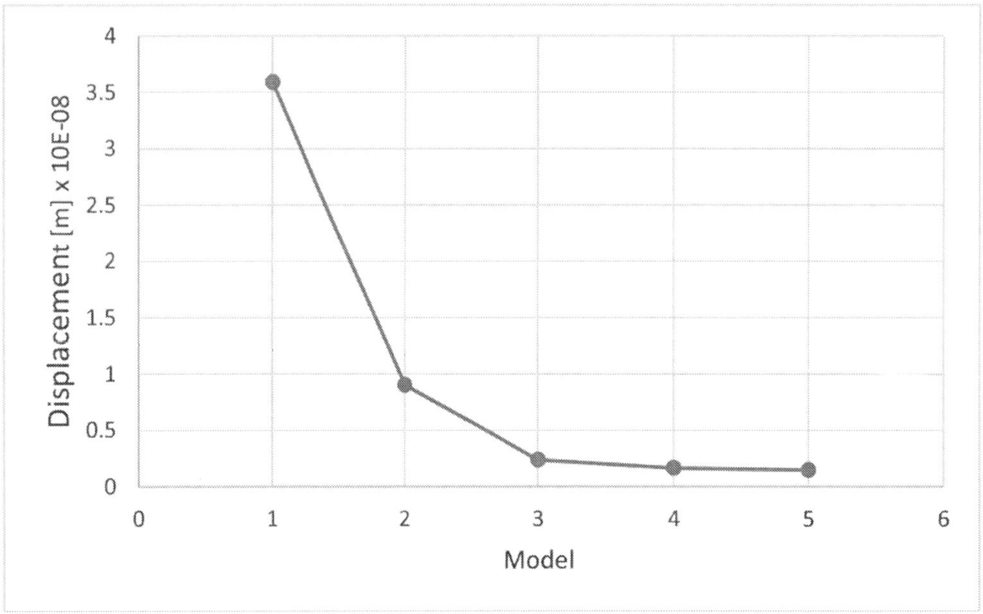

Fig. 6. Maximum Displacement

Fig. 6 shows that the displacement of stiffened plate is reduced from basic model ($3.59x10^{-8}$m) to regular optimization ($9.02x10^{-9}$m). The number decreased further for suggested design after topography optimization that are, $2.389x10^{-9}$m (1st optimization), $1.658x10^{-9}$m (2nd optimization) and $1.489x10^{-9}$m respectively. It means, when the displacement is reduced, the bending that occurs on the stiffened plate which can lead to fracture will also be reduced. In terms of tension and compression in Fig. 4, results show that second optimization design is the best optimization with lowest value of $1.05x10^{-3}$ Pa. The results was reinforced by the maximum value of von Mises Stress in Fig. 5 that proved second optimization as the best optimization for the stiffened plate with the value of $1.31x10^{-3}$ Pa. Besides, second optimization has the lowest mass per diagram than other types of optimization. From the result in Table 1, second optimization can be considered as the best optimization in giving the best performance to the stiffened plate.

Other than using single type optimization, Belblidia et al has introduced in their work integrated design optimization by combining topology optimization and sizing optimization as suitable method in producing lightweight stiffened plate [16]. Thus, combination of topology optimization and topography optimization is the line of research arising from this work which could be continued.

4 Conclusions

Based on this research, it shows clearly that the stiffened square plate can be optimized and the performance of stiffened square plate can be improved using the topography optimization. The value of maximum displacement can be accepted for all design because it represents a very small number which can be ignored. However for maximum tension, maximum compression and maximum von Mises Stress, second design optimization represents the best design because it has the lowest value for these three properties.

The team would like to thank the Ministry of Higher Education (MOHE) and Universiti Technical Malaysia Melaka (UTeM) for the financial support in publishing this paper under grant PJP/2015/FTK(34C)S01459.

References

1. O.C. Zienkiewicz, R.L. Taylor, David Fox, "*The Finite element Method for Solid and Structural Mechanics*", 7th Edition, Butterworth- Heineman Publisher (2014)
2. R. Ramkumar, V.R. Lenin, A. Krishnaraju, M. Senthilkumar, International Journal of Engineering Trends and Technology (IJETT), **21**, 12 (2015)
3. H.W. Leheta, S.F. Badran, A.S. Alhanafi, Thin-Walled Structures, **94,** 545 (2015)
4. Y.V. Satish Kumar, M. Mukhopadhyay, Applied Ocean Research, **22**, 361 (2000)
5. A.S. Anumod, S. Harinarayanan, S.Usha, Int. J. Eng. Res. Ind. Appl., 47-54 (2015)
6. W.D. Callister, D.G. Rethwisch, *Material Science and Engineering - An Introduction.* 7th Edition, John Wiley & Sons (2007)
7. Y. Liu, G. Yang, J. Lian, J. Mater. Process. Technol., **31**, 327 (1992)
8. P. Plotnikov, J. Sokolowski, "*Compressible Navier-Stokes Equation. Theory and Shape Optimization*", Springer, Basel (2012)
9. R. Chakravarty, *Study of Topography Optimization on Automotive Body Structure.* SAE Technical Paper. (2009)
10. L. Naesung, S. Kazuhiro, Journal of Mechanical Design, **127**, 170, (2005)
11. F. Behrooz, A. Ali, Int. J. Solids Struct., **47**, 2508 (2010)
12. E. Adel, Mech. Adv. Mater. Struct., **20**, 764 (2013)
13. S. Timoschenko, S. Woinowsky-Krieger, Theory of Plates and Shells. McGraw-Hill Classic Textbook, 2nd Edition, USA (1959)
14. Y. Okumoto, Y. Takeda, M. Mano, T. Okada, *Design of Ship Hull Structures - A Practical Guide for Engineers.* Springer-Verlag Berlin Heidelberg (2009)
15. W. Jiang, G. Jianming, C. Hu, T. Shandong, T. Shan Tung, Acta Metall. Sin., **43**, 539 (2007)
16. F. Belblidia, S.M.B. Afonso, E. Hinton, G.C.R., Engineering Computations. **16**, 934 (1999)

MATEC Web of Conferences 97, 01072 (2017) DOI: 10.1051/matecconf/20179701072
ETIC 2016

Development of Magnetorheological Engine Mount Test Rig

Mohd Razali Md Yunos[1,*], *Mohamad Hafiz* Harun[1], *M. Z.* Sariman[2], and *A. K.* Mat Yamin[2]

[1] Faculty of Engineering Technology, Universiti Teknikal Malaysia Melaka, Malaysia
[2] Faculty of Mechanical Engineering, Universiti Teknikal Malaysia Melaka, Malaysia

Abstract. Ride comfort is an important factor in any road vehicle performance. Nonetheless, passenger ride comfort is sometimes affected by the vibrations resulting from the road irregularities. Vehicle ride comfort is also often compromised by engine vibration. Engine mount is one of the devices which act as vibration isolator from unwanted vibration from engine to the driver and passengers. This paper explains the development of the test rig used for laboratory testing of Magnetorheological (MR) engine mount characterization. MR engine mount was developed to investigate the vibration isolation process. An engine mount test machine was designed to measure the displacement, relative velocity and damper force with respect to current supply to characterize the hysteresis behavior of the damper and as force tracking control of the MR engine mount.

1 Vehicle engine mounting system

Engine performance is measured based on power-to-weight ratio value in which the higher value represents high performance. Hence, car manufacturer focus on development of small-sized engine to reduce fuel consumption and emission gasses. Additional power recovery devices such as turbocharger and supercharger are implemented to compensate the loss power. As a result, greater engine vibration will be caused by the unbalance force generated by reciprocating piston movement and the combustion pressure. Thus, engine mounts is a proper device used to isolate the vibration from the engine to the chassis.

Active system consists of spring and actuator, where actuator act as external power supplied to overcome the force act on the system. Semi-active system is known the intermediate ways to counter these problems due to the limitation of passive system to provide the flexible damping force in isolating various frequencies from engine and due to the drawback of complexity of active system. Engine excitation is extreme at idle condition within frequency range 5~20 Hz while at driving condition the frequency range about 20~250Hz. Therefore a variable engine mount that can be operated within both conditions is needed [1]. Thus a semi-active engine mount is used to satisfy the variable range of frequency problem. Car makers put much effort to satisfy both condition by tuning the

* Corresponding author: razali.yunos@utem.edu.my

stiffness of each engine mount. Design of Experiment DOE methods which is using ADAMS software is a specific method to optimize those parameters as one of the commonly used technique [2]. Besides that, different approach was proposed by repositioning the engine mounts which also affects the dynamic stiffness of the engine mount [3]. In addition, the simulation approaches were used widely for testing and parameter estimation of the engine mounts in order to reduce cost and time [4]. It is undeniable that the semi active damper in engine mounts is step up approach in isolating variable range of frequency.

2 MR engine mount as semi-active damper

Various kinds of semi-active devices are currently available [5]. There are semi-active devices that can generate damping force from viscous/visco-elastic-plastic fluids [6]. MR damper falls into this category and currently become the subject of many researches. MR dampers contain controllable fluids that can change their properties when exposed to magnetic fields. By controlling the current to an electromagnetic coil inside the piston of the damper, the MR fluids viscosity can be changed, resulting in continuously variable real time damping system. The application MR damper in engine mount in vehicle was first implemented on 2011 by Porsche. They implements the magnetorheological (MR) technology for the engine mounts of its seventh-generation 911 Turbo which is part of a strategy to bring added smoothness and refinement to the power train, as well as contributing to enhanced cornering and traction capabilities. The system uses magnetically "soft" iron particles, suspended in a synthetic base fluid, held within a cavity in the rubber body of the mount. When a magnetic field is applied by a coil, the particles become aligned, increasing the shear stress and therefore the resistance to flow. The stronger the magnetic field, the higher the dynamic rate of the system. An MR damper is reasonably a latest damping device, in which the magnitude of the resisting force upon a mechanical structure can be adjusted in real time. Adjustment takes place by altering the amount of current passing through wires fixed in the damper. Its characteristics have been studied through both numerical simulation and laboratory tests.

3 Work principle of MR engine mount

A schematic of the fabricated engine mounts is shown as Fig.1. The MR engine mounts consists of a four orifice, wire coil, piston, piston rod, oil seals, inner chamber, upper cover, bottom cover, bracket, ball joint and 250ml MR fluid. The chamber is the working area for piston passing through the MR fluids. The upper cover and bottom cover serve to prevent the MR fluids flow out from the inner chamber. While, the oils seals used at the moving piston area to prevent the MR fluid flow out. Mounting bracket is designed on the bottom cover. Moreover, spring is using as to support the static and dynamics load from the engine

The ball joint is used as the main component and take responsible to transfer forces from the engine to the piston in one direction. It moves linearly during compression and extension. The socket which is connecting to the piston provides free working space for the ball joint to move. Four orifices with diameter 3mm drilled on the piston to let the MR fluids pass through. Coiled copper wire around the piston meant to generate magnetic field when current supplied on it. A total of 1200 winding wires were wrapped around the circumference of the piston. This is the maximum number of turns that can be made on the provided piston area. The circumference had been insulated wire with araldite to avoid reaction between MR fluids and the winding cooper wire. Furthermore, the maximum

DOI: 10.1051/matecconf/20179701072

current to the electromagnet in the magnetic choke is 1.0 Ampere and the coil resistance is 25 Ohm and the maximum stroke produce is 10 mm.

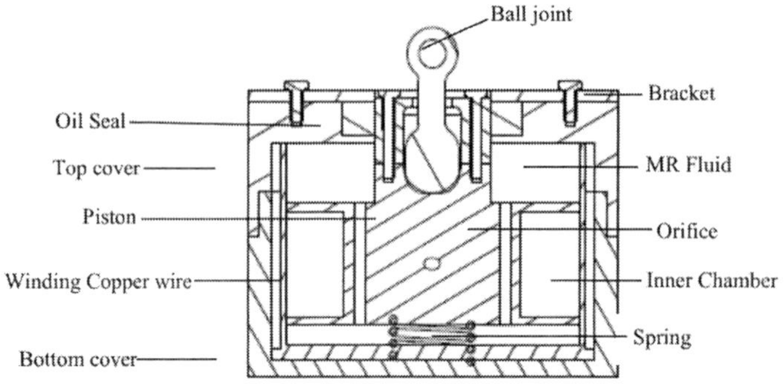

Fig. 1. Schematic of MR engine mounting design

4 Design of MR engine mount test rig structure

The key design specification for the MR damper test rig is sufficiently stiff in all frequency ranges of operation. The MR damper test rig structure constructed as stiff as possible in order to absorb the unwanted structural vibration. Several efforts were performed to make the structure as stiff as possible in the most critical directions particularly in vertical plan of the towering structure, side to side towering of the vertical structure and stiffening the vertical sliding shaft. To protract side to side towering of the MR engine mount test rig, diagonal braces were welded to form a diagonal structure with the vertical frames. These additional frames are useful for maintaining torsional rigidity of the structure as well as to prevent torsional bending caused by the unwanted rotational vibration. Stiffening the sliding shaft was performed by holding the shaft edges on the vertical structure and the basement rectangular frame. The solid plate was joined with a sliding bush or well known as sliding bearing at the two main posts as shown in Figure 2.

Fig. 2. Sliding bearing.

5 Test rig actuation system

A common test procedure for the MR damper characterization requires a number of actuators such as pneumatic cylinders [7], a hydraulic cylinder [8] and the slider crank mechanism. In an effort to generate a sinusoidal profile, pneumatic and hydraulic actuators requires additional equipment such as compressors, pumps, pipes and control devices. The slider crank mechanism has been chosen in this study because of the cheaper cost and easier to fabricate. Movement of the slider crank mechanism consists of the slide crank assembly, reducer gearbox, 3-phase motors and three phase motor drive. In an effort to identify the criteria of MR engine mount, a moving slider crank mechanism is mounted to the MR engine mount test rig. Reducer gear box consists of a pair of worm gears with a power capacity of 3.7 kW, sufficient to test the MR engine mount. In this project, the gear reducer with a transmission ratio of 10:1 was selected.

Reducer gear can be installed onto the flange, foot or shaft mounting solutions. In this test rig, foot mounting have been employed because of its simplicity and easy installation. Gear reducer covered by casing made of cast iron provides maximum strength for maximum durability, high accuracy, high overload capacity and stable transmission with lower vibration and noise. Efficient heat dissipation and lubrication for longer operating life provided by a major oil reservoirs. Meanwhile, lubrication pipe plugs allow easy fill. Worm gear reducer coupled with 3-phase motor by the B-type V-belt with a girth of 43 inches. MS-series 3-phase motor made by DELTA ELE is used to authorize on the amount of 5 HP or 3.7 kW. The maximum speed of the motor can produce about 1500 rpm. Mount the foot which has a frame 112 M4 was selected because of the simplicity in installation.

While performing experiment on MR engine mount, some frequencies have been used to produce different experimental conditions. By using 3-phase motor drive, motor speed can be regulated depending on the requirements. In addition, the main reason to use the variable drive speed is to reduce costs and extend the motor life itself. The motor drive working principle is as diverse driving frequencies. It is an electronic controller that adjusts the speed of electric motors regulates the frequency of electric voltage. Variable frequency drive also reduces the mechanical and electrical stresses on the motor, so it will prolong motor life. 3-phase motor drives used in this study. There are two types of sensor used in this test which are Linear Variable Displacement Transducer (LVDT) to measure the displacement and relative velocity of the damper while the Load Cell sensor to measure the damper force. The National Instrument (NI) card device provides signal processing of the transducers and excitation signals of the slider crank actuator system. These signals are digitally processed and stored in a personal computer using MATLAB Simulink 2009B. Control signals to the MR damper are converted to analogue signals by NI card device. Then, the voltage signals are passed through current driver and sent to the MR damper.

6 Conclusion

A MR engine mount test rig has been developed for laboratory testing of MR damper characterization. The limitation of the small peak-to-peak distance of the MR damper is challenging. The sliding mechanism is self-fabricated according to the limitation have solved the problem. The sliding component installed on the structure is aligned properly in resulting smooth running during the experiment. The well software set up give promising results from the sensors used. The MR engine mount testing was successfully done by applying a cyclic motion between the upper and lower ends of the MR engine mount for different values of applied currents to the dampers coils.

Figure 3 above shows the harmonic motion graph which are the results from the up and down movement of the piston. The increasing force in each different current is effect from the generated magnetic field increase as the current increase. The increasing magnetic field leads the increasing of the viscosity of MR damper. Therefore, high viscosity generates large resistance force to passing through the orifice resulting in high damping force generation.

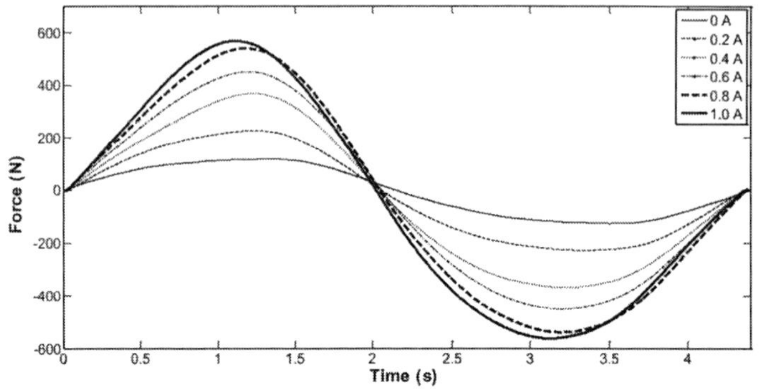

Fig. 3. Measured forces for five constant current levels

This work was financially supported by Short Term Research Grant from Universiti Teknikal Malaysia Melaka (UTeM).

References

1. C. Ciocanel, T. Nguyen, C. Schroeder, M.H. Elahinia, *Performance Evaluation of a Semi-Active Magnetorheological Mount.* SAE Technical Papers. 10.4271/2008-01-0429, (2008)
2. A. Keshavarz, M. Bayani Khaknejad, S. Azadi, *Improving Vehicle NVH Behavior via Tuning the Engine Mount Stiffness Using DOE Method.* SAE Technical Paper 2011-01-1510, (2011)
3. Z. Mohd Ripin, L. E. Ooi, *Dynamic Characterization of Engine Mount at Different Orientation Using Sine Swept Frequency Test.* Regional Conference on Mechanical and Aerospace Technology, Bali, (2010)
4. P.A.G. Zavala, M.G. Pinto, R. Pavanello, J. Vaqueiro, *Experimental and Computational Simulation Approaches for Engine Mounting Development and Certification.* SAE Technical Paper 2000-01-3239, (2000)
5. J. Wang, G. Meng, Proc. Inst. Mech. Eng., Part L., **215**, 165-174, (2001)
6. M. Elahinia, C. Ciocanel, T.M. Nguyen, S. Wang, Smart Mater. Res., **2013**, 21, (2013)
7. K. Hudha, H. Jamaluddin, P. Samin, R. Rahman, International journal of vehicle autonomous systems, **3**, 230 (2005)
8. F.D. Goncalves, J.D. Carlson, J. Phys.: Conf. Ser., **149**, (2009)

Process Optimization of EDM Cutting Process on Tool Steel using Zinc Coated Electrode

Hanizam H[1,*], *N.* Mohamad[2], *U.A.A.* Azlan[1], and *M.* Qamaruddin[1]

[1]Faculty of Engineering Technology, Universiti Teknikal Malaysia Melaka (UTeM), Hang Tuah Jaya, 76100 Durian Tunggal, Melaka, Malaysia
[2] Faculty of Manufacturing Engineering, Universiti Teknikal Malaysia Melaka (UTeM), Hang Tuah Jaya, 76100 Durian Tunggal, Melaka, Malaysia

Abstract. In WEDM machining process, surface finish quality depends on intensity and duration of spark plasma. Electrode wire diameter has significant effect on the spark intensity and yet the studies on this matter still less. Therefore, the main objectives of this studies are to compare the different diameters of zinc coated and uncoated brass electrode on H13 tool steel surface roughness. The experiments were conducted on Sodick VZ300L WEDM and work piece material of tool steel AISI H13 block. Electrode of zinc coated brass with diameters of 0.1 mm, 0.2 mm, 0.25 mm and uncoated brass 0.2 mm were used. The surface roughness of cutting was measured using the SUR-FTEST SJ-410 Mitutoyo, surface roughness tester. The results suggest that better surface roughness quality can be achieved through smaller electrode wire diameter. The zinc coated improves flushing ability and sparks intensity resulting in better surface finish of H13 tool steel. New alloys and coating materials shall be experimented to optimized the process further.

1 Introduction

In die making industries, H13 tool steel which has high wear resistance and thermal shock is commonly used as die material and requires high precision cutting process. Hence, wire electrical discharge machining (WEDM) has been adopted to tackle this requirement. WEDM uses a thin single-strand metal wire to cut a conductive work piece that submerged in dielectric fluid. It uses electro-thermal energy to cut with high dimensional accuracy and surface finish. The high quality surface improves wear resistance, corrosion resistance and fatigue strength of work piece [1]. Moreover, brass is common material used as wire compared to copper, molybdenum and tungsten as more cost effective electrode material. However, in recent years, the development of alloys and coating technologies have led to more robust and improved materials, such as, zinc coated brass wire [2]. The coated wire also helps to protect the brass from thermal shock and broken [3]. In addition, the coated wire able to increase machine productivity by 70% over uncoated brass wire [4].

* Corresponding author: hanizam@utem.edu.my

DOI: 10.1051/matecconf/20179701073

The electrode wire performances are normally measured on two machining responses, material removal rate (MRR) and surface roughness (Ra) and the corresponding input parameters studies by most were current, pulse on time, pulse off time, wire feed and wire tension [3,5,6]. Observations made by several works showed that Ra of H13 tool steel increases when the electrode wire tension, pulse off time and servo voltage increased, and decreases with the increases of electrode wire feed, pulse on time and peak current [7,8]. Furthermore, the MRR shall increase when pulse on time and peak current increases [9]. Since the surface finish quality depends on intensity and duration of spark plasma, another important input parameter such as electrode wire size is always neglected. The studies conducted on wire diameter to surface finish is still less. Therefore, the main objectives of this study are to compare the different diameters of zinc coated and uncoated brass electrode on H13 tool steel surface roughness.

2 Research methodology

The experiments were conducted on Sodick VZ300L WEDM and work piece material of tool steel AISI H13 block of 100mm x 50mm x 20mm. The tool steel melting point and tensile strength were 2600OF and 1990 MPa, respectively. Two type of electrode wires were investigated, brass with tensile strength and thermal conductivity of 1000 N/mm2 and 109 Watt/mK, respectively and with diameters of 0.2mm. Zinc coated brass of 883 N/mm2 and 20 Watt/mK with diameters of 0.1mm, 0.2mm, 0.25mm [5]. Process parameters applied for all samples were current <20A, electrode feed rate 420mm/sec, voltage 120V and submerged into dielectric of distilled water [10]. The surface roughness of cutting was measured using the SUR-FTEST SJ-410 Mitutoyo, surface roughness tester. Analysis data were carried out using Minitab17 software.

3 Result & discussion

3.1 Surface roughness comparison of zinc coated brass electrode diameters

Based on 95% of confident level, two-sample T-test has been conducted to compare the 3 pairs of surface roughness taken at 0.1 mm, 0.2 mm and 0.25 mm diameters. The P-values (< 0.05) obtained from each pair shown significant different of the means and also shown in the boxplot diagram in Fig.1:

DOI: 10.1051/matecconf/20179701073

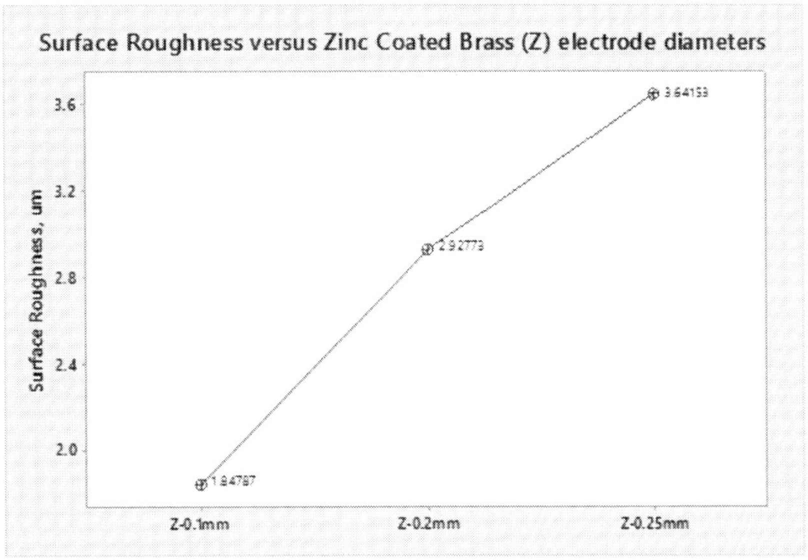

Fig. 1. Surface Roughness of H13 material using different wire diameters

The surface roughnesses of cutting area were affected by the diameter of electrode wires. The results showed that the surface roughness increases almost linearly with electrode diameter. The SR value increases from 1.847 um to 3.64 um respectively. Similar observations have been made by other work on mild steel workpiece material cut using two brass wire diameters of 0.3 mm and 0.2 mm [11]. The 0.2 mm diameter's surface roughness reduced by 40% compared to 0.3 mm diameter from 2.5 um to 3.5 um, respectively.

The surface roughness might be influenced by the heat affected zone sparks during cutting process. The larger the diameter causing higher spark intensities, therefore resulting in more penetration and large amount of material being removed [12].

3.2 Surface roughness comparison between brass and zinc coated brass electrode

The comparison has been made on similar 0.2mm diameter of both electrode materials. Two-sample T-test have been conducted and the result as follow;

Electrode	N	Mean	StDev	SE Mean
B-0.2mm	15	3.375	0.123	0.032
Z-0.2mm	15	2.928	0.0735	0.019

Difference = μ (B-0.2) - μ (Z-0.2)
Estimate for difference: 0.4477
95% CI for difference: (0.3709, 0.5245)
T-Test of difference = 0 (vs ≠): T-Value = 12.09 P-Value = 0.000 DF = 22

The P-values (< 0.05) obtained shown significant different of the means and also shown in the boxplot diagram in Fig.2:

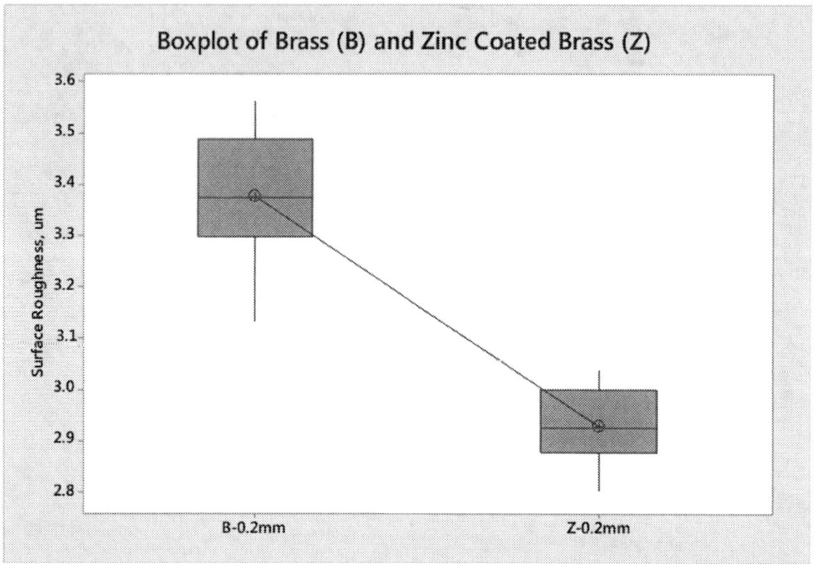

Fig. 2. Surface roughness comparison between coated and uncoated electrode wire

The tool steel machined using zinc coated brass produced better surface quality compared to brass wire and similar to other work [12]. It is due to the formation of surface profile, decreased dielectric flushing ability and zinc coated also acts as coolant to minimize surface modification. The ineffective flushing result in poor surface finishes [3].

As a proven method for the result, the value of surface roughness for both wires is compared graphically by time series plot graph (Figure 2). By refer to the graph; the surface roughness value for zinc coated brass is definitely lower than brass wire electrode. Similar conclusion made by other author when comparing the SR of INCONEL 600 material cut using 0.25 mm brass and zinc coated brass wires [13]. Zinc coated brass wire also given the optimum result when comparing brass, zinc coated brass and molybdenum wires.

4 Conclusions

This study set out to examine the relationship between the surface roughness and electrode diameters. These data suggest that better surface roughness quality can be achieved through smaller electrode wire diameter on top of electrode wire tension, pulse off/on time, servo voltage, electrode wire feed, and peak current. Indirectly, it might be useful to the die making industry to save some cost on the wire. Another important finding, in general, it seems that the zinc-enriched outer coating of brass improves flushing ability and sparks intensity resulting in better surface finish of H13 tool steel.

The authors would like to acknowledge Universiti Teknikal Malaysia Melaka (UTeM) and Ministry of Higher Education Malaysia for sponsoring this study under a Short Term Research Grant – Scheme (PJP/2014/ FTK (13B)/S01328).

DOI: 10.1051/matecconf/20179701073

References

1. J. Singh, S. Sharma, Int. J. of Current Eng. and Tech., 1852-1857, (2013)
2. M. Manjaiah, S. Narendranath, S. Basavarajappa, V.N. Gaitonde, Precis. Eng., **41**, 68 (2015)
3. J. Kapoor, S. Singh, J.S. Khamba, Proc. World Congr. Eng., **2**, 1 (2010)
4. D. Thomas, R. Kumar, G.K. Singh, P. Sinha, S. Mishra, Materials Today: Proceedings. **2**, 3520 (2015)
5. B.K. Lodhi, S. Agarwal, Procedia CIRP., **14**, 194 (2014)
6. V. Singh, S.K. Pradhan, Procedia Eng., **97**, 1597 (2014)
7. M.T. Antar, S.L. Soo, D.K. Aspinwall, D. Jones, R. Perez, Procedia Eng., **19**, 3 (2011)
8. J.R. Mevada, Int. J. Sci. Eng. Res., **4**,1398 (2013)
9. Ashok Kumar Choudhary, Prof. K.K. Chhabra, Int. J. Res. Dev. Eng. Sci. & Manage., **1**, 72 (2014)
10. D. Ghodsiyeh, A. Golshan, J.A. Shirvanehdeh, Indian J. Sci. Technol., **6**, 4128 (2013)
11. A.A. Khan, M.B.M. Ali, N.B.M. Shaffiar, J. Appl. Sci., **6**, 2317 (2006)
12. K. Kanlayasiri, P. Jattakul, Precis. Eng., **37**, 556 (2013)
13. G. Amitesh, K. Jatinder, Int. J. Adv. Eng. Technol., **3**, 170 (2012)

A Design Method of Metasurface using CCSR

A.A.M Ezanuddin[1,*] , and *A.H* Ismail[1]

[1]Fakulti Teknologi Kejuruteraan, Universiti Malaysia Perlis (UniMAP), Aras 1, Blok S2, Kampus UniCITI Alam, 02100 Sungai Chucuh, Padang Besar Perlis, Malaysia.

Abstract. This research paper describes the steps used to design an alternative metasurface which is based on Circular Complementary Split Ring (CCSR) formation at 5.8 GHz. All the 3D modelling was conducted using specialized EM simulation software that utilizes Eigen algorithm to solve different frequency modes that can resonate within the CCSR structure itself. The final design will be then fabricated using a precise laser CNC machine to achieve a better dimension as closely as possible to the simulation dimension. By using laser CNC, this step will able to overcome generic PCB design issue related to via's placement, signal integrity, power integrity and multilayer designs. Completed CCSR metasurface will be able to mitigate or block unwanted EM mode from flowing to its neighbouring sensitive area which can affect the entire component performance due to unseen harmonics.

1 Introduction

The systematic design of unit cells for a CCSR based surface, a particular class of metasurface, is presented here. The design of these unit cells is based on transmission-line theory and waveguide. The CCSR unit cells presented here are machined on a typical microwave laminate instead of many stacked, interspaced layers [1]. This simplifies fabrication and enables the scaling down of the metasurfaces to shorter wavelengths and are capable of synthesizing the full range of impedances required. In contrast to frequency-selective surfaces (FSSs) and traditional resonator arrays, which are on the order of a wavelength thick, these designs are only $\lambda/10$ thick while incurring minimum reflections losses.

Figure 1 displays the basic Maxwell equations used to perform the required calculation There are four steps used in the simulation to produce a final result in the from Maxwell Grid equations. Whereas in Figure 2, it illustrates the basic material (Dielectric value of PCB) equations involved and it's corresponding mathematical variables used in calculating the transient solver for every 3D mesh grid of the proposed unit cell and full array unit cells. The transient purpose is to conduct integration calculations in iteration form until it can produce a stable and real signal waveform in scattering parameter format.

* Corresponding author: mezanuddin85@gmail.com

 DOI: 10.1051/matecconf/20179701074

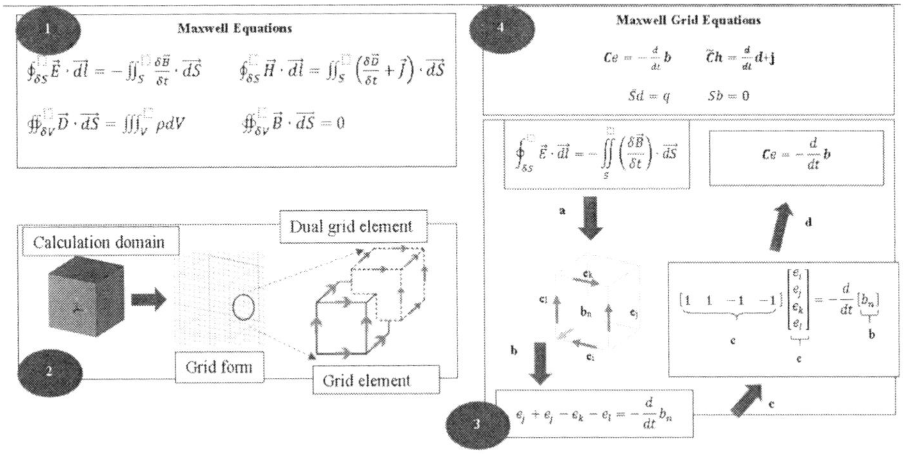

Fig. 1. Finite integration technique used within the 3D EM simulation technique and its steps

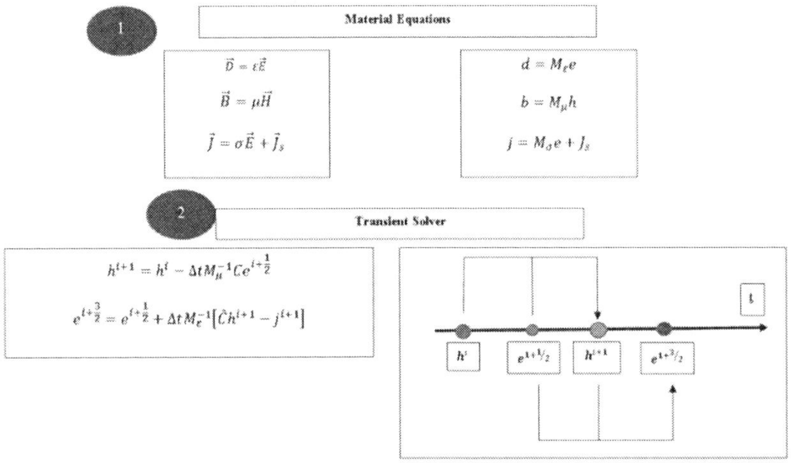

Fig. 2. 3D EM Transient solver for general S-Parameter simulation steps before moving to Eigen algorithm

2 Initial simulation

Provided that the unit cell is sub-wavelength, the CCSR unit cell can be approximated by an electric loop. In addition to this, the orientations of the CCSR unit cell must correspond with the polarizations of the input and output EM fields. This reflects the anisotropies of the impedances required to enforce the surface currents. Keeping this issue in mind, unit cells proposed here were synthesized using commercial full-wave 3D electromagnetics solver to obtain the required electric and magnetic impedances. Here infinite array analysis was used. This result makes sense in view of the transmission-line description of infinite array analysis [2].

Figure 3(a) and (b) show the proposed CCSR unit cell and one period of the simulated unit cell. Note that the 1D refraction is in the horizontal plane, and the unit cell is compatible with the field polarizations stipulated in these problems. The input and output electric field polarizations are vertically oriented. The input and output EM field polarizations are in the horizontal plane as can be seen in Figure 4. The structure was simulated using known waveguide model WR159 (operational between 4.90 to 7.05 GHz) This first unit cell was designed as an entirely printable metasurface, and are implemented

DOI: 10.1051/matecconf/20179701074

on a FR4 substrate [ε_r = 4.7, tan(δ) = 0.0025]. Final result shows that this CCSR can resonate well at 5.8 GHz.

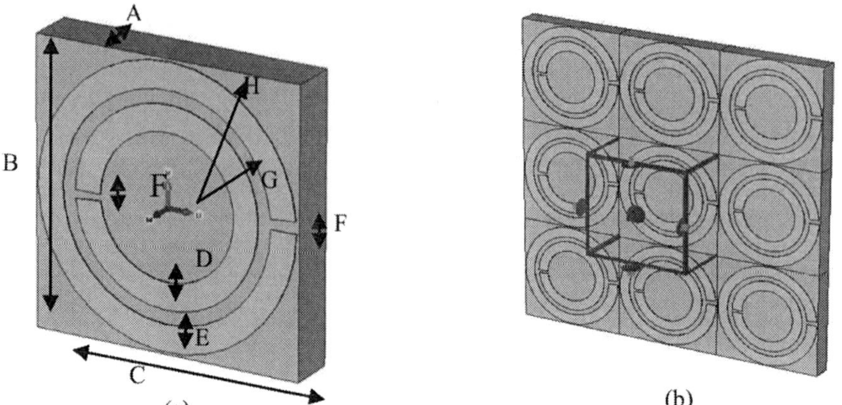

(a) (b)

Fig. 3. CCSR unit cell a) the unit cell configuration are A = 1.6 mm, B = C = 9 mm, D = E = 0.85 mm, F = 3550 (Angle), G = 2.71 mm (radius) and H = 4 mm (radius) and b) unit cell with its boundary condition, two port setup was used here

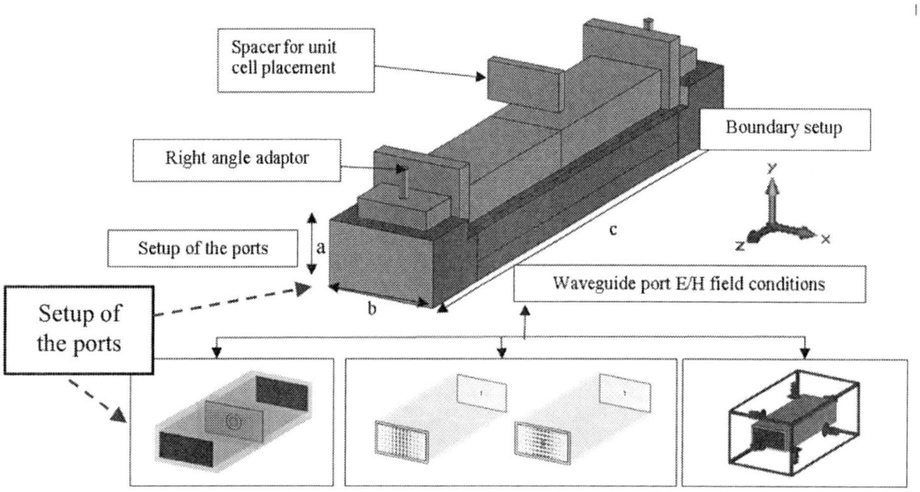

Fig. 4. Simulation of the CCSR unit cell inside a WR 159 waveguide to characterise its S-parameters from 5.85 to 8.2 GHz

3 Fabrication and near-field measurement

Now that all of the issues with the layout design have been resolved, the next step is to actually transfer to out the circuit layout and fabricate it. Layout editor was used to create a Gerber file, Gcode and Gerbtools™ to reorganize every gerber files into a printable dimension. The board was fabricated using the combination of laser CNC Figure 5(a) and wet-etching Figure 5(b). Laser CNC uses open-source software for its layout instructions and commands to control laser intensity level upon material used. The final layout and the assembled PCB are shown below in Figure 5(c) below.

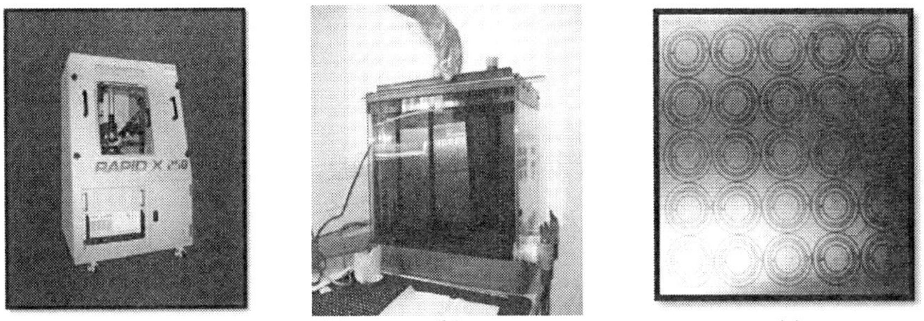

(a) (b) (c)

Fig. 5. (a) Resonetics Rapid X 250 laser CNC micromachining machine which allows laser cutting for any material with high precision and accuracy, (b) The outer PCB area was done with wet etching technique, and (c) the final output

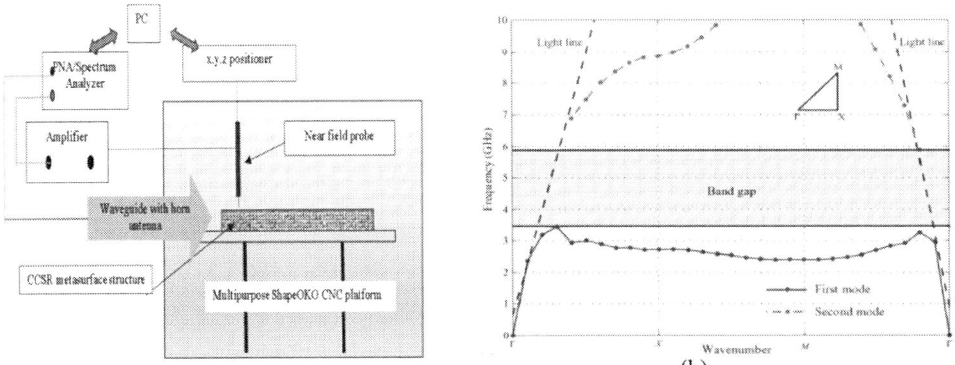

(a) (b)

Fig. 6. Diagram of the experimental setup, (a) the LANGER EMV-Tehcnik™ near field probe, and (b) Shows the expected dispersion diagram of the CSAR structure. The vertical axis shows the frequency and the horizontal axis represents the values of the transverse wavenumbers (β_x, β_y). Three specific points are: Γ, X and M

Agilent E5071C network analyzer was used to measure the phase shift (angle of S_{21}), insertion loss (magnitude of S_{21}),signal delay and return loss (magnitude of S_{11}).The PNA can be replaced with a much simpler equipment such as the Spectrum Analyzer since it will be easier to map E-field and H-field and plot it back using Excel™. Figure 6(a) shows the diagram of the intended platform for measuring purposes. The dispersion of a metasurfaces are tested and compared in terms of measurement, simulation and the theoretical model. The distance of the waveguide is set to d = 10 mm from the CCSR metasurface.

4 Expected dispersion results

The expected results for each dispersion method are shown in Figure 6(b). The measured dispersion [3] will closely match with measured, simulation and theory. Only the lowest mode can be measured, and the results stop at 10 GHz. Above this frequency, data is difficult to obtain (due to lack of equipment and facility). The measurement probe is scanned 5 mm above the surface, and if significant wave power is located below this height, it becomes difficult to obtain an accurate reading.

When two or more modes are presented, it becomes difficult to determine the phase velocity [4] of any single mode due to interference between the modes. In simulation and theory, single mode operation can be strictly enforced, and the dispersion curves can be extended to higher frequencies that can be measured experimentally.

From Figure 6(b), the first two modes are shown as dashed lines for the theoretical solutions, and as dots for simulation. The first theory curves uses ideal impedance theory to obtain impedance at intermediate angles and the second theory curve uses simulation to get these values. A black dashed line is the dispersion of light in a vacuum. Guided modes must lie below the dispersion relation of the outer region, and this is shown as a black line. Simulated and theoretical curves are shown for the first two guided modes. The second mode has a cutoff when it intersects the dispersion curve of the outer region. The outer region dispersion represents the highest phase velocity that can be achieved in the waveguide, as this outer region has a lower index than the interior region.

5 Expected near-field measurement

Two-dimensional near field measurement testbed were built using the setup described in previous section. The final dimension of the CCSR surface produced was 75 x 75 mm. The measured sample is smaller than the total table board size to allow space for the feed and absorbers around the edge of the surface. The excitation source, shown before in Figure 6(a), does not couple perfectly into the guided modes at all frequencies, and some power is excited directly to the outer region. A vertical probe was swept 5 mm above the surface along a 2 mm grid. The probe was made to scan higher above the surface for two-dimensional field measurements because the surface was not completely flat and scanning lower could cause contact between the probe and surface at the edges.

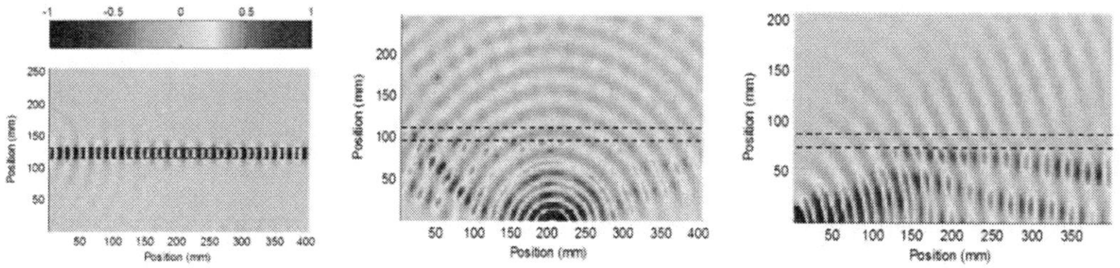

(a) (b) (c)

Fig. 7. Expected measured fields. The plot show normalized vertical electrical field at 5.8 GHz for (a) guided mode, (b) orthogonal mode, and (c) angled mode. Fields in (a), (b), and (c) are not plotted to the same scale

A guided mode is displayed in Figure 7(a). The mode is bound near the guide and does not leak to the outer region. The measured bandwidth for guided mode is from 5 to 8 GHz. below 5 GHz the measured mode is regular surface wave. At 5 GHz, as shown in Figure 7(a) almost all of the fields are in the waveguide mode and minimal fields are present in the outer region. Above 8 GHz, the source excites more measurable field in the outer region than as guided mode. This is because higher frequency modes have high impedance[5].

Figure 7(b) shows a two-dimensional field plot for a surface with a feed located in the exterior region. The feed was a vertically-oriented horn antenna that excited electromagnetic normal phase fronts on the surface. The guiding and outer regions have identical surface impedance for surface wave incident normal to the guide. Therefore, as seen in Figure 7(b), waves at normal incidence pass through the guide as if it was not present, and transparency is confirmed. In this case, the wave front is only orthogonal directly above the source. At off-normal angles, some reflections and distortion can be seen off the guide. In Figure 7(c) a trapezoidal feed is used to launch a surface wave toward the waveguide at a small grazing angle. In this case, the higher impedance guiding region affects a significant portion of the incident power.

The authors are grateful to the financial support by Project No 9003-00449 from Fundamental Research Grant Scheme (FRGS) 2014 from Ministry of Education (MOE), Malaysia.

References

1. J.Q. Liu, M.D. He, S. Chen, C.P. Huang, L. Zhou, Y.Y. Zhu, J. Phys.: Condens. Matter., **23**, 215303 (2011)
2. O. Abu Safia, L. Talbi, K. Hettak, Magnetics, IEEE Transactions., **49**, 968 (2013)
3. Y. Hao, R. Mittra, *FDTD modeling of metamaterials: theory and applications*. Artech house (2008)
4. H.X. Xu, G.M. Wang, M.Q. Qi, Z.M. Xu, Appl. Phys. Lett., **101**, 173501 (2012)
5. D. Seetharamdoo, R. Sauleau, K. Mahdjoubi, A.C. Tarot, J. Appl. Phys., **98**, 063505 (2005)

Study on Surface Integrity of AISI 1045 Carbon Steel when machined by Carbide Cutting Tool under wet conditions

N. Fauzi Tamin[1,*] , *U.A.A.* Azlan[1] , *M.* Hadzley[2] , *N.S.* Remle[1] , *M. Hairizal* Osman[1] , *M. Rahimi* Yusoff[1] , and *N.Z.* Noriman[3]

[1]Faculty of Engineering Technology, Universiti Teknikal Malaysia Melaka (UTeM), Hang Tuah Jaya, 76100 Durian Tunggal, Melaka, Malaysia
[2]Faculty of Manufacturing Engineering, Universiti Teknikal Malaysia Melaka (UTeM), Hang Tuah Jaya, 76100 Durian Tunggal, Melaka, Malaysia
[3]Faculty of Engineering Technology, Universiti Malaysia Perlis (UniMAP), Aras 1, Blok S2, Kampus UniCITI Alam Sungai Chuchuh, Padang Besar, 02100 Perlis, Malaysia

Abstract. This paper presents the evaluation of surface roughness and roughness profiles when machining carbon steel under wet conditions with low and high cutting speeds. The workpiece materials and cutting tools selected in this research were AISI 1045 carbon steel and canela carbide inserts graded PM25, respectively. The cutting tools undergo machining tests by CNC turning operations and their performances were evaluated by their surface roughness value and observation of the surface roughness profile. The machining tests were held at varied cutting speeds of 35 to 53 m/min, feed rate of 0.15 to 0.50 mm/rev and a constant depth of cut of 1 mm. From the analysis, it was found that surface roughness increased as the feed rate increased. Varian of surface roughness was suspected due to interaction between cutting speeds and feed rates as well as nose radius conditions; whether from tool wear or the formation of a built-up edge. This study helps us understand the effect of cutting speed and feed rate on surface integrity, when machining AISI 1045 carbon steel using carbide cutting tools, under wet cutting conditions.

1 Introduction

Carbon steel is an engineering material that is generally contained with steel and carbon. The properties of carbon steel depend on the carbon contained within their composition, ranging from low (0.05-0.2%), medium (0.30-0.60%) and high (0.7-2.5%) [1]. Carbon steel possesses several important characteristics, such as being difficult to weld, bend, machine and cut; depending on its properties. When heat treatment is applied, the properties of carbon steel can be altered whereby it can change the formability, machinability and weldability of the products [2]. Carbon steel is considered to be an important alloy used in industry, since it accounts for over 90% of all construction materials, automotive, petroleum and chemical industries [3]. One of the dominant processes used to produce the

* Corresponding author: norfauzi@utem.edu.my

parts made from carbon steel is machining. Carbon steel is still being machined, as this material possesses good characteristics during material removal and is considered soft and rigid when engaged with a cutting tool. During workpiece-cutting tool engagement, contact at the tool-chip and tool-workpiece interfaces generates heat; and therefore, wet cutting is generally applied to enhance machining performance. The use of wet cutting enables the machining process to obtain maximum metal removal and accurate dimensions; thus achieving an adequate surface quality.

Despite the advantages of machining carbon steel in wet conditions, there are some difficulties machining this material; especially when the cutting parameters are not applied correctly. Existing industry cutting tools usually facilitate guidance parameters for general machining practices. The use of specific cutting tools for specific materials; especially in wet conditions, still need to be explored - as it can enhance the machinist's knowledge to observe the behaviour of material during machining and apply parameters correctly. This paper investigates the surface roughness when machining carbon steel AISI 1045 in wet conditions at different cutting speeds, feed rates and a constant depth of cut. In this study, a 3 axis CNC lathe was used for machining trials. The surface integrity of carbon steel was evaluated using a surface roughness tester. The data obtained from this study will be useful as a reference for machinists planning their machining, in order to produce maximum surface quality, reduce processing steps and minimize operational costs.

2 Experimental procedures

This work aims to identify the surface roughness and roughness profile of carbon steel at several cutting speeds and feed rates in wet conditions. The experiments were carried out using a CNC Machine (DMG-CTX 310). The medium, carbon steel of AISI 1045, was selected as the workpiece material. The types of carbide tools used in this work were made by SANDVICK-Coromat (as shown in Fig. 1). The tool holder used was a DCLNL 2020K (as shown in Fig. 2). Pre-tech mineral oil was used as a coolant. A surface roughness tester (Mitutoyo SJ-410) was used to determine the surface roughness value and roughness profiles. Table 1 show the cutting parameters used in this study.

Table 1. Cutting conditions for machining tests (Depth of cut: constant 1 mm)

No.	Cutting speed (m/min)	Feed rate (mm/rev)
1	35	0.15
2	41	0.15
3	47	0.15
4	53	0.15
5	35	0.50
6	41	0.50
7	47	0.50
8	53	0.50

Fig. 1. SANDVICK Coromat cutting tool

Fig. 2. Tool Holder DCLNL 2020K

3 Results and discussion

Fig. 3 shows a plot of surface roughness (Ra) based on the determined cutting speed and feed rates. It shows the line graph of the surface roughness, Ra versus cutting speed and surface roughness data, Ra in wet conditions at different cutting speeds with a constant feed rate of 0.15 mm/rev. At a cutting speed of 35 m/min, the value of surface roughness was 3.085 µm, increasing to 3.948 µm at 47 m/min; but decreasing to 3.283 µm at 53 m/min. Fig. 4 shows the 0.50 mm/rev with different cutting speeds. The line graph shows that surface roughness, Ra decreased slightly for the following cutting speeds; 35 m/min, 41 m/min, 47 m/min and decreased even more at a cutting speed of 53 m/min. Surface roughness values following these cutting speeds were 16.327 µm, 15.774 µm, 15.022µm and 14.923µm, respectively. Thus, surface roughness decreased with an increased cutting speed. It should be noted that the Ra recorded at lower feed rate of 0.15 mm/rev (3.085-3.948 µm) are so much lower than higher feed rate of 0.5 mm/rev (14.923-16.327 µm). This shows that the feed rate is the most influencing parameter affecting on surface roughness. The results consistent with the review from [4] where the surface finish obtained is good when the lower feed rate is applied as compared to higher feed rate. Low surface roughness enables to produce accurate dimension and increase the fatigue life of the machined components [5].

Fig. 5 shows the surface roughness profile at feed rate 0.15 mm/rev with different cutting speeds. It shows that at the lower cutting speed of 35 m/min, the surface roughness profile showed uniform feed marks according to the nose radius engagement of the cutting tool at the upper machined surface. As the cutting speed increased to 41 m/min and 47 m/min, the distance between feed marks became closer; hence, resulting in a higher surface roughness value (as shown in Fig. 3). However, as the cutting speed increased to 53 m/min, the surface roughness profile exhibited an almost flat appearance. It is suspected that at the highest cutting speed of 53 m/min, the cutting tool was already worn, which had altered the nose radius of the cutting tool, thus resulting in an alteration of feed marks to an almost flat appearance on the roughness profile. Another possible mechanism was the presence of a built-up edge that altered the nose radius of the cutting tool. A built-up edge is the formation of molten metal that adheres to the nose radius due to high temperatures generated at the tool-chip interface [6]. The existence of a built-up edge may indicate that the coolant did not penetrate the cutting zone due to the higher cutting speed or feed rate that splashing away the coolant.

Fig. 6 shows the surface roughness profile for a feed rate of 0.50 mm/rev. It shows that every different cutting speed had different surface roughness profiles. The surface profile of the presented figures shows a larger distance between feed marks (compared to Figure 5) due to the larger distance of feed rate. The surface profile exhibited fine feed marks at a lower cutting speed of 35 m/min. As the cutting speed increased to 41 m/min and 47 m/min, the feed marks exhibited finer distance of peak and valley, resulting in a lower surface roughness (as shown in Fig. 4). It should be noted however, as the cutting speed

DOI: 10.1051/matecconf/20179701075

increased to 53 m/min, the surface roughness profile demonstrated smaller gaps between the feed marks, may be due to several aspects that need to further investigated.

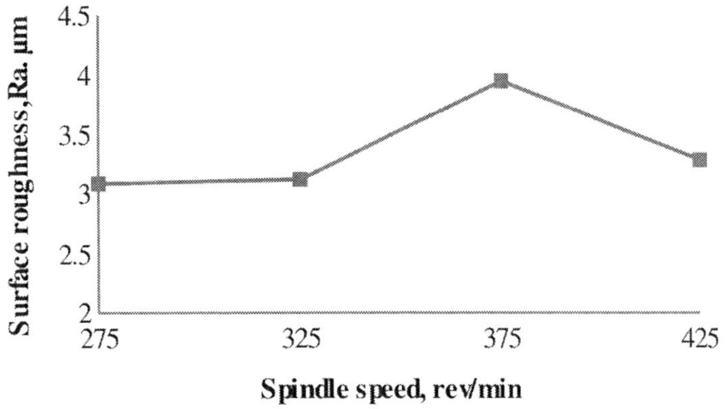

Fig. 3. Surface roughness vs. cutting speed using feed rate 0.15 mm/rev

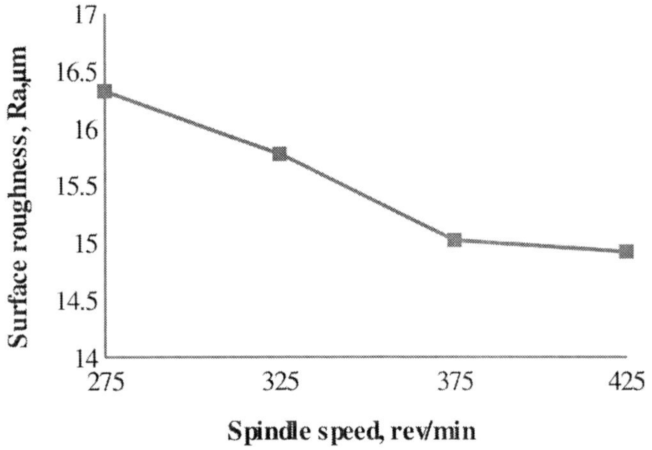

Fig. 4. Surface roughness vs. cutting speed using feed rate 0.50 mm/rev

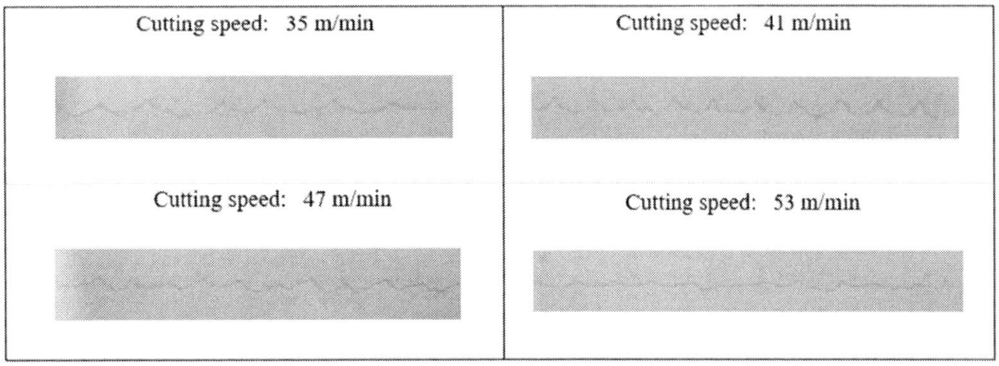

Fig. 5. Surface roughness profile at feed rate 0.15 mm/rev

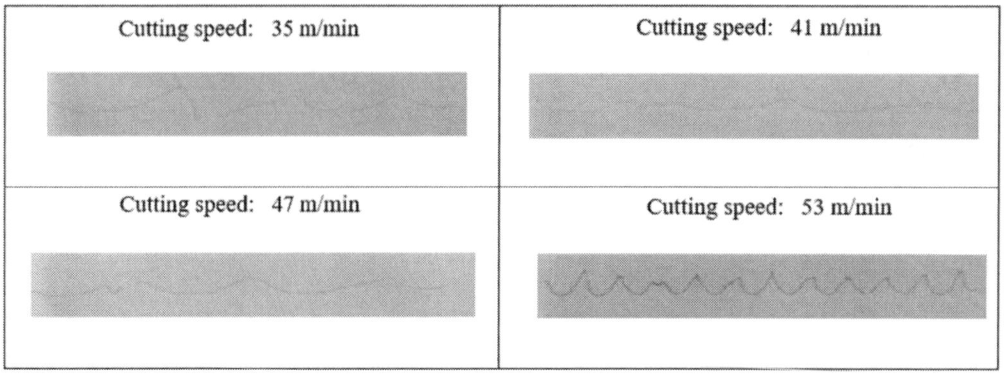

Fig. 6. Surface roughness profile at feed rate 0.50 mm/rev

4 Conclusion

This paper presents the experimental of surface roughness of Carbon Steel AISI 1045 using wet cutting CNC turning. From the machining data collected, and based on the experimental findings, the following conclusions were made:

i. The surface roughness increased when the cutting speed was increased at a feed rate of 0.15 mm/rev.

ii. The surface roughness decreased when the cutting speed was increased at a constant feed rate with a value of 0.50 mm/rev.

iii. The use of lower feed rates is suitable for wet cutting.

iv. This study helps us to understand the effects of cutting speed and feed rate on the surface roughness of Carbon Steel AISI 1045 under wet cutting conditions.

The author would like to thank Universiti Teknikal Malaysia Melaka and Ministry of Higher Education under the Fundamental Research Grants Scheme (FRGS) number: FRGS/1/2014/TK04/FTK/02/F00207 for funding this project.

References

1. J.K. Odusote, T.K. Ajiboye, A.B. Rabiu, J. Miner, Mater. Charact. Eng., **11** (2012)
2. D. Gandy, Carbon Steel Handbook. Paolo Alto: Electric Power Research Institute. (2007)
3. S.O. Seidu, B.J. Kutelu, J. Miner, Mater. Charact. Eng., **1** (2013)
4. S.B. Chikalthankar, R.B. Kakade, V.M. Nandedkar, Int. J. Eng. Res. Technol., **3**, 10 (2014)
5. A.B. Mohd Hadzley, M.R. Nurul Fatin, R.A. Raja Izamshah, N.I.S. Hussein, A. Siti Sarah, Sivarao, Appl. Mech. Mater., 465-466 (2014)
6. A.B. Mohd Hadzley, A. Siti Sarah, R.A. Raja Izamshah, A.A. Mohd, M.S. Kasim, M.A. Sulaiman, M.R. Nurul Fatin, Appl. Mech. Mater., 465-466 (2014)

Observation of Built-up Edge Formation on a Carbide Cutting Tool with Machining Aluminium Alloy under Dry and Wet Conditions

U.A.A. Azlan[1,*] , *M.* Hadzley[2] , *N. Fauzi* Tamin[1] , *F.M.* Noor[2] , *A.A.* Azhar[2] , *M. Rahimi* Yusoff[1] , *N.Z.* Noriman[3]

[1]Faculty of Engineering Technology, Universiti Teknikal Malaysia Melaka (UTeM), Hang Tuah Jaya, 76100 Durian Tunggal, Melaka, Malaysia
[2]Faculty of Manufacturing Engineering, Universiti Teknikal Malaysia Melaka (UTeM), Hang Tuah Jaya, 76100 Durian Tunggal, Melaka, Malaysia
[3]Faculty of Engineering Technology, Universiti Malaysia Perlis (UniMAP), Aras 1, Blok S2, Kampus UniCITI Alam Sungai Chuchuh, Padang Besar, 02100 Perlis, Malaysia

Abstract. This paper presents the morphology of built-up edge (BUE) formation under wet and dry conditions with low and high cutting speeds. The workpiece materials and cutting tools selected for this work were aluminium alloy and canela carbide inserts graded PM25. The cutting tools underwent turning operation machining tests and their performance was evaluated by the flank wear and observation of the tool wear area. The machining tests were conducted at different spindle speeds and feed rates while the cut depth was kept constant. The analysis showed that formation of the BUE was dominant at low cutting speeds in dry conditions, but in wet conditions at high cutting speeds, a better performance was exhibited in terms of wear analysis.

1 Introduction

Aluminium alloy is a lightweight material that has been widely used in manufacturing industries, especially in the automotive and aeronautics sectors. This is because of its superior mechanical behaviour; exhibiting good corrosion resistance, low density, high strength and low-cost material [1]. Despite the advantages of this aluminium, there are some difficulties machining this material especially when the cutting parameter is not applied correctly. In any machining process, apart from obtaining accurate dimensions, achieving a good surface quality and maximizing metal removal are also important criteria. Machining aluminium is normally applied in dry conditions whereby the processes are normally applied to finishing and cleaning environments. Therefore, machining aluminium sometimes produces quick tool wear due to high friction at the cutting interface [2].

There are other techniques such as wet cutting that are also applicable to cutting aluminium. The wet cutting technique reportedly produces longer tool life due to lower friction and a cooler environment [3]. However, using coolant will be hazardous to the

* Corresponding author: umar@utem.edu.my

environment, and the availability of the material raises the question of which technique is better to machine this material. There are cases where machining aluminium may produce the best surface finish and the lowest tool wear in dry conditions when the cutting parameters are applied correctly. There are also cases where machining aluminium gave the worst surface finish and the highest tool wear in wet conditions when improper cutting parameters were applied [4]. Therefore, it is necessary to make a comparison between wet and dry conditions in order to evaluate the best surface finish and tool wear when machining Aluminium 2024. This paper investigates the tool wear of carbide cutting tools when machining aluminium alloy on wet and dry conditions at different cutting speeds. In this research, the evaluation of the machining performance of the cutting tools mentioned above depends on tool wear and tool life and wears mechanisms.

2 Experimental procedures

The aim of this work is to discover the performance value of the tool life of carbide cutting tools at several different cutting speeds on dry and wet conditions. The experiments were carried out on a MOMAC SM200X manual lathe machine. Since the manual lathe machine was used, 12 bars of workpiece materials were prepared at consistent cutting lengths of 127 mm long and 19 mm diameter for each cutting trial. The types of carbide tools used in this work were made from Canela insert grade PM25, and held by a PSBNR16-4R174.3-2525-12 tool holder. Mineral oil was used as the coolant for wet conditions. An optical microscope and a scanning electron microscopy (SEM) were used to determine tool wear and make observations of the edge tool profile. For this experiment, cutting speeds and feed rates were varied according to the machine settings, while the cut depth was kept constant at 1 mm. Table 1 shows the cutting parameters used in this study.

Table 1. Cutting Conditions for Machining Tests (Depth of Cut: Constant 1 mm, Cutting Conditions: Dry and Wet

No.	Cutting speed (m/min)	Feed rate (mm/rev)
1	10,000	45.72
2	15,000	66.04
3	22,000	96.52
4	32,000	142.24
5	47,000	210.82
6	71,000	307.34

3 Results and discussion

Fig. 1 shows a comparison plot of wet and dry cutting conditions in terms of tool wear value. For dry cutting, the flank wear was considered high at the cutting speed of 10,000 m/min and it decreased steadily between 22,000 m/min to 71,000 m/min. For wet cutting, the graph shows that low flank wear was recorded at a starting point of 10,000 m/min, increased tremendously at 15,000 m/min, gradually decreased at 22,000 m/min and slightly increased at the end of 71,000 m/min. It clearly shows that the flank wear in wet cutting was lower than in dry cutting for most of the cutting conditions, except at the higher cutting speed of 71,000 m/min.

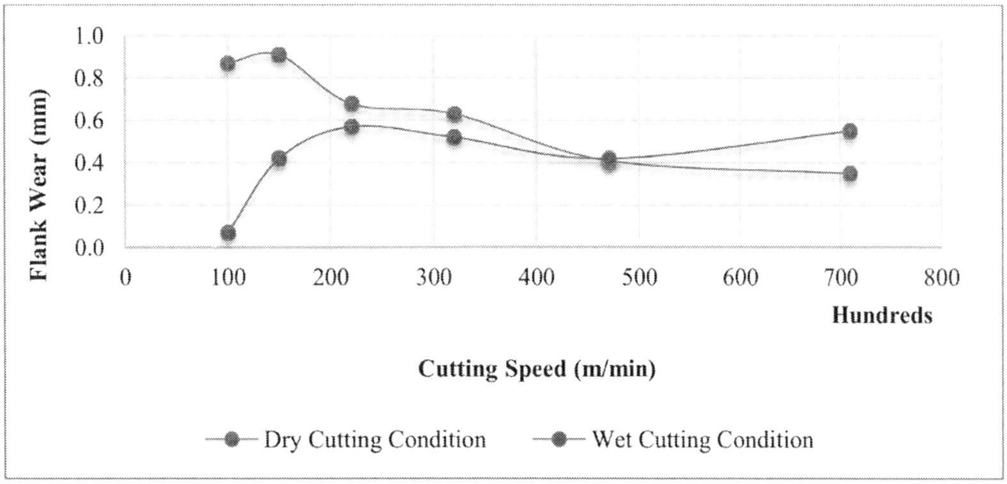

Fig. 1. Comparison between Wet and Dry Cutting Conditions

Fig. 2 and Fig. 3 show the images of flank wear of the cutting tool taken by SEM in dry and wet conditions respectively. The images clearly illustrate the formation of BUE at some locations at the edge of the tool wear area. This was observed especially during machining in dry conditions. BUE can be described as the machined material building up on the cutting edge due to the material having a tendency to anneal itself to the cutting edge of a tool. It occurs most frequently on softer metals with lower melting points, and it can be prevented by increasing cutting speeds and using lubricant. During machining, high temperatures generated at the tool-chip interface resulted in some of the material melting. As a result, this material gradually attached to the cutting tool as a permanent structure in the form of bulk layer [5].

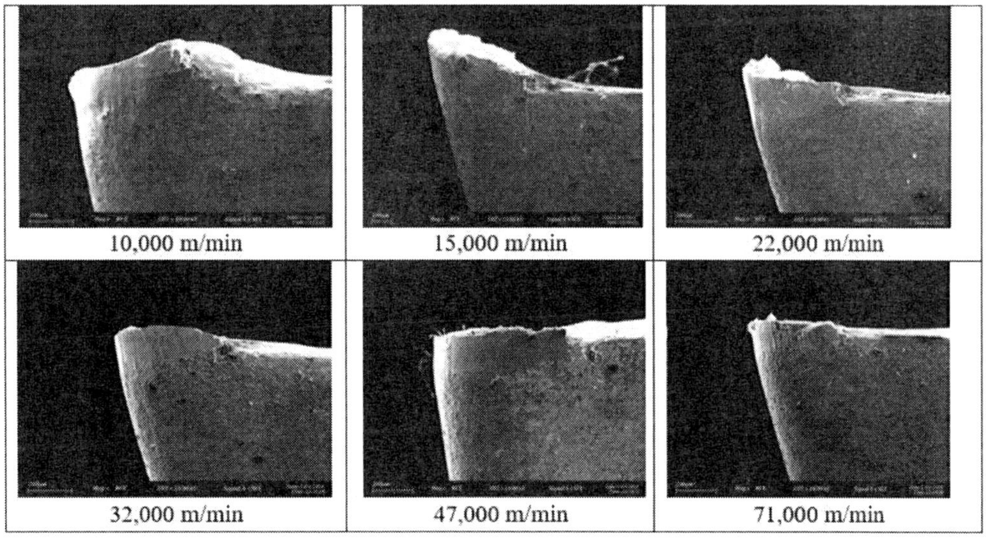

Fig. 2. Image of flank wear of the cutting tool in dry cutting conditions

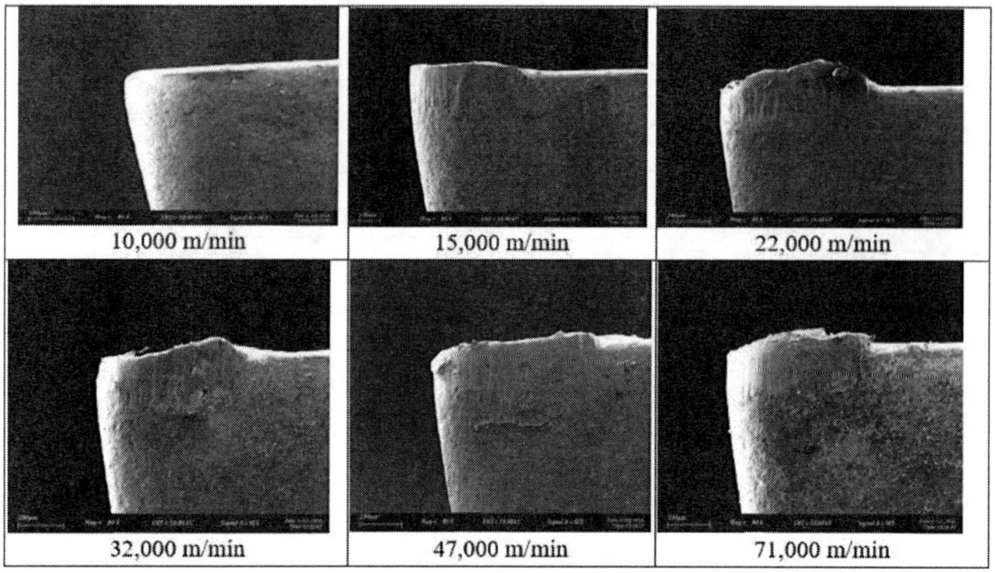

Fig. 3. Image of flank wear of the cutting tool in wet cutting conditions

The existence of BUE is highly dependent on spindle speed and feed rate. The additional appearance of BUE altered the tool nose radius, resulting in changes in the tool-chip contact as well as cutting force. Changes in the tool-chip contact may result in surface finish of the machined component while change in the cutting force may increase the risk of tool breakage [6]. It should be noted that, in dry cutting conditions, the formation of BUE was only observed at low cutting speeds. For instance, BUE significantly appeared at cutting speeds of 10,000 m/min, 15,000 m/min and 22,000 m/min only. Conversely, the BUE formation was not clearly observed at the high cutting speeds of 32,000 m/min, 47,000 m/min and 71,000 m/min. This phenomenon occurred due to the sliding action of the chip during high cutting speed. As the cutting speed increased, the sliding velocity of the chip that formed during machining increased tremendously. As a result, the force from the sliding chip slid the BUE immediately, leading to a gradual decrease in the BUE structure [7].

The appearance of BUE was observed to be minimized when wet cutting was performed, as shown in Fig. 3. The use of cutting fluids enabled heat to be dissipated away from the cutting zone, and at the same time, the fluid reacted as a lubricant to reduce the friction between the cutting tool and workpiece material. Additionally, the force from the overhead cutting fluids removed the chip formation which helped dispose of the chips that became entangled in the machining zone, decreasing the contact length between chip and control chip formation. This consequently had a positive effect on improved machining performance.

4 Conclusions

This paper presents the results in terms of tool performance when turning Aluminium Alloy 2014 in wet and dry conditions. The process appears to offer an economic advantage over completely dry machining current aluminium alloys. Based on the experimental findings the following conclusions were made:

- Flank wear is higher at lower cutting speeds and lower feed rates and slightly decreased when the speed is increased in dry cutting conditions. Conversely, flank wear in wet cutting conditions is lower until 47,000 rev/min and constantly uniform from then onwards.

- The tool wear on wet cutting is lower than on dry cutting due to the presence of cutting fluid that reduced the friction and dissipated heat away from the contact zone.
- BUE is significant in both cutting conditions, but it is most effective in dry cutting.
- The presence of cutting fluid can reduce the formation of BUE at the cutting zone.

Special thanks to Ministry of Higher Education Malaysia (MOHE) (FRGS/1/2014/TK04/FTK/02/F00207) and Universiti Teknikal Malaysia Melaka for their financial support.

References

1. S. Kalpakjian, S.R. Schmid. 4th ed. New Jersey: Prentice-Hall (2001)
2. A.B. Mohd Hadzley, A. Siti Sarah, R.A. Raja Izamshah, A.A. Mohd, M.S. Kasim, M.A. Sulaiman, M.R. Nurul Fatin, Appl. Mech. Mater., 465-466 (2014)
3. M.S. Kasim, C.H. Che Haron, J.A. Ghani, M.A. Azam, R. Izamshah, M.A. Md Ali, M.S.A. Aziz, J. Sci. Ind. Res. India., 73 (2014)
4. G. List, M. Nouari, D. Gehin, S. Gomez, J.P. Manaud, Y. Le Petitcorps, F. Girot, Wear., **259**, 7(2015)
5. S. Atlati, B. Haddag, M. Nouari, A. Moufki, Tribol. Int., **90** (2015)
6. A. Gomez Parra, M.Alvarez Alco, J. Salguero, M. Batista, M. Marcos, Wear., **302**, 1 (2015)
7. J. Kummel, J. Gibmeier, E. Muller, R. Schneider, V. Schulze, A. Wanner, Wear., **311**, 1 (2014)

Numerical Simulation of Different Models of Heat Pipe Heat Exchanger Using AcuSolve

Nurul Amira Zainal [1,*], *Saiful Naim* Sulaiman[1], *Mohd Afdhal* Shamsudin[1], and *Iskandar* Waini[1]

[1] Faculty of Engineering Technology, Universiti Teknikal Malaysia Melaka, Hang Tuah Jaya, 76100 Durian Tunggal, Melaka, Malaysia

Abstract. In this paper, a numerical simulation of heat pipe heat exchanger (HPHE) is computed by using CFD solver program i.e. AcuSolve. Two idealized model of HPHE are created with different variant of entry's dimension set to be case 1 and case 2. The geometry of HPHE is designed in SolidWorks and imported to AcuSolve to simulate the fluid flow numerically. The design of HPHE is the key to provide a heat exchanger system to work proficient as expected. Finally, the result is used to optimize and improving heat recovery systems of the increasing demand for energy efficiency in industry.

1 Introduction

The heat exchanger is a device used to transfer heat between two or more fluid streams at different temperature. Heat exchangers find widespread application in space heating, refrigeration, air-conditioning, power stations, petroleum refinery, and automotive application. For efficiency purposes, heat exchangers are designed to maximize the surface area of the wall between the two fluids while minimizing resistance to fluid flow through the exchanger. There are several common heat exchanger designs. These include heat pipe heat exchanger (HPHE). The heat pipe heat exchanger is playing a considerable role in different fields in the industry. HPHE is a heat-transfer device that combines the principles of both thermal conductivity and phase transition with high effective thermal conductivity. Concisely, heat pipe is an evaporation-condensation device for transferring heat in which latent heat of vaporization is utilized to transfer heat over a long distance with a corresponding small temperature difference.

The aim of augmenting heat transfer is to accommodate high heat fluxes. An advantage of a heat pipe over other conventional methods to transfer heat is that HPHE can have an extremely high thermal conductance in steady state operation. Firouzfar [1] stated that heat pipe with liquid metal working fluids can have a thermal conductance of a thousand or even tens of thousands of times greater than the best solid metallic conductors, silver or copper. However, there are limitations on the ability of HPHE to transfer heat; commonly known as the sonic limit, the capillary limit, the viscous limit, the entrainment limit and the boiling

* Corresponding author: nurulamira@utem.edu.my

limit. Recently, HPHE found a wide variety of applications and research studies. Yau [2] reviewed on the application of horizontal HPHE air conditioning systems in tropics. It shows that on the basis of results obtained, the application of horizontal HPHE in terms of energy saving, dehumidification enhancement and condensate drainage is recommended for the tropics. Ma [3] studied heat pipe assisted heat exchanger used for industrial waste heat recovery experimentally. The results indicated that the exergy efficiency (also known as the second-law efficiency or rational efficiency) increased with the increment of waste water mass flow rate at constant fresh water mass flow rate, while the effectiveness decreased at the same operation condition. Meanwhile, investigation of HPHE integrated with a water spray foe the heat recovery from bile exhaust gas was studied by Yuan [4]. The study presents a thermodynamic analysis and a numerical simulation of a heat pipe heat exchanger which recovers both sensible and latent heat from the exhaust gases of a boiler.

Heat pipe technology has found increasing demands in enhancing the thermal performance of heat exchange for energy efficiency, especially in domestic appliances and industrial systems. Heat transfer efficiency in such systems is the primary factor for efficient performance of the whole systems [5]. Therefore, the aim for continuously introducing and improving heat recovery systems is highly needed. Based on the nature of industrial demands, understanding the physics of the processes, introducing adequate simplifications and establishing an appropriate design are essential factors for obtaining reasonable results and correct thermal design [6]. It is clearly that the key to provide a heat exchanger system to efficiently work as expected is the design of the heat recovery systems with heat pipe units. Heat pipe is unable to transport enough heat and may function poorly as thermal conductor in the systems without correct design. Lin [7] suggested that Computational Fluid Dynamics (CFD) modelling is able to predict thermal performance of the dehumidification solution with heat pipe heat exchangers. In the meantime, CFD modelling of flow and heat transfer in a thermosiphon was determined by Alizadehdakhel [8]. The studied concluded that CFD is a useful tool to model and explain the complex flow and heat transfer in a thermosyphon.

CFD is an effective and appropriate method that uses finite volumes to solve the processes that consist of transport phenomena. It is one of the robust methods for fluid flow simulation that analyses the systems consists of momentum transfer, heat transfer and mass transfer. In fact, CFD is actually not only for simulation work, but it is also a good guide for engineers, because it reinforces designers having good sights about the operations and processes in the system [9]. Moreover, CFD is a time-saving approach, cost effective and may reduce the overall needs for man-power compared to real experimental work. Referring to the advantages of the above mentioned, this paper aim to analyse fluid conditions on the tube bundle in the HPHE by using CFD programme packages i.e. AcuSolve. CFD modelling was carried out to show the ability of this approach as well as to investigate the flow pattern in HPHE.

2 Mathematical formulation

CFD uses numerical analyses and algorithms to solve and analyze problem that involves fluid flows. The fundamental basis of almost all CFD problems is the Navier-Stokes equations. Principally, the equations are a set of coupled differential equation and could be solved for a given flow problem mathematically. However, in practice these equations are too difficult to solve analytically. Thus CFD was employed in order to solve approximations to the equations using a variety of techniques including finite element method. The Navier-Stokes equation consists of conservation of mass (Eq. 1), conservation of momentum (Eq. 2) and conservation of energy (Eq. 3).

Conservation of Mass:

$$\frac{\partial \rho}{\partial t} + \nabla \cdot (\rho \vec{V}) = 0 \tag{1}$$

Conservation of Momentum:

$$\rho \frac{D\vec{V}}{Dt} = -\nabla P + \rho \vec{g} + \mu \nabla^2 \vec{V} \tag{2}$$

Conservation of Energy:

$$\rho \left(\frac{\partial h}{\partial t} + \nabla \cdot (h\vec{V}) \right) = -\frac{Dp}{Dt} + \nabla \cdot (k \nabla T) + \phi \tag{3}$$

To solve the above equations, a turbulence model is needed. This study employed Spalart-Allmaras model. Spalart-Allmaras incorporates some of the recent advances in turbulence modelling. It regularly shows that it has equal or superior accuracy for nearly all classes of flows. In addition, the Spalart- Allmaras model is more computationally efficient compared to other model. Javaherchi [10] reviewed the Spalart- Allmaras model and its modification. The model is presented in Eq. (4).

Spalart- Allmaras model :

$$\frac{\partial}{\partial t}(\rho \vec{v}) + \frac{\partial}{\partial x_i}(\rho \vec{v} u_i) = G_{\upsilon} + \frac{1}{\sigma_{\vec{\upsilon}}} \left[\frac{\partial}{\partial x_j} \left\{ \left(\mu + \rho \vec{\upsilon} \right) \frac{\partial \vec{\upsilon}}{\partial x_j} \right\} + C_{b2\rho} \left(\frac{\partial \vec{\upsilon}}{\partial x_j} \right)^2 \right] - Y_{\upsilon} + S_{\vec{\upsilon}} \tag{4}$$

3 Design of HPHE

Two idealized models of HPHE are designed in this study with different variants of entry's dimension as shown in Fig. 1 and Fig 2. The model dimensions were taken from previous studies which based on the basic design and with attention to needs and process constraints on the operation unit gas field [9]. The geometry of HPHE was designed in SolidWorks. SolidWorks is a CAD & CAE program and utilizes a parametric feature-based approach to create models and assemblies. The model is then imported to CFD Solver i.e AcuSolve to simulate the fluid flow numerically.

Fig. 1. Schematic model of HPHE Case 1

DOI: 10.1051/matecconf/20179701077

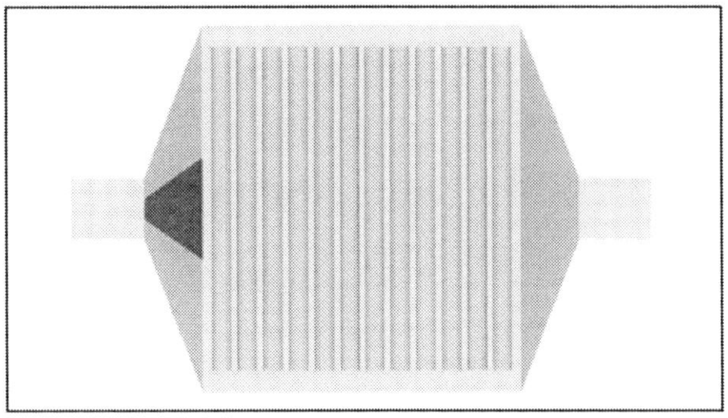

Fig. 2. Schematic model of HPHE Case 2

4 Simulation model of HPHE

There are several components that need to be established before the simulation process begins. The basic components include wall as the outermost component of the body, tube inside HPHE, a fluid that represents the volume in which the gas will flow, and inlet-outlet flow represents the area where the gas will flow in and out. Then, the model is meshed in Hypermesh with total number of elements is 1782371 relatively and exported to AcuSolve for further analysis hereafter. Fig 3 shows the complete meshed model; meanwhile Fig. 4(a) shows the side-view of HPHE. Meanwhile, the top-view of HPHE is shown in Fig 4(b).

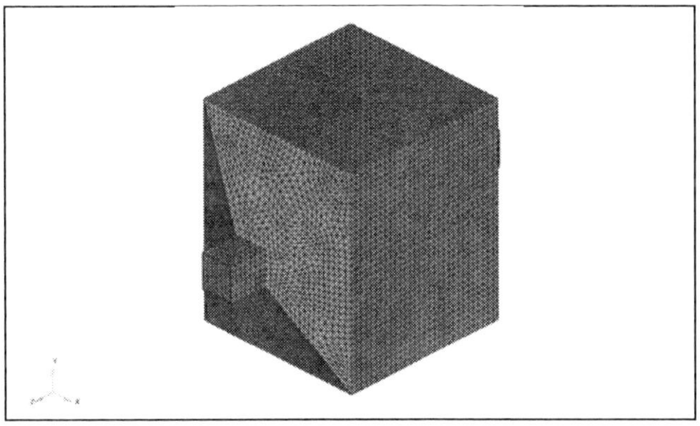

Fig. 3. Complete meshed model

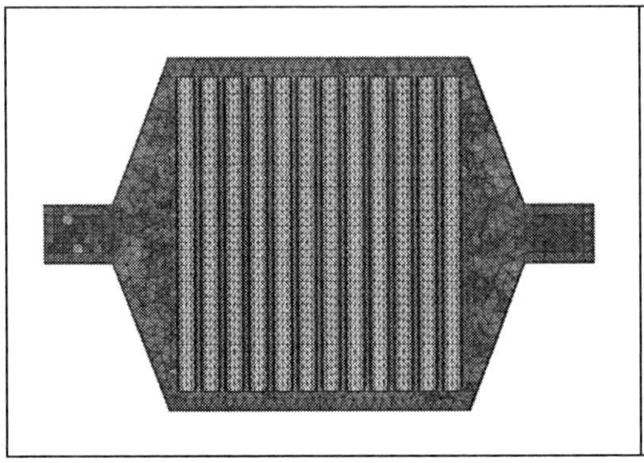

Fig. 4(a). Side view (cut) of HPHE after meshed.

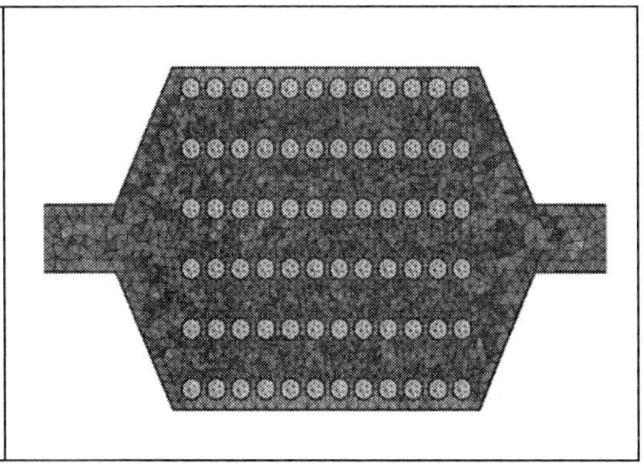

Fig. 4(b). Top view (cut) of HPHE after meshed.

The boundary condition specification for each component is given in Table 1 and the value implies to both cases.

Table 1. Boundary condition specification of HPHE

Components	Modelling details
Fluid	- Medium: Fluid - Material model: Air
Inflow	- BC type: Inflow - Inflow type: Mass flux - Mass flux: 3.75 kg/s - Temperature: 793 K
Outflow	- BC type: Outflow
Tube	- BC type: Wall - Temperature BC type: Flux - Heat flux : -2714.56 W/m2 (-156568 / 72 tubes)
Wall	- BC type: Wall

In case 2, the simulation model was derived from case 1 model, means that the elements are all the same. However a component is added to the model, which is a cone located right after the inlet. The property of the meshed cone is given below in Table 2.

Table 2. The property of the meshed cone in case 2

Components	Modelling details
Cone	- Modelled as a wall - Element size : 50mm - No of elements: 3616 - Element type: TRIA3 (tria elements with 3 nodes)

Henceforth, the total number of elements in case 2 is 1 868 865. On the other hand, the material modeling and CFD analysis for case 1 and case 2 was set as in Table 3 and Table 4 below.

Table 3. Material modeling of HPHE

Parameters	Value
Material Type	Air
Density	0.88 kg/m3
Viscosity	2 e-5 kg/ms
Conductivity	95 W/mK

Table 4. CFD analysis of HPHE

Parameters	Type
Analysis Type	Steady state
Temperature equation	Advective Diffusive
Turbulence model	Spalart Allmaras
Convergence tolerance	0.001
Temperature at initial condition	793 K

5 Results and analysis

As mentioned earlier, the main objective of the present study is to explore the ability of this CFD modelling approach i.e AcuSolve, as well as to investigate the flow pattern in HPHE. Fig.5 shows the velocity distribution of HPHE in case 1 meanwhile, Fig. 6 illustrates the velocity distribution of HPHE in case 2.

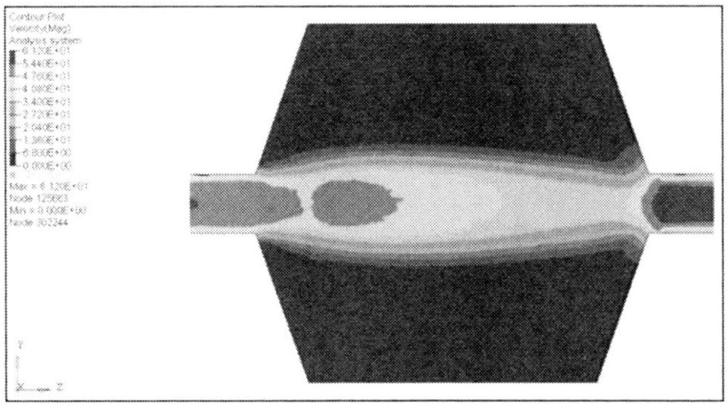

Fig. 5. Velocity distribution of HPHE in case 1

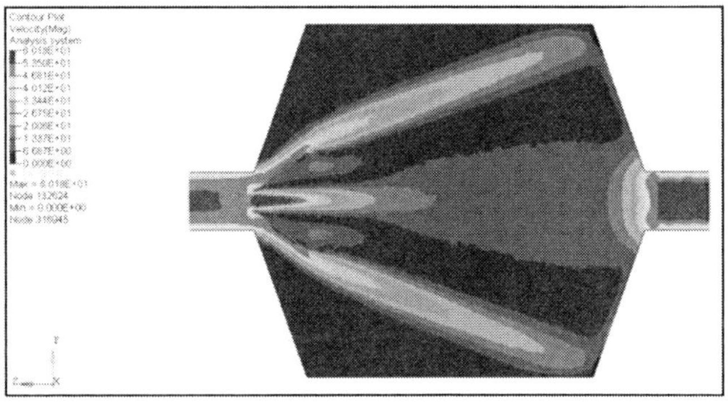

Fig. 6. Velocity distribution of HPHE in case 2

In case 1, the gas flow approaches the inlet at an average velocity of 54.15 m/s, whereas the gas flows out from the outlet with an average velocity of 61.05 m/s. In case 2, the gas enters the inlet and exits from the outlet at an average velocity of 54.28 m/s and 56.61 m/s respectively. From the results of the simulation, the distribution flow of velocity in case 2 shows a better dissemination compared to case 1. The flow is perfectly distributed along the center of the HPHE and also along the right and left side of the heat pipe. In contrast, the flow distribution of velocity in case 1 is only focused at the center of heat pipe which is not fully allocated over the whole tube bundle completely. Therefore, it can be concluded that in case 1, the basic design of HPHE was not fully utilized since some parts of the tube bundle did not expose to the gas flow thus the efficiency of HPHE to transfer the heat is low. Comparing this two model design, this study would recommend case 2 which has a better distribution flow that increases thermal efficiency.

As for temperature distribution, the result in both cases is shown in Fig. 7 and Fig. 8. In case 1, the average temperature at the inlet of HPHE is 793K, while the average temperature at the outlet is 779.6K. Fig. 7 illustrates the temperature distribution where the heat of the gas flow did not affect the areas of HPHE very much. This probably because of some parts of the tube bundle has not exposed really well to the heat. Thus, the heat transfer is low in case 1, which leads to inappropriate heat recovery. However, there are slightly

DOI: 10.1051/matecconf/20179701077

significant differences shown in Fig. 8 for case 2. In case 2, the average temperature at the inlet is 793K and the average temperature at the outlet is 774.8K. The temperature distribution in case 2 is considerable compared to case 1. The development of dispersion in temperature distribution in case 2 shows the best distribution flow between both cases that had been studied earlier. Therefore, from this analysis it can be concluded that case 2 has the highest rate of temperature distribution that leads to higher heat transfer and boost the thermal efficiency in HPHE.

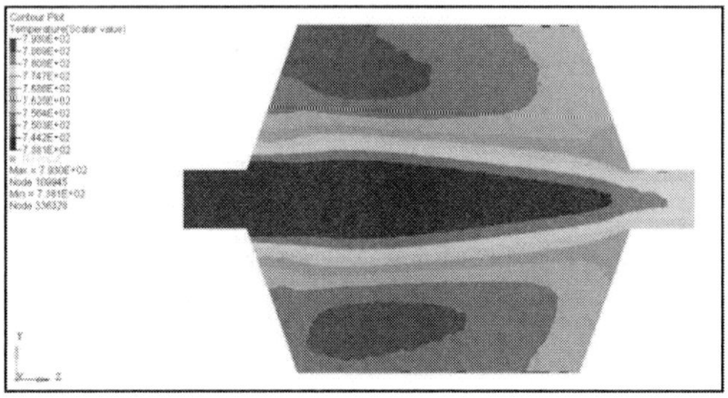

Fig. 7. Temperature distribution of HPHE in case 1

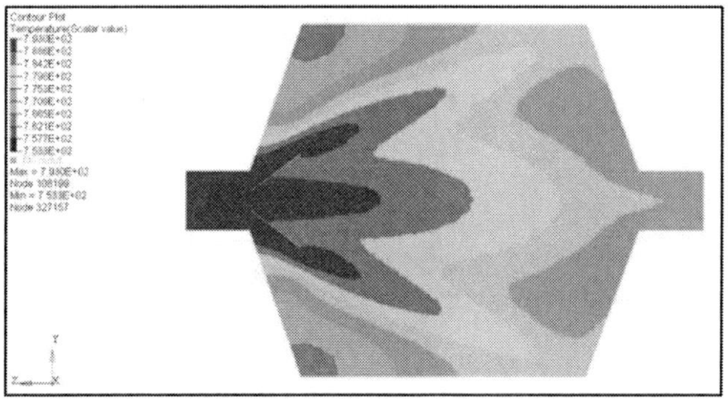

Fig. 8. Temperature distribution of HPHE in case 2

6 Results and analysis conclusion

The aim of this paper is to analyse fluid conditions on the tube bundle in the HPHE by using CFD packages i.e. AcuSolve. CFD modelling was carried out to show the ability of this approach as well as to investigate the flow pattern in HPHE. Two idealized models of HPHE were designed in this study with different variants of entry's dimension set to be case 1 and case 2. From the present investigation, the following conclusions can be drawn:

(1) Cone inserted in case 2 are very effective and can increase the performance of HPHE. An appropriate development of HPHE with differing size of inserting cone should be discussed in further study.

(2) CFD solver program that has been used in this study are acceptable and suitable to simulate and analyse the HPHE performance.

(3) The design of HPHE is the key to provide a heat exchange system to work proficiently as expected. The result is used to optimize and improving heat recovery systems of the increasing demand for energy efficiency in industry.

The authors would like to acknowledge Universiti Teknikal Malaysia Melaka (UTeM) for financial support under UTeM short term grant PJP/2015/FTK(1C)/S01394 and PJP/2015/FTK(8C)/S01411.

References

1. E. Firouzfar, M. Attaran, Int. Sch. Sci. Res. Innov., **2**, 266 (2008)
2. Y. H. Yau, M. Ahmadzadehtalatapeh, Appl. Therm. Eng., **30**, 77 (2010)
3. H. Ma, L. Yin, X. Shen, W. Lu, Y. Sun, Y. Zhang, N. Deng, Appl. Energy., **169**, 177 (2016)
4. Y. Yuan, Y. Lu, H. Bao, Y. Wang, W. Wang, A.P. Roskilly, Energy Procedia, **61**, 2141 (2014)
5. I. Sauciuc, A. Akbarzadeh, P. Johnson, Heat Recover. Syst. CHP., **15**, 631 (1995)
6. M. Cruchaga, D. Celentano, Int. J. Comut. Fluid Dyn., **16**, 247 (2002)
7. S. Lin, J. Broadbent, R. McGlen, Appl. Therm. Eng., **25**, 127 (2005)
8. A. Alizadehdakhel, M. Rahimi, A.A. Alsairafi, Int. Commun. Heat Mass Transf., **37**, 312 (2010)
9. M. Saber, H. Ashtiani, in Proceedings of 5th IASME/WSEAS International, (2010)
10. T. Javaherchi, *Wind Engineering*, (University of Washington, 2010)

MATEC Web of Conferences 97, 01078 (2017) DOI: 10.1051/matecconf/20179701078
ETIC 2016

Aligned Magnetic Field Effects on Flow and Heat Transfer of the Upper-Convected Maxwell Fluid over a Stretching/Shrinking Sheet

Iskandar Waini[1,*], *Nurul Amira* Zainal[1] , and *Najiyah Safwa* Khashi'ie[1]

[1] Faculty of Engineering Technology, Universiti Teknikal Malaysia Melaka, Hang Tuah Jaya, 76100 Durian Tunggal, Melaka, Malaysia

Abstract. In this paper, the effect of aligned magnetic field towards the flow and heat transfer of the upper-convected Maxwell (UCM) fluid over a stretching/shrinking sheet is numerically studied. The governing partial differential equations are reduced into a system of ordinary differential equations using a similarity transformation, which are then solved numerically using the shooting method. The skin friction and heat transfer coefficients, the velocity, as well as the temperature profiles of the fluid are presented and discussed. Results indicate that an increase in the aligned angle strengthens the applied magnetic field which decrease the velocity and increase the temperature profiles of the fluid. This implies that an increase in the aligned angle increases the skin friction coefficient and decreases the heat transfer coefficients.

1 Introduction

The fluid dynamics due to a stretching sheet are important in extrusion processes. The production of sheeting material arises in a number of industrial manufacturing processes and includes both metal and polymer sheets like the cooling of an infinite metallic plate in a cooling bath, paper production, glass blowing, etc. In view of these applications, Sakiadis [1] first investigated the boundary layer flow on a continuous solid surface moving at constant speed. Due to the entrainment of the ambient fluid, this boundary layer flow is quite different from the Blasius flow past a flat plate. Sakiadis's theoretical predictions for Newtonian fluids were later corroborated experimentally by Tsou *et al.* [2].

Since the pioneering study by Crane [3], who presented an exact analytical solution for the steady two-dimensional flow due to a stretching surface in a quiescent fluid, many authors have considered various aspects of this problem and obtained similarity solutions. Gupta and Gupta [4] investigated the heat and mass transfer corresponding to the similarity solutions for the boundary layer over a stretching sheet subjected to suction or blowing. Chen and Char [5] analyzed the heat transfer characteristics over a continuous stretching surface with variable surface temperature. Research involving the steady two-dimensional

* Corresponding author: iskandarwaini@utem.edu.my

flow due to a stretching surface in a quiescent fluid, many authors has considered various aspects of this problem and obtained similarity solutions (see [6-9]).

Investigations of magnetohydrodynamic (MHD) flow are very important industrially and have applications in different areas of research such as petroleum production and metallurgical processes. The properties of the end product depend greatly on the rate of cooling involved in these processes and the desired properties of the end product can be controlled by the use of electrically conducting fluids and application of magnetic field. Besides, some electrically conducting fluids such as nuclear fuel slurries, liquid metals, mercury amalgams, biological fluids, lubrication oils and certain greases have many applications in the absence as well as in the presence of a magnetic field [10]. The problem involving the MHD boundary layer has been investigated by the number of authors (see [11-17]) by considering various effects.

However, some interest has been given to the study of non-Newtonian fluid which is viscoelastic fluid due to the numerous applications in the industry. Birds *et al.* [18] discussed that the viscoelastic fluid models used in previous works are simple models such as second-order model and/or Watler's B model which are known to be good only for weakly elastic fluids subject to slow and/or slowly-varying flows. Though, the models can describe the aforementioned problems adequately, but physically it is violating certain rules of thermodynamics. Furthermore, a non-Newtonian second grade fluid does not give meaning full results for highly elastic fluids (polymer melts) which occur at high Deborah numbers. Therefore, as far as polymer industry is concerned, the significance of the results reported in the above works is limited. Obviously, for the theoretical results to become of any industrial significance, more realistic viscoelastic fluid models such as upper-convected Maxwell model or Oldroyd-B model should be invoked in the analysis [19]. The boundary layer flow and heat transfer due to stretching sheet for the upper-convected Maxwell fluid has been considered by many researchers (see [20-29]) in various ways.

Recently, Raju *et al.* [30] have been reported the effects of aligned magnetic field on steady two-dimensional flow over a vertical stretching surface. They found that increase in aligned angle decreases the velocity profiles and increases the temperature and concentration profiles of the fluid. In addition, the effect of the aligned magnetic field and cross-diffusion on a nanofluid over an exponentially stretching surface in a porous medium has been studied by Sulochana *et al.* [31]. The researchers observed that an increase in the aligned angle decreases the velocity profiles and increases the temperature and concentration profiles of the nanofluid.

In this present study, we investigate the effect of aligned magnetic field towards the flow and heat transfer of the upper-convected Maxwell fluid over a stretching/shrinking sheet. The numerical results are tabulated and shown graphically to illustrate the influence of the various parameters on the velocity and temperature profiles, as well as on the skin friction and the heat transfer coefficients. As a contribution, this present study will gain a better understanding in terms of the fundamental scientific knowledge of the flow and heat transfer phenomena-related subjects. Besides, the theoretical predictions and numerical solutions obtained in this study will help to explain and verify experiments for the engineers.

2 Mathematical formulation

A steady two-dimensional boundary layer flow of an upper-convected Maxwell fluid over a stretching/shrinking sheet at $y = 0$ is considered as shown in Figure 1. The stretching/shrinking velocity of the sheet is $u_w(x) = ax$, where a is a positive constant

($a>0$). Here, the aligned magnetic field B_0 is applied to the flow with acute angle γ along y direction. This magnetic field acts like a transverse magnetic field at $\gamma=\pi/2$. The induced magnetic field due to the motion of the electrically conducting fluid is negligible. It is also assumed that the external electric field is zero and the electric field due to polarization of charges is negligible [13].

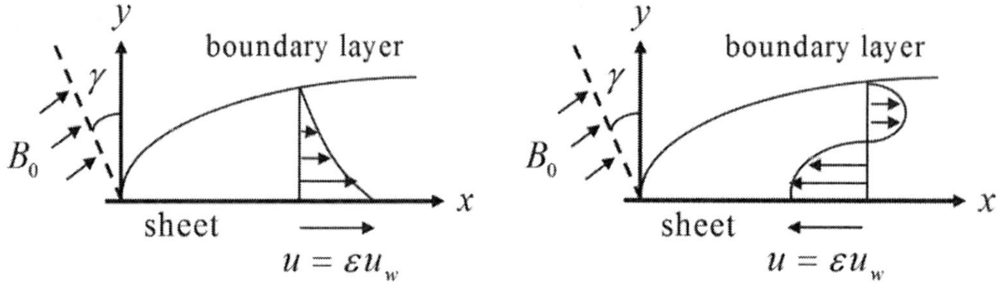

Fig. 1. Physical model and coordinate system.

By implementing the boundary layer approximations and applying the order of magnitude analysis, the governing equations for the upper-convected Maxwell fluid can be written as follows (see Abel *et al.* [19]):

$$\frac{\partial u}{\partial x}+\frac{\partial v}{\partial y}=0 \tag{1}$$

$$u\frac{\partial u}{\partial x}+v\frac{\partial u}{\partial y}+\lambda\left(u^2\frac{\partial^2 u}{\partial x^2}+v^2\frac{\partial^2 u}{\partial y^2}+2uv\frac{\partial^2 u}{\partial x\partial y}\right)=\nu\frac{\partial^2 u}{\partial y^2}-\frac{\sigma B_0^2}{\rho}\sin^2(\gamma)u \tag{2}$$

$$u\frac{\partial T}{\partial x}+v\frac{\partial T}{\partial y}=\frac{k}{\rho c_p}\frac{\partial^2 T}{\partial y^2} \tag{3}$$

where u and v are the velocity components along the x and y axes, respectively, and T is the temperature of the fluid, ρ is the fluid density, σ is the electrical conductivity of the fluid and λ is the relaxation time parameter of the fluid, while ν is the kinematic viscosity of the fluid, c_p is the specific heat at a constant pressure, k is the thermal conductivity of the fluid, and γ is the aligned angle. The boundary conditions for the problem are

$$u=\varepsilon u_w,\ v=v_0,\ T=T_w(x)=T_\infty+bx^2\ \text{at}\ y=0$$
$$u\to 0,\ T\to T_\infty\ \text{as}\ y\to\infty \tag{4}$$

where b is a some constant, and v_0 is the mass flux velocity. Introducing the stream function ψ defined by $u=\partial\psi/\partial y$ and $v=-\partial\psi/\partial x$, equation (1) is satisfied identically. Then, Eqs. (2) and (3) can be written as

$$\begin{aligned}&\left(\frac{\partial\psi}{\partial y}\right)\left(\frac{\partial^2\psi}{\partial x\partial y}\right)-\left(\frac{\partial\psi}{\partial x}\right)\left(\frac{\partial^2\psi}{\partial y^2}\right)\\&+\lambda\left[\left(\frac{\partial\psi}{\partial y}\right)^2\left(\frac{\partial^3\psi}{\partial y\partial x^2}\right)+\left(\frac{\partial\psi}{\partial x}\right)^2\left(\frac{\partial^3\psi}{\partial y^3}\right)-2\left(\frac{\partial\psi}{\partial y}\right)\left(\frac{\partial\psi}{\partial x}\right)\left(\frac{\partial^3\psi}{\partial x\partial y^2}\right)\right]\\&=\nu\left(\frac{\partial^3\psi}{\partial y^3}\right)-\frac{\sigma B_0^2}{\rho}\sin^2(\gamma)\left(\frac{\partial\psi}{\partial y}\right)\end{aligned}\tag{5}$$

$$\left(\frac{\partial\psi}{\partial y}\right)\left(\frac{\partial T}{\partial x}\right)-\left(\frac{\partial\psi}{\partial x}\right)\left(\frac{\partial T}{\partial y}\right)=\frac{k}{\rho c_p}\left(\frac{\partial^2 T}{\partial y^2}\right)\tag{6}$$

Further, we seek for a similarity solution of Eqs. (5) and (6) subject to the boundary conditions (4) by introducing the following similarity transformation:

$$\eta=\left(\frac{a}{\nu}\right)^{1/2}y,\quad \psi=(a\nu)^{1/2}xf(\eta),\quad \theta(\eta)=\frac{T-T_\infty}{T_w-T_\infty}\tag{7}$$

where η is the similarity variable, $f(\eta)$ is the dimensionless stream function, and $\theta(\eta)$ is the dimensionless variables for the temperature.

Then, by substituting similarity transformation (7) into Eqs. (5) and (6), we obtain the following nonlinear ordinary differential equations:

$$f'''+ff''-f'^2-M\sin^2(\gamma)f'+\beta(2ff'f''-f^2f''')=0\tag{8}$$

$$\frac{1}{\Pr}\theta''+f\theta'-2f'\theta=0\tag{9}$$

and the boundary conditions (4) becomes

$$\begin{aligned}&f(0)=S,\ \ f'(0)=\varepsilon,\ \ \theta(0)=1\ \text{ at }\ \eta=0\\&f'(\eta)\to 0,\ \ \theta(\eta)\to 0\ \text{ as }\ \eta\to\infty\end{aligned}\tag{10}$$

Here primes denote differentiation with respect to η, ε is the stretching/shrinking parameter with $\varepsilon>0$ for stretching and $\varepsilon<0$ shrinking, β is the dimensionless Maxwell parameter, M is the magnetic parameter, Pr is the Prandtl number, and S is the suction/injection parameter with $S>0$ for suction and $S<0$ for injection, which are defined as

$$\beta=\lambda a,\quad M=\frac{\sigma B_0^{\ 2}}{a\rho},\quad \Pr=\frac{\mu c_p}{k},\quad S=-\frac{v_0}{\sqrt{a\nu}}\tag{11}$$

The physical quantities of interest are the skin friction coefficient C_f and the local Nusselt number Nu_x which is defined as

$$C_f = \frac{\tau_w}{\rho u_w^2(x)/2}, \quad Nu_x = \frac{x q_w}{k(T_w - T_\infty)} \tag{12}$$

where $\tau_w = \mu(\partial u/\partial y)_{y=0}$ and $q_w = -k(\partial T/\partial y)_{y=0}$ are the shear stress, and heat flux at the surface, respectively. By using the similarity transformation (7), we obtain

$$\frac{1}{2} C_f Re_x^{1/2} = f''(0), \qquad Nu_x / Re_x^{1/2} = -\theta'(0) \tag{13}$$

where $Re_x = u_w(x)x/\nu$ is the local Reynolds number based on the stretching velocity $u_w(x) = ax$ whereas the skin friction coefficient is represented by $f''(0)$ while the heat transfer coefficient is represented by $-\theta'(0)$, respectively.

3 Numerical solution

The nonlinear equations (8) and (9) are transformed into a system of five first order differential equations as follows:

$$\begin{aligned}
&\frac{df_0}{d\eta} = f_1, \\
&\frac{df_1}{d\eta} = f_2, \\
&\frac{df_2}{d\eta} = \frac{(f_1)^2 + M\sin^2(\gamma)f_1 - f_0 f_2 - 2\beta f_0 f_1 f_2}{1 - \beta f_0^2}, \\
&\frac{d\theta_0}{d\eta} = \theta_1, \\
&\frac{d\theta_1}{d\eta} = \Pr\left[2 f_1 \theta_0 - f_0 \theta_1\right]
\end{aligned} \tag{14}$$

Subsequently the boundary conditions (10) take the form,

$$\begin{aligned}
&f_0(0) = S, \;\; f_1(0) = \varepsilon, \;\; f_1(\infty) = 0, \\
&\theta_0(0) = 1, \;\; \theta_0(\infty) = 0
\end{aligned} \tag{15}$$

Here $f_0 = f(\eta)$ and $\theta_0 = \theta(\eta)$. The aforementioned boundary value problem is first converted into an initial value problem (IVP) by appropriately guessing the missing slopes $f_2(0)$ and $\theta_1(0)$. The resulting IVP is solved by shooting method for a set of parameters appearing in the governing equations with a known value of $f_2(0)$ and $\theta_1(0)$. The convergence criterion largely depends on fairly good guesses of the initial conditions in the shooting technique. The iterative process is terminated until the relative difference between the current iterative values of $f_2(0)$ matches with the previous iterative value of $f_2(0)$ up to a tolerance of 10^{-6}.

4 Results and discussions

The governing partial differential equations of flow and heat transfer are transformed into a system of non-linear ordinary differential equations by using the similarity transformation. Eqs. (8) and (9) subjected to the boundary conditions (10) are solved numerically by using a shooting method. The accuracy of the above mentioned numerical method is validated by direct comparisons with earlier published results in some limiting cases which are presented in Table 1 and Table 2. It is noted that the present results are in good agreement with the existing results.

Table 1. Comparison of skin friction coefficient $f''(0)$ for different values of Maxwell parameter β when $\Pr = 1$, $M = 0$, $S = 0$ and $\varepsilon = 1$ (stretching case).

β	$f''(0)$			
	Sadeghy *et al.* [22]	Mamaloukas *et al.* [29]	Abel *et al.* [19]	Present results
0	-1.0000	-0.999962	-0.999962	-1.00000005
0.2	-1.0549	-	-1.051948	-1.05188989
0.4	-1.10084	-1.101850	-1.101850	-1.10190327
0.6	-1.0015016	-	-1.150163	-1.15013734
0.8	-1.19872	-1.196692	-1.196692	-1.19671125
1.2	-	-1.285257	-1.285257	-1.28536326

Table 2. Comparison of skin friction coefficient $f''(0)$ for different values of suction parameter S when $\beta = 0$, $\Pr = 1$, $M = 2$ and $\varepsilon = -1$ (shrinking case).

S	$f''(0)$		
	Muhaimin *et al.* [16]	Bhattacharyya [17]	Present results
2	2.414214	2.414300	2.41421357
3	3.302776	3.302750	3.30277563
4	4.236068	4.236099	4.23606797

Figure 2 and 3 shows the effect of aligned angle and magnetic parameter on velocity profiles for the stretching case, respectively. It can be observed that the increase in aligned angle and magnetic parameter decreases the velocity profiles of the fluid for the stretching case. The reason behind this behavior is increased in aligned angle causes to strengthen the magnetic field. Due to enhanced magnetic field, it generates opposite force to the flow which is called Lorentz force. This force declines the velocity boundary layer thickness.

Figure 4 and 5 demonstrate the effect of the aligned angle on the velocity and temperature profiles for the shrinking case, respectively. It is observed that an increase in the aligned angle enhances the velocity profiles and depreciates the temperature profiles for the shrinking case. We observed that the Lorentz force helps to enhance the flow in shrinking case.

DOI: 10.1051/matecconf/20179701078

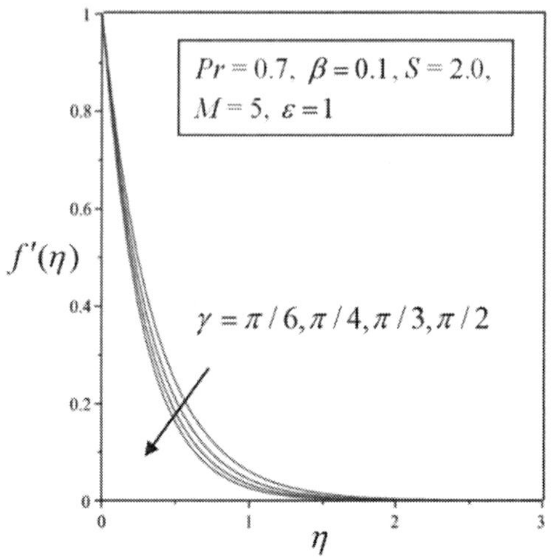

Fig. 2. Effect of the aligned angle γ on the velocity profiles.

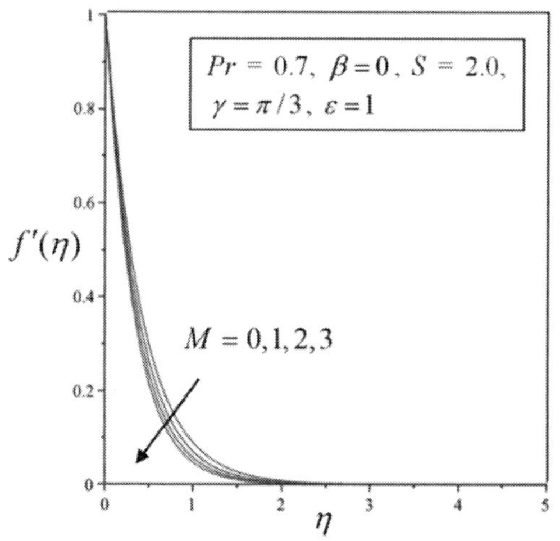

Fig. 3. Effect of the magnetic parameter M on the velocity profiles.

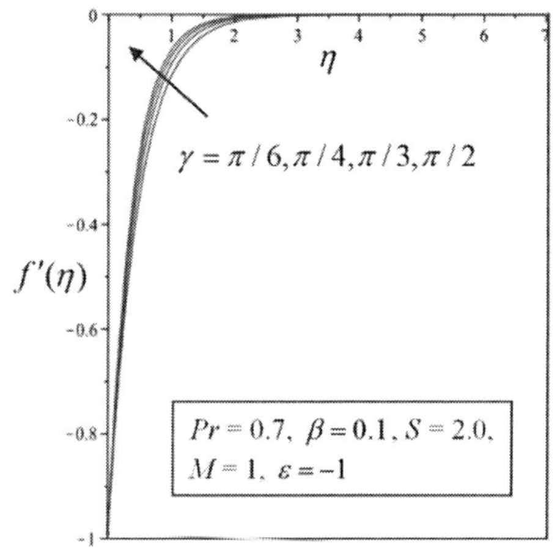

Fig. 4. Effect of the aligned angle γ on the velocity profiles.

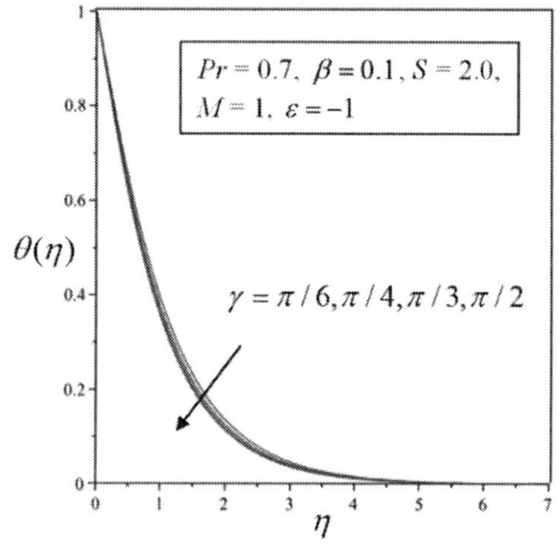

Fig. 5. Effect of the aligned angle γ on the temperature profiles.

Meanwhile, Figure 6 and 7 are depicted to show the effect of magnetic parameter on velocity and temperature profiles in shrinking sheet. As expected, an increase in the magnetic parameter leads to increase the velocity and decrease the temperature of the fluid.

The effect of the Maxwell parameter on velocity and temperature profiles displayed in Figure 8 and 9. An increase in the Maxwell parameter is noticed to increase the fluid velocity and decrease the fluid temperature in shrinking sheet.

Figure 10 and 11 illustrate the effect of the aligned angle and magnetic parameter on the skin friction and heat transfer coefficients. It is found that the increasing of aligned angle and magnetic parameter has the capability to increase the value of the skin friction and heat transfer coefficients in shrinking sheet. This behavior also can be observed in Table 3.

DOI: 10.1051/matecconf/20179701078

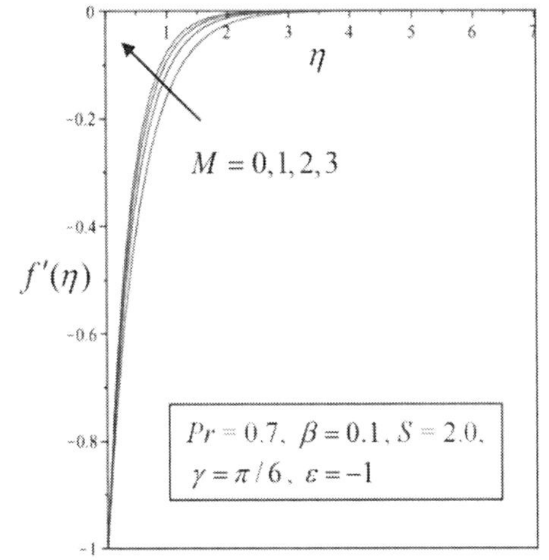

Fig. 6. Effect of the magnetic parameter M on the velocity profiles.

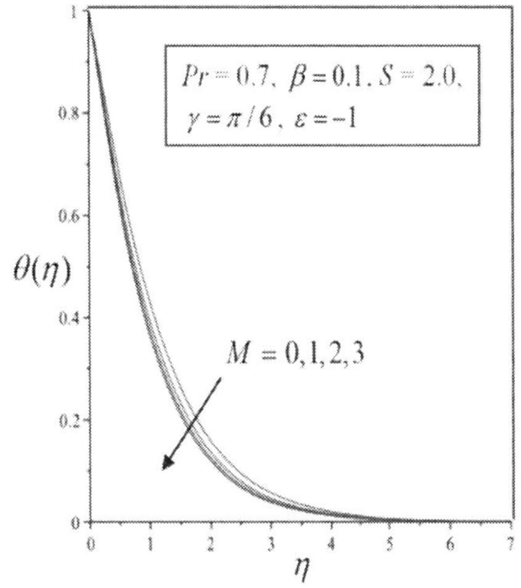

Fig. 7. Effect of the magnetic parameter M on the temperature profiles.

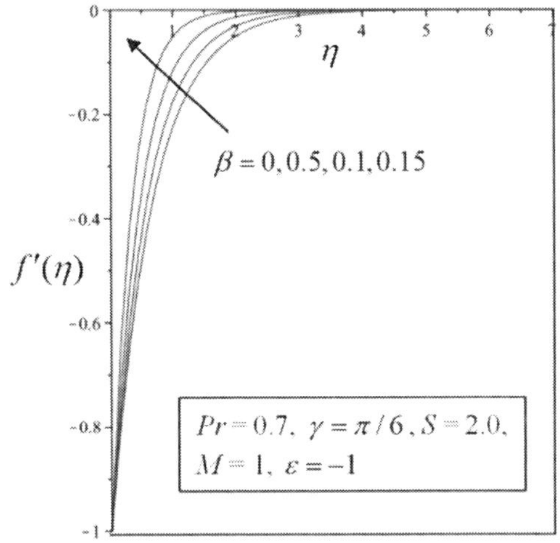

Fig. 8 Effect of the Maxwell parameter β on the velocity profiles.

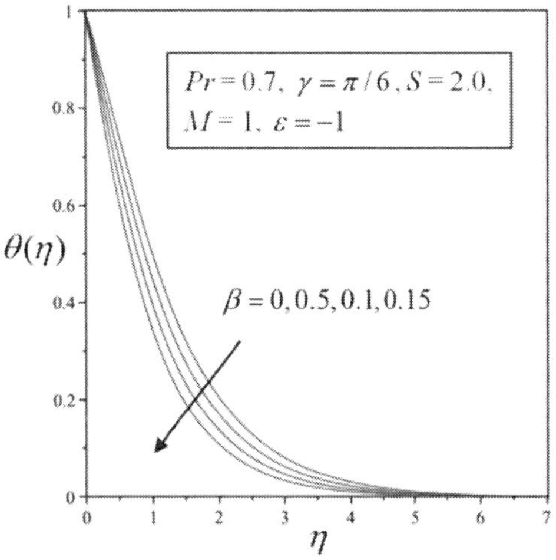

Fig. 9 Effect of the Maxwell parameter β on the temperature profiles.

Table 3. Values of the skin friction $f''(0)$ and heat transfer $-\theta'(0)$ coefficients for different values of aligned angle γ and the magnetic parameter M when $\Pr = 0.7$, $\beta = 0.1$, $S = 2.0$, and $\varepsilon = -1$.

γ	M	$f''(0)$	$-\theta'(0)$
$\pi/6$	0	1.81868494	0.59941014
	1	2.13643677	0.70462676
	2	2.36184718	0.76153317
	3	2.54657746	0.80073211
$\pi/4$	0	1.81866870	0.59795336
	1	2.36184718	0.76153317
	2	2.70691873	0.83005752
	3	2.98177503	0.87421087

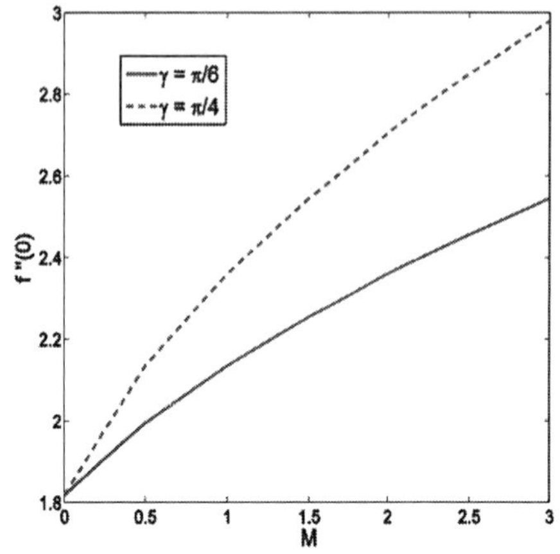

Fig. 10. Effect of the aligned angle and magnetic parameter on the skin friction coefficient when Pr = 0.7, $\beta = 0.1$, S = 2.0, and $\varepsilon = -1$.

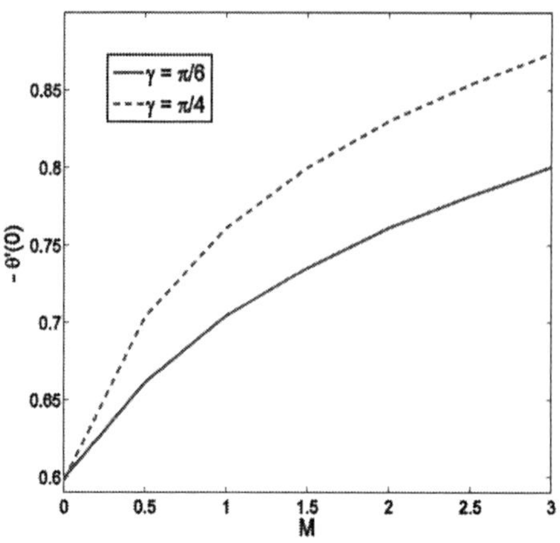

Fig. 11. Effect of the aligned angle and magnetic parameter on the heat transfer coefficient when Pr = 0.7, $\beta = 0.1$, S = 2.0, and $\varepsilon = -1$.

5 Conclusion

This paper presents a numerical solution for aligned magnetic field towards the flow and heat transfer of the upper-convected Maxwell (UCM) fluid over a stretching/shrinking sheet. By means of similarity transformation, the governing mathematical equations are reduced into ordinary differential equations which are then solved numerically using a shooting method. From the present investigation, the following conclusions can be drawn:

1. An increase in the aligned angle strengthens the magnetic field parameter and depreciates the velocity profiles in stretching case. However, increases in aligned angle enhances the fluid velocity for the shrinking case. Meanwhile, the fluid temperature decrease with the increase of aligned angle.
2. The increasing of magnetic parameter produces a decrease in the fluid velocity for the stretching case, but the opposite trend is observed in shrinking case. Besides, the fluid temperature decrease with the increase of magnetic parameter.
3. An increase in the Maxwell parameter is noticed to increase the fluid velocity and decrease the fluid temperature for the shrinking case.
4. The increasing of aligned angle and magnetic parameter has the capability to increase the value of the skin friction and heat transfer coefficients in shrinking sheet.

The authors would like to acknowledge Universiti Teknikal Malaysia Melaka (UTeM) for financial support under the UTeM short term grants PJP/2015/FTK(8C)/S01411 and PJP/2015/FTK(1C)/S01394.

References

1. B.C. Sakiadis, Am. Inst. Chem. Engineers J., **7**, 26 (1961)
2. F. Tsou, E. Sparrow, R. Goldstein, Int. J. Heat Mass Transfer, **10**, 219 (1967)
3. L.J. Crane, Z. Angew, Math. Phys., **21**, 645 (1970)

4. P.S. Gupta, A.S. Gupta. Can., J. Chem. Eng., **55**, 744 (1977)
5. C.K. Chen, M. Char, Journal of Mathematical Analysis and Applications, **135**, 568 (1988)
6. M.E. Ali, Warme Stoffubertragung, **29**, 227 (1994)
7. M.E. Ali, International Journal of Heat and Fluid Flow, **16**, 280 (1995)
8. R.S.R. Gorla, I. Sidawi, Appl. Sci. Res., **52**, 247 (1994)
9. E.M.A. Elbashbeshy, J. Phys. D: Appl. Phys., **31**, 1951 (1998)
10. T. Sarpkaya, AIChE J., **7**, 324 (1961)
11. I.C. Liu, Int. Commun. Heat Mass Transfer, **32**, 1075 (2005)
12. A. Chakrabarti, A.S. Gupta, Quart. Appl. Math. **37**, 73 (1979)
13. T.C. Chiam, Int. J. Engng Sci., **33**, 429 (1995)
14. S.P. Anjali Devi, M. Thiyagarajan, Heat and Mass Transfer, **42**, 671 (2006)
15. T. Javed, Z. Abbas, M. Sajid, N. Ali, International Journal of Heat and Mass Transfer, **54**, 2034 (2011)
16. Muhaimin, R. Kandasamy, Azme B. Khamis, Applied Mathematics and Mechanics, 1309 (2008)
17. K. Bhattacharyya, Chemical Engineering Research Bulletin, **12** (2011)
18. R.B. Bird, R.C. Armstrong, O. Hassager, *Dynamics of polymeric liquids*, **1**, Wiley, New York (1987).
19. M.S. Abel, J.V. Tawade, M.M. Nandeppanavar, Meccanica, **47**, 385 (2012)
20. M. Renardy, J. Non Newton, Fluid Mech., **68**, 125 (1997)
21. K. Sadeghy, A.H. Najafi, M. Saffaripour, Int J Non-Linear Mech., **40**, 1220 (2005)
22. K. Sadeghy, H. Hajibeygi, S.M. Taghavi, Int J Non-Linear Mech., **41**, 1242 (2006)
23. T. Hayat, Z. Abbas, M. Sajid, Phys Lett A., **358**, 396 (2006)
24. A. Alizadeh Pahlavan, V. Aliakbar, F. Vakili Farahani, K. Sadeghy, Commun Nonlinear Sci Numer Simul., **14**, 473 (2009)
25. V. Aliakbar, A. Alizadeh Pahlavan, K. Sadeghy, Commun Nonlinear Sci Numer Simul., **14** (3), 779 (2009)
26. A. Alizadeh Pahlavan, K. Sadeghy, Commun Nonlinear Sci Numer Simul., **14** (4), 1355 (2009)
27. T. Hayat, Z. Abbas, M. Sajjid, Chaos Solitons Fractals., **39**, 840 (2009)
28. B. Raftari, A. Yildirim, Computers and Mathematics with Applications, **59**, 3328 (2010)
29. C. Mamaloukas, M.S. Abel, J.V. Tawade, U.S. Mahabaleswar, International Journal of Pure and Applied Mathematics, **66**, 1 (2011)
30. C.S.K. Raju, N. Sandeep, C. Sulochana, V. Sugunamma, M. Jayachandra Babu, Journal of the Nigerian Mathematical Society, **34**, 169 (2015)
31. C. Sulochana, N. Sandeep, V. Sugunamma, B. Rushi Kumar, Applied Nanoscience, **1** (2015)

The Influence of Tool Geometry towards Cutting Performance in Machining Aluminium 7075

Jumali Muhammad Syafik [1,*], *Muhamad* Mohd Razali [2], and *Sundi* Syahrul Azwan [1]

[1] Faculty of Engineering Technology, Universiti Teknikal Malaysia Melaka, Malaysia
[2] Faculty of Manufacturing Engineering, Universiti Teknikal Malaysia Melaka, Malaysia

Abstract. Aerospace industries often use Computer Numerical Control (CNC) machining in manufacturing aerospace parts. Aluminium 7075 is the most common material used as aircraft components. This research aims to produce end mill with optimum geometry in terms of the helix angle, primary radial relief angle and secondary relief angle. End mills with different geometry parameters are tested on Aluminium 7075 and data on surface roughness and tool wear were collected. The results were then analysed to determine which parameters brought the optimum result with regards to surface roughness and tool wear.

1 Introduction

Machining is the most common material removal process being practiced in the metal fabrication industry such as in the aerospace and automotive industries. This is due to the flexible nature of machining in that it can be used to produce almost any design features of metal components. The capability of machining is hugely improved through the introduction of advanced technologies such as Computer Numerical Control (CNC) and Computer-Aided Manufacturing (CAM).

In aerospace industry where volume is low and flexibility is high, machining is one the most profitable and viable method of production. Conventional machining method using CNC machines is often used to produce aerospace parts. Variable parameters in machining are spindle speed, feed rate and depth of cut. These three parameters partly determine whether any parts that are produced have good surface finish while improving cutting tool performance and having longer tool life during machining. However, in modern CAM software, higher cutting speed is obtained by manipulating different cutting strategies in order to fully utilize the capabilities of CNC machines.

In order to investigate the relationship between cutting speed with surface finish and tool wear, a combination of varied primary and secondary clearance angles and also the flute helix angle of end mill will be used to determine the optimum tool geometry to cut Aluminium 7075. Aluminium 7075 is commonly used for aircraft structures because it has high strength and also the best machinability among Aluminium Alloys. Many researchers studied the effects of helix angle to vibrations. Omar et al. (2007) found that cutting forces

* Corresponding author: syafik@utem.edu.my

decrease as helix angle increases. Zatarain et al. (2006) found that helix angle reduces the importance of higher order harmonics vibrations. The effect of helix angle on surface roughness is studied by Buj-Corral et. al (2011) who pointed out that for helix angles other than zero, the same roughness pattern appears on the workpiece's surface along each helix pitch regardless of helix angle. However the research was done by cutting steel. Vobrouček (2015) suggested for changing surface quality, friction and toughness of mill, we must change the clearance angle on the tool; the research was done by cutting Akrylonitrilbutadienstyrene (ABS) plastic. Figure 1 shows end mill nomenclature.

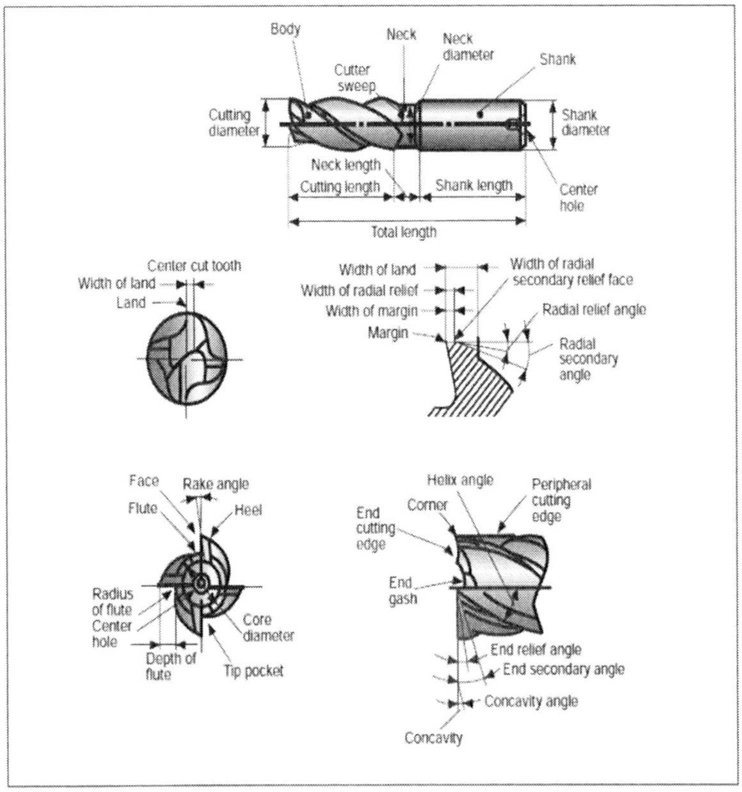

Fig. 1. End mill Nomenclature

2 Experimental setup

Three critical parameters are considered in the fabrication of the end mills. The parameters are helix angle (30°-60°), primary relief angle (5°-9°) and secondary radial relief angle (14°-17°). The helix angle of 30°-60° is selected because these are the angles typically found in the market and recommended by tool manufacturers. The primary and secondary relief angles are also angles recommended by tool manufacturers that are commonly used in end mill manufacturing.

These angles were used in Design of Experiment (DOE) using Full Factorial method to generate the matrix for experimental specimens. The matrix is generated using Minitab 16 software. The software was also used later on to analyse the results after all experiments are completed. Based on the Full Factorial method, there are eight specimens of end mill produced. One replication was added which makes the total number of specimens 16 pieces. Table 1 shows the matrix of parameters that are used in producing the end mill.

Table 1. Experimental setup

Run	A:Helix Angle [Degree]	B:Primary Relief Angle [Degree]	C:Secondary Relief Angle [Degree]
1	60	9	14
2	30	5	17
3	60	5	17
4	60	9	17
5	30	5	14
6	60	9	14
7	30	9	17
8	60	5	14
9	60	5	14
10	60	9	17
11	30	9	14
12	30	9	17
13	30	9	14
14	30	5	17
15	60	5	17
16	30	5	14

3 Experimental results and analysis

The following results were obtained after machining.

Table 2. Experimental results

Run	A:Helix angle [Degree]	B:Primary Relieve Angle [Degree]	C:Secondary Relieve Angle [Degree]	Axial Surface Roughness [μm]	Radial Surface Roughness [μm]	Tool Wear [μm]
1	60	9	14	0.48	0.38	83.9
2	30	5	17	0.24	0.32	78.54
3	60	5	17	0.29	0.44	89.46
4	60	9	17	0.58	0.57	114.83
5	30	5	14	0.37	0.24	84.34
6	60	9	14	0.67	0.5	194.06
7	30	9	17	0.26	0.45	59.84
8	60	5	14	0.47	0.58	49.39
9	60	5	14	0.38	0.43	43.64
10	60	9	17	0.49	0.34	101.93
11	30	9	14	0.31	0.26	64.05
12	30	9	17	0.33	0.35	69.55
13	30	9	14	0.37	0.33	60.42
14	30	5	17	0.38	0.4	82.91
15	60	5	17	0.46	0.51	75.92
16	30	5	14	0.37	0.32	76.29

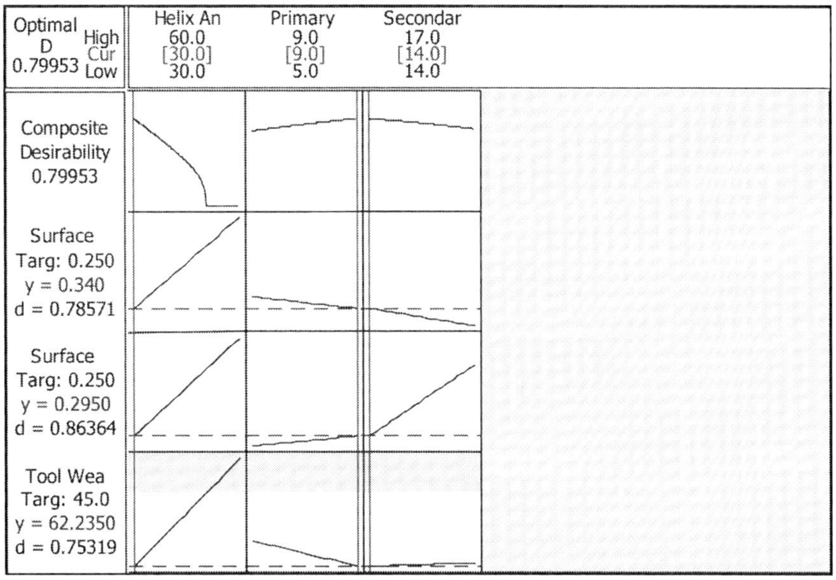

Fig. 2. Response Optimization Graph

Figure 2 shows response optimization graph. Response optimization is an analysis done to get the optimum values for the combination of all parameters. This analysis will provide the prediction of which combination angles will provide the best trade-off between axial and radial surface roughness and also tool wear. From this graph, it can be seen that the recommended values are as follows;

a) Helix Angle: 30°
b) Primary Radial Relief Angle: 9°
c) Secondary Relief Angle: 14°

By using this combination of angles, based on software calculation, the response will be as follows;

a) Axial Surface Roughness: 0.340 μm
b) Radial Surface Roughness: 0.295 μm
c) Tool wear: 62.2350 μm

4 Conclusion

From the results and analysis done the following conclusions can be deduced:

a) Higher primary radial relief angle gives higher tool wear. Higher secondary radial relief angle gives lower tool wear.
b) Higher helix angle gives higher tool wear.
c) Higher primary radial relief angle gives higher axial surface roughness but lower radial surface roughness. Higher secondary radial relief angle gives lower tool wear and vice versa.

The author would like to acknowledge the Ministry of Science, Technology and Innovation and Universiti Teknikal Malaysia Melaka for the support and funding throughout this study.

References

1. M. Alauddin, M.A El Baradie, M.S.J Hashmi, J. Mater. Process Tech., **71**, 456 (1997)
2. D. Begic Hajdarevic, A. Cekic A, M Kulenovic, Procedia Engineering, **69**, 291 (2014)
3. I. Buj Corral, J. Vivancos Calvet, H. Gonzalez Rojas, Int. J. Mach. Tool Manu., **51**, 889 (2011)
4. T. Dursun, C. Soutis, Materials & Design, **56**, 862 (2014)
5. J.A. Ghani, C.H. Che Haron, S.H. Hamdan, A.Y.M. Said, S.H. Tomadi, **39**, 4449 (2013)
6. ISO 4287, Geometrical Product Specifications (GPS) –Surface texture: Profile method-Terms, definitions and surface texture parameters, (1997)
7. ISO 8688-2, Tool Life Testing In Milling – Part 2: End Milling, (1989)
8. S. Kalidas, P. Palanisamy, J. Adv. Manu. Sys., **13**, 1 (2014)
9. S.M. Lou, C.J. Chen, M.C. Li, J. Ind. Tech., **5**, 1 (1999)
10. D.C. Montgomery, *Design and Analysis of Experiments,* (John Wiley & sons, Inc, Hoboken, 2005)
11. O.E.E.K. Omar, T. El Wardany, E. Ng, M.A. Elbestawi, Int. J. Mach. Tool. Manu., **47**, 1263 (2007)
12. J. Vobroucek, Procedia Engineering, **100**, 1556 (2015)
13. M. Zatarain, J. Munoa, G. Peigne, T. Insperger, (2006), CIRP Annals of Manu. Tech., **55**, 365 (2006)

The Management of Humidifying Treatment for Low Contamination Risks During Indoor Cultivation of Grey Oyster Mushroom (*Pleurotus pulmonarius*)

Md. Tariqul Islam[1,*], *Zarina* Zakaria[2], *Nasrul* Hamidin[3], *Mohd Azlan* Bin Mohd Ishak[4], and *Chong* Shi Fern[5]

[1]School of Bioprocess Engineering, Universiti Malaysia Perlis (UniMAP), 02600 Arau, Perlis, Malaysia.
[2, 5]Faculty of Engineering Technology, Universiti Malaysia Perlis (UniMAP), 02100 Padang Besar, Perlis, Malaysia.
[3]School of Environmental Engineering, Universiti Malaysia Perlis (UniMAP), 02600 Arau, Perlis, Malaysia.
[4]Universiti Technology MARA (UiTM), Perlis Campus, 02600 Arau, Perlis, Malaysia.

Abstract. In this study, grey oyster mushroom (*Pleurotus pulmonarius*) was cultivated in indoor controlled environment to seeking out the possible risks of contamination and ways of treatment to avoid the contamination. For this, mushroom was cultivated in providing artificial humidifying and ventilation system to ensure optimum humidity (80-90%) and fresh air recirculation in different ways of treatment. The ways of treatment were included as in position of humidifier, frequency of humidifying, plastic cork of bags opening part and cleaning of humidifier water container. Maximum percentages of bag contamination (2.5-25.30%), cap contamination (5.6-30.75%), stalk contamination (4.75-23.25%) and root contamination (2.6-18.45%) were found in front to front humidifier position, long humidifying with long interval frequency, without plastic cork, without cleaning and bi-monthly cleaning of humidifier water container treatment but no diseases and pest infection was found. Whereas, very low percentages of contamination (0.1-0.5%) were found in surrounding humidifying position, short humidifying duration with short interval frequency, with plastic cork and weekly cleaning of humidifier water container treatment.

1 Introduction

Mushrooms of *Pleurotus spp.* are commonly known as oyster mushrooms which occupy the second position among most cultivated edible mushrooms worldwide due to their nutritional and medicinal values [1–3]. Pleurotus spp. were able to produce a maximum

* Corresponding author: zarinaz@unimap.edu.my

yield when cultivated on sterilized sawdust substrate in a control environment [4]. The mushroom requires low temperature and high relative humidity to produce fruiting bodies [5, 6], whereas, high temperatures caused deformed fruiting body and consequencely low mushroom yield [6]. Kong (2004) reported that *P. ostreatus*, *P. florida*, *P. sajor-caju* can reach their optimum growth at temperature of 25-30^{0}C [7].Chang (2008) reported a humidity at 80–95% was optimum for formation of fruiting bodies and at relative humidity lower than 80%, the fruiting bodies may dry out easily [8]. Jang et al. (2003) suggested that, 80% or more relative humidity is suitable for the formation of highest cap size and individual weight of the *P. ostreatus* and the cap size and stipe thickness started to decrease at lower than 80% and decreased severely at 60% or less relative humidity [9]. Yang *et al.* (2013) reported that, during the fruiting period, the suitable air relative humidity can be as high as 85-90% [10].

High humidity in mushroom cultivation surrounding in the other way can cause disease from the microbial contamination [11]. Upadhyay *et al.* (2011) reported that humidity should be kept low (60–70 %) to prevent microbial contamination when the protecting plastic sheets are removed [13, 14]. Contamination in mushroom cultivation always require the growers to establish new facilities to stop the spreading out of contaminants and therefore increase the production cost [15, 16]. So, from above all literature it is clear that high humidity is the most essential factor for mushroom cultivation and for indoor cultivation an artificial humidifying system is needed to achieve the favourable humidity. Since then, there is no study yet on the possible risks of contamination for mushrooms under controlled humidifying condition and their managements to avoid such situation. Therefore this study had conducted to investigate the possible risks of contamination and their management during indoor cultivation of *Pleurotus pulmonarius* which helps to avoid the contamination that can ensure proper quality and yield of mushrooms.

2 Materials and methods

2.1 Preparation of mushroom growing bags

Substrate's for preparing mushroom bags were prepared by established mushroom cultivators. The substrates were formulated by the mixture of saw dust, rice bran and agricultural lime with the ratio of 100:10:1. The mixtures were packed in autoclavable polypropylene bags (6×9 inch) and sterilized at 20 psi pressure and 121^{0}C temperatures for 6 hours to remove spore of bacteria, fungi and other microorganisms [16]. After sterilization, it was cooled at room temperature and injected with spawn prepared through tissue culture technique. Then the opening parts of the bags were closed by plastic cork, paper and rubber band to protect loss of moisture content from inside the bags and left it vertically for 30-35 days in dark for mycelium colonization. The mycelium colonization was completed when the bags were fully covered by white colour mycelium. Then the bags with completed colonization were transferred into a growing room inside a building. Cultivation period was counted from the first day of bags were transferred and arranged in the cultivation room.

2.2 Design of indoor cultivation room

The indoor cultivation room was measured at 5.8 L×4.57 W×2.74 H m^3 in size. The wall was made by bricks at one side and gypsum board on the rest of the walls. The room was installed with four rows of racks; each rack contained four iron bars distance by 15 cm from each other and 55 cm distance between the racks. The room was also installed with roof ventilation system and light in order to facilitate fresh air and maintain proper light intensity respectively. The bags were placed horizontally through the racks by using rope and made it tight by cable tie and hanged vertically. Twelve bags were arranged in one column of rope by 3 bags in together in 4 parts. A total of 100 columns of rope were used to arrange 1200 bags. The columns of rope were arranged at 25 cm distance from each other in every rack. The distance of the top end bag from the ceiling was 76 cm and the down end bag from the floor was 30 cm.

2.3 Position of humidifier

After complete colonization the bags were transferred into the growing room and arranged as mentioned above. The humidifiers were placed inside the growing room at two positions, 'Faced to Faced' and 'Surrounding'. In 'Faced to Faced', the humidifier was placed between the two columns of bags where the distance from the column to humidifier was 15cm in both side and the opening parts (mouth to come out fruitbody) of the bags were faced to the humidifier. In the 'Surrounding', the humidifier was placed also between the two columns of bags but the distance between the opening part of the bags (front view) and humidifier were 25-30 cm, between side view to humidifier were 20-25 cm and back view to humidifier 10-15 cm (Fig. 1a). In both positions, the humidity was maintained at 80-90 % using the humidifiers. The data was collected on the basis of percentages of bags, stalk, cap and root contamination.

2.4 Humidifying frequency

Two ways of humidifying frequency was conducted; a long humidifying with long interval (LHLI) and short humidifying with short interval (SHSI). In LHLI, two units of humidifier were placed at 3.5-4 m distance from each other and it was applied 30 minutes with 45 minutes interval from two positions to maintain 80-90% indoor humidity. Whereas in SHSI, four units of humidifiers were placed at 2-2.5 m distance from each other's and it was applied 15 minutes with 15 minutes interval from four positions to maintain 80-90% indoor humidity. The humidifiers in both treatments were configured as followed by timer:

LHLI:- 30H1+45I+30H2+45I+30H1+45I+30H2+45I+.......................
SHSI:- 15H1+15I+15H2+15I+15H3+15I+15H4+15I+15H1+15I+15H2+15I+15H3+.....

*H1,H2,H3,H4 = Humidifier; 1,2,3,4. I= Interval

Mushroom was cultivated in this both humidifying system with continuous ventilation. The data was collected on the basis of percentages of bags, stalk, cap and root contamination.

2.5 Plastic cork of bags opening part

In this method the bags were cultivated with and without the plastic cork that was used in the opening part during substrate packing. After arranging the completed colonization bags in the growing room, 50% plastic cork were removed and another 50% were not removed (Fig. 1b). Thus all the bags were cultivated in same environment by providing artificial humidifying and ventilation to maintain 80-90% humidity and proper air circulation. The data was collected on the basis of percentages of bags, stalk, cap and root contamination.

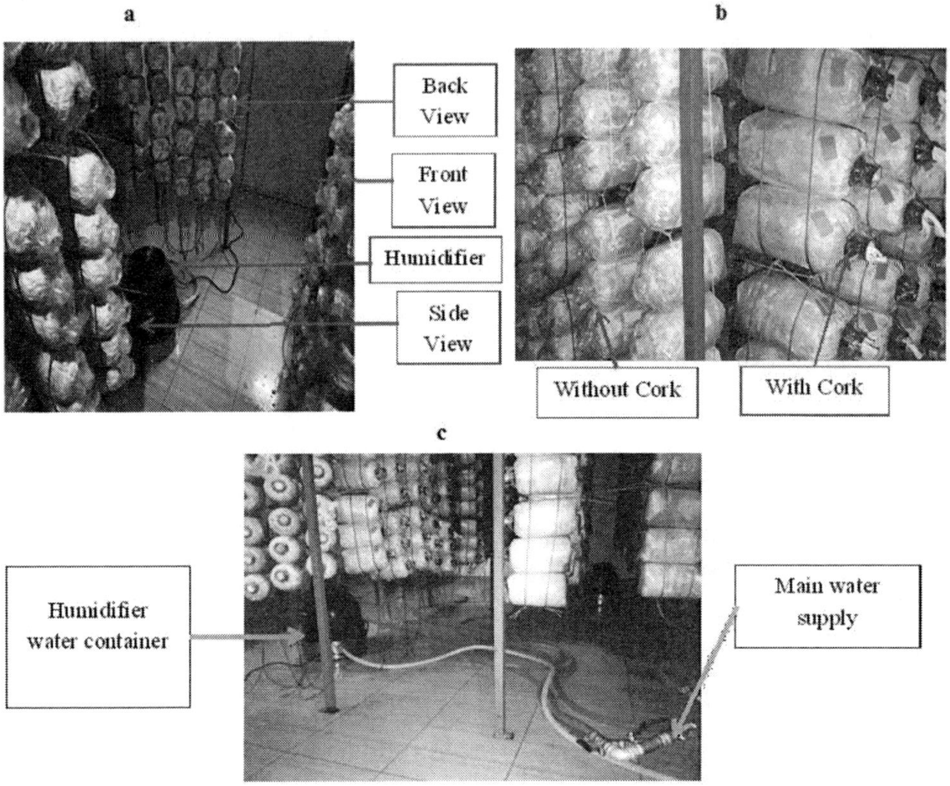

Fig. 1. Different ways of management for indoor mushroom cultivation (a: Position of humidifier; b: Plastic cork; c: Humidifier water container)

2.6 Cleaning of water container

The water container which contained water sources for humidifying was subjected to three ways of cleaning treatment (Fig. 1c). First treatment was the control, where no cleaning was conducted throughout the cultivation period. The second treatment was bi-monthly cleaning and the third was weekly cleaning. The data was collected on the basis of percentages of bags, stalk, cap and root contamination.

2.7 Data collection and analysis

The percentages of different contamination including bags, caps, stalk and root was calculated as follows:

$$Contamination(\%) = \frac{Total\ (Bag\ or\ cap\ or\ stalk\ or\ root)\ Contaminated}{Total\ numbers\ of\ bag\ or\ cap\ or\ stalk\ or\ root} \times 100. (1)$$

The collected data were analyzed as descriptive statistic and one way ANOVA test where the significant level was considered lower than 0.05 at p-value. All of the statistical calculations were executed using SPSS 10.0.

3 Results and discussion

The highest percentage of contaminated bags was recorded as 25.30%, 20.33%, 15.5%, 5.6% and 2.5% in LHLI, front to front, without cork, without cleaning and bi-monthly cleaning applications respectively. On the other hand, the lowest was found as 0.25%, 0.4% and 0.5% in SHSI, with cork, weekly cleaning and surrounding humidifying treatments (Table 1; Fig. 2).

Table 1. Percentages of contaminations in different ways of treatment during indoor mushroom cultivation.

		Types of Contamination			
Methods	**Treatments**	Bag (%)[1]	Stalk (%)[2,4]	Cap (%)[2,3,6]	Root (%)[5]
Position of humidifier	Front to Front	20.33±1.1	10.45±1.6	25.65±2.5	15.75±1.9
	Surrounding	0.50±.2	0.50±.2	0.25±.05	0.25±.1
Humidifying Frequency	LHLI	25.30±1.9	20.56±2.4	30.75±3.2	18.45±2.9
	SHSI	0.25±.13	0.20±.1	0.23±.2	0.23±.1
Plastic Cork of Bags Opening Part	Without Cork	15.50±1.9	4.75±2.4	5.60±2.1	13.45±1.8
	With Cork	0.40±.3	0.20±.1	0.32±.1	0.33±.1
Cleaning of Water Container	Without	5.60±1.9	23.25±2.7[4]	30.40±2.0[6]	7.56±2.8
	Bi-monthly	2.50±1.2	7.20±1.9[4]	10.70±2.9[6]	2.40±.6
	Weekly	0.40±.2	0.25±.1	0.50±.2	0.22±.1
*[1,2,3,4,5,6]: Picture of different way of contamination in Fig.-2					

The highest percentages of contaminated stalk was found as 23.25%, 20.56%, 10.45%, 7.2% and 4.75% in without cleaning, LHLI, front to front, bi-monthly cleaning and without cork application, whereas the lowest were 0.2%, 0.25% and 0.5% as found in with cork, SHSI, weekly re-cleaning and surrounding humidifying treatments respectively (Table 1; Fig. 2). The highest contaminated cap of 30.75%, 30.4%, 25.65%, 10.7% and 5.6% was found in LHSI, without cleaning, front to front, bi-monthly cleaning and without cork treatment respectively. On the other hand, the lowest of 0.1%,0.25% and 0.5% was found in SHSI, surrounding humidifying, with cork and weekly cleaning treatment respectively. Finally, the highest contaminated root of 18.45%, 15.75%, 13.45%, 7.56% and 2.4% was found in LHLI, front to front, without plastic cork, without cleaning and bi-monthly cleaning treatment respectively. Instead, the lowest of 0.1%, 0.25% and 0.33% was found in SHSI, weekly cleaning, surrounding humidifying and with cork application respectively (Table 1; Fig. 2). All the types of contamination in different ways of treatment for indoor cultivation were found statistically significant in one-way ANOVA test at .000 significant value.

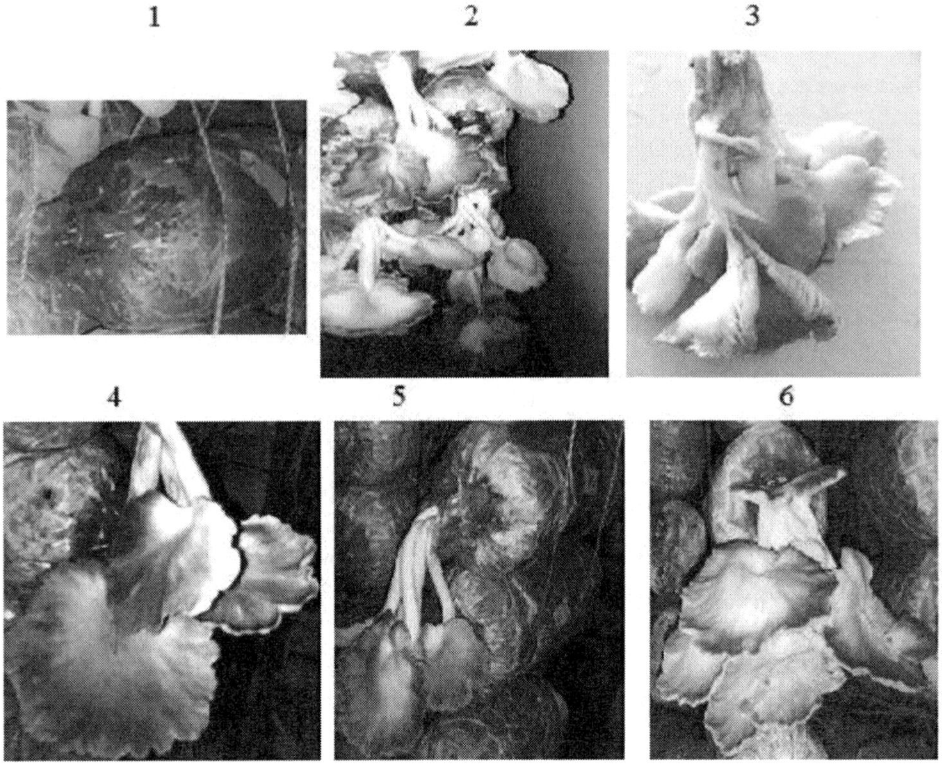

Fig. 2. Contaminated bags (1), caps (2,3,6), stalk (2,4) and root (5) in different ways of treatment for indoor cultivation.

The results showed that the humidifying in front of the bags opening part with short distance, LHSI, without plastic cork, without and bi-monthly cleaning of humidifier water container application were applications identified to impose high risk for contamination. Those applications provided easy access of contaminated vapour into the bags and fruting bodies. Moreover, contamination was also found in without and bi-monthly cleaning application because in both applications the water container became dirty and produced bad smells which indication of water contamination. The contaminated water was sprayed as vapor by humidifier and consequently contaminated the bags and fruitbodies. Deacon, (2013) reported that, excessive water can cause disease, contamination and poor mushroom quality [11]. In this present study, no disease was found but contaminated bags and fruiting bodies were found in different ways of application.

On the other hand, surrounding humidifying, short humidifying with short distance, with plastic cork of the bags mouth and weekly cleaning were found had caused low risks of contamination. Surrounding humidifying provided a proper distance between bags and humidifier so as not excessive vapour had penetrated directly into the bags opening and into the fruitbodies. Other treatments such as the plastic cork, weekly cleaning of water container also ensured fresh water and vapour surrounding the cultivation room. Therefore this study suggested that management of humidifying conditions may reduce certain percentage of water-caused contamination and to ensure the present of water will give positive effect and not vice versa.

4 Conclusion

This study found that humidifying application could be imposed contamination for certain percentages of bags and fruitbodies. Therefore, the management of humidifying conditions as suggested in this study may reduce the risks of disease from water-caused contamination.

References

1. M.A. Khan, M. Tania, S.M.R. Amin, N. Alam, M.N. Uddin, Bangladesh J. Mushroom, **2**, 17 (2008).
2. S.A. Ahmed, J.A. Kadam, V.P. Mane, S.S. Patil, M.M.V. Baig, Nat. Sci., **7**, 44 (2009).
3. U. Lindequist, T.H J. Niedermeyer, W.D. Julich, Evidence-Based Complement. Altern. Med., **2**, 285 (2005).
4. E.S. Dias, Ciencia e Agrotecnologia, **34**, 795 (2010).
5. P. Patel, R. Trivedi, Int. J. Sci. Res., **4**, 1126 (2015).
6. N. Klomklung, S.C. Karunarathna, K.D. Hyde, E. Chukeatirote, Acta Biol. Szegedlensis, **58**, 39 (2014).
7. W.S. Kong, Oyster mushroom Cultiv. Part II. Oyster mushrooms. Seoul Heineart Inc., **54** (2004).
8. S.T. Chang, Mushrooms as Funct. foods, **1** (2008).
9. K.Y. Jang, C.S. Jhune, J.S. Park, S.M. Cho, H.Y. Weon, J.C. Cheong, S.G. Choi, J.M. Sung, Mycobiology, **31**, 145 (2003).
10. J. Yang, J. Zhao, H. Yu, Y. Wang, R. Wang, L. Tang, *Computer and Computing Technologies in Agriculture VI*, (Springer, China, 2013).
11. J. Deacon, *Fungal Biology*, (John Wiley & Sons, UK, 2013).
12. R.C. Upadhyay, M. Singh, *Industrial Applications*, (Springer, Germany, 2011).
13. A.W. Chen, *Mushroom Grower's Handbook1*, (Springer, Netherlands, 2004).
14. T. Hyunjong Kwon, *Mushroom Grower's Handbook 1*, (Mush World, China, 2004).
15. S.W. Kang, *Mushroom Grower's Handbook 1*, (MushWorld, China, 2004).
16. J. Christopher, D. Custodio, *Mushroom Grower's Handbook 1*, (MushWorld, China, (2004).

Development of Solar Biomass Drying System

Samson Mekbib Atnaw[1,*], *Che Ku Mohammad Faizal* Bin Che Ku Yahya[1] , and *Abduaziz* Jama Oumer[2]

[1] Faculty of Engineering Technology, University Malaysia Pahang, Tun Razak Highway, 26300 Kuantan, Pahang, MALAYSIA.
[2] Faculty of Industrial Management, University Malaysia Pahang, Tun Razak Highway, 26300 Kuantan, Pahang, MALAYSIA.

Abstract. The purpose of this paper focuses on the experimental pre-treatment of biomass in agricultural site using solar energy as power source and contribution of common use and efficiency solar dryer system for consumer. The main purpose of this design for solar cabinet dryer is to dry biomass via direct and indirect heating. Direct heating is the simplest method to dry biomass by exposing the biomass under direct sunlight. The solar cabinet dryer traps solar heat to increase the temperature of the drying chamber. The biomass absorbs the heat and transforms the moisture content within the biomass into water vapour and then leaves the chamber via the exhaust air outlet. This problem however can be solved by adopting indirect solar drying system. High and controllable temperatures can be achieved as a fan is used to move the air through the solar collector. This project has successfully created a solar cabinet dryer that combines both direct and indirect solar drying systems and functions to dry biomass as well as crops effectively and efficiently with minimal maintenance. Hence, it is indeed a substitution for conventional dryers which are affordable to local farmers.

1 Introduction

Biomass is a biological material that is abstracted from living organism [1]. Malaysia is a country that has abundant source of biomass such as woods, sugar canes, organic matter and so on. In order to convert this biomass to electricity, it can undergo the gasification process in a gasification reactor [2]. For the gasification process it requires relatively low moisture input. In order to ensure an effective and complete combustion of biomass, it should undergo the pre-treatment process called drying or moisture removal [3]. Geographical location shows that Malaysia is located near the equator [4]. The scope of study for this work is to fabricate a solar cabinet dryer to dry biomass. The current study will cover the development of drying mechanism and heat transfer calculations and analysis in order to ensure efficient utilization of the solar heat inside the drying chamber. Equilibrium moisture content (EMC) for the biomass will be experimentally determined for

* Corresponding author: mekbibsams@um p.edu.my

the specific biomass type to be studied. Therefore, the efficiency of this system can be estimated according to the EMC result.

2 Background

In many parts of the world there is a growing awareness that renewable energy have an important role to play in extending technology to the farmer in developing countries to increase their productivity [5]. Solar thermal technology is a technology that is rapidly gaining acceptance as an energy saving measure in agriculture application. It is preferred to other alternative sources of energy such as wind and tidal energy, because it is abundant, inexhaustible, and non-polluting. Solar air heaters are simple devices to heat air by utilizing solar energy and employed in many applications requiring low to moderate temperature below 80°C, such as crop drying and space heating [6]. In addition to moisture removal of biomass, solar drying process plays an important role in the preservation of agricultural products. It is defined as a process of moisture removal due to simultaneous heat and mass transfer [7].

In order to improve traditional drying, solar dryers which have the potential of substantially reducing the above-mentioned disadvantages of open air drying have received considerable attention over the past 20 years. Though forced convection type of solar dryers could be effectively used, however they require electricity in order to operate the extraction fans. This may significantly increase the operation cost and also limit the use of dryer in rural areas where there is no elecricity. Even when electricity exists, the potential users of the dryers are unable to pay for it due to their very low income. Forced convection dryers are for this reason not going to be readily applicable on a wide scale in many developing countries [8-9]. Natural convection dryers circulate the drying air without the aid of a fan. They are therefore, the most applicable to the rural areas in developing countries.

Table 1. Merits and Demerits

The Applied Drying Technique	Advantages	Disadvantages
Open air drying	• No investment is required. No fuel is required.	• The product is exposed to risks, such as dust, animals, ants, rain, etc.
Firewood/fuel drying	• Faster than the open air type	• Cost of labor and building the dryer box.
Solar Drying	• Cheap, Environmentally friendly and ease of maintenance	• The difficult dryer operation.

Shown in Table 1 is a summary of the merits and demerits of various drying methods. As shown in Table 1, there is a clear indication, that the solar drying technique is the most suitable for the small scale farmers to cover their needs, during the processing of cash crops. It is also suitable for drying the available fruits and vegetables.

2.1 Solar Drying

In solar drying the heat is created by the sun radiation. The heat could be circulated by a ventilator powered by PV array. The three first mentioned techniques can be replaced by the environmental friendly drying techniques. The Solar Dryer Technology and Utilization The description of the solar drying technologies will be limited to few types, which are very common and well known. Low Cost Solar Dryer: The low cost dryer has different varieties

DOI: 10.1051/matecconf/20179701081

such as the cabinet solar dryer, the flat solar dryer, etc. The low cost dryer is the most suitable even for the small farmer, who have a very limited income. It will be possible for them to cover the investment cost. This is produced from locally available materials, and by local carpenter. Solar Dryer: This type developed for drying only food crops, cash crops and fruits. The dryer consists of: Sun collector panel for air heating, Drying chamber, Fan for the hot air circulation, Moisture and temperature sensor and PV- panel for powering the fan.

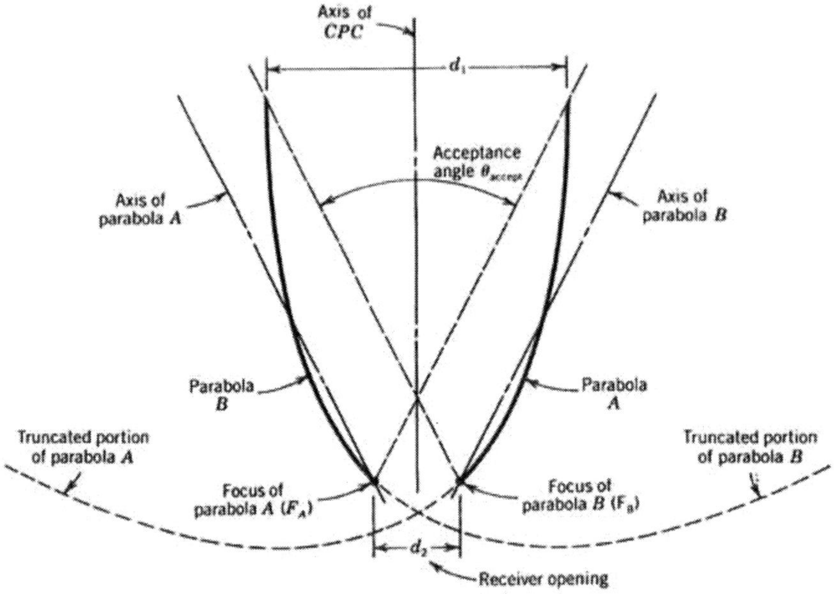

Fig. 1. Compound parabolic concentrator (CPC)

Initially, a mechanism that uses a compound parabolic concentrator (CPC) to heat the biomass was considered. CPC is comprised of two parabolic mirror segments with different focal points as indicated. It is capable of collecting direct sunlight as well as diffused sunlight without the need of a solar tracking system. However, it can attain a very high temperature up to 300 °C that makes it inappropriate for this drying process. The basic shape of the compound parabolic concentrator is illustrated in Fig 1.

The second type of design considered was a solar dryer based on the conventional solar dryer which uses direct drying or indirect drying. The solar dryer combines both drying methods so as to improve its efficiency. The design is illustrated in Fig 2. The biomass is laid out on trays, which sit behind a transparent glass. Sun beams heat up the incoming fresh air and force the hot air to move upwards. The hot air is then discharged via air outlet, taking the moisture from the biomass.

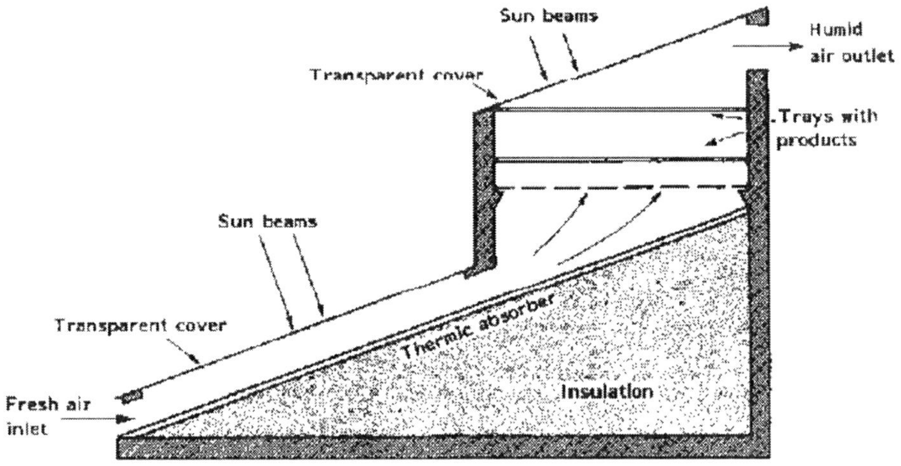

Fig. 2. Combined direct and indirect solar dryer

The moisture content of the biomass depends on the relative humidity and temperature of the air surrounding. Moisture content express the amount of water present in a moist sample. Moisture content can be expressed on wet basis or dry basis.

1. Moisture content(MC) on dry basis is the amount of water per unit mass of dry solids in the sample :

$$\text{MC for dry basis} = \frac{initial\ mass\ of\ sample - final\ mass\ of\ sample}{final\ mass\ of\ sample} \text{x } 100\%$$

2. Moisture comtent(MC) on wet basis is the amount of water per unit mass of moist or wet sample :

$$\text{MC for wet basis} = \frac{initial\ mass\ of\ sample - final\ mass\ of\ sample}{initial\ mass\ of\ sample} \text{ x } 100\%$$

3 Methodology

The main purpose of this design for solar cabinet dryer is to dry biomass via direct and indirect heating. Direct heating is the simplest method to dry biomass by exposing the biomass under direct sunlight. The solar cabinet dryer traps solar heat to increase the temperature of the drying chamber.

Fig 3. shows the schematic representation of proposed solar drying system. The transparent glass is fixed at an inclination of 21° to achieve a maximum absorption of solar radiation. The wall of the drying chamber ismade up of three different layers, i.e. fibreglass, wood and aluminium. Fibreglass is used as the outer layer because it makes a good heat insulator that prevents heat loss to the surrounding[10-12]. Wood isused as the middle layer due to its low cost. The inner layer is made of aluminium which has high resistivity against rust and corrosion[11]. As shown in Fig 3., the design of the drying chamber issimilar to a drawer cabinet that has two drawers. The drawer compartments are made up of aluminium and they are designed in such a way that they allow a maximum flow of hot air through the biomass.

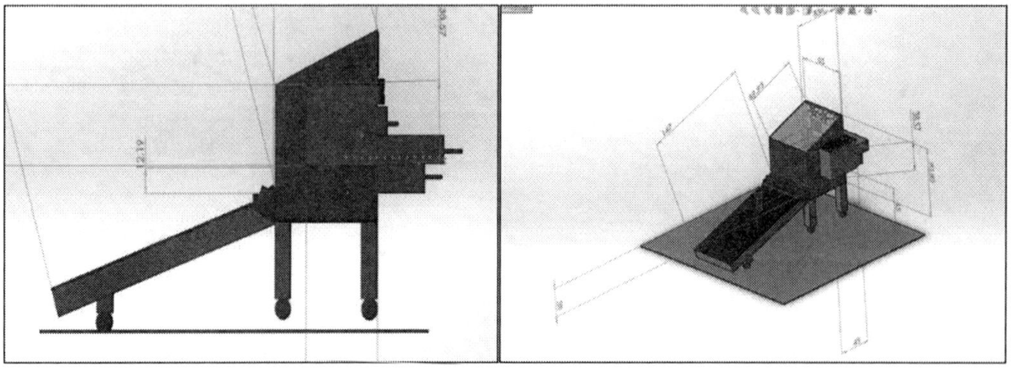

Fig. 3. Solar Cabinet Dryer

The whole ventilation system is driven by direct current (DC) generated from solar panels (PV).The air flow rate of the ventilation fan is 70 cubic feet per minutes (CFM). The temperature of position parts of the drying chamber is measured using type K thermocouple.The bottom is made up of aluminium and painted in black to increase the radiation absorptivity and emissivity. The top of the absorber is covered by a dark glass. The black surface will absorb the radiation and changes it into heat energy to heat up the incoming fresh air

This paper is conducted based on research and experiences into risk management its potential to prevent and eliminate risk and learning from performance team in industry which matched by an application in project refurbishment in Batam manufacturing Indonesia.

4 Results and discussion

After the course of study, the solar cabinet dryer is found to be more efficient on drying the coconut coir as compared to natural drying. In 3 days of continuous drying under the same climatic condition, a total of 31.46% moisture content was removed from sample 1, 2 and 3 whereas only 21.99% of moisture content was removed from sample 4, 5 and 6. Moisture content (MC) of the coconut coir will be calculated by using dry basis method.

$$\text{MC for dry basis} = \frac{initial\ mass\ of\ sample - final\ mass\ of\ sample}{final\ mass\ of\ sample} \text{ x } 100\%$$

Table 2. Drying Coconut Coir Experiment via Solar Cabinet Dryer

Sample	**Initial mass (g)**	**Final mass (g)**	**MC (%)**
1	23.4875	15.6826	33.23
2	27.4722	19.5932	28.68
3	19.1241	12.9164	32.46
Average			31.46

After full drying at 107°C for 24 hours in the furnace, the moisture content of coconut coir is 0%, implies that the sample 7, 8 and 9 do not containany water. The moisture loss of full drying process is 39.02%. It indicates the total moisture content in coconut coir sample for experiment of solar cabinet dryer and natural drying process is 39.02%.

Table 3. Drying Coconut Coir Experiment via Natural Drying Process

Sample	Initial mass (g)	Final mass (g)	MC (%)
4	22.9496	17.7476	22.67
5	22.1241	17.3972	21.37
6	17.9478	14.0103	21.94
Average			21.99

The purpose of full drying experiment is to analyze the ability of sample to absorb moisture at room condition to achieve equilibrium moisture content (EMC). The sampleswere exposed in room condition for 3 days and the result was analyzed to determine the EMC of coconut coir.

Table 4. Drying Coconut Coir Experiment via Furnace for the Purpose of EMC

Sample	Initial mass (g)	Full drying mass(g)	MC (%)
7	4.4584	2.5721	42.31
8	2.2811	1.8913	17.09
9	4.6336	1.9613	57.67
Average			39.02
Sample	**Full drying mass(g)**	**Final mass(g)**	**EMC (%)**
7	2.5721	2.9119	11.67
8	1.8913	2.2351	15.38
9	1.9613	2.2510	12.87
Average			13.31

5 Conclusion

Direct solar drying allows direct absorption of solar radiation by the biomass which is the most effective way of converting solar radiation into useful heat for drying. The final dried quality of biomass is also enhanced by direct exposure to solar radiation. However, because the biomass absorbs solar radiation directly, its temperature is difficult to control. This problem however can be solved by adopting indirect solar drying system. High and controllable temperatures can be achieved as a fan is used to move the air through the solar collector. This project has successfully created a solar cabinet dryer that combines both direct and indirect solar drying systems and functions to dry biomass as well as crops effectively and efficiently with minimal maintenance. Hence, it is indeed a substitution for conventional dryers which are affordable to local farmers.

The authors would like to acknowledge the support from University Malaysia Pahang and Malaysian Ministry of Higher Education FRGS grant (RDU140109).

References

1. P.Basu, *Biomass Gasification, Pyrolysis and Torrefaction (2nd Edition) Practical Design and Theory,* (Academic Press, 2013)
2. R.L. Earle, *Unit operations in food processing*, (Oxford Pergamon Press, 2003)
3. *The Free Encyclopedia.Geography of Malaysia* (Wikipedia, 2015)

DOI: 10.1051/matecconf/20179701081

4. J.A. Waewsak, Int. J. Sci. Technol., **11**, 14 (2006)
5. I. Kurtbas, E. Turgut, Int. J. Sci. Technol., **1**, 75 (2006)
6. C. Ertekin, O. Yaldiz, J. Food Eng., **63**, 349 (2004)
7. Y.H. Hui, *Food Drying Science and Technology: Microbiology, Chemistry, Applications*, (DEStech Publications 2008)
8. A. Gatea, Journal of Agricultural Biotechnology and Sustainable Development, **2**, 39 (2010)
9. S.M. Atnaw, S.C. Kueh, S.A. Sulaiman, Sci. World J., 8 (2014)
10. *Thermal Insulation Association of South Africa Thermal insulation-the invisible energy saver: The Guide to Energy Efficient Thermal Insulation Buildings,* (Midrand, AAAMSA administered 2010)
11. S.A. Sulaiman, M.R.T. Ahmad, S.M. Atnaw, AIP Proceedings, 981 (2012)
12. S.M. Atnaw, S.A. Sulaiman, S. Yusup, Scientific World J., (2014)

In-situ Transesterification of *Jatropha curcas* L. Seeds for Biodiesel Production using Supercritical Methanol

M.A.M. Ishak[1,*], *K.* Ismail[2], *W.I.* Nawawi[1], *Ali* H. Jawad[1], *A.Y.* Ani[1] , and *Z.* Zakaria[3]

[1]Faculty of Applied Sciences, Universiti Teknologi MARA, Perlis Branch, 02600 Arau, Perlis, Malaysia
[2]Faculty of Applied Sciences, Universiti Teknologi MARA, 40450 Shah Alam, Selangor, Malaysia
[3]Faculty of Engineering Technology, Universiti Malaysia Perlis, 02100 Padang Besar, Perlis, Malaysia

Abstract. *In-situ* supercritical methanol transesterification for production of biodiesel from *Jatropha curcas* L. (JCL) seeds was successfully being carried out via batch-wise reactor system, under varying temperatures of 180 - 300 °C, pressures of 6 - 18 MPa, reaction time of 5 - 35 min and seeds-to-methanol ratio of 1:15 - 1:45 (w/v). In this study, the extracted oil obtained showed the presence of FAME referring as biodiesel, indicating that transesterification reaction had occurred during the extraction process. The results showed that the biodiesel yield was obtained at optimum conditions of 280 °C, 12 MPa, 30 min and 1:40 (w/v) were 97.9%.

1 Introduction

Vegetable oil methyl esters, commonly referred to as biodiesel are prominent candidates as alternative to diesel fuels [1]. Biodiesel is technically competitive with or offers technical advantages compared to conventional petroleum diesel fuel. The vegetable oils, as alternative engine fuels, are all extremely viscous with viscosities ranging from 10 to 20 times greater than that of petroleum diesel fuel [2]. Fortunately, inedible vegetable oils, mostly produced by seed-bearing trees and shrubs can provide as alternative for biodiesel sources. Among the various renewable energy choices, seed oil crops especially the *Jatropha curcas L.* has a high potential to meet with the increasing requirements of petroleum and its products. With no competing food uses, this characteristic turns attention to *Jatropha curcas*, which grown in tropical and subtropical climates across the developing world [3]. It is significant to point out that, the inedible vegetable oil of *Jatropha curcas* has the requisite potential of providing a promising and commercially viable alternative to diesel oil since it has desirable physicochemical and performance characteristics comparable to diesel.

Recently, oil contents, fatty acid methyl esters (FAME) compositions and energy values of *Jatropha* species were investigated [1]. The conventional method of extracting the oils

* Corresponding author: azlanishak@perlis.uitm.edu.my

from *Jatropha* seeds is through mechanical screwed pressing technique. However, much of the un-extractable oil still remains in the kernel; hence better ways of extracting the oils are needed. Among the extraction techniques reported in the literature include the use of ultrasound-assisted system [4], enzymes extraction [5] and the use of catalytic materials [6]. Some of these extraction methods, however, required a longer extraction time and multiple steps to achieve. However, the used of *in-situ* supercritical methanol extraction on *Jatropha* seeds has not been reported thus far. Supercritical fluid extraction using polar solvent such as methanol as extraction solvent is highly potential extraction technique to be used whereby high yield of oil can be achieved within a shorter time. Further, at supercritical state, the solvent solubility increased dramatically, and the extracted oil is relatively low in impurities. The objectives of this study is to determine the effect of parameters i.e. temperatures, pressures, reaction times and ratios on the percent of biodiesel yields from the *in-situ* supercritical methanol transesterification.

2 Materials and methods

2.1 Materials

Matured JCL fruits were obtained from Plantation Unit of Universiti Teknologi MARA Perlis, Malaysia. The fruits were cleaned and de-hulled to obtain the seeds. The seeds were then dried in an oven at 105 °C for 30 min in order to remove surface moisture and ground to particle size of less than 1 mm using grinder. All of the chemicals used were of analytical reagent grade.

2.2 *In-situ* supercritical methanol transesterification

A batch type reactor at supercritical methanol was used for *in-situ* conversion of JCL seeds and into biodiesel. A schematic diagram of the system employed is shown in Fig. 1. A 1 liter stainless steel reactor (PAAR Model 4571) was charged with the seeds to methanol ratio (w/v) of 1:40. The reaction vessel was heated with a heater to the desired reaction temperatures (180 – 300 °C), pressures (6 – 18 MPa), reaction times (5 – 35 min) and seeds-to-methanol ratios (1:15 – 1:45 w/v). After each reaction the vessel was removed from the heater and placed into a cold water bath to quench the reaction. The reaction products were discharged from the reactor and were evaporated to remove the methanol. The products were allowed to settle and separated into two phases. The top phase consists of the methyl ester (biodiesel) and the lower phase consists of glycerol with a minor components.

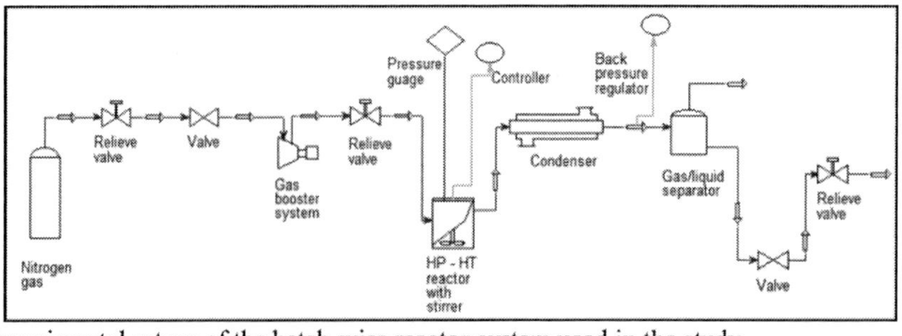

Fig. 1. Experimental set-up of the batch-wise reactor system used in the study.

DOI: 10.1051/matecconf/20179701082

3 Results and discussion

3.1 Effect of temperature

The effect of temperature on percent of biodiesel yield from JCL seeds were investigated. These experiments were carried out at 12 MPa of pressure, 1:40 of seeds-to-methanol ratio, 30 min of reaction time, 450 rpm of stirring speed and at varying temperatures of 180, 200, 240, 280 and 300 °C. The results of *in-situ* supercritical methanol transesterification on percent of biodiesel yield from JCL seeds at various temperatures are shown in Fig. 2. From Fig. 2, the results indicate that the percent of biodiesel yield obtained at temperatures of 180 to 300 °C were in the range of 51.2 – 97.9%. At 180 °C, which is the lowest temperature of investigation, low yield of biodiesel (51.2%) were obtained. This observation might be due to the subcritical state of methanol or the instability of the supercritical state of methanol [7-9]. Apparently, by increasing the reaction temperature from 200 to 280 °C, the conversion increases significantly with biodiesel yield increased from 76.0 – 97.9%. The higher conversions observed in the supercritical state can be attributed to the formation of a single phase between alcohol and oil [8]. Under supercritical conditions, the solubility parameter of alcohol reduces and was close to the solubility parameter of oil [10]. At the 300 °C of temperature, the percent of biodiesel (90.9%) yields were slightly decreased. This observation was due to the decomposition of polyunsaturated methyl esters and unreacted triglycerides in postcritical methanol at severe high temperature [11]. As a result, the effect of temperature from *in-situ* supercritical methanol transesterification was optimized at 280 °C with biodiesel yield at 97.9%.

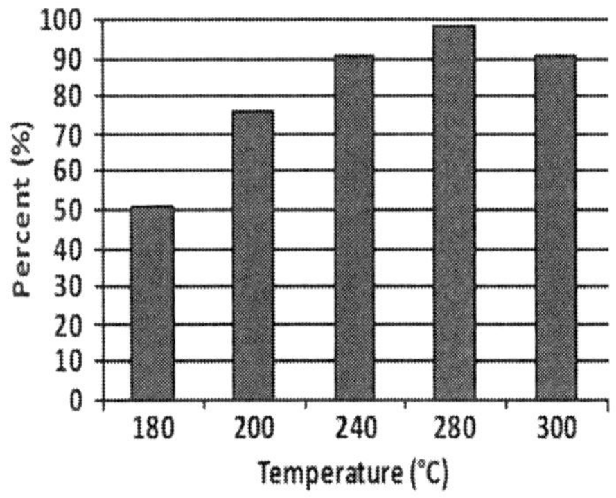

Fig. 2. Effect of temperatures on percent of biodiesel yield.

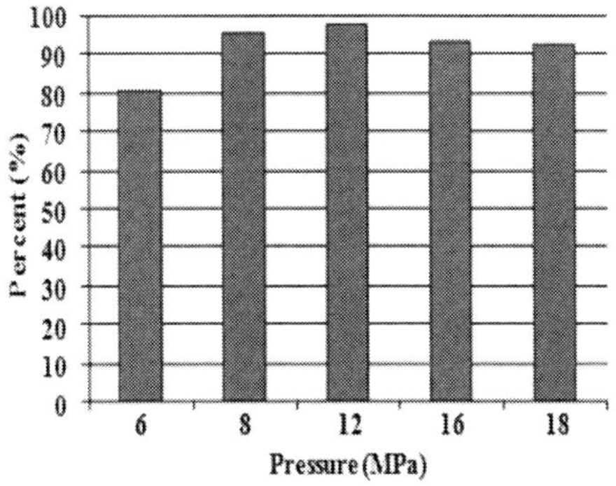

Fig. 3. Effect of pressure on percent of biodiesel yield

DOI: 10.1051/matecconf/20179701082

3.2 Effect of pressure

The results of in-situ supercritical methanol extraction and transesterification *in-situ* on percent of extraction (seed basis) and FAMEs yield from JCL seeds at various pressures are shown in Fig. 3. The temperature was fixed at 280 °C based on the maximized yield conditions from the previous experiment. Meanwhile the seeds-to-methanol and reaction time was fixed at 1:40 w/v and 30 min, respectively with stirring speed of 450 rpm. From Fig. 3, the results indicated that the percent of biodiesel yield obtained at temperatures of 280 °C and pressures of 6-18 MPa was in the range 80.6 – 97.9%, with maximum yields at 12 MPa. At the lowest pressure of 6 MPa, the biodiesel yield are only 80.6%, but increases to 97.9%, when the pressure are increased to 12 MPa. The high biodiesel yield achieved at 12 MPa, which is slightly above the critical pressure of methanol (8.09 MPa), might be due to the increase in solvent power of methanol with increasing pressure. At highest pressure (18 MPa), the biodiesel yield decreases slightly to 92.5%. After the pressure increased to a specific level, the increase of pressure does not cause an obvious improvement in the FAME yield [12]. This phenomenon might be due to the maximum solubility and/or hydrogen donor ability of the solvent that has been achieved regardless of high pressure employed. Thus, a 12 MPa of pressure was chosen in order to maximize the biodiesel yield up to 97.9%.

3.3 Effect of reaction time

Fig. 4 shows the effect of reaction time on percent biodiesel yield from JCL seeds using *in-situ* supercritical methanol transesterification. The reaction conditions were fixed based on maximum yields at optimized conditions discussed previously, i.e. 280 °C of temperature, 12.7 MPa of pressure and 1:40 seeds-to-methanol ratios at different reaction times. From Fig. 4, the results indicated that the percent of biodiesel yield obtained at reaction time of 5 – 35 min was in the range of 88.4 – 97.9%, with maximum yields at 30 min. From the results, it can be seen that the percent of biodiesel yield were only 88.4%, at 5 min of reaction time. According to Saka and Kusdiana [13], in the common method, the reaction is initially slow because of the two-phase nature of the methanol/oil system, and slows even further because of the polarity problem even with the help of an acid or an alkali catalyst. It can be seen that the conversion was increased in the reaction time ranges between 5 and 30 min with the percent of biodiesel yield showed a slight increase in the range of 88.4 – 97.9%. Thus, the percent of biodiesel yield showed an increased with increasing reaction time. The results indicated that an extension of the reaction time from 30 to 35 min had leads to a reduction in the biodiesel yield (93.1%). This is because longer reaction time enhanced the hydrolysis of esters (reverse reaction of transesterification), resulted in loss of esters as well as causing more fatty acids to form soap [14]. Hence, for this process, there is no reason to prolong the reaction time beyond 30 min. Thus, the reaction time of 30 min can be considered as the economic reaction time by considering the percent of biodiesel yield being achieved.

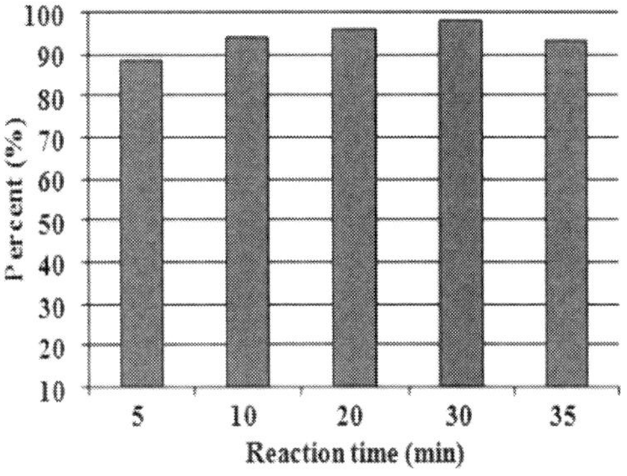

Fig. 4. Effect of reaction time on percent of biodiesel yield.

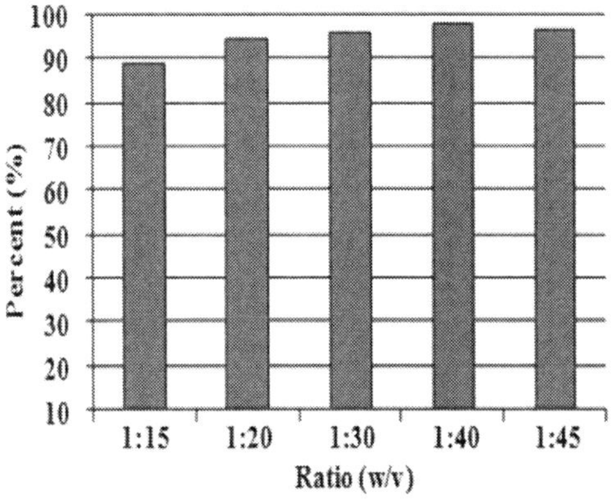

Fig. 5. Effect of ratio on percent of biodiesel yield.

3.4 Effect of seeds-to-methanol ratio

Fig. 5 shows the effect of seeds-to-methanol ratio on percent of biodiesel yield from JCL seeds using *in-situ* supercritical methanol transesterification. The reaction conditions were fixed based on maximized yields at optimized conditions discussed previously, i.e. 280 °C of temperature and 12.7 MPa and 30 min of reaction time at different ratios. The results indicated that the percent of biodiesel yield obtained at temperatures of 280 °C, pressures of 12.7 MPa, reaction time of 30 min and at various seeds-to-methanol ratio (1:15 – 1:45 w/v) was in the range of 89.0 – 97.9%, with maximum yields at 1:40 w/v. At the lowest seeds-to-methanol ratio of 1:15, the percent of biodiesel yield were relatively low (89.0%) and increased with increasing seeds-to-methanol ratio. In this reaction, an excess of methanol was used in order to shift the equilibrium in the direction of the products [15]. It can be seen that an increment in seed-to-methanol ratio can enhance biodiesel yield due to higher contact area between methanol and triglycerides. However, when the ratio is beyond 40, the yield of biodiesel begins to decrease substantially. This might be due to the restriction of the reaction equilibrium and difficulties in separating excessive methanol from methyl esters and glycerol, which subsequently lowered the yield of biodiesel [11]. Maximum percent of biodiesel yield were obtained at a 1:40 of seeds-to-methanol ratio.

4 Conclusions

Based on the findings, it can be concluded that temperature is an important property in this *in-situ* process. As the temperature increased, the extraction and FAMEs yields also increased. The loss was caused by thermal decomposition, dehydrogenation and other side reactions. For the effect of pressure, the extraction and FAMEs yield increased with increasing pressure. For the effect of reaction time, it can be seen that the conversion was increased in the reaction time ranges between 5 and 30 min, and thereafter reduced as a representative of the equilibrium conversion. The excess reaction time did not promote the conversion but favors the reverse reaction of transesterification which resulted in a reduction in the ester yield. For the effect of seeds-to-methanol ratio, it can be seen that an increment in seed-to-methanol ratio can enhance biodiesel yield due to higher contact area between methanol and triglycerides. However, when the ratio is beyond 40, the yield of biodiesel begins to decrease substantially.

This research work was funded by the Ministry of Education, Malaysia under the FRGS grant scheme. The authors also thank the management of Universiti Teknologi MARA for supporting the research work.

References

1. W.N.M.W. Ghazali, R. Mamat, H.H. Masjuki, G. Najafi, Renew. Sustainable Energy Rev., **51**, 585 (2015)
2. A. Demirbas, Prog. Energy Combust. Sci., **31**, 466 (2005)
3. H.J. Berchmans, S. Hirata, Bioresour. Technol., **99**, 1716 (2008)
4. E. Fayyazi, B. Ghobadian, G. Najafi, B. Hosseinzadeh, R. Mamat and J. Hosseinzadeh, Ultrason. Sonochem., **26**, 312 (2015)
5. E.C.G. Aguieiras, E.D.C. Oliveira, D.M.G. Freire, Fuel, **159**, 52 (2015)
6. M.R. Avhad, J.M. Marchetti, Renew. Sustainable Energy Rev., **50**, 696 (2015)
7. W. Cao, H. Han, J. Zhang, Fuel, **84**, 347 (2005)
8. G. Madras, C. Kolluru, R. Kumar, Fuel, **83**, 2029 (2004)
9. K. Bunyakiat, S. Makmee, R. Sawangkeaw, Ngamprasertsith, Energ. Fuel, **20**, 812 (2006)
10. H. Han, W. Cao, J. Zhang, Process Biochem., **40**, 3148 (2005)
11. K.T. Tan, K.T. Lee, A.R. Mohamed, Biomass Bioenergy, **33**, 1096 (2009)
12. H. He, T. Wang, S. Zhu, Fuel, **86**, 442 (2007)
13. S. Saka, D. Kusdiana, Fuel, **80**, 225 (2001)
14. T. Eevera, K. Rajendran, S. Saradha, Renewable Energy, **34**, 762 (2009)
15. A. Demirbas, Energy Conversion and Management, **48**, 937 (2007)

Spreadsheet simulation-based optimization for personnel capacity requirement in breakdown maintenance

Tan Hauw Sen[1,*], *Wan Hasrulnizzam* Wan Mahmood[1] , and *Siti Shahara* Dolah[1]

[1]Dept. of Manufacturing Engineering Technology, Faculty of Engineering Technology, Universiti Teknikal Malaysia Melaka, Hang Tuah Jaya, 76100 Durian Tunggal, Melaka, Malaysia

Abstract. This study's objective is to find out the optimum number of personnel in breakdown maintenance activities. Spreadsheet simulation was used for that purpose considered spreadsheet is relatively not an expensive simulation software, familiar to many people, easy to use, yet powerful enough to do the simulation. Input data included in the simulation are time to repair and time between failures which are generated randomly following normal and uniform distribution respectively. The distribution identified by using the histogram and chi-squared test for distribution to find the best-fit distribution. The simulation results show that, the optimum number of personnel obtained when utilization of maintenance personnel is maximized while the number and/or time machines/equipment waiting in queued for maintenance personnel was minimized. In that condition, the cost of breakdown maintenance will be optimum.

1 Introduction

The maintenance function plays an important role in a manufacturing system. The high-performance manufacturing system is achieved not from its operational design only, but also from the capability to maintain the condition of the system [1]. This refers to maintenance function that includes all actions to keep machines/equipment in the operating condition and daily problem solving to bring them back in operating condition after breakdown [2, 3].

Although maintenance function is important to support manufacturing system, it is often considered as a cost centre rather than a competitive advantage [4]. Commonly, the annual cost of maintenance runs between 7–15 % of the investment [3]. U.S industry spent over $300 billion per year for maintenance cost [2]. Chelson et al. [5] discovered that in manufacturing, maintenance cost could consume 2-10 % of the company's revenue. Most of the cost is breakdown maintenance cost due to production stop, productivity loss, miss delivery schedule, etc. Breakdown maintenance can cost 50 % to 200 % more than the

* Corresponding author: tanhauwsr@utem.edu.my

 DOI: 10.1051/matecconf/20179701083

preventive maintenance cost [6]. Therefore, optimum maintenance is achieved when all actions were taken at a minimum cost [7].

The ultimate goal of breakdown maintenance is to make the machine/equipment running again immediately. Ensuring resources, especially maintenance personnel, available when breakdown happened is critical to minimize breakdown maintenance cost. For that reason, Nyman and Levitt [8] suggested a division of work in the maintenance function with one of the group focus to urgent or emergency job which is breakdown maintenance.

It is not easy to determine an exact number of personnel required since the occurrences of breakdown are uncertain and random. Stochastic simulation is the most suitable tool to analyse system with uncertainty in the form of random variables with specific probability distributions [9, 10]. Atlason et al. used Analytic Center Cutting Plane Method (ACCPM) simulation to optimize call centre staffing [11]. Mjema used SIMPLE++ simulation from AESOP GmbH to analyse personnel capacity requirement in the maintenance department [12]. However, simulation can be difficult, complicated, time-consuming and expensive because it requires special knowledge, skill, training and software. Therefore, spreadsheet simulation could be alternative for simulation analysis. Spreadsheet simulation is relatively simple, familiar by many people and relatively low cost. Although Spreadsheet is simple, it provides an almost complete set of tools for simulation [13]. Microsoft Excel is used throughout this study.

The main objective of this study was to find an optimum number of breakdown maintenance personnel that will maximize utilization of manpower and minimize cost caused by production loss due to production stoppage. The specific objectives of this study were to identify the probability distribution of time between failures (TBF) and time to repair (TTR) as the input of simulation in order to find out the optimum number of breakdown maintenance personnel required.

2 Research methodology

Operationally, breakdown maintenance activity is triggered by an event, which is a failure. The nature of this failure is random following a certain probability distribution. Fig. 1. show, when the failure occurs, and there are any available personnel, the breakdown maintenance activity can start immediately. But, if all the personnel busy, the machine/equipment will queue for repaired.

Fig. 2. show, when a repair is completed, the personnel state will change. If there is no machine/equipment waiting to be repaired, personnel status changes to idle; if there is one or more machines/equipment waiting, the personnel will start repairing the next machine/equipment.

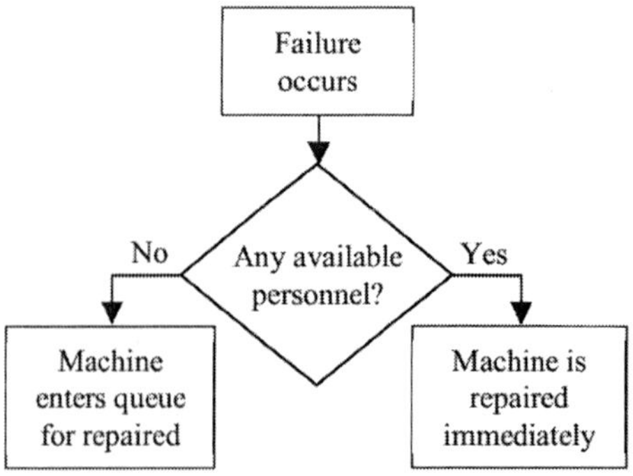

Fig. 1. Failure event flow diagram

DOI: 10.1051/matecconf/20179701083

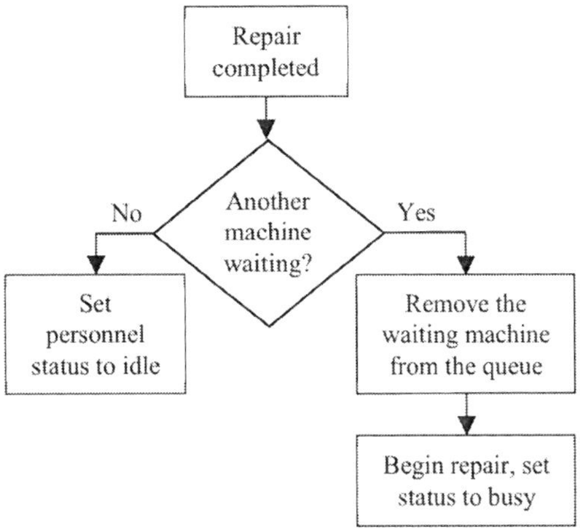

Fig. 2. Personnel state flow diagram

The real-life system used for the study was a Plastic Injection Moulding Department of a manufacturing company. The company belongs to the so-called a fully integrated contract manufacturing service provider of IT-related electronics and mechanical assemblies. The Department had 29 injection moulding machines with different tonnage that operate 3 shifts, 24 hours a day and 7 days a week. During the study, there were six breakdown maintenance personnel per shift who will respond if there was a failure.

TTR and TBF data used as input for simulation to optimize the number breakdown maintenance personnel required for the department. TTR is elapsed time needed to perform a maintenance activity that consists of repairing and replacing. TTR is random, can be a short-time period or long-time period, following certain probability distribution. TBF is a time between the first failure occurred, continued by a maintenance call to the next failure or next maintenance call. In another word, TBF is failure interarrival time. This TBF is also random following certain probability distribution.

The histogram was constructed from TTR and TBF data collected as shown in Fig. 3 and Fig. 4. Based on the figures, hypothesized distribution for TTR was a normal distribution and hypothesized distribution for TBF was a uniform distribution.

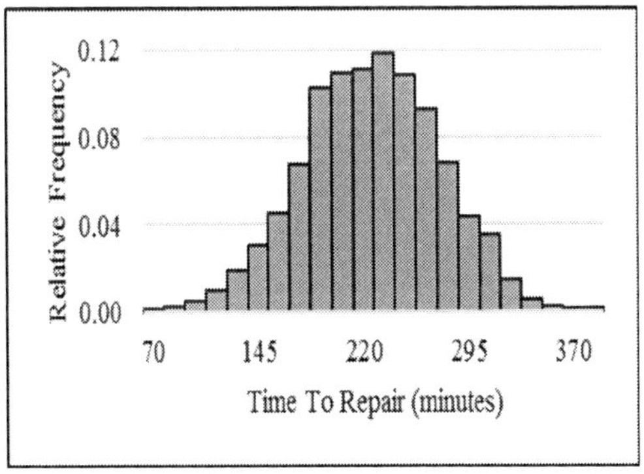

Fig. 3. Histogram of time to repair

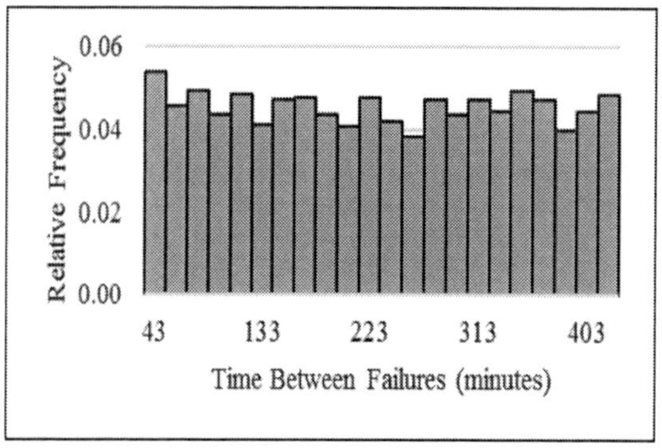

Fig. 4. Histogram of time between failures

The chi-square test for distribution was performed for both data, TTR data and TBF data, for fitting with hypothesized distribution under 0.05 significance level. Table 1 shown the results for TTR & TBF data. As P-value for TTR and TBR were more than the significance level, hypothesized distribution was accepted for both data or in another word TTR data fit normal distribution and TBR data fit to uniform distribution.

Table 1. Chi-square test for distribution of TTR & TBF data with significance level = 0.05

TTR		TBF
0.000857	Square Error	0.000308
14	Degree of Freedom	21
15.9	Test Statistic	9.35
0.332	Corresponding P-value	0.771
Accept	Decision	Accept

3 Results and discussions

The simulation results in this study are focused on three main performance measures, which are:

1. Maximum number of maintenance personnel "busy"
2. Utilization of maintenance personnel
3. Number and time machines/equipment waiting in queued for maintenance personnel

The breakdown maintenance system simulation used TTR and TBF as random input data produced random results. To interpret results accurately, the simulation should be run in several replications and results will be accumulated to get statistical performance results. Twelve replications represent a year of 43200-minutes (1 month) simulation were performed in this study. Three simulation scenarios were performed to find the optimum number of breakdown maintenance personnel.

The first simulation performed for the current condition with six personnel for breakdown maintenance. The maximum number of personnel "busy" during the simulation time shown in Fig. 5 and utilization of maintenance personnel shown in Fig. 6.

The maximum number of personnel "busy" at the same time is five personnel and only happened 1 time. Four personnel "busy" at the same time happened 5 times, three personnel "busy" at the same time happened 32 times, two personnel "busy" at the same time happened 85 times and one personnel "busy" happened 89 times. This brought the utilization of the personnel become low with average 18.09 %. With six breakdown maintenance personnel available, no machine/equipment found waiting for personnel in the queue to be repaired. In summary, it can be said six breakdown maintenance personnel is more than required that may increase the cost of breakdown maintenance in labour cost. From the results, it could be hypothesized only three or two personnel required.

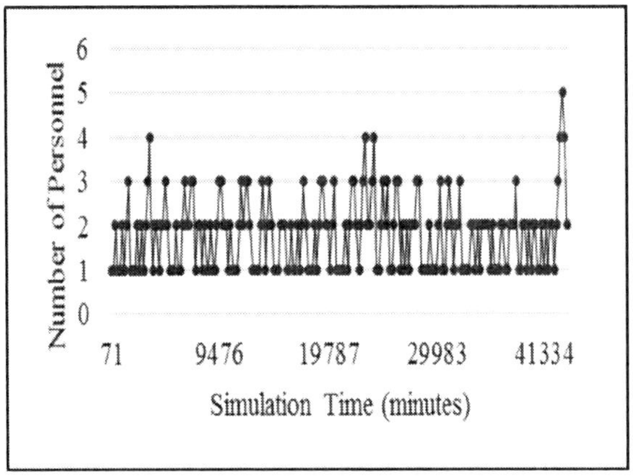

Fig. 5. Number of personnel "busy" over time in 6-personnel scenario simulation

Fig. 6. Personnel utilization over time in 6-personnel scenario simulation

The second simulation carried out for optimization with 3 personnel for breakdown maintenance. The maximum number of personnel "busy" during simulation time shown in Fig. 7 and utilization of maintenance personnel shown in Fig. 8.

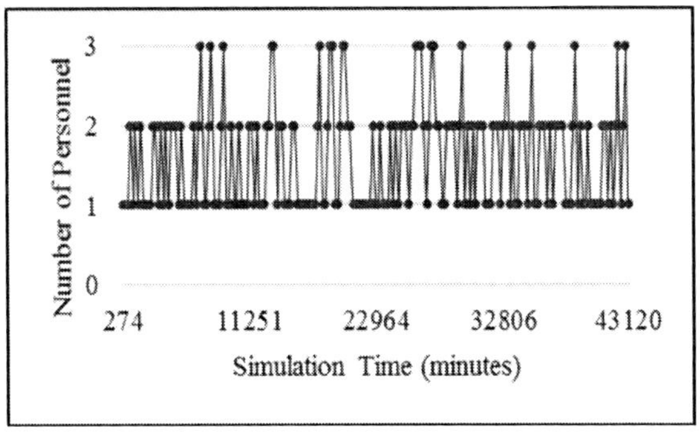

Fig. 7. Number of personnel "busy" over time in 3-personnel scenario simulation

 DOI: 10.1051/matecconf/20179701083

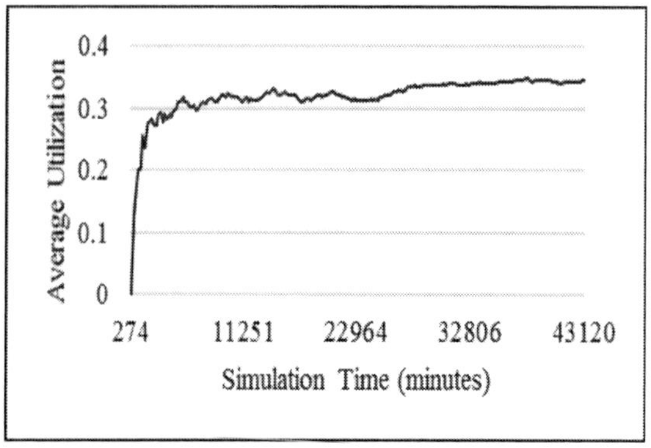

Fig. 8. Personnel utilization over time in 3-personnel scenario simulation

In the 3-personnel scenario, found that three or all of the personnel "busy" at the same time happened 21 times, two personnel "busy" at the same time happened 80 times and 1 personnel "busy" 100 times. The utilization of the personnel increases to 34.60 % in average. The good news was still not found machine waiting for personnel in the queue to be repaired. With three personnel, the labour cost will be less compared to six personnel in breakdown maintenance.

The third simulation carried out for optimization with 2 personnel for breakdown maintenance. The maximum number of personnel "busy" during simulation time shown in Fig. 9 and average utilization of maintenance personnel shown in Fig. 10.

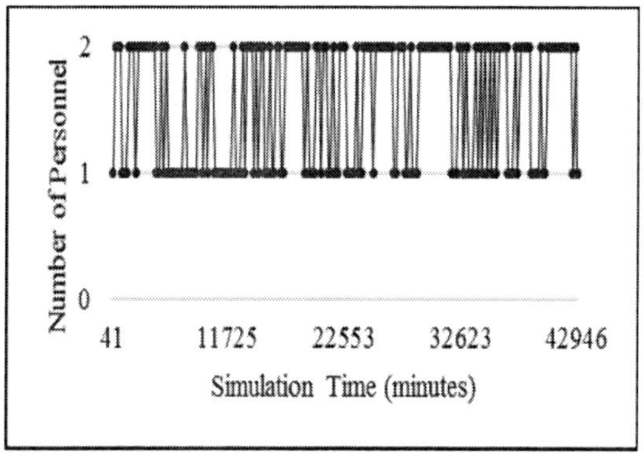

Fig. 9. Number of personnel "busy" over time in 2-personnel scenario simulation

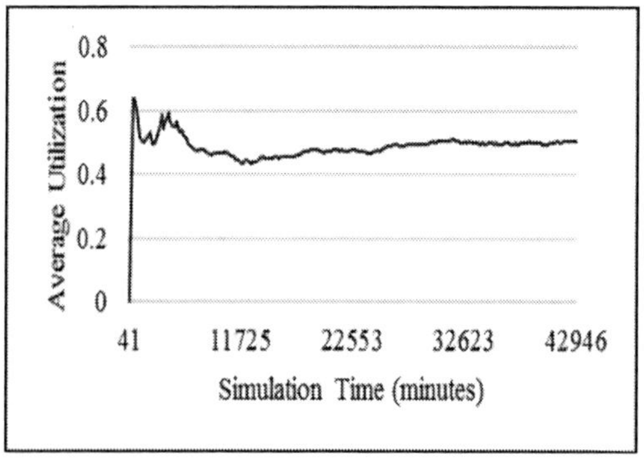

Fig. 10. Personnel utilization over time in 2-personnel scenario simulation

In the 2-personnel scenario, found that two or all of the personnel "busy" at the same time happened 112 times and 1 personnel "busy" 89 times. Now, the utilization of the personnel increase to 50.61 % on average, which means that it was likely there will be always while one technician was repaired, the other personnel available and stand by for next maintenance call-in. But, this resulting machine/equipment waiting for personnel in the queue to be repaired.

There were 31 occurrences where machine/equipment waiting for personnel in the queue to be repaired with averages waiting time 184.58 minutes. There was one occurrence that machine/equipment must wait for 515 minutes for personnel to repair the problem. Therefore, although the labour cost will decrease in the 2-personnel scenario compares to the 3-personnel scenario, the company need to calculate the cost due to production loss and miss scheduled caused by machine/equipment stopped waiting for repaired carefully. Considered number of machines/equipment and the waiting time, it seemed 2-personnel scenario was not preferred.Performance measures results for each simulation presented in Table 2.

Table 2. The 43200-minutes simulation results for breakdown maintenance system

Performance Measure	6-Personnel Simulation	3-Personnel Simulation	2-Personnel Simulation
Number of failures	212	201	201
Average TBF (minutes)	202.66	214.53	213.66
Average TTF (minutes)	219.97	222.66	216.27
Personnel Utilization	0.1809	0.3460	0.5061
Max number personnel "busy"	5 of 6	3 of 3	2 of 2
Number of machines queue	0	0	31
Average waiting time in queue (minutes)	0	0	184.58
Max waiting time in queue (minutes)	0	0	515

4 Conclusions

The main finding from the simulation results is the TTR and TBF are the dominant factors to the number of breakdown maintenance personnel required. Higher TTR will result to more personnel required while higher TBF will result to less personnel required. The efforts to reduce the workforce level in breakdown maintenance should be focused on how to reduce TTR and how to increase TBF. TTR could be reduced by improving the skill of the personnel. TBF could be reduced by improving preventive maintenance and maintenance engineering.

Maintenance cost directly correlated with the number of personnel available for breakdown maintenance. To reduce the cost of labour in breakdown maintenance, the company could reduce the number of personnel. However, the company should analyse the effects carefully since if it is done recklessly, it will cause machine/equipment stop operating waiting for personnel to be repaired. That situation can lead to bigger cost such as production loss and miss scheduled delivery. This spreadsheet simulation is a good tool for that kind of analysis.

This result is only valid when material and tools required for breakdown activity are always available. In addition, the skill of the personnel also assumed same. Further study could include those variables in the simulation to get more accurate results for decision making.

References

1. W.R. Blischke, D.P. Murthy in: Introduction and Overview, *Case Studies in Reliability and Maintenance*, (John Wiley & Son, USA, 2003)
2. B.S. Dhilon: *Engineering Maintenance: A Modern Approach*, (CRC Press, Florida, 2002)
3. R.K. Mobley in: Introduction to the Theory and Practice of Maintenance, *Maintenance Engineering Handbook*, **7**th ed, (McGraw-Hill, USA, 2008)
4. A. Salonen, M. Deleryd in: Cost of Poor Maintenance, JQME, **17(1)**, 63 (2011)
5. A.E. Haroun, S.O. Duffuaa in: Maintenance Organization, *Handbook of Maintenance Management and Engineering*, edited Ben-Daya et. Al., (Springer, London, 2009)
6. J.D. Campbell, A.K. Jardine: *Maintenance Excellence: Optimizing Equipment Life-Cycle Decisions*, (CRC Press, Florida, 2001)
7. A. Kelly: *Managing Maintenance Resources*, (Butterworth-Heinemann, UK, 2006)
8. D. Nyman, J. Levitt: *Maintenance Planning, Coordination and Scheduling*, (Industrial Press, New York, 2010)
9. J. Banks, J.S. Carson II, B.L. Nelson, D.M. Nikol: *Discrete-Event System Simulation*, **5**th ed, (McGraw-Hill, USA, 2010)
10. H.K. Al Fares, S.O. Duffuaa in: Maintenance Forecasting and Capacity Planning, *Handbook of Maintenance Management and Engineering*, edited Ben-Daya et. Al., (Springer, London, 2009)
11. J. Atlason, M.A. Epelman, S.G. Henderson, Management Science, **54(2)**, 295 (2007)
12. E.A.M. JQME, **8(3)**, 253 (2002)
13. J.P. Pecherska, Y. Merkuryev, *Proceedings 19th European Conference on Modelling and Simulation*, (2005)

Isolation and Molecular Characterization of Local Goat Milk Casein for Nutraceutical Value

Mohd Akmal Azhar[1,*], and *Norshafiqa* Salim[2]

[1]Faculty of Engineering Technology, UMP Gambang, Lebuhraya Tun Razak, 26300, Kuantan, Pahang, Malaysia
[2]Politeknik Nilai, Kompleks Pendidikan Bndr Enstek, 71760 Bandar Enstek, Negeri Sembilan,Malaysia

Abstract. Bioactive peptide from casein play a very important role in biological functionalities such as antioxidant and antimicrobial activities. Casein is the main protein that derived from goat milk which consists of alpha (α), beta (β) and kappa (κ) casein. Dietary protein such as casein from animal can provide rich source of bioactive peptide. However, the macromolecular protein such as cow milk can cause allergic response to certain individuals. On the other hand, goat milk have been known for its hypoallergenic and therapeutic properties in human nutrition and health. The purpose of this study is to extract casein from local breed goat milk and identify the molecular characterization of casein for nutraceutical value. The casein was successfully extracted using extraction method. Extraction is a common technique used to separate a desired substance when it is mixed with other components. The average percentage of casein obtained was 24.25%. Then, the casein was analysed by running it in the SDS-Page. The major fraction is β-casein and the minor is α-casein that can be seen between 20kDa and 30kDa respectively. There is no contaminated protein appear in the purified α-amylase. The result obtained in this study indicates that isolated casein from Malaysian goat milk was pure and can be used as bioactive peptide for nutraceutical value.

1 Introduction

Nutraceutical food may provide numerous benefits to the consumers because of the active components of the food. Nowadays, A lot of scientific research indicate health are mostly related to food consumption resulting to increase a concern regarding functional food [1]. Although several plant and animal can be the source of bioactive peptide, however milk protein have become the main source of bioactive peptide for functional food [2]. Bioactive peptide from casein play a very important role in biological functionalities such as antioxidant, antimicrobial, antihypertensive, opioid and mineral carrying activities. Researcher for the last decade has provided that biologically active peptide release from casein contain 3 to 20 amino acid per molecule [3]. The increasing knowledge of this bioactive compound have lead the demand of obtaining of this component in enriched purified form. Casein fraction have been prove to be biologically active and also good source

* Corresponding author: akmalba@ump.edu.my

 DOI: 10.1051/matecconf/20179701084

of different bioactive peptide. It is known to have good nutritive value such as amino acids, calcium, phosphate and several trace elements. Dietary protein such as casein from animal can provide rich source of bioactive peptide. However, the macromolecular protein such as cow milk and whey protein can cause allergic response to certain individuals. On the other hand, goat milk have been known for its hypoallergenic and therapeutic properties in human nutrition and health. The demand of goat milk was increase dramatically worldwide when people start to notice that goat milk are better than bovine milk in term of nutrient, flavour and digestion. Goat milk is vary in term of nutritional properties, composition and biological stability depending on the genetic factor and goat farming practices [4]. The interest of goat milk as a functional food has become the current trend to developed country because the advantages of the nutritional properties compared to cow's milk. The lower allergenicity especially toward children increases the popularity of goat milk as the functional food. The aims of this study are to isolate and purified the protein casein from local breed goat milk which is the Boer and Saanen species. This study also aims to identify the molecular characterization of the isolated casein.

2 Materials and methods

2.1 Sample collection

Goat milk sample was obtained from Tanjung Agrofarm at Nilai, Negeri Sembilan. The milk was brought back to the laboratory of Nilai Polytechnic. The container was placed in a sterile condition and kept in the chiller at 5 °C for further use.

2.2 Preparation of whole casein

The whole casein preparation was done using acid precipitation method [5]. The procedure involves the following steps. The fresh goat milk was centrifuge at 4000rpm for 20 minutes at 20°C. The upper layer is discarded and supernatant skim milk was collected. The pH of supernatant was adjusted to 4.6 with 1M HCl by slow stirring. The mixture was centrifuge for 15 minutes at 4000 rpm. The pellet was washed using ethanol and acetate buffer for three times and analysed by SDS-PAGE electrophoresis.

2.3 Purification of α-casein

The α-casein was purified by urea fractionation method [6]. The procedure involves the following steps. Casein was dissolve in 6.6M urea. The molarity of the urea was reduced to 4.6 by diluted with water slowly and stirred. The solution was centrifuge for 20 minutes at 4000 rpm at 20°C. This pellet was dissolved in 6.6 M urea containing 0.15 M NaCl and diluted to 4.6 M urea. The pellet was separated by centrifugation and analysed by SDS-PAGE electrophoresis.

2.4 Protein molecular weight determination

Prepared proteins were separated on polyacrylamide gel electrophoresis as describe by [7]. The purified caseins were dissolved with 1M NaOH before mixed with an equal volume of sample buffer. After gel polymerization, 20μg proteins were applied to each lane in the gel. The electrophoresis was performed using mini-protein Tetra system with 10% acrylamide gel at 150V for 40 minutes. The proteins were visualized on the gel coomassie blue staining (0.1% coomassie blue brilliant blue in 50% methanol and 10% acetic acid).

DOI: 10.1051/matecconf/20179701084

3 Results and discussions

The average percentage of casein extracted from goat milk was 24.25%. This figure was obtained by calculate the total dry weigh of casein extracted by acid precipitation method. It is agree with the result obtained from [8] where it state that the casein protein extracted from whole milk was around 21-27%. Figure 1 shows the casein extracted from local goat milk which was observed to be white in colour. The casein was in semi-solid form after the drying method was done.

Fig. 1. Isolated casein

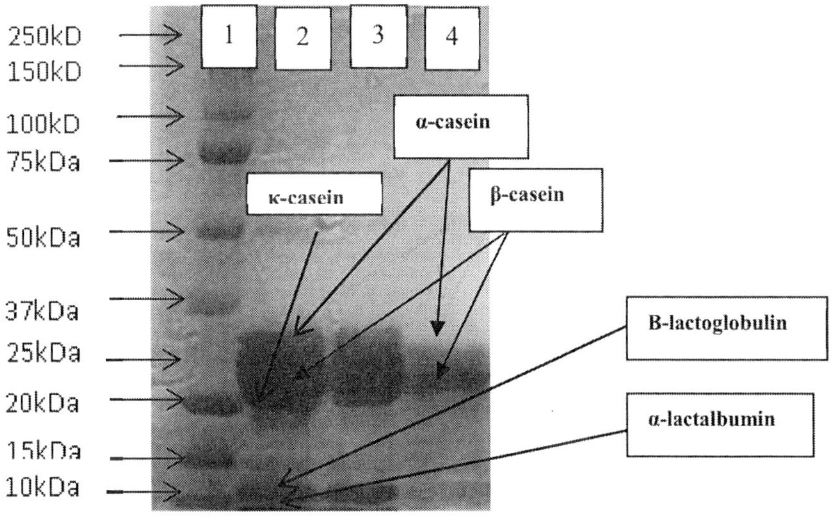

Fig. 2. SDS page of goat milk. Lane 1: Standard protein marker. Lane 2 and 3: Whole goat milk. Lane 4: whole casein protein.

The SDS-PAGE electrophoresis patterns of protein from goat milk are presented in figure 2. Lane 1 shows the whole goat milk that contain three major band which were α-casein, β-casein and κ-casein. Another two minor band were identified as β-lactoglobulin and α-lactalbumin. Lane 2 shows the whole casein extraction. The major fraction is β-casein and the minor is α-casein that can be seen between 20kDa and 30kDa at lane 4. No major band of contaminating cross proteins were observed in whole goat milk as well as whole casein extraction. Both were reasonably pure because only relevant bands were obtained in the SDS-PAGE. However, it was difficult to differentiate β-casein and κ-casein in the whole casein fraction. Although the molecular weight of casein fractions fall in the range 19–25 kDa. This result agrees with other reported data [9], where it state that the dominant fraction of casein is β-casein which is 70.2% while α-casein is the minor which is 29.8%.

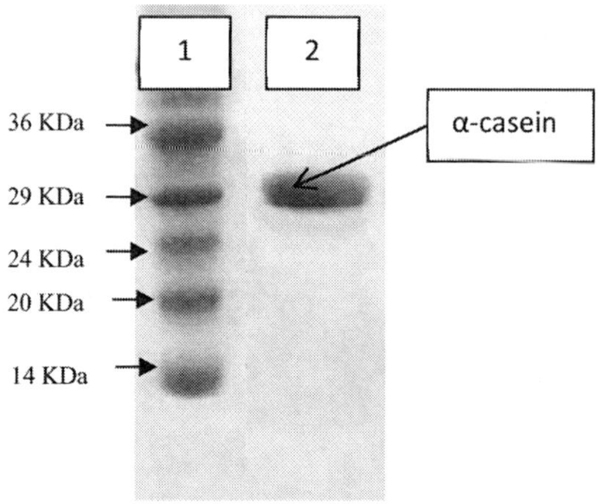

Fig. 3. SDS page of goat milk. Lane 1: Standard protein marker. Lane 2: purified α-casein

Figure 3 shows that one single band without any protein contamination for α-casein after purified using fractionation method. It was stated by [10] that the molecular of α-casein and β-casein were estimated at 30.2kDa and 23.9kDa respectively. This result shown that the isolated casein has the molecular weight at the range of previous study. In general, the protein in milk from different species have different characteristic therefore affected the protein proportion. However, many of them shared approximately the same molecular weight because the similarity amino acid composition.

4 Conclusion

Goat milks protein from local breed goat (Malaysia) was extracted by using acid precipitation method and purified using urea fractionation. The source of whole milk was validate as pure goat milk based on the absence of other contaminated protein and the presence of five major types of protein. The whole casein preparation and purified α-casein can be validate as pure casein and pure α-casein because it was identified to have the dominant proportion which are α- and β- casein in the whole casein while only α-casein band in purified α-casein based on the SDS-Page identification. Hence, the result suggests that isolated casein from Malaysian goat milk was pure and can be used as bioactive peptide for nutraceutical value.

This work was financially supported by grant from UMP (RDU 1403146). The authors are grateful to Tanjung Agrofarm, Nilai for donating the milk samples.

References

1. A.C. Ribeiro, S.D.A. Ribeiro, Small Rumin. Res., **89**, 225 (2010)
2. J. Dziuba, P. Minkiewicz, D. Nalecz, Polish J. Food Nutr. Sci., **849**, 3 (1999)
3. Y.W. Park, Bioactive components in milk and dairy products. (Wiley-Blackwell, USA, 2009)

4. F. Yangilar, J. Food Nutr. Res., **1**, 68 (2013)
5. S. Srinivas, V. Prakash, Int. J. Pept. Res. Ther., **16**, 7 (2009)
6. R. Aschaffenburg, J. Dairy Res., **30**, 259 (1963)
7. U.K. Laemmli, Nature, **227**, 680 (1970)
8. N.K. Kamizake, M.M. Goncalves, C.T.B. Zaia, D.A. Zaia, J. Food Compos. Anal., **16**, 507 (2003)
9. S.A. Salem, E.I. El Agamy, F.A. Salama, N.H. Abo Soliman, Trop. Subtrop. Agroecosystems, **11**, 29 (2009)
10. E.I. El Agamy, Z.I. Abou Shloue, Y.I. Abdel Kader, Third Alexandria Conference on Food Science and Technology, (1997)

Application of Electroporation Technique in Biofuel Processing

Abu Yousuf[1,*], *Maksudur Rahman* Khan[2], *Amirul* Islam[2], *MdMinhaj Uddin* Monir[1], *Zularisam* Ab Wahid[1], and *Domenico* Pirozzi[3]

[1]Faculty of Engineering Technology, University Malaysia Pahang, Malaysia
[2]Faculty of Chemical and Natural Resources Engineering, University Malaysia Pahang, Malaysia
[3]Department of Chemical Engineering, Materials and Industrial Production, University Naples Federico II, Italy

Abstract. Biofuels production is mostly oriented with fermentation process, which requires fermentable sugar as nutrient for microbial growth. Lignocellulosic biomass (LCB) represents the most attractive, low-cost feedstock for biofuel production, it is now arousing great interest. The cellulose that is embedded in the lignin matrix has an insoluble, highly-crystalline structure, so it is difficult to hydrolyze into fermentable sugar or cell protein. On the other hand, microbial lipid has been studying as substitute of plant oils or animal fat to produce biodiesel. It is still a great challenge to extract maximum lipid from microbial cells (yeast, fungi, algae) investing minimum energy.Electroporation (EP) of LCB results a significant increase in cell conductivity and permeability caused due to the application of an external electric field. EP is required to alter the size and structure of the biomass, to reduce the cellulose crystallinity, and increase their porosity as well as chemical composition, so that the hydrolysis of the carbohydrate fraction to monomeric sugars can be achieved rapidly and with greater yields. Furthermore, EP has a great potential to disrupt the microbial cell walls within few seconds to bring out the intracellular materials (lipid) to the solution. Therefore, this study aims to describe the challenges and prospect of application of EP technique in biofuels processing.

1 Introduction

Biofuel (biodiesel, bioethanol, biogas, bio-H_2) is a promising alternative to conventional fossil fuels since it does not have the same environmental impact associated with fossil fuels. Lignocellulosic biomass is considered as a reliable renewable source for the production of biodiesel, bioethanol, biogas and bio-H_2[1-3]. The conversion of lignocellulosic biomass to biofuels requires the following common steps: hydrolysis of cellulose and hemicellulose, sugar fermentation, lignin residue separation and finally recovery and purification of biofuels in order to fulfill fuel specifications . The

*Corresponding Author: ayousufcep@yahoo.com,ayousuf@ump.edu.my

hydrolysis of cellulose and hemicellulose is technically difficult because the digestibility of these two components are hindered by physico-chemical complexities in biomass. These structural complications are the reason why a pretreatment step is necessary for obtaining various chemicals and fermentable sugars during the hydrolysis step. The reason for the implementation of pretreatment is to open up the biomass structure and break down the lignocellulosic bonding in order to promote enzymatic accessibility to cellulose and hemicellulose for hydrolysis[4-6]. Therefore, the pretreatment is a rate limiting step for the conversion of biomass to fermentable sugars, but it has been viewed as one of the most expensive processing steps. The benefits with pretreatment are the great potential for improvement of efficiency of lignocellulosic biomass processing and it can reduce the economic costs as well as facilitate the total biofuel process[4, 6].On the other hand, microbial oils are considered as substitute of plant oils to synthesis biodiesel. Microalgae and yeast are the main source of microbial oils but extraction of oils from those is a challenging step. Existing processes are associated withhigh cost, large volume of solvent and time consuming.

Electroporation (EP) is a microbiologicaltechnique, which involves application of a short burst of high voltage to a sample placed between two electrodes. EP treatment has serious effects on the structure of biological tissues [7]. When a high-intensity, external electric field is applied, a critical electric potential is induced across the cell membrane, which leads to rapid electrical breakdown and local structural changes of the cell membrane and the cell wall[8]. The electric field results in a dramatic increase in mass permeability and, in some cases, mechanical rupture of the plant, animal or microbial tissues. Application of high-intensity electric field pulses from nanoseconds to microseconds in duration leads to the permeabilization of biological membranes. This technology is massively using in food industry and genetic engineering. The fractionating characteristics of EP and its high efficiency can overcome current processing difficulties for biofuel applications [7]. This study describes the pathway of EP application to enhance the biofuel production focusing its limitations and prospects.

2 EP in biogas technology

There is a large potential of biogas production from different kinds of organic residues as well as from different crops, farmland residues or organic solid waste. These residues are abundantly available and is in many cases seen as a problem instead of what they really are, a resource. Currently biogas plays a minor, but steadily growing its role in renewable bioenergy sector. However, the technology for biogas production is still developing and not yet optimized and thus not fully cost-effective. To be fully commercially competitive with other types of fuels, efficiency improvements of the process is urgent. Recently, commercial biogas is producing from agriculture (energy crops) and forest wastes, i.e., from lignocellulosic biomass. Although, the source of lignocellulosic biomass is abundant in almost every country, their contribution in RE sector is insignificant. Lignocellulosic biomass is composed of carbohydrate polymers (cellulose and hemicellulose) and lignin. Typically, LCB contains 35–55% cellulose, 20–40% hemicellulose and 10–25% lignin [2]. The pre-hydrolysis, often referred as pretreatment, is required to alter the structure of lignocellulosic biomass to make cellulose more accessible to the enzymes that convert the carbohydrate polymers to fermentable sugars. Structural change of lignocellulosic biomass is the rate-limiting step for the conversion of lignocellulosic biomass to biogas. This change will reduce the retention time and increase the production rate. EP technique can be applicable to disrupt or deform the structure of cellulosic biomass. The mechanism of EP is illustrated in the Fig. 1., where biomass is placed between two electrodes with a short burst

DOI: 10.1051/matecconf/20179701085

of high voltage. As a result, biomass structure is deformed or destroyed by the high electric field.

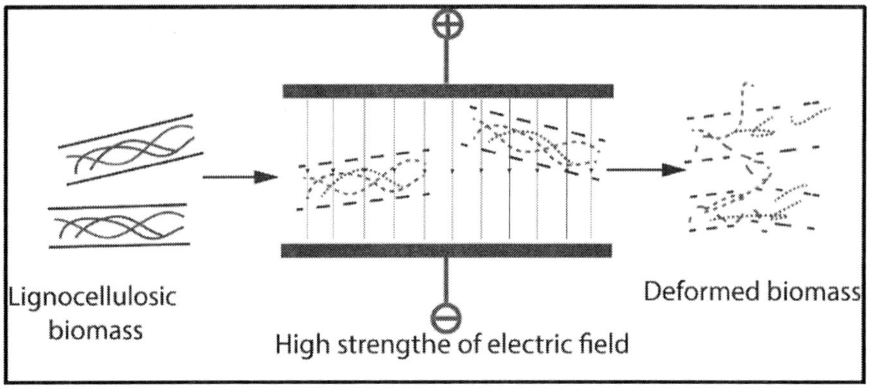

Fig. 1. Conceptual effect of EP on lignocellulosic biomass.

Garoma and Shackelford [9] applied EP as a pretreatment method for algal biomass used as feedstock for anaerobic digestion. They were able to increase bio-CH_4 production as high as 110%. Lindmark [10] studied the effect of EP on lay crop silage to enhance the biogas output and the data from the electroporation experiments show an increased gas production of over 100 % during 36 days of fermentation. Though, EP has shown its potentiality over the biological cell destruction, its application in biofuel sector is still limited.

In recent year, dry fermentation or solid-state fermentation (SSF) has attracted the attention due to its various advantages over wet fermentation, such as it requires smaller volume of digester, no water addition, no mechanically stirred, no dewatering or drying of the effluent. The commercial biogas plant, where single type substrate (energy crop, forest residue) is used, EP technique may be applicable (Fig. 2.) to boost up the biogas production and reduce the retention time and processing cost.

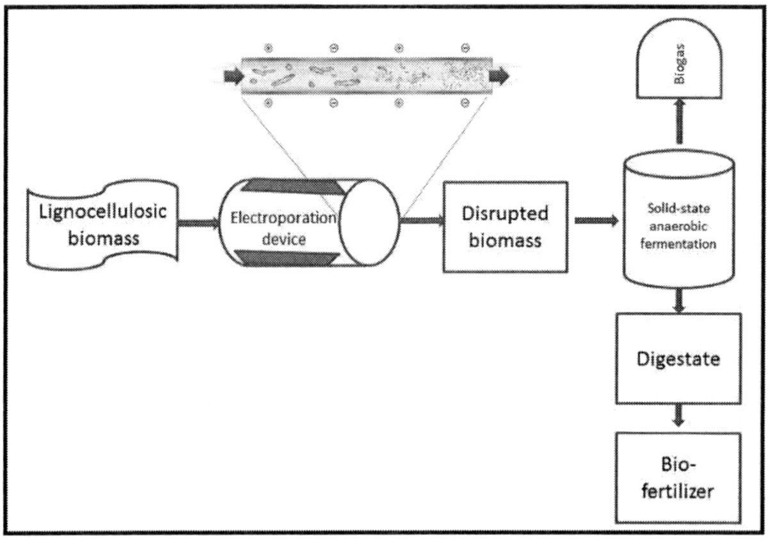

Fig. 2. Simplified process diagram of EP and SSF.

3 EP in biodiesel processing

In recent times, biodiesel production from microbial oils or single cell oil (SCO) has become very widespread. Most attractive candidates for this purpose are microalgae and yeasts. The challenging part of the biodiesel synthesis from SCO is the extraction of microbial oils from intracellular compartments. EP can be applied in the stage of oil/lipid extraction process. In traditional way, microbial biomass is dried, added solvent and applied physical forces (glass bed, ultrasonic sound etc.) to disrupt cell wall to extract lipid. Afterward lipid is separated and carried out esterification to synthesis biodiesel. However, implication of EP can simplify the process offering financial and environmental benefits, since the technique works for nano- to micro-second, it consumes lower amount of energy.

Applying EP, transesterification can be done in situ, also called direct transesterification, where reaction is carried out in a single step (Fig.3). EP technology enhances the transesterification process producing pores in the microbial cells by exposing the substrate to electrical fields. During the treatment, the cells are damaged and releases nutrients that can lead to an increase in the biodiesel production rate.

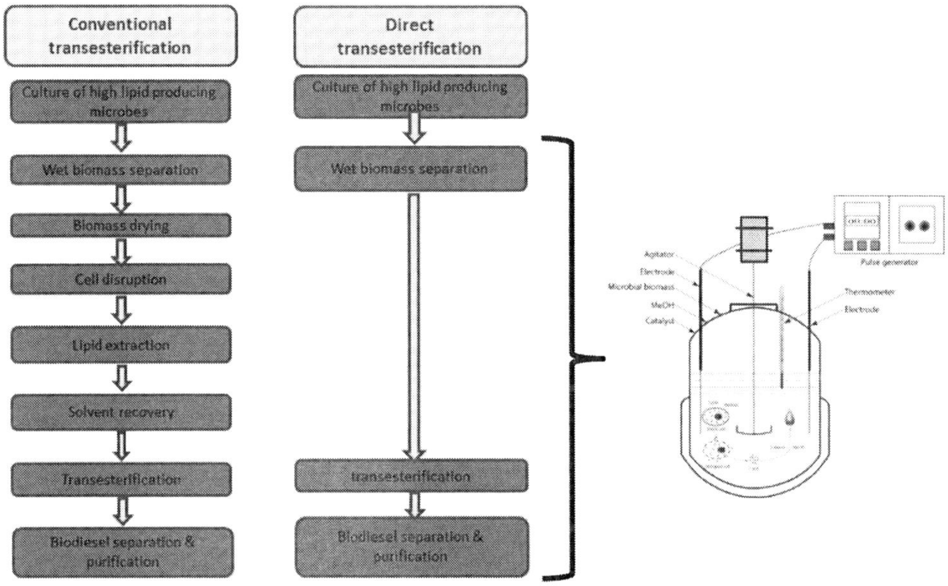

Fig. 3. Application of EP for direct transesterification of microbial lipid to synthesis biodiesel.

A feasibility study of electroporation as a pretreatment method has been conducted by Garoma and Shackelford [9] for algal biomass used as feedstock for anaerobic digestion. Electroporation of cells results in a significant increase in cell conductivity and permeability caused due to the application of an external electric field. They also reported that pretreating algal biomass with electroporation significantly improved the soluble COD (SCOD), increasing it to more than 830% at 28 kWh/m^3 treatment intensity (TI). Addition to TI, culture conditions also influenced the performance of the electroporation process. According to Lindmark [10], the reaction container, electric current, the distance between the electrodes and the number of pulses have also effect of electroporation on the treated material. The cell disruption can be maximized using EP technique by keeping the lowest level of TI and solvent volume [11].

DOI: 10.1051/matecconf/20179701085

4 Conclusion

In summary, integration of electroporation technique in biogas and biodiesel processing has offered promising results for lowering the costs and environmentaleffects of the solvent-based extraction step. The optimal strength of the electrical field depends on the type, size and geometry of the cell tissue. In addition, the intensity of the field, the frequency of the field and the geometry of the pulses probably has an impact on the result.

The University Malaysia Pahang has supported this research under the project RDU1403148.

References

1. N. Ren, W. Guo, B. Liu, G. Cao, J. Ding, Curr. Opin. Biotechnol., **22**, 365 (2011)
2. A. Yousuf, Waste Manage., **32**, 2061 (2012)
3. S. Chaikitkaew, P. Kongjan, S. O Thong, Energy Procedia, **79**, 838 (2015)
4. D. Chiaramonti, M. Prussi, S. Ferrero, L. Oriani, P. Ottonello, P. Torre, F. Cherchi, Biomass Bioenerg., **46**, 25 (2012)
5. P. Kumar, D. M. Barrett, M. J. Delwiche, P. Stroeve, Ind. Eng. Chem. Res, **48**, 3713 (2009)
6. J. Ariunbaatar, A. Panico, G. Esposito, F. Pirozzi, P. N. Lens, Appl. Energy, **123**, 143 (2014)
7. T. Kotnik, W. Frey, M. Sack, S. Haberl Meglic, M. Peterka, D. Miklavcic, Trends Biotechnol., **33**, 480 (2015)
8. R. Joshi, K. Schoenbach, Phys. Rev. E, **62**, 1025 (2000)
9. T. Garoma, T. Shackelford, Bioresource Technol., **169**, 778 (2014)
10. J. Lindmark, *The Biogas Optimization Project*, (Malardalen University, Vasteras, Sweden, 2013)
11. J. Sheng, R. Vannela, B. E. Rittmann, Environ. Sci. Technol, **45**, 3795 (2011)

Improved Sound Absorption Performance of Nonwoven Fabric using Fabric Facing and Air Back Cavity

Ahmad Yusuf Ismail[1,*], *Muhamad Rafi* Omar[1] , *Muhamad Arif* Anuar Hashim[1] , and *Muhamad Ridzuan* Md Radzi[1]

[1]ViBRO Research Group, Centre for Advanced Research on Energy (CARe)
Faculty of Engineering Technology, Universiti Teknikal Malaysia Melaka, Durian Tunggal, 76100, Melaka, Malaysia

Abstract. This paper presents the improvement methods to increase sound absorption performance of polyethylene based nonwoven fabric (PNF). The methods are placing a woven fabric in front of the sample as well as providing air cavity behind the sample. The samples were experimentally tested in an impedance tube based on ISO 10354-2:2001 whereby two microphones are used and the transfer matrix methods are employed. From the results, it can be seen that placing front woven fabric effectively increases sound absorption performance. Moreover, introducing air cavity gap behind the sample is also found to be more significant to increase sound absorption.

1 Introduction

The nonwoven industry has changed and grown rapidly since its beginning decades ago. The growth is including expansion with many other technologies and applications in textile industry [1]. In automotive industry, particularly, nonwoven fabric has been widely applied for interior cabin materials [2 – 3]. This is due to the good acoustical performances of the nonwoven fabric which provides good insulation from noise sources.

Researchers have been investigated the nonwoven fabric particularly for sound insulation applications. Lee and Joo (2003) investigate the sound absorption performance of recycled polyester nonwoven fibrous materials. It is found that the absorption can be significantly changed by controlling the fibre and web properties [4]. Parikh et.al. (2006) fabricate nonwoven fabric from natural waste fibres for automotive floor applications. Kenaf, jute, waste cotton as well as flax are mixed with polypropylene and polyester. It is found that their sound insulation performance is good and relatively better than commercial cotton pad [5]. Tascan and Vaughn (2006) study the effect of total surface area and fabric density of a needle punch nonwoven fabric on the acoustical performance. It is found that both parameters are proportional with the sound insulation obtained [6]. Kucuk and

* Corresponding author: ahmadyusuf.ismail@utem.edu.my

Korkmaz (2012) also study the effect of physical parameters i.e. thickness, weight per unit area, air permeability on the sound absorption performance. By increasing the thickness and reducing the air permeability, sound absorption performance is significantly increased [7]. Liu et.al. (2014) use general regression neural network for predicting sound absorption of sandwich nonwoven fabric. Parametric study is also conducted in the research. It is found that the prediction is reliable and efficient to be used [8]. Tiuc et.al. (2016) creatively improve sound insulation performance of polyurethane (PU) foam by mixing the PU with nonwoven textile waste. The result shows that by certain composition of the nonwoven waste fabric and PU foam, the sound insulation can be improved [9]. Shimizu and Koizumi (2015) conduct a compatibility study of a nonwoven fabric that is installed as gap and slits cover. Although it is thin, but the result shows that the nonwoven cover provides good sound insulation without distracting flow resistivity [10]. Recently, Ghorbani et.al. (2015) investigate the sound insulation performance of polypropylene needle punch nonwoven fabric. It is found that the most influential factor to increase insulation is the web mass. By increasing the mass, the sound insulation performance is improved significantly [11].

From previous researches, one can note that most of the factor that gives significant effect to increase sound insulation is the physical parameters. This paper, therefore, investigates the effect of another physical parameter i.e. air back cavity as well as additional front woven fabric cover purposely to improve the sound absorption performance of nonwoven fabric.

2 Materials and method

2.1 Material preparation

In this research, polyethylene based nonwoven fabric (PNF) as seen in Fig. 1 was chosen and used as a sample due to its wide application in industry [2 – 3]. The thickness of the sample was set to be 1, 2 and 3 cm. While the air cavity gap behind the sample was set to be 0.5, 1 and 1.5 cm. A commercial woven fabric was used to cover the PNF in the experiment.

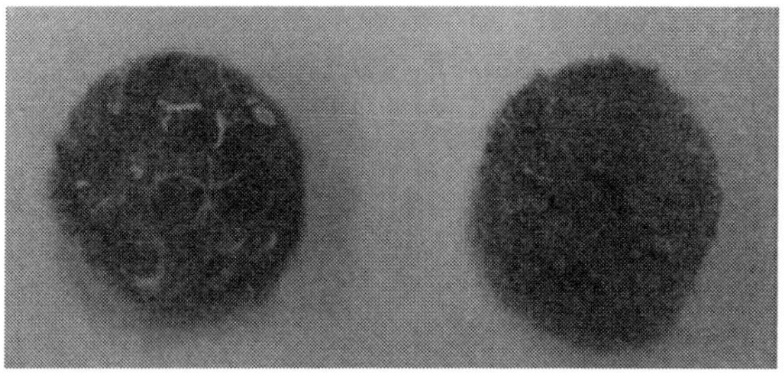

Fig. 1. Nonwoven fabric (PNF).

2.2 Experimental setup

The measurement setup is shown in Fig. 2. This uses the impedance tube method according to ISO 10534-2:2001 [12]. The sample was placed against rigid backing and the loudspeaker at the other tube end. Two acoustic microphones, The G.R.A.S ½ inch free-field type 40AE were located in the front of the sample to save the incident and reflected acoustic signals. ACUPRO software and spectrum analyser was used as the data acquisition system.

The diameter of the tube was 29.4 mm and the tube was capable to provide frequency range between 100 – 5000 Hz. This corresponds to the diameter of the tube as well as the reliability of the software. The recorded signal was processed in the ACUPRO software and all the result was presented in one-third octave band frequency.

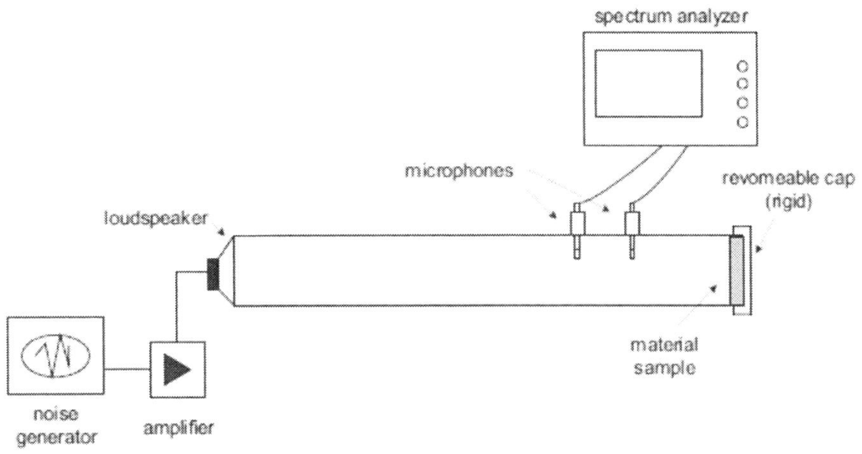

Fig. 2. Experimental Setup.

3 Results and discussion

Fig. 3 shows the initial observation of sound absorption performance from PNF. It is seen that the thickness of the fibre plays a major role in increasing absorption. As the thickness increases, the sound absorption performance is increased as well. This is similar with previous research results [7, 11] whereby the sample thickness increasing the porosity and mass density. This, in turn, makes more absorption of sound energy.

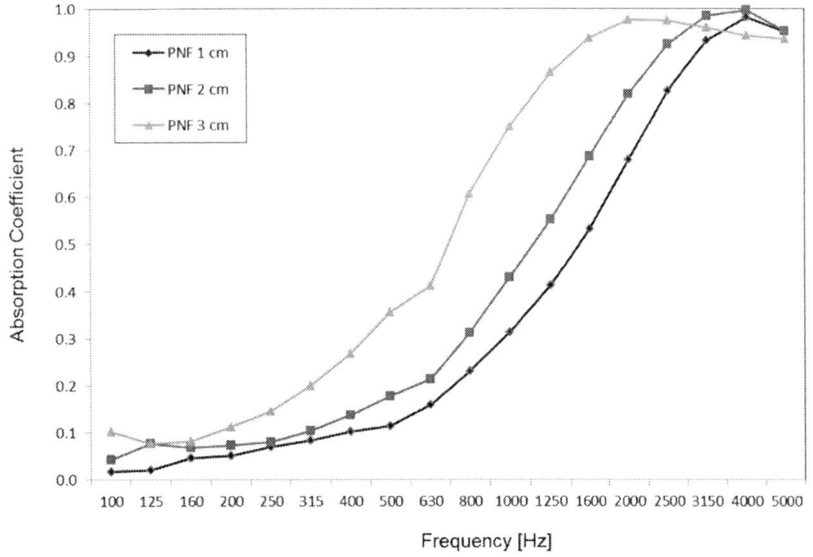

Fig. 3. Sound absorption of PNF in several thicknesses.

While Fig. 4 and 5 shows the effect of placing front woven fabric on the sound absorption performance. It can be seen that even thin woven fabric facing (FF) may

 DOI: 10.1051/matecconf/20179701086

increase the absorption performance. From 250 to 2000 Hz the increment is obtained roughly by 0.05. At around 500 – 650 Hz, the increment achieves its peak for almost reaching 1 increment. This is interesting since the use of front woven fabric often far from discussion. The even fibre matrix from the front woven facing effectively dissipates most of incident sound energy into the PNF where eventually lean the absorption processes in the nonwoven porosity. This dissipation is still effective even if the thickness of the PNF is increased as seen in Fig. 5.

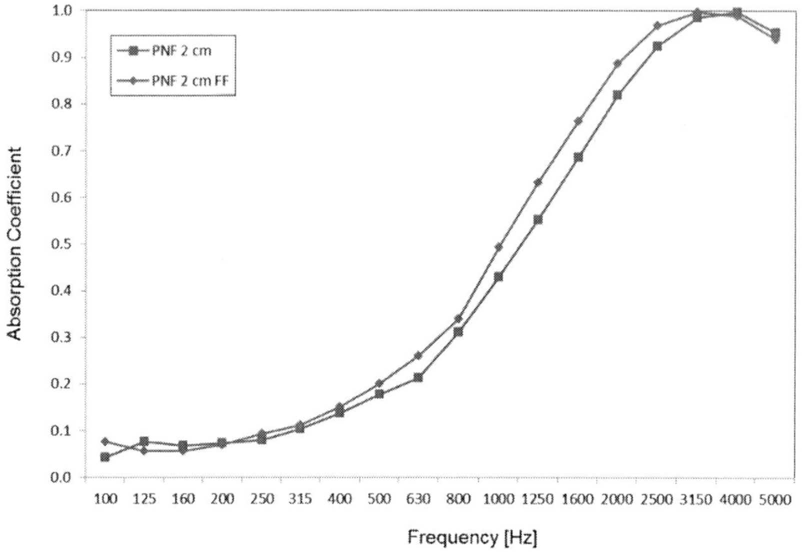

Fig. 4. The effect of front woven fabric facing (FF) on the sound absorption.

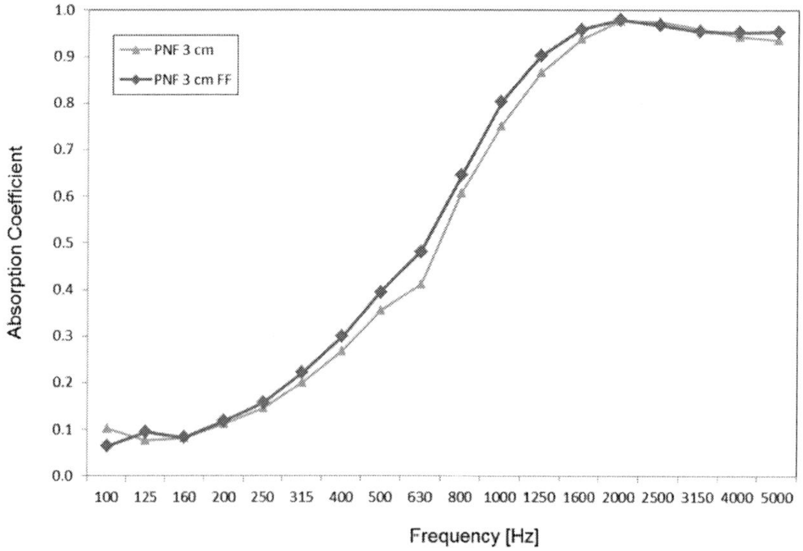

Fig. 5. The effect of front woven fabric facing (FF) on the sound absorption.

Fig. 6 shows the effect of air back cavity on the sound absorption. It is clearly seen that adding air back cavity increases the sound absorption significantly. The bigger the gap, the higher absorption obtained. The increment is obtained almost for all frequency range. It can also be said that the increment of air cavity gap shifts the absorption performance to the

 DOI: 10.1051/matecconf/20179701086

lower frequency range. From the figure as well, it is also seen that the peak absorption performance is shifted to the lower frequency while the cavity gap is increased. This is very useful when the noise source is dominant at low frequency region.

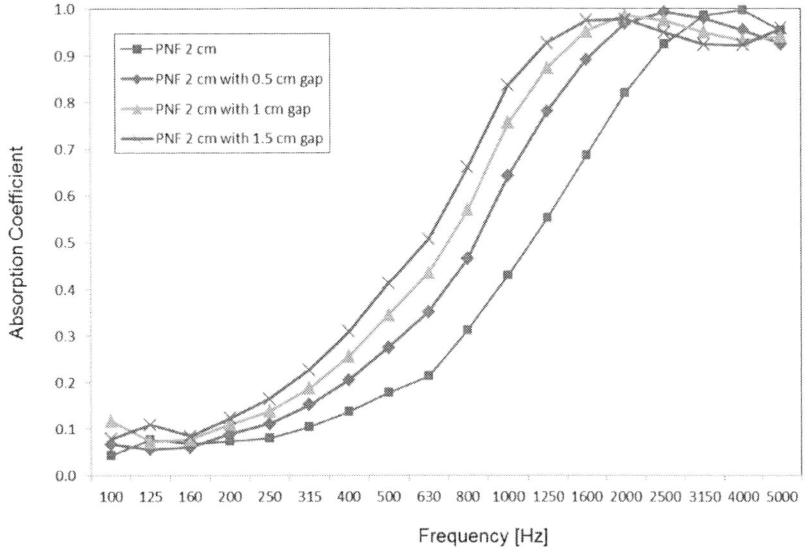

Fig. 6. The effect of air back cavity on the sound absorption.

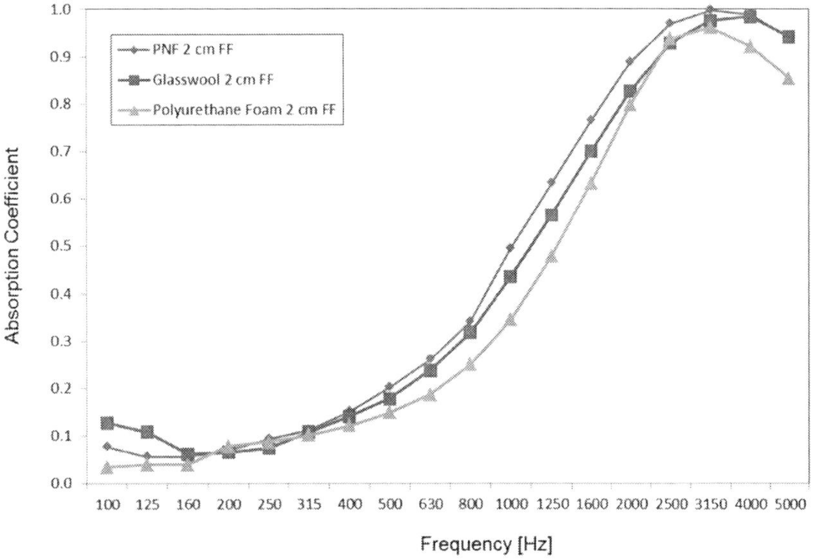

Fig. 7. Comparison between PNF absorption with other commercial acoustic absorbers.

Lastly, Fig. 7 shows a comparison between PNF absorption performances with other acoustic absorbers. With the same thickness, it can be seen that PNF absorber with front woven facing shows its dominance among others materials. The highest sound absorption performance is obtained from 500 – 4000 Hz. The performance difference is even up to 0.1 at around 1300 Hz whereas the effective sound quality range is placed. From this information, it can be concluded that PNF fibres with front woven fabric facing is effective to be employed in any mechanical or even building applications.

 DOI: 10.1051/matecconf/20179701086

4 Summary

The effect of woven fabric facing and air back cavity on the sound absorption performance has been investigated experimentally. It is found that both parameters give significant effect of increasing the acoustical performance. This result can possibly expands the application of nonwoven fabric usage particularly for area that needs air cavity gap or aesthetic look of the sound absorber materials without reducing the sound quality performance for example building, environment and ventilation.

This work was financially supported by Faculty of Engineering Technology, Universiti Teknikal Malaysia Melaka (UTeM).

References

1. INDA, *INDA Nonwoven Glossary*, **69** (2002)
2. N.A. Kalebek, Fib. and Text. in East. Eur., **24 (1),** (2016)
3. D, Gon, P. Paul, K. Das, The Ind. Text. J. 9 (2013)
4. Y. Lee, C. Joo, AUTEX Res. J. **3 (2),** (2003)
5. D.V. Parikh, Y. Chen, L. Sun, Text. Res. J. **76 (11),** (2006)
6. M. Tascan, E.A. Vaughn, Text. Res. J. **78 (4)**, (2008)
7. M. Kucuk, Y. Korkmaz, Text. Res. J. **82 (20)**, (2012)
8. J. Liu, W. Bao, L. Shi, B. Zuo, W. Gao, Appl. Acoust. **76** (2014)
9. A. Tiuc, H. Vermesan, T. Gabor, O. Vasile, En. Proc. **85** (2016)
10. T. Shimizu, S. Koizumi, Build. Env. **94** (2015)
11. K. Ghorbani, H. Hasani, M. Zarrebini, R. Saghafi, Alex. Eng. J. **8** (2015)

Preliminary Study on Non-Fatal Occupational Injury among Operational Workers in Malaysia Palm Oil Mill

Rumaizah Ruslan[1,*], *Ishak* Baba[1] , and *Abdul Mutalib* Leman[1]

[1]Faculty of Engineering Technology, Universiti Tun Hussein Onn Malaysia, 86400 Parit Raja Batu Pahat, Johor, Malaysia

Abstract. Non-fatal occupational injury had becoming major global concern and its consequences to safety and health would be heavily burdening. The aim of this preliminary study was to investigate the distribution of non-fatal occupational injury among specific group of workers in palm oil mill and to acknowledge potential factors of injury causation. A questionnaire survey was designed to assess injury involvement during the employment period among operational workers of palm oil mill located in Southern Peninsular Malaysia. Thirty three (n= 33) workers volunteered and completed the questionnaire. Prevalence of injury among palm oil mill workers was 39.4% with sprain and burn were the common types of injury reported. Press Plant workers reported to have high cases of injuries. Majority of workers (78.8%) stated noise was the main hazard in the palm oil mill, followed by heat hazard. Press Plant was identified as one of the specific risk group in palm oil mill based on the rate of injury occurrences reported by the workers. Exposure to noise hazard was estimated to be one of the potential factors of injury causation and further study should be made to measure the risk of hazard.

1 Introduction

Non-fatal occupational injury is defined by as any condition that resulted in 4 hours or more of restricted activity; for example a person could not perform work or other normal duties, missed work, or a condition that required professional medical treatment [1]. For the purpose of this study, non-fatal occupational injury could be defined as a damaged to human body resulting from working activities, which involve with or without sick leave, ranges from negligible injury to permanent disability among workers of palm oil mill industry. The importance of injuries in low- and middle-income contexts is expected to increase, and by 2020 injury will rank as the second major contributor to disability adjusted life year (DALY) loss [2]. Compared to other health conditions, the physical, psychological and social impacts of injuries are a relatively new area of study. However, initial accounts suggest that the impact of injuries is staggering [2]. Injuries could lead to the loss of the

* Corresponding author: rumaizah@uthm.edu.my

DOI: 10.1051/matecconf/20179701087

individual's future social and economic contribution, and the permanent loss of the usually considerable material and human resources already invested in education and development [3].

In Malaysia, it was noted the highest annual average incidence rate of non-fatal occupational injuries in the agricultural sector (24.1/1000) in the year 2011. The main types of injuries reported were superficial injuries (17%), followed by sprains and strains (10%) and fractures (5%). Falling from heights (31%) was the main cause of accidents followed by being struck by objects (25%). The main accident agent was identified as working environment (45%) [4]. Research on health and safety related to palm oil workers in Malaysia have vigorously been done based on published articles in recent years [5 - 11]. The findings had proven that health and safety issue among palm oil workers is required full attention and commitment. Neglecting health and safety aspects which are included for social wellbeing, may affect the sustainability of palm oil industry. However, lack attention had been paid for occupational injuries issues in local palm oil mill due to narrow references of related research that focusing in milling factory. Hence, this paper was aimed to explore the distribution of nonfatal occupational injuries among palm oil mill workers by estimating the prevalence rate of injury, types of injury based on their workstations, to identify workers' perception of workplace hazards and to determine high risk group of worker based on injury by workstations.

2 Methods

2.1 Subject and study population

This preliminary research had took place in one of the palm oil mill in Southern Peninsular Malaysia, with 33 operational workers were volunteered in assisted questionnaire session. All workers involved in this research were Malaysian.

2.2 Questionnaire

Set of questionnaires were distributed to the respondents for gaining workers' information and potential risk factors, perceived workplace hazards and injury due to hazard exposure at the workplace. The questionnaire session was carried out with the assistance of the mill supervisor for the purpose of gathering information from the respondents. The management personnel of the palm oil mill were adequately informed and were aware of the purpose of the study. They were also made to understand that participation in the study was voluntary and under a strict confidentiality.

2.3 Data analysis

This pilot study was to explore the preliminary results of nonfatal occupational injury among workers in palm oil mills. Hence, statistical analysis involved in this study will only contain descriptive data outcome. This is due to recommendation by other authors in past literature [12, 13] which stated any type of pilot study should be mainly descriptive. Although the outcome from a pilot study provided with relevant information, but hypothesis testing should be interpreted with caution. The two main reason of this are because of the limited state of knowledge about the methods and the smaller proposed sample size [14].

DOI: 10.1051/matecconf/20179701087

In this paper, the preliminary results were accumulated from profiling data of respondent's epidemiological data consists of socio-demographic information, working profile, lifestyle, health status and injuries information. Details of the profiling data used in this study were explained in authors' previous paper [15].

3 Results

3.1 Socio-demographic data and work profile

All respondents who volunteered to participate in this pilot study were Malaysian male operational workers in palm oil mill. There were about 33 feedbacks were gained in form of complete answered questionnaire. For reliability test, these questionnaires were tested with Cronbach's α value of 0.827. The accepted alpha value more than 0.70 reveals that the inter-item for measurement items in a construct is consistent [16-18].

The respondents' age ranges were around 18 to 57 years old. Most of them (69.7%) have received and completed formal education until secondary schools. They are working six (6) days a week from Sunday to Friday with an average of 8 hours of working a day. All of them are reported to have an overtime period of working in the mill, with majority (63.6%) claimed to have more than 5 times overtime in a month. The workstations of the respondents varies from Loading Ramp, Sterilizer, Press Plant, Kernel Plant, Boiler Room, Engine Room, Oil Room, lab workers and grader workers.

3.2 Non-fatal occupational injury information

Non-fatal occupational injury information was obtained through an active survey and included in the research's questionnaire. The information of injuries was adapted from the WHO Injury Surveillance Guideline [19]. Out of 33 respondents, thirteen (13) of them claimed that they had experienced injury while working in the mill. The prevalence rate for injury occurrences among operational workers in palm oil mills was 39.4% as in Table 1. The highest rate of injury occurrences was among workers who work at press plant and other workstations such as Engine Room, Oil Room, lab and grader workers.

Table 1. Number of injured workers based on workstations.

Workstations	Respondents	Injured workers n (%)	Noise hazard reported
Loading ramp	9	0 (0%)	4 (12.1%)
Sterilization	1	0 (0%)	1 (3%)
Press plant	5	5 (15.2%)	4 (12.1%)
Kernel plant	3	1 (3.0%)	3 (9.1%)
Boiler room	2	2 (6.1%)	2 (6.1%)
Others (engine room, oil room, lab and grader workers)	13	5 (15.2%)	12 (36.4%)
Total	33	13 (39.4%)	26 (78.8%)

Average age of injured workers was around 25 years old. The common type of injuries were sprain (30.8%), burn (30.8%) and followed by major cut and scratches (15.4%). Most of workers claimed that injuries were caused by falling (46.2%) and sharp object (30.8). Lower body was reported to be main body part affected due to the injuries (30.8%). Severity and frequency of non-fatal occupational injury were measured based on the respondents' self-report. Majority of them experienced moderate injury (30.3%) and there was no repetition of the same injury at the same year which was indicated by less than one injury occurrences in a year. Table 2 summarized the results.

Table 2. Socio-demographic of injured workers and characteristics of injuries.

Variables	Characteristics	Frequencies (n)	Proportion (%)	Mean ± SD	Range
Age		-	-	25.46	7.253
Work Profile	*Month of employment*	-	-	19	11.762
	Month of employment in the work station	-	-	22	11.080
Type of injury	*Sprain*	4	30.8	-	-
	Burn	4	30.8	-	-
	Major cuts and scratches	2	15.4	-	-
	Minor cuts and scratches	1	7.7	-	-
	Bruise	1	7.7	-	-
	Uncertain	1	7.7	-	-
Mechanism of injury	*Falling*	6	46.2	-	-
	Sharp object	4	30.8	-	-
	Bang/ beaten by object	1	7.7	-	-
	Fire/Fire smoke	1	7.7	-	-
	Others	1	7.7	-	-
Body part effected	*Lower part of body*	4	30.8	-	-
	Upper part of body	3	23.1	-	-
	Head	3	23.1	-	-
	Face	1	7.7	-	-
	In various part of body	1	7.7	-	-
Severity	*First aid*	2	6.1		
	Moderate injury	10	30.3		
	Major injury	1	3.0		
Frequency of injury	*Less than one in a year*	10	30.3		
	Once in a month	1	3.0		
	Once in a week	1	3.0		
	2-3 times a week	1	3.0		

n=13

3.3 Reported self-perceived hazards at the workplace

Hazard at workplace had been determined based on perception of workers while they are working at the mill. Those hazards were rank based on the frequency and the result is shown in the Fig. 1. Majority of workers reported noise (78.8%) was the main hazards in their workplace. The distribution of self-perceived hazards was further analysed by focusing on injured workers. Results as illustrated in Fig. 1 showed that majority of injured workers claimed that they are exposed by noise hazard (92.3%), followed by dust exposure and slippery floor (61.5%, respectively). Majority of workers reported noise (78.8%) was the main hazards in their workplace.

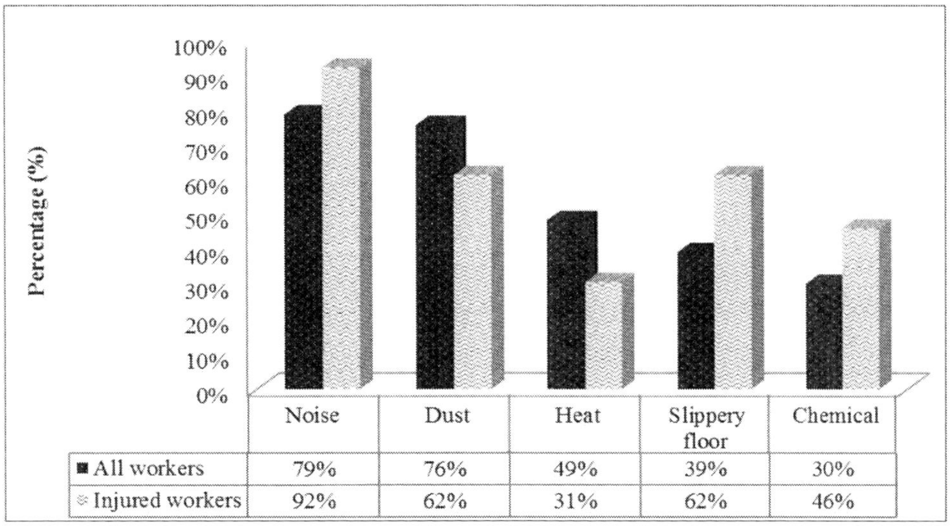

Fig. 1. Distribution of self-perceived hazards at workplace.

3.4 High risk group of nonfatal occupational injuries based on different workstation in palm oil mill

High risk group was determined based on the descriptive distribution of nonfatal occupational injury cases with regards to the respondents' workstation. According to the result as illustrated in Table 1, respondents worked in the press plant and others section had high percentage of injury reported (15.2%). Self-perceived noise hazard for Press plant was 12.1% and for others section was 36.4%. Meanwhile workers of boiler room revealed the same percentage for both parameters of injury (6.1%) and reported noise hazard (6.1%).

4 Discussions

This study had found out that majority of injured workers were around 25 years old. As stated in previous literature review [20], workers who aged under 25 years old is consider as young workers. He added that the majority of studies on nonfatal injuries showed that young workers had a higher injury rate than the overall rate. This result is align with other recent study which stated workers aged less than 30 years old is commonly reported to have injury [21, 22].

Prevalence of injury was 39.4% with workers in Press plant reported high incidents of injuries. The findings estimated out of 100 operational workers in palm oil mill, 39 of them had injured during work. Handling the press and digester machine were among the job tasks that had to be accomplished by Press Plant workers. Their work intensively involved in hand movement and walking. Common injury was determined to be sprain (30.8%), burn (30.8%), major cut and scratches (15.4%). These findings had proven that workers in palm oil mills are heavily engaged with physical movement, walking and working at height, which could cause sprain injury. Palm oil mill workers exposed to heat due to job process in palm oil production where mainly the processes were in contact with hot machinery. The reason had explained the cause of burn injury that commonly reported by the workers in this study. However, the findings contradict with previous research conducted in Malaysia where the main types of injuries reported were superficial injuries (17%), followed by sprains and strains (10%) and fractures (5%) [4]. This might be due to different scope of industry, where this study specifically focused on palm oil mill workers rather than general workers in various agricultural industries.

The determinants of injury are relying on descriptive analysis, which the analysis can be used to generate hypotheses of association between exposures and outcomes. In this study, majority of workers reported to be interfered by noise during work. Moreover, greater figure of injured workers claimed to have exposed to noise compared to other hazards. Thus the direction of future study will be carried out by having an initial hypothesis, which is to measure the impact of noise exposure to occupational injuries. These had been supported by past research that stated risk of workplace injury correlated with noise exposure based on a dose-response relationship [23].

5 Conclusions

Younger workers which aged around 25 years old had a higher risk of non-fatal occupational injury in palm oil mill. Press Plant was identified as one of the risk group in palm oil mill based on the rate of injury occurrences reported by the workers. The common injuries were sprain and burn. Exposure to noise hazard was estimated to be one of the potential factors of injury causation and further study should be made to measure this risk of hazard.

This study was funded by the Office of Research, Innovation, Commercialization and Consultancy Management (ORICC) Universiti Tun Hussein Onn Malaysia under Multidicplinary Research Grant Scheme (Project No: U099). The authors would like to thank the management and respondents of palm oil mill who involved in the study.

References

1. Information on http://www.cdc.gov/niosh/programs/expa/A.V. Niekerk, N. Duncan, ASP, 1-4 (2002)
2. P. Barss, G. Smith, S. Baker, D. Mohan: Injury Prevention: An International Perspective, (Oxford University Press, New York, 1998)
3. L. Abas, A.R.B.M. Said, M.A.B.A. Mohammed, N. Sathiakumar, Int. J. Occup. Environ. Health, **17** (2011)
4. S.G. Normala, K.M. Dileep, S.R. Subrahmanium, Asian Social Science, **10** (2014)
5. Y.G. Ng, S.B.M. Tamrin, W.M. Yik, I.S.M. Yusoff, I. Mori, Industrial Health, **52** (2014)
6. Y.G. Ng, S.B.M. Tamrin, Y.I. Shah, I. Mori, Z. Hashim, J. Occup. Health, **55** (2013)
7. E.H. Sukadarin, B.M. Deros, J.A. Ghani, A.R. Ismail, M.M. Mokhtar, D. Mohamad, Advanced Engineering Forum, **10** (2013)
8. N.D.M. Yusof, K. Karrupiah, S.B.M Tamrin, Adv. Environ. Biol., **8** (2014)
9. H.M. Hadi, S.B.M. Tamrin, K. Karrupiah, Adv. Environ. Biol., **8** (2014)
10. R.L. Naeini, S.B.M. Tamrin, Asian J. Med. Pharm. Res., **4** (2014)
11. A. Bauhofer, B. Stinner, U. Plaul, A. Torossian, M. Middeke, I. Celik, M. Koller, W. Lorenz, Br. J. Surg., **88** (2001)
12. G. Lancaster, S. Dodd, P. Williamson, J. Eval. Clin. Pract.,**10** (2004)
13. A. Leon, L. Davis, H. Kraemer, J. Psychiatr. Res., **45** (2011)
14. R. Ruslan, I. Baba, A. M. Leman, T. L. Wai, ARPN Journal of Engineering and Applied Sciences, **11** (2016)

15. M. F. Ahmad: Basic Statistical Analysis: Step by Step Using SPSS (Penerbit UTHM, Johor, 2016)
16. J.F. Hair: Multivariate Data Analysis (Pearson Prentice Hall, New York, 2010)
17. J.C. Nunnally: Psychometric Theory (McGraw-Hill, New York, 1978)
18. Y. Holder, M. Peden, E. Krug, J. Lund, G. Gururaj, O. Kobusingye: Injury Surveillance Guideline (World Health Organization, Geneva, 2001)
19. S. Salminen, J. Safety Res., **35** (2004)
20. S.G. Kumar, A. Dharanipriya, Indian J. Occup Environ Med, **18** (2014)
21. S.G. Kumar, U. Rathnakar, H. Harsha Kumar, Indian J. Community Med., **35** (2010)
22. J.H. Yoon, J.S. Hong, J. Roh, C.N. Kim, J.U. Won , Noise Health, **17** (2015)

Notched Strength of Woven Fabric Kenaf Composites with Different Fiber Orientations

Hilton Ahmad[1,*], and *Safwan* Rahim [1]

[1]Faculty of Civil and Environmental Engineering, Universiti Tun Hussein Onn Malaysia, 86400 Parit Raja, Batu Pahat, Johor, Malaysia.

Abstract. The awareness of implementing sustainable materials in construction industry is gaining good attention among engineers worldwide. Kenaf fibers are local renewable materials to combine with epoxy polymers matrix in producing lightweight composite materials which may replace imported synthetic fiber composites especially in developing countries. Other benefits of using kenaf fiber composites are relatively cheap, less abrasive and less hazardous during fabrication handling. Current study investigates parametric study on notched strength of woven fabric kenaf composite plates with different fiber orientations and circular hole sizes. Stress concentration occurred at the notch edge promotes to micro-damage events such as matrix cracking and fiber fracture as remote tensile loading applied. Current study showed that 0° fiber orientation gives optimum notched strength, plates with larger fiber tilting than 0° are associated with further strength reduction. Kenaf fibers give an alternative to material designers to opt woven fabric kenaf composites in low and medium load bearing applications.

1 Introduction

Materials engineers and scientists had started to use natural fibres in composite materials fabrications due to comparable specific strength (and stiffness) to synthetic fibre composites (e.g. GFRP). Kenaf fibre has excellent strength and modulus properties [1], thus potentially used as fibre reinforcement in composite system. Large amount of research works were reported to improve the mechanical properties of this materials (such as adding composite fillers) to demonstrate the applicability of kenaf composites system as an advanced engineering materials [1-2]. The drawback of hydrophilic nature of kenaf fibres may provide poor adhesion with polymer, but less significant effect were exhibited when combined with thermo-set polymer than thermoplastic polymer counterparts [2].

Woven fabric fibre composites are another class of reinforcing fibre system in composite materials. It has lower strength than unidirectional fibre composites due to existence of crimping region. However, woven fabric composite system has good drape-ability for manufacturing processes, superior resistance against fatigue and impact loading. However, due to large arrays of woven fabric composite combinations (lay-up types,

* Corresponding author: hilton@uthm.edu.my

weaving architectures, fiber orientations, etc.), different failure mechanisms may exhibit in different fibre reinforcing/binders combination systems.

Most structures assembly are joined by bolted joints, which requires introduction of notch in composite plates. It is important to study the behaviour of the notch plate prior to implementation of composite materials in bolted joints system. Due to discontinuity of plate geometry, stress raisers are exhibited at the vicinity of the notch edge. The critical length from the notch edge showed complex damage mechanisms, comprised of matrix cracking, splitting, delamination and ultimately 0° fibre fracture. The critical length is associated with stable self-similar fashion damage prior to catastrophic failures along net-tension plane, and widely used as strength predictive parameter in notched composite plates [3-7]. Experimental works on woven fabric GFRP and CFRP system is slightly different in terms of critical length exhibited in both systems, however, it is suggested that the failure mechanisms showed similar damage fashion throughout all layers in a composite plate. This work can be further expanded to study structures response subjected to other stress raisers such as welded joints [8] and floor connections [9].

The behaviour of natural fibres composite system with notched plate is relatively very few as compared to synthetic fibre composites, and effect of fibre orientations on notched strength is remains vague. Current project aim to study effect of fiber orientations and hole size in woven fabric kenaf composite plates in more precise manner.

2 Experimental frameworks

Kenaf fiber yarns used in current project are processed fibers through drying stage, fiber grading stage (to conform uniformity) and formation of fiber yarn or string (by fibre twisting). Kenaf fiber threads available are in various diameter sizes, ranging between 0.75 mm to 3 mm diameter size. Current experimental work used kenaf fiber yarn with a diameter size of 0.75 mm, weaved using handloom with extra precision. Extra care should be taken during hand lay-up weaving process as kenaf thread are sometimes easy to break apart (due to non-uniformity of kenaf yarn during yarn formation stage) to in-plane layer size of 250 mm x 250 mm.

Composite material strength is attributed to fiber-matrix adhesion. The poor adhesion can be avoided if the natural fibers are pre-treated beforehand prior to composite fabrications. However, it was reported that hydrophilic nature in kenaf composites is negligible by combining with thermoset polymers (e.g. epoxy resin). Current work use epoxy resin as matrix binder in composite fabrication stage. Amount of woven fabric kenaf layers and matrix binders (i.e., epoxy resin and hardener) are calculated based on materials weight.

Table 1. Testing series investigated in current work.

Weave type	Lay-up	Designation	Estimation thickness (mm)	No. of testing coupons (hole size)		
				2.5 mm	5 mm	10 mm
Plain weave (PW)	0°	X0	0.7	3	3	3
	15°	X15	0.7	3	3	3
	30°	X30	0.7	3	3	3
	45°	X45	0.7	3	3	3

The testing series involves various fibre orientations and notch sizes, three testing coupons are prepared for each testing series and notch sizes respectively as shown in Table 1. Material preparations and composite fabrication works were carried out in Textiles gauge length and plate width of 25 mm with notch sizes (i.e., hole diameters), *d*, of 2.5 mm, 5.0 mm, and 10.0 mm giving *d/W* ratios of 0.1, 0.2 and 0.4 respectively. End tab is prepared for surface treatment to get good adhesion with aluminium strip to provide better gripping and to avoid testing coupon slippage during mechanical testing.

3 Results and discussions

Notched strength of testing coupons are calculated by using notched stress at failure expression by dividing peak load, P_{max} with its cross-section (i.e., a product of plate thickness,t and plate width, *W*) or can be mathematically expressed as, $\sigma_N = P_{max}/Wt$. There are four testing series as stipulated in Table 1 of different plate orientations and hole sizes. A number of orthogonal tows crossing over are given as "n_g" which larger "n_g" demonstrates higher degree of drapeability that is easier to form into complex shapes but reduced stability during manufacturing. However smaller "n_g" improves impact resistance and tighter nesting for energy absorption. Current work used plain weave system (n_g=2) and due to more crimping region, these fabric system may exhibited less strength as compared to other fabric systems with less crimping region such as harness satin (n_g =8) or twill satin (n_g=3). However, this fabric system is favorable due its stability during weaving process.

The failure modes of all testing coupons subjected to remote tensile testing was net-section failures associated with stress concentration. Net-section failure occurred at net-tension plane, where cracks were initiated and propagated along net-tension plane. It was observed that crack was initiated at hole edge perpendicularly to applied load and progressively propagated to plate edge. Damage associated within damage zone comprised of not only limited to matrix cracking and splitting, but also fiber tow fracture. From Fig. 1, initially the notched plate of woven fabric kenaf composite displayed a linear elastic behavior until sudden fracture and brittle failure type were demonstrated. It is assumed that plastic deformation in tensile testing of notched plate is negligible, as found in other composite system (Belmonte [4, 5] and Manger [6]). Experimental observation found that a screeching sound can be heard after few minutes of testing machine started, associated with matrix cracking and fiber-matrix breakage occurred at notch edge. Thereafter, fiber fracture was taken place to indicate the reaching of peak load prior to catastrophic failures.

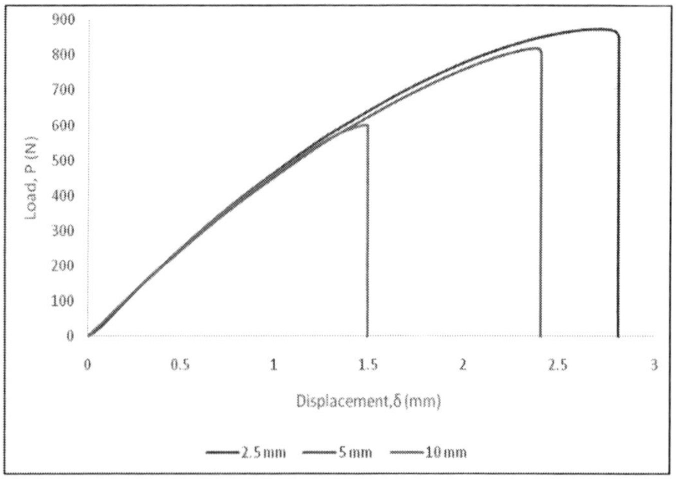

Fig. 1. Graph Load against Displacement of 0° orientation notch for 2.5, 5 and 10 mm notch size.

As remote tensile load is applied at the plate edge, crack initiation was created at net-section plane perpendicular to the applied remote loading. As the crack size reaching a critical length (associated with process zone length), catastrophic failure occured along net-tension plane. At the process zone length, three types of micro damage events are expected to occur, i.e matrix cracking, delamination and fiber fracture at the vicinity of the hole edge. The size of damage zone length is approximately a size of hole radius as reported by previous researchers (Belmonte [4, 5] and Manger [6]). The notched strength decreased as hole size increased due larger remaining unloaded plate width also associated with smaller area of damage zone length developed at the hole vicinity prior to catastrophic failures as reported by Belmonte [5].

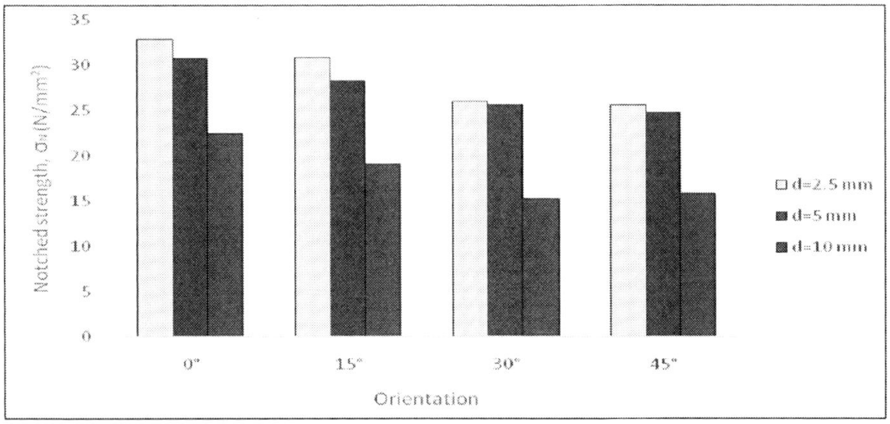

Fig. 2. Graph of notched stress at failures of various fibre orientations

Fig. 2 shows notched strength of all testing series with different fibre orientations. As expected, increased notched strength were observed with testing coupon of fiber orientation close to 0° remote loading direction. On the other hand, decreased notched strength was seen in coupons with fiber orientation tilting further from 0° fibre direction. Notice that the highest notched strength were given with coupon X0 with circular hole size of 2.5 mm to give notched stress at failure about 33 N/mm² and decreased accordingly as fiber orientation gets more inclination from 0° fiber orientation. These trend is shown in fiber orientation of 15°, 30° and 45° to give 30.84 N/mm², 26 N/mm², and 25.64 N/mm² respectively. Similar trends were demonstrated to the other hole sizes. This is expected as fiber orientations are tilting further than remote applied loading direction, effective fiber direction to withstand ultimate loading capacity is further decreased. Bear in mind that a woven composite layer is equivalent to two layers of unidirection plies, therefore 0° woven fabric layers has an orientation of (0/90) ply.

The effect of hole sizes to notched strength were also been carried out. All testing composite coupons was drilled using high speed drilling jig to produce a circular hole of different hole size (i.e., 2.5 mm, 5 mm, and 10 mm) for each fiber orientations plates understudied. The largest value of maximum load, P_{max} was given by 2.5 mm hole size coupons followed by 5 mm and 10 mm hole size respectively. The strength of 0° fibre orientation is given as 32.9 N/mm² for 2.5 mm circular hole size followed by 30.7 N/mm² in 5 mm hole diameter and 22.51 N/mm² in 10 mm hole diameter. Increased notched strength in smaller hole size testing coupons are expected due to larger unloaded region from the vicinity of the notch as fracture energy (or strain energy) released is exceeded. It is also associated with the formation of damage zone length which formed perpendicularly to damage formation at the hole vicinity can help to explain lower strength of testing coupons with larger hole sizes. Belmonte (2004) conducted interrupted test (at 90% of peak load) to confirm the occurrence of matrix cracking, delamination and numbers of tow fracture under scanning electron microscope (SEM) investigations. However, interrupted test and determination of damage zone length is outside the scope of the current work.

4 Conclusions

Experimental observations showed that crack initiation was taken place at the vicinity of the hole edge and propagated to the plate edge along net-tension plane. The micro-damage formations are localized within damage zone length adjacent to the notch edge as results of matrix cracking, delamination and fiber fracture. It was found that 0° fibre orientation and smaller hole size give optimum notched strength. 0° fibre orientation is the direction of remote applied loading, therefore these direction is effective to resist applied load. Notched strength decreased gradually with larger hole size due to larger unloaded region width prior to catastrophic failures.

The author would like to thank Universiti Tun Hussein Onn Malaysia (Vot U245) to support this project work.

References

1. H.M. Akil, M.F. Omar, A. Mazuki, S. Safiee, Z. Ishak, A. Abu Bakar, Mater. Des., **32** (2011)
2. O. Faruk, A.K. Bledzki, H.P. Fink, M. Sain, Prog. Polym. Sci., **37** (2012)
3. H. Ahmad, A.D. Crocombe, P.A. Smith, Adv. Mat. Res., **845** (2014)
4. H.M.S. Belmonte, C.I.C. Manger, S.L. Ogin, P.A. Smith, R. Lewin, Comp. Sci. Tech., **61** (2001)
5. H.M.S. Belmonte, S.L. Ogin, P.A. Smith, R. Lewin, Comp. Pt. A, **35** (2004)
6. C.I.C. Manger, *Failure of Notched Woven GFRP Composites: Damage Analysis and Strength Modelling.* PhD Thesis, (Uni. of Surrey, 1999)
7. H. Ahmad, A.D. Crocombe, P.A. Smith, Plas. Rub. Comp., **42(3)**, (2013)
8. N. Abd Rahman, W. Tizani, Adv. Mat. Res. 1025, (2014)
9. N. Abd Rahman, Z. Jaini, N.A. Abd Rahim, S.A. Razak, InCEIC2014 (2014)

Roof Box Shape Streamline Adaptation and the Impact towards Fuel Consumption

M.F. Abdul Latif[1,*], *M.N.* Hashim [1], *M.Z.*A Rashid [1], *M.A.* Azhari [1], and *M.N.* Othman[1]

[1]Faculty of Engineering Technology, Universiti Teknikal Malaysia Melaka, Hang Tuah Jaya 76100, Malaysia.

Abstract. The fuel price hike is currently a sensational national issue in Malaysia. Since the rationalization of fuel subsidies many were affected especially the middle income family. Vehicle aerodynamic were directly related to the fuel consumption, were extra frontal area result a higher drag force hence higher fuel consumption. Roof box were among the largest contributor to the extra drag, thus the roof box shape rationalization were prominent to reduce the extra drag. The idea of adopting water drop shape to the roof box design shows prominent result. The roof box has been simulated using MIRA virtual wind tunnel modelling via commercial computational fluid dynamic (CFD) package. This streamline shape drastically reduce the drag force by 34% resulting to a 1.7% fuel saving compare to the conventional boxy roof box. This is an effort to reduce the carbon foot print for a sustainable green world.

1 Introduction

The effort of drag reduction mainly focused on the design of the vehicle where designer added an aerodynamic flavour to their car design right after the First World War [1]. However the aerodynamic flavour to the car design were never a successful story due to the design limitation and the sole purpose of the vehicle design itself as a transportation device for human where space were a big issue [2]. For a bluff body vehicle segment having a streamline body were just a non-desirable idea. Most of them were boxy and bulky. The main issue of the bluff body vehicle like MPV was the lag of storage space and many end up having an extra boxy storage box on top of their vehicle.

The box definitely add up extra surface area that face the wind direction and directly affected the total drag of the vehicle which likely contribute to an extra fuel consumption [3]. Recently, designer have shown an increased interest in drag reduction effort to the box design. Somehow an idea of adopting streamlined shape like water drop shape to the roof box design were less explored. The streamlined shape to a car design has been adopted during the post war era. During that era most German aeronautical engineer were forbidden designing an airplane so most end up being a car designer [4]. As their passion on aeronautic most designer like Paul Jarray add up the aeronautical flavor to his car design

* Corresponding author: faruq@utem.edu.my

 DOI: 10.1051/matecconf/20179701089

where the main focus was to have the lowest air drag as possible [5]. During that era fuel consumption was a big issue due to the gloomy postwar economy. The most famous Jarray design was the zeppelin airship.

The design was simply based on the most aerodynamic basic shape of an air foil where the drag coefficient were about 0.04. The design was a success in the perspective of aerodynamic but in the perspective of consumer Jarray design has several set back. Base on the fact that the water drop shape was the most aerodynamic shape this research were carried out to explore the potential of such concepts [6].Recent developments in the field of aerodynamic have led to a renewed interest in automotive drag reduction design. Most studies in roof box aerodynamic have only been carried out in a small number of focus areas. Several studies have produced estimates of roof box contribution to the total aerodynamic drag but there is still insufficient data for roof box shape optimization [7]. A. Howell has depicted the automotive accessory contribution to the total drag of the vehicle where the roof box add up around 0.15 to 0.175 [3]. A typical multi-purpose vehicle has a drag coefficient around 0.45, installation of roof box increase the total drag from 25% to 28%. These figure contribute to 2.27% of total fuel consumption calculated using the formula adapted by Zambri Harun [8]. Elaborating the 2.27% increase of total fuel consumption by taking the average fuel price in Malaysia between year 2013 to 2016 which is about Malaysian Ringgit (RM) 2.00 per litter of fuel; A 2.0 litter engine MPV consume about 50 L of fuel a week for average daily drive of 20 km. The MPV consume about 2400 L per year. Installing the roof box add up about RM 109 to the total of one year of fuel cost. There are variety choices of roof box design available in the market depending on the storage requirement. This paper focused on the commonly used roof box design in Malaysia which is for luggage storage.

2 Roof box modelling

The commonly used roof box model in Malaysia were selected as the reference model in this study. In this case model Thule 667ES Excursion were referred. The 3D model were modelled using CATIA V5 software based on the measurement made on the actual model as displayed in Fig. 1 below. This model has the capacity of 350 L. This 3D model were built in single piece where upper cover and the lower tray were merge to simplify the CAD data. It is assumed to be fully solid with no wall thickness as this study only focus on the external aerodynamic.

Fig. 1. Common roof box 3D model developed using CATIA V5.

DOI: 10.1051/matecconf/20179701089

Table 1. Measured Drag Coefficient.

Wind direction →									
Shape 	Sphere	Half-sphere	Cone	Cube	Angled Cube	Long Cylinder	Short Cylinder	Streamlined Body	Streamlined Half-body
Drag Coefficient	0.47	0.42	0.5	1.05	0.8	0.82	1.15	0.04	0.09

This paper focus on the idea of streamlined shape adaptation and its impact to the total aerodynamic drag. Table 1 exhibit the drag coefficient of various shape. It is easy to identify that the streamlined body has the lowest drag coefficient. Based on this fact the author has adapted the shape onto the roof box design. The details of the design adaptation as illustrated in Fig. 2.

Fig. 2. Streamline roof box 3D model developed using CATIA V5.

3 Pre-processing Configuration

MIRA wind tunnel model of an open back with a rectangular sectional dimension of the test chamber as displayed in Fig. 3 were use in this study as it produce excellence result with optimum calculation time [9]. The meshing setting was set as proposed by Gmbh [10]. The details of the meshing configuration as exhibit in Table 2. The setup chosen is Pressure-Based to monitor pressure. The time chosen is steady state which the iteration will be captured after the flow of air is steady. The details of setup configuration as exhibit in Table 3.

Table 2. Meshing Configuration.

Criteria	Setup
Tunnel Mesh size	Course
Smoothing	Medium
Mesh Extreme size	250mm
Base size	1mm
Growth rate	1.2
Surface mesh size	10mm
Refinement zone size	15mm
Inflation	1.2

Table 3. Setup Configuration.

Criteria	Setup
Velocity formulation	absolute
Turbulence model	Realizable k-epsilon
Wall function	non-equilibrium [11]
Backflow turbulent intensity	5%
Backflow viscosity ratio	10% [12]
Air density	1.225
Temperature	302.15 K
pressure-velocity coupling	COUPLE

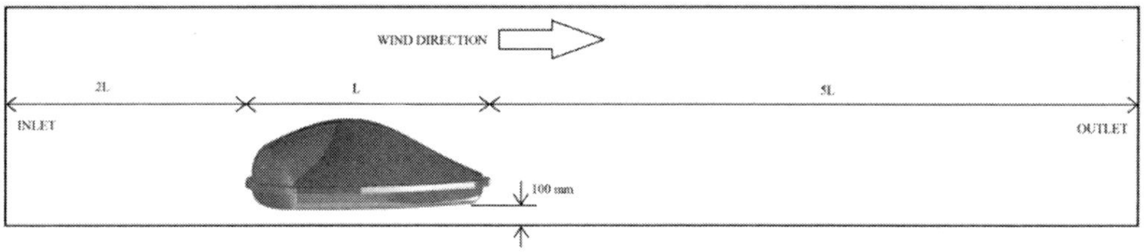

Fig. 3. MIRA Wind tunnel configuration.

4 Result and discussion

Based on the result acquire from CFD analysis, the drag coefficient Cd value for both roof boxes can be obtain after the iteration value has converged. Converging means the iteration value has not changed after ten iterations. Common roof box converging after 800 iterations and 700 for optimise roof box. The velocity contour plot of both common design and optimize design roof box as shown in Fig. 4. Strong evidence of boundary layer separation size reduction was found at the rear region of the optimize design. There was a significant difference between the two designs and there was a significant positive correlation between reduction of boundary layer separation size and the drag coefficient. The different in value as reflected in Table 1.

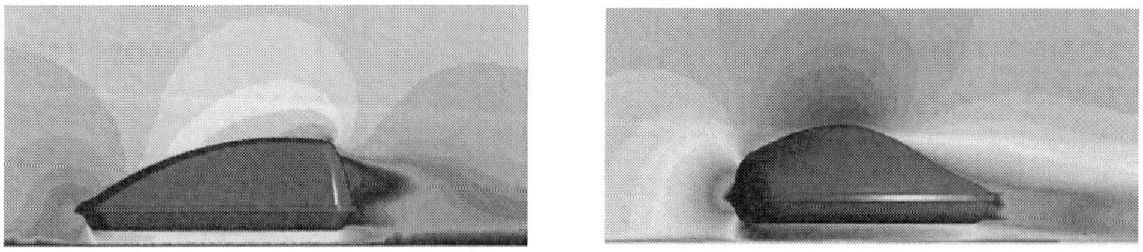

Fig. 4. Velocity contour of common and optimized roof box.

The most striking result to emerge from the data is that by streamlining the box shape resulted a significant drag reduction by 34%. This contribute to a total of 1.7% fuel consumption reduction when calculated using fuel consumption formula adapted from Zambri et all [8] as illustrated on Table 4.

Table 4. Fuel consumption contribution of both design.

	Common roof box	Optimise roof box
Total energy	8834.25 w	8297.463 w
Percentage of fuel consumption	3.616 %	1.9088 %

Fig. 5 exhibit that back area of common roof box has a bigger flow layer separation compare to optimize roof box. According to Tsubokura [13] this phenomena contribute the value of drag coefficient. This separation contribute to a pressure difference over the roof box body where the larger pressure difference the higher the drag will be.

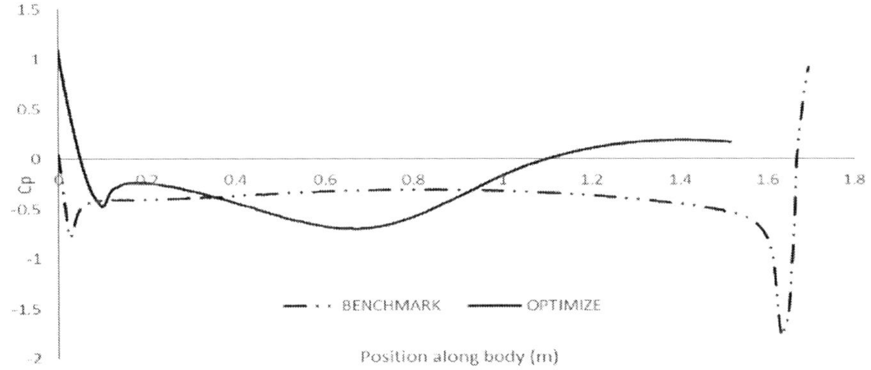

Fig. 5. Cp along upper body of both roof box

This phenomenon was clearly shown in the graph of pressure coefficient (Cp) versus the distance along the upper body in Fig. 5. As can be observed the curve of the common model shows sudden change pressure at the front where a large boundary layer as expected at frontal area where there is the first contact of air flow to the body and it is normal to has such patent. This trend were also experience by the optimize roof box. Then the curve of common roof box again shows significantly sudden change in pressure when it reach at the rear end. This trend were not observe in the optimize roof box curve.

5 Conclusion

In conclusion the improvement of drag coefficient directly affected the fuel consumption. The total saving by installing optimize roof box is 1.7% reduction compare to common roof box. Incorporating this value onto the assumption made in the introduction the optimize roof box contribute a total of RM 70 of saving annually. This data may appear relatively small but by looking at the big picture of global carbon foot print this data has significant contribution toward efforts of greener world.

This work was financially supported by Universiti Teknikal Malaysia Melaka UTeM under the grant PJP/2014/FTK(20C)/S01355.

DOI: 10.1051/matecconf/20179701089

References

1. G. P. Boretos, Technol. Forecast. Soc. Change, **76** (2009)
2. B. Khalighi, S. Jindal, and G. Iaccarino, J. Wind Eng. Ind. Aerodyn., 107–108 (2012)
3. A. Howell, Gaylard, Mot. Ind. Res. Assoc. Tech. Pap., 2006.
4. M. Takagi, J. Wind Eng. Ind. Aerodyn., **33** (1990)
5. P. Jaray, Patent No. 1631269, (1927)
6. F. Alam, H. Chowdhury, H. Moria, R. L. Brooy, A. Subic, 17th Australasian Fluid Mechanics Conference, (2010)
7. F. Alam, H. Chowdhury, H. Moria, S. Watkins, "Effects of Vehicle Add-Ons on Aerodynamic Performance,(2 010)
8. Z. Harun, M. F. Abdul Latif, Int. J. Veh. Syst. Model. Test., **9** (2014)
9. D.Brown, M. Baxendale, Hickman, *Second MIRA International Conferences on Vehicle Aerodynamics Conventry*, (1998)
10. F. D. Gmbh, Best practice guidelines for handling Automotive External Aerodynamics with FLUENT, **2**, (2005)
11. L. Sapienza, Technology (2002)
12. C. Hakansson, M. J. Lenngren, CFD Analysis of Aerodynamic Trailer Devices for Drag Reduction of Heavy Duty Trucks, (2010)
13. N. Tsubokura, M. Kobayashi, T. Nakashima, T. Nouzawa, T. Nakamura, T. Zhang, H. Oshima, J. Comput. Fluids, **38** (2009)

Study of Solid-Liquid Ratio of Fly Ash Geopolymer as Water Absorbent Material

Fandi Angga Prasetya [1,3,*], *Ndaru* Candra Sukmana [2,3], and *Ufafa* Anggarini [2,3]

[1]Departement of Engineering Management, Universitas Internasional Semen Indonesia, Indonesia
[2] Departement of Chemical Engineering, Universitas Internasional Semen Indonesia, Indonesia
[3]Konsorsium Riset Geopolimer Indonesia (KORIGI), Lab Beton dan Bahan Bangunan ITS, Surabaya, Indonesia

Abstract. Geopolymer has been synthesized from fly ash to be applicated as water absorbent material. This research conducted to determine the ability of geopolymer to abrsop water by variation of solid – liquid ratio at optimum molarity of NaOH; 3 M. In this research, the synthesis of geopolymer was conducted at the variation of solid-liquid ratio; 60:40, 65:35, 70:30, and 75:25. Result of the treatment were characterized by XRD and SEM to compare the geopolymer structure. Water absorption capacity was measured by immersing the geopolymer specimens in water for 24 hours. Based on the result, solid – liquid ratio with maximum water absorbed was 70:30 with 13,04 wt%.

1 Introduction

Urban farming is an activity utilizing the open space that is not productive (vacant land and green land) into an open space that is productive in quality and quantity. Planting medium is essential if urban farming is done in a narrow open space, with the right media plants can grow well. Planting medium as a place to grow plants must have several properties, one of which is capable of storing water [1]. Ceramic based porous material is one type of material that has the potential to be used as water absorbent. Water-absorbing material is especially useful as a planting medium in the narrow land or areas with limited land. Porous ceramics prepared from clay minerals indicate the adsorption capacity of the high water, high adsorption rate and slow water release rate associated with their unique microporous structure in the material [2, 3].

Geopolymer is one of the best candidates in preparing a porous ceramic material. Geopolymer can be synthesized and undergo a process of hardening at room temperature with the formation of gel alkali aluminosilicate framework structure. Geopolymer synthesis is generally carried out by reacting the alkali silicate solution with a solid aluminosilicate such as metakaolin, fly ash and so forth under alkaline conditions [4, 5]. Microstructure and mechanical properties of the material is influenced by the chemical composition of the starting materia [6]. Duxson et al., [7] explains that the mechanical properties of geopolymer produced increased with increasing ratio of SiO2 / Al2O3 is used in the synthesis process, while the porosity of the material increases with increasing ratio H2O /

* Corresponding author: fandi.prasetya@uisi.ac.id

SiO2. In this research, the synthesis of geopolymer was conducted at the variation of solid-liquid ratio; 60:40, 65:35, 70:30, and 75:25 at optimum molarity of NaOH; 3 M. Result of the treatment were characterized by XRD and SEM to compare the geopolymer structure.

2 Experimental section

2.1 Material and Instrumentation

The materials used in this research were fly ash. The fly ash refers to industrial waste collected from a power plant in Paiton area which is used as the aluminosilicate source. Na_2SiO_3, NaOH pellet (99%, Merck) as alkali activator, distilled water and KCl. X-ray Diffraction (XRD) was conducted to identify the crystal phase of geopolymer by using Cu-Kα radiation. Scanning Electron Microscope (SEM) images were used to identify the morphology of geopolymer. The water storage capacities of all geopolymer samples were defined by immersing geopolimer matrix in distilled water for 24 hour at 25°C.

2.2 Procedure

This material was prepared by shieving with sieve no. 325 and dried at 85°C for 24 hours. NaOH solution was prepared using NaOH pellet with distilled water. The solution was stored for 24 hours. Alkali solution was prepared by mixing sodium silicate with NaOH solution. The synthesis of geopolymer was conducted at the variation of solid-liquid ratio; 60:40, 65:35, 70:30, and 75:25 at optimum molarity of NaOH; 3 M .Fly ash, as the starting material, was mixed with alkali solution, then KCl salt was added into geopolymer paste. This mixture was casted and cured at 85°C for 24 hours.

2.3 Measurement of water absorption capacity

Geopolymer specimen was weighed and determined as initial weight (w_0). These samples were then immersed in destilled water for 24 hours at 25°C. The saturated dry surface was weighed as final weight(w_f). Water absorption capacity was calculated and reported as percentage increase in weight.

3 Result and discussion

Geopolymer has been synthesized using fly ash and alkali solution. Abdullah, et al [8] stated that geopolymer containing silica and alumina will react to produce chain and ring that contain three dimensional polymer. The existance of three dimensional ring in geopolymer can be used for water and nutrient storage for plantation aplication. Okada et al, [9] stated that geopolymers are good candidates for water absorbers due to their microstructure.

3.1Effect of Solid/Liquid Ratio Toward the Structure

The crystal phase of geopolymer samples were investigated by XRD. Fig. 1 showed that XRD pattern of Geopolymer with solid-liquid ratio 60:40 (a), 65:35 (b) 70:30, (c) , and 75:35(d) at NaOH 3M respectively. All variation of solid-liquid ratio has formed Geoplymer phase. Geopolymerization reaction can form phases, such as quartz, mullite, magnetite and sodalite. Fig. 1 show the XRD pattern of fly ash and geopolymer samples.

 DOI: 10.1051/matecconf/20179701090

While quartz was the phase most widely found at (2θ) of 26.72°, 39.49° and 50.49°. The other phases found in fly ash sample was mullite shown from peak at (2θ) of 16.48°, 23.47°, 31.15°, 33.2°, 35.29°, 40.89°, 42.55°, 49.52°, 53.66° and 57.56°. The diffractogram of geopolymer samples was synthesized from fly ash exhibit transformation from fly ash diffractogram. X-Ray diffraction pattern of geopolymer show the presence of hump peak at (2θ) of 11°-40° indicating the formation of geopolymer phase which has been done. However, the quartz phase could be found in geopolymer samples at (2θ) of 27° that confirms partly fly ash that unreacted.

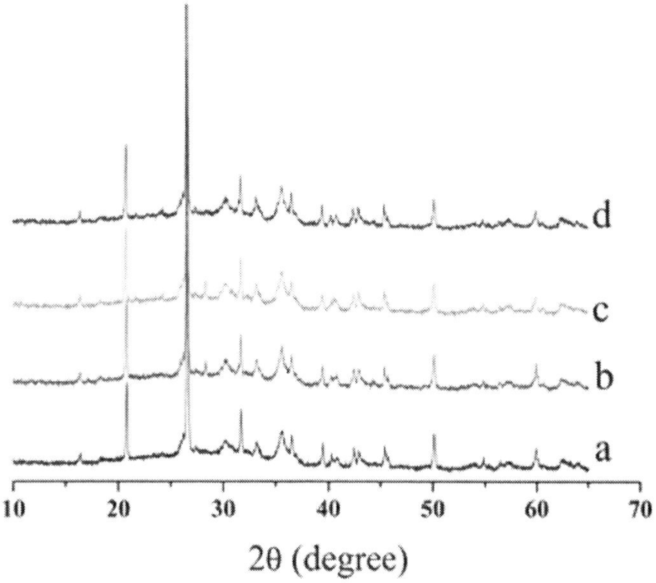

Fig. 1. The XRD pattern of Geopolymer with solid-liquid ratio 60:40 (a), 65:35 (b) 70:30, (c) , and 75:35(d) at NaOH 3M.

3.2 Effect of solid/liquid weigt ratio toward the morphology of geopolymer

Fig. 2 shows the morphology of geopolymer that synthesis with variation of solid/liquid weight ratio. Lloyd and Rangan (2010) in their experiement reveal that the smallest ratio generates geopolymer with higher workability and dense material. SEM images of geopolymer with solid/liquid weight ratio of 70:30 show the smallest particle size, so the material produce high specific area and high water abrsorption capacity.

 DOI: 10.1051/matecconf/20179701090

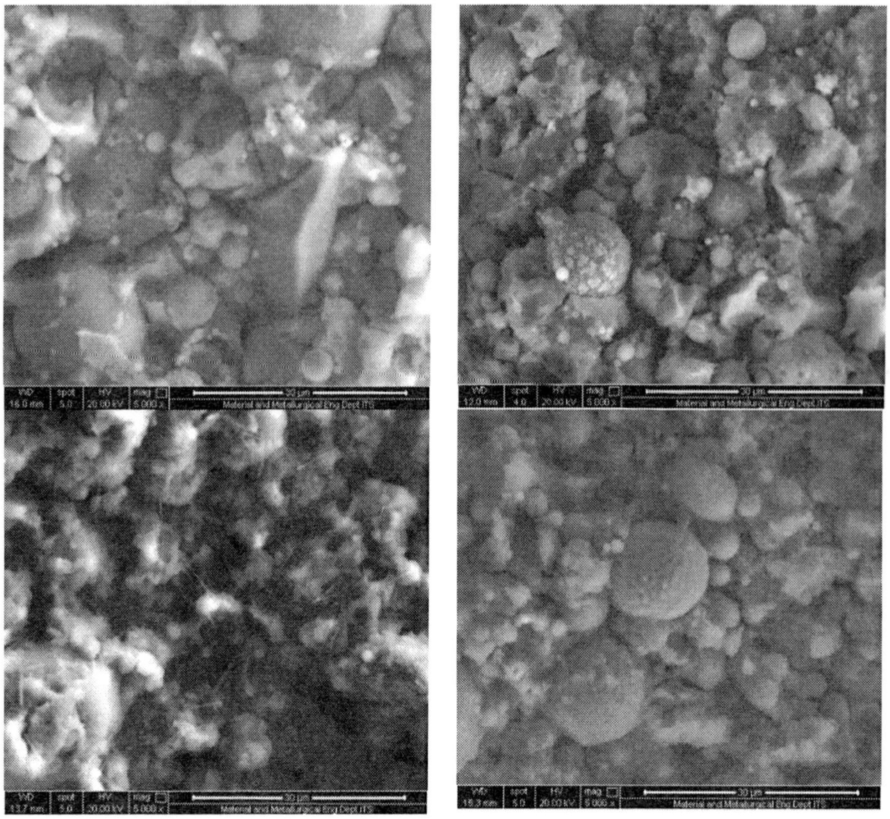

Fig. 2. SEM images of geopolymer with variation solid/liquid weight ratio of (a) 60:40, (b) 65:45, (c) 70:30, (d) 75:35.

3.3 Water absorption capacity measurement

Water absorption capacity of geopolymer samples with variation of solid-liquid ratio depicted in Fig. 3. Solid-liquid ratio of 60:40 resulted 11,10% water absorption. Meanwhile 65:35 resulted 10,89%. Maximum capacity of water absorption is 13,04% with 70:30 ratio of salod-liquid. Solid phase contain fly ash and KCl. Liquid phase contain Na_2SO_3, NaOH 3M. After reached optimum condition that caused Geopolymer has maximum ability to absorp water. Increasing of liquid ratio at 75% decrease ability of geopolymer to absorp water significantly at 4,39%.

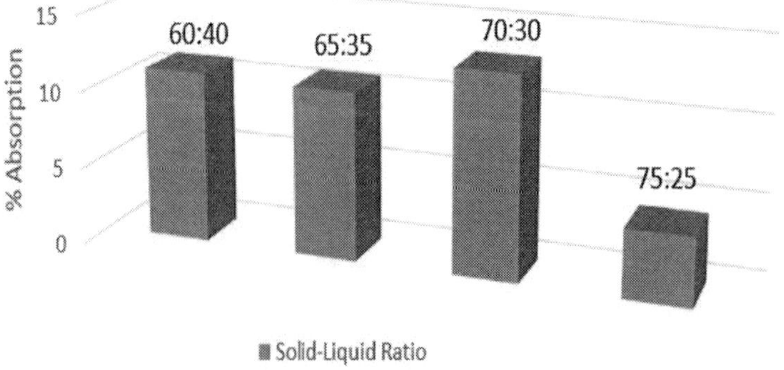

Fig. 3. Water absorption capacity of geopolymer samples with variation of solid-liquid ratio.

4 Summary

Geopolymer has been synthesized from fly ash as an aluminosilicate source with variations of solid-liquid 60:40, 65:35, 70:30, and 75:25 with 3M of NaOH. The XRD confirmed that all variation resulted geopolymer phase. SEM images of geopolymer with solid/liquid weight ratio of 70:30 show the smallest particle size, so the material produce high specific area and high water abrsorption capacity. It is proved Water Absorption Capacity Measurement that confirmed with 70:30 ratio of solid-liquid resulted 13,04%wt.

The authors wish to thank LPPM UISI for providing *Hibah Riset Bersaing UISI*, contract number 03/KP.01/10-01/02.16 date February 19th, 2016

References

1. B. Cohen B., Technol. Soc., **28** (2006)
2. K. Okada, S. Matsui, T. Isobe, Y. Kameshima, A. Nakajima, Ceram. Intern., **34** (2008)
3. J. Davidovits, Therm. Anal., **37** (1991)
4. S. Thokchom, P. Ghosh, S. Ghosh, ARPN Journal of Engineering and Applied Sciences, **4**, (2009)
5. U. Rattanasak, P. Chindaprasirt, Min. Engine., **22** (2009)
6. Skvara, Frantisek, Kopecky, Lumboir, Nemecek, Jiri, Bittnar, Zdenek, Microstructure of Geopolymer Materials Based on Fly Ash, Original Paper, Ceram-Silikaty, **50** (2006)
7. P. Duxon, A. Ferandez Jimenez, J.L. Provis, G.C. Luckey, A. Palomo, J.S.J. Van Deventer, J. Mater. Sci, **42** (2007)
8. M.M.A. Abdullah, K. Hussin, M. Bnhussain, K.N. Ismail, W.M.W. Ibrahim, Int. Journal of Pure and App. Sci and Tech, **6** (2011)
9. K. Okada, A. Ooyama, T. Isobe, Y. Kameshima, A. Nakajima, K.J.D. MacKenzie, J Eur Ceram Soc., **29** (2009)

Tool Wear Analysis on Five-Axis Flank Milling for Curved Shape Part – Full Flute and Ground Shank End Mill

Sundi Syahrul Azwan[1,*], *Muhamad* Mohd Razali[2] , *Kasim* Mohd Shahir[2] , and *R.Abdullah* R. Izamshah[2]

[1]Faculty of Engineering Technology, Universiti Teknikal Malaysia Melaka, Hang Tuah Jaya, 76100 Durian Tunggal, Melaka, Malaysia
[2]Faculty of Manufacturing Engineering, Universiti Teknikal Malaysia Melaka, Hang Tuah Jaya, 76100 Durian Tunggal, Melaka, Malaysia

Abstract. This paper is a study on full flute (extra-long tool) and ground shank end mill wear analysis by utilizing five-axis CNC to implement flank milling strategy on curved shape part. Five-axis machining eases the user to implement variations of strategy such as flank milling. Flank milling is different from point milling. Point milling cuts materials by using the tip of the tool whereas the flank milling uses the cutting tool body to cut material. The type of cutting tool used was end mill 10 mm diameter with High Speed Steel (HSS) material. One factor at a time was utilized to analyze the overall data. Feed rate and spindle speed were the two main factors that been set up equally for both full flute and ground shank end mill. At the end of this research, the qualitative analysis based on tool wear between full flute and ground shank end mill is observed. Generally, both types of cutting tools showed almost the same failure indication such as broken edge or chipped off edge, formation of pinned hole on the surface and serration formation or built-up edge (BUE) on the primary flute. However, the results obtained from the enlarged images which were captured by Optical Microscope indicated that, the ground shank end mill is better than the full flute end mill.

1 Introduction

There are a lot of parts that are difficult to machine especially in industry mentioned as aerospace, molds and dies, automotive, and heavy machinery. Usually, these industries need to produce complex structure and hard materials. Therefore, it is extremely need an analysis on tool wear. Most of the studies are concentrating on the strategy of machining such as tool path algorithm, ruled surface, and cutting stability to reduce tool wear while do machining. All these research are good to be implemented but the study on tool wear need to be taken into account before it. Hou, Y. [1] stated that the tool wear is a key factor

* Corresponding author: syahrul.azwan@utem.edu.my

affecting the machining quality. They also agreed that the need for tool wear analysis is important in the machining. In the previous studies, Choudhury and Rath [2] proposed a milling tool flank wear assessment approach based on the relationship of the average cutting force coefficients and the tool flank wear. Sarhan [3] et al. proposed an approach for tool wear detection based on the variation of cutting force components in frequency domain. Bhattacharyya [4] et al. used the average signal energy and signal mean square deviation as the tool wear state indicators for the single-tooth milling in same conditions. Kuljanic and Sortino [5] defined the normalized cutting force (NCF) indicator and the torque-force distance (TFD) indicator by cutting forces, and characterized the tool wear state through these two indicators. In this research, tool wear analysis on flank milling was done and it slightly different from the others studies. This research analysis approaches the qualitative measurement of the tool named full flute and ground shank end mill after the machining by observing the conditions. Basically, full flute is extra-long (long series) tool which is very useful in machining deeper contour or profile where is not accessible by normal or standard series of tool. In the other hand, ground shank end mill is a normal or typical series cutting tool with standard flute length. There is minor adjustment made to the shank diameter by grinding out a few millimeters to ensure that the shank will not be rubbing up the work piece while machining.

2 Methodology

This research activity started with preparation of the part modelling which was assisted by Computer Aided Design (CAD) software. The modified part modelling was then gone through a process called machining programming which was assisted by Computer Aided Manufacturing (CAM) software. The completed program was post processed and transferred to the Computer Numerical Control (CNC) machine center to perform the actual machining. Fig. 1 illustrates the CAD model with different view point namely as isometric, top, and side views.

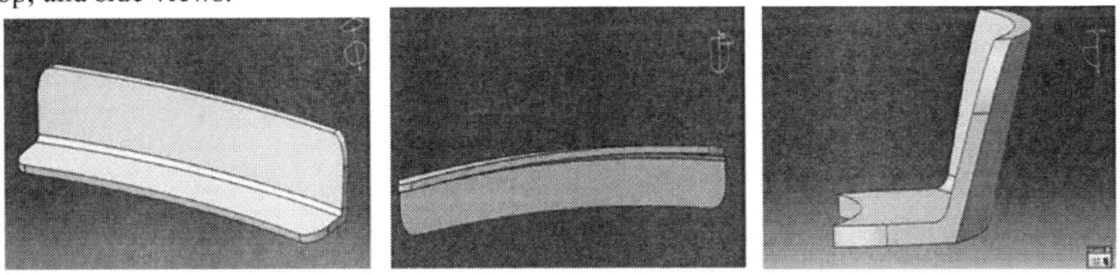

Fig. 1. Part or CAD model selected in isometric, top and side views.

At the present moment, not every industry awares of the advantages of using ground shank tool especially in aerospace industry. In general, ground shank tool is very useful in cutting at the deeper and shallow area. There are number of methods available in grinding the tools' shank. In shank grinding process, the even roundness diameter is the most critical criteria in order to ensure the balancing and stability of the tool during the machining. In machining, the stability and the balancing of the tool plays a very important role in giving the best machining result especially the surface roughness and the dimensional accuracy. Imbalanced motion can create high chattering and vibration phenomenon. Ground shank tool can be seen clearly as shown in Fig. 2 below. The reachable flute length for both tool was differentiated by 10mm. Feed rate and spindle speed remained constant for both physical machining.

Optical Microscope was utilized to observe the condition of both tools closely. The magnifying power used to analyze the tool wear images in this study was 3500 times. The accuracy of the measurement taken is relying on the user's skill to plot and perform the manual measurement using the tools provided. In this research the tool wear analysis was categorized into a few areas namely front face of the flutes, side or body of the tool and slot area or secondary flutes. For the side or the body of the tool, there were another three other different sections were taken into consideration such as bottom, middle and top portion. In general, the analysis done was only concentrated on physical comparison. The discussion and conclusion made were based on the overall appearance and physical images of the tool's sections only.

Fig. 2. The physical comparison of the long series/full flute and ground shank end mill.

3 Result and discussion

At the end of the machining, the analysis on tool wear is thoroughly observed. For this research, the tool wear of the tools were investigated by looking at the physical appearance of the tool using the Optical Microscope. The wear of the tool was only observed by looking at the overall tool's geometry captured without any specific physical measurement taken. An interpretation was made from the differences observed to the both end mills. In other words, the approach of analyzing the tool wear in this research was qualitative approach or first level interpretation analysis. In order to obtain better overview on which geometry of cutting tool shall resulting better outcomes in tool wear, a fair comparison needs to be done.

Therefore, both setups were having the same parameters. The spindle speed applied was 4500 rpm whilst 1000 mm/min for the feed rate. There were three specimens or machined part examples machined for every type of cutting tool. The most significant result obtained was the chipped off edges. It happened in full flute but not in ground shank end mill which clearly indicated by Fig. 3. In addition, there were much more serration or built up edge (BUE) formed and pinned holes tabulation were higher and bigger in size for the full flute end mill to be compared to the ground shank tool as illustrated in Fig. 4.

This failures happened were probably because of high cutting forces occurred during machining. The full flutes end mill has longer cutting flute length which encourages the creation of high vibration during machining [1]. On the other hand, there was different situation happened for ground shank tool. Since the amount of the contact area with the material was less, this automatically reduces the tendency of vibration occurrence. The main tool failure mode was catastrophic chipping of the cutting edge for all cutting operations. In five-axis milling, the tool life would be expected to increase due to the variation of the contact area [2]. Probably bull nose or radius type end mill can be considered in replacing the flat end mill in order to overcome this chipped off edges problem in machining of the curved shape profile.

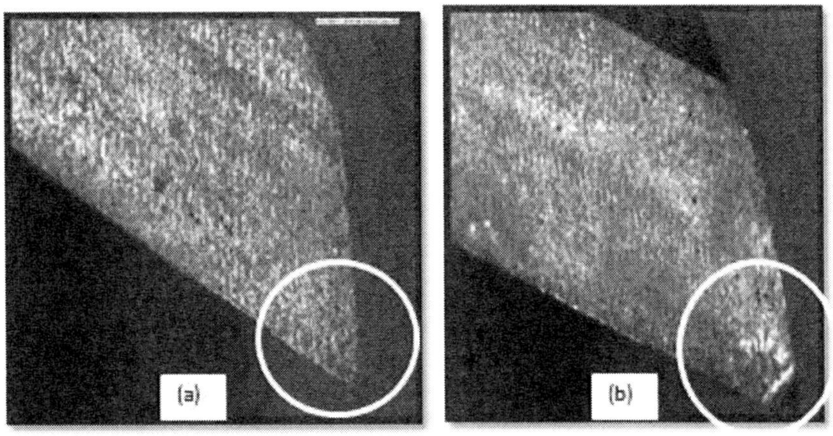

Fig. 3. Comparison of chipped off edges between (a) ground shank and (b) full flute end mill.

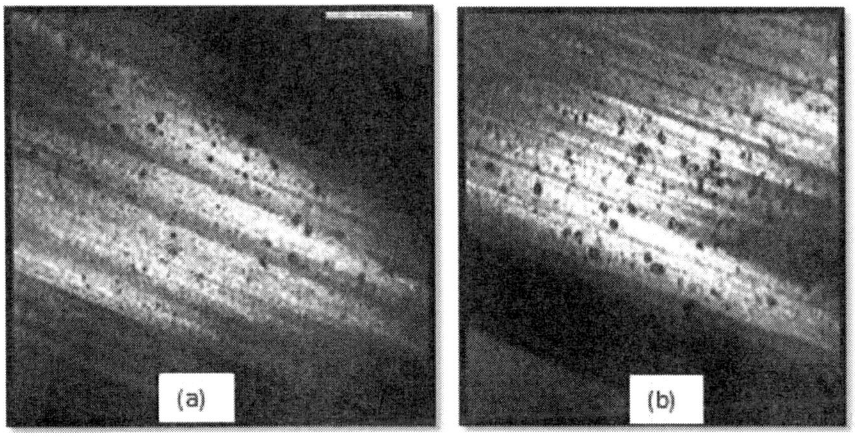

Fig 4. Comparison of pinned holes formation between (a) ground shank and (b) full flute end mill.

4 Conclusion

As for the conclusion, both types of cutting tools showed almost the same failure indication such as broken edge or chipped off edge, formation of pinned holes on the surface and serration formation or built-up edge (BUE) on the primary flute. However, the results obtained from the ground shank end mill was better than the full flute end mill. In the specific comparison made, the ground shank resulted no chipped off edges with minimum pinned holes and built-up edge (BUE) formation at constant machining parameters. Ultimately, in flank cutting of curved shape profile the ground shank end mills is preferably to be used rather than full flute end mill as it is more economical.

Author acknowledges, with gratitude, authors' debt of thanks to Universiti Teknikal Malaysia Melaka (UTeM) for having sufficient facilities and equipment along this study was conducted. Special thanks to the funding body; Ministry of Higher Education Malaysia for the research grant (RAGS/1/2014/TK01/UTEM/B00079) that support most of the involved cost to complete this study.

References

1. Y. Altintas, S. Engin, E. Budak, Journal of Manufacturing Science and Engineering, **121** (1999)
2. R. Sarhan, A. Sayed, A. Nassr, R. M. El Zahry, J. Mater. Process. Tech. J., **109**, (2001)
3. C.E. Becze, P. Clayton, L. Chen, T.I. El-Wardany, M.A. Elbestawi, International Journal of Machine Tools & Manufacture, **40** (2000)
4. E. Kuljanic, M. Sortino, Int. J. Mach. Tool. Manu. J., **45** (2005)
5. I. Korkut, M. A. Donertas, The influence of feed rate and cutting speed on the cutting forces, surface roughness and tool–chip contact length during face milling, (2007)
6. P. Bhattacharyya, D. Sengupta, S. Mukhopadhyay, Mech. Syst. Signal Process, **21** (2007).
7. F.H. Ramy, G. Hu, B. Alain, Comput. Aided Des., **45** (2013)
8. S. K. Choudhury and S. Rath, J. Mater. Process. Tech. J., **99** (2000).
9. S.A. Sundi, M.R. Muhamad, B.A. Bakar, R.I.R. Abdullah, Science International (Lahore), **26** (2014)
10. Y. Hou, D. Zhang, M. Luo, B. Wu, *2014 IEEE/ASME International Conference on Advanced Intelligent Mechatronics*, (2014)

3D Miniature Antenna Design for RFID Applications in IoT Environment

A.Nasir Mohamed [1,*], *S.N* Azemi[2], *S.A* Suhaimi[2], and *A.A.M.* Ezanuddin[2]

[1] School of Computer and Communication Engineering, Universiti Malaysia Perlis (UniMAP), Perlis, Malaysia.
[2] Advanced Communication Engineering Centre (ACE), UniMAP, Perlis, Malaysia

Abstract. The antenna application for internet of things (IoT) in urban infrastructure is introduced. Here, an effective design of microstrip 3D antenna for a RFID system in IOT which can also be used in narrow-band wireless communication is proposed. The antenna is combination of meander line techniques implemented with 3D structure resulted to miniature size of antenna that will be useful in many application. The critical parameters analysis and simulation results demonstrate the validity of the antenna with high gain and impedance matching are discussed as well.

1 Introduction

The Internet of Things (IoT) is an environment that has capability to transfer data over a network without needing human-to-human or human-to-computer interaction. IoT has developed from the convergence of wireless technologies, micro-electromechanical systems (MEMS) and the Internet. Furthermore is a global social network infrastructure based on normal and communication interoperable protocols where real and virtual "things" have unique characteristics, real quality, and particular personalities [1]. Recently, IoT are predictable to become active contributors in business, social processes and information where they are allowed to cooperate and communicate between themselves and with the surroundings by replacing information and data senses about the environments, while responding independently to the "real /physical world" events influence it by consecutively processes that start create services and actions with or not including direct human being interference [2]. IoT will extend abilities of conventional and localized exploitation of automatic identification and data capture and other interfacing technologies. As prospective new technology IoT has great potential applications in the administration of urban infrastructure [3].

With the Internet of Things (IoT), everything will be wireless. The physical connection for many IoT hubs leverages legacy wired system found in homes, workplaces, schools, industrial facilities and other areas. All the 'things' such as bag, cloth, shoes, glasses, could be connected to the internet and will communicate. Nevertheless, in order to become the ultimate mobile information sensing platform, a powerful and robust antenna is required.

*Corresponding author: nasir121080079@gmail.com

Every wireless connection needs at least two antennas which is one for transmitter (Tx) and one for receiver (Rx).

RFID (Radio frequency identification) is a word that is used to describe a system that transmits the identity of an object, goods or person wirelessly or using radio waves [4]. Technology of radio communication was rapid growth in last 10 years and improves their performance in order to make it smaller, cheaper and reliable. A basic component of RFID is RFID tags, readers or antenna and transceiver. Small antenna will be easier to apply in any application of the latest technologies which is demanded by other sector and attraction from antenna designer. There are several technique to reduce the antenna size such as meander line and fractal design beside, using specific higher permittivity substrate will also decrease the antenna size as well as the bandwidth due to the attentiveness involving the actual power field in the higher permittivity locations. However, by reducing the size of the antenna will be increased unwanted radiation because of the unbalance current which appear on the small ground plane. There are mostly occurring from the ground plane and the input connector part.

Nowadays antenna miniaturization becomes one of the research focuses, owing to the limitation of size of RFID card. In the present paper, Hilbert fractal theory and technique are employed to achieve the goal of reduction dimension of tag antenna. Additionally, in order to improve antenna efficiency, the impedance matching of the antenna must be considered [5]. It is known that the input impedance of IC chip varies from manufacturer to manufacturer. Generally speaking, the tag antenna is requested to directly match to IC chip. This is to say, the input impedance of tag antenna can be adjusted flexibly to realize impedance matching. To solve this problem, meander line technique is applied in our work. Simulations indicate that the value of input impedance of tag antenna can be changed readily by tuning the high line.

Here, microstrip meander line 3D antenna operating at 2.45GHz for IoT environment is described. Recent technology 3D antenna is believed to overcome some of the drawbacks identified in conventional antennas where it is required for certain application. The design and simulation processes are done by using CST Microwave Studio. The simulation result is optimized to achieve the best performance of the antenna. All manuscripts must be in English, also the table and Fig. texts, otherwise we cannot publish your paper.

2 Antenna design and analysis

Antenna structure shown in Fig. 1 until Fig. 4 shows the proposed antenna designed. The antenna is mounted in the middle of a rectangular ground plane with the dimension of 21.6mm x 21.6mm. Originally, the size of antenna was determined by $\lambda/4$. The width of the antenna was calculated by:

$$\lambda = \frac{c}{f} \tag{1}$$

Where;

λ=wave length, c=speed of light, f= *frequency*

Calculation:

$$\lambda = \frac{3x10^8}{2.45GHz} = 0.1224m$$

With the meander line technique implemented with 3D structure, each arm of the antenna L_2 is 4.3mm. The antenna size is reduced to $\frac{\lambda}{30}$. Table 1 shows the antenna size and parameter.

 DOI: 10.1051/matecconf/20179701092

Table 1. 3D Antenna's Parameters.

Parameters	Description	Value [mm]
L	Length of the ground plane	21.6
W	Width of the ground plane	21.6
L_1	Length of the antenna	4.3
L_2	Length of the antenna	2.15

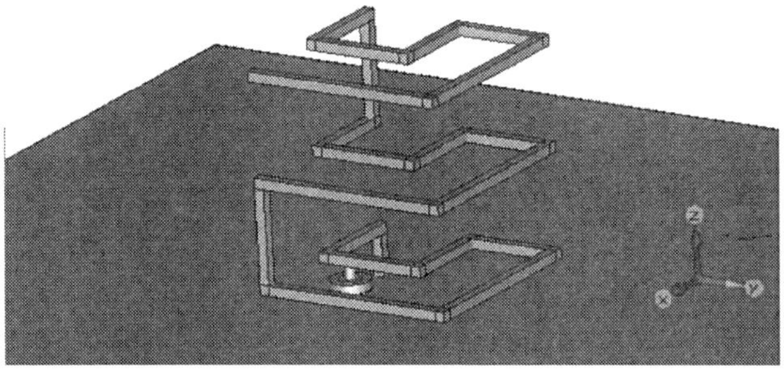

Fig. 1. Perspective Top View of 3D Antenna.

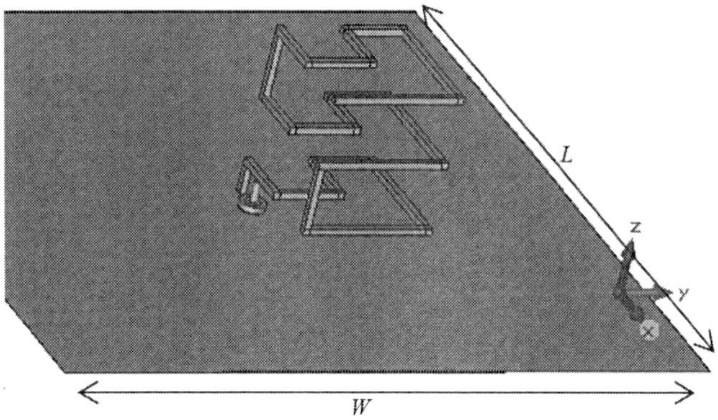

Fig. 2. The proposed 3D Antenna.

DOI: 10.1051/matecconf/20179701092

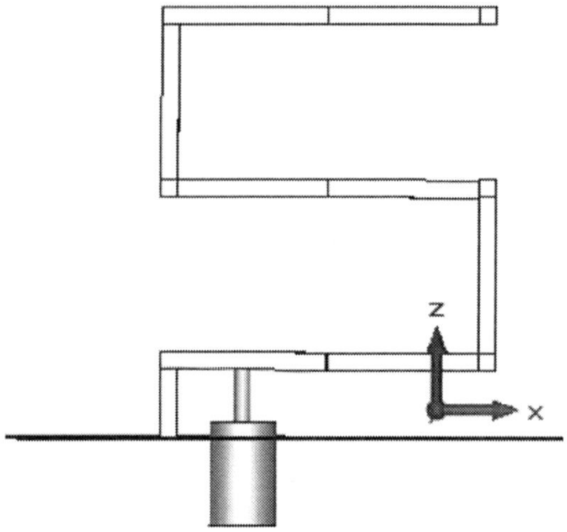

Fig. 3. Side View of the 3D Antenna.

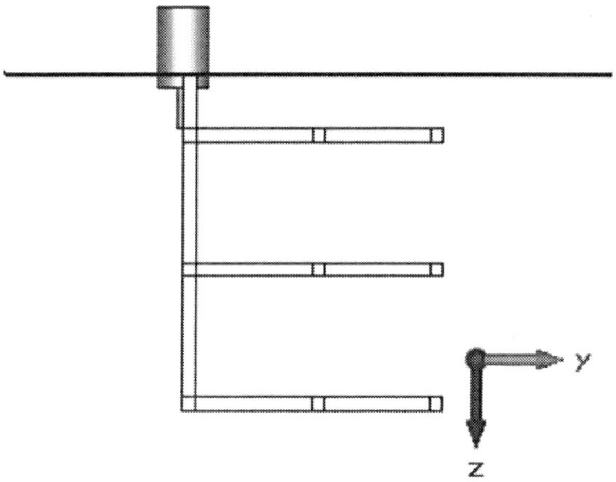

Fig. 4. Bottom View of 3D Antenna.

3 Simulation results

The return loss and fabricated antenna for the optimized dimension as given in Table 1 is shown in Fig. 5. The Fig. shows the minimum return loss is -20 dB operating at the frequency of 2.45 GHz. From the Fig., it is clearly observed that the proposed 3D meander line antenna covers the frequency range from 2.447GHZ to 2.459 GHz for return loss -10 dB. Hence, the impedance bandwidth of this antenna is 0.012 GHz that is Equivalent to 0.48% at the center frequency.

Other importance performance parameters of the antenna are radiation patterns. There are 3 information that can be retrieve from the radiation pattern which is efficiency, gain and directivity. Shown in Fig. 6 until 8 are the radiation pattern results in 3D as well in 2D. As can be seen, at frequency 2.45HGz, the radiation pattern indicates an Omni directional radiation pattern and the far field gain of the antenna are 2.69 dB. Main lobe direction of the direction is 103 degrees. The angular width at the 3 dB points of the main lobe is 95.3 degrees that a wider and suitable for good performance antenna.

Efficiency can be measured by bringing in the entire power flux. The efficiency of this antenna are -0.5dB/89%. Antenna gain is the main performance which combined the antenna directivity and electrical efficiency. Directivity is probably the standard antennas in fact it is way of measuring type of radiation or directional pattern. The directivity of this antenna is 1.28dB at 2.45GHz.

Fig. 5. (a) Fabricated 3D miniature antenna.

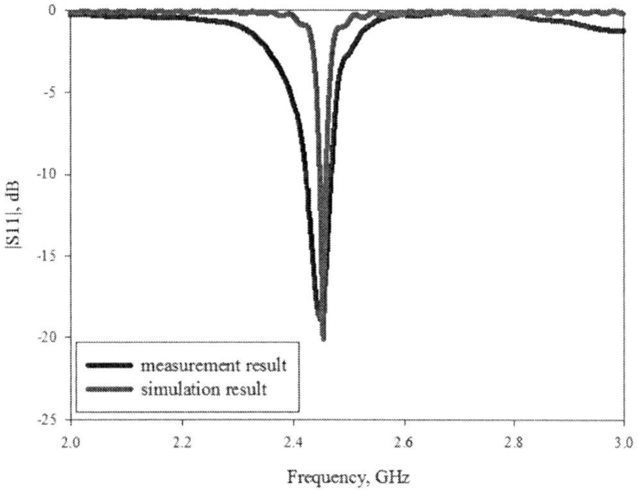

Fig. 5. (b) $|S_{11}|$ Antenna results.

DOI: 10.1051/matecconf/20179701092

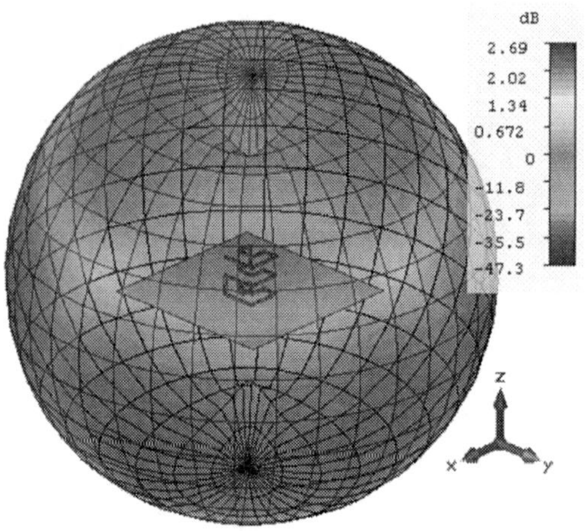

Fig. 6. 3D Antenna Radiation.

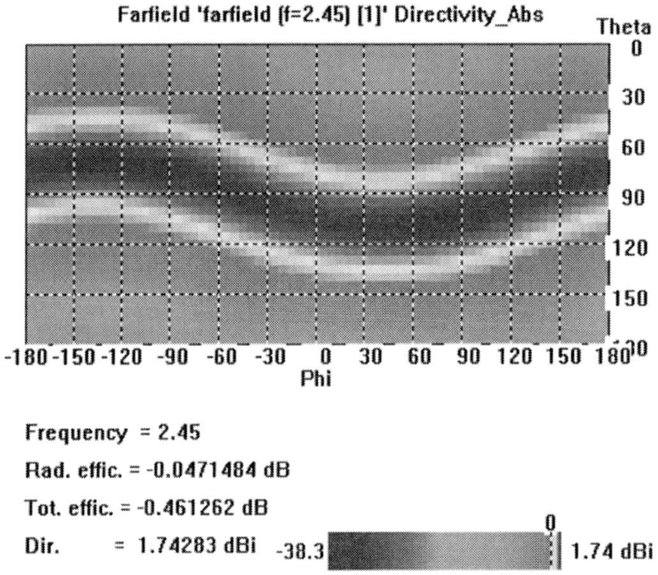

Fig.7. 2D Antenna Radiation Pattern.

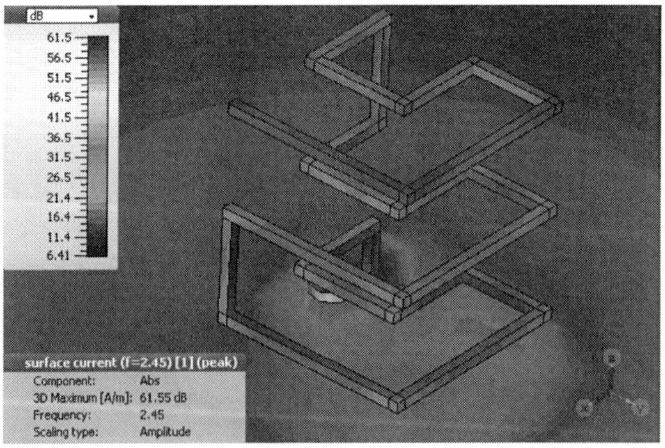

Fig. 8. Surface Current of the Antenna.

4 Conclusion

In this paper, we proposed a miniature 3D antenna operating at 2.45GHz that consists of meander line technique with perfect electric conductor (PEC). By implementing meander lines techniques in 3D structure, the antenna enable to reduce it size up to 1/30λ. The final design occupies a small area of 4.3 mm x 4.3 mm. It is not only small in size, but yet has < 90% efficiency with relatively gain. With these entire properties make this antenna is a good choice in compact antenna application for IoT in urban infrastructure. Miniaturized scale – able to be integrated with WLAN Access points, 3G Routers (either internally or externally), mobile/smart phones, mobile devices (PDA, laptop, Tablets etc).

References

1. Y. Chen, D. Zeng, T. Tang, Antenna Design for IOT in Urban Infrastructure, Journal of Convergence Information Technology, **8**, (2013)
2. S. Erumal, N.M. Norwawi, V. Raman, In Digital Information Processing and Communications (ICDIPC), *2015 Fifth International Conference*, (2015)
3. Y. Chen, Applied Mechanics and Materials, (2013)
4. V.D. Hunt, A. Puglia, M. Puglia, RFID: A Guide To Radio Frequency Identification, John Wiley & Sons, (2007)
5. Z. Katbay, S. Sadek, R. Lababidi, A. Perennec, M. Le Roy, *In New Circuits and Systems Conference (NEWCAS), IEEE 13th International*, (2015)

Gravitational Wave & Relativity Impact Electronic Communication & Engineering

Shahrudin Zakaria[1,*], *Norazlina* Abd Razak[2] , and *Zikri Abadi* Baharudin[1]

[1]Faculty of Engineering Technology, UTeM, Malaysia
[2]Faculty of Electronics and Computer Eng, UTeM, Malaysia

Abstract. About a few months ago (Feb 11, 2016), the LIGO (Laser Interferometer Gravitational-Wave Observatory) scientist team researchers made an announcement that they had confirmed the gravitational wave already detected on Sept 14, 2015 (by LIGO's twin detectors in Livingston, Louisiana and Hanford, Washington). The wave was predicted by Einstein back in 1916 with his theory of General Relativity. This paper is about gravitational wave and relativity theory that may contribute to the field of Telecommunication and other engineering as well.

1 Introduction

Indirect evidence of Gravitational wave was detected by Hulse and Taylor, they were awarded a Nobel Prize for their work in 1993. The way it was detected was by measuring the timing of radio pulses by Neutron star. They found the orbital period slow down at just the rate predicted by the theory [1]. 2016 is the year of confirming the predicted wave existence. This is an interesting year!

When the x-ray wave was found, nobody knows what to do with it. Later it found it is something to do with medicine world. As the new wave discovered (gravitational wave), again we might ask what to do with this discovery. Whenever they asked the scientist, all answer given probably directly to the goal of science's dream. This is natural for any scientists. If you love science and engineering application, this is perhaps your favorite paper.

2 What is the hope for telecommunication & others?

What analogy of this phenomenon? For Electronics communication's communities, this kind of wave is just like other kind of other wave. It can be produced and detected. For now it is only generated from the sky as nature resource. What happen in this scenario? The answer is the same goes as in electronic communication. Electromagnetic wave transmits and receives. In optic fiber as well: light transmits through fibre optic is then received by phototransistor.

* Corresponding author: shahrudinzakaria@utem.edu.my

There is no optical engineering work standalone in this arena; they normally pack together with electronic communication engineering. That makes such Faculty of Communication engineering significant; important key player. The purpose of this paper is also to alert the community of Telecommunication of higher learning; the higher prospect believes to affect the world of communication and other engineering as well.

So what is gravitational wave? It is actually the space-time wave predicted by Einstein in 1916 in General theory of Relativity. For many years, many countries and scientist try to prove the existence of the prediction. Now it is happening to 2016 announcement that the US laboratory (LIGO) confirming the founding (Happened to be about 100 years ago, one decade when once predicted).

The space is proven, that it can be distorted. Right now the highest speed of internet or other communication channel can reach from one point to another point in fiber optic is about almost the speed of C (C is Speed of light). If interconnect with an electronic device, the connection must slower than that. From the engineering viewpoint, we have limitation or constraint of the speed of signal carrier. The speed of signal carrier cannot exceeded the speed of light.

If space-time wave that reached by LIGO about 0.5C (theoretically about speed of light; C), therefore this speed of gravitational waves, make it possible for optical information reach will about 1.5C (hypothesizes). The maximum speed of light remains C, the only make possible to gain more speed is throughout the 'medium' of 'space wave' speed. This time round regarding telecommunication, the space-time not bending or distort or slowing down but acting as super carrier.

To generate space-time wave is not easy, no technology today can help us, and the natural wave is only hopeful from the sky. However, like the light wave from the sun, we cannot replicate that, unless we know technically about how to produce it from electrical current. Same goes to the space-time generator, for Mechanical Engineering, this kind of challenge will come closer to their community (Since the kind of wave has mass-space-time relationship property).

There might have several kinds of gravitational wave predicted, one that detected by LIGO, it is such like ripples that generated exactly like a mass being dropped in water (Fig. 1(a)). It is like omnidirectional antenna, which transmitting wave in all directions. The characteristic detected by LIGO: low frequency at beginning, the end up with higher frequency with higher amplitude, then settling to rest. The other type perhaps does not come from merging black hole, but might come from single axis, which could come from rotating star too, with steady frequency, which is, more-less like come from a directional antenna transmitter. The illustration of gravitational wave is as in Fig. 1(b).

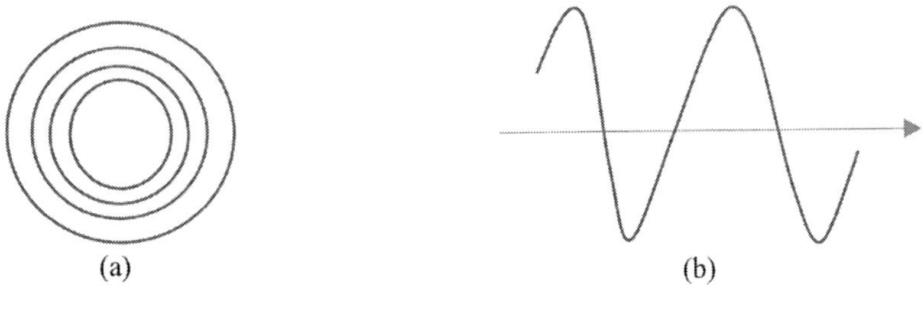

(a) (b)

Fig. 1. (a) Sphere propagation. (b) Monochrome of propagation wave.

It is postulated: (i) light can travel in empty space at the speed of light in all reference frames with the same speed of C, (ii) a wave of space-time in directional wave can bind space-time as a reference frame along with all physic law apply normally in its space-time dimension. If the light is transmitted from one point through the beam of the space-time wave as directional wave pointed to the second point; predicted the receiver will receive the signal correctly. Therefore, (i) Communication data can be transmitted faster than the speed of light (light speed still constant C in its measurement), (ii) Communication condition (data such as '1' or '0') will remain normal with all physic law, regardless with or without space-time waving as a carrier.

Revising Einstein concept of relativity, which contrast with Newton, gravity, the force is not pulling but 'push' from the effect of bending space-time curvature, therefore any spacecraft, or else will react to the space-time curvature accordingly. Remember, this hypothesis contrast with wormhole, where spacetime is stretch out.

Anything moves along with very strong directional wave propagation perhaps to have properties like a screw like shape, the space curvature will react to it (as bending space-time effect from the Einstein theoretical principle of gravity). In this paper, we think Lorenz transformation (mathematical tool) need to be updated in the version of spiral axis in terms of spacetime relationship.

The problem with the theory of relativity is that, reader tends to get confused on what actually relativity wants to deliver to reader about speed limits. Actually the limitation of speed travel by light is C explained by the theory is the speed of light measured by its own observer (measured in their relative frame only). For an example, you cannot measure two points between wormholes as a distance travel by light (inside wormhole) at outside the wormhole. So light predicted to deliver earlier when it conveys by this sense with its own of space-time behaviour. Any attempt to increase its speed limit measurement beyond C will be governed by this factor, as in Eq.1 below.

$$\frac{u + v}{1 + uv/c^2} \tag{1}$$

The other possibility, which this kind of wave can give its special properties, is that, it can be used to 'hide' the information, unseen by the third party. It can be battle/war secret information, the electromagnetic wave that used to transfer the data perhaps hiding inside the 'shield' wave. Prof Chen was once governed to lead TEKO-100 Gravitational wave detector at UTM around the year 2000's had mention the possibility of future use as new wave technology. As compared to conventional electromagnetic wave/radio, it was well known on how to manipulate the signal. The development of the TEKO-100 in Malaysia was based on support from ISAS (Institute of Space and Astronomical Science), Japan [10].

3 Current issues

To give reader a deep understanding about how speed and time in space-time is really an issue, see the problems that currently influence our own telecommunication system, where it becomes the reality now, which of time dilation really happens in our GPS satellites.

All total 24 GPS satellites that work around the GPS system have to adjust their clock, just to make sure they all remain synchronized relative to each other. The engineering done on that system did adjust their clock, whether relativity has accepted as truth or they still feel skeptical about it. The effect of velocity is represented by the equation Eq.2, and the effect of acceleration/gravity [3], as in the Eq.3 (all variables and constants are in standard form, assume to be known).

$$t' = \frac{t}{\sqrt{(1 - v^2/c^2)}} \tag{2}$$

$$t' = \frac{t}{\sqrt{(1 - 2GM/rc^2)}} \tag{3}$$

What is the way to demonstrate on how this situation actually happens? To the extent of convincing, so academician in the area of field will realize how important it is regarding our satellites or any different gravity or whenever acceleration system take place. There were many experiments already justify the theory [11], for example the time dilation existence when something moving faster relative to another.

Indeed, there is a strong requirement of relativistic coordinate in translating all the problems in the future design. As the current situation, the GPS was designed by Newtonian system coordinate (Euclidean space and absolute time) with post cutting edge technology of relativistic effect adjustment [4].

4 Recommendations

The issue of speed and time of space-time bender is really the issue. Should the relativity study included as a pair to EMT (electromagnetic theory) subject? Especially those in specific program field such as satellite telecommunication system engineering, bachelor or master program or courses. It is clear how electromagnetic become compulsory to electrical bachelors program, and now relativity theory should also be clear as compulsory knowledge to telecommunication engineering bachelor program.

We fully recommend the Special Relativity to be taught and basic of General Relativity is to be understood in related engineering applications but not in detail. The borderline is, is it not so suitable dealing in larger scale of the universe, for example, inside a black hole, expending of the universe, or some others, unexplained observation. Left side of the Field equation may need adjusting to take account of some of observed problems [8].

Anyway, there are many pros and cons about Einstein's relativity, which never agree at the bottom theory of physic (Quantum Mechanic) for so long. But many experiments done agree with the theory [11], perhaps maybe because of status of race (Jews). Today, the problems of GPS will spread to the user of smart phone worldwide. Soon or later reader will judge fairly when the problem real reach all of us in everyday life.

Here is the example of GPS problems: relativistic effected of velocity; -7mirosecond/day, affected by acceleration/gravity give +45microsecond/day, total timing relativistic effect: +38micro second/day, so the GPS system on the satellite has to adjust -38microsecond/day to remain intact synchronous with GPS system on the ground [5]. If the system not corrected in a day of operation, it can give error more than 11km dragging error in the receiver's system [4, 5]. Therefore, because of this clear evidence, it is hard for us to reject the theory of general relativity easily.

The rule of thumb is, Newtonian system and application engineering is the best approximation for on Earth application and when the system is relatively slow. (Whenever Newtonian system move to outer space and with fast system relative to one another, Newtonian seem to have an error clear, for instant the GPS system). While Einsteinian system and application engineering is the best approximation for outer space application and when the system is relatively faster.

For application that need faster communication like precise detection of the event of lighting (anything related to speed of light), it is recommended to have Einsteinian system. For an example to take account of its relativity consideration in its measurement.

5 Conclusions

We hope this delivery of the science application message is successful. In the medium of engineering perspective. Wide opportunity then open by the gravitational wave window.

 DOI: 10.1051/matecconf/20179701093

The new wave application, which probably may enhanced the transmission quality, which may speed up information travel through space or can shield the information of being transmitted. This predictive hypothesis perhaps could come true one day no matter how weird its look like. Even the weirdest thing in mechanic quantum, now turn into public test on quantum computer in 2016! We do not fully understand about mechanic quantum physical phenomenon, but it is work! You can test it online.

Then probably, this will spark the mind of all those academicians in the related area. With GPS have already shown a close relationship between speed, acceleration and time; give all relativistic effect in real action. Real effect. Perhaps they could get involved in this area such as proposing new exciting projects. As well as thinking of some new theories related in the near future.

References

1. B.P. Abbott, Rep. Prog. Phys., **72**, 25 (2009)
2. E.S. Phinney, Astrophysics, (2009)
3. E.G. Haug, International Journal of Astronomy and Astrophysics, **6**, 206 (2016)
4. J.F.P. Sanchez, ArXiv, (2006)
5. D. Pope, *Everday Einstein: GPS and Relativity* (Perimeter Institute for Theoretical Physics, Canada, 2010)
6. A.G. Polnarev, D. Baskaran, Phys. Rev. D., **77**, 12 (2008)
7. M. Cavaglia, *International Year of Astronomy ASP Conference Series*, (2008)
8. C. Corda, Astrophysics, (2009)
9. S.V. Dhurandhar, A.Vecchio, Astrophysics, (2010)
10. Y.T. Chen, *Proceedings of the MGIX MM Meeting*, (2000)
11. C.W. Misner, K.S. Thorne, J.A. Wheeler, *Gravitation* (W.H. Feeman and Company, England, 1973)

DOI: 10.1051/matecconf/20179701094

Preparation and characterization of form-stable paraffin/polycaprolactone composites as phase change materials for thermal energy storage

M.S. Aludin[1,*], and *S. Saidatul* Akmal[1]

[1]Faculty of Engineering Technology, Universiti Teknikal Malaysia Melaka, Hang Tuah Jaya, 76100, Durian Tunggal, Melaka, Malaysia.

Abstract: Paraffin is Phase Change Materials (PCM) that possesses desirable properties such as high thermal energy storage and thermal stability to make it suitable for thermal energy storage applications. However, paraffin has been reported to leak out during the melting process. In this study, composites were prepared by dissolving paraffin and polycaprolactone (PCL) at varied mass percent compositions in chloroform and then purified through precipitation techniques. The leakage test was conducted by placing the composite samples on a set of four-layer filter papers and left in a furnace at 90°C for 1 hour. By incorporating PCL into paraffin phase, the leakage mass percentage was drastically reduced. The PCL polymer matrix in the composites may have trapped the paraffin molecules during melting process thus prevent it from leaking.

1 Introduction

Thermal energy storage (TES) functions as temporary storage of thermal energy in the forms of cold or heat energy. TES can be generally classified as sensible heat storage and latent heat storage [1]. Latent heat storage is the most preferred forms of energy storage because it can provide high energy storage density and nearly isothermal heat storage or retrieval processes [2]. The latent heat can be discharged or charged based on demand and applications. One of prospective techniques of storing thermal energy is the application of phase change materials (PCM). PCM are latent heat storage materials that can absorb, store and release latent heat of fusion to the environment during melting and freezing processes [3]. Solid-liquid phase change process is illustrated in Fig. 1.

* Corresponding author: aludin@utem.edu.my

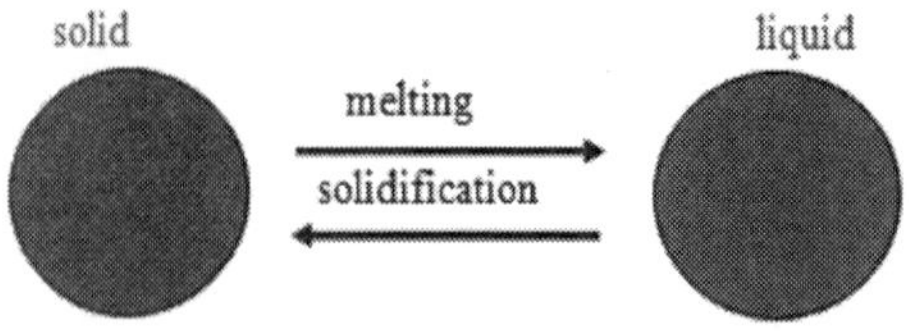

Fig. 1. Solid-liquid phase change process

Paraffin (saturated hydrocarbons with C_nH_{2n+2} formulas) not only can absorb and store heat but it also can release large amount of heat over a large number of phase change cycle which make it useful as an excellent PCM [4]. Paraffin is also considered non-toxic. However, the major problem of using paraffin is leakage during melting process [5, 6]. As a result, in recent years, much attention has been paid to form-stable paraffin PCM which consists of solid-liquid paraffin and matrix that provides supportive function [2]. One of the selected matrixes is polymer. In this study, paraffin/polycaprolactone (PCL) composites were prepared at varied mass percent compositions in order to obtain form-stable PCM composites thus possibly eliminates the leakage problem. PCL was selected as the polymer matrix since it has good mechanical properties and excellent miscibility with a large range of materials and it is one of the few commercially available biodegradable polymers [7].

2 Methodology

2.1 Materials

Technical grade paraffin wax with melting temperature 56°C, supplied by Aldrich Co., Inc. PCL with average molecular weight 80,000 and melting temperature 61°C was also purchased from Aldrich Co., Inc. Chloroform and ethanol were purchased from Kanto Chemical Co., Inc. Both of the chemicals had a purity of 99 wt% and were used as received.

2.2. Preparation of paraffin/PCL composites

The composites were prepared by dissolving paraffin and PCL at varied mass percent compositions in chloroform and then purified through precipitation techniques by using ethanol solution. The composites were then dried in a fume hood.

2.3 Thermal properties of paraffin/PCL composites

The melting temperature and the latent heat of fusion of the composites were measured by Differential Scanning Calorimeter (DSC). The DSC measurements were performed in the temperature range of 0 to 80°C with a heating rate 10°C per minute and under a constant stream of nitrogen at atmospheric pressure.

2.4 Leakage test

The leakage test was conducted by placing 1.0 g of each composite sample on a set of four-layer filter papers and left in a furnace at 90°C for 1 hour. Leakage test set-up in a furnace is illustrated in Fig. 2. The leakage percentage was calculated based on the change of sample mass before and after the leakage test.

DOI: 10.1051/matecconf/20179701094

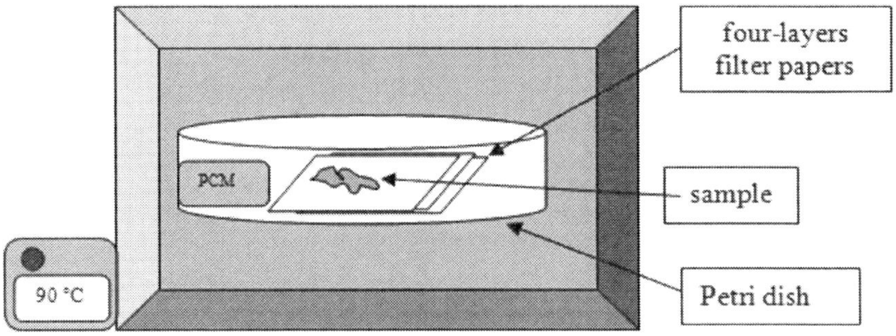

Fig. 2. Leakage test set-up

3 Results and discussions

The leakage test results show that the leakage mass percentage of the composites is significantly lower than that of pure paraffin (leakage percentage 78.4%). The composite with percentage mass composition of paraffin/PCL 40/60 shows almost no leakage during the leakage test. Table 1 tabulates the overall leakage test results.

Table 1. Leakage test results

Sample No.	Paraffin / PCL (Mass%)	Leakage Mass	
		(g)	(%)
PCM1	100/0	0.784	78.4
PCM2	80/20	0.251	35.1
PCM3	60/40	0.197	29.7
PCM4	50/50	0.305	30.5
PCM5	40/60	0.001	0.10
PCM6	20/80	0.182	18.2
PCL	0/100	0.000	0.00

The resistance of PCL polymer matrix to higher temperature than the melting temperature of the paraffin makes the composites able to maintain the solid morphology in macro level even the paraffin has melted. The photographs indicating the remaining PCM on the filter papers after the leakage tests are shown in Fig. 3. As can clearly be seen in the photos, the composite sample PCM5 maintains it shape in the solid state without seepage.

DOI: 10.1051/matecconf/20179701094

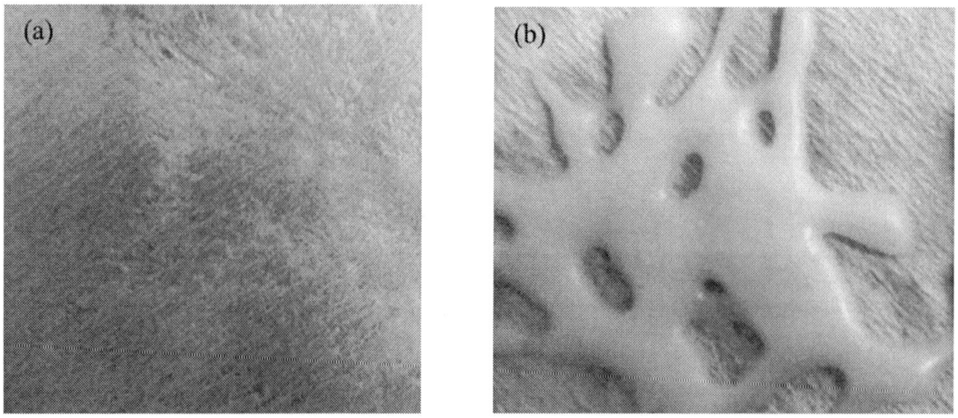

Fig. 3. Sample appearance after leakage test: (a) PCM1 (b) PCM5

The melting temperature and latent heat of fusion of the composites that were measured by DSC are shown in Table 2. The latent heat of fusion of the composites decreased at PCL mass compositions 40% to 60%. The melting temperatures are also higher by about 4°C at this range of mass compositions. Fig. 4 shows the DSC thermograph of PCM5 (paraffin 40%/PCL 60%).

Table 2. DSC results of paraffin/PCL composites

Sample No.	Paraffin/PCL (Mass %)	Melting temperature (°C)	Latent heat of fusion (kJ/kg)
PCM1	100/0	55.6	139.2
PCM2	80/20	56.6	129.1
PCM3	60/40	64.8	85.2
PCM4	50/50	63.6	84.3
PCM5	40/60	65.2	81.4
PCM6	20/80	56.5	96.7
PCL	0/100	60.6	72.1

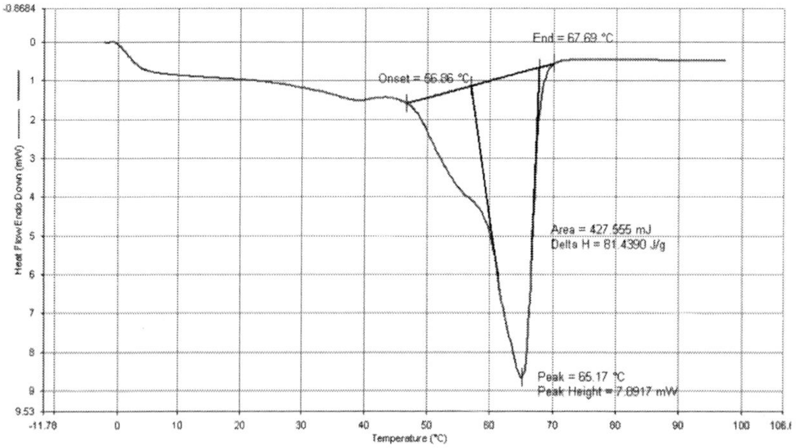

Fig. 4. DSC thermograph of PCM5

 DOI: 10.1051/matecconf/20179701094

The lower of heat of fusion and the slight increase in the melting temperatures of the composites with PCL mass compositions 40% to 60% indicated that paraffin phase and PCL matrix phase were homogeneously mixed. Therefore, the PCL polymer matrix in the composites may have trapped the paraffin molecules during melting process thus prevent it from leaking, as illustrated in Fig. 5.

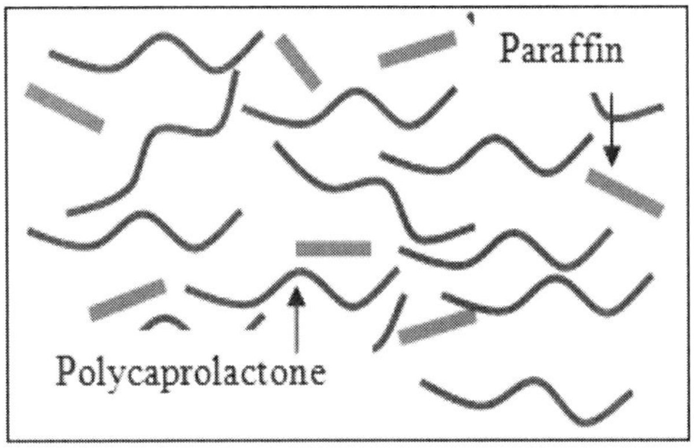

Fig. 5. Paraffin molecules in PCL matrix

4 Conclusions

The leakage test results indicated that by incorporating polycaprolactone (PCL) into paraffin phase, the leakage problem that associated to the pure paraffin can be prevented. The leakage characteristic of the composites was somehow related to the heat of fusion and melting temperatures of the composites. The main advantage of these composites is that its can maintain the shape and prevent paraffin leakage during phase change process thus can be used without extra encapsulation because the paraffin has been encapsulated in the polymer matrix structure. These form-stable PCMs would potentially become novel candidates of PCM for thermal energy storage applications.

The authors would like to thank Universiti Teknikal Malaysia Melaka and Ministry of Higher Education under Short Term Grant number: PJP/2014/FTK(12A)/S01322 for funding this project.

References

1. I. Dincer, M.A. Rosen, *Thermal energy storage: systems and application* (John Wiley & Sons, Canada, 2012)
2. K. Pielichowska, K. Pielichowski, Prog. Mater. Sci., **65**, 67 (2014)
3. A. Sharma, V.V. Tyagi, C.R. Chen, D. Budhi, Renewable and Sustainable Energy Reviews, **12**, 318 (2009)
4. D. Zhou, C.Y. Zhao, Y. Tian, Appl Energy, **92**, 593 (2012)
5. F. Chen, M. Wolcott, Solar Energy Materials and Solar Cells, **137**, 79 (2015)
6. A. Sari, Energy Conversion and Management, **45**, 2033 (2004)
7. H. Yasuda, M.S. Aludin, N. Kitamura, M. Tanabe, H. Shirahama, Macromolecules, **32**, 6047 (1999)

MATEC Web of Conferences **97**, 01095 (2017) DOI: 10.1051/matecconf/20179701095
ETIC 2016

Validation of Vehicle Model Response with an Instrumented Experimental Vehicle

Mohamad Hafiz Harun[1,*], *Mohd Razali* Md Yunos[1] , and *Muhamad Azwar* Azhari[1]

[1] Faculty of Engineering Technology, Universiti Teknikal Malaysia Melaka, Malaysia

Abstract. A steering aid system called active steering is evaluated by simulating different kinds of driving events. The main purpose of the steering system is to allow the driver control the vehicle independently. A full car vehicle model is simulated in Matlab/Simulink with 14 degree of freedom of equations which include the ride vehicle model and also the handling model. The steering angle is the input of the vehicle model that should be focused on. The angle of the steering system between the tires when turning the vehicle is taken in consideration. Simulations are made on different road conditions effect and also side wind disturbances. Different values are applied to the simulation to reduce the effect of the driving events. Therefore, these simulations results to provide a better improvement to the steering system. The aim for this work is to validate the vehicle response with an instrumented experiemental vehicle. Specific driving events in these simulations are the road adhesions and lateral side wind disturbances.

1 Introduction

In an age of advanced technology, automobile is a life's necessity. It also acts as a transportation to travel for a distance no matter a short or long journey. Therefore, automobile had become more and more convenience nowadays. In this competitive field, vehicle must be improved in comfort and safety. By doing so, development of vehicles should be done to ensure new technology is implanted into the driving system to enhance driver pleasure.

A vehicle consists of many systems to maintain each of the vehicle parts function well. The steering system acts as one of the important system in the vehicle. For the past few decades, the development of steering system has tremendously experienced many stages and the steer by wire system is the latest technology of steering system for passenger cars. Steer by wire system are still not fully trusted by public consumers. On top of that, with the permission issues by the state regulations, AFS has become a customer choice. Active front steering system is the latest technology for passenger cars which implements an electronically controlled superposition of an angle to the hand steering wheel angle that is

* Corresponding author: mohamadhafiz@utem.edu.my

prescribed by the driver. However, the permanent mechanical connection between steering and road remains [1, 2].

There are a lot studies focusing about the active steering system are carried out. But most of the studies focus on the stability of vehicle [3-8]. In active steering system, the driver will experience the variable steering gear ratio function at first and perceive the improvement of steering portability. AFS enables continuous and situation dependent variations of the steering ratio according to the vehicle's motion state, therefore AFS improve the manoeuvrability of the vehicle at low speed and the stability at high speed. The performances of stability improvement with active suspension system depend upon the quality of variation of the steering ratio to a certain extent. A controller of variable steering ratio for Active Front Steering system is designed, and virtual road tests are made in CarMaker driver vehicle-road simulation environment. The results of simulation tests validate the controller performance and the advantage of variable steering ratio function, also show that the driving comfort is improved at low speed especially, due to the Active Front Steering system alters the steering ratio according to the driving situation [9, 10].

The proposed AFS control structure is implemented on a validated full vehicle model. The full vehicle model can be approximately described as a 14 DOF system subject to excitation from braking and throttling inputs. It consists of 7-DOF vehicle ride model and 7-DOF vehicle handling model coupled with Calspan tire model [11]. The time used in this software was less compared to commercial program. Therefore, a more accurate result can be obtained.

This paper is organized as follows: The first section contains the introduction and the review of some related works, followed by the methodology of development of active front steering with affect of crossing wind test in the second section. The third section presents the result and discussion of the validation performance. The last section contains some conclusion.

2 Methodology

Computer simulations are carried out to verify the development of the active steering system in this study. For a basic start up, a model of the main component was developed. Vehicle Model of Mercedes E-Class was chosen as the simulation vehicle model. The characteristics of the Mercedes E-Class vehicle model was studied and simulated by Matlab/Simulink software. The stability test of the Mercedes E-Class was tested using crosswind test. Crosswind test is a wind test which is perpendicular or lateral to the vehicle itself. This affects the stability of the vehicle model when simulated in Carsim software.

In the analysis, crosswind test was applied on the vehicle model and changes of the vehicle are observed. The stability of the vehicle is focused and discussed. The driving stability is affected by the aerodynamic characteristics as well as the vehicle dynamic characteristics. This includes the yaw, lateral acceleration and longitudinal acceleration of the vehicle model. The vehicle dynamic characteristics is then plotted into graphs and compared. Different vehicle model will provide different graphs pattern. The other factors that affect the vehicle dynamic characteristics are the velocity of the wind.

The three vehicle dynamics characteristics act as an important role in the simulation. Longitudinal acceleration refers to the acceleration produces by the vehicle in a straight line. The lateral acceleration created when a vehicle tends to push a vehicle sideway and yaw rotation is the movement around the y axis. Stability test focused on these dynamic characteristics to improve the stability of the vehicle.

The validation of the ride and handling model is done by Matlab/Simulink software. The responses of the 14 DOF Mercedes E-Class vehicle models is compared with the

DOI: 10.1051/matecconf/20179701095

instrumented experimental vehicle data using the same input conditions and data value. The dynamic response characteristics to be measured include lateral acceleration, longitudinal acceleration and yaw. Validation is done to obtain the graph result by two software which are the Carsim and Matlab software.

3 Results and discussion

Figure 1 below shows the lateral acceleration versus time. The dotted line represents the vehicle model simulation while the dark line depicts the Carsim model simulation. Both lines display similar pattern with different values. The duration of the simulation is set at 8 second. Each data was recorded at interval of 0.025 second. From Figure 1, data after 8 second is unrecorded as the lateral acceleration remains constant.

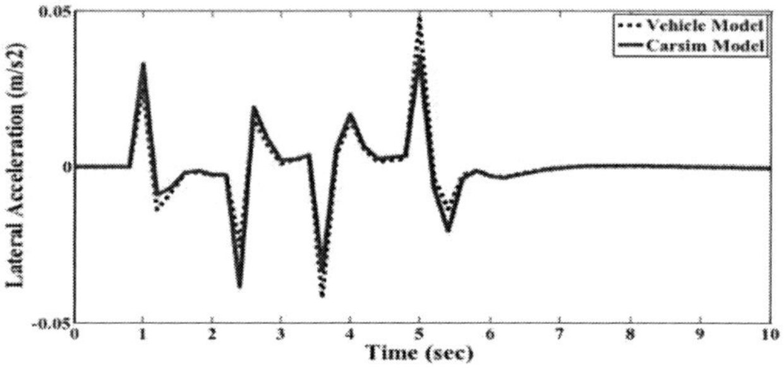

Fig. 1. Validation of lateral acceleration

The lateral acceleration is created when the vehicle is cornering. The acceleration tends to push the vehicle outwards. This is due to the centrifugal force created on the vehicle. Therefore, little acceleration is applied before reaching the apex of the curve. This is done to pull the vehicle through the curve. Therefore, the best way to take a sharp corner is to slow down before reaching the curve. Then, accelerate at the apex to turn the vehicle.

In this simulation, a vehicle model Mercedes E-Class was driven on a straight track with lateral side wind applied on it. Simulations showed that the vehicle speed with non-stationary wind excitation plays an important role in the crosswind stability of vehicle model. Therefore, the values obtained are slightly different compared to the actual values. As in real life, the speed of the wind and the wind turbulence will influence the crosswind stability. Actual measurement is hard to obtain from the vehicle model as simulation of vehicle model in Matlab/Simulink represents the vehicle model driven in real life. As for Carsim simulation, the speed of the wind is set ideally at both sides. It means that equal forces are applied side to side of the Mercedes E-Class vehicle model. As a result, perfect line graph can be obtained as true value of measurement is inserted to the Carsim software.

Figure 2 below illustrates the validation of longitudinal acceleration versus time. The range of the longitudinal acceleration is between 0.03 m/s^2 to -0.04 m/s^2 and the maximum simulation time is 8 seconds.

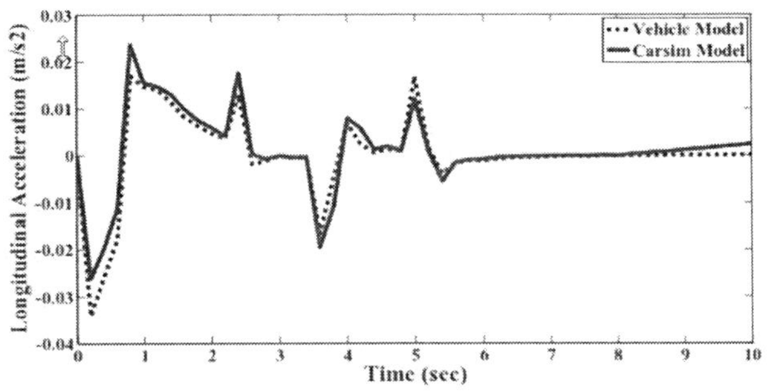

Fig. 2. Longitudinal acceleration

The longitudinal acceleration is the speed of a vehicle accelerates in a straight path. The positive value indicates acceleration whereas the negative value is for braking. A force felt in acceleration or braking movement around the horizontal axis of the vehicle is named as pitch.

First, the inconsistent pattern of the graph indicates the stability of the vehicle model. Pressure applied laterally to the vehicle will affect the movement of the vehicle. This also depends on the pressure value. A higher pressure will cause the stability of the vehicle to decrease. There are some factors that caused the value of both graphs to be asymmetry. Speed of the vehicle is one of the important factors to be discussed. Result cannot be compared between simulations and in real-life. As mentioned above, Carsim data is correct and cannot be denied. Therefore, the asymmetry data of Matlab/Simulink simulation is caused by the inconsistent speed driven on the vehicle. For example, an 80km/h and 100km/h speed value will produce different data when simulated with same lateral side wind condition. The value of 100km/h will higher than the value of 80km/h and the point in the graph will show significantly. Hence, speed of the vehicle may affect the stability of the vehicle even the weight of the Mercedes E-Class is one of the heaviest vehicles among luxury cars.

Figure 3 below represents the validation of yaw versus time on a same vehicle, Mercedes E-Class. The pattern of the graph goes smoothly compared to the previous graphs. The range of yaw degree is between 0.1 degree and -0.1 degree. The value remains constant zero for 0.0825 seconds.

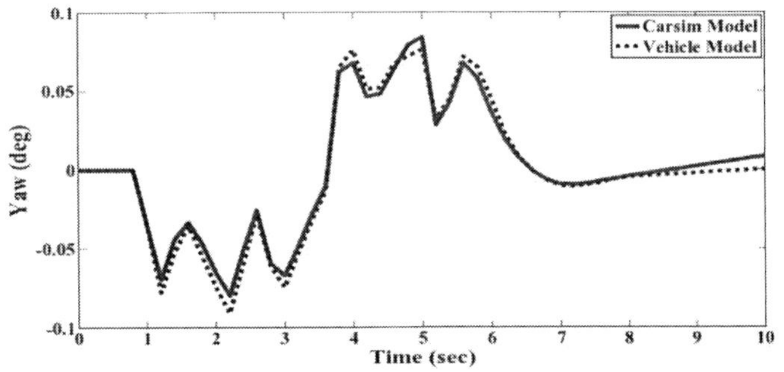

Fig. 3. Validation of Yaw

Yaw can be explained briefly as the ability to turn or rotate. A yaw rotation is the movement around the yaw axis of a rigid body. It changes the pointed direction either to the left or right of its direction of motion. Besides, yaw rate control is one of the main targets of electronic stability control systems that function to maintain the yaw rate of the vehicle.

DOI: 10.1051/matecconf/20179701095

This is similar to the purpose of this analysis as the focus point is the stability of the vehicle.

The lateral side wind conditions show the actual stability of the vehicle. The vehicle model with 0.087007 m/s^2 positive yaw moment generates high yaw rate thus creates lower stability. The effect of surrounding also will affect the yaw rate, thus affect the stability of the vehicle. Higher wind force applied to the vehicle model will results the higher yaw rate. A rise and drop of the values runs by Matlab/Simulink software is due to the wind forces and side wind effect of the surroundings. Besides that, tire and suspension system of the vehicle act an important factors on the stability of the vehicle. Tire model can be categorized into many types. Each type may have different pressure and air density. Thus, different volume density of the tire may affect the stability result of the vehicle. Furthermore, flat tire do counted a part of it. The suspension system is the main stability system that stabilise the vehicle while having side conditions on roads. Nowadays, the Mercedes E-Class is using Airmatic air suspension system to improve the handling stability by reducing the vibrations and quieter actuation noises. With the advanced technology, stability may be improved.

4 Conclusions

The vehicle model is created by Matlab software. 14 DOF mathematical equations were developed in blocks. The 14 DOF includes the ride model and handling model. Both models acting the important part in this project to improve the vehicle stability due to the only factor lateral side wind effect. Therefore, validation graphs are compared in Matlab/Simulink simulations.

The author would like to acknowledge the Ministry of Science Technology and Innovation and Universiti Teknikal Malaysia Melaka for the support and funding throughout this study.

References

1. P. Koehn, M. Eckrich, *Active steering-the BMW approach towards modern steering technology* (SAE Technical Paper, 2004)
2. W. Klier, W. Reinelt, *Active Front Steering (Part 1): Mathematical Modelling and Parameter Estimation* (SAE Technical Paper, 2004)
3. C. Deling, Y. Chengliang, China Mechanical Engineering, **18**, 3019 (2007)
4. X. Sun, J. Zhao, Tractor and Farm Transporter, **35**, 91 (2008)
5. J. Guo, J. Li, Y. Li, Automobile Technology, 4 (2007)
6. Z. Yu, Z. Zhao, H. Chen, China Mechanical Engineering, **31**, 652 (2005)
7. Q. Li, G. Shi, Y. Lin, W. Zhao, Automotive Engineering, **31**, 629 (2009)
8. B. Lee, A. Khajepour, K. Behdinan, *SAE 2006 World Congress*, (2006)
9. J. Huang, J. Ahmed, A. Kojic, J.P. Hathout, *Proceeding of the 2004 American Control Conference Boston*, (2004)
10. Z. Gao, J. Wang, D. Wang, Procedia Eng., **15**, 1030 (2011)
11. Z.A. Kadir, K. Hudha, M.Z.M. Nasir, M.R. Said, *Proceedings of the International Conference on Plant Equipment and Reliability*, (2008)

Preliminary Application of a Helicopter Toy for Environment Monitoring

Abdul Kadir[1,*]

[1]Faculty of Engineering Technology, Universiti Teknikal Malaysia Melaka, 7600 Durian Tunggal, Melaka, Malaysia

Abstract. This paper presents a result of an experimental study by using a helicopter toy for monitoring several environment parameters such as temperature, humidity, and air quality. A microcontroller board named Arduino was embedded in the helicopter. This board controlled various sensors. The data captured by the sensors was sent to the receiver by using a serial communication provided by a pair of XBee Pro.

1 Introduction

Researches in using aero vehicles for many purposes have been done by several researches. For example, the application of unmanned aero vehicles (UAV) has been studied and implemented in [1]. Other researches focussed on how to control the aero vehicles, such as used Android to control the quadcopter. Hanafi, et al. [2] highlighted the use of graphical user interface (GUI) to control the quadcopter. Whilst, Varshnev [3] concentrated on how to handle gesture controlled aero vehicle and obstacle avoidance occurred in its path.

This research was actually the part of an integral project for utilizing UAV to do monitoring several environment parameters. However, to ease in getting results for further development of the whole system, the implementation of monitoring environment was the main priority. This goal was to get the proper hardware needed so the problem weight of the hardware could be known in advance.

2 Material and methods

To implement this project, an experimental project was conducted. The experiment was accomplished by embedding the sensors in a helicopter toy. Figure 1 shows the block diagram of the system developed in this project. The outputs of sensors were read by the Arduino and sent to the receiver through XBee Pro. In this case, the graphical LCD was used to display the information gathered from the helicopter. Some materials were used to implement this project. Therefore, the explanations of all materials will be given first.

* Corresponding author: akadir64@gmail.com

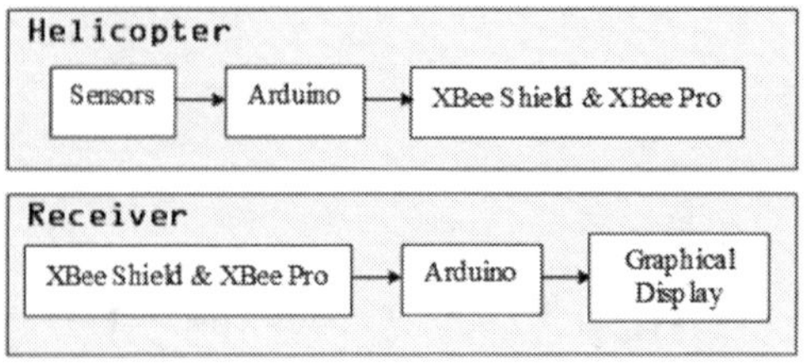

Fig. 1. Block diagram of the system.

2.1 The helicopter

To accelerate the result achievement in this research, a helicopter toy was used. The consideration of choosing this aero vehicle was the cost. This toy was relatively cheap compared to a quadcopter. Besides, because its size was big enough, it was possible to embed many devices, including the microcontroller board, the sensors, and other equipment.

2.2 The arduino board

The Arduino board was chosen as the main controller because it was powerful, cheap, and easy to program. Besides, many sensors were available for this board. Arduino board is an open source singleboard-based microcontroller [4]. Actually, there are many kinds of Arduino boards. In this project, Arduino Uno was used. This board was selected because it provided the various shields. One of the requirements in this system was XBee Pro, which needs a special shield as shown in Figure 2.

2.3 The XBee Pro

The XBee and XBee-PRO OEM RF Modules were engineered to meet IEEE 802.15.4 standards and support the unique needs of low-cost, low-power wireless sensor networks [5]. It can be used indoor and support up to 100 m. For outdoor application, the XBee Pro with 60mW output can handle communication up to 1,500 m.

Fig. 2. Arduino, XBee shield and XBee Pro.

2.4 The humidity and temperature sensors

For measuring the humidity and the temperature of the environment, a sensor named DHT11. A module consisting this sensor is shown in Figure 3(a). This sensor can measure the humidity between 20 and 90o RH with the accuracy of +5% RH. Measurement for temperature is between 0 and 50°C with the accuracy of +2°C. Communication between this sensor and the Arduino is accomplished by using one-wire serial bus.

2.5 The dust sensor

The dust sensor named GP2Y1010AU0F (Sharp Optical Dust Sensor) was used for measuring air quality. It will report the dust density in mg/m^3 Figure 3(b) shows this sensor.

2.6 The MQ-4 sensor

MQ-4 (Figure 3(c)) is one type of MQ family sensor. The MQ-4 sensor is useful for detecting methane and compressed natural gas (CNG). It can detect gas concentration from 200ppm to 10,000ppm.

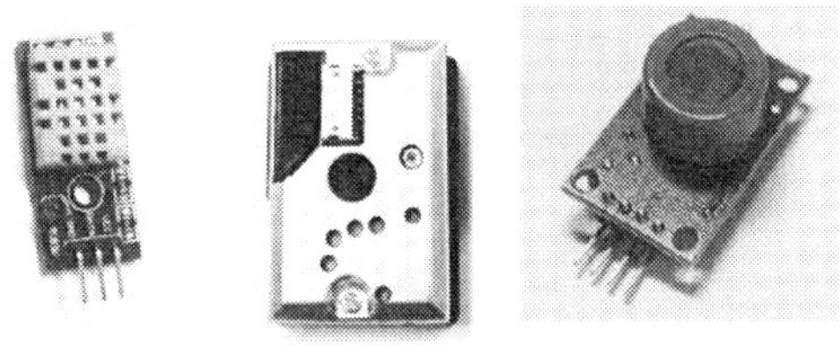

(a) DHT11 (b) GPY2Y1010AU0F (c) MQ-4

Fig. 3. Three sensors used in the experiment.

2.7 The CLCD

A graphical liquid crystal display (GLCD) is useful for displaying charts and also texts. The display shown in Figure 4 has 64x128 pixels.

Fig. 4. The GLCD.

3 Experimental research and discussion

The helicopter toy and the receiver used in the experiment were shown in Figure 5. All sensors were located in the lower part of the helicopter.

(a) Transmitter (b) Receiver

Fig. 5. The helicopter toy used in the experiment.

Figure 6 shows how to connect all sensors to the Arduino board. The dust sensor is obtained by reading the pin A0. The pin 2 is used to control the LED pin of the sensor.

DOI: 10.1051/matecconf/20179701096

How to program this sensor can be achieved at [6-7]. According to [7], the dust density can be obtained by using the following command:

density = 0.172 * voltage + 0.0999;

In this case, the voltage is the value on the pin A0.The DHT sensor was connected to the Arduino by utilizing the analog pin A1. Whilst, the output of the MQ-4 is monitored on the pin A2 by using the following command:

mq_value = analogRead(A2);

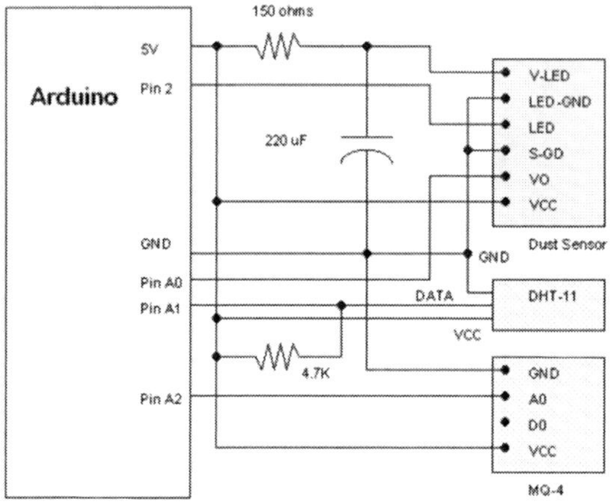

Fig. 6. The connection between the DHT sensor and the Arduino board.

To ease in programming the DHT-11 sensor, a library named DHT is needed. This library is available at www.hobbyist.co.nz/?q=documentations/wiring-up-dht11-temp-humidity-sensor-to-your-arduino. This library has a class named dht. The fragment of code to get the temperature and the humidity is as below:

```
dht DHT;
DHT.read11(A1);
float humidity = DHT.humidity;
float temperature = DHT.temperature;
```

Sending the sensor data is handled by the XBee Pro of the transmitter part. The data has a format as follows:

T=value,H=value,Q=value,G=value

In this case, T represents the temperature, H represents the humidity, D represents the air quality, and G represents the gas concentration.

On the receiver part, the XBee shield will read those data every two seconds and then the data is parsed and displayed to the GLCD. To program the GLCD, a library named U8glib is used. This library can be downloaded freely at https://github.com/olikraus/U8glib_Arduino.

Testing was done several times by flying the helicopter manually using a remote control to the targeted area and monitoring the output of sensors in the GLCD. The maximum distance was tested not more than 0.5 km although XBee could handle until 1.5 km. At the range, it was easy to see the helicopter. The results shown that all sensors could function as expected on that range.

Based on the results, the helicopter can be used to monitor the environment parameters remotely. It is useful for ad hoc measurements in certain places such as in agriculture area or in the dangerous area for human such as the mining to detect certain parameters.

4 Conclusions

The communication used in monitoring the environment monitor by using XBee Pro gave satisfied results. Therefore, it is expected that the communication channel can be used for handling some additional sensors. Besides, the helicopter is still able to bring some sensors because the total weights of all equipment needed for monitoring the environment parameters are tolerable. Based on the performance, it is possible also to equip the helicopter with a light camera and GPS module sensors as well. Therefore, for next future implementation, the use of a camera to capture the view of environment and a GPS module to record the geographical position in latitude and longitudes will be the priorities. An SD card is also possible to include to record the data.

Thanks to students of Engineering Technology Faculty, UTeM, who participated in testing this project.

References

1. C.R. Verschoor, "Integrating vision-based algorithms on an parrot ar drone to autonomously follow linear shaped structures in a landscape," Faculty of Science, University of Amsterdam (2012)
2. D. Hanafi, M. Qetkeaw, R. Ghazali, M.N.M. Than, W.M. Utomo, R. Omar, Int. J. Communications, **6**, 52 (2013)
3. V. Varshney, M. Wilson, S. Sivaraman, Inter. J. Eng. Adv. Technol, **3**, 222 (2014)
4. J. Bayle, *C programming for Arduino* (Packt Publishing, Birmingham, 2013)
5. Information on http://www.sparkfun.com/datasheets/Wireless/Zigbee/XBee-Manual.pdf
6. Information on https://www.sparkfun.com/datasheets/Sensors/gp2y1010au_e.pdf.
7. Information on http://www.howmuchsnow.com/arduino/airquality/ (2012)

Applications of Foamed Lightweight Concrete

Kamarul Aini Mohd Sari[1,*], and *Abdul Rahim* Mohammed Sani[2]

[1,2]Faculty of Engineering Technology, Universiti Tun Hussein Onn Malaysia, 86400 Batu Pahat, Johor.

Abstract. Application of foamed concrete is increasing at present due to high demand on foamed concrete structures with good mechanical and physical properties. This paper discusses on the use of basic raw materials, their characteristics, production process, and their application in foamed lightweight concrete with densities between 300 kg/m^3 and 1800 kg/m^3. It also discusses the factors that influence the strengths and weaknesses of foamed concrete based on studies that were conducted previously.

1 Introduction

The introduction of polymers into cementitious materials to improve the bond adhesion, flexibility and workability of resultant composite was first made in the 1930s with natural rubber. However, the main difficulty encountered when using this natural polymer material is the inability to alter the chemical structure and tailor the physical properties to those required for a particular application (Miller 2005).

The primary reason for introducing polymer materials into concrete and cement-based compounds is to improve the physical and mechanical characteristics of the resultant cementitious material to be lightweight, self compacting, high workability, good thermal insulation, and can be cut and nail/screw easily (Koh 2008). One of the important characteristics of lightweight concrete is light or low-density foamed. El-Reedy, 2009 has state the classification of lightweight concrete in Fig. 1.

Foamed concrete is formed when air-voids are entrapped in mortar by suitable foaming agent. It possesses high flow ability, low self-weight, minimal consumption of aggregate, controlled low strength and excellent thermal insulation properties. Foamed concrete with densities between 400 kg/m^3 and 1600 kg/m^3 is obtained for the purpose of structural, partition, insulation, and filling grades (Ramamurthy 2009).

An improvement in the workability and bond adhesion will also increase the flexural and tensile strength. Polymer modified cementitious materials also showed an increase in durability and water resistance. Hence, the addition of the polymer into cementitious system enabled its use in more complex structures and situations (Ohama 1998).

* Corresponding author: aini@uthm.edu.my

Fig. 1. Classification of lightweight concrete.

No-fines Concrete	**Lightweight Aggregate Concrete**	**Aerated Concrete**	
		Chemical	**Foaming Mixture**
Gravel Crushed stone	Clinker Foamed slag	Aluminium powder Hydrogen peroxide and bleaching powder	Performed foam Air-entrained foam
Coarse clinker Sintered pulverized fuel ash	Expanded clay Expanded shale		
Expanded clay or shale Expanded slate	Expanded slate Sintered pulverized fuel ash		
Foamed slag	Exfoliated vermiculite Expanded perlite Pumice Organic aggregate		

The initial use of polymers in concrete systems is achieved using a liquid additive, which was added to the cementitious material during mixing. However, in certain circumstances it has become more advantageous nowadays to include the polymer in the actual cementitious powder, which polymerizes when in contact with water during the mixing process. This allows the quantity of polymer throughout the mix to be controlled more precisely, reducing the errors associated with the mixing of the dispersion directly with the cement-based product. This can have the advantage of lowering the cost of manufacturing and construction process.

2 Materials and method

2.1 Materials

Main materials to produce foam concrete are mortar or cement grout and foam. Mortar is normally fresh concrete which contains a mix of cement, fine sand and water. Additional mix is foam protein-based or synthetic to reduce the weight of the concrete. Hydrolyzed proteins have higher strength performance, while synthetic based foaming agents are easier to handle, cheaper and requires less energy and can be stored longer.

2.2 Production methods

Foaming admixtures is based on BS8443:2005. Barnes 2009 and Ramamurthy 2009 discussed the method of producing foam, known as the wet method and the dry method.

(i) Wet foam has a large loose bubble structure. Although stable, is not recommended for the production of foamed concretes with densities below 1000 kg/m^3. It involves spraying a solution of the agent and water over a fine mesh, leading to a foam with bubbles of size between 2 mm and 5 mm.

(ii) Dry foam is extremely stable. Stability is a characteristic that has become increasingly important as the density of the foamed concrete reduces. It is produced by forcing a solution of foaming agent and water through restrictions whilst forcing compressed air

 DOI: 10.1051/matecconf/20179701097

into the mixing chamber. The resulting bubble size is smaller than wet foam, i.e. less than 1 mm in diameter.

After the foam is stable, it can be mixed into the concrete using two method (Barnes 2009), i.e. the inline or pre-foam methods.

2.1.1 Inline

Cement, fine sand and water are placed into a unit where they are blended together with the foam. The mixing process is more controlled and greater quantities can be easily produced. There are two types of processes, depending on whether the wet method or dry method is used.

In wet method, the base materials are the same as those used in the pre-foam system, but are generally wetter. The base material and the foam are fed through a series of static inline mixers where the two are mixed together. The foam and the base materials are blended together and continuously monitored by an on-board density monitor. The output volume is not governed by the size of the pre-mix concrete truck, but by the density of the foamed concrete, where one 8 m^3 delivery of base material can produce 35 m^3 of a 500 kg/m^3 foamed concrete.

2.1.2 Pre-foam

In pre-foam method, pre-mix concrete truck delivers the base materials to site and then the pre-formed foam is injected directly into the back of the truck whilst the mixer is rotating. The advantage of this method is that, relatively small quantities can be ordered. However, it does rely on the mixing action of the concrete truck. Densities in the range of 300 kg/m^3 to 1200 kg/m^3 can be achieved. Typically foamed air in this method is in the range of 20% to 60% of air.

3 Properties

Ramamurthy, 2009 listed types of foam concrete properties as:

(i) fresh state properties to be familiar with the consistency and stability,
(ii) physical properties to recognize the drying shrinkage, air-void system and density,
(iii) mechanical properties to identify the compressive strength, flexural and tensile strengths, and modulus of elasticity,
(iv) thermal properties to classify the thermal conductivity, thermal expansion, specific heat, and flammability,
(v) strength prediction models,
(vi) durability of foamed concrete such as permeation characteristics, and resistance to aggressive environment, and
(vii) functional characteristics for instance thermal insulation, acoustical properties, and fire resistance.

Table 1 summarizes the applications of foamed concrete based on its density (Aldridge 2005, Liew 2005, Jones 2005, Barnes 2009). Common density of foamed concrete to be developed is between 1000 kg/m^3 to 1500 kg/m^3, which generally is used for housing applications, prefabrications and cast-in place wall, either load bearing or non-load bearing structures.

Table 1. Summary of foamed concrete applications based on density

Density (kg/m^3)	Application
300-600	Replacement of existing soil, soil stabilization, raft foundation.
500-600	Currently being used to stabilize a redundant, geotechnical rehabilitation and soil settlement. Road construction.
600-800	Widely used in void filling, as an alternative to granular fill. Some such applications include filling of old sewerage pipes, wells, basement and subways.
800 - 900	Primarily used in production of blocks and other non-load bearing building element such as balcony railing, partitions, parapets, etc.
1100-1400	Used in prefabrication and cast-in place wall, either load bearing or non-load bearing and floor screeds.
1100-1500	Housing applications.
1600-1800	Recommended for slabs and other load bearing building element where higher strength required.

Apart from the density of the aggregates, the density of the concrete also depends on the grading of the aggregates, their moisture content, mix proportions, cement content, water to binder ratio, and chemical and mineral admixtures. Besides characteristics of the materials, it also depends upon the method of mixing, compaction, and curing condition (Osman Unal 2007).

Barnes (2009) reported that there is a correlation between the density and compressive strength of foamed concrete. Comparing the wet density of the foamed concrete at the point of discharge, with the wet density determined during development testing gives an estimate of likely strength at a given age.

The cellular structure of foamed concrete contributes to good thermal insulating properties and low thermal conductivity values. Table 2 shows that the thermal conductivity of foamed concrete ranges between 0.1 W/mK and 0.7 W/mK for dry densities of 600 kg/m^3 to 1600 kg/m^3. These thermal conductivities of foamed concrete are typically 5% to 30% of those measured on normal weight concrete. Thermal conductivities reduces with decreasing densities.

Table 2. Typical properties of foamed concrete based on British Concrete Association (BCA)

Dry Density (kg/m^3)	Compressive strength (N/mm^2)	Thermal Conductivity (W/mK)	Modulus of Elasticity (kN/mm^2)	Drying Shrinkage (%)
400	0.5 – 1.0	0.10	0.8 – 1.0	0.30 – 0.35
600	1.0 – 1.5	0.11	1.0 – 1.5	0.22 – 0.25
800	1.5 – 2.0	0.17 – 0.23	2.0 – 2.5	0.20 – 0.22
1000	2.5 – 3.0	0.23 – 0.30	2.5 – 3.0	0.15 – 0.18
1200	4.5 – 5.5	0.38 – 0.42	3.5 – 4.0	0.09 – 0.11
1400	6.0 – 8.0	0.50 – 0.55	5.0 – 6.0	0.07 – 0.09
1600	7.5 – 10.0	0.62 – 0.66	10.0 – 12.0	0.06 – 0.07

Foamed concrete depends on its many characteristics for its different applications in building construction. These are some guidelines on production and processes to produce the best foamed concrete.

The strength of foamed concrete are increase by :

(i) decreasing porosity (less foam),

(ii) using finer sand, less than 5 mm (Nambiar 2006),

(iii) introducing small air bubbles (between 0.1 mm and 1.5 mm) with uniform distribution (Brady 2001, Kunhanandan Nambiar 2006),

DOI: 10.1051/matecconf/20179701097

(iv) using fly ash and silica fume for pozolanic reaction,
(v) air curing (compared to sealed/ water curing) (Kearsley 1998,), and
(vi) usage of polypropylene.

4 Conclusions

Applications and properties of foamed lightweight concrete have been discussed in this paper. In extension of this paper, the production of lightweight concrete for insulation purposes with target density below 1500 kg/m^3 will be carried out. The natural material considered is the palm kernel oil-based polyurethane, which will be mixed with concrete for the production of building materials converging on wall provisions.

References

1. D. Aldridge, *Introduction to Foamed Concrete: What, Why and How? Use of Foamed Concrete in Construction* (Thomas Telford Publishing, UK, 2005)
2. Information on https://www.thenbs.com/PublicationIndex/documents/details?Pub=BCA&DocId
3. R.A. Barnes, *Foamed Concrete: Application and spesification. Excellence in Concrete Construction through Innovation* (Taylor & Francis Group, London, 2009)
4. E.K. Kunhanandan, K.R. Nambiar, Cement Concrete Comp., **28**, 475 (2006)
5. M.A.E. Reedy, *Advance materials and techniques for reinfoorced concrete structure* (Taylor and Francis Group, Egypt, 2009)
6. H.B. Koh, Y.L. Lee, *Foamed Concrete - A Potential Civil Engineering Materials* (UniMAP, Perlis, 2008)
7. K.C. Brady, M.R. Jones, *Specification for Foamed Concrete* (Highway Agency, UK, 2001)
8. K. Ramamurthy, E.K.K. Nambiar, G.I.S. Ranjani, Cement Concrete Comp., **31**, 388 (2009)
9. E.P. Kearsley, P. Booysens, Concrete Beton, **91,** 9 (1998)
10. A. Liew, *New Innovaative Lightweight Foam Concrete Technology. Use of Foamed Concrete in Construction* (Thomas Telford Publishing, UK, 2005)
11. M.R. Jones, A. McCarthy. *Behaviour and Assessment of Foamed Concrete for Construction Applications. Use of Foamed Concrete in Construction* (Thomas Telford Publishing, UK, 2005)
12. M. Miller, *Polymers in Cementitious Materials* (Rapra Technology Limited, UK, 2005)
13. Y. Ohama, Cement Concrete Comp., **20**, 189 (1998)
14. Osman Unal, T. Uygunoglu, A. Yildiz, Build. Environ., **42**, 584 (2007)

Smart Garbage Monitoring System for Waste Management

Norfadzlia Mohd Yusof[1,*], *Aiman Zakwan* Jidin[1], and *Muhammad Izzat* Rahim[1]

[1]Faculty of Engineering Technology, Universiti Teknikal Malaysia Melaka, Malaysia

Abstract. Piles of rubbish are one of the major problems faced by most people in Malaysia, especially those who live in flats, as the number of bins is limited and shared among all residents. It may cause pollutions, which may lead to sanitary issues and diseases. This project presents the development of a smart garbage monitoring system in order to measure waste level in the garbage bin in real-time and to alert the municipality, in particular cases, via SMS. The proposed system is consisted by the ultrasonic sensor to measure the waste level, the GSM module to send the SMS, and an Arduino Uno which controls the system operation. It supposes to generate and send the warning messages to the municipality via SMS when the waste bin is full or almost full, so the garbage can be collected immediately. Furthermore, it is expected to contribute to improving the efficiency of the solid waste disposal management.

1 Introduction

Nowadays, there are tons of flats and apartments which have been built in the rapid urbanization area. This is due to high housing demands which have been drastically risen as a result of migration from villages to cities to find works. In order to accommodate the growing population in the urban area, the government has built flats, apartments or condominiums, to provide shelter for them. Besides, flats become popular mainly in the United State, Europe and other developing countries [1, 2].

There are several issues faced by the inhabitants of the flats. One of them is the issue of the domestic solid waste disposal, which cause pollutions [3]. Unlike landed houses, the flats' waste disposal bins are shared among all residents which live in the same building, and thus, the bins tend to be filled very quickly. Thus, an unsystematic and inefficient disposal waste management may cause the bins to be always full with of garbage, and further littering from the residents will cause the garbage piles to be scattered outside the bins. Therefore, there will be a question of sanitary as those garbage piles may become the root cause of illness and diseases like dengue, diarrhea, and cholera.

Besides, there are also problems regarding the attitudes of each inhabitant of the flats. There are cases where some irresponsible residents, who normally live at the higher levels of the building, littered or simply threw their domestic waste directly from the floor which

* Corresponding author: norfadzlia@utem.edu.my

they live into the bins. This may cause pollutions if the garbage thrown fell outside the bins or injuries to innocent people downstairs if they fell onto them.

The waste disposal can be managed more properly and efficiently by constantly monitoring the bin status and the garbage level [4]. In addition, the municipality can be alerted when the bin is full or almost full, thus promoting dynamic scheduling and routing of the garbage collection. By comparing to the conventional static scheduling and routing, this dynamic scheduling and routing are said to allow operational cost reduction, by reducing the number of trucks, the manual labour cost and the transport mileage savings [5, 6, 7].

This paper presents an alternative in managing domestic waste especially in flat areas via a smart garbage monitoring system, which is developed based on Arduino Uno. This system will automatically monitor the garbage level at each bin and will alert the municipality in the case where the bins are almost full.

2 Research methodology

2.1 System architecture

Fig. 1 describes the block diagram of the monitoring system. It consists of several ultrasonic sensors, the Arduino Uno microcontroller, and the GSM module. The ultrasonic sensors are used to detect the level of garbage in each bin, and it will send this information to the Arduino Uno, which acts as the system controller. In the case where the bins are already full or almost full, then it will generate a warning message which will be sent to the municipality via SMS, by using the GSM module. Moreover, this system is also connected to several LEDs, in green or red colours, in order to alert all the residents regarding the bin status.

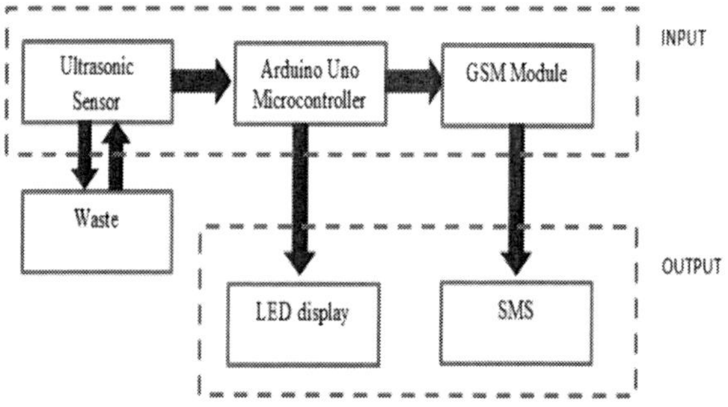

Fig. 1. Waste Management Monitoring System Block Diagram

2.2 System process flow

At first, the level or the height of the garbage in each bin is measured by using the ultrasonic sensor. This information is then received and processed by the Arduino Uno. It will determine whether the garbage level has been surpassing the threshold level or not. For this research purpose, there are two threshold levels sets: the first threshold is at 70% of the bin height, and the second threshold is set at 90% of the bin height.

If the garbage level in the bin is crossing the first threshold level, then the first warning message is generated and sent to the municipality. Besides, the green LEDs will be turned ON in order to alert all the residents at every floor. Next, if the garbage level in the bin is crossing the second threshold level, then the second warning message is generated and sent to the municipality. In this case, all the residents will be alerted when the red LEDs are turned ON.

3 Results and discussions

Fig. 2 shows the hardware connection setup for the system. An Arduino Uno board is connected to the HC-SR04 ultrasonic sensor via digital I/O pin. Besides, the SIM900A GSM Module is serially connected to the Arduino Uno board, where the TX port of the GSM Module is connected to the RX port (PIN 2) of the Arduino Uno, while RX port of the former is tied to the TX port (PIN 3) of the latter. Moreover, a green LED and a red LED are also connected to the Arduino Uno.

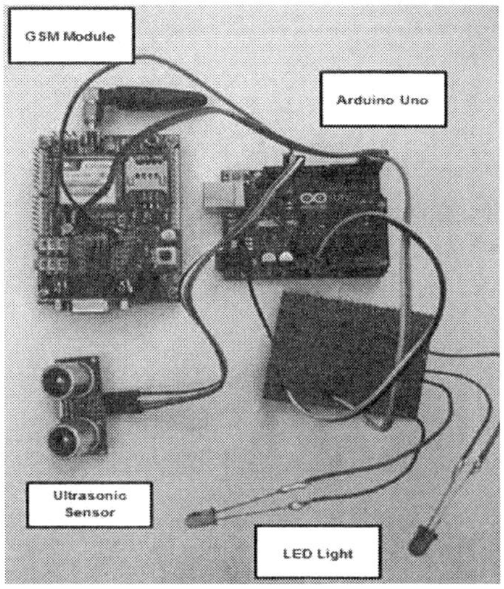

Fig. 2. All connection component of hardware in this system

The first test conducted is the situation where the garbage bin is empty or its garbage level is very low. As can be observed in Fig. 3, neither the green LED nor the red LED is turned on.

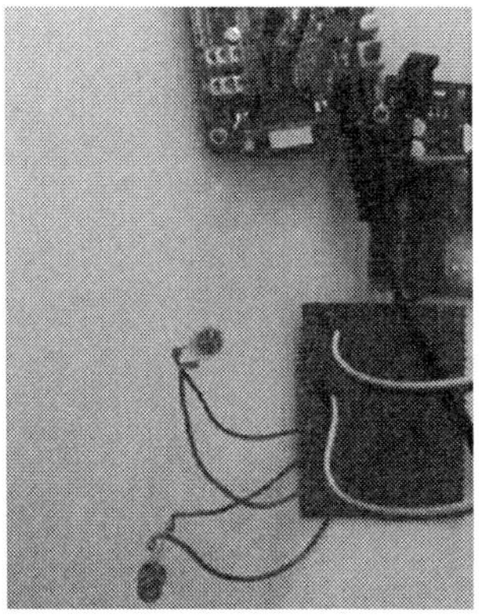

Fig. 3. Both LEDs are turned OFF.

Then, the bin is filled with more garbage until its level has surpassed the first threshold value, which is set to 70%. Fig. 4 shows that the green LED is turning ON, and then the first warning SMS is being sent, as depicted in Fig. 5.

Fig. 4. GREEN LED is turned ON once the waste reaches the level of 70% full.

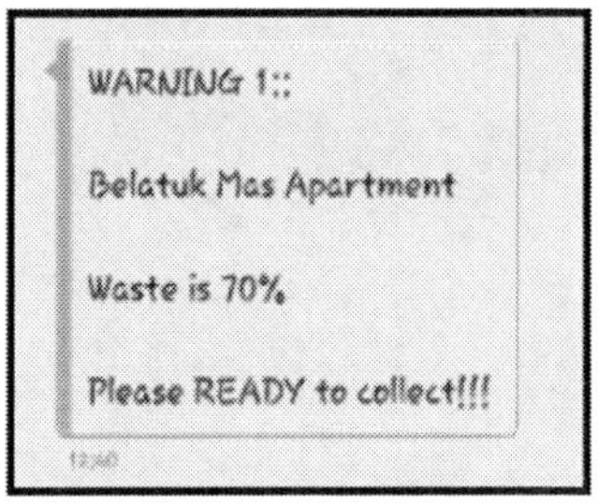

Fig. 5. The first notification SMS sent by the system, once the waste reaches the level of 70% full.

Next, the bin is fully filled and thus, the garbage level is now exceeding the second threshold level, which is set to 90% of the bin height. As the consequences, the red LED is now turning ON while the green LED is turned OFF, and the second warning SMS is sent, telling the municipality to immediately collect the waste disposal. This is shown in Fig. 6 and Fig. 7, respectively.

Fig. 6. Red LED is turned ON once the waste reaches the level of 70% full.

Fig. 7. The second notification SMS sent by the system, indicating that bin is at least 90% full and the garbage needs to be collected immediately.

4 Conclusions

This paper has described the development of a smart garbage monitoring system, which is based on Arduino Uno microcontroller. It is very useful in improving the efficiency of solid waste disposal management especially in the flat residential areas, where the garbage piles at the bins are one of the residents' major concerns owing to its ability to continuously measure the garbage level in the bin and alerting the municipality for immediate collection. The outputs from the conducted tests show that all the functionality of the system has performed correctly. The proposed system is suitable to be implemented in all flat residential areas, due to its practicality, reliability and reasonable cost.

The authors wish to acknowledge Universiti Teknikal Malaysia Melaka (UTeM) for the financial funding and providing instrumentation devices support of this project.

References

1. P. Sukholthaman, K. Shirahada, *Proceedings of PICMET '14 Conference: Portland International Center for Management of Engineering and Technology; Infrastructure and Service Integration*, (2014)
2. C. K.M. Lee, T. Wu, International Conference on Industrial Engineering and Engineering Management, 798 (2014)
3. A.F. Thompson, A.H. Afolayan, E.O. Ibidunmoye, *Information Science, Computing and Telecommunications*, 206 (2013)
4. M.A. Hannan, M. Arebey, R.A. Begum, H. Basri, Waste Manage., **32**, 2229 (2012)
5. M.A.A.Mamun, M.A. Hannan, A. Hussain, H. Basri, IEEE Sensors Journal, **15**, 561 (2015)
6. H. Krikke, I.L. Blanc, M. van Krieken, H. Fleuren, Int. J. Prod. Econ., **111**, 209 (2008)
7. O. M. Johansson, Waste Manage., **26**, 875 (2006)

MATEC Web of Conferences **97**, 01099 (2017) DOI: 10.1051/matecconf/20179701099
ETIC 2016

Location tracking system using wearable on-body GPS antenna

Thennarasan Sabapathy[1,*], *Mohd Amirudin* Mustapha[2], *Muzammil* Jusoh[1], *Shakhirul* Mat Salleh[3], *Ping Jack* Soh[4]

[1] Bioelectromagnetic Research Group (BioEM), School of Computer and Communication Engineeering, Universiti Malaysia Perlis, 02600 Arau, Malaysia
[2]Department of Electronic Industry, Industrial Training Institute Kepala Batas, 13200 Pulau Pinang, Malaysia
[3] Faculty of Engineering Technology, Universiti Malaysia Perlis, 02600 Arau, Malaysia
[4] Advanced Communication Engineering Centre (ACE), School of Computer and Communication Engineeering, Universiti Malaysia Perlis, 02600 Arau, Malaysia

Abstract. An on-body location tracking system is developed and integrated with a wearable GPS antenna. Such system is beneficial in human location tracking of patients and elderly within a radius of 1 km. The system consists of a wearable antenna, a GPS module, a low cost microcontroller, two RF modules and a local monitoring system. A user equipped with the GPS antenna, GPS module and a RF transmitter is able send his/her location to the local monitoring system via a RF receiver. The proposed wearable antenna is validated to be safe for human use in terms of specific absorption rate (SAR). This antenna was then incorporated into the complete prototype and tested. Several suggestions for future improvements are also proposed and discussed.

1 Introduction

During the past decade, location tracking/monitoring systems have been actively developed for tracking children, elderly, disabled [1-4], vehicles and etc. Specifically, human location tracking systems are important to ensure human safety and to efficiently initiate search and rescue in the event of emergencies. Such location monitoring can be developed using different technologies such as global positioning system (GPS) [2, 5], radio frequency identification (RFID) [4, 6] and ZigBee [3]. Their selection depends on the application requirements such as coverage, cost and accuracy.

A majority of previously developed tracking systems focused on aspects such as security and privacy [4, 7], accuracy improvement [8], and body effects to the antenna used [2]. The need of developing low cost tracking system is addressed in [1]. However, these works have yet to address the issue of implementing location tracking system on?? Human body. Ref. [2] described the effect of body towards the performance of the system but did

* Corresponding author: thennarasan@unimap.edu.my

not discuss the effect of the system towards the human body. Apart from that, none of the aforementioned works adopted the use of wearable antennas, despite their attractive features and suitable properties for on-body application [9].

In this work, the issue of implementing a body-wearable antenna onto a low cost location tracking system is addressed. The wearable antenna is first designed for operation before being studied in terms of specific absorption rate (SAR) level to ensure safety for operation on body. . The rest of this paper discusses the system description and results obtained through simulations and measurements.

2 System overview

The overall location tracking system consists of a wearable, circularly-polarized GPS antenna, a microcontroller, GPS module and RF module to enable its functionality.

2.1 Antenna design

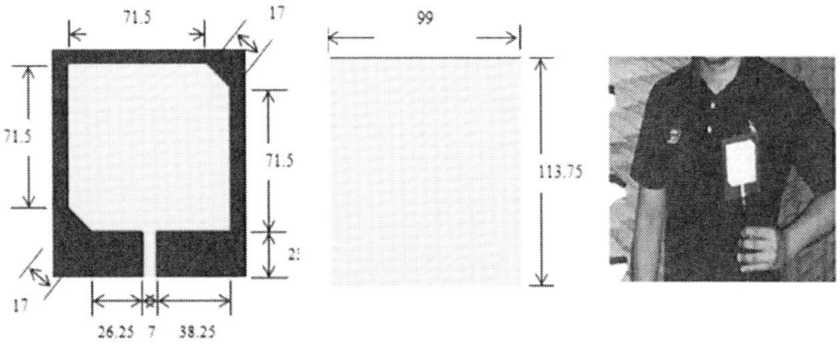

(a) (b) (c)

Fig. 1. Structure of the proposed wearable CP Antenna (a) top view of the antenna (dimensions in mm) (b) Bottom view of the antenna (dimensions in mm) (c) Fabricated antenna during on-body measurements.

As presented in [9], it is demonstrated that the existence of the full ground plane on the reverse side of the patch antenna is expected to minimize the effect of the body, resulting in very low SAR levels. Based on this, the selected circularly polarized (CP) antenna topology for this system adopts a full ground plane. It is based on a square shaped topology with corner truncations, as illustrated in Fig. 1. It operates at 1.575 GHz with a right-handed circular polarization (RHCP). Textile materials are used for the fabrication of the antenna. The substrate used is Felt with a thickness of 1.5 mm and a dielectric constant of 1.22. Similar to the work in [10], the conductive textile known as SheildIt super is used to form its patch and ground plane.

2.2 System design

The developed antenna was then incorporated with a GPS module and Arduino Mega microcontroller. The microcontroller is programmed to capture the GPS data and send it to the local monitoring system through RF modules which operate at 2.4 GHz band. The graphical user interface (GUI) for the monitoring system is developed using Visual Basic software. The overall system is illustrated in Fig. 2.

DOI: 10.1051/matecconf/20179701099

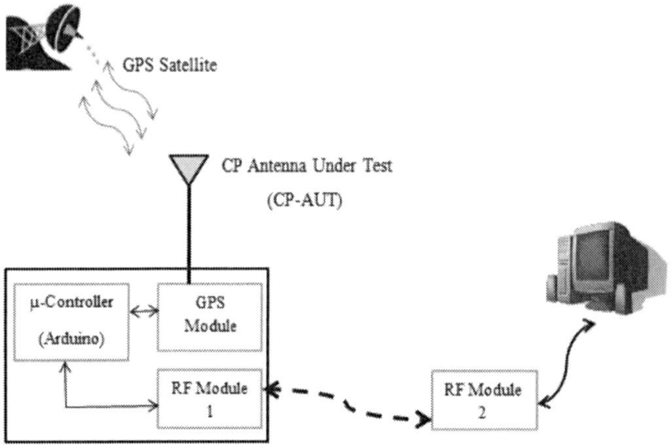

Fig. 2. Location tracking system using wearable CP antenna

3 Results

3.1 Antenna performance

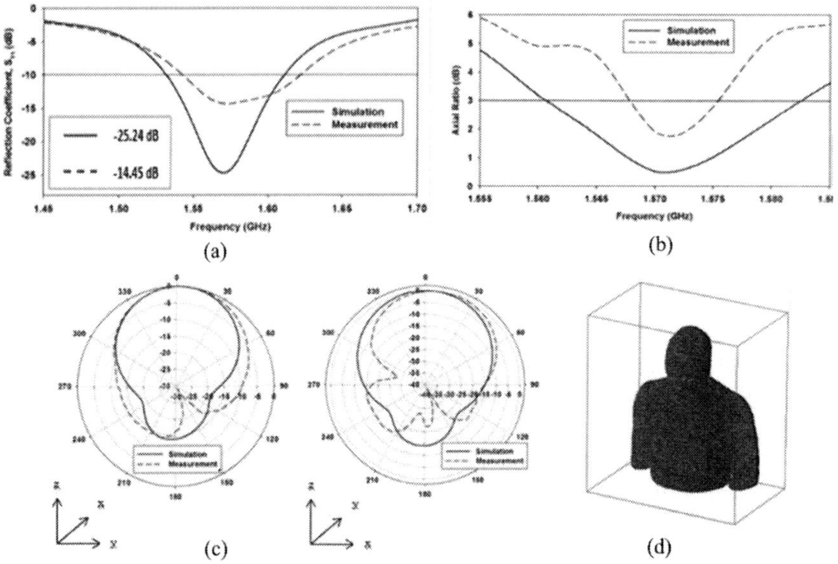

Fig. 3. Results of the antenna. (a) Reflection coefficient. (b) Axial ratio. (c) Radiation pattern (d) SAR analysis

All evaluated antenna parameters are satisfactory, as shown in Fig. 3.. Measured 10 dB impedance bandwidth indicated operation from 1.538 GHz to 1.622 GHz, while its 3 dB axial ratio from 1.568 GHz to 1.577 GHz. The antenna gain is approximately 3 dBi with a unidirectional radiation pattern. The SAR analysis conducted in proximity of a HUGO human body model resulted in values of 1.478 W/kg for averaged over a 1g of tissue, and 0.356 W/kg for 10g tissue-mass averaging.. Both values do not exceed the limit regulated by the ICNIRP guidelines [9].

3.2 GPS data processing and transmission

In this section, the developed system is tested to validate its practical usability. The designed antenna was capable in receiving the GPS signal. Meanwhile, the developed processing architecture was able to process the data and send it to the local monitoring system/data centre as depicted in Fig. 4.

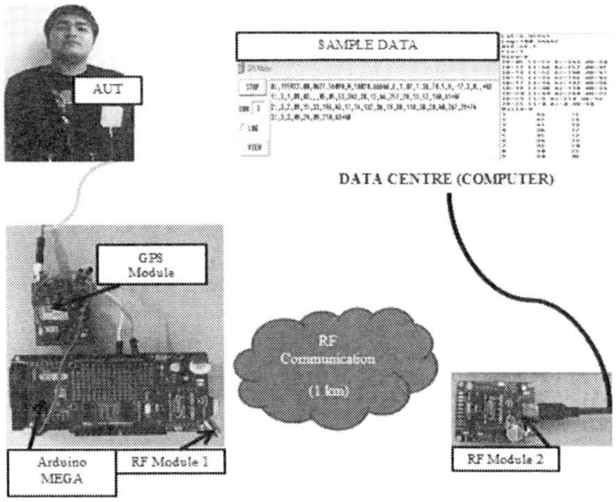

Fig. 4. Practical outdoor measurement setup using the developed location tracking system

The outdoor measurement is conducted in a free space environment. Fig. 4 shows the location tracking system in two parts, firstly, the microcontroller module and secondly, the interface module located in the data centre. Each module has its own functionality. The microcontroller module is attached to the on-body GPS antenna and GPS module to receive Global Positioning System Fixed Data (GGA) and GNSS Satellites in View (GSV) values from available GPS satellites.). Sample of such received data is shown below:

: $GPGSV,3,1,11,10,63,137,17,07,61,098,15,05,59,290,20,08,54,157,30*70
: $GPGSV,3,2,11,02,39,223,19,13,28,070,17,26,23,252,,04,14,186,14*79
: $GPGSV,3,3,11,29,09,301,24,16,09,020,,36,,,*76

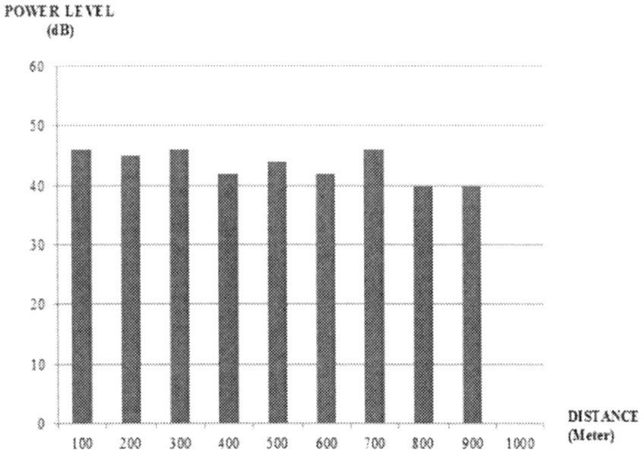

Fig. 5. Power level (dBHz) vs distance (meter).

The received power level is then extracted from the data and is illustrated in Fig. 5. It indicates the power level versus distance between the microcontroller on the human body and data centre. It can be observed that from distance 100 meter to 900 meter, the power

DOI: 10.1051/matecconf/20179701099

level ranges from 40 dBHz to 46 dBHz. However, beyond the distance of 1000 meter, communication between RF module 1 and RF module 2 has ceased to exist. This result validates that the specification RF modules are able to operate up to a maximum distance of 1 km.

4 Conclusions

A low cost location tracking system using wearable GPS antenna is successfully developed and tested. The designed wearable antenna operates at 1.575 GHz with acceptable impedance and axial ratio bandwidths. The antenna is validated to be safe for on body use with SAR values of 1.478 W/kg (averaged over 1g of tissue) and 0.356 W/kg (averaged over 10g of tissue). These values do not exceed the regulatory limit specified in the ICNIRP guidelines. The system can be operated for coverage of up to 1 km of radial distance, and can be improved with the use of high gain antenna at RF modules.

References

1. D. Punetha, V. Mehta, 2014 International Conference on Advances in Computing, Communications and Informatics, 2349 (2014)
2. X.L. Chen, N. Kuster, Y.C. Tan, and N. Chavannes, 6th European Conference on Antennas and Propagation, (2012)
3. C. Park, J. Kim, 2011 Seventh International Conference, (2011)
4. J. Lee, K.E. Khatib, 2010 IEEE Second International Conference, (2010)
5. R.I. Rajkumar, P.E. Sankaranarayanan, G. Sundari, 2013 International Conference on Advanced Electronic Systems, (2013)
6. H. Rajagopalan, Y.R.Samii, 2010 IEEE Antennas and Propagation Society International Symposium, (2010)
7. C.Y. Chow, M.F. Mokbel, T. He, IEEE Transactions on Mobile Computing, 10, 94 (2011)
8. D. Rosner, R. Tataroiu, C. Ilas, A.L. Beciu, O.M. Alexandru,19th International Conference on Control Systems and Computer Science, (2013)
9. P.J. Soh, G. Vandenbosch, F.H. Wee, A.v.d. Bosch, M.M. Vazquez, D. Schreurs, IEEE Antennas and Propagation Magazine, 57, 229 (2015)
10. M.S. Shakhirul, M. Jusoh, A.H. Ismail, M.R. Kamarudin, R. Yahya, M.N.M. Yasin, IEEE Asia-Pacific Conference on Applied Electromagnetics, (2014)

Effective Design for Optical CDMA Based on Radio over Fiber (RoF) Technique

Rashidi C. B. M[1,*], *Aljunid* S. A[1] , *Rahman* A.K[1] , *Anuar* M.S[1] , and *Yaakob* S[2]

[1] Advanced Communication Engineering, Centre of Excellence (ACE-CoE), School of Computer and Communication Engineering, Universiti Malaysia Perlis,Kangar, Perlis,
[2] TM Research and Development, Cyberjaya, Selangor Darul Ehsan, MALAYSIA

Abstract. In this paper, the performance of OCDMA coding systems utilizing the radio over fiber (RoF) technique is presented. It has been done by means of conventional OptiSystem simulation tools, where the propagation of radio signals up to 50 km using standard single mode fiber (SMF) was investigated. The analysis was made based on the performance of eye diagram, bit rate, bit error rate and optical received power.

1 Introduction

As the integration of wireless and optical communications, radio-over-fiber (RoF) technology has triggered tremendous interest in being considered as the powerful solution for future high-speed wireless access networks [1]. By using a fiber to connect central-station (CS) with multiple remote simplified base-stations (BS), it can considerably increase the network capacity, coverage and mobility and furthermore serve both fixed and mobile end users [2]. Recently, many investigations have been done about the RoF system and most of them are mainly focused on the overall architecture design, the RF signal generation or the crosstalk reduction [3-5]. Since security is one of the most important performance indexes of a system, it is of great urgency and significance to improving the security of the RoF system.

Optical code-division multiple access (OCDMA) is a promising multiple-access method for high secure access networks [6]. It combines the large bandwidth of fiber with the flexibility of CDMA technique, so it can achieve high-speed connectivity and guarantee the security against interception along the fiber link. Furthermore, OCDMA also has other attractive features like all optical processing, asynchronous transmission, bandwidth efficiency, soft capacity on demand, protocol transparency, simplified network control and so on [7-8]. In consequence, OCDMA technique can be introduced into the RoF system to form the OCDMA-RoF system and it can solve the security problem of the traditional RoF system.

* Corresponding author: rashidibeson@unimap.edu.my

 DOI: 10.1051/matecconf/20179701100

2 Essential of radio over fiber technique

RoF presents the distribution of radio frequency (RF) signals over optical fiber links from a central office to RAU. Radio over fiber-scheme for both down and up-link paths are demonstrated based on digitized RF over fiber technique. The digitized RF-over fiber technique is able to improve spectral efficiency and minimize hardware count, while the performance is limited by inter-channel interference as a result of spectra aliasing of the regenerated IF signals which limits the signal-to-noise-ratio [8]. There are many techniques which are used to transport the RF signal to end-users such as; RF-over-fiber, intermediate frequency (IF) over fiber, and baseband (IF) over fiber. The first two former schemes are examples of analog photonic links and the last is a digital photonic link.

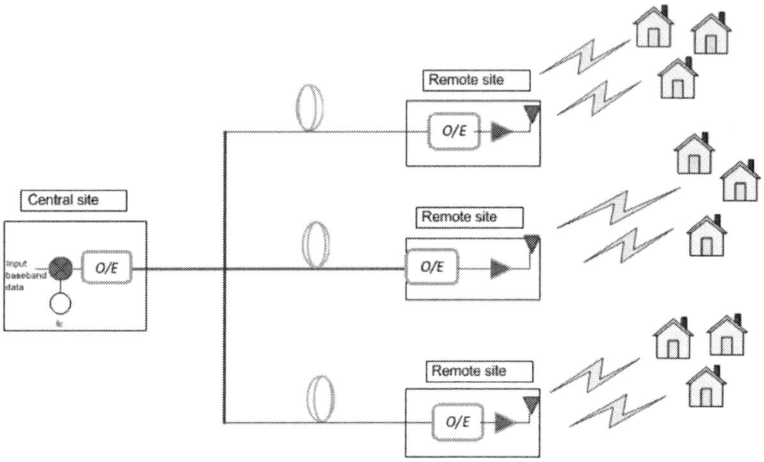

Fig. 1. Basic of RoF Architecture [10]

Figure 1 shows the RF signal is modulated at the central site and combined with continuous wave (CW) laser by external Mach–Zehnder Modulator (MZM), which converts the signal to optical with direct transmission through the fiber to the BSs. At the BS, the signal is detected by the photo-detector (PD) without any need for frequency translation at the remote BS. RF-over fiber transport has the advantage of realizing simple base-station designs with additional benefits of centralized control, independently of the air-interface and also enabling multi wireless band operation. However, one of its major drawbacks is the requirement for high-optical modulation techniques that have the ability to generate speed modulated optical signals as well as high-speed photo-detection schemes that directly convert the modulated optical signals back to signals in the RF domain, Using the external modulator, the dispersion effect can be minimized; thus, this is preferred most in any roof system [9-10].

3 Optical CDMA coding algorithm

Optical codes are family of *K* (for *K* users) binary [0, 1] sequences of length *N*, code weight *W* (the number of "1" in each codeword) and the maximum cross-correlation, λ_{max}. In OCDMA system, to allow receivers to distinguish each of the possible users, to reduce channel interference and to accommodate large number of users, optical codes should have large values of *W* and the size *K*.

Step 1: The set optical code consists of (N, W, λ_{max}) FCC code for K users. The $K \times N$ code matrix is here called the **Tridiagonal Code Matrix.** These sets of codes are then represented by [9];

$$A_K^W = \begin{bmatrix} a_{11} & a_{12} & a_{13} & 0 & 0 & \cdots & 0 \\ a_{21} & a_{22} & a_{23} & a_{24} & 0 & \cdots & \vdots \\ 0 & a_{32} & a_{33} & a_{34} & a_{35} & 0 & \vdots \\ 0 & 0 & a_{43} & a_{44} & a_{45} & a_{46} & \vdots \\ \vdots & \ddots & \ddots & \ddots & \ddots & \ddots & \vdots \\ 0 & 0 & \cdots & \cdots & \cdots & \cdots & a_{KN} \end{bmatrix} = \begin{bmatrix} A_1 \\ A_2 \\ A_3 \\ \vdots \\ \vdots \\ A_K \end{bmatrix} \quad (1)$$

where,

$$\begin{aligned} A_1 &= a_{11}, a_{12}, a_{13} \ldots a_{1N} \\ A_2 &= a_{21}, a_{22}, a_{23,} a_{24} \ldots a_{2N} \\ A_3 &= a_{31}, a_{32}, a_{33,} a_{34} \ldots a_{3N} \\ &\vdots \\ A_K &= a_{K1}, a_{K2}, a_{K3} \ldots a_{KN} \end{aligned}$$

The rows of A_1, A_2 and A_k represent the K codeword and it is assumed that, the code weight of each of the K codeword is to be W.

Step 2: After the K codes, represented by the K rows of the $K \times N$ code matrix A_K^W in Eq. (1), are to represent a valid set of K codeword with in phase cross-correlations λ_{max} and code weight W; it must satisfy the following conditions;

1. The elements $\{a_{ij}\}$ of A_K^W must have values "0" or "1"

$$a_{ij} = \text{"0" or "1" for } i = 1,2,..K, j = 1\ 2,..N \quad (2)$$

2. The in phase cross-correlation λ_{max}, between any of the K code words (K rows of the matrix A_K^W) should not exceed code weight W. That is,

$$X_i X_j^T = \begin{cases} \leq \lambda_{max} & for\ i \neq j \\ = W & for\ i = j \end{cases} \quad (3)$$

3. The code weight of each codeword should be equal to W where,

$$\sum_{j=1}^{N} a_{ij} = W, \ i = 1,2 \ldots K \quad (4)$$

4. From Eq. (3), it is seen that the $W = X_i X_i^T$ is the in phase auto-correlation function of codes. $X_i Y_j^T$ is the out of phase cross-correlation between the ith and the jth codes. It follows that $X_i X_i^T$ should be greater than $X_i Y_j^T$. In other words, $W > \lambda_{max}$.

5. All K rows of A_K^W should be linearly independent because each codeword must be uniquely different from other code words. That is to say the rank of the A_K^W code matrix should be K. Moreover, for A_K^W to have rank K, thus codes $N \geq K$.

Step 3: From the five conditions above in **Step 2**, one of the matrices binary sequences as shown in Eq. (1), whose the first i^{th} row for the first K user is given by;

$$A_i = \overbrace{0 \ldots 0}^{r(i-1)} \ \overbrace{1 1 .. 1}^{W} \ \overbrace{0 \ldots 0}^{r(K-i)} \quad (5)$$

DOI: 10.1051/matecconf/20179701100

The length N of the codes which is the length of the rows of the $K \times N$ code matrix is given by;

$$N = WK - \lambda_{\max}(K-1) \tag{6}$$

It can be seen that the length N is minimum under the assumed conditions.

4 Result and discussion

The performance of OCDMA-RoF was demonstrated by using *OptiSystem* software from *Optiwave*^{TM} and the simulation parameters are shown in Table 1. The simulation parameters such as attenuation, dispersion effects and non-linear effects such as four wave mixing are activated during simulation and specified according to the typical industry values to simulate the real environment as close as possible. The performances of the system were characterized by referring to the fiber length and effective optical received power, ORP for the OCDMA-RoF.

Table 1. Typical parameters for the simulation analysis.

Parameters	Values
Data rates	155, 622 Mbps
Operating wavelength	λ=1550 nm
Number of users, K	3, 9
Number of code weights, W	2, 4
Attenuation	0.2 db/km
Thermal noise	1.0 x 10^{-22} W/Hz
Dispersion	16.75 ps/nm-km
Fiber length (km)	10-80
Fiber loss	0.2 dB
Pulse width	0.8 nm

Fig 2 shows the plots between fiber length and system performance BER for OCDMA-RoF for bit rates of 155 Mbps and 622 Mbps when the number of users fixed at nine, (K=9). It can be seen that, at system performance BER of 10^{-9}, bit rate of 155 Mbps, the OCDMA-RoF code definitely has better system performance BER where it is capable of achieving an acceptable BER threshold adequately for 45 km fiber length. In contrast the OCDMA-RoF code can perform adequately for 10 km fiber length within bit rate of 622 Mbps and system performance BER of 10^{-9}. It is clearly observed that, from the eye diagrams bit rate of 155 Mbps having a large eye-opening between the top and bottom level. The height of the eye-opening at the specified time integral shows the noise margin or immunity to noise. While, at a bit rate of 622 Mbps shows the more the eye close the more difficult it is to differentiate between the ones and zeros in the signal. From this observation, OCDMA-RoF at bit rate of 155 Mbps can successfully eliminate and suppressed the effects of PIIN and MAI for the SAC-OCDMA coding system.

 DOI: 10.1051/matecconf/20179701100

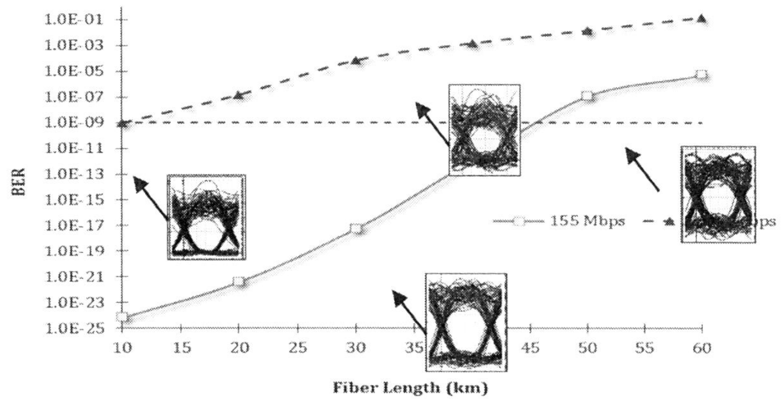

Fig. 2. Effect of bit rates on fiber length for OCDMA-RoF.

Fig 3 shows the plots of optical received power, ORP versus system performance BER for OCDMA-RoF based on AND subtraction detection technique for OCDMA coding system. This evaluation is to see the effect of bit rates of 155 Mbps, 622 Mbps and 1 Gbps, respectively on ORP. From this figure, at a bit rate of 155 Mbps, the OCDMA-RoF coding system can achieve an optimum value of optical received power ORP = - 25 dBm without requiring any amplification. At system performance BER of 10^{-9}, while at at bit rates of 622 Mbps and 1 Gbps can accommodate minimum optical received power ORP = - 19 dBm and - 18 dBm, respectively. There are about - 36 dB margin between bit rate of 155 Mbps and 622 Mbps, where - 37 dB between bit rate of 155 Mbps and 1 Gbps effective received power margins for optical transmission.

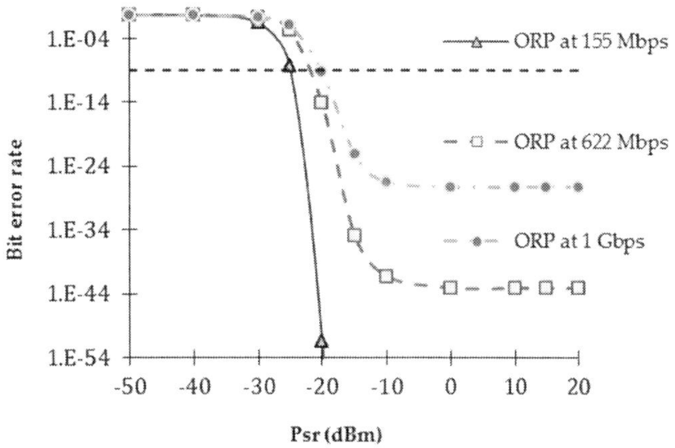

Fig. 3. Effect of various bit rates on optical received power (ORP).

5 Summary

The performances of the OCDMA coding systems using Radio over Fiber (RoF) have been presented. The results indicated that the OCDMA-RoF achieves long transmission up to 50 km at minimum effective optical received power ORP = -25 dBm. In addition to that, the architecture of OCDMA-RoF also presented with a MAI cancellation property. Without a doubt, the maximum out of phase auto-correlation and flexibility in phase cross-correlation values, respectively with different combination of (N, W, λ_{max}); by increasing W and increasing λ_{max}will improve the system performance with shorter, N. This will give an offer in OCDMA-RoF for better quality of service (QoS), fast asynchronous transmission, variable bit rate traffic and higher bandwidth in optical access networks.

References

1. Z. Jia, J. Yu, G. Chang, IEEE Photon. Technol. Lett., **18(16)**, 1726 (2006)
2. Y. Hsueh, Z. Jia, H. Chien, A. Chowdhury, J. Yu, G.K. Chang, J. Lightw. Technol., **29(8)**, 1105 (2011)
3. D. Wake, A. Nkansah, J. Gomes, J. Lightw. Technol., **28(16)**, 2456 (2010)
4. Abdullah O. Aldhaibania, S. Yaakobb, R.Q. Shaddada, S.M. Idrusa, M.Z. Abdul Kadirb, A.B. Mohammada, Optik, 3678 (2013)
5. S. Yaakob, N. M. Samsuri, R. Mohamad, N.E. Farid, I.M. Azmi, S.M. Mohd. Hassan, N. Khushairi, S.A. Enche Ab Rahim, A.I. A Rahim, A. Rasmi, A.K. Zamzuri, S.M. Idrus, S.H. Fan, *The 3rd International Conference on Photonics*, (2012)
6. S. Yaakob, M.A. Ismail, R. Mohamad, M.R. Yahya, A.F.A. Mat, M.R. Mokhtar, H.A.A Rashid, *IEEE International Conference On Semiconductor Electronics*, (2006)
7. C. Zhang, L. Wang, K. Qiu, Opt. Express, **19(15)**, 13957 (2011)
8. B. Zhang, Y. Lu, J. Zhang, B. Yang, (Book Chapter FKE, UTM, 2007)
9. B. Hraimel, X. Zhang, Y. Shen, *International Conference in Microwave and Millimeter Wave Technology (ICMMT)*, (2012)
10. C.B.M. Rashidi, S.A. Aljunid, F. Ghani, H.A. Fadhil, M.S. Anuar, Procedia Eng., **53**, 420 (2013)
11. S.M. Idrus, I.N. Salleh, N.S. Sahedan, S.Y. Cheng, V.O.S. Yee, (Book Chapter FKE, UTM, 2008)

Evaluation of Antioxidant Activity from Different Plant Parts of Senduduk Herb: Extraction Conditions Optimization of Selected Plant Part

Nor Helya Iman Kamaludin[1,*], and *Nor Syahida Adila* Jaafar[1]

[1]Department of Chemical Engineering Technology, Faculty of Engineering Technology, Universiti Malaysia Perlis (UniMAP), Perlis, Malaysia.

Abstract. This work reports a study on evaluation of antioxidant properties from flower of Senduduk herb. Natural occurring antioxidant was mostly preferred due to their little or no toxicity compared to the synthetic antioxidants which posses carcinogenic effects. Extraction was done on selected plant parts of Sendududk herb including leaves, stem, flower and berry parts to evaluate their antioxidant potentiality. Flower part of Sendudk herb extracted using acetonic solvent promotes highest antioxidant activity which is 93.97 ±1.38 % as compared to leaves (92.20 ± 0.20 %), stem (47.94 ±1.42%) and berry (92.88 ±0.63%) using the same extracting solvent. Thus, Senduduk flower was chosen to be continued with screening and optimization process. Single factor experiment using the one factor at a time (OFAT) method was done to study the effect of each extraction parameter that was solid to solvent ratio, temperature and solvent concentration. The extraction condition in each parametric study which results in highest antioxidant activity was used as the middle level of optimization process using Response Surface Methodology (RSM) coupled with Central Composite Design (CCD). The optimum condition was at 1:20 solid to solvent ratio, 64.61°C temperature and 80.24% acetone concentration which result in antioxidant activity of 96.53%. The verification of RSM showed that the model used to predict the antioxidant activity was valid and adequate with the experimental parameters.

1 Introduction

The requirement or demand towards natural antioxidants has increases in recent years especially in food and supplements industry. Most of people prefer something natural as their antioxidant source in order to maintain healthy life. Therefore, this situation has given rise to interest in studying herbal as natural antioxidants. This is because herbal are easy to obtain as they can be found abundantly in our country and also they have more economic value as compared to synthetic antioxidants. Furthermore, synthetic antioxidants may contain undesirable side effects such as toxics and carcinogenic properties when being used as food additives, therefore their uses being limited [1].

* Corresponding author: helya@unimap.edu.my

Senduduk herbs with scientific name *Melastoma malabathricum L.* (Melastomataceae) is one of the herbal that have high potential to be commercialized as natural antioxidant sources. This herb has attained as one of the herbs status in the Malay folklore belief among Melastomatacae family in our country [2]. Potential as antidiarrhoic drugs, antiperglycemic, antihypertensive, antihyperglycemic, hemostatic, and antioxidants have been reported among numerous species of this family in conventional folk medicine [3,4].

Solvent extraction is one of extraction method that was routinely used for the isolation of antioxidants from plant material. The antioxidant activity of extracts are strongly dependent on various extraction parameters such as solid to solvent ratio, temperature and solvent concentration. Therefore, the aim of this research is to evaluate the potentiality of different part of Senduduk Herb as a source of antioxidants. In fact, the optimum operating conditions was being examined as well for better extraction efficiency.

2 Materials and method

This section explains the sequences of steps required in this research activities. The first step involves the collection and preparation of sample. Next, the extraction using specific solvents was conducted in two screening steps including screening of Senduduk plant parts and screening of extraction parameters prior proceed to optimization studies using Design of Experiment (DOE) software.

2.1 Sample Preparation

Flower of Senduduk herb was collected from Pekan, Pahang. The herbal plant then was washed under tap water to removes any impurities and dirts. The samples were dried using conventional oven at 50°C for 24 hours and later was grind into smaller size using domestic blender. The ground plant parts are then kept in the separate closed container for later used.

2.2 Solvent extraction

The extraction was done using specific solvent in thermal-control water bath shaker. 2.5 g of samples were placed in 250 ml conical flask and 50 ml of solvent was added and then placed in thermostatic water bath shaker at 40 °C for 4 hours and shook at 150 rpm. The extracts were then filtered using Whatman No.1 filter paper to separate the liquid extract from solids. The solvent used was then removed by using rotary evaporator prior being analyzed.

2.3 Screening of senduduk plant parts and solvent types

Screening of plant part and type of solvent were performed by extracting different plant parts of Senduduk herbs including leaves, flower, stem and berry with different types of solvent (distilled water, methanol, ethanol and acetone) for evaluation of antioxidant activity. The plant part and solvent type that contribute to the highest antioxidant activity will be selected for further study of their extraction parameter.

2.4 Screening of extraction parameters of selected plant part using OFAT

Screening of extraction parameters was done using classical One-Factor-at-a-Time (OFAT) method. Using this approach, effect of temperature, extraction time and concentration of solvent on antioxidant activities of selected plant parts of Senduduk herbs were studied by evaluating each parameter while the others parameter are keep constant.

2.4.1 Effect of Solid to Solvent Ratio on Antioxidant Activity

To study the effect of solid to solvent ratio on antioxidant activity, the flower of Senduduk herb was extracted with solid to solvent ratio of 1:05, 1:10, 1:20, 1:30 and 1:40 at 40 °C, 150 rpm for 2 hours at concentration of 100%. The best solid to solvent ratio was chosen according to the highest antioxidant activity (%).

2.4.2 Effect of Temperature on Antioxidant Activity

To study the effect of temperature on antioxidant activity, the best solid to solvent ratio selected at previous step was used to extract Senduduk flower at different temperatures (45, 55, 65, 75, 85) °C by fixing the concentration of 100% for 2 hours. The best temperature was chosen according to the highest antioxidant activity (%).

2.4.3 Effect of Concentration of Solvent on Antioxidant Activity

To study the effect of concentration of solvent on antioxidant activity, the best solid to solvent ratio and temperature selected at previous steps were used at different concentration of solvent (20%, 40%, 60%, 80% 100%). The best concentration of solvent was chosen accordingly to highest antioxidant activity (%).

2.5 Optimization using Response Surface Methodology (RSM)

Optimization of extraction process was conducted by using Design-Expert software (Version 7.1.5). Response Surface Methodology coupled with Central Composite Design (CCD) at three level (-1, 0 and +1) was employed. The factors that were optimized include solvent to solid ratio, temperature and solvent concentration while the response variable was the antioxidant activity. A second order equation for two factors was developed by multiple regression procedure using the general equation (1):

$$Y = \beta_0 + \sum_{i-1}^{3} \beta_i X_i + \sum_{i-1}^{3} \beta_{ii} X^2{}_i + \sum_{i-1}^{2} \sum_{j-1+1}^{3} \beta_{ij} X_i J_i + \varepsilon \quad (1)$$

Where Y is the predicted response (% DPPH), β_0 is the model constant, β_i, β_{ii}, and β_{ij} are model coefficients (linear, squared and interactive effects), and ε is the error. Analysis-of-Variance (ANOVA) was performed for comparison of means. Regression analysis was performed and independent variables 3D surfaces were plots for analysis.

2.6 Analysis: determination of antioxidant activity

The 2,2-diphenyl-1-picrylhydrazyl (DPPH) radial-scavenging activity was determined by the method described by Jamal *et al.* [5] with slight modification. 250 µl of the extract was diluted with distilled water to 10 ml. Next, 200µl of the aliquots were mixed with 2 ml of 100 µM DPPH methanolic solution. Then, this mixture was placed at room temperature for 60 minutes in dark environment. The absorbance reading of the mixture was measured at 520 nm. The following equation (2) was used to calculate the ability to scavenge of DPPH radical.

$$\text{DPPH scavenging effect (\%)} = \frac{A_0 - A_1}{A_0} \text{x } 100 \quad (2)$$

where,
A_0 = the absorbance of the control at 60 min
A_1 = the absorbance of the sample at 60 min

3 Result and discussion

This section discusses on the results obtained through several analysis steps. There are three main sub-sections starting with discussion on the screening results of Senduduk plant parts and extraction parameters from selected plant part followed by analysis on optimization studies.

3.1 Screening of senduduk plant parts and solvent types

The result of DPPH scavenging activity among different plant parts of Senduduk herbs are shown in Table 1. In general, the percentage of DPPH inhibition among different plant parts ranging from 34.90±1.00 to 93.97±1.38%. In comparison to the various parts of Senduduk herbs, methanolic stem extract showed least effectiveness in DPPH scavenging activity with only 34.90±1.01% of radical inhibition. Meanwhile, acetonic extracts of three plant parts of Senduduk flower, berry and leaves possess the highest antioxidant activity among the extracts with percentage of inhibition were 93.97±1.38, 92.88±0.63 and 92.20±0.20%, respectively. Thus, it can be concluded that acetone was the best solvent for the extraction of antioxidant activity of Senduduk herb. This is in agreement with Martin *et al.*, [6] which states that acetone can be considered as a quite efficient solvent for the extraction of phenolic substances which responsible for antioxidant activity.

Table 1. Screening of different plant parts of Senduduk herb and solvent types on antioxidant activity

Plant Part Extract	Type of Solvent (pure)	Antioxidant Activity (%)
Leaves	Distilled water	87.58 ±0.24
	Methanol	85.61 ±0.45
	Acetone	92.20 ± 0.20
	Ethanol	86.68 ±0.23
Stem	Distilled water	88.00 ± 0.76
	Methanol	34.90 ± 1.01
	Acetone	47.94 ±1.42
	Ethanol	86.42 ±2.09
Flower	Distilled water	85.79 ±0.70
	Methanol	79.93 ±0.57
	Acetone	93.97 ±1.38
	Ethanol	67.53 ±1.04
Berry	Distilled water	88.20 ±0.37
	Methanol	83.31 ±0.11
	Acetone	92.88 ±0.63
	Ethanol	86.78 ±0.28

Remarks:
Lowest AA
Highest AA

3.2 Screening of extraction parameters of senduduk flower using one factor at a time (OFAT)

3.2.1 Effect of Solid to Solvent Ratio on Antioxidant Activity

The antioxidant activity of flower of Senduduk herb was significantly affected by solid to solvent ratio as the probability P-values is higher than 0.05 ($P>0.05$). From Figure 1, the best solid to solvent ratio were obtained at 1:20 with the highest DPPH scavenging activity of 81.89±1.87%. According to Tan *et al.*, [7] higher solid to solvent ratio could result in the higher antioxidant activity. This was due to the increase in concentration gradient between solid and the solvent that acts as the driving force for the mass transfer. This phenomenon was observed at a range of 1:05 to 1:20, where antioxidant activity showed significantly increased from 73.74±1.86 to 81.89±1.87%. However, further increased in the ratio up to 1:40 resulted in the reduction of antioxidant activity. This could be caused by saturation of solvent occurred at 1:20 and thus the final equilibrium were achieved [8]. Therefore, 1:20 solid to solvent ratio was selected as the best solid to solvent ratio to be used for further study of other effect in OFAT approach.

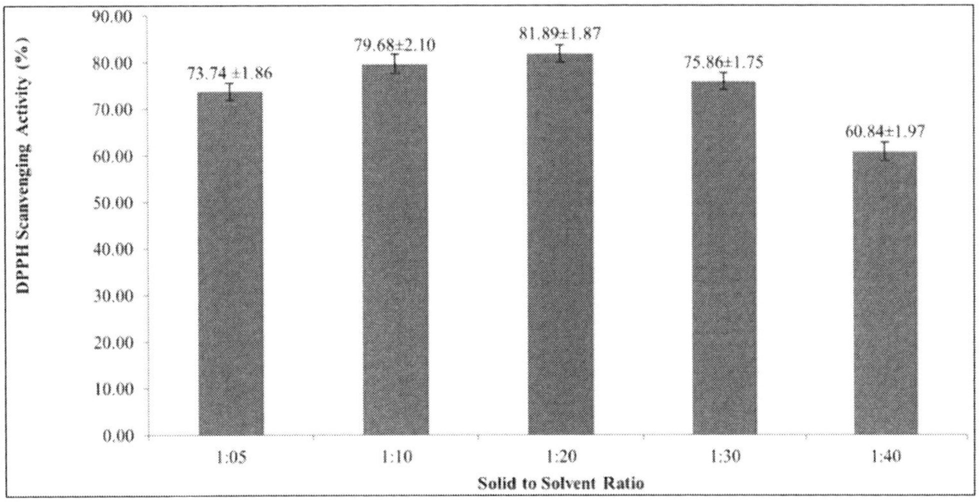

Fig. 1. Effect of solid to solvent ratio on antioxidant activity from flower of Senduduk herb.

3.2.2 Effect of temperature on antioxidant activity

Meanwhile, the effect of temperature on antioxidant activity of Senduduk flower is shown in Figure 2. A significant increased of antioxidant activity was observed from 79.03±2.17% to 95.88±2.50% as the temperature increased from 45°C to 65°C. This was due to at higher extraction temperature, the solubility of analyte would increase and thus promotes higher rates of mass transfer. Hence, the extraction efficiency could be improved and thus allow the solvent reach to the sample matrices easily. However, beyond 65°C, the antioxidant activity starts to decrease. These might due to the degradation of thermo-sensitive phenolic compound such us flavonoids that was heat sensible [9]. Therefore, 65°C was selected as the best temperature to be used for further study of other effect in OFAT approach.

DOI: 10.1051/matecconf/20179701101

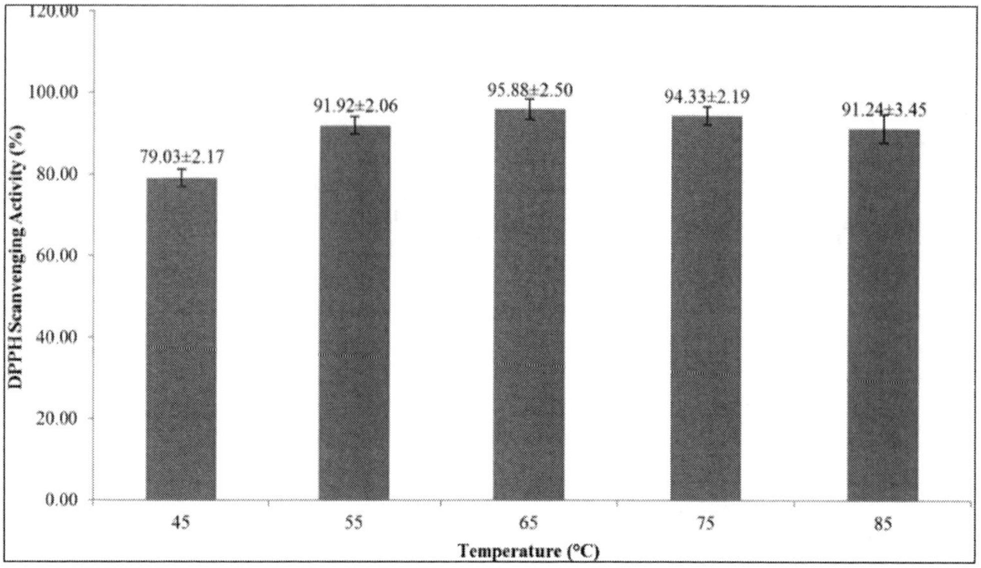

Fig. 2. Effect of temperature on antioxidant activity from flower of Senduduk herbs.

3.2.3 Effect of Solvent Concentration on Antioxidant Activity

The antioxidant activity as a function of solvent concentration was shown in Figure 3. Results indicated that there is a significant increase in antioxidant activity from 84.74±0.58% to 87.56±0.23% with the increasing of acetone concentration from 20% to 80%. However, the further increase in solvent concentration up to 100% results in the decrease of antioxidant activity. This is in agreement with Fatma et al., [10] that stated the extraction using a mixture of organic solvent and water could promotes higher recovery of antioxidant phenolic compounds as compared using of pure organic solvent due to the polarity effects. Therefore, acetone with concentration of 80% was selected as the best concentration of solvent to extract antioxidant activity from flower of Senduduk herb.

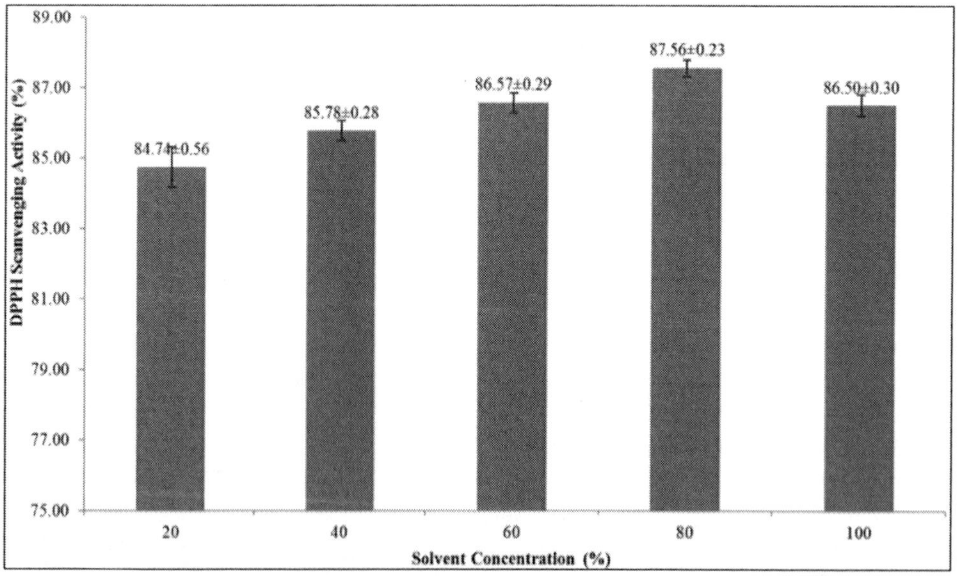

Fig. 3. Effect of solvent concentration on antioxidant activity of flower of Senduduk herb.

3.3 Optimization studies

Central composite design (CCD) with three factors and five levels were used in this study to optimize the DPPH scavenging activity of Senduduk herb flowers. Twenty run of extraction condition with eight factorial points, six axial points and six repetition of central point was given in Table 2. The result indicates that the highest antioxidant activity (97.05%) of Senduduk herb flower was shown by Run 8 which lies within the central point at 1:20 solid to solvent ratio, 65°C temperature and 80% acetone concentration. The residual error obtained was only 0.61% for both experimental and predicted values. Based on the results in Table 2, second order equation was developed as shown in equation (3) where X_1 is solid to solvent ratio, X_2 is the temperature and X_3 is solvent concentration.

$$\text{Antioxidant activity} = +96.44-0.67X_1-0.31X_2+0.083X_3-0.020X_1X_2-0.092X_1X_3 + 0.019X_2X_3-1.39X_1^2-1.94X_2^2-1.8403X_3^2 \quad (3)$$

Table 2. The experimental and predicted design of CDD from flower of Senduduk herb

Run	Solid to solvent ratio (g/ml)	Temperature (°C)	Solvent concentration (%)	Antioxidant activity (%)		
				Experimental	Predicted	Residual (%)
1	0.05	65.00	63.18	90.78	91.09	0.31
2	0.05	65.00	80.00	95.57	96.44	0.87
3	0.05	65.00	96.82	90.03	91.37	1.34
4	0.04	70.00	90.00	92.24	92.01	-0.23
5	0.05	56.59	80.00	90.25	91.47	1.22
6	0.05	65.00	80.00	96.41	96.44	0.03
7	0.04	70.00	70.00	91.84	91.29	-0.55
8	0.05	65.00	80.00	97.05	96.44	-0.61
9	0.05	65.00	80.00	95.48	96.44	0.96
10	0.07	60.00	90.00	91.34	90.73	-0.61
11	0.08	65.00	80.00	90.48	91.37	0.89
12	0.04	60.00	90.00	93.53	92.21	-1.32
13	0.05	65.00	80.00	97.04	96.44	-0.60
14	0.03	65.00	80.00	92.86	93.62	0.76
15	0.04	60.00	70.00	92.36	92.23	-0.13
16	0.07	70.00	90.00	91.48	90.44	-1.04
17	0.05	73.41	80.00	90.00	90.43	0.43
18	0.05	65.00	80.00	97.35	96.44	-0.91
19	0.07	60.00	70.00	92.05	92.05	0.00
20	0.07	70.00	70.00	89.94	90.09	0.15

Note: Run 2, 6, 8, 9, 13, 18 represents the centre points.
Condition shows the highest antioxidant activity
Condition shows the lowest antioxidant activity

3.4 Model analysis

In fact, in order to investigate the interactive effect of independent variables on the optimization of extraction condition from Senduduk flower, three dimensional (3D) plots were examined for the estimated response. Figure 4 represents the combination effect of two variables on the response. As shown in Figure 4 (a), a response surface plot for interaction effect of temperature and solid to solvent ratio at a fixed 80% solvent concentration on antioxidant activity was illustrated. From the figure, it can be observed that solid to solvent ratio and temperature has a simultaneous effect on antioxidant activity. The scavenging rate increased up to 97.35% with the increasing of the temperature from 60°C to 65°C at solid to solvent ratio of 1:20. However, beyond 65°C the scavenging effect starts to decrease. This could be caused by degradation of phenolic compound that contribute to reduction in antioxidant activity as they are easily oxidized [11]. Cacae and Mazza [12] also stated that the decreased of antioxidant activity at higher temperature was due to degradation of thermo-sensitive phenolic compound that was heat sensible.

The interaction effect of solid to solvent ratio and solvent concentration on antioxidant activity at fixed extraction temperature of 65°C was shown in Figure 4 (b). The result revealed that the antioxidant activity increased up to 97.35% with the increament of solvent concentration from 70 to 80 % and solid to solvent ratio of 1:20. The increasing trends of antioxidant activity with respect to solid to solvent ratio was demonstrated by the mass transfer principle. At higher solid to solvent ratio, the concentration gradient between the solid and the liquid solvent also was higher and thus could promote the leaching of the bioactive compound to the solid matrix, and hence better extraction rate could be obtained [13][14]. However, further increased in solid to solvent ratio and solvent concentration beyond 1:20 and 80%, respectively gave negative effects on antioxidant activity. This could be caused by saturation of solvent occurred at 1:20 and thus achieved their final equilibrium [8].

Figure 4 (c) demonstrated the interactive effect between solvent concentration and temperature at constant solid to solvent ratio of 1:20. The result indicates that an increase in solvent concentration from 70% to 80% and temperature from 60°C to 65°C resulting an increase in antioxidant activity up to 97.35%. However, the declined trends were observed in antioxidant activity when the solvent concentration was kept increase to 90%. This might be due to a decrease in polarity of acetone as more polar solvent will facilitates the extraction of phenolic compound [10].

 DOI: 10.1051/matecconf/20179701101

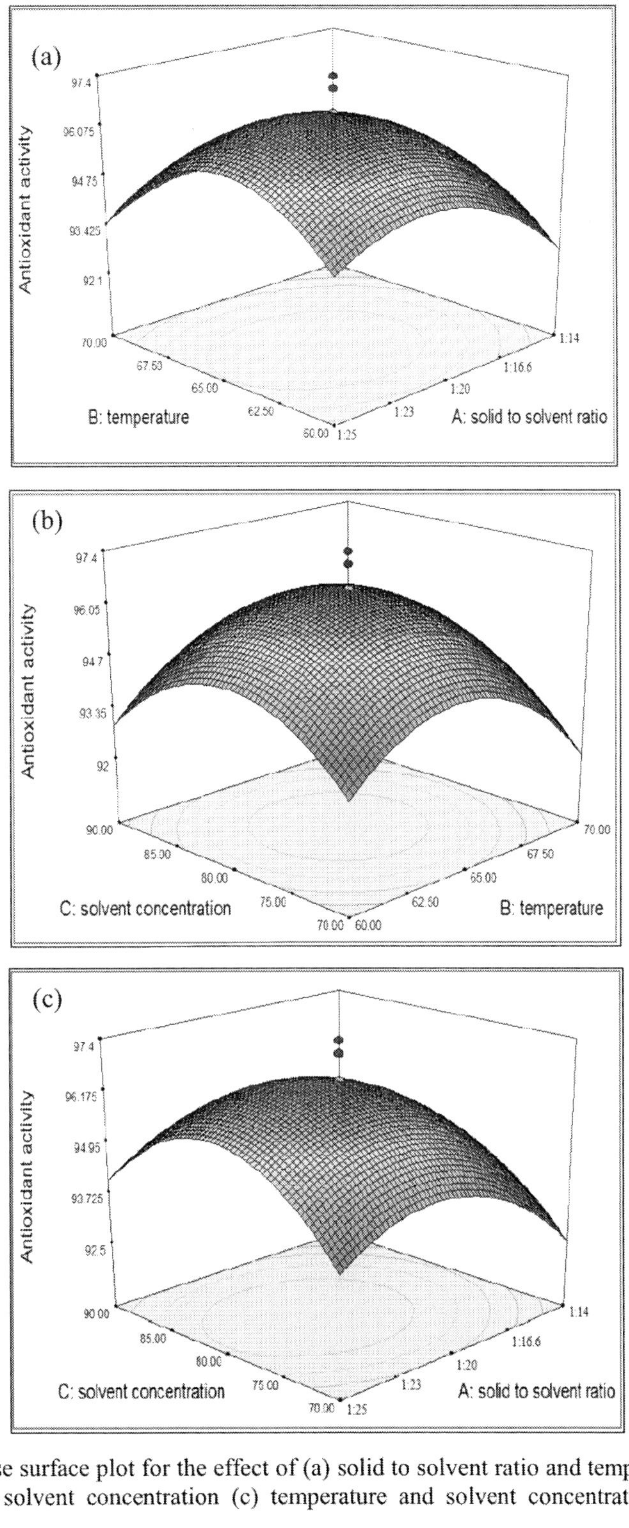

Fig. 4. 3D response surface plot for the effect of (a) solid to solvent ratio and temperature (b) solid to solvent ratio and solvent concentration (c) temperature and solvent concentration on antioxidant activity

4 Conclusion

For overall view, the antioxidant activities was detected exist in Senduduk herbal plants. Flower part with acetonic extracts showed the highest antioxidant activity of 93.97±1.38% as compared to the other three plant parts (leaves, stem and berry) of Senduduk herb tested. Through OFAT approach, the antioxidant activities of Senduduk flower was significantly affected by various extraction parameters studied including solid to solvent ratio, temperature and solvent concentration. The extraction was continued in optimization by using Response Surface Methodology (RSM) with Central Composite Design (CCD). The optimum process parameters for maximum antioxidant activity were determined and a second order polynomial model for predicting the response was obtained. The highest antioxidant activity of 96.53% was obtained at optimum extraction conditions at 1:20 solid to solvent ratio, 64.61°C temperature and 80.24% acetone concentration. The verification of RSM showed that the model was valid and adequate for the optimum extraction conditions of antioxidant activity. Therefore, CCD and RSM enabled the determination of optimal condition for the extraction of antioxidant activity from flower of Senduduk herb.

The authors would like to express their appreciations to Faculty of Engineering Technology, Universiti Malaysia Perlis (UniMAP) for facilitating this project.

References

1. S.C. Jo, K.C. Nam, B.R. Min, D.U. Ahn, S.H. Cho, W.P. Park, S.C. Lee, Int. J. Food Sci. Tech., **41**, 15 (2006).
2. K. Meyer, Journal of Plant Taxonomy and Plant Geography, **46**, 351 (2001)
3. D. Susanti, H.M. Sirat, F. Ahmad, R.M. Ali, N. Aimi, M. Kitajima, Food Chem., **103**, 710 (2007)
4. J.T. Cheng, F.L. Hsu, H.F. Chen, Planta. Med., **59**, 405 (1993)
5. P. Jamal, Z.M. Idris, M.Z. Alam, Food Chem., **124**, 1595 (2010)
6. S. Martins, C.N. Aguilar, J.A. Teixeira, S.I. Mussatto, Sep. Purif. Technol., **88**, 163 (2012)
7. P.W. Tan, C.P. Tan, C.W. Ho, Int. Food Res. J., **18**, 557 (2011)
8. E.M. Silva, H. Rogez, Y. Larondelle, Sep. Purif. Technol., **55**, 381 (2007)
9. J.E. Cacace, G. Mazza, J. Food Eng., **59**, 379 (2003)
10. F. Kallel, D. Driss, F. Chaari, L. Belghith, F. Bouaziz, R. Ghorbel, S.E. Chaabouni, Ind. Crops. Prod., **62**, 34 (2014)
11. Y.Y. Thoo, S.K. Ho, J.Y. Liang, C.W. Ho, C.P. Tan, Food Chem., **120**, 290 (2010)
12. J.E. Cacace, G. Mazza, J. Food Eng., **59**, 379 (2003)
13. M.A. Al Farsi, C.Y. Lee, Food Chem., **108**, 977 (2008)
14. P.W. Tan, C.P. Tan, C.W. Ho, Int. Food Res. J., **18**, 557 (2011)

 DOI: 10.1051/matecconf/20179701102

Sulphuric Acid Resistant of Self Compacted Geopolymer Concrete Containing Slag and Ceramic Waste

I. Shafiq [1,*], *M.* Azreen[2], and *M.W.* Hussin[1]

[1]Faculty of Civil Engineering, Universiti Teknologi Malaysia, 81310 UTM Johor Bahru, Johor, Malaysia.
[2]Forensic Engineering Centre, Institute for Smart Infrastructure Innovative Construction, Faculty of Civil Engineering, Universiti Teknologi Malaysia, 81310 UTM Johor Bahru, Johor, Malaysia.

Abstract. Malaysia is a one of the developing countries where the constructions of infrastructure is still ongoing, resulting in a high demand for concrete. In order to gain sustainability factors in the innovations for producing concrete, geopolymer concrete containing granulated blast-furnace slag and ceramics was selected as a cement replacement in concrete for this study. Since Malaysia had many ceramic productions and uses, the increment of the ceramic waste will also be high. Thus, a new idea to reuse this waste in construction materials have been tested by doing research on this waste. Furthermore, a previous research stated that Ordinary Portland Cement concrete has a lower durability compared to the geopolymer concrete. Geopolymer binders have been reported as being acid resistant and thus are a promising and alternative binder for sewer pipe manufacture. Lack of study regarding the durability of the geopolymer self-compacting concrete was also one of the problems. The waste will be undergoing a few processes in the laboratory in order to get it in the best form before undergoing the next process as a binder in geopolymer concrete. This research is very significant in order to apply the concept of sustainability in the construction field. In addition, the impact of this geopolymer binder is that it emits up to nine times less CO_2 than Portland Cement.

1 Introduction

Higher demand for concrete from the construction field has almost peaked in every country in the world including Malaysia that is developing country, leading to the search for a new solution that focuses more on recycle and reuse of waste product in order to generate new construction material that can achieve a sustainable concept [1].

OPC production is considered an environmental problem because it requires large quantities of raw materials and fuel and contributes 8% of global CO_2 emission [2, 3]. Furthermore, OPC is not chemically stable, deteriorates through sulfate, sulfuric acid

* Corresponding author: mohdazreen@utm.my

attacks and elevated temperature exposure [4, 5]. The use of pozzolon to replace part of OPC has become more attractive and important. Geopolymers (GP) are one of the alternatives to overcome all these issues [6].
Previous research has shown that OPC concrete has a very low durability against sulphuric acid attack compared to the geopolymer concrete [7]. Durability refers to the ability of concrete to resist deterioration from the environment or service in which it is placed. Properly proportioned concrete that is properly placed, finished, and cured should endure without significant distress throughout its service life [8]. This is because the penetration of sulfates in solution causes change in the composition and microstructure of the concrete. These changes may vary in type or severity, but commonly include extensive cracking, expansion and loss of bond between the cement paste and aggregate. Alteration of paste composition results in monosulfate phase converting to ettringite and, in later stages, gypsum formation. The necessary additional calcium is provided by the calcium hydroxide and calcium silicate hydrate in the cement paste. The effect of these changes is an overall loss of concrete strength [9].

In addition, geopolymer binders have been reported as being acid resistant and thus are a promising alternative binder for sewer pipe manufacture compared to Ordinary Portland Cement Concrete [10]. Lack of study about high strength of self-compacting geopolymer concrete against sulphuric acid attack is also one of the problems. The previous research studied the normal strength of self-compacting geopolymer concrete.

In this research, slag and ceramics will be used as the replacement to OPC. Slag contains high calcium, which is suitable for aggressive environment [11], while the ceramic that contain high silica can make the concrete more durable and more binder gel [12]. There is also pressure to provide a better quality concrete to the construction sector and to preserve natural resources and the air quality, and it is expected that this geopolymer self-compacting concrete (GSCC) can solve all the issues described in OPC [13].

The aim of this research is to study the durability of self-compacting geopolymer concrete against sulphuric acid attack which mainly use Ground Granulated Blast-Furnace Slag (GGBFS) and Ceramics waste as a binder. In order to achieve this aim, several objectives are identified and measured to ensure that the fundamental parameters needed in the design are evaluated. These parameters include determining the optimum mix design for geopolymer concrete containing Granulated Blast-Furnace Slag and Ceramics as a binder, the durability performance of Granulated Blast-Furnace Slag and Ceramics concrete against sulphuric acid attack and the effect of geopolymer concrete after immersion in sulphuric acid solution.

This research has been carried out to investigate the durability performance of Granulated Blast-Furnace Slag and Ceramics concrete against sulphuric acid attack. The specimens were prepared in 100 mm x100 mm x100 mm cubes. Type of curing for this concrete was ambient temperature. Compressive strength test is conducted to study the engineering properties of slag concrete. The cubes have been tested for compressive test after being immersed in sulphuric acid solution for 28 days or more. The concentration of sulphuric acid solution has a pH of 3.0.

2 Methodology

2.1 Preparation of materials

Materials used in this research basically consist of Granulated Blast-Furnace Slag, Ceramics, fine (Sand) and coarse aggregates (Gravel 10mm) and alkaline solution. The grading of the fine and coarse aggregates followed the ASTM standard C33 (Standard Specification for Concrete Aggregates) [14].

The Granulated Blast-Furnace Slag was collected from slag factories in parts of southern Malaysia while the ceramic waste was collected from a ceramic factory. After collection, the ceramic will be crushed by a crusher jaw to make it smaller and easier to sieve. The ceramic waste that has been crushed will then be sieved by a mechanical sieve shaker for a sufficient period and will be grounded in Los Angeles Abrasion Machine for at least 8 hours for each 4 kg using a steel bar of 18mm in diameter and 800mm long. Only the finest ceramic will be used in the mixing. Then the fine ceramic will be kept tightly secured in a container. The coarse aggregates, which are gravel, were collected from the Structure and Materials Laboratory at Universiti Teknologi Malaysia (UTM), Malaysia. The gravel will be sieved by a mechanical sieve shaker and only the gravel that passed through 10 mm sieve were collected and used in the mixing.

For the alkaline solution, 8M and 12M of sodium hydroxide and sodium silicate were used because previous research stated that 8M of NaOH has a better workability compared to 12M of NaOH [15]. Davidovits (2002) suggested that it is preferable to mix the sodium silicate solution and the sodium hydroxide solution together at least one day before adding the liquid to the solid constituents [16].

2.2 Mix Design

2.2.1 Cube Sample

The mix proportion of this geopolymer self-compacting concrete was shown in Tables 1 and 2 for the control sample mix proportion.

Table 1. Mix Proportion of GSCC.

	Slag(kg)	Ceramic(kg)	F.A(kg)
GSCC	1.8	1.8	2.58

C.A (kg)	SP(kg)	Na_2O_3Si(kg)	NaOH(kg)
1.11	0.26	1.18	0.47

Extra Water(kg)
0.41

Table 2. Mix proportion of OPC concrete.

	W/B ratio	Water (kg/m^3)	OPC (kg/m^3)	FA (kg/m^3)	CA (kg/m^3)
OPC(100%)	0.59	192	384	898	861

DOI: 10.1051/matecconf/20179701102

2.2.2 Testing procedure of GSCC

In this research, the slump flow and L-Box test were performed to ascertain the self-compacting capabilities. All those tests are in accordance with European Federation of National Associations representing producers and applicators of specialist building products for Concrete (EFNARC) guidelines [17]. The slump flow test was used to assess the horizontal free flow of SCC in the absence of obstruction. On lifting the slump cone filled with concrete, the concrete begins to flow. The time T_{50cm} is a secondary indication of flow. It measures the time taken in seconds from the instant the cone is lifted to the instant when horizontal flow reaches a diameter of 500 mm.

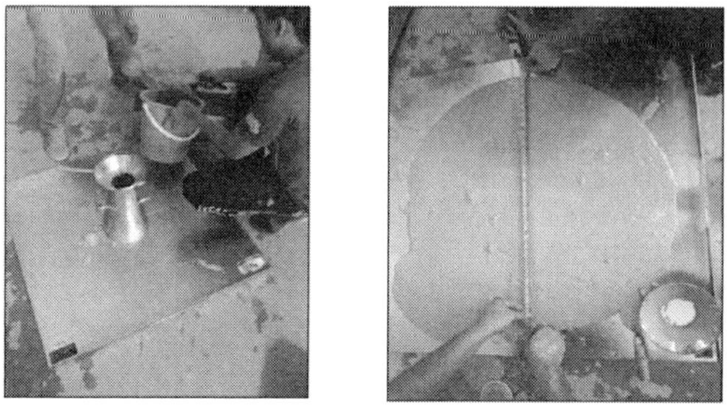

Fig 1. Procedure for Slump-Flow Test.

The passing ability is determined by the L-box test as shown in Figure 2. The vertical section of the L-box is filled with concrete, and then the gate lifted to let the concrete flow into the horizontal section. The height of the concrete at the end of the horizontal section is expressed as a proportion of that remaining in the vertical section (H_2/H_1). This is an indication of passing ability.

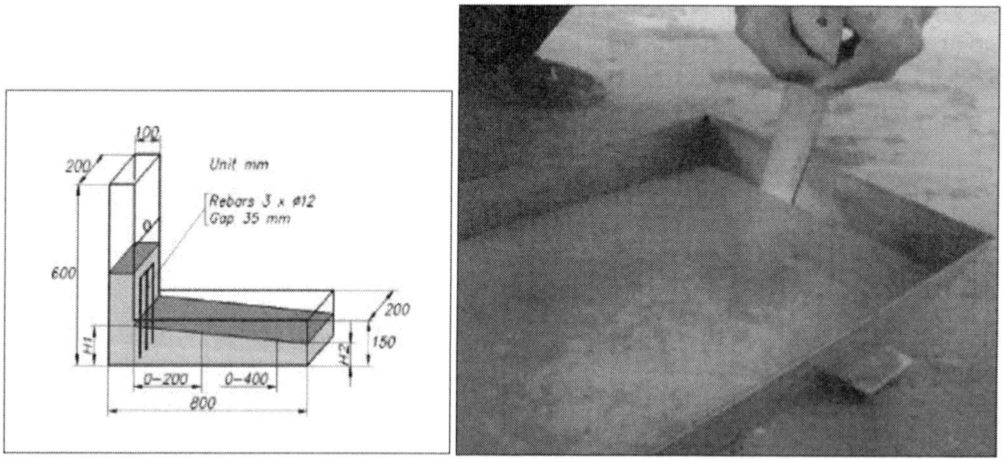

Fig 2. Procedure for L-Box.

3 Results

In this chapter, all 100 mm x100 mm x100 mm samples are used for compressive testing. These results were presented in the form of table, graph and images where all comparisons of the results will be shown. In addition, the test methods for self-compacting, which is Slump flow, T_{50cm} slump flow and L-box will be shown, including the results.

3.1 Compressive Test

The most valuable concrete property is the concrete compression strength because it gives the overall definition of the quality of concrete in terms of concrete strength in relation to the hydrated cement paste. Only compressive testing of the hardened concrete has been conducted. For the compression test, 3.50 N/mm^2 was set according the laboratory standard [14].

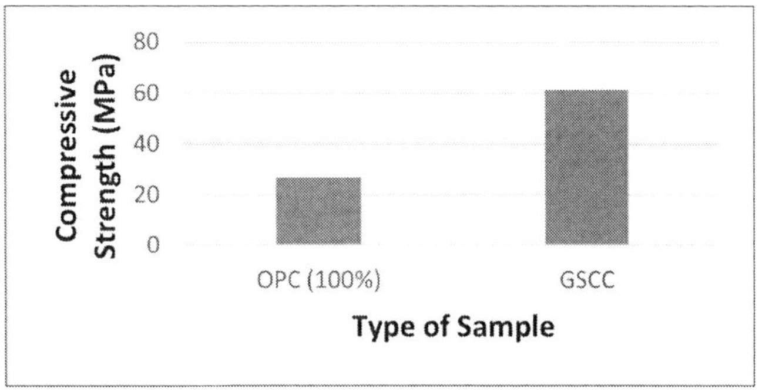

Fig 3. Compressive Strength against Type of Sample.

From Figure 3, it can be seen that Geopolymer Self-Compacting Concrete blended with slag (50%) and ceramics (50%) have a higher compressive strength than normal concrete (100% OPC). The strength improvement of Self-Compacting Concrete blended with slag contributed to the early strength of a concrete, thus making the concrete more durable and better than normal SCC (100% OPC). Also, it is clear that the role of slag and ceramics that contain high calcium and silica increase the compressive strength of the concrete.

3.2 Testing for Self-Compacting Concrete (SCC)

A concrete mix can only be considered as SCC if the three characteristics for workability are satisfied [17]. The three fresh concrete characteristics mandatory for SCC are filling ability, passing ability and resistance to segregation. Only Slump flow test, T_{50cm} slump flow and L-box test have been conducted.

3.3 Fresh Properties of GSCC

The slump flow values of SCC with slag and ceramics are shown in Table 3. From the results, the slump flow values were calculated at 780 mm. According to the values recommended by EFNARC for fresh state properties of SCC, the mixtures fulfilled the filling ability.

The L-box test results are shown in Table 3. From the results, the blocking ratio for trial 6 is 1.0. A satisfactory blocking ratio was observed for this GSCC (50% slag and 50% ceramics). In a previous study, Felekoglu et al. concluded that a blocking ratio of 0.6 to 1 is acceptable for SCC to obtain the satisfactory filling ability. In this regard, this concrete mix was satisfactory.

Table 3. Workability Test Results.

Workability Test	Slump Flow Test (mm)	T_{50cm} slump flow (sec)	L-Box (H_2/H_1)
GSCC	780	4	1.0

3.4 Durability test for sulphuric acid

It is accepted that concrete is not fully resistant to acids. All acids will have an impact on concrete. The rate of speed of action may differ, but they finally disintegrate the concrete. Six cube specimens of size 100 mm x100 mm x100 mm were prepared and immersed in sulphuric acid solution (pH 3) for 28 and 42 days.

3.4.1 Compressive strength comparison between OPC concrete and Geopolymer concrete after acid immersion after 28 and 42 days.

Figure 4 shows the compressive strength comparison between OPC concrete and geopolymer concrete (GSCC) after acid immersion for 28 days. GSCC has a better strength compared to the OPC concrete after 28 and 42 days. The high percentage of calcium in the geopolymer binder (slag) produces a better concrete in terms of durability and strength. The composition and microstructure of the OPC concrete have changed due to the penetration of sulfate in the sulphuric acid solution, thus gypsum is formed. The effect of these changes is an overall loss of concrete strength.

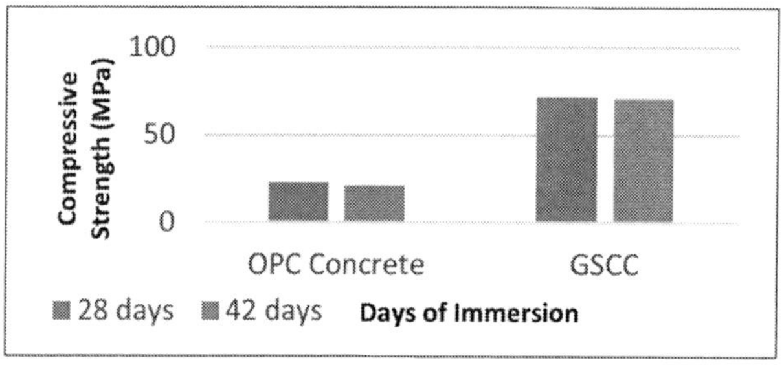

Fig 4. Compressive Strength against Days of Immersion.

4 Conclusion

All conclusions drawn upon here are based on the objectives of the study and observations done throughout the whole course of this study. The conclusions that can be drawn in studying geopolymer self-compacting concrete against sulphuric acid attack are that the most optimum mix design for granulated blast furnace slag and ceramic powder as a cement replacement is 50 % for slag and 50% for ceramic powder. In addition, 8M of Sodium Hydroxide (NaOH) gives a better workability as compared to 12M of Sodium Hydroxide (NaOH). Slag also has a low workability because it has a low relative density that causes an increase in paste volume. Thus, more water is needed for the self-compacting concrete, which can lead to low compressive strength. Lastly, the sulphuric acid solution had no effect on the strength of the concrete and the weight after being immersed in sulphuric acid solution for 28 and 42 days.

The authors would like to acknowledge UTM for providing the Research University Grant (RUG), VOT No.Q.J130000.2522.13H01 from Ministry of Education (MOE) of Malaysia for the financial supports.

References

1. Tarun R Naik, Rakesh Kumar. *Third International Conference on Sustainable Construction Materials and Technologies* 1 (2013)
2. K.L.Scrivener, R.J. Kirkpatrick, Cem. Concr. Res. **38**, 128 (2008)
3. J.G.J. Olivier, G.Janssens-Maenhout, J.A.H.W. Peters, *Trends in global CO_2 emissions report* (Rep., PBL Neth. Environ. Assess. Agency, The Hague, 2012)
4. M.A.M Ariffin, M.A.R. Bhutta, M.W. Hussin, M. Mohd Tahir, Nor Aziah, Constr. Build. Mater. **43**, 80 (2013).
5. M.W. Hussin, M.A.R Bhutta, M. Azreen, P.J. Ramadhansyah, J. Mirza, Materials and Structures **48**, 709 (2015)
6. P. Duxson, J.L. Provis, G.C. Lukey, J.S.J Deventer, *Cem.* Concr. Res. **37**, 1590 (2007)
7. H.S. Shankar, R. B. Khadiranaikar, Journal of Civil and Structural Engineering **3**, 396 (2012)
8. Peter Taylor, *Durability of Concrete. Transportation Research Board of the National Academies* (2013)
9. C.F. Ferraris, P.E. Stutzman, K.A. Snyder, *Sulfate Resistance of Concrete: A New Approach, R&D Serial No. 2486, Portland Cement Association* (Skokie, Illinois, USA, 2006)
10. X.J. Song, M. Marosszeky, M. Brungs, R. Munn, *10DBMC International Conference On Durability of Building Materials and Components* (Lyon, France, 2005)
11. Md. Moinul Islam, Md.Saiful Islam, Bipul Chandra Mondal, Mohammad Rafiqul Islam, Journal of Civil Engineering (IEB) **38**, 129 (2010)
12. Paratibha Aggarwal, Rahul Pratap Singh and Yogesh Aggarwal, Civil and Environmental Engineering **2**, 1 (2015)
13. Srinivas Vasam, K. Jagannadha Rao, M.V. Seshagiri Rao. International Journal of Research in Engineering and Technology **4**, 1 (2015)
14. American Society for Testing and Materials, *Standard Specification for Concrete Aggregates* (ASTM C33 – 03)
15. M. Fadhil Nuruddin, Samuel Demie, M. Fareed Ahmed, Nasir Shafiq, International Journal of Civil and Environmental Engineering **3**, 122 (2011)
16. P. Ganapati Naidu, A.S.S.N. Prasad, S. Adiseshu, P.V.V. Satayanaraya. International Journal of Engineering Research and Development **2**, 19 (2012)
17. European Federation of national trade associations representing producers and applicators of specialist building products, *Specification and Guidelines for self-compacting concrete* (Hampshire, U.K, 2002)

Utilization of Organic Fertilizer on Sweet Corn (*Zea mays* saccharata Sturt) Crop at Shallow Swamp Land

Midranisiah[1,*], *Neni* Marlina[2], *S.E* Rahim[2], and *Erni* Hawayanti[3]

[1]STIPER Sriwigama of Palembang, Jl. Demang IV Demang Lebar Daun Lorok Pakjo IB 1Palembang Indonesia
[2]Faculty of Agriculture, Palembang University, Jl.Darmapala No 1A Bukit Lama Palembang Indonesia
[3]Faculty of Agriculture, Muhammadiyah University, Palembang, Jl Jend. A.Yani 13 Ulu Palembang Indonesia

Abstract. Shallow lowland swamp area has significant potential for cultivation of sweet corn crop. This lowland swamp has rich natural resources such as organic fertilizers from chicken dunk, cow dunk, oil palm fresh bunches and legume cover crops (LCC) that are not maximally utilized yet by farmers. These organic fertilizers can be utilized to increase the growth and production of sweet corn crop. The research objective was to determine organic fertilizer types that capable to increase the growth and production of sweet corn crop at shallow lowland swamp area. This research had been conducted from January to April 2015 in Pulau Semambu Village, North Indralaya Subdistrict, Ogan Ilir District, South Sumatra Province. The design used in this research was non-factorial Randomized Block Design (RBD) with four treatments of organic fertilizer types with six replications for each treatment. The treatments were consisted of organic fertilizers from chicken dunk, cow dunk, oil palm fresh bunches and legume cover crops (LCC). The results showed that treatment of organic fertilizers from chicken dunk could increase the growth and production of sweet corn at shallow lowland swamp area with yield level of 4.37 kg.plot $^{-1}$.

1 Introduction

Area of lowland swamp in Indonesia is estimated about 13.28 million ha which consisted of 4.167 millions ha of shallow lowland swamp, 6.075 million ha of intermediate lowland swamp and 3.038 million ha of deep lowland swamp. It is estimated that lowland swamp which has potential for agricultural area was about 10.19 million ha, but only about 1.55 million ha that had already been opened and about 0.729 million ha had been utilized for agricultural enterprise. This lowland swamp generally is located in Sumatra, Kalimantan and Papua. From the magnitude of this utilized lowland swamp, only about 694.294 ha was

[*] Coresponding Author: marlina002@yahoo.com

planted with rice crop and only about 62.844 ha was planted with rice crop having two times planting per year. Therefore, very extensive area of lowland swamp is available to be utilized for agriculture enterprise [1].

Lowland swamp can be classified as shallow lowland swamp, intermediate lowland swamp and deep lowland swamp. Shallow lowland swamp generally has better soil fertility level than that of intermediate lowland swamp and deep lowland swamp because of enrichment from mud sediment carried by river water outflow. Shallow lowland swamp has high potential for agricultural enterprise of food crops and horticulture crops, especially rice in wet season and second crops as well as seasonal vegetables and fruits [2]. One of this potential crop is sweet corn.

Sweet corn crop will not produce maximum yield if its nutrients requirement are insufficient. Fertilization can increase harvest yield quantitatively and qualitatively. Lingga and Marsono [8] stated that fertilizer is key point of soil fertility because fertilizer contains one or more nutrients to substitute the exhausted nutrients absorbed by crop.

Fertilizers used in this research was organic fertilizers which consisted of organic fertilizers from chicken dunk, cow dunk, oil palm fresh bunches and legume cover crops (LCC). According to Suriadikarta and Simanungkalit [23], organic fertilizer is fertilizer that mostly or all of its components is consisted of organic matters originated from crops or animals that had been processed, in solid or liquid form which can be used to supply organic matters for improvement of physical, chemical and biological properties of soil. Improvement of physical, chemical and biological properties of soil can provide better growth environment for roots and addition of macro and micro nutrients from organic fertilizers can increase production of sweet corn crop.

Some previous studies had been done in term of organic fertilizers usage (such as plus organic fertilizer, bio-fertilizer, rice straw compost enriched with microbe or bio-organic fertilizer, chicken dunk fertilizer, swamp grass compost) on crops such as food crops (rice, corn and nuts), vegetable crops (sweet corn, cucumber, tomato and others) in shallow and deep lowland swamp areas as well as in tidal lowland of C-type flooding. Results of study by Marlina *et al.* [10-15] and [26] showed that plus organic fertilizer at dose of 750 kg/ha combined with organic pesticides (animal and vegetation sources), bio-organic fertilizer at dose of 225 kg/ha to 300 kg/ha and 0 % to 75 % recommended dose of inorganic fertilizer was capable to increase rice production in shallow lowland swamp with magnitude of 2.16 kg/plot.

Bio-organic fertilizer at dose of 300 kg/ha and 75% recommended dose of inorganic fertilizer gave rice production of 93.67 g/pot. Bio-organic fertilizer at dose of 200 kg/ha could increase production of hybrid corn in shallow lowland swamp. Combination of 75% bio-organic fertilizer and 25% recommended dose of inorganic fertilizer could increase rice production with magnitude of 48.56 g/pot. Addition of formula 9 plus organic fertilizer could increase rice production with magnitude of 5 kg/plot. Grass swamp compost could increase production of red pepper and cucumber with magnitude of 20.48 kg/raft and 21.44 kg/raft, respectively. The use of chicken dunk fertilizer at dose of 10 ton/ha was capable to increase peanut production with magnitude of 2.73 kg/plot.

2 Research method

This research had been conducted from January to April 2015 in Pulau Semambu Village, North Indralaya Subdistrict, Ogan Ilir District, South Sumatra Province. The design used in this research was non-factorial Randomized Block Design (RBD) with four treatments of organic fertilizer types with six replications for each treatment. The treatments were

DOI: 10.1051/matecconf/20179701103

consisted of organic fertilizers from chicken dunk, cow dunk, oil palm fresh bunches and legume cover crops (LCC).

2.1 Land preparation

Land or area that had been measured was subsequently cleaned from weeds and the existing crop residues. Land clearing was done manually by using equipments such as chopping knife, mattock and others and followed by plot development with dimension of 1 m x 3 m consisting of 36 plots with inter plot distance of 1 m. Land preparation was conducted 2 weeks before planting operation.

2.2 Soil tillage

Soil was tilled by using mattock until loose in order to improve soil structure, air circulation within soil and facilitate soil microbe activity.

2.3 Fertilizing

Fertilizing of crop was done according to the treatments in which the used fertilizers were consisted of organic fertilizers from chicken dunk, cow dunk, oil palm fresh bunches and legume cover crops (LCC) with respective dose of 5 ton/ha (1.5 kg/plot), given one week before planting which was uniformly distributed and subsequently covered with soil. Inorganic fertilizer as base fertilizer in form of combined NPK fertilizer at dose of 150 kg/ha (45 g/plot) was given at planting time.

2.4 Planting

Seeds were planted by putting 2 seeds per planting hole with planting distance of 75 cm x 25 cm. Each plot consisted of 12 plants.

2.5 Maintenance

Crop maintenance was consisted of thinning out, irrigation, weeds control, and heaping. Thinning out was done two weeks after planting in which one planting hole only consisted of one plant. Irrigation was done every day, especially during vegetative growth stage or during plant ages of one up to eight weeks. Weed controls was done according to requirement and condition in land itself, whereas heaping was done in order to strengthen the plant upright.

2.6 Harvesting

Harvesting was done when the plant ages was about 60 days after planting (DAP) which characterized by appearance of brown hairs of corn, corn ears were still soft and fully fill up.

2.7 The observed parameters

The observed parameters were consisted of soil's chemical properties before planting, analyses of C-organic, N, P and K within respective organic fertilizers, crop height (cm), leave numbers (sheet), stem of an ear of corn length (cm), stem of an ear of corn diameter (cm), stem of an ear of corn weight per crop (g), stem of an ear of corn weight per plot (kg) and dry matter weight (g).

3 Results and discussion

Based on analysis results of soil chemical properties conducted before research implementation and criteria [18] as well as [22], soil used in this study can be classified as acid soil (pH H_2O = 4.81) with low level of cation exchange capacity (13.53 me/100g), medium level of C-organic content of 2.67 %, medium level of N-total content of 0.22 %, very high level of available P (180.37 ppm), exchanged base such as Ca-dd (1.04 me/100g) was very low, Mg-dd (0.28 me/100g) was very low, K-dd (0.21 me/100g) was very low, Na-dd (0.53 me/100g) was very low, base saturation (15.23 %) was very low, Al-dd was 1.96 me/100g with soil texture consisting of 62.42 % sand, 17.00 % loam and 20.00 % clay so that this soil can be classified as sandy loam.

Sandy loam soil which contains higher percentage of sand than that of loam and clay has porous characteristics so that addition of organic fertilizer will bind soil particles firmly which in turn capable to store water containing more nutrients and subsequently available for sweet corn crop.

Although C-organic content of soil before planting was 2.67 % (classified as medium), organic fertilizer is still needed for sweet corn cultivation because nutrients content within soil will decrease along with time so that it is expected that addition of organic matter in form of chicken dunk, cow dunk, oil palm fresh bunches and legume cover crops (LCC) will improve soil fertility which related to physical, chemical and biological properties of soil in the long term. This is in accordance to statement from [6] and [3] which showed that organic matter is important element for development of soil fertility. Organic matter in general can improve soil physical, chemical and biological properties. Organic matter improves soil physical property through development of loose and crumbly condition of soil particles which provide better aeration so that crop's root can easily penetrate into soil. Organic matter in sandy soil will increase the bonding amongst soil particles and improve water holding capacity. Soil chemical property is improved due to the increment of cation exchange capacity and nutrients availability, whereas soil biological property is improved due to addition of energy required for soil microorganism life [24].

Nutrients content within soil will decrease along with time because they are consumed by plant. This nutrients depletion condition should be recovered by addition of nutrients in order to prevent low level of crop growth and production. Lack of nutrients required by crop can be solved by fertilizer addition [25].

Analysis results of organic fertilizers used in this research was shown in Table 1. These organic fertilizers had diverse C/N ratio content. C/N ratio value of chicken dunk organic fertilizer was lower (21.58) than C/N ratio values of organic fertilizers from cow dunk, oil palm fresh bunches and legume cover crops (LCC). This fact showed that chicken dunk organic fertilizer had higher decomposition rate and capable to provide balance and sufficient nutrients of N, P and K for sweet corn crop. This is in accordance to statement from [9] which showed that optimum growth and production of corn crop require sufficient nutrients, especially N, P and K.

In addition, lower value of C/N ratio of chicken dunk organic fertilizer than that of other organic fertilizers used in this study showed the occurrence of mineralization process (the change from organic state to inorganic state). Mineralization process will release N nutrient and others resulting in increase of N nutrient availability within soil. This is in accordance to statement from [27] which showed that chicken dunk organic fertilizer had relatively faster decomposition rate than other organic fertilizers based on the same weight unit.

Table 1. N, P and K nutrients content and C/N ratio of organic fertilizer types.

Organic fertilizer types	N total (%)	P total (%)	K total (%)	C/N ratio
Chicken dunk	1.07	3.61	0.91	21.58
Cow dunk	2.27	1.81	2.47	25.53
Oil palm fresh bunches	2.58	0.55	5.59	22.57
LCC	1.54	2.55	2.34	27.31

Results of study showed that treatment of chicken dunk organic fertilizer on sweet corn had produced higher leave numbers, longer stem of an ear of corn and higher stem of an ear of corn diameter than that of other organic fertilizers as well as had significantly different effect (Table 2). This was due to the fact that chicken dunk organic fertilizer had provided N, P and K nutrients in sufficient and balance condition resulting in increase of crop height and leave numbers. Crop height affects leave numbers. Leave numbers on treatment by using chicken dunk organic fertilizer were higher than that of other organic fertilizers which in turn increase photosynthesis capacity. High photosynthesis capacity will increase dry matter weight of crop.

This dry matter is subsequently used for vegetative and generative growths. Jumin [7] had stated that dry matter is photosynthate storage in cells and tissues. Photosynthate or net yield of photosynthesis is the energy reduction due to respiration. Higher dry matter magnitude will produce higher photosyntate storage which in turn will determine photosynthate distribution magnitude (transfer of dry matter) into economic part of crop (stem of an ear of corn) indicated by stem of an ear of corn weight per crop and stem of an ear of corn weight per plot. Gardner *et al.* [4] showed that the higher photosynthate yield, the higher was the food supply storage transferred into seeds by assuming that other factors such as light, water, temperature and nutrients are in optimum condition.

Table 2. The effect of organic fertilizer types on vegetative growth of sweet corn crop.

Organic fertilizer types	Crop height (cm)	Leave numbers (helai)	Stem of an ear of corn length (cm)	Stem of an ear of corn diameter (cm)
Chicken dunk	187.67 c	11.33 c	22.15 d	4.21 b
Cow dunk	177.25 b	10.75 b	20.78 c	4.05 a
Oil palm fresh bunches	137.41 a	9.83 a	17.44 a	3.99 a
LCC	166.21 b	10.08 a	18.83 b	3.96 a
HSD 0.05 =	17.64	0.54	1.22	0.13

Remarks: Number followed by the same letter in the same column showed not significantly different based on HSD test at $\alpha = 5$ %

In addition, average production of sweet corn per crop and per plot as well as dry matter weight were higher on treatment by using chicken dunk organic fertilizer than that of organic fertilizers of cow dunk, oil palm fresh bunches and LCC (Table 3). Dry matter weight produced on treatment by using chicken dunk organic fertilizer was higher and significantly different than that of treatment by using organic fertilizers of cow dunk, oil palm fresh bunches and LCC. This indicates that crop having higher dry matter weight will produce higher stem of ear corn weight per crop and per plot. This is in accordance to statement from [17] which showed that higher dry matter weight will produce higher stem of an ear of corn length, stem of an ear of corn diameter and husked stem of an ear of corn weight.

Table 3. The yield and dry matter weight of sweet corn crop related to several organic fertilizers treatment

Organic fertilizer types	**Production**		**Dry matter weight (g)**
	per plant (g)	**per plot (kg)**	
Chicken dunk	284.52 c	4.37 c	65.39 d
Cow dunk	219.49 b	3.78 b	55.47 c
Oil palm fresh bunches	140.20 a	2.97 a	34.82 a
LCC	184.16 b	3.23 a	45.41 b
HSD 0.05 =	42.16	0.48	9.06

Remarks: Numbers followed by the same letter in the same column showed not significantly different based on HSD test at α = 5 %

During seed filling period, nitrogen and photosynthate transport from leaves section is very high in which nitrogen regulates phosphorus usage that stimulate flowering and seed formations. During this growth period, it is assumed that chicken dunk organic fertilizer is continuously decomposed and nitrogen as well as other nutrients becomes more available when plant is entering into flowering and seed formations phases. According to [21], nitrogen has function to accelerate the growth, to increase ear corn size, to increase protein content on cereal crops and to regulate the use of phosphorus, potassium and other nutrients. According to [8], nitrogen has function to accelerate the whole growth and to develop leave chlorophile which are very important for photosynthesis process. In addition, nitrogen also has function for the formation of protein, lipid and other organic compounds.

Application of chicken dunk organic fertilizer not only capable to increase sweet corn production, but also had positive impacts in term of improvement of soil physical, chemical and biological properties. Improvement of soil physical properties is achieved because organic matter (organic fertilizer) has function as adhesive of loose soil particles or aggregate stabilizer compound [5,19]. Improvement of soil chemical properties is achieved because organic matter (organic fertilizer) facilitates crop's root to penetrate deeper within soil so that it can absorb more water and nutrients, improve rhizosphere environment that can maintain nutrients, improve exudation by crop's roots that capable to increase soil organic matter decomposition as well as N mineralization [16]. Improvement of soil biological properties is achieved because organic matter provides energy source for most of soil's microorganisms [20].

Fertilizer of oil palm fresh bunches had produced the lowest growth and production than that of chicken dunk organic fertilizer. This was due to the fact that fertilizer of oil palm fresh bunches had imbalance nutrients which was shown by higher K availability than that of N and P. According to [28], availability of sufficient and balance nutrients had effect on metabolism process in crop tissues. Metabolism process is a process of development and degradation of nutrients and organic matters within crop. The lack of certain nutrients within crop can produce the negative effect, whereas excessive quantity of certain nutrients within crop can damage the crop's growth itself.

4 Conclusion

Treatment of chicken dunk organic fertilizer was capable to increase the growth and production of sweet corn crop in shallow lowland swamp with magnitude yield of 4.37 kg/plot.

References

1. T. Alihamsyah, Izzudin Noor, *Tidal Lowland: Supporting of Food Tenacity and Agribusiness Growth Source,* (Council of Agricultural Research and Development, 2003)
2. T. Alihamsyah, *Development of Lowland Swamp Area for Agricultural Enterprise*, (Council of Agricultural Research and Development, 2005).
3. D. Budianta, D. Ristiani. *Soil Fertility Management*, (Sriwijaya University, Press. Palembang, 2013)
4. F.T. Gardner, R.B. Pearce, R. L. Mitchell, Physiology of Crops., (2008)
5. R.F. Gonzales, L.R. Cooperband, Compost. Sci. Util., **10**, 226 (2002)
6. N. Hakim, M.Yusuf, A.M. Lubis, N. Sutopo, M.R. Saul, M.A. Diha, Go Ban Hong, H.H. Bailey, *Fundamentals of Soil Science*, (Lampung University, 1986)
7. H. B. Jumin, *Fundamentals of Agronomy*, (Raja Grafindo Perkasa, Jakarta, 2005)
8. P. Lingga, Marsono, *Guidance of Fertilizer Utilizaton*, (Penebar Swadaya, Jakarta, 2008)
9. S.L. Margaretha, Ningsih, W. Subandi, Zubachtiroddin, *Corn Response to N, P and K Fertilizers at Dry Land and Dry Climate* (Cereal Crop Research, East Lombok, 2014)
10. N. Marlina, E.A. Saputro, N. Amir, Journal of Suboptimal Land, **1(2)**, 138 (2012)
11. N. Marlina, N. Gofar, N. Amir, B.A. Putra, *7th National Seminar Proceeding of Indonesia Soil Conservation Society*, (Palembang, 2013)
12. N. Marlina, N. Gofar, A. Halim, A.M. Rohim, Journal of Agricultural Science Agrivita, **36(1)**, 48 (2014)
13. N. Marlina, Syafrullah, *Proceeding of National Seminar of Suboptimal Land*, (Palembang, 2014)
14. N. Marlina, R.I.S. Syafrullah, Aminah, Gusmiatun, Rosmiah, Midranisiah, Y. Purwanti, Gribaldi, International Journal of Engineering Research Science & Technology, **4(2)**, 100 (2015)
15. N. Marlina, R.I.S. Aminah, Rosmiah, L.R. Setel, Journal of Biosaintifika, **7(2)**, 136 (2015)
16. J.A.W. Morgan, G.D. Bending, P.J. White, J. Exp. Bot., **56**, 1729 (2005)
17. A. Navioside, Y. Sugito, M. Dewani, Jurnal Agrivita, **24(2)**, 136 (2002)
18. Soil Research Center, *Terms of Reference Type*, (P3TT, Bogor, 1983)
19. H. Riley, R. Pommeresche, R. Eltun, S. Hansen, A. Korsaeth, Agric. Ecosyst. Environ., **124**, 275 (2008)

20. A. Saviozzi, R. Cardelli, P. Nkou, R. Levi Minzi, R. Riffaldi, Compost. Sci. Util., **14**, 54 (2006)
21. G. Soepardi, *Soil Properties and Characteristics*, (Soil Science Department, 1983)
22. Soil Research Council, *Analyses of Sol Chemistry, Plant, Water and Fertilizer*, (Department of Agriculture, Jakarta, 2005)
23. R.D.M. Suriadikarta, D.A. Simanungkalit, *Organic Fertlzer and Bofertilizer*, (Agricultural Research and Development Council, Bogor, 2006)
24. R. Sutanto, *Organic Framing*, (Kanisius, Yogyakarta, 2002)
25. Y. Sutoro, Soeleman, Iskandar, *Corn Crop Cultivation*, (Research and Development Center for Food Crops, Bogor, 1988)
26. N. Syafrullah, Marlina, Gribaldi, R.I.S. Aminah, Midranisiah, G.A. Nasser, Gusmiatun, International Journal of Engineering Research Science & Technology, **4(3)**, 1 (2015)
27. L.R. Widowati, Sri Widati, U. Jaenudin, W. Hartatik, *The Effect of Organic Fertilizer Compost Enriched with Minerals and Biofertilizer on Soil Properties* (Soil Research Council, 2005)

Warpage Optimisation on the Moulded Part with Conformal Cooling Channels using Response Surface Methodology (RSM) and Glowworm Swarm Optimisation (GSO)

M.H.M. Hazwan[1,2,*], *Z.* Shayfull[1,4], *S.* Sharif[3], *S.M.* Nasir[1,4], *M.M.* Rashidi[5]

[1]School of Manufacturing Engineering, Universiti Malaysia Perlis, Kampus Tetap Pauh Putra, 02600 Arau, Perlis, Malaysia.
[2]School of Engineering Technology, Universiti Malaysia Perlis, Kampus UniCITI Alam, Sungai Chuchuh, 02100 Padang Besar, Perlis, Malaysia.
[3]Faculty of Mechanical Engineering, Universiti Teknologi Malaysia, 81310 UTM Skudai, Johor, Malaysia.
[4]Green Design and Manufacture Research Group, Centerof Excellence Geopolymer and Green Technology (CEGeoGTech), Universiti Malaysia Perlis, 01000 Kangar, Perlis, Malaysia.
[5]Faculty of Mechanical Engineering, Universiti Malaysia Pahang, 26600 Pekan, Pahang, Malaysia.

Abstract. Today, there are many of optimisation method have been explored by previous researchers to find the appropriate processing parameters setting for the injection moulding process. From the previous researches, it has been proven that the optimisation work improved the quality of the moulded part. However, the application of optimisation work in conformal cooling channels still lacks. Therefore, in this study, the application of optimisation work to improve warpage of front panel housing with conformal cooling channels moulds have been explored. By choosing cooling time, coolant temperature, packing pressure and melt temperature as the variable parameters, design of experiment (DOE) has been defined by using the rotatable central composite design (CCD) approach. Response Surface Methodology (RSM) was performed to determine the mathematical model. The mathematical model then will be used in Glowworm Swarm Optimisation (GSO) method in order to obtain the optimal processing parameters setting which will optimise the warpage condition. Based on the results, cooling time is the most significant factor contributes to the warpage condition and warpage have optimised by 38.7% after optimisation using the proposed approach. This finding shows that the application of conformal cooling channels with optimisation work will produce better quality of the moulded part.

1 Introduction

Injection moulding process consists into four main stages which are filling, packing, cooling and ejecting processes [1, 2]. Due to the complexity of injection moulding process, it is

[*]Corresponding author: hazwanhanid@unimap.edu.my

difficult for injection moulding industries in order to produce the best quality of the moulded part. Most common defects in injection moulding is a warpage [3]. Warpage is difficult to avoid due to design complexity and numerous of influence factors which affect the assembly process because of uneven clearance or interference problems. With an appropriate setting of injection moulding processing parameters, warpage condition can be reduced [4]. Previously, most of injection moulding industries have used try-and-error approach in order to obtain an appropriate processing parameters setting which consume time and production cost [5]. Today, with the advancement of computer technology, an appropriate combination of processing parameters can be obtained by simulation analysis software with and aid of optimisation approach which is highly accurate, shorter time, and cost effective [3]. These days, many researchers came with various proposals of optimisation approaches to determine an optimal processing parameters in order to optimise the quality of injection moulded parts.

For instances, Wang et al. [6], studied in minimising warpage on cellular phone cover made from PC and ABS material. Expected improvement (EI) optimisation method has been proposed in this research with the aid of Kriging surrogate model and genetic algorithm (GA). At first, the moulded part have been simulated to determine the initial condition. Then, Kriging surrogate model was employed to obtain the mathematical model. Next, the EI method with the combination of GA was performed to find the optimum processing parameters and to obtain the significant factor that contribute to warpage. The results show that warpage reduced up to 28.24% and packing pressure has detected as the significant factor contributed to warpage.

Chen et al. [7], explored the sequential optimisation methods which are Taguchi's method, hybrid back-propagation neural network and genetic algorithm (BPNN-GA) and hybrid back-propagation neural network and particle swarm optimisation (BPNN-PSO) to find an optimum length and to reduce warpage. ADSL modem cover made of PA-765 material was used as a specimen in this research. At first, Taguchi's method was conducted to investigate the initial processing parameters. Next, by implementing ANOVA, packing pressure and packing time have been found as the significant factors that contributed to the length and warpage condition. Then, BPNN-GA was conducted to obtain the first stage of optimal processing parameters setting. From the first stage analysis, BPNN-PSO was performed to obtain the second stage of optimal processing parameters and the results show that warping have been reduced by 38,6%, from 0.2202mm to 0.1351mm and optimal length achieved by 123.997mm from target value which is 124mm.

Gang Xu and Zhitao Yang [8], investigated the sequence of optimisation methods consist of Taguchi's method, back-propagation neural networks (BPNN), grey correlation analysis (GCA) and particle swarm optimisation (PSO) in order to improve the part weight, flash and volumetric shrinkage on the thin-walled part which made of PP material. At first, Taguchi's method was used to identify significant processing parameters. Then, GCA was performed to establish the hidden neuron numbers of BPNN. Next, the BPNN was used to generate the optimisation algorithm. Then, this algorithm has been used by PSO to get optimal outcomes. The results of local optimisation show that part weight was improved by 30%, flash improved by 46% and volume shrinkage reduced by 53%. Meanwhile, global optimisation results show that part weight was improved by 28%, flash improved by 48%, and volume shrinkage reduced by 41%.

Based on the literatures, moulded part qualities can be enhanced by the application of optimisation work. However, most of previous researches focused an optimisation work on conventional straight-drilled cooling channels. Nowadays, there are some researchers have introduced the various of conformal cooling channels design which have been proven improved quality and productivity of the moulded part.

For instance, Gao et al. [9], analysed the cooling effect of conformal cooling channels (CCC). By using a plastic cup as a moulded part, the performance of conformal cooling channels has been carried out and compared with conventional cooling channels by using Autodesk Moldflow Insight software. The simulation results show that cooling time have been reduced from 11.34s to 8.284s compared with conventional cooling channels. Average temperature also have been decreased which is from 75.56°C to 58.67°C and warping deformation by using the proposed conformal cooling channels have been reduced from 0.6417mm to 0.5752mm.

Saifullah and Masood [10], evaluated the performance of conformal cooling channels compared with conventional cooling channels of a moulded circular plastic bowl made from PP. The cooling channel diameter was 12mm and have been located as close as possible with the surface of core and cavity that in contact with the melted plastic. ANSYS thermal analysis software has been performed to evaluate the performance of conformal cooling channels and compared it with conventional cooling channels. The results show that conformal cooling channels give more uniformity of temperature distribution in the moulding process and cooling time has been reduced about 20% of the total cycle duration compared to conventional cooling channels.

Based on the previous conformal cooling channels researches, the application of optimisation approach in order to determine an appropriate processing parameters setting are still lacking. Therefore, this study will be introduced an alternative optimisation approach to improve warpage on front panel housing part made of Acrylonitrile-Butadiene-Styrene (ABS) with Milled Groove Square Shape (MGSS) conformal cooling channels mould. Based on input processing parameters, design of experiment (DOE) will be generated by using full factorial Design (FFD) with an aid of rotatable central composite design of experiment (CCD). Then AMI 2013 software will be used to analyse warpage condition for each run. Response surface methodology (RSM) will be performed in order to obtain the mathematical model function and analysis of variance (ANOVA) will be used to determine the significant factors influencing the warpage condition. The mathematical model obtained from RSM will be used in glowworm swarm optimisation (GSO) method as an objective function to determine the optimal processing parameters which will optimise the warpage of the moulded part.

2 Response surface methodology

Response surface methodology (RSM) is a classical optimisation approach. It was used to demonstrate the relationship between variable parameters which influence the response condition in two or three-dimensional hyperbolic surface[11]. The mathematical model function will be obtained by using the second-order polynomial regression model in this study which will be used as the objective function in GSO. The necessary information to construct the response model was accumulated by the simulation analysis [12, 13]. Figure 1 shows the RSM flowchart in this study.

DOI: 10.1051/matecconf/20179701104

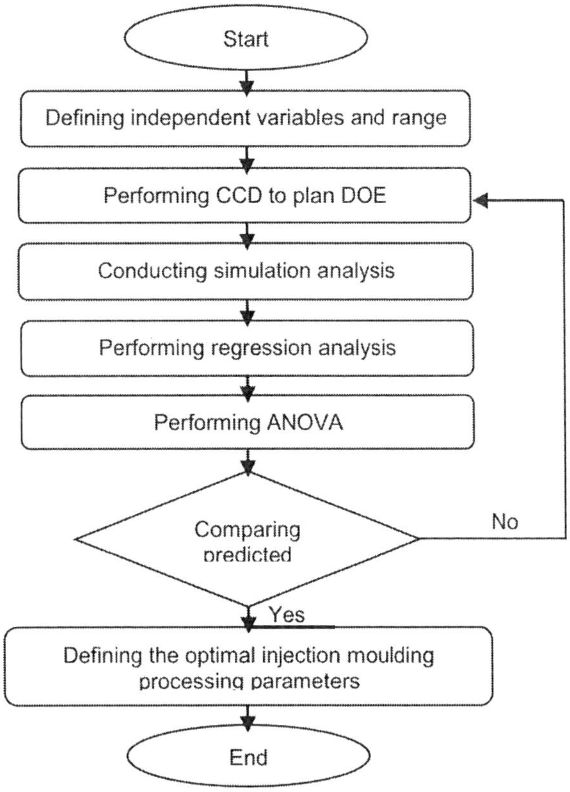

Fig. 1. Response Surface Methodology (RSM) flowchart

2.1 Design of experiment setup

In this study, coolant temperature, melting temperature, packing pressure and cooling time have been selected as variable parameters and the range of each parameter was shown in Table 1. At first, full factorial design (FFD) with four centre points was selected as an experimental design to evaluate the model and main effects contribute to the warpage condition by using Design Expert 7.0 software. In order to obtain the significant curvature which is important in the RSM regression analysis, the augment of rotatable central composite design (CCD) was added. Therefore, 30 runs of specified condition have been generated and each run will be set in AMI 2013 simulation software to evaluate the warpage condition of the moulded part.

Table 1. Variable parameters and levels

Factors	Level	
	Minimum	**Maximum**
Coolant inlet temperature, A (°C)	25	65
Melt temperature, B (°C)	220	260
Packing pressure, C (MPa)	46.74	56.74
Cooling time, D (s)	10	25

2.2 Finite element analysis setup

The CAE software, Autodesk Moldflow Insight (AMI) 2013 have been used to simulate the injection moulding process and to evaluate the warpage condition of front panel housing moulded part which has 2.5mm of average thickness and made from ABS material with trade name as Toray/Toyolac 700-314 with MGSS conformal cooling channels mould as shown in Fig. 2. The moulded part, gating system and cooling channels 3D data was created and imported into the software for meshing process. In order to obtain more precise results, the mould insert material made of P20 steel and Nissei NEX1000, 80 tonne injection moulding machine specification have been set in the AMI software. Then, the Cool (FEM)+Fill+Pack+Warp analysis has been performed to evaluate warpage condition for each run which have been generated by DOE.

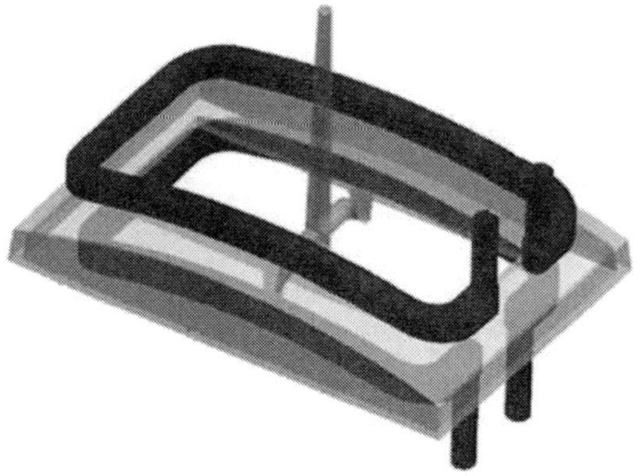

Fig. 2. Front panel housing with MGSS conformal cooling channels

2.3 RSM regression analysis

The warpage results for each run which obtained from the simulation analysis will be used in the RSM regression analysis. By using Design Expert 7.0 software the regression analysis was performed with the backward quadratic model in order to obtain the mathematical model function which representing the relationship between variable parameters and response. The software will calculate using second-order polynomial regression model in the statistically manner and the results will be verified with ANOVA.

2.4 Analysis of variance (ANOVA)

The result of the quadratic model obtained from the RSM regression analysis will be verified by analysis of variance (ANOVA) to determine either the mathematical model was statistical significant or otherwise. In the same analysis, the significant factors that contribute to the warpage condition will be defined.

3 Glowworm swarm optimisation

The operators in GSO are recognised as glowworms [14] that carry a luminescent amount which have been called as *luciferin* [15]. Every glowworm utilises an artificial proportional luciferin to transmit the fitness of its current location and evaluate based on the objective model function to its neighbours. The glowworms rely on the variable of their neighbourhood which is based on the radial sensor range boundary in order to identify their neighbours and evaluate their movements. Each glowworm, use the probabilistic mechanism by selecting a neighbour that contain the higher luciferin amount than its own and moves toward it. In other words, it will be attracted to the neighbours which is glowing brighter. Finally, the movement of the glowworms will be based on the local information and particular neighbour interactions which empower the majority swarm to form multiple optima of a given multimodal function.

In this study, GSO method will be carried out to obtain the optimal processing parameters of injection moulding process based RSM mathematical model function. MATLAB R2014a software will be used to conduct GSO analysis. Figure 3 shows the flowchart of GSO in this study.

3.1 Defining mathematical model function

The objective function used in GSO is the mathematical model function which have been obtained from RSM regression analysis. The formulated objective function has been taken in this study is a warpage function.

3.2 Initialisation of glowworms' parameters

The initial glowworms' parameters were set based on the selected injection moulding variable parameters limit in this study to create the solution space. Then, the GSO agents will be deployed randomly in the solution space by setting the initial glowworms population size and maximum iteration based on the research requirement. In this study 30 numbers of glowworms and 40 iterations have been set.

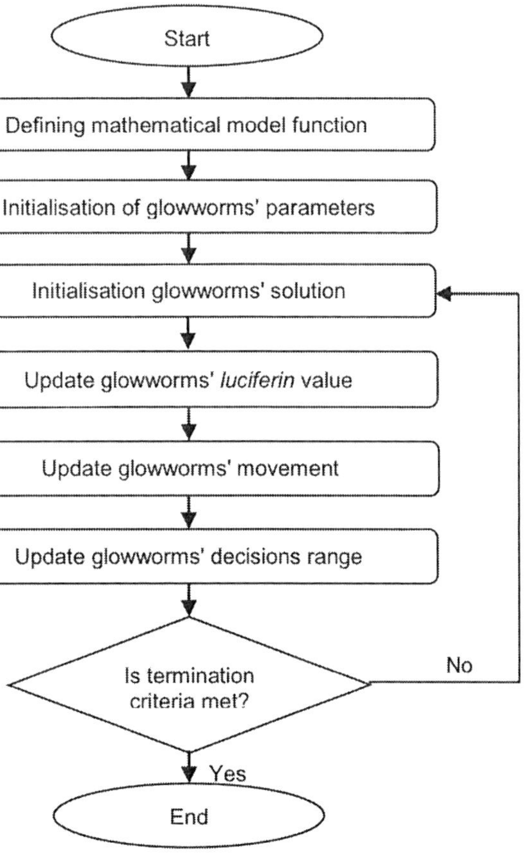

Fig. 3. Flowchart of Glowworm Swarm Optimisation (GSO)

3.3 Initialisation glowworms' solution

The initial solution of initial *luciferin* value (l_o), *luciferin* decay constant (ρ), *luciferin* enhancement constant (γ), beta (β), step size (s), neighbourhood range $r_d^i(t)$ and parameter used to control the number of neighbours (n_t) was set based on the research requirement where each glowworm contained the same *luciferin* value and sensor range in this phase. In this study, GSO control parameters optimal setting are: $\beta = 0.08, \rho = 0.4, \gamma = 0.6$ and $r_d^i(t) = 3$.

3.4 Update glowworms' luciferin value

At the beginning, each glowworm contains the same *luciferin* value. Influenced by the objective function value of their current location, the *luciferin* value will be changed. The rule of *luciferin* update was given by standard Equation (1) where $li(t)$ is the *luciferin* level for glowworm i at time t, ρ is the *luciferin* decay constant $(0 < \rho < 1)$, and $Ji(t)$ indicates the objective function at agent i's location at time t.

$$li(t+1) = (1-\rho)li(t) + \gamma Ji(t+1) \quad (1)$$

3.5 Update glowworms' movement

In the movement-update phase, each glowworm will move toward neighbour that contained the higher *luciferin* value than its own using a probabilistic mechanism. The probability of agent i move to agent j was given by a standard Equation (2) where $l_i(t)$ is the *luciferin* value for glowworm i, $d(i,j)$ is the Euclidean distance between agent i and j.

$$P_{ij}(t) = \frac{l_j(t) - l_i(t)}{\sum_{k \in N_i(t)} l_k(t) - l_i(t)} \quad (2)$$

Then, the glowworm i movement can be expressed by Equation (3) where s is the step size.

$$x_i(t+1) = x_i(t) + s\left[\frac{x_j(t) - x_i(t)}{\|x_j(t) - x_i(t)\|}\right] \quad (3)$$

3.6 Update glowworms' local decisions range

Local decision range update phase will be used to determine multiple peaks in a multimodal functional landscape in order to obtain the optimal variable parameters which optimise warpage value effectively. When the number of neighbours changed, the local decision domain needs to update in each of iteration. The rule is given by Equation (4) where $r_d^i(t+1)$ is the local decision domain of glowworm i in the $t+1$ iteration, β is the constant parameter that affected the rate of change of the neighbour domain and n_t is the threshold used to control the number of neighbours.

$$r_d^i(t+1) = \min\{r_s, \max\left\{0, r_d^i(t) + \beta(n_t - |N_i(t)|)\}\right\} \quad (4)$$

4 Results and discussions

4.1 Injection Moulding Process Simulation

Warpage values obtained from simulation analysis are shown in Table 2. The results tabulate the warpage value for each run with the specified variable parameters condition which obtained from the DOE. The specified variable parameters condition was set and simulated in the AMI 2013 software.

Table 2. Injection moulding process simulation results.

Standard Order	Data Source	Variable parameters for injection moulding simulation				Response
		Coolant temperature (°C)	Melt temperature (°C)	Packing Pressure (MPa)	Cooling time (s)	Warpage (mm)
1	DOE	25	220	46.74	10	0.115
2		65	220	46.74	10	0.095
3		25	260	46.74	10	0.165
4		65	260	46.74	10	0.125
5		25	220	56.74	10	0.110
6		65	220	56.74	10	0.090
7		25	260	56.74	10	0.155
8		65	260	56.74	10	0.130
9		25	220	46.74	25	0.095
10		65	220	46.74	25	0.085
11		25	260	46.74	25	0.155
12		65	260	46.74	25	0.115
13		25	220	56.74	25	0.085
14		65	220	56.74	25	0.090
15		25	260	56.74	25	0.140
16		65	260	56.74	25	0.115
17	Centre	45	240	51.74	17.5	0.125
18		45	240	51.74	17.5	0.125
19		45	240	51.74	17.5	0.125
20		45	240	51.74	17.5	0.125
21	Axial	5	240	51.74	17.5	0.130
22		85	240	51.74	17.5	0.090
23		45	200	51.74	17.5	0.050
24		45	280	51.74	17.5	0.155
25		45	240	41.74	17.5	0.120
26		45	240	61.74	17.5	0.130
27		45	240	51.74	2.5	0.140
28		45	240	51.74	32.5	0.110
29		45	240	51.74	17.5	0.125
30		45	240	51.74	17.5	0.125

4.2 Response surface methodology (RSM) regression analysis

From regression analysis, the determination coefficient, R2 that fits the model is 0.9774. The adjusted determination is 0.9655, which shows the regression model is significant with an adequate precision is more than 4, which is 38.862. The model is more significant if R2 value is closer to 1 [16]. The standard deviation for this model is $4.683\times10\text{-}3$ and this value indicate that it is relatively lower than 0.05 which is good for better precision and reliability of the experiment [17]. The polynomial regression model which related to the warpage with all input parameters (coolant inlet temperature (A), melt temperature (B), packing pressure (C) and cooling time (D) is established by Design Expert 7 software represented in Equation 5. This mathematical model function will be used in GSO as an objective function.

DOI: 10.1051/matecconf/20179701104

$$Warpage = -0.979 + (1.763 \times 10^{-3}A) + (8.342 \times 10^{-3}B) - (1.391 \times 10^{-3}C) \\ -(1.573 \times 10^{-3}D) - (1.328 \times 10^{-5}AB) + (2.8125 \times 10^{-5}AC) \\ +(1.458 \times 10^{-5}AD) - (9.082 \times 10^{-6}D^2) - (1.377 \times 10^{-5}B^2) \quad (5)$$

4.3 Analysis of Variance (ANOVA)

From the ANOVA table as shown in Table 3, the F calculated value is bigger than F tabulated. This condition shows that the mathematical model obtained from regression analysis was significant. The ANOVA also indicated that three from four selected variable parameters give significant effect on the warpage condition.

Table 3. ANOVA of response surface model.

	Sum of Squares	df	Mean Square	F (calculated)	R^2	RA^2	F (tabulated)
SSR	0.0180	10	0.0018	82.1355	0.9774	0.9655	2.37
SSE	0.0004	19	0.0000				
Total	0.0184	29					

The results show that cooling time was the most significant factor contributed to the warpage condition. This result was inline with previous researchers which found out the same significant factor which influenced the warpage condition [18, 19]. Then, it follows by cooling temperature and melt temperature. Packing pressure was the least significant factor contributed to the warpage condition for MGSS conformal cooling channels as shown in Figure 4. In addition, this result proves that MGSS conformal cooling channels give better uniformity of thermal distribution in the mould insert where the temperature is no longer become the most significant factor contribute to the warpage condition.

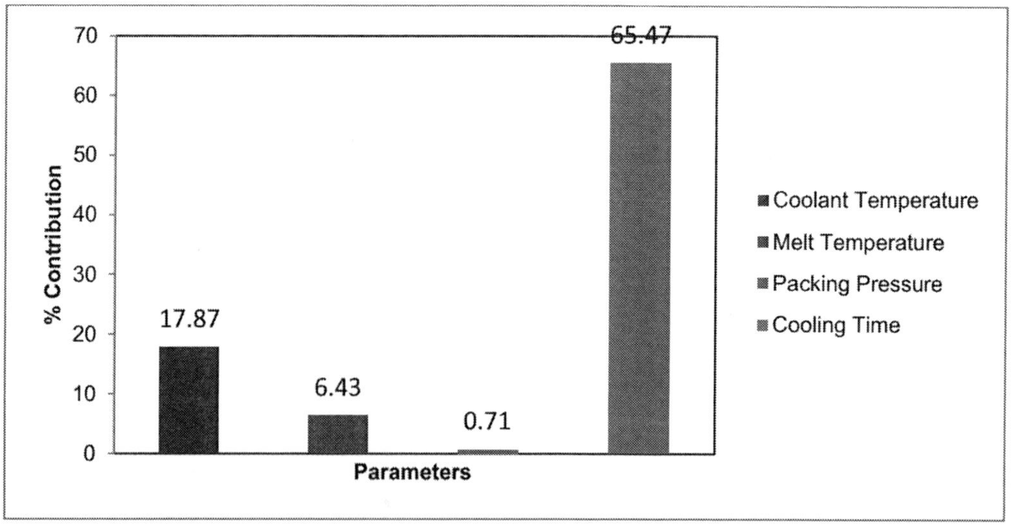

Fig. 4. Contribution for each parameters in percentage.

Equation 5 was applied to calculate the prediction warpage values of the polynomial model and the results is summarised in Figure 5. From the comparison, it shows that the simulation and predicted value are very close to each other which indicated the mathematical response function have given a good prediction value in estimating warpage.

DOI: 10.1051/matecconf/20179701104

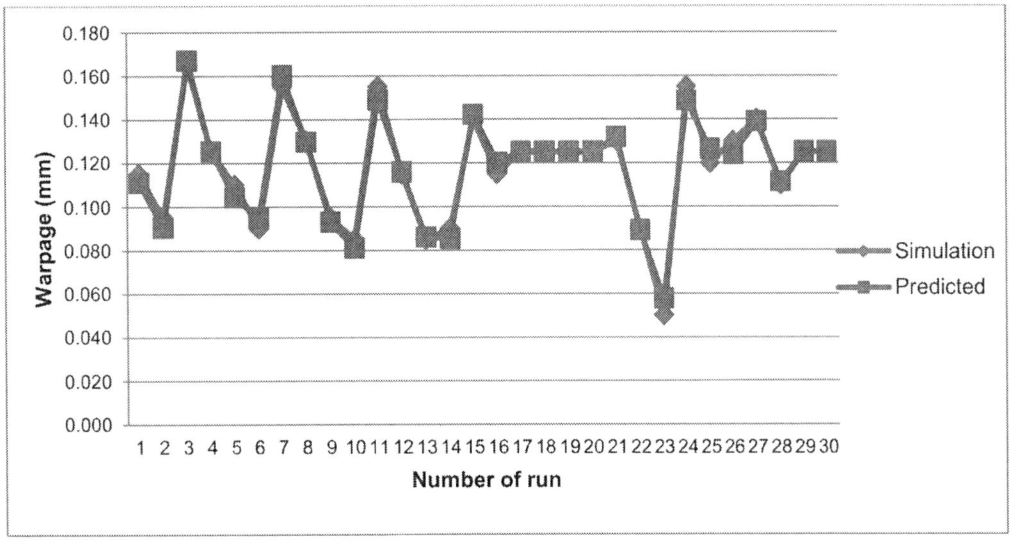

Fig. 5. Simulation and predicted warpage comparison

4.4 Glowworm Swarm Optimisation (GSO) analysis

The GSO algorithm has been conducted on injection moulding processing parameters. The algorithm was tested on four selected variable parameters. By using the mathematical model function obtained from RSM, the optimal results were shown in Table 4 and the warpage variation for each glowworm was shown in Figure 6.

Table 4. Recommended simulation results versus GSO Optimised results

Factors	Recommended simulation results	GSO optimised results
Coolant inlet temperature, A (°C)	25	43.01
Melt temperature, B (°C)	240	222.54
Packing pressure, C (MPa)	46.74	50.33
Cooling time, D (s)	10.55	24.85
Warpage, (mm)	0.155	0.095

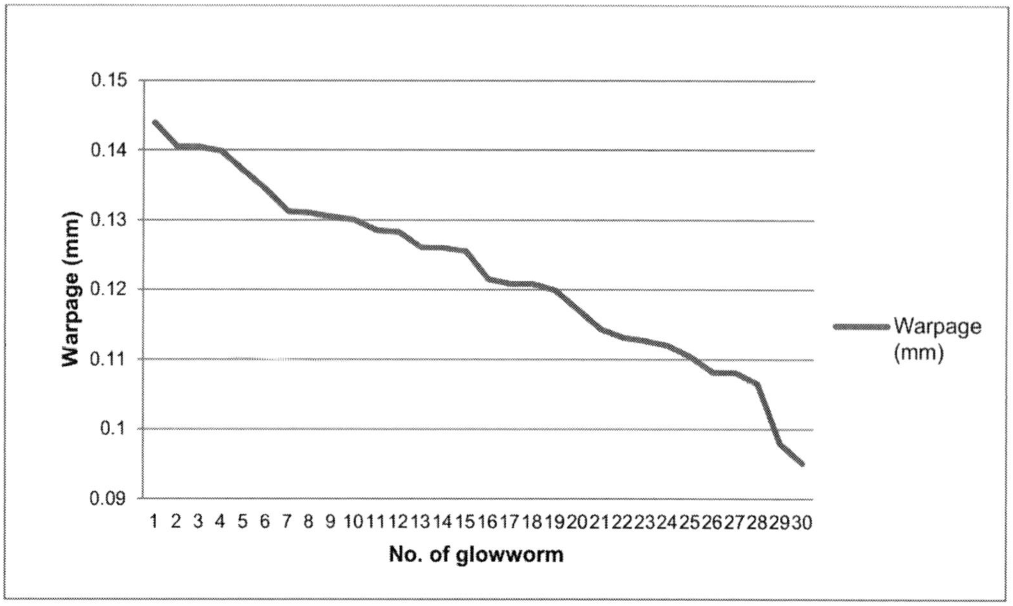

Fig. 6. Warpage variation for each glowworm

5 Conclusion

This study is definitely helpful in enhancing the quality of moulded parts produced where the objective is to optimise warpage of the front panel housing moulded part with MGSS conformal cooling channels mould have been achieved. Based on the results, the warpage has been optimised by using an alternative approach of response surface methodology (RSM) and Glowworm Swarm Optimisation (GSO). The results also show that:

- By using RSM, the significant mathematical model function can be obtained in order to predict warpage value with reasonable accuracy.
- From the ANOVA results, cooling time is the most significant factor influencing the warpage condition on the moulded part, follow by coolant temperature and melt temperature.
- The optimal processing parameter obtained from a GSO optimising warpage by 38.7% which is from 0.155mm from the simulation result to 0.095mm.

In this study, it proved that the application of conformal cooling channels with the proposed optimisation approach has enormous potential in order to obtain better quality of moulded parts.

References

1. W. Michaeli, G. Potsch, Injection molding: an introduction, München: Hanser, Fachbuchverlag, 1995.
2. L. Sors, I. Balazs, Design of plastic moulds and dies, 1989.
3. S.S. Teklehaimanot, Simulation and design of a plastic injection Mold, 2012.
4. D. Annicchiarico, J.R Alcock, Mater. Manuf. Process, **29**, (6), 662, (2014).
5. Y.C. Lam, L.Y. Zhai, K. Tai, S.C. Fok, Int. J. Prod. Res., **42**, (10), 2047, (2004).
6. X. Wang, J. Gu, C. Shen, X. Wang, Int. J. Adv. Manuf. Tech., **78**, (1-4), 177, (2015).
7. W.C Chen, P.H. Liou, S.C. Chou, Int. J. Adv. Manuf. Tech., **73**, (9-12), 1465, (2014).
8. G. Xu, Z. Yang, Int. J. Adv. Manuf. Tech., **78**, (1-4), 525, (2015).

9. G.Q. Gao, Y.S. Liu, Y.J. Cai, Analysis and Manufacturing of Conformal Cooling Channels in Injection Mold, (Trans. Tech. Publ.), 2014.
10. A. Saifullah, S. Masood, Finite element thermal analysis of conformal cooling channels in injection moulding, (Engineers Australia), 2007.
11. D. Montgomery, Design and Analysis of Experiments, (6th edn.,) John Wiley and Sons', New York, NY, 2005.
12. H. Oktem, T. Erzurumlu, H. Kurtaran, J. Mater. Process. Technol., **170**, (1), 11, (2005).
13. B. Ozcelik, T. Erzurumlu, Int. Commun. Heat Mass, **32**, (8), 1085, (2005).
14. E. Bonabeau, M. Dorigo, G. Theraulaz, Swarm intelligence: from natural to artificial systems, (Oxford university press), 1999.
15. R. Brits, A.P. Engelbrecht, F. Van den Bergh, A niching particle swarm optimizer, (Singapore: Orchid Country Club), 2002.
16. Y.H. Wang, J.T. Feng, Q. Zhang, X. Zhang, J. Appl. Microbiol, **104**, (3), 735, (2008).
17. M.S. Tanyildizi, D. Ozer, M. Elibol, Process Biochem, **40**, (7), 2291, (2005).
18. W.C. Chen, D. Kurniawan, Int. J. Precis. Eng. Manuf., **15**, (8), 1583, (2014).
19. W.C. Chen, M.H. Nguyen, W.H. Chiu, T.N. Chen, P.H. Tai, Int. J. Adv. Manuf. Technol., **83**, (9-12), 1873, (2016).

Warpage Optimisation on the Moulded Part using Response Surface Methodology (RSM) and Glowworm Swarm Optimisation (GSO)

M.H.M. Hazwan[1,2,*], *Z.* Shayfull[1,4], *S.* Sharif[3], *S.M.* Nasir[1,4], *M.M* Rashidi[5]

[1]School of Manufacturing Engineering, Universiti Malaysia Perlis, Kampus Tetap Pauh Putra, 02600 Arau, Perlis, Malaysia.
[2]School of Engineering Technology, Universiti Malaysia Perlis, Kampus UniCITI Alam, Sungai Chuchuh, 02100 Padang Besar, Perlis, Malaysia.
[3]Faculty of Mechanical Engineering, Universiti Teknologi Malaysia, 81310 UTM Skudai, Johor, Malaysia.
[4]Green Design and Manufacture Research Group, Centerof Excellence Geopolymer and Green Technology (CEGeoGTech), Universiti Malaysia Perlis, 01000 Kangar, Perlis, Malaysia.
[5]Faculty of Mechanical Engineering, Universiti Malaysia Pahang, 26600 Pekan, Pahang, Malaysia.

Abstract. Nowadays, there are various of optimisation methods that have been explored by many researchers to find the appropriate processing parameters setting for the injection moulding process. From the previous researches, it was reported that the optimisation work has improved the moulded part quality. In this study, the application of optimisation work to improve warpage of the front panel housing have been explored. By selecting cooling time, coolant temperature, packing pressure and melt temperature as the variable parameters, design of experiment (DOE) have been constructed by using the rotatable central composite design (CCD) approach. Response Surface Methodology (RSM) was performed to obtain the mathematical model. This mathematical model then will be used in Glowworm Swarm Optimisation (GSO) method in order to determine the optimal processing parameters setting which will optimise the warpage condition. Based on the results, melt temperature is the most significant factor contribute to the warpage condition and warpage have optimised by 39.1% after optimisation. The finding shows that the application of optimisation work offers the best quality of moulded part produced.

1 Introduction

Injection moulding process consists into four main stages which are filling, packing, cooling and ejecting processes [1, 2]. Due to the complexity of injection moulding process, it is difficult for injection moulding industries in order to produce the best quality of the moulded part. Most common defects in injection moulding is a warpage [3]. Warpage is difficult to prevent due to design complexity and numerous influencing factors which affected the assembly process because of uneven clearance or interference problems. With an appropriate

*Corresponding author: hazwanhanid@unimap.edu.my

setting of injection moulding processing parameters, warpage condition can be reduced [4]. Previously, most of injection moulding industries have used try-and-error approach in order to obtain an appropriate processing parameters setting which consume time and production cost [5]. Today with the advancement of computer technology, an appropriate combination of processing parameters can be obtained by simulation analysis software with an aid of optimisation approach which is highly accurate, shorter time, and much more cheaper [3]. These days, many researchers come with various proposals of optimisation approaches to determine an optimal processing parameters in order to optimise the quality of injection moulded parts.

For instances, Annicchiarico et al. [6], proposed the half fractional factorial design of experiment optimisation method to investigate the processing parameter influence on the shrinkage in order to improve part quality. Squared specimen made from polyoxymethylene (POM) was used in this research. The preliminary screening was performed to identify high and low values for each selected parameter. Then, the significant factors have determined by conducting statistical analysis. The result shown that mould temperature are the most critical factor influenced the shrinkage for parallel and normal flow direction.

Yin et al. [7], explored Taguchi's method with multi-objectives optimisation approach to find an optimal processing parameters in order to reduce warpage and birefringence. An optical lens made of PMMA material was used in this research. At first, Taguchi's orthogonal array with three level was created to evaluate the initial process condition. S/N ratio was calculated and analysis of mean (ANOM) was performed as a multi-objectives optimisation approach in order to obtain an optimal processing parameter setting and to determine the most critical factor to warpage and birefringence. Experimental verification was conducted in this research by using optimal processing parameters that obtained from the optimisation work. The results show that the birefringence and warpage have reduced, which were smaller than minimal value obtained from the Taguchi's method. Melt temperature was found as the most significant factor to give effect to both warpage and birefringence.

Chen et al. [8], explored the optimisation work using Taguchi method, response surface methodology (RSM) and hybrid of genetic algorithm-particle swarm optimisation (GA-PSO) in order to find an optimum length and to reduce warpage. In this research, printer rear cover was used as specimen which made from PBT-2100 material. At first, Taguchi's method was conducted to investigate the initial processing parameters. By implementing ANOVA, the significant factors have been determined which are packing time and cooling time. Next, the RSM was conducted to generate mathematical models and the combination of RSM-GA was performed to determine an optimal processing parameters. This first stage optimal processing parameters will be optimised again by using hybrid GA-PSO method. The results of the second stage optimisation show that the optimum length achieved was 170.483mm which was claimed to be the closest target value (170.5mm) and warpage was reduced from 0.092 to 0.025mm.

In other research, Chen et al. [9], proposed the sequence of optimisation methods which consist of Taguchi's method, back-propagation neural network (BPNN), genetic algorithm (GA) and hybrid particle swarm optimisation and genetic algorithm (PSO-GA) to get an optimum length and to reduce warpage. The same printer rear cover made from PBT-2100 material was used as research specimens. For initial, Taguchi's method was performed to obtain the best combination of processing parameters. Next, ANOVA was carried out to identify the significant factors contributed to the responses and the result are the packing time and cooling time. Afterward, BPNN was conducted to plot the relationship between variables and responses. Then, the first stage optimisation was employed by using GA to obtain the initial optimal processing parameters. Next, the second stage optimisation which used hybrid PSO-GA approach to determine an optimal processing parameters setting was performed.

The results show that the optimal length achieve was 170.52mm from the target length which was 170.5mm and warpage was reduced from 0.198mm to 0.096mm.

Based on the literatures, moulded part qualities can be enhanced by the application of optimisation work. Therefore, in this study an alternative optimisation approach has been introduced to improve warpage on front panel housing part made of Acrylonitrile-Butadiene -Styrene (ABS). Based on input processing parameters design of experiment (DOE) will be generated by using full factorial Design (FFD) with an aid of rotatable central composite design of experiment (CCD). Then AMI 2013 software will be used to analyse warpage condition for each run. Response surface methodology (RSM) will be performed in order to obtain the mathematical model function and analysis of variance (ANOVA) will be used to define the significant factors influencing on the warpage condition. The mathematical model obtained from RSM will be used in glowworm swarm optimisation (GSO) method as an objective function to determine the optimal processing parameters which will optimise warpage of the moulded part.

2 Response surface methodology

Response surface methodology (RSM) is a classical optimisation approach. It was used to demonstrate the relationship between variable parameters which influence the response condition in two or three-dimensional hyperbolic surface[10]. The mathematical model function will be obtained by using the second-order polynomial regression model in this study which will be used as the objective function in GSO. The necessary information to construct the response model are generally accumulated by the simulation works [11, 12]. Figure 1 shows the RSM flowchart in this study.

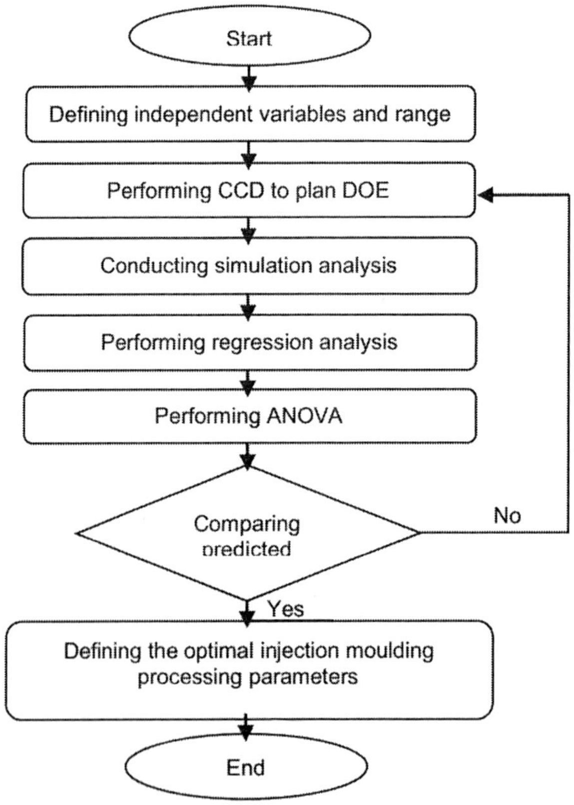

Fig. 1. Response Surface Methodology (RSM) flowchart

2.1 Design of Experiment Setup

In this study, coolant inlet temperature, melting temperature, packing pressure and cooling time have been selected as variable parameters and the range of each parameter as shown in Table 1. At first, full factorial design (FFD) with four centre points was selected as an experimental design to evaluate the model and main effects contribute to the warpage condition by using Design Expert 7.0 software. In order to obtain the significant curvature which is important in the RSM regression analysis, the augment of rotatable central composite design (CCD) was performed. Therefore, 30 runs of specified condition have been generated and each run will be set in AMI 2013 simulation software to evaluate the warpage condition of the moulded part.

Table 1. Variable parameters and levels

Factors	Level	
	Minimum	Maximum
Coolant inlet temperature, A (°C)	25	65
Melt temperature, B (°C)	220	260
Packing pressure, C (MPa)	46.74	56.74
Cooling time, D (s)	20	35

2.2 Finite element analysis setup

The CAE software, Autodesk Moldflow Insight (AMI) 2013 have been used to simulate the injection moulding process and to evaluate the warpage condition of front panel housing moulded part which has 2.5mm of average thickness and made from ABS material with trade name as Toray/Toyolac 700-314 as shown in Fig. 2. The moulded part, gating system and cooling channels 3D data was created and imported into the software for meshing process. In order to obtain more precise results, the mould insert material made of P20 steel and Nissei NEX1000, 80 tonne injection moulding machine specification have been set in the AMI software. Then, the Cool (FEM)+Fill+Pack+Warp analysis has been performed to evaluate warpage condition for each run which have been generated by DOE.

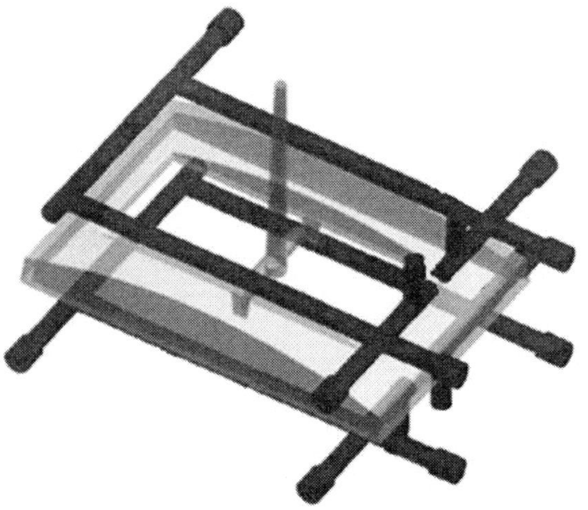

Fig. 2. Front panel housing with straight drilled cooling channels

DOI: 10.1051/matecconf/20179701105

2.3 RSM Regression Analysis

The warpage results for each run which obtained from the simulation analysis will be used in the RSM regression analysis. By using Design Expert 7.0 software the regression analysis was performed with the backward quadratic model in order to obtain the mathematical model function which representing the relationship between variable parameters and response. The software will calculate using second-order polynomial regression model in a statistical manner and the results will be verified with ANOVA.

2.4 Analysis of Variance (ANOVA)

The result of the quadratic model obtained from thc RSM regression analysis will be verified by analysis of variance (ANOVA) to determine either the mathematical model was statistically significant or otherwise. In the same manner, the significant factors that contribute to the warpage condition will be defined.

3 Glowworm swarm optimisation

The operators in GSO are recognised as glowworms [13] that carry a luminescent amount which has been called as luciferin [14]. Every glowworm utilises an artificial proportional luciferin to transmit the fitness of its current location and evaluate based on the objective model function to its neighbours. The glowworms rely on the variable of their neighbourhood which is based on the radial sensor range boundary in order to identify their neighbours and evaluate their movements. Each glowworm using the probabilistic mechanism by selecting a neighbour that contain the higher luciferin amount than its own and moves toward it. In other words, it will be attracted to the neighbours which are glowing brighter. Finally, the movement of the glowworms will be based on the local information and particular neighbour interactions which empower the majority swarm to form multiple optima of a given multimodal function.

In this study, GSO method will be carried out to obtain the optimised processing parameters of injection moulding process based RSM mathematical model function. MATLAB R2014a software will be used to conduct GSO analysis. Figure 3 shows the flowchart of GSO for this study.

3.1 Defining mathematical model function

The objective function used in GSO is the mathematical model function which has been obtained from the RSM regression analysis. The formulated objective function has been taken in this study is a warpage function.

3.2 Initialisation of glowworms' parameters

The initial glowworms' parameters were set based on the selected injection moulding variable parameters limit in this study to create the solution space. Then, the GSO agents will be deployed randomly in the solution space by setting the initial glowworms population size and maximum iteration based on the research requirement. In this study 30 numbers of glowworms and 40 iterations have been set.

DOI: 10.1051/matecconf/20179701105

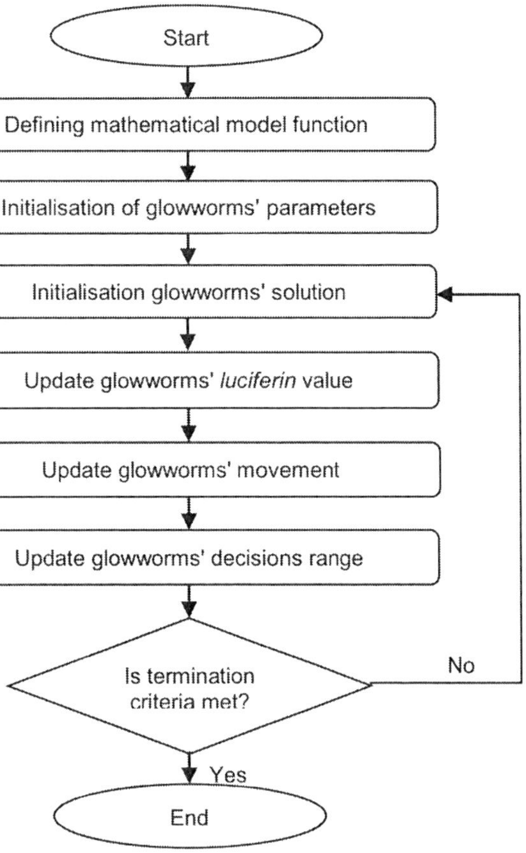

Fig. 3. Flowchart of Glowworm Swarm Optimisation (GSO)

3.3 Initialisation glowworms' solution

The initial solution of initial *luciferin* value (l_o), *luciferin* decay constant (ρ), *luciferin* enhancement constant (γ), beta (β), step size (s), neighbourhood range $r_d^i(t)$ and parameter used to control the number of neighbours (n_t) was set based on the research requirement where each glowworm contained the same *luciferin* value and sensor range in this phase. In this study, GSO control parameters optimal setting are: $\beta = 0.08, \rho = 0.4, \gamma = 0.6$ and $r_d^i(t) = 3$.

3.4 Update glowworms' luciferin value

At the beginning, each glowworm contains the same *luciferin* value. Influenced by the objective function value of their current location, the *luciferin* value will be changed. The rule of *luciferin* update was given by standard Equation (1) where $li(t)$ is the *luciferin* level for glowworm i at time t, ρ is the *luciferin* decay constant $(0 < \rho < 1)$, and $Ji(t)$ indicates the objective function at agent i's location at time t.

$$li(t+1) = (1-\rho)li(t) + \gamma Ji(t+1) \quad (1)$$

3.5 Update glowworms' movement

In the movement-update phase, each glowworm will move toward neighbour that contained the higher *luciferin* value than its own using a probabilistic mechanism. The probability of agent i move to agent j was given by a standard Equation (2) where $l_i(t)$ is the *luciferin* value for glowworm i, $d(i,j)$ is the Euclidean distance between agent i and j.

$$P_{ij}(t) = \frac{l_j(t)-l_i(t)}{\sum_{k \in N_i(t)} l_k(t)-l_i(t)} \quad (2)$$

Then, the glowworm i movement can be expressed by Equation (3) where s is the step size.

$$x_i(t+1) = x_i(t) + s\left[\frac{x_j(t)-x_i(t)}{\|x_j(t)-x_i(t)\|}\right] \quad (3)$$

3.6 Update glowworms' local decisions range

Local decision range update phase will be used to determine multiple peaks in a multimodal functional landscape in order to obtain the optimal variable parameters which optimise warpage value effectively. When the number of neighbours changed, the local decision domain needs to update in each of iteration. The rule is given by Equation (4) where $r_d^i(t+1)$ is the local decision domain of glowworm i in the $t+1$ iteration, β is the constant parameter that affected the rate of change of the neighbour domain and n_t is the threshold used to control the number of neighbours.

$$r_d^i(t+1) = \min\{r_s, \max\{0, r_d^i(t) + \beta(n_t - |N_i(t)|)\}\} \quad (4)$$

4 Results and discussions

4.1 Injection moulding process simulation

Warpage value obtained from simulation analysis was shown in Table 2. The results tabulate the warpage value for each run with the specified variable parameters condition which obtained from the DOE. The specified variable parameters condition was set and simulated in the AMI 2013 software.

4.2 Response surface methodology (RSM) regression analysis

From regression analysis, the determination coefficient, R^2 that fitted the model is 0.9710. The adjusted determination is 0.9618, which indicated that the regression model is significant with an adequate precision is more than 4, which is 37.564. The model is more significant if R^2 value is closer to 1 [15]. The standard deviation for this model is 8.807×10-3 and this value indicated that it is relatively lower than 0.05 which is better for precision and reliability of the experiment [16]. The polynomial regression model which relates to the warpage with all input parameters (coolant inlet temperature (A), melt temperature (B), packing pressure (C) and cooling time (D) is established by Design Expert 7 software and represented in Equation 5. This mathematical model function will be used in GSO as an objective function.

$$\begin{aligned} Warpage = {} & -0.251 + (3.43 \times 10^{-3}A) + (2.212 \times 10^{-3}B) + (7.917 \times 10^{-4}C) \\ & -(2.806 \times 10^{-3}D) - (1.328 \times 10^{-5}AB) - (1.832 \times 10^{-5}D^2) \end{aligned} \quad (5)$$

Table 2. Injection moulding process simulation results.

Standard Order	Data Source	Variable parameters for injection moulding simulation				Response
		Coolant temperature (°C)	Melt temperature (°C)	Packing Pressure (MPa)	Cooling time (s)	Warpage (mm)
1	DOE	25	220	46.74	20	0.220
2		65	220	46.74	20	0.155
3		25	260	46.74	20	0.305
4		65	260	46.74	20	0.225
5		25	220	56.74	20	0.215
6		65	220	56.74	20	0.170
7		25	260	56.74	20	0.310
8		65	260	56.74	20	0.235
9		25	220	46.74	35	0.175
10		65	220	46.74	35	0.130
11		25	260	46.74	35	0.255
12		65	260	46.74	35	0.185
13		25	220	56.74	35	0.175
14		65	220	56.74	35	0.135
15		25	260	56.74	35	0.245
16		65	260	56.74	35	0.190
17	Centre	45	240	51.74	27.5	0.215
18		45	240	51.74	27.5	0.215
19		45	240	51.74	27.5	0.215
20		45	240	51.74	27.5	0.215
21	Axial	5	240	51.74	27.5	0.240
22		85	240	51.74	27.5	0.140
23		45	200	51.74	27.5	0.170
24		45	280	51.74	27.5	0.270
25		45	240	41.74	27.5	0.205
26		45	240	61.74	27.5	0.240
27		45	240	51.74	12.5	0.260
28		45	240	51.74	42.5	0.180
29		45	240	51.74	27.5	0.215
30		45	240	51.74	27.5	0.215

4.3 Analysis of variance (ANOVA)

From the ANOVA table as shown in Table 3, the F calculated value is bigger than F tabulated. This condition shows that the mathematical model obtained from regression analysis was significant. The ANOVA also indicated that three from four selected variable parameters give significant effect on the warpage condition.

Table 3. ANOVA of response surface model.

	Sum of Squares	df	Mean Square	F (calculated)	R^2	RA^2	F (tabulated)
SSR	0.0571	7	0.0082	105.2280	0.9710	0.9618	2.46
SSE	0.0017	22	0.0001				
Total	0.0588	29					

The results show that melt temperature was the most significant factor contributed to the warpage condition. This result was inline with previous researchers which found out the same significant factor which influencing the warpage condition [7, 17-21]. Then it follows by cooling temperature and cooling time. Packing pressure was the least significant factor contributed to the warpage condition as shown in Figure 4. This result shows the uneven

DOI: 10.1051/matecconf/20179701105

volumetric shrinkage occurred to the poor uniformity of thermal distribution in the mould [22]. According to Subramanian et al. [23] and Huang et al. [24] the temperature difference between upper and lower surfaces in the cavity and core can give effect to the shrinkage condition. This behaviour will cause either warpage or residual stress, depending on the mechanical stiffness of the moulded part design.

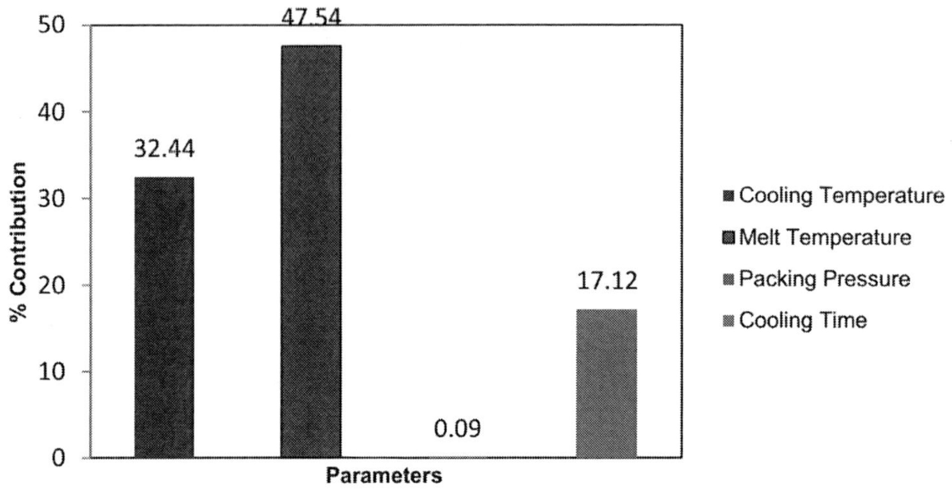

Fig. 4. Contribution for each parameters in percentage.

Equation 5 was applied to calculate the prediction warpage values of the polynomial model and the results is summarised in Figure 5. From the comparison, it shows that the simulation and predicted value are very close to each other which indicated the mathematical response function have given a good prediction value in estimating warpage.

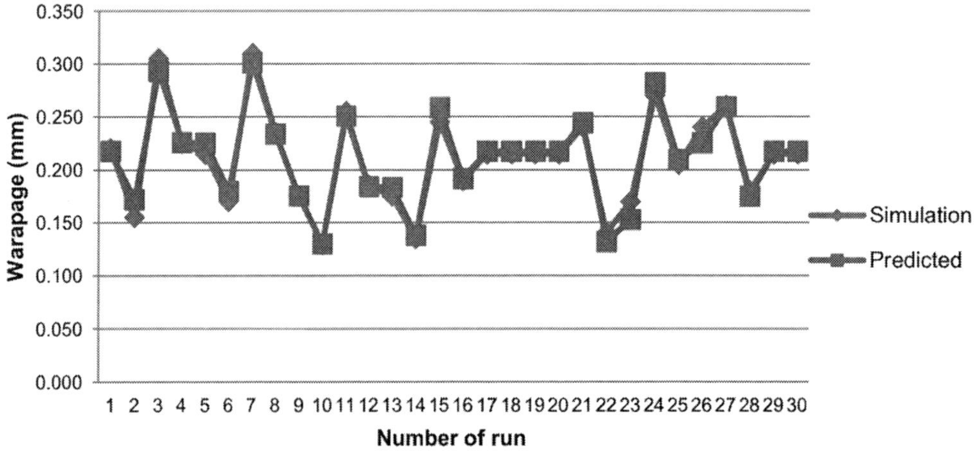

Fig. 5. Simulation and predicted warpage comparison

DOI: 10.1051/matecconf/20179701105

4.4 Glowworm swarm optimisation (GSO) analysis

The GSO algorithm has been conducted on injection moulding processing parameters. The algorithm was tested on four selected variable parameters. By using the mathematical model function obtained from RSM, the optimal results were shown in Table 4 and the warpage variation for each glowworm was shown in Figure 6.

Table 4. Recommended simulation results versus GSO Optimised results

Factors	Recommended simulation results	GSO optimised results
Coolant inlet temperature, A (°C)	25	64.27
Melt temperature, B (°C)	240	232.51
Packing pressure, C (MPa)	46.74	52.01
Cooling time, D (s)	30	31.77
Warpage, (mm)	0.2650	0.1614

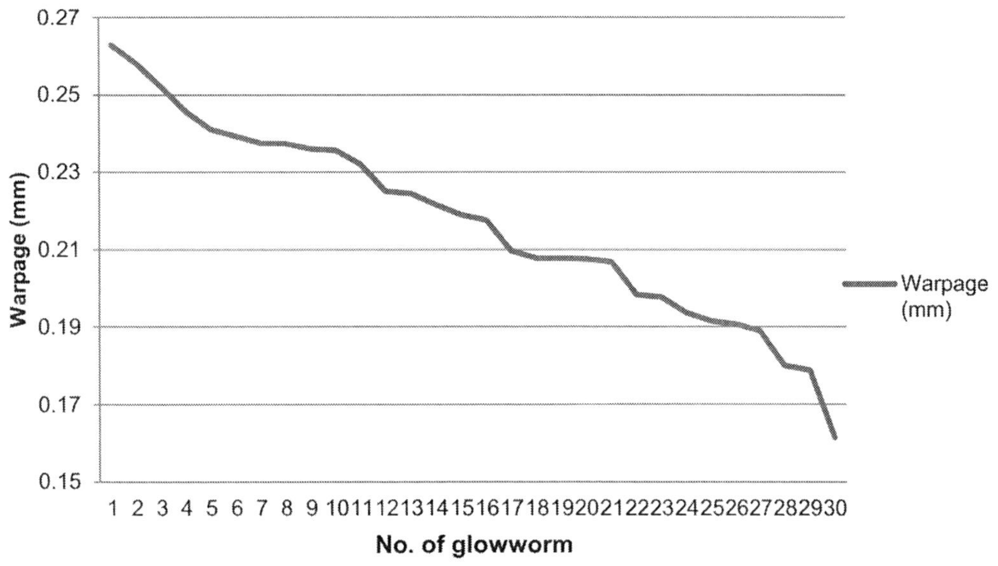

Fig. 6. Warpage variation for each glowworm

5 Conclusion

This study is definitely helpful in enhancing the quality of moulded parts produced where the objective is to optimise warpage of the front panel housing moulded part have been achieved. Based on the results, the warpage has been optimised by using an alternative approach of response surface methodology (RSM) and Glowworm Swarm Optimisation (GSO). The results also show that:

- By using RSM, the significant mathematical model function can be obtained in order to predict warpage value with reasonable accuracy.

DOI: 10.1051/matecconf/20179701105

- From the ANOVA results, melt temperature is the most significant factor influencing the warpage condition on the moulded part, follow by coolant temperature and cooling time.
- The optimal processing parameter obtained from GSO has optimised warpage by 39.1% which is from 0.2650mm from the simulation result to 0.1614mm.

In this study, it proved that the proposed optimisation approach has enormous potential in order to obtain better quality of the moulded part.

References

1. W. Michaeli, G. Potsch, *Injection molding: an introduction*, (Munchen, Hanser, Fachbuchverlag, 1995)
2. L. Sors, I. Balazs, *Design of plastic moulds and dies*, (1989)
3. S.S. Teklehaimanot, *Simulation and design of a plastic injection Mold*, (2012
4. D. Annicchiarico, J.R Alcock, Mater. Manuf. Process, **29(6)**, 662 (2014)
5. Y.C. Lam, L.Y. Zhai, K. Tai, S.C. Fok, Int. J. Prod. Res., **42(10)**, 2047 (2004)
6. D. Annicchiarico, U.M. Attia, J.R. Alcock, Polym. Test, **32(4)**, 769 (2013)
7. X. H. Yin, C. Yang, X.P. Li, Polym. Plast. Technol., **54(17)**, 1772 (2015)
8. W.C. Chen, M.H. Nguyen, W.H. Chiu, T.N. Chen, P.H. Tai, Int. J. Adv. Manuf. Technol., **83(9-12)**, 1873 (2016)
9. W.C. Chen, D. Kurniawan, Int. J. Precis. Eng. Manuf., **15(8)**, 1583 (2014)
10. D. Montgomery, *Design and Analysis of Experiments*, (John Wiley and Sons, New York, NY, 2005)
11. H. Oktem, T. Erzurumlu, H. Kurtaran, J. Mater. Process. Technol., **170(1)**, 11 (2005)
12. B. Ozcelik, T. Erzurumlu, Int. Commun. Heat Mass, **32(8)**, 1085 (2005)
13. E. Bonabeau, M. Dorigo, G. Theraulaz, *Swarm intelligence: from natural to artificial systems*, (Oxford university press, 1999)
14. R. Brits, A.P. Engelbrecht, F. Van den Bergh, *A niching particle swarm optimizer*, (Singapore, Orchid Country Club, 2002)
15. Y.H. Wang, J.T. Feng, Q. Zhang, X. Zhang, J. Appl. Microbiol., **104(3)**, 735 (2008)
16. M.S. Tanyildizi, D. Ozer, M. Elibol, Process Biochem., **40(7)**, 2291 (2005)
17. M.A. Barghash, F.A. Alkaabneh, Qual. Eng., **26(3)**, 319 (2014)
18. W. Guo, L. Hua, H. Mao, Z. Meng, J. Mech. Sci. Technol., **26(4)**, 1133 (2012)
19. S. Kitayama, R. Onuki, K. Yamazaki, Int. J. Adv. Manuf. Tech., **72(5-8)**, 827 (2014)
20. H. Oktem, Int. J. Adv. Manuf. Tech., **61(5-8)**, 519 (2012)
21. R. Wang, J. Zeng, X. Feng, Y. Xia, J. Macromol. Sci. B, **52(1)**, 206 (2013)
22. D. Lee, W.A. Chen, T.W. Huang, S.J. Liu, Int. Polym. Proc., **28(2)**, 221 (2013)
23. N.R. Subramanian, L. Tingyu, T.A. Seng, Mechatronics, **15(1)**, 111 (2005)
24. M.C. Huang, C.C. Tai, J. Mater. Process. Technol., **110(1)**, (2001)

Organic and Inorganic Fertilizers Application on NPK Uptake and Production of Sweet Corn in Inceptisol Soil of Lowland Swamp Area

Neni Marlina[1,*], *Nurbaiti* Amir[2], *R. Iin Siti* Aminah[2], *Gamal Abdul* Nasser[1], *Yani* Purwanti[1], *Laili* Nisfuriah[1], and *Asmawati*[1]

[1]Faculty of Agriculture UNPAL Palembang, Jl. Darmapala No 1A Bukit Lama Palembang Indonesia
[2]Faculty of Agricultre UM Palembang, Jl. A. Yani 13 Ulu Palembang Indonesia

Abstract. This study objective was to determine the dose of organic and inorganic fertilizers which can increase N, P and K nutrients uptake as well as the growth and yield of sweet corn on inceptisol soil of lowland swamp. Inceptisol soil has low soil fertility and relatively low to moderate levels of organic matter content. Application of organic fertilizer on inceptisol soil of lowland swampis expected capable to increase N, P and K nutrients as well as yield of sweet corn. This research was conducted from April to July 2014 at Experimental Farm Area of Pulau Semambu Village, Indralaya Utara Subdistrict, Ogan Ilir District, South Sumatra Province. The method used in this research was randomized block design consisting treatments as follows: 75% inorganic fertilizer + 5 ton.ha^{-1}organic fertilizer, 50% inorganic fertilizer + 5 ton.ha^{-1}organic fertilizer, 25% inorganic fertilizer + 5 ton.ha^{-1} organic fertilizer, 0% inorganic fertilizer + 5 ton.ha^{-1}organic fertilizer with six replications. The recommended dose of inorganic fertilizerswas 200 kg.ha^{-1} urea, 100 kg.ha^{-1} SP-36 and 100 kg. ha^{-1} KCl. The results showed that treatment of 75% of inorganic fertilizer + 5 ton.ha^{-1}organic fertilizer had produced N, P and K nutrients uptake with magnitude of 1.850, 0.418 and 2.374 g.plant^{-1} respectively as well as good growth and yield of sweet corn with magnitude of 356.36 g. plant^{-1}or 15.21 ton ha^{-1}.

1 Introduction

Sweet corn(*Zea mays saccharata*) seasonal crop that has high economic value because its seed has higher sugar content than that of other corn varieties. Sugar content of sweet corn is 14 to 18 % which is close to sugar content of sugarcane crop with magnitude of 19%, whereas sugar content of local sweet corn is only 9 to 11 %. Sweet corn has energy of 96 cal, Protein of 3.5 g, Lipid of 1.0 g, Carbohydrate of 22.8 g, Calcium of 3.0 mg, Phosphorusof 111 mg, Iron of 0.7 mg, Vitamin A of 400 SI, Vitamin B of 0.15 mg,

* Coresponding author: marlina002@yahoo.com

Vitamin C of 12.0 mg andwater of 72.7 g [12].Sweet corn crop can also cultivated on suboptimal land such as lowland swamp.

Lowland swamp has very high potential to be developed in order to increase crop production such as sweet corn crop. Lowland swamp area in South Sumatra is relatively extensivewith magnitude of 368,685 hectares and most area was cultivated for rice crop and small area was cultivated for other crops such assweet corn crop [23].

The main problem in lowland swamp area is its physical constraint in form of water logging and chemical constraint in form of high soil acidity, Al and Fe cations which bind phosphorus and nutrients defficiency [13]. In addition, according to [3], soil fertility level in lowland swamp area is very low to medium so that fertilization effort is important to increase the productivity of lowland swamp.

Fertilizer is agricultural input that must be used in order to achieve maximum yield of sweet corn cultivated on lowland swamp area. However, the real problem experienced by farmers to increase productivity of lowland swamp through fertilization was frequently impeded by high price of inorganic fertilizer and its unavailability in site or location. Therefore, an alternative technology should be developed in order to increase crops yield and to decrease the dependenceon inorganic fertilizer (chemical fertilizer) at suboptimal land. This alternative technology should be economically feasible and technologically easy to be used by farmers, for instance the use of organic fertilizer (manure of chicken dunk). Ogan Ilir District is one of organic fertilizer producer (manure of chicken dunk) which can be utilized for sweet corn cultivation and capable to decrease the transportation cost.

Several study results in term of organic fertilizer on dry land had been published. Application of organic fertilizer in form of chicken dunk manure always produce the best crop response in the first season. This is due to the fact that organic fertilizer is relatively faster to be decomposed and has sufficient level of nutrients compared to other manures at the same weight unit [22]. According to Balittanah [25], nutrients content of chicken dunk manure is as follows: 1.70 % N, C/N ratio of 10.80, P of 2.12 % and K of 1.45 %. In addition, according to Adimihardja [2], application of chicken dunk manure with dose of 5 ton.ha^{-1}on dry land (Ultisol in Jambi) had significant effect on increasing C-organic content and yield of corn and soybean.

The use of chicken dunk manure on soil can decrease environmental pollution, decrease heavy metals content which are toxic to crops [27] and can increase stem of ear cornweight of sweet corn with magnitude of 247.10 g.plant^{-1} (11.57 ton.ha^{-1}) [12].It is expected that organic fertilizer combined with inorganic fertilizer if applied to lowland swamp can increase N, P and K nutrients uptake as well as the growth and yield of sweet corn crop. This is in accordance to Permentan No 40/2007 which recommend the development of organic matter or application of organic fertilizer combined with inorganic fertilizer in order to improve soil condition and fertility all at onceto improve the usage efficiency of inorganic fertilizer [26].

2 Methodology

This research was conducted from April to July 2014 at Experimental Farm Area of Pulau Semambu Village, Indralaya Utara Subdistrict, Ogan Ilir District, South Sumatra Province. The method used in this research was randomized block design consisting treatments as follows: 75% inorganic fertilizer + 5 ton.ha^{-1} organic fertilizer, 50% inorganic fertilizer + 5 ton.ha^{-1} organic fertilizer, 25% inorganic fertilizer + 5 ton.ha^{-1} organic fertilizer, 0% inorganic fertilizer + 5 ton.ha^{-1} organic fertilizer with six replications. The recommended dose of inorganic fertilizerswas 200 kg.ha^{-1} urea, 100 kg.ha^{-1} SP-36 and 100 kg. ha^{-1} KCl. Organic fertilizer dose (chicken dunk manure) was 5 ton.ha^{-1}.The further test was

 DOI: 10.1051/matecconf/20179701106

conducted by using Least Significant Different (LSD) at 5% level, whereas regression and correlation analysis was used to determine the relationship between N, P and K nutrients uptake and stem of ear cornweight of sweet corn. Statistical analysis was done by using SAS 9.1.3 Portable program.

2.1 Land preparation

Land was cleared from the existing vegetation two weeks before planting and two tillages operation were conducted until soil become loosefollowed by development of 24 plots.

2.2 Planting

Planting was done by planting two seeds per planting hole with dibblemethod. The planting distance was 75 cm x 25 cm.

2.3 Fertilizing

Addition of manure was conducted at the same time with soil tillage period. Urea fertilizer was applied two times, i.e. 1/3 part during planting and 2/3 part at period of 4 weeks after planting, whereas SP-36 and KCl fertilizers were given during the planting period.

2.4 Crop maintenance

Crop maintenance was consisted of watering (irrigation), replacement, thinning out, heapingand weeds clearing. The watering operation was done every day in morning and afternoon period. Replacementwas done one week after planting. Thinning outwas done one week after planting. Heaping and weeds clearingwas done two weeks after planting.

2.5 Harvest

Harvest was done after crop was at period of about 60-70 days after planting which was indicated by come out ofcorn hairs having brown color and corn seeds were still soft and fully filled.

The observed parameters were soil analysis before planting, crop height (cm), leave numbers (sheets), stem of ear corndiameter (cm), stem of ear cornlength (cm), uptake of N, P and K nutrients, stem of ear cornweight per plant (g) and stem of ear cornweight per hectare (tons).

3 Results and discussion

3.1 Results

3.1.1 Soil characteristics before treatment

The observed characteristics of soil chemical properties were consisted of soil pH, C-organic content, N total, P-available, K-dd, Na-dd, Ca-dd, Mg-dd, Cation Exchange

Capacity, H-dd, Al-dd andC/N ratio.Results of soil chemical analysis can be seen in Table 1.

Based on research criteria according to [17] and [25], soil used in this research was classified as acid soil (pH H_2O=4.20) with high cation exchange capacity (29.23cmol$_{(+)}$ kg^{-1}), C-organic content classified as medium (2.67 %), C/N ratio classified as medium (8.34), N-total content classified as low (0.32 %), P-available classified as high (216.62 mg kg^{-1}), exchangable bases such as Ca-ddclassified as low (4.03cmol$_{(+)}$ kg^{-1}), Mg-dd classified as high (0.26 cmol$_{(+)}$ kg^{-1}), K-dd classified as low (0.27 cmol$_{(+)}$ kg^{-1}), Na-ddclassified as low (0.61cmol$_{(+)}$kg^{-1}), base saturation classified as very low (17.63) and Al-dd of 1.27 cmol $_{(+)}$ kg^{-1}.

Table 1. Soil chemical analysis resultsprior to planting.

No.	Analysis type	Analysis results	Evaluation*
1.	pH H_2O (1:1)	4.20	Very acid
2.	C/N ratio	8.34	Low
3.	C-organic (%)	2.67	Medium
4.	N-total (%)	0.32	Low
5.	P Bray I (µg g^{-1})	216.62	High
6.	Ca-dd (cmol$_{(+)}$ kg^{-1})	4.03	Low
7.	Mg-dd (cmol$_{(+)}$ kg^{-1})	0.26	Very low
8.	K-dd (cmol$_{(+)}$ kg^{-1})	0.27	Low
9.	Na-dd (cmol$_{(+)}$ kg^{-1})	0.61	Low
10.	Base saturation (%)	17.63	Very low
11.	Al-dd (cmol$_{(+)}$ kg^{-1})	1.27	
12.	Al saturation	1.14	Very low

Remarks: Analysis results from Nubika Laboratory, Bogor (2014)
*Soil Research Center (1983) and Soil Research Council (2005)

3.1.2 Vegetative Growth of Sweet Corn Crop

Results of variance analysis showed that organic fertilizer and inorganic fertilizer treatments had significant effect on leave numbers, stem of ear corn diameter, N, P and K nutrients uptake, but had no significant effect on other parameters. The summary of variance analysis results wer shown in Table2.

Table 3 showed that the highest value of average crop height and the highest leave numbers was found on treatment of 75 % inorganic fertilizer + 5 ton ha^{-1}organic fertilizer with magnitude of 217.60 cm and 17.24sheets, respectively; whereas the lowest value of average crop height and the lowest leave numbers was found on treatment of 0 % inorganic fertilizer + 5 ton ha^{-1} organic fertilizer with magnitude of 209.08 cm and 17.08 sheets, respectively. Treatment of 75 % inorganic fertilizer + 5 ton ha^{-1} organic fertilizer was significantly different than treatments of 50 %, 25 % and 0 % inorganic fertilizer + 5 ton.ha^{-1}organic fertilizer in term of N, P and K nutrients uptake. There was very close relationship between doses of organic ferilizer + inorganic fertilizer and N, P and K nutrients uptake (see Figure 1, 2 and 3).

Table 2. Calculated value of F for the observed parameters.

Source of Variation	Crop height (cm)	Leave numbers (sheets)	N uptake ($g.plant^{-1}$)	P uptake($g.plant^{-1}$)	K uptake ($g.plant^{-1}$)
Inorganic fertilzer + organic fertilizer	Not significant	Not significant	Significant	Significant	Significant

Table 3. The effect of inorganic and organic fertilizers on crop height, leave numbers and N, P and K nutrients uptake.

Inorganic fertilzer + organic fertilizer(%)	Crop height (cm)	Leave numbers (sheets)	N uptake ($g.plant^{-1}$)	P uptake ($g.plant^{-1}$)	K uptake ($g.plant^{-1}$)
75+po 5 $ton.ha^{-1}$	217.60	17.24	1.850 c	0.418 d	2.374 d
50+po5 $ton.ha^{-1}$	216.36	17.08	1.790 b	0.368 c	1.922 c
25+po5 $ton.ha^{-1}$	214.88	17.08	1.638 a	0.348 b	1.594 b
0+po 5 $ton.ha^{-1}$	209.08	17.08	1.628 a	0.338 a	1.208 a
HSD 0.05=	Ns	ns	0.030	0.006	0.001

Remarks: Numbers followed by the same letters in the same column are represent not significantly different.

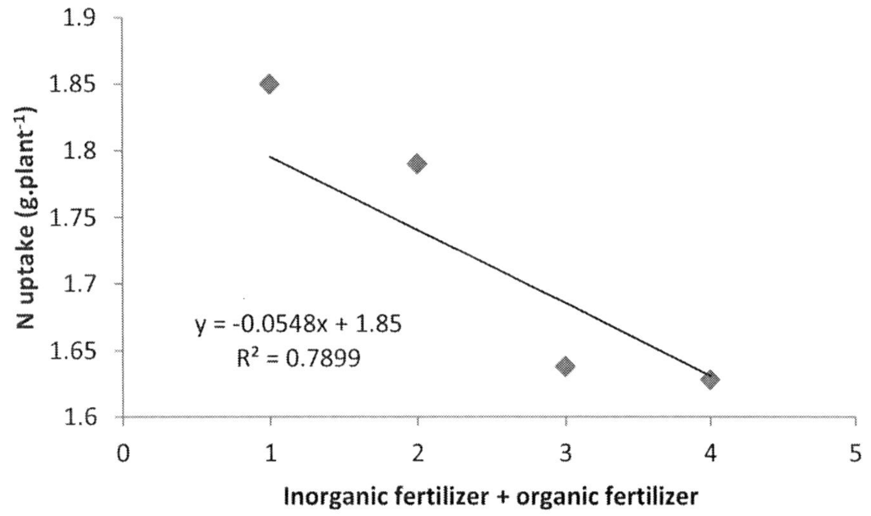

Fig. 1. Relationship between inorganic and organic fertilizers and N nutrient uptake

DOI: 10.1051/matecconf/20179701106

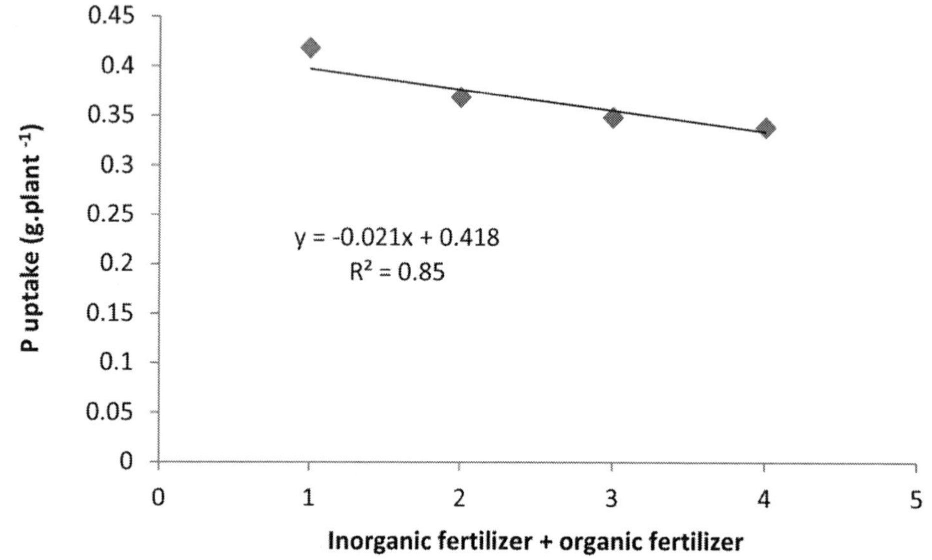

Fig. 2. Relationship between inorganic and organic fertilizers and P nutrient uptake

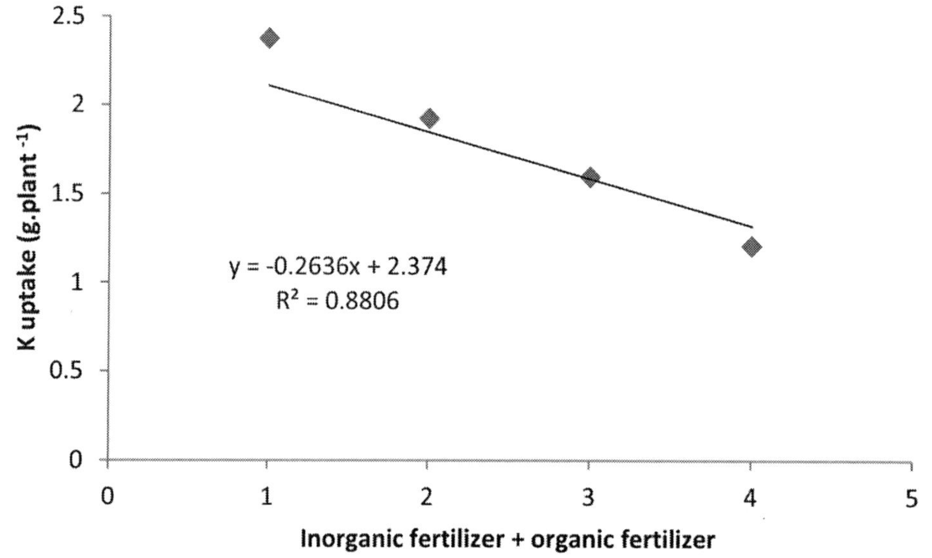

Fig. 3. Relationship between inorganic and organic fertilizers and K nutrient uptake

Figure 1, 2 and 3 showed that there was very close relationship amongst inorganic fertilizer + organic fertilizer and N, P and K nutrients uptake which was shown by R^2values of 0.789, 0.850 and 0.880 as well as r values of 0.890, 0.920and 0.940. It means that every addition of organic fertilizer to lower dose of inorganic fertilizer would results in decrease of N, P and K nutrients uptake with magnitude of 0.054 x, 0.021 x and 0.263x, respectively.

3.1.3 Production of sweet corn crop

Results of variance analysis showed that application of organic and inorganic fertilizers had no significant effect on stem of ear corn length, stem of ear corn weight per plant and stem of ear corn weight per hectare. The summary of variance analysis results for the observed parameters can be seen in Table 4.

Table 5 showed that the highest average values of stem of ear corn length, stem of ear corn weight per plant and stem of ear corn weight per hectare was found on treatment of 75 % inorganic fertilizer + 5 $ton.ha^{-1}$ organic fertilizer with magnitude of 20.76 cm, 356.36 $g.plant^{-1}$ and 15.21 $ton.ha^{-1}$, whereas the lowest average values of stem of ear corn length, stem of ear corn weight per plant and stem of ear corn weight per hectare was found on treatment of 0 % inorganic fertilizer + 5 $ton.ha^{-1}$ organic fertilizer with magnitude of 17.08 cm, 333.68 $g.plant^{-1}$ and 14,24 $ton.ha^{-1}$, respectively.

Table 4. Values of calculated F for the observed parameters.

Source of Variation	Stem of ear corn length (cm)	Stem of ear corn weight per plant (g)	Stem of ear corn weight per hectare (ton)
Inorganic fertilizer + organic fertilizer	Not significant	Not significant	Significant

Table 5. The effect of inorganic fertilizer and organic fertilizer on stem of ear corn length, stem of ear corn weight per plant and stem of ear corn weight per hectare

Inorganic fertilizer + organic fertilizer (%)	Stem of ear corn length (cm)	Stem of ear corn weight per plant (g)	Stem of ear corn weight per hectare (ton)
75+po	20.76	356.36	15.21
50+po	20.58	354.96	15.14
25+po	20.56	348.52	14.78
0+po	17.08	333.68	14.24
BNJ 0,05=	ns	ns	ns

Remarks: Numbers followed by the same letters in the same column are represent not significantly different.

3.2 Discussion

3.2.1 Soil characteristics before treatment

Soil used in this study had low pH and very acid as well as low N-total content. Although availability of P is high, but available P was also low because most of it was bindedby Al which produce Al-P so that P nutrient can not be absorbed by sweet corn crop.

Low value of pH H_2O in this soil was due to the fact that this soil had been used for years for agricultural enterprise and used chemical fertilizer, especilly urea. One molecule of urea can contribute4 mol ions of H^+into soil which subsequently produce soil acidity. Soil in this research location also had higher quantity of metal ions such as Fe, Cu, Mn and Zn when hydrolized would contributeH^+ions into soil.

Soil in this lowland swamp has low fertility level which is showed by very acid soil pH and this is oen ciri of inceptiol soil. According to [1], inceptiol soilhas low fertility level and acid soil pH. This condition is exageratedby limited use of organic fertilizer, especially on seasonalfood crops. Inceptisol soil is characterized by loamy textural class, soil reaction in the range of relative acid to relative alkaline, nutrient content and nutrient supply were relatively medium and cation exchange capacity in the range of medium to high.

Soil condition used in this research was not provide good support for the growth of sweet corn crop so that this soil should be added with inorganic fertilizer and organic fertilizer (chicken dunk manure). This organic fertilizer has function as source of organic acids which capable to control metals precipitation in soil or as nutrients source for crops. Decomposition result might be in form of organic acids which can chelate metal ions so that P which bounded by metal ions (Al-P and Fe-P) can be released and become available again. This available P can be utilized by sweet corn crop for good growth and production.

Organic acids available in organic fertilizer can chelate toxic elements (metals) in soil so that soil condition is not harmful for crop growth [22] and [7]. Organic acids are capable to decrease phosphate quantity which fixedby Fe and Al through chelating mechanisms so that P is become available for crops (Barker dan Pilbeam, 2007). Pengikatan mechanisms of Fe and Al was done by functional groupof organic component due to availability of carboxyl groupwhich close to and react with metal ions.

3.2.2 Vegetative growth of sweet corn crop

Results of the study showed that organic fertilizer application was very helpful in providing nutrients which is highly required by sweet corn crop either in addition with inorganic fertilizer or without addition of inorganic fertilizer.

There was an interesting phenomenon in which application of 5 ton.ha^{-1} organic fertilizer followed by application of 0% to 75% recommended dose of inorganic fertilizers had produced good growth of crop height, leave numbers and good nutrients uptake of N, P and K for sweet corn crop. This was due to the fact that soil analysis results before planting showed low availability of N, P and K nutrients as well as low base saturation. Therefore, sweet corn crop had good positive response to the addition of organic and inorganic fertilizers. This positive response was estimated due to excessive organic fertilizer that capable to increase soil absorption to water and to help nutrients absorbption from inorganic fertilizer. This organic fertilizer was quickly decomposed so that nitrate production rate was readily available for crops. Nitrate absorbed by sweet corn crop has a role in increasing the crop growth and formation of leave numbers.

Organic fertilizer has a role in improving soil fertility. Nutrients content in organic fertilizer is not high, but it has specific characteristics, i.e. improving soil physical properties such as soil permeability, soil porosity, soil structure, water holding capacity and soil cations [9], [6]. N nutrient from urea has a role for leave development, but this nutrient is easily leached so that organic matter is required to increase water holding capacity and soil cations.

Nutrients content within organic fertilizer, soil and nutrients addition from chemical fertilizer which are absorbed by crops will be translocated on vegetative parts of crop. These nutrients are highly required especially during meddle ingenerative growth of crop for flowering and seed development processes. According to [21], organic fertilizer (manure) is nitrogen source that will give the fastest and significant effects on crop growth compared to other elements.

Crop height affects leave numbers. The higher the crop height, the more was the leave numbers. Increase of leave numbers results in increase of photosynthesis process which is indicated by the increase of N, P and K nutrients uptake (Figure 1, 2 and 3). The increase of N, P and K nutrients uptake can be used by crop to increase its vegetative parts such as crop height and development of leave numbers.

3.2.3 Production of Sweet Corn Crop

Application of 5 ton.ha^{-1}organic fertilizer + 75 % inorganic fertilizer had produced higher magnitude of stem of ear corn length, stem of ear corn weight per plant and stem of ear corn weight per hectare than that of application of 5 ton.ha^{-1}organic fertilizer + 50 %, 25 % and 0 % inorganic fertilizers. This results were estimated due to the fact that organic fertilizer application has highly significant role in improving environmental condition of roots zone for sweet corn and in helping nutrients availability from unavailable to available condition for crop. If the growth environment of roots is better, then nutrients will be absorbed by crops either from organic fertilizer or from inorganic fertilizer to support photosynthesis process and cell formation or cell enlargement which has direct effect in increasing nutrients uptake of N, P and K, crop growth and yield of sweet corn crop. According toAgromedia (2007), crops require the proper nutrients in term of type and quantity for their growth and production processes.

In addition, production of sweet corn crop (tongkol weight per plant and tongkol weight per hectare) is highly affected by the growth of crop height, leave numbers and nutrients uptake of N, P and K by crop itself. Transportation of nitrogen, phosphorusand potassiumas well as photosyntate from leaves was very greatduring seed filling process. Nitrogen controls the use of phosphoruswhich stimulate flowering and fruit formation processes. It is estimatedthat organic fertilizer during this period had been continously decomposed and nitrogen as well as other nutrients became more available when crops meddle inflowering and seed filling processes. According to Soepardi (1983) and Hanafiah (2005), nitrogen has function to stimulate the growth and to increase protein content of cereal crops as well as to regulate the use of phosphorus, potassiumand other nutrients. According to Lingga and Marsono (2008), nitrogen has a role in stimulating the overall growth and leave chlorophyleformation which is very important in photosynthesis process.In addition, nitrogen has a role in formation of protein, lipid and other organic cmpounds. This is supported by study results of Nurdin *et al.* (2009) which showed that appication of NPK fertilizer could produced corn crop height of 175.25 cm.

P elementhas a role in flower, fruit and seed formations. Availability of P elementwithin soil is very low. Most of P elementis in unavailable formthat can not be absorbed by crop and fixation of P by Al is occurred within acid soil or fixation of P by Ca is occurred within alkaline soil (Hardjowigeno, 2003). Application of manure can neutralize acid soil and application of NPK fertilizer can provide P unsur requirement for crops.

The sweet taste on sweet corn is estimateddue to availability of potassiumnutrients. Potassiumis absorbed in form of K^+ ion. Salisbury and Ross (1992) had stated that K+ has a role in starch formation as sintetase starch enzyme activator. This is one of reason why K+ is important for crop and why sugar instead of starch which is heapedwithin crop that experience potassiumdefficiency. The study results from Maruapey and Faesal (2010) showed that application KCL fertilizer dose of 100 kg.ha^{-1} had produced the highest value of biomass weight with average value of 11.083 kg.ha^{-1} and high amylopectine content.

Tongkol length and diameter showed the form of tongkol itself which can affect tongkol weight. Higher values of tongkol length and tongkol diameter will produce higher magnitude of seed numbers in tongkol which in turn produce higher magnitude of tongkol weight (Table 5).

Results of this study in overall showed that sweet corn crop require good environment for roots growth ad balance nutrients if we want to apply balance fertilizing either by using organic fertilizer or inorganic fertilzer. Balance fertilizing is location specific nutrients management which depend on local environment, especially soil. According to

Dobermann *et al.* (2003) *in* [20], concept of location specific nutrients management should take into account soil capability to provide nutrients naturally and nutrients recovery that had been previously used by crops.The similar concept is also used for new fertilizing recommendation for corn crop with specific emphasize on understanding of potential yield and gapyield as a base for recommendation improvement of location specific nutrient management.Nutrients management of location specific is an effort to provide nutrients for crops properly in term of quantity, types and application time by considering crop requirement and land capacity in providing nutrients for crops.

4 Conclusion

It can be concluded that application of 5 ton.ha^{-1}organic fertilizer + 75 % inorganic fertilizer could increase N, P and K nutrients uptake (1.850, 0.418 and 2.374 g.plant^{-1}) as well as good growth and yield of sweet corn crop with magnitude of 356.36 g.plant^{-1} or 15.21 ton.ha^{-1}.

The writers want to deliver acknowledgement to Bambang Tulistiono which had provided signficant help in the field work.

References

1. A. Abdurachman, A. Dariah, A. Mulyani, J. Agrie. Res., **27(2)**, 43 (2008)
2. A. Adimihardja, I. Juarsah, U. Kurnia, *Proceeding of National Seminar of Soil Resources, Climate and Fertilizer,* (Book II. Lido Bogor, 1999)
3. T. Alihamsyah, Ar Riza, *Potential and Technology of Lowland Swamp Utilization for Agriculture. Main Paper of National Workshop for Lowland Swamp Development*, (Regional Government of Hulu Sungai District and Agricultural Council of South Kalimantan Province, Kandangan, 2004)
4. Agricultural Research and Development, *The Map of Inorganic Fertilizer Saving Potential and Organic Fertilizer Development on Paddy Field in Indonesia*, (Agriculture Ministry, Indonesia 2010)
5. A.V Barker, D.I. Pilbeam, *Handbook of Plant Nutririon*, (CRC Press, New York, 2007)
6. D. Budianta, D. Ristiani, *Soil Fertility Management to Support Conservation of Land Resource and Environment*, (UNSRI Press, Palembang 2013)
7. Dubey, India Soil Sci., **45**, 506 (1997)
8. K.A. Hanafiah, *Principles of Soil Science*, (RajaGrafindo, Jakarta 2005)
9. S. Hardjowigeno, *Soil Science*, (Akademika Pressindo, Jakarta 2003)
10. W. Hartatik, L.R. Widowati, Manure, *Organic Fertilizer and Biofertilizer. Research and Development Council of Agriculural Land Resource*, (Agricultural Research and Development Council, Bogor 2005)
11. D. Iskandar, *National Seminar of Teknologi Negeri*, *Pusat Teknologi Negeri, Vol. II*, (Jakarta, 2003)
12. P. Lingga, Marsono, *Fertilizer Use Guidance*, (Penebar Swadaya, Jakarta 2008)
13. A. Maruapey, Faesal, *Proceeding of National Cereal Day*, (2010)
14. A. Maruapey, *Cereal National Seminar*, 23 (2011)
15. M. Noor, *Lowland Swamp: Technology, Utilization and Its Development*, (Rajawali Pers., Jakarta, 2007)
16. P. Nurdin, Z. Maspeke, Ilahude, F. Zakaria, Journal of Tropical Soil., **14(1)**, 49 (2009)

 DOI: 10.1051/matecconf/20179701106

17. Soil Research Center, *Terms of Reference Type*, (P3TT, Bogor, 1983)
18. Redaksi Agromedia, *Fertilizing Guidance*, (Agromedia Pustaka, Jakarta, 2007)
19. F.B. Salisbury, W.C. Ross, *Crop Physiology: Part 2*, (ITB Press, Bandung, 1992)
20. M.P. Sirappa, *Increasing of Corn Productivity Through Application of N, P and K Fertilizers as well as Manure in Dry Land*, (Agricultural Technology Research Council, Maluku, Ambon, 2010)
21. G. Soepardi, *Properties and Characteristics of Soil*, (IPB, Bogor, 1983)
22. F.J. Stevenson, *Humus Chemistry, Genesis, Composition, Reaction*, (John Wiley and Son Inc., New York, USA, 1984)
23. H. Subagyo, *Lowland Swamp*, (Research and Development Council of Agricultural Land Resource Development, Bogor, 2006)
24. L.R. Widowati, U. Sri Widati, Jaenudin, W. Hartatik, *The Effect of Compost Enriched with Minerals and Biofertilizer on Soil Properties, Nutrients Uptake and Organic Vegetable Production*, (Soil Research Council, 2005)

Effect Of Flow Resistance In Open Rectangular Channel

Noor Aliza Ahmad [1,*] , *Saiful Iswal* Saiful Bahry[2] , *Zarina* Md Ali[2] , *Azra Munirah* Mat Daud[2] , and *Sabariah* Musa[2]

[1]Faculty of Civil Engineering and Environmental, Department of Water Engineering and Environment, University Tun Hussein Onn Malaysia,86400 Parit Raja,Batu Pahat,Johor,Malaysia
[2]Faculty of Civil Engineering and Environmental, Department of Water Engineering and Environment, University Tun Hussein Onn Malaysia,86400 Parit Raja,Batu Pahat,Johor,Malaysia

Abstract. The determination of flow resistance for open channels remains a challenge in practices. In this research, an experimental study was carried out to investigate the hydraulic roughness characteristics in an open channel. The experimental flume (10 m length, 0.30 m width and 0.46 height) was carried out with two conditions ; contain gravel bed and without gravel bed (as controller). Blocks of concrete foam were fabricated and laid on the bed surface for providing uniformly roughened along the open channel. A velocity flow meter was used to quantify the average velocity, and other parameters such as cross-section, hydraulic radius, wetted perimeter, and channel slopes were also calculated. Finally, the flow resistance, *n* obtained using the flume with gravel bed surface is higher than the flume without gravel bed surface. The flow conditions were declared as subcritical as the Froude number is less than 1 for both conditions in the flume. As a conclusion, it was identified that the hydraulic roughness *n*, was influenced by the type bed roughness, flow rate and channel slope.

1 Introduction

Open channel can be classified as either natural or artificial channel. The terminology natural channel refers to all channels which have been developed by natural processes and have not been significantly improved by humans. [1] An important characteristic of open channel flow is that it has a free surface at atmospheric pressure. [2]

The Manning formula has been widely used to determine the roughness coefficient, *n* for open channels. The variation of roughness coefficient, n occurs due to many contributing factors such as surface roughness, channel irregularity, channel alignment, silting, vegetation, obstruction and flow rate [2]. Although the *n* values have been practiced in hydraulic and hydrology analysis, the determination of n is becoming a challenge to engineers and researcher because the values cannot be computed equally for all types of open channels [3].

* Corresponding author: aliza@uthm.edu.my

In analyzing the flow through open channels of regular sectional shape and hydraulic roughness, in general, it is sufficientto use the overall hydraulic radius as the parameter, which characterizes the properties of the cross section [4]. It is then possible to calculate the discharge through the channel from one range of well-known formulas in term of the channel roughness, slope, and depth. Manning's equation was based on data from flume studies and developed for uniform flow conditions in which the water-surface slope, friction slope, and energy gradient are parallel to the stream bed, and the cross-sectional area, hydraulic radius, and depth remain constant throughout the reach.

The influence of bed roughness gives some effect to flow rate and the roughness characteristics [5]. Focusing on the experimental investigations for hydraulic roughness, [6], it used the particle image velocimetry to study the effects of roughness on the flow structure in a gravel bed channel. They established a new formula to calculate the friction factor for the gravel bed , expressed in terms of the Reynolds number and the surface geometric parameters.

This study has been carried out to analyze the effects of bed roughness to the velocity and determine the hydraulic roughness at roughened rectangular bed in open channel. All data were collected in a laboratory work at the Hydraulic Laboratory of the Faculty of Civil Engineering and Environmental, Universiti Tun Hussein Onn Malaysia (UTHM), Malaysia.

2 Experimental apparatus and method

2.1 Sample material and model set-up

Two blocks of lightweight concrete, which are formed by form agent with cement mixture was used in this experiment. The samples of the block with size 400 mm x 295 mm were prepared by adding the gravel and without gravel. Approximately, 50% of each gravel was submerged in the concrete foam. Three grades of gravel were used (5-10 mm, 20-30 mm, and 50-70 mm) to create a single geometric and hydraulic roughness as shown in Figure 1. The blocks without gravel were functioning as a controller.

Fig.1. Gravel bed as a bed roughness in flume

The experiment was carried out in the Hydraulic Laboratory of the Faculty of Civil Engineering and Environmental, Universiti Tun Hussein Onn Malaysia (UTHM), Malaysia. It was accompanied in a flume of 0.30 m wide and 0.46 m deep with an overall 10 m long as in Figure 2. The flume was set to the slopes 1:500(0.002), 1:400(0.0025), and 1:300(0.0033) respectively. Block samples: lightweight concrete foam with gravel and without gravel, were placed on the bed along the open channel to provide a uniform

DOI: 10.1051/matecconf/20179701107

roughness. Water was pumped into the flume from a circulation storage tank and each bed material was tested with 4 fixed discharges; 0.020 m^3/s, 0.025 m^3/s, 0.030 m^3/s, and 0.035 m^3/s by adjusting the water pump valve, which can be opened and closed by a steel wheel. The flow depth and velocity were measured at three different points (Section I, II and III) by the current meter.

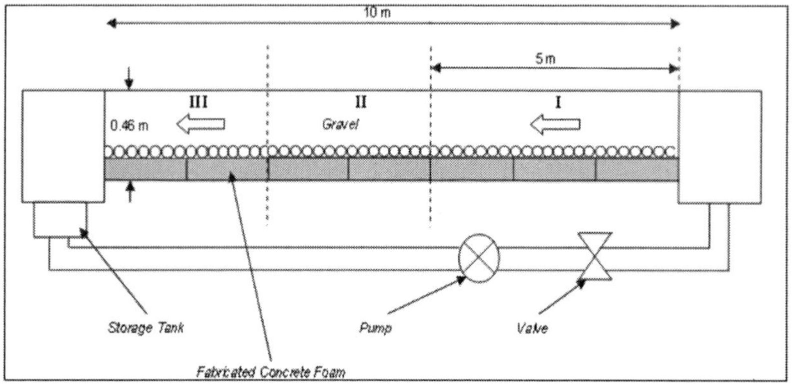

Fig. 2. Schematic diagram of flume with gravel

3 Results and analysis

3.1 The discharge and the flow resistance coefficient

Roughness coefficients represent the resistance to flows in channels and flood plains. The results of Manning's formula, an indirect computation of flow, have applications in flood plain management, in flood studies and in the design of bridges and highways across flood plains [7]. A channel roughness can be described through Manning roughness coefficient, n as stated in the equation (1) [2]:

$$n = \frac{1}{Q} A R^{2/3} S^{1/2} \tag{1}$$

$$Q = AV \tag{2}$$

Where; Q = fixed discharges (m^3/s), A = cross sectional area of flow (m^2), n = Manning coefficient of roughness, R = hydraulic radius (m), S = fixed bed slope, V = mean velocity (m/s)

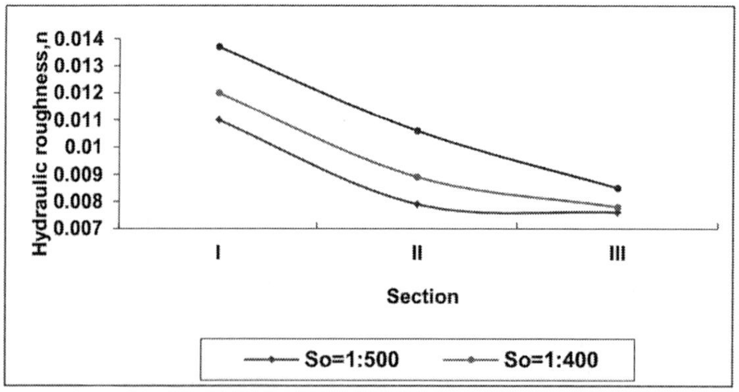

Fig. 3. Hydraulic roughness coefficient along three section with fixed discharges

0.025 m^3/s with gravel bed.

Figure 3 and 4 present the flow resistance coefficient along the section I, II and III with gravel bed and without gravel bed and fixed discharges of 0.025 m^3/s respectively. All data were performed in three different slopes; 1:300, 1:400, and 1:500. It showed the relation between the hydraulic roughness, n under fixed discharges and slopes for the tested bed materials (with gravel bed). Also, it was noticed that the maximum and minimum hydraulic coefficients were found in the case of gravel bed material. Consequently, it can be concluded that the hydraulic roughness coefficient inversely proportionally to the discharge, slopes and fixed bed material.

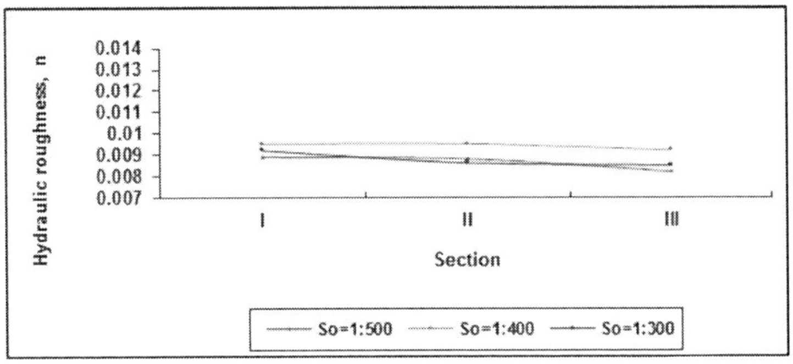

Fig. 4. Hydraulic roughness coefficient along three section with fixed discharges 0.025 m^3/s in flume without gravel.

Table 1 showed the summary between hydraulic roughness, n according to fixed discharges and slopes along three sections of flume with gravel bed and without gravel bed. The average roughness coefficient, n yield from gravel bed ranging from 0.0090 to 0.0119 while the value of *n*, for flume without gravels range from 0.0082 to 0.0096. It means that, the material of bed surface; gravel gave a resistance force to the flow as stated in [2].

Table 1. The average summary of roughness coefficient n, according to fixed discharges and slopes (gravel bed)

Slope	Controlled Flow Rate (m^3/s)	*n* values (gravel bed)			*n* values (without gravel bed)		
		Section I	Section II	Section III	Section I	Section II	Section III
1:500 (0.002)	0.020	0.0092	0.0101	0.0094	0.0092	0.0088	0.0084
	0.025	0.0093	0.0099	0.0092	0.0089	0.0088	0.0082
	0.030	0.0093	0.0092	0.0086	0.0088	0.0087	0.0084
	0.035	0.0090	0.0094	0.0087	0.0088	0.0087	0.0082
1:400 (0.0025)	0.020	0.0102	0.0105	0.0102	0.0096	0.0094	0.0095
	0.025	0.0101	0.0105	0.0102	0.0095	0.0095	0.0092
	0.030	0.0100	0.0103	0.0097	0.0096	0.0092	0.0089
	0.035	0.0100	0.0101	0.0098	0.0092	0.0093	0.0088
1:300 (0.0033)	0.020	0.0116	0.0117	0.0107	0.0093	0.0087	0.0088
	0.025	0.0113	0.0119	0.0110	0.0092	0.0086	0.0085
	0.030	0.0112	0.0113	0.0107	0.0086	0.0086	0.0082
	0.035	0.0107	0.0108	0.0106	0.0087	0.0087	0.0083

3.2 Froude Number, Fr

To classify the type of flow in the channel, the mean velocity for a given discharge must be determined first. For calculation of Froude number, F_R is given from equation (2):

$$F_R = \frac{v}{\sqrt{gD}} \tag{3}$$

With V for mean velocity (m/s), g is gravitational acceleration (m/s^2) and D is the hydraulic depth. Table 2 shows tabulated Froude number calculated from an average depth of each section in the flume and average velocity measured from experiment according to flow rates and bed slope of 1:500, 1:400 and 1:300 respectively. The flow measurement on the gravel and without gravel are in a subcritical state with Froude number, Fr = 0.781 to 0.888 and Fr = 0.799 to 0.902. Froude number increased when slope increases.

Table 2. Froude number calculated from two different bed surfaces

Slope	$Q_{experiment}$ (m^3/s)	Fr (gravel)	Fr (without gravel)
1:500 (0.002)	0.020	0.781	0.799
	0.025	0.788	0.799
	0.030	0.802	0.782
	0.035	0.819	0.775
1:400 (0.0025)	0.020	0.836	0.813
	0.025	0.826	0.821
	0.030	0.835	0.818
	0.035	0.852	0.822
1:300 (0.0033)	0.020	0.799	0.888
	0.025	0.828	0.879
	0.030	0.882	0.902
	0.035	0.888	0.874

4 Conclusion

The experimental work was accompanied to study the hydraulic roughness characteristics in an open channel with two bed surfaces. As conclusion, flow rate and roughness coefficient were influenced by bed roughness. For fixed bed material (with gravel and without gravel), the hydraulic roughness coefficient was influenced by the discharges. Bed surface without gravel also affects the differences between Q_{theory} and $Q_{experiment}$ because of the bed material. It was found that gravel has more effect at steeper slope although this study was done in a small scale.

The authors gratefully acknowledge to University Tun Hussein Onn Malaysia and Ministry of Higher Education for sponsoring this research (FRGS Vot: 1414).

References

1. R.H. French, *Open Channel Hydraulics,* (Mc Graw-Hill Book Company,New York 1987)
2. V.T. Chow, *Open Channel Hydraulics*, (Mc Graw-Hill Book Company, New York 1959)

3. J.M. Buffington, D.R. Montgomery, Water Resour. Res., **35(11)**, 3507 (1999)
4. K. Subramaya, *Flow in Open Channel 2nd Ed.,* (New Delhi. Mc Graw-Hill Companies Ltd. 1997)
5. B.J. McKeon, *Proceeding of the Summer Program, Center for Turbulent Research*, (2008)
6. Wang Xi, Sun Yi, Weizhen Lu, Wang Xiekang, J. Hydrology. Engr., **16(9)**, 710 (2011)
7. J.T. Limerinos, *Dept. of Water Resources*, (1970)

Corn Oil Concentrations Detection for Food Industry Research Development by Using Application of Fiber Optic Liquid Sensor Concept

Md Ashadi Md Johari[1,*], *Aziean* Mohd Azize[1], *Rahaini* Mohd Said[1], *Nor Azazi* Ngatiman[1], *Noradila* Zaine[1]

[1] Faculty of Engineering Technology, Universiti Teknikal Malaysia Melaka, Hang Tuah Jaya, 76100 Durian Tunggal, Melaka, Malaysia

Abstract. This paper purposed application of fiber optic as liquid sensor for detection of various concentrations of corn oil. The final result would an additional understanding about fiber optic sensor and could be used in related field such as food industry sector. They are 20 samples of corn oil with different concentrations between 0% until 100% are tasted. For every concentration, would experience a dipped process together with fiber optic before measuring process. Each of concentrations would present different results in line graph. The experiment results would be discussed in listed values of sensitivity, correlation and coefficient of determination of graft which are totally dependent on concentration of corn oil and the light source. With a different value of concentration and 1550nm wavelength of light source, the maximum sensitivity of 0.00072 the coefficient of determination 94.0% is obtained at third slope, respectively. Fiber optic is successfully to be corn oil liquid sensor and would be in high performance by using 1550 nm wavelength of light source and measuring concentration more than 85%.

1 Introduction

Fiber optic has become one of the most influence technologies in communication engineering recently [1]. Fiber optic cable already tasted which able to convey large bundle of data transmission in one time, causing to be use in other field rather than communication [2 - 3]. Originally, fiber optic cable was known as bundle of glass which able to transmit modulated light wave as massage data. Its capability has been tasted over years which are more advantage than metal cable in data transmissions [5]. The biggest advantage of optical fiber is the fact it is the most cost effective means of transporting information. Fiber can transport more information longer distances in less time than any other communication medium. The bandwidth and distance capability of fiber means that fewer cables, fewer

* Corresponding author: ashadi@utem.edu.my

repeaters, less power and less maintenance are needed. In addition, fiber is unaffected by the interference of electromagnetic radiation which makes it possible to transmit information and data with less noise and less error. These perhaps would be evidences for fiber as a best medium in data transmission. However, this paper would present implementation of fiber optic cable as liquid sensor. The light travelled inside fiber optic would be interfered by existing of liquid molecules around fiber cable. This would be the result of experiment because each of concentrations would consist of different molecules and different value of light source travelled.

Corn oil is oil extracted from the germ of corn. Its main use is in cooking, where its high smoke point makes refined corn oil valuable frying oil. Corn oil is also a feedstock used for biodiesel. Other industrial uses for corn oil include soap, salve, paint, rust proofing for metal surfaces, inks, textiles and insecticides. It is sometimes used as a carrier for drug molecules in pharmaceutical preparations. However, this experiment is to analyze sensitivity of fiber optic sensor in determine the concentration of corn oil. It is because only right concentration can be used in certain purpose such as in chemical industry.

There a various tremendous experiment conducted recently regarding liquid concentration measurement [4][6]. Be that as it may, without precedent for this trial, it was involve the fiber optic link as a fundamental segment in the olive oil detecting action. This examination maybe could help another strategy to detecting fluid focus for different scientists. Decent ramifications behind this trial would be to recognize the ability of fiber optic in fluid fixation detecting action.

The final result from this experiment would produce few different graphs. Each of graphs would be analyzed for correlation R value, coefficient of determination (R^2) value and also sensitivity value. Correlation (R) value would represent a relationship between corn oil concentrations with voltage value. The value would be perfect if it is close to 1.0, which is mean that the changes in oil concentrations would affect the voltage value. Coefficient of determination (R^2) value is a percentage of corn oil will affect the final voltage value. In other words, the higher percentage value could cause larger influence of corn oil towards final voltage value. These three values would determine the best performance of fiber optic as liquid sensor towards different concentrations of coil oil.

2 Experiment setup

The experiment schematic showed in Fig.1.1, consist of light source, silicon detector, amplifier and work station (computer). The main component in this experiment is fiber optic cable which is separated into two parts. The fiber optic cable is made from plastic with core diameter is 0.25 and NA is 0.35. The first part is connected from light source to container full with corn oil liquid and the second part is from container to the silicon detector. The importance area between of these 2 parts is in middle of fiber cable; which is a sensor activity will happen at this area. It is a same fiber cable; however it has been stripped using appropriate stripper to make it thinner than other part of fiber. This procedure is to make that part more sensitive to liquid particles. In other words, the more thinner of diameter, the more sensitive fiber optic would be and its capability to sense particle of corn oil liquid would be grater. Light source used would produce several different of wavelengths which are 850 nm, 1300 nm, 1310 nm and 1550 nm. However due to high demand of corn oil, only 1310 nm and 1500 nm would be use as main light sources. The middle stripped fiber will be dipped inside a container full with coil oil liquid in several times and normally more than an hour. This procedure is to ensure that fiber optic cable could react to the coil oil liquid particle. However, only one concentration will be use in one time. In other word, this experiment will be repeated by using different concentrations. Before change to another concentration, fiber would be taken out from container and would be clean using an alcohol liquid.

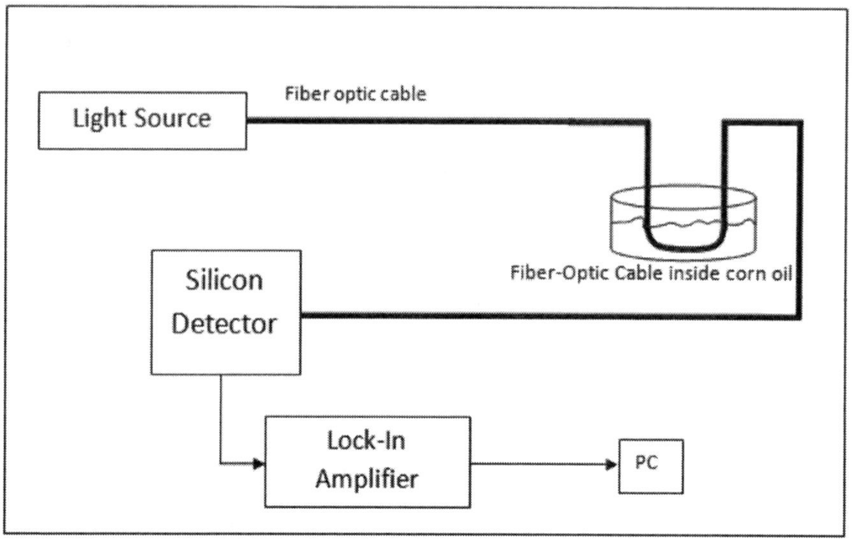

Fig. 1. Schematic diagram of the proposed fiber-optic sensor

The modulated light source would be transmitted inside fiber optic cable from beginning until silicon detector. However, during light travelling, it will experience uneven situation right after reached at stripped area which is inside corn oil liquid. This is because; the light travelling process will be disturbed by coil oil particle which are presence around stripped area. Each concentrations of corn oil will give different impact to the light source. Light travelled finally reached to silicon detector for filtering process and transfer to amplifier for amplification process. The final result would be in decibel (dB) and will be plotted into line graph in the computer.

3 Results and discussion

Fig.2 shows the overall result from this experiment. There are two different line in this graph representing two types of light source (different wavelength) used in this experiment. Coil oil liquid has been sampled into 20 different concentrations started from 0% of concentration until 100%. The line graphs are plotted based on different concentrations. In other words, every concentration would produce different final result. The output in this experiment is in decibel; converted into output voltage. The first experiment started with 1310nm of wavelength. Begun with 0% of concentration, the line graph showed down trend until it reached to 65% of concentration. However, the line graph showed up trend from that point until 100% of concentration. This experiment is continued with 1550 nm wavelength used as main light source. There are three different trend showed in the graph which are down trend from beginning until 40% of concentration, uptrend from 50% to 80% of concentrations and finally back to down trend from concentration 85% until 100%. Based on the results, by used 1310 nm as light source, there would be two slope in the line graph would be sampled for further analysis. However for 1550 nm of light source, three different slopes would be used for analysis. This analysis will determine which of the slopes give major contribution to the performance of fiber optic as corn oil liquid sensor.

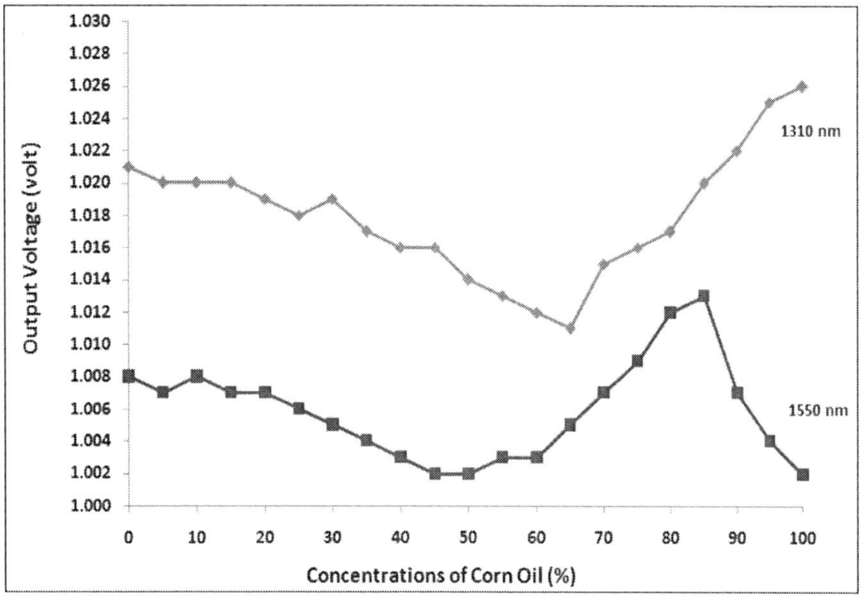

Fig. 2. Fiber-optic sensor responses to different value of corn oil concentrations

The lines graph analysis is recorded in Table 1. These analysis are been separated into several information which are linear range, correlation R, coefficient of determination (COD) and sensitivity. Linear range is a concentration range used from the graph for analysis. For an example, 1310nm of light source is separated into two slopes which are 0% to 65% of concentration in the first slope and 65% to 100% of concentration for the second slope. Correlation R is represented relation between concentrations value with output voltage. COD is show the percentage of corn oil that affected the voltage result. Sensitivity is show sensitive level capture by fiber optic sensor toward different value of concentrations.

The first analysis is based on slopes for light source 1310 nm used. Table 1 showed comparisons between two sample slopes with different range of concentrations. It is show that second slope received higher correlation, COD and sensitivity value. From here it is show that fiber optic would be in higher performance. Different concentrations would definitely give an impact to the output voltage. This experiment continues with 1550 nm of light source and received three slopes analysis. However, the third slope is showed the higher performance of fiber optic as liquid sensor. The third is choose to represent the best result from light source 1550 nm is because it received higher sensitivity value of the bunch. There is one similarity between these results which is fiber optic sensor could definitely in high performance for high concentration value. However, from both of final results, the third slope of 1550 nm wave length is choose to be the best result from others. The reason because this slope received correlation R almost perfect (equal to 1), 94% of corn oil particles interacted with light source traveled through it and sensitivity value is higher among others. In other words, fiber optic sensor is in high performance for corn oil concentrations detection by using 1500 nm as light source and operated for concentrations above 85%.

DOI: 10.1051/matecconf/20179701108

Table 1. Summary of the performance for the fiber-optic sensor.

Light Sources	1310nm		1550nm		
	1st Slope	2nd Slope	1st Slope	2nd Slope	3rd Slope
Linear range (%)	0 ~ 65	65 ~ 100	0 ~ 50	50 ~ 85	85 ~ 100
Correlation R	0.97	0.99	0.96	0.98	0.97
COD (%)	94.2	98.1	92.2	95.6	94.0
Sensitivity (volt/ %)	0.00015	0.00042	0.00013	0.00034	0.00072

4 Conclusion

Fiber-optic sensor experimentally demonstrated using intensity modulation technique in different of liquid concentrations. The results show the different value of voltage output for each of different concentrations and light sources. The result could be observed in the graft which contains two different slopes (for each of light sources) in the voltage response. In this experiment, the liquid concentrations sensing in the range of 20 different oil concentrations using 1310nm and 1550nm wavelength for the main input light sources. The sensitivity correlation R and COD are totally depends on these concentration, condition of fiber-optic and light sources. With the 1550nm light source, the maximum sensitivity of 0.00072 and COD (linearity) of 94.0% had been obtained in third slope of fiber-optic sensor, respectively.

Our appreciations and gratitude to Universiti Teknikal Malaysia Melaka (UTeM) for providing the necessity in completing this project. This work was financially supported by UTeM short term grants PJP/2015/FTK(25B)/S01446. To our group members for their great teamwork and our family and friends whom without their support this project would not be a success.

References

1. S. Binu, V.P. Mahadevan Pillai, N. Chandrarasekaran, Opt. Laser Technol., **39(8)**, 1537 (2007)
2. G.Y. Chen, M. Ding, T. Newson, G. Brambilla, The Open Optics Journal, **7(1)**, (2013)
3. Jingyi Lou, Yipei Wang, Limin Tong, Sensors, **14(4)**, 5823 (2014)
4. Y.S. Chiam, K.S. Lim, S.W. Harun, S.N. Gan, S.W. Phang, Sensor Actuat. A-Phys., **205**, 58 (2014)
5. G.Y. Chen, T.P. Newson, G. Brambilla, Opt. Fiber Technol., **19(6)**, 802 (2013)
6. Y. Zhao, Z.Q. Deng, Q. Wang, Sensor Actuat. B-Chem., **192**, 229 (2014)

Development of Acetone Liquid Concentration Detection Sensor by Using Fiber Optic for Diabetic Level Detection

Md Ashadi Md Johari[1,*], *Aziean* Mohd Azize[1], *Rahaini* Mohd Said[1], *Nor Azizi* Ngatiman[1], *Carvyn* Blaise[1]

[1] Faculty of Engineering Technology, Universiti Teknikal Malaysia Melaka, Hang Tuah Jaya, 76100 Durian Tunggal, Melaka, Malaysia

Abstract. The design of current fiber optic sensor technology especially when related to the chemical still requires a certain degree of the quantitative reliability. The acetone is one of chemical found in the bodily fluid for diabetic person and it would be obvious during fasting. The acetone level could be monitored which is indicate the severity of the diabetes. In this experiment, 4 samples of acetone are used with different concentrations between 20% until 80%. It is represented four different level of diabetic. Fiber optic would be dipped in every concentration before measuring process. Each concentration presented in different line graph and analyze for sensitivity value using statistical method. By using 1550nm wavelength of light source, the maximum sensitivity of 1.064 is obtained at second slope for 40% of concentration, respectively. This is showed that fiber optic sensor could be use as diabetic level sensor. However this sensor would be in high performance at 40% concentration of acetone.

1 Introduction

In recent years, fiber optic sensor received more considerable research efforts as the need for more sensitive and reliable sensors to measure a large of physical, chemical and biomedical quantities [1 – 5]. These research efforts are done due to the fact that fiber optic sensor has a potential sensitivity, detection speed and abilities to the widely assay condition [6]. Acetone is an organic compound with the formula $(CH_3)CO$. It is a colorless, volatile, flammable liquid, and is the simplest ketone. The term was named by IUPAC (International Union of Pure and Applied Chemistry) which is propan-2-one. Along with acetoacetate and β-hydroxybutyrate, acetone is produced in periods of glucose deficiency or insulin insufficiency as an alternative energy source. All these three substances are commonly known as ketone body. Measurement of ketone in urine and blood are widely used in the management of patients with diabetes as adjuncts for both diagnosis and ongoing

* Corresponding author: ashadi@utem.edu.my

monitoring of Diabetic Ketoacidosis (DKA). Overall this project is mainly focusing on the fiber optic sensor of acetone detection development.

A precise and highly sensitive sensor is highly preferred when it's come to measurement and analysis a chemical substance. In medical laboratories for example, a blood sample is obtained from a diabetic ketoacidosis patient and thus it requires accurate and precise acetone concentration information for further evaluation. Various kind of sensor has been developed nowadays. In general, the chemical sensors are broadly classified into gas, liquid, and solid particulate sensors based on phases of the analyte. They can be further categorized as optical, electrochemical, thermometric, and gravimetric sensor according to the operation principle of the transducer. The design of chemical sensor also requires appreciation of needed degree of quantitative reliability (precession and accuracy). An electrochemical sensor for example, has been also introduced to determine the chemical concentration. Nonetheless, a cross sensitivity could happen because of the chemical that is used might require a very active working electrode catalyst and high operating potential for its oxidation. In this case a chemical substance which is more easily oxidized such as alcohol and carbon monoxide could interference the sensitivity of the sensor. Thus, this fiber optic sensor in purposed to overcome these problems.

2 Experiment setup

The experiment setup show in Fig. 1, consist of light source, a silicon detector and lock-in amplifier. A fiber-optic cable is used as a main component in this experiment. It would be connected to be a transmitting and receiving fiber which is made from plastic material with 0.25 of core diameter and have a NA of 0.35. The end of each fiber-optic cable is connected to the light source and silicon detector using pig-tail connector. The middle of the fiber-optic cable would be scraped from its cladding and be used as a tasted area. This experiment used the light source of 1550 nm. The tasted fiber-optic cable is positioned in the middle and will be immersed in the container full with acetone liquid for every experiment. Three different acetone concentrations are used and only one of concentration value is used for every session. The fiber-optic would be undergo immersing process which is take 2 to 30 minutes of duration time, in a way to reinsure that all the results taken would be same and accurate. This will contribute to have accurate results. There are 15 values of final voltage output be taken for every 2 minutes time interval. The fiber-optic cable would be taken out from the container and be cleaned after finished.

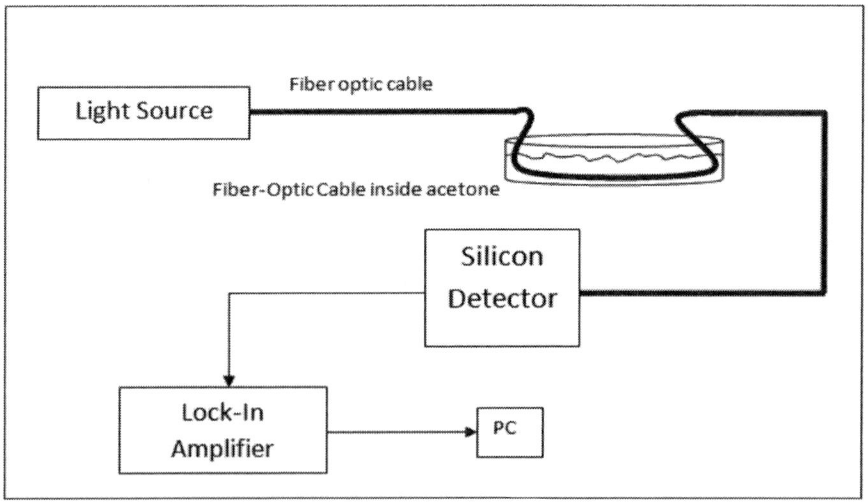

Fig. 1. Schematic diagram of the proposed fiber-optic sensor

The modulated light source is transmitted inside fiber-optic cable through pig-tail connection until the end of connection. At the olive oil liquid container, this light source would experience uneven transmission and some of the light particle will be lost. Each of concentrations used will produce different final results. The light source reached inside silicon detector and be filtered before regenerated inside lock in amplifier which then translate the result into decibel (dB) numbers. A lock-in amplifier is connected with the detector to reduce the dc drift voltage due to an ambient light.

The experiment begins with the soaking process of fiber-optic cable inside olive oil liquid for more than 2 minutes. The result will be taken since that moment until the time reached 30 minutes in total. The soaking process is importance which is to ensure olive oil particle would folded around microfiber which than producing differences in final result.

By the time experiment conducted, there are several situation would happen which influence the final result. One of uneven situation could be thermal noise which is normally caused by equipment used or certain unpredictable circumstances. An action should be considered to overcome this matter such as by repeating the same experiment more than once in different atmospheres. However, this would not reduce thermal noise disturb, but it could help to get batter result. Consideration to use Continuous Wavelet Transform off-line technique to de-noise during data collection also could help to reduce uncertainty during data collection.

3 Results and discussion

Fig. 2 shows the output power value collected for different concentrations of acetone in 30 minutes time durations. There are 15 collections of data which represented 2 minutes time interval to complete 30 minutes in overall. The date collected in 15 output power value. During experiment, fiber optic sensor would be disregarded in acetone with different concentration for 30 minute. The result will be taken every 2 minutes. This step would be handled carefully because of time consummation which is too short.

This experiment started with 20% concentration of acetone and ended with 80% concentration. As overall, each grafts showed different pattern. For concentrations between 20% and 40%, it showed the same pattern which is increase at beginning and finally became constant until the end. However, for concentration of 60% and 80%, the graft started increased at early stage and finally decreased until final. The pattern which is not same for these grafts became the main reason to separate the graft into two slopes for sensitivity analysis.

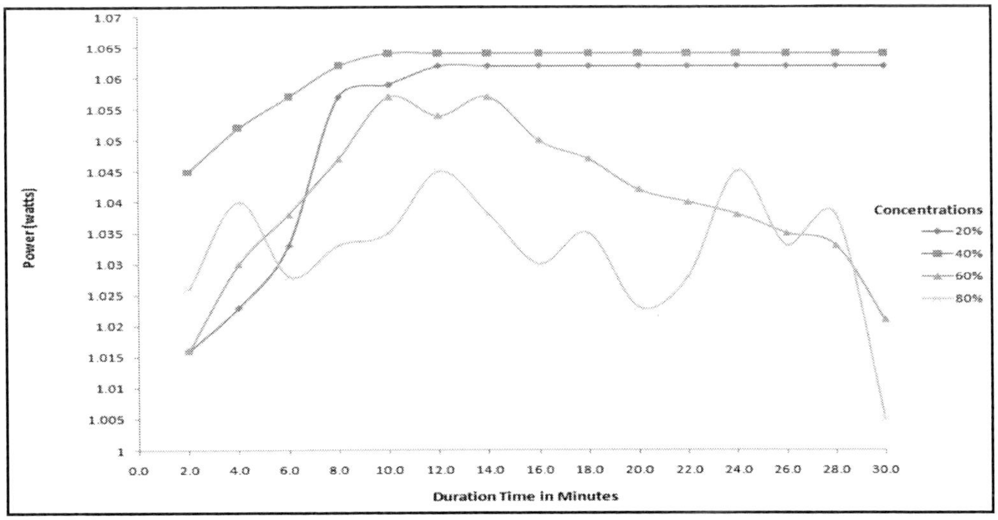

Fig. 2. Fiber-optic sensor responses to different value of olive oil concentrations in time duration

The analysis of these slopes recorded in Table.1. The graft divided into 2 different slopes which is the first slope is between 2 minutes until 14 minutes time interval and the second slope is between 16 minutes up to 30 minutes time interval. This separation would help to understand the pattern of the graft and easy for linearity analysis. The concentrations of acetone are 20%, 40%, 60% and 80% only. Due to the liquid analysis using fiber optic, these ranges of concentration is the best to produce accurate result [6].

The first concentration used is 20%, which is showed highest sensitivity value 0.008. However, sensitivity value decreased when the concentration changed to 40% and remained the same for concentration 60%, which is 0.003. Sensitivity value dropped to 0.001 for 80% concentration. In the second slope, the trend showed as same as before, which is decreased when the concentration of acetone is increase. Started from 20% concentration, sensitivity value is 1.062. However, sensitivity value was increasing a little to be 1.064 when the concentration changed to 40%. The sensitivity value only reached 0.003 for 60% and 0.001 for 80% concentration. From the overall results, light source of 1550nm showed highest sensitivity which is 1.064 for 40% concentration, which is almost perfect. In other words, in every changes of time would definitely affect the output power value. This is due to the natural behavior of liquid with low concentration contained active particles which is affecting the final result.

4 Conclusion

Fiber-optic sensor experimentally demonstrated experimentally using intensity modulation technique in different concentrations of liquid. At the end of this experiment, results show variation value of power output for each of different concentrations. The result could be observed in the graft which contains two different slopes in power response. In this experiment, the liquid concentrations sensing in the range of 4 different acetone concentrations using 1550nm wavelength for the main input light sources. The sensitivity and linearity are totally depends on these concentration, condition of fiber-optic and light sources. With the 1550nm light source, the maximum sensitivity of 1.064 had been obtained in second slope of fiber-optic sensor with 40% of concentration, respectively. This is an evidence that fiber optic could be a sensor for acetone level monitoring for diabetic purpose. Fiber optic sensor would be in high performance condition at 40% of acetone concentration. The development of fiber optic sensor on acetone detection could be a milestone in the health and sensor field.

Table 1. Summary of the performance for the fiber-optic sensor.

Concentration of Acetone (%)	Sensitivity of the 1st Slope (watt/min)	Linear range (minutes)	Sensitivity of the 2nd Slope (watt/min)	Linear range (minutes)
20%	0.008	2 ~ 14	1.062	16 ~ 30
40%	0.003	2 ~ 14	1.064	16 ~ 30
60%	0.003	2 ~ 14	0.003	16 ~ 30
80%	0.001	2 ~ 14	0.001	16 ~ 30

Our appreciations and gratitude to Universiti Teknikal Malaysia Melaka (UTeM) for providing the necessity in completing this project. This work was financially supported by UTeM short term grants

PJP/2015/FTK(25B)/S01446. To our group members for their great teamwork and our family and friends whom without their support this project would not be a success.

References

1. D. Li, J. Wu, P. Wu, Y. Lin, Y. Sun, R. Zhu, J. Yang, J. Yang, K. Xu, Sensor Actuat. B-Chem., (2015)
2. H.Z. Yang, X.G. Qiao, D. Luo, K.S. Lim, W. Chong, S.W. Harun, Meas. J. Int. Meas. Confed., **48**, 333 (2014)
3. Y. Zhao, L. Cai, X.G. Li, F.C. Meng, Sensor Actuat. B-Chem., **196**, 518 (2014)
4. S.S. Patil, P.B. Buchade, D. Shaligram, Sensors Actuat. A-Phys., **20**, 79 (2013)
5. Y.G. Lee, Y.Y. Kim, C.G. Kim, Rev. Sci. Instrum., **83**, 1 (2012)
6. H.A. Rahman, M. Yasin, S.W. Harun, H. Ahmad, 3 (2012)

DOI: 10.1051/matecconf/20179701110

Anchor Bolt Position in Base Plate In Terms Of T and J Anchor Bolt

Mohamad Hairi b Osman[1,*], *Luqmanalhakim* bin Imran[1], *Hannifah* binti Tami[2], *Nor Azira* bt Abdul Rahman[3], and *Salman* bin Salim[4]

[1]Faculty of Engineering Technology, UTHM, 86400 Parit Raja, Batu Pahat, Johor, Malaysia
[2]Faculty of Civil and Environmental Engineering, UTHM, 86400 Parit Raja, Batu Pahat, Johor, Malaysia
[3]POLIMAS, Bandar Darulaman, 06000 Jitra, Kedah Darul Aman, Malaysia
[4]Centre for Diploma Studies, UTHM, 86400 Parit Raja, Batu Pahat, Johor, Malaysia

Abstract. Generally, L anchor bolt system has been used for a long period of time in construction industry as one of the distributing load structures. However, there are some weaknesses in L anchor bolt which may straighten and pullup when charged with tensile load. Current practices prefer to use other types of anchor bolt systems, such as headed studs anchor bolt system to replace the L anchor bolt design. There has been lack of studies to prove that it is more effective in terms of performance. A new T anchor bolt which was basically modified from headed studs anchor bolt was proposed in this study to compare its performance of tensile loading in concrete failure to typical L design. This study aims to determine whether the T anchor bolt system gives better performance as compared to an L anchor bolt system. The performance was rated based on tensile loading on concrete failure pattern. A pullout test was conducted on two different anchor bolt systems, namely L and T. The anchor bolt embedded depth, h in concrete were varied according to their hook or bend radius. Each sample was repeated twice. There were totally eight samples. The hook or bend radius used were 50 mm and 57.5 mm for sample L1 and L2, respectively. 90-degree bend were used on sample T1 and T2. Based on test results, it can be seen that the performance of concrete failure pattern under tensile load on both L and T anchor bolt design samples with 200 mm embedment depth was better than deeper embedment depth of 230 mm. But the L anchor bolt design gives the best results as compared to T design. Although T anchor bolt design shows higher resistance before first bond failure to the concrete sample. T anchor bolt was analysed and needed deeper embedment depth to allow formation of cone pull-out shape to acquire better performance.

* Corresponding author: mhairi@uthm.edu.my

1 Overview of anchor bolt

Anchor bolt on column base plate plays important role to transfer loads (i.e. compression, tension and shear) from column to foundation. The main purpose of the anchor bolt is to hold down the column by transferring tensile loads to the corresponding foundation. These loads may appear in form of pure tension or tension in one side of the column caused by a bending moment. There are various types of anchor bolts, as shown in Fig. 1, and they should be chosen carefully according to appropriate conditions.

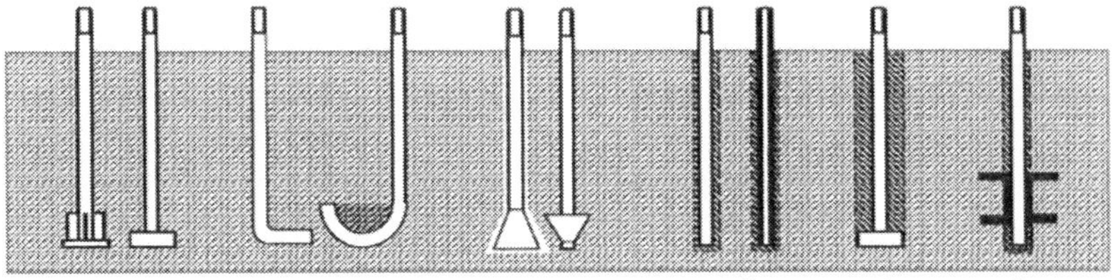

Fig. 1. Types of anchor bolts a) cast in-situ anchor bolts b) hooked bars c) undercut anchor bolts d) bonded anchor bolts e) grouted anchor bolt f) anchoring to grillage beams

There have been a few studies carried out for T anchor bolt to prove that it performs better on tensile loading and in concrete failure as compared to J anchor bolt system. This study aims to design a new anchor bolt that produces better performance on tensile load. The proposed design is T shape which is similar to the design of a headed anchor bolt and is commonly used in industry. A proper experiment was conducted to compare the two different anchor bolt designs. The main objectives of this research are to study the performance using Tensile Strength Test on "T" anchor bolt in concrete under tensile load compare to "J" anchor bolt.

Cast in-situ bolt is the simplest anchor bolt available today. There are many types of cast in-situ anchor bolts. The end of the bolt has some hooked or bend-like-hook, as shown in Fig.2 below. Cast in-situ anchor bolts are the strongest type of fastener, but unfortunately the casting for in-situ anchor bolt is difficult. They are usually used for heavy machines mounted on poured floor or foundations. Another application for this anchor bolt is to connect concrete foundation of a building to its walls.

Fig. 2. Cast in situ

2 Experimental works

Tensile strength test has been carried out to obtain capacity load for anchor bolt. The experiment used 8 samples having same concrete block dimensions of 150mm x 150mm x 300mm. Portland Cement was used for the samples and have been casted with an anchor bolt each and the anchor bolt embedded in certain depth from the surface for sample L anchor bolt (200mm and 230 mm) and "T" anchor bolt (200 mm and 230 mm). The anchor bolt that has been used are "T" shaped anchor bolt that cast in-situ for 4 samples and "L"

 DOI: 10.1051/matecconf/20179701110

shaped anchor bolt that cast in-situ for the remainder 4 samples. Both "T" and "L" anchor bolts were bought from Metal Shop at Taman Maju, Parit Raja. Concrete mix has been designed for the grade 30 concrete based on DOE method. The depth of concrete block for "L" and "T" are shown in Table 1 and 2.

Table 1. "L" anchor bolt Tensile Strength Tests

Samples	Tensile Strength	
	300 mm Block Depth	400 mm Block Depth
L1A	√	
L1B	√	
L2A		√
L2B		√

Table 2. "T" anchor bolt Tensile Strength Test

Samples	Tensile Strength	
	300 mm Block Depth	400 mm Block Depth
T1A	√	
T1B	√	
T2A		√
T2B		√

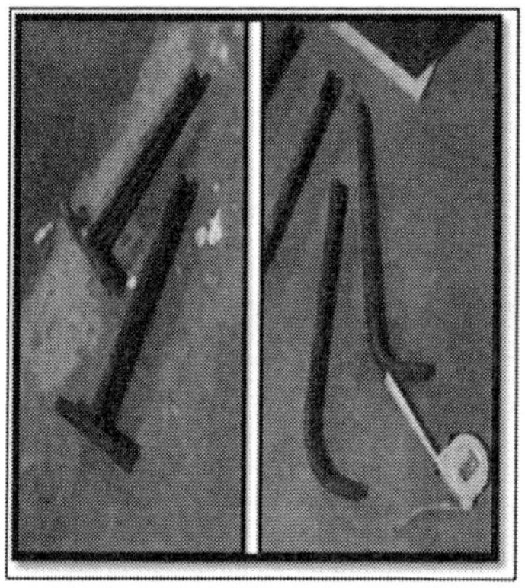

Fig. 3. T and L Anchor Bolts

The samples positioned inside a rig and have been tested using Universal Testing Machine (UTM) as shown in Fig.5 to measure and record tensile strength for each anchor bolt that casted in in-situ concrete block. While gathering the results for tensile strength for each block, the concrete blocks were examined while being tested. The concrete blocks have been recorded using camera for further observation and references for data analysis.

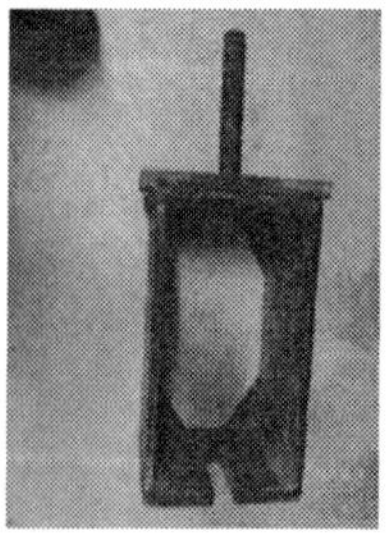

Fig. 4. Design of cast in-situ anchor bolt with rig

Fig. 5. Sample for pullout test using UTM machine

2.1 Pull-out Test

UTM has been used to perform the pull-out test. It consists of two solid bars and two rail solid bars for attaching upper and lower holders as shown in Fig.5 above. The samples are placed in the middle between the upper and lower holders and it is attached at lower holder. The machine pulls or pushes the sample until it reaches maximum length.

3 Results and discussions

The results of Pullout test is shown in Fig. 6 and 7 below.

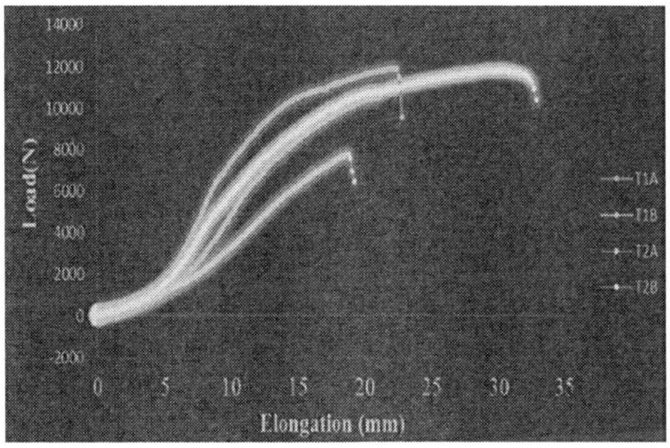

Fig. 6. Relationship between Load and Elongation of Sample T for 28 days

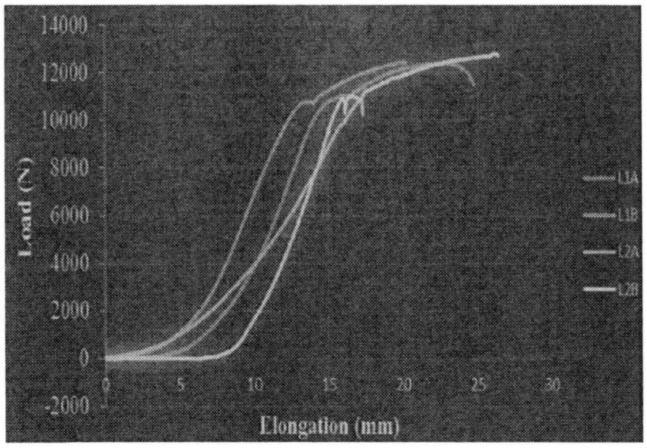

Fig. 7. Relationship between Load and Elongation of Sample L for 28 days

DOI: 10.1051/matecconf/20179701110

Based on Fig. 6, the relationship between load and elongation of anchor bolt sample 'T' for 28 days had been recorded and it has been found that sample T1A gave highest tensile load reading at 11.702 kN, on the contrary, sample T2A gave the lowest reading of tensile load. The graph shows that the shorter the embedded depth, the higher the tensile load to be received by sample structure. The lowest load was archived by sample T2A due to early failure of the anchor bolt grip. The surplus bolt length of sample T2A was less than 150 mm when installed in the machine. The load was not transferred completely to the concrete in the rig plate. The graph also shows the pattern of 'T' sample, where the load decreasing as concrete failure occur and starts to form a cone failure with an angle. The same bond failure occurred between the concrete and anchor bolts. The structure then resist from failure by frictional bond of the concrete cone and the anchor bolt until ultimate tensile load was reached. Fig.7 shows the relationship between Load and Elongation for Sample 'L' which was named as 'L1' and 'L2' with embedded depth of 200 mm and 230 mm, respectively. The highest tensile load failure occurred at sample L2B at 12.583 kN, while the lowest tensile load reading before failure occurred at L2A at 10.625 kN. The same graph also shows the pattern of 'L' sample where the load decreases as the concrete failure occur and starts to form a cone failure with an angle. This is where the bond failure between the concrete and anchor bolts occurred. The structure then resisted from failure by frictional bond of the concrete cone and the anchor bolt until the ultimate tensile load had been reached.

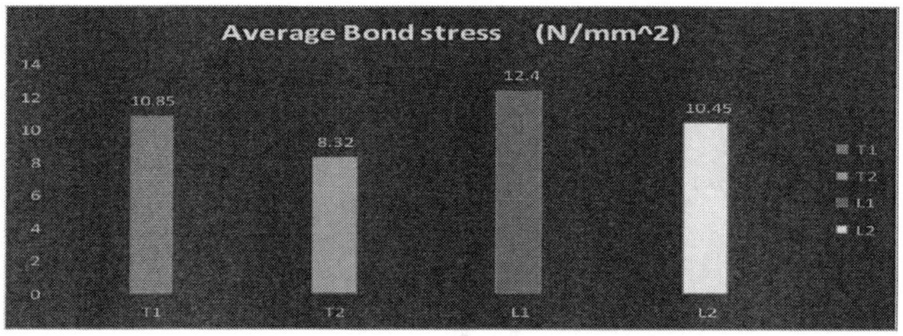

Fig. 8. Relationship between different Anchor Bolts and Average Bond Stress (N/mm^2)

Fig.8 shows the same ranking of samples' performance according to anchor bolt sample and design. The highest bond stress fell on sample L1 at 12.4 N/mm^2, while the lowest bond stress was held by T2. This could due to failure of anchor bolt in concrete and loss connection of transferring loads. The results above show that L performs better and this is because the value of tensile strength depends on steel surface area on concrete. The friction that occurred on the surface of steel contributed the value of tensile strength.

4 Conclusions

Based on the results above, the samples with lower embedment at 200 mm give better results in tensile loading. It can be concluded that the optimum depth in this study is found to be 200 mm and sample L gives higher performance. But it does not tally with the hypothesis that with shallower embedment depth, the better is the performance in tensile loading. This is because the T samples show higher resistance to tensile loading before occurrence of bond failure between anchor bolt and concrete. The T samples require deeper embedment to prevent fracture failure. When insufficient embedment depth is used, T sample could distribute load to the concrete in standard cone shape which causes the anchor bolt fail without affecting the concrete structure. Thus, there wasn't any support of friction bond to resist more tensile loading. The anchor bolt design T is still relevant to be used in

DOI: 10.1051/matecconf/20179701110

the industry but further research need to be conducted to find the optimum embedment depth for the same design.

References

1. P.M. Amaral, *Steel Column Bases under Biaxial Loading Conditions*, (Master's Thesis, University of Porto, 2014)
2. British Standard Institution, *Specification for Aggregates from Natural Sources for Concrete,* (London, BS 882, 1992)
3. British Standard Institution, *Structural Use of Concrete*, (London, BS 8110, 1997)
4. British Standard Institution, *Structural Use of Steelwork in Building*, (London, BS 5950, 1990)
5. M.T.G. Barbosa, S.S. Filho, GJRE, **13(3)**, 54 (2013)
6. A.Chanakya, *Design of Structural Element: Concrete, Steelwork, Masonry and Timber Designs to British Standards and Eurocodes, 3rd Ed.*, (New York, Taylor & Francis, 2009)
7. E.A. Dechant, *Relationship Between Support Structure Forces And Anchor Bolt Stresses*, (Master's Thesis, Lehigh University, 1996)
8. S. Edward, *Anchor Bolt Design for Industrial Structures*, (Master's Thesis, Brigham Young University, 2009)
9. European Committee for Standardization, *Eurocode 3: Design of Steel Structures,* (London, EN 1993-1-8, 2005)
10. European Committee for Standardization, *Eurocode 2: Design of Concrete Structures,* (London, EN 1992-1-1, 2004)
11. European Committee for Standardization, *Eurocode 2: Design of Concrete Structures,* (London, EN 1992-1-1, 2004)

MATEC Web of Conferences **97**, 01111 (2017) DOI: 10.1051/matecconf/20179701111
ETIC 2016

Design and Implementation of Battery Management System for Electric Bicycle

Muhammad Ikram Mohd Rashid[1,*], and *James Ranggi* Anak Johnny Osman[2]

[1]Sustainable Energy & Power Electronics Research Group, Fakulti Kejuruteraan Elektrik & Elektronik, Universiti Malaysia Pahang, Malaysia
[2]Sustainable Energy & Power Electronics Research Group, Fakulti Kejuruteraan Elektrik & Elektronik, Universiti Malaysia Pahang, Malaysia

Abstract. Today the electric vehicle (EV) has been developed in such a way that electronic motor, battery, and charger replace the engine, tank and gasoline pump of the conventional gasoline-powered [1]. In other word, instead of using fossil fuel to move the vehicle, in this case we used a pack of batteries to move it. The global climate change and the abnormal rising international crude oil prices call for the development of EV [2]. To solve these problems, a new energy needs to be developed or optimized in order to replace the current energy which is fossil fuel. A clean and green energy [2]. Because of this, it is very important to make sure that the battery that being used is reliable as the fossil fuel. Thus, the design of the battery management system plays an important role on battery life preservation and performance improvement of EV [3]. The BMS also performs many tasks including the measurement of system voltage, current and temperature, the cells' state of charge (SOC), state of health (SOH), remaining useful life (RUL) determination, controlling and monitoring the charge / discharge characteristics and cell balancing [3]. For this project, 18650 Lithium-Ion battery is used to develop battery management for 144V 50Ah. As lithium-ion batteries have high value of specific energy, high energy density, high open circuit voltage, and low self-discharge, they are a proper candidate for EVs among other cell chemistries [4].

1 Introduction

The purpose of this paper is to provide a review of past research efforts related to Battery Management System (BMS) for electric vehicle and 18650 Lithium-Ion battery. Some articles discuss on the importance of BMS in electric vehicle while others focusing on the balancing of the State of Charge (SOC) of the battery. The articles discussed on how the BMS help in prolonged the life span of the battery used through cell balancing. The review of the articles showed that the BMS provided the users various operating parameters that

* Corresponding author: mikram@ump.edu.my

 DOI: 10.1051/matecconf/20179701111

very important to take the EV to next level, reliability enough to replace current vehicle that used fossil fuel to power up the vehicle.

2 Battery management systems

The battery need to be protected from high voltage or critical operation conditions. High voltage management need to be developed according to the safety standard of EV such as unbalance cell of the battery [5]. The system also monitors and manages vehicle battery modules, and provides early warning protection [2]. That why BMS are important subsystem for hybrid and electric vehicle, which also ensures that the batteries are operating within the specified conditions [3].

3 Cell balancing

Cell balancing is one of the main functions of BMS. A battery is balanced when all of the cells are exactly at the same SOC. The battery may be unbalanced in multiple ways including; SOC, self-discharge current, internal resistance, and capacity. The balancing topologies can be categorized as passive and active balancing [3].

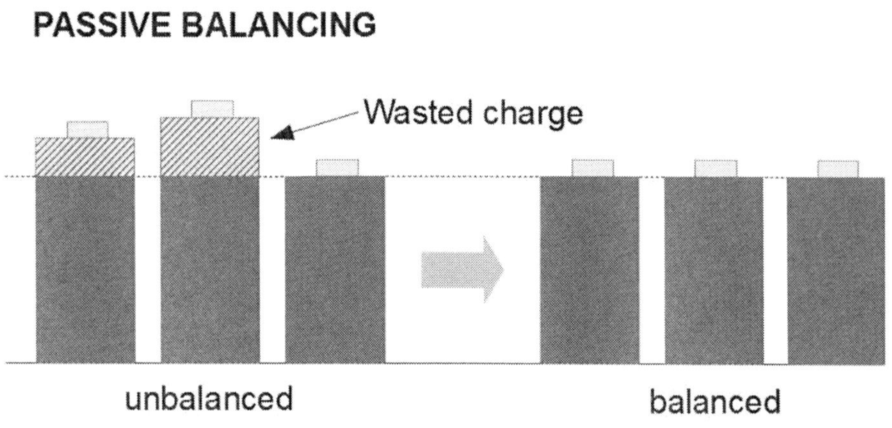

Fig. 1. Passive balancing

Passive cell balancing balance the battery by sampling all cells voltages, levelling all voltages to the lowest among the cells by removing charge from these higher potential cells [3].

DOI: 10.1051/matecconf/20179701111

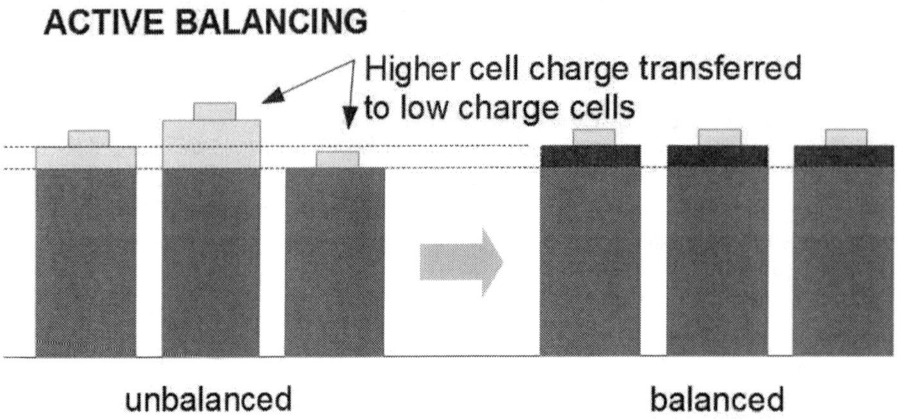

Fig. 2. Active balancing

Active cell also samples all voltages but in different way, by calculating the mean value of the cells then transferring the charge from the cells over this mean value to the cells under the mean value. No waste[3].

4 Protection

This system will cut off the relay that measure the voltage between positive and negative terminal of the battery if it detects overcharge, over discharge, overload and short circuit[6]. Place the figure as close as possible after the point where it is first referenced in the text. If there is a large number of figures and tables it might be necessary to place some before their text citation.

5 18650 Lithium-ion Battery

This project focuses on the management of secondary Lithium Ion cells because of their energy density compared to other types as shown in the graph below Figure 2. Panasonic NCR18650 Lithium-Ion battery was used for this project. The target voltage for this voltage is 58.8V while the target current is 15maH. In order to archive the target values, amount of cell needed are 70. 14 series and 5 parallels. The arrangement of the battery is shown in Figure 6.

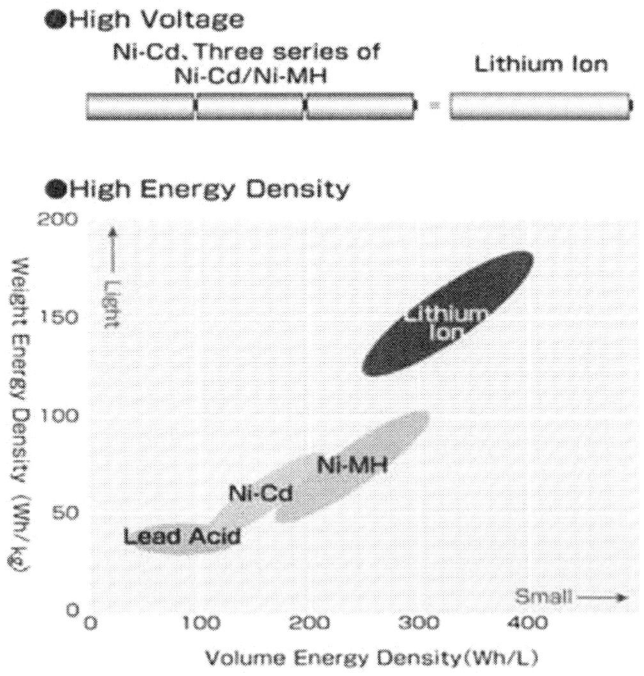

Fig. 3. Comparison of various battery types

This type of battery also has more density, long life cycle and more safety for many applications [7]. This battery also more cheap compare to other rechargeable battery.

Table 1. Specifications of battery

Rated capacity	Min. 2700 mAh
Capacity	Min. 2750 mAh Typ. 2900 mAh
Nominal voltage	3.6 V
Charging	CC-CV, Std. 1925 mA, 4.20 V, 3.0 hrs
Weight (max.)	46.5 g
Temperature	Charge: 0 to +45 °C Discharge: -20 to +60 °C
Energy density	Volumetric: 577 Wh/l Gravimetric: 214 Wh/kg

DOI: 10.1051/matecconf/20179701111

Fig. 4. Spot weld battery in progress

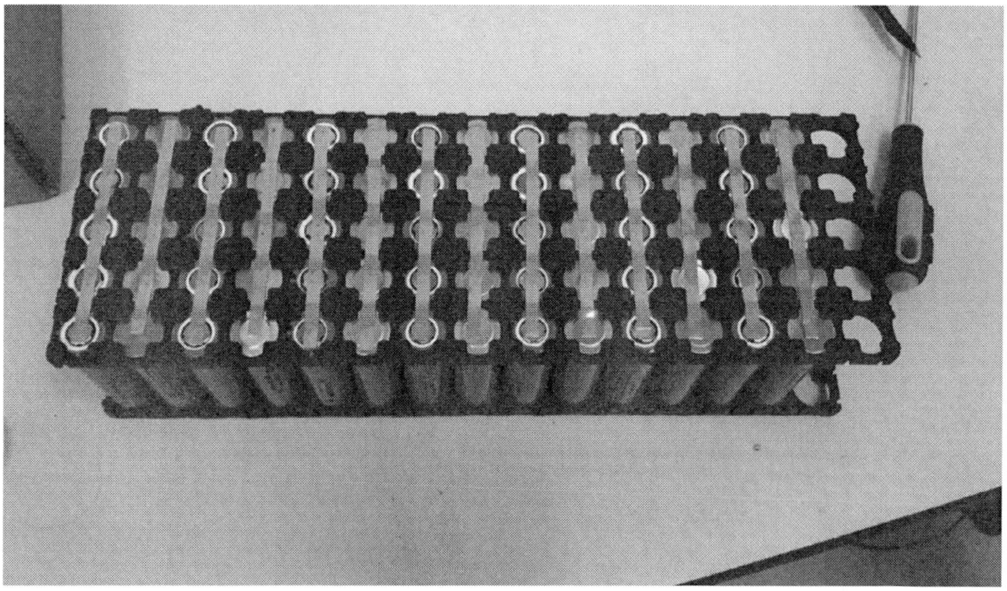

Fig. 5. After spot weld completed

 DOI: 10.1051/matecconf/20179701111

Fig. 6. Battery with battery management system

6 Simulation results

Fig. 7. LTSpice Simulation Charging of cells

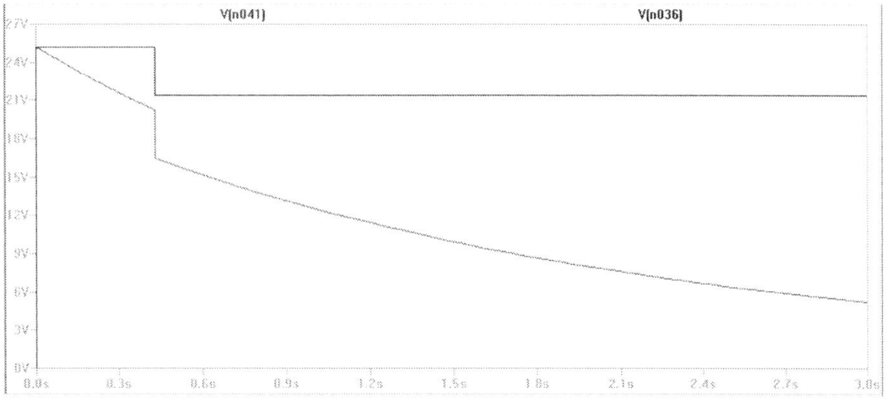

Fig. 8. LTSpice Simulation

DOI: 10.1051/matecconf/20179701111

The graph obtained from the simulation shows the charging and discharging of the battery. From Figure 8, the graph keeps increasing with time until it reached 25.2 V. The voltage for each cell used in this simulation is 4.2 V. So total voltage for 6 cells are 25.2 V. Once the voltage for all 6 cells reached it optimum voltage, the cells will stop charging. Red line in the graph represents the charging and discharging of the battery pack. While the Blue line represents the total voltage for 6 cells. The graph shown in Figure 9 and Figure 10 is only for one IC. In order to achieve the total voltage needed for this project, three IC were needed. After the battery is fully charged, it will start to discharge. From the simulation, one of the main functions of BMS can is proven, protection from overcharged.

7 Summary

After implementation of battery management system, the lifespan of the battery is much longer than the one without it and to ensure the battery operating within safe operating system.

This project was financially supported by Universiti Malaysia Pahang with RDU 150340.

References

1. C.L. Wey, P.C. Jui, *International Conference on Connected Vehicles and Expo (ICCVE)*, 903 (2013)
2. C.H. Lin, H.Y. Chao, C.M. Wang, M.H. Hung, *International Conference Proceedings/TENCON*, 863 (2011)
3. M. Bowkett, K. Thanapalan, T. Stockley, M. Hathway, J. Williams, *19th International Conference on Automation and Computing (ICAC)*, 1 (2013)
4. P. Amiribavandpour, W. Shen, A. Kapoor, *IEEE Vehicle Power and Propulsion Conference (VPPC)*, 117 (2013)
5. J. Qiang, L. Yang, G. Ao, H. Zhong, *International Conference on Vehicular Electronics and Safety (*ICVES), 134 (2006)
6. N. Haq, E. Leksono, M. Iqbal, F.X.N. Soelami, D. Kurniadi, Int. Conf. Electr. Eng. Comput. Sci., 203 (2014)
7. K.J. Chung, C.C. Hsiao, *International Conference on Computer, Consumer and Control (IS3C)*, **2(3)**, 930 (2012)

The Effect of Surface Finish on Sculptured Shape Utilizing Scanned Data – Reversed Engineering (CATIA V5 & DELCAM)

Sundi Syahrul Azwan[1,*], *Jumali* Muhammad Syafik[1], *Attan* Hassan[1] and *R.Abdullah* R. Izamshah[2]

[1]Faculty of Engineering Technology, Universiti Teknikal Malaysia Melaka, Hang Tuah Jaya, 76100 Durian Tunggal, Melaka, Malaysia
[2]Faculty of Manufacturing Engineering, Universiti Teknikal Malaysia Melaka, Hang Tuah Jaya, 76100 Durian Tunggal, Melaka, Malaysia

Abstract. The ultimate aim of this study was to investigate the effect of surface finish for a machined part which was programmed by two popular CAD/CAM software namely Catia V5 and Delcam using scanned data obtained from one of the Reverse Engineering methods namely Three-Dimensional (3D) scanning process. A gaming mouse was chosen as the physical part to be scanned and machined because of its sculptured shape and wavy looks. In this study, simultaneous three-axis machining programs were created and machined using three-axis Computer Numerical Control (CNC) Milling machine. Initially, the physical gaming mouse gone through the first process called scanning process using a 3D scanner; model 700 CX in order to capture the 3D CAD data in points cloud form. The raw scanning data then gone through editing process to obtain better surfaces. Moreover, the stable and edited CAD model then undergone CAD/CAM programming process for both mentioned software respectively. There were ten random points chosen to be analyzed for the surface roughness analysis. The arithmetic mean value (Ra) was used as the roughness value reference. The results reveal that the average surface roughness for the ten points was 0.232 μm for Catia V5 whilst for Delcam was around 0.274 μm. Ultimately, Catia V5 producing better surface finish compared to Delcam software concluded from the analysis carried out as presented in this paper.

1 Introduction

CAD/CAM system is commonly used in daily engineering tasks. CAM may be defined as the use of computer systems to plan, manage, and control the operations of a manufacturing plant through either direct or indirect computer interface with the production resources of the plant. The geometric model generated during the CAD process forms the basis for the CAM process. Various activities in CAM may require different types of information of the

* Corresponding author: syahrul.azwan@utem.edu.my

CAD process [5]. As it is developed through time and widely used, it needs the best decision to choose the best software to be owned. In the process of product design and manufacture, this role is becoming increasingly important as competitive pressures call for improvements in product performance and quality, and for reductions in development time scale [4].

The process of discovering the technological principles of a device, object, or a system through analysis of its structure, function and operation is known as Reverse Engineering (RE). Reverse engineering has become a viable method to create a 3D virtual model of an existing physical part for use in 3D CAD, CAM, CAE or other software [9]. The measured data alone, usually represented as a point cloud, lacks topological information and is therefore often processed and modeled into a more usable format such as a triangular-faced mesh, a set of NURBS surfaces, or a CAD model. A digital 3D model must be constructed using a device that can analyze a real-world object or environment to collect data on its shape and possibly its appearance for example the color. This device is known as 3D scanner. Nowadays, many different technologies can be used to build 3D-scanning devices, but each of the technology comes with its own limitations, advantages and costs. Some of the limitations of this technology that still present are optical technologies encounter many difficulties with shiny, mirroring or transparent objects [3].

The quality of the surface plays a very important role in the performance of milling as a good-quality milled surface significantly improves fatigue strength, corrosion resistance, or creep life. Surface roughness also affects several functional attributes of parts, such as contact causing surface friction, wearing, light reflection, heat transmission, ability of distributing and holding a lubricant, coating, or resisting fatigue. An increase in cutting forces would degrade or reduce the surface finish [7]. Therefore, the desired finish surface is usually specified and the appropriate processes are selected to reach the required quality. The machining quality was assessed through the comparison of surface roughness, surface texture and dimensional control parameters.

2 Methodology

Scanning process was the first step done in this research. A gaming mouse is chosen as the physical part to be scanned and machined because of its sculptured shape. The scanning process was done by using 3D scanner model 700 CX. In 3D scanning, it was very hard to obtain fine surfaces from the raw data. Typical problems normally happened in the raw scanned data such as opened, overlapped and unfinished surfaces. Hence, it is really important to perform editing process in order to resolve all the problems mentioned. The editing process or smoothing process normally involves with sewing unattached surfaces, re-closing opened surfaces and re-creating overlapping surfaces. Geomagic studio was the software used to perform the editing process. After editing process completed, the file must be saved in igs / IGES or stp / STEP file format so that it could be read in Catia V5 and Delcam software for CAM preparation.

For the physical machining, 3-Axis CNC milling machine model DMC 635 Ecoline DMG Mori machine was the machine selected. The size of raw materials or stocks prepared were 150 mm x 100 mm x 50 mm. The material was Aluminum 1000 series. Cutting tool of End Mill D16 High Speed Steel (HSS) four flutes is used to perform the roughing process. Meanwhile, Ball Nose Mill R6 (HSS) cutting tools were chosen to perform the machining on the surfaces which were to be analyzed in surface roughness analysis. In Catia V5, Advanced Machining workbench is utilized to create the CAM program. Machining process offered by Catia V5 for surface machining called "Sweeping" process. On the other hand, in Delcam software it was automatically selected by the software once the specific geometries of the 3D CAD model are defined. There were setups or settings which need to be set for both software. It is realized that some different terms used carry the same meanings such as "depth of cut" used in Catia V5 whereas "Z increment" in Delcam. Table 1 indicates the settings have been applied for both software in preparing the CAM program whilst Fig. 1 shows the completed CAM programs for both software. Completed programs were then post processed using integrated post processor and finally transferred to the CNC machine for physical machining.

Surface Roughness is a measurable surface characteristic quantifying high frequency deviation from an ideal surface. It is usually described by the arithmetic mean value (Ra), based on the mean of the normal deviations from a nominal surface over a specified "cut off" length and normally measured in micrometers (μm). This process was done using Mitutoyo surface roughness tester in metrology laboratory. Moreover, this analysis used a contact surface roughness tester to measure the surface finish. In other words, this method involves dragging a measurement stylus in traverse direction on the surface of the sample. The surface features were measured relatively to the drive unit reference surface. Hence, this accurately measures waviness and finally stepped features in addition to the surface roughness.

Table 1. Settings / parameters applied for both software in preparing the CAM program

Catia V5 Software			Delcam Software	
Spindle speed	7000 Rpm		Spindle speed	7000 Rpm
Offset on part	0.5mm		Finishing allowance	0.5mm
Machining tolerance	0.1mm		Machining tolerance & Stepover	0.1mm
		ROUGHING		

Catia V5 Software			Delcam Software	
Spindle speed	8000 Rpm		Spindle speed	8000 Rpm
Machining tolerance	0.1 mm		Machining tolerance	0.1mm
Maximum distance between pass	0.05 mm		Stepover	0.05mm
Offset on part	0 mm		Finishing allowance	0mm
Cut of depth	0.01 mm		Z increment	0.01mm
SWEEPING			FINISHING	

DOI: 10.1051/matecconf/20179701112

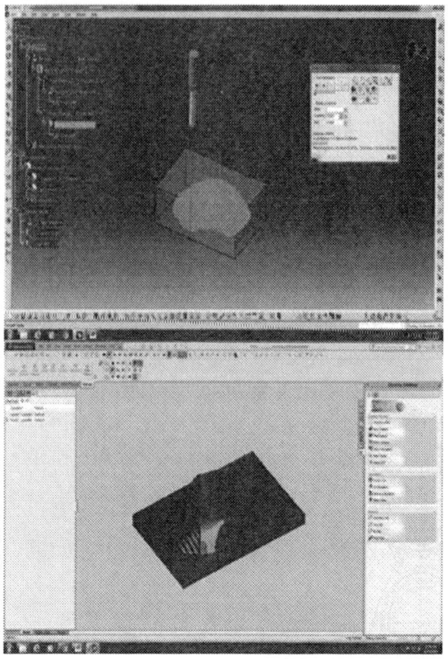

Fig. 1. Completed CAM program in Catia V5 (top) and Delcam (bottom)

3 Result and discussion

In measuring the surface finish, there were ten points randomly picked on the top surface of the machined parts. Various locations have been chosen to be analyzed. Eight points were taken on the main part. Meanwhile, two points were on the bottom of the machined part. All points taken for both machined parts were approximately on the same locations. Furthermore, three readings are taken for each identified point and the average value is calculated. The stylus traverse distance was set constantly at 5 mm at every single reading. Fig. 2 indicates the location of the points taken for the surface analysis.

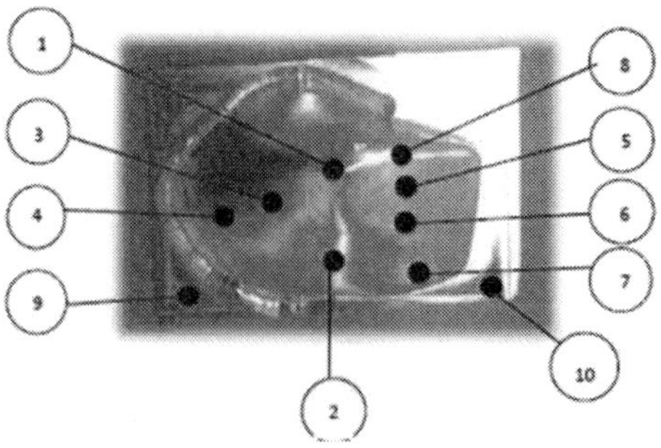

Fig. 2. Points taken for surface roughness analysis for both machined parts

CATIA V5 SOFTWARE DELCAM SOFTWARE

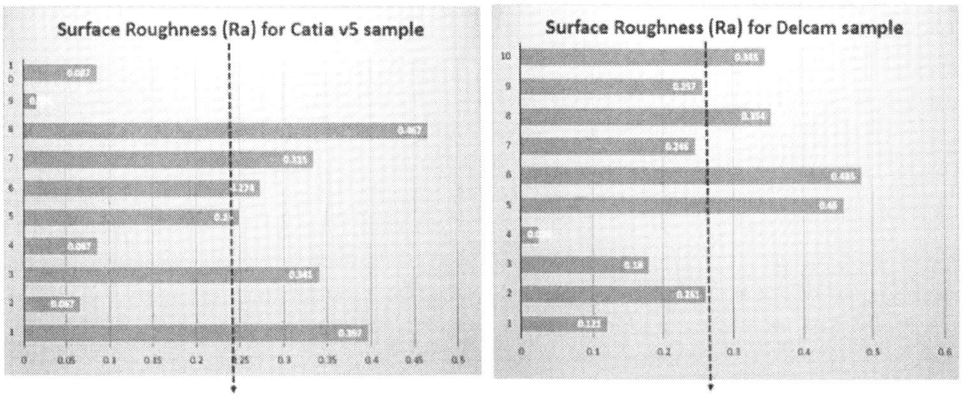

Average=0.232 Average=0.274

Fig. 3. Comparison of surface roughness average values for the 10 points taken

From the result illustrates above, it is obviously shown that the highest average value for Catia V5 is 0.467 μm at the 8th point whilst for Delcam's is 0.485 μm representing the 6th point. On the other hand, the best surface roughness value falls at the point marked as 9 for Catia V5 with the value of 0.016 μm. Meanwhile, for Delcam it falls at the 4th point with 0.026 μm. The location of the 9th point was located on the bottom side. However, the location for the 4th point for the Delcam was located on the main part itself. Nevertheless, it appears the average value obtained for Catia V5 is 0.232 μm shows slightly better result to be compared to the Delcam's which is 0.274 μm. This finding highlights that the same scanned 3D CAD model programmed by Catia V5 resulted slightly better surface finish in average than programmed by Delcam.

The main factor contributes to the result obtained was possibly caused by the different way of preparing the program. Although the same 3D CAD model were used and the same parameters applied, somehow the differences in determining the important settings whilst preparing the CAM program such as tool path direction and machining processes. In Catia V5 users were given options in determining the direction of tool path as this was not being offered in Delcam. In this study, the direction of the tool path selected was along the length of the part to be machined. However, in Delcam the tool path was automatically calculate which was opposite direction from the one selected in Catia V5 and at certain areas the tool paths looked undefined movement. Furthermore, in Delcam there were too many shortcuts to simplify the machining program compared to Catia V5. Machining parameters were also automatically calculated from the library referring to the range set by the users as not in the practice of Catia V5. Catia V5 emphasizes details in each step and process compared to Delcam. There were two machining processes chosen in Catia V5 namely roughing and sweeping process. Sweeping process was chosen because of its ability in generating tool paths for curved or sculptured surfaces in point milling. On the other hand, in Delcam machining process were defined automatically by the software once all necessary geometries were chosen. The differences mentioned are presented in Fig. 4.

Ultimately, Catia V5 offers higher flexibility in determining the way of CAM program to be created than Delcam. Nevertheless, simplification and automatic defined by the software makes programming time in Delcam much faster than Catia V5. Importantly, both software are able to produce the desired machined part.

DOI: 10.1051/matecconf/20179701112

4 Conclusion

In this study, the effect of surface finish of machined parts which were programmed by Catia V5 and Delcam applying one of Reverse Engineering methods namely 3D Scanned Data have been investigated. The result reveals that the surface finish of machined part programmed by Catia V5 slightly better than Delcam in average although the same 3D CAD data is used. Important conclusion drawn from this work is that; differences in preparing the CAM program especially on the tool path direction as well as the details of machining processes involve will also somehow effect the surface finish of the machined part even though the same 3D CAD model is used as the reference. In particular, further research should be conducted on study of dimensional accuracy in producing machined part from scanned data to be compared to the actual part for both meant software.

Author acknowledges, with gratitude, authors' debt of thanks to Universiti Teknikal Malaysia Melaka (UTeM) and to the funding body; Ministry of Higher Education Malaysia for the research grant (RAGS/1/2014/TK01/UTEM/B00079).

References

1. Abbott, Steven, *SPE (Surface Profile Explorer), Abbott Apps,* (Steven Abbott TCNF Ltd. 2014)
2. R. Baptista and J.F. Antune Simoes, J. Mater. Process. Technol., **103**, 398 (1999)
3. F. Bernardini, H.E. Rushmeier, *The 3D Model Acquisition Pipeline*, (Compute Graphics Forum 2002)
4. Chris McMahon, Jimmie Brownie, *CAD/CAM Principle Practice and Manufacturing Management, 2nd Edition,* (Prentice-Hall, 1999)
5. R. Alavala Chennakesava, *CAD/CAM Concept and Application*, (Prentice-Hall India. 2008)
6. R. Dubovska, J. Jambor, J. Majerik, Procedia Eng., **69**, 638 (2014)
7. S.A. Sundi, M.R. Muhamad, B.A Bakar, R.I.R. Abdullah, Sci. Int. (Lahore), **26**, 1869 (2014)
8. Y.W. Sun, D.M. Guo, Z.Y. Jia, J. Mater. Process. Technol., **180**, 74 (2006)
9. T. Várady, R.R. Martin, J. Cox, Comput. Aided Des., 253 (1997)

Investigation on Heavy Precipitation Effects over FSO Link

M.F. Talib[1,*], *A.K.* Rahman[1], *M.S.* Anuar[1], *C.B.M.* Rashidi[1], and *S.A.* Aljunid[1]

[1]Advance Communication Engineering, Centre of Excellence (ACE-CoE), School of Computer & Communication Engineering University Malaysia Perlis (UniMAP)

Abstract. This paper has examined the performance of free space optic system and scattering effects over the laser propagation. Analysis result based on Optisystem simulation software and rain data collected from Malaysia Meteorological Department Perlis, which covering beneath the tropical rainforest weather. The measurement in a simulation using the real parameter from Laserbit system which using wavelength 785 nm ,0.5 – 15mrad beam divergence, 70mW laser power and range between 100m – 2500m. The result can estimate the intensity of rain to impair the link and alternative method to overcome the drawback system.

1 Introduction

In telecommunications, Free Space Optics (FSO) are an optical communication technology that uses light propagating in free space to transmit data between two points. This technology is new in Malaysia and not widely use. It might relate to environmental issues that familiar with this system. FSO also well known as "free space photonic" or "optical wireless" which is classified as unguided beam. Various applications can be deployed in FSO system such as Bridging Wireless Local Area Network (WLAN), LAN to LAN Building Connectivity, Education Campus Network, Gigabit Ethernet, Access and Backhaul. Among the advantage of FSO is not required license from Federal Communication Commission (FCC) for installation. The latest technology in FSO which is design to deploy behind window has eliminating the need for costly rooftops rights. In terms of communication security, FSO provides far more secure than other wireless-based band transmission, such as uses narrow laser beam light making harder to intercept and crack [1].

Unfortunately, FSO system is vulnerable to atmosphere phenomenon such as attenuation and scintillation which can reduce laser power and limit the path length for a given link availability. Basically, in FSO the impairments link is caused by air turbulence that produces a beam spreading, beam wander, scintillation and degradation of the coherence of the wave front [2] and hydrometers (rain and snow) and suspended water particles (fog represents the worst meteorological situation for FSO) that yield high losses due to scattering [3,7,8].

* Corresponding author: karahman@unimas.my

DOI: 10.1051/matecconf/20179701113

2 Theoretical approaches

2.1 Scattering coefficient

Rain is liquid precipitation. It is the condensation of atmospheric water vapour into drops heavy enough to fall, often making it to the surface and the most frequent type of atmospheric precipitation in temperate climates. Rain is under non-selective or geometrical scattering where it occurs in the lower portion of the atmosphere when the particles are much larger than the incident radiation. The scattering coefficient can be calculated by using Stroke Law [3].

$$\beta_{rainscat} = \pi a^2 N_a Q_{scat}\left(\frac{a}{\lambda}\right) \tag{1}$$

Where:

a = radius of raindrop (0.001cm to 0.1cm)
N_a = Rain drop distribution
Q_{scat} = scattering efficiency
λ = wavelength

2.2 Atmospheric attenuation

The attenuation of laser power in the atmosphere is described by Beer's Law [5]:

$$\tau(R) = -\frac{P(R)}{P(O)} = e^{-(\beta)R} \tag{2}$$

$\tau(R)$ = transmittance at range R
$P(R)$ = laser power at R
$P(O)$ = laser power at the source
β = scattering coefficient
R = link range (km)

Equation (2) can be converted to logarithms scale using Equation (3)

$$= -10\log\frac{P(R)}{P(O)}$$

$$= -10\log e^{-\beta R}$$

$$= 10\log e^{-\beta R} \tag{3}$$

3 Result & discussion

Fig 1 shows the performance of the scattering coefficient against intensity rain. The x-axis represents the intense rain varies from 0 – 0.00045 cm/sec and y-axis represent the scattering coefficient in km^{-1}. The curves were plotted based on Equation (1).

DOI: 10.1051/matecconf/20179701113

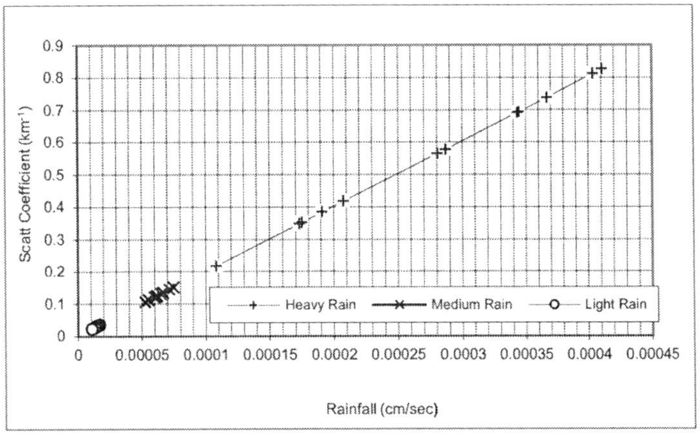

Fig. 1. Scattering Coefficient against Rainfall rate.

The graph also indicates the performance of three different rain i.e. The heavy rain, medium rain and light rain. Scattering for light rain varies from 0.022 km^{-1} to 0.035 km^{-1} while for medium rain varies from 0.112 km^{-1} to 0.151 km^{-1} and for heavy rain varies from 0.218 km^{-1} to 0.827 km^{-1}.

From the analysis the heavy rain exhibits a high scattering coefficient effect comparable to the other rain type. The heavy rain rate varies from 0.0001 cm/sec to 0.00045 cm/sec. The graph also shows that the scattering coefficient is linearly proportional with rainfall intensity. The highest precipitation rate will produce a high scattering coefficient. In this context scattering coefficient mean a measure of the light attenuation due to scattering of light as it traverses a medium containing scattering particles. Thus the rise of scattering coefficient will yield high atmospheric attenuation with respect of precipitation rate and this relationship will be elaborate more in Fig 2.

The x-axis represents rainfall rate and y-axis correspond to atmospheric attenuation. Fig 2 proves that the atmospheric attenuation is linearly proportional with rainfall rate. The curve obtained from Equation (2) from Beer's Law [4] which explain that laser power at receiver over laser power at transmittance is equivalent with exponential scattering coefficient times link range of laser transmission. Link range between transmitter and receiver is within 0.5 kilometer. As the rainfall rate increases, the attenuation will increase as well.

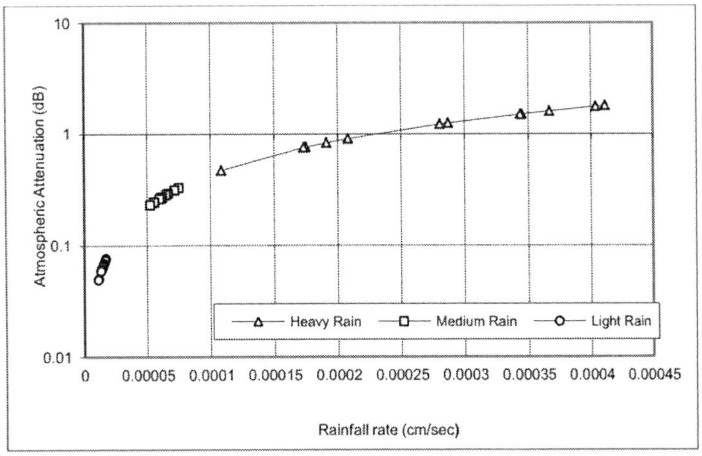

Fig. 2. Atmospheric Attenuation due to intensity rain.

Basically, the graph shows three different types of rain. The high atmospheric attenuation is measure at 1.7949 dB and lowest attenuation at 0.0485dB. Mostly severe rain (heavy rain) yields 0.4730 dB – 1.7939dB and medium rain and light rain below 1dB changing from 0.3274dB – 0.2292dB and 0.0728dB - 0.0485dB respectively. As explain in Fig 1 the high scattering contributes high atmospheric attenuation. However, these attenuation not enough to impair the link but as the link range increases the atmospheric attenuation begin to increase. All of the can observe in Fig 3 which show the relation laser power at receiver with link distance.

Fig 3 shows the performance of the heavy rain due to link distance. The powers at receiver laser in range 20 dBm to -120 dBm and distance about 5 km. The curve is obtained by simulation using Optisystem software. The equation that use to simulate this curve is

$$P_{received} = P_{transmit} \frac{d_R{}^2 R}{(d_T + \theta R)^2} \times 10^{-\left(\frac{\alpha R}{10}\right)} \tag{4}$$

d_R = Receiver Aperture (cm)

d_T = Transmitter Aperture (cm)

θ = Beam divergence (mrad)

α = Atmospheric attenuation (dB/km)

From the equation, the input parameter is based on laser bit equipment. Receiver aperture = 20cm, transmitter aperture = 5cm, beam divergence = 8 mrad , wavelength = 785nm and power consumed = 70mW.

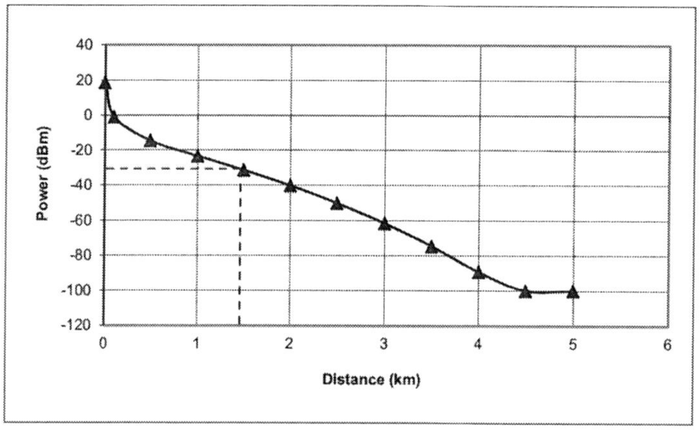

Fig. 3. The changing laser power at receiver due to increasing link distance.

The graph shows that power laser is inversely proportional with link distance. As the distance increases, the power at receiver decreases. This condition depict that when power laser transmit through heavy rain it experiences weaken power. The blue dot lines indicate the minimum sufficient laser power at receiver for full modulation transmission about -31.438 dBm with range 1.5 km. When power continue to decreases, the link begin to impair and can drop it worst condition totally error. The minimum sufficient laser power is vital because represent the link performance under heavy rain but if the receiver receives below minimum power big possibility data error in transmission is high. This result based on Optisystem simulation in Bit Error Rate (BER) Analyser that has been done.

The solution for this problem is to setup short range between transmitter and receiver. As we can see in Fig 3 above, the distance between 0 less than 1.5 km is best range and above minimum power require at receiver. Thus can optimize line transmission without add any cost to upgrade the link performance.

Another alternative solution is to use narrow laser beam. This can be implementing by using small beam divergence. Previous parameter use beam divergence = 8mrad. Fig 4 represents power laser correspond with link distance by using beam divergence = 1mrad.

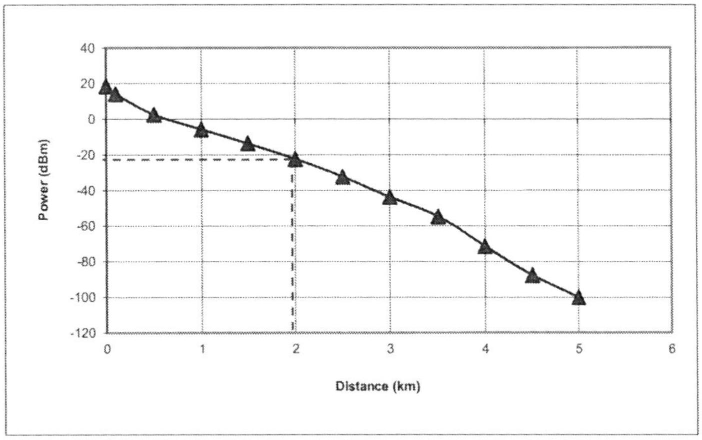

Fig. 4. Relation performance of laser power with link distance (using divergence = 1 mrad).

Based on the graph, laser powers receive at receiver in scale 20 to -100 dBm and minimum sufficient power at receiver is -22.341 dBm at range 2 km. Compare to beam divergence a 8mrad, the minimum sufficient power at range 1.5 km with distance different about 0.5 km. In laser communication, the more capable system to go further is better because can reduce the usage of laser at short range and moreover add wide coverage communication.

The beam divergence usually provides by vendors. Consumption of wide beam is to make sure enough beam receive at detector and it is to ensure automatically stay align even if small shifting point happens. However, this scheme is acceptable for low data rate only. The advantage of narrow beam can generate high data rates plus add more distance to transmit.

The relationship between beam divergences with distance is illustrated in equation (5) below [5]:

$$W = R\theta \tag{5}$$

W = width (or diameter beam area)
R = link distance
θ = beam divergences

Therefore to avoid larger spread beam it better to use narrow beam such 1mrad [6,7].

4 Conclusions

Rain is the main major effect in the design of FSO system operating in high rainfall regions. The usage of laserbit at 785nm wavelength can optimize the line transmission with setup short distance between transmitter and receiver. It can reduce the atmospheric attenuation and fulfil the minimum required laser power at transmitter for full modulation at range below 1.5 km. Other than that the use of narrow beam is suitable in high rainfall region because produce high power at receiver for better transmission. It also gives the advantage in data rates transfer. Since FSO bandwidth can up into gigabit it can carry a tremendous amount of information and narrow beam is suitable for high data rate transfer. Therefore a small beam divergence is much better compared to big beam divergence.

This work has been supported by a grant from the Fundamental Research Grant Scheme (FRGS) in FSO research area under ACE - CoE School Communication and Computer, University Malaysia Perlis.

References

1. M. Achour, *Proceedings of SPIE*, (2002)
2. D. Deirmendjian, J. Appl. Meteorol., **14**, 15584 (1975)
3. J.P.V Poiares Baptista, *Precipitation, Clouds and other relate non- refractive effect*, (ESA Publication Division, Netherlands)
4. H.Weichel, *Laser beam Propagation in the Atmosphere* (SPIE Press, Bellingham, 1990)
5. Pedrotti, L. Frank, S. Pedrotti Leno, *Introduction To Optics, 2nd edition*, (Prentice Hall International Editions, USA, 1993)
6. A.K. Rahman, M.S. Anuar, S.A. Aljunid, M.N. Ismail, B.S. Naimullah, *International Conference on Electronic Design*, (2008)
7. A.K. Rahman, M.S. Anuar, S.A. Aljunid, M.N. Junita, *Proceedings of IEEE 2008 6th National Conference on Telecommunication Technologies*, (2008)

Multi responses optimization of decarboxylation process of Free Fatty Acid (FFA) into biodiesel using Grey Relational Analysis (GRA)

Erfina Oktariani[1,*], and *Rita* Istikowati[1]

[1]Department of Polymeric Chemical Engineering, Polytechnic of STMI Jakarta, 10510, Jakarta, Indonesia
[2]Department of Automotive Industrial Engineering, Polytechnic of STMI Jakarta, 10510, Jakarta, Indonesia

Abstract. Grey Relational Analyses (GRA) was utilized to determine optimal setting parameters of Free Fatty Acid (FFA) conversion from into hydrocarbon chain by decarboxylation process. The considered setting parameters were voltage and acetic acid (CH_3COOH) concentration. Decarboxylation process occurred by adding acetic acid as the ion contributor for the dimerization process. In order to facilitate FFA well ionized, electrolysis was employed during the decarboxylation with voltage varied from 10 Volt to 25 Volt. The concentration of Acetic Acid varied from 0.5M to 1.5M for an optimum condition with a high yield. Decarboxylation reaction took place at 130°C for 1 hour. The experiments were carried out by using full factorial design and multi responses considering % Yield, specific gravity (spgr), °API, viscosity, and Net Heating Value (NHV). Grey Reasoning Grade (GRG) was used to analysis multi responses. The setting parameters of sequence seven (A2 B3) has the highest GRG. Analysis of variance (ANOVA) results indicated that acetic acid concentration was contributing parameter. The optimal level for acetic acid concentration based on polynomial regression model was found to be 1.325 M.

1 Introduction

Biodiesel is commonly known as short-chain mono-alkyl esters of long-chain fatty acids derived from renewable lipid feedstocks, such as vegetable oil or animal fats [1]. If all of the produced edible oils are transesterified, the petrol-diesel global consumption can be replaced until 3%.

Biodiesel has more advantages than petrol-diesel due to its shorter carbon chain, this results biodiesel is more flammable. It has no sulfur content which sulfur may damage the environment as a pollutant. Biodiesel is more available as a renewable energy than fossil fuel. Furthermore, it can be directly employed in existing engines, vehicles and

* Corresponding author: erfina@kemenperin.go.id

infrastructure [2]. It also has a better energy balance than petrol-diesel because it contains 3.2 times the amount of energy it takes to produce it [3].

There are some properties that classified a good biodiesel i.e: contains Carbon (wt) 80%, Hydrogen (wt) 13%, 0.8 specific gravity, Cetane number at 40-45, viscosity at 3 mm^2/s [4]. However, biodiesel which consists of fatty acid alkyl esters has some drawbacks. Biodiesel has a low fluidity, high oxygen content, poor stability of thermal oxidation and polymerization of ester fuel molecule [3]. Decarboxylation of Free Fatty Acid (FFA) is regarded as one of solutions to solve these problems, that can directly convert FFA into hydrocarbon chains.

There are several factors that influence Free Fatty Acid (FFA) decarboxylation process into hydrocarbon chain [5]. The selection of the right setting parameter for each factor is needed to have an optimal quality characteristics. A full-factors design of experiments (DOE) can be proposed as a method to select the right setting parameter. DOE methods can evaluate each factor's effect on average and variability in quality characteristics[6]. Although DOE has the advantage, this method has limitations in handling multi responses of quality characteristics.

Free Fatty Acid (FFA) decarboxylation process into hydrocarbon chain has some responses as quality characteristics considering % Yield, Net Heating Value (NHV), specific gravity (spgr), °API and viscosity. Therefore, it needs a method to solve multi responses problem. There are several methods for solving multi responses problem such as engineering judgement, regression analysis, Principal Component Analysis (PCA) and Grey Relational Analysis (GRA) [7]. The knowledge and experience of engineers are commonly used for engineering judgement which results its weakness. This method is only suitable for experienced engineers. Regression analysis is basically a technique-based approach to regression for each quality characteristic. Using regression analysis for solving multi response can increase complexity in the calculation and possibility of a correlation between quality characteristics not adequately considered. PCA transforms multi response into key components that are not correlated. PCA methodbhas difficulties in determining a feasible solution if there are more than one component that has eigen value more than one or there are more than one main component selected. GRA converts multi responses to the value of Grey Relational Grade (GRG) and optimizes the parameters by analyzing the GRG. GRA is based on the theory of Grey System can provide weighting factors that influence processes and can be applied to solve problems of complex multi responses of quality characteristics [8]. DOE based on the Grey Relational Analysis (GRA) offers an approach transforming single response optimization to multi response optimization processes.

2 Literature review

Catalytic decarboxylation of benzoic acid was performed by Takemura et.al. (1985) at the temperature range of 370–440°C in a batch autoclave with Y type zeolite. From the three rare earth metal-exchanged Y zeolite catalysts, NdHY resulted the highest activity, on which the benzene yield was 75.6% under nitrogen at 390°C [9]. Watanabe et.al. (2005) proposed a reaction mechanism based on the decarboxylation experiment of stearic acid (C17H35COOH) using batch reactor with supercritical water (SCW) of 400°C and 0.17 g/cm^3 for 30 min. Furthermore, it was found that the addition of alkali hydroxide (NaOH and KOH) in the SCW reaction could enhance the decarboxylation of C17-acid with the main products being CO_2 and C_{17} alkane [10]. Kolbe electrolysis of potassium salts of coconut fatty acids and acetic acid using platinum electrodes has produced a clear liquid product (58% theoretical yield) with very good diesel-fuel properties through the studies of

Sumera et.al. (1990). The analysis of the liquid product showed that it contained 83% straight chain hydrocarbons[5].

Grey Relational Grade was used by Deng (1989) in Grey System to measure the level of approximation among sequences[11]. Recently, the optimization of process parameters for multi responses using GRA have been investigated. Taguchi method is commonly used instead of full factorial DOE in performing experiment. Kalsi et. al, (2013) used Taguchi based GRA for optimizing in turning process [12]. There are some researches that comparing GRA and other method for optimizing multi responses. Johny et. al, (2016) compared Taguchi based GRA and Taguchi based PCA as an optimization method [13]. Jeyapaul et al. (2005) compared engineering judgement, regression analysis, Principal Component Analysis (PCA), and Grey Relational Analysis (GRA) methods on optimizing multi responses experiment [7].

3 Methods

3.1 Experiment

Crude Palm Oil (CPO) was first hydrolysed at 120°C using Sulfuric Acid as the catalyst with excess water for 2 hours. The volume ratio between water and CPO was 16:1. Products obtained in two layers after settling for 3 hours which upper layer was FFA and the lower layer was glycerol with water. FFA then separated to be employed as feed in the decarboxylation. 12 L of FFA was flowed into a reactor with 600 ml CH3COOH as the catalyst in the concentration range 0,5-1,5 M. Electrolysis was occured in the voltage range 10-25 volt at 130°C. Two electrodes were employed i.e. silver (Ag) and copper (Cu). This decarboxylation with the aid of electrolysis took 2 hours to produce hydrocarbon chain.

3.2 Optimization

There are factors that influenced Free Fatty Acid (FFA) decarboxylation process into hydrocarbon chain such as temperature, pressure, voltage and electrolyte concentration [5]. Two control factors were determined based on screening experiments, namely voltage and electrolyte concentration (acetic acid). Each parameter was designed to have different level, three levels of acetic acid concentration and four levels for voltage which are shown in Table 1.

Table 1. Experiment level factor

Factor			Level			
Control factor	Coded	Uncoded	1	2	3	4
	A	Acetic acid concentration	0,5M	1M	1,5M	
	B	Voltage	10V	15V	20V	25V

According to factorial design, an experimental design with two factors with three and four levels required twelve experiments. In this process, each experiment was conducted with three repetitions (M1, M2, M3). Replication was used to measure of variance that arises. Sequence number of experiments was employed from the Minitab as shown in Table 2. % Yield, NHV, spgr, %API, and viscosity, were measured as quality characteristics from each experiment. Response variables data from each experiment are summarized in Table 3.

DOI: 10.1051/matecconf/20179701114

Table 2. Sequence numbers of experiments

Sequence Number	Control factor		Repetitions		
	A	B	M1	M2	M3
1	1	1	8	13	24
2	1	2	10	15	16
3	1	3	22	25	26
4	1	4	6	31	32
5	2	1	2	12	33
6	2	2	5	9	28
7	2	3	17	20	35
8	2	4	4	7	29
9	3	1	3	14	27
10	3	2	11	19	23
11	3	3	1	18	30
12	3	4	21	34	36

Table 3. Experiment's result

Seq	% Yield			spgr			°API			Viscosity			NHV		
	M1	M2	M3	M1	M2	M3	M1	M2	M3	M1	M2	M3	M1	M2	M3
1	47	48	48	0.95	0.97	0.96	16	16.5	16.3	0.26	0.25	0.26	17774.4	17780.1	17770.0
2	49.8	50	51	0.97	0.965	0.97	18.3	18.6	18.9	0.21	0.2	0.23	17750.0	17751.3	17751.3
3	57.2	58	58	0.96	0.96	0.96	23	23.5	23.2	0.17	0.18	0.17	17776.8	17776.5	17776.3
4	56.8	57	58	0.87	0.86	0.86	22	22.6	23	0.23	0.26	0.27	18282.0	18282.3	18282.0
5	67.9	68	69	0.97	0.97	0.96	23.5	23.3	23.6	0.92	0.93	0.93	17740.5	17740.0	17740.1
6	66.5	66	67	0.86	0.86	0.87	33	32.4	32.6	1.14	1.17	1.18	18303.6	18304.0	18303.0
7	67.9	68	68	0.86	0.87	0.87	32.8	33.1	33.2	1.28	1.25	1.27	18305.2	18304.9	18305.0
8	73	74	74	0.86	0.85	0.87	32.7	32.8	32.6	1.22	1.25	1.27	18317.1	18317.1	18317.0
9	81	80	83	0.92	0.9	0.93	19	19.3	18.9	1.75	1.73	1.76	18041.5	18045.0	18041.0
10	82.7	83	83	0.88	0.87	0.88	28	28.3	28.4	1.8	1.76	1.81	18205.3	18205.2	18206.0
11	85	86	85	0.885	0.87	0.89	28.3	28.4	28.3	1.09	1.08	1.08	18204.2	18204.1	18204.5
12	84.5	85	85	0.86	0.86	0.88	26	27	26.6	1.01	1.02	1	18298.0	18297.8	18297.6

The GRA is principally measure relational analysis. This approach is suitable for process with incomplete information or uncertain model. The result of the relational analysis is a set of discreet number. Normalization of experimental data, range within 0 to 1, was conducted and called grey relational generating. Quality characteristic can be divided into three criteria, 'larger the better', 'smaller the better', and 'nominal the best'. Criteria of quality characteristics for Free Fatty Acid (FFA) decarboxylation process into hydrocarbon chain are 'larger thebetter' for % Yield, °API, NHV, and 'nominal the best' for spgr, viscosity. 'larger-the-better' responses were normalized using equation (1) and 'nominal the best' were normalized using equation (2)

$$x_i^*(k) = \frac{x_i^o(k) - \min x_i^o(k)}{\max x_i^o(k) - \min x_i^o} \quad (1)$$

$$x_i^*(k) = 1 - \frac{|x_i^o(k) - x_0^o(k)|}{max(\max x_i^o(k) - x_0^o(k), x_0^o(k) - \min x_i^o(k))} \quad (2)$$

where $x_i^*(k)$is the normalized value, $x_i^o(k)$ is mean value of each response, $x_0^o(k)$ is target value, and $max\ x_i^o(k)$is the largest mean response,$min\ x_i^o(k)$is the smallest mean response, i = 1,2,3,...,12. All sequences after application the data preprocessing through Eq. (1) and (2) are shown in Table 4.

Table 4. Normalized value

Running	% Yield	Spgr	°API	Viscosity	NHV
1	0.0000	0.0847	0.0000	0.0570	0.0600
2	0.0658	0.0000	0.1392	0.0274	0.0185
3	0.2633	0.0847	0.4155	0.0000	0.0630
4	0.2536	0.9322	0.3738	0.0547	0.9394
5	0.5463	0.0169	0.4294	0.5154	0.0000
6	0.5009	0.9322	0.9781	0.6773	0.9765
7	0.5365	0.9661	1.0000	0.7480	0.9791
8	0.6895	0.8983	0.9801	0.7343	1.0000
9	0.8977	0.5254	0.1670	0.9236	0.5240
10	0.9422	0.9322	0.7137	0.8940	0.8066
11	1.0000	0.8814	0.7197	0.6226	0.8045
12	0.9884	0.9661	0.6123	0.5724	0.9666

The larger normalized results should be expected, therefore; best normalized result should be equal to one. The relation between the best (ideal) and each normalized value can be shown by grey relational coefficient (GRC). GRC wascalculated using the following equation:

$$\xi_i(k) = \frac{\Delta\min + \xi\Delta\max}{\Delta_{0i}(k) + \xi\Delta\max} \quad (3)$$

where, $\Delta_{0i}(k)$is an absolute value of the difference between$x_0^*(k)$ and$x_i^*(k)$,$\Delta_{0i}(k) = |x_0^*(k) - x_i^*(k)|$, Δmin is the minimum value of the absolute differences of all comparing sequences. Δmax is the maximum value of the absolute differences of all comparing sequences. ξ is the identification coefficient and its value is between 0 and 1. This research used 0.5 for coefficient. Grey relational grade (GRG) was calculated to express level of correlation and an average of the GRC. It was calculated using the following equation:

$$\bar{y_j} = \frac{1}{k}\sum_{i=1}^{m} \xi_i(k) \quad (4)$$

All sequences after application the data preprocessing for coefficients and grades using Equation (3) and (4) are listed in Table 5.

4 Result and discussion

Table 5 demonstrates that the parameters setting of sequence seven (A2 B3) has the highest GRG. In addition, the average of the GRG for each level of the parameters setting is summarized and given in Table 6. Furthermore, the total average of the GRG for the twelve sequences experiment was calculated to be 0.6359 as shown in Table 6.

Table 5. GRC and GRG

Sequence	GRC					GRG
	% Yield	**spgr**	**°API**	**Viscosity**	**NHV**	
1	0.3333	0.3540	0.3333	0.3476	0.3472	0.3431
2	0.3486	0.3333	0.3674	0.3401	0.3375	0.3454
3	0.4043	0.3540	0.4610	0.3333	0.3479	0.3801
4	0.4011	0.9344	0.4440	0.3470	0.8919	0.6037
5	0.5243	0.3373	0.4670	0.5308	0.3333	0.4385
6	0.5004	0.9344	0.9581	0.6522	0.9552	0.8001
7	0.5189	1.0000	1.0000	0.7245	0.9600	**0.8407**
8	0.6169	0.8769	0.9618	0.7093	1.0000	0.8330
9	0.8301	0.5229	0.3751	1.0000	0.5123	0.6481
10	0.8963	0.9344	0.6359	0.9397	0.7211	0.8255
11	1.0000	0.8507	0.6408	0.6054	0.7189	0.7632
12	0.9774	1.0000	0.5633	0.5680	0.9374	0.8092

Table 6. Response table for GRG.

Factor		Level				Delta	Rank
Coded	Uncoded	1	2	3	4		
A	Acetic acid concentration	0.4181	0.7281	0.7615		0.3434	1
B	Voltage	0.4766	0.6570	0.6613	0.7486	0.2720	2

DOE based GRA can provide optimal parameters for multi responses of quality characteristics. Additionally, analysis of variance (ANOVA) on the grey relational grade was conducted to determine the effect of process parameters andidentifythe significant factors. ANOVA analysis was performed with a 95% confidence level and 5% significance level. As listed in Table 7, ANOVA analysis was carried out to see the effects of the experimental parameters on the result. The results of the ANOVA in Table 7 indicates that voltage has no statistically significance on the GRG values.

Table 7. Anova for GRG

Control Factors	Degree of freedom	Sum Square	Adj SS	Adj MSE	F	P_{value}
A	2	0,28684	0,28684	0,14342	14,06	0,005
B	3	0,11754	0,11754	0,03918	3.84	0,076
Error	6	0,06119	0,06119	0,01020		
Total	11	0,46557				

Considering that voltage has no statistically significance on the GRG, regression model is built in software Minitab. Voltage will give a significance effect if its level difference more than 15 Volt [10]. Regression is conducted to analyze between output parameter (GRG) and one input parameter (acetic acid concentration). As shown by Fig 1, regression model fits into the following equation:

$$GRG = -0{,}1685 + 0.7248A - 0.1383A^2 \quad (5)$$

The optimal level for acetic acid concentration (A) based on polynomial regression model is found to be at level 2.62 or 1.325 M.

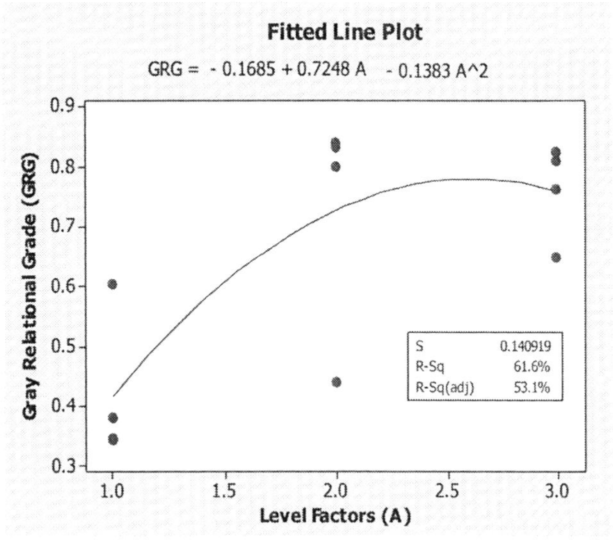

Fig. 1. Regression level factors on GRG

5 Conclusions

This paper focuses on application of DOE based GRA to optimize the parameters Free Fatty Acid (FFA) decarboxylation process into hydrocarbon chain such as acetic acid concentration and voltage. Decarboxylation process has multiple quality characteristics, % Yield, NHV, spgr, %API, and viscosity. The application of DOE based GRA directly combined the multiple responses into a single performance characteristic called grey relational grade (GRG). The multi responses optimization results demonstrated that the parameters setting of sequence seven (A2 B3) has the highest GRG, combination of 1 M values acetic acid concentration and 20 V voltage. They were the recommended levels of the controllable parameters concurrently considered. Analysis of variance (ANOVA) was utilized to determine the effect of factors on GRG. ANOVA results demonstrated that voltage had no statistically significance on the GRG values. Polynomial regression model was developed to know the optimal parameter for acetic acid concentration which had statistically significance on the GRG values. It was found out to be 1.325 M.

This work was partially supported by Polytechnic STMI Jakarta, Ministry of Industry, Republic of Indonesia.

References

1. L.P. Christopher, A. Kumar, V.P. Zambare, Appl. Energy, **119**, 497 (2014)
2. M. Aghbashlo, A. Demirbas, Biofuel Research Journal, **10,** 379 (2016)
3. J. Van Gerpen, B. Shanks, R. Pruszko, D. Clements, G. Knothe, *Biodiesel Production Technology, Golden,* (National Renewable Energy Laboratory, USA, 2004)
4. S. K. Hoekman, B. Amber, R. Curtis, C. Eric, N. Mani, Renew. Sustainable Energy Rev., **16**, 143 (2012)
5. F.C. Sumera, S. Sadain, Philipp. J. Sci., **119(4),** 333 (1990)
6. I. Soejanto, Desain Eksperimen Dengan Metode Taguchi, (Graha Ilmu, Yogyakarta, 2009)
7. R. Jeyapaul, P. Shahabudeen, K. Krishniah, Int. J. Adv. Manuf. Tech., **26**, 1331 (2005)

8. Y.M. Chiang, H.H. Hsien, Computer and Industrial Engineering Journal, **56(2)**, 648 (2009)
9. Y. Takemura, A. Nakamura, H. Taguchi, Ind. Eng. Chem. Prod. Res. Dev., **24(2),** 213 (1985)
10. M. Watanabe, T. Iida, H. Inomata, Energ, Convers, Manage., **47**, 3344 (2016)
11. J.L. Deng, J. Grey. Syst., **1**, 1 (1989)
12. N.S. Kalsi, R. Sehgal, V.S. Sharma, International Journal of Strategic Decision Sciences, **4(2),** 79 (2013)
13. S.M. Johny, C.R. Sri Sai, V.R. Rao, B.G. Singh, Int. Res. J. Eng. Tech., **3**, 680 (2016)

Drough tolerance estimation of several rainfed rice (*Oryza sativa* L.) of lokal veriety using *polyethilene glycol* (PEG)

Susanti Diana[1,*], Novriani[1], Firnawati Sakalena[1], Dora Fatma Nurshanti[1], and Supli Effendi Rahim[2]

[1]Agrotechnology Study Program, Faculty of Agriculture, Baturaja University Jl. Ratu Penghulu Karang Sari No.20301, OKU Sumatera Selatan Indonesia
[2]Agrothecnolgy Faculty of Agriculture, University of Palembang, Palembang

Abstract. The yield of rice crop was highly affected by water availability and drought condition . The effort was done to seek local varieties of rainfed rice which tend to be tolerance to drought condition. This research objective was to seek some local varieties of rainfed rice that tolerance to several drought conditions. This research was conducted at Farm Exprimentation Site of Agrotechnology Study Program, Baturaja University from November 2015 to April 2016. Exprimental design used in this research was Split Plot Design with three replications. The main plot was variety treatments (varieties of Henik, Henik Pisang, Serenik, Remaja, Semester, Selanggar, Duku, Serendah Putih, Agai and Tinta), whereas the subplot was *Polyethilene glycol* (PEG) treatments (concentrations of 0%, 20%, 25% dan 30%). The results showed that varieties of Henik, Henik Pisang, Serenik, Semester, Duku, Serendah Putih and Tinta had produced high yield at drought condition. Rainfed rice could produced high yield at treatments of 20% and 25% PEG, respectively. Treatment combination of Duku variety and 30% PEG had produced the highest weight of unhulled rice/clump and fully unhulled rice/clump, respectively. The highest weight of unhulled rice per 1000 grains was found on treatment combination of Duku variety and 25% PEG.

1 Introduction

Food increment need to be done in accord to the increase of population numbers. The effort to increase food production was done through dry land empowerment. According to Pusdatin (2013), dry land area in Indonesia that was not yet utilized was 14,262,383 ha. This dry land resource has potential to support food self sufficiency. Rainfed rice

* Corresponding author: susa12j@yahoo.com

DOI: 10.1051/matecconf/20179701115

cultivation can be applied at dry land area. Ogan Komering Ulu District of South Sumatra Province had extensive dry land area and local varieties of rainfed rice that have potential to be developed in order to increase food self sufficiency. Danial and Nurbani (2015) had stated that rainfed rice development had received less attention from the government than that of paddy field rice. Productivity of rainfed rice is lower than that of paddy field rice. Yuniarti (2015) had stated that one factor which cause low productivity of rainfed rice is drought condition. According to Rahayu and Harjono (2010), rainfed rice crop that tolerance to drought would have high prolin content in its leaves. The crop height and tiller numbers of rainfed rice will decrease at drought condition.

The growth and yield of rice crop are highly affected by water availability. Unfavourable climate condition due to climatic change has high impact on crop yield or production. Short wet season period coupled with long dry season will results in crop management difficulty due to water availability which solely available from rainfall water. The impact of El Nino phenomenon in 2015 was extreme drought condition and long dry season.

Drought causes accmulation of absisate acid, decrease in transpiration and decrease of CO_2 assimilation (Razak *et al.*, 2013). Drought condition will decrease rice yield level (Quampah *et al.*, 2011). Drought tolerance characteristics of a rice genotype is always related to the changes of morphology and physiology as adaptation way toward drought condition so that a rice genotype can be classified as drought tolerance. Crop characteristics either morphology or physiology can be used as assessment base of tolerance properties toward drought condition. One of substance that can be used to select drought tolerance rice crop is PEG (*Polyethylene Glycol*). PEG concentration of 18.1% or equivalent to – 6.24 bar of water potential can be used to determine the tolerance of variety to drought stress (Halimursyadah *et al.*, 2013). PEG 6000 solution at 25% concentration is relatively effective concentration level to produce drought stress on rice crop. The used of PEG 6000 solution at 25% concentration at seedling phase can detect hybrid rice genotype which tolerance to drough condition (Afa *et al.*, 2012).

This research was conducted to analyze the growth and yield of local varieties of rainfed rice originated from Ogan Komering Ulu District, South Sumatra Province, Indonesia at drought condition in order to determine the adaptive local varieties of rainfed rice at drought condition.

2 Material and Method

This research was conducted at Farm Exprimentation Site of Agrotechnology Study Program, Agricultural Faculty, Baturaja University from November 2015 to April 2016. Exprimental design used in this research was Split Plot Design with three replications. Variety treatments as the main plot were consisted of 10 varieties as follows: V_1: Henik, V_2: Henik Pisang, V_3: Serenik, V_4: Remaja, V_5: Semester, V_6: Selanggar, V_7: Duku, V_8: Serendah Putih, V_9: Agai and V_{10}: Tinta). *Polyethylene glycol* (PEG) concentrations as the sub-plot were consisted of four levels as follows: G_1:0%, G_2: 20%, G_3:25% and G_4:30%.

The first activity was planting media preparation in which yellowish red podsolic soil (PMK) was refined and sieved followed by putting it into polybag with size of 5 kg. Soil was then sprinkled with water until field capacity condition and subsequently was left over for one week. The seeds were soaked in beaker glass according to teatments for 12 hours and then were germinated within germination box. The next step was that seeds having 14 days old were transplanted into the previously prepared polybags in which each polybag was contained three stalks of seed. Crop maintenance was consisted of weeds cultivating, watering was done once per day and rice bug pest control was conducted by using demolish spraying. The observed parameters were crop height, crown dry weight, root dry weight, tiller emerging time, maximum tiller numbers, productive tiller numbers, flowering time, unhulled rice weight/clump, fully unhulled rice weight/clump and weight of 1000 grains unhulled rice. Data analysis was done by using Analysis of Variance with statistical package of SAS 6.12. If analysis results were significantly different, then it was followed by Duncan test using 5% level.

3 Results and Discussion

3.1 Results

Results of F-test (Table 1) showed that variety treatments had significant effect on parameters of crop height, crown dry weight, root dry weight, maximum tiller numbers, productive tiller numbers, flowering time, unhulled rice weight/clump, fully unhulled rice weight/clump and weight of 1000 grains unhulled rice. Variety treatments had no significant effect on tiller emerging time. The PEG treatments had significant effect on parameters of crown dry weight, productive tiller numbers and weight of 1000 grains unhulled rice, but had no siginificant effect on other parameters. Treatments interaction had significant effect on parameters of root dry weight, productive tiller numbers and weight of 1000 grains unhulled rice, but PEG treatments had no significant effect on other parameters.

3.2 Growth Parameter

The results of Duncan test showed that crop height of Henik variety (V_1) was not significantly different than that of V_2, V_4, V_7 and V_8. Crown dry weight of V_1 was not significantly different than that of V_3, V_7 and V_{10}. Root dry weight of V_1 was not significantly different than that of V_4 and V_8. Faster emerging time of tillers were found on treatments of V_1, V_3, V_4, V_6 and V_9. Maximum tiller numbers and productive tiller numbers were found on treatments of V_6. Tiller emerging time for V_3 treatment was faster than that of other varieties (Table 2). The comparison of PEG treatments and without PEG treatment (G_0) showed that G_3 treatment had taller crop height, tiller emerging time was not significantly different than that of G_1 and G_2 treatments as well as had faster flowering time. The G_1 treatment showed better results in increasing of crown dry weight, root dry weight, maximum tiller numbers and productive tiller numbers (Figure 1 to Figure 3). Results of treatment combination of variety and PEG showed that the highest values of crop height, crown dry weight and root dry weight were found on $V_{10}G_3$ treatment combination. The fastest tiller emerging time was found on V_9G_1 treatment combination. The highest values of maximum tiller numbers and productive tiller numbers were found on V_6G_3 treatment combination than that of other treatment combinations (Table 3).

DOI: 10.1051/matecconf/20179701115

3.3 Yield Parameter

Faster flowering time was found on V_6 treatment. Treatment of V_1 had unhulled rice weight/clump that was not significantly different than that of V_2, V_3, V_4, V_5, V_7, V_8 and V_9. Fully unhulled rice weight for V_7 treatment was not significantly different than that of V_1, V_3, V_5 and V_8. The 1000 grains weight of unhulled rice for V_1 treatment was not significantly different than that of V_3, V_7 and V_{10} (Table 2). The G_1 treatment had produced the highest value of 1000 grains weight of unhulled rice. The weight of unhulled rice/clump and fully unhulled rice/clump for G_2 treatment were higher than that of other PEG treatments (Figure 4 and Figure 5). Treatment combination of V_6G_2 had produced faster flowering stage. Treatment combination of V_7G_3 had higher values of unhulled rice weight/clump and fully unhulled rice weight/clump than that of other treatment combinations. The highest value of 1000 grains weight of unhulled rice was found on V_7G_2 treatment combination (Table 4).

Table 1. Variance analysis (F test) of variety treatments (V) and PEG (G)on all variables.

No	Variable	F value V	Pr > F V	F value G	Pr > F G	F value VxG	Pr >F VxG
1	Crop height (cm)	39.38*	0.0001	1.69^{ns}	0.1778	1.00^{ns}	04828
2	Crown dry weight (g)	2.45*	0.0501	5.83*	0.0015	2.34^{ns}	0.0032
3	Root dry weight (g)	7.26*	0.0002	1.86^{ns}	0.1454	1.76*	0.0351
4	Tiller emerging time (dap)	2.14^{ns}	0.0804	1.88^{ns}	0.1426	1.44^{ns}	0.1212
5	Maximum tiller numbers (stalk)	24.65*	0.0001	2.58^{ns}	0.0622	0.98^{ns}	0.5072
6	Productive tiller numbers (stalk)	13.68*	0.0001	9.23*	0.0001	3,17*	0.0001
7	Flowering time (dap)	2.96*	0.0238	1.77^{ns}	0.1626	1.07^{ns}	0.3997
8	Unhulled rice weight per clump (g)	3.99*	0.0060	0.56^{ns}	0.6466	1.37^{ns}	0.1541
9	Fully unhulled rice weight per clump (g)	4.40*	0.0036	0.40^{ns}	0.7510	1.37^{ns}	0.1560
10.	Weight of 1000 grains (g)	2.45*	0.0501	5.83*	0.0015	2.34*	0.0032

Remarks: *= significant; ns = not significant.

Table 2. Average results of variety (G) as the main plot for all variables

No	Variable	Variety										Dun-can Test
		V_1	V_2	V_3	V_4	V_5	V_6	V_7	V_8	V_9	V_{10}	
1	Crop height (cm)	104,2 Ab	104 ab	98.1 bc	110.7 a	99.6 b	63.9 d	108.6 a	109.7 a	91.9 c	100.9 b	57.1
2	Crown dry weight (g)	96.20 a	68.5 b	78.0 ab	75.2 b	61.2 b	62.8 b	77.58 ab	68.83 b	69.92 b	77.75 ab	487.9
3	Root dry weight (g)	31.3 A	23.08 bc	21.58 bc	34.08 a	24.08 b	19.17 bc	24.4 b	31.7 a	21.5 bc	16.9 c	54.1
4	Tiller emerging time (DAP)	17.6	18.5	17.2	17.7	18.0	17.3	18.4	18.3	17.4	19.2	-
5	Maximum tiller numbers (stalk)	11.74 d	13.4 cd	13.94 cd	13.0 cd	13.9 cd	26.4 a	12.1 d	12.1 d	20.7 b	15.5 c	10.7
6	Productive tiller numbers (stalk)	9.02 cde	8.57 cde	11.27 cb	8.08 de	11.8 b	11.2 a	7.4 e	8.6 cde	10.6 bcd	11.2 cb	8.5
7	Flowering time (dap)	95.09 bc	111.94 ab	99.73 bc	101.41 bc	93.18 bc	85.95 c	94.64 bc	94.38 bc	116.09 ab	134.59 a	838.6
8	Unhulled rice weight per clump (g)	37.9 a	26.9 abc	38.4 a	34.9 ab	36.1 ab	16.0 c	39.9 a	37.9 a	37.4 ab	23.4 bc	206.1
9	Fully unhulled rice weight per clump (g)	35.18 ab	23.99 bcd	34.74 ab	31.16 abc	32,75 ab	14.29 d	37.65 a	35.58 ab	31.50 abc	20.55 d	157.9
10	Weight of 1000 grains (g)	96.2 a	68.5 b	78.0 ab	75.2 b	61.7 b	62.8 b	77.6 ab	68.8 b	69.9 b	77.7 ab	487.9

Remarks: Numbers followed by the same letters are not significantly different.
V_1: Henik, V_2: Henik Putih, V_3: Serenik, V_4: Remaja, V_5: Semester, V_6: Selanggar V_7: Duku, V_8: Serendah Putih, V_9: Agai and V_{10}: Tinta. DAP = Days After Planting.

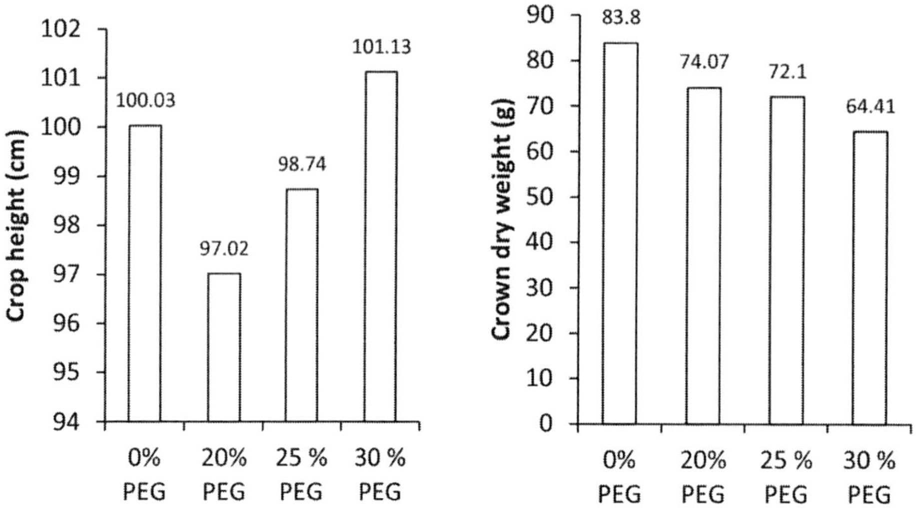

Fig. 1. Average crop height and crown dry weight at PEG treatment.

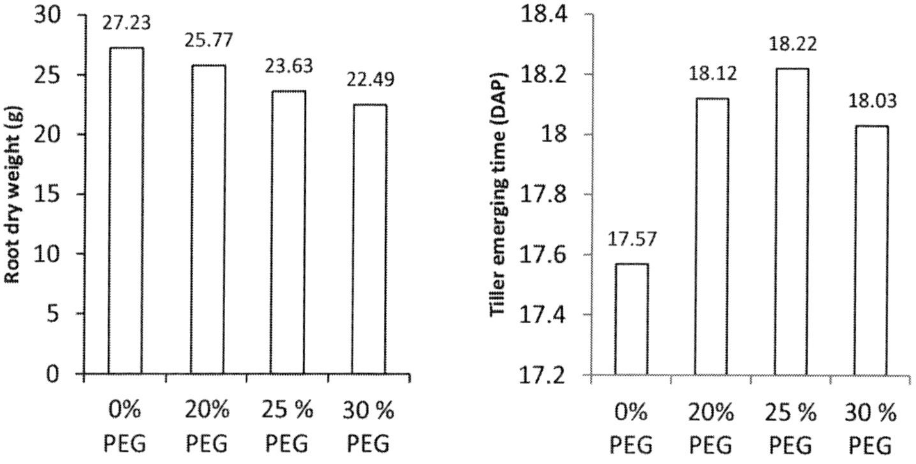

Fig. 2. Average of root dry weight (g) and tiller emerging time (DAP) at PEG treatment.

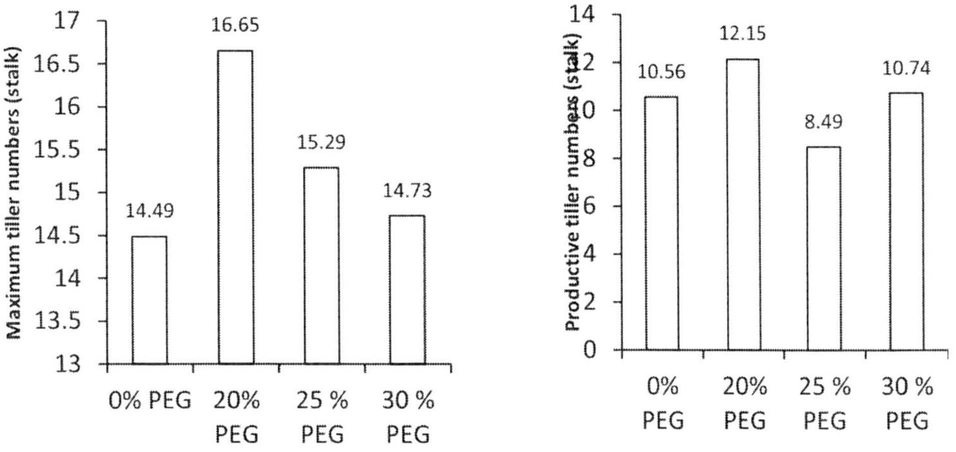

Fig. 3. Average of maximum tiller numbers (stalks) and productive tiller numbers (stalks) at PEG treatment.

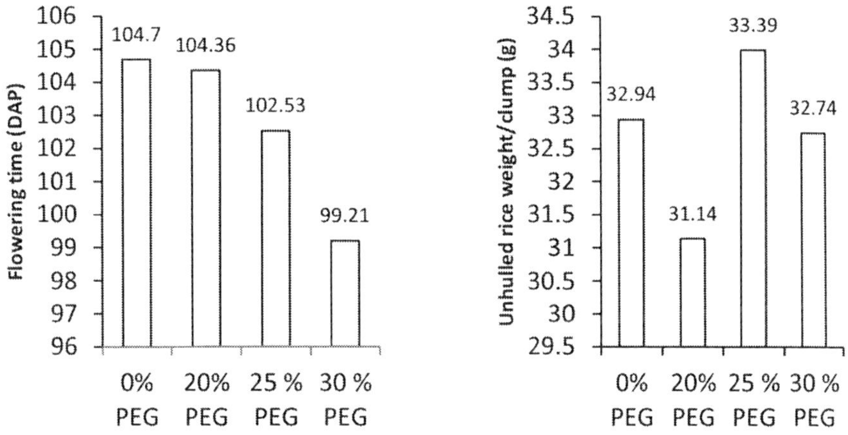

Fig. 4. Average flowering time (DAP) and unhulled rice weight/clump (g) at PEG treatment

DOI: 10.1051/matecconf/20179701115

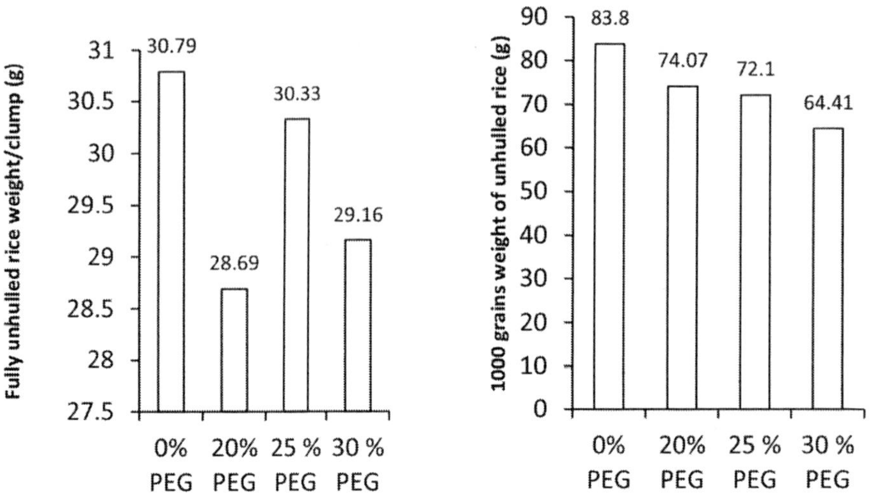

Fig. 5. Average weight of fully unhulled rice/clump (g) and 1000 grains weight of unhulled rice (g) at PEG treatment

Table 3. Average growth and yield of several local varieties of rainfed rice (V) at PEG treatment (G).

Variety	PEG	Crop height (cm)	Crown dry weight (g)	Root dry weight (g)	Tiller emerging time (DAP)	Maximum tiller numbers (stalk)
V_1	G_0	104.67	34.89	11.63	15.3	10.77
V_1	G_1	101.63	33.88	11.29	18.33	11.9
V_1	G_2	101.53	33.84	11.28	18.87	13.2
V_1	G_3	108.9	36.3	12.1	17.97	11.1
V_2	G_0	102.3	34.1	11.37	18.43	12.9
V_2	G_1	99.8	33.27	11.09	19.43	13.9
V_2	G_2	106	35.33	11.78	18.23	12.23
V_2	G_3	110.33	36.78	12.26	18.1	14.87
V_3	G_0	104.13	34.71	11.57	16.97	14.87
V_3	G_1	93.8	31.27	10.42	17.67	12.33
V_3	G_2	100.23	33.41	11.17	16.43	16.13
V_3	G_3	94.33	31.44	10.48	17.67	12.43
V_4	G_0	112.7	37.57	12.52	17.67	14.33
V_4	G_1	111.37	37.12	12.37	17.57	14
V_4	G_2	108.23	36.08	12.02	18.1	11.47
V_4	G_3	110.43	36.81	12.27	17.57	12.53
V_5	G_0	99.57	33.19	11.06	17.67	13.3
V_5	G_1	99.233	33.08	11.02	18.43	14.8

V_5	G_2	101.1	33.7	11.23	18.23	13.9
V_5	G_3	98.57	32.86	10.95	18	13.7
V_6	G_0	64.37	21.46	7.15	17.23	22.1
V_6	G_1	60	20	6.66	17	27.9
V_6	G_2	61.23	20.411	6.80	17.23	25.47
V_6	G_3	70.1	23.37	7.79	18	30.23
V_7	G_0	113.7	37.9	12.63	17.43	12.57
V_7	G_1	106.1	35.37	11.79	18	13.47
V_7	G_2	106.2	35.4	11.8	18.77	11.47
V_7	G_3	108.53	36.18	12.06	19.33	11
V_8	G_0	113.67	37.89	12.63	18.57	10.9
V_8	G_1	109	36.33	12.11	18.8	14.57
V_8	G_2	111.23	37.08	12.36	17.83	10.33
V_8	G_3	104.87	34.96	11.65	18	12.43
V_9	G_0	94.47	31.49	10.49	17	18.77
V_9	G_1	89.87	29.96	9.98	16.13	25.67
V_9	G_2	91.03	30.34	10.11	18.97	20.03
V_9	G_3	92.2	30.73	10.24	17.57	18.2
V_{10}	G_0	90.77	30.26	10.09	19.43	14.37
V_{10}	G_1	99.433	33.14	11.05	19.87	18
V_{10}	G_2	100.57	33.52	11.17	19.53	13.1
V_{10}	G_3	113	37.67	12.56	18.1	16.43

Table 4. Average growth and yield of several local varieties of rainfed rice (V) at PEG treatment (G).

Variety	PEG	Productive tiller numbers (stalk)	Flowering time (DAP)	Unhulled rice weight/clump (g)	Fullly unhulled rice weight/clump (g)	Weight of 1000 grains (g)
V_1	G_0	8.67	92.33	36.1	33.77	27.32
V_1	G_1	11.37	98.43	42.23	40.6	25.02
V_1	G_2	6.13	98.23	42.63	40.27	25.85
V_1	G_3	9.9	91.37	30.83	26.07	26.02
V_2	G_0	7	115.23	25.23	24	20.27
V_2	G_1	10.9	116.53	24.43	21.6	22.91
V_2	G_2	10.23	106.57	31.6	24.67	22.74
V_2	G_3	6.13	109.43	26.7	25.7	24.66
V_3	G_0	14.7	101.1	42.8	40.03	26.06

 DOI: 10.1051/matecconf/20179701115

V_3	G_1	9.2	102.57	29.07	25.3	23.91
V_3	G_2	11.2	98.9	44.13	43.73	26.73
V_3	G_3	9.97	96.33	37.77	29.9	24.92
V_4	G_0	8.23	100.53	36.87	34.57	25.69
V_4	G_1	9.13	100.1	29.93	27.57	24.35
V_4	G_2	9.2	103.53	39.63	29.67	24.29
V_4	G_3	5.77	101.47	33.03	32.83	23.49
V_5	G_0	12.43	94.13	32.77	30.93	25.74
V_5	G_1	14.57	91.67	42.1	38.27	25.36
V_5	G_2	10.23	92.57	38.83	35.2	25.31
V_5	G_3	10.3	94.33	30.63	26.6	26.24
V_6	G_0	18.63	87.33	30.33	18.43	23.38
V_6	G_1	17.87	86.33	15.7	13.97	21.52
V_6	G_2	16.2	86.47	11.37	9.63	21.05
V_6	G_3	20.3	83.67	16.9	15.13	21
V_7	G_0	5.87	91.77	34.77	33.37	28.24
V_7	G_1	11.13	94.57	37.83	34.5	25.42
V_7	G_2	4.57	94.67	37.63	35.23	27.4
V_7	G_3	8	97.57	49.3	47.5	25.69
V_8	G_0	8.87	98.33	47.67	44.57	28.67
V_8	G_1	9.67	96.2	38.37	37.1	26.75
V_8	G_2	7.97	88.87	30.7	28.07	27.98
V_8	G_3	7.9	94.1	35.37	32.6	27.71
V_9	G_0	8.9	115.67	31.57	28.8	22.78
V_9	G_1	15.2	116.1	30.47	29.77	26
V_9	G_2	1	117.03	32.63	29.63	23.04
V_9	G_3	17.3	115.58	46.88	37.8	25.26
V_{10}	G_0	12.3	150.57	21.6	19.4	23.34
V_{10}	G_1	12.4	141.13	21.27	18.2	22.11
V_{10}	G_2	8.17	138.43	30.8	27.17	20.77
V_{10}	G_3	11.8	108.23	19.97	17.43	23.46

4 Discussion

The results of F-test showed that all varieties used in this study had significant effect on all parameters except for tiller emerging time parameter. This showed that all varieties used in this study had different responses. These different responses were estimated due to

 DOI: 10.1051/matecconf/20179701115

differences in genetic characteristics of each variety. According to Farooq *et al.* (2009), drought tolerance varieties would produce differences in genetic characteristics althouh they are cultivated in the same location. These differences were due to genetic control from each variety.

The PEG treatment had significant effect on crown dry weight, productive tiller numbers and 1000 grains weight of unhulled rice. The effect of PEG on growth and yield was related to crown dry weight in which crown dry weight is the accumulation of photosyntate yield. Interaction of varieties and PEG had significant effect on root dry weight, productive tiller numbers and 1000 grains weight of unhulled rice. The PEG treament had produced drought stress condition in which roots can not absorb more water. Water absorbtion has effect on photosynthesis process of crop. The photosynthesis yield is used for productive tillers development and is accumulated in seeds. Biomass of crop is highly affected by water availability and photosynthesis yield. Ahadiyat (2011) had stated that tolerance of rainfed rice crop at drought stress condition is affected by physiological capability in controlling water distribution during its growth which is tend to be controlled by genetic characteristics. This is in accordance to the statement from Zain et al. (2014) which showed that rice variety that capable to use water efficiently would be more tolerance to water stress condition.

Results of further test by using Duncan test showed that V_1 (Henik), V_3 (Serenik), V_5 (Semester), V_6 (Selanggar), V_7 (Duku) and V_{10} (Tinta) were varieties that capable to increase growth and to produce high yield at PEG treatments of G_1 (20% PEG) and G_2 (25% PEG). Treatment combinations that had produced high yield were V_7G_2 (Duku +25 % PEG) and V_7G_3 (Duku + 30% PEG). There was six varieties that gave high yield at drought stress condition out of ten varieties tested in this study. These varieties had different characteristics and can be adaptive to water shortage environment. Adaptive capability from each variety is controlled by internal properties. The adaptive and tolerance variety is the one that capable to increase water use efficiency.

Treatment of 20% PEG was capable to increase crown dry weight, root dry weight, maximum tiller numbers, productive tiller numbers and 1000 grains weight of unhulled rice. The highest weight values of unhulled rice/clump and fully unhulled rice/clump were found on treatment of 25% PEG. This was estimated that rainfed rice at drought condition could still absorb water which was subsequently used for growth and production, whereas rice crop no longer could absorb water at 30% PEG condition. According to Afa *et al.* (2012), the higher the PEG concentration results in higher decrease of water potential so that water can not be absorbed by crop. The 1000 grains weight of unhulled rice for 20% PEG treatment was higher than that of 25% PEG treatment which was estimated due to water availability factor. Water availability is highly affected by soil temperature and soil water content. According to Ji et al. (2012), water is the main factor that affect rice production. Water shortage affects stomata closure and subsequently impede photosynthesis process.

Treatment combinations of variety and PEG which capable to increase rice growth was found on $V_{10}G_3$, V_9G_1, V_6G_2 and V_6G_3. Treatment combinations of V_7G_2 and V_7G_3 were superior in increasing rice production than that of V_9G_3. The capability of V_6 (Selanggar), V_7 (Duku), V_9 (Agai) and V_{10} (Tinta) varieties in increasing average values of growth and production parameters were lower than that of control treatment (G_0: without addition of PEG). Selanggar, Agai and Tinta varieties had capability in absorbing water and they could only used water to increase the growth, whereas Duku variety was more efficient in utilizing water and this water could be used to increase rice production with average value higher than that of control treatment. The weight of unhulled rice/clump for V_7G_0 was 33.37 g and for V_7G_3 was 47.50 g, respectively. The weight of fully unhulled rice/clump for V_7G_0 was 34.77 g and for V_7G_3 was 49.30 g, respectively. Agai variety also had higher average values in term of weight of unhulled rice/clump (V_9G_3:46.88 g) and weight of fully unhulled rice/clump (V_9G_3: 37.8 g) than that of control treatments (V_9G_0: 31.53 g and V_9G_0: 28.8 g, respectively).

Duku and Agai varieties in combination with PEG 30% had higher yield than that of other treatment combinations. These varieties were more tolerance to drought stress than that of other varieties. Duku variety was capable to cope with critical period during seedling phase which was shown by higher value of crown dry weight found in combination treatment of Duku variety and PEG than that of combination treatment of Agai variety and PEG. However, average value of crown dry weight found in these two treatment combinations was not significantly different than that of control treatment. This showed the capability in coping with critical phase so that crop can absorb water and nutrients which were important for the on-going of photosynthesis process. Results of the study by Akram *et al.* (2013) showed that drought stress could decrease the crown dry weight of crop.

5 Conclusion

The results showed that varieties of Henik, Serenik, Semester, Selanggar, Duku and Tinta were capable to produce high yield at drought treatment condition. Treatment of 20% PEG and 25% PEG had significant effect on growth and yield of rice crop. Treatment combination which had high yield was Duku +25 % PEG and Duku + 30% PEG, respectively.

This research was self funding research and we want to deliver our acknowledment to the Dean of Agricultural Faculty, Baturaja University, Head of Laboratory as well as Head of Farm Exprimentation Site of Agrotechnology Study Program that had already gave facilities for this research.

References

1. Pusdatin, *Statistics of Agricultural Land in Year of 2008- 2012", Secretary General of Agricultural Ministry*, (The Government of Indonesia Republic, Jakarta, 2013)
2. D. Danial, Nurbani, *Proceeding of National Seminar of Indonesia Biodiversity Society*, **1(4)**, **9**10 (2015)
3. S. Yuniarti, *Proceeding of National Seminar of Indonesia Biodiversity Society*, **1(4)**, 848 (2015)
4. Y.A. Rahayu, T. Harjono, J. Akta Agrosia., **13(1)**, 40 (2010)
5. A.A. Razak, M.R. Ismail, M.F. Karim, P.E.M. Wahab, S.N. Abdullah, H. Kausar, Aust. J. Crop. Sci., 7, 1582 (2013)
6. A. Quampah, R.M. Wang, I.H. Shamsi, G. Jilanil, G. Zang, S. Hua, H. Xu, Int. J. Agri. Bio., **13**, 1560 (2011)
7. Halimursyadah, A.I. Hereri, A. Hafnizar, J. Floratek, **8**, 73 (2013)
8. O.L. Afa, S.P. Bambang, J. Ahmad, H. Oteng, S.D. Iswari, J. Agrivigor., **11(2)**, 292 (2012)

9. M. Farooq, A. Wahid, N. Kobayashi, D. Fujita, S.M.A. Basra, Agron. Sustain. Dev., **29**, 185 (2009)
10 Y.R. Ahadiyat, Journal Agrin., **15(1)**, 1 (2011)
11. M.A.N. Zain, R.I. Mohd, M. Maziah, P. Adam, H.I. Mohd, Molecules, **19**, 1795 (2014)
12. K. Ji, W. Yangyang, S. Weining, L. Qiaojun, M. Hanwei, S. Shihna, C. Hui, J. of Plant, **169(4)**, 336 (2012)
13. M.H. Akram, A. Ali, A. Sattar, H.S.U. Rehman, A. Bibi, J. Anim. Plant. Sci., **23(5)**, 1415 (2013)

Assessment of Functional and Dysfunctional On Implant Stability Measurement for Quality Of Life

Norshahrizan Nordin[1,*], *Mohd Mustafa* Al Bakri Abdullah[2,3], *Fauziah* Mat[4], *Muhammad* Abdullah[1], *Razli* Che Razak[5], *Abdul Hamid* Adom[4]

[1]School of Business Innovation and Technopreneurship (PPIPT), Universiti Malaysia Perlis (UniMAP), Kangar, Perlis, Malaysia
[2]Faculty of Engineering Technology, Universiti Malaysia Perlis (UniMAP), Kangar, Malaysia
[3]Center of Excellence Geopolymer & Green Technology (CeGeoGTech), School of Material Engineering, Universiti Malaysia Perlis (UniMAP), Perlis, Malaysia
[4]School of Mechatronic Engineering, Universiti Malaysia Perlis (UniMAP), Pauh Putra, Perlis, Malaysia
[5]Faculty of Business and Entrepreneurship, Universiti Malaysia Kelantan (UMK), PengkalanChepa, Kelantan, Malaysia

Abstract. This study was conducted to investigate the effect of an implant wearer comprising among orthopedic patients as well as the use of implant dentistry in Northern Malaysia. A total of 100 questionnaires were distributed and 70 questionnaires can be used to record, analyze and test hypotheses. Data for all variables were collected through a questionnaire administered alone and analyzed by using SmartPLS V3. A total of four (4) hypotheses have been formulated and the results show that the hypothesis is supported. The results show that: (1) limit the functionality and quality of life was significantly (0.904) in connection with the implant wearer, (2) physical pain was significantly (0.845) relating to the quality of life, (3) physical discomfort was significantly (0.792) in connection with quality of life, and also (4) social discomfort is significant as well (0.809). This finding suggests that there are positive effects on the implant wearer who through life routine. The results of the study may also serve as a basis for reliable decisions related to quality of life and for the implementation of awareness campaign that increase how the need for humanity in the field of quality involvement.

1 Introduction

Bone structure is the most powerful and strong structure in a body. It is harder than iron and stronger than stone. Bone consists of bonds of minerals such as calcium, phosphorus, taurine and so on. This mineral molecular bond strength determines the bone strength.

* Corresponding author: norshahrizan@uthm.edu.my

Surgical services include general surgery, orthopedics, ophthalmology, otorhinolaryngology, urology, neurosurgery, plastic surgery, and various other sub-specialties [1].

General surgery and orthopedic services can be found in nearly all hospital specialists. Certain services such as ophthalmology surgical specialties and sub-specialties such as vascular surgery provide network services.

2 Implant

The implant is medical device designs specifically for medical replace a function or structure of a part of a human body. In the field of orthopedic implants as a bone substitute is the way in restoring bone fractures. Similarly, dental implants are a way to replace missing teeth so that the function of mastication, aesthetics and comfort. A dental implant is a device that is surgically implanted into the jaw bone or soft tissue so that it can serve as a substitute root to hold the denture or bridge [2].

The restoration of dental implants is very similar to natural teeth because of embedded in the network so that it can support in terms of aesthetics, protection and development of neighboring teeth confidence. Table 1 explained about the implant is implanted in principle requires an acceptable network material body, strong enough and can work together with the restoration of the prosthesis on it [2][3]. Providing that the implants are as follows:

Table 1. Functional materials of implant [2].

Functional materials of implant
Biocompatible is defined as non-toxic, non-allergenic, non-carcinogenic, non-destructive and disruptive tissue healing around and not corrosive
Strong enough to support the weight of mastication
High resistance to thermal and corrosion
Elasticity is the same or almost the same as the surrounding tissue
Can be made in various forms

3 Orthopedic implant

Orthopedic implants are medical devices that are used to replace or provide fixation of bone or to replace the joint surface whether cracks, fractures or looseness. In other words, the use of orthopedic implants is used to replace damaged or problematic joints [3].

Implant surgery is also performed only by surgeons specialized and trained. For each implant surgical procedure involves removing the damaged joint replacement prostheses and artificial. These implants are mostly made of stainless steel and titanium alloy for strength and filled with plastic to act as bone substitutes or artificial [4].

However, orthopedic implants are designed differently by different manufacturers. This means that manufacturers use different design theory to develop an implant for each specific application [5]. Orthopedic implants do not only restore the quality of life but also help in mobility and reduce pain.

About restoration surgery, infection, and damage that may be part of the disadvantages of orthopedic implants. Each designed to correct orthopedic implants affected joints so it withstand movement and associated pressure and to increase mobility and reduce pain. In general, orthopedic implants available for hip, knee, shoulder and elbow [6].

The suitability of implants is referred to as the does the implant in accordance with the wearer, the implant design meets the standards or standard material used to make the implants. However, if the implants have a defect it will be a worse situation for the implant wearer. The Table 2 shows the implant that does not fit to the specific location of the bodies.

Table 2. Dysfunctional Materials of Implant [7].

Types of dysfunctional	Explanation
Infection	It may occur after surgery in the early stages or after a few years. Minimal infection can be treated with antibiotics. A long internal infection and may require surgery and may change the bone
Blood Clots	It may be caused by several factors, including the lack of movement of venous blood causes a slow journey. It will cause swelling and pain in the affected limbs. If this happens orthopedic surgeon will perform as a physical examination and laboratory services and will talk to patient about treatments and methods of prevention of blood clots
Looseness Implants	This can occur on the surface of the implant and the bone and cause pain. If looseness is a problem for patients, surgery is needed to convert a fake bone to another
Transition Fake Bones	Sometimes the implant in the bone may be moved from the original. The implant can be inserted into the original place without surgery. If the transition is often and cause problems for patients, surgery may be done to change the new fake bones
Wear	Wear implant will occur in all bones false. Significant wear will require surgery.
Implant Damage	Damage such as metal or plastic fracture in the bone fakes rare. If it happens revision surgery should be done

4 Dental implant

Dental implants are made of metal objects metal shaped like a tooth and usually have a thread on the outside, installed by planted in the jaw bone (upper or lower jaw) serves to replace lost natural teeth in the mouth [8]. This system is suitable in technological developments, particularly in the field of dentistry to restore beauty, masticatory function, and the function of the pronunciation when talking. Dental implants are usually on the outer surface received additional materials that aim to get the most out of the relationship or union jawbone known as Osseo integration. Research proves this material is safe to use, can be accepted by the human body in the sense of not being rejected by the body as a foreign object or give a negative reaction, and has a maximum adaptability to the bone and soft tissue such a gums in the mouth . The process of making a dental implant was by melting the metal at very high temperatures in a vacuum or air.

Dental implants were originally designed to help patients who have lost teeth due to aging or those who have no teeth at all. However, now the dental implant treatment is given to people of all ages who have lost teeth due to various reasons, such as is involved in an accident or gum disease. However, it is rarely done on children because their jaws are not yet fully mature.

Tooth implant was shaped like a screw, produced from titanium metal (the same substance for the treatment of bone fractures) grown in the jawbone at the original place of missing teeth. It acts as a root for dentures. This method can be used to replace one tooth or several teeth. There is also a bar-shaped implants; also implanted into the jawbone. However, now more than 90% of implants available in the market around the world is shaped like a screw. Table 3 showed that dental implant does not just allow someone to eat better, but it also aims to: [10]

DOI: 10.1051/matecconf/20179701116

Table 3. Application of dental implants [Nickenig, Wichmann, Andreas, &Eitner (2008)].

Dental Implants Application
Replace one or more missing teeth without affecting adjacent teeth
Provide support to its line of denture, making it more sturdy and
No longer need to use removable dentures
Help maintain the shape of the face due to missing teeth eventually cause the jaw to shrink and make the bottom face looks also narrowed Allowing someone to talk comfortably Help restore confidence because many feel inferior and ashamed to open my mouth when loss of teeth, especially the incisors. It is also important to show the image of an attractive and convincing, especially for individuals who are looking for work

Dental implants can be thought of as permanent teeth because it is grown in the jawbone. When teeth come off, the jawbone shrinks slowly and it affects the soft tissue of the gums. This is why people who wear dentures have a problem matching, discomfort, often dated and difficult to chew after a certain period [8].

In Malaysia, the problem of the absence of teeth occurs at a younger age - 3% of adults aged 35 to 44 years do not have teeth, and this figure rose to 40% (65 to 74 years) and reached 70% for senior citizens over 75 years. Furthermore, it is more than 70% of cases of tooth loss due to trauma or accident. Among individuals with no teeth, only 52% wearing dentures even know the absence of teeth can affect their appearance [11].

5 Theoretical framework.

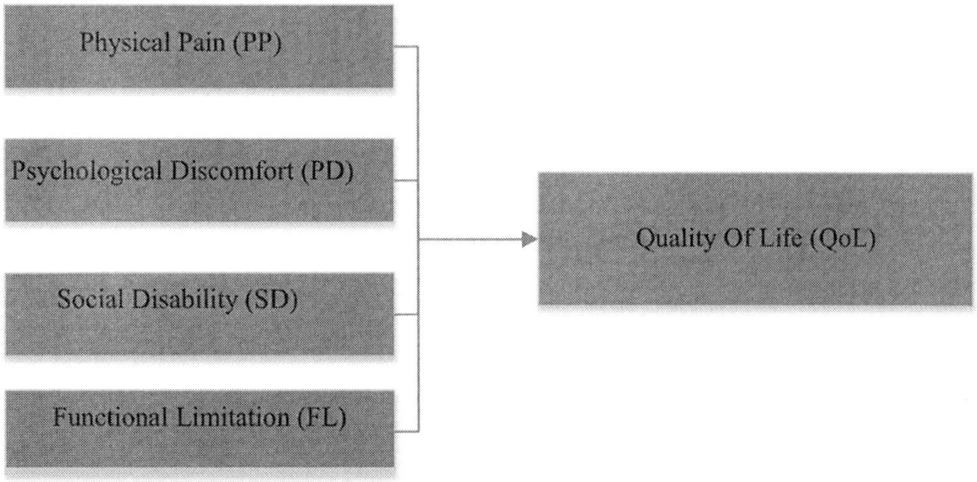

Fig.1. Theoretical Framework.

The model in Fig.1 shows a schematic diagram of the theoretical framework creates for this study. There are four independent variables in this study which are functional limitation, physical pain, psychological discomfort and social disability. All of this is referred to as the independent variable in quality of life for users implant factors that can affect the way of life of a patient in any place either outside the home, the workplace or in the home.

While the dependent variable is the effect the way they perform their daily activities with the use of implants is whether it is positive or negative in nature. This Figure 1 also shows that the independent variables have a positive correlation with the dependent variable.

6 Composite reliability

Likewise composite reliability, all instruments shown above 0.8 except Physical Pain that performance nearly 0.8, 0.792 stated in Table 4 below:

Table 4. Composite Reliability.

Functional Limitation	0.904
Physical Disability	0.845
Physical Pain	0.792
Quality Of Life	0.821
Social Discomfort	0.809

Table 4 assessed the composite reliability for functional limitation is 0.904, physical disability is 0.845, and social discomfort is 0.809. These results are all above 0.8, unlike physical pain that only 0.792. All this result indicate to all variables to quality of life that shown 0.821, still above 0.8. These finding showed that the determinant factors for service design in implant stability measurement particularly on non-linear relationship should be considered for more accurate evaluation of patient needs [12]. The needs and demands of the patients must be fulfilled for them to spread a positive word of mouth. In the present scenario, a positive word of mouth plays an important role in promoting services and product [13].

7 Average variance extracted (AVE)

Table 5. Average Variance Extracted (AVE).

Functional Limitation	0.653
Physical Disability	0.579
Physical Pain	0.508
Quality Of Life	0.537
Social Discomfort	0.523

Table 5 displayed Average Variance Extracted (AVE) of each instrument are above 0.5 as expected. The result indicated that for Functional Limitation, Physical Disability, Physical Pain, and Social Discomfort will be 0.653, 0.579, 0.508, and 0.523. Then, all instruments were proved that it is applicable to this model.

8 Discriminant reliability towards quality of life (QoL)

Discriminant validity value represents the cross loading factor that is to know whether the construct has sufficient discrimination quality of life that is by comparing the loading on the intended constructs must be greater than the value of the loading with other constructs.

This is to ensure the reliability quality of life instrument is reflected to relationship has shown acceptable or not acceptable as Table 6 below.

Table 6. Discriminant Reliability.

	FL	PD	PP	QOL	SD
FL	**0.80800**				
PD	0.33600	**0.76100**			
PP	0.50600	0.59900	**0.71300**		
QOL	0.31900	0.58900	0.60500	**0.73300**	
SD	0.47700	0.50200	0.54500	0.59100	**0.72300**

Table 6 above shown comparing the square root of the average variance extracted (AVE) correlation between each construct with other constructs in the model, if the square root of the average variance extracted (AVE) constructs larger than the correlation with all other lantern constructs then said to have good discriminant validity. Recommended the measurement must be greater than 0.50. The Functional Limitation (0.808), Physical Disability (0.761), Physical Pain (0.713), and Social Discomfort (0.723), there are every side of instruments is all above than 0.5.

9 Conclusion

The implications that further increasing fulfilment of the literature on medical biostatistics terms of quality of life practices can be derived from the results of this study. The finding indicated that all the factors discussed in this study are relevant and important in predicting quality of life for implant wearer. This study also can give the authority to the management of organization in taking care of them to the quality of life for implant wearer. Likewise require attention to satisfy and motivate implant wearer in fulfilling their obligations and assisting the achievement of the life goals. Furthermore, such findings may also contribute to the additional literature for researchers to pursue and improve on, particularly in the area of humanity of medical institutions.

References

1. A. Preciado, J. Dent., **41(9)**, 753 (2013)
2. Y.H. Pan, T.M. Lin, C.H. Liang, Biomedical Journal, **37**, (2014)
3. M. Manasas, K.Oslakovic, C. Sultan, J. Hamilton, D. Ingber, C. Sammarco, J. Kummailil, D. Skinner, U.S. Patent No. 20020128714 A1, (2001)
4. C.R. Arciola, Y.H. An, D. Campoccia, M.E. Donati, L. Montanaro, Int. J. Artif. Organs, **28**, (2005)
5. C.T. Laurencin, A. Ambrosio, M. Borden, J. Cooper Jr, Annu. Rev. Biomed. Eng., **1(1)**, 19 (1999)
6. S. Pike, J. Van Diepenbos, J. Hazard. Mater., **47**, (2003)
7. J. Schroeder, S. Goodman, K.J. Kim, Google Patents, (2006)
8. Y. F. Chen, T.H. Yang, J.H Chen, H.E. Lee, Y.C. Lin, J. Ebinger, T.M. Chou, Journal of Dental Sciences, **7(3)**, 289 (2012)

DOI: 10.1051/matecconf/20179701116

9. H.J. Nickenig, M. Wichmann, S.K. Andreas, S. Eitner, J. Craniomaxillofac. Surg., **36**, (2008)
10. N. Yunus, H. Abdullah, F. Hanapiah, J. Prosthet. Dent., **85**, 6 (2001)
11. N. Nordin, R.C. Razak., Advanced Review on Scientific Research, **4**, 1 (2014)
12. N. Nordin, A.A. Yaacob, R.C.Razak, W.N.W. Radzi, U.N. Saraih, Journal of Advanced Research in Business and Management Studies, **4**, 1 (2016)

MATEC Web of Conferences **97**, 01117 (2017) DOI: 10.1051/matecconf/20179701117
ETIC 2016

Design of Electric Vehicle Racing Car Chassis using Topology Optimization Method

Mohd Suffian bin Ab Razak[1,*], *Mohd Hudairee* bin Hasim[1], and *Nor Azazi* bin Ngatiman[1]

[1]Faculty of Engineering Technology (FTK), Technology Campus, Universiti Teknikal Malaysia Melaka, Hang Tuah Jaya, 76100 Durian Tunggal, Melaka, Malaysia.

Abstract. The goal of this project is to improve the design of space frame chassis of electric vehicle which meet the rules and regulations of Formula Varsity (FV) Malaysia. The chassis was designed using topology optimization method and analyzed for its structural performance using various loading analyses and RULA assessment.

1 Introduction

FV Malaysia is a biennial racing car design competition organized by Universiti Teknikal Malaysia Melaka (UTeM), Malaysia, which is similar to the international Formula SAE. The electric vehicle category was introduced in 2015 to cultivate students' interest in electric vehicle technology. A number of initiatives has been applied in order to develop a light, fast, energy-efficient vehicle such as battery downsizing, use of advanced materials and design optimization to reduce excess weight. A study showed that by reducing 10 percent of the vehicle weight, energy efficiency could be increased by 5 to 8 percent. On top of that, reducing weight would improve the vehicle dynamics. Chassis, one of the most important components in the electric vehicle racing car must be designed optimally to reduce its weight, thus enhancing overall vehicle performance. All components and design must comply with the rules and regulation specified in the FV Malaysia technical specifications [1].

The objective of this project is to design an electric vehicle (EV) chassis for upcoming FV Malaysia competition that focus in weight reduction whilst not compromising on chassis strength and ergonomics (for the driver).

2 Research methodology

The research begins with material selection, generating conceptual design using topology optimization and preliminary Rapid Upper Limb Assessment (RULA). The design used ASTM 1 inch, A500 hollow steel with 1mm wall thickness as it is cheaper than the higher grade A1085 and readily available in local market.

* Corresponding author: suffian@utem.edu.my

 DOI: 10.1051/matecconf/20179701117

Table 1. Material properties of the chassis

Properties	**ASTM A500 steel grade B**
Young's Modulus, E [Mpa]	200E+03
Poisson's Ratio, ν	0.3
Material density, ρ [t / mm^3]	7.85E-09

Chassis was initially designed in a monocoque body made of steel plate which is later converted to space frame. This novel method allows hollow structural section steel (HSS) to be modelled and simulated in shell meshes in order to distribute the load freely on a two-dimensional space (plate). By using size optimization, the exact same size of hollow tube and plate (1-meter long, 1-inch width) is fixed on one side and 100N force was applied on the other side to observe the maximum displacement and von mises stress, as shown in Figure 1 and 2. Plate thickness is varied using Hyperstudy in order to have similar performance with HSS tube in term of displacement and von mises stress. Optimization details are listed in Table 2.

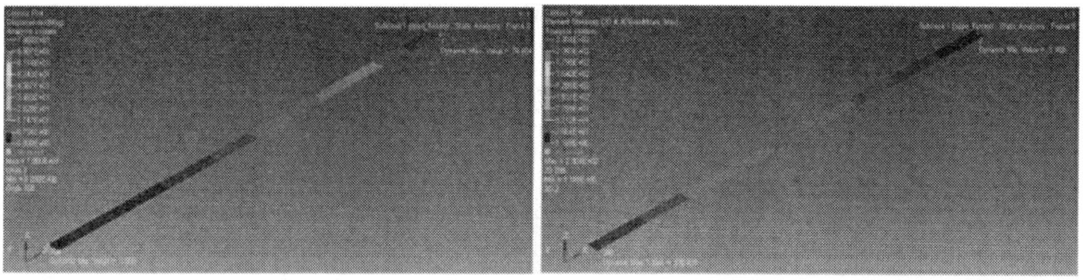

Fig. 1. Displacement of the plate (left) and von mises stress of the plate (right)

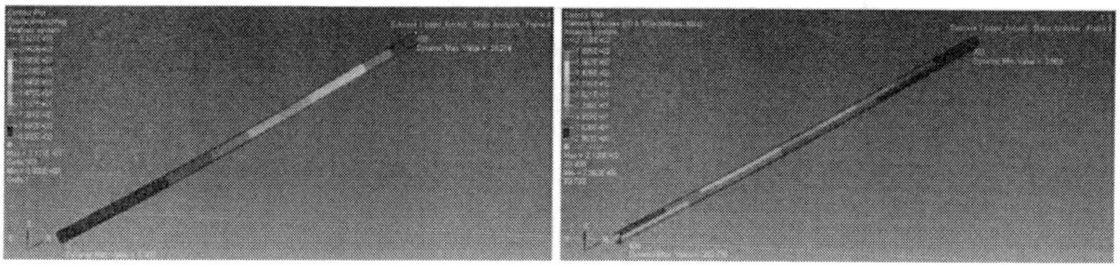

Fig. 2. Displacement of the tube (left) and von mises stress of the tube (right)

Table 2. Size optimization parameters

Objective	Minimize mass
Constraint	Displacement plate ≤ tube (33.214mm) Von mises stress plate ≤ tube (22.778Mpa)

Optimum plate thickness of 13.3034 mm obtained from Altair Hyperstudy software will be used for material thickness of the plate throughout this study.

Next, benchmarking on racing car chassis design was conducted and an Asian-size mannequin had been used to design a monocoque chassis which meet the FV Malaysia design regulations [1, 2].

The suspension mounting is located at the same location with 2015 EV model in order to maintain the suspension design and configurations. While the suspension mounting points and main hoop are set as non-design space, other area is set as design space where

 DOI: 10.1051/matecconf/20179701117

the material is allowed to be distributed within the defined space. The design of chassis is shown in Figure 3.

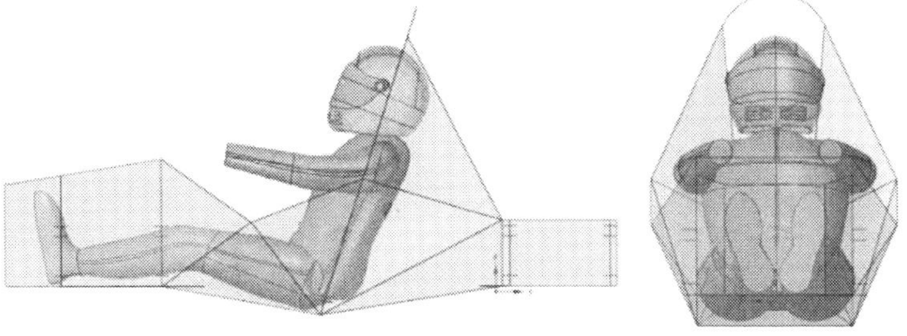

Fig. 3. Conceptual design of the chassis side view (left) and front view (right)

Symmetrical plane on Y-axis is applied at the centre. Support is assigned at rear mountings while combine load case for torsional (vertical, opposite direction), lateral (horizontal, same direction) loading and vertical bending load are applied at the front mounting and driver seat mounting [2]. Meshing setup and boundary conditions are shown in Figure 4.

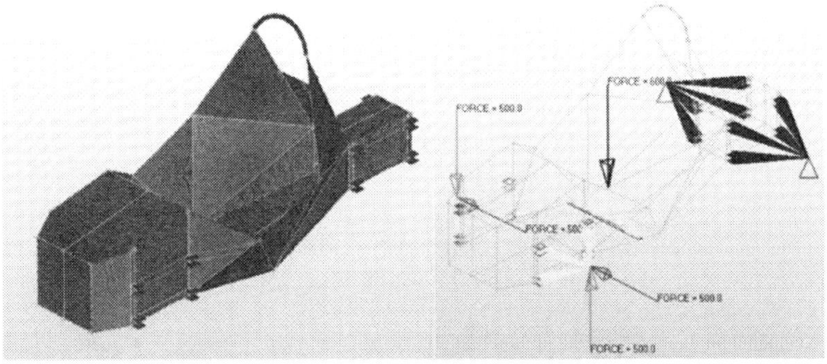

Fig. 4. Meshing setup (left) and boundary conditions (right)

Each force (Fn) acted on each mounting is calculated based on following equation:

$$Fn = \frac{(Vm \cdot g)}{mn} \tag{1}$$

Where *Vm* is gross vehicle mass, *g* is gravitational acceleration and *mn* is no. of suspension mountings. Altair Optistruct is a structural analysis solver in Hyperworks which allows user to develop a structurally efficient conceptual design that is right at the first time, thus reducing cost and shortening development time. Parameters for topology optimization are shown in Table 3. Generated optimized conceptual design (Figure 5) is then exported to CAD software in STL file format.

Table 3. Topology optimization optimization parameters

Design Variable	Monocoque body (design space)
Objective	Minimize weighted compliance
Constraint	Maximum 30% of design space volume

Next, the chasis is designed with topology result as a reference in a hybrid modelling environment in CAD software, Ansys SpaceClaim. Direct modelling capability in SpaceClaim allows user to design a CAD model in the presence of STL file within same environment.

 DOI: 10.1051/matecconf/20179701117

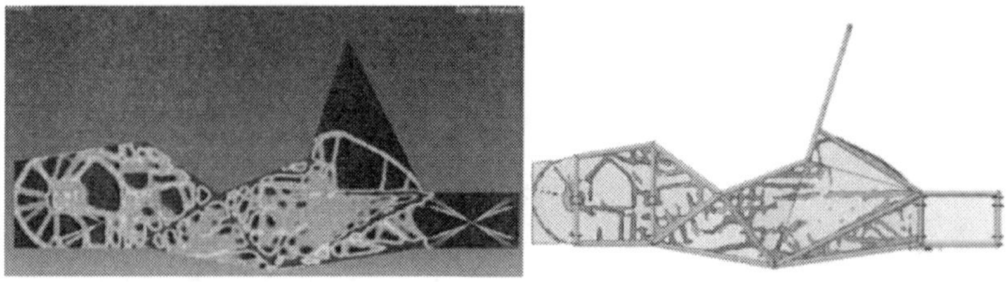

Fig. 5. Topology Optimization result in Altair Hyperview (left) and in Ansys SpaceClaim (right)

Topology optimization process will undergo few design iterations and optimization result will be easier to interpret compared to the initial design iteration.

Once the design is complete, the space frame chassis have to be analyzed for torsional, lateral and vertical loading. The performance is then compared to the 2015 EV chassis.

Lastly, the space frame chassis design is evaluated based on FV Malaysia regulations and assessed on vehicle ergonomics while driving using Rapid Upper Limb Assessment (RULA). RULA assessed driver's comfort to avoid fatigue and muscle disorder. The correct driving posture, pedals, seat and steering location will be determined based on RULA assessment.

3 Result and discussion

Figure 6 shows the displacement and von mises stress result of the structure. Maximum deformations of 23.8mm are at the front suspension mountings where the forces were applied. Since maximum von mises stress of 720Mpa are observed at seat mounting area, brackets should be installed to support excessive load. The chassis is considered safe as the driver's head is positioned 50 mm below the main hoop [3].

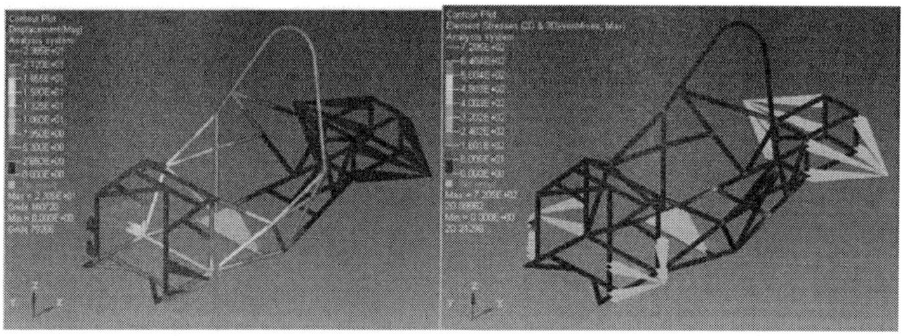

Fig. 6. Displacement result (left), von mises stress result (right) of the new model

The summary of improvement of the current design compared to the previous design is shown in Table 4. New EV chassis achieved mass and stress reduction, while structure deformation is similar to 2015 EV model (Figure 7) with 1mm difference.

Table 4. Comparison of performance

Result	**New model**	2015 EV model	Difference
Mass (chassis only) [kg]	**16.87**	18.97	-11.07%
Max. displacement [mm]	23.8	**22.8**	+4.38%
Max. Von Mises Stress [Mpa]	**720**	910	-21.74%

 DOI: 10.1051/matecconf/20179701117

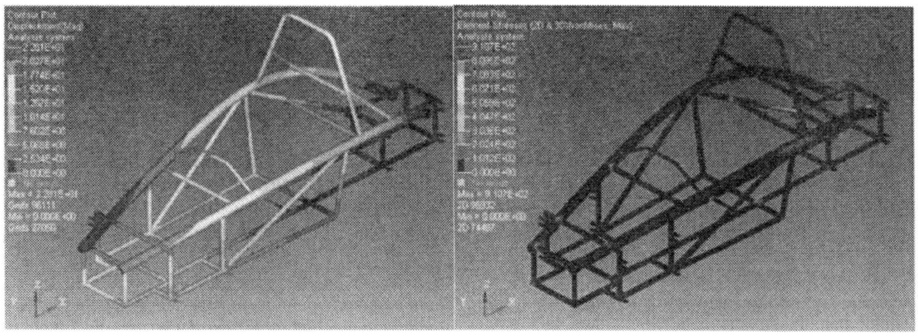

Fig. 7. Displacement result (left), von mises stress result (right) of 2015 EV model

To ensure comfort for the driver, posture plays an important role in terms of ergonomic during racing. RULA result in Figure 8 showed a final score of 2 (acceptable). This means the driver will not feel fatigue and will be comfortable around the upper-body.

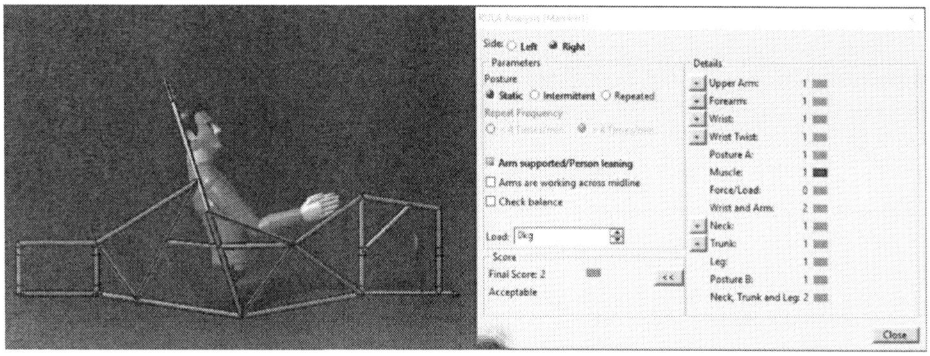

Fig. 8. RULA analysis

4 Conclusions

The objective to design an electric vehicle (EV) racing car chassis for FV Malaysia that focused in ergonomics and performance was achieved. The final score of RULA is 2 and the comfort level is within acceptable level. Topology optimization method can be applied to improve the vehicle performance.

The author would like to express appreciation to ACTIVE Research Group, Centre of Research on Energy (CARe) and Universiti Teknikal Malaysia Melaka (UTeM) for funding this project under PJP/2014/FTK(19C)/S01348.

References

1. Efa Riani, Ali. *Vehicle Technical Specifications,* (Universiti Teknikal Malaysia Melaka, 2015)
2. S. Kakria, D. Singh, *CAE Analysis,* SAE Technical Paper, 2015
3. M.H. Mat, A.R.A Ghani, Procedia Eng., **41**, 1756 (2012)

DOI: 10.1051/matecconf/20179701118

A Novel Study on Using Vietnam Rice Hush Ash and Cullet as Environmental Materials

PHAM Trung Kien[1,*], *TRAN* Thi Thien Ly[2], *PHAM* Thi Lan Thanh[3], *TRAN* Pham Quang Nguyen[1], *NGUYEN* Hoc Thang[4], and *Mohd Mustafa Al Bakri* ABDULLAH[5]

[1] Faculty of Materials Technology, Ho Chi Minh City University of Technology (HCMUT), Vietnam National University (VNUHCM) 268 Ly Thuong Kiet, District 10, Ho Chi Minh City, Vietnam
[2] Hai Anh Equipment and Technology Co. Ltd. 127/33 Hoang Hoa Tham, District Tan Binh, Ho Chi Minh City, Vietnam
[3] Faculty of Pharmacy, Lac Hong University, 10 Huynh Van Nghe, Dong Nai, Vietnam
[4] Faculty of Chemical Technology, Ho Chi Minh City University of Food Industry, 140 Le Trong Tan, Tan Phu District, Ho Chi Minh City, Viet Nam
[5] Center of Excellence, Geopolymer & Green Technology (CeGeoGTech), School of Material Engineering, Universiti Malaysia Perlis (UniMAP), Perlis Malaysia

Abstract. This work introduces novel method to prepare ecomaterials wollastonite ($CaSiO_3$) using the hydrothermal treatment technique of Rice hush ash (RHA), glass cullet and CaO at 180°C for 24 hour so as the Ca/Si molar ration of 1.0. The sample after hydrothermal treatment at 180°C for 24 hours is calcined at 1000°C to obtained wollastonite $CaSiO_3$. The obtained wollastonite has many applications in biomaterials and thermal insulator. This research also aim to improve the quality of RHA, thus reduce negative impact to environment.

1 Introduction

Vietnam is an agriculture country, in which produce lot of rice and its by product is rice husk ash (RHA). The RHA is consider as waste of agriculture industry, and treated by burn in the open air. This process cause air pollution, thus attracted Vietnamese researcher to find alternative method to reduce the impact of rice husk ash to environment. The research group in Department of Ceramic Materials aim to reuse rice husk ash as source of Silica (SiO_2). It need to emphasized that the main content of rice husk ash is silica. In addition, glass cullet also contain up to 70% of Silica. This research report new technology to mixing RHA with glass cullet and CaO with the Ca/Si molar ratio of 1.0, in order to synthesize Wollastonite ($CaSiO_3$) as environmental materials. Compare with the other method to synthesize wollastonite using chemical reaction between CaO and SiO_2 at 1500°C [1-7], the hydrothermal treatment technique require lower temperature. The advantage of our study is utilize the Vietnam RHA and reduce the

* Corresponding author: phamtrungkien@hcmut.edu.vn

2 Materials and methods

2.1 Preparation of RHA

Rice husk is burn at 600°C with the heating rate of 10°C/min (Naberthem 1400, Nabertherm, Germany), then soaking at 2 hour for complete burning. The phase composition of obtained RHA is characterized using Xray Diffraction and Fourier transform infrared spectroscopy (FTIR).

2.2 Preparation of glass cullet

Glass cullet is supplied by Coca Cola Vietnam. The cullet is pre-milled by hammer milling, then heating at 700°C using electric furnace (Nabertherm 1400, Nabertherm, Germany) with the heating rate of 10°C/min, soaking at 30 min, quench by tap water to increase the milling ability. The cullet is milled in ball milling machine for 20 mins, and pass the sieve 125 μm. The chemical composition of obtained RHA is characterized using XRay Flouresence (XRF) method.

2.3 Preparation of CaO

CaO is supplied by Xilong Chemical in bottle 500g.

2.4 Preration of mixture

The mixture of RHA, glass culled and CaO is mixing with the weight ratio of CaO: RHA: glass cullet is 1.43: 1.00: 1.00 so that the Ca/Si molar ratio of the mixture is 1.0 according to the chemical composition of wollastonite (CaSiO3). The obtained mixture is pressing at 25kG/cm^2 to form the disk the diameter of 14mm and height of 5mm, following by hydrothermal treatment at 180°C for 6, 12 and 24 hours. The hydrothermal treated disk is calcination at 1000°C (Nabertherm 1400, Nabertherm, Germany) to form wollastonite.

2.5 Phase analysis

The powder Xray Diffraction (XRD) patterns of disk samples were recorded with a vertically mounted diffractometer system (Bruker-AXS: D8 ADVANCE, Germany) using Ni filtered CuKa generated at 15 kV.

2.6 The morphology of samples using Scanning Electron Microscope (SEM)

The surface of samples were observed using a scanning electron microscope (SEM) (JSM 5400LV, JEOL Co. Ltd., Japan) under an accelerating voltage of 20 kV after being coated with gold.

2.7 Chemical bonding of sample is analyzed by Fourier Transformer Infrared (FTIR)

The sample is mixing with KBr with the ratio 1: 200 and put in the FTIR machine with the waveband vary from 400-4000cm-1.

2.8 Chemical composition of sample is analyzed by Xray-Fluorescence (XRF)

The sample is energy using Xray (MESA-50, Horiba, Japan) and measure the secondary beam to analyze the chemical composition of sample.

3 Results and discussions

The chemical composition of RHA is shown in Table 1:

Table 1. The chemical composition of RHA (weight percent)

SiO_2	K_2O	CaO	P_2O_5	MgO	Al_2O_3	MnO	Fe_2O_3	SO_3	other	LOI	Total
92,7	3,16	1,33	0,596	0,466	0,306	0,291	0,242	0,126	0,153	0,63	100

The phase analysis of RHA is shown in Fig. 1. The main phase compostion of RHA is Cristoballite (SiO_2) at 2theta = 22°, as shown in Fig. 1

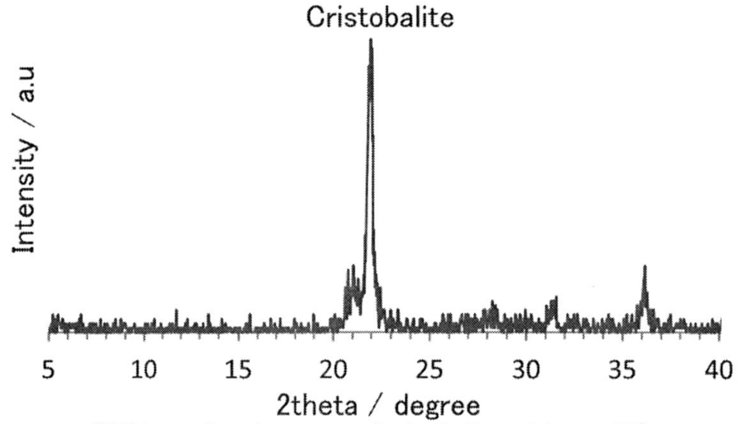

Fig. 1. XRD pattern of RHA, confirm the present of cristobalite at 2theta = 22°

The chemical bonding of RHA is shown in Fig. 2. This data confirm that the main chemical bonding of RHA is Si-O bonding.

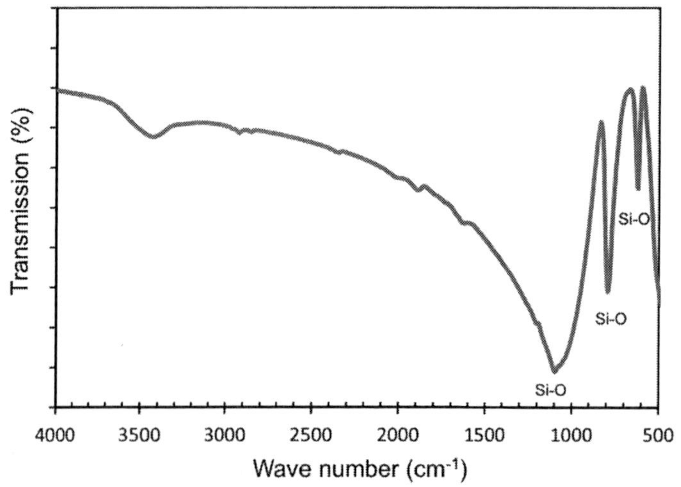

Fig. 2. FTIR of RHA. The bonding of Si-O is display on the figure at 600, 700 and 100 cm^{-1}

The chemical composition of glass cullet is shown in Table 2:

Table 2. The chemical composition of glass cullet (weight percent)

CaO	SiO_2	Al_2O_3	Fe_2O_3	MgO	Na_2O	K_2O	P_2O_5	TiO_2	LOI	Total
10,48	70,2	1,75	0,11	0,75	13,71	0,58	0,03	0,28	2,11	100

The SEM of sample hydrothermal treatment at different treatment time (0h, 6h, 12h and 24h) is shown in Fig. 3. We can observe new crystal interlock together at Fig. 3 (c) and (d) to enhance the setting property of sample.

Fig. 3. SEM morphology of mixture hydrothermal treatment at 180°C for (a) before (b) 6h (c) 12h and (d) 24h.

The XRD of sample hydrothermal treatment at different treatment time (0h, 6h, 12h and 24h) is shown in Fig. 4. After hydrothermal treatment, we can observed the peal of Tobermorite at 2theta = 29°.

 DOI: 10.1051/matecconf/20179701118

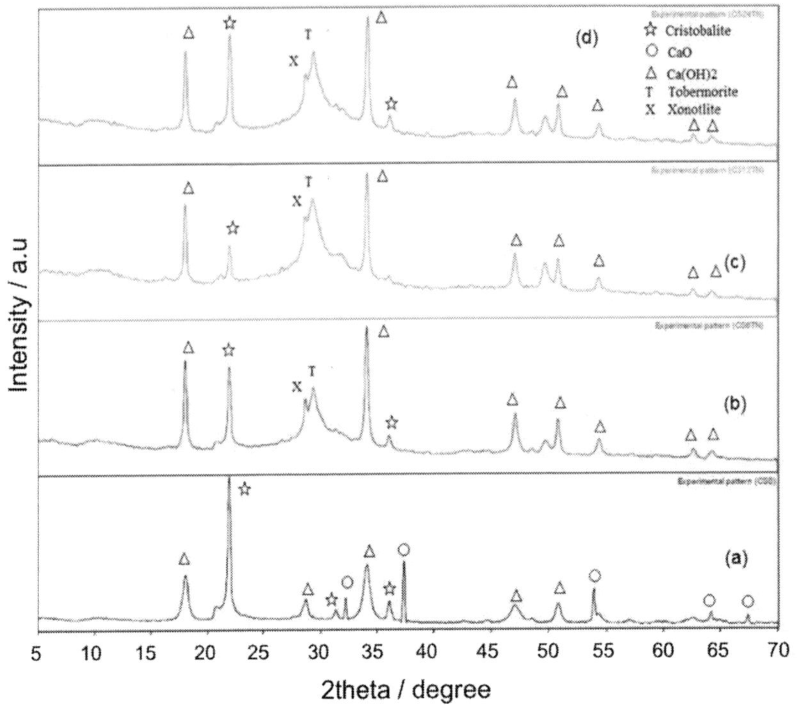

Fig. 4. XRD pattern of sample hydrothermal treatment at 180°C for different treatment time: (a) before; (b) 6h; (c) 12h and (d) 24h

The sample hydrothermal treatment at 180°C for 24h is calcinated at 1000°C to obtain the wollastonite. XRD of wollastonite obtained is shown in Fig. 5.

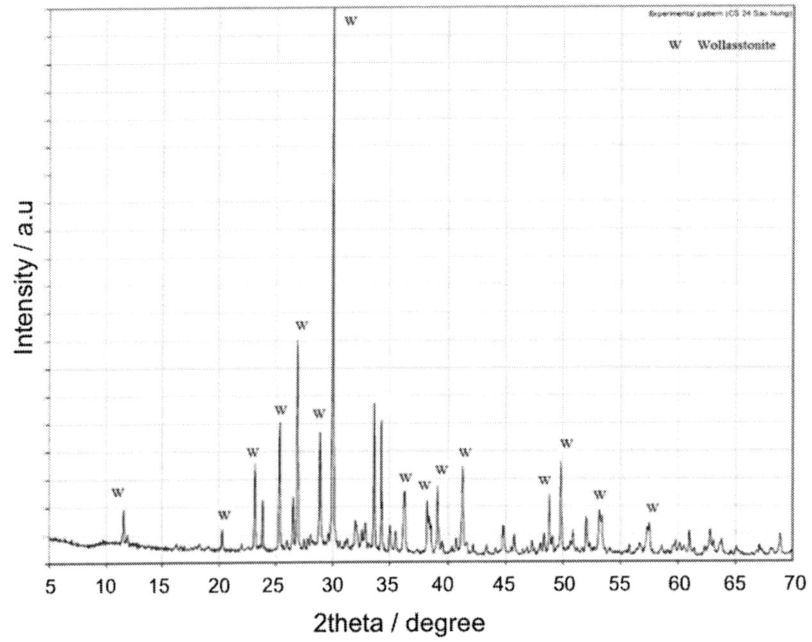

Fig. 5. XRD pattern of sample hydrothermal treatment at 180°C for 24h then calcination at 1000°C to obtain wollastonite (W:Wollastonite).

Figure 5 shows that we can obtain wollastonite by calcination hydrothermal treatment sample at 1000°C. The meaning of the study need to put in to the context of our facility. By using hydrothermal treatment of RHA, glass cullet and CaO, following by calcination, we can obtain the wollastonite at 1000°C. This temperature is much lower than the conventional method. For example, the solid reaction between CaO and SiO_2 to form wollastonite requires the temperature of 1500°C or above to happen. The reason that we can lower temperature to form wollastonite from 1500°C to 1000°C is we utilize the hydrothermal treatment technique. In the other hand, this research is contributing to improve the quality of RHA, thus reduce the negative impact to environment.

4 Conclusions

By using hydrothermal treatment technique of mixture of RHA, glass cullet and CaO, following by calcination at 1000°C, my research group can obtain wollastonite (CaSiO3) at low temperature. The obtained wollastonite has many applications in biomaterials and thermal insulator. This research also aim to improve the quality of RHA, thus reduce negative impact to environment. Further application of wollastonite in biomaterials is still under investigated by my group.

This research is funded by Vietnam National University Ho Chi Minh City University of Technology (VNU-HCMUT) under grant number T-CNVL-2016-12". The first author also thankful to Lac Hong University and Center of Excellence, Geopolymer & Green Technology (CeGeoGTech), School of Material Engineering, Universiti Malaysia Perlis (UniMAP) for helping on material facility.

References

1. Tasci, J. Aust. Ceram. Soc., **50**, 43 (2014)
2. R. Mathur, A.K. Misra, P. Goel, J. Sci. Ind. Res., **66**, 1029 (2007)
3. X. Liu, C. Ding, P. K. Chu, Biomaterials, **25**, 1755 (2004)
4. M. H. Tiggemann, D. Tomacheski, F. Celso, V. F. Ribeiro, S. M. B. Nachtigall, Polym. Test., **32**, 1373 (2013)
5. A.M. Soliman, M.L. Nehdi, Cem. Concr. Compos., **46**, 81 (2014)
6. S. Saadaldin, A. Rizkalla, Dent. Mater., **30**, 364 (2014)
7. T. Pham, D. Kieu, M. Do, H. Nguyen, H. Ly, T. Pham, Journal of Science and Technology, **52**, 198 (2014)

Lightweight Heat Resistant Geopolymer-based Materials Synthesized from Red Mud and Rice Husk Ash Using Sodium Silicate Solution as Alkaline Activator

Nguyen Hoc Thang[1,*], *Pham* Trung Kien[2], and *Mohd Mustafa Al Bakri* Abdullah[3]

[1]Faculty of Chemical Technology, Ho Chi Minh City University of Food Industry, 140 Le Trong Tan Str, Tan Phu Dist., Ho Chi Minh City, Viet Nam
[2]Faculty of Materials Engineering, Ho Chi Minh City University of Technology, 268 Ly Thuong Kiet Str, Dist. 10, Ho Chi Minh City, Viet Nam
[3]Center of Excellence, Geopolymer & Green Technology (CeGeoGTech), School of Material Engineering, Universiti Malaysia Perlis (UniMAP), Perlis Malaysia

Abstract. Geopolymer is an inorganic polymer composite with potentials to replace Ordinary Portland Cement (OPC)-based materials in the future because of its lower energy consumption, minimal CO_2 emissions and lower production cost as it utilizes industrial waste resources. Hence, geopolymerization and the process to produce geopolymers for various applications like building materials can be considered as green industry. Moreover, in our study, the raw materials we used are red mud and rice husk ash, which are are industrial and agricultural wastes that need to be managed to reduce their impact to the environment. The red mud and rice husk ash combined with sodium silicate (water glass) solution were mixed to form geopolymer materials. Moreover, the geopolymer specimens were also tested for heat resistance at a temperature of 1000°C for 2 hours. Results suggest high heat resistance with an increase of compressive strength after exposed at high temperature.

1 Introduction

Geopolymer, as coined by Davidovits [1], is an inorganic polymer composite which has a potential sustainable construction material because of its lower energy and carbon footprint as compared to Portland cement-based materials [2]. Such material uses industrial waste such as red mud as raw material which reduce its cost of production. Red mud (RM) is the waste generated during aluminum production from bauxite ore. It depends on the quality of bauxite mine which could affect the output of aluminum production. To produce 1 ton of metallic aluminum, this Bayer process generates from 0.6 – 2.5 tonnes of red mud [3-4]. Globally, it is estimated that there has been over 70 million tonnes of this red mud (RM) waste being discharged every year [4, 5]. RM waste is fine powder and thus easy to disperse into the

* Correspoding author : thangnh@cntp.edu.vn

surrounding environment which may lead to ecological problems. Currently, RM is contained in dams but these use up a lot of land area. Moreover, these dams pose risk to surrounding residential areas. In fact, in 2010, an accident wherein the failure of red mud dam occurred in Hungary caused serious environmental consequences and health impact to people [3,7]. Therefore, it is imperative to manage these red mud wastes properly to reduce its impact to environment and community.

In the other hand, rice husk ash (RHA) is high–silica material made from burning rice husk. Rice husk takes up about 20% weight of rice and its compositions include 15 – 20% SiO_2, the others are cellulose and lignin [8, 9, 10]. After the burning process, total weight of obtained RHA is about 20% weight of rice husk. Hence, the total estimated RHA reserves are over 28 million metric tons every year [9].

2 Materials and methods

In this research, both red mud (RM) and rice husk ash (RHA) from Vietnam were used to produce the geopolymer mixture. They are the raw materials in geopolymerization reactions to create the mechanical strength for product. All of the raw materials were characterized before carrying out the mixing to form the geopolymer mortar. The red mud contains high alumina in amorphous or semi-crystalline structure which serves as one of the reactants in the geopolymerization process. RM waste came from Tan Rai Bauxite Plant, Lam Dong Province, Vietnam and its chemical composition is summarized in Table 1. On the other hand, rice husks from Mekong delta, South of Viet Nam were burned at 700°C for 1 hour at Laboratory of Ceramic Materials to obtain rice husk ash with chemical compositions described in the same table.

Table 1. Chemical composition of raw materials from Vietnam

Raw materials	Chemical composition [wt.%]								
	SiO_2	Al_2O_3	Fe_2O_3	CaO	Na_2O	K_2O	TiO_2	Other	L.O.I
Red mud	4.52	18.98	49.90	0.87	2.06	0.05	5.62	1.48	16.52
Rice husk ash	90.90	1.12	0.54	1.41	-	4.66	-	0.06	0.77

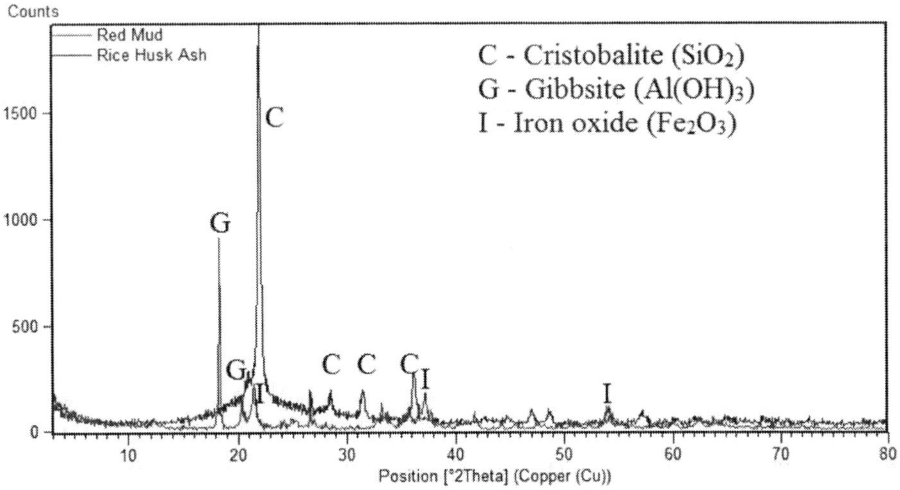

Fig. 1. XRD patterns of red mud and rice husk ash

As shown from the XRD of red mud and RHA, evidence of amorphous silica and alumina was observed as described by the noise at the bottom hump of the XRD spectrums (see Figure.1). There also have been some sharp peaks associated with crystalline minerals of gibbsite ($Al_2O_3.3H_2O$) at 18° and iron oxide (Fe_2O_3) at 20° in XRD pattern of RM (Figure.1) and cristoballite (SiO_2) in XRD spectrum of RHA (Figure.1). Note that the amorphous compositions are necessary conditions to conduct geopolymerization for RM based geopolymer materials. Water glass solution was then used as alkaline solution to increase pH value and supply the sodium hydroxide reactant for the geopolymer mixtures.

Table 2 summarizes the mix proportions used in this experiment. The schematic diagram of the experimental process is shown in Figure 2. RM, after being dried for 24 hours, was ground in 30 minutes (Ceramic Instrument) and then passed through 90 µm- mesh (Retsch AS-200 Seive Shaker). Rice husk was burned at 700°C for one hour (NABERTHERM), and then also ground in 30 minutes and passed through 90µm-mesh. All raw materials were mixed for 20 minutes with 20% water glass solution (Lab cement mixer) and water is added to get a pH value around 12. Geopolymer mixtures were molded in 50x50x50mm cube-cement mold and cured at room temperature (28°C, 80% humidity). The 28-day geopolymer specimens were tested for compressive strength, volumetric weight, and heat resistance. Then, the geopolymer specimen with the highest value of compressive strength was characterized with X-ray diffraction.

Table 2. Mix proportions of raw materials

	Proportion (wt.%)		
Sample	Red mud	Rice husk ash	Water glass solution (WGS)
A	20	60	20
B	40	40	20
C	60	20	20

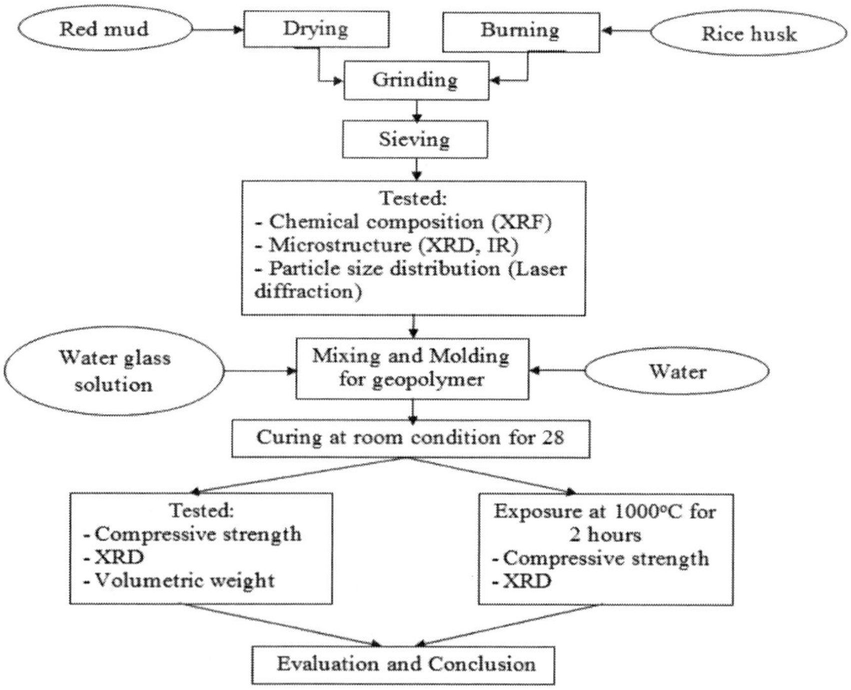

Fig. 2. The experimental process to produce geopolymer-based materials from red mud and RHA.

3 Results and discussions

3.1 Compressive strength and volumetric weight

Table 3. Properties of geopolymer produced from red mud and rice husk ash

Sample	Mix Proportions (wt.%)			Volumetric weight (kg/m^3)	Compressive strenghth (MPa)		Change of compressive strength (%)
	RM	RHA	WGS		Room condition	After exposed at 1000°C	
A	20	60	20	1270	15.50	67.83	337.63
B	40	40	20	1460	8.33	30.17	262.00
C	60	20	20	1590	6.83	35.33	417.07

The volumetric weight of geopolymer specimens were found to be from 1270 to 1590 kg/m^3 (see Table 3). The geopolymer samples were also tested for 28 day- compressive strength and the results showed that the values range from 6.83 MPa to 15.50 MPa. In addition, sample A has the highest value at 15.50 because of the high silica in geopolymer mixture that reacted with alumina of red mud to form the geopolymer networks (see Table 3). These values can be compared with concrete brick in ASTM C55 and C90 [13-14]

3.2 Heat resistance

All geopolymer samples were tested for heat resistance by keeping the specimen at a furnace at the temperature at 1000°C for 2 hours, with a heating rate of 5°C/min and a natural cooling process to room temperature. This procedure is similar to test done by Kong et al., (2007) [11] and Pan et al., (2012) [12]. After exposure with this high temperature, the samples were tested for compressive strength. Results indicate that the compressive strength significantly increased from 262% to 417% as shown in Figure 3 and summarized in Table 4. This could be due to sintering that may be occurred at high temperature in geopolymer samples, similar to what is happening in ceramic process [11, 12].

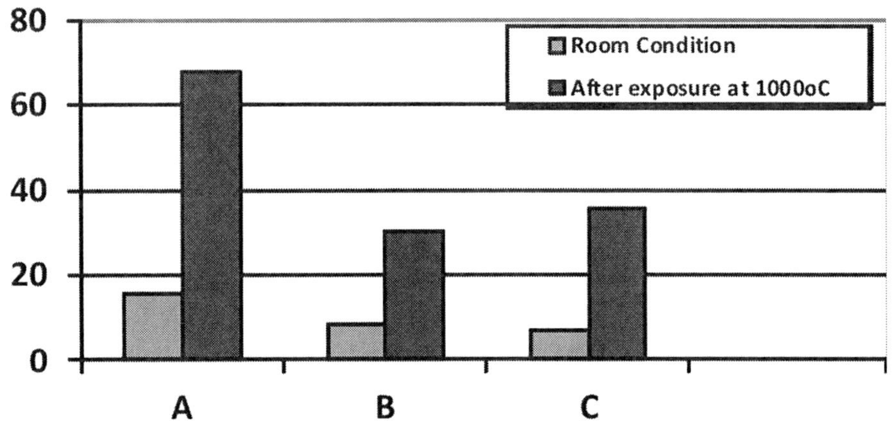

Fig. 3. Compressive strength of geopolymer before and after exposed at 1000°C.

3.3 Microstructure

Geopolymer sample A has the highest compressive strength and was therefore selected for microstructure characterization by XRD (see Figure 4).

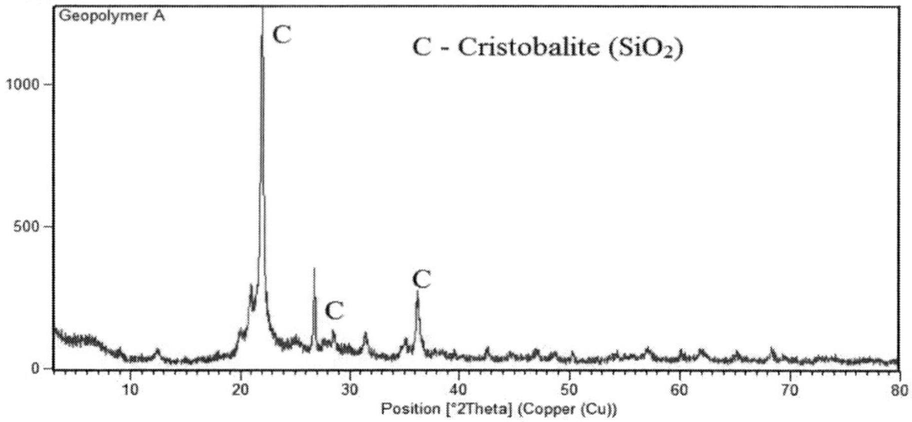

Fig. 4. (Top) XRD pattern of geopolymer sample A

The main compositions in the crystalline structure of geopolymer are silicon oxide (at 22° - 2Theta angle) and some minerals such as sodium aluminium silicate (at 21.5° - 2Theta angle), iron silicon oxide (36° - 2Theta angle). Amorphous phase in its microstructure has also been observed. Two peaks of gibbsite and iron oxide at 18° - 2Theta angle and 20° - 2Theta angle, respectively are typically associated with red mud raw material (Figure 1) disappeared in this XRD spectrum because there was reaction of geopolymerization among reactants resulting to an increase of mechanical strength of geopolymer specimens.

4 Conclusions

Red mud is an industrial waste that can effect negatively the environment if not properly managed. It is necessary to have more research in utilizing this as raw materials for other applications. In this research, red mud in combination with rice husk ash in high alkaline condition can be geopolymerized to form products that can be compared with concrete brick in ASTM C55 and C90 in terms of mechanical strength and volumetric weight. More specially, the geopolymer products have high compressive strength even if exposed at high temperature. That can be evaluated as heat-resistant materials. Future work will be done to characterize further the microstructure and optimize the properties for a lightweight heat-resistant geopolymer material.

Our research group would like to give deep appreciation to Ho Chi Minh City University of Technology, De La Salle University for the provided facility. Also, we would like to thank Tokyo Institute of Technology for the discussions and reference provided. The first author also acknowledge the support of AUNSEED-Net in the conduct of this collaborative research.

References

1. J. Davidovits, *Geopolymer Chemistry and Application* (Institute Geopolymere, France, 2011)

2. Z. Li, Z. Ding, Y. Zhang, *Proceedings of the International Workshop on Sustainable Development and Concrete Technology* (2004)
3. J. Benedek, *Causes and Lessons from the Red Mud Disaster* (Greens / European Free Alliance Parliamentary Group in the European Parliament and LMP, Cypress Ltd, 2011)
4. W. Liu, J. Yang, B. Xiao., Int. J. Miner. Process., **93**, 220 (2009)
5. S. Amritphale, A. Anshul, N. Chandra, N. Ramakrishnan., J. Eur. Ceram. Soc., **27**, 1845 (2006)
6. G. Zhang, J. He., *Developments in Strategic Materials and Computational Design II*, (John Wiley & Son, United States, 2011)
7. S. Chandra, *Waste Materials Used in Concrete Manufacturing* (Noyes, USA, 1997).
8. R. Siddique, M. Khan. *Supplementary Cementing Materials* (Springer-Verlag Berlin Heidelberg, New York, 2011)
9. B.K.Y. Foo, B.H. Hameed, Adv. Colloid Interface Sci., **152(1)**, 39 (2003)
10. D. Kong, J. Sanjayan, K.S. Crentsil, Cem. Concr. Res., **37(12)**, 1583 (2007)
11. Z. Pan, G. Jay, G. Kong, Constr. Build. Mater., **36**, 365 (2012)
12. Standard ASTM C55 99, Standard specification for concrete brick.
13. Standard ASTM C90 99a, Standard specification for load bearing concrete masonry units.

Effect of Grinding Treatment of Fly Ash on Compressive Strength of Hardened Geopolymers using Warm Press Method

Shinobu Hashimoto[1,*], *Kosuke* Kubota[1], *Kotaro* Ando[1], *Yusuke* Diko[1], *Sawao* Honda[1], and *Yuji* Iwamoto[1]

[1]Nagoya Institute of Technology, Department of Life Science and Applied Chemistry, Gokiso-cho, Showa-ku , 466-8555 Nagoya, JAPAN

Abstract. In this study, starting mixed wet powders were uniaxially compressed at 200 MPa for 40 min and simultaneously heated at 200 °C to form hardened bodies. Three different starting slurries composed of 100 g of fly ash powders with particles of different sizes after grinding treatment, 15 g of reagent-grade sodium, and 33.3 g of fresh water were considered. In this time, the uniaxial pressure was fixed at 200 MPa. This method is called the warm press method because the heating temperature is not as high as that in the conventional hot press sintering method used for ceramics. When ground fly ash with a diameter of 6.8 μm was used, the average compressive strength of the hardened bodies reached approximately 120 MPa. As the heating temperature and duration were increased at the fixed uniaxial pressure of 200 MPa, the compressive strength of the geopolymers increased. When ground fly ash with a diameter of 6.8 μm was used and the heating temperature and duration were 280 °C and 60 min, respectively, the compressive strength of the geopolymers reached approximately 150 MPa.

1 Introduction

Generally, geopolymer hardened bodies are fabricated by curing a mold of starting slurry consisting of amorphous aluminosilicate, such as fly ash exhausted from a power plant or metakaolin obtained by heating kaolinite clay at approximately 700 °C, and alkali solution [1–3]. Alkali silicate solution is most often added to the starting slurry as a reaction accelerator, but the use of alkali silicate solution is not necessary when geopolymers are fabricated from volcanic ash [4]. Generally, the volume fraction changes easily after the slurry have hardened, causing some cracks to form in the bodies when the curing conditions are not good for the hardening process. To avoid changes in the volume fraction and the formation of cracks inside the hardened bodies, we developed a warm press method in a previous study [5]. In this method, geopolymer starting slurry is first dried at more than

* Corresponding author: hashimoto.shinobu@nitech.ac.jp

100 °C for several hours to vaporize any water it contains. The resultant hardened sample is ground into particles of less than 500 μm in diameter. The particles of the dried sample are placed in a metal mold, and the mold is then set in a warm press device. The mold is uniaxially pressed at 200–300 MPa and simultaneously heated to 300 °C. This method is called the warm press method because the heating temperature is lower than that of conventional ceramic heating using the hot press method. Recently, this warm press method was used to fabricate both geopolymer and calcium hydroxide hardened bodies [6].

So far, the grinding treatment of the starting fly ash has positively influenced the compressive strength of the resultant geopolymers [7–9]. In this study, the effect of the grinding treatment of the starting fly ash on the compressive strength of the resultant geopolymers fabricated using the warm press method was clarified. The relationship between the compressive strength and the relative density of the geopolymer hardened bodies fabricated using the warm press method was examined. Under the warm press conditions of a singular uniaxial pressing at 200 MPa, the effect of the heating temperature and duration on the relative density and compressive strength were investigated.

2 Experimental procedure

Fly ash exhausted from a Japanese power plant (Chubu Electric Power Co., Inc., Japan) was used as the starting material in this study. The chemical composition of the fly ash is given in Table 1. This is a standard-grade fly ash produced by Japanese coal-fired power generation. The fly ash was ground by a tower mill device (Takahama Industry Co., Ltd., Japan). After tower milling, both the ground power and the fine powder collected by a dust collector were used as the starting coal fly ash material. As a reference, fly ash without tower milling treatment was also used.

Compound	SiO_2	Al_2O_3	Fe_2O_3	CaO	K_2O	TiO_2	SO_3	MgO	Na_2O	P_2O_5
Mass %	54.3	27.1	7.9	3	2.2	2.1	1	0.8	0.5	0.5

Table 1. Chemical composition of the fly used in this study

For each of the three types of fly ash, 100 g of fly ash, 15 g of reagent-grade sodium hydroxide, and 33.3 g of fresh water were mixed to form starting slurry. The slurry was heated at 130 °C for 2 h to vaporize the water. Finally, the dried hardened sample was ground by a conventional mill into particles of less than 500 μm in diameter. The obtained particles were placed in a cylindrical steel mold with a diameter of 15 mm, and the mold was set in the warm press device. Beforehand, a thin carbon sheet was used to cover the cylindrical surface of the mold. Fig. 1 shows the warm press device and cylindrical metal mold used in this study.

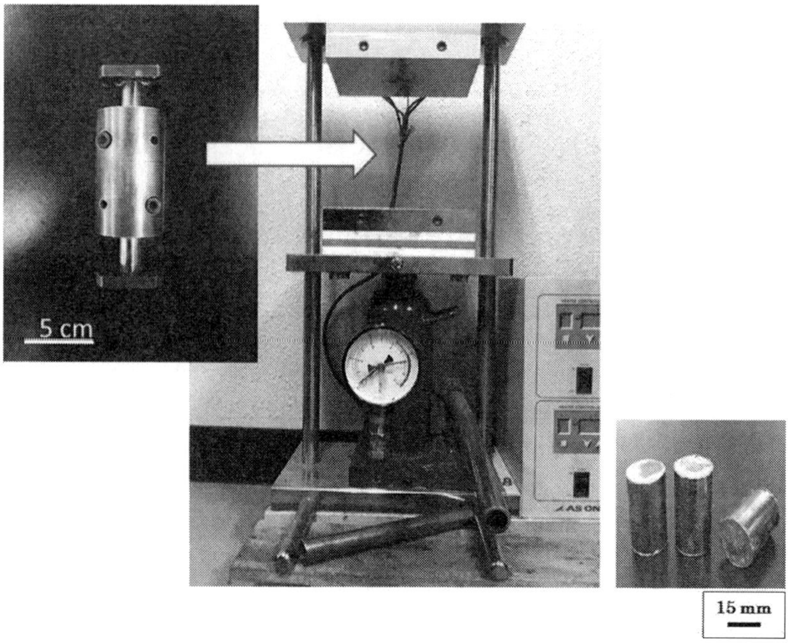

Fig. 1. Photographs of the warm press device, cynlindrical metal mold, and resultant hardened bodies

During the warm press process, temperatures ranged from room temperature to 280 °C at a constant uniaxial pressure of 200 MPa. Starting at room temperature, this fixed uniaxial pressure was applied to the steel mold holding the powder, and the mold was then heated to the desired temperature as measured at the surface of the mold by a contact-type thermocouple sensor. After a fixed molding time, starting from when the mold reached the target temperature, the mold was released from the warm press device, and the hardened sample was removed from the mold. The compressive strength of each of the hardened bodies was determined using an Instron 5582 test apparatus (Instron Co., Ltd., Japan). At least three hardened bodies were used for the compressive strength examination under each set of warm press conditions. The crystal phases of the resultant bodies were analyzed by powder X-ray diffraction (XRD; X'Pert-MPD, Philips). The micromorphology of the sample particles was observed before and after the warm press treatment by scanning electron microscopy (SEM; JMS-6010LA, JEOL). Finally, the concentrations of aluminum and silicon ions eluted into the sodium hydroxide solution were analyzed by inductively coupled plasma mass spectrometry (ICP-MS; ICPS-7000, Shimadzu). During the ICP-MS analysis, 100 g of distilled water, 5 g of sodium hydroxide, and 5 g of fly ash were mixed well while stirring with a rotor for 60 min. After filtration, the remaining water was evaluated.

3 Results and discussion

3.1 Grinding treatment of fly ash

Fig. 2 shows SEM photographs of the starting fly ash with and without tower milling treatment and their particle size distribution. This figure also shows the fine particles collected by the dust collector. With the grinding treatment, the average size of the fly ash particles decreased from 10.4 to 6.8 μm, and most particles were not spherical. This fly ash

 DOI: 10.1051/matecconf/20179701120

was used as the starting material to form the starting slurry for the geopolymer hardened bodies created by warm press.

Fig. 3 shows the variation in the compressive strength and bulk density of the hardened bodies fabricated by warm press (temperature: 200 °C, uniaxial pressure: 200 MPa) with respect to the average size of the fly ash. When ground fly ash particles with a diameter of 6.8 μm were used, the average compressive strength of the hardened bodies reached to approximately 120 MPa. When fine spherical fly ash particles gathered by the dust collector were used, the average compressive strength of the

Fig. 2. SEM photographs of the starting fly ash samples with and without tower milling treatment and their particle size distrubutions. As a refecence, data for the original fly ash are also shown

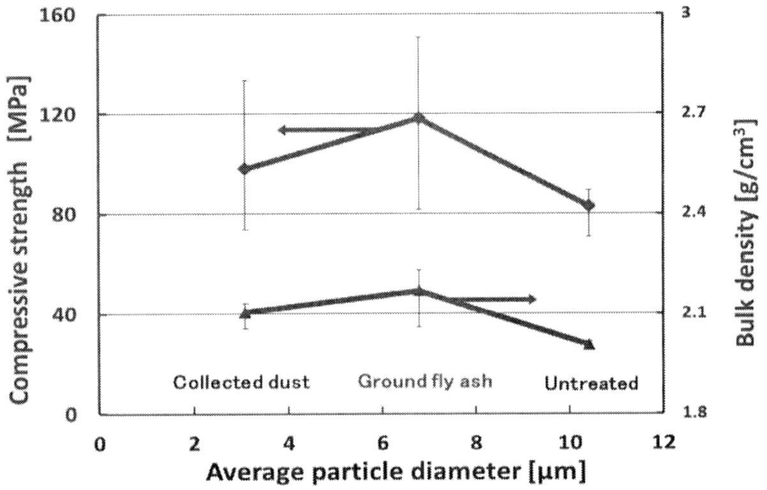

Fig. 3. Compressive strength and bulk density of hardened bodies fabricated by warm press (temperature: 200 °C, uniaxial pressure: 200 MPa) plotted against the average diameter of the fly ash

hardened bodies was approximately 100 MPa, which is lower than that of the ground fly ash-derived hardened bodies. Furthermore, after warm press treatment, analysis of magic angle spinning nuclear magnetic resonance (MAS-NMR) spectra confirmed that geopolymerization occurred inside the hardened bodies [5]. These results indicate that the grinding treatment of fly ash produces an excellent starting material for geopolymerization.

To clarify the increase mechanism of mechanical strength of geopolymers, the elusion behavior of the three types of fly ash particles in the sodium hydroxide solution was analyzed using ICP-MS. Fig. 4 shows the amount of aluminum and silicon ions eluted from the three types of fly ash immersed in the sodium hydroxide solution. Silicon ions escaped easily from the fly ash particles into the solution. In particular, when the ground fly ash was used, the amount of eluted silicon ions was the largest compared to that for the other types of fly ash, which increased the reaction activity of geopolymerization inside the body. Therefore, the ground fly ash produced the geopolymer products with the largest average compressive strength.

3.2 Effects of temperature and duration

Fig. 5 shows the variation in the compressive strength and bulk density of the hardened geopolymer products derived from ground fly ash under a uniaxial pressure of 200 MPa for 40 min with respect to the heating temperature. The compressive strength of the geopolymer hardened bodies increased with increasing heating temperature. When the heating temperature was 280 °C, the average compressive strength reached approximately 120 MPa. However, after heating at 280 °C, the bulk density of the hardened body was the same as that of the body heated at 200 °C. Therefore, as a result, after heating at 200 °C, the compressive strength almost showed the same value.

Fig. 6 shows the variation in the compressive strength and bulk density of the geopolymer hardened bodies under a uniaxial pressure of 200 MPa and a heating temperature of 280 °C with respect to the heating duration. The average compressive strength increased with increasing heating duration. When the duration was 60 min, the average compressive strength reached approximately 150 MPa. In contrast, after 40 min, the bulk density of the hardened geopolymer body was almost the same as that of the hardened body after 60 min. Therefore, if the compressive test sample number increases, the compressive strength remains almost the same.

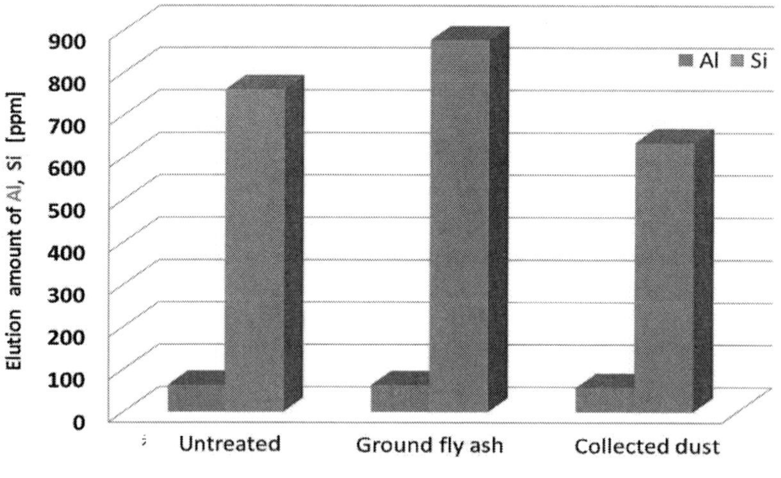

Fig. 4. Amount of aluminum and silicon ions eluted from three types of fly ashes immersed in sodium hydroxide solution

 DOI: 10.1051/matecconf/20179701120

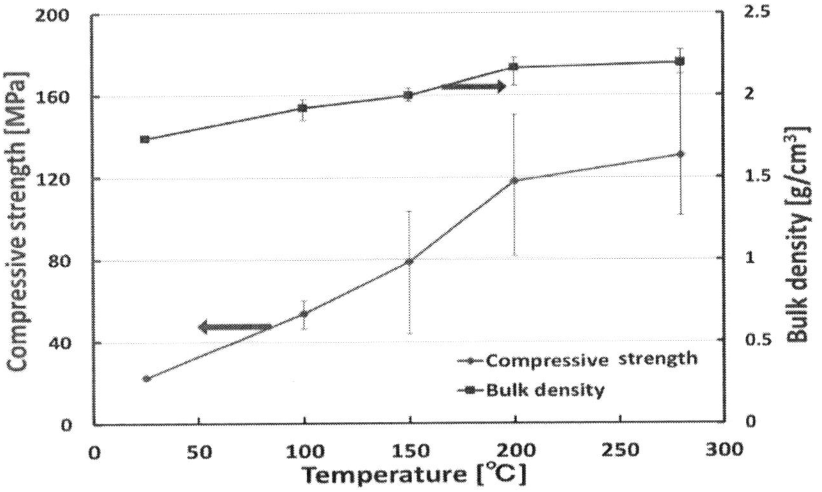

Fig. 5. Compressive strength and bulk density of hardened geopolymer products derived from ground fly ash under a uniaxial pressure of 200 MPa for 40 min plotted against heating temperature

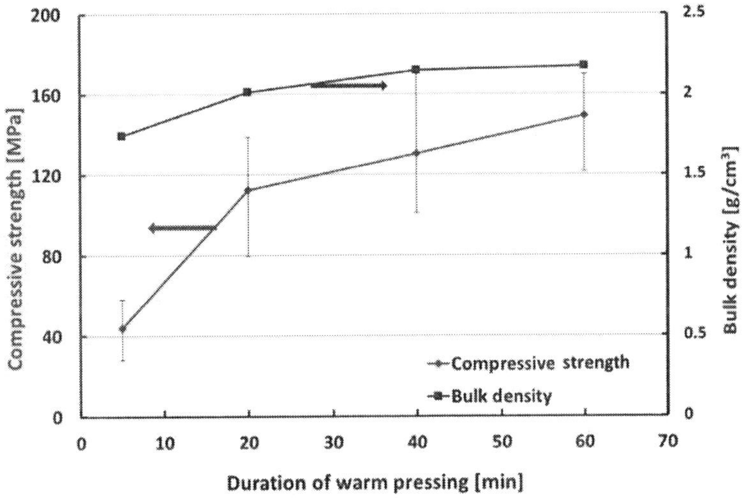

Fig. 6. Compressive strength and bulk density of geopolymer hardened bodies under a uniaxial pressure of 200 MPa and heating at 280 °C plotted against duration

3.3 Microstructure of hardened bodies

Fig. 7 shows SEM photographs of the samples before and after warm pressing at 280 °C for 60 min. After warm pressing, the microstructure became dense. Uniaxial pressing during heating was thought to decrease the pore volume, leading to the formation of a dense body. Our previous paper discussed the effect of warm pressing on the pores of geopolymer bodies [5].

 DOI: 10.1051/matecconf/20179701120

Fig. 7. SEM photographs of samples before and after warm pressing at 280 °C for 60 min

This warm press method is a very unique and convenient method of fabricating geopolymer products because a geopolymer hardened body can be obtained after only several hours and a sample product with high dimensional accuracy can be fabricated. Furthermore, because the product has a high density and the mechanical strength of the hardened body is very high, this method may be useful as a novel fabrication method of high-performance geopolymer products.

4 Summary

Fly ash was ground by a tower mill device into particles with an average diameter of 6.8 μm. Fine fly ash powder gathered by a collection filter after grinding was also used as a starting fly ash. As a reference, original fly ash was considered as well. Starting slurries consisting of 100 g of fly ash, 15 g of potassium hydroxide, and 33.3 g of fresh water were dried at 1300 °C for 2 h. The resultant dried samples were ground into particles with diameters of 500 μm. Subsequently, a cylindrical steel mold containing the particles was set into a warm press device. When the ground fly ash was used among the three types of fly ash, hardened bodies with the highest compressive strength could be fabricated. When the ground fly ash with an average diameter 6.8 μm was used and the heating temperature and duration were 280 °C and 60 min, respectively, the compressive strength of the geopolymers reached approximately 150 MPa.

References

1. J. Davidovits, J. Therm. Anal., **37**, 1633 (1991)
2. H. Takeda, S. Hashimoto, H. Yokoyama, S. Honda, Y. Iwamoto, Materials, **6**, 1767 (2013)
3. H. Takeda, S. Hashimoto, T. Iwata, S. Honda, Y. Iwamoto, J. Mater. Cycles Waste., **14**, 403 (2012)
4. H. Takeda, S. Hashimoto, H. Kanie, S. Honda, Y. Iwamoto, Ceram. Int., **40**, 4071 (2014)
5. H. Takeda, S. Hashimoto, H. Matsui, S. Honda, Y. Iwamoto, Constr. Build. Mater., **50**, 82 (2014)

6. S. Hashimoto, W. Shimoda, H. Takeda, Y. Daiko, S. Honda and Y. Iwamoto, Constr. Build. Mater., **110**, 65 (2016)
7. G. Mucsi, S. Kumar, B. Csoke, R. Kumar, Z. Molnar, A. Racz, F. Madai, A. Debreczeni, Int. J. Miner. Process., **143**, 50 (2015)
8. A.M. Mustafa Al Bakri, H. Kamarudin, M. Bnhussain, J. Liyana, C.M. Ruzaidi, Appl. Mech. Mater., **313/14**, 169 (2013)
9. G. Mucsi, Z. Molnar, S. Kumar, Acta Phys. Pol. A., **126**, 994 (2014)

Utilization of Rice Husk Ash as partial replacement with Cement for production of Concrete Brick

Le Tuan Minh[1,*], and *Nguyen Xuan Thanh* Tram[1]

[1]Department of Silicate Materials, Faculty of Materials Technology, Ho Chi Minh city University of Technology, VNU-HCM

Abstract. Rice husk ash (RHA) obtained from agricultural by-product could be a potential source for partial cement replacement to produce concrete brick. In this study, primary RHA and RHA heat-treated at 500°C were used to substitute cement in raw material mixtures. Preliminary analyses were conducted to evaluate some properties of RHA such as loss of ignition (LOI), morphology, chemical and phase compositions. X-ray Powder Diffraction (XRD) and X-Ray flourescence (XRF) results indicated that silica (SiO_2) is the major component of RHA. Transmission Electron Microscopy (TEM) images showed that RHA particles were in polygonal shapes with diameter approximately from 300 nm – 1 μm. Cement was replaced with RHA by weight at 10 %wt. The effects of RHA on the physical and mechanical properties of the aggregate cement bricks were investigated. Results demonstrated that the compressive strength of concrete bricks increased with age at curing and decreased as the percentage of RHA content increased. It is recommended that further studies should be carried out to achive optimum parameters for production of concrete brick in economic aspects.

1 Introduction

Currently, sustainable development is one of the most interests all over the world. Historically, sintered-clay bricks have been used as construction materials, however, the production of them created energy and environmental concerns. There is an economical benefit in using concrete bricks over sintered-clay bricks. Concrete bricks comprise of cement, fine and coarse aggregates. They are bound together tightly into a compact whole without sintering since cement has ability to be cohesive and adhesive when mixed with water. For that reason, energy and environmental issues could reduce significantly in mass production. Cement is the most expensive component among the raw materials of concrete bricks. Alternative materials have been considered in order to reduce the manufacturing cost [1-3]. Every year million tons of paddies are produced in Vietnam. This results in large amount of agricultural waste materials that are not adequately managed and utilized. Rice husk, generally considered as agricultural by-products, is mainly used for fertilizer and energy production. Rice husk ash (RHA) is obtained from the burning of rice husk. Previous

* Corresponding author: letuanminhsx@gmail.com

studies showed that the predominant component of RHA is silica (SiO_2). It could chemically react with calcium hydroxide ($Ca(OH)_2$) at room temperature to achieve adhering properties and combine materials firmly. Therefore, reactive silica is also known as pozzolanic additive commonly used in cement production and the manufacture of concrete [4-6]. Despite research on RHA utilization in construction materials is not new, the use of RHA as a partial replacement to cement will solve the enviromental problem of argicultural by – product as well as contribute to reduce the manufacturing cost of concrete bricks. Therefore in this study, primary RHA and RHA heat-treated at 500°C were used to substitute cement in raw material mixture. The effects of RHA on the physical and mechanical properties of the aggregate cement bricks were investigated. It is expected that the cheaper concrete bricks are produced, the lower cost of civil construction could be achieved.

2 Materials and methods

Primary RHA was collected from Sepzone – Linh Trung 1 (Freetrend Industrial Limited). RHA was sieved using 125-µm sieve to remove impurities. Then RHA was dried in an oven at 100°C for 24h. The loss of ignition (LOI) test was done by placing crucible with 1 g of dried RHA in furnace and igniting at 900 °C for 30 min to achieve a constant mass. The loss of ignition is calculated as mass percentage based on the formula (eq. 1):

$$LOI = \frac{M_0 - M_1}{M_0} \times 100\% \tag{1}$$

Where: M_o is the mass of the starting sample and M_l is the mass of the sample after ignition.

The chemical composition of RHA was determined by X-ray fluorescence spectrometry (XRF, ARL ADVANT'X, Thermo Fisher, USA). The phase composition of RHA was analyzed by X-ray diffraction (XRD) using diffractometer system (D2 Phaser, Bruker AXS GmbH, Karlsruhe, Germany). Particle size and shape of RHA were evaluated using Transmission Electron Microscope (TEM, JEM-1400, JEOL, Japan).

The raw materials of concrete bricks were mixed according to the ratios shown in table 1 [7]. Three mixtures differing in the RHA contents were prepared. Control was the batch using cement with 0%wt of RHA. Primary RHA and RHA-500°C were the batches using cement with 10%wt of primary RHA and RHA heat-treated at 500°C, respectively. Mixing mixtures were conducted using concrete mixers for 7 -10 minutes. Then mixtures were poured into mold (4x8x16cm). Simultaneously vibration was applied to increase homogeneity of mixture and density of finished product. After 7 and 14 days of curing with 100% humidity, concrete bricks were characterized in term of compressive strength and bulk density. Diameter and height of each specimen were measured. Then a load was applied to crush each specmen using compression machines 3000KN (C071, Matest, Italy) at a crosshead speed of 0.6 KN/s.

Table 1. Mixtures of raw materials

Batch/Materials	Cement (g)	Coarse Agg. (g)	Fine Agg. (g)	RHA (g)	Water (ml)
Control	300	2450	1350	0	300
RHA Primary	270	2450	1350	30	300
RHA 500	270	2450	1350	30	300

DOI: 10.1051/matecconf/20179701121

3 Results and discussion

Loss of ignition (LOI) of rice husk ash (RHA) is frequently due to the presence of moisture, unburned residues and unstable mineral phases at high temperature. In this study, LOI of primary RHA and RHA heat-treated at 500°C were 5.27 %wt and 4.13 %wt, respectively (Table 2). Difference of LOI values could be atributed to the different amount of unburned residues and moitures [8].

Table 2. Loss of Ignition (LOI) of primary RHA and RHA heat-treated at 500°C

Sample	LOI (%)
RHAprimary	5.27 ± 0.57
RHA 500^{0}C	4.13 ± 0.55

Table 3 showed the chemical composition of primary RHA and RHA heat-treated at 500°C. Basically, silica (SiO_2) is the major component of RHA (90 %wt). Besides, chemical component of RHA also included K_2O and CaO (4 %wt) as well as some minor content of MgO, MnO and Al_2O_3 (less than 1%wt).

Table 3. Major chemical composition of raw materials (wt%)

Composition	SiO_2	K_2O	CaO	MgO	MnO	P_2O_5	Al_2O_3	Fe_2O_3
RHAprimary	90.30	4.17	1.30	0.50	0.47	0.37	0.17	0.15
RHA 500^{0}C	87.70	4.04	1.16	0.52	0.48	0.40	-	0.17

Figure 1 showed XRD patterns of primary RHA and RHA heat-treated at 500°C. The results indicated that cristobalite that is one of SiO_2 polymorphs is the main crystal phases of RHA. Quartz was also detected as minor phase. The broad peak between 21-37^0 implied the presence of amorphous phase of carbon and silica (as shown in Fig 1).

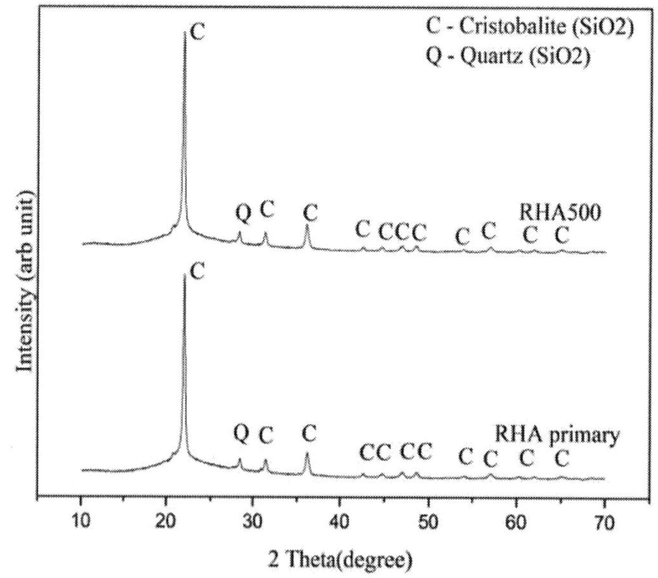

Fig. 1. XRD patterns of primary RHA and RHA heat-treated at 500°C.

Crystal size is calculated by using the Scherrer's equation as follows (eq. 2): [10]

$$D = \frac{K\lambda}{B\cos\theta} \quad (2)$$

With B, K, λ and D, respectively half peak width (FWHM) in radians, Scherrer constant (K=0.9), x-rays wavelength (1.5406 Å) and the crystal diameter (nm). Crystal size values of primary RHA and RHA heat-treated at 500°C were shown in Table 4.

Table 4. Crystal size of rice husk ash

	2 Theta (deg)	FWHM (deg)	D (nm)
RHA primary	22.0564	0.2165	
	31.4861	0.2362	342.2199 ±102.3046
	36.1971	0.2755	
RHA 500	21.9239	0.2165	
	31.3095	0.1378	607.1219 ± 284.8269
	35.9967	0.0984	

TEM images showed that RHA particles were in polygonal shapes with diameter is approximately from 300 nm – 1 μm (as shown in figure 2). Particle size of RHA heat-treated at 500°C was larger than that of primary RHA. This could be explained due to the coagulation of RHA particles and development of crystal size (as shown in table 3) as being treated at high temperature.

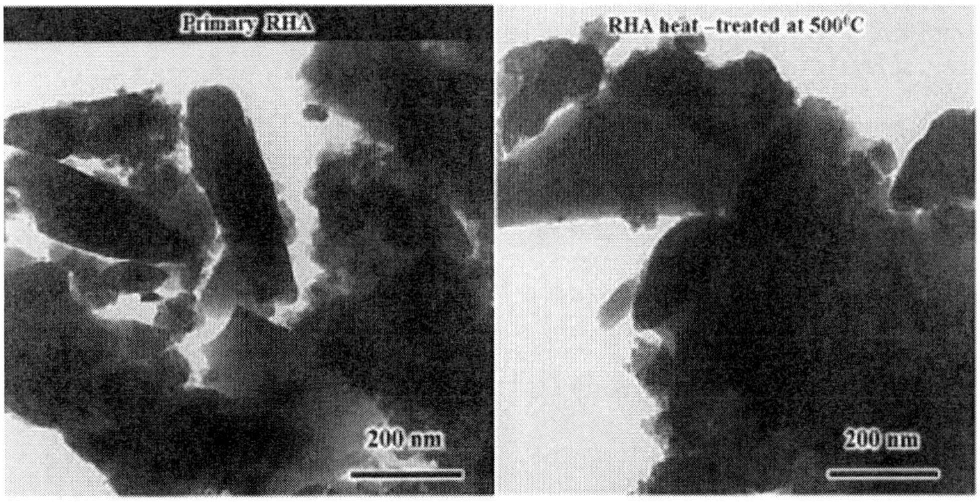

Fig. 2. TEM images of primary RHA and RHA heat-treated at 500°C

The standard consistency of hydraulic cement was determined using Vicat Apparatus. The basic aim is to find out the water content required to produce a cement paste of standard consistency as specified by ASTM C 187 [10]. Results were showed in Table 5. The replacement of RHA changed the normal consistency of cement. The required water for normal consistency was increased as 10% replacement of RHA by weight of cement. Comparison between primary RHA and RHA500, higher quantity of water needed as using RHA heat-treated at 500°C.

Table 5. The normal consistency of hydraulic cement.

Sample	Water (ml)	Slump of Vicat (mm)
Cement (400g)	130	6 ± 0.5
Cement (360g)+ RHA Primary (40g)	145	5 ± 0.5
Cement (360g) + RHA500 (40g)	155	6.5 ± 0.5

Table 6 showed compressive strength and bulk density of concrete bricks after 7 and 14 days of curing with 100% humidity. Results demonstrated that the compressive strength of concrete bricks increased with age at curing and decreased as the percentage of RHA content increased. RHA can replace cement from 10 – 40 %wt, therefore, it is suggested that various RHA substitution contents could be investigated in the next studies [11].

Table 6. Compressive strength and bulk of concrete bricks after 7 and 14 days of curing with 100% humidity.

Days	Sample	Bulk Density (kg/m^3)	Compressive strength (MPa)
7	Control	2248.46 ± 17.81	3.75 ± 0.10
	RHA Primary	2103.05 ± 48.64	3.60 ± 0.11
	RHA500	2192.52 ± 26.65	3.65 ± 0.05
14	Control	2175.31 ± 75.01	4.97 ± 0.31
	RHA Primary	2300.55 ± 25.34	3.93 ± 0.12
	RHA500	2216.10 ± 31.18	3.77 ± 0.09

Fig 4 showed the practical images of specimens before and after mechanical testing. It can be seen that pores on the surface of RHA specimen was higher than that of control specimen. After mechanical load test, control specimen was cracked in one direction whereas RHA specimen was destroyed in multi-direction.

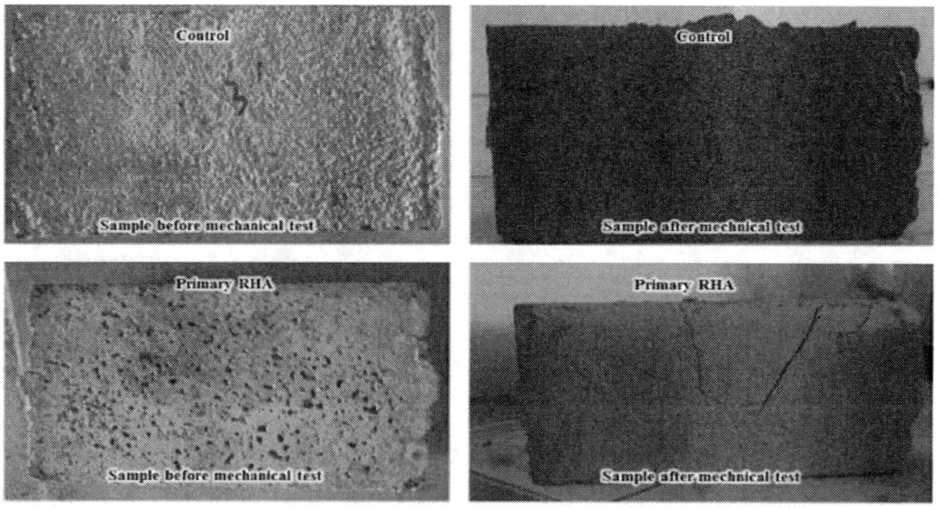

Fig. 4. Specimens before and after mechanical testing.

DOI: 10.1051/matecconf/20179701121

4 Conclusion

The rice husk ash (RHA) could be used to replace cement for production of concrete bricks to eeduces material cost and emission of CO_2 due to less cement utilization. However, the content of RHA needs to modify to achive optimum parameters for production of concrete brick. Rice husk ash after heat treatment is advatageous on reducing organic component that may contribute to chemical resistance of concrete brick.

The authors would like to express appreciation for the financial support of the scientific research grant of Ho Chi Minh City University of Technology for undergraduate (SVCQ -2015-CNVL-126).

References

1. G.A. Habeeb, H.B. Mahmud, Mater. Res.,**13(2)**, 185 (2010)
2. S. Singh, D. Kumar, Int. J. Innov. Res. Adv. Eng., **1(6)**, 214 (2014)
3. A. Kachwala, A. Pamnani, A. Raval, Int. Res. J. Eng. Technol., **12(5)**, 175 (2015)
4. S.H. Sathawane, V.S. Vairagade, K.S. Kene, Procedia Eng., **51**, 35 (2013)
5. A. Shukla, C.K. Singh, A.K. Sharma, Int. J. Earth Sci. Eng., **4(6)**, 965 (2011)
6. I.O. Obilade, Int. J. Eng., **5(4)**, 8269 (2014)
7. S.H. Kosmatka, B. Kerkhoff, W.C. Panarese, *Design and control of concrete mixtures* (Portland Cement Association, USA, 2002)
8. C. Santasnachok, W. Kurniawan, H. Hinode, J. Life Sci., **10(3)** (2015)
9. E.M. Ginting, B. Wirjosentono, N. Bukit, H. Agusnar, Chem. Mater. Res., **6(7)**, 14 (2014)
10. ASTM C 187 – 98 , Standard Test Method for Normal Consistency of Hydraulic Cement, ASTM C 187 - 98 in: Annual Book of ASTM Standards, Vol 04.01, (1995)
11. ASTM C 109 - 88 , Standard Test Method for Compressive Strength of Hydraulic Cement Mortars, ASTM C 109 - 88 in: Annual Book of ASTM Standards, Vol 04.01, (1995)

 DOI: 10.1051/matecconf/20179701122

Strategies to Reduce Water Footprint in Palm Oil Production: A Case of PTP Mitra Ogan, Baturaja, South Sumatra

Herda Sabriyah Dara Kospa[1,*], *Kris R.D.* Lulofs[2], *Chay* Asdak[3] and *Supli* E. Rahim[4]

[1]Urban and Regional Planning Department, Indo Global Mandiri University, Palembang, 30129, Indonesia
[2]Master of Environmental and Energy Management, University of Twente, Leeuwarden, The Netherlands
[3]Master of Environmental Science, University of Padjadjaran, Bandung, 40132, Indonesia
[4]University of Palembang, Palembang, South Sumatra, 30139, Indonesia

Abstract. The massive expansion of palm oil industry in Indonesia has triggered environmental issues including water-related problems which have become an important concern. Regarding the issues, sustainable practice standard has been set up as a requirement for palm oil to enter global market. Inevitably, water consumption in this sector is very crucial to be analyzed. One of the methods that can be used as a tool for sustainable appropriation of fresh water resources is water footprint analysis. The primary aim of this study was to formulate the strategies to reduce the water footprint in the palm oil production based on the best practice criteria. Both quantitative and qualitative research was conducted to get the value blue water (volume of surface or groundwater evaporated) and grey water (dilution volume to dilute pollutants according to agreed water quality standards). The values of water footprint in palm oil production obtained were used to represent the existing water use and were utilized as the basis for formulating strategies in reducing water use in the palm oil milling processes which was compared with the best practice criteria. The result showed that the blue water of CPO was 109.6 m^3/ton and the grey water was 537.7 m^3/ton, while the blue water of palm kernel was 62,4 m^3/ton and grey water was 306,2 m^3/ton. The value indicated that there was an inefficient use of water in the production of palm oil. The use of steam accumulator has been proposed to reduce the use of blue water by optimizing the steam supply. Besides, the reuse of water from fat-pit pond for pressing purposes, or recovery of condensate water as dilution water in the press unit which will affect the amount of wastewater discharged can be done as the strategies in reducing both blue and grey water, as well as reuse of cooling water turbines.

* Corresponding Author: darabantet@rocketmail.com

1 Introduction

Nowadays, palm oil is a strategic sector in Indonesia because the trend of its export in the global market is increasingly higher, reaching 23.5 million tons per year (RSPO, 2013).The palm oil industry in Indonesia has been developing rapidly. It was twice higher from 4.2 million ha in 2000 to be 8.0 million ha in 2010.

As one of the key elements of Indonesian economic development, the development of palm oil sector is predictably higher. It is because the Government of Indonesia has targeted to export Crude Palm Oil (CPO) at 52 million tons in 2020. Consequently, the oil palm plantation in Indonesia was projected to reach 13 hectares by 2020.

In realizing the target, all stakeholders, including companies should ensure that the palm oil industry is sustainably organized. It is because sustainable practices of palm oil industry become one of the requirements for a palm oil industry to enter the global market. In Indonesia, however, palm oil sector is facing sustainability issues due to the standards set up by RSPO. The issues include deforestation, loss of biodiversity and habitats, soil erosion, air pollution (GHG emissions), water-related problems, and social conflicts (WWF, 2012).

Among those issues, water-related problems have become an important concern. In plantation, for instance, it needs relatively high water use. According to Bergert (2000), after 11 years from plantation, palm trees with a population of 150 trees/ ha require annual rainfall of 2000 mm which is evenly distributed throughout the year. During the dry period with a relatively low rainfall, the freshwater consumption will be increasingly higher.

The aforementioned condition potentially causes conflicts in water utilization with other sectors, especially in dry season. For instance, villagers in Central Kalimantan and other provinces in Indonesia living nearby palm oil plantations faced water problems because of the palm oil plantation activities (SEI, 2012). The problems include reduced water debit and diverted water flow due to the constructions of channels and dams developed for the palm oil irrigation.

Meanwhile, in mills, large amount of water is also used for the palm fruit extraction. As much as 50 % of water used in the extraction process becomes palm oil mill effluents (POME). From 1 ton of CPO, 5 to 7.5 tons of water in the form of POME is produced (Muyibi et al., 2014). If the effluent is directly discharged to the stream, it can cause freshwater pollution, affect the people and downstream biodiversity (WWF, 2012).

Those impacts indicate the urgency of applying sustainable water resource management in palm oil industry, which takes the efficiency and the renewability of sources into account. One of the methods that can be used as a tool for sustainable appropriation of freshwater resources is water footprint analysis. According to Hoekstra et al. (2011), water footprint can be used as a communication tool between stakeholders in better understanding on how water management can support social, economic and environmental outcomes. (Chapaign & Tickner, 2012). For companies, the results of water footprint in their activities will provide insight, information and knowledge to consider the strategies to reduce the use of water. In this case, reducing water can give benefits for companies and for environment in general. Water reduction benefits for the companies include saving on water and sewage disposal costs, saving energy costs (through reducing appliance and hot water use) and enhancing the company reputation by being proactive in protecting environment. Meanwhile, for environment, the benefits include conserving groundwater and protecting surface water from pollution (Sappi, 2012).

Water footprint analysis is geographically, technologically, and temporally dependent. There are many studies related to the water footprint in palm oil sector throughout the world. In Indonesia, however, there is no sufficient detailed water footprint information in that sector. Hence, this research was conducted to investigate water footprint of palm oil production based on the case of PTP Mitra Ogan in South Sumatra, Indonesia. The primary aim of this study is to formulate the strategies to reduce the water footprint in the palm oil production based on the best practice criteria.

2 Methods and data

The unit of analysis use was PTP Mitra Ogan in Peninjauan District, Ogan Komering Ulu Regency, South Sumatra, Indonesia. It was one of the largest plantation companies in the South Sumatera whose land is about 65.678,95 ha in 2013. This company had two processing mills and had developed an integrated palm oil production from cultivation to processing CPO since 1988 at a processing capacity of each 30 tons per hour. This study identified the nucleus plantation area of 655,64 ha.

2.1 Palm Oil Production

The stages of palm oil production include cultivation of oil palm and palm oil milling process. The processes in the mill consists of several steps, namely transportation of oil palm fruits (known as fresh fruit bunches, FFB), reception of FFB in the mill, FFB processing, waste treatment, and waste reuse.

Water is very important in the activities of a palm oil mill. It plays a role as the feed for boiler in generating power and for product treatment (Naibaho, 1996). Water used for the boiler in producing steam is equal to 0.903 tons of FFB or 2.77 m^3 / ton of production.

Besides as the feed for boiler, water is also used in two hot water tanks which functions to supply water to some production machines, i.e. one machine located at the press station press and other machines in the clarification station. The stages that involve water in the palm oil production machines are illustrated in Figure 1.

In addition, water is also used in the purification process in the clarification station. The use of water in this station is very important because the purification process will produce a liquid at a desire quality level. A process of refining palm oil into CPO is intended to eliminate the water content in it reaching 0.2%, so a good quality of CPO is obtained.

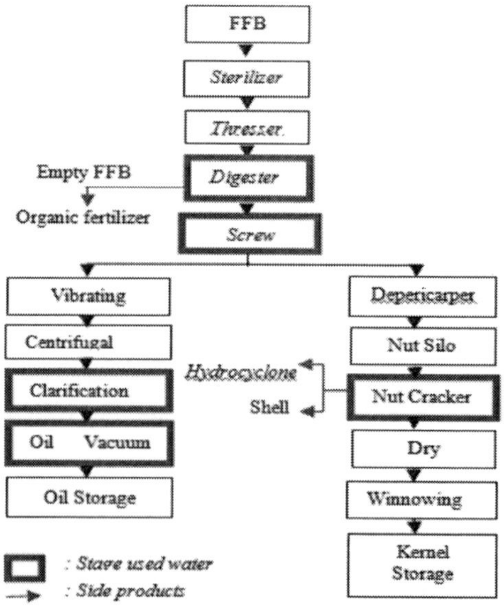

Fig. 1. CPO and Kernel Processing

DOI: 10.1051/matecconf/20179701122

In the clarification station, the water is conducted to the hot water tank and to the pump of vacuum drier for further processing. Water in the hot water tank will then be used by the machines in the refining process. The flow of water use in the clarification station is illustrated in Figure 2.

The process of dilution required in the refining process aims to separate oil efficiently and to obtain the quality palm oil. The amount of water recommended is proportional to the amount of oil contained in the liquid. The amount of water required in the process of purification is 0.114 m^3 / ton of FFB.

In addition, the use of water in the processing of palm oil also occurs in the kernel station at the machine of clay baths (separation of seeds). This separation is based on differences in the density between the kernel (ρ = 1.07 kg/m^3) and shell (ρ = 1.3 kg/m^3). The mixture between the kernel and the shell is mixed with the liquid clay (ρ = 1.2 kg/m^3) that is free of sand and calcium carbonate ($CaCO_3$), so that the kernel will be floating and shells will be sinking. This process is performed in a conical tank equipped with a pump to circulate the liquid clay. The addition of water in this process is as much as 0.012 m^3 / ton of FFB which is done twice by mixing water and $CaCO_3$.

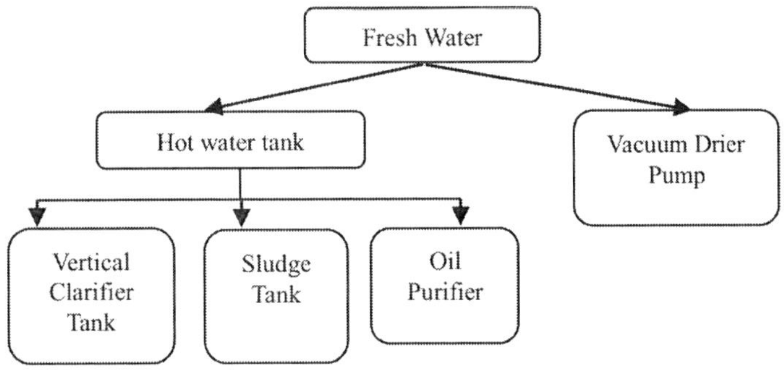

Fig. 2. The process of water use in the clarification station

2.2 Methodology of Water Footprint for Palm Oil Production

In general, water footprint of palm oil production in this study focuses on palm oil milling processes. In the palm oil production, the calculation of water footprint component in the industrial stage only includes the components of blue and grey water because the phases mostly use surface water or groundwater. The calculation of blue water in industrial process refers to the amount of water consumed in each stage of production (Hoekstra et al., 2011).

According to Hoekstra et al. (2011), the water footprint of a product can be calculated by using water requirement mechanism in each step of processing product. The calculation of water footprint involves inputs of product, process, product fraction and value fraction (Hoekstra et al., 2011).

The amount of blue water in this stage can be known by calculating the amount of water that is lost through evaporation mechanism at every phase of production. According to Hoekstra et al. (2011), there are difficulties to quantify the amount of evaporation in the production process directly. Therefore, the calculation of water consumption (Consumptive Water Use) for each phase using water balance approach in the industrial stage can be used to calculate the blue water. In analysing this data, a value of product fraction was used. It was a value representing the fraction of water lost during the process in each step in the mill.

Finally the grey water in the industrial stage is an indicator that refers to the pollution of water bodies due to the production activity. Meanwhile, the grey water footprint in the palm

DOI: 10.1051/matecconf/20179701122

oil mill was estimated based on data of the volume of waste water having quality lower than the ambient water quality. In determining whether the volume of waste water was calculated or not, data of on-site water quality, effluent quality, ambient of water quality standards, effluent quality standards and the yields were used.

According to Hoekstra et al. (2009), the value of grey water can be assumed to be equal to the amount of the waste if the industry has applied the waste treatment before it is discharged to the environment. As a result, the concentration of pollutants in the effluent has the same value or less than the standard of the ambient water quality value.

The values of water footprints in the mill of palm oil production obtained in this research were then used to represent the existing water use in the palm oil company. Information of existing water use was utilized as the basis for formulating strategies in reducing water footprint in the palm oil company. In formulating the strategies of water reduction in the palm oil company, field notes obtained from the interview were studied and compared with the best practise criteria. In this study, the certified palm oil company, PTP London Sumatra, was identified to get information related to the reduction strategies which have been implemented by the company. The strategies could be the best practise criteria having possibilities to be implemented in the palm oil company based on existing conditions in term of technical and technological, managerial, human resource, and economic aspects.

3 Results and discussions

3.1 Palm Oil Milling Processes in PTP Mitra Ogan

Figure 3 shows the mass balance of palm oil production in PTP Mitra Ogan milling process.

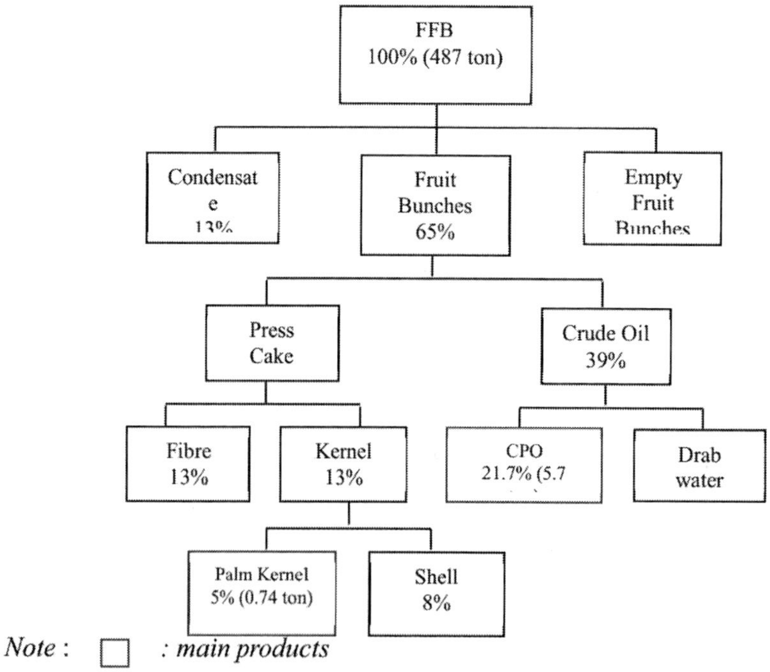

Source: Company Documentation

Fig. 3. Palm Oil Processing in PTP Mitra Ogan

The average production of FFB produced by the central estate of PTP Mitra Ogan per year is 14.81 tons per hectare. In every hectare of FFB harvested, Crude Palm Oil (CPO) and Palm Kernel produced are about 5.7 ton and 0.74 ton by the mill of PTP Mitra Ogan. With the processing capacity of 30 ton FFB per hour, it can process for about 487 ton of FFB during 22 operating hour in a day.

3.2. Water Footprint of CPO and Kernel

3.2.1. Blue Water Footprint of CPO and Kernel

Total water use in all stages of the production of CPO and kernel in the PKS unit of PTP Mitra Ogan is 1.36 m^3 / ton of FFB. It is equal to 5.08 m^3 / ton of production (Table 1). Detailed calculation is presented in Table 5. Based on the average number of processing issued Naibaho (1996: 154) and H-Kittikun et al (2000: 7), the national standard of water use in the PKS unit is equal to 1.1 to 1.2 m^3 / ton of FFB. In addition, the amount of blue water in the extraction of CPO is larger than the value blue water that has been analyzed by Jefferies, et al. (2012), i.e. 2.9 m^3 / ton of production. This value shows that there is a wasteful use of water in the processing of palm oil at the study site.

Table 1. The Values Of Blue Water Component For Cpo And Kernel

No.	Processes	Water Quantity (m^3/ton FFB)	Water Quantity (m^3/ton Prod)
1	Boiler	0.90	3.38
2	Hot water tank (press station)	0.18	0.68
3	Clay bath Mixing (Kernel Station)	0.01	0.043
4	Hot Water Tank (Clarification Station)	0.11	0.35
5	Vacuum Drier Pump (Clarification Station)	0.15	0.55
	Total	1.36	5.08

Hoekstra et al., (2009) stated that generally blue water has no effect or greater impacts on the environment than green water does. Blue water represents the need of water obtained from surface water (river) or groundwater. Blue water used will be "lost" through the process of evapotranspiration, whereas it should be a "reserve" of water resources in the ground or river.

Hence, a more efficient use of water for an industrial process becomes important. In addition to minimizing the impact on the environment, reducing the use of water in the treatment process will also reduce the discharging of waste. This means that efforts to reduce water usage during the process potentially reduce the level of pollution caused by wastewater of palm oil. The implementation of clean production technologies and reuse of wastewater produced can be an alternative in reducing the value of blue water in the palm oil processing.

3.2.2. Grey Water Footprint of CPO and Kernel

In processing FFB into palm oil and palm kernel, a palm oil mill has a wastewater treatment plant. Wastewater generated during the production process is treated before it is discharged into the environment and meets quality standard (Ceff = Cmax). This means that the value of the grey water footprint component in the production process will be equal to the volume of waste water generated due to the capacity of a water body to assimilate contaminants/ pollutants. The presence of wastewater in a water body changes Cnat into Cmax (Hoekstra et al., 2009).

Based on observations in the study location, the waste water from palm oil extraction in the PTP Mitra Ogan is treated before it is discharged into the environment. So, it was assumed that the concentration of pollutants in the effluent is within the standard of the ambient water quality value. As a result, according to its classification, the grey water footprint of the process is the same as the volume of wastewater generated. Due to the limited access to primary data, the data volume of waste water produced was estimated based on the documentation from preliminary studies in PTP Mitra Ogan done by the Estate Departement of Baturaja.

Liquid waste the PKS unit of PTP Mitra Ogan is one of the by products derived from the condensate of the sterilization process, the water from the clarification process, and hydrocyclone (clay bath) water. Based on these results it is known that in the palm oil extraction process based on the material balance and the secondary data of the company. In one operating day, the PKS unit will produce wastewater of approximately 306.81 m^3 / ton. This amount is equivalent to 0.63 m^3 / ton of FFB processed or 2.36 m^3 / ton of waste per ton of production (Table 2).

Table 2. The Value Of Grey Water Component For Cpo And Kernel

Source of Liquid waste	**Persentage (%)***	**WFproses,grey (m3/ton)**
Condensate Water	13	63.31
Clarification Station	45	219.15
Hydrocyclone (*Clay Bath*)	5	24.35
Total	63	306.81
Average / ton TBS processed		0.63
Average /ton production		2.36

**Source: Company document*
Note: - Total amount of CPO & kernel produced per day: 130.03 ton

Handling of palm oil waste includes changing complex organic substances into a simple one microbiologically and anaerobically. Biological processes that occur in each treatment pond are: wastewater is collected in effluent ponds, before being released into the water, the water is treated anaerobically in a digester. Generally, waste water treatment plant (WWTP) consists of the following units:

a. Fat pit

Wastewater produced from washing apparatus, condensate sterilizer, sludge separator, and purifier is directly stored in the pit fat ponds. Wastewater in a pit fat sewer can separate oil and water on its density. The top layer of waste is conducted back into the cleaning (clarified) part. Furthermore, waste water containing less oil is conducted into the ponds.

b. Cooling Tower

This cooling tower is used to cool the hot waste water, so that the waste water temperature is lower than 40°C for supporting the life of anaerobic bacteria in the next ponds. The towers are arranged in series (sequence). Furthermore, the waste discharged into a seeding pond in which bacteria are bred.

c. Anaerobic Pond

An anaerobic pond consists of two ponds with a total volume of wastewater of 126.720 m^3. Basically, the system used in the multiple feeding is an anaerobic system, namely a system that does not require oxygen in the process of changing organic materials. This change requires anaerobic bacteria to enable this process. After the process in WWTP has been done, waste water is conducted into an area through a piping system designed based on the field conditions.

3.3 Strategies of Reducing Water Footprint during Milling Process

During the processing of palm oil in the mill, it is known that it is possible to use water in a smaller amount to obtain the same yields. The amount of blue water footprint of palm oil in the PKS I of PTP Mitra Ogan is 1.36 m^3 / ton of FFB. This amount of water indicates an uncontrolled and inefficient water use. Reducing the value of the use of surface water can be done by recycling wastewater as recommended in the manual of Environmental Impact Control Technology Palm Oil Industry (BAPEDAL, 1988 in Fitria 2003).

In addition, based on a comparative study conducted by the researcher to the PKS of PTP London Sumatra (Lonsum) located in Banyuasin, saving water use can be done by using a steam accumulator. However, it requires further study to be applied in PTP Mitra Ogan such as the technical and financial analysis of the company.

3.3.1 The Use of Steam Accumulator

Based on a study conducted by the researcher in the PKS of PTP London Sumatra in Banyuasin in August 2014, the use of blue water in the mill can be reduced by the use of steam accumulator. Besides, a research conducted by Shah et al. (2009) has proposed steam accumulator to stabilise the steam pressure fluctuation in palm oil mil especially at the demand sites. However, additionally study needs to be done to ensure the possibility of installing a steam accumulator in the mill in which the size of Back Pressure Vessel (BPV) and the capacity of steam consumed vary.

According to an interview conducted with the manager of mill, the PKS will use a steam accumulator when increasing the production capacity at 45 tonnes of FFB / hour to 60 tons of FFB / hour. Despite an increase in capacity, the amount of steam that flows will remain the same as the initial capacity, so the efficient use of water will automatically happen in the boiler. The steam accumulator is used in order to improve the efficiency of the steam supply from BPV (Back Pressure Vessel) to be conducted into the process. This means that the steam accumulator can save about 40% of water use by the boiler. However, further study need to be done so that it can be applied properly accordance with the technical, financial and environmental aspects of PTP Mitra Ogan mills.

The work principle of a steam accumulator is when the boiler is supplying steam to the turbine, the remaining steam coming from the turbine is used for other processes in the palm oil mill. This remaining steam is stored in the BPV tank and will be distributed to some of the equipment like sterilizer in the oil mill at a tank pressure of 3 bar (44.1 psi). When the steam consumption is too much, the tank pressure decreases. It causes the process in the sterilizer takes a relatively long time (95 minutes). To save time and water use in the boiler, a steam accumulator is added to improve the quality and efficiency of the supply of steam to the processing areas in the oil mill. In order to maintain tank the BPV pressure at 3 bars, when the pressure of BPV tank is reduced by less than 3 bars, then the valve (PCV-b) will open slowly to supply steam to the BPV tank from the steam accumulator.

Meanwhile, when the pressure of the BPV tank has reached 3 bars, the valve (PCV-b) will close. This process is continuous if there is a change in pressure between the BPV tank and steam accumulator under or above 3 bars. As result, a steam accumulator accelerates

 DOI: 10.1051/matecconf/20179701122

the process in the sterilizer becoming 40 minutes. The steam accumulator is illustrated in Figure 4.

3.3.2 Reuse of Condensate Water for Diluting Press and Vibro

In addition, saving water in processing palm oil can also be done in the pressing station. The reuse of wastewater is done based on the recommendations of the manual of Environmental Impact Control Technology Palm Oil Industry (BAPEDAL, 1988 in Fitria 2003), key informant (machinist head assistant) and the amount of discharged water, as well as previous studies conducted by Fitria (2003) in the POM of Sei Mangkei, North Sumatra. The strategy based on the study in POM of Sei Mangkei can also be applied in PTP Mitra Ogan because both companies have similarities in term of production capacity, production process and the types of machines used. However, further studies which consider the suitability of the financial, technical, and environmental aspects of the company should be done to apply this strategy.

The efficiency is done by maximizing the water condensate in the process of sterilization or TBS boiling. TBS boiling process in the sterilizer produces liquid

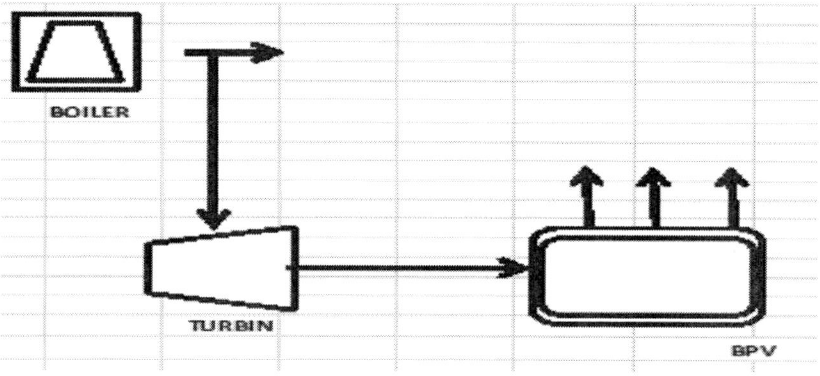

The flow scheme before applying a steam accumulator

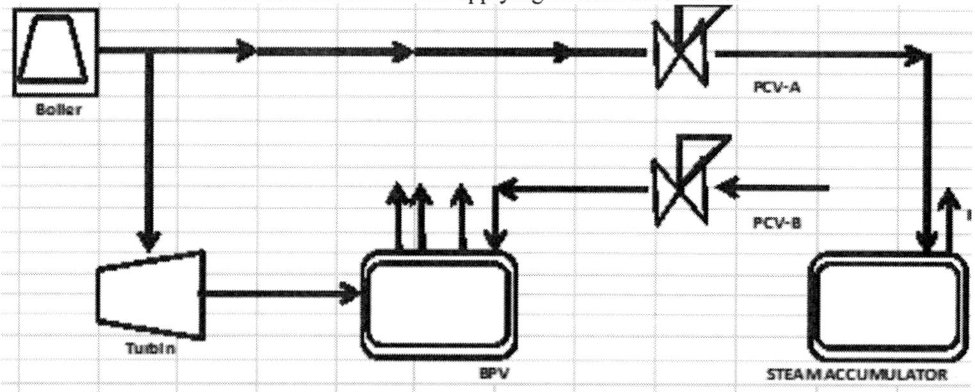

The flow scheme after applying a steam accumulator

Note: PVC = Pressure Valve Control

Fig. 4. The Flow Diagram of Before and After the Application of Steam Accumulator

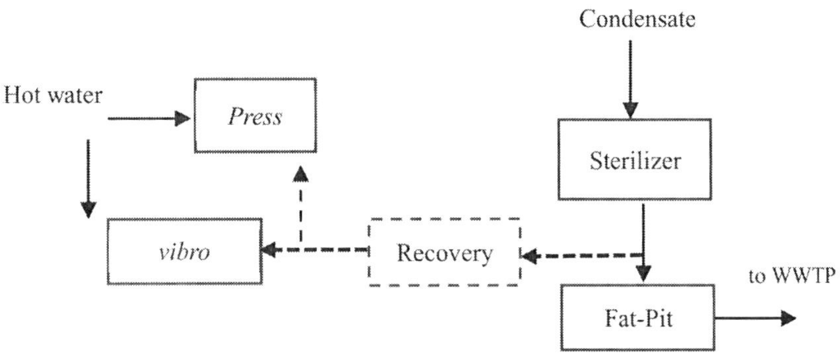

Fig. 5. The flow diagram of Reuse of Condensate Water for Diluting Press and Vibro

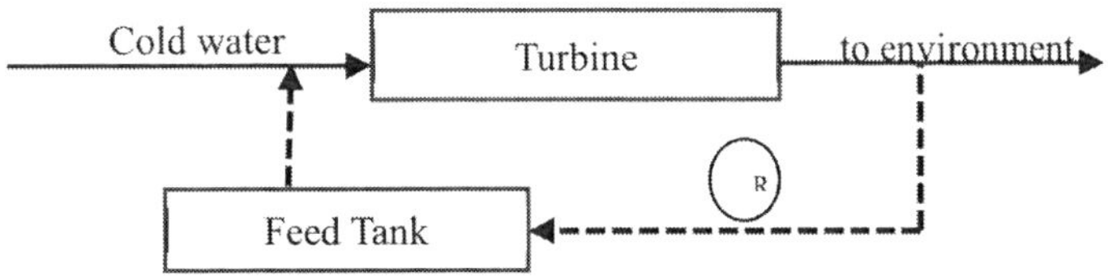

Fig. 6. The Flow Diagram of Recycling of Water from Turbine

waste in the form of condensate generally disposed to the receiver after being processed in the fat pit and WWTP. By considering the demand of water for dilution in the press unit, the temperature for dilluting press is about of 90 - 95 °C (Standardization Team of Palm Oil Processing. Directorate General of Estate, 1997 in Fitria, 2003).This aims to facilitate the separation of the oil in the press and clarification units. In addition, the function of hot water is to ease the process of the separation of palm kernel and seed, and maintain the calorific value of fiber (pulp press) as the boiler fuel.

The vibro screen unit acts as the filter for filtering solids from the oil coming from the press unit. In addition to requiring water for diluting oil in the press unit (if the viscosity is too high), water is also required for the dilution process to facilitate the flow of solids (sludge) as a result of the separation of waste that must be disposed. For the purposes, the water used does not require any special requirements, so the use of condensate water can be done.

The use of water as shown in Figure 5 is the process of water recovery from boiled waste water discharged into the water body after being processed in the fat-pit and WWTP. The water will be reused for dilution purposes, extra water (dilution) in the unit of press and a small portion for vibro waste, so that the high polluted wastewater does not negatively affect the environment. A decrease in the use of water by reusing wastewater can reach 10% where it will affect the amount of wastewater discharged. Approximately 25% of high polluted wastewater can be reduced by applying this method (Fitria, 2003).

3.3.3 Recycling of Water from Turbine

Based on the recommendations of the machinist head assistant, the recycling of water from turbine can also reduce the use of blue water in the mill. The concept of water recycling in Figure 6 is the process of reusing the cooling water coming from the turbine, so the water can be recovered for cooling after being collected in the collector and then returned to the hot well tank for the purposes in the clarification unit. Water as much as 0.20 m^3 / ton of FFB is directly discharged into the environment during this process. Figure 6 illustrates that the water used for cooling the turbine that is low pollutant can be reused without any special treatments. The water is then collected in a collector tub for being reused. Based on the results of a research conducted by Fitria (2013) in the POM of Sei Mangkei, North Sumatra, reusing (recycling) water for cooling turbine can reduce water consumption by 16% accompanied by a decrease in the discharge of liquid waste by 100%, which is low pollutant. Based on the analysis, POM of Sei Mangkei uses similar specification of turbine and has similar power plant production flow, so it is assumed that it is also applicable in PTP Mitra Ogan.

For saving time of using water for cooling the turbine, a cooling tower can also be used in which a fan that will speed up the cooling process in the turbine. Decision making in using water efficiently depends on the costs, technical and environmental considerations of the mill.

4 CONCLUSIONS

The average value of palm oil water footprint in PTP Mitra Ogan showed that there is an inefficient use of water. The increase use of the blue water, the potential contamination by the waste of palm oil will be larger, so that reducing strategy of the water footprint is very crucial. Waste water reuse strategies and application of clean technology are expected to minimize the inefficient use of water by the company, which will reduce the amount of liquid waste.

However, additionally study based on more exact data is required in the assessment of the water use impacts because there were not enough detailed and specific data available in this research. Besides, further studies in quantifying the affectivity and efficiency aspects of the strategies proposed are also required to be done.

The authors are grateful to Beasiswa Unggulan Scholarship Indonesia and University of Twente Scholarship, Leeuwarden, Netherland, for supporting this work and to PTP Mitra Ogan and PTP Lonsum, South Sumatra, Indonesia, for making data available.

REFERENCES

1. L.D. Bergert, (Thesis Master, Michigan Technology University 2000)
2. A.K. Chapaign, D. Ticker, Water Footprint: Help and Hidrance in Water Alternatives, **5(3)**, 565 (2012)
3. A. Fitria, (Thesis Master, Universitas of Indonesia 2003)
4. A.H. Kittikun, Environmental Management for Palm Oil Mill. Integreted Bio-system Network. Tokyo. 12 (2000). http :www.ias.unu.edu/proceeding/icibs/ic.unfal.index.
5. Hoekstra et al., 2009. Water Footprint Manual: State of the Art. Netherland: Water Footprint Network (2009). Available online at: www.waterfootprint.org. Accessed on April 19th, 2014.
6. Hoekstra, A.Y., A.K. Chapagain, M.M. Aldaya, M.M. Mekonnen. The Water footprint Assessment Manual : Setting the Global Standard. London: Earthscan (2011). Media release: (http://www.waterfootprint.org).
7. A.Y. Hoekstra, A.K. Chapagain, M.M. Aldaya, M.M. Mekonnen, Water footprint manual: State of the art (2009)
8. D. Jefferies, I. Munoz, J. Hodges, V.J. King, M. Aldaya, A.E. Ercin, L.M. Canals, A.Y. Hoekstra, J. Clean. Prod., **32**, 155 (2012)
9. S.A. Muyibi, T. Tajari, M.S. Jami, M.K. Amosa, Adv. Environ. Biol., 590 (2014)
10. M.P. Naibaho, *Teknik Pengolahan Kelapa Sawit Pusat Penelitian Kelapa Sawit,* Medan, 306 (1996)

DOI: 10.1051/matecconf/20179701122

11. I. Pahan, *Panduan Lengkap Kelapa Sawit,* Bogor, Penebar Swadaya (2006)
12. Sappi, (Sustainable Palm Oil: Conversation and Debate 2013)
13. Sappi, Water Use and Treatment in the Pulp and Paper Industry. Boston: Fine Paper North America vol.5 (2012).
14. Stockholm Environment Institute (SEI), (2012)
15. M.H.S. Ismail, A. Aziz, M. Kamal, N.A. Morad, Eur. J. Sci. Res., **37(4)**, 628 (2009)
16. World Wild Fund (WWF). Palm Oil: Minimizing Water Use (2013). Available online at: wwf.panda.org. Accessed on April 17th, 2014.

Effects of fertilization and submersion on morphological changes and production of two rice varieties with submersion stress conditions

Gribaldi[1,*], *Nurlaili*[1], and *Supli* Effendi Rahim[2]

[1] Agrotechnology Study Program Faculty of Agriculture, University of Baturaja, Jl. RatuPenghulu No. 02301 Karang Sari Baturaja 32115, South Sumatra, Indonesia.
[2] Agrotechnology Study Program Faculty of Agriculture, University of Palembang, Jl. Dharmapala No. 1A, Bukit Lama, Palembang City 30139, South Sumatra, Indonesia.

Abstract. This study aimed at studying the effects fertilization and submersion on morphological changes and production of two rice varieties with submersion stress conditions. The experimental design used in this research was randomized block design with three replications. The first factor is: Varieties (V): IR 64 (V_1), Inpara 5 (V_2). The second factor is the treatment (P): P_0 = Without submersion, P_1 = Submersion 7-14 dap and fertilizer N, P and K 7 days after being submerged, P_2 = Submersion 7-14 dap and fertilizer N, P and K + PPC Micro 7 days after being submerged P_3 = Submersion 7-14 dap and 28-35 dap and fertilization N, P and K 7 days after the first submersion, P_4 = submersion 7-14 dap and 28-35 dap and fertilization N, P and K + PPC Micro 7 days after the first submersion. Results of the study showed the fertilization treatment after being submerged affected the morphological changes of rice crop with submersion stress conditions. Treatment of the fertilization N, P and K + PPC Micro 7 days after the first submersion give the best effect on the growth and yield of rice under submergence stress conditions.

1 Introduction

Submerged rice plants result in reduced oxygen supply. According to [1], the diffusion of oxygen in the water 10^4 times slower than in the air. This decrease in gas diffusion results in poor growth, metabolism, and tolerance of plants [5]. [1] revealed that the roots of oxygen deficient environment greatly inhibit ion uptake by roots and ion transport to the header, so the concentration of nutrients like N, P and K in leaves decreased. In addition the plant is submerged under stress shortly after being submerged. For subsequent metabolic processes and internal repair plant, required the availability of sufficient carbohydrate reserves once submerged. Efforts can be made to promote tolerance and availability of plant nutrients after being submerged can be through the fertilization treatment after being submerged. The results of the research in rice conducted by [2] states that, fertilization after the stress submerged conducted one week after the completion of submerged stress

[*] Corresponding author: gribaldi64@yahoo.co.id

DOI: 10.1051/matecconf/20179701123

and fertilizers are given around 30-50 kg N, 20-30 kg of P2 O5 and 20 -30 kg K ha-1 can speed up the recovery of rice plants.

Recovery after being submerged require sufficient nutrients and can be quickly absorbed by plants. Fertilization through leaves is the additions and improvements fertilization through the soil or roots in certain circumstances where the absorption of the roots of the important elements is reduced. One effort to overcome the problem by granting Complementary Liquid Fertilizer (PPC) as an additional micro nutrients Si and Zn through the leaves, with a recommended dosage for the paddy crop is 2 cc / l of water. Furthermore [6] states that giving treatment *"Plant Phytoregulator"* and Nitrogen can help rice plants speed recovery after being submerged. This study aimed at studying the effects fertilization and submersion on morphological changes and production of two rice varieties with submersion stress conditions.

2 Materials and methods

This experiment was conducted at the experimental garden plastic Faculty of Agriculture, University of Baturaja. This experiment used randomized block design arranged as factorial with two factors treatments and three replications. The experimental treatment consists of: The first factor is: Rice Varieties (V) consists of: = IR 64 Variety (V_1), Inpara 5 Variety (V_2). The second factor is the treatment (P), comprising: P_0 = Without submersion + basic fertilizer, P_1 = Submersion 7-14 dap and fertilizer N, P and K 7 days after being submerged, P_2 = Submersion 7-14 dap and fertilizer N, P and K + PPC Micro 7 days after being submerged P_3 = submersion 7-14 dap and 28-35 dap and fertilizer N, P and K 7 days after the first submersion, P_4 = submersion 7-14 dap and 28-35 dap and N, P and K + PPC Micro 7 days after the first submersion.

Seed of Inpara 5 and IR 64 varieties were incubated for 3 days, after germinating seedling in a medium plastic tub with length of 40 cm, width 30 cm and in 13 cm were filled valley marshland as much as 15 kg, in which the soil previously treated for N, P, K, Si and Zn, and manure, each with a dose of 60, 40, 40, 30 and 20 kg ha-1 and 10 ton ha-1 [7]. Seeds that have been aged 21 days in seedbed tubs lifted and planted three seedlings of rice plants into each black polybag filled valley marshlands each 5kg and has flooded approximately 30 days. Submersion treatment tailored to each treatment, for treatment of P_1 and P_2 7-14 submerged dap and for the treatment P_3 and P_4 submerged 7-14 dap and 28-35 dap 7 days. High water depth of at least 15 cm from the surface of the plant.

Basic fertilization (Urea 100, SP36 128 and KCl 100 (kg ha-1)) is performed at the time of planting and fertilizing once submerged done 21 days after planting with fertilizer dose used is for fertilizer N, P2O5, K2O each with a dose of 50, 30 , 30 (kg ha-1) as well as micro PPC 2 ml / l of water. PPC fertilizer application of micro-interval of 7 days until the phase of primordia, whereas inorganic fertilizers administration by means embedded into the ground as deep as 10 cm.

Observations conducted in this study included: 1) plant height (cm), 2) The number of tillers (tillers), 3) dry weight of plants (g), 4) The ratio of root / shoot, 5) Age flowering (day), 6) Grain yields per clump, and (g), 7) Weight of 100 grains (g).

This experiment was conducted at the experimental garden plastic Faculty of Agriculture, University of Baturaja. This experiment used randomized block design arranged as factorial with two factors treatments and three replications. The experimental treatment consists of: The first factor is: Rice Varieties (V) consists of: = IR 64 Variety (V1), Inpara 5 Variety (V2). The second factor is the treatment (P), comprising: P0 = Without submersion + basic fertilizer, P1 = Submersion 7-14 hst and fertilizer N, P and K 7 days after being submerged, P2 = Submersion 7-14 hst and fertilizer N, P and K + PPC Micro 7 days after being submerged P3 = submersion 7-14 hst and 28-35 hst and fertilizer N, P and K 7 days after

 DOI: 10.1051/matecconf/20179701123

the first submersion, P4 = submersion 7-14 hst and 28-35 hst and N, P and K + PPC Micro 7 days after the first submersion.

Seed of Inpara 5 and IR 64 varieties were incubated for 3 days, after germinating seedling in a medium plastic tub with length of 40 cm, width 30 cm and in 13 cm were filled valley marshland as much as 15 kg, in which the soil previously treated for N, P, K, Si and Zn, and manure, each with a dose of 60, 40, 40, 30 and 20 kg ha-1 and 10 ton ha-1 (Suwignyo et al., 2012). Seeds that have been aged 21 days in seedbed tubs lifted and planted three seedlings of rice plants into each black polybag filled valley marshlands each 5kg and has flooded approximately 30 days. Submersion treatment tailored to each treatment, for treatment of P1 and P2 7-14 submerged hst and for the treatment P3 and P4 submerged 7-14 hst and 28-35 hst 7 days. High water depth of at least 15 cm from the surface of the plant.

Basic fertilization (Urea 100, SP36 128 and KCl 100 (kg ha-1)) is performed at the time of planting and fertilizing once submerged done 21 days after planting with fertilizer dose used is for fertilizer N, P2O5, K2O each with a dose of 50, 30 , 30 (kg ha-1) as well as micro PPC 2 ml / l of water. PPC fertilizer application of micro-interval of 7 days until the phase of primordia, whereas inorganic fertilizers administration by means embedded into the ground as deep as 10 cm. Observations included: 1) plant height (cm), 2) The number of tillers (tillers), 3) dry weight of plants (g), 4) The ratio of root / shoot, 5) Age flowering (day), 6) Grain yields per clump (g), 7) Weight of 100 grains (g).

3 Result and discussion

3.1 Result

Results of variance analysis in Table 1 showed that for every parameter between varieties affect no significant effect unless the number of productive tillers while between fertilization treatments significantly affect except grain weight of 100 grains. The interaction between varieties and fertilizer treatment effect is not obvious unless the dry weight of the plant.

Table 1. Results of variance analysis influence fertilization treatment after being submerged in both rice varieties for all the observed variables.

No	Observed parameters	Variety	Fertilization	Interactions
1	Percentage of survived plant	Ns	*	Ns
2	Plant height	Ns	*	Ns
3	The number of productive tillers/plant	*	*	Ns
4	Plant dry matter weight	Ns	*	*
5	Percentage of filled grains per panicle	Ns	*	Ns
6	Number of grains per panicle	Ns	*	Ns
7	100 grains weight	Ns	Ns	Ns
8	Grain yield per clump	Ns	*	Ns

Notes: * = significant effect
Ns = no significant effect

3.1.1 Percentage of survived plant

F test results showed fertilization treatment give significant effect, but the varieties of interaction affect no significant effect on the percentage of survived plants. The percentage of plants in the twice submersion (P_3 and P_4) tend to be lower than those experienced once submersion (P_1 and P_2), at various fertilization treatment for all varieties tested (Table 2). In the twice submersion conditions, Inpara 5 varieties have live plants percentage higher than IR 64 varieties at various fertilization treatment, which amounted to 50 percent. While the provision of fertilizer N, P and K + PPC Micro 7 days after the first submersion (P_4) on IR 64 varieties tend to be higher than the percentage of plants P_3 fertilization, which is 45.83 percent, an increase of 22.2 percent.

Table 2. Effect of fertilization after being submerged in both rice varieties on the percentage of survived plants stress submersion conditions in turbidwater.

Treatment	P_0	P_1	P_2	P_3	P_4
IR 64	100.00	100.00	100.00	37.50	45.83
Inpara 5	100.00	100.00	100.00	50.00	50.00

Notes: $\mathbf{P_0}$ = Without submersion + basic fertilizer, $\mathbf{P_1}$ = Submersion of 7-14 dap and fertilizer N, P and K 7 days after being submerged, $\mathbf{P_2}$ = Submersion of 7-14 dap and fertilizer N, P and K + PPC Micro 7 days after being submerged $\mathbf{P_3}$ = submersion of 7-14 and 28-35 dap and fertilizer N, P and K 7 days after the first submersion, $\mathbf{P_4}$ = submersion of 7-14 dap and 28-35 dap and fertilizer N, P and K + PPC Micro 7 days after the first submersion.

3.1.2 Plant height

F test results showed significant effect on fertilization treatment, but the varieties of interaction affect no significant effect on plant height. Higher plants by the twice submersion (P_3 and P_4) tend to be lower than that experienced once submersion (P_1 and P_2), at various fertilization treatment for all varieties tested (Table 3). IR 64 varieties treated with N, P, K + PPC micro 7 days after being submerged (P_2) tend to have the highest plant which the height is 112.17 cm.

Table 3. Effect of fertilization after being submerged in both rice varieties to the height of plant (cm) stress submersion conditions in turbidwater.

Treatment	P_0	P_1	P_2	P_3	P_4
IR64	106.02	108.83	112.17	70.50	78.67
Inpara 5	114.00	101.95	110.17	81.50	86.12

Notes : P_0 = Without submersion + basic fertilizer, P_1 = Submersion 7-14 dap and fertilizer N, P and K 7 days after being submerged, P_2 = Submersion 7-14 dap and fertilizer N, P and K + PPC Micro 7 days after being submerged P3 = submersion 7-14 dap and 28-35 dap and fertilizer N, P and K 7 days after the first submersion, P4 = submersion 7-14 dap and 28-35 dap and N, P and K + PPC Micro 7 days after the first submersion.

3.1.3 The number of productive tillers per plant

F test results showed fertilization treatment and varieties give significant effect, but the interaction effect was not significant on the number of productive tillers. The provision of fertilizer treatment on varieties undergo twice submersion (P3 and P_4) inclined tiller number less than that experienced once submersion (P_1 and P_2) (Fig. 1). Inpara 5 varieties which given various treatments of N, P and K + PPC Micro 7 days after being submerged tend to have the highest number of productive tillers, which are 25.17 stem.

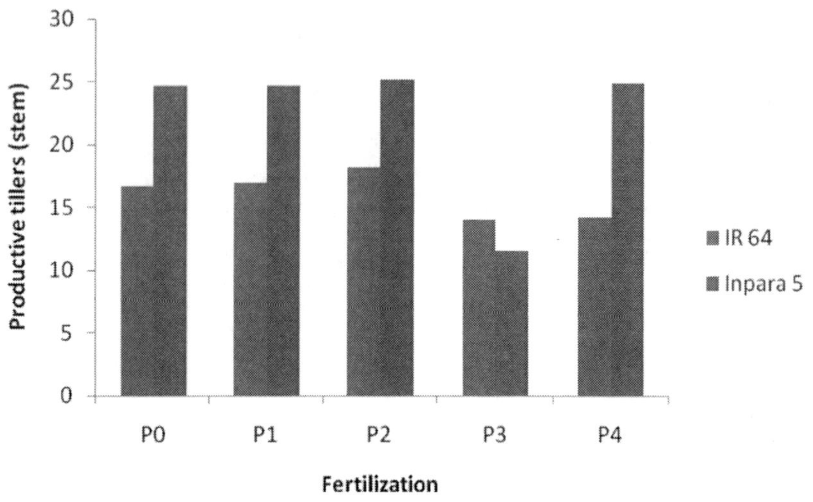

Fig. 1. Number of tillers 42 dap, on each of the varieties of rice and fertilizer treatment in a stress submerged conditions. $\mathbf{P_0}$ = Without submersion + basic fertilizer, $\mathbf{P_1}$ = Submersion 7-14 dap and fertilizer N, P and K 7 days after being submerged, $\mathbf{P_2}$ = Submersion 7-14 dap and fertilizer N, P and K + PPC Micro 7 days after being submerged $\mathbf{P_3}$ = submersion 7-14 dap and 28-35 dap and fertilizer N, P and K 7 days after the first submersion, $\mathbf{P_4}$ = submersion 7-14 dap and 28-35 dap and fertilizer N, P and K + PPC Micro 7 days after the first submersion.

3.1.4 Plant dry matter weight

F test results showed varieties treatment give no significant effect, but fertilization and their interactions significantly affected on the plant dry matter weight. The provision of fertilizer treatment on varieties undergo twice submersion (P3 and P4) the dry matter weight of plants tend to be lower than the varieties experiencing once submersion (P1 and P2) (Fig. 2). IR 64 varieties treated fertilizer of N, P and K + PPC Micro 7 days after being submerged tend to have the highest plant dry matter weight, that is 63.1 g/clump for once submersion, while at the twice submersion, Inpara 5 varieties given fertilization treatment for N, P and K + PPC Micro 7 days after being submerged tend to have the highest plant dry weight, which is 40.9 g / clump.

DOI: 10.1051/matecconf/20179701123

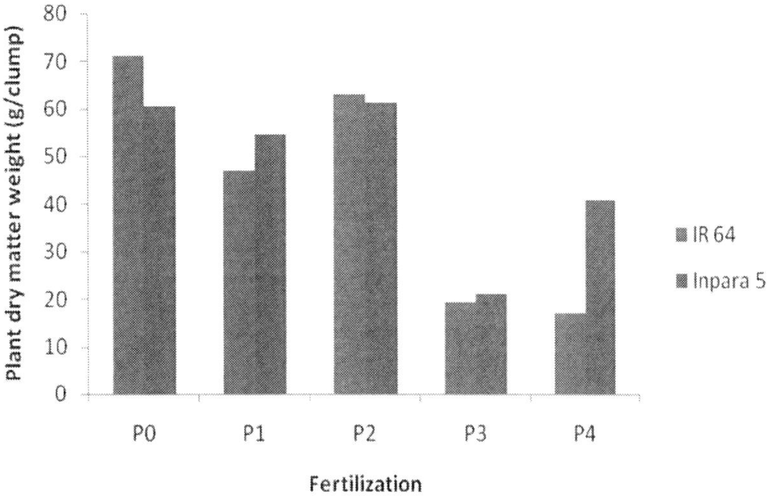

Fig. 2. The dry weight of the plant life of 42 dap, on each of the varieties of rice and fertilizer treatment in a stress submerged conditions. $\mathbf{P_0}$ = Without submersion + basic fertilizer, $\mathbf{P_1}$ = Submersion 7-14 dap and fertilizer N, P and K 7 days after being submerged, $\mathbf{P_2}$ = Submersion 7-14 dap and fertilizer N, P and K + PPC Micro 7 days after being submerged $\mathbf{P_3}$ = submersion 7-14 dap and 28-35 dap and fertilizer N, P and K 7 days after the first submersion, $\mathbf{P_4}$ = submersion 7-14 dap and 28-35 dap and fertilizer N, P and K + PPC Micro 7 days after the first submersion.

3.1.5 Yield and yield components

F test results showed significant effect on fertilization treatment, but the variety and interaction give no significant effect on grain yield per clump.

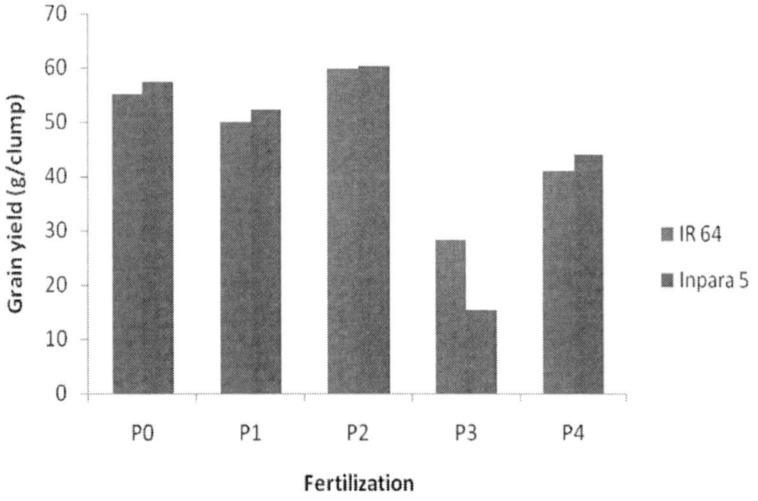

Fig. 3. Grain yields in each variety of rice and fertilizer treatment in a stress submerged conditions.

P0 = Without submersion + basic fertilizer, **P1** = Submersion 7-14 hst and fertilizer N, P and K 7 days after being submerged, **P2** = Submersion 7-14 hst and fertilizer N, P and K + PPC Micro 7 days after being submerged **P3** = submersion 7-14 hst and 28-35 hst and fertilizer N, P and K 7 days after the first submersion, **P4** = submersion 7-14 hst and 28-35 hst and fertilizer N, P and K + PPC Micro 7 days after the first submersion.

The provision of fertilizer treatment on varieties undergo twice submersion (P_3 and P_4) grain yield per clump tended to belower than varieties which experience once submersion (P_1 and P_2). Inpara 5 varieties and IR 64 were treated for N, P, K + PPC micro 7 days after being submerged (P_2) tend to have the highest grain yield, 60.32 and 59.67 g / clump and grain yield exceeded that did not experience submersion (P_0), that is 57.37 and 40.83 g per clump (Fig. 3).

Table 4. Effect of fertilization after being submerged in both rice varietiesto some components of the results of stress submersion conditions in the turbid water.

Treatment	Inpara 5			IR64		
	1	2	3	1	2	3
P0	141.67	94.33	2.22	145.85	93.11	2.18
P1	153.67	94.91	2.24	140.67	91.94	2.22
P2	157.17	95.32	2.32	141.50	95.68	2.21
P3	117.00	84.06	2.20	117.50	81.59	2.19
P4	127.33	83.80	2.32	119.83	82.34	2.23

Remarks: **1**: The number of grains per panicle (grains), **2**: Percentage of grain pithy (%), **3**: grain weight of 100 grains (g). **P0** = Without submersion + basic fertilizer, **P1** = Submersion 7-14 hst and fertilizer N, P and K 7 days after being submerged, **P2** = Submersion 7-14 hst and fertilizer N, P and K + PPC Micro 7 days after being submerged **P3** = submersion 7-14 hst and 28-35 hst and fertilizer N, P and K 7 days after the first submersion, **P4** = submersion 7-14 hst and 28-35 hst and fertilizer N, P and K + PPC Micro 7 days after the first submersion.

Yield components in both rice varieties tend to decrease in submersion twice compared with once submersion (Table 4). Treatment of N, P and K + PPC Micro 7 days after being submerged in both rice varietiesthe components of grain yield tend to be higher compared to other fertilizer treatment, even at Inpara5 varieties tend to be higher than without submergence at once submersion.

4 Discussion

Giving fertilization treatment after being submerged can improve the recovery of rice plants to stress submerged, it can be seen from the changes in plant height, number of tillers, plant dry weight and grain yield per clump that has a value approaching and even exceeding the changes in rice plants that are not submerged (P0), the ability of plants to quickly recover from stress submerged is very important. Plants that are tolerant with stress submerged are able not only to survive but also to grow back after reeling submerged. [3] research results mention that the recovery was good for rice crops experiencing stress submerged characterized by rapid out new tillers so that at the end of growth, the plant has a number of tillers almost same or even exceed compared with plants that do not undergo submersion.

Twice submersion treatment (P3 and P4) in both rice varieties showed a greater drop than once submersion (P1 and P2) on all the observed variables. This is due to the longer and the frequent the plant having a submerged condition, so the damage suffered by the plants is greater because the metabolism of plants is hampered by the lack of oxygen. According to [1] revealed that, the lack of oxygen greatly inhibit ion uptake by roots and ions transport to penetrate, so the concentration of N, P and K in leaves decreased. More [4] stated that the lack of oxygen due to gripped by submersionis a limiting factor of growth and productivity.

To increase crop productivity growth and recovery efforts are needed to the treatment of plant once gripped by submersion, through fertilization after being submerged in order that thenutrients neededby plants are enough and can be quickly absorbed by plants. The provision of combined fertilization through roots and leaves is one attempt to address these conditions. [6] states that the giving of treatment *"Plant Phytoregulator"* and Nitrogen can help rice plants speed the recovery after being submerged. The results of the research in rice conducted by [2] states that the provision of fertilizer after stress submerged conducted one week after the stress submerged completedand fertilizers is given around 30-50 kg N, 20-30 kg P2O5 and 20-30 kg K2O ha-1 can speed up the recovery of rice plants.

Giving fertilization treatment after being submerged with N, P and K + PPC micro 7 days after being submerged (P_2) is the best fertilizer treatment to improve recovery of rice plant, it is seen changes in almost all the observed variables. Inpara 5 varieties treated for N, P and K + PPC micro 7 days after being submerged (P_2) produces highgrain yield per clump, due to the growth of this variety is better than the otherfertilizer treatment, it can be seen from the changes in plant height, number of tillers, and plant dry weight. The dry weight of the high plant in P2 treatment is due to the high plant height and number of tillers that many lead to better plant growth so as to produce a high grain anyway.

5 Conclusion

Based on results and discussion in the study, it can be concluded that: (1) Fertilization after being submerged in both rice varieties were experiencing stress submersion in the turbid water affect the growth and production of rice, (2) Provision of fertilizer N, P and K + PPC Micro 7 days after being submerged best effect on the growth and production of rice, and (3) The longer the submersion of the rice variety, the greater the effect on growth and rice production.

We especially wish to acknowledge the Ministry of Research, Technology and Higher Education that had already provided funding for this research through Competitive Research Grant (*Hibah Bersaing*) for fiscal year of 2015-2016.

References

1. W. Armstrong, M.C. Drew, *Plant roots: the hidden half* **3**, 729 (2002)
2. IRRI, The Research and Development Council of Food Crops, *Submergence Tolerant Rice* (Food Crops Research and Development Center, 2009)
3. A.K. Makarim, E. Suhartatik, G.R. Pratiwi, Ikhwani, *Assembly of rice production technology on swamp land and flooding prone swamp (>15 days) for minimum productivity of 7 t ha-1* (Indonesian Rice Research Institute, Indonesian Agricultural Research and Development, 2009)
4. R.K. Sairam, D. Kumutha, K. Ezhilmathi, Curr. Sci., **96**, 674 (2009)
5. R.K. Sarkar, J.N. Reddy, S.G. Sharma, A.M. Ismail, Curr. Sci., **91**, 899 (2006)
6. R.A Suwignyo, J. TanamanTropika., **8**, 45 (2005)
7. R.A. Suwignyo, A. Wijaya, H. Sihombing, Gribaldi, J. Lahan Suboptimal., **1**, 1 (2012)

AUTHOR INDEX

AUTHOR INDEX

AUTHOR INDEX

AUTHOR INDEX